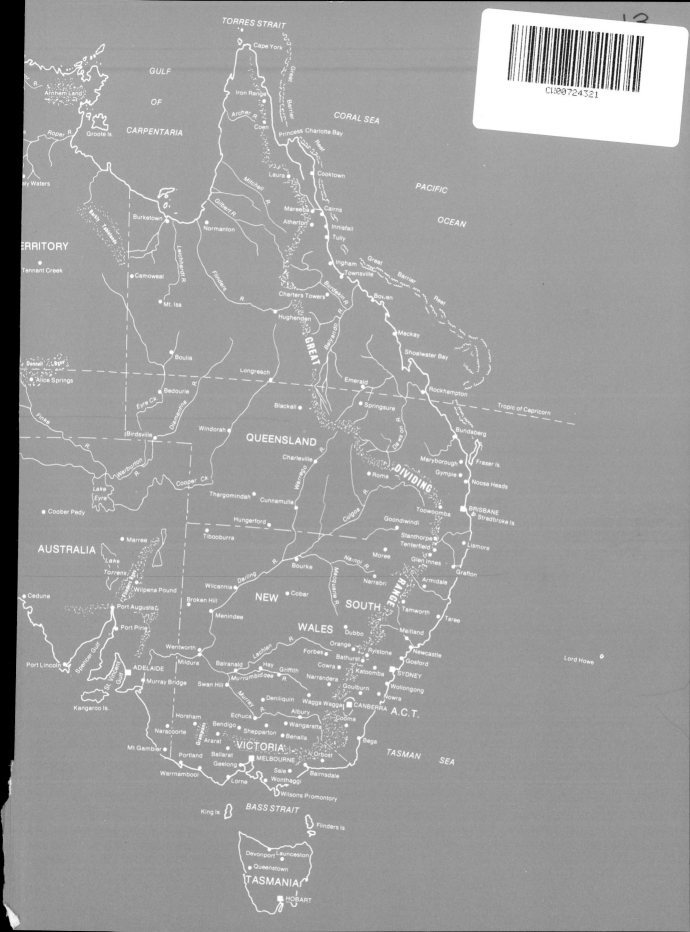

ENCYCLOPAEDIA OF AUSTRALIAN PLANTS

suitable for cultivation

VOLUME SEVEN

ENCYCLOPAEDIA OF AUSTRALIAN PLANTS

suitable for cultivation

VOLUME SEVEN

W. Rodger Elliot
David L. Jones
B.Ag.Sc. Dip. Hort.

Line drawings by Trevor L. Blake

Lothian
B O O K S

Thomas C. Lothian Pty Ltd
11 Munro Street, Port Melbourne, Victoria 3207

First published 1997

National Library of Australia
Cataloguing-in-Publication data:

Elliot, W. Rodger (Winston Rodger), 1941– .
Encyclopaedia of Australian plants suitable for
cultivation. Volume 7, N–Po.

Includes bibliographies and indexes
ISBN 0 85091 634 8 (v. 7)
ISBN 0 85091 148 6 (set)

1. Wild flower gardening. 2. Wild flower gardening —
Australia. 3. Plants, Ornamental — Australia.
4. Botany — Australia. I. Jones, David L. (David Lloyd),
1944 – . II. Title.

635.96760994

Produced by Publishing Solutions
Printed in Singapore

Frontispiece: *Pomaderris lanigera* at Karwarra Gardens, Victoria   W. R. Elliot

Contents

This volume is dedicated to the memory of Dr James Hamlyn Willis AM (1910–1995), known as Jim to his many friends and admirers.

During his long and extremely productive life, Jim Willis inspired countless numbers of us to appreciate better the wonders of the natural world.

Preface

Indeed, a further volume is published! This volume covers over 1300 species and cultivars from more than 260 genera. The following pages describe many fascinating plants and many horticulturally interesting genera such as *Nemcia, Nephrolepis, Nicotiana* (Australia has many species), *Nothofagus, Nymphaea, Olearia, Orthrosanthus, Ozothamnus, Pandanus, Pandorea, Passiflora, Persoonia, Petrophile, Phebalium, Pileanthus, Pimelea, Pittosporum, Platycerium, Podocarpus* and *Pomaderris*.

In recent times, government policy-makers have decided that botanical research does not warrant as much support as it has received in the past. This is an extremely short-sighted view if we are to try to understand the value of Australia's rich floral heritage in its multiplicity of applications for habitat, medicine, food, regeneration, horticulture, landscape and fuel. In many cases, we don't even know what exists in some of our natural areas. Two recent events dramatically highlight our current lack of knowledge. First, scientists in NSW discovered the very primitive Wollemi Pine, *Wollemia nobilis*, a member of the Araucaria family in NSW. This news hit the newspapers of the world in an extremely short period of time and created a frenzy in botanical and horticultural circles. Second, the discovery and description of the endemic monotypic genus, *Eidothea*, from north-eastern Qld has shed more light on the relationships of the family Proteaceae. It is possible that the family Platanaceae, which includes Liquidambars and Plane trees, is very closely allied.

On a global scale, we are still trying to source plants which could benefit the human race in terms of their food and medicinal value.

At a recent United Nations endorsed conference held in Newcastle, NSW ('Pathways to Sustainability'), Dr Tim Flannery, Principal Research Scientist at the Australian Museum, stressed the importance of cultivating native grasses, trees and shrubs in gardens and streets of urban areas. The planting of indigenous flora would increase biodiversity and should therefore decrease the possibility of certain species facing extinction and also provide increased habitat for Australian wildlife.

Australian plants have a very important ongoing role to play in amenity horticulture especially with the increasing interest in the value of cultivating indigenous plants of known provenance. Plants which are suited to the local conditions can only help to provide valuable habitat in urban areas for our native wildlife.

Since the publication of Volume 6 of this Encyclopaedia it has been marvellous to have Volumes 16, 28, 49 and 50 of the *Flora of Australia* published. Also Volumes 2 and 3 of the *Flora of Victoria* reached publication. These tomes are important in helping to enrich our understanding of our Australian plant life.

Once again we have tried to keep apace of botanical studies and revisions but it is difficult to be completely up to date as the time between finishing writing and actual publication can mean changes to nomenclature or new species described, and inevitably this information cannot be included.

Preface

We always welcome comments and criticism which can help to improve the standard of the Encyclopaedia, so do not hesitate to correspond with us via the publishers. As for previous volumes, we are happy to accept responsibility for errors and omissions.

The input of information by many enthusiasts and scientists is greatly appreciated and is given recognition in the Acknowledgements.

The light is starting to get much brighter as we approach the end of this project. Now there are only two more volumes to complete the series. It has certainly been a mammoth task for all involved, but from the comments received we do believe it has also been very worthwhile. Thank you all for your continued support of the project.

Explanation of Text

Examples of the layout used for genus and species in this book are shown below together with explanations of relevant terms and headings.

Example of Genus Layout

Generic name and author citation

An author citation is placed at the end of the botanical name in recognition of the person or persons who named and described the taxon. One of the fascinating aspects of author citations is the wide range of nationalities of botanists who have studied, or are studying, plants of the Australian flora.

Author citations are extremely important because they can help to provide us with the naming history of plants.

Over recent years, an accepted standard format for author citations has been implemented. This has resulted in changes to previously accepted citations, and has affected their application in the *Encyclopaedia of Australian Plants*. The *Authors of Plant Names* edited by R. K. Brummitt & C. E. Powell, published by the Royal Botanic Gardens, Kew, England, is the accepted standard reference. In the Supplement to this Encyclopaedia there is a listing of author citations which have been used in the Encyclopaedia. It provides full name, nationality and date of birth and death (where applicable) for each author.

(derivation of plant name)
Plant family
Simplified botanical description, including:

> form of plant — shrub, tree, etc,
> leaves — shape and form,
> flowers — arrangement, size, colour, often including calyx, petals, stamens,
> fruit — type, size and colour.

Information about the distribution of the genus and the number of species.
Cultivation information, plus propagation notes.
N.B. In the case of a genus being monotypic (ie. with only one species in the genus), no generic description is included.

Example of Species Layout

Name and author citation

Sometimes there can be a number of botanists whose author citations are attached to a genus, species, variety or subspecies name. These author citations provide us with the history of name changes as the taxon has undergone botanical studies. It may be that a single author is responsible for the name of the plant or in some cases it may be two or three people who work together on a botanical or taxonomic study and then the two or three

Explanation of Text

names are used in the author citation. Bear in mind that it is always the last listed author citation which refers to the most recent name change. An example is *Angophora leiocarpa* (L. A. S. Johnson ex G. Leach) K. Theile & Ladiges. Before it was given its present species status, it was known as *Angophora costata* ssp. *leiocarpa* L. A. S. Johnson ex G. Leach and before that it was *Angophora costata* var. *leiocarpa* L. A. S. Johnson.

On some occasions the same botanical name is used but with a different author citation. For instance, *Chorizema varium* has been known as *Chorizema varium* Benth. ex Lindl. but is now known as *Chorizema varium* Paxton. We know from the differing author citations they are almost sure to be different taxa. Usually one of these taxa has been reduced to synonymy with another taxon. However there always seem to be exceptions to the rules, eg. *Calochilus grandiflorus* Rupp and *Calochilus grandiflorus* (Benth.) Domin are both the very same species.

(derivation of plant name)

Distribution within Australia   Vernacular (common) name

Growth dimensions, height × width   Flowering time

Simplified botanical description including:

    form of plant — herb, shrub, tree etc.,

    bark — texture, colour,

    branches — form, colour and other characteristics,

    branchlets — as above,

    leaves — dimensions, type, shape, colour and other characteristics

        (eg. margins, apex, nerves etc),

    inflorescence — type and size,

    flowers — size, colour, arrangement and other differentiating characteristics,

    fruit — type, dimensions, shape, colour and other characteristics.

Specific information on natural habitat, when available.

Requirements for soil and climatic conditions, and tolerances.

Specific uses. Propagation notes.

Reference to related species and their differences.

Information about previous names, confusion of names and allied facts.

In some cases, further subspecies and varieties are included, with differentiating characteristics. When this is done, the species description always refers to the typical form, eg. *Acacia pulchella* var. *pulchella*. The other varieties, eg. *A. pulchella* var. *reflexa* appear below the typical form.

Cultivars

Cultivars are included with relevant information. The names of cultivars are designated by single quotation marks, eg. *Callistemon* 'Burgundy'.

Synonym Entries

Major and recent entries are included to explain recent name changes, and to facilitate the use of the book.

Group Entries

Group entries are included to cover plant families, major plant groups such as ferns or orchids, interesting features of plants such as aromatic flowers or foliage, and specialised features or groups such as bog plants, cushion plants, annuals etc.

Terms Used in Text

1 For any botanical terms, see glossary.

2 Distribution
States are given only when records are listed by the various herbaria, or are known by personal observation. There is only minor emphasis given to overseas distribution of a species, subspecies or variety since such information is sometimes difficult to find, and can be inaccurate.

3 Vernacular name (common name)
Only those names that are in common use have been included, as there is no Australia-wide convention for such names.

4 Growth dimensions
These are given as height × width. They should be taken as a guide only because plant growth can be variable due to differing soil and climatic conditions. In cultivation, some plants may be vigorous with dense foliage, whereas in nature the same species will be of open habit, and leggy, straggly or misshapen. In some case, the latter habits of growth add to their beauty.
HEIGHT GROUPINGS
Dwarf shrub — 0–1 m
Small shrub — 1–2 m
Medium shrub — 2–4 m
Tall shrub — 4–6 m
Small tree — 6–12 m
Medium tree — 12–25 m
Tall tree — over 25 m

5 Flowering time
This can be a most variable aspect of plants. The periods given are derived from different sources such as records of dried specimens in herbaria, publications, and from observations of cultivated and wild plants by ourselves and others. As far as possible they relate to all areas of Australia.

Cultivation often results in plants flowering at times which differ from those recorded in the natural habitat. This is often due to artificial watering, and applications of fertiliser which stimulate new growth, resulting in the production of flowers outside the normal season.

In many cases, the flowering time given, eg. Sept–Feb, does not mean that the flowers are displayed for the full length of that period, but rather that flowering occurs during that time. In general, flowering in tropical and coastal areas occurs within the earlier part of the time period provided for each entry.

When a flowering time such as 'Sept–Feb; also sporadic' is given, this indicates that limited flowering often occurs outside the period of Sept–Feb. Sometimes the phrase throughout the year is used, and this refers to a more-or-less continual flowering period.

6 Soils are generally described under three headings:
a) light — including sands, gravels,

b)   medium — including loams,

c)   heavy — including clay-loams and clay soil types.

For further information, see Soils, page 53, Volume 1.

7   Drainage

The terms related to drainage are as follows:

a)   well-drained or freely draining; water does not accumulate and has a free passage through the soil,

b)   relatively well-drained; water is retained in the soil for very short periods,

c)   poorly drained; refers to soils in which water is retained for varying periods,

d)   periodic inundation; soils retain water above ground level for short periods, ie. 1–30 days,

e)   waterlogged; heavy soils where the retention of water is maintained at a maximum level for extended periods. This can also result in inundation.

8   Degree of light

a)   full shade; receives no direct sunlight,

b)   semi-shade; receives very little sunlight,

c)   dappled shade or filtered sun; refers to areas having an overhead canopy of trees or shrubs, allowing some sunlight to penetrate throughout the day,

d)   partial sun; can be an open situation, but only receiving full sunlight for less than half of the day,

e)   full sun; an open position, with virtually no protection from direct sunlight through-out the day.

9   Cultivation

For more detailed information, see page 47, Volume 1.

10 Maintenance

For details about terms used and other relevant information, refer to page 79, Volume 1.

11 Propagation

Procedures in relation to propagation methods are dealt with in detail on page 187, Volume 1.

12 Illustrations and colour plates

We have tried to include a representation of as many genera as possible. The ideal would be an illustration of every species, including photographs showing growth habit, close-ups of flower and foliage, and other diagnostic features. This is not possible because of practical and economic restraints, and a selection has had to be made. The black-and-white line drawings can be used as an aid to identification, and such illustrations are often more useful than colour photographs.

Acknowledgements

It is with pleasure and gratitude that we acknowledge the assistance of the many people and institutions who willingly provide information, or help in myriad ways, because their input adds tremendously to the value of this volume.

The support of Australian botanical and horticultural institutions has continued to be invaluable as they make their facilities available as well as answering various queries we have as authors. We sincerely thank the Directors and Staff of the Herbaria from Queensland, New South Wales, Victoria, Tasmania, South Australia, Western Australia, the Northern Territory and the Centre for Plant Biodiversity Research in Canberra.

We thank Judy Crafter for making available the slide collection of the late Brian Crafter, from which a number were chosen for inclusion in this volume.

The gardens of Neil and Jane Marriott, Gordon Paterson and Royce and Jeanne Raleigh have provided us with specimens for illustration. We thank you.

Botanists who have been of special assistance include Neville Walsh (*Pomaderris*) and John Williams (*Parsonsia*). We also sincerely thank Lindy Cayser (Pittosporaceae), Barry Conn, Lyn Craven, Mike Crisp (Fabaceae), John Dowe (Pandanaceae), Clyde Dunlop, Don Foreman (*Petrophile*), Paul Forster (Asclepiadaceae, Euphorbiaceae, Piperaceae and *Plectranthus*), Gordon Guymer, John Jessup, Nicholas Lander (*Olearia*), Brendan Lepschi (*Porana*), Brian Molloy (Podocarpaceae), Brian Morley (*Negria*), Christopher Puttock (*Ozothamnus*), Estelle Ross, Roger Spencer, Dr Judy West (Portulacaceae and *Polyscias*) and Peter Weston (*Persoonia*).

We wish to acknowledge the help of technical staff at various Botanic Gardens and Herbaria, especially Helen Cohn, Jill Thurlow and Catherine Coles at the Royal Botanic Gardens, Melbourne, Catherine Jordan of the Australian National Botanic Garden and Ray Cranfield from Perth.

Mark Richardson was very helpful when he was at the Australian National Botanic Garden and Iain Dawson of the Australian Cultivar Registration Authority also provided information.

The experiences of Australian plant enthusiasts is never underestimated by us and the following people are all sincerely thanked for their assistance: Robert Anderson, John Armstrong, John Arnott, Judy Barker, Dick Burns, Ross and Joyce Cowling, James and Annette Frew, Marilyn Gray, Peter Jones, John and Sue Knight, Alan Lacey, Dean Lewis, Peg McAllister, Max MacDowall, Ian Mitchell, Jenny Rejske, Helen Richards, the late Fred Rogers and June Rogers, Maureen and Vic Schaumann, Brian and Diana Snape, Paul Thompson, J. S. and D. E. White and Glen Wilson.

Jeff Irons is an indefatigable grower of Australian plants in England and is forever passing on cultural information. Ray Collett (Director) and Brett Hall (Manager) of the Arboretum of the University of California, Santa Cruz USA, are always eager to share their experiences.

Acknowledgements

There are also some extremely important people who have much more than a passing interest in this endeavour. Trevor Blake is an illustrator with a wealth of knowledge about Australian plants combined with a wonderful sense of humour, which is just as well with the demands often placed upon him. He also contributes many photographs and makes valued comments and judgements which help to make this a better publication.

Our best friends and wives, Gwen and Barbara, continue in their inimitable manner with their tremendous input into the project in countless ways. It is always easy to say thanks in printed word but we would like to add in public that your contribution is thoroughly appreciated and completely invaluable.

Thank you very much, one and all.

Abbreviations

aff.	with affinity
alt.	alternately
approx.	approximately
°C	degrees Celsius
cm	centimetres
comb. nov.	*combinatio nova*, new combination
cv.	cultivar
km	kilometres
m	metres
m^2	square metres
mm	millimetres
pH	hydrogen-ion concentration
sp.	species (singular)
sp. aff.	species with affinity
spp.	species (plural)
var.	variety
Qld	Queensland
NSW	New South Wales
Vic	Victoria
Tas	Tasmania
SA	South Australia
WA	Western Australia
NT	Northern Territory
NZ	New Zealand
UK	United Kingdom
USA	United States of America
Jan	January
Feb	February
Aug	August
Sept	September
Oct	October
Nov	November
Dec	December

Nymphaea violacea

N

NABLONIUM Cass.
(from the Greek *nable*, a ship; a possible allusion to its island and coastal habitats)
Asteraceae (alt. *Compositae*)
A monotypic genus endemic in Tasmania. Some botanists prefer to transfer *Nablonium* to *Ammobium* R. Br.

Nablonium calyceroides Cass.
(similar to the genus *Calycera*)
Tas
0.05–0.1 m × 0.2–1 m   Nov–April
Perennial **herb**, stoloniferous, often forming small colonies; **leaves** about 2.5 cm × 0.3–0.7 cm, linear or lanceolate, in a basal rosette, green and faintly hairy above, white woolly-hairy below, margins recurved, apex pointed; **scape** to about 10 cm tall, 1 or, rarely, 2, shorter or taller than leaves, erect, leafless or with 1–2 short bracts; **flowerheads** to about 1 cm across, solitary, daisy-like, surrounded by pointed green bracts; **flowers** small, tubular, erect, white with prominent purple anthers; **fruit** a cypsela, flattened.
A dainty perennial herb that inhabits wet soils on the north-eastern Bass Strait Islands and on the west coast. Rarely cultivated, it needs moist but well-drained soils and tolerates plenty of sunshine. Hardy to moderate frosts. It has potential for use in general planting as well as for rockeries, miniature gardens and containers. Propagate from seed or by division of stolons.

NAJADACEAE Juss.   Water Nymphs
A family of monocotyledons consisting of about 50 species in the solitary genus *Najas*. They are annual or perennial aquatic plants which grow in fresh or brackish water. The genus is widely distributed around the world with 5 or 6 species occurring in Australia. The Australian species have little horticultural merit although they may provide food and shelter for fish and waterfowl.

NAUCLEA L.
(from the Greek *naus*, ship; *cleio*, to confine; referring to the boat-shaped fruits)
Rubiaceae (alt. *Naucleaceae*)
Trees; **leaves** opposite, large, pinnately veined; **stipules** free, terminal pair large, persistent, lower pairs often deciduous; **inflorescence** terminal or axillary pedunculate heads; **flowers** bisexual, tubular, cylindrical to funnel-shaped, crowded, 4–5-lobed; **stamens** 4–5; **calyx lobes** 4–5; **fruit** a indehiscent syncarp, fleshy; **seeds** flattened, not winged, numerous.

A tropical genus of about 10 species, distributed in Africa, Asia and with 1 endemic species in Australia.

Nauclea orientalis (L.) L.
(eastern)
Qld, WA, NT   Leichhardt Tree; Canary Cheesewood
10–20 m × 6–15 m   May–Feb
Small to medium **tree** with hairy, bronze-red young growth; **bark** tessellated or corky, pale grey; **branches** spreading, often providing a layered effect; **branchlets** more or less terete, faintly hairy or glabrous; **leaves** 12–25 cm × 7–15 cm, elliptical to ovate-elliptical, opposite, dark green, glabrous or faintly hairy above, faintly hairy below, venation prominent, apex rounded; **flowerheads** 2–4 cm across, globular, terminal, on short branchlets; **stipules** 1.5–4 cm long; **flowers** about 1 cm long, yellow with slightly spreading lobes and exserted white style, profuse; **fruit** 2–4 cm across, globular, becoming succulent when ripe, edible but bitter.
A distinctive tree which inhabits vine thickets and riparian forest of the lowlands in northern Australia and extends south to near Mackay. Often retained during clearing of other vegetation. Trees are generally fast-growing and are suitable for tropical and warm subtropical regions. They require good drainage and

Nablonium calyceroides × .6

3

Nauclea orientalis

D. L. Jones

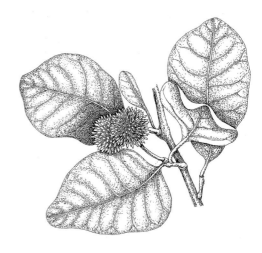

Nauclea orientalis × .25

tolerate a sunny site but must have plenty of water. Applications of organic mulch are beneficial and regular light applications of fertilisers are helpful. Plants are very sensitive to frost. Does well as a container plant while young. Propagate from fresh seed.

NAUCLEACEAE (DC.) Wernham
A small family of dicotyledons consisting of about 200 species in 10 genera. They are often included in the family Rubiaceae, but can be recognised by their large, entire, stipulate leaves, and their flowers arranged in dense globose heads. They also have minute seeds. Mainly tropical in distribution, they consist of trees, shrubs or climbers. Australian genera: *Nauclea*, *Neonauclea*.

NEEDHAMIA R. Br. = *NEEDHAMIELLA* L. Watson

NEEDHAMIELLA L. Watson
(after Rev. John T. Needham, 18th-century cleric and microscopist, and from the Greek *ella*, diminutive form)
Epacridaceae
A monotypic genus endemic in south-western WA.

Needhamiella pumilio (R. Br.) L. Watson
(dwarf, pygmy)
WA
0.1–0.2 m × 0.2–0.6 m   July–Nov
Dwarf **shrub**; **stems** prostrate to ascending or erect; **leaves** about 0.2 cm × up to 0.1 cm, linear to narrowly ovate, opposite or alternate, sessile and more or less appressed, crowded, concave, green; **upper leaves** with ciliate margins; **lower leaves** glabrous, apex pointed; **flowers** about 0.3 cm long, tubular, red with white lobes, solitary in upper axils forming leafy spikes; **fruit** a small drupe.

An attractive species from the southern Darling and Eyre Districts where it occurs in mallee heath and heathlands, on winter-wet sites in sand and sandy clay soils. *N. pumilio* is evidently not cultivated but it should prove adaptable in well-drained soils with a sunny or semi-shaded aspect. It has potential for use in gardens, including rockeries, as well as for container cultivation. Hardy to moderate frosts. Should respond well to pruning. Propagate from seed, which may be slow to germinate, or from cuttings of fairly soft young growth.

Previously known as *Needhamia pumilio* R. Br., which was an invalid name.

NEGRIA F. Muell.
(after Professor Christopher Negri, 19th-century Italian diplomat and academic)
Gesneriaceae
A monotypic genus which is endemic on Lord Howe Island.

Negria rhabdothamnoides F. Muell.
(similar to the NZ genus *Rhabdothamnus*)
NSW (Lord Howe Island)   Pumpkin Flower
4–9 m × 3–6 m   sporadic all year
Medium to tall **shrub** or small **tree** with hairy young growth; **bark** soft, yellow-grey; **branches** spreading to ascending, thick; **branchlets** initially hairy; **leaves** 4–15 cm × 3–9.5 cm, ovate to ovate-elliptical, in whorls of 3–4 at ends of branchlets, base rounded or tapering, green above, paler below, hairy to glabrous and glossy above, hairy below, often densely so on raised veins, apex bluntly or finely pointed; **inflorescence** an axillary cyme, 3–flowered; **flowers** about 2 cm long, tubular, curved, yellow with orange-red spots inside throat, prominently lobed, fleshy, hairy; **calyx** to 1.4 cm long, with erect narrow teeth; **stamens** 4; **capsules** to 1.6 cm long, beaked.

This species is confined to Mt Gower and Mt Lidgbird in the southern region of Lord Howe Island. It grows in closed evergreen forest which is subjected to much wind and rain. On Lord Howe Island, plants rarely reach more than 5 m in height unless they grow in very sheltered sites. Rarely cultivated, this attractive species deserves wider recognition. It is best suited to tropical and subtropical regions but may succeed in temperate regions if grown in a warm protected site. A sheltered location with very well-drained soils that are rich in organic matter and are moist for most of the year should be the ingredients for successful cultivation. Responds well to pruning, which is often needed regularly. Plants are damaged by frosts. It is an excellent container plant. Flowers are attractive to nectar-feeding birds. Propagate from seed or from cuttings of firm young growth, which strike readily.

NEISOSPERMA Raf.
(from the Greek *neiso*, ?spherical; *sperma*, seed)
Apocynaceae

Trees with milky sap; **leaves** opposite or whorled, with transverse veins; **inflorescence** of dichotomous or trichotomous cymes; **flowers** tubular, with twisted lobes overlapping in bud, stalked, strongly fragrant; **anthers** not exserted; **calyx** shortly lobed; **fruit** a drupe, red.

A genus of about 20 species which extends from the Seychelles to the western Pacific region. In Australia, there are 2 species that are confined to eastern Qld and NSW.

Rarely cultivated, they have potential for greater use due to their attractive foliage, fragrant flowers and colourful fruits. They are best suited to tropical and subtropical regions but could adapt to temperate areas if grown in protected sites. Propagate from fresh seed.

Previously included in *Ochrosia* Juss.

Neisosperma kilneri (F. Muell.) Fosberg & Sachet
(after Mr F. Kilner, 19th-century phycologist)
Qld
4–10 m × 3–8 m   Jan–April

Tall **shrub** or small **tree** with a dense rounded canopy; **bark** bitter tasting, does not exude sap when cut, like fruit does; **leaves** 6–14 cm × 3-5 cm, obovate, in whorls of 4–5, narrowed to base, with stalk to about 2.5 cm long, glossy green above, pale green below, nerves spreading, apex blunt; **panicle** a terminal trichotomous cyme; **flowers** to about 1.5 cm wide, white, shortly tubular with spreading lobes, strongly fragrant; **fruit** 5–9 cm long, ovoid, bright red, smooth.

An attractive, endemic rainforest tree from the North and South Kennedy Districts. Although rarely cultivated, it has potential for growing in tropical and subtropical areas. Probably prefers moist well-drained soils rich in organic matter. Supplementary watering during extended dry periods is beneficial. May need protection from hot sun when young. Plants may become brittle with age. Propagate from fresh seed.

Previously known as *Ochrosia kilneri* F. Muell.

Neisosperma poweri (F. M. Bailey) Fosberg & Sachet
(after Mr R. D. Power)
Qld, NSW   Milkbush
4–10 m × 3–6 m   Jan–April

Tall **shrub** to small **tree**, with slightly milky sap; **leaves** 4–15.5 cm × 2–6.5 cm, obovate to elliptical, opposite, narrowed to base, with stalk 0.3–1.2 cm long, dark green and glossy above, paler below, veins not prominent, apex bluntly pointed; **panicle** a terminal dichotomous cyme; **flowers** to about 1.2 cm across, white, cream or yellow, tubular with spreading lobes, strongly fragrant; **fruit** about 4 cm × 1.5 cm, bright red, smooth, usually paired.

This decorative endemic species extends from north-eastern Qld to north-eastern NSW. It grows in rainforests and usually develops as an open shrub. Plants are worthy of wider recognition for cultivation. They need a shaded site and moist well-drained soils for best results. May need protection from hot sun when young and supplementary watering during extended dry periods. Propagate from fresh seed.

Previously known as *Ochrosia poweri* F. M. Bailey and *O. newelliana* F. M. Bailey.

Nelitris ingens F. Muell. ex C. Moore = *Acmena ingens* (F. Muell. ex C. Moore) Guymer & B. Hyland

NELSONIA R. Br.
(after David Nelson, 18th-century English gardener and botanical collector)
Acanthaceae

Herbs, with many hairs; **leaves** opposite; **floral bracts** leaf-like; **inflorescence** axillary or terminal spike with overlapping bracts; **calyx lobes** 4, unequal; **corolla** tubular with 5 spreading more or less equal lobes; **stamens** 2, included or slightly exserted; **fruit** a capsule.

Nelsonia is a small tropical genus of about 5 species. It has a wide distribution and is found in Africa, Asia, America and Australia. One species occurs in Australia and it extends to New Guinea. Propagation is from seed or cuttings.

Nelsonia campestris R. Br.
(of the fields or open plains)
Qld, WA, NT
0.1 m × 0.5–1 m   May–Aug

An annual **herb** with densely hairy young growth; **stems** prostrate, hairy, often self-layering at nodes; **leaves** to about 6 cm × 3.5 cm, lowest leaves largest,

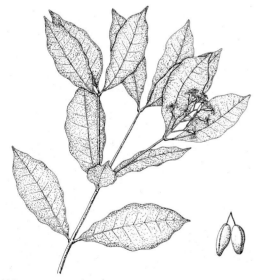

Neisosperma poweri × .4

ovate, spreading to ascending, green to greyish-green, hairy, apex pointed; **spikes** 1–6 cm × about 1 cm, axillary or terminal; **flowers** to about 0.5 cm long, white, often profuse; **capsules** to 0.5 cm long, glabrous.

Widespread throughout the tropical region, where it is usually associated with melaleuca woodlands on river or creek banks and on edges of swamps. It grows in sandy soils or clay loams. This species lacks highly ornamental flowers but some plants have an attractive silvery-grey appearance provided by a dense coverage of hairs. It is possibly useful for stabilising loose soils on slight slopes because of the capability of stems to self-layer at nodes. Plants are cold-sensitive and therefore best suited to tropical and subtropical regions. They prefer plenty of sunshine but will tolerate a semi-shaded site.

Propagate from seed or by division of the layered stems.

NELUMBO Adans.
(from a native Sri Lankan name)
Nelumbonaceae

Large aquatic **herbs**; **rhizomes** creeping, thick, fleshy, nodes prominent; **leaves** large, peltate, circular, floating or more usually held above water, long-stalked; **flowers** large, showy, solitary on long thick scapes, yellow, pink or white; **sepals** 4 or 5, overlapping, deciduous; **petals** many; **stamens** many, with long petal-like filaments; **carpels** many, immersed singly in large spongy top-shaped receptacle; **fruit** a nut, oblong to globular.

A tropical genus of 2 species with one extending to Australia.

Nelumbo nucifera Gaertner
(nut-bearing)

Qld, NT	Pink Lotus-lily, Pink Waterlily
Aquatic herb	March–Dec

Perennial aquatic **herb** with a submerged creeping rootstock; **leaves** 20–90 cm across, roundish, peltate, on erect, thick, prickly stems 0.5–1.5 m long, held well above water, blue-green, waxy, concave, venation radiating from the centre; **flowers** 15–25 cm across, petals and sepals pink, purplish-pink, pinkish-white or rarely white, with many yellow stamens, sweetly fragrant, solitary on erect stems to about 2 m long, most conspicuous; **receptacle** 6–12 cm across, top-shaped, woody, flat-topped or slightly convex, with sunken cavities for fruits; **fruits** about 2 cm long, oblong, nut-like, hard, smooth.

A magnificent aquatic plant of sheltered freshwater lagoons and billabongs in tropical lowland regions extending from the Top End, NT, to southern Qld. Also occurs in southern Russia and Asia. In India it is known as the Sacred Lotus-lily. Although best suited to tropical and subtropical regions there are now cold-tolerant Australian selections offered which will succeed as far south as Tasmania. A selection with pink-tipped creamy-white flowers has proved adaptable. Plants do well in deep water in northern Australia but they do best in southern Australia in water of about 15 cm depth which heats up more readily than deeper water. In southern regions they are best planted in late spring to allow them to develop over summer

before the onset of cold periods which can retard their growth. Plants are best grown in pots with a typical water-lily mix which contains about one-quarter cow manure and three-quarters good quality lime-free loam plus a half to one handful of blood and bone per pot and a final mulch of coarse sand or gravel.

N. nucifera is a valuable food source as most parts are edible. Sliced roots are excellent raw or steamed and are sometimes part of Asian meals. The seed is highly nutritious before or after roasting but is often difficult to extract from the hard outer covering. Leaves are used to wrap food before steaming. Aboriginals of northern Australia ground the seed for flour, and the juice from leaf stalks was traditionally used for treatment of diarrhoea and other sicknesses. Dry fruit receptacles are popular with floral artists.

Propagation is by division of the jointed stems (especially for named selections) or from seed, which usually germinates between 25–30°C. The hard shell needs to be cracked or filed to allow entry of moisture. Seed is renowned for its lengthy viability and in fact a 237-year-old seed is known to have germinated.

NELUMBONACEAE Dumort.   Lotus Family
A small family of dicotyledons consisting of 2 species in the genus *Nelumbo*. Although few in number these large, impressive water plants are distributed in America, Asia, various Pacific islands and Australia. As they are regarded by many cultures as sacred plants, being prized for their large leaves and colourful flowers, this may have aided in their distribution. The seeds and rhizomes are eaten by people in many countries including the Australian Aborigines.

NEMATOLEPIS Turcz.
(from the Greek *nema, atos*, thread; lepis, *scale*; referring to the long hairs at the base of the style)
Rutaceae

A monotypic genus endemic in south-western WA.

Nematolepis phebalioides Turcz.
(similar to the genus *Phebalium*)

WA	
0.6–2 m × 0.6–2 m	July–Nov; also sporadic through the year

Dwarf to small **shrub**; **branches** spreading to erect, covered in silvery scales when young; **leaves** 1–2.5 cm × 0.8–2 cm, oblong to nearly orbicular, alternate, simple, stalked, mid-to-dark green, glabrous, leathery, glandular, apex blunt; **flowers** to 2 cm × 0.7 cm, tubular, pendant, bright red with yellow-green tips, solitary, sometimes profuse and very conspicuous; **fruit** 2-valved cocci.

An ornamental bird-pollinated species from the Eyre and Roe Districts where it grows in a range of soils including sand, clay, clay loam and gravel. Limestone is present in some regions. An extremely adaptable species, best suited to temperate and semi-arid regions. Needs fairly well-drained soils with a sunny or semi-shaded aspect. Hardy to moderate frosts and withstands extended dry periods. Plants may become leggy if not pruned and they respond very well to pruning. The small flowers are long-lasting and attractive for indoor decoration. A number of

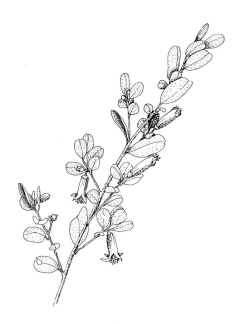

Nematolepsis phebalioides × .5

selections with differing leaf shapes are cultivated. They are useful for general or accent planting in gardens and containers. Propagate from cuttings of firm young growth, which strike readily. Seed propagation is difficult due to the presence of inhibiting agents. Lengthy periods of leaching may be helpful. For further information see Treatments and Techniques to Germinate Difficult Species, Volume 1, page 205.

NEMCIA Domin
(after Bohumil Nemec, 19–20th-century plant physiologist, Prague, Czechoslovakia)
Fabaceae (alt. *Papilionaceae*)

Shrubs; **leaves** opposite or in whorls of 3–4, narrow to broad, simple, leathery, stiff, margin entire or lobed, non-poisonous; **stipules** bristle-like; **inflorescence** axillary or terminal clusters, condensed corymbs or racemes; **flowers** pea-shaped, yellow to orange, yellow to orange with red, or reddish; **stamens** free; **calyx** 5-lobed, 2 upper lobes joined; **fruit** a hairy pod, small, inflated.

A genus of about 28 species which is endemic in south-western WA. Recent revision has enlarged the genus, and some species previously included in *Callistachys, Gastrolobium* and *Oxylobium* are now placed in *Nemcia*. (The genus *Callistachys* will be described in the supplementary volume. The *Gastrolobium* species that have moved into *Nemcia* are described in this volume as they have had revision since their description in Volume 4).

Nemcia is widespread in the region and species occur on a range of soil types. They are found at low altitudes as well as on mountain slopes, such as in the Stirling Range.

Nemcias are not common in cultivation, undoubtedly because a number are typical eggs-and-bacon pea flowers of which there are a multitude in the Australian flora.

Most nemcias are small to medium shrubs. Some plants are spreading and compact while others are tall and somewhat open with stiff, few-leaved branches. Some species have spectacular floral displays, eg. *N. leakeana*. There is a range of foliage shapes, colours and textures. Some species have narrowly pointed leaves, while others are broad, greyish or hairy. There are also lobed leaves.

Nemcias generally require freely draining soils which are acidic to slightly alkaline, and a sunny or semi-shaded site. Most species tolerate light to moderate frosts. Plants usually grow well without regular fertilising and watering.

Some plants can become leggy, especially the large-leaved species, and to overcome this tendency it is best to begin tip pruning from the planting stage to promote lateral growth as hard pruning may be detrimental. Small-leaved species tolerate regular pruning.

Pests and diseases are not prime problems, with caterpillars and borers among the most common pests, but usually they do not cause much damage. Root rot can be a problem if plants are growing in poorly drained soils which are wet for extended periods.

Propagation is from seed or cuttings. Seeds reach maturity during the hot weather and are expelled from the pods when ripe. The seed can be collected just before the pods mature (when they are green but fully inflated). The hard-coated mature seeds need pre-sowing treatment, see Volume 1, page 205. Cuttings of firm young growth usually strike readily and applications of rooting hormones can be beneficial.

Nemcia acuta (Benth.) Crisp
(acute)
WA
0.3–1 m × 0.3–0.6 m   Aug–Sept

Dwarf **shrub** with soft, silky-hairy young growth; **branches** ascending to erect, few, hairy; **leaves** 1.2–2.2 cm × 0.3–0.9 cm, ovate-elliptical to oblong, in whorls of 3 or scattered, becoming glabrous, stiff, venation reticulate, apex pungent-pointed; **stipules** bristle-like, persistent; **flowers** pea-shaped, about 1.2 cm across, yellow and reddish, in short, loose, axillary clusters, often profuse and very conspicuous; **standard** yellow with red or brown; **wings** and **keel** red or brown; **calyx** about 0.7 cm long, silky-hairy; **pods** to about 0.9 cm long, leathery, pointed.

A dwarf, floriferous species from the Darling District where it grows mainly in heavier sandy loams. Evidently not cultivated now, but was grown in England during the mid 19th-century. Plants require relatively good drainage in a site which is sunny or semi-shaded. Probably will do best in temperate regions. Pruning at a young stage will promote lateral growth. Hardy to moderate frosts. Suitable for gardens or containers. Propagate from seed, which germinates readily, or from cuttings of firm young growth.

Gastrolobium acutum Benth., *Oxylobium acutum* (Lindley) Benth. and *Callistachys acuta* (Benth.) Kuntze are synonyms.

N. epacridoides differs in having ovate leaves but lacks stipules. Its calyx is only faintly hairy.

Nelumbo nucifera

T. L. Blake

Nemcia atropurpurea (Turcz.) Domin =
N. leakeana (Drumm.) Crisp

Nemcia atropurpurea (Turcz.) Domin var. **minorifolia**
Domin = *N. rubra* Crisp

Nemcia axillaris (Meisn.) Benth.
(axillary)
WA
0.6–1.2 m × 0.5–1 m   Aug–Oct

Dwarf to small **shrub**, often with silky-hairy young growth; **branches** ascending to erect, stiff, slender; **leaves** 1–2.5 cm × 0.6–2 cm, more or less elliptical to nearly orbicular, mostly opposite, spreading to ascending, glabrous above, minutely silky-hairy below, margins often wavy, reticulate venation barely prominent, apex blunt or acute with a small point; **flowers** pea-shaped to about 1 cm across, yellow, with red or maroon, in axillary clusters or terminal heads, often profuse; **standard** yellow with red base; **wings** and **keel** red or maroon; **pods** to about 0.6 cm long, with long hairs.

N. axillaris is from the northern Darling and Irwin Districts where it grows in sand, clay loam and gravelly soils. Plants prefer a sunny site but they also tolerate semi-shade. Soils should not become waterlogged. Should do best in semi-arid and warm temperate regions. Hardy to moderate frosts and extended dry periods. Propagate from seed, which germinates readily, or from cuttings of firm young growth.

Gastrolobium axillare Meisn., *Nemcia reticulata* (Meisn.) Domin var. *axillaris* (Meisn.) Domin and *Oxylobium reticulatum* Meisn. var. *gracile* Benth. are synonyms.

Nemcia brownii (Meisn.) Crisp
(after Robert Brown, 18–19th-century British botanist)
WA
0.5–2 m × 0.6–1.5 m   Sept–Nov

Dwarf to small **shrub** with softly hairy young growth; **branches** ascending to erect, hairy; **leaves** 0.8–2.5 cm × 0.3–0.7 cm, obovate or oblong, usually broadest above the middle, or linear cuneate, ascending to erect, leathery, glabrous, venation reticulate, apex rounded or very blunt ending in a short sometimes pungent point; **flowers** pea-shaped, about 0.8 cm across, yellow and red, in clusters in upper axils, often profuse and very conspicuous; **standard** orange-yellow; **wings** and **keel** red to reddish-brown or purplish-red; **calyx** softly hairy to nearly glabrous; **pods** small, silky-hairy.

A showy species from the southern Darling District where it is usually associated with granitic soils, and often grows on sandy stream banks. Well suited to temperate regions with freely draining soils which are not dry for extended periods. A situation with dappled shade is probably preferable but plants should also tolerate plenty of sunshine. Hardy to light frosts. Pruning from planting should promote bushy growth. Propagate from seed, which germinates readily, or from cuttings of firm young growth.

Gastrolobium brownii Meisn. is a synonym.

N. hookeri is allied but is a spreading dwarf shrub with shorter leaves.

Nemcia capitata (Benth.) Domin
(in heads)
WA   Bacon and Eggs
0.2–1 m × 0.6–1.3 m   May–Nov

Dwarf **shrub** with silky-hairy young growth; **branches** prostrate to ascending, often straggly, usually hairy; **leaves** 1.5–5 cm × 0.2–0.8 cm, variable, usually linear to oblong, lower ones sometimes obovate, spreading to ascending and sometimes reflexed, glabrous and reticulate above, glabrous or faintly silky-hairy below,

Nemcia capitata × .5, leaves and fruit × .75

8

midrib prominent, apex ending in a recurved rigid point; **flowers** pea-shaped, to about 1.2 cm across, yellow to orange with red or reddish-brown, in axillary clusters, sometimes forming a leafy terminal head, often profuse and very conspicuous; **standard** yellow to orange; **wings** and **keel** red to reddish-brown; **pods** to about 1 cm long, ovoid, hairy.

Plants are striking when in flower and the species is worthy of greater cultivation. It occurs in the Avon, northern Darling and Irwin Districts where it often grows in damp or swampy, sandy or gravelly soils, which are sometimes alkaline. It has had limited cultivation. Adapts to a range of soil types and a sunny or semi-shaded site is suitable. Hardy to light frosts. Tip pruning is beneficial from an early stage to promote bushy growth. Introduced into England in 1837 as *Oxylobium capitatum*. Propagate from seed, which usually begins to germinate 10–35 days after sowing. Cuttings of firm young growth are worth trying.

Callistachys capitata (Benth.) Kuntze and *Oxylobium capitatum* Benth. are synonyms.

Nemcia carinata Crisp
(keeled)
WA
0.4–1 m × 0.5–1 m   Aug–Oct
Dwarf **shrub** with woolly-hairy young growth; **branches** spreading to erect, thick, somewhat woolly-hairy; **leaves** 0.4–1 cm × 0.2–0.3 cm, narrowly ovate to oblong, in irregular whorls of 3, crowded, spreading to recurved, leathery, glabrous, venation reticulate, apex blunt; **flowers** pea-shaped, about 1 cm across, orange and brown, in axillary clusters forming irregular whorls near ends of branches, often profuse and very conspicuous; **standard** orange; **wings** and **keel** reddish-brown to brown; **pods** to about 0.7 cm long, nearly globular, hairy.

From the Dale, Eyre and Roe Districts where it grows in a range of sandy soils which may have a high content of clay or lateritic gravel. It displays its flowers well and deserves to be grown more widely. Should adapt to most soils in temperate regions. Prefers a sunny or semi-shaded site with moderate to well-drained soils. Hardy to moderate frosts and extended dry periods. Propagate from seed, which germinates readily, or from cuttings of firm young growth.

Eutaxia reticulata Meisn. and *Gastrolobium reticulatum* (Meisn.) Benth. are synonyms.

N. punctata (Turcz.) Crisp is allied but has longer and narrower leaves with very thick veins and recurved apices, as well as slightly longer flowers.

Nemcia coriacea (Smith) Domin
(leathery)
WA
1–2.5 m × 1–2.5 m   Sept–Jan
Small to medium **shrub** with hairy young growth; **branches** many, stiff; **branchlets** angled when young, hairy; **leaves** 2–5.5 cm × 1.5–2.7 cm, oblong to ovate, mostly opposite, stalked, mainly spreading, somewhat leathery, glabrous and reticulate above, usually silky-hairy below, apex blunt, truncate or emarginate with a very small point; **flowers** pea-shaped, about 1.2 cm across, yellow to orange and red, in crowded terminal

Nemcia coriacea   W. R. Elliot

clusters or racemes, sometimes also axillary, profuse and very conspicuous; **standard** yellow to orange; **wings** and **keel** red; **pods** about 0.8 cm long, ovoid, densely hairy.

A floriferous species which occurs in the southern Darling and western Eyre Districts where it is found in gravelly or peaty sands. Plants tolerate winter-wet sites and often have to cope with extended dry periods over summer. Rarely cultivated, it has potential for greater use in temperate gardens. Sunny or semi-shaded sites would be suitable and it should adapt to a range of acidic soils. Hardy to moderate frosts. Pruning from an early stage should promote bushy growth. Thought to be introduced into England as *Oxylobium retusum* in 1823. Propagate from seed, which usually begins to germinate about 25–40 days after sowing, or from cuttings of firm young growth.

Synonyms include *Gastrolobium ovalifolium* (Meisn.) Lemaire, *Oxylobium coriaceum* (Smith) C. Gardner, *Oxylobium ovalifolium* (Meisn.) and *Oxylobium retusum* R. Br. ex Lindl.

N. coriacea has been confused with *N. vestita*, which differs in having much larger leaves.

Nemcia crenulata (Turcz.) Crisp
(with small convex teeth)
WA
1–2m × 1–2 m   Sept–Nov
Small **shrub** with hairy young growth; **branches** angled; **leaves** 2–6 cm × 1.5–3 cm, obovate, usually in whorls of 3, stalked, glabrous, margins revolute and crenulate, apex notched with a small point; **flowers** pea-shaped, about 1 cm across, yellow to orange-yellow with deep red, in few-flowered axillary racemes longer than the leaves; **standard** yellow to orange-yellow; **wing** deep yellow; **keel** maroon; **calyx** hairy; **pods** small.

A decorative species which occurs in the Eyre District and appears to be confined to the Barren Ranges. Apparently rarely cultivated, it should adapt

9

to a range of acidic soils in temperate regions. Although appreciating sunshine it will also tolerate a semi-shaded aspect. Good drainage is needed for optimum growth. Hardy to light frosts. Tip pruning from an early stage should be beneficial. Propagate from seed, which germinates readily, or from cuttings of firm young growth.

Gastrolobium crenulatum Turcz. is a synonym.

N. pyramidalis (T. Moore) Crisp is allied but has larger flowers and flowerheads, as well as broadly ovate to oblong leaves.

Nemcia cuneata (Benth.) Domin =
   N. dilatata (Benth.) Crisp

Nemcia cuneata (Benth.) Domin var. **drummondii**
   (Meisn.) Domin = *N. retusa* (Lindley) Crisp

Nemcia dilatata (Benth.) Crisp
(enlarged or widened)
WA
0.5–1.5 m × 1–2 m   July–Nov

Dwarf to small **shrub**, with silky-hairy young growth; **branches** spreading to erect, stiff; **branchlets** angular when young; **leaves** 2–5.5 cm × 0.6–2.5 cm, narrowly cuneate, spathulate to triangular, mostly opposite or in whorls of 3–4, narrowed to the base, somewhat folded lengthwise, leathery, glabrous, sometimes faintly silky-hairy below, very faint reticulate venation, apex truncate; **flowers** pea-shaped, to about 1.2 cm across, yellow to orange and dark red to purple, in crowded axillary or terminal clusters or racemes, often profuse and very conspicuous; **standard** yellow to orange; **wings** and **keel** dark red to purple; **pods** about 1 cm long, ovoid.

A very floriferous species from the Avon, Darling and Eyre Districts where it is usually associated with gravelly granitic or lateritic soils. Introduced into England in 1840 as *Oxylobium cuneatum* but it is rarely grown in Australia. It is well suited to temperate and possibly semi-arid regions with well-drained soils and a sunny or semi-shaded site. Hardy to moderate frosts and extended dry periods. Tip pruning promotes bushy growth. Propagate from seed, which usually begins to germinate 25–55 days after sowing, or from cuttings of firm young growth.

Nemcia cuneata (Benth.) Domin, *Oxylobium cuneatum* Benth. and *Oxylobium dilatatum* Benth. are synonyms.

N. emarginata is allied but differs in having deeply notched leaves.

Nemcia dorrienii Domin =
   N. emarginata (S. Moore) Crisp

Nemcia effusa Crisp & Mollemans
(loosely spreading)
WA
0.6–1.2 m × 0.5–1.3 m   July–Aug

Dwarf to small **shrub** with greyish-hairy young growth; **branches** spreading to ascending, stiff; **branchlets** spreading, greyish-hairy; **leaves** 1–2.5 cm × about 0.4 cm, narrowly oblong-elliptic, in whorls of 3, spreading, with very short hairy stalk, deep green, reticulate venation prominent, apex pointed and slightly

recurved; **stipules** slender, persistent; **flowers** pea-shaped, to about 1 cm across, in condensed axillary racemes; **standard** apricot with reddish-maroon markings on front and reddish-maroon back; **wings** apricot with reddish-maroon; **keel** maroon; **calyx** silky-hairy; **pods** not known.

This poorly known species was described in 1993. It is known only from one locality in the western Roe District and is thought to be rare. It inhabits gravelly soil among mallee and shrub vegetation. Warrants cultivating as part of a conservation strategy. Plants should adapt to a range of freely draining acidic soils in semi-arid and warm temperate regions. A sunny or semi-shaded site will be suitable. Plants tolerate moderate frosts and extended dry periods. Tip pruning from an early stage will promote bushy growth. Propagate from seed or cuttings.

N. stipularis is similar but lacks spreading leaves. *N. punctata* has shorter leaves and insignificant stipules.

Nemcia emarginata (S. Moore) Crisp
(notched at apex)
WA
1–2 m × 1–2 m   Sept–Nov

Small **shrub** with greyish-hairy young growth; **branches** becoming glabrous, terete; **leaves** 0.8–2 cm × 0.6–1.2 cm, obcordate-oblong, in whorls of 3 or opposite, short-stalked, leathery, greyish-green, becoming glabrous above, hairy below, margins revolute, apex prominently notched; **flowers** pea-shaped, about 1 cm across, yellow and red, in crowded elongated heads, in upper axils and terminal, profuse and very conspicuous; **calyx** with yellow-brown hairs; **pods** small.

This showy species is not well known and deserves to be planted more widely. It occurs in the Avon, central Darling and Eyre Districts where it grows in sandy or gravelly soils. It can be a somewhat open shrub and should adapt to a range of soils which drain at least moderately well. Plants will need a sunny or semi-shaded site. Hardy to moderate frosts and extended dry periods. Propagate from seed, which germinates readily, or from cuttings of firm young growth.

Oxylobium emarginatum S. Moore and *O. emarginatum* var. *major* S. Moore are synonyms.

Nemcia epacridoides (Meisn.) Crisp
(similar to the genus *Epacris*)
WA
0.3–1.5 m × 0.5–1.5 m   Aug–Sept

Dwarf to small **shrub**, with hairy young growth; **branches** slender, twiggy, hairy; **leaves** 1–2.5 cm × 0.7–1 cm, ovate, usually in whorls of 3, sessile, spreading, many, crowded, rigid, glossy, apex ending in a pungent point; **stipules** absent; **flowers** pea-shaped, about 1 cm across, in loose axillary clusters; **standard** and **wings** yellow and purple or reddish-brown; **keel** purple or reddish-brown; **calyx** faintly hairy; **pods** about 0.7 cm long, ovoid, hairy.

A rare species from the northern Darling District where it is usually found on lateritic soils. Rarely cultivated, it needs to be grown for conservation purposes. Good drainage and plenty of sunshine are important for optimum growth, but plants should tolerate semi-shaded sites. Plants may suffer damage from moderate frosts but are hardy to extended dry periods. Tip pruning from an early stage promotes lateral growth. Propagate from seed, which usually begins to germinate 16–25 days after sowing, or from cuttings of firm young growth.

Previously known as *Gastrolobium epacridoides* Meisn.

N. acuta is allied but differs in having stipules, elliptical leaves and a densely hairy calyx.

Nemcia hookeri (Meisn.r) Crisp
(after Sir William J. Hooker, 19th-century Director of Royal Botanic Gardens, Kew, England)
WA
0.3–0.6 m × 0.3–1 m   July–Nov

Dwarf **shrub** with densely hairy young growth; **branches** spreading to erect, many; **branchlets** densely hairy; **leaves** 0.8–2 cm × 0.3–1 cm, obovate or elliptical, opposite, sessile, sparsely hairy to glabrous above, hairy below, margins wavy, apex recurved and pungent-pointed; **stipules** bristle-like; **flowers** pea-shaped, about 0.8 cm across, yellow and red, in small axillary clusters, often profuse and very conspicuous; **standard** yellow; **wings** and **keel** red; **calyx** hairy; **pods** to 0.6 cm long, ovoid, hairy.

A somewhat variable species from the Avon, central Darling, Irwin and Roe Districts where it occurs in a wide range of soil types. Worth growing in temperate and semi-arid areas, in gardens and containers. Probably prefers a sunny aspect but will also adapt to a semi-shaded site. Soils should be acidic to neutral and moderately well drained. Pruning from an early stage promotes bushy growth. Plants are hardy to moderate frosts and extended dry periods. Propagate from seed, which germinates readily, or from cuttings of firm young growth.

Gastrolobium hookeri (Meisn.), *Gastrolobium stewardii* S. Moore and *Gastrolobium tricuspidatum* var. *subinerme* Meisn. are synonyms.

Nemcia ilicifolia (Meisn.) Crisp
(leaves similar to the genus *Ilex*)
WA
1.5–4 m × 1–3 m   Aug–Oct

Small to medium **shrub** with densely hairy young growth; **branches** ascending to erect, thick; **branchlets** densely hairy; **leaves** 1.8–6.5 cm × 1.6–3.3 cm, somewhat triangular to broadly ovate or elliptical, opposite or in whorls of 3, sessile to short-stalked, leathery, flat to folded lengthwise, glabrous and dark green above, paler below, entire or sometimes lobed, upper margins pungent-toothed, faint reticulate venation; **flowers** pea-shaped, to about 1 cm across, yellow to orange with red to purple, in crowded axillary clusters, often profuse and very conspicuous; **standard** and **wings** yellow to orange; **keel** red to purple; **calyx** silky-hairy; **pods** to about 0.7 cm × 0.4 cm, ovoid to nearly globular, hairy.

A floriferous, poorly known species with interesting, distinctive foliage. It occurs in the Avon, northern Darling and southern Irwin Districts and usually is found on sandy or lateritic soils. Deserves wider cultivation and should adapt to a range of acidic to slightly alkaline soils in temperate and semi-arid regions.

It tolerates a sunny or semi-shaded aspect. Plants are hardy to moderate frosts and extended dry periods. Useful for foot traffic control, as well as for ornamental applications. Pruning from an early stage promotes a bushy framework. Propagate from seed, which germinates readily, or from cuttings of firm young growth.

Gastrolobium ilicifolium Meisn., along with its var. *lobatum* Benth. and *Gastrolobium verticillatum* Meisn. are synonyms.

Nemcia leakeana (Drumm.) Crisp
(after Robert B. Leake, 19–20th-century pharmaceutical chemist)
WA   Mountain Pea
0.5–2 m × 0.5–1.5 m   Aug–Dec

Dwarf to small **shrub** with softly hairy young growth; **branches** ascending to erect, thick; **branchlets** angled, hairy; **leaves** 3.5–6 cm × 1–2.8 cm, elliptical to broadly elliptical, opposite, decussate, stalked, spreading to ascending, stiff, dark green, glabrous above, glabrous to silky-hairy below, faint venation, apex blunt to prominently notched; **flowers** pea-shaped, 2.5–2.2 cm long, bright to deep red or rarely yellow to orange with red, sometimes pendant, in crowded axillary clusters, often profuse and extremely conspicuous; **standard** and **wings** red or, rarely, yellow to orange with red; **keel** red; **calyx** densely silky-hairy; **pods** 0.9 × 0.4 cm long, silky-hairy.

An outstanding species in flower, which is quickly noticed in its habitat of the higher rocky slopes in the Stirling Range. Occasionally cultivated, it can become upright and leggy, but pruning from an early stage can promote bushy growth. This spectacular pea plant requires very good drainage but appreciates moist soils. Should do best in a protected sunny or semi-shaded site in temperate regions. Plants are hardy to moderate frosts but dislike extended dry periods. Propagate from seed, which germinates readily, or from cuttings of firm young growth.

Nemcia atropurpurea (Turcz.) Domin, *Nemcia luteifolia* Domin, *Gastrolobium leakeanum* Drumm. and *Oxylobium atropurpureum* Turcz. are synonyms.

Nemcia rubra Crisp is closely allied but has narrowly oblong elliptical leaves to 10 cm long which are usually in 3s and have a cordate base and a scarcely notched apex. The flowers are also slightly larger.

Nemcia lehmannii (Meisn.) Crisp
(after Johann G. C. Lehmann, 19th-century German botanist)
WA   Cranbrook Pea
0.5–1.5 m × 0.5–1.5 m   Sept–Oct

Dwarf to small **shrub** with softly hairy young growth; **branches** ascending to erect; **branchlets** hairy; **leaves** 1.5–5 cm × 0.7–1.4 cm, oblong, short-stalked, ascending to erect, rounded at base, leathery, glabrous above, densely hairy below, margins thickened, apex very blunt or notched with a small point; **flowers** pea-shaped, about 0.8 cm across, yellow and purple, in axillary clusters; **standard** yellow; **wings** and **keel** purple; **calyx** silky-hairy; **pods** longer than calyx.

This apparently extinct species is recorded from the southern Avon and south-eastern Darling Districts. The most recent collection was from and area between

Nemcia leakeana   D. L. Jones

Cranbrook and the Stirling Range in 1918. It may be rediscovered and, if so, it should be introduced into cultivation as part of a conservation strategy. Should require good drainage and a semi-shaded or sunny site. Propagate from seed, which germinates readily, or from cuttings of firm young growth.

Previously known as *Gastrolobium lehmannii* Meisn.

Nemcia luteifolia Domin = *N. leakeana* (Drumm.) Crisp

Nemcia obovata (Benth.) Crisp
(broadest above the middle)
WA   Boat-leaved Poison
0.2–1 m × 0.6–1.5 m   Aug–Oct

Dwarf **shrub** with woolly-hairy young growth; **branches** slender, spreading; **branchlets** hairy; **leaves** to 2.7 cm × 1.2 cm, obovate to rhomboid, scattered, spreading, somewhat folded lengthwise, leathery, glabrous, reticulate venation, apex tapering to a fine pungent point; **flowers** pea-shaped, about 1 cm across, yellow to orange and red, in somewhat loose axillary clusters, often profuse and conspicuous; **standard** yellow to orange; **wings** and **keel** bright to dark red; **calyx** silky-hairy; **pods** to 0.4 cm long.

This dwarf species has potential for wider cultivation in temperate and semi-arid regions as a garden or container plant. It occurs in the Avon, Eyre, Irwin and Roe Districts where it grows in sandy soils which are often incorporated with clay loam and lateritic gravels. Moderate drainage is essential and a sunny or semi-shaded site should be suitable. It is hardy to moderate frosts and should withstand extended dry periods and waterlogging for limited periods. Tip pruning from planting promotes bushy growth. Propagate this species from seed, which germinates readily, or from cuttings of firm young growth. Although known as Boat-leaved Poison it is not toxic.

Previously known as *Gastrolobium obovatum* Benth. and its var. *verticillatum* Meisn. is also a synonym.

N. pauciflora is allied but has flat leaves and 1–3 flowers per axil.

Nemcia pauciflora (C. A. Gardner) Crisp
(few-flowered)
WA
0.5–1 m × 0.5–1 m   Sept–Oct

Dwarf **shrub** with hairy young growth; **branches** ascending to erect; **branchlets** spreading; **leaves** to about 2 cm × 0.8–1 cm, broadly oblanceolate to obovate, opposite, short-stalked, stiff, flattish, glaucous, glabrous, prominent reticulated venation, apex pointed; **stipules** bristle-like; **flowers** pea-shaped, about 1 cm across, yellow and red, 1–3 per axil, sometimes profuse; **standard** yellow with red veins; **wings** and **keel** red; **calyx** silky-hairy; **pods** not seen.

This glaucous-foliaged species is from the northern Avon and northern Irwin Districts where it grows in sandy, gravelly and clay loam soils. Well suited for cultivation in temperate and semi-arid regions, it should adapt to a range of soils but will need a warm to hot aspect. Hardy to moderate frosts and extended dry periods. Propagate from seed, which germinates readily, or from cuttings of firm young growth.

Previously known as *Gastrolobium pauciflorum* C. A. Gardner.

Nemcia plicata (Turcz.) Crisp
(pleated or folded)
WA
0.3–2 m × 0.3–1.5 m   Sept–Oct

Dwarf to small **shrub** with silky-hairy young growth; **branches** rigid; **branchlets** covered in silky hairs when young; **leaves** about 2.5 cm long, opposite, obovate-cuneate, folded lengthwise, concave above, often glaucous, glabrous, leathery, apex usually blunt with a small recurved point; **stipules** long; **flowers** pea-shaped, about 1.2 cm across, in axillary clusters; **standard** yellow; **wings** yellow with red; **keel** reddish; **calyx** about 0.6 cm long, with long hairs; **pods** about 0.8 cm long, hairy.

N. plicata is distributed in disjunct populations in the Avon and Roe Districts of south-western WA where it usually grows in sandy soils containing lateritic gravel. Evidently not cultivated but its sizeable flowers and often glaucous foliage make it an attractive species. Needs good drainage and plenty of sunshine. Should tolerate extended dry periods and light to medium frosts. Propagate from seed or cuttings of firm young growth.

Previously known as *Gastrolobium plicatum* Turcz.

Nemcia pulchella (Turcz.) Crisp
(beautiful)
WA
1–2 m × 1–2 m   Aug–Sept

Small **shrub**; **branches** slender, with silky hairs; **leaves** to 2.5 cm long, oblong, rounded at base, leathery, glabrous, apex blunt or notched, with a short recurved point; **flowers** pea-shaped, about 0.5 cm across, in upper axillary clusters of terminal heads, not usually exceeding the leaves; **standard** yellow; **wings** red; **keel** reddish; **calyx** to 0.6 cm long, with silky hairs; **pods** small.

This species is thought to be confined to the northern Darling and southern Irwin Districts. It is recorded as growing in gravelly or clayey soils. Most likely not cultivated but worth trying in temperate and semi-arid regions. Plants would need well-drained soils and a sunny situation. They may be susceptible to frost damage but will tolerate extended dry periods. Propagate from seed or from cuttings of firm young growth.

Previously known as *Gastrolobium pulchellum* Turcz.

Nemcia punctata (Turcz.) Crisp
(dotted)
WA
0.3–1 m × 0.5–1.5 m   July–Oct

Dwarf **shrub** with silky-hairy young growth; **branches** ascending to erect, many; **leaves** 1–3.5 cm × 0.3–1 cm, oblong-elliptical, falcate to incurved, in whorls of 3 or opposite, mainly sessile, ascending to erect, somewhat crowded, glabrous, appearing dotted due to very close reticulate venation, apex pointed and strongly recurved; **flowers** pea-shaped, to about 1.2 cm across, yellow and purplish-red, in upper axillary clusters or terminal racemes, often profuse and very conspicuous; **standard** and **wings** yellow; **keel** purplish-red; **pods** to about 0.7 cm long.

This dwarf shrub is from the Eyre and Roe Districts of south-western WA where it grows in sandy and gravelly soils. A sunny or warm semi-shaded aspect with well-drained soils should be suitable. Plants are hardy to moderate frosts and extended dry periods. Propagate from seed, which germinates readily, or from cuttings of firm young growth.

Previously known as *Eutaxia punctata* Turcz. and *Gastrolobium reticulatum* (Meisn.) Benth. var. *recurvum* E. Pritzel.

N. carinata Crisp is allied, but differs in having ovate leaves with margins flat or only slightly recurved, and more open, reticulate venation.

Nemcia pyramidalis (T. Moore) Crisp
(in the shape of a pyramid; referring to the growth habit)
WA
1.5–2 m × 1–1.5 m   Sept–Nov

Small erect **shrub**; **branches** and **branchlets** densely hairy; **leaves** 2.5–6.5 cm × 3–5 cm, mostly in whorls of 3, oval to rotund, stalked, spreading initially, densely hairy, becoming glabrous above; **stipules** about 2 cm long, bristly, brown ageing to black, initially hairy, deciduous; **flowers** pea-shaped, about 1.5 cm across, in dense, globular, axillary racemes, profuse; **standard** orange-yellow, reddish-brown at base; **wings** deep yellow; **keel** maroon; **calyx** with shaggy hairs; **pods** hairy.

N. pyramidalis is confined to north-east of Albany, where it occurs on the eastern part of the Stirling Range or near the coast, usually in gravelly soils. This species was described as a very handsome evergreen shrub when first introduced to cultivation in England, as *Gastrolobium pyramidale*, during the early 1850s. Initially it was grown as a container plant in glasshouses but was found to be semi-hardy and placed outside during summer. At present it is rare in cultivation. Plants require very well-drained sandy or loamy soils. They need plenty of sunshine but will tolerate a semi-shaded situation. May suffer damage from heavy frosts. Pruning young plants and older plants after flowering is recommended if dense growth is desired. Grows well in containers. Propagate from seed or cuttings of firm young growth.

Gastrolobium polycephalum Turcz. and *Gastrolobium pyramidale* T. Moore are synonyms.

Nemcia reticulata (Meisn.) Domin
(netted veins))
WA
0.3–1 m × 0.5–1.5 m   July–Oct

Dwarf **shrub** with silky-hairy young growth; **branches** ascending to erect, many; **leaves** 1–3.5 cm × 0.3–1.5 cm, mostly opposite or in whorls of 3–4, ovate-oblong to obovate, somewhat crowded, ascending to erect, becoming glabrous above and below, margins often crenulate, venation reticulate, apex blunt or notched, with or without a small point; **flowers** pea-shaped to 1.2 cm across, yellow to orange with red or reddish-brown, in axillary clusters or terminal racemes, often profuse and very conspicuous; **standard** yellow to orange; **wings** and **keel** red to reddish-brown; **pods** to about 0.7 cm long, ovoid, hairy.

N. reticulata occurs in the north-western Avon, northern Darling and Irwin Districts where it grows in acidic or alkaline sands and sandy clay soils. This floriferous species should do well in temperate or semi-arid regions. A sunny or semi-shaded site with moderately good drainage is desirable. Hardy to moderate frosts and tolerates dry periods. Pruning may promote bushy growth. Propagate from seed, which germinates readily, or from cuttings of firm young growth.

Callistachys oxyloboides Meisn., *C. reticulatum* Meisn., *Oxylobium nervosum* Meisn. and *O. reticulatum* Meisn. are synonyms.

Nemcia retusa (Lindl.) Crisp
(notched at apex)
WA
0.5–1 m × 0.5–1 m   Aug–Nov

Dwarf **shrub** with hairy young growth; **branches** ascending to erect; **leaves** about 1.2 × 0.5 cm, nearly linear to ovate-oblong, opposite or in whorls of 3 or 4, stalked, somewhat flexible, glabrous above, silky-hairy below, margins usually recurved, apex very blunt and often notched; **flowers** pea-shaped, about 1.2 cm across, yellow and red, in crowded terminal clusters or racemes, sometimes also axillary, often profuse and conspicuous; **standard** yellow with red; **wings** and **keel** yellow; **pods** about 0.9 cm long, ovoid, hairy.

Plants are found in the Avon, Darling and western Eyre Districts. They inhabit a range of sandy and gravelly soils. A sunny or warm semi-shaded site with relatively good drainage should be suitable. Evidently rarely cultivated but was introduced into England in 1830 as *Gastrolobium retusum*, where it was regarded as a 'pretty greenhouse plant'. Hardy to moderate frosts and dry periods. Best suited to temperate and semi-arid regions. Propagate from seed, which germinates readily, or from cuttings of firm young growth which strike readily.

Gastrolobium retusum Lindl., *Nemcia cuneata* (Benth.) Domin var. *drummondii* (Meisn.) Domin, *Oxylobium cuneatum* Benth. var. *emarginatum* Benth, *O. drummondii* Meisn., *O. melinocaule* E. Pritzel and *O. virgatum* Benth. are synonyms.

Nemcia rubra Crisp
(red)
WA
0.6–2 m × 0.5–1.5 m   Aug–Dec

Dwarf to small **shrub** with softly hairy young growth; **branches** ascending to erect, thick; **branchlets** angled, hairy; **leaves** 4–10 cm × 1–1.5 cm, narrowly oblong-elliptical, often in whorls of 3, stalked, spreading to ascending, stiff, dark green, glabrous above, silky-hairy below, apex blunt or faintly notched; **flowers** pea-shaped to 2.5 cm long, bright to deep red, in crowded axillary clusters, often profuse and very conspicuous; **calyx** densely silky-hairy; **pods** not seen.

A very ornamental species when in full flower. It is from the Stirling Range where it grows on the rocky slopes. *N. rubra* has similar requirements to the closely allied *N. leakeana*, which differs in its broader and notched, opposite leaves and slightly smaller flowers. Propagate from seed, which germinates readily, or from cuttings of firm young growth.

Previously known as *Nemcia atropurpurea* (Turcz.) Domin var. *minorifolia* Domin.

Nemcia spathulata (Benth.) Crisp
(spoon-shaped; referring to the leaves)
WA
0.2–1 m × 0.3–1 m   Aug–Nov

Dwarf, often rounded **shrub**; **branches** twiggy, erect; **branchlets** slightly hairy; **leaves** to 1.5–3 cm × up to about 1 cm, spathulate, crowded, folded lengthwise, concave above, glabrous, leathery, venation reticulate, apex blunt or notched, with a short recurved point; **flowers** pea-shaped, small, about 0.7 cm across, in axillary clusters or racemes shorter than the leaves, often profuse and conspicuous; **standard** yellow to orange with dark red base and reverse; **wings** orange; **keel** deep red; **calyx** about 0.5 cm long, with silky hairs; **pods** small.

This dwarf species occurs in the Avon, northern Darling and Irwin Districts where it grows in gravelly soils which can be clayey. Evidently not well known in cultivation but was grown in greenhouses in England during the mid 19th-century. Plants should adapt to warm or cool temperate regions. They need fairly good drainage and plenty of sunshine. May suffer damage from heavy frosts. Propagate from seed, which usually begins to germinate 12–20 days after sowing, or from cuttings of firm young growth.

Gastrolobium spathulatum Benth. and its var. *latifolium* Benth. are synonyms, but further studies may determine the variety to be distinct.

Nemcia stipularis (Meisn.) Crisp
(with stipules; referring to the long stipules at the base of leaves)
WA
0.6–1 m × 0.6–1 m   Oct–Nov

Dwarf **shrub**; young growth hairy; **branches** erect; **branchlets** hairy; **leaves** 2.5–4 cm × up to 0.2 cm, opposite or irregularly whorled, narrow-linear, ascending to erect, crowded, rigid, becoming glabrous, margins revolute, venation prominent below, apex ending in a small point; **stipules** to 1.2 cm long, bristly, black; **flowers** pea-shaped, about 0.6 cm across, in axillary clusters; **standard** mostly yellow to orange; **wings** yellow to orange; **keel** red-brown to purplish; **calyx** about 0.6 cm long, with silky hairs; **pods** about 0.8 cm long, with brown hairs.

This species is now considered very rare in nature. It is restricted to a small area in the southern Avon District of south-western WA. Plants usually occur in yellow sand over laterite. They need to be introduced into botanic and private gardens as a conservation strategy. *N. stipularis* requires well-drained soils and partial or full sun. Should be tolerant of light to medium frosts. Propagate from seed or cuttings of firm young growth.

Previously known as *Gastrolobium stipulare* Meisn.

Nemcia tricuspidata (Meisn.) Crisp
(three-pointed)
WA
0.7–1.3 m × 0.7–1.5 m   Sept–Oct

Dwarf to small **shrub**; **branches** rigid, erect to spreading; **branchlets** with soft hairs; **leaves** 2–4 cm long, opposite or in whorls of 3, wedge-shaped, folded lengthwise and concave above, leathery, initially hairy, becoming glabrous, apex truncate or with 3 short lobes ending in pungent points; **flowers** pea-shaped, about 0.6 cm across, yellow and red, in axillary clusters; **standard** yellow with red to reddish-brown; **wings** yellow; **keel** reddish-brown to brownish-purple; **calyx** about 0.7 cm long, hairy; **pods** enclosed in the calyx.

N. tricuspidata is found among sandheath vegetation in the Avon, Darling and western Roe Districts, and usually grows in sandy or gravelly soils. It is not well known and is not thought to be in cultivation. Plants need well-drained soils and should grow best in sunny positions. Propagate from seed or cuttings of firm young growth.

Nemcia truncata (Benth.) Crisp
(cut off abruptly; referring to the leaf apex)
WA
prostrate × 0.8–1.5 m   Sept–Nov

Spreading dwarf **shrub**; new growth hairy; **branches** many, slender, often rooting at nodes; **branchlets** brownish, hairy; **leaves** 1–1.8 cm × 0.7–1.2 cm, opposite or irregular, somewhat orbicular, dark green, concave and glabrous above, hairy below, margins wavy, apex truncate, with fine central point; **flowers** pea-shaped, about 0.8 cm across, profuse, in axillary clusters; **standard** yellow; **wings** yellow with red; **keel** brownish-red; **calyx** about 0.3 cm long, slightly hairy; **pods** to 0.7 cm long.

This ground-covering species occurs naturally in the southern Avon, Darling and western Roe Districts. It grows in heavy clay soils which may be wet for extended periods. A showy plant for cultivation although the massed floral display may only last for a few weeks. Hardy under a wide range of conditions in temperate zones but should also adapt to subtropical conditions. Needs relatively well-drained soils. Grows well in semi-shade to full sun. It is hardy to most frosts and extended dry periods. Suited to container cultivation including hanging baskets. Also useful for soil erosion control in small areas. Propagate from seed, cuttings or by division of layered stems. Transplanting of layered stems directly in a garden situation has proved successful during late winter and early spring.

Previously known as *Gastrolobium truncatum* Benth. and *G. crispifolium* Domin is also a synonym.

Nemcia vestita Domin
(covered with hairs)
WA
1–2 m × 1–2 m   Sept–Oct

Small **shrub** with densely hairy young growth; **branches** ascending to erect, hairy; **branchlets** slender, hairy; **leaves** 2–7 cm × 1.2–3.5 cm, broad ovate to oblong, usually in whorls of 3, ascending to erect, sometimes spreading, darker and glabrous above, paler and hairy below, margins crenulate and slightly recurved, fine reticulate venation, apex blunt or notched, usually with a very small mucro; **flowers** pea-shaped, to about 1 cm across, yellow with deep orange, in terminal and axillary racemes, can be profuse and conspicuous; **standard** yellow to orange; **wings** and **keel** bright red; **calyx** densely silky-hairy; **pods** about 1.2 × 0.5 cm, enclosed by calyx, densely hairy.

An attractive species from the eastern Eyre District where it is rare and usually occurs in mallee heath. Evidently not cultivated it should be introduced as part of a conservation strategy. Plants require well-drained soils with a sunny or semi-shaded aspect. They can become leggy and pruning from an early stage and after flowering will promote bushy growth. Hardy to light frosts but may not appreciate extended dry periods. Propagate from seed, which germinates readily, or from cuttings of firm young growth.

N. coriacea differs in having smaller leaves.

NEOALSOMITRA Hutch.
(from the Greek *neos*, new; and the genus *Alsomitra*, which is named for the resemblance of its fruits to a mitre)
Cucurbitaceae

Climbing perennial **herbs**, often with a woody root-stock, dioecious or sometimes monoecious, glabrous or pubescent; **tendrils** simple or 2-branched; **leaves** simple and palmately 3–5-lobed or compound and 3–5-foliolate; **male flowers** small, in axillary panicles; **calyx** 5-lobed; **corolla** deeply 5-lobed; **stamens** 5; **female flowers** in smaller panicles or racemes than males or, rarely, one female among a cluster of males; **styles** 3; **fruit** a capsule, somewhat 3-angled or cylindrical, dehiscent, 3-valved.

A genus of about 15 species which occurs from south-eastern Malesia to north-eastern Australia, where there are 4 species.

Although not highly ornamental, plants will provide variety and texture in a garden with their growth habit and foliage. They are rarely encountered in cultivation. Once the woody rootstock becomes well developed, plants may respond successfully to hard pruning each year. See species descriptions for other requirements.

Propagate from fresh seed, which apparently does not retain viability for long. Cuttings of firm young growth are worth trying, as other cucurbits usually produce roots readily.

Neoalsomitra capricornica (F. Muell.) Hutch.
(after the Tropic of Capricorn)
Qld
Climber   Feb–March; possibly also sporadic

Slender or bushy **climber**; **stems** to 3 m or longer; with simple or branched tendrils; **leaves** 3–7 cm × 4–9 cm, broadly ovate in outline, on long stalks, deeply 5-lobed, and with coarsely toothed margins, thin-textured, faintly hairy; **male flowers** to about 0.5 cm across, cream to pale yellow, in few-flowered sparsely branched racemes to about 1.5 cm long; **female flowers** to about 0.2 cm across, cream to pale yellow, solitary on long stalk; **fruit** 1.8–2.5 cm long, cylindrical, yellowish, glandular-hairy, ripe mainly July–Sept.

Plants are found in north-eastern Qld and possibly extend to New Guinea. This species occurs in mountainous rainforest and along the coast. Useful as a background climber due to its moderate ornamental appeal. As it is frost sensitive, plants are best suited to tropical and subtropical regions and require well-drained loam soils. Judicious pruning should rejuvenate plants. Propagate from seed, or from cuttings of firm young growth.

Neoalsomitra suberosa (Bailey) Hutch. =
        Nothoalsomitra suberosa (Bailey) I. Telford

Neoalsomitra trifoliolata (F. Muell.) Hutch.
(three leaflets)
Qld
Climber   March–April; possibly also sporadic

Slender or bushy **climber**; **stems** up to 3 m or more long, with simple or 2-branched tendrils; **leaves** trifoliolate, with stalk to 3 cm long; **leaflets** 5–12 cm × 3–7 cm,

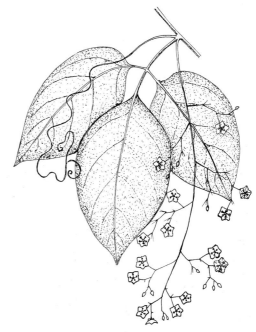

Neoalsomitra trifoliolata × .5

15

Nemcia rubra

W. R. Elliott

lanceolate to ovate, stalked, dark green, thin-textured, sparsely hairy or becoming glabrous, margins entire, apex pointed; **male flowers** to about 0.5 cm across, pale yellow, in much-branched broad, pendant panicles to about 40 cm long; **female flowers** not seen; **fruit** 3–5 cm long, cylindrical to bell-shaped, mainly ripe Aug–Sept.

A climber with decorative foliage, it occurs in the rainforests of the mountains and coastal regions in north-eastern Qld and New Guinea. Rarely seen in cultivation, it requires well-drained loamy soils in filtered sun. May need protection from hot sun when young. Plants are damaged by frosts and are best grown in tropical and subtropical areas. Should respond well to supplementary watering and mulching in extended dry periods. Propagate from seed, or from cuttings of firm young growth.

NEOASTELIA J. B. Williams
(from the Greek *neo*, new; and *Astelia*; referring to similarity to the genus *Astelia*)
Asteliaceae (alt. *Liliaceae*)
A monotypic genus endemic in north-eastern NSW.

Neoastelia spectabilis J. B. Williams
(remarkably spectacular)
NSW
0.5 - 1.5 m × 0.6 - 2.5 m   Nov–Dec

Tufted, rhizomatous, perennial **herb**; **leaves** 0.6–1.65 m × 0.3–0.6 cm, linear-lanceolate, somewhat lax, green above, silvery white and keeled below, parallel veined; **inflorescence** a panicle, 24–70 cm long, much-branched with many flowers, open arrangement, on thick stalk to 40 cm long, with usually a long leaf near top, panicle branches subtended by spathes;

flowers 1.5–1.8 cm across, whitish, profuse, somewhat membranous; **fruit** a somewhat globular berry, about 1.25 cm across, pale green, grooved at the top.

A very decorative lily which was described in 1987. It is from the New England region in north-eastern NSW where it is known from only a few populations. It occurs in rocky crevices beside waterfalls and is continually exposed to water spray and moist conditions. Likely to be highly suitable for cultivation in containers and in gardens. Plants require very good drainage in moisture-retentive soils and a sheltered site, although they are able to tolerate some sunshine. Supplementary watering during dry periods may be essential. Hardy to moderate frosts. Propagate from seed, which may respond to pre-sowing treatment, (See Volume 1, page 205), or from division of clumps.

NEOBASSIA A. J. Scott
(from the Greek *neo*, new; and the allied genus *Bassia*, from which it has been segregated)
Chenopodiaceae
Small **shrubs**; **branches** brittle, hairy, often reddish; leaves alternate, narrow, sessile; **flowers** bisexual, small, solitary, axillary; **perianth** 5-lobed, fused nearly to apex; **stamens** 5; **fruiting perianth** cylindrical, with 5 spines from base of lobes.

A small endemic genus of 2 species. It is closely allied to *Dissocarpus*, which differs in having 2 or more flowers per leaf axil.

Propagation is from seed, which does not require pre-sowing treatment, or from cuttings of firm but not too old growth.

Neobassia astrocarpa (F. Muell.) A. J. Scott
(star-shaped fruits)
WA, NT
0.6–1.5 m × 1–2 m   Nov–Feb; possibly also sporadic

Dwarf to small **shrub**; **stems** spreading to ascending, silky-hairy; **leaves** 0.5–1.5 cm × about 0.15 cm, semi-terete, somewhat succulent, silky-hairy, often S-shaped or apex recurved; **flowers** solitary, axillary; **perianth** shortly lobed, silky-hairy; fruit about 0.3 cm × 0.1 cm, cylindrical, crustaceous to thinly woody, spines to three quarters length of tube.

This species is usually found in saline soils. It occurs along the WA coast from Shark Bay to Broome and extends eastward inland to western NT. It has limited potential but may be useful for reclaiming saline soils in subtropical and semi-arid regions within its natural range. May be worth trying in gardens or containers too. Hardy to light frosts. Propagate from seed, or from cuttings of firm, young growth.

Previously known as *Bassia astrocarpa* F. Muell. or *Sclerolaena astrocarpa* (F. Muell.) Domin.

Neobassia proceriflora (F. Muell.) A. J. Scott
(very tall flowers)
Qld, NSW, NT   Soda Bush
0.3–0.5 m × 0.5–0.6 m   Dec–Feb

Dwarf annual or short-lived perennial **subshrub**; **stems** ascending to erect, often reddish, hairy; **leaves** 1–2 cm × about 0.15 cm, linear to terete, somewhat succulent, with long hairs or nearly glabrous; **flowers** solitary, axillary; **fruit** about 0.8 cm × 0.25 cm, cylindrical, glabrous to densely hairy, becoming spongy, spines short and erect.

A wide-ranging but uncommon species in semi-arid regions. Usually occurs on saline soils. It has very limited potential for general cultivation, although may do well in containers. Needs plenty of sunshine. Could be useful for reclamation of saline soils within its natural range. Edible to stock but is recorded as poison-ous for very hungry stock or if other feed is not readily available. Propagate from seed, or from cuttings of firm young growth.

Previously known as *Threlkeldia proceriflora* F. Muell.

NEOBYRNESIA J. A. Armstr.

(from the Greek *neo*, new; and after Norman B. Byrnes, contemporary Australian botanist and original collector)

Rutaceae

A monotypic genus endemic in Northern Australia.

Neobyrnesia suberosa J. A. Armstr.

(corky)

NT

0.2–0.5 m × 0.2–0.5 m   Feb–July;possibly sporadic

Dwarf **shrub**, young growth reddish-brown, hairy; **bark** corky, creamy-white; **branches** slender, upright to pendulous, with whitish to creamy hairs; **branchlets** covered in reddish-brown hairs; **leaves** 1.3–4 cm × 0.6–1.6 cm, oblong-lanceolate to ovate, opposite, decussate, prominently stalked, entire, dark green and oil-dotted above, white to cream with reddish brown hairs below, margin sometimes recurved, apex blunt with small mucro; **flowers** about 0.25 cm across, white, in 3–7-flowered axillary cymes, somewhat inconspicuous; **petals** 4; **stamens** 4; **calyx** 4-lobed, with cream and reddish-brown hairs; **fruit** of 1–4 basally joined cocci, reddish-brown.

N. suberosa is known only from skeletal soils in rocky crevices of sandstone cliff faces in the East Alligator River area. Evidently not cultivated, it is suitable for tropical regions and possibly may adapt to subtropical areas. Plants need excellent drainage. Although not highly ornamental it is deserving of cultivation. It may make an interesting container plant because of its hairy foliage and stems. Propagate from seed, which may contain inhibitors and require pre-sowing treatment, see Volume One, page 205. Cuttings of firm young growth would be worth trying.

NEOFABRICIA Joy Thomps.

(from the Greek *neo*, new, and the genus *Fabricia*)

Myrtaceae

Shrubs or small **trees**; **bark** fibrous; **leaves** alternate, spirally arranged, simple, entire, with prominent oil dots; **flowers** bisexual, open-petalled, tea-tree-like, sessile or very short-stalked, usually solitary, axillary or appearing terminal; **petals** 5, white, cream or yellow, glabrous; **stamens** many, in several irregular rows; **fruit** a many-celled woody capsule; **seeds** flattened and winged.

A small, endemic, tropical genus of 3 species which is confined to the Cape York Peninsula in northeastern Qld. This genus is closely allied to *Leptospermum* but differs in a number of characteristics, including the staminal arrangement, many-celled fruits and winged seeds.

All species are attractive ornamental plants and *N. myrtifolia* is cultivated irregularly in tropical and subtropical regions.

Plants are propagated from seed, which does not require pre-sowing treatment, or from cuttings of firm young growth.

Neofabricia mjoebergii (Cheel) Joy Thomps.

(after Dr E. Mjöeberg, original collector)

Qld

4–10 m × 3–6 m   Aug–Oct

Tall **shrub** or small **tree**; with hairy young growth; **bark** grey to greyish-brown, flaky; **branchlets** becoming glabrous; **leaves** 0.6–1.4 cm × 0.2–0.6 cm, elliptic to obovate, sessile, base tapered or wedge-shaped, ascending to erect, glabrous or faintly hairy, apex pointed; **flowers** about 1 cm across, white or cream, solitary or rarely in 3s in upper axils; **capsules** to about 0.4 cm across, with a conical base, initially with long hairs, becoming glabrous; **seeds** shed anually.

Occurs in the central inland areas of Cape York Peninsula in open forest and woodland where it is often associated with other shrubs. Plants are often found growing in white sands. This uncommon species is rarely cultivated. It is suitable for tropical and subtropical regions and generally develops a somewhat sparse cover of foliage. Appreciates moist but well-drained soils. It will tolerate sunshine but is also suitable for a semi-shaded site. Propagate from seed, or from cuttings of firm young growth.

Neofabricia myrtifolia (Gaertn.) Joy Thomps.

(leaves similar to the genus *Myrtus*)

Qld   Black Tea-tree; Untarra

1–10 m × 1–5 m   March–Aug; also sporadic

Small to tall **shrub** or rarely a small **tree**; young growth reddish hairy; **bark** hard, dark grey-brown; **branchlets** glabrous or hairy; **leaves** 2–5.5 cm × 0.4–1.3 cm, oblanceolate to narrowly obovate, sessile or short-stalked, spreading to ascending, deep green, glabrous, margins often hairy when young, apex blunt with a small point; **flowers** to about 2.5 cm across, pale to deep yellow, solitary in upper axils, sometimes profuse and very conspicuous; **calyx lobes** hairy; **capsules** to 1 cm across, hairy, with a cup-shaped base, 8–12-celled; **seeds** shed annually.

N. myrtifolia is from the northern and eastern Cape York Peninsula, where it occurs in rocky and sandy sites. This very decorative species can be a dwarf windshorn shrub of coastal heathlands, or in sheltered sites it can develop as a tree-like tall shrub. Generally plants from southern populations have flowers of deeper colour and one cultivar, *N. myrtifolia* 'Gold', is available. This species is becoming popular in cultivation in tropical and subtropical regions and needs fairly good drainage with plenty of sunshine for best results. Plants can become open with age and often they do not respond well to hard pruning. Tip pruning at planting and during early development can help to promote a bushy framework. Heavy frosts can damage plants. They appreciate mulching and supplementary

Neofabricia sericisepala

Neofabricia myrtifolia × .8

water during extended dry periods. They do well in large containers or as general garden plants and are recommended for coastal sites. They respond well to slow-release fertilisers. Aborigines used the reddish gum for gluing skins to their drums. Propagate from seed, or from cuttings of firm young growth, which can be slow to strike.

N. sericisepala is closely allied but differs in having narrower leaves, smaller flowers and smaller 5–8-celled capsules.

Neofabricia sericisepala J. Clarkson & Joy Thomps.
(silky sepals)
Qld
2–6 m × 1–4 m   April–Aug
Medium to tall **shrub** with hairy young growth; **bark** hard, grey; **leaves** 1.5–4 cm × 0.2–0.5 cm, narrow oblanceolate to narrow elliptic, often slightly falcate, sessile, spreading to ascending, initially silky-hairy, becoming somewhat glabrous, apex tapering to a sharp point; **flowers** to about 1.5 cm across, yellow, solitary in upper axils, sometimes profuse and very conspicuous; **calyx** silky-hairy; **capsules** to about 0.6 cm across, somewhat silky-hairy, base cup-shaped, 5–8-celled; **seeds** usually shed Oct–Dec.
An attractive species from the central and southern inland regions of Cape York Peninsula where it grows in shallow sandy, often gravelly soils in eucalyptus woodland. Regarded as rare in nature, it should adapt to seasonally dry tropical and subtropical regions. Requires good drainage and should do best in a sunny or semi-shaded site. Frost can damage plants. Should respond well to pruning and supplementary watering during extended dry periods. Propagate from seed, or from cuttings of firm young growth.

N. myrtifolia is allied, but has larger, non-prickly leaves and larger flowers and 8–12-celled capsules.

Neogoodenia minutiflora C. A. Gardner & A. S. George
= *Goodenia neogoodenia* Carolin

NEOGUNNIA Pax & K. Hoffm. = *GUNNIOPSIS* Pax

NEOLITSEA Merr.
(from the Greek *neo*, new; *Litsea* a related genus in the same family)
Lauraceae
Shrubs or **trees**, dioecious; young growth at first limp and hairy, later glabrous; **leaves** simple, entire, spirally arranged or clustered, 3-veined, with minute oil dots; **inflorescence** of clustered axillary umbels; **flowers** small, dull-coloured, unisexual, 4–5-merous; **staminodes** 5–9; **fruit** a hard drupe not enclosed in a swollen receptacle.
A genus of about 80 species distributed in Asia, South-East Asia, the Pacific Region, New Guinea and in Australia. There are 3 species in Australia and 2 of these are endemic. Even though the Australian species have handsome foliage and attractive flushes of pale-coloured, drooping new growth, they are still uncommonly grown. Birds such as fruit-eating pigeons are attracted to the ripe fruit. Trees of this genus are unisexual and male and female plants are necessary for fruit production. Propagation is from seed, which has a limited period of viability and for best results should be sown while fresh. Cuttings of some species have proved successful.

Neolitsea australiensis Kosterm.
(Australian)
Qld, NSW   Grey Bollywood
12–30 m × 10–18 m   March–May
Either a small **tree** with a slender crown or developing as a large **tree** with a narrow crown; young growth hairy, becoming glabrous; **leaves** 6.5–13.5 cm × 2–5 cm, elliptical or lanceolate, dark green and glossy above, paler, dull green and waxy beneath, acuminate; **umbels** very small, clustered, bearing 4 or 5 flowers; **flowers** about 0.4 cm across, cream or pale brown, pleasantly fragrant; **drupes** 1.2–2 cm × 1–1.4 cm, ellipsoid to globular, reddish ripening black, mature Feb–May.
Distributed from south-eastern Qld to south-eastern NSW, this species is a common component of rainforests ranging from sea-level to the mountains up to an altitude of about 1000 m. An attractive, usually slender tree with a dense canopy and ornamental foliage. Often slow-growing and may require some protection from direct sun when young. Best in well-drained, loamy soils and the use of mulches and watering during dry periods is beneficial. Tolerant of light to moderate frosts. Propagate from fresh seed, which takes one to two months to germinate. Cuttings would be worth trying.

N. australiensis has been erroneously recorded as *N. cassia* (L.) Kosterm., a species which does not occur in Australia.

Neolitsea brassii Allen
(after Leonard Brass, American botanist and leader of
the Archbold expedition)
Qld, NT
15–20 m × 8–15 m   Jan–April
   Medium to tall **tree** with a slender canopy; young
growth with white hairs, later glabrous; **leaves**
6.5–15.5 cm × 2.5–7 cm, elliptical to lanceolate, dark
green above, green or glaucous beneath, leathery,
acuminate; **umbels** 0.4–0.6 cm long, in axillary clusters, bearing 3–5 flowers; **flowers** about 0.5 cm across,
cream or greenish, pleasantly fragrant, opening widely; **drupes** 1–1.2 cm × 0.75–1 cm, globular, red, mature
Aug–Nov.
   Widely distributed from north-eastern to south-
eastern Qld and western areas of the NT. Also occurs
in New Guinea. Usually found growing in rainforests
but also extends to moist gullies and wet sclerophyll
forests. A slender bushy tree worth growing for its
foliage and small colourful fruit which are attractive to
birds. Best in a semi-shaded position in well-drained
soil. Responds to mulches and watering during dry
periods. Tolerant of light to moderate frosts.
Propagate from fresh seed which takes 40–170 days to
germinate. Cuttings would be worth trying.

Neolitsea cassia (L.) Kosterm. —
                                    see under *N. australiensis*

Neolitsea dealbata (R. Br.) Merr.
(whitewashed, referring to the white undersurface of
the leaves)
Qld, NSW   Hairy-leaved Bolly Gum
8–20 m   March–June
   Small, medium or tall **tree** with a rounded canopy;
young shoots with pale-coloured, limp, hairy leaves;
leaves 8–22 cm × 3.5–8.5 cm, broadly elliptical, ovate
or lanceolate, dark green above, white or glaucous
beneath with brown hairs, stiff-textured, venation
prominent, acuminate; **umbels** very short, in dense
axillary clusters, bearing 3–5 flowers; **flowers** about
0.5 cm across, cream, yellowish or brown, pleasantly
fragrant; **drupes** 0.9–1.1 cm × 0.8–1.1 cm, globular,
red, ripening to black, mature Feb–May.
   Widely distributed from north-eastern Qld to south-
eastern NSW, this is a common tree of rainforests,
stream banks and moist areas in open forests. It usually
grows as a small tree but occasionally larger specimens
are encountered. Often colonises disturbed sites with
young plants germinating and growing rapidly in conditions of high light. Birds are attracted to the ripe
fruit. Flushes of new growth hang limply and are a
highly decorative feature. Plants are suitable for a wide
range of climates and can be successfully grown in
southern Australia. The young leaves are susceptible
to wind damage and for best results trees should be
planted in a situation sheltered from strong winds.
Young plants are often straggly but become bushy with
age. Tolerant of light to moderate frosts. Requires
well-drained soil and responds to mulches, watering
and light applications of fertiliser. Propagate from
fresh seed, which usually germinates in 1–3 months
but can germinate sporadically for up to a year.
Cuttings of firm young growth can be successful.

NEONAUCLEA Merr.
(from the Greek *neo*, new; and the allied genus
Nauclea, from which it was segregated)
Rubiaceae (alt. *Naucleaceae*)
   Trees or **shrubs**; **leaves** opposite, venation pinnate;
stipules large, deciduous, ovate, elliptical to obovate
or sometimes linear-oblong to narrowly triangular;
flower-heads globular, terminal, 1–3 or, rarely, 5–7,
surrounded by 2 ovate bracts, many-flowered; **flowers**
5-merous, more or less sessile, on hairy receptacle;
corolla funnel-shaped, 5-lobed, interior glabrous; **style**
exserted; **fruit** a capsule; **seeds** many, flattened,
winged.
   A genus of about 70 species with one endemic in
north-eastern Australia. Plants are rarely encountered
in cultivation. Propagate from fresh seed.

Neonauclea gordoniana (F. M. Bailey) Ridsdale
(after Mr P. R. Gordon, 19–20th-century Chief In-
spector of Stock, Qld)
Qld   Hard Leichhardt Tree
8–12 m × 5–8 m   June–Aug
   Small **tree**; **bark** thin, scaly; **leaves** 10–15 cm × 4–6 cm,
lanceolate to somewhat elliptical, opposite, tapering to
stalk at base, dark green, glabrous, apex bluntly pointed; **stipules** prominent, blunt, glabrous; **flowerheads**
to about 2.5 cm across, globular, terminal, solitary or 3
together; **flowers** small, white, glabrous, in dense
globular heads; **fruit** to about 3 cm across, burr-like.

Neolitsea dealbata × .7

19

An uncommon tree of north-eastern Qld rainforests. It has potential for cultivation due to its large leaves and intriguing flowerheads. Reputedly slow-growing, it is best suited to tropical and subtropical regions. Moist well-drained soils, rich in organic matter, and a protected site should provide conditions for good growth. Hot sun can damage young plants and plants are also sensitive to frost. The wood is close-grained and has been used for timber framing and lining boards. Propagate from fresh seed.

NEOPAXIA O. Nilsson

(from the Greek *neo*, new; and the genus *Paxia* after 19–20th-century German botanist Ferdinand A. Pax)
Portulacaceae

A monotypic genus which is confined to Australia and New Zealand.

Neopaxia is closely allied to *Calandrinia* which differs in its variable number of stamens and to *Montia* which has erect stems, alternate leaves and unequal petals.

Neopaxia australasica (Hook. f.) O. Nilsson.

(Australian)

NSW, Vic, Tas, SA, WA — White Purslane
prostrate × 1–2 m — Aug–April

Perennial; **stems** creeping, rooting at nodes, glabrous; **leaves** 3–10 cm × 0.2–1 cm, linear-lanceolate, opposite, winged at base and surrounding stem, glabrous, apex blunt; **inflorescence** a loose terminal cyme of 1–6 flowers, borne on stalks to 3.5 cm long; **flowers** about 1.5–2 cm wide, spreading equal petals, white or pale pink, fragrant; **stamens** 5; **capsules** about 0.3 cm long.

A widely distributed species that occurs at most altitudes, usually found on moist soils, and on occasions as an aquatic. It has adapted very well to cultivation, tolerating most soils, but with a preference for moisture throughout the year. It has the capacity to be dormant during dry periods and regrow when conditions are suitable. Grows best in partial or full sun. Frost tolerant. Propagate from seed, cuttings or by division.

There is some disagreement amongst botanists regarding the genus to which this species belongs, some including it in *Claytonia* and others in *Montia*. It is now generally regarded as best included in *Neopaxia*.

NEORITES L. S. Sm.

(from the Greek *neo*, new, latest; after the genus *Orites*)
Proteaceae

A monotypic genus endemic in north-eastern Qld.

Neorites kevediana L. S. Sm.

(from the given names of Kevin J. White and H. Edgar Volck, original collectors in 1952)

Qld — Fishtail Silky Oak
15–25 m × 6–12 m — April–May

Medium **tree**; young growth light green with pinkish to rusty-brown hairs; **trunk** often fluted on mature plants; **bark** dark brown with many whitish lenticels; **juvenile leaves** 20–35 cm long, pinnate, stiff, initially hairy becoming bright glossy green; **leaflets** up to 15, to 16 cm × 7 cm, margins crenately toothed, apex blunt to pointed; **adult leaves** 6–16 cm × 2–7 cm, simple, crenately toothed, apex blunt; **inflorescence** to about 7 cm long, panicles of short axillary spikes; **flowers** to about 0.6 cm long, cream, usually profuse, pairs of flowers initially enclosed by broad, striate bracts; **fruit** a curved, soft-walled follicle to about 7 cm long, with 6–8 flat, winged seeds.

An ornamental species from the Cook District where it grows as a tree of the rainforest canopy. It has proved to be fast-growing in tropical regions but can sometimes struggle in subtropical areas. Requirements include well-drained organically rich soils which are moist for much of the year. Plants dislike exposed sites when young and are damaged by frosts. Mulching and supplementary watering are usually necessary. Worthy

Neopaxia australasica, natural carpet — T. L. Blake

Neopaxia australasica × .75

Neoroepera banksii × .8

Neosepicaea jucunda   D. L. Jones

of growing as an outdoor or indoor container plant. Cut foliage is long-lasting for indoor decoration. Propagate from fresh seed.

NEOROEPERA Müll. Arg. & F. Muell.
(from the Greek *neo*, new, latest; and after the allied genus *Roeperia*)
Euphorbiaceae

Glabrous **shrubs**; **leaves** alternate, short-stalked, entire, leathery; **inflorescence** axillary clusters; **flowers** monoecious, small; **male flowers** perianth divided to base, 5 or 6 petal-like segments; **stamens** 5 or 6, exserted; **female flowers** perianth deeply divided into 6 narrow lobes; **styles** 3; **fruit** a globular capsule, separating into 3, 2-valved cocci.

An endemic genus of 2 species which is confined to eastern Qld.

Neoroepera banksii Benth.
(after Sir Joseph Banks, 18–19th-century English botanist and patron of the natural sciences)
Qld
0.6–1.5 m × 0.6–1 m   Aug–Oct

Dwarf to small, open-branched **shrub**, with light green young growth; **branches** ascending to erect; **leaves** 1–2 cm × 0.3–0.7 cm, cuneate-oblong, short-stalked or nearly sessile, spreading to ascending, mid green, glabrous, margins flat, apex blunt or sometimes notched; **inflorescence** few-flowered axillary clusters; **male flowers** small, greenish-white; **female flowers** small, on pendulous slender stalks to about 1.2 cm long, with 3 short broad styles; **capsules** to about 0.8 cm, globular.

A dweller of moist sandy soils in coastal scrubland in the Cook and Kennedy Districts. Of interest mainly to enthusiasts in tropical and subtropical areas. It will

need light soils which drain freely and a sunny or semi-shaded site should be satisfactory. Propagate from seed, or from cuttings of firm young growth.

Neoroepera buxifolia Müll. Arg. & F. Muell.
(leaves similar to the genus *Buxus*)
Qld
0.6–1.5 m × 0.6–1 m   May–Oct

Dwarf to small **shrub**; **branches** ascending to erect; **leaves** 1.5–2.5 cm × 0.3–1.2 cm, elliptical-oblong, short-stalked, spreading to ascending, mid green, glossy, glabrous, leathery, margins flat, apex blunt; **inflorescence** of numerous male flowers with one female flower per cluster; **male flowers** about 0.2 cm long, greenish-cream, on stalks of about 0.8 cm long; **female flowers** on longer and thicker stalks than male flowers, with 3 club-shaped styles; **capsules** small, globular.

This species is from the Port Curtis District where it grows in moist sites in heathland and forests along the coast and on adjacent islands. It has similar needs to *N. banksii*. Propagate from seed or from cuttings of firm young growth.

NEOSEPICAEA Diels
(from the Greek *neo*, new, latest; and after the allied genus *Sepikea*)
Bignoniaceae

Woody **climbers**; **leaves** opposite, frequently digitate; **inflorescence** terminal panicles or racemes; **flowers** tubular, somewhat funnel-shaped, with shortly flared lobes, interior of throat hairy; **calyx** short, 4–5-toothed; **stamens** 4, included in the corolla; **style** 2-lobed; **fruit** a capsule.

A tropical genus of 4 species with 2 occurring in far north-eastern Australia.

21

Neosepicaea jucunda (F. Muell.) Steenis
(pleasant, agreeable)

Qld   Jungle Vine
Climber   July–Sept

Vigorous twining **climber**, with glabrous young growth; **stems** long; **leaves** trifoliolate, prominently stalked; **leaflets** to 12 cm × 5 cm, elliptical, stalked, glabrous, concave, light-to-mid green, venation prominent, apex pointed; **racemes** axillary, many-flowered; **flowers** to about 2.5 cm long, tubular, with 5 flared lobes, pink to reddish-pink, most conspicuous; **capsules** about 5 cm long, woody, ripe Dec–March; **seeds** winged.

A showy climber which can grow to great heights and cover large areas. It occurs mainly in the lowland rainforest of the Cook and North Kennedy Districts, but is also found on the Atherton Tableland. Plants do best in moist well-drained soils with a semi-shaded aspect. They are appreciative of organic mulches and respond well to light applications of slow-release fertiliser. Plants are damaged by frosts and they dislike extended dry periods. They may need regular pruning to curb their spread. Propagate from fresh seed or from cuttings of firm young growth, which strike readily.

Neosepicaea viticoides Diels
(like the genus *Vitex*)
Qld
Climber   Sept

Climber; **bark** light grey; **stems** thick; **leaves** 3–5 times foliate, prominently stalked; **leaflets** 10–18 cm × 4–7 cm, broadly oblanceolate, stalked, glabrous, glossy, apex pointed; **panicles** axillary, many-flowered; **flowers** to about 1.5 cm long, tubular with 5-flared lobes, exterior pale brown, hairy, interior with reddish-brown longitudinal stripes; **capsules** about 10 cm × 2 cm, woody; **seeds** winged.

N. viticoides occurs in the Cook District and in New Guinea. It inhabits rainforest. Plants are not well known and apparently not cultivated. Should do best in freely draining soils which are moist for most of the year, and with a semi-shaded aspect. Its other requirements are similar to those of *N. jucunda*. Propagate from fresh seed or from cuttings of firm young growth.

NEOSTREARIA L. S. Sm.
(from the Latin *neo*, new; and *Ostrearia*, a closely allied genus)
Hamamelidaceae
A monotypic genus endemic in Qld.

Neostrearia fleckeri L. S. Sm.
(after Dr Hugo Flecker, 20th-century medical practitioner)
Qld
6–12 m × 4–8 m   May–Aug

Small **tree** with stellate-hairy or scaly young growth; **branches** spreading to ascending; **branchlets** stellate-hairy or with small scales; **leaves** 8–18 cm × 2–7 cm, lanceolate to oblong-lanceolate, prominently stalked, glabrous, glossy and dark green above, greyish below, aromatic, venation prominent, apex pointed; **stipules** to 0.4 cm, bristle-like, stellate-hairy; **inflorescence** loose terminal spike to 10 cm long, on stalk to about

3.5 cm long, pendulous beneath the leaves; **flowers** about 1.5 cm long; **petals** 5, to 1.2 cm × 0.2 cm, white to cream; **stamens** 5; **calyx** 3-lobed; **fruit** a 2-celled capsule, 1–1.5 cm long, obovoid, flattened, becoming glabrous.

This member of the witch-hazel family occurs near waterways of lowland rainforest in north-eastern Qld. It has ornamental foliage which is aromatic when crushed. Evidently not cultivated, it should be best suited to tropical and warm subtropical regions, in moist shaded sites which drain freely. This species will need protection from excessive exposure while young. Application of slow-release fertiliser and mulching should be beneficial. Plants are likely to require supplementary watering during dry periods. Heavy frosts can damage plants. Propagate from fresh seed and possibly from cuttings.

NEPENTHACEAE Dumort.   Pitcher Plants
A family of dicotyledons consisting of about 70 species in the solitary tropical genus *Nepenthes*. A single widespread species, *N. mirabilis*, extends to north-eastern Qld. Plants of this genus are woody herbs which climb with the aid of tendrils that develop from an apical extension of the midrib. The end of the midrib expands into a pitcher which traps and drowns insects and other small animals. As the creatures decay their useful products are absorbed by the plant.

NEPENTHES L.
(after a plant of Greek literature which could diminish sorrow; here used in reference to its supposed medicinal properties)
Nepenthaceae

Dioecious **shrubs** or **vines**, often using their leaves to climb; **leaves** alternate, sessile or stalked, adult leaves often develop colourful, pendant, pitcher-like appendages with recurved fluted rim and a lid at the end of coiled tendrils; **inflorescence** terminal raceme or panicle; **flowers** with 3–4 tepals in 2 whorls, nectar-producing; **male flowers** with 2–24 stamens joined to form a column; **female flowers** with 3–4 carpels, style 1; **fruit** an elongated leathery dehiscent capsule.

This genus of a monogeneric family contains about 68 species and is represented in Madagascar, southern China, Indomalaysia, New Guinea, New Caledonia and with one species in north-eastern Australia.

These fascinating insectivorous plants secrete digestive enzymes into water-filled pitchers which can then digest drowned insects. The nutrients from the insects are absorbed by the plant.

Nepenthes mirabilis (Lour.) Druce
(wonderful)
Qld   Tropical Pitcher Plant
0.5–5 m × 0.5–1.5 m   Nov–March; also sporadic

Dwarf to small **shrub** or loose **vine** with glabrous or hairy young growth; **leaves** 10–40 cm × 2–8 cm, oblong to lanceolate, base tapering to a thick stalk up to 25 cm long, faintly papery to faintly leathery, concave to prominently folded, bright green, apex blunt or pointed and extending to a coiled tendril on which a green or green with reddish spotted or veined pitcher of up to 18 cm long is appended; **male flowers** green to purplish-black, with tepals to 0.7 cm long, in racemes to 30 cm long, with stalk to about 15 cm long, strong mouse odour; **female flowers** with tepals to 0.5 cm long, raceme similar but shorter than male, strong mouse odour; **capsule** 1.5–3 cm long, angular.

This intriguing species occurs mainly in eastern Cape York Peninsula, with isolated populations as far south as Innisfail. It is often found on swamp margins. Generally grows as a straggly vine in permanently wet areas but takes on the form of a low shrub in drier sites. Best suited to tropical and subtropical areas and needs a heated growing area if temperatures reach below 10°C. Plants grow in a range of moderately heavy to light soils. They prefer bright sunshine, which helps to give colour to the pitchers. They grow well as container plants as long as humidity is maintained in their growing area. Highly successful in hanging baskets. An infertile growing medium is recommended, such as sand with live sphagnum peat or ground peat, or sand with pine bark. Do not use fertilisers. Tip pruning of young plants promotes lateral growth. Propagate from seed or from cuttings of firm wood.

N. mirabilis has been used to a limited degree for breeding programmes with other species.

N. kennedyana F. Muell. is synonymous as are the following names given by F. M. Bailey to what are now regarded as minor variants within the species: *N. albo-lineata, N. alicae, N. ambrustae, N. bernaysii, N. cholmondeleyi, N.garrawayae, N. jardinei, N. moorei, N. pascoensis* and *N. rowanae.*

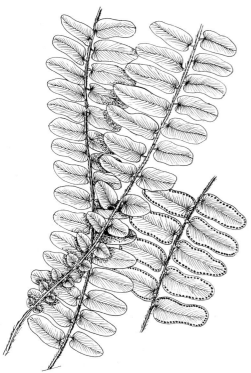

Nephrolepis acutifolia × .5, section of fertile frond (at right) × 1

NEPHROLEPIDACEAE Alston

A family of ferns consisting of about 40 species in 3 genera, distributed mainly in tropical regions, particularly Asia. They are terrestrials, epiphytes or climbers with erect or creeping rhizomes and some species spread by stolons. Their pinnate fronds have jointed leaflets and free veins and the sori are covered by kidney-shaped indusia. Australian genera: *Arthropteris, Nephrolepis.*

NEPHROLEPIS Schott

(from the Greek *nephros,* kidneys; *lepis,* a scale; referring to the shape of the indusia which cover the sporangia)

Nephrolepidaceae (alt. *Oleandraceae*)

Terrestrial or **epiphytic ferns** with long, slender, scaly runners; **rhizome** erect, covered with old woody stipes; **fronds** pinnate, often long, erect, arching or pendulous; **leaflets** entire, crenate or lobed; **sori** covered with a kidney-shaped or crescentic indusium.

A genus of about 40 species widely distributed in tropical regions. There are 7 species occurring in Australia; 2 of these are endemic. A couple of species have proved to be hardy and adaptable and are widely grown in gardens or containers for their ornamental fronds. Large containers of *Nephrolepis* have for many years been used to decorate hotels or halls for important functions. Some species also make excellent basket plants. Propagation is from spores, plantlets produced on runners or by division of clumps.

Nephrolepis acutifolia (Desv.) Christ.
(with pointed leaflets)
Qld, NT
0.6–1.2 m tall

Perennial fern forming dense clumps; **stipes** 15–30 cm long, grey-brown, densely scaly; **fronds** 60–120 cm × 12–16 cm, erect, arching or pendulous, pale green to yellowish-green; **leaflets** 5–8 cm × 1–1.5 cm, linear to linear-oblong, leathery, falcate, blunt; **fertile leaflets** longer than sterile leaflets and with an acute apex; **sori** close to margin.

A widely distributed species which extends from Asia to tropical Australia where it occurs in Arnhem Land and north-eastern Qld, usually growing among rocks. A hardy species which withstands considerable exposure to sun, especially if well watered. Requires well-drained soil. It is cold sensitive and in temperate regions requires the protection of a heated glasshouse. Propagate from spores and by plantlets produced in runners.

Nephrolepis arida D. L. Jones
(arid; referring to the prevailing climate)
NT
0.8–1.7 m tall

Perennial fern forming small, spreading colonies; **stipes** 30–40 cm long, pale brown, scaly at base; **fronds** 80–170 cm × 3–11 cm, erect or arching, dark green; **leaflets** 1.5–5.5 cm × 0.7–1.4 cm, oblong-lanceolate, widely spaced, falcate, bluntly toothed near apex, with a prominent basal auricle; **sori** well within margin.

N. arida is endemic in Australia where it is found growing in wet crevices and ledges of sheltered sand-

Nepenthes mirabilis   D. L. Jones

Nephrolepis auriculata   D. L. Jones

stone gorges in Central Australia, parts of the Top End and the Wessell Islands. Experience in cultivation is limited but plants grow readily in a pot of well-drained, friable mixture. Requires bright light and regular watering. Tolerates moderate frosts. Propagate from spores and by plantlets produced on runners.

Nephrolepis auriculata (L.) Trimen
(with a small ear-like lobe)
Qld, WA, NT   Giant Fishbone Fern
0.5–2 m tall
   Perennial fern forming dense colonies; **stipes** 30–50 cm long, dark brown; **fronds** 50–200 cm × 30–40 cm, erect to arching, dark green and glossy; **leaflets** 15–20 cm × 1.5–2.5 cm, oblong-lanceolate, straight or falcate, well-spaced, margins crenate, base often with a small lobe, apex acuminate; **sori** well away from the margin.
   Widely distributed throughout the tropics and in Australia where it occurs in northern areas. Grows in rainforest, near swamps and close to streams. Usually grows as a terrestrial, on rocks, or occasionally as an epiphyte in trees. A coarse, large-growing fern which needs ample room to develop. Excellent in large tropical and subtropical gardens. Requires abundant moisture and plants grown in shade have best appearance. Frost tolerance is not high. Propagate from spores and plantlets produced on runners.
   Previously known as *N. biserrata* (Sw.) Schott

Nephrolepis biserrata (Sw.) Schott =
                    N. auriculata (L.) Trimen

Nephrolepis cordifolia (L.) Presl
(with cordate leaflets)
Qld, NSW, NT   Sword Fern, Fishbone Fern
0.3–1.2 m tall
   Perennial fern forming dense colonies; **runners** often bearing fleshy tubers to 1 cm long; **stipes** 10–30 cm long, light brown; **fronds** 30–120 cm × 2–7 cm, narrowly lanceolate to oblong, yellowish-green to dark green, erect or arching; **leaflets** 1–3.5 cm × 0.4–1 cm, oblong-elliptical, straight or falcate, not crowded, margins bluntly crenate, apex blunt; **sori** close to margin.
   Widely distributed in tropical Asia and in Australia extending well south into temperate regions. It grows as an epiphyte or terrestrial in a wide range of habitats from rainforest to sheltered areas of open forest. Can be weedy and may naturalise suitable sites. Excessive spread may need to be curtailed. An excellent fern for containers and adapts to a wide range of garden situations from shade to sun. Tolerant of light to moderate frosts. Propagate from spores or by plantlets produced on runners.
   The cultivar 'Kimberley Queen' is hardly distinguishable from the typical species.

Nephrolepis hirsutula (G. Forst.) Presl
(hairy)
Qld, WA, NT   Hairy Sword Fern
0.3–1 m tall
   Perennial fern forming dense colonies; **stipes** 10–40 cm long, pale brown, covered with hairy scales; **fronds** 30–100 cm × 10–20 cm, lanceolate to elliptical, yellowish-green, erect or arching; **leaflets** 5–10 cm ×

24

1–2 cm, linear, straight or falcate, base with a narrow, triangular lobe, margins crenate, hairy; **fertile leaflets** narrower than sterile; **sori** close to margin.

Widely distributed in tropical Asia and Pacific islands. In Australia, it occurs in tropical regions and grows as a terrestrial or epiphyte in open forest and on the fringes of rainforest. An adaptable fern which is commonly cultivated in tropical regions. Tolerant of considerable exposure to sun especially if well-watered. Frost tolerance is not high. Propagate from spores or by plantlets produced on runners.

Nephrolepis obliterata (R. Br.) J. Sm.
(weakly developed)
Qld, WA, NT
1–1.5 m tall

Perennial fern forming dense colonies; **stipes** 10–30 cm long, brown, shiny; **fronds** 100–150 cm × 15–28 cm, lanceolate to elliptical, green, erect or arching; **leaflets** 8–14 cm × 1–2.5 cm, linear, straight or slightly falcate, well spaced, glabrous, apical margins toothed; **fertile leaflets** narrower than sterile; **indusia** close to margin.

Occurs in New Guinea and Australia where it is widely distributed across the north growing on rocks and trees in and close to rainforest. A robust species which has similar cultural requirements to *N. hirsutula*. Frost tolerance is not high. Propagate from spores or by plantlets produced on runners.

Nephrolepis radicans (Burm. f.) Kuhn
var. **cavernicola** Domin
(with aerial roots; growing in caves)
Qld
0.3–1 m tall

Perennial fern with long-creeping climbing runners bearing at intervals short lateral branches; **stipes** light brown, clustered; **fronds** 30–100 cm × 4–8 cm, narrowly elliptical, dark green, thin-textured; **leaflets** 2–4 cm × 0.8–1 cm, oblong, crowded, blunt, margins minutely crenate; **sori** close to margin.

This variety is endemic in Australia. It is known only from the vicinity of Chillagoe in north-eastern Qld where it grows close to caves. Plants have an unusual scrambling growth habit. Rare in cultivation. Suitable for tropical and subtropical regions. Requires good drainage in semi-shade or filtered sun. Tolerates moderate frosts. Propagate from spores or by division.

NEPTUNIA Lour.
(after Neptune, the Roman god of the sea, as plants often grow in or near water)
Mimosaceae

Dwarf to tall perennial **herbs**, or small to medium **shrubs**; **leaves** bipinnate; **stipules** membranous; **flowerheads** globular, many-flowered, yellow, solitary on axillary stalks, upper heads fertile, lower heads sterile with petal-like staminodes; **stamens** usually 5 or 10; **pods** oblong, flattened, with 1 to several seeds; **seeds** transverse.

Neptunia is a tropical and subtropical genus of 11 species. Five species occur in Australia where they are distributed across the northern region.

They are only moderately ornamental in flower but some species have decorative grey-green foliage. They are rarely encountered in gardens but they are of interest. Their leaves react to touch by closing the leaflets upwards, returning to their previous position after a period of time. A number of species are valued as fodder plants for stock. They all should respond well to pruning.

Propagation is from seed, which will germinate better if pre-soaked in very hot water for 1–12 hours. For other treatments and techniques see Volume 1, page 205. Cuttings of firm young growth are worth trying.

Neptunia amplexicaulis Domin
(stem-clasping)
Qld
0.2–1 m × 0.8–2 m   Nov–May

Perennial **herb**; stems to about 1 m long, glabrous; **leaves** bipinnate, on stalks to about 2.5 cm long, grey-green to blue-green, glabrous; **pinnae** 2–3 pairs, to 7 cm long; **leaflets** 0.7–1.9 cm × 0.3–0.7 cm, 7–14 opposite pairs, apex rounded; **stipules** 1.2–2.2 cm × 0.4–1.1 cm, persistent; **stipels** leaf-like at base of lowest pair of pinnae; **upper flowerheads** about 0.8 cm across, yellow stamens with green petals, on axillary stalks to 5.5 cm long; **lower flowerheads** to about 2.5 cm across, yellow with petal-like staminodes; **pods** 1.5–2.3 cm × 0.9–1.3 cm, flattened, dark brown, 2–4-seeded.

The typical form is from the Burke District where it grows only on alkaline soils. It is not highly regarded in its native region as grazing fodder because of selenium poisoning. Plants are moderately ornamental and are worth trying in containers or as garden plants in tropical and subtropical regions. Needs good drainage and plenty of sunshine.

The forma *richmondii* Windler occurs in the Burke and Moreton Districts and differs in having all parts of plants covered in long hairs, and it lacks stipels.

Propagate both forms from seed, which germinates readily, or from cuttings of firm young growth.

Neptunia dimorphantha Domin
(two distinct forms of flowers)
Qld, SA, WA, NT   Sensitive Plant
prostrate–0.3 m × 0.6–1.5 m   Dec–April

Neptunia dimorphantha   D. L. Jones

Perennial **herb** with glabrous or hairy young growth; **stems** to about 0.6 cm long, spreading to ascending, slender, glabrous or hairy; **leaves** bipinnate, on stalks to 1.5 cm long, sensitive to touch; **pinnae** usually 2–4 pairs; **leaflets** 0.3–1.2 cm × 0.1–0.3 cm, 7–25 opposite pairs, blue-green, glabrous except for hairy margins; **stipules** to 1 cm long, persistent, tapering to a fine point; **upper flowerheads** about 0.7 cm across, yellow stamens with green petals, on axillary stalks to 4.5 cm long; **lower flowerheads** to about 1.8 cm across, with yellow petal-like staminodes; **pods** 0.7–1 cm × about 0.6 cm, flattened, hairy, brown, 1–2-seeded.

A glaucous-foliaged species with a wide distribution. Usually occurs in open sites on sandy or clay soils. Suitable for seasonally dry tropical and subtropical regions, where it may be useful as a groundcover for gardens and for soil erosion control. Also worth trying in containers. Propagate from seed, which germinates readily, or from cuttings of firm young growth.

Neptunia gracilis Benth.
(slender)
Qld, NSW, NT   Native Sensitive Plant
prostrate–0.3 m × 0.6–1 m   Nov–May
Perennial **herb**; **stems** slender, to about 0.5 cm long, often reddish, glabrous or hairy; **leaves** bipinnate, on stalks to 3 cm long; **pinnae** 2–6 pairs; **leaflets** 0.2–1.1 cm × up to 0.25 cm, 7–33 opposite pairs, grey-green, glabrous or faintly hairy; **upper flowerheads** to about 1 cm across, yellow stamens with green or reddish petals, on slender axillary stalks to about 3.5 cm long; **lower flowerheads** to about 2.5 cm across, with yellow petal-like staminodes; **pods** 1.5–2.3 cm × 0.5–0.8 cm, flat, glabrous, brown to black, 3–8-seeded.

A widespread species of open forest, usually found on heavy soils where it can become weedy. Also extends to New Guinea and the Philippines. It is valued as stock fodder and can be cultivated for that purpose. Plants have little appeal for gardens. They appreciate plenty of moisture and should adapt to most soils. Not hardy to frosts.

The forma *glandulosa* Windler occurs in northern NT and differs in having 1 or 2 obscure glands at the base of the lowest pair of leaflets.

Propagate both forms from seed, which germinates readily, or from cuttings of firm young growth.

The var. *major* Benth. has been given species status.

Neptunia major (Benth.) Windler
(larger, taller)
Qld, WA, NT
1.5–3 m × 1–2 m   Dec–May
Smooth to medium perennial **herb** or **subshrub**; stems somewhat woody, glabrous; **leaves** bipinnate, on stalks to 2.5 cm long; **pinnae** 2–4 or more pairs; **leaflets** 0.6–1.3 cm × up to 0.25 cm, 22–40 opposite pairs, grey-green, glabrous; **upper flowerheads** about 1 cm across, yellow with green petals, on axillary stalks to 7 cm long; **lower flowerheads** to about 2 cm across, yellow with petal-like staminodes; **pods** 1.4–2.6 cm × 0.9–1.2 cm, stalked, flat, glabrous, brown, 3–6-seeded.

This wide-ranging species occurs in seasonally dry tropical and subtropical regions, where it is often found near rivers and streams, especially on coastal plains. It is of limited interest for cultivation. It is not regarded as useful for stock fodder. Plants should adapt to a range of soil types in a sunny situation. Propagate from seed, which germinates readily, or from cuttings of firm young growth.

Previously known as *N. gracilis* var. *major* Benth.

Neptunia monosperma Benth.
(single seed)
Qld, WA, NT
0.5–1 m × 0.4–1.5 m   Nov–May
Perennial **herb**; **stems** ascending to erect, glabrous, slender; **leaves** bipinnate, on stalks to 1.5 cm long; **pinnae** 1–3 pairs; **leaflets** 0.3–1.3 cm × up to 0.25 cm, 20–39 opposite pairs, grey-green to blue-green, glabrous; **upper flowerheads** to about 0.8 cm across, yellow with green petals, on axillary stalks to 2 cm long; **lower flowerheads** to about 1.4 cm across, yellow with petal-like staminodes; **pods** 0.7–1 cm × about 0.8 cm, flat, dark brown, faintly hairy.

A grassland species which occurs on a range of soil types. It is palatable to stock but not highly regarded for grazing. The glaucous foliage makes this an interesting species and it has potential for seasonally dry tropical and subtropical areas, as a garden plant or in containers. It needs maximum sunshine for optimum growth. Pruning promotes compact growth. Propagate from seed, which germinates readily, or from cuttings of firm young growth.

NERTERA Banks and Sol. ex Gaertner
(from the Greek *nerteros*, lowly; referring to the creeping growth habit of species)
Rubiaceae

Perennial **herbs**; **branches** terete to somewhat quadrangular; **leaves** opposite, stalked; **stipules** joined to form a sheath around node; **flowers** bisexual, small, solitary, terminal or sometimes axillary, 4–5-merous; **corolla** somewhat funnel-shaped; **fruit** a drupe.

A genus of about 12 species which is mainly distributed in the southern hemisphere (excluding Africa), but also in China and Taiwan. Australia has one species which has limited potential for cultivation. Propagation is from seed, which does not require pre-sowing treatment, or from cuttings, which strike readily.

Nertera depressa Gaertner =
                              N. granadensis (Mutis) Druce

Nertera granadensis (Mutis) Druce
(from Granada)
NSW, Vic, Tas, SA   Matted Nertera
prostrate × 0.1–0.3 m   Nov–Feb
Perennial mat-forming **herb**; **stems** slender, much-branched to about 10 cm long, often rooting at nodes; **leaves** 0.3–0.5 cm × 0.3–0.45 cm, ovate to orbicular, with stalks to 0.4 cm long, glabrous, green, apex usually blunt; **flowers** to 0.2 cm long, tubular with spreading lobes, yellowish-green, solitary, terminal, glabrous; **fruit** small, yellow to orange-red drupe.

A dweller of wet, sheltered sites, such as mossy rocks and cliff faces or moist-to-wet soils in open forests. More of interest than having strong ornamental qualities, but useful for cool shaded spots such as around pools or amongst ferns or other shade-loving plants. Also suitable for miniature gardens and terrariums. Hardy to moderate frosts. Plants will need supplementary watering during dry periods. Propagate from seed or from cuttings.

Nertera reptans (F. Muell.) Benth. =
$\qquad$ *Leptostigma reptans* (F. Muell.) Fosb.

NERVILIA Comm. ex Gaudich.
(from the Latin *nervus*, a vein; referring to the prominent veins in the leaves)
Orchidaceae
$\quad$ Terrestrial orchids with subterranean tubers; **leaf** solitary, appearing after the inflorescence, entire or with undulate or scalloped margins, erect or prostrate, with raised, radiating veins; **inflorescence** a raceme, often fleshy; **flowers** dull-coloured, short-lived, opening tardily to widely, often more or less pendant; **labellum** enclosing the column at the base.

$\quad$ A genus of about 65 species with 5 or 6 occurring in tropical parts of Australia. All species can be cultivated but require fairly specific conditions and are often shy of flowering. They must be grown in a mix suitable for terrestrial orchids and require a heated glasshouse in temperate regions. Plants should be kept moist while in active growth over summer and dry when dormant during winter. Propagation is from seed sprinkled around the base of adult plants or by natural increase of tubers.

Nervilia aragoana Gaudich.
(a native name)
Qld, NT
10–35 cm tall $\hfill$ Sept–Dec
$\quad$ **Terrestrial orchid** growing in loose colonies; **leaf** 8–14 cm across, appearing circular, bright green with radiating veins, margins wavy; **racemes** 10–25 cm tall, erect, fleshy, bearing 2–8 flowers; **flowers** 3–4 cm across, creamish-green, with purple streaks on the labellum; **sepals** and **petals** to 3.5 cm long; **labellum** white with purple or green veins, hairy.

$\quad$ A widely distributed species which extends to Cape York Peninsula and the Top End. Plants are found in or close to monsoonal rainforest and in vegetation close to streams. The large distinctive leaves are very ornamental. Plants can be grown readily in a pot of terrestrial orchid mix, but require a heated glasshouse in temperate regions. They need regular watering when in active growth over summer but should be kept dry when dormant. Annual repotting is best. Propagate from seed or by natural increase.

Nervilia crociformis (Zoll. & Moritzi) Seidenf.
(resembling the genus *Crocos*)
Qld
2–8 cm tall $\hfill$ Nov–Dec
$\quad$ **Terrestrial orchid** growing in loose colonies; **leaf** 3–7 cm across, orbicular, about 10 veins radiating, dull grey-green, margins scalloped; **racemes** 2–8 cm tall, erect, fleshy, bearing a single flower; **flowers** 2–2.5 cm across, green with a white labellum; **sepals** and **petals** to 1.5 cm long; **labellum** with crisped and fringed margins.

$\quad$ Occurs in the Philippines, Indonesia and New Guinea. In Australia it is restricted to north-eastern Qld where it grows in large open colonies along the edge of rainforest. Can be grown in a pot of terrestrial orchid mix with conditions and treatment as for *N. aragoana*. Propagate from seed or by natural increase.

Nervilia discolor (Blume) Schltr. =
$\qquad$ *N. plicata* (Andr.) Schltr.

Nervilia holochila (F. Muell.) Schltr.
(with an entire lip)
Qld, WA, NT
15–25 cm tall $\hfill$ Nov–Dec
$\quad$ **Terrestrial orchid** growing in loose colonies; **leaf** 10–20 cm × 5–7 cm, broadly ovate, petiolate, erect, bright green, with 7 prominent veins; **racemes** 15–25 cm tall, erect, fleshy, bearing 1–6 flowers; **flowers** about 2.5 cm across, pale green to brownish-mauve with a purple and white labellum; **labellum** with wavy margins.

$\quad$ Occurs in New Guinea and is widely distributed across tropical Australia growing on rainforest margins and in moist areas of open forest. Plants have a distinctive, erect, ornamental leaf. They can be grown readily in a pot of terrestrial orchid mix with conditions and treatment as for *N. aragoana*. Propagate from seed or by natural increase.

Nervilia peltata B. Gray & D. L. Jones
(shield-shaped with a central stalk)
Qld, NT
6–10 cm tall $\hfill$ Dec–Feb
$\quad$ **Terrestrial orchid** growing in loose colonies; **leaf** 2–4.5 cm across, orbicular, peltate, ground-hugging, pale grey-green, somewhat granular, margins irregular; **racemes** 6–10 cm tall, erect, fleshy, bearing 1–3 flowers; **flowers** to about 3 cm across, pale green with a white labellum; **sepals** and **petals** to 1.6 cm long; **labellum** with crisped and fringed margins.

$\quad$ A recently described (1994) species which occurs in the Top End, NT, and is distributed in north-eastern Qld, from Torres Strait to Hinchinbrook Island. It grows in open forest and woodland. Plants have grown and flowered successfully in a pot of terrestrial orchid mix with conditions and treatment as for *N. aragoana*. Propagate from seed or by natural increase.

Nervilia plicata (Andr.) Schltr.
(pleated, folded)
Qld, NT
10–18 cm tall $\hfill$ Nov–Dec
$\quad$ **Terrestrial orchid** growing in loose colonies; **leaf** 8–12 cm across, orbicular, ground-hugging, grey-green, reddish or purplish, hairy, with numerous, prominent radiating veins; **racemes** 10–18 cm tall, pink, fleshy, bearing 1–4 flowers; **flowers** about 2.5 cm across, pale green or brown with a purple and white labellum; **labellum** apex notched.

$\quad$ Occurs in Philippines, Indonesia and New Guinea. In Australia it is widely distributed from Cape York Peninsula to near Yeppoon in Qld and in the Top End of the NT. It grows in open forest and monsoonal thickets. An attractive species which can be grown

Nervilia plicata, leaf

D. L. Jones

readily in a pot of terrestrial orchid mix with conditions and treatment as for *N. aragoana*. Propagate from seed or by natural increase.

Previously known as *N. discolor* (Blume) Schltr.

Nervilia uniflora (F. Muell.) Schltr.
(with a single flower)
Qld, NT
10–20 cm tall   Nov–Jan
   Terrestrial orchid growing in loose colonies; **leaf** 2–3.5 cm across, heart-shaped, dark green above, reddish beneath, margins irregular; **racemes** 10–20 cm tall, bearing a single flower; **flowers** 3–3.5 cm across, pink or mauve with a darker labellum; **sepals** and **petals** to 2 cm long; **labellum** abruptly recurved near middle.
   N. uniflora is apparently endemic in Australia where it is distributed in tropical parts. Plants can be grown readily in a pot of terrestrial orchid mix with conditions and treatment as for *N. aragoana*. Propagate from seed and by natural increase.

Nestegis ligustrina (Vent.) L. Johnson =
Notelaea ligustrina Vent.

NEURACHNE R. Br.
(from the Greek *neuron*, nerve; *achne*, husk, glume; referring to the many-nerved glumes)
Poaceae (alt. *Graminae*)   Mulga-grasses
   Densely tufted perennial **grasses**, with shortly spreading rhizomes; **leaves** many, mostly at base, linear, flattish, midrib not prominent; **ligule** a dense tuft of hairs; **stems** erect, branching or non-branching; **inflorescence** spike-like raceme; **spikelet** with 1 bisexual floret and a male or sterile floret below, bearded; **glumes** 2, longer than florets, many-nerved; **lemmas** 1–7-nerved, margins translucent and sometimes hairy; **stamens** 3.
   This endemic genus of 6 species occurs in semi-arid and arid regions, mainly in southern Australia. Rarely encountered in cultivation, plants have value as stock fodder and species such as *N. alopecuroidea* have ornamental characteristics. It is important to be alert to the possibility of these grasses escaping to become problem weeds. Propagation is from seed, which is thought to have a dormancy period of 3–9 months to allow for after-ripening. Division of rhizomes is also worth trying.

Neurachne alopecuroidea R. Br.
(similar to the genus *Alopecurus*)
Vic, SA, WA   Foxtail Mulga-grass
0.2–0.5 m × 0.2–0.6 m   Aug–Dec
   Tufting perennial **grass**, with short spreading rhizomes; **leaves** 1.6–8 cm × to about 0.3 cm, mostly basal, flat or inrolled, glabrous or lower leaves slightly hairy, bluish-grey; **sheaths** exterior not hairy; **spikelets** to about 1.3 cm long, bearded at base; **glumes** 3–11-nerved.
   Foxtail Mulga-grass usually occurs on white sands in Mallee heath and low woodlands of north-western Vic, south-eastern SA and south-western WA. It is decorative in flower and has potential for use in gardens and containers. Plants will need very good drainage and plenty of sunshine. Hardy to most frosts and extended dry periods. Propagate from seed or by division of clumps. Small divisions can be difficult to establish.
   N. lanigera S. T. Blake is a poorly known species from north-western SA and the Austin District in WA. It is closely allied but differs in having densely woolly leaf-sheaths and less hairy glumes.
   N. minor S. T. Blake, from the Austin and Avon Districts of WA, also has some similarities to *N. alopecuroidea*, but it differs in having a hairy interior of the leaf sheaths and shorter spikelets.

Neurachne lanigera S. T. Blake —
see *N. alopecuroidea* R. Br.

Neurachne minor S. T. Blake —
see *N. alopecuroidea* R. Br.

Neurachne munroi (F. Muell.) F. Muell.
(after Mr Munro)
Qld, NSW, SA, WA, NT   Window Mulga-grass;
                                          Slender-headed Mulga-grass
0.2–0.4 m × 0.2–0.3 m   May–Oct; March–May
   Tufting perennial **grass**; with short creeping rhizomes; **stems** to about 40 cm tall, many, stiff, erect, branching or non-branching, nodes 2–3 and hairy; **leaves** to 12 cm × up to 0.35 cm, mainly basal, mostly inrolled, semi-erect, nearly glabrous, prickly pointed; **raceme** 2.5–5.5 cm long, silky; **spikelets** to 0.7 cm long, not in groups; **glumes** 7-nerved.
   Usually found in stony soils or deep red sands which can be alkaline in inland regions. Due to its palatability to stock, populations of plants have been greatly reduced by grazing. It is worth cultivating within its natural range. In nature, plants respond well to seasonal rains, so supplementary watering during extended dry periods should be beneficial. Also has potential for garden planting in a well-drained sunny site. Hardy to most frosts. Propagate from seed or by division of clumps.
   N. queenslandica S. T. Blake, from the Mitchell and Warrego Districts in Qld is closely allied but differs in having stems which are silky-hairy below the racemes, and its spikelets are in groups.

Neurachne alopecuroidea   W. R. Elliot

mainly to comparative unavailability and strict cultivation requirements. Most species need very good drainage and are best suited to sandy or light loam soils. For optimum growth an open, sunny site is necessary. Some species have a tendency to develop into erect shrubs but with regular tip pruning a bushy framework may be promoted. Generally plants do not suffer damage from light or moderate frosts and they withstand extended dry periods once well established.

A number of species have potential for use in floriculture and trials for this purpose in semi-arid and warm temperate areas would be helpful.

Propagation is from seed or cuttings. Seed is rarely available commercially. A limited number of species have been germinated without pre-sowing treatment, with germination occurring about 30–70 days after sowing. There is need for further experimentation in order to ascertain the best seed-raising procedures. Cuttings of firm young growth usually provide best results but they may be slow to form roots. Application of rooting hormones can be beneficial. Excessive moisture on the hairy stems and leaves of cuttings is often troublesome.

Newcastelia bracteosa F. Muell.
(conspicuous bracts)
SA, WA, NT
0.3–1 m × 0.5–1.5 m   Aug–Dec; also sporadic

N. tenuifolia, S. T. Blake, from southern NT and the Darling Downs, Qld, has a tussocking growth habit, similar to that of spinifex-grass. Its stems have glabrous nodes. The strongly inrolled leaves are very slender.

Neurachne queenslandica S. T. Blake —
see *N. munroi* (F. Muell.) F. Muell.

Neurachne tenuifolia S. T. Blake —
see *N. munroi* (F. Muell.) F. Muell.

NEWCASTELIA F. Muell.
(after Henry P. Clinton, 5th Duke of Newcastle, Secretary of State for the Colonies, 1852–54)
Chloanthaceae
Small perennial **shrubs**, densely covered with cottony or woolly hairs; **leaves** decussate or in whorls of 3, simple, sessile, entire, margins often recurved, venation reticulate; **inflorescence** terminal, solitary spikes or with 1 or more lateral pairs of spikes, or in semi-globular heads, usually with crowded flowers, rarely loosely arranged; **flowers** bisexual, more or less sessile, 2–3 per axil of bract; **calyx** 5–6-lobed; **corolla** bell-shaped, usually 5–6-lobed, exterior glabrous, interior hairy; **stamens** 5–6, inserted between the lobes; **fruits** drupe-like, indehiscent, 1–2-seeded.

Newcastelia is a small endemic genus which now has about 9 species. Recent botanical study resulted in the transfer of 3 species to *Physopsis*. Plants are mainly distributed in arid regions. They vary from dwarf to tall shrubs and are very conspicuous due to the dense covering of hairs on stems and leaves, as well as the ornamental, terminal, woolly-hairy inflorescences. Even though they have desirable ornamental characteristics, to date they are not popular in cultivation. This is due

Newcastelia cephalantha var. *cephalantha*   T. L. Blake

Dwarf **shrub**; **branches** spreading to ascending, densely covered with yellowish-brown woolly hairs; **leaves** 1.5–2.5 cm × 0.3–0.7 cm, oblong to oblong-lanceolate, decussate, spreading to ascending, sessile, widely spaced, initially densely greyish to yellowish-brown, margins recurved, apex bluntly pointed; **inflorescence** of 1 or more cylindrical spikes, 4–8 cm × 0.5–0.7 cm, woolly, purplish-grey, bracts prominent and deciduous as flowers open; **flowers** to about 0.25 cm long, bell-shaped, purple-violet, finely lobed; **calyx** 5–6-lobed, exterior glandular and densely woolly-hairy.

N. bracteosa occurs in north-western SA, southern NT and in the Carnegie, Giles, Helms and Coolgardie Districts of WA. Plants are found in deep red sandy soils. This very attractive species must have excellent drainage in a very warm to hot aspect. It is best suited to semi-arid and arid regions, but is also worth trying in warm temperate areas. It is hardy to moderate frosts. Plants have limited potential for floriculture because of they have relatively short branches. Propagate from seed or from cuttings of firm young growth.

N. elliptica Munir is a synonym.

Newcastelia cephalantha F. Muell.
(flowers in heads)
Qld, SA, WA, NT
0.5–1.3 m × 0.6–1.5 m   July–Oct; also sporadic

Dwarf to small **shrub**; **branches** ascending to erect, densely covered with brownish-grey woolly hairs; **leaves** 0.8–2.3 cm × 0.2–0.8 cm, ovate to oblong-lanceolate, decussate, sessile, spreading to ascending, greyish woolly-hairy, margins recurved, venation visible on undersurface, apex pointed; **spikes** somewhat globular, 0.8–1.8 cm × 0.8–1.8 cm, terminal, sessile, greyish-woolly; **flowers** to 0.5 cm long, bell-shaped, purple-violet, usually 5-lobed, pale yellow anthers prominent; **calyx** 5- or, rarely, 4-lobed, exterior glandular and densely woolly-hairy; **fruit** to about 0.3 cm across, globular, glabrous, glandular.

The typical variant of this variable species occurs in dry inland regions where it often grows in moisture-retentive sandy loams. It has ornamental flowers and foliage which makes it worthwhile for cultivation. A hot, well-drained site is required. Plants withstand moderately heavy frosts. May adapt best to container cultivation in warm temperate regions.

There are 2 other varieties:

var. *oblonga* Munir occurs in south-western Qld, the Austin and Helms Districts in WA, and in southern NT. It differs in having leaves 1–2.8 cm × 0.4–0.8 cm, and oblong flower-spikes.

var. *tephropepla* Munir is from north-western SA, the Ashburton, Austin and Carnegie Districts in WA, and the Central Australian region of NT. It has whitish-grey woolly hairs and often crowded leaves of 0.6–1.5 cm × 0.2–0.6 cm. The flowerheads are very fluffy.

Propagate all varieties from seed, or from cuttings of firm young growth.

N. insignis is allied but differs in having greyish-woolly flowerheads.

Newcastelia chrysophylla C. A. Gardner =
Physopsis chrysophylla (C. A. Gardner) Rye

Newcastelia chrysotricha F. Muell. =
Physopsis chrysotricha (F. Muell.) Rye

Newcastelia cladotricha F. Muell.
(branched hairs)
WA, NT
0.3–1 m × 0.3–1 m   May–Sept

Dwarf **shrub**; **branches** densely covered in greyish-rusty hairs; **leaves** 1–3 cm × 0.3–1.5 cm, oblong, decussate, spreading to ascending, more or less sessile, rounded at base, hairy all over, reticulate venation below, margins slightly recurved, apex pointed; **inflorescence** usually a solitary spike to 12 cm long, woolly purplish-grey, initially crowded when young, becoming interrupted and lax, often prominently arched, sometimes profuse and very conspicuous; **flowers** to 0.6 cm long, tubular, purple-violet, with 5 pointed lobes, solitary in axil of each bract; **stamens** and **style** not exserted; **calyx** 5-lobed, densely covered in purplish-grey hairs; **fruit** to about 0.3 cm across, globular, glabrous, glandular.

A fairly widespread species from arid regions of northern WA and western NT. Needs very well-drained soils and a hot aspect. Best suited to semi-arid and arid regions but may prove adaptable in warm temperate areas. Hardy to light or moderate frosts. Tip pruning while young should promote bushy growth. Propagate from seed or from cuttings of firm young growth.

N. interrupta is allied, but differs in having strongly recurved leaves and more crowded spikes. *N. velutina* has 3 flowers per axil of each bract.

Newcastelia elliptica Munir = *N. bracteosa* F. Muell.

Newcastelia hexarrhena F. Muell.
(six stamens)
WA   Lamb's Tails
0.3–1.3 m × 0.3–1.5 m   July–Oct

Dwarf to small **shrub**; **branches** densely covered with greyish-woolly hairs; **leaves** 3–7 cm × 0.8–1.8 cm, lanceolate to oblong-lanceolate, decussate, ascending to erect, sessile, thick, covered in greyish-woolly hairs, margins recurved to revolute, apex pointed; **inflorescence** solitary, spike to 14 cm × 1.5 cm, or with 1 or more lateral pairs of spikes, white woolly-hairy, with deciduous stalked bracts, most conspicuous; **flowers** to 0.65 cm long, bell-shaped, purple-lilac with 6 pointed lobes; **stamens** 6; **calyx** 6-lobed, densely covered with long white-woolly hairs; **fruit** to about 0.25 cm across, globular, woolly-hairy.

This very ornamental species from central WA deserves greater recognition. The woolly foliage and long flower-spikes make it desirable for use in semi-arid and possibly warm temperate regions. It requires excellent drainage and plenty of sunshine. Plants are hardy to moderate frosts. It has potential for floriculture and is also worth trying as a container plant. Propagate from seed or from cuttings of firm young growth.

N. bracteosa is similar, but has thicker and larger leaves, as well as longer and thicker flower-spikes.

N. hexarrhena is sometimes confused with *Lachnostachys verbascifolia*, which differs in having very short corolla lobes.

Newcastelia insignis E. Pritzel
(remarkable, distinguished)
WA
0.3–1 m × 0.3–1 m   Sept–Nov

Dwarf **shrub**; **branches** densely covered with greyish-yellow hairs; **leaves** 3–7 cm × 0.5–1.2 cm, linear-lanceolate, decussate, mostly erect, sessile, somewhat overlapping, densely covered with greyish-yellow hairs, margins recurved, apex pointed; **inflorescence** terminal, corymb of sub-globular to sub-oblong heads, 1–3 cm × 1–1.4 cm, bright yellow, woolly-hairy, very conspicuous; **flowers** to 0.55 cm long, bell-shaped, pale yellow to yellow, with 6–8 narrow lobes; **calyx** 6–8-lobed near apex, covered in yellow-woolly hairs, **fruit** not seen.

A goldfield species from the Comet Vale–Coolgardie Region. It has attractive characteristics which give it potential for cultivation. Plants require very good drainage and plenty of sunshine. It will be best suited to semi-arid and warm temperate regions and should tolerate moderate frosts. May grow well in containers in warm to hot protected sites in cool temperate regions. Tip pruning from an early age should promote bushy growth. Propagate from seed or from cuttings of firm young growth.

Newcastelia interrupta Munir
(interrupted)
Qld
0.5–1 m × 0.6–1.5 m   Aug–Feb

Dwarf **shrub**; **branches** densely covered in short greyish hairs; **leaves** 1–3.5 cm × 0.3–0.6 cm, oblong-lanceolate, decussate or in whorls of 3, especially below the flowerheads, ascending to erect, sessile, covered in greyish hairs, margins strongly recurved, apex pointed; **inflorescence** terminal, solitary spike 4–18.5 cm × 0.6–0.8 cm, or branched spikes, initially short then becoming long and interrupted, greyish-woolly, peduncles purplish; **flowers** to 0.4 cm long, tubular, purple-violet, with 5 or, rarely, 6 narrow lobes; **stamens** and **style** not exserted; **calyx** 5- or, rarely, 6-lobed, densely covered with greyish-woolly hairs: **fruit** to 0.15 cm across, obovoid, sparsely hairy above and glabrous below.

One of the few species which occurs in Qld, it is distributed in the Darling Downs and Maranoa Districts of southern Qld. It inhabits open forest, where it can be dominant, and is usually found in sandy soils. Plants have potential for cultivation in semi-arid and warm temperate regions. They will need good drainage and maximum sunshine. Hardy to moderate frosts. May prove suitable for floriculture. Tip pruning from a young stage is recommended if bushy growth is desired. Propagate from seed or from cuttings of firm young growth.

N. cladotricha is closely allied, but differs in having oblong, densely hairy leaves with strongly recurved margins.

Newcastelia roseoazurea Rye
(pink and blue)
WA
0.3–1 m × 0.6–1.5 m   July – Sept

Dwarf **shrub**; **branches** spreading to ascending, densely covered in whitish to rusty hairs; **leaves** 0.8–3.8 cm × 0.4–1.9 cm, oblong to broadly ovate, decussate, ascending to erect, sessile or very short-stalked, somewhat rough and grey-green with whitish-grey hairs above and rusty hairs near apex, below with white to rusty hairs, margins recurved, apex blunt; **inflorescence** of 1 or more interrupted spikes to about 10 cm × 1–1.5 cm, pink and blue, terminal, cylindrical, most conspicuous; **flowers** to about 0.4 cm long, bell-shaped, deep pink to red in bud, opening blue; **calyx** 5- or rarely 6-lobed, exterior densely covered in bright pink hairs; **fruit** not seen at maturity.

An eyecatching somewhat mounding species which is found in the Rudall River region of the Eremaean Botanical Province. It grows in association with spinifex on red sandy soils. Evidently not cultivated, it has high potential for use in semi-arid and warm temperate regions. Plants will need excellent drainage and plenty of sunshine. Should be hardy to moderate frosts. Tip pruning from an early stage should help to promote bushy growth. Propagate from seed or from cuttings of firm young growth.

N. spodiotricha is closely allied but it has leaves with a densely white-hairy undersurface, and whitish-hairy flower-buds.

Newcastelia spodiotricha F. Muell.
(ash-grey hairs)
QLD, SA, WA, NT
1–2 m × 1–2 m   Sept–Dec; also sporadic

Small **shrub**; **branches** densely covered in grey hairs; **leaves** 3–6.5 cm × 1.2–3 cm, broadly ovate, decussate, ascending to erect, sessile or short-stalked, covered in whitish-grey hairs, reticulate venation visible on under-surface, margins somewhat revolute, apex pointed; **inflorescence** terminal, rarely solitary, usually 1–3 lateral pairs of spikes, 3–8.5 cm × 1–1.5 cm, densely greyish-hairy, very conspicuous; **flowers** to 0.8 cm long, tubular, bluish, with 5 narrow lobes; **stamens** and **style** exserted; **calyx** 5-lobed, glandular and densely hairy; **fruit** to 0.3 cm across, somewhat globular, glabrous, sparsely glandular.

An impressive, grey-foliaged, arid region species from far south-western Qld, northern SA, northern WA and mainly in southern NT. Plants require excellent drainage and maximum sunshine, and should tolerate moderate frosts. Potentially useful in semi-arid and arid regions, and it may prove adaptable in warm temperate zones. Plants may be suitable for floriculture. Propagate from seed, or from cuttings of firm young growth.

Newcastelia velutina Munir
(velvety)
Qld
not known   not known

Erect **shrub**; **branches** densely covered in woolly hairs; **leaves** to 4 cm × 1.2 cm, broadly lanceolate, decussate, spreading to ascending, sessile, densely covered in greyish hairs, margins recurved, apex pointed; **inflorescence** terminal branched spikes, 6–17 cm long, elongated, densely greyish woolly-hairy; **flowers** to about 0.4 cm long, somewhat bell-shaped, with 5

Newcastelia spodiotricha   T. L. Blake

pointed narrow lobes, 3 in axil of each bract; **stamens** and **style** not exserted; **calyx** 5-lobed, exterior glandular and densely hairy; **fruit** not seen.

This species has soft velvety foliage. It is known only from one collection in the Burnett District, and needs to be introduced to cultivation as a conservation measure. Should grow well in semi-arid and warm temperate regions. Good drainage and plenty of sunshine seem likely to be a prerequisite. Probably hardy to moderate frosts. Propagate from seed (when it is collected), or from cuttings of firm young growth.

N. cladotricha is allied, but differs in having ovate leaves. *N. interrupta* has some similarities but it usually has leaves in whorls of 3.

Newcastelia viscida E. Pritzel =
Physopsis viscida (E. Pritzel) Rye

NICOTIANA L.
(after Jean Nicot, 16th-century French ambassador to Portugal, who sent tobacco plant seeds to France in 1560)
Solanaceae   Wild Tobaccos

Annual or perennial **herbs**, **shrubs** or small **trees**; **leaves** entire, alternate, small to large, glabrous or hairy, odourous; **inflorescence** panicles, or rarely racemes or flowers, solitary; **flowers** bisexual, tubular with 5 spreading lobes, white but often with green or purple or yellow; **stamens** 5; **style** slender; **stigma** bilobed; **fruit** a dehiscent capsule, 4-valved or rarely 2-valved, ovoid or ellipsoid.

Nicotiana comprises about 65 species, with its main representation in North and South America. It also occurs on some Pacific Islands and extends to Australia where there are about 18 species, all of which are endemic, with the exception of the introduced

weedy *N. glauca*. The most famous (or infamous) species in this genus is *N. tabacum* which is commonly known as tobacco.

Nicotiana is represented in all states, with the majority of species occurring in semi-arid and arid regions. Although they are prevalent in drier areas, plants usually occupy niches which are sheltered by rock outcrops and tall trees and they grow in soils which are usually moisture-retentive.

The flowers of this genus are extremely interesting, as they close in sunlight and only open in very dull daylight or during the night. They also have strong and sometimes sweet perfumes, which are noticeable when the flowers open. This phenomenon is undoubtedly to attract pollinating insects.

Aborigines found many species valuable as chewing tobacco before the arrival of Europeans. *N. benthamii*, *N. excelsior* and *N. gossei* were highly regarded for this purpose. Others were rarely used for chewing, eg. *N. megalosiphon*, *N. occidentalis* and *N. rosulata* ssp. *ingulba*.

Exotic *Nicotiana* species have been used for many years as short-term bedding plants and there is no reason why some of the Australian species cannot be used for the same purpose. They have potential for use in informal drifts as well as for massing or growing singly in containers. Although generally regarded as annuals, some plants are short-lived perennials in the wild. Plants trials may reveal how to extend their usefulness in gardens. Some species branch near the base and removal of the tip bud on young seedlings may produce bushy plants, as well as possibly helping to prolong their life span.

Most species appreciate a sunny aspect and they should tolerate a fair degree of shade, as long as the sites are not too cold. Plants should respond well to slow-release and quick-release fertilisers. Supplementary watering may be needed during dry periods, as they appreciate moist soils.

Nicotiana species also have potential for regeneration projects within their natural range because they grow very quickly to provide soil cover.

Propagation is from seed, which does not require pre-sowing treatment and usually germinates within 15–50 days after sowing. Cuttings are rarely used, but soft wood or semi-hardwood, especially the young branchlets from the base of plants, should strike readily.

Nicotiana amplexicaulis N. T. Burb.
(stem-clasping)
Qld
0.6–1.3 m × 0.3–0.6 m   sporadic

Annual **herb**; **stems** 1 to a few, hairy; **leaves** to 30 cm × 1–19 cm, lanceolate to broadly elliptical, often stem-clasping at base, sometimes with winged stalks to 8 cm long, green, hairy, margins entire or irregularly lobed, apex pointed; **inflorescence** an elongated, few-branched panicle on leafless stem; **flowers** to 2 cm × 0.35 cm, white, with blunt or notched lobes, closing in sunlight, fragrant; **calyx** to 1.4 cm, lobes pointed; **capsules** 0.5–0.9 cm, ellipsoid to ovoid.

Restricted to the Leichhardt District where it grows at the bases of sandstone cliffs often with protection from overhanging rocks. *N. amplexicaulis* has limited potential. Its scented flowers open at night. Plants are worth trying in semi-arid and seasonally dry subtropical regions. Needs freely draining soils and a semi-shaded site. Propagate from seed.

N. grossei is allied, but it is a more robust species with larger flowers.

Nicotiana benthamiana Domin
(after George Bentham, 19th-century English botanist and author of *Flora Australiensis*)
Qld, WA, NT
0.4–1.3 m × 0.3–0.6 m   Jan–July; also sporadic

Annual **herb** with sticky and hairy young growth; **stems** often branched near base, glandular, hairy; **leaves** 0.7–23 cm × 0.3–15 cm, narrowly to broadly ovate, lowest ones broadest and stalked, upper ones becoming narrower and sessile, hairy, soft, green, margins entire, shallowly lobed or toothed, apex pointed; **flowers** to about 5.5 cm × 3 cm, white, closing in sunlight, fragrant, solitary in axils of small leaves along upper parts of stems, often profuse; **calyx** 0.8–1.7 cm, with loose and spreading lobes, hairy; **capsules** 0.6–1.1 cm long, ovoid to ellipsoid.

A sweetly fragrant widely distributed species with disjunct populations. It usually occurs on sheltered, rocky hillsides in western Qld, northern WA and throughout NT. Cultivated to a limited degree it has potential for greater use and may adapt to temperate regions. A sunny, moist location is required. Tolerates alkaline soils. Worth trying in containers which can be moved to a different location during flowering. This species was highly prized by Aborigines as a chewing tobacco. Propagate from seed.

Nicotiana occidentalis ssp. *obliqua*   D. L. Jones

Nicotiana burbidgeae Symon
(after Nancy T. Burbidge, 20th-century Australian botanist)
SA
0.5–1 m × 0.3–0.6 m   June–Sept

Annual or short-lived perennial **herb** with sticky and hairy young growth; **stems** branched at base, lower ones woody; **leaves** to 10 cm × 7 cm, subcordate to obovate, sessile at the broad base, thick, somewhat fleshy, glandular-hairy, apex pointed; **flowers** 4–7 cm × up to 5 cm, white, fragrant, closing in sunlight, solitary in axils of leaves along upper parts of stems, forming a leafy spike, very conspicuous; **calyx** to 2 cm long; **capsules** to 1.2 cm long, ovoid to oblong.

A recently described (1984) species, known only from an isolated population in the far northern region of the Lake Eyre District where it grows on sandy soils of creek beds or gibber plains. A very ornamental species with relatively large flowers which may prove successful in semi-arid and warm temperate regions. It may adapt to container cultivation in cooler climates. Tip pruning of young plants may promote bushy growth. Propagate from seed.

Nicotiana cavicola N. T. Burb.
(inhabiting cavities or holes)
WA
0.5–1 m × 0.3–0.6 m   July–Dec

Annual **herb** with sticky and hairy young growth; **stems** branched at base, glandular-hairy; **leaves** 1–20 cm × 1–12.5 cm, broadly ovate to broadly cordate, basal and along stems where they are smaller with winged stalk, glandular-hairy, soft, margins shallowly lobed, undulate or toothed, apex tapering to a point; **inflorescence** an elongated raceme; **flowers** 2.2–4.5 cm × 1–4 cm, white, closing in sunlight, fragrant, very conspicuous; **calyx** 0.7–2 cm long, with narrow lobes; **capsules** 0.6–1.2 cm long, ovoid to ellipsoid, hairy.

Although a dweller of arid regions in mid-western WA, it is always found in sheltered sites among rocks or cliffs and breakaways where it grows in gravelly or sandy soils. It has potential for cultivation due to its

Nicotiana cavicola × .5

fragrant flowers which open at night. Needs very well-drained soils and ample moisture. Should be hardy to moderate frosts and may adapt to temperate regions. Propagate from seed.

Nicotiana debneyi Domin
(after G. L. Debney)
Qld, NSW
0.6–1.5 m × 0.4–1 m   Aug–March; also sporadic
   Annual or short-lived perennial **herb; stems** solitary or branched near base, hairy; **leaves** 1.5–25 cm × up to 13.5 cm, linear to elliptical, lower ones broadest and stalked, upper ones becoming narrower and sessile to stem clasping, light green, hairy, margins entire or very shallowly lobed, apex blunt to pointed; **inflorescence** an elongated much-branched, glandular-hairy panicle; **flowers** 1–2.3 cm × 0.6–1.3 cm, white, with broad lobes, closing in sunlight; **calyx** to 1 cm long, light green, narrowly lobed, hairy; **capsules** 0.5–1.1 cm long, ovoid to ellipsoid.
   The typical subspecies occurs south of Cairns in north-eastern Qld and extends to the Central Coast of NSW and Lord Howe Island. It is found in a wide range of habitats and should adapt to most soils in sub-tropical regions. May prove successful in temperate areas. It will tolerate a sunny or semi-shaded site and needs relatively good drainage, although appreciating an ample supply of moisture.
   The subspecies *monoschizocarpa* P. Horton is known only from the Daly River and Reynolds River region in north-western NT where it is found in clay soils beside rivers. It differs in having narrow corolla lobes and 2-valved capsules.
   Propagate both subspecies from seed.

Nicotiana excelsior (J. M. Black) J. M. Black
(taller, more noble)
SA, ?WA, NT
0.5–1.5 m × 0.3–0.8 m   April–Sept; also sporadic
   Annual or short-lived perennial **herb; stems** usually solitary, more or less glabrous; **leaves** 1.5–25 cm × up to 14 cm, elliptical or ovate or lanceolate, mainly on stem where they are sessile and decurrent, basal leaves have winged stem-clasping stalks, more or less glabrous, margins entire or shallowly lobed or sometimes finely toothed, apex pointed; **inflorescence** a few-branched, faintly hairy panicle; **flowers** 4–7 cm × 2–5 cm, white with blunt lobes which can be purplish on underside, closing in sunlight; **calyx** 1.5–3 cm long, narrowly lobed; **capsules** 1.2–2 cm long, ovoid to ellipsoid.
   A vigorous species which is confined mainly to the mountains of north-western SA and southern NT, where it grows in rocky gullies and creeklines. Although recorded in WA it is thought to have been spread there by Aborigines, who valued it for chewing. Rarely cultivated, it needs maximum temperatures and excellent drainage. Should be best suited to arid and semi-arid regions but may succeed in a warm, protected site in cooler areas. Hardy to moderate frosts. Propagate from seed.

Nicotiana exigua H. -M. Wheeler = *N. suaveolens* Lehm.

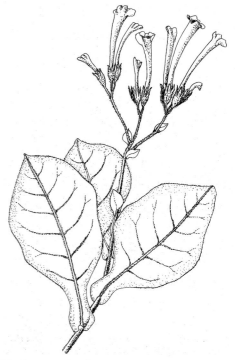

Nicotiana gossei × .5

Nicotiana goodspeedii H. -M. Wheeler
(after Thomas H. Goodspeed, 20th-century Californian botanist who specialised in *Nicotiana*)
NSW, Vic, SA, WA   Small-flowered Tobacco;
                                    Smooth-flowered Tobacco
0.5–1 m × 0.3–0.6 m   throughout the year
   Annual or short-lived perennial **herb; stems** many, leafless or with a few leaves, usually glabrous; **leaves** 0.5–19 cm × up to 5.5 cm, narrowly elliptical to spathulate basal leaves, stalked and largest upper leaves more or less sessile and narrow, glabrous, margins entire or very shallowly lobed, apex pointed; **inflorescence** an elongated few-branched panicle; **flowers** 1–2 cm × 0.6–1.2 cm, white to creamy-white, with blunt or notched lobes, closing in sunlight, fragrant; **calyx** 0.4–1.2 cm long; **capsules** 0.5–1 cm long, ellipsoid.
   A wide-ranging species of the drier regions, where it grows in open sites on alkaline soils. Not very ornamental in appearance but its fragrant flowers are delightful at night. Best suited to semi-arid and warm temperate regions. Plants will need good drainage and maximum sunshine. Hardy to moderate frosts. Propagate from seed, which usually begins to germinate 10–20 days after sowing.

Nicotiana gossei Domin
(after William C. Gosse, 19th-century explorer of Central Australia)
SA, NT   Native Tobacco
0.6–2 m × 0.3–1.5 m   July–Sept; also sporadic
   Annual or short-lived perennial **herb** with densely hairy young growth; **stems** 1–few, leafy, hairy, often

Nicotiana maritima × .45

woody at base; **leaves** 5–35 cm × 2–16 cm, broadly elliptical to lanceolate, few basal with broadly winged stem-clasping stalks, stem leaves smaller but still stem-clasping, soft, hairy, margins entire to shallowly lobed, sometimes undulate, apex blunt but with a distinctive point; **inflorescence** a few-branched panicle; **flowers** 3–6.5 cm × 1.5–3.5 cm, white, sometimes with purplish tonings, lobes blunt, closing in sunlight, fragrant; **calyx** 1.2–3.1 cm long, very narrow lobes; **capsules** 0.8–1.6 cm long, ellipsoid to ovoid.

N. gossei occurs in fertile loams or sandy loams, among rocks, often in shady crevices of the mountain range in north-western SA and southern NT. A robust species with potential for semi-arid and warm temperate areas. Needs a warm protected site. Its fragrant flowers open in the evening. Highly regarded for chewing by Aborigines who harvested leaves carefully without endangering the life of plants. Propagate from seed. Leaf extracts show potential for insect control.

N. amplexicaulis is allied, but it is not as vigorous and has smaller flowers.

Nicotiana maritima H. -M. Wheeler
(growing by the sea)

Vic, SA	Coast Tobacco
0.4–1 m × 0.3–0.6 m	Aug–Dec; also sporadic

Annual or short-lived perennial **herb** with greyish-hairy young growth; **stems** 1–few, white to grey woolly hairs at base; **leaves** 2–22 cm × up to 14 cm, spathulate to linear, mostly basal, with long slightly stem-clasping stalks, upper ones more or less sessile, hairy, margins usually shallowly lobed and sometimes undulate, apex pointed; **inflorescence** an elongated panicle; **flowers** 1.3–3 cm × 0.7–2.4 cm, white, with blunt or notched lobes, closing in sunlight; **calyx** 0.6–1.6 cm long; **capsules** 0.5–1.2 cm long, ellipsoid to broadly ellipsoid.

Coast Tobacco extends from far western Vic to the Eyre Peninsula, where it grows mainly on and near the coast, in sand or gravelly soil among rocks. It lacks strong ornamental features although it has fragrant flowers which open at night. Needs good drainage and a sunny or semi-shaded site. Propagate from seed.

Nicotiana megalosiphon Van Heurck & Müll-Arg.
(large tube)

Qld, NSW	Long-flowered Tobacco
0.5–1 m × 0.2–0.4 m	March–Nov; also sporadic

Annual or short-lived perennial **herb**; with hairy young growth; **stems** 1–few, branched at base; **leaves** 1–17 cm × up to 9 cm, elliptical or ovate to linear, broadest and largest near base, stalked, dark green above, paler below, hairy, margins entire or shallowly lobed, apex pointed; **inflorescence** an elongated few-branched panicle; **flowers** 3.4–9 cm × 1.3–3.5 cm, white with notched lobes, closing in sunlight, fragrant; **calyx** 1–2.1 cm long; **capsules** 0.7–1.6 cm long, ovoid-ellipsoid.

The typical subspecies occurs in south-eastern mid-central Qld and mid-northern and north-eastern NSW. Plants are often found in disturbed sandy loam or clay soils. It has well displayed fragrant flowers and may have potential for cultivation. Prefers a warm to hot open site but will tolerate some shade. Soils need to be moderately drained. Hardy to most frosts.

The subspecies *sessiliflora* P. Horton differs in having stems with sessile leaves. It occurs in western Qld and central to southern NT. Plants are most commonly found along streams and dry creeks.

Propagate from seed.

Nicotiana occidentalis H. -M. Wheeler
(western)

WA	
0.5–1.5 m × 0.3–0.8 m	July–Oct; also sporadic

Annual or short-lived perennial **herb**; with sticky and hairy young growth; **stems** 1–several, glandular-hairy; **leaves** 2–20 cm × up to 9 cm, ovate, elliptical, lanceolate or linear, basal ones broadest and with stalks to 16 cm long, upper ones becoming sessile and narrow, glandular-hairy, margins entire or shallowly lobed, apex pointed; **inflorescence** an elongated few-branched panicle; **flowers** 2.6–5.4 cm × 1–2.5 cm, white, with notched lobes, closing in sunlight, fragrant; **calyx** 0.5–1.4 cm long, glandular-hairy; **capsules** 0.7–1.4 cm, ovoid to narrowly ellipsoid.

The typical subspecies occurs in the Ashburton, Carnarvon and Fortescue Districts where it is most common on or near the coast and on offshore islands. It has well displayed fragrant flowers and will be best suited to semi-arid, warm temperate and seasonally dry subtropical regions. It prefers some shade and needs good drainage.

This variable species has a further 2 subspecies.

The subspecies *hesperis* (N. T. Burb.) P. Horton has much smaller flowers, to 1.8 cm long, and is a coastal species, mainly in the same area as ssp. *occidentalis*.

Nicotiana rosulata

The subspecies *obliqua* N. T. Burb. occurs in Qld, NSW, SA, WA and NT. It has flowers of 1.5–4 cm long. It is found usually in sheltered locations in sandy or rocky sites.

Propagate all subspecies from seed, which generally begins to germinate 11–20 days after sowing.

Nicotiana rosulata (S. Moore) Domin
(a small rosette)
SA, WA
0.5–1 m × 0.2–0.5 m   Aug–Sept; also sporadic
Annual or short-lived perennial **herb**; with hairy young growth; **stems** 1–few, glabrous except for base; **leaves** 3–21 cm × up to 11.5 cm, spathulate to linear, with stalks to 8 cm long, upper leaves narrowest and more or less sessile, soft, hairy, margin entire or shallowly lobed, apex blunt or pointed; **inflorescence** an elongated few-branched panicle; **flowers** 1.5–4.5 cm × 0.7–2.5 cm, creamy-white, with notched lobes, closing in sunlight, fragrant; **calyx** 0.6–1.5 cm long; **capsules** 0.6–1.6 cm long.

A variable species with the typical subspecies occurring mainly along tree-lined creek beds, over a wide range in central WA and western SA. It commonly grows in sandy soils which are sometimes gravelly or stony. Needs good drainage and a warm site. Should do best in semi-arid and warm temperate regions. Hardy to most frosts.

The subspecies *ingulba* (J. M. Black) P. Horton differs in having mainly glabrous stems and leaves, and its flowers are usually larger, from 2–6.4 cm long. It occurs in southern NT and in WA near the NT/SA border.

Propagate both subspecies from seed.

Nicotiana rotundifolia Lindl.
(round leaves)
0.4–1 m × 0.2–0.6 m   Aug–Dec; also sporadic
Annual or short-lived perennial **herb**; with sticky and hairy young growth; **stems** 1–few, hairy; **leaves** 1.5–25 cm × up to 16 cm, broadly elliptical to linear, mainly basal with stalks to 14 cm long, hairy, margin entire to shallowly lobed, apex blunt to pointed; **inflorescence** an elongated few-branched panicle; **flowers** 1.3–1.8 cm × 0.5–1.6 cm, white, often with deeply notched lobes, closing in sunlight, fragrant; **calyx** 0.5–1.3 cm long; **capsules** 0.4–0.8 cm long.

A somewhat insignificant species with fragrant flowers, from south-western WA where it grows in a wide range of habitats, but generally with protection from boulders and trees. Needs good drainage and should adapt to semi-arid and temperate regions. Hardy to moderate frosts. Propagate from seed, which usually begins to germinate 10–50 days after sowing.

Nicotiana simulans N. T. Burb.
(similar)
Qld, NSW, SA, WA, NT
0.5–1.2 cm × 0.2–0.6 cm   May–Oct; also sporadic
Annual or short-lived perennial **herb**; young growth hairy ; **stems** 1 to many, hairy; **leaves** 2–23 cm × up to 12 cm, broadly spathulate to lanceolate, sometimes mainly basal, usually with stem-clasping stalks to 9 cm long, hairy, margins usually entire, apex usually

Nicotiana velutina   T. L. Blake

pointed; **inflorescence** an elongated few-branched panicle; **flowers** 2.2–4.2 cm × 0.6–2.4 cm, white and usually notched, closing in sunlight, fragrant; **calyx** 0.5–1.7 cm long; **capsules** 0.5–1.3 cm long, usually shorter than the calyx.

N. simulans occurs over a wide range in semi-arid and arid regions. It is usually found in sandy or rocky soils and often grows in the shelter of boulders and trees. Worth trying in semi-arid and warm temperate regions. Plants will need very good drainage and a warm but sheltered and protected site. Hardy to moderate frosts. Propagate from seed.

N. megalosiphon is similar but has smaller flowers.

Nicotiana suaveolens Lehm.
(sweetly scented)
NSW, Vic, Tas   Austral Tobacco
0.5–1.5 m × 0.3–0.6 m   Sept–May; also sporadic
Annual or short-lived perennial **herb**; **stems** 1–few, somewhat woody and hairy at base; **leaves** 2–31 cm × up to 19.5 cm, ovate to linear, both basal and on stems, broadest near base and sometimes hairy on stalks to 16 cm long, upper ones narrow and glabrous, margins entire to shallowly lobed, apex pointed; **inflorescence** an elongated few- to much-branched panicle; **flowers** 1.7–5.5 cm × 1.4–3.7 cm, white to creamy-white, with blunt or notched lobes, closing in sunlight, strongly fragrant; **calyx** 0.6–2.6 cm long; **capsules** 0.7–1.2 cm long, ovoid.

Renowned for its sweet fragrance, this species deserves to be more widely cultivated. It occurs in eastern and southern NSW and over a wide area of Vic, in a range of sandy and stony soils along watercourses and on rocky slopes. It may also be found as an understorey plant in woodland and shrublands. Best suited to temperate regions and will need relatively good drainage with some protection. Hardy to most frosts. Can be used in gardens or containers. Introduced into England in 1800. Propagate from seed.

Nicotiana umbratica N. T. Burb.
(living in shade)
WA
0.5–0.8 m × 0.3–0.5 m   April–June; possibly also
                                                         sporadic
Annual or short-lived perennial **herb**, with sticky and hairy young growth; **stems** 1 or, rarely, more, glandular-hairy; **leaves** 0.7–12 cm × up to 10 cm, broadly cordate to lanceolate or linear, with narrow stalks to 11 cm long, mostly on stems, hairy, margins entire or shallowly lobed, apex pointed; **inflorescence** an elongated few-branched panicle; **flowers** 2.5–6.6 cm × 2–3.5 cm, white, with pointed or slightly blunt lobes, closing in sunlight; **calyx** 0.6–1.5 cm long; **capsules** 0.6–1 cm long, ovoid-ellipsoid.

N. umbratica is from the western Pilbara region where it grows in shaded sites among rocky outcrops. Plants may prove useful in semi-arid and warm temperate regions. Needs freely draining soils in a warm protected site. It has potential for gardens and containers. Hardy to moderate frosts. Propagate from seed.

N. cavicola has affinity but differs in having winged leaf stalks.

Nicotiana velutina H. Wheeler
(velvety)
Qld, NSW, Vic, SA, NT   Velvet Tobacco
0.5–1.5 m × 0.3–0.8 m   April–Nov; also sporadic
Annual or short-lived perennial **herb**; with velvety hairy young growth; **stems** few to many, usually hairy, sometimes becoming glabrous; **leaves** 1–28 cm × up to 12 cm, spathulate to linear, mainly basal with stalks to 15 cm long, upper ones more or less sessile, woolly-hairy, margins entire or shallowly lobed, apex pointed; **inflorescence** an elongated few- to many-branched panicle; **flowers** 1.1–3.5 cm × 0.7–3 cm, white to creamy-white, sometimes with greenish or purplish tonings, with lobes blunt or notched, closing in sunlight, honey fragrance.

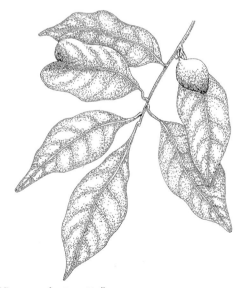

Niemeyera chartacea × .5

Nitraria billardierei   T. L. Blake

A very widespread species which is found almost exclusively on sandy soils of semi-arid and arid regions. The soils are often alkaline or sometimes saline. It is capable of invading disturbed areas and becoming weedy. May be useful for regeneration in eroded and saline regions within its own natural range. Tolerates plenty of sunshine and needs good drainage. Hardy to moderate frosts. Worth trying in gardens and containers. Propagate from seed.

NIEMEYERA F. Muell.
(after Dr Felix Niemeyer, 19th-century German professor of medicine)
Sapotaceae
A monotypic genus endemic in the rainforests of eastern Australia.

Niemeyera chartacea (Bailey) C. White
(papery)
Qld, NSW
10–20 m × 6–10 m   Dec–March
Small to medium **tree**, with rusty-hairy young growth and milky sap; **leaves** 5–15 cm × 1.5–4 cm, lanceolate to oblong-obovate, alternate, base wedge-shaped, stalked, spreading to ascending, thin-textured, becoming glabrous, margins entire and wavy, apex pointed; **flowers** small, to about 0.5 cm across, white, tubular with 5 recurved lobes, 4 or more per axillary cluster; **calyx** about 0.2 cm long, 5-lobed; **fruit** 2–2.5 cm × 1–1.3 cm, ovoid, purplish-black.

A poorly known but fairly widespread species from the Cook, Port Curtis, South Kennedy, Wide Bay and Moreton Districts of eastern Qld, and extending to north-eastern NSW. Suitable for parks and large gardens in tropical and subtropical regions. May also adapt to more temperate climes. Needs organically rich, freely draining soils and most likely protection from hot sun and drying winds when young. Frosts can damage plants. Should respond well to supplementary

37

watering during dry periods. Propagate from fresh seed. Cuttings may prove difficult to strike.

Some botanists place this species in *Chrysophyllum* L.

Niemeyera prunifera (F. Muell.) F. Muell. =
Amorphospermum whitei Aubrév
A. whitei will be described in the Supplement to the Encyclopaedia.

Nigromnia globosa Carolin =
Scaevola globosa (Carolin) Carolin

NITRARIA L.
(from Greco-Latin *nitrum*, saltpetre; referring to the saline plains where the first species was collected)
Zygophyllaceae
Shrubs; **branches** often thorny; **leaves** simple, spiral or clustered, wedge-shaped base, somewhat succulent; **stipules** very small, often deciduous; **inflorescence** a cyme with alternate branches not in the same plane; **flowers** small; **calyx** 5-lobed; **petals** 5 or rarely 6; **stamens** 10–15; fruit a fleshy drupe.

A small genus of about 9 species which extends from the Sahara to Eastern Siberia. One species occurs in Australia. Plants are found in deserts and coastal sand dunes and usually on saline and/or alkaline soils.

Propagation is from seed, which does not require pre-sowing treatment, or from cuttings of firm young growth.

Nitraria billardierei DC.
(after Labillardiere, 18–19th-century French naturalist)
Qld, NSW, Vic, SA, WA, NT   Nitrebush;
Dillon Bush; Wild Grape
1–2 m × 3–5 m   Sept–Nov; also sporadic
Small **shrub**, with hairy young growth; **branches** spreading widely, often arching; **branchlets** often spiny and tangled; leaves 1–4 cm × 0.1–0.3 cm, oblanceolate, alternate or clustered, spreading to ascending, thick, somewhat succulent, green to glaucous, usually hairy, apex blunt; **flowers** about 0.5 cm across, white, in small clusters along the branches; **fruit** 1–2 cm × 0.5–1 cm, oblong, fleshy, golden, purple or red when ripe.

A common plant of overgrazed areas in semi-arid regions. Plants inhabit alkaline or saline clay and loam soils, which can be subject to flooding. Also found on coastal sand dunes. Thought to be introduced into Qld. Often forms large impenetrable mounds and is excellent for soil erosion control and reclamation and also as a fauna refuge. Plants are drought tolerant and hardy to most frosts. Suitable for semi-arid and warm temperate regions. They respond well to supplementary watering during extended dry periods and although unlikely to be required they can be pruned heavily. Stock will graze plants when there is a shortage of other food. The colourful fruits are edible but slightly astringent. A food source for Aborigines, as well as for emus, smaller birds, lizards and marsupials. Propagate from seed, which usually begins to germinate 22–50 days after sowing, or from cuttings of firm but not too young growth.

In the past, this species has been confused with *N. schoberi* L., which occurs from European Russia to Central Asia.

Nitraria schoberi L. — see *N. billardierei* DC.

NOAHDENDRON P.K. Endress, B. Hyland & Tracey
(after Noah Creek and from the Greek *dendron*, a tree)
Hamamelidaceae
A monotypic genus endemic in Northern Qld.

Noahdendron nicholasii P.K. Endress, B. Hyland & Tracey
(after Mr Nicholas)
Qld
3–10 m × 3–6 m   Sept–Nov
Medium to tall **shrub** with stellate-hairy young growth; **trunk** usually erect, with few branchlets; **branchlets** short, slightly pendulous; **leaves** to 30 cm × 10 cm, alternate, oblong to elliptic, narrowing to prominent stalk, arranged in 2 rows, dark green, glabrous, glossy, venation prominent, apex pointed; **stipules** to 2 cm × 1 cm, ovate, some persistent; **inflorescence** a terminal spike to about 7 cm long, on a stalk to about 5 cm long; **flowers** small, red or purple, with inrolled glabrous petals; **stamens** red; **fruit** a 2-celled capsule, about 1 cm × 1 cm, woody; **seeds** about 0.7 cm long.

This species has shiny leaves and slightly pendulous foliage. It has a very limited distribution near Cape Tribulation where it grows in the understorey of lowland rainforests. Plants occur in shallow rocky soils which are moist for most of the year and are covered with organic matter. Rarely cultivated, it is suited to tropical and warm subtropical regions and will require good drainage. Should respond well to copious organic mulch, slow-release fertilisers, and supplementary watering during dry periods. Plants are damaged by frosts. Worth trying as an indoor plant. Propagate from fresh seed and possibly from cuttings.

NOMISMIA Wight & Arn.
(derivation unknown)
Fabaceae (alt. *Papilionaceae*)
Twiners, faintly hairy to hairy; **leaves** pinnate with 3 leaflets, prominently stalked, minute glands on undersurface; **inflorescence** axillary racemes, often on young shoots; **flowers** pea-shaped; **calyx** hairy, deeply 2-lipped, with 5 recurved lobes; **stamens** 10, with 1 free; **fruit** a pod, orbicular, stalked, flattened, 1–2-seeded.

A small tropical genus, with 1 species endemic in northern Australia. Some authors include *Nomismia* in *Rhynchosia* Lour.

Nomismia rhomboidea (Benth.) Pedley
(diamond-shaped)
WA, NT
Climber or trailer   Dec–April
Lightly climbing or trailing **perennial** with hairy young growth which is often sticky; **rootstock** woody; **stems** to about 1.5 m long; **leaves** trifoliolate; **leaflets** to 2.5 cm × 2 cm, rhomboid to nearly circular, slightly hairy, leathery, green, margins flat, apex blunt or slightly pointed; **flowers** pea-shaped, about 0.5 cm long, yellow, sweetly fragrant, in small axillary racemes; **pods** to 1.5 cm × 1 cm, flat, thin valves, reticulate venation.

A poorly known species from far northern WA and

the Victoria River District in NT. Plants inhabit sandy soils of tropical woodlands which are subject to long seasonally dry periods. Although lacking highly decorative qualities, the flowers emit a sweet fragrance. Plants die back to the woody rootstock after flowering. A warm, frost-free area in well-drained soil is required. Regular pruning after flowering may be beneficial in promoting vigour. Propagate from seed, or from cuttings of firm, young growth.

Previously known as *Rhynchosia rhomboidea* Benth.

NORMANBYA Becc.
(after the Marquis of Normanby)
Arecaceae (alt. *Palmae*)

A monotypic genus endemic in north-eastern Qld.

Normanbya normanbyi (W. Hill) L. H. Bail.
(after the Marquis of Normanby)

Qld	Black Palm; Dowar
15–22 m tall	Oct–Jan

Solitary **palm**; **trunk** 15–20 m tall, slender except for enlarged base, light grey; **crownshaft** 0.5–1 m long, mealy-white, enlarged near base; **fronds** 2–2.5 m long, gracefully arching, somewhat feathery; **rhachis** stout, with a covering of mealy-white hairs; **pinnae** 30–45 cm × 2–3 cm, tapering to base, arranged in whorls to give a plumose appearance, dark green above, silvery below, apex jagged; **spathes** 15–30 cm long; **inflorescence** a panicle, 15–35 cm long; **flowers** about 1.5 cm across, cream to pinkish with pink anthers, profuse; **drupes** 3–3.5 cm long, deep pink to scarlet, ovoid to pear-shaped, ripe March–Aug.

An outstanding, majestic palm from coastal regions of north-eastern Qld, where it is found on swamp margins. It is best suited to cultivation in tropical and subtropical areas but also adapts to warm temperate regions where it is slow growing. Needs well-drained and friable, rich loamy soil which is moist for most of the year. Sun scorch can be a problem for young plants, so they need shelter for the first 3–5 years. They respond well to organic mulches and light applications of slow-release fertilisers. Plants are damaged by frost and they dislike extended cool periods as well as extended dry periods. Supplementary watering may be necessary. Plants are excellent for indoors, provided there is constant humidity. The wood of the trunks is almost black with cream streaks and is highly prized because of its long grain which polishes beautifully. It is used for walking sticks and Aborigines made spears from it. Propagate from fresh seed which may begin to germinate after 2 months, but results are usually very sporadic and some seeds take over 12 months to sprout.

NOTELAEA Vent.
(from the Greek *notos*, the south and *elaia*, olive; referring to southern hemisphere members of the olive family)
Oleaceae   Mock Olives

Shrubs or **trees**; **leaves** opposite, simple, usually entire or rarely crenulate; **inflorescence** axillary racemes or sometimes sessile clusters; **bracts** deciduous; **flowers** bisexual, small; **calyx** 4-lobed; **corolla** 4 free petals or joined in 2 pairs by stamens; **stamens** 2;

style 2-lobed, short; **fruit** a drupe with solitary seed.

A small endemic genus of about 11 species. It is distributed in the eastern states extending from north-eastern Qld to western Vic and with one species in Tas. Identification is often difficult because of the propensity of some species to hybridise in the wild.

Notelaea johnsonii P. S.Green
(after Lawrence A. S. Johnson, former Director, Royal Botanic Garden, Sydney and contemporary botanist)

Qld, NSW	Veinless Mock Olive
4–8 m × 3–6 m	April–July; also sporadic

Tall **shrub** or small **tree**; with hairy young growth; **branches** spreading to ascending; **branchlets** faintly hairy or glabrous; **leaves** 4–12 cm × 1–5 cm, usually elliptic to narrowly elliptic, opposite, glabrous, dark green, somewhat glossy, stiff, margins entire or slightly crenulate, main veins usually visible, apex pointed; **racemes** 4–8 cm long, axillary; **flowers** very small, bluish-black; **drupes** 1.8–2 cm × about 1 cm, elliptic, bluish-black.

This attractive species is endemic in central-eastern Qld and north-eastern NSW. It occurs in rainforest. Plants deserve to be better known in cultivation and should do well in tropical and subtropical regions as well as in warm protected sites in temperate areas. A sunny or semi-shaded aspect with well drained soils is recommended. Plants are slow-growing and may be straggly in early years but they respond well to pruning and can develop bushy foliage as they mature. They prefer soils which are moist for most of the year and are hardy to light frosts. Birds are attracted to ripe fruits. Propagate from fresh seed.

Notelaea ligustrina Vent.
(similar to the genus *Ligustrum*)

NSW, Vic, Tas	Privet Mock Olive
6–8 m × 4–6 m	Aug–Sept

Small **tree** often with whitish or grey young growth; **branches** spreading to ascending; **leaves** to 7 cm ×

Notelaea ligustrina × .45

1.5 cm, lanceolate, opposite, short-stalked, glabrous, dull green, somewhat thick-textured, margins entire, veins inconspicuous, apex bluntly pointed; **racemes** short, more or less clustered, short, axillary, often on long stalks; **flowers** small, white; **drupes** to about 1.2 cm long, ovoid to nearly globular, pink to deep purple-black.

A widespread plant in mountain forests of temperate regions, extending from the Southern Tablelands in NSW to western Vic. It is also prevalent over much of Tas. Plants adapt well to cultivation although they can be slow to develop. Sites which receive some sunshine or which are shaded for most of the day are suitable. They prefer freely draining loamy soils which are moist for most of the year but will tolerate limited periods of dryness. Hardy to most frosts. Responds well to tip pruning. Usually requires little attention once well established. Useful for underplanting of tall trees and for screening. Introduced into England from Tas during 1807. The fruits are eaten by birds. Propagate from seed or cuttings of firm young wood.

Some botanists refer to this species as *Nestegis ligustrina* (Vent.) L. Johnson.

Notelaea linearis Benth.
(linear)
Qld, NSW
1–2 m × 1–2 m   Sept–Dec
Small **shrub**, with glabrous or faintly hairy young growth; **branches** spreading to ascending; **leaves** 2–11.5 cm × up to 0.7 cm, linear or rarely narrowly ovate, opposite, short-stalked, thick textured, dark green, glabrous, margins thickened and entire, venation conspicuous, apex pointed; **racemes** to about 1 cm long, glabrous to faintly hairy, axillary, sessile; **flowers** about 0.2 cm across, white to cream; **drupes** to 0.7 cm × about 0.5 cm, ovoid, white or blue.

An interesting shrubby species from south-eastern Qld and north-eastern NSW. It usually occurs in valleys where it inhabits sandy or stony clay soils in eucalypt forests. Well worth growing in subtropical and temperate regions. Needs well-drained soils in a slightly sunny or semi-shaded site. Suitable for general planting. Should respond well to pruning. Probably tolerates light frosts and limited dry periods. Worth trying as a container plant. Propagate from fresh seed and possibly from cuttings of firm young growth.

Notelaea lloydii Guymer
(after Lloyd H. Bird, 20th-century botanical collector, Qld)
Qld
1–4 m × 1–3 m   June–Aug
Small to medium, multi-stemmed **shrub** with hairy young growth; **bark** pale grey, smooth; **branches** spreading to ascending; **branchlets** initially faintly hairy, becoming glabrous; **leaves** 7–14 cm × up to 0.6 cm, linear, slightly falcate, opposite, spreading to ascending, dark green, glabrous except for usually faintly hairy base, margin entire, dotted above and below, venation conspicuous, apex pointed; **racemes** to about 1 cm long, 1 or 2 per axil; **flowers** small, white or cream; **drupes** to about 0.8 cm long, more or less globular, dark blue, ripe Oct–Dec.

A recently described species which is vulnerable in nature. It occurs west to south west of Brisbane, where it is found on the margins of eucalypt forests and vine thickets. It is deserving of cultivation for its ornamental qualities and also as a conservation strategy. Plants grow well in subtropical and warm temperate regions, in semi-shaded, well-drained sites, and are suitable for underplanting of taller trees. Being multistemmed, plants also respond well to pruning. Heavy frosts will damage plants. Birds are attracted to the fruits. Propagate from fresh seed or possibly from cuttings of firm young growth.

Notelaea longifolia Vent.
(long leaves)
Qld, NSW   Large Mock Olive
6–10 m × 4–6 m   May–Oct; also sporadic
Small **tree** with hairy young growth; **branches** spreading to ascending or often somewhat pendant; **branchlets** often pendant, hairy; **leaves** 5–15 cm × up to 8 cm, lanceolate to broadly ovate, opposite, prominently stalked, dull green, hairy, stiff-textured, venation prominent and reticulate, apex pointed; **racemes** to about 2.5 cm long, axillary; **flowers** small, pale yellow, often profuse; **drupes** about 1.5 cm long, ovoid, black, fleshy, very conspicuous and often profuse, ripe Nov–Feb.

The typical form occurs in south-eastern Qld and along the coast of NSW as well as on the Central Tablelands, Southern Tablelands and Central Western Slopes. This species inhabits a range of sands and loams. It is cultivated mainly by enthusiasts but deserves better recognition as an ornamental small tree. Plants often produce copious quantities of black fruits, which are excellent for birds. *N. longifolia* adapts to most freely draining acidic soils and does well in sites which receive plenty of sun or are semi-shaded. Hardy to moderate frosts, it also withstands extended dry periods once well established. In odd seasons, pruning may be needed to lighten the load of fruits on the branches. Excellent for planting beneath established tall trees.

Notelaea longifolia   D. L. Jones

There are 2 other recognised forms which have similar attributes and cultivation needs to the typical form:

forma *glabra* P. S. Green (previously known as *N. longifolia* var. *decomposita* Domin) differs in having glabrous stems and leaves. It occurs in south-eastern Qld and the North Coast District of NSW.

forma *intermedia* P. S. Green is from the Northern Coast and Central Western Slopes of NSW, and has shortly hairy stems and leaves.

Propagate all forms from fresh seed or possibly from cuttings of firm young growth.

Notelaea longifolia Vent. var. **decomposita** Domin =
N. longifolia forma *glabra* P. S. Green

Notelaea longifolia Vent. var. **ovata** (R. Br.) Domin =
N. ovata R. Br.

Notelaea longifolia Vent. var. **pedicellaris** Domin =
N. venosa F. Muell.

Notelaea longifolia Vent. var. **velutina** F. M. Bailey =
N. microcarpa var. *velutina* (F. M. Bailey) P. S. Green

Notelaea microcarpa R. Br.
(small fruits)
Qld, NSW, NT   Velvet Mock Olive
6–10 m × 4–6 m   May–Sept; also sporadic

Small **tree** with furry young growth; **trunk** often crooked; **bark** rough, grey; **branches** spreading to ascending; **branchlets** hairy when young; **leaves** 2–15 cm × 0.3–3 cm, linear ovate to ovate, opposite, stalked, dull green and usually glabrous above, or with scattered hairs above and below, venation prominent and reticulate, apex acute or blunt with a small point; **racemes** 0.5–2 cm long, axillary, hairy; **flowers** small, cream, often profuse; **drupes** 0.7–1 cm × 0.5–8 cm, ovoid, dark purple to black, often profuse, ripe July–Sept.

The typical variety has a wide distribution extending from the Darwin and Gulf District to the Southern Tablelands in NSW. Often occurs in rocky soils and has proved adaptable in cultivation as it grows in sun or semi-shade on soils which have good drainage. Suited to tropical, subtropical and temperate regions. Plants may be damaged by heavy frosts but they withstand extended dry periods. Worth trying as a container plant. Birds avidly feed on the fruit.

The var. *velutina* (F. M. Bailey) P. S. Green differs in having densely velvety hairy leaves. It occurs in south-eastern Qld and north-eastern NSW.

Propagate both varieties from fresh seed or possibly from cuttings of firm young growth.

Hybrids between *N. microcarpa* and *N. ovata* occur in south-eastern Qld, and it is doubtful if these are in cultivation.

Notelaea neglecta P. S. Green
(neglected)
NSW
4–10 m × 3–6 m   Sept–Oct

Tall **shrub** to small **tree**, with softly hairy young growth; **branches** spreading to ascending; **branchlets**

Notelaea microcarpa var. *microcarpa*   D. L. Jones

becoming glabrous; **leaves** 4–10 cm × 0.5–0.7 cm, lanceolate to narrow-elliptical, opposite, stalked, narrow base, softly hairy when young, becoming more or less glabrous, dull green, venation inconspicuous, apex pointed; **racemes** 1–2 cm long, axillary, or sometimes forming leafy spikes; **flowers** small, cream, often profuse; **drupes** about 0.5 cm long, globular, blackish, often profuse.

A poorly known species which is endemic in the Central Tablelands. It usually inhabits open forest and occurs in sandy soils. This species warrants cultivation because of its floriferous nature (although the flowers are minute). Fruit is produced in large quantities because of the profuse flowering. Should adapt to a wide range of well-drained soils and a sunny or semi-shaded site in temperate and subtropical regions. Hardy to moderate frosts and extended dry periods. Worth trying as a container plant. Propagate from fresh seed or possibly from cuttings of firm young growth.

Notelaea ovata R. Br.
(ovate, egg-shaped)
Qld, NSW
4–8 m × 3–6 m   Jan–March; also sporadic

Tall **shrub** or small **tree**; usually with hairy young growth; **bark** greyish, flaky; **branches** spreading to ascending; **branchlets** softly hairy or glabrous; **leaves** 3–12 cm × 2–5 cm, mostly ovate, opposite, base cordate and blunt or with short stalk, dull green, leathery, glabrous or softly hairy, margins wavy and crenulate, venation conspicuous and reticulate above but obscure below, apex pointed; **racemes** to about 2 cm long, prominently stalked, usually attached just above the axil; **flowers** small, cream, often numerous; **drupes** to about 1.5 cm long, globular to ovoid, purplish-black, sometimes numerous.

This species usually occurs in rainforest margins and open forest where it grows in sandy soils which are often associated with sandstone. Evidently rarely cultivated, it should be suited to tropical and subtropical regions in freely draining soil and a sunny or semi-shaded aspect. Useful for growing beneath taller vegetation and as a background plant. May suffer damage from frost. Introduced into England in 1824. Birds eat the fruits. Propagate from fresh seed and possibly from cuttings.

N. longifolia is similar but differs in having leaves with entire margins while *N. venosa* has prominent venation on both surfaces of leaves.

Notelaea punctata R. Br.
(dotted)
Qld
0.6–2 m × 0.6–2 m   sporadic all year

Dwarf to small, often multi-stemmed **shrub**, with glabrous or faintly hairy young growth; **branches** spreading to ascending; **branchlets** glabrous or faintly hairy, spreading; **leaves** 3.5–13 cm × 1–5 cm, narrowly ovate to elliptical, opposite, prominently stalked, dull green, glabrous and somewhat shiny, dotted above and below, margins more or less entire, apex pointed; **racemes** 1–3 cm long, glabrous, axillary; **flowers** small, cream, often profuse; **drupes** 0.7–1 cm × about 0.6 cm, ovoid, purple-black.

N. punctata is widespread in eastern Qld. It occurs on the mountains and plateaux, in open forest and woodland, where it is found on sandy soils. Suitable for sunny or semi-shaded sites in tropical, subtropical and temperate regions. It needs good drainage. Provenances originating from higher altitudes should be more tolerant of frosts, while those from lower elevations may be prone to frost damage. This shrubby species is capable of reshooting from its woody rootstock. Suitable for underplanting of trees and tall shrubs. Worth trying as an indoor and outdoor container plant. Introduced into England during 1826. Birds are attracted to the fruits. Propagate from fresh seed and possibly from cuttings of firm young growth.

Young plants of *N. microcarpa* may be confused with *N. punctata*, but *N. microcarpa* differs in having hairy racemes. Hybrids between these two species occur in the wild.

Notelaea pungens Guymer
(sharp, prickly-pointed)
Qld
0.15–0.6 m × 0.3–1 m   May–July

Dwarf **shrub** with hairy young growth; **branches** spreading to ascending; **branchlets** initially faintly hairy, becoming glabrous; **leaves** 1.1–3.2 cm × 0.4–0.5 cm, lanceolate, opposite, short-stalked, dark green, glabrous except for base, leathery, margins thickened and slightly recurved, apex prickly-pointed; **flowers** about 0.2 cm across, pale yellow to pale green, 1 or 2 per axil; **drupes** 0.7–1 cm × 0.5–0.9 cm, ovoid to globular, dark blue, ripe Nov–March.

This recently described species occurs in open forests in the Darling Downs District, and is known only from a small area near Chinchilla where it grows in sandstone derived soils. It has attractive fruits and is worthy of cultivation in subtropical and warm temperate regions. Also needs to be grown as a conservation strategy. Soils need to drain freely and a sunny or semi-shaded site is suitable. Plant trials in containers may be worthwhile. Container-grown plants may also succeed in cool temperate regions. Heavy frost may damage plants. Birds eat the fruits. Propagate from fresh seed and possibly from cuttings of firm young growth.

Notelaea venosa F. Muell.
(many veins)
Qld, NSW, Vic   Veined Mock Olive; Smooth
                                                        Mock Olive
1.5–8 m × 1.5–6 m   Oct–March

Small to tall **shrub** or small **tree**, with faintly hairy young growth; **branches** spreading to ascending; **branchlets** faintly hairy or glabrous; **leaves** 4–17 cm × 1–6 cm, narrowly ovate to elliptical, opposite, prominently stalked, spreading to ascending, dull green, paler below, becoming glabrous, stiff-textured, thickish, margins entire or minutely crenulate, venation reticulate and prominent above and below, apex pointed; **racemes** 1–7 cm long, axillary, glabrous; **flowers** small, greenish, often profuse; **drupes** 1–1.5 cm × 0.8–1 cm, ovoid, dark purple to blackish.

A bushy species from moist sheltered gullies of closed forest. Its distribution extends from southeastern Qld, through eastern NSW and eastern Vic. It is suitable for subtropical and temperate regions and should adapt to a range of freely draining soils in sites which are semi-shaded or have full shade all day. Plants are tolerant of moderate frosts and they may need supplementary watering during extended dry periods. Birds eat the decorative fruits. Propagate from fresh seed and possibly from cuttings of firm young growth.

N. longifolia var. *pedicellaris* Domin is a synonym.

NOTHOALSOMITRA I. Telford
(from the Greek *nothos*, false; and the Malesian genus *Alsomitra*)
Cucurbitaceae

A monotypic genus endemic in south-eastern Qld.

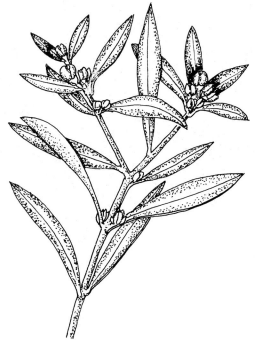

Notelaea pungens × .9

Notelaea venosa × .45

Nothoalsomitra suberosa (F. M. Bailey) I. Telford
(corky)
Qld   Corky Cucumber
Climber   Dec–March; possibly also sporadic
   A vigorous dioecious **climber**; **bark** with corky
flanges at base of old stems; **stems** to 3 m or more
long, with 2-branched tendrils; **leaves** trifoliolate, with
stalks 2.5–6 cm long; **leaflets** 3–11 cm × 1.5–5 cm,
more or less equal, short-stalked, bright green, entire
or bluntly toothed, margins wavy, apex shortly
pointed; **male flowers** about 1 cm across, broadly bell-
shaped, woolly white exterior, pale yellow interior, in
racemes to about 5 cm long, with stalks to about
2.5 cm long; **female flowers** about 1.2 cm across, cream,
solitary, on stalks to 2.5 cm long; **fruit** to 12 cm × 5 cm,
oblong, green and white to yellow, fragrant yellow
flesh, ripe mainly March–May.
   An ornamental climber from the moist forests and
rainforests of south-eastern Qld where it grows in
loamy soil. It has had limited cultivation and responds
well to moist, well-drained soils in a site with filtered
sun. Best cultivated in tropical and subtropical
regions, unless frost protection can be provided (but
may not suffer damage from light frosts). The fruits
are edible but are not very palatable.
   Propagate from fresh seed during warm periods.
Cuttings strike readily.
   Previously known as *Alsomitra suberosa* F. M. Bailey.

NOTHOFAGUS Blume
(from the Greek *nothos*, false; and Latin *fagus*, beech;
referring to the close relationship with European
beech)
Fagaceae

Evergreen or deciduous **shrubs** or **trees**; **leaves** alternate, with stipules, entire, toothed or lobed, penninerved; **inflorescence** catkins or small clusters; **flowers**
unisexual, solitary or in groups of 3; **male flowers** bell-
shaped, short-stalked, in catkins; **female flowers**
solitary or few together, sessile or short-stalked, covered by overlapping scales; **fruit** a nut, angled or
winged, enclosed in a prickly 4-winged, husk-like envelope of bracts.
   A genus of about 35 species, many of which are very
ornamental. Restricted to the southern hemisphere,
the genus is distributed in New Caledonia, New
Guinea, New Zealand and temperate South America,
as well as in Australia. There are 3 species in Australia
and they are confined to the eastern states. They occur
in areas of high rainfall and are all cultivated to a
limited extent. Propagation is from seed, cuttings
(which may be difficult to strike if taken from mature
trees), and possibly aerial layering.

Nothofagus cunninghamii (Hook.) Oerst.
(after Alan Cunningham, 19th-century botanist)
Vic, Tas   Myrtle Beech; Myrtle; Beech
6–50 m × 3–20 m   Nov–Jan
   Small to tall **tree**; with brilliant rose-pink to orange-
red young growth; **bark** hard, dark brown; **branchlets**
rusty-hairy; **leaves** 0.6–2.5 cm × 0.4–1.8 cm, ovate to
nearly round or triangular, wedge-shaped at base,
fairly crowded, dark green, glossy and glabrous above,
paler green below, margins bluntly toothed, venation
prominent, apex blunt; **flowers** small, unisexual,
greenish, axillary; **fruit** a very small nut, winged, ripe
March–May, expelled from prickly woody husk which
is about 0.6 cm long.
   In nature, this species can be a stately tree or tall

Nothofagus cunninghamii × .75

43

bushy shrub. It inhabits sheltered valleys and gullies at an altitude of 100–1200 m and is usually associated with eucalypt forests. In cultivation it generally develops a bushy habit and will take many years to reach 15 m in height. It is a fairly adaptable species which prefers sheltered sites but also tolerates exposure to sunshine. For best growth this species must have well-drained moisture-retentive soils which are organically rich. Plants are hardy to most frosts and withstand snowfalls. The new leaves are a very attractive feature of this species. Best suited to large gardens and parks, but can be grown as a container plant for many years, and is highly regarded for bonsai. Also useful for hedging, as it tolerates regular clipping. Plants respond well to slow-release fertilisers and supplementary watering during dry periods. Cut foliage is long-lasting and exploited to a very limited degree by the floriculture trade. Introduced into England in about 1843. The reddish to pale brown timber is valued for its hardiness and attractive grain and is used for flooring, furniture and wood-turning. Propagate from seed, or from cuttings of firm young growth.

Nothofagus gunnii (Hook. f.) Oerst.
(after Ronald C. Gunn, 19th-century Tasmanian botanical collector)
Tas   Tanglefoot; Dediacuous Beech
2–10 m × 2–6 m   Nov–Jan

Deciduous medium **shrub** or small **tree**; **branches** spreading to ascending, sometimes slender and wiry; **leaves** 1–2 cm × 0.5–1.5 cm, broadly ovate to nearly circular, alternate, short-stalked, bright green in spring and summer, turning golden-brown to reddish before falling, deeply impressed venation above, hairy and venation prominent below, margins bluntly toothed, apex blunt; **flowers** small, unisexual, creamy green; **fruit** very small nut, winged, ripe mid March to May, expelled from the prickly woody husk which is about 0.8 cm long.

An outstanding plant which grows from altitudes of 900–1250 m where it creates a dramatic effect in autumn as Tasmania's only deciduous tree. It has proved difficult to tame in cultivation and seems to prefer cool temperate climates above altitudes of 330 m. Needs cool and protected, moist but freely draining sites. Plants are very slow-growing. They are hardy to most frosts and snowfalls but, although occurring naturally in harsh areas, they don't seem to tolerate extended low temperatures, such as those that are experienced in parts of Europe and northern America. It can do well as a container plant in an acidic potting mix and is used by bonsai enthusiasts. Propagate from fresh seed, which is best stratified at 2–5°C for 4 weeks before sowing. Temperatures of about 15–20°C seem suitable for germination. Cuttings of leafless hardwood stem 2–6 years old, taken in autumn or early spring, are usually successful.

Nothofagus moorei (F. Muell.) Krasser
(after Charles Moore, late 19th-century Superintendent of Royal Botanic Gardens, Sydney)
Qld, NSW   Niggerhead Beech; Antarctic Beech
6–50 m × 4–25 m   Aug–Oct

Small to tall **tree**, with deep reddish to bronze young growth; **trunk** rarely straight, massive and gnarled in old plants; **bark** hard, dark brown, scaly; **branchlets** slender, reddish-brown, hairy; **leaves** 4–8 cm × up to 3 cm, ovate-lanceolate to ovate, usually slightly oblique at base, short-stalked, stiff, dark green above and below (turning to orange before falling), upper margins finely toothed, venation prominent, apex pointed; **flowers** small, unisexual, greenish, axillary; **fruit** a very small nut, winged, ripe Dec–Feb, expelled from the prickly woody husk which is about 0.8 cm long.

A majestic dweller of cool temperate rainforests at altitudes above 600 m. It occurs in the south-eastern corner of Qld and extends to the nearby North Coast and Northern Tablelands of north-eastern NSW. Generally not renowned as fast-growing trees they can nevertheless develop fairly quickly in some areas. Plants need cool moist conditions with well-drained, organically rich loam soils. Best for large gardens and parks, but can be kept for many years growing in a container. They are suitable for bonsai. Plants are prone to having crooked sculptural trunks and branches, which often adds to their appeal. They tolerate fairly heavy frosts and need supplementary watering during extended dry periods. Cut stems of foliage are excellent for indoor decoration. Propagate from fresh seed, or from cuttings of firm young growth.

NOTOCHLOE Domin
(from the Greek *notos*, south; *chloe*, grass; therefore a southern grass)
Poaceae (alt. *Graminae*)
A monotypic genus endemic in the Blue Mountains, west of Sydney NSW.

Notochloe microdon (Benth.) Domin
(minutely toothed)
NSW
0.3–0.6 m × 0.3–0.6 m   Sept–Dec

A tufting perennial **herb**; **leaves** to about 30 cm × 0.2–0.3 cm, flat or inrolled, glabrous, green; **panicles** 6–10 cm long, loose, usually with 5–6 spikelets, erect becoming pendulous; **spikelets** to 2.5 cm × 0.3–0.5 cm, with 8–14 bisexual florets; **glumes** to about 0.7 cm long, narrow, glabrous, awnless; **lemmas** slightly larger than glumes, with 3 minute terminal awns.

N. microdon is confined to higher altitudes of the Blue Mountains where it occurs in swamps. This species has potential for cultivation but it may become weedy in some areas. Plants withstand plenty of sunshine or an aspect which is shady for part of the day. It needs moisture-retentive soils but may dislike heavy clay soils. Propagate from seed, which may have a pre-germination period and is best sown 6–12 months after ripening. Division of clumps is also worth trying but the divisions should not be too small.

Triraphis microdon Benth. is synonymous.

Notoxylinon australe (F. Muell.) Lewton =
                           Gossypium australe F. Muell.

NUYTSIA R. Br.
(after Pieter Nuyts, 17th-century member of Council of the Dutch East Indies, who was on *Gulde Zeepard* in

1627, one of the first ships to explore the south and west coast of WA)
Loranthaceae
A monotypic genus endemic in south-western WA.

Nuytsia floribunda (Labill.) R. Br. ex Fenzl
(abundant flowers)
WA   Christmas Tree; Mudja
3–8 m × 2–6 m   mainly Oct–Jan
Parasitic medium **shrub** to small **tree**; **bark** hard, grey; **branches** spreading to ascending, brittle; **leaves** 4–10 cm × 0.3–0.8 cm, linear, alternate, sessile, olive-green, thick, brittle, glabrous, apex pointed or blunt; **racemes** terminal, of 3-flowered stalked clusters, crowded; **flowers** about 1.2 cm long, bisexual, gold to orange, tubular, with 6–8 spreading narrow petals, sessile, sweet honey fragrance, profuse and most conspicuous; **calyx** to 0.5 cm long; **fruit** about 1.5 cm long, with 3 short thick **wings**.

One of the world's few tree-sized mistletoes. *N. floribunda* is an outstanding sight in full flower, and surpasses most other plants for its dramatic impact. In nature it extends from near Israelite Bay to the Murchison River region. Plants were often retained during clearing of vegetation for farming activities. It is usually associated with sandy soils.

This species has had limited cultivation because of its parasitic nature. Little is understood of its host requirements but seedlings have been grown successfully by having grasses and banksias in the same container. It is a very slow-growing and long-lived species that can take 15–30 years before the first flowering occurs.

Good drainage and a warm to hot sunny site is needed. Plants are hardy to moderate frosts and if

Nymphaea gigantea   T. L. Blake

young plants are damaged by heavy frost they usually reshoot from the base near ground level. Mature plants also reshoot if cut off at ground level. Nectar-feeding birds and many insects delight in the flowers.

Propagate from seed, which usually germinates 22–90 days after sowing. May be best if planted in pots to alleviate the shock of transplanting at the seedling stage. Some growers place seed in moist sawdust or a folded hessian bag and, when first signs of germination are noted, seedlings are transplanted to desired location, whether it be in a pot or in the ground. There has been limited success from cuttings, which may take a number of years to become vigorous.

NYCTAGINACEAE Juss.
A family of dicotyledons consisting of about 300 species in 30 genera mostly developed in tropical America. They are herbs, shrubs or trees with opposite or entire leaves which are often unequal. The flowers, which are usually small, are subtended by bracts which are often large and colourful. The fruit are ridged and sometimes glandular. In Australia the family is represented by 4 genera and about 8 species. Australian genera: *Boerhavia, Ceodes, Heimerliodendron, Pisonia.*

NYMPHAEACEAE Salisb.   Waterlilies
A family of dicotyledons consisting of about 35 species in 6 genera with about 12 species in 2 genera occurring in Australia. They are aquatic plants with creeping rhizomes, floating leaves and floating or emergent, often colourful, flowers. The fruit is a spongy berry which has a layer of mucilaginous material surrounding the seeds. Many exotic species and hybrids of *Nymphaea* are grown in ponds and dams for their colourful flowers. Australian genera: *Nymphaea, Ondinea.*

NYMPHAEA L.
(from the Greek, *nymphaia*; the waterlily)
Nymphaeaceae   Waterlilies
Emergent aquatic **perennials**; **rhizomes** submerged, creeping; **leaves** submerged or floating, rounded to

Nuytsia floribunda × .55

broadly ovate, with cordate or cleft base, with long petiole, glossy above, margins entire to crenately toothed; **flowers** solitary on thick stalks, often large and very conspicuous, white, yellow, pink, red or blue; **petals** 6–50; **sepals** 3–5; **stamens** many; **carpels** united; **fruit** a spongy berry, ripening under water.

A widely distributed genus which is confined mainly to tropical regions and comprises about 60 species. Australia has 11 species of which 4 are endemic, 4 are shared with New Guinea and 2 (*N. mexicana* and *N. nouchali* var. *zanzibarensis*) are naturalised introductions which have become weedy in warmer parts of Australia. The remaining species extends to Asia.

Cultivars of introduced species are very popular for cultivation in ponds and dams but do not seem to pose a threat to become naturalised in temperate regions.

Australian *Nymphaea* species are cold-sensitive and do not survive for very long in temperate regions unless grown in a heated structure or pool. The exotic, blue-flowered *N. stellata*, which tolerates low temperatures, is often marketed wrongly as an Australian species.

Aborigines found that most parts of Australian waterlilies were valuable as a source of food. Before eating the rhizomes it is best to cook them to remove the moderate poison 'nupharine'. Raw or cooked seeds and leaf stems were a regular part of Aboriginal diets and they were also ground and used as a flour.

Plants are vigorous and need plenty of space and nutrients to develop to their potential. Generally for best growth pools should be 0.3–1 m deep. Cultivation of young waterlilies is best undertaken by planting them in pots of 20–50 cm diameter. The potting mixture should contain 3–4 parts of friable humus-rich soil, 1 part of aged cow manure and the addition of 30–40 g of blood and bone per 30–40 litres of potting mix. To prevent the potting mix from becoming dislodged it is best to cover the mixture with a layer of coarse sand or gravel of 2–5 cm deep.

Once plants have become well established they can, if desired, be transplanted to a permanent position in the pool. If retained in pots it is best to divide and transplant waterlilies every 3 years. This action helps to promote better flower production.

Plants do best in pools without too much water movement. They prefer a sunny aspect but will tolerate limited periods of shade apparently without affecting flower production.

Flowers are long-lasting when cut and are excellent for interior decoration.

Propagation is from seed or by division of rhizomes. Seed is best sown fresh and the bog method or use of capillary beds are suitable for germination. See Volume 1, page 204, for further information.

Divisions of rhizomes should contain at least one crown with a bud and as many roots as possible should be retained. Tissue culture has been very successful with this genus.

Nymphaea atrans S. W. L. Jacobs
(darkening; referring to the flower)
Qld   Waterlily
Aquatic   July–Nov
Perennial aquatic **herb**, with a short vertical swollen rhizome; **leaves** to about 40 cm across, elliptical to orbicular, with a radial slit, floating, stalk winged, green, margins with short sparse teeth; **flowers** 5–20 cm across, solitary, on stems to 40 cm above the water, most conspicuous; **sepals** 4 to 8 cm long, exterior green with purple streaks, margins light pink ageing to deep pink, apex blunt; **petals** to 33, white with pink, darkening with maturity; **stamens** many; **berry** 4 cm across, globular.

Apparently this recently described (1992) species is restricted to Cape York Peninsula where it grows in billabongs, lakes and dams. Its flowers open in the morning and possibly at night. It is suited to tropical and subtropical areas. Propagate from seed or by division of rhizomes.

Allied to other species in the 'gigantea' group, but differs in its sepals and petals darkening with maturity and its apparent nocturnal flowering.

Natural hybrids between *N. atrans* and *N. immutabilis* produce pink flowers which do not darken. Some of these may be in cultivation.

Nymphaea brownii F. M. Bailey = *N. violacea* Lehm.

Nymphaea casparyi Rehnelt & Henkel ex Henkel, Rehnelt & Dittman = *N. violacea* Lehm.

Nymphaea dictyophlebia Merr. & Perry = *N. macrosperma* Merr. & Perry

Nymphaea elleniae S. W. L. Jacobs
(after Ellen, daughter of S. W. L. Jacobs)
Qld   Waterlily
Aquatic   April–December
Perennial aquatic **herb** with an erect rhizome; **leaves** to 22 cm × 18 cm, elliptical, with radial slit, floating, green above and often reddish below, margins entire to slightly sinuate; **juvenile leaves** often retained for a number of years and are usually arrowhead-shaped and reddish; **flowers** 5–10 cm across, solitary on stems to 20 cm above water, very conspicuous; **sepals** 4, 7 cm long, exterior green with purple spots, margins white, apex usually pointed; **petals** to 25, white, narrow; **stamens** many; **berry** about 2.5 cm across, globular.

A recently described (1992) species which occurs on Cape York Peninsula and extends to New Guinea. Generally grows in permanent water which can be to 5 m deep. Suited to tropical regions. The flowers open during the day but close at night. Propagate from seed or by division of the rhizome.

N. hastifolia is closely allied but differs because it can be an annual or perennial. It grows in ephemeral pools or water to about 1 m deep. It lacks purple spots on sepals and has larger berries.

Nymphaea gigantea Hook.
(gigantic)
Qld, NSW, NT   Giant Waterlily
Aquatic   most of the year
Perennial aquatic **herb** with a globular rhizome; **leaves** to about 80 cm across, orbicular, with a radial slit, floating, green, margins with prominent evenly spaced teeth to about 0.5 cm long; **flowers** to about

25 cm across, solitary, on stems to about 50 cm above water, a distinct gap between petals and stamens, very showy and conspicuous; **sepals** 4, to 11 cm long, exterior green often with small purple stripes; **petals** to 32, white, pink or blue, fading to near white with maturity; **stamens** numerous; **berry** about 5 cm across, globular.

This magnificent waterlily occurs mainly in coastal areas south of the Tropic of Capricorn to north-eastern NSW and with scattered populations in southern Qld extending just into NT. Plants usually grow in permanent water but are also found in ephemeral pools. Flowers open during the day and close at night. Suitable for tropical and subtropical areas. Over the years many plants have been offered for sale wrongly as this species. Propagate from seed or by division of rhizomes.

The var. *neorosa* K. Landon is synonymous.

Plants from northern Australia previously referred to as variants of *N. gigantea* are now known by the following names: *N. atrans*, *N. immutabilis* and *N. macrosperma*.

Nymphaea hastifolia Domin
(spear-shaped leaves)

WA, NT   Waterlily
Aquatic   Jan–April

Annual or perennial **herb** with a globular rhizome; **leaves** to 20 cm × 15 cm, elliptical, with radial slit, floating, green above with purplish below, margins slightly sinuate, arrowhead-shaped juvenile leaves often retained for a number of years; **flowers** 5–9 cm across, solitary, on stems to 30 cm above water, very conspicuous; **sepals** 4–5, to 6 cm long, exterior green, apex pointed; **petals** to 30, white, narrow, pointed; **stamens** to about 100; **berry** about 4.5 cm across, globular.

A near-coastal species of the Kimberley and Darwin and Gulf Regions where it grows in ephemeral pools or shallow water. Suitable for tropical regions. Flowers open during the day and close at night. Propagate from seed or by division of rhizomes.

N. elleniae is closely allied but differs in being only perennial. It grows in deep water and has sepals with purple spots, and much smaller berries.

Nymphaea holtzei Rehnelt & Henkel ex Henkel,
Rehnelt & Dittman = *N. violacea* Lehm.

Nymphaea immutabilis S. W. L. Jacobs
(not changing)

Qld, WA, NT   Waterlily
Aquatic   March–Nov

Annual or perennial aquatic **herb**, with an upright globular rhizome to about 8 cm long; **leaves** to about 70 cm across, orbicular, with a radial slit, floating, green, margins with prominent evenly spaced teeth about 0.45 cm long; **flowers** to 12 cm across, solitary, on stems to about 50 cm above the water, distinct gap between petals and stamens, very conspicuous; **sepals** 4, to about 5 cm long, exterior green with purple spots; **petals** to 34, mostly white, outer ones with blue tint or uncommonly all white, or all blue and not fading with maturity; **stamens** numerous; **berry** about 5 cm across, globular.

The typical subspecies has a widespread distribution across northern Australia where it inhabits ephemeral or permanent waterholes. Its flowers open during the day and close at night. This attractive waterlily is only suitable for tropical regions.

N. lotus var. *australis* F. M. Bailey is a synonym.

The ssp. *kimberleyensis* S. W. L. Jacobs occurs in the Kimberley region where it is known only from one ephemeral lagoon. Its flowers are blue with white at base. The sepals are to 12.5 cm long and the leaves have shorter toothed margins. The flowers are most likely produced during March to June.

Propagate both subspecies from seed or by division of rhizomes.

Nymphaea lotus L. var. australis F. M. Bailey =
N. immutabilis S. W. L. Jacobs

Nymphaea macrosperma Merr. & Perry
(large seed)

Qld, NT   Waterlily
Aquatic   March–Oct

Perennial aquatic **herb** with a globular rhizome; **leaves** to 55 cm across, elliptical to globular, with a radial split, floating, green, margins with prominent evenly spaced teeth; **flowers** to 15 cm across, solitary, on stems to 30 cm above water, very conspicuous, sweetly fragrant; **sepals** 4, to 6.5 cm long, exterior green with purple stripes which are often dominant, apex blunt; **petals** to 22, white, deep blue or pink, can fade slightly with maturity, distinct gap between petals and stamens; **stamens** many; **berry** about 4 cm diameter, globular.

N. macrosperma is a very attractive waterlily which inhabits water to about 3 m deep in near-coastal areas of far north Qld and NT, and also extends to New Guinea. Its flowers open during the day and close at night. Suitable for tropical regions. Propagate from seed or by division of rhizomes.

N. dictyophlebia Merr. & Perry is a synonym.

Nymphaea minima F. M. Bailey
(smallest)

Qld, NT   Waterlily
Aquatic   Nov–March

Annual or perennial aquatic **herb** with small tuberous rhizome; **leaves** 10–23 cm across, oblong to orbicular, floating, with smooth slender stalk, margins slightly sinuate; **flowers** to about 7 cm across, barely emergent from water, very conspicuous; **sepals** 4, usually to 3 cm long but sometimes to 7 cm long, exterior green with purple streaks commonly near margins, apex pointed; **petals** to about 10, blue, pink or white, distinct gap between petals and stamens; **stamens** about 20; **berry** 1.5–4 cm across, globular.

This decorative species occurs in north-eastern Qld, northern NT and extends to New Guinea. Generally found near the coast where it grows in ephemeral swamps and pools. It is cold-sensitive and best suited to tropical and subtropical regions but may succeed in heated aquaria in temperate regions. Propagate from seed or by division of rhizomes.

N. minima has been confused with *N. nouchali*

Burm.f. (an introduced species), *N. pygmaea* Ait. and *N. tetragona* Georgi (both not recorded in Australia).

Nymphaea pubescens Willd.
(downy, slightly hairy)

NT   Waterlily
Aquatic   Jan–July

Annual or perennial aquatic **herb** with a short erect rhizome; **leaves** to about 43 cm across, elliptical to orbicular, with a radical slit, floating, smooth and green above, usually hairy below, margins toothed; **flowers** to 15 cm across, solitary on stems to 10 cm above water, most conspicuous; **sepals** 4–5, to 9 cm long, green; **petals** to 19, white usually with pink tonings, apex blunt; **stamens** to about 60; **berry** about 5 cm across, globular.

N. pubescens occurs mainly on the coastal flood plains of the Darwin and Gulf District where it grows in water to about 2 m deep. Also extends to New Guinea, India and south-eastern Asia. The flowers of this decorative waterlily open in the morning and possibly at night but usually close by midday. It is cold-sensitive and best grown in tropical and subtropical areas. Propagate from seed or by division of rhizomes.

Further botanical studies are likely to create changes to this species in the near future.

Nymphaea violacea Lehm.
(violet)

Qld, WA, NT   Waterlily
Aquatic   Jan–Aug

Annual or perennial aquatic **herb** with a globular rhizome; **leaves** 10–30 cm × 10–30 cm, elliptical to orbicular, with a radial slit, floating, green, margins entire to sinuate; **flowers** 5–16 cm across, solitary on stems to about 60 cm above water, most conspicuous; **sepals** 4, to 11.5 cm long, apex blunt or pointed; **petals** to 45, blue, mauve, pink and fading minimally or white; **stamens** many; **berry** about 5 cm across, globular.

A highly attractive and variable species which extends across northern Australia from Cape York to the Kimberley and also occurs in New Guinea. It is found in billabongs, swamps, creeks and rivers and usually grows in water to about 2 m deep. The flowers open during the day and close at night. It is popular in cultivation in tropical regions. It is cold-sensitive. Aboriginals placed great importance on the species because of its value as a food source. The rhizomes, stalks and fruit were eaten raw or cooked and the seeds were sometimes ground for flour. Propagate from seed or by division of rhizomes.

The following are synonyms: *N. brownii* F. M. Bailey, *N. casparyi* Rehnelt & Henkel ex Henkel, *N. gigantea* var. *violacea* (Lehm.) Conard, *N. holtzei* Rehnelt & Henkel ex Henkel and *N. violacea* var. *coerulea* Lehm.

Over the past years, the exotic species *N. stellata* has been confused with *N. violacea*.

NYMPHOIDES Hill
(from the Greek, *nymphaea*, the waterlily; and *oides*, like, similar)

Menyanthaceae   Marshworts

Perennial or annual aquatic or marsh **herbs**; **stems** spreading or erect, often forming roots at nodes of perennial species; **leaves** usually floating on water or prostrate on mud, ovate to rounded, alternate or clustered at base, prominantly cordate, with short to long petiole; **inflorescence** a cyme or reduced to a few flowers or in clusters of up to 20 flowers, held above water on slender stalks; **flowers** bisexual, short-lived; **petals** 4–5, yellow or white, rarely pink, flimsy, often winged or hairy on inner surface; **stamens** same number as petals; **stigmas** 2–5; **fruit** an irregularly dehiscing capsule which is submerged at maturity; **seeds** smooth or rough.

A cosmopolitan genus of about 30 species with 20 species occurring in Australia, of which about 17 are endemic. They are found in all states of Australia where they grow in pools or slow-flowing water in depths of up to 2 m. Some species also grow as emergents in drying mud.

They are decorative plants and are likely to become desirable in cultivation. The production of flowers over a long period makes them very useful.

Growers need to be aware that some species may become weedy when they are introduced to areas outside their natural range which have similar climatic conditions.

Planting can be direct into clay edges of pools or in pots which are submerged in the water. Potting ingredients are similar to those described under *Nymphaea*.

Although best growth is achieved in pools and dams, some species such as *N. geninata* grow successfully in pots which have the base submerged in water.

Nymphoides species are usually unsuccessful as aquarium plants because they require very bright light conditions.

Propagation is from seed or by division for species which have stoloniferous stems. Seeds need to be sown while fresh and the bog method or capillary beds are suitable (see Volume 1, page 204). Species with stoloniferous stems are readily propagated by cutting the stems into lengths which have roots growing from 2–3 nodes, and placing them into propagating mixture or the potting mixture as given for *Nymphaea* species. The pots with stems should then be placed in a container with enough water to cover the divisions until the plants become sufficiently established and can be placed permanently in pools or dams.

Limnanthemum S. G. Gmel. is a synonym.

Nymphoides aurantiaca (Dalzell) Kuntze
(orange-coloured)

Qld, WA, NT   Orange Fringe; Marshwort
Aquatic   April–Sept

Perennial aquatic **herb**; **stems** to about 1.4 m long, slender, stoloniferous; **basal leaves** 0.7–13.5 cm × 0.7–10 cm, more or less circular, floating with stalk to 1.1 m long, deeply cordate at base, glossy green above, dotted below, margins usually entire; **stem leaves** similar; **inflorescence** terminal or with distinct branches to 32 cm long; **flowers** 1.1–4.5 cm across, rich yellow to deep orange, margins and base of lobes fringed, on paired stalks to 15 cm long, often profuse and very conspicuous; **capsules** 0.4–0.85 cm long, ellipsoid.

A tropical species from northern Australia which

also occurs in southern India, Moluccas, New Guinea, Sri Lanka, Taiwan and Thailand. This brilliantly flowered species grows in permanent or ephemeral pools, swamps and streams where water is usually 1–2 m deep and the base is sandy, gravelly or clayey sand. Often associated with paperbarks. Only suited to tropical or subtropical regions unless grown in a heated pool.

Propagate from seed or divisions of stoloniferous stems.

Limnanthemum hydrocharoides (F. Muell.) F. Muell. ex Benth. and *Nymphoides hydrocharoides* (F. Muell.) Kuntze are synonyms.

Nymphoides beaglensis Aston
(after Beagle Bay)

WA	Marshwort
Aquatic	April–Aug

Apparently an annual aquatic **herb**; **stems** 4–20 cm long; **basal leaves** 2–5 cm × 2.2–4.7 cm, broadly ovate to circular, floating, with stalk 5–30 cm long, deeply cordate base, dark green above, paler green with purple or completely deep maroon-purple below, margins entire; **inflorescence** often clustered, tightly arranged, subtended by leaf; **flowers** 1.8–2.2 cm across, white or very pale pinkish-mauve with deep maroon-mauve throat, on slender stalks 2.5–8 cm long, margins fringed, clusters of hairs on each side of base of stamens, often profuse and conspicuous; **capsules** to about 0.6 cm long, ellipsoid.

A recently described (1987) species which is known only from two areas in the Kimberley region. Plants grow on the edges of shallow freshwater pools or in claypans which receive seasonal rains. Evidently not cultivated. It has potential for use in tropical and subtropical pools and is worth trying in bog gardens. Propagate from seed.

Nymphoides crenata (F. Muell.) Kuntze
(with convex teeth)

Qld, NSW, Vic, SA, WA, NT	Wavy Marshwort; Yellow Fringe
Aquatic	Sept–May

Perennial aquatic herb; **stems** to about 2 m long, more or less floating, stoloniferous; **basal leaves** 3–15 cm long, broadly ovate to orbicular, on long stalks, deeply cordate at base, glossy green and somewhat waxy above, dull and dotted below, thick-textured, margins crenate; **stem leaves** smaller, on stolons, short-stalked; **inflorescence** few to many-flowered, subtended by upper leaves; **flowers** to about 3.5 cm across, yellow, 4–5-lobed, margins and midline fringed, very conspicuous; **capsules** 0.5–1 cm long, shortly beaked.

Wavy Marshwort is a very widely distributed and attractive species. It usually inhabits slow-moving waterways in water to about 1.5 m deep. Plants take root in the muddy base. Although preferring constant water, plants will survive in moist mud for extended periods. This showy aquatic does best in warm climates and likes plenty of sunshine. Can colonise readily but is not usually of pest proportions.

There are a number of selections available including a large-leaved variant from the Murray River which has very prominent serrations, and 'Purple Mosaic'

which is thought to have originated from south-eastern Qld. It has reddish-purple leaf markings and is sometimes sold as 'Variegata'. In cool climatic regions it usually dies back each winter.

Propagate from seed or by division of stoloniferous stems.

Nymphoides disperma Aston
(two seeds)

WA	Marshwort
Aquatic	May–Aug

Annual or possibly perennial **herb** with a slender rootstock; **stems** to 50 cm long, slender, flexuose, floating, simple or once-forked; **basal** and **stem leaves** 1.5–4 cm × 1.4–4.5 cm, usually more or less circular, floating, with a stalk to about 30 cm long, deeply cordate at base, green above, green to deep purplish-maroon below, thin-textured, margins entire; **inflorescence** terminal or sometimes on side shoot, with spaced pairs of stalks; **flowers** 1.5–2.5 cm across, yellow to orange-yellow, margins fringed, glabrous except for one or two clusters of hair near base of lobes, often profuse and very conspicuous; **capsules** about 0.4 cm long, oblong.

N. disperma is known only from the northern and north-western Kimberley region where it occurs in freshwater creeks and pools to about 0.7 m deep. Apparently not cultivated, it is worth trying in tropical and possibly subtropical regions. Propagate from seed and possibly from cuttings of stems.

Nymphoides elliptica Aston
(elliptic)

Qld	Marshwort
Aquatic	Jan–July

Apparently an annual aquatic **herb**; **stems** 8–80 cm long, arising from rootstock, many, slender, flexuose; **juvenile leaves** to about 2.2 cm long, ovate to somewhat triangular, thin-textured; **adult leaves** 1.5–8.5 cm × 1–6 cm, narrowly elliptic to ovate-elliptic, deeply cordate at base, green above, somewhat purplish below, margins entire; **inflorescence** 7–22 stalks per cluster, subtended by leaf; **flowers** 1.4–2.5 cm across, white or white-tinged with very pale pink or mauve-pink, with yellow throat, glabrous except for hairs at base of lobes, upper margins toothed, sometimes profuse and very conspicuous; **capsules** 0.4–0.6 cm long, ellipsoid.

N. elliptica is an ornamental species which is known only from the Cape York Peninsula. It grows in fresh water to a depth of 0.6 m in seasonally flooded swamps and ephemeral streams of *Melaleuca* woodland. Plants usually take root in sandy soils. It has potential for cultivating in tropical and subtropical regions. Plants are ideally suited for shallow pools. Propagate from seed.

N. triangularis Aston occurs in the same area as *N. elliptica* and is closely allied to that species but differs in having triangular leaves, prominently fringed flower lobes and much smoother seeds. Cultivation requirements are similar to those for *N. elliptica*.

Nymphoides exigua (F. Muell.) Kuntze
(small, insignificant)

Tas	Marshwort
Aquatic	Oct–Feb

Nymphoides exiliflora

Perennial aquatic **herb**, often forming an entangled mat of slender stems from a somewhat stout rootstock; **leaves** 0.4–1 cm × 0.3–0.7 cm, ovate to almost circular, base tapering to slender stalk 0.1–0.3 cm long, mid green above, pale green below, glabrous, margins entire; **flowers** about 0.6 cm across, pale yellow, solitary on stalks less than 2 cm long; **capsules** about 0.3 cm across, ovoid.

This species is widespread in Tas and is often locally plentiful ranging from sea-level to about 800 m altitude. It grows in shallow, fresh or brackish water and constantly wet mud. Cultivated to a very limited degree but should adapt well in cool temperate regions. *N. exigua* is worth trying in pools, bog gardens and as a container plant with its base immersed in water. Propagate from seed or by division of clumps.

Previously known as *Limnanthemum exiguum* F. Muell.

Nymphoides exiliflora (F. Muell.) Kuntze
(small or thin flowers)
Qld, NT   Marshwort
Aquatic   Jan–May; also sporadic

Perennial aquatic **herb**; **stems** more or less floating, stoloniferous; **leaves** 1–2.5 cm × 0.6–2.5 cm, ovate, orbicular or kidney-shaped, base blunt to cordate, tufted, on long stalks, glossy and mid green above, margins entire; **inflorescence** of solitary flowers or paired or clustered, on slender stems to about 3 cm long at stem nodes; **flowers** to about 1.5 cm across, yellow, with faintly fringed lobes and bearded near base of interior; **capsules** 0.2–0.4 cm long, ovoid; **seeds** velvety.

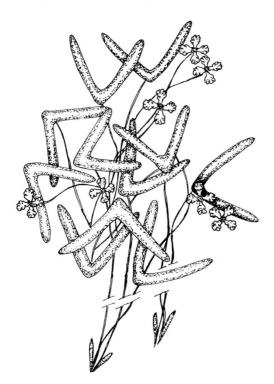

Nymphoides furculifolia × .75

N. exiliflora is a dweller of shallow water or wet mud in the Burke, Cook, Moreton, North Kennedy, Port Curtis and Wide Bay Districts of Qld, and also occurs in the Darwin and Gulf region of NT. and in New Guinea. It is suitable for pools in tropical and subtropical areas and is also worth trying in bog gardens, as well as in containers standing in saucers of water. Plants grow well in drying mud but rarely flower. Propagate from seed or by division of stoloniferous stems.

Nymphoides furculifolia Specht
(deeply forked leaves)
NT   Marshwort
Aquatic   Dec–May

Annual or perennial aquatic **herb**; **stems** 20–40 cm long, very slender, erect from tuft of basal leaves; **basal leaves** to about 2.5 cm × 0.2 cm, very narrowly obovate; **stem leaves** 1.1–1.4 cm × 1–1.5 cm, V-shaped, solitary at end of stem, with narrow lobes, forward pointing lobe shortest; **inflorescence** clusters of short-stalked flowers, below stem leaf; **flowers** about 0.25 cm across, white, never profuse; **capsules** about 0.2 cm long.

An uncommon species with intriguingly-shaped leaves. It is known only from the Darwin and Gulf District of northern NT where it grows in shallow water. It is plentiful in Arnhem Land. Evidently not cultivated, it has potential as an horticultural oddity, for tropical and subtropical pools and possibly worth growing as a container plant placed in a saucer of water. Propagate from seed.

Nymphoides geminata (R. Br.) Kuntze
(in pairs)
Qld, NSW, Vic   Entire Marshwort; Star Fringe
Aquatic   Sept–April

Tufted annual or perennial aquatic **herb**; **stems** usually absent; **leaves** 3–8 cm across, more or less circular, with deeply cordate base, semi-erect or floating, on stalks to about 20 cm long, mid green and glossy above, dull green and dotted below, somewhat thick-textured, margins entire or slightly crenate; **inflorescence** terminal, stalks in pairs or apparently clustered, on stems 3–23 cm long; **flowers** about 3 cm across, bright yellow, margins prominently fringed, very conspicuous, sometimes profuse; **capsules** 0.5–0.8 cm long; **seeds** globular or slightly flattened.

A widespread species which usually inhabits permanent water of up to 2.5 m depth. Plants grow in pools, rivers and streams, but also tolerate semi-exposure by growing in saturated mud at the end of dry seasons. This excellent pond plant is cultivated to a limited degree and it adapts well in tropical, subtropical and temperate regions where it does very well in permanent water. Also successful in pots which are sitting in a saucer of water. Propagate from seed or division of clumps.

N. geminata has been confused with *N. montana*, which differs in having long, floating, stoloniferous stems, and the seeds are prominently flattened.

Nymphoides hydrocharoides (F. Muell.) Kuntze =
N. aurantiaca (Dalzell) Kuntze

Nymphoides indica　　　　　　　　　D. L. Jones

Nymphoides minima　　　　　　　　　D. L. Jones

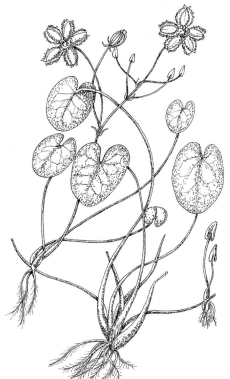

Nymphoides geminata × .5

Nymphoides indica (L.) Kuntze
(from India or the Far East)
Qld, NSW, WA, NT　　　Water Snowflake; White Fringe
Aquatic　　　　　　　　　　Aug–April; also sporadic
　　Perennial aquatic **herb**; **stems** to several metres
long, floating, stoloniferous; **leaves** to about 30 cm
across, more or less circular, deeply cordate base, floating, short-stalked, thick-textured, mid-to-light green,
margins entire; **inflorescence** terminal at base of leaf,
many-flowered; **flowers** 1–2.5 cm across, white or white
with yellow or orange base, prominently bearded, on
stalks 2–8 cm long, very conspicuous; **fruit** to about
0.5–0.8 cm long; **seeds** straw-coloured.
　　A widely distributed species which occurs from
northern WA to near Newcastle in NSW, and extends
throughout the tropics of Africa, America and Asia.
Grows in water to about 2 m deep and is found in still
or flowing water of dams, pools, creeks and rivers.
Popular in cultivation and is best suited to tropical and
subtropical regions. Frosts can damage or kill plants.
It can become weedy and restrict waterflow in channels. Propagate from seed or by division of stolons.

Nymphoides minima (F. Muell.) Kuntze
(smallest)
Qld, WA, NT
Aquatic　　　　　　　　　　　May–Sept; also sporadic
　　Tufting, apparently annual aquatic **herb**; **stems** to
80 cm long; **leaves** 0.6–3 cm × 0.6–2.7 cm, elliptical
to kidney-shaped, base often deeply cordate, floating,
smooth above and somewhat spongy below, margins
entire; **inflorescence** clusters of 3–9 flowers on separate stalks to about 2 cm long; **flowers** 0.8–1.5 cm
across, white with yellow base, glabrous except for

51

basal fringe; **petals** 4–6; **fruit** about 0.25 cm long; **seeds** cream to light brown.

A tropical species which inhabits flowing fresh water in creeks and pools and also in wet shallow soils among rocks and boulders. Evidently not cultivated, it will not tolerate cold and in temperate regions it may be worth trying in heated aquaria. May also succeed in bog gardens. Propagate from seed.

M. parvifolia (Griseb.) Kuntze occurs in Australia in north-eastern Qld and northern NT, and extends to Asia and New Guinea. It has some similarities but differs in having slightly larger leaves which lack the spongy undersurface.

Nymphoides montana Aston
(of the mountains)
NSW, Vic
Aquatic   Nov–May

Perennial aquatic **herb**; **stems** to about 2 m long, floating or embedded in mud, stoloniferous; **basal** and **stem leaves** to about 11 cm across, more or less circular, base deeply cordate, with slender stalks, smooth, margins entire or faintly crenate; **inflorescence** usually up to 11-flowered, terminal, loosely arranged, floating, with paired stalks 2–10 cm long; **flowers** to about 3.8 cm across, lemon-yellow to bright yellow, margins prominently fringed, very conspicuous; **fruit** to 0.9 cm × 0.5 cm, ellipsoid; **seeds** blackish, smooth.

N. montana is found between altitudes of 600–1400 m in eastern NSW and eastern Vic. It usually grows on the edges of freshwater creeks, rivers, pools and swamps and has a preference for water of up to about 1 m deep. This species tolerates cold conditions and is well suited to growing in temperate regions in pools or in bog gardens. Propagate from seed or by divisions of stolons.

N. subacuta Aston, from the Darwin and Gulf region, NT, is an allied species which is generally an annual but is possibly a perennial in permanent water. It differs in having pointed leaf blades with flattened stalks and broader flower lobes with finely fringed wings. It is suitable for tropical areas.

The allied *N. spinulosperma* differs in having spiny seeds. The leaf stalks also have a rosy-pink spot.

Nymphoides parvifolia (Griseb.) Kuntze —
see *N. minima* (F. Muell.) Kuntze

Nymphoides planosperma Aston
(flat seeds)
NT
Aquatic   Dec–May

Annual or possibly perennial aquatic **herb**; **stems** slender, arising from base; **leaves** 0.8–1.7 cm × 0.9–1.6 cm, somewhat boomerang-shaped with broadly cordate base, floating, smooth above, spongy below, margins entire; **inflorescence** terminal clusters at base of leaf, many-flowered; **flowers** to about 1 cm across, white with yellow base, on stalks to about 1.8 cm long, conspicuous; **fruit** to about 0.25 cm long, ellipsoid; **seeds** flattened, black at maturity.

A somewhat quaintly foliaged species from the Kakadu region of northern NT where it grows in seasonal freshwater pools or water trapped in rocky escarpments. Evidently not cultivated, it is suitable for tropical and subtropical regions. Propagate from seed.

Nymphoides quadriloba Aston
(four-lobed)
Qld, WA, NT
Aquatic   Dec–Sept; possibly also sporadic

Annual or possibly perennial aquatic **herb**; **stems** 7–85 cm long, very slender, arising from the base; **leaves** 1–11 cm × 0.8–8 cm, somewhat triangular to horseshoe-shaped, floating, glossy green above, not spongy below, margins entire; **inflorescence** terminal cluster at base of leaf, loosely arranged, many-flowered; **flowers** 0.6–1.9 cm across, white or very pale pink or pale mauve-pink with yellow base, lobes fringed, conspicuous; **fruit** to about 0.5 cm long, ellipsoid; **seeds** creamish, brownish-black or black when mature.

N. quadriloba is from northern tropical regions where it usually occurs in shallow, still, freshwater lagoons and periodic swamps. Evidently not cultivated. It is cold-sensitive. Propagate from seed.

Nymphoides spinulosperma Aston
(seeds with small spines)
Qld, NSW, Vic
Aquatic   Sept–Feb

Perennial aquatic **herb**; **stems** to about 1.5 cm long, floating or often producing roots at nodes in mud, forking at nodes; **leaves** 2.5–12 cm × 2–11 cm, broadly ovate to nearly circular in outline, base deeply cordate, on stalks to about 50 cm long with rosy-pink spot at base of blade, upper surface often mottled with deep green, yellow-green, maroon-brown and/or brown-tan, lower surface green to whitish-green, margins entire to slightly crenate; **inflorescence** terminal, loosely arranged, floating, many-flowered with paired flowers at each node; **flowers** to about 5.5 cm across, yellow, prominently fringed, very conspicuous; **fruit** to about 11.5 cm × 0.7 cm, ellipsoid to ellipsoid-ovoid; **seed** with many small spines.

This recently described (1997) species is found in freshwater pools or seasonal swamps in disparate locations such as southern Qld, North Western Plains of NSW and in western Vic. Very uncommon in cultivation but it adapts well and is suitable for subtropical and temperate regions. Plants will tolerate cold temperatures. The distinctive mottling of leaves is not as evident during summer months. Propagate from seed or by division of stolons.

The closely allied *N. montana* does not have the rosy-pink spot on leaf stalks and its seeds are rarely spiny.

Nymphoides spongiosa Aston
(spongy)
NT
Aquatic   April–July; possibly sporadic

Apparently an annual aquatic **herb**; **stems** 3–90 cm, very slender, flexuose, arising from base; **leaves** 1.5–5 cm × 0.8–4.5 cm, somewhat kidney-shaped, floating, terminal, glossy green above, whitish and spongy but smooth below, margins entire; **inflorescence** terminal cluster at base of leaf, loosely arranged; **flowers** 0.7–2 cm across, white with yellow base, lobes prominently winged, glabrous except for basal hairs, on slender stalks 1–3 cm long, conspicuous; **fruit** to 0.4 cm long, somewhat globular; **seeds** creamish to pale brownish-grey at maturity.

N. spongiosa occurs in the Darwin and Gulf District where it is usually found in still, fresh water to about 60 cm deep, such as in swamps and lagoons or billabongs. Plants are sensitive to cold. Propagate from seed.

Nymphoides stygia (J. M. Black) H. Eichler =
Villarsia reniformis R. Br.

Nymphoides subacuta Aston — see *N. montana* Aston

Nymphoides triangularis Aston — see *N. elliptica* Aston

NYPA Steck
(from the Moluccan vernacular name of *Nipa*)
Arecaceae (alt. *Palmae*)

A monotypic genus which has its distribution from South-East Asia and the Pacific Islands to northern Australia.

Nypa fruticans Wurmb
(shrubby or bush-like)
Qld, NT   Mangrove Palm
4–9 m × 3–6 m   June–Aug

Clumping maritime **palm**; **trunk** prostrate, much-branched, subterranean, usually below water; **fronds** 4–9 m long, pinnate, feathery, with a powdery, thick stalk 1–1.4 m long; **pinnae** 0.6–1.3 m × 5–8 cm, 100–120 per frond, glossy, stiff, dark green above, powdery white below; **spathes** many, small and paper-like, with one large and strap-like; **inflorescence** 1–2 m long, arising from the underground trunk; **male flowers** bright yellow, arranged in catkins on lateral part of inflorescence; **female flowers** arranged in terminal spherical clusters 20–25 cm across; **fruit** 10–15 cm × 5–8 cm, angular, fibrous, reddish-brown, in a globular head of about 30–45 cm diameter.

The mangrove palm is an uncommon component of the margins of coves, bays and coastal river estuaries in northern Australia. It occurs in scattered localities in mud which is regularly inundated by brackish or sometimes fresh water. It is valuable for its contribution to soil stabilisation and ecology in its natural habitat. The fruits are carried by sea currents to new locations. This palm is best suited to tropical areas and as plants are very cold-sensitive they will need to be grown in heated structures in cooler climes. Successful cultivation has been achieved in tidal mudflats, salt-marsh and freshwater swamps. Generally plants are slow-growing. Fronds can be used for building and thatching and the soft unripened seed can be eaten raw. Propagation is from fresh seed which is usually best germinated in brackish soils. Propagation by division is difficult but has been carried out in some tropical countries

The var. *neameana* F. M. Bailey is a synonym.

Ozothamnus alpinus   W. R. Elliot

OBERONIA Lindl.
(after Oberon, King of the Fairies)
Orchidaceae

Epiphytes or **lithophytes** usually forming small clumps; **growths** flattened, often iris-like; **leaves** flattened, curved, overlapping and sheathing at the base; **inflorescence** a slender, terminal raceme; **flowers** tiny, arranged in whorls, often red or green, sometimes fragrant; **tepals** entire or toothed, often reflexed; **labellum** 3-lobed.

This large genus of about 340 species is widely distributed in the world's tropics. About 5 species occur in Australia; one on Norfolk Island and the others in Qld and NSW. They are generally popular with orchid growers. Propagation is from seed which must be sown on sterilised media.

Oberonia attenuata Dockrill
(attenuate, drawn out)
Qld
10–15 cm tall   May–Sept
Epiphytic orchid forming small clumps of pendulous, iris-like growths; **leaves** 4–7 per growth in a fan-like arrangement, 10–15 cm × 0.5–0.8 cm, dark green, tapered from base to apex, acuminate; **racemes** 10–15 cm long, pendulous, slender, densely flowered; **flowers** about 0.16 mm across, pale brown; **labellum** about 0.1 cm × 0.1 cm, mid-lobe deeply notched.

Recorded from Mossman Gorge, north-eastern Qld but apparently not relocated since the original collection. Grows near streams in lowland rainforest. Cultural requirements are largely unknown but are probably the same as for other species in the genus. Propagate from seed.

Oberonia carnosa Lavarack
(fleshy, succulent)
Qld
1–3 cm tall   Feb–June
Epiphytic orchid forming very small clumps, often consisting of a single growth; **leaves** 4–6 per growth, 1–2.5 cm × 0.8 cm, light green, fleshy, curved, tapered to the blunt apex; **racemes** 4–6 cm long, arching to pendulous, slender, sparsely flowered; **flowers** about 0.1 cm across, orange; **petals** toothed; **labellum** about 0.1 cm × 0.07 cm, margins irregularly toothed.

This little orchid is restricted to central parts of Cape York Peninsula where it grows in lightly vegetated, humid scrubs, sometimes close to streams. Plants are easy to grow on a small slab of treefern or

weathered hardwood. They need warm, airy, humid conditions and the protection of a heated glasshouse in cold climates. Propagate from seed.

Oberonia complanata (A. Cunn.) M. A. Clem. & D. L. Jones
(flattened in one plane)
Qld, NSW
10–15 cm tall   Aug–Oct
Epiphytic orchid forming small, dense clumps of erect, iris-like growth; **leaves** 3–8 per growth, in a fan-like arrangement, 10–15 cm × 1–1.5 cm, flattened, yellowish-green, widest near the base, apex blunt; **racemes** 10–20 cm long, erect to arching, slender, densely flowered; **flowers** about 0.25 cm across, greenish-cream; **petals** and **sepals** reflexed back against the ovary; **labellum** about 0.15 cm × 0.15 cm, margins toothed.

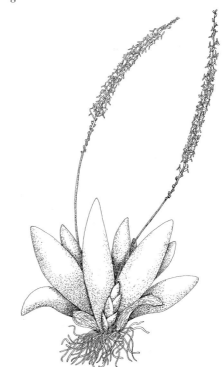

Oberonia complanata × .4

57

Oberonia muelleriana

This orchid is distributed from south-eastern Qld to north-eastern NSW. It grows on trees and rocks, sometimes in very exposed, sunny situations. Plants are very easily grown on a slab of treefern, weathered hardwood or cork. They require bright light, humidity and adequate air movement. An unheated glasshouse may be necessary in cold climates. Propagate from seed. *Oberonia muelleriana* Schltr. is a synonym.

Oberonia muelleriana Schltr. =
O. *complanata* (A. Cunn.) M. A. Clem. & D. L. Jones

Oberonia neocaledonica Schltr. =
O. *titania* (Endl.) Lindl.

Oberonia palmicola F. Muell.
(growing on palms)
Qld, NSW
5–8 cm tall   Feb–May
Epiphytic orchid forming small, dense clumps of erect to semi-pendulous, iris-like growth; **leaves** 5–10 per clump, in a fan-like arrangement, 5–8 cm × 0.5–0.9 cm, flattened, pale green to pinkish-green, widest near the base, tapered, acute; **racemes** 10–16 cm long, erect to arching, slender, densely flowered; **flowers** about 0.12 cm across, bright reddish-brown; **labellum** about 0.8 cm × 0.6 cm, mid-lobe about the same length as width.

This orchid occurs in southern Qld and northern NSW between Gympie and the Hastings River. It grows on rocks and trees in rainforest. It is popular in cultivation and grows well on a small slab of treefern if given warm, humid conditions with ample air movement. Propagate from seed.

O. *titania* is similar but it is smaller growing with paler flowers and the width of the labellum mid-lobe is less than the length.

Oberonia titania (Endl.) Lindl.
(after Titania, Queen of the Fairies)
Norfolk Island   Soldier's Crest Orchid
3–5 cm tall   Feb–May
Epiphytic orchid forming small dense clumps of erect to semi-pendulous, iris-like growths; **leaves** 3–7 per growth, in a fan-like arrangement, 3–5 cm × 0.4–0.7 cm, flattened, pale green or pinkish, widest near the base, tapered, acute; **racemes** 5–12 cm long, erect to arching, slender, densely flowered; **flowers** about 0.1 cm across, pink to greenish-red; **labellum** about 0.7 mm × 0.5 mm, the length of the mid-lobe is greater than the width.

Occurs on Norfolk Island and New Caledonia, growing on trees in rainforest. Rarely cultivated but easily grown on a small slab of treefern. Requires warm, humid conditions, bright light and plenty of air movement. Propagate from seed.

Oberonia neocaledonica Schltr. is a synonym.
Similar to O. *palmicola* which is larger growing, has brighter red flowers and the labellum mid-lobe is about equal in length and width.

OCHNACEAE DC.

A family of dicotyledons consisting of about 600 species in 40 genera distributed principally in America,

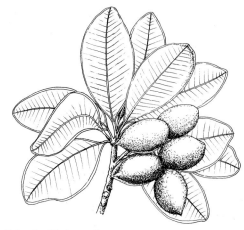

Ochrosia elliptica × .5

Africa and the Pacific islands. It is very poorly represented in Australia with only a single taxon in the genus *Brackenridgea*. An introduced species, *Ochna serrulata* , is frequently a weed in subtropical areas. An interesting feature of these plants is that the fruit develops on a swollen, persistent and often colourful calyx.

OCHROCARPUS A. Juss. = *MAMMEA* L.

OCHROSIA A. Juss.
(from the Latin *ochraceus*, yellow-brown)
Apocynaceae
Shrubs or **trees** with milky sap; **leaves** simple, entire, opposite or in whorls; **inflorescence** cymose, terminal or axillary; **flowers** small, tubular, 4–5-lobed; **stamens** 5; fruit a **drupe**, often compressed, divided into 2 hollow portions containing the seeds.

A small genus of 21 species distributed from Mauritius to Polynesia, with 3 species occurring in Australia. They are rarely encountered in cultivation. Propagation is from seed which has a short period of viability and must be sown while fresh.

Ochrosia elliptica Labill.
(elliptic)
Qld   Mangrove Ochrosia
5–9 m × 3–8 m   Oct–Feb
Tall **shrub** to small **tree** with a spreading habit; young growth shiny; **leaves** in whorls of 3 or 4, 8–20 cm × 4–8 cm, obovate to elliptical, dark green, glossy, leathery, blunt to acute; **cymes** short, in the upper axils; **flowers** about 0.8 cm long, crowded, white fragrant; **drupes** 3–6 cm × 1–2 cm, scarlet, pointed, often in opposite pairs; **seeds** about 1 cm across, round, with a narrow wing, mature Aug–Dec.

A common strand plant in the Pacific region extending to north-eastern and central-eastern Qld where it grows amongst mangroves and coastal scrub. This species has an open habit and large colourful fruit. The fruits are interesting and showy but they are poisonous. Plants bleed white sap copiously from damaged tissue. Although rarely cultivated this species has potential for use in beach stabilisation projects.

Seedlings may be slow to establish. Frost tolerance is very low. Propagate from fresh seed.

Ochrosia moorei (F. Muell.) F. Muell. ex Benth.
(after Charles Moore, former Government Botanist, NSW)
Qld, NSW　　　　　　　　　　　　Southern Ochrosia
4–8 m × 1–3 m　　　　　　　　　　　　Dec–Feb

Medium to tall **shrub** or small bushy **tree**; **bark** dark brown to blackish; young growth dark green to purplish, shiny; **leaves** 8–20 cm × 1.5–3.5 cm, obovate to lanceolate, bright green and glossy on both surfaces, numerous veins prominent, tapered to the base, apex long-acuminate; **cymes** short, dense, terminal; **flowers** about 0.8 cm long, tubular, white; **drupes** 4–8 cm × 2–2.5 cm, scarlet, shiny, mature Dec–Feb.

Distributed from extreme south-eastern Qld to the Richmond River in northern NSW, growing in rain forest. Plants are ornamental with attractive foliage and large colourful fruit but they have proved to be very slow-growing. They are suitable for tropical and subtropical regions and require a sheltered location, well-drained moist soil and mulches. Tolerates light to moderate frosts. Propagate from fresh seed and possibly also by cuttings.

OCHROSPERMA Trudgen
(from the Greek *ochros*, pale; *sperma*, seed; referring to the pale straw-coloured seeds)
Myrtaceae

Dwarf to small, glabrous **shrubs**; **leaves** opposite, entire, spreading to appressed, short-stalked; **flowers** small, axillary, solitary or paired on a common stalk; **petals** 5, white or flushed with pink; **stamens** 5, one opposite each calyx lobe, terete; **fruit** a 3-valved capsule which opens widely; **seeds** pale straw-coloured, with a white aril.

This genus, described in 1987, comprises 3 species and is confined to the eastern states. Two species previously referrable to *Baeckea* were transferred to this genus.

Propagation is from seed, which does not require pre-sowing treatment, or from cuttings of firm young growth, which usually produce roots readily.

Ochrosperma citriodorum (A. R. Penfold & J. L. Willis) Trudgen
(lemon-scented)
NSW
0.1–0.4 m × 0.3–0.7 m　　　　Nov–Feb; also sporadic

Dwarf **shrub**; **branches** spreading to ascending, twiggy; **leaves** 0.1–0.3 cm × up to 0.15 cm, narrowly to broadly elliptic or oblong, crowded, thick, glabrous, margins thin, apex blunt, strong citrus odour when crushed; **flowers** to about 0.3 cm across, white, solitary or, rarely, paired, in upper axils; **capsules** small.

Apparently confined to coastal areas of northern NSW where it occurs in isolated small patches, from near Lismore to the vicinity of Port Macquarie. It is readily recognised by the powerful lemon or citrus odour released from crushed leaves. Though not common in cultivation, it has potential for use in containers, rockeries or for general garden use. Tolerates short periods of waterlogging, but frosts may damage plants. Responds well to pruning. Propagate from seed, or from cuttings of firm young growth.

Previously known as *Baeckea citriodora* A. R. Penfold & J. L. Willis

Ochrosperma lineare (C. T. White) Trudgen
(linear)
Qld, NSW　　　　　　　　　　　　Straggly Baeckea
0.3–2 m × 0.3–2.5 m　　　　　　Aug–Dec; also sporadic

Dwarf to small **shrub**; young growth often pinkish; **branches** spreading to ascending, often spindly, twiggy, tips usually weeping; **leaves** 0.4–1.1 cm × up to 0.1 cm, linear, closely appressed to the stems, green and often attaining reddish tones in autumn and winter, glabrous, thickened towards the pointed apex, aromatic when crushed; **flowers** to about 0.3 cm across, white or faintly flushed either with pink or, rarely, green, solitary or paired in upper axils, often profuse; **capsules** small.

An ornamental species which extends from south of Bundaberg, Qld, to near Taree, NSW. It grows in a range of habitats including wallum scrub, heath, open forest, sand dunes as well as on the edges of swamps. Plants adapt to most acidic soils and will tolerate periodic waterlogging. They prefer a semi-shaded site but also tolerate plenty of sunshine. Heavy frosts may damage plants. They can become straggly but respond well to pruning although heavy pruning can deprive plants of their attractive pendulous growth habit. Propagate from seed, or from cuttings of firm young growth.

Previously known as *Baeckea linearis* C. T. White.

Ochrosperma monticula Trudgen
(mountain dweller)
NSW
0.2–0.5 m × 1.5–2 m　　　　　　　　　Sept–Nov

Dwarf **shrub**; **branches** many, spreading to ascending; **leaves** 0.25–0.55 cm × to about 0.2 cm, elliptic to broadly elliptic, spreading to recurved, more or less flat, green and glabrous above, paler below, apex bluntly pointed; **flowers** to about 0.5 cm across, white, solitary or paired in upper axils, often profuse; **capsules** not seen.

O. monticola occurs mainly on rocky ridges and rock outcrops of the Great Dividing Range, north-west of Sydney, with an outlying population west of Nowra. Evidently not cultivated, it will need good drainage and a sunny or semi-shaded site. Plants tolerate moderate frosts. The low spreading growth will appeal and they have potential for low shrubberies, rockeries and for general planting. Propagate from seed, or from cuttings of firm young growth.

OCIMUM L.
(from the Greek *ocimon*, an aromatic herb, a name that was possibly used for basil)
Lamiaceae

Herbs or soft woody **shrubs** with aromatic foliage; **leaves** simple or toothed, opposite, decussate, with numerous oil dots; **flowers** tubular, in false whorls of 6–10, arranged in terminal racemes; **corolla tube** 2-lipped, upper lip of 4 lobes, lower lip entire; **stamens** 4; **fruit** a nut enclosed in an enlarged calyx.

A large genus of about 150 species. They are widely

distributed in the tropics and across Africa. Basil (*Ocimum basilicum* L.) is the best known member of this genus which includes many aromatic herbs. One widespread species extends to Australia. The taxonomy of *Ocimum* in Australia is uncertain and needs clarification. *O. sanc-tum* L. and *O. sanctum* var. *angustifolium* Benth. are recorded as being naturalised weeds, but whether they are indigenous or not needs further study. Propag-ation is from seed, cuttings of firm young wood or by removal of suckers.

Ocimum tenuiflorum L.
(with slender flowers)
Qld, NT   Sacred Basil
0.3–0.6 m × 0.5–2 m   May–Dec

Dwarf **shrub** with a sprawling often suckering habit; **branches** numerous, soft and pithy; **leaves** 1–5 cm × 0.5–1.5 cm, oblong to lanceolate, hairy, dull green, margins entire or more usually with a few coarse teeth, strongly aromatic when crushed; **flowers** about 1 cm long, 2-lipped, purple to white, in false whorls of 6, forming erect leafy racemes; **stamens** shortly exserted; **nuts** small, smooth.

Widely distributed from tropical Asia to northern Australia where it grows in heavy clay soils in inland districts. It often forms colonies, apparently spreading by suckers as well as from seed. The leaves have a strong spicy perfume although the fragrance is variable between clones. Crushed leaves have been used by Aborigines to prepare a drink for treating fevers. This same species is commonly planted in Asia and is regarded as a sacred plant by the Hindus. Propagate from seed, by cuttings or suckers.

OCTARRHENA Thwaites
(from the Greek *octa*, eight, *arrhen*, male; referring to to the 8 pollinia)
Orchidaceae

Small **epiphytic orchids**, solitary or clumping; **stems** short, branched; **leaves** entire, fleshy, sessile; **inflorescence** a slender raceme; **flowers** small, green to cream; **sepals** and **petals** similar; **labellum** nearly entire; fruit a capsule.

A small genus of about 35 species distributed in Malaysia, Polynesia and New Guinea. One species, which is endemic in Australia, is grown by orchid enthusiasts. Propagation is from seed which must be sown under sterile conditions.

Octarrhena pusilla (F. M. Bailey) M. A. Clem. & D. L. Jones
(small, puny, weak)
Qld
Epiphyte   Sept–Nov

Tiny **epiphyte** forming small clumps; **stems** to 4 cm long, short-branched, often covered with thread-like roots; **leaves** 3–6 per growth, 2–3 cm × 0.2–0.25 cm, cylindrical, fleshy, curved, dark green to yellowish; **racemes** 2–3 cm long, erect, thread-like; **flowers** 5–20, about 1.5 mm across, cream to white; **tepals** widely spreading, blunt; **labellum** obscurely lobed.

Locally common in highland rainforests of northeastern Qld between Cooktown and Townsville, growing in groups on the trunks and branches of trees. A tiny species which is of interest only to ardent orchid enthusiasts. Plants are easily grown on a small slab of treefern. They require cool, shady, humid conditions with free air movement. Propagate from seed sown under sterile conditions.

Phreatia pusilla (F. M. Bailey) Rolfe and *P. baileyana* Schltr. are synonyms.

ODIXIA Orchard
(transposition of letters in the genus *Ixodia*, in which the species were previously included)
Asteraceae (alt. *Compositae*)

Perennial **shrubs** wih densely hairy young growth; **leaves** alternate, simple, entire, glabrous and sticky above, densely matted hairs and yellow resin below; **branches** initially somewhat angled, becoming smooth; **inflorescence** umbel-like, with crowded small slender flowerheads; **involucral bracts** linear, golden-brown; **fruit** a cylindrical achene.

A small endemic genus of 2 species confined to Tasmania. They were previously included in the genus *Ixodia*. Plants are cultivated to a limited degree. Propagation is from seed, which does not require pre-sowing treatment, or from cuttings.

Odixia achlaena (D. I. Morris) Orchard
(lacking scales)
Tas
1–2 m × 0.6–1.5 m   Oct–Jan

Small **shrub**, with sticky golden-yellow to yellow-green young growth; **branches** many, spreading to

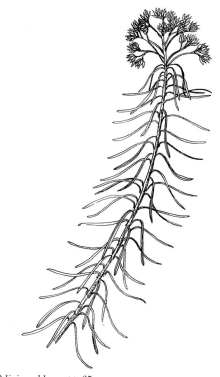

Odixia achlaena × .65

ascending, hairy with sticky yellow resin; **leaves** 1–2.5 cm × about 0.1 cm, narrow-linear, alternate, spreading to recurved, sticky, glabrous and becoming dark green above, margins strongly recurved and partially concealing hairy undersurface, apex pointed; **flowerheads** about 0.4 cm across, daisy-like, white, short-stalked, profuse and arranged in terminal umbel-like heads to about 3.5 cm across, very conspicuous.

An attractive and interesting species with yellowish stems and clustered flowerheads. It is known only from a limited area near the coast in north-eastern Tas where it grows in wet sclerophyll forest. Uncommon in cultivation but it adapts well in temperate regions to a range of moderate to well-drained soils with a semi-shaded aspect. Plants are hardy to most frosts and although withstanding dry periods they respond well to supplementary watering. Suitable for general use and as a container plant. They have a tendency to become leggy but tip pruning from an early stage and pruning after flowering usually controls the habit. Flowers are long-lasting and excellent for indoor decoration. They could prove to be suitable for floriculture. Propagate from seed or from cuttings of firm young growth.

Previously known as *Ixodia achlaena* D. I. Morris.

Odixia angusta (N. A. Wakef.) Orchard
(narrow)
Tas
1–3 m × 0.6–2 m   Nov–Dec
Small to medium **shrub**; young growth hairy and often yellowish-green; **branches** spreading to ascending, hairy while young; **leaves** 0.3–1 cm × 0.2–0.5 cm, obovate, alternate, sessile, spreading to ascending, crowded especially on young branches, dark green, glabrous and shiny above, greyish-hairy below, margins slightly recurved, apex blunt with a minute point; **flowerheads** to about 0.4 cm across, daisy-like, white, short-stalked, profuse, in rounded, terminal, umbel-like heads to about 2.5 cm across, very conspicuous.

This coastal species extends from near Coles Bay to Eaglehawk Neck in south-eastern Tas. Plants are fairly common in the heaths and open woodlands. Evidently rarely cultivated, it should adapt to a wide range of well-drained soils in temperate climates. A sunny or semi-shaded site is suitable. Plants respond well to regular light pruning which promotes bushy growth. Hardy to moderate frosts. Useful for general application and may be suitable for floriculture. Propagate from seed, or from cuttings of firm young growth.

Previously known as *Ixodia angusta* (N. A. Wakef.) N. T. Burb. and *Helichrysum angustum* N. A. Wakef.

OECEOCLADES Lindl.
(from the Greek *oiceos*, private; *clados*, branch; referring to Lindley's interpretation of their unique features)
Orchidaceae
Terrestrial orchids; stems swollen as pseudobulbs, crowded, erect; **leaves** 2 or more, terminal on the pseudobulb; **inflorescence** a raceme which arises from the base of a pseudobulb; **flowers** usually opening widely, colourful petals shorter and broader than the sepals; **labellum** usually 3-lobed, with a short basal spur.

Odixia angusta × .8, flower × 4

A genus of about 30 species distributed mainly in Africa and Madagascar. One widespread species extends to Australia. Propagate from seed which must be sown under sterile conditions.

Oeceoclades pulchra (Thouars) M. A. Clem. & Cribb
(beautiful)
Qld
0.3–0.45 m tall   April–June
Terrestrial orchid forming small clumps; **pseudobulbs** 10–15 cm × 1.5–2 cm, cylindrical, crowded, fleshy, tapered, covered with sheathing bracts when young; **leaves** 2, apical, 20–30 cm × 8–10 cm, lanceolate, dark green, 3 veins prominent; **raceme** to 80 cm tall, 10–20-flowered; **flowers** about 1.2 cm long, opening tardily if at all, green with brown markings; **petals** broader than the sepals.

This species is widely distributed in Asia and extends to New Guinea as well as Australia where it is found in central Cape York Peninsula. The flowers are self-pollinating and they usually remain closed. This orchid is very rarely cultivated. It can be grown in a pot of well-drained terrestrial mixture fortified with leaf mould. Requires the protection of a heated glasshouse in areas with a cold climate. Propagate from seed.

OENOTRICHIA Copel.
(from the Greek *oenos*, wine; *thrix, trichos*, hair; apparently referring to the colour of the hairs on one species)
Dennstaedtiaceae

Oenotrichia tripinnata

D. L. Jones

Terrestrial ferns with finely dissected fronds; **rootstock** creeping, young parts covered with short hairs; **stipes** hairy; **lamina** 2–4-pinnate; **sori** terminal on veins, indusiate; **indusium** round or reniform.

A small genus of five species, two in New Caledonia, one in New Guinea and two endemic in Australia. They are attractive ferns with restrictive growing requirements and are mainly for the fern specialist. Propagation is from spores which germinate readily when fresh, but the sporelings are slow growing.

Oenotrichia dissecta (C. T. White & Goy) S. B. Andrews
(finely cut, dissected)
Qld   Lace Fern
30–60 cm tall
Rootstock short-creeping, hairy; **fronds** 30–60 cm long, erect or semi-erect in a dense tuft; **stipes** to 30 cm long, densely hairy; **lamina** ovate-triangular in outline, to 30 cm × 20 cm, tripinnate, dark green above, glaucous beneath, main veins hairy; **ultimate segments** to 1 cm long, deeply divided into several linear or falcate lobes; **sori** small; **indusium** reniform.

This species is confined to north-eastern Qld where it is extremely rare. It grows in rainforests at high altitudes. Plants are not known to be in cultivation but probably have similar requirements to *O. tripinnata*. Propagate from spores.

Oenotrichia tripinnata (F. Muell. ex Benth.) Copel.
(with tripinnate fronds)
Qld   Hairy Lace Fern
20–50 cm tall
Rootstock short-creeping, hairy; **fronds** 20–50 cm long, erect or semi-erect in a dense tuft; **stipes** to 20 cm long, slender, hairy, yellowish-green; **lamina** ovate-triangular in outline, to 20 cm × 12 cm, tripinnate, dark green above and below, main veins densely hairy; **ultimate segments** to 0.7 cm long, deeply divided into 2–4 blunt lobes; **sori** small; **indusium** reniform.

These beautiful ferns are endemic in north-eastern Qld. Plants grow in cool, dark situations in humid rainforests at moderate to high altitudes. They commonly grow in small colonies and are often found on rocks and boulders. It can be a difficult fern to grow. Requires moderate to dense shade and high humidity with air movement. Plants will not tolerate drying out and succeed best in an organically-rich soil with plenty of organic mulch. They can be grown in a container but choose a container which will barely accommodate the root system. Propagate from spores.

OLACACEAE Juss.
A small family of dicotyledons consisting of 250 species in 25 genera; two genera and about 13 species are found in Australia. The bulk of the family occurs in America, Africa and Asia. They are shrubs, climbers or trees with tough, leathery, alternate leaves and small bisexual flowers borne in axillary inflorescences. The fruit is fleshy, often red, and may be subtended by a swollen calyx. Australian genera: *Olax, Ximenia.*

OLAX L.
(from the Latin *olax*, with a strong bad odour; referring to the scent of some species)
Olacaceae

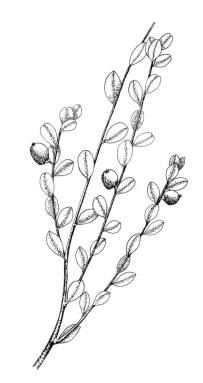

Olax aurantia × .5

Shrubs or small **trees**; **branchlets** often angular, slightly flexuose; **leaves** simple, entire, alternate, sparse to crowded, grey-green or yellow-green; **flowers** small, solitary in the axils, bisexual or unisexual; **calyx** short, cup-shaped; **petals** 5, partly joined; **fruit** a drupe, the hypanthium enlarged when ripe.

This genus includes about 60 species distributed from tropical Africa and Madagascar to New Guinea with 11 species endemic in Australia. The ornamental appeal of most native species is extremely limited and they are rarely grown. Propagation is from seed and possibly also from cuttings.

Olax aurantia A. S. George
(orange)
WA
1–2 m × 0.6–1.5 m   Aug–Oct
Small **shrub** with glabrous growth and an open habit; **branchlets** angular, striate; **leaves** 5–18 cm × 0.3–1 cm, lanceolate, more or less sessile, dark green, leathery; **flowers** about 0.7 cm long, creamish to pale green; **petals** about 0.4 cm long; **drupe** 1–1.4 cm long, ellipsoid, orange.

This species occurs mainly in the northern Irwin District with an isolated occurrence in the northern Carnarvon District. It grows in coastal scrub and forests, usually on well-drained sandy soil. Frost tolerance is unlikely to be high. Requires excellent drainage and some protection. Propagate from fresh seed and possibly also from cuttings.

Olax obcordata A. S. George
(reverse heart-shaped; referring to the leaf shape)
SA
0.3–0.6 m × 0.2–0.5 m   Aug–Dec
Dwarf **shrub**, slender to branched; **branches** flexuose, granular; **leaves** 0.4–1 cm × 0.5–1.2 cm, broadly obcordate, leathery, concave, shallowly notched, sessile; **flowers** about 0.5 cm across, greenish-cream; **drupe** about 0.4 cm long, reverse pear-shaped.

This species occurs on Kangaroo Island and southern parts of the Eyre Peninsula, growing among shrubs in coastal and near-coastal districts. It is unknown in cultivation and has limited ornamental appeal. Propagate from seed and possibly also from cuttings.

Olax pendula L. S. Smith
(pendulous, weeping)
Qld, NT
2–4 m × 1–2 m   Aug–Oct
Small to medium **shrub** with an open habit; **bark** greenish, smooth, mottled; **branchlets** pendulous, striate; **leaves** 2.5–9 cm × 0.5–1.5 cm, lanceolate, dark green, leathery; **flowers** about 0.5 cm long, pale green; **petals** about 0.4 cm long; **drupe** 0.8–1.1 cm long, ellipsoid, enlarged red hypanthium.

An interesting shrub which occurs on Cape York Peninsula, western Arnhem Land and on some northern islands of the Great Barrier Reef. It grows in rainforest and coastal scrubs, usually on well-drained sandy soil. Plants have graceful arching to pendulous branches. Best suited to tropical regions and may be a useful coastal plant. Frost tolerance is unlikely to be

high. Requires excellent drainage and some protection. Propagate from fresh seed and possibly also from cuttings.

Olax retusa Benth.
(blunt and shallowly notched)
Qld, NSW
0.3–0.6 m × 0.3–0.5 m   July–Dec
Dwarf **shrub**, slender, often single-stemmed; **leaves** 0.4–1.2 cm × 0.2–0.3 cm, oblong to cuneate, appressed against the stem, yellowish-green, apex truncate or notched; **flowers** about 0.5 cm across, creamy white; **drupe** 0.5–0.7 cm long, ellipsoid.

Distributed in south-eastern Qld and north-eastern NSW, growing in heath, woodland and coastal scrubs. Plants have a woody rootstock and branch more profusely after fires. Because it has limited ornamental appeal this species is rarely grown. Propagate from seed and possibly cuttings.

Olax stricta R. Br.
(upright, erect)
Qld, NSW, Vic
1–2 m × 0.5–1.5 m   all year
Small **shrub**, erect, often much-branched; **branchlets** striate; **leaves** 0.3–1 cm × 0.1–0.2 cm, linear to oblong, yellowish-green, blunt or hooked; **flowers** about 0.5 cm across, white to cream; **drupe** 0.4–0.8 cm × 0.5 cm, obovoid.

Widespread in open forest and woodland, often in rocky areas and also sometimes prominent in coastal dunes. A nondescript shrub which has limited ornamental appeal. Requires well-drained soil in a partially protected or exposed position. Propagate from seed.

OLEA L.
(from the Greek *elaia,* the olive tree)
Oleaceae
Shrubs or **trees**; **leaves** simple, entire, opposite, lower surface often scaly; **inflorescence** a terminal or axillary panicle; **flowers** small, bisexual; **calyx** 4-lobed; **corolla** tubular, 4-lobed; **stamens** 2; **fruit** a drupe.

A genus of about 40 species distributed in North America, Europe, Africa, Asia and New Zealand with one species occurring in Australia. It is rarely cultivated but could be used in reclamation projects. Propagation is from seed, which has a short period of viability, and for best results it should be sown fresh.

Olea paniculata R. Br.
(in panicles)
Qld, NSW   Australian Olive; Native Olive
15–25 m × 5–12 m   Oct–Dec
Medium to tall bushy **tree**; young growth shiny; **leaves** 5–10 cm × 1.5–6 cm, ovate to elliptic, dark green and glossy above, paler beneath, with small, hollow domatia along the midvein, margins entire, apex acuminate; **panicles** 4–10 cm long, terminal or in the upper axils, dense; **flowers** 0.3–0.4 cm across, greenish-white; **drupes** 0.8–1.2 cm × 0.6–0.8 cm, ovoid, bluish-black, mature May–Sept.

Widely distributed from Cape York Peninsula to north of Newcastle and extending to Lord Howe Island and New Caledonia. A common tree which

Oleaceae

Olea paniculata × .5

grows along streams and in dry rainforests and littoral rainforests. The close-grained, hard wood is attractively marked and has been used for carving, turning and inlays. A wide range of birds feed on the fruit. Native Olive is not widely grown but it is an excellent tree for planting in reclamation projects, especially on moist slopes near streams. Plants are quick growing and adaptable. Requires well-drained soil, some protection when young and responds to mulching and light fertiliser application. Tolerates light frosts. Seedlings frequently germinate on the ground beneath roosting birds. Propagate from fresh seed.

OLEACEAE Hoffmanns. & Link   Olive Family
A family of dicotyledons consisting of about 650 species in 24 genera. It is widely distributed in temperate and tropical parts of the world. They are trees, shrubs or climbers with opposite (rarely alternate), simple or compound leaves. The flowers have a corolla tube containing 2 or 4 stamens and commonly 4 sepals and petals (some species have up to 10 or more). This is a notable horticultural family because it contains many genera of important cultivated plants including *Jasminum, Fraxinus, Ligustrum, Olea* and *Syringa*. The family is represented in Australia by 5 genera and about 30 species. Australian genera: *Jasminum, Ligustrum, Linociera, Notelaea, Olea*.

OLEANDRA Cav.
(alluding to the fronds resembling an oleander leaf)
Oleandraceae

    Terrestrial, **lithophytic** or **epiphytic ferns** growing in clumps; **rhizome** densely scaly, bearing aerial roots; **fronds** simple, entire, scattered along the rhizome or in tufts; **sori** borne on lateral veins, often arranged close to the midrib; **indusia** reniform.

This genus of about 50 species is widely distributed in the tropics with a single widespread species extending to Australia. It has a very interesting growth habit and is mainly cultivated by enthusiasts. Propagation is mostly by division of the clumps.

Oleandra neriiformis Cav.
(like a *Nerium* (Oleander); referring to the leaves)
Qld   Stilt Fern
15–40 cm tall
    Terrestrial, **lithophytic** or **epiphytic fern** forming untidy clumps; **rhizome** wiry, about 0.5 cm thick, densely scaly, glaucous, supported on long, stilt-like wiry roots; **fronds** 20–60 cm × 2–4.5 cm, thin, simple, entire, light green, tapered to both ends, apex acute, **veins** conspicuous; **sori** in an irregular row, close to the midrib; **indusium** reniform.
    Occurs in north-eastern Qld and is widely distributed in Polynesia and Asia. An interesting fern which adapts very well to cultivation. Requires bright light, warmth, humidity and air movement. It can be grown in a pot or on a slab but is best in a basket of coarse, porous potting mixture. Needs protection from frosts. Propagate by division or from spores.

OLEANDRACEAE Ching ex Pichi-Serm.
A small family of tropical ferns consisting of about 50 species in the genus *Oleandra*. One species occurs in Australia. These ferns have scales attached peltately to the rhizome. The rhizomes are supported by wiry aerial roots and simple fronds which are jointed to short stalks on the rhizome.

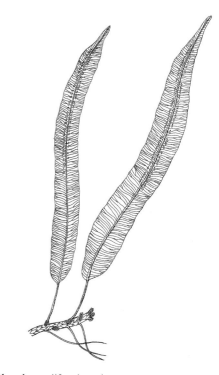

Oleandra neriiformis × .4

OLEARIA Moench
(probably named after Adam Olearins, 17th-century German botanist)
Asteraceae (alt. *Compositae*)   Daisy Bushes

Dwarf to tall **shrubs** or rarely small **trees** with glabrous, hairy or sometimes resinous growth; **leaves** alternate or opposite, sessile or stalked, simple, margins entire, toothed or lobed; **flowerheads** often daisy-like, terminal, axillary, sessile or stalked, solitary, arranged in terminal corymbs or panicles or leafy racemes; **involucral bracts** in several series, soft; **ray florets** female, petal-like or slender, white, cream, blue, violet, mauve, purple or pink; **disc florets** tubular, usually yellow but can be white, cream, blue, mauve, violet or purple; **fruit** a hairy or glabrous achene.

Olearia occurs exclusively in the southern hemisphere. The genus comprises about 180 species; Australia has about 130 species which are all endemic.

Olearia species have typical daisy-type flowerheads but generally they are easily distinguished from other members of the daisy family because they have woody stems and branches. Except for a couple of species all of them are shrubby in form. The two exceptions, *O. grandiflora* and *O. rhizomatica*, are low-growing and may sucker.

Olearia occurs in all states but the greatest representation is in south-eastern Australia. In the small state of Tasmania there are about 23 species of which 8 are endemic. In Western Australia, several new species have been described — currently there are about 33 named species occurring in WA.

In eastern states, daisy bushes are more common in sheltered sites where the soils are moist for the greater percentage of the year. They are found mainly on acidic soils in heathland, shrubland, woodland, and in wet and dry sclerophyll forests, from sea-level to about 1500 m altitude. In WA and the NT, as well as in parts of western Vic and SA, daisy bushes are usually found in drier habitats, such as in mallee and woodland where soils are often alkaline.

Olearias range from dwarf multi-stemmed shrubs, for example *O. grandiflora*, to solitary-stemmed tall shrubs. Some plants may even reach small tree proportions, for instance *O. argophylla*, which is the tallest member of the genus. Most species are quick-growing and some such as *O. flocktoniae* are pioneer plants in nature which thrive in disturbed soils although they live for perhaps only 3–4 years. Conversely, others are slow-growing and renowned as long-lived in cultivation, for example *O. argophylla*.

Foliage is one of the most decorative features of some olearias. The young growth is a highlight of species such as *O. rugosa* and *O. pannosa*. Many species have greyish foliage including much-admired selections of *O. phlogopappa*. Frequently, the underside of leaves is covered with silvery or sometimes rusty-brown hairs, and it is most appealing to see these leaves buffeted by gentle or bold winds. Foliage shape varies from small narrow leaves, for example *O. ciliata*, to sizeable lobed leaves eg. *O. pannosa* or *O. grandiflora*.

Flower production is usually excellent. Some species have outstanding displays of daisy flowers. Most species produce their flowers at or near the ends of branches or branchlets. Selections of *O. phlogopappa* provide a most pleasing range of colours including white, blue (of various hues), mauve, purple or pink (of various shades). On occasions it is difficult to see the foliage because of the profusion of flowerheads. Flowerhead diameter within the genus can range from about 0.5 cm (eg. *O. axillaris*), to over 7 cm (eg. *O. grandiflora*).

To date there has been very little horticultural breeding and development of *Olearia*. In most instances, the available cultivars are usually raised from selected flower colour variants. *O. phlogopappa* has provided the most variation. *O. stellulata* has one registered cultivar, 'Olwyn Barnett', which has a massed display of purple-pink flowerheads during Nov–Jan.

Potentially, breeding programmes could produce a stunning range of hybrids for horticulture.

Olearias are mainly used in amenity horticulture for home gardens. Although olearias are not regarded as highly desirable commercial cut flowers, they are very popular for home decoration (like many other daisies). Flowering stems will last 10–12 days or sometimes longer.

This genus is not considered medicinally valuable nor is it a food source for humans, however, flowering plants are an important source of nectar and pollen for insects. *O. pimeleoides* is regarded as a good honey and pollen producer. Butterflies and hover flies are among the most common of the myriad insects attracted to olearias.

Little is known about the value of olearias for forage grazing purposes but records show that *O. ramulosa*, which has low palatibility, and *O. magniflora* have been grazed.

Olearias are not important for forestry but the tallest species, *O. argophylla*, Musk Daisy Bush, produces mottled timber. During the late 1800s this timber — prized for wood-turning and cabinet-making — slabs of up to 36 inches (about 0.9 m) wide were available. The gnarled butts and roots were also eagerly sought after.

O. argophylla is also recorded as producing a brilliant sap-green dye in the 1800s. More recently dyeing wool with this species has produced a yellow colour when alum is used as a mordant.

Cultivation

Possibly *O. tomentosa* was the first species to be cultivated and it is recorded as being introduced into England in 1793. Other species cultivated in England and Europe during the early 1800s include *O. argophylla*, *O. myrsinoides*, *O. phlogopappa*, *O. ramosissima* and *O. stellulata* (then grown as *O. lirata*).

The range of species utilised so far has been limited. *O. argophylla*, *O. ciliata*, *O. erubescens*, *O. floribunda*, *O. minor* (often sold as the allied *O. iodochroa*), *O. microphylla*, *O. myrsinoides*, *O. phlogopappa* (with its many variants), *O. ramulosa*, *O. stellulata*, *O. teretifolia* and *O. tomentosa* are among the most popular.

O. phlogopappa deserves special mention. It has always received acclaim for its floral display and is often described as the most beautiful member of the genus, but this comment should be open to question because there are so many other exquisite species.

Much of the attention given to *O. phlogopappa* came through collections made by the English botanical explorer and gardener, Harold Comber. During 1929–1930 he visited Tasmania and made collections of flower-colour variants which were sent to England. In his field notes he wrote, 'every plant should be saved, as I consider this to be the best find that I have made so far in Tasmania'.

Another daisy bush which has proved reliable in Europe and to a lesser degree in Australia is usually marketed as *O. scilloniensis*. This plant is now thought to be a selection of *O. phlogopappa*. The cultivar arose as a seedling in the Tresco Abby Garden in the Scilly Isles off the Cornwall coast in south-western England.

Other olearias which deserve wider recognition include *O. astroloba*, *O. calcarea*, *O. frostii*, *O. grandiflora*, *O. magniflora*, *O. muelleri*, *O. obcordata*, *O. pannosa* (including ssp. *cardiophylla*), *O. persoonioides*, *O. picridifolia*, *O. quercifolia*, *O. rhizomatica*, *O. speciosa*, *O. stuartii* and *O. tasmanica*.

In cultivation, plants usually do best in freely draining acidic soils with a semi-shaded aspect, but some species tolerate alkaline soils, eg. *O. calcarea*, *O. magniflora* and *O. muelleri*.

Nearly all species tolerate light frosts. Plants which come from subalpine areas, such as *O. frostii*, *O. persoonioides*, *O. phlogopappa* (var. *flavescens* and var. *subrepanda*) and *O. tasmanica*, are much more cold tolerant.

The cool temperate species have a higher tolerance of high humidity than those from semi-arid and warm temperate regions.

In general, olearias respond well to low applications of slow-release complete fertilisers but in many cases plants do well without supplementary feeding. Regular or excessive fertilising can force growth on plants resulting in top-heavy foliage growth which can lead to splitting of branches and stems (see notes on pruning, below).

As many olearias occur in moist shaded sites in nature, these species may need supplementary watering during extended dry periods. But there are also many species which come from semi-arid areas and they are capable of growing well with minimum or no extra water.

Many daisy bushes develop open and somewhat leggy growth. It is usually beneficial to start pruning from an early stage. Tip pruning of seedlings or rooted cuttings of most species may stimulate lateral branching. Young plants also benefit from pruning. Gardeners should make sure they prune plants during flowering (when the cut stems can be used for indoor decoration), or immediately after flowering. Usually it is best not to cut into leafless wood, although good results have been achieved on vigorously growing *O. phlogopappa* plants. Prune very hard and new growth appears within 4–6 weeks. Most species respond very well to regular clipping and some have potential for hedging.

Pests and diseases do not cause major problems. One of the most common pests is stem borer which can be a regular minor problem on the more woody species, eg. *O. phlogopappa*. The larvae of light brown apple moth hide between leaves during the day and will feed on new growth. They are rarely troublesome — instead, they are best viewed as beneficial support for a pruning programme. Slugs and snails can attack young plants. See Volume 1, page 143 for control methods of these chewing pests. Collar rot can occur if too much mulch is placed around the trunk or stems of plants.

Propagation

Olearias are propagated from seed or from cuttings.

The seed is shed as soon as it is ripe, which may be a very short time after flowering. The ripe, swollen but small seed is attached to a hairy pappus (typical of daisy plants). Seed does not usually need any treatment before sowing but seed of species from subalpine areas can be stratified at 4°C for 4–6 weeks before sowing. See Treatment and Techniques to Germinate Difficult Species, Volume 1, page 205.

Germination time from sowing is variable. Species such as *O. axillaris* have been known to germinate in 8–10 days although they usually take much longer. It is not uncommon for seed to take up to 50 days to germinate, or longer. *O. teretifolia* may take about 6 weeks to germinate.

Very little is known about the reaction of *Olearia* seed to different pre-germination treatments and there is a need for further detailed germination research.

Striking cuttings is the most common method of propagating *Olearia*. Generally semi-woody young growth of about 5–10 cm long gives good results. The application of root inducing hormones can be beneficial but cuttings of many species will readily form roots without them. Choose cuttings with plenty of healthy nodes because the nodes may be damaged or destroyed by slugs, snails or caterpillar larvae.

Olearia muelleri in its natural habitat near Norseman, WA   W. R. Elliot

Olearia phlogopappa at Mt Buffalo,
north-eastern Victoria   W. R. Elliot

Olearia adenophora × .65

Root cuttings have been used to a very limited
degree and this method should be successful with
O. grandiflora and *O. rhizomatica.*

Aerial layering could be used with some of the
larger species, eg. *O. phlogopappa, O. argophylla* and
O. stellulata.

Olearia adenolasia (F. Muell.) F. Muell ex Benth.
(glandular hairs)
SA, WA
0.3–1.5 m × 0.4–1.5 m   Aug–Oct

Dwarf to small **shrub** with glandular-hairy young
growth; **branches** ascending to erect, glandular-hairy;
branchlets with sticky woolly hairs; **leaves** 0.7–1.4 cm ×
about 0.1 cm, more or less linear, sessile, spreading to
ascending, crowded, deep green and covered with soft
glandular hairs above, densely covered with non-glandular cream hairs below, apex blunt; **flowerheads**
daisy-like, to about 2 cm across, terminal; **ray florets**
white to bluish-violet; **disc florets** yellow; **achenes** flattish.

A floriferous species from the Eyre Peninsula, SA,
and from the Austin, Avon, Coolgardie and Irwin
Districts in WA. Plants occur in freely draining sandy
soils in heath and woodland. They require excellent
drainage and should adapt to well-drained sands and
loams in semi-arid and temperate regions. A sunny or
semi-shaded site is suitable. Hardy to moderate frosts
and extended dry periods. Tip pruning of young
plants promotes bushy growth. Propagate from seed,
which usually begins to germinate 20–30 days after
sowing, or from cuttings of firm young growth.

O. homolepis is closely allied but differs in having only
glandular hairs and blue flowers.

Olearia adenophora (F. Muell.) F. Muell. ex Benth.
(gland-bearing)
Vic   Scented Daisy Bush; Forest Daisy Bush
1–2 m × 1–1.5 m   Sept–Nov

Small **shrub** with resinous and hairy young growth;
branches many, spreading to ascending; **branchlets**
hairy and resinous; **leaves** 2–4.5 cm × 0.1–0.3 cm,
linear, alternate, sessile, spreading to ascending,

Olearia alpicola   D. L. Jones

67

crowded, sweetly aromatic, resinous, mid-to-deep green, hairy and rough above, hairy below, margins entire, revolute, apex pointed; **flowerheads** daisy-like, to about 4 cm across, on stalks shorter than the leaves, solitary or in few-flowered clusters, often profuse and very conspicuous; **ray florets** 20 plus, pale to deep blue, purple or white; **disc florets** yellow; **achenes** silky-hairy.

This poorly known species from eastern Victoria grows in disjunct areas on rocky mountain slopes in dry sclerophyll forest. Cultivated to a limited degree, it adapts well to most freely draining acidic soils in a semi-shaded site. Hardy to heavy frosts. Plants may be leggy but respond well to pruning from an early stage. Propagate from seed, or from cuttings of firm young growth.

In the 1970s plants were grown as *O. adenophora* but in most cases these were a compact selection of *O. tomentosa*, which has pale mauve flowerheads about 3 cm across.

Olearia aglossa (Maiden & Betche) Lander
(without a tongue; referring to florets)
NSW, Vic
2–3 m × 1.5–2.5 m   mainly Jan–Feb
Small **shrub** with greyish-hairy young growth; **branches** spreading to erect; **branchlets** hairy; **leaves** 3–10 cm × 0.8–3 cm, ovate, opposite, prominently stalked, spreading to ascending, scattered, green to glabrous above, densely grey-hairy below, margins revolute, venation reticulate, apex pointed; **flowerheads** daisy-like, to 2 cm across, in terminal flat-topped clusters, often profuse and conspicuous; **ray florets** white; **disc florets** yellow; **achenes** glabrous or with silky apex.

O. aglossa is confined to mountainous areas of the Southern Tablelands in south-eastern NSW and in the Victorian Alps. It occurs in dry sclerophyll forest. Evidently rarely cultivated, it should adapt to a wide range of freely draining acidic soils in temperate regions and will probably do best in a semi-shaded site. Plants are hardy to most frosts. Tip pruning at an early stage promotes a bushy framework. Propagate from seed, or from cuttings of firm young growth.

Previously known as *O. alpicola* var. *aglossa* Maiden & Betche.

Olearia algida N. A. Wakef.
(cold)
NSW, Vic
0.4–1 m × 0.5–1 m   Oct–April
Dwarf **shrub** with woolly-hairy young growth; **branches** many, spreading to erect; **branchlets** short, hairy; **leaves** to 0.3 cm × 0.1 cm, elliptic, alternate, spreading to ascending, crowded, sessile, dark green and glabrous above, grey-woolly below, margins revolute, apex pointed or rounded; **flowerheads** daisy-like, to about 1.2 cm across, solitary, axillary, often profuse and conspicuous; **ray florets** white; **disc florets** cream; **achenes** silky-hairy.

This species is found in cold habitats in south-eastern NSW and north-eastern Vic. It grows in swampy heathland and grassland. Cultivated to a limited degree, it has potential for greater use in revegetation of the Australian subalpine region. Plants require

freely draining moist sites but will tolerate short periods of dryness. Hardy to frost and snowfalls. Plants respond well to pruning. Propagate from seed or from cuttings of firm young growth.

O. ramulosa is allied but differs in having minutely glandular leaves and prominently stalked flowers.

Olearia allenderae J. H. Willis
(after Marie J. Allender, 20th-century botanical assistant, Victoria)
Vic   Promontory Daisy Bush
1–2 m × 1–1.5 m   Sept–Nov
Small **shrub** with whitish-hairy young growth; **branches** ascending to erect, purplish, ribbed; **branchlets** hairy; **leaves** 1–3 cm × 0.5–1.2 cm, ovate-lanceolate to oblong, alternate, very short-stalked, spreading to ascending, scattered, shiny deep green and wrinkled above, densely whitish-hairy below, margins irregularly toothed and revolute, apex blunt; **flowerheads** daisy-like, to about 2.5 cm across, on stalks to 1 cm long and usually solitary in upper axils, forming leafy clusters to about 8 cm long, often profuse and conspicuous; **ray florets** 10–12, white; **disc florets** usually purplish, becoming yellow; **achenes** glabrous.

A rare but not threatened species from eastern Victoria, where it is known from two disjunct locations. It occurs in swampy acidic soils. Evidently rarely cultivated, it should adapt well in temperate regions and to most acidic soils which are moisture-retentive. Plants tolerate moderate frosts. They need pruning to promote bushy growth. Propagate from seed or from cuttings of firm young growth.

O. stellulata is allied but has stellate hairs, leaves with flat regularly toothed margins, and a pointed apex.

Olearia alpicola (F.Muell.) F. Muell. ex Benth.
(growing in the alps)
NSW, Vic   Alpine Daisy Bush
1.5–3 m × 1–2.5 m   Dec–March
Small to medium **shrub** with rusty-hairy young growth; **branches** spreading to erect; **branchlets** hairy; **leaves** 3–14 cm × 0.3–2.3 cm, ovate, opposite, stalked, spreading to ascending, scattered, green and glabrous above, greyish-hairy below, margins revolute, apex ending in a fine point; **flowerheads** daisy-like, to about 3 cm across, in terminal clusters, often profuse and conspicuous; **ray florets** 6–7, white; **disc florets** yellow; **achenes** glabrous.

This floriferous, cold-tolerant species is from the mountains of eastern NSW and eastern Vic. Plants are found in sclerophyll forest. They produce lovely flushes of rusty new growth. *O. alpicola* has potential for greater use in a range of temperate regions. Plants grow best in well-drained acidic soils with semi-shade or dappled shade. This species tolerates frosts and light snowfalls. Pruning promotes bushy growth. Propagate from seed or from cuttings of firm young growth.

O. alpicola var. *aglossa* Maiden & Betche is referrable to *O. aglossa* (Maiden & Betche) Lander.

Olearia alpina (Hook. f.) W. M. Curtis =
O. tasmanica (Hook. f.) W. M. Curtis

Olearia archeri Lander
(after William Archer, 19th-century Tasmanian botanical collector)
Tas
1.5–2.5 m × 1–2 m   Sept–March
   Medium **shrub** with bronze-, yellowish-, or pinkish-hairy young growth; **branches** ascending to erect; **branchlets** hairy, brownish to yellowish; **leaves** 1.5–10 cm × 0.3–1.5 cm, narrowly elliptic, opposite near apex, alternate, short-stalked, spreading to ascending, scattered, shiny above, deep green and glabrous above, densely covered with brownish to yellowish hairs below, margins entire, flattish, apex bluntly pointed; **flowerheads** daisy-like, to about 3 cm across, on stalks to about 8 cm long, in crowded terminal clusters, profuse and very conspicuous; **ray florets** 7–9, white; **disc florets** yellow; **achenes** glabrous except for hairy apex.
   A handsome daisy bush which occurs from sea-level to an altitude of 650 m on the east coast. It inhabits dry sclerophyll forests where it favours shady sites on hillsides and stony slopes. This species is cultivated to a limited degree but it deserves to be better known. Plants should adapt to a wide range of acidic soils, which drain freely, and a semi-shaded aspect. Hardy to heavy frosts. Pruning from an early stage promote bushy growth. Propagate from seed or from cuttings of firm young growth.
   Previously known as *O. lanceolata* (Benth.) D. I. Morris and *O. persoonioides* var. *lanceolata* Benth.

Olearia argophylla (Labill.) F. Muell. ex Benth.
(silvery leaves)
NSW, Vic, Tas   Musk Daisy Bush; Native Musk; Silver Shrub
3–10 m × 2–6 m   Sept–April
   Medium to tall **shrub** or small **tree** with silvery young growth; **trunk** solitary or multiple; **bark** ribbony, greyish; **branches** spreading to ascending; **branchlets** silvery-hairy; **leaves** 2.5–20 cm × 1–9 cm, broadly elliptic, alternate, prominently stalked, scattered, spreading to ascending, dark green and shiny above, silvery-hairy below, margins toothed or entire, venation reticulate, apex pointed; **flowerheads** daisy-like, 1.2–2.7 cm across, terminal, in large crowded heads, usually profuse and very conspicuous, faintly spicy-scented; **ray florets** 3–8, white; **disc florets** cream to pale yellow; **achenes** glabrous or silky.
   This tallest member of the genus is a handsome species which inhabits fern gullies and other moist areas of wet sclerophyll forests. The delightful contrasting leaf surfaces are readily observed on windy days. It adapts well to cultivation and was introduced into England in 1834. Plants prefer sheltered, moist sites but success has been achieved in exposed positions. Soils need to be acidic and drain freely. Plants are hardy to most frosts and light snowfalls. They respond well to hard or light pruning. Supplementary watering may be required during extended dry periods. This species is excellent for background planting and for windbreaks in areas of high rainfall. Wood from mature trees is sometimes used for sculpture and woodturning. Propagate from seed or from cuttings of firm young growth.

Olearia argophylla × .2

Olearia arguta (R. Br.) Benth.
(sharply pointed)
WA, NT
0.3–0.6 m × 0.3–1 m   sporadic throughout the year
   Dwarf **shrub** with hairy, often sticky young growth; **branches** spreading to ascending; **branchlets** hairy; **leaves** 0.3–1 cm × 0.1–0.25 cm, narrowly elliptic to obovate, alternate, sessile, ascending to erect, green to greyish, hairy, margins prickly, apex pointed; **flowerheads** daisy-like, to about 3.5 cm across, solitary on terminal stalks, often profuse and very conspicuous; **ray florets** about 10–20, white, blue or purple; **disc florets** yellow; **achenes** hairy.
   The typical form of this interesting dwarf species occurs in the Kimberley Region of WA and in the Top End of the NT. It is found in a range of rocky soil types. Evidently not cultivated, it is best suited to seasonally dry tropical regions and is worth trying in subtropical areas. Plants need excellent drainage and a semi-shaded site.
   The var. *lanata* Benth. differs in having densely greyish-hairy stems and leaves with entire margins which usually become glabrous with age. It occurs in Qld, WA and NT.
   Propagate both variants from seed or from cuttings of firm young growth.

Olearia arida E. Pritzel
(arid, dry)
SA, ?WA
1–2 m × 1–2 m   Aug–Oct
   Small **shrub** with resinous woolly-hairy young growth; **stems** erect, woody, with sticky ribs; **branches** spreading to ascending, hairy; **leaves** 0.7–2 cm × 0.1–0.25 cm, linear to narrowly oblanceolate, sessile, spreading to ascending, scattered, viscid and glabrous above, white-woolly below, margins revolute, apex blunt; **flowerheads** daisy-like, to 2 cm across, terminal on short branchlets; **ray florets** white; **disc florets** yellow; **achenes** about 0.2 cm long, terete.
   A poorly known species which dwells on sandy hills. It occurs in the North Western and Lake Eyre Districts

69

of SA, and possibly in south-eastern WA. Best suited to arid, semi-arid and warm temperate regions, in excellently drained acidic soils. Plants tolerate exposure to sunshine and are hardy to moderate frosts. Pruning from a young stage promotes bushy growth. Propagate from seed or from cuttings of firm young growth.

Olearia aspera W. Fitzg. =
        Minuria macrorhiza (DC.) Lander.

Olearia asterotricha (F. Muell.) F. Muell. ex Benth.
(with starry or stellate hairs)
NSW, Vic   Rough Daisy Bush
1–2 m × 1.5–2 m   Oct–July
    Small **shrub** with stellate-hairy young growth; **branches** spreading to ascending, hairy; **branchlets** stellate-hairy, often arching; **leaves** 0.6–2.5 cm × 0.2–0.8 cm, narrowly obovate, alternate, scattered to crowded, sessile, soft and flexible, deep green and glossy with stellate hairs above, paler green and hairy below, margins prominently lobed, toothed or entire, apex rounded; **flowerheads** daisy-like, 2–4 cm across, solitary or clustered, terminal, often profuse and very conspicuous; **ray florets** 15–21, pale blue, mauve or, rarely, white; **disc florets** yellow; **achenes** silky-hairy.
    This appealing daisy bush has a distribution which extends from the Blue Mountains to western Vic. Generally it occurs in moist soils of heath or dry sclerophyll forests and extends from the coast to some distance inland. Although uncommon in cultivation, it deserves to be better known. Plants do best in semishade or dappled shade and prefer well-drained but moist acidic soils. They are hardy to moderate frosts and tolerate short dry periods. Plants often have an open growth habit but they respond very well to hard pruning which promotes fuller growth. Useful as a cut flower. Propagate from seed or from cuttings of firm young growth.
    The var. *parvifolia* Benth. is regarded as a synonym by some botanists.

Olearia astroloba Lander & N. G. Walsh
(stellate-hairy lobes; referring to the disc florets)
Vic
0.4–1 m × 0.5–1.5 m   June–Nov
    Dwarf, often spreading **shrub** with pale green, hairy young growth; **branches** spreading to ascending; **branchlets** hairy; **leaves** 0.5–2 cm × 0.2–1 cm, spathulate, alternate, sessile, ascending to erect, crowded, greyish-green and hairy above, paler and hairy below, margins thick and revolute, toothed in upper parts, apex bluntly toothed; **flowerheads** daisy-like, to about 3.2 cm across, sessile, solitary, scattered to profuse, very conspicuous; **ray florets** about 20, bluish-violet; **disc florets** purple; **achenes** silky-hairy.
    An extremely rare and vulnerable species, recently described in 1989 and known only from an area of about 40 ha in the subalpine region of eastern Gippsland. This highly decorative species occurs on steep slopes and cliff faces in skeletal loams derived from marble. It has adapted well to cultivation and is best suited to freely draining soils in a semi-shaded or dapple-shaded site, but it also tolerates plenty of sunshine. Plants may develop spreading branches which

Olearia astroloba × .8

are best pruned early to promote bushy growth. Hardy to heavy frosts. Propagate from seed or from cuttings of firm young growth.

Olearia axillaris (DC.) F. Muell. ex Benth.
(axillary; referring to flowerheads)
NSW, Vic, Tas, SA, WA   Coast Daisy Bush
1.5–3 m × 1–2.5 m   Nov–July
    Small to medium **shrub** with whitish- to greyish-hairy young growth; **stems** woody; **branches** ascending to erect, many; **branchlets** whitish to greyish, hairy; **leaves** 1–4.5 cm × 0.15–0.6 cm, linear to narrowly oblanceolate, sessile, spreading to ascending, somewhat crowded, deep green and more or less glabrous above, densely white-hairy below, margins revolute, apex blunt; **flowerheads** about 0.5 cm across, axillary, more or less sessile; **ray florets** 2–5, white to cream, somewhat insignificant; **disc florets** yellow; **achenes** flattened to terete.
    A common plant in some coastal areas where it grows on sand dunes. It is an important component of shrubland and heathland. This species is adaptable and fast-growing in acidic or slightly alkaline soils. Plants are suitable for semi-arid and temperate regions. In coastal sites, plants are excellent soil-binders, and they readily form windbreaks. They tolerate salt spray, moderate frosts and extended dry periods. Plants respond very well to pruning and can be used for hedging as foliage is the strongest ornamental feature. There is ample opportunity for selection of variants which are superior for horticultural purposes.

The var. *eremicola* Diels, Inland Daisy Bush, is a small shrub to about 1.5 m × 1.5 m. It has somewhat terete leaves and white to purplish ray florets.

Propagate both varieties from seed, which usually begins to germinate by 24 days after sowing. Cuttings of firm young growth strike readily.

O. revoluta is closely allied but has prominent ray florets.

Olearia ballii (F. Muell.) Hemsl.
(after Lieutenant Henry Lidgbird Ball, discoverer of Lord Howe Island in 1788)
NSW (Lord Howe Island)   Mountain Daisy
1–2 m × 1–2 m   mainly Nov–March; also sporadic

Small **shrub** with white-hairy young growth; **branches** spreading to ascending; **branchlets** short, usually spreading; **leaves** 0.5–1.2 cm × up to 0.15 cm, linear, alternate, more or less sessile, spreading to ascending, often crowded on branchlets, shiny dark green and glabrous above, white-hairy below except on midrib, apex bluntly pointed; **flowerheads** daisy-like, to about 2 cm across, solitary, terminal, sessile, often profuse and very conspicuous; **ray florets** 20–30, white with purple tip; **disc florets** purplish; **achenes** brown, glandular.

This admirable species deserves to be better known in cultivation. It is restricted to Lord Howe Island where it grows above 400 m altitude. Plants adapt well to cultivation in subtropical and temperate regions. They need freely draining acidic soils which are moist for most of the year. Probably best suited to semi-shade

Olearia ballii   W. R. Elliot

Olearia brachyphylla, long-leaved variant   D. L. Jones

Olearia axillaris × .65, flower × 3

71

or dappled shade conditions. Moderate to heavy frosts will damage them. They respond well to pruning. Propagate from seed or from cuttings of firm young growth.

Olearia brachyphylla (F. Muell. ex Sond.) N. A. Wakef.
(short leaves)
SA, WA
0.6–1.5 m × 0.6–2 m   July–Feb
  Dwarf to small **shrub** with whitish-hairy young growth; **stems** woody; **branches** many, whitish-hairy, spreading to ascending; **leaves** 0.2–0.3 cm × about 0.1 cm, oblong, sessile, spreading to ascending, incurved, moderately crowded, flat, green and more or less glabrous above, hairier below, margins revolute, apex blunt; **flowerheads** somewhat daisy-like, to about 0.4 cm across, solitary, terminal, sessile, usually profuse and most conspicuous; **ray florets** 2–3, white; **disc florets** yellow; **achenes** glabrous to faintly hairy.
  A very floriferous species which is common in mallee and heath communities . It has a disjunct distribution with representation in south-eastern SA and the Eyre District in WA. Rarely cultivated, it is suited to semi-arid and temperate regions. Plants require freely draining acidic soils in a sunny or semi-shaded site. It is hardy to most frosts and tolerates dry periods. Responds well to pruning. Propagate from seed or from cuttings of firm young growth.
  O. pimeleoides var. *minor* Benth. was considered synonymous but is now separated as *O. minor* (Benth.) Lander. It has longer leaves and larger flowerheads to about 2.2 cm across.
  O. ramulosa differs in having longer leaves and it has ray florets which are much longer than the styles.

Olearia burgessii Lander
(after Rev. Colin Burgess, 20th-century botanical collector)
NSW
1.5–2.5 m × 1.5–2 m   July–Nov
  Small to medium **shrub** with hairy young growth; **branches** spreading to ascending; **branchlets** initially greyish-hairy, becoming nearly glabrous; **leaves** 0.3–1.5 cm × 0.1–0.4 cm, elliptical or obovate, alternate, scattered, more or less sessile, spreading to ascending, green and glabrous or greyish-woolly and glandular above, densely greyish-hairy below, apex pointed or rounded; **flowerheads** daisy-like, to about 1.6 cm across, terminal on stalks to about 1.2 cm long, solitary or in loose clusters forming leafy panicles, often profuse and conspicuous; **ray florets** 9–13, white; **disc florets** yellow; **achenes** silky-hairy.
  A recently described (1991) species which has two disjunct populations; one on the Central Coast and the other on the Central and Southern Tablelands. It is poorly known and needs to be cultivated as part of a conservation strategy. Plants grow in dry sclerophyll forest in sandy loam over limestone or shale. They should prove adaptable in freely draining acidic soils in temperate regions. Best grown in a semi-shaded or dapple-shaded site. Hardy to moderate frosts. Propagate from seed or from cuttings of firm young growth.

Olearia calcarea F. Muell. ex Benth.
(growing in limestone)
NSW, Vic, SA, WA   Limestone Daisy Bush
0.6–1.5 m × 0.6–2 m   May–Oct
  Dwarf to small **shrub** with resinous young growth; **stems** woody; **branches** spreading to ascending; **branchlets** glabrous, ribless; **leaves** 0.2–1.5 cm × 0.2–1 cm, broadly cuneate to nearly circular, alternate, sessile, ascending to erect, crowded on branchlets but otherwise scattered, green, glabrous, glandular, margins toothed, apex rounded; **flowerheads** daisy-like, 3–4.5 cm across, solitary, terminal, sessile or short-stalked, very conspicuous; **ray florets** white; **disc florets** yellow; **bracts** green, sticky; **achenes** silky-hairy, terete.
  A very ornamental species which inhabits alkaline sandy and loamy soils in woodland and mallee in low rainfall regions. It should be very useful in arid, semi-arid and warm temperate regions. Plants will grow in alkaline and acidic soils which drain freely and sites which are sunny or semi-shaded. Hardy to moderate frosts and extended dry periods. It should respond well to pruning. Worth trying as a container plant. Propagate from seed or from cuttings of firm young growth.
  The closely allied *O. muelleri* has smaller flowerheads and entire or toothed leaves. It may intergrade with *O. calcarea.*

Olearia canescens (Benth.) Hutch.
(hoary)
Qld, NSW
1.5–2.5 m × 1.5–2 m   Feb–Aug
  Small to medium **shrub** with densely grey-woolly young growth; **branches** ascending to erect; **branchlets** densely grey-woolly hairy; **leaves** 1–6 cm × 0.3–2 cm, elliptical or ovate, alternate, spreading to ascending, more or less sessile, scattered, flat, densely covered with greyish-woolly hairs above and below, apex pointed; **flowerheads** daisy-like to about 2 cm across, in terminal panicles, on stalks to about 2 cm long, often profuse and conspicuous; **ray florets** 8–15, white; **disc florets** yellow; **achenes** silky-hairy.
  A quick-growing and often short-lived ornamental daisy bush from south-eastern Qld and north-eastern NSW. It grows in sclerophyll forest and woodland. Rarely cultivated, it has potential to become more popular with its greyish appearance. Plants need good drainage, acidic soils and a semi-shaded or dapple-shaded site. May suffer damage from heavy frosts. Old plants may need rejuvenation by thinning out mature growth. Propagate from seed or from cuttings of firm young growth.
  O. stellulata is similar but it lacks the dense covering of greyish-woolly hairs and it has cream disc florets.

Olearia cassiniae (F. Muell.) Benth. —
                    see *O. ramulosa* (Labill.) Benth.

Olearia chrysophylla (DC.) Benth.
(golden leaves)
Qld, NSW
2–3 m × 1.5–2.5 m   Nov–Jan

Medium **shrub** with hairy young growth; **branches** ascending to erect; **branchlets** with pale brown hairs; **leaves** 2–13 cm × 0.4–4 cm, elliptic, opposite, stalked, spreading to ascending, scattered, flat, yellow-green to mid green and glabrous above, yellowish or pale buff to light brown hairs below, margins variably toothed to entire, venation reticulate, apex pointed; **flowerheads** daisy-like, to about 3 cm across, on stalks to about 5 cm long, forming terminal panicles, often profuse and very conspicuous; **ray florets** 4–7, white; **disc florets** yellow; **achenes** glabrous.

A distinctive and decorative species with interesting foliage and well-displayed flowers. It inhabits mountains in south-eastern Qld and in the Central and Northern Tablelands and the Central Coast of NSW. Plants grow in sclerophyll forest. Worthy of greater cultivation, it should adapt well to most freely draining but moisture-retentive acidic soils, in sites which are semi-shaded or with dappled shade. Plants are hardy to moderate frosts. May need pruning to promote bushy growth. Propagate from seed or from cuttings of firm young growth.

This species has been confused with *O. oppositifolia* (F. Muell.) Lander, which has regularly toothed leaves with a greyish-hairy undersurface.

Olearia ciliata (Benth.) F. Muell. ex Benth.
(fringed with fine hairs)
Qld, Vic, Tas, SA, WA   Fringed Daisy Bush
0.15–0.4 m × 0.2–0.6 m July–Dec; also sporadic

Olearia ciliata ssp. *ciliata* × .55

Dwarf twiggy **shrub** with slightly hairy young growth; **branches** few, spreading to ascending; **branchlets** thick, hairy; **leaves** 0.5–2.5 cm × up to 0.2 cm, linear, alternate, sessile, recurved to ascending, crowded, bright to deep green and more or less glabrous above except for the fringed revolute margins, hairy below, apex slightly prickly-pointed; **flowerheads** daisy-like, to about 3 cm across, solitary on slender terminal stalks 2–15 cm long, scattered to profuse but always very conspicuous; **ray florets** 20–35, usually pinkish-mauve, lilac or blue, rarely white to pink; **disc florets** yellow; **achenes** glabrous to slightly hairy.

An outstanding but variable dwarf daisy bush which occurs in a range of soils and habitats in temperate regions. Good drainage is an essential requirement for successful cultivation. Plants do well in alkaline or acidic soils and will grow in full sun although they usually do better in a semi-shaded site. They tolerate heavy frosts and extended dry periods. Tip pruning of young plants and again during or immediately after flowering will quickly create a bushy framework. Excellent for use in gardens or containers.

The var. *squamifolia* (F. Muell.) Benth. is confined to Kangaroo Island, SA, where it grows in lateritic sandy soils. It differs in having smaller, strongly decurved leaves which are arranged in crowded, rosette-like clusters.

Propagate both varieties from seed, which usually begins to germinate 11–25 days after sowing, or from cuttings of firm young growth. The leaves of cuttings may rot if they are wet for extended periods.

Olearia cordata Lander
(heart-shaped)
NSW
1.5–2.5 m × 1.5–2 m   Nov–April

Small to medium **shrub** with hairy young growth; **branches** ascending to erect; **branchlets** hairy; **leaves** 1–4 cm × 0.2–0.8 cm, linear to narrow-lanceolate, with a cordate base, alternate, spreading to ascending, scattered, green with dense covering of simple hairs and glandular hairs above and below, margins revolute, apex pointed or somewhat rounded; **flowerheads** daisy-like, to about 3.5 cm across, solitary, terminal on stalks to about 4 cm long, may be profuse and conspicuous; **ray florets** 10–18, deep blue to mauve; **disc florets** yellow; **achenes** silky-hairy.

This rather handsome daisy bush is very rare in nature. It grows in a reserve but it needs to be cultivated as part of a conservation strategy. The species occurs on sandstone north-east of Gosford in shrubland and dry sclerophyll forest. Plants need excellent drainage, acidic soil and a site with semi-shade or dappled shade but they may tolerate an open sunny site too. Hardy to light frosts. May need pruning early to promote bushy growth. Propagate from seed or from cuttings of firm young growth.

O. tenuifolia is allied but it has crowded, less hairy, or glabrous leaves with a narrow base.

Olearia covenyi Lander
(after Robert G. Coveny, contemporary botanical collector, Royal Botanical Gardens, Sydney)
NSW
3–5 m × 2–3.5 m   Nov–Jan

Medium to tall **shrub** with densely hairy, yellowish-brown young growth; **branches** ascending to erect; **branchlets** densely hairy; **leaves** 3–15 cm × 1.2–5 cm, ovate, opposite, stalked, spreading to ascending, scattered, green and glabrous above, yellowish-brown hairy below, margins entire, venation reticulate, apex blunt or rounded, ending in a very short point; **flowerheads** to about 2.2 cm across, prominently stalked, forming terminal compound clusters, often profuse and conspicuous; **ray florets** 2–3, white; **disc florets** yellow; **achenes** glabrous.

Described in 1991, this species has a wide distribution range over the North Coast, Northern Tablelands and North Western Slopes. It inhabits wet sclerophyll forests of the mountainous areas. Plants need moist but freely draining acidic soils in semi-sheltered or sheltered sites. May need supplementary watering during extended dry periods. Hardy to moderate frosts. Judicious pruning promotes bushy growth. Propagate from seed or from cuttings of firm young growth.

O. megalophylla has similarities but its flowerheads are larger with 5–9 ray florets and flowering is usually during Dec–March.

Olearia cydoniifolia (DC.) Benth.
(leaves similar to those of the quince, *Cydonia*)
Qld, NSW
3–4.5 m × 2–4 m   Oct–Nov
Medium to tall **shrub** with densely hairy young growth; **branches** spreading to ascending; **branchlets** felty-hairy; **leaves** 2.5–13 cm × 0.9–5.5 cm, elliptical, alternate, spreading to ascending, scattered, prominently stalked, green and glabrous above, densely silver-hairy below, flat, margins entire, apex rounded or pointed; **flowerheads** daisy-like, to about 2 cm across, in loose few-flowered terminal clusters, often profuse and conspicuous; **ray florets** 6–10, white; **disc florets** yellow; **achenes** silky-hairy.

A shrubby species which occurs on the margins of rainforest and in the sclerophyll forest of south-eastern Qld and north-eastern NSW. It grows in rocky gorges, on escarpments and on steep slopes. Plants should do well in warm temperate and subtropical regions in freely draining acidic soils. Best suited to a semi-shaded site. Hardy to light frosts. Tip pruning from an early stage promotes bushy growth. Propagate from seed or from cuttings of firm young growth.

O. gravis is allied but has crowded hairs on the upper surface of the sometimes toothed, shorter leaves.

Olearia decurrens (DC.) Benth.
(base of leaf extending down stem as wings)
NSW, Vic, SA, WA   Clammy Daisy Bush
1–2 m × 1–2 m   Dec–June; also sporadic
Small **shrub** with resinous young growth; **branches** spreading to ascending; **branchlets** resinous; **leaves** 0.7–5 cm × 0.1–0.5 cm, oblanceolate, alternate, spreading to ascending, more or less sessile, scattered, glabrous, deep green, resinous, margins entire or toothed near the blunt apex; **flowerheads** daisy-like, to 2 cm across, on stalks about 1 cm long in terminal panicles, often profuse and conspicuous; **ray florets** 3–5, white; **disc florets** yellow; **achenes** about 0.2 cm long, silky-hairy.

A dry area species which inhabits mallee and mulga scrubland and woodland over a wide range. This long-flowering daisy needs excellent drainage in acidic or slightly alkaline soils. It appreciates plenty of sunshine but also tolerates semi-shade. Plants are hardy to most frosts and they are drought tolerant. They can develop spindly growth but pruning should promote a bushy framework. Propagate from seed or from cuttings of firm young growth.

O. passerinoides differs in having 5–8 ray florets and narrower leaves.

Olearia dentata Moench =
                                    O. tomentosa (H. L. Wendl.) DC.

Olearia elaeophila (DC.) F. Muell. ex Benth.
(marsh-loving)
WA
0.4–0.75 m × 0.5–1 m   Jan–June; also sporadic
Dwarf **shrub** with glabrous to faintly hairy young growth; **branches** spreading to erect, slender; **branchlets** twiggy; **leaves** 1–3 cm × about 0.1 cm, narrowly linear, alternate, sessile, spreading to ascending, scattered on upper parts while often clustered on lower parts, green, glabrous or glandular-hairy, margins revolute, apex pointed; **flowerheads** daisy-like, to about 1.5 cm across, on slender stalks, arranged in somewhat loose terminal clusters, often profuse and conspicuous; **ray florets** about 12–15, white, pink, mauve or lilac; **disc florets** yellow; **achenes** silky-hairy.

This somewhat dainty species occurs in the forests of the Darling District where it grows in moist to swampy soils. Although rarely cultivated, it may prove to be an excellent garden plant. Plants should adapt well to a range of moisture-retentive acidic soils in cool temperate regions. A semi-shaded site would be best. Plants tolerate light frosts. Pruning from an early stage should promote bushy growth. Suitable for gardens or containers and should be worth trying as a cut flower.

The var. *major* Benth. has a more woolly-hairy appearance. Its leaves are larger and the lowest ones may be toothed.

Propagate both varieties from seed, which usually begins to germinate 20–30 days after sowing. Cuttings of firm young growth should strike readily.

Previously known as *O. heliophila* (F. Muell.) Benth.
O. muricata has been confused with this species but differs in having rough, hairy leaves.

Olearia elliptica DC.
(elliptical)
Qld, NSW   Sticky Daisy Bush
0.6–2.5 cm × 1–2 cm   Nov–May
Dwarf to medium **shrub** with resinous young growth; **branches** spreading to ascending; **branchlets** somewhat resinous; **leaves** 2.5–15 cm × 0.5–4 cm, narrowly to broadly elliptic, alternate, stalked, ascending to erect, scattered, green and resinous above, pale green and glabrous below, margins faintly revolute, apex pointed to rounded, strongly aromatic; **flowerheads** daisy-like, 1–2.6 cm across, in terminal clusters; **ray florets** 8–25, white; **disc florets** yellow; **achenes** faintly hairy.

The typical form occurs in south-eastern Qld and

Olearia ferresii   T. L. Blake

eastern NSW. It grows in a range of mountainous habitats including heath, woodland and dry sclerophyll forest. Lacks strong ornamental characteristics but adapts well to cultivation in well-drained acidic soils. A sunny, semi-shaded or dapple-shaded site is suitable. Hardy to moderate frosts.

The ssp. *praetermissa* P. S. Green is restricted to rocky ledges of the taller mountains on Lord Howe Island. It has obovate-elliptic leaves of about 4 cm × 1.5 cm and flowers about 0.8 cm across which bloom during May–Oct.

Propagate both forms from seed or from cuttings of firm young growth.

Olearia eremaea Lander
(occurs in the desert)
WA
1–1.5 m × 1–1.5 m   July–Aug

Small **shrub** with resinous, sometimes reddish young growth; **branches** spreading to erect; **branchlets** reddish when young, resinous; **leaves** 0.6–1.6 cm × 0.2–0.5 cm, elliptic to obovate, alternate, sessile, spreading to ascending, scattered, pale green with hairs above and below, resinous, flat, margins toothed, apex pointed; **flowerheads** daisy-like, about 1.5–4 cm across, solitary, on terminal stalks to about 2 cm long, scattered to profuse, conspicuous; **ray florets** 13–22, white; **disc florets** yellow; **achenes** silky-hairy.

This recently described (1990) species is regarded as rare but not threatened in nature. Nevertheless it could be cultivated as part of a conservation strategy. It occurs in the Ashburton and Giles Districts of the Eremaean Botanical Province where it grows in shallow soils on lateritic breakaways. It will be best suited to semi-arid and warm temperate regions, in well-drained acidic soils with a sunny or semi-shaded aspect. Hardy to moderate frosts. Propagate from seed or from cuttings of firm young growth.

Olearia ericoides

O. calcarea has thicker, broader (sometimes nearly circular) leaves, while *O. muelleri* has leaves which are toothed near the apex.

Olearia ericoides (Steetz) N. A. Wakef.
(resembles the genus *Erica*)
Tas
0.6–1.5 m × 0.6–1.5 m   Dec–Feb

Dwarf to small **shrub** with hairy young growth; **branches** many, spreading to ascending; **branchlets** spreading to ascending, slender; **leaves** 0.3–0.5 cm × about 0.1 cm, narrowly linear, alternate, sessile, spreading to ascending, scattered or clustered, green and slightly rough above, densely hairy below, margins revolute, apex pointed; **flowerheads** daisy-like, to about 1.8 cm across, solitary, terminal, forming leafy racemes, often profuse and very conspicuous; **ray florets** about 24, pale bluish-purple or, rarely, white; **disc florets** purple; **achenes** hairy.

An attractive, floriferous, twiggy daisy bush from the south and south-east where it grows on dryish hillsides in sclerophyll forest. Not commonly cultivated but adapts well to freely draining acidic soils in semi-shaded sites. Tolerates most frosts and limited dry periods. Pruning from an early stage is beneficial in promoting bushy growth. Propagate from seed or from cuttings of firm young growth.

O. hookeri is similar but it has thick, nearly terete leaves and a much sparser floral display.

Olearia ericoides × .75

75

Olearia erubescens (Sieber ex DC.) Dippel
(turning red)
NSW, Vic, Tas, SA   Silky Daisy Bush; Moth
                                    Daisy Bush
1–2 m × 1–2 m   Sept–March; often sporadic
    Small **shrub** with often reddish-hairy young growth;
branches spreading to ascending; **branchlets** often
reddish when young, hairy; **leaves** 2–13.5 cm ×
0.3–2 cm, elliptical, alternate, spreading to ascending,
stalked, scattered, thick, dark green and glabrous
above, densely white-hairy below, margins flat and
prickly-toothed, apex pointed; **flowerheads** daisy-like,
to about 3 cm across, in terminal clusters, often profuse and conspicuous; **ray florets** 4–7, white; **disc
florets** yellow; **achenes** ribbed, glabrous.
    A showy olearia with interesting young growth. It
grows in dry sclerophyll forest in temperate regions.
Plants do well in most freely draining acidic soils with
a semi-shaded aspect but will tolerate plenty of sunshine. Tolerates most frosts and lengthy dry periods.
Healthy plants respond very well to hard pruning,
which needs to be done regularly. Cultivated in
Ireland and the UK. Propagate from seed or from cuttings of firm young growth.
    The closely allied *O. myrsinoides* differs in having
longer-stalked flowerheads with only 2–4 ray florets.

Olearia exiguifolia (F. Muell.) F. Muell. ex Benth.
(small, narrow leaves)
SA, WA   Small-leaved Daisy Bush
1.5–2.5 m × 1–2 m   July–Feb
    Small to medium **shrub** with densely hairy young
growth; **branches** many, ascending to erect; **branchlets**
densely white-hairy; **leaves** 0.2–0.6 cm × 0.2–0.4 cm,
broadly ovate to wedge-shaped, tapering to base, alternate, spreading to ascending, scattered, deep green and
becoming glabrous above, with cream to rusty hairs
below, margins 3-lobed near apex and sometimes
lobed towards base, apex blunt; **flowerheads** daisy-like,
to about 1.2 cm across, solitary, terminal, on short axillary branchlets and often forming leafy panicles, often
profuse; **ray florets** 5–6, white; **disc florets** yellow; **achenes** hairy.
    This species inhabits sand dunes in semi-arid
regions. It deserves to be much better known. Plants
require excellently drained soils with a warm to hot
sunny aspect. They are best suited to semi-arid and
warm temperate regions. Hardy to moderate frosts.
Prone to becoming leggy if not pruned from an early
age. Propagate from seed or from cuttings of firm
young growth.

Olearia exilifolia (F. Muell.) Benth. =
    O. brachyphylla (F. Muell. ex Sond.) N. A. Wakef.

Olearia ferresii (F. Muell.) F. Muell. ex Benth.
(after Mr J. Ferres)
SA, WA, NT
1–2 m × 1–2 m   June–Sept
    Small **shrub** with resinous young growth; **branches**
ascending to erect; **branchlets** glabrous, resinous;
leaves 4–10 cm × 0.5–2 cm, elliptical, alternate, spreading to ascending, often crowded on branchlets, sessile,
green, resinous, strongly aromatic, flat, margins finely

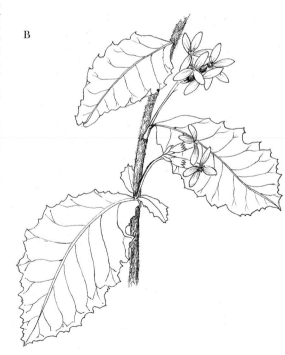

Olearia erubescens variants; (A) × .85, (B) × .7

toothed, apex pointed; **flowerheads** daisy-like, to about 3 cm across, in terminal clusters, often profuse and very conspicuous; **ray florets** 15–20, white; **disc florets** yellow or orange; **achenes** ribbed, silky-hairy.

This distinctive and handsome daisy occurs in arid areas on ephemeral watercourses of Central Australia. Plants are rare in cultivation. They are best suited to semi-arid and warm temperate regions with well-drained soils. Heavy frosts can damage plants. Pruning from an early stage should promote bushy growth. Propagate from seed or from cuttings of firm young growth.

Olearia flocktoniae Maiden & Betche
(after Margaret L. Flockton, 19–20th-century botanical illustrator)
NSW Dorrigo Daisy Bush
1–2.5 m × 1–2 m Feb–March
Small to medium **shrub** with hairy young growth; **branches** spreading to ascending; **branchlets** finely hairy; **leaves** 2–9 cm × 0.1–0.5 cm, linear, alternate, spreading to ascending, crowded, often clustered, more or less sessile, green, finely hairy, margins revolute, apex pointed; **flowerheads** daisy-like, to about 2.5 cm across, on stalks to about 5 cm long, in terminal clusters, often profuse and very conspicuous; **ray florets** 30–48, white, sometimes with violet tonings; **disc florets** yellow; **achenes** silky-hairy.

This quick-growing and often short-lived daisy bush was rediscovered recently after it was thought to be extinct. It is confined to the Dorrigo region of north-eastern NSW where it occurs in warm temperate rainforest and dry sclerophyll forest. It is a pioneer species which thrives in disturbed sites. Plants need good drainage in acidic soil and should do best in semi-

Olearia floribunda × .7, leaf × 2

shaded sites. It is cultivated to a limited degree. Hardy to moderate frosts. Pruning from an early age can slightly improve its longevity. Propagate from seed or from cuttings of firm young growth.

Olearia floribunda (Hook. f.) Benth.
(abundant flowers)
NSW, Vic, Tas, SA Heath Daisy Bush
1–2.5 m × 1–2.5 m July–Feb
Small to medium **shrub** with greyish-hairy young growth; **branches** spreading to ascending, often arching; **branchlets** hairy; **leaves** to about 0.3 cm × 0.1 cm, elliptical to ovate, alternate, ascending to erect, sessile, crowded, deep green and glabrous above, greyish-hairy below, margins revolute, apex pointed to blunt; **flowerheads** daisy-like, 1–2 cm across, solitary, terminal, sessile, on short branchlets in leafy panicles, often profuse and very conspicuous; **ray florets** 3–7, white to pale pink or, rarely, lavender; **disc florets** cream to yellow; **achenes** densely hairy.

An ornamental species which is distributed over a wide area in temperate regions. It is often found in woodland, open forest and near swamps. Plants adapt very well to cultivation in acidic sandy and clay loam soils with a sunny or semi-shaded aspect. They are hardy to most frosts and tolerate limited periods of dryness. Plants respond well to pruning, which promotes bushy growth. Propagate from seed or from cuttings of firm young growth.

This species is often confused with *O. ramulosa*, which differs in having longer and slightly tuberculate, linear leaves.

O. pimeleoides var. *minor* Benth. has been regarded by some botanists as a variant of *O. floribunda* but it was recently (1991) raised to species status as *O. minor* (Benth.) Lander.

O. floribunda var. *lanuginosa* J. H. Willis is now known as *O. lanuginosa* (J. H. Willis) N. A. Wakef.

Olearia fluvialis Lander
(riverine habitat)
WA
0.3–0.6 m × 0.3–0.6 m April–May
Dwarf **shrub** with hairy young growth; **branches** ascending to erect; **branchlets** with long hairs; **leaves** 0.2–1 cm × 0.1–0.2 cm, narrowly ovate, alternate, sessile, ascending to erect, scattered, green and nearly glabrous to covered with long hairs, flat, margins entire, apex pointed; **flowerheads** daisy-like, to about 1.3 cm across, solitary on terminal stalks to about 8 cm long, scattered to profuse and often conspicuous; **ray florets** 12–14, white or mauve; **disc florets** yellow; **achenes** densely silky-hairy.

This vulnerable species is known only from a small area, in a National Park in the Fortescue District, where it grows in the iron-rich alluvium of the creek and river beds. It needs to be cultivated as part of a conservation strategy. Plants should be best suited to semi-arid or warm temperate regions in excellently drained acidic soils. They tolerate moderate frosts. Pruning may help to promote a bushy framework. Propagate from seed or from cuttings of firm young growth.

Olearia frostii (F. Muell.) J. H. Willis
(after Charles Frost, 19th-century botanist)
Vic   Bogong Daisy Bush; Hoary Daisy Bush
0.6–1 m × 0.8–1.5 m   Dec–Feb

Dwarf **shrub** with woolly-hairy young growth; **branches** spreading to ascending; **branchlets** densely woolly-hairy; **leaves** 2–4 cm × 0.4–1 cm, obovate, alternate, sessile, spreading to ascending, scattered, pale green, densely woolly-hairy above and below, flat, margins entire except for toothed or notched apex; **flowerheads** daisy-like, to 3 cm across, usually solitary, terminal, scattered to profuse, always very conspicuous; **ray florets** usually 40–50, white to deep mauve or purple; **disc florets** yellow to orange-yellow; **achenes** glabrous.

A spectacular, cold-tolerant species which is restricted to the Victorian alpine region. It grows in the open high plains in freely draining acidic soils. This highly desirable daisy bush is best suited to cool temperate regions. It can be short-lived in cultivation. Plants require good drainage and plenty of sunshine to do well. Pruning and shaping at an early stage promotes bushy growth. *O. frostii* is suitable for use in gardens or containers. It has performed moderately well in English gardens. Propagate from seed or from cuttings of firm young growth.

Hybrids of *O. frostii* and *O. phlogopappa* with intermediate characteristics are sometimes encountered in cultivation.

Olearia frostii × .7

Olearia glandulosa   D. L. Jones

Olearia glandulosa (Labill.) Benth.
(bearing many glands)
NSW, Vic, Tas, SA, WA   Swamp Daisy Bush
1–2 m × 1–2 m   Nov–June

Small **shrub** with glandular, resinous young growth; **branches** ascending to erect; **branchlets** glandular, resinous; **leaves** 1–6 cm × 0.1–0.2 cm, narrow-linear, alternate, spreading to erect, scattered, channelled above, convex below, dark green, glabrous, glandular, margins finely crenate, apex pointed; **flowerheads** daisy-like, to about 1.2 cm across, stalked, in terminal clusters, often profuse and conspicuous; **ray florets** 15–22, white or blue; **disc florets** yellow; **achenes** silky-hairy.

A lover of moist sites, this species is distributed from south of Sydney to south-western WA. It occurs in woodland, wet heath, along waterways, in moist gullies, swamps and other low-lying sites. Cultivated to a limited degree, it requires moist acidic soils in sunny or semi-shaded sites. Hardy to moderate frosts. May become leggy but should respond well to pruning from an early age. Propagate from seed or from cuttings of firm young growth.

Olearia glutinosa (Lindl.) Benth.
(covered with a sticky secretion)
Vic, Tas, SA   Sticky Daisy Bush
1–3 m × 1–2.5 m   Dec–Feb

Small to medium **shrub** with sticky young growth; **branches** ascending to erect; **branchlets** sticky, appearing glabrous; **leaves** 1–4 cm × 0.1–0.2 cm, narrow-linear, alternate, sessile, spreading to erect, crowded, dark green to slightly glabrous above and below, resinous, margins entire and revolute, apex pointed; **flowerheads** daisy-like, to about 2 cm across, solitary, on terminal stalks about same length as leaves, forming leafy clusters, often profuse and conspicuous; **ray florets** white to pale blue; **disc florets** violet; **achenes** glandular-hairy.

This robust, widespread species is usually found on coastal sand dunes. It is excellent for use in coastal gardens and does well in acidic or slightly alkaline, freely draining soils in a sunny or semi-shaded site. Plants are hardy to moderate frosts and tolerant of extended dry periods. Usually develops as a densely foliaged plant but pruning at an early stage promotes bushy growth and can be useful for rejuvenating mature plants. It was introduced into England in the 19th century. Propagate from seed or from cuttings of firm young growth.

Olearia gordonii Lander
(after David M. Gordon, 20th-century Australian plant enthusiast)
Qld
0.3–0.6 m × 0.3–1 m   Jan–July

Dwarf **shrub** with hairy young growth; **branches** spreading to erect; **branchlets** green becoming brown, hairy; **leaves** 1–7.5 cm × 0.2–0.8 cm, linear to narrowly elliptic, tapering to base, alternate, spreading to ascending, scattered, pale green with hairs above and below, margins toothed, apex pointed; **flowerheads** daisy-like, 2–3 cm across, solitary, on terminal or axillary stalks to about 8 cm long, forming terminal clusters, often profuse and very conspicuous; **ray florets** 13–33, blue; **disc florets** yellow; **achenes** densely silky-hairy.

This showy daisy bush was described in 1989. It occurs in the Maranoa and Warrego Districts of south-eastern Qld where it grows in shallow soils on lateritic slopes and ridges. Rarely cultivated, it should do best in semi-arid and temperate regions. Plants need acidic soils with very good drainage and a sunny or semi-shaded aspect. It is worth trying this species as a container plant and it should be useful in gardens. Pruning of young plants is likely to be beneficial. Propagate from seed or from cuttings of firm young growth.

Olearia grandiflora Hook.
(large flowers)
SA   Mt Lofty Daisy Bush
0.3–1 m × 0.3–1.5 m   Nov–Dec

Dwarf **shrub** which spreads by suckering and has hairy young growth; **stems** ascending to erect, with woody bases; **branchlets** few, woolly-hairy; **leaves** 3–12 cm × 2–6 cm, ovate, alternate, short-stalked, spreading to ascending, scattered, shiny above, dark green and glabrous above, densely whitish- to rusty-hairy below, flat, margins toothed, apex pointed; **flowerheads** daisy-like, 6–7 cm across, solitary, on terminal stalks 20–30 cm long, often profuse and very conspicuous; **ray florets** 15–25, white; **disc florets** yellow; **achenes** silky-hairy.

A very handsome dwarf daisy bush which is confined to the southern Mt Lofty Ranges near Adelaide. It grows in sheltered moist sites in woodland and dry sclerophyll forest. It is best suited to cool temperate regions in well-drained but moisture-retentive acidic soils with a semi-shaded or shaded aspect. Hardy to moderate frosts. Plants can be rejuvenated by hard pruning. Propagate from seed, cuttings of firm young growth, or root suckers.

O. pannosa is similar but it is a taller shrub with entire leaves.

Olearia gravis (F. Muell.) F. Muell. ex Benth.
(heavy, weighty)
Qld, NSW
0.8–2 m × 1–2 m   Aug–Nov

Dwarf to small **shrub** with greyish-hairy young growth; **branches** ascending to erect; **branchlets** hairy; **leaves** 2–6 cm × 0.5–1.5 cm, elliptical to ovate, alternate, spreading to ascending, stalked, scattered, dark green with greyish hairs above, hairy and pale green to yellowish-brown below, flat, margins coarsely toothed or entire, apex pointed; **flowerheads** daisy-like, 3–4 cm across, terminal, solitary, on stalks to about 6.5 cm long, often profuse and very conspicuous; **ray florets** 20–22, white; **disc florets** yellow; **achenes** glabrous.

A highly desirable daisy bush which inhabits dry sclerophyll forests in mountainous regions of south-eastern Qld and north-eastern NSW. Rarely cultivated, plants require freely draining acidic soils and should do well in subtropical and temperate regions. A semi-shaded site will probably be best but in southern Australia plants may need more sunshine. Hardy to light frosts. Tip or light pruning helps promote bushy growth. Propagate from seed or from cuttings of firm young growth.

O. cydoniifolia is closely allied but differs in having longer leaves which are glabrous on the upper surface.

Olearia gravis   D. L. Jones

Olearia gunniana (DC.) Hook. f. ex Hook. =
O. *phlogopappa* (Labill.) DC.

Olearia heliophila (F. Muell.) Benth. =
O. *elaeophila* (DC.) F. Muell. ex Benth.

Olearia heterocarpa S. T. Blake
(unequal fruits)
Qld, NSW
3–5 m × 2–3 m   May–Dec
Medium to tall **shrub** with stellate-hairy young growth; **branches** ascending to erect; **branchlets** stellate hairy; **leaves** 2–12 cm × 0.4–1.8 cm, narrowly elliptical to oblanceolate, alternate, spreading to ascending, short-stalked, scattered, dark green and more or less glabrous above, with grey to yellowish stellate hairs below, margins toothed, venation reticulate, apex pointed; **flowerheads** daisy-like, to about 1.5 cm across, in leafy terminal clusters, often profuse and very conspicuous; **ray florets** 4–7, white; **disc florets** yellow; **achenes** about 0.15 cm long, hairy.

A somewhat rare species from the southern Moreton District of Qld and the North Coast of NSW. It grows in wet sclerophyll forest, woodland and on the margins of mallee scrubland. This floriferous daisy bush is well worth cultivating. Plants prefer freely draining acidic soils and a sunny or semi-shaded site. Hardy to light frosts. Propagate from seed or from cuttings of firm young growth.

Olearia homolepis (F. Muell.) F. Muell. ex Benth.
(with scales of one type)
WA
0.6–1.25 m × 0.6–1.25 m   Aug–Oct

Olearia hookeri × 1

Dwarf to small **shrub** with densely hairy young growth; **branches** ascending to erect; **branchlets** rough, with many woolly hairs; **leaves** 1–2 cm × 0.2–0.3 cm, linear, alternate, sessile, spreading to ascending, crowded, green and rough with woolly hairs, margins revolute, apex pointed; **flowerheads** daisy-like, to about 2.5 cm across, solitary or 2–3 together, on short terminal stalks, often profuse and very conspicuous; **ray florets** usually more than 20, white and sometimes becoming pinkish; **disc florets** yellow; **achenes** silky-hairy.

O. *homolepis* is distributed over a wide area of the south-west. It occurs in the Avon, Coolgardie, Northern Darling, Irwin and Roe Districts where it grows in sandy and loam soils in mallee and woodland. This charming daisy bush is best suited to semi-arid and warm temperate regions but it may succeed in cool temperate zones. Plants need good drainage and tolerate a sunny or semi-shaded aspect. They are hardy to moderate frosts. Pruning from early stages should help to promote a bushy framework. Propagate from seed or from cuttings of firm young growth.

The var. *pilosa* Ewart is synonymous.

O. *homolepis* is closely allied to O. *adenolasia*, which differs in having only glandular hairs and it usually has blue flowerheads.

Olearia hookeri (Sond.) Benth.
(after Sir William Hooker, 19th-century Director of Royal Botanic Gardens, Kew)
Tas
0.6–1.5 m × 0.5–1 m   Sept–Dec
Dwarf to small **shrub** with more or less glabrous young growth; **branches** ascending to erect, slender; **branchlets** becoming glabrous; **leaves** 0.3–0.5 cm × about 0.1 cm, narrow-linear, alternate, sessile, spreading to ascending, crowded, thick, mid green and glabrous above and below, margins entire, apex bluntly pointed; **flowerheads** daisy-like, to about 1.8 cm across, solitary, sessile, terminal, scattered to profuse, very conspicuous; **ray florets** 20–30, initially white then usually acquiring crimson to bluish-purple tonings; **disc florets** yellow; **achenes** hairy.

A quick-growing and decorative daisy bush from eastern and south-eastern Tas. It inhabits open woodland and dryish slopes. Rarely cultivated, it has potential for greater use in gardens in temperate regions. Plants need well-drained acidic soils and a semi-shaded or somewhat sunny site. They are hardy to heavy frosts. Pruning can promote bushy growth. Propagate from seed or from cuttings of firm young growth.

O. *ericoides* is closely allied but differs in having thinner, somewhat hairy leaves and leafy racemes of flowers.

Olearia humilis Lander
(low, small)
WA
0.6–1.3 m × 0.5–1 m   July–Nov
Dwarf to small **shrub** with slightly sticky, hairy young growth; **branches** few, ascending to erect, slender; **branchlets** hairy and slightly sticky; **leaves** 2–3 cm × 0.1–0.3 cm, linear to narrowly obovate, alternate, sessile, spreading to ascending, scattered or clustered,

green and faintly hairy above and below, sometimes curved, margins entire to faintly toothed and recurved, apex pointed; **flowerheads** daisy-like, 1.5–3 cm across, solitary, terminal on stalks to about 4 cm long, often profuse and conspicuous; **ray florets** 10–20, pink to mauve or purple; **disc florets** yellow; **achenes** silky-hairy.

A small, elegant daisy bush which occurs in the Avon and Austin Districts where it grows on a range of acidic soils in open woodland and shrubland. Uncommon in cultivation, it should be best suited to semi-arid and warm temperate regions. Plants cope with sandy, loam or clay loam soils and a sunny or semi-shaded site. Hardy to moderate frosts and extended dry periods. Pruning from a very early age should be beneficial. Worth trying as a container plant as well as in gardens. Propagate from seed or from cuttings of firm young growth.

O. stuartii is allied but differs in having hairy, prominently toothed leaves and more ray florets per flowerhead.

Olearia hygrophila (DC.) Benth.
(moisture-loving)
Qld
1.5–2.5 m × 1–2 m   Sept–Nov

Small to medium **shrub** with faintly hairy young growth; **branches** spreading to ascending; **branchlets** faintly hairy; **leaves** 1.5–7.5 cm × up to 0.6 cm, linear-elliptical, alternate, short-stalked, recurved to spreading, scattered, green and often nearly glabrous above, hairy below, margins recurved, apex pointed; **flowerheads** daisy-like, to about 3 cm across, on slender stalks, forming loose to fairly dense terminal clusters, often profuse and very conspicuous; **ray florets** 12–20, white; **disc florets** yellow; **achenes** glabrous

An endangered ornamental species which is confined to Stradbroke Island where it grows in swampy acidic soils. This species is cultivated to a very limited degree. Plants should adapt well to moisture-retentive but freely draining acidic soils with a semi-shaded aspect in subtropical and temperate regions. May suffer frost damage. Pruning from early stages, and during or after flowering, should be beneficial in promoting bushy growth. Propagate from seed or from cuttings of firm young growth.

Olearia imbricata (Turcz.) Benth.
(overlapping)
WA   Imbricate Daisy Bush
0.4–1 m × 0.6–1 m   Sept–Nov

Dwarf **shrub** with somewhat sticky young growth; **branches** many, spreading to erect; **branchlets** slightly resinous; **leaves** 0.2–0.5 cm × about 0.1 cm, linear-cuneate, alternate, sessile, spreading to ascending, crowded, thick, often concave above, green and glabrous, slightly resinous, apex blunt; **flowerheads** daisy-like, about 1 cm across, solitary, on short terminal stalks, scattered to profuse, often conspicuous; **ray florets** 15–20, pinkish-white to lilac; **disc florets** yellow; **achenes** hairy.

This dwarf daisy bush occurs in the Eyre and Roe Districts where it grows in sand heath communities. Evidently rarely cultivated, it is suited to semi-arid and temperate regions. Plants need very well-drained acidic soils and a sunny or semi-shaded site. They tolerate moderate frosts and extended dry periods. Pruning may not be needed to promote bushy growth because branches may naturally grow entangled. Propagate from seed or from cuttings of firm young growth.

O. passerinoides is allied but it has less ray florets per flowerhead and its leaves have revolute margins.

Olearia incana (D. A. Cooke) Lander
(grey, hoary)
NSW, Vic, SA, WA
1–1.5 m × 1–1.5 m   June–Nov

Small **shrub** with greyish-hairy young growth; **branches** spreading to ascending; **branchlets** greyish-hairy; **leaves** 0.2–1.8 cm × 0.1–0.4 cm, narrowly elliptical or wedge-shaped, alternate, sessile, spreading to ascending, scattered, greyish-hairy, margins entire or toothed, recurved, apex blunt; **flowerheads** daisy-like, to about 4.5 cm across, on stalks to 9 cm long, terminal, solitary or clustered, often profuse and very conspicuous; **ray florets** 11–21, white; **disc florets** pale yellow; **achenes** hairy.

A charming daisy bush from semi-arid and warm temperate regions. The distribution extends from west of Dubbo, NSW, to the Carnarvon District in WA. Plants must have excellent drainage in soils which can be acidic or slightly alkaline. They tolerate a sunny or semi-shaded site. Hardy to moderate frosts and drought tolerant. Pruning from an early stage promotes bushy growth. Useful as fresh cut flowers. The greyish foliage reflects light at night very well. Propagate from seed or from cuttings of firm young growth.

Previously known as *O. pimeleoides* ssp. *incana* D. A. Cooke. *O. pimeleoides* has leaves which are dark green above, and it has solitary flowerheads.

Olearia incondita Lander
(unkempt)
WA
0.6–1.3 m × 0.6–1.5 m   Jan–April

Dwarf to small **shrub** with hairy young growth; **branches** spreading to erect; **branchlets** initially hairy, becoming glabrous; **leaves** 0.4–1.5 cm × 0.1–0.3 cm, narrowly elliptical, alternate, sessile, spreading to ascending, scattered, thick with small blisters, green and glabrous above, hairy below, margins revolute, apex blunt; **flowerheads** daisy-like, 2.5–3.5 cm across, solitary, terminal, often profuse and very conspicuous; **ray florets** 7–10, white to pink; **disc florets** yellow; **achenes** hairy.

This recently described (1990) species is vulnerable. It is restricted to the Avon and Roe Districts where it occurs in small populations around granite outcrops and on margins of ephemeral lakes. It needs to be cultivated as part of a conservation stratgegy. Plants are probably best suited to semi-arid and warm temperate regions. They need good drainage and a sunny or semi-shaded site. Hardy to moderate frosts and extended dry periods. May need pruning from an early stage to promote bushy growth. Propagate from seed or from cuttings of firm young growth.

Olearia iodochroa

Olearia imbricata

W. R. Elliot

Olearia iodochroa (F. Muell.) F. Muell. ex Benth.
(blue-coloured)
NSW, Vic   Violet Daisy Bush
0.8–1.5 m × 1–1.5 m   Sept–Dec; also sporadic

Dwarf to small **shrub** with resinous young growth; **branches** many, spreading to ascending; **branchlets** hairy and resinous; **leaves** 0.2–1.2 cm × up to 0.4 cm, obovate, alternate, spreading to ascending, more or less sessile, crowded to scattered, dark green and glabrous above, cream to greyish woolly-hairy below, margins revolute, apex blunt, faintly curry-scented; **flowerheads** daisy-like, 1–2 cm across, terminal, solitary or clustered, forming leafy sprays, often profuse and very conspicuous; **ray florets** white, mauve or violet; **disc florets** cream, yellow or blue; **achenes** silky-hairy.

A delightful, floriferous daisy bush which occurs in dry sclerophyll forest or woodland, often in rocky sites. It is found in south-eastern NSW and eastern Vic. Plants adapt very well to cultivation and will grow in acidic sand, clay loam and gravels. Although tolerant of full sunshine, this species prefers a semi-shaded site. Withstands most frosts and extended dry periods. Responds very well to pruning. Propagate from seed or from cuttings of firm young growth.

Plants in cultivation attributed to *O. iodochroa*, from the Brisbane Ranges, west of Melbourne, are usually referrable to *O. minor*.

Olearia laciniifolia Lander
(fringed or narrowly lobed leaves)
WA
0.6–1.3 m × 0.5–1 m   June–Nov

Dwarf to small **shrub** with very hairy young growth; **branches** ascending to erect; **branchlets** hairy, initially yellow becoming purplish; **leaves** 0.6–3.5 cm × 0.1–1 cm, oblong, alternate, sessile, spreading to ascending, scattered, greyish-green, hairy above and below, margins with many short narrow lobes, slightly revolute, apex pointed; **flowerheads** daisy-like, 2.5–3.5 cm across, solitary, on terminal hairy stalks to 2.5 cm long, often profuse and very conspicuous; **ray florets** 35–43, lilac; **disc florets** yellow; **achenes** silky-hairy.

Fairly recently described (1990), this attractive but vulnerable species is known only from the Roe District. It is found in white sand amongst mallee and shrubs on the margins of ephemeral lakes. It should adapt well to cultivation in semi-arid and temperate regions. Plants need excellent drainage and a warm to hot sunny or semi-shaded site. They are hardy to moderate frosts and extended dry periods. Pruning from an early stage promotes bushy growth. Propagate from seed or from cuttings of firm young growth.

O. rudis has been confused with this species. It differs in having flowerheads (with up to 75 ray florets) in compound clusters.

Olearia lanceolata (Benth.) D. I. Morris =
                                    O. archeri Lander

Olearia lanuginosa (J. H. Willis) N. A. Wakef.
(with long cottony hairs)
Vic, SA, WA   Woolly Daisy Bush
1–2 m × 1–1.5 m   Oct–March

Small **shrub** with densely woolly young growth; **branches** many, ascending to erect, woolly-hairy; **branchlets** white to greyish woolly-hairy, short, thick, stiff; **leaves** 0.1–0.3 cm × about 0.1 cm, ovate to broadly ovate, alternate, ascending to erect, curved, sessile, clustered and often overlapping on young branchlets, faintly hairy and greyish-green above, densely white woolly-hairy below, margins revolute, apex blunt; **flowerheads** daisy-like, to about 1.5 cm across, terminal and sessile on very short branchlets, often profuse and conspicuous; **ray florets** 4–6, white; **disc florets** mauve; **achenes** hairy.

This compact species occurs in mallee and heathland on sandy soils, in semi-arid and warm temperate regions of western Vic, SA, and in the Austin District of WA. Plants require excellent drainage in a sunny or semi-shaded site. They are hardy to moderate frosts and are drought tolerant. May do well in cool temperate regions if grown in a protected, warm to hot sunny site. Propagate from seed or from cuttings of firm young growth.

O. lepidophylla is similar but has longer leaves which are reflexed against the branches, while *O. ramulosa* has spreading, linear leaves to about 1.5 cm long.

Olearia lasiophylla Lander
(hairy leaves)
NSW
0.6–1.3 m × 0.6–1.5 m   Nov–Feb

Dwarf to small **shrub** with yellowish-hairy young growth; **branches** ascending to erect, hairy; **branchlets** densely stellate-hairy, becoming brownish; **leaves** 0.8–6 cm × 0.4–1.6 cm, elliptical, alternate, spreading

Olearia magniflora   W. R. Elliot

to ascending, short-stalked, scattered, dark green and faintly stellate-hairy above, densely stellate-hairy below, flat, margins entire to toothed, venation reticulate, apex bluntly pointed; **flowerheads** daisy-like, to about 2.7 cm across, on stalks to about 3.5 cm long, terminal (often on very short axillary branchlets), solitary or clustered, often profuse and conspicuous; **ray florets** 14–15, white; **disc florets** yellow; **achenes** silky-hairy.

A handsome and floriferous but poorly known species which is restricted to mountainous slopes in dry sclerophyll forests west of Mt Kosciuszko. It should adapt well to cultivation because it is closely allied to *O. stellulata* (see below). Plants need acidic, freely draining soils and a semi-shaded or dapple-shaded site. Hardy to frost and light snowfalls. Pruning from an early stage will promote bushy growth. Propagate from seed or from cuttings of firm young growth.

This species has often been identified as *O. stellulata*, which differs insofar as its leaves are glabrous above, or nearly so, and lightly greyish-hairy below. *O. stellulata* flowerheads are held in leafy panicles.

Olearia ledifolia (DC.) Benth.
(leaves similar to the genus *Ledum*)

Tas   Mountain Daisy Bush
0.3–1 m × 0.5–1 m   Nov–Jan

Dwarf **shrub** with rusty-hairy young growth; **branches** ascending to erect; **branchlets** thickish, brownish-hairy; **leaves** 1–3 cm × about 0.1 cm, oblong-linear, alternate, sessile, spreading to ascending, thick, crowded, deep green and glabrous above, silvery to rusty hairs below, margins entire and revolute, apex blunt; **flowerheads** daisy-like, to about 2 cm across, solitary, axillary, on rusty-hairy stalks about as long as

Olearia megalophylla   T. L. Blake

the leaves, forming leafy racemes, often profuse and conspicuous; **ray florets** 20–24, white; **disc florets** yellow; **achenes** glabrous or with hairs near apex.

A floriferous species from the mountainous regions of Tasmania where it often grows in the protection of boulders. Cultivated to a limited extent, it is best suited to temperate regions. Plants need freely draining acidic soils and an open sunny or semi-shaded aspect. They are hardy to most frosts and snowfalls. Responds well to pruning and has potential for use as a dwarf hedge. Propagate from seed or from cuttings of firm young growth.

Olearia lepidophylla (Pers.) Benth.
(scaly leaves)

Qld, NSW, Vic, Tas, SA, WA   Clubmoss Daisy Bush
0.6–1.5 m × 0.6–1.5 m   Dec–June

Dwarf to small **shrub** with grey-woolly young growth; **branches** many, ascending to erect; **branchlets** woolly-hairy; **leaves** to about 0.15 cm × 0.1 cm, ovate, alternate, spreading to reflexed, clustered, deep green, shiny and more or less glabrous above, with dense pale greyish hairs below, margins revolute, apex blunt; **flowerheads** daisy-like, to about 1.5 cm across, solitary, sessile, terminal, often profuse and conspicuous; **ray florets** 4–7, white; **disc florets** yellow; **achenes** silky-hairy.

An inhabitant of mallee sandhills across the southern half of Australia. It is cultivated to a limited degree. Plants adapt well to warm and cool temperate regions with acidic or slightly alkaline soils, in a sunny or semi-shaded position. Hardy to moderate frosts and extended dry periods. Plants produce bushy growth if pruned during or just after flowering. Propagate from seed or from cuttings of firm young growth.

83

Olearia lirata

O. lanuginosa has densely woolly stems and leaves which are not reflexed. *O. ramulosa* has spreading linear leaves to about 1.5 cm long.

Olearia lirata (Sims) Hutch. =
O. stellulata (Labill.) DC.

Olearia macdonnellensis D. A. Cooke
(from the Macdonnell Ranges)
NT
0.6–1.2 m × 0.5–1 m   Aug–Sept
Dwarf to small **shrub** with resinous young growth; **branches** many, spreading to ascending; **branchlets** resinous; **leaves** 1.2–3 cm × 0.6–1.5 cm, oblong to broadly ovate, alternate, short-stalked, spreading to ascending, scattered, green, varnished, margins nearly entire to shallowly lobed, apex truncate; **flowerheads** daisy-like, to about 3 cm across, on stalks 2–8 cm long, in terminal clusters of usually 2–5 flowerheads, often profuse and very conspicuous; **ray florets** 8–14, white; **disc florets** yellow; **achenes** silky-hairy.
This poorly known species occurs in a couple of gorges in the Macdonnell Ranges. It grows at the base of steep rocky slopes. Plants need to be cultivated as part of a conservation strategy. They are probably best suited to semi-arid and warm temperate regions. Plants will require a semi-shaded site which is excellently drained. They should tolerate light frosts. Propagate from seed or from cuttings of firm young growth.
O. calcarea is allied but it has prominently toothed leaves and more-or-less sessile flowerheads. *O. muelleri* has entire to toothed leaves and more-or-less sessile flowerheads.

Olearia magniflora (F. Muell.) F. Muell. ex Benth.
(bearing large flowers)
NSW, Vic, SA, WA   Splendid Daisy Bush
1–2 m × 1–1.5 m   July–Jan
Small **shrub** with faintly hairy young growth; **branches** spreading to ascending, glabrous; **branchlets** more or less glabrous; **leaves** 0.3–2.6 cm × 0.15–1 cm, narrowly obovate, alternate, spreading to ascending, more or less sessile, scattered, green, more or less glabrous above and below, margins entire or with 3–5 teeth near the blunt apex; **flowerheads** daisy-like, 3.5–6 cm across, solitary, terminal, more or less sessile, often profuse and extremely conspicuous; **ray florets** 12–24, violet to purple; **disc florets** yellow; **achenes** to 0.5 cm long, glabrous.
An outstanding and distinctive daisy bush which deserves wider cultivation. *O. magniflora* is distributed in arid and semi-arid regions of southern Australia. It grows in mallee and eucalypt woodland. Plants are best suited to semi-arid and warm temperate regions but may do well in cooler areas if grown in a protected warm to hot site. Soils may be acidic to slightly alkaline but must have excellent drainage. Plants are hardy to moderate frosts. Pruning from an early age promotes bushy growth. Propagate from seed or from cuttings of firm young growth.
Hybrids with *O. magniflora* and *O. muelleri* as parents occur in nature. Plants usually have smaller pale mauve flowers.

Olearia megalophylla (F. Muell.) F. Muell. ex Benth.
(large leaves)
NSW, Vic   Large-leaf Daisy Bush
1.5–2.5 m × 1–2 m   Dec–March
Small to medium **shrub** with pale brownish- to reddish-hairy young growth; **branches** ascending to erect; **branchlets** densely covered in pale brownish hairs; **leaves** 2.5–15 cm × 0.6–7 cm, elliptic to narrowly ovate, alternate or opposite, stalked, spreading to ascending, scattered, dark green and glabrous above, densely covered with pale brownish to rusty hairs below, margins flat, apex pointed to rounded; **flowerheads** daisy-like, to about 3 cm across, on stalks to 5.5 cm long, in terminal clusters, often profuse and conspicuous; **ray florets** 5–12, white; **disc florets** yellow; **achenes** glabrous.
A relatively common plant in wet sclerophyll forests, extending from near Orange, NSW, to Matlock, Vic. It adapts well to cultivation in temperate regions in acidic, well-drained soils with a semi-shaded or dappleshaded aspect. Plants are hardy to moderate frosts and light snowfalls. Pruning may help to promote bushy growth. Supplementary watering may be necessary during extended dry periods. Propagate from seed or from cuttings of firm young growth.

Olearia microdisca J. M. Black
(small disc)
SA   Small-flowered Daisy Bush
1–1.5 m × 0.6–1 m   mainly Feb–May; also sporadic;
Small **shrub** with resinous hairy young growth; **branches** ascending to erect; **branchlets** finely hairy, resinous; **leaves** 0.15–0.3 cm × about 0.1 cm, oblong, alternate, sessile, ascending to erect, crowded to overlapping, green and glabrous above, woolly-hairy below, margins revolute, apex bluntly pointed; **flowerheads** somewhat daisy-like, to about 1 cm across, solitary and terminal on short erect branchlets, often profuse and conspicuous; **ray florets** 2–5, white; **disc florets** yellow; **achenes** hairy.
An erect-branched, small-flowered species which is an endangered endemic on Kangaroo Island where it grows in mallee, scrubland and woodland on a range of soils. Although in limited cultivation, it should adapt well to warm and cool temperate regions. Prefers freely draining soils in a semi-shaded or dapple-shaded site but also tolerates plenty of sunshine. Hardy to light frosts and responds well to moderate pruning. Propagate from seed or from cuttings of firm young growth.

Olearia microphylla (Vent.) Maiden & Betche
(small leaves)
Qld, NSW
1–2.5 m × 0.8–2 m   June–Oct
Small to medium **shrub** with hairy young growth; **branches** ascending to erect; **branchlets** hairy and glandular; **leaves** 0.2–0.7 cm × 0.1–0.3 cm, narrowly to broadly spathulate, alternate, more or less sessile, spreading to ascending, crowded, dark green and glandular above, grey-woolly hairs below, margins strongly revolute, apex rounded to pointed; **flowerheads** daisy-like, to about 1.7 cm across, terminal,

solitary, sessile, often in leafy racemes, may be profuse and very conspicuous; **ray florets** 6–8, white or, rarely, mauve-blue; **disc florets** yellow; **achenes** glandular and sometimes sticky.

This dainty and floriferous species occurs from south-eastern Qld to north of Canberra, in heathland and dry sclerophyll forest. It is often associated with rock outcrops. Cultivated to a limited extent, it is worthy of greater interest. Does well in subtropical and temperate regions on acidic soil which drains freely. Plants do best in a semi-shaded site. They tolerate moderate to heavy frosts and respond well to moderate pruning. Propagate from seed or from cuttings of firm young growth.

O. ramulosa is allied but it lacks glandular leaves and its achenes are hairy.

Olearia minor (Benth.) Lander
(smaller)
NSW, Vic, SA, WA
1–1.5 m × 1–1.5 m   June–Nov
Small **shrub** with greyish-woolly young growth; **branches** ascending to erect; **branchlets** greyish woolly-hairy; **leaves** 0.15–0.8 cm × 0.1–0.25 cm, elliptical to obovate, alternate, more or less sessile, spreading to ascending, scattered, green and nearly glabrous above, greyish-woolly hairs below, margins revolute, apex rounded to bluntly pointed; **flowerheads** daisy-like, to about 2.2 cm across, terminal, solitary or clustered, sessile, often profuse and conspicuous; **ray florets** 7–12, white; **disc florets** yellow; **achenes** glandular.

A wide-ranging species which extends from southern NSW to south-eastern WA. It occurs in mallee and shrubland, growing in loam or sand, on dunes or rocky slopes. Suitable for semi-arid and temperate regions. Plants adapt to acidic or alkaline freely draining soils with a sunny or semi-shaded aspect. Hardy to moderate frosts and extended dry periods. Responds well to moderate pruning. Propagate from seed or from cuttings of firm young growth.

Plants marketed as *O. iodochroa* from the Brisbane Ranges, Vic, are referrable to *O. minor*.

Previously known as *O. pimeleoides* var. *minor* Benth.

O. brachyphylla differs in having incurved leaves and flowerheads to about 0.4 cm across. *O. floribunda*, with which *O. minor* has been confused, has crowded leaves with flat margins, and flowerheads with 3–7 ray florets.

Olearia montana Lander
(of the mountains)
NSW
1.5–2.5 m × 1–2 m   Sept–Oct
Small to medium **shrub** with woolly-hairy young growth; **branches** ascending to erect, hairy, yellowish-brown becoming greyish-brown; **branchlets** hairy; **leaves** 1.5–5 cm × 0.7–1.8 cm, elliptical, alternate, short-stalked, spreading to ascending, scattered, dark green and more or less glabrous above, pale brownish woolly-hairy below, flat, margins toothed, venation reticulate, apex blunt; **flowerheads** daisy-like, to about 2.7 cm across, terminal, on short stalks, forming panicles, often profuse and very conspicuous; **ray florets** 17–28, mauve; **disc florets** purple; **achenes** glabrous.

A highly desirable and rare species which inhabits

Olearia minor × .5

dry sclerophyll forest and woodland in a limited region of the Southern Tablelands. It grows on rocky slopes in dry shallow soils over granite. Plants have been cultivated as *O. stellulata* (see below). They adapt to a range of freely draining acidic soils in a semi-shaded or dapple-shaded site. Hardy to heavy frosts and light snowfalls. Responds well to pruning. Propagate from seed or from cuttings of firm young growth.

O. montana has also been known as *Olearia* species 2. *O. stellulata* differs in having flowerheads with 11–16 ray florets.

Olearia mooneyi (F. Muell.) Hemsl.
(after Thomas Mooney, a 19th-century resident of Lord Howe Island)
NSW (Lord Howe Island)   Pumpkin Bush
1–4 m × 1–3 m   Nov–Jan
Small to tall **shrub** with woolly young growth; **stems** thick; **branches** thick, spreading to ascending; **branchlets** short, hairy; **leaves** 4–10 cm × 1.5–3.5 cm, narrowly elliptic to obovate, stalked, alternate, somewhat crowded, shiny green and glabrous above, deep cream woolly-hairy below, flat, margins entire, apex blunt to pointed; **flowerheads** daisy-like, about 0.8 cm across, arranged in a crowded terminal corymb, usually profuse and very conspicuous; **ray florets** about 12, white; **disc florets** pale yellow; **achenes** bristly-hairy.

Pumpkin Bush is one of three endemic olearias on Lord Howe Island. It occurs on mountains above 750 m altitude and is one of the dominant species in the high regions. Rarely cultivated, it should do well in subtropical and temperate regions. It needs acidic soils which have excellent drainage and plenty of air movement. Plants will do well in a sunny or semi-shaded site. They may suffer damage from frosts. Judicious pruning will promote bushy growth. Propagate from seed or from cuttings of firm young growth.

Olearia mucronata

Olearia microphylla   W. R. Elliot

Olearia mucronata Lander
(ending in a short, sharp point)
WA
0.6–1 m × 0.5–1 m   Aug–Jan

Dwarf **shrub** with hairy young growth; **branches** ascending to erect, becoming reddish; **branchlets** hairy; **leaves** 1.3–4.6 cm × 0.1–0.5 cm, linear or, rarely, obovate, alternate, sessile, ascending to erect, crowded, dark green and lightly hairy above and below, margins entire or sparingly toothed, apex pointed; **flowerheads** daisy-like, to about 1.4 cm across, on stalks to about 2.3 cm, solitary, terminal, forming leafy clusters, often profuse and conspicuous; **ray florets** 9–12, white; **disc florets** yellow; **achenes** silky-hairy.

A recently described (1990) species. It is vulnerable and confined to a small region of the Fortescue District where it inhabits hills composed of schist. It needs to be cultivated as part of a conservation strategy. Plants are best suited to semi-arid and warm temperate regions and require excellent drainage with plenty of sunshine. They are hardy to light frosts. Plants may need regular pruning to maintain bushy growth. Propagate from seed or from cuttings of firm young growth.

O. stuartii is allied but has smaller flowerheads with more ray florets (21–64).

Olearia muelleri (Sond.) Benth.
(after Baron Sir Ferdinand von Mueller, first Government Botanist of Victoria)
NSW, Vic, SA, WA, NT   Mueller's Daisy Bush
0.6–1.5 m × 0.6–1.5 m   July–Jan

Dwarf to small, often rounded **shrub** with glabrous young growth; **branches** many, spreading to ascending; **branchlets** glabrous; **leaves** 0.5–1.5 cm × 0.2–1 cm, obovate, alternate, more or less sessile, spreading to ascending, scattered, deep green and glabrous above and below, dotted, margins entire or toothed, apex rounded; **flowerheads** daisy-like, 1.3–3.1 cm across,

solitary, terminal, more or less sessile, often profuse and extremely conspicuous; **ray florets** 7–13, white or very pale mauve; **disc florets** yellow; **achenes** densely silky-hairy.

A very showy olearia which is widely distributed in the semi-arid and warm temperate regions. It can be found growing in woodland, mallee and spinifex communities on sand and clay loams. For best results, plants need excellent drainage in acidic or slightly alkaline soils, as well as plenty of sunshine, but they tolerate semi-shaded sites. Hardy to moderate frosts and extended dry periods. Responds well to light pruning. It has potential for use as a container plant and in general planting. Propagate from seed, which usually begins to germinate between 20–45 days after sowing, or from cuttings of firm young growth.

O. calcarea is allied but differs in having larger flowerheads and prominently toothed leaves. Intergrading variants of these two species occur in nature.

Olearia muricata (Steetz) Benth.
(short, sharp points on the surface)
WA   Rough-leaved Daisy Bush
0.3–1.3 m × 0.5–1 m   Dec–May

Dwarf to small **shrub** with densely hairy young growth; **branches** many, spreading to ascending; **branchlets** roughly hairy; **leaves** 0.4–1 cm × about 0.1 cm, linear, alternate, sessile, ascending to erect, scattered, green with stiff glandular hairs, margins revolute, apex blunt; **flowerheads** daisy-like, about 1 cm across, sessile, terminal, sometimes profuse and conspicuous; **ray florets** 8–10, white, lilac or blue; **disc florets** yellow; **achenes** hairy.

A somewhat open but entangled dwarf shrub which occurs in the Avon, Eyre and Roe Districts where it grows in sandy soils. Rarely cultivated, it should adapt well to semi-arid and temperate regions. Needs acidic and well-drained soils and a sunny or semi-shaded site is suitable. Hardy to moderate frosts. Plants will need pruning to promote a bushy growth habit. Propagate from seed, which usually begins to germinate 25–40 days after sowing, or from cuttings of firm young growth.

O. elaeophila has been confused with this species but it lacks rigid hairs on its leaves and it has larger flowerheads.

Olearia myrsinoides (Labill.) F. Muell. ex Benth.
(similar to the genus *Myrsine*)
NSW, Vic, Tas   Blush Daisy Bush; Silky Daisy Bush
0.6–1.5 m × 0.8–2 m   Nov–March

Dwarf to small **shrub** with faintly hairy young growth; **branches** spreading to ascending or sometimes arching; **branchlets** greyish-hairy; **leaves** 0.5–3.5 cm × 0.3–2.2 cm, elliptical to obovate, alternate, short-stalked, spreading to ascending, scattered to somewhat crowded, dark green and becoming glabrous above, silky-hairy below, margins finely toothed, venation reticulate, apex pointed to rounded; **flowerheads** somewhat daisy-like, to about 2 cm across, in terminal leafy panicles, often profuse and very conspicuous, sweetly scented with a somewhat coconut-like aroma; **ray florets** 2–4, white; **disc florets** violet or pale yellow; **achenes** small, glabrous.

This wide-ranging species is distributed from near the Qld/NSW border along the coast or hinterland, over most of Victoria (except for the north-west) and to Tas. It is often associated with mountains and hills, in sites which can be rocky to swampy. Adapts very well in cultivation to acidic soils which drain relatively freely and withstands extended dry periods. A semi-shaded or partially shaded site is suitable. Responds well to pruning and old plants can be rejuvenated by removal of woody stems. Propagate from seed or from cuttings of firm young growth.

It is sometimes confused with *O. erubescens* which has reddish young growth and larger flowerheads with 4-7 ray florets.

An olearia with pale purple flowers was introduced into England during 1835 as *O. myrsinoides*, but the accuracy of this record is not confirmed.

Olearia nernstii (F. Muell.) F. Muell. ex Benth.
(after Joseph Nernst, 19th-century gardener)
Qld, NSW
1–2.5 m × 1–2 m   July–Nov
Small to medium **shrub** with rusty- to whitish-hairy young growth; **branches** spreading to ascending; **branchlets** with rusty to whitish hairs; **leaves** 2.5–12 cm × 0.8–3.5 cm, ovate to elliptical, alternate, stalked, spreading to ascending, green and more or less glabrous above, rusty-hairy below, margins toothed and sometimes prickly, apex pointed; **flowerheads** daisy-like, to about 3 cm across, in loose terminal clusters, often profuse and conspicuous; **ray florets** 9–20, white; **disc florets** yellow; **achenes** glabrous.
O. nernstii is a short-lived species which occurs in the Moreton and Wide Bay District of south-eastern Qld, in the Central Coast and Northern Coast, and Central and Northern Tablelands of NSW. It is found in dry sclerophyll forest and woodland. Rarely cultivated, but it should do well in subtropical and cool temperate regions. Plants need well-drained acidic soil with a semi-shaded or dapple-shaded site. Hardy to light frosts. Pruning from an early stage promotes bushy growth. Ideal for coastal areas. Propagate from seed or from cuttings of firm young growth.
O. gravis is similar, but it usually has larger flowerheads and the leaves may be entire or coarsely toothed.

Olearia obcordata (Hook. f.) Benth.
(inversely heart-shaped)
Tas
0.5–1 m × 0.5–1 m   Dec–Feb
Dwarf **shrub** with faintly hairy young growth; **branches** many, spreading to erect; **branchlets** white-hairy; **leaves** 0.7–1.2 cm × 0.6–1 cm, obcordate-cuneate, alternate, sessile, spreading to ascending, scattered to crowded, deep green and glabrous or faintly hairy above, densely white-hairy below, margins entire except for 3–5 blunt lobes at tips; **flowerheads** daisy-like to about 2 cm across, solitary on short upper branchlets, forming leafy clusters, often profuse and conspicuous; **ray florets** 10–15, white; **disc florets** yellow; **achenes** glabrous.
This floriferous daisy bush occurs above 900 m altitude where it is usually subjected to frost and heavy

Olearia obcordata × .6

snowfalls. It should be cultivated more widely. Best suited to temperate regions. Plants require well-drained acidic soils and a site which is semi-shaded or sunny but not extremely hot. Pruning from an early stage promotes bushy growth. Propagate from seed or from cuttings of firm young growth.

Olearia occidentissima Lander
(most western)
WA
0.1–0.3 m × 0.3–0.5 m   Sept
Dwarf **shrub** with woolly, white-hairy young growth; **branches** spreading to erect; **branchlets** white, woolly-hairy, becoming greyish; **leaves** 0.6–2.5 cm × 0.3–0.6 cm, narrowly elliptic, alternate, stalked, spreading to ascending, scattered, greyish-green and hairy above, white-woolly hairs below, margins entire, revolute, apex pointed; **flowerheads** daisy-like, 2.5–3 cm across, sessile, solitary, terminal, often profuse and conspicuous; **ray florets** 10–12, white to pink; **disc florets** white and often tinged with violet; **achenes** silky-hairy.
A recently described (1990) species which is regarded as vulnerable in the alkaline soils of its coastal cliff top habitat on an island in the southern Carnarvon District. Needs to be cultivated as a conservation strategy. Plants are suited to semi-arid and warm temperate regions. They should tolerate acidic or alkaline soils but must have excellent drainage. An open or semi-shaded site should be suitable. Withstands coastal exposure very well. In protected sites it may reach larger dimensions. Propagate from seed or from cuttings of firm young growth.
O. pimeleoides is allied but it is a much smaller shrub with leaves that have a dark green upper surface.

Olearia oliganthema F. Muell. ex Benth.
(bearing few flowers)
NSW
Unknown   unknown

   Shrub; **branches** ascending to erect; **leaves** 3–8 cm ×
1.6–4 cm, alternate, stalked, spreading to ascending,
glabrous above, faintly silvery-hairy below, flat to somewhat wavy, margins entire to faintly toothed, apex
pointed to rounded; **flowerheads** to 1.8 cm across, in
crowded terminal clusters; **ray florets** 1 or 2, probably
white; **disc florets** probably yellow.

   This species from the Blue Mountains is presumed
extinct. It was last collected in 1866. If it is ever
rediscovered it will probably require freely draining
acidic soils with a semi-shaded aspect. Plants may be
hardy to moderate frosts. Propagate from seed or from
cuttings of firm young growth.

Olearia oppositifolia (F. Muell.) Lander
(opposite leaves)
NSW
1.5–2.5 m × 1–2.5 m   Nov–Jan

   Small to medium **shrub** with faintly hairy young
growth; **branches** ascending to erect; **branchlets** pale
brown, becoming greyish; **leaves** 2–12.5 cm × 0.6–4 cm,
ovate to elliptical, opposite or in whorls of 3, stalked,
spreading to ascending, scattered, green and more or
less glabrous above, greyish- to brownish-hairy below,
margins usually toothed, slightly revolute, apex
pointed or rounded; **flowerheads** daisy-like, to about
2–3 cm across, on long stalks, in terminal clusters,
often profuse and conspicuous; **ray florets** 4–6, white;
disc florets yellow; **achenes** glabrous or faintly hairy.

   This species is an inhabitant of dry and wet sclerophyll forests in the Northern Coast and Northern
Tablelands Districts of north-eastern NSW. Plants
require acidic, freely draining soils and a semi-shaded
or dapple-shaded site. They should withstand moderate frosts. Suitable for subtropical and temperate
regions. May require supplementary watering during
extended dry periods. Prune during or after flowering
to promote bushy growth. Propagate from seed or
from cuttings of firm young growth.

   O. chrysophylla is closely allied and has been confused with this species. It usually has entire leaves and
larger flowerheads.

Olearia pannosa Hook.
(felt-like)
Vic, SA   Silver-leaved Daisy; Velvet Daisy Bush
1–2 m × 1–1.5 m   Aug–Oct

   Small **shrub** which often suckers; young growth
covered in white-woolly hairs; **branches** ascending
to erect; **branchlets** densely woolly-hairy; **leaves** 3–
10.5 cm × 1.5–5 cm, ovate to elliptic, alternate, stalked,
cordate at the base, spreading to ascending, scattered,
deep green and shiny above, with dense white to very
pale rusty brown woolly hairs below, margins entire
and flat, venation reticulate, apex bluntly pointed;
flowerheads daisy-like, 4–5.5 cm across, solitary, on terminal stalks to about 30 cm long, scattered to profuse,
very conspicuous; **ray florets** 12–25, white or, rarely,
pale mauve; **disc florets** yellow; **achenes** hairy.

   The typical variant is confined to the south-eastern

region where it grows in open forest, woodland and
mallee in a range of soil types. This showy daisy bush is
best suited to temperate regions. It needs well-drained
acidic or slightly alkaline soils and a semi-shaded site
for best growth. Plants are hardy to moderate frosts
and they withstand extended dry periods. Tip pruning
of young plants promotes a bushy framework.
Introduced into England during the 1850s.

   The ssp. *cardiophylla* (F. Muell.) D. A. Cooke occurs
in scattered locations in Vic, and in the Southern Lofty
and Flinders Ranges Regions in SA. It has leaves with
cordate bases and a buff to rusty-brown hairy undersurface. This subspecies has been previously marketed
as *O. pannosa* or *O. species* aff. *pannosa*.

   Propagate both variants from seed or from cuttings
of firm young growth.

   O. grandiflora has similarities, but it is a dwarf shrub
with toothed leaves.

Olearia passerinoides (Turcz.) Benth.
(similar to the genus *Passerina*)
NSW, Vic, SA, WA   Slender Daisy Bush
1.5–2.5 m × 1–2 m   mainly Nov–April

   Small **shrub** with resinous young growth; **branches**
many, ascending to erect; **branchlets** resinous; **leaves**
0.25–0.7 cm × up to 0.15 cm, linear, alternate, more or
less sessile, ascending to erect, scattered, green and
resinous above and below, margins entire and flat to
revolute, apex rounded or pointed; **flowerheads** daisy-
like, to 2–3 cm across, terminal, solitary, may be profuse and conspicuous; **ray florets** 5–8, white to pale
mauve; **disc florets** yellow; **achenes** silky-hairy.

   The typical subspecies has a distribution which
extends from central NSW to the Coolgardie and Eyre

Olearia passerinoides ssp. *glutescens* × .75

Districts of WA. Rarely cultivated, plants need well-drained acidic or slightly alkaline soils. Although they tolerate high temperatures they would prefer a semi-shaded site. Hardy to moderate frosts. Pruning helps promote bushy growth.

The ssp. *glutescens* (Sond.) D. A. Cooke has spreading to ascending leaves, and flowerheads in small clusters which bloom for most of the year. It occurs in NSW, Vic and SA.

Propagate both subspecies from seed or from cuttings.

O. imbricata differs in having concave leaves and more ray florets per flowerhead.

Olearia paucidentata (Steetz.) Benth.
(with few teeth)
WA   Hispid Daisy Bush
1–1.5 m × 0.6–1.2 m   April–Jan
  Small **shrub** with hairy young growth; **branches** ascending to erect; **branchlets** hairy; **leaves** 0.3–2.5 cm × up to 1 cm, linear to narrowly obovate, alternate, sessile, narrowed to base, spreading to ascending, scattered or clustered, green, hairy, margins revolute with 1–6 teeth or lobes on each side, apex pointed; **flowerheads** daisy-like, to about 2 cm across, solitary, terminal, usually stalked but can be sessile, forming leafy clusters, often profuse and conspicuous; **ray florets** 10–20, white, pale violet, blue or mauve; **disc florets** white, violet or blue; **achenes** hairy.

An attractive daisy bush which occurs in the Avon, Darling and Eyre Districts where it grows in a range of soils which are often damp to wet for extended periods. This species is cultivated to a limited degree. Plants are best suited to temperate regions. They need freely draining but moisture-retentive acidic soils and a sunny or semi-shaded site. Hardy to light frosts. Pruning from an early stage should promote a bushy framework. Propagate from seed, which usually begins to germinate 20–40 days after sowing, or from cuttings of firm young growth.

O. muricata has been confused with this species but it has rigid hairs over all its parts.

Olearia persoonioides (DC.) Benth.
(similar to the genus *Persoonia*)
Tas   Mountain Daisy Bush
1–3 m × 0.6–2 m   Dec–Feb
  Small to medium **shrub** with whitish to pale brownish hairs on young growth; **branches** ascending to erect; **branchlets** with pale brown to whitish hairs; **leaves** 2–6 cm × 0.6–2.2 cm, elliptical or obovate, alternate, short-stalked, spreading to ascending, scattered, deep green and shiny glabrous above, white to pale brown silky-hairy below, margins entire and recurved, apex bluntly pointed; **flowerheads** daisy-like, to about 2 cm across, on slender stalks, 3–5 together, in upper axils, forming leafy terminal clusters, often profuse and very conspicuous; **ray florets** usually 3–6 fully developed, white; **disc florets** yellow; **achenes** hairy.

A handsome species occurring on plateaux and rocky slopes between about 600–1000 m altitude. It grows in subalpine scrub woodland, open forest and rainforest, in clay loam and peaty soils. It was introduced into England during the 1800s, but is not

Olearia persoonioides × .5

common in cultivation now. Plants are suited to temperate regions and they tolerate heavy frosts and snowfalls. They require good drainage and grow well in acidic sandy or loamy soils in a semi-shaded site. Tip pruning promotes bushy growth. Propagate from seed or from cuttings of firm young growth.

The var. *lanceolata* Benth. is a synonym of *O. archeri* Lander.

O. tasmanica is smaller in size and has much smaller leaves.

Olearia phlogopappa (Labill.) DC.
(with a *Phlox*-like pappus)
NSW, Vic, Tas
1–2.5 m × 1–2.5 m   Oct–March
  Small **shrub** with hairy young growth; **branches** ascending to erect; **branchlets** usually covered in short greyish hairs; **leaves** 0.5–5 cm × 0.2–1 cm, elliptic to obovate, alternate, spreading to ascending, deep green to greyish-green or bluish-green, becoming glabrous above, greyish- to creamy-hairy below, margins entire or toothed, apex pointed or rounded; **flowerheads** daisy-like, 1.7–2.7 cm across, terminal, often on very short branchlets, solitary or clustered, sessile or on stalks to about 4 cm long, usually profuse and very conspicuous; **ray florets** usually 16–30, white, pink, mauve, purple-blue or blue; **disc florets** yellow; **achenes** silky-hairy or glandular.

This extremely variable and often extremely decorative species is currently undergoing revision. It is usually found at high elevations in woodland, open forest, closed forest and in subalpine woodland but it also occurs in heath at lower elevations. Plants do best in a semi-shaded site with freely draining acidic soils but will tolerate plenty of sunshine if grown in soils which are moist for most of the year. Plants grow

Olearia phlogopappa var. *subrepanda* × .9

Olearia phlogopappa, pink   W. R. Elliot

quickly and may lose their lower foliage but regular pruning during or after flowering often promotes new growth from bare lower branches.

Olearia phlogopappa, blue-purple   W. R. Elliot

Many selections of flower colour variants have been made and a number are usually available. During 1929–1930 the English collector, H. F. Comber collected some forms in Tasmania and in his field notes wrote 'this is quite the best find that I have made so far in Tasmania'. The selections 'Comber's Blue' and 'Comber's Pink' are still available in England, and other selections known as 'Splendens Blue', 'Splendens Lavender' and 'Splendens Pink' are also grown there. These variants may need clarification as some could possibly be referrable to *O. stellulata* (see below). *O.* cv Scilloniensis originated in England and is now regarded as a variant of *O. phlogopappa*. It has white flowers.

Plants are very useful for mixed shrub plantings, as well as for accent in other areas. Flowering plants attract many butterflies and cut stems are excellent for indoor decoration.

A further 2 varieties which occur in subalpine areas are included. The var. *flavescens* (Hutch.) J. H. Willis differs in having thick, usually longer leaves which have a creamy-yellow hairy undersurface, and flower-heads on long, somewhat thicker stalks. The var. *subrepanda* (DC.) J. H. Willis has short obovate leaves, usually under 1.2 cm long, and often solitary flower-heads on very short stalks.

Propagate all varieties from fresh seed, or from cuttings of semi-hardwood current year's growth.

The var. *angustifolia* (Hutch.) W. M. Curtis is no longer regarded as warranting varietal status.

Olearia picridifolia   W. R. Elliot

Olearia picridifolia (F. Muell.) Benth.
(leaves similar to the genus *Picris*)
Vic, SA   Rasp Daisy Bush
0.5–1 m × 0.5–1.5 m   June–Dec; also sporadic
   Dwarf **shrub** with densely hairy young growth;
branches few, spreading to ascending; **branchlets**
hairy; **leaves** 2–6 cm × 0.2–1 cm, narrowly oblanceolate, alternate, sessile, spreading to erect, scattered,

grey-green and very hairy above and below, tapering to
base, flat, margins entire, apex pointed; **flowerheads**
daisy-like, to about 3.5 cm across, on hairy stalks, solitary or in clusters, often profuse and very conspicuous;
ray florets 15–30, white, blue or violet; **disc florets** yellow; **achenes** glabrous.
   A highly desirable, ornamental species which occurs
in western Vic and south-eastern SA extending to the
Eyre Peninsula. It is found on alkaline soils in mallee
and heath. It deserves to be better known in cultivation. Grows well in freely draining acidic or alkaline
soils in a sunny or semi-shaded site. Hardy to moderate frosts. Plants may be short-lived but pruning from
an early stage promotes bushy growth and usually
increases the longevity of plants. Propagate from seed
or from cuttings of firm young growth.

Olearia pimeleoides (DC.) Benth.
(similar to the genus *Pimelea*)
Qld, NSW, Vic, SA, WA   Mallee Daisy Bush;
                                  Pimelea Daisy Bush
1–2 m × 0.6–1.5 m   May–Oct
   Small **shrub** with greyish-hairy young growth;
branches ascending to erect; **branchlets** greyish-hairy;
leaves 0.3–3 cm × 0.1–0.7 cm, linear, elliptical or obovate, alternate, more or less sessile, spreading to
ascending, scattered, dark green and nearly glabrous
above, greyish woolly-hairy below, margins entire or
irregularly toothed, revolute, apex pointed to
rounded; **flowerheads** daisy-like, to about 3.7 cm
across, on stalks to about 8.5 cm long, terminal, solitary or clustered, often profuse and very conspicuous;
ray florets 8–25, white; **disc florets** yellow; **achenes**
silky-hairy.
   An appealing species which is common over a wide
region of semi-arid, southern Australia. It grows in
shrubland, mallee, open woodland and dry sclerophyll
forest on sandy and clay soils which are often alkaline.

Olearia pimeleoides   D. L. Jones

Olearia pimeleoides × .5

Olearia pinifolia

This species has had limited cultivation. It adapts well in semi-arid and warm temperate regions in sites which have freely draining acidic or alkaline soils. Needs a warm to hot site for best results. Hardy to moderate frosts and extended dry periods. Responds well to pruning. Plants are very sensitive to *Phytophthora cinnamomi*, Cinnamon Fungus. Propagate from seed or from cuttings of firm young growth.

Two varieties have been raised to species level (see *O. incana* (D. A. Cooke) Lander, and *O. minor* (Benth.) Lander).

Olearia pinifolia (Hook. f.) Benth.
(leaves similar to those of the genus *Pinus*)
Tas
0.6–3 m × 0.8–2.5 m   Dec–Feb
Dwarf to medium **shrub** with silky-hairy young growth; **branches** spreading to ascending; **branchlets** silky-hairy; **leaves** 1.5–4.5 cm × to about 0.2 cm, narrowly linear, alternate, sessile, spreading to erect, crowded, deep green and nearly glabrous above, silky-hairy below, margins strongly revolute, apex pungently pointed; **flowerheads** daisy-like, about 2 cm across, solitary on upper axillary stalks, forming leafy terminal clusters, often profuse and very conspicuous; **ray florets** 8–10, white; **disc florets** yellow; **achenes** glabrous.

An interesting, pine-like shrub from mountains and subalpine areas above 750 m. It grows in heath, shrubland and subalpine woodland in freely draining acidic sandy loam soils which are moist for most of the year. Cultivated to a limited degree, it adapts well to temperate regions in well-drained soils with a semi-shaded site. Hardy to frost and snowfalls. Responds well to pruning and has potential as a hedging plant. May be used in bonsai as well as for general planting. Propagate from seed or from cuttings of firm young growth.

Olearia plucheacea Lander
(similar to the genus *Pluchea*)
WA
1–1.5 m × 1–1.5 m   Aug–Oct
Small **shrub** with somewhat resinous young growth; **branches** spreading to erect; **branchlets** resinous, initially yellow, becoming brownish; **leaves** 1.2–4.5 cm × 0.1–0.5 cm, filiform to narrowly linear, alternate, sessile, ascending to erect, scattered, green, faintly to densely hairy, margins entire to toothed, strongly revolute, apex pointed; **flowerheads** daisy-like, to about 1.4 cm across, in crowded and slender-stalked more or less flat-topped terminal clusters, often profuse and very conspicuous; **ray florets** 5–7, white; **disc florets** yellow; **achenes** hairy.

This recently described (1990) species is regarded as vulnerable in nature. It occurs in the Ashburton and Austin Districts where it grows in stony soils of sandstone breakaways, in woodland or shrubland. Needs to be cultivated as a conservation strategy. Best suited to semi-arid or warm temperate regions. Prefers freely draining soils and a sunny or semi-shaded site. Hardy to moderate frosts and extended dry periods. Tip pruning may be necessary to promote bushy growth. Propagate from seed or from cuttings of firm young growth.

Olearia pinifolia, broad-leaf variant × .7

Olearia propinqua S. Moore
(related to, or resembling)
WA
0.6–1.3 m × 0.6–1.3 m   Aug–Oct
Dwarf to small **shrub** with faintly woolly-hairy young growth; **branches** spreading to ascending; **branchlets** initially white-hairy, becoming glabrous; **leaves** 1.5–2 cm × about 0.1 cm, narrowly oblong to obovate, alternate, sessile, tapering to base, spreading to ascending, scattered, green to grey-green and glabrous or hairy above, densely whitish-hairy below, more or less flat, margins faintly toothed, apex blunt; **flowerheads** daisy-like, to about 1.5 cm across, usually solitary, sessile or short-stalked, terminal, often profuse and very conspicuous; **ray florets** about 14, white, sometimes with pale blue undersurface; **disc florets** yellow; **achenes** silky-hairy.

This olearia occurs over a wide area in the southwest. It is found in the Austin, Avon, Coolgardie, Helms and Roe Districts where it grows in sandy loams of mulga country. Rarely cultivated, it is best suited to semi-arid and warm temperate regions. A sunny or semi-shaded site with good drainage is essential. Hardy to moderate frosts and extended dry periods. May need tip pruning to promote bushy growth. Propagate from seed, which usually begins to germinate 20–40 days after sowing, or from cuttings of firm young growth.

O. pimeleoides is closely allied but differs in having leaves with recurved entire margins.

Olearia quercifolia Sieber ex DC.
(leaves similar to the genus *Quercus*)
NSW
1.5–2 m × 1.5–2 m   July–Dec
Small **shrub** with yellowish-hairy young growth; **branches** ascending to erect; **branchlets** yellowish-hairy; **leaves** 1.5–6 cm × 0.7–2.5 cm, elliptical to

obovate, alternate, short-stalked, spreading to ascending, scattered, green and rough above, yellowish-hairy below, margins shortly lobed, venation reticulate, apex pointed to rounded; **flowerheads** daisy-like, to about 3 cm across, on erect stalks to about 5 cm long, axillary, solitary or clustered, often profuse and conspicuous; **ray florets** 7–15, white; **disc florets** yellow; **achenes** glabrous.

A rare but not threatened species from the Blue Mountains where it grows in moist or swampy soils. Evidently rarely cultivated, it prefers moist but freely draining acidic soils in a semi-shaded site. Should adapt well in temperate and subtropical regions. Hardy to moderate frosts. Pruning promotes bushy growth. May need supplementary watering during extended dry periods. Propagate from seed or from cuttings of firm young growth.

Olearia ramosissima (DC.) Benth.
(much-branched)
Qld, NSW, WA
0.6–1.5 m × 0.6–1.5 m   June–Sept

Dwarf to small **shrub** with woolly-hairy young growth; **branches** many, spreading to ascending; **branchlets** twiggy, hairy; **leaves** to 0.5 cm × 0.2 cm, elliptical, obovate or triangular, alternate, sessile, reflexed, crowded, green and glabrous or slightly rough above, greyish-woolly hairs below, margins revolute, apex pointed; **flowerheads** daisy-like, to about 2.7 cm across, sessile, terminal, solitary, often profuse and conspicuous; **ray florets** 8–13, blue to mauve; **disc florets** yellow; **achenes** silky-hairy.

A handsome daisy bush which warrants greater cultivation. It has a disjunct distribution with populations in the Darling Downs District of Qld, mainly north of Narrabri in NSW, and in the Avon, Coolgardie and Eyre Districts in WA. Plants should do well in semi-arid and temperate regions in well-drained soils with semi-shade or dappled shade. Hardy to moderate frosts. Pruning from a young stage will promote a bushy framework. Introduced into England during 1818. Propagate from seed or from cuttings of firm young growth.

O. ramulosa differs in having stalked flowerheads to 2 cm across and longer leaves, while *O. floribunda* has sessile flowerheads to about 2 cm across with 3–7 ray florets.

Olearia ramulosa (Labill.) Benth.
(with many very small branches)
Qld, NSW, Vic, Tas, SA   Twiggy Daisy Bush
0.5–2.6 m × 0.6–2 m   Oct–May

Dwarf to medium **shrub** with greyish-hairy young growth; **branches** many, ascending to erect; **branchlets** twiggy, greyish-hairy; **leaves** 0.2–1.2 cm × up to 0.25 cm, linear to narrowly obovate, alternate, sessile, spreading to ascending or recurved, scattered, green and slightly rough above, greyish woolly-hairy below, margins entire and revolute, apex rounded, sometimes curry-scented; **flowerheads** daisy-like, 1–2 cm across, on stalks to 1.5 cm long, solitary, axillary or terminal, forming leafy spikes, often profuse and very conspicuous; **ray florets** 2–15, white, pale pink, pale blue or mauve-blue; **disc florets** yellow; **achenes** hairy.

An extremely variable species which comprises many intergrading variants. Over its wide distribution it occurs in dry sclerophyll forest, woodland, mallee and coastal scrub. Adapts well to cultivation in temperate and semi-arid regions. Grows in acidic or slightly alkaline soils in semi-shaded or dapple-shaded sites. Hardy to moderate frosts. May need supplementary watering during extended dry periods. Responds very well to pruning and can be used for low hedging. Propagate from seed or from cuttings of firm young growth.

The allied *O. ramosissima* has larger flowerheads but shorter leaves. The rarely cultivated *O. cassiniae* (F. Muell.) Benth. from the southern Darling District, WA, is also closely allied. It has narrow-linear leaves and its white flowerheads usually have 2–3 ray florets.

Olearia revoluta F. Muell. ex Benth.
(margins rolled backwards)
WA
0.6–1.5 m × 0.6–1.5 m   Feb–April; also sporadic

Dwarf to small **shrub** with whitish-hairy young growth; **branches** many, spreading to erect; **branchlets** whitish-hairy; **leaves** 0.5–1.5 cm × up to 0.2 cm, linear to oblong-cuneate, alternate, sessile, spreading to ascending, somewhat crowded, deep green and more or less glabrous above, densely white-hairy below, margins strongly revolute, apex bluntly pointed; **flowerheads** daisy-like, about 1 cm across, axillary, more or less sessile; **ray florets** 4–8, white; **disc florets** yellow; **achenes** usually hairy.

A widespread coastal species which extends virtually all the way from the Fortescue District down to the Coolgardie District but is not found in the northern Darling District. Not highly ornamental but nevertheless invaluable for coastal planting and revegetation projects. Must have good drainage but

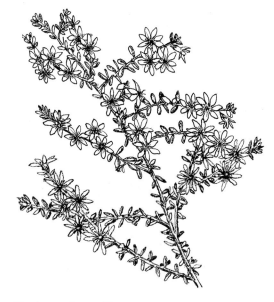

Olearia ramulosa × .95

tolerates acidic or slightly alkaline soils. Hardy to moderate frosts. Propagate from seed or from cuttings of firm young growth.

The allied *O. axillaris* has more-or-less insignificant ray florets.

Olearia rhizomatica Lander
(with a rhizome)
NSW
0.2–0.5 m × 0.6–1.5 m   Dec–March

Dwarf spreading **shrub** with rhizomatous roots and resinous young growth; **branches** spreading; **branchlets** resinous; **leaves** 0.5–1.2 cm × 0.1–0.4 cm, narrowly cuneate, infolded, alternate, sessile, spreading to erect, crowded, dark green above, glabrous and resinous above, glabrous and somewhat blistered below, margins entire except for 3-lobed apex; **flowerheads** daisy-like, to about 3 cm across, solitary, on stalks to about 4–5 cm long, terminal, often profuse and very conspicuous; **ray florets** about 30, blue; **disc florets** yellow with bluish tips; **achenes** small, silky-hairy.

A rare and beautiful species from the Southern Tablelands where it grows in alpine herbfields and snow gum woodlands. Evidently rarely cultivated, it should adapt well to temperate regions if grown in freely draining acidic soils with a sunny or semi-shaded aspect. Tolerates snowfalls and heavy frost. Plants have excellent potential for use in gardens, especially rockeries, and in containers. Propagate from seed, or stem cuttings, and it may be worth trying root cuttings.

Olearia rosmarinifolia (DC.) Benth.
(leaves similar to the genus *Rosmarinus*)
Qld, NSW
1–2 m × 1–2 m   Oct–Dec

Small **shrub** with hairy young growth; **branches** ascending to erect; **branchlets** finely hairy; **leaves** 1.2–9 cm × 0.15–0.4 cm, linear, alternate or opposite, more or less sessile, spreading to ascending, scattered, green and more or less glabrous above, greyish woolly-hairy below, margins entire and revolute, apex pointed; **flowerheads** daisy-like, to about 2.6 cm across, on stalks to 5.5 cm long, in loose terminal clusters, often profuse and conspicuous; **ray florets** 5–7, white; **disc florets** yellow; **achenes** glabrous.

This narrow-leaved, shrubby species is found from south-eastern Qld to near Cooma in south-eastern NSW. It usually inhabits rocky sites such as in gorges and waterways. Cultivated to a limited extent. It needs well-drained but moisture-retentive acidic soils and a semi-shaded or dapple-shaded site. Hardy to moderate frosts. Pruning from an early stage promotes bushy growth. Propagate from seed or from cuttings of firm young growth.

Olearia rudis (Benth.) F. Muell. ex Benth.
(rough)
NSW, Vic, SA, WA   Azure Daisy Bush
0.5–1.5 m × 0.3–1 m   June–Oct; also sporadic

Dwarf to small **shrub** with resinous young growth; **branches** ascending to erect, brittle; **branchlets** resinous when young; **leaves** 1.5–12 cm × 0.6–4 cm, elliptical, ovate or obovate, alternate, sessile, spreading to ascending, scattered, green and hairy on both

Olearia rudis × .5

surfaces, margins entire or toothed, apex pointed to rounded; **flowerheads** daisy-like, to about 4 cm across, on stalks to 4 cm long, terminal, usually in loose clusters, often profuse and very conspicuous; **ray florets** 40–75, pale blue, mauve or purple; **disc florets** orange; **achenes** glabrous.

An extremely showy species which inhabits semi-arid regions of southern Australia where it grows in usually alkaline soils of mallee and woodland. Limited in cultivation, it often becomes leggy and has proved somewhat short-lived. Regular pruning from planting promotes a bushy growth habit and may improve its longevity. Plants require good drainage and plenty of sunshine. In less sunny conditions, they must have a warm to hot, protected site. Suitable for acidic or alkaline soils. Hardy to moderate frosts and drought tolerant. Useful as a cut flower. Propagate from seed, which usually begins to germinate 15–45 days after sowing, or from cuttings of firm young growth.

O. laciniifolia differs in having solitary flowerheads which usually have up to 43 ray florets.

Olearia rugosa (F. Muell. ex Archer) Hutch.
(wrinkled, corrugated)
Vic, Tas   Wrinkled Daisy Bush
1–2.5 m × 1–2 m   Oct–Dec

Small to medium **shrub** with brownish-hairy young growth; **branches** ascending to erect; **branchlets** densely hairy; **leaves** 2.5–6 cm × 1–3 cm, ovate-lanceolate to broadly ovate, alternate, short-stalked, spreading to ascending, scattered, wrinkled, deep green above, rough and hairy above, densely brownish-hairy below, margins coarsely toothed or lobed, apex blunt; **flowerheads** daisy-like, to about 2.5 cm across, on stalks 1–3 cm long, in leafy terminal clusters, often profuse and conspicuous; **ray florets** 8–16, white; **disc florets** white to cream; **achenes** hairy.

The wrinkled daisy bush occurs in the mountains of eastern Vic and in Tas. It grows in woodland and dry sclerophyll forest. This species is best suited to cool temperate regions. Cultivated to a limited degree, it adapts well to acidic soils which drain freely but are moisture-retentive. A semi-shaded site is preferred. Plants are hardy to most frosts and tolerate snowfalls. Pruning can help to promote bushy growth. Propagate from seed or from cuttings of firm young growth.

O. stellulata is allied but differs markedly in its larger leaves which have entire or very shortly toothed margins.

Olearia cv **Scilloniensis** = *O. phlogopappa* (Labill.) Benth.

Olearia species affinity **lanuginosa** (Mornington Peninsula) —
      see *O. lanuginosa* (J. H. Willis) N. A. Wakef.

Olearia species affinity **pannosa** (Anglesea) =
      O. pannosa ssp. *cardiophylla* (F. Muell.) D. A. Cooke

Olearia speciosa Hutch.
(showy, beautiful)
Vic   Netted Daisy Bush
0.6–2 m × 0.6–1.5 m   Oct–Dec

Dwarf to small **shrub** with hairy young growth; **branches** ascending to erect; **branchlets** whitish-hairy; **leaves** 4–7.5 cm × 1–3 cm, narrowly oblong to ovate-lanceolate, attenuate, short-stalked, spreading to ascending, deep green and becoming glabrous above, with dense whitish to buff hairs below, margins finely to coarsely toothed or, rarely, entire, apex pointed; **flowerheads** daisy-like, to about 2.5 cm across, in long-stalked terminal clusters, often profuse and very conspicuous; **ray florets** 5–6, white to cream; **disc florets** yellow to brownish-yellow; **achenes** nearly glabrous.

This showy daisy bush has a scattered distribution from the north-east to the south-west of the state. It is often associated with rocky soils and is found on mountain slopes in dry sclerophyll forest as well as on outlying sandy slopes. Although cultivated to a limited extent it is worthy of greater attention and should adapt well to temperate regions. Prefers acidic, freely draining soils with semi-shade or dappled shade. Pruning promotes bushy growth. Plants withstand moderate to heavy frosts and light snowfalls. This species was introduced into England in 1888 with seed supplied by Royal Botanic Gardens, Melbourne. Propagate from seed or from cuttings of firm young growth.

O. erubescens differs in having shorter, oblong leaves with toothed or lobed margins and it lacks long-stalked flowers.

Olearia stellulata (Labill.) DC.
(resembling small stars)
Qld, NSW, Vic, Tas   Snow Daisy Bush
2–3 m × 2–3 m   Aug–Jan

A medium **shrub** with creamish-hairy young growth; **branches** ascending to erect; **branchlets** creamy-hairy; **leaves** 1–17.5 cm × 0.5–3.5 cm, elliptic to obovate,

A

Olearia stellulata and compact variant (A) × .6

alternate, short-stalked, spreading to ascending, scattered, pale to deep green and more or less glabrous above, greyish-hairy below, flat, margins entire or toothed, apex pointed, slightly aromatic when crushed; **flowerheads** daisy-like, to about 2–5 cm across, on stalks to about 3.5 cm long, in leafy terminal clusters, often profuse and very conspicuous; **ray florets** 8–16, white or, rarely, purple, violet or pink; **disc florets** cream; **achenes** silky-hairy.

A common and widespread species which occurs in wet and dry sclerophyll forest. It is usually found in soils which are moist for extended periods. Adapts very well to cultivation in temperate regions and is usually fast-growing. Should do well in acidic sandy to clay soils which drain freely. Prefers a semi-shaded or dapple-shaded aspect. Plants respond well to pruning, which helps promote bushy growth. Hardy to most frosts. May require supplementary watering during extended dry periods. Use of organic mulch is recommended. A white-flowered selection was introduced into England during 1812 and a purple-flowered selection from Tas was introduced during 1823. Propagate from seed or from cuttings of firm young growth.

O. lirata (Hutch.) Sims is now included in this species and the cultivar 'Olwyn Barnett' is now known as as *O. stellulata* 'Olwyn Barnett'. This cultivar grows to about 2 m × 1.5 m and produces lovely purple-pink daisies of about 2 cm across during Nov–Jan.

O. phlogopappa has been confused with this species but it differs in having smaller leaves, more ray florets

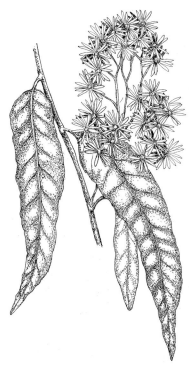

Olearia stellulata, long-leaved variant × .55

per flowerhead and flowerheads produced in non-leafy clusters. Hybrids with parentage of *O. phlogopappa* and *O. stellulata* are in cultivation.

O. lasiophylla is sometimes misidentified as *O. stellulata.* The former differs in having leaves with a densely hairy upper surface.

Olearia stilwelliae Blakely
(after Miss Sylvia Stilwell, co-discoverer of this species, 20th-century botanical collector)
NSW
0.3–0.6 m × 0.6–1.5 m   June–Oct
Dwarf **shrub** with cream- to rusty-hairy young growth; **branches** ascending to erect; **branchlets** cream- to rusty-hairy; **leaves** 3–16 cm × 1.6–3.5 cm, elliptical to ovate, alternate, stalked, spreading to ascending, scattered, green and glabrous greyish-hairy above, beige to rusty-hairy below, flat, margins entire or toothed, apex pointed to rounded; **flowerheads** daisy-like, to about 5 cm across, on stalks to over 30 cm long, solitary, terminal, very conspicuous; **ray florets** 8–12, pale blue, or white with mauve tinges; **disc florets** yellow; **achenes** silky-hairy.

A highly decorative, rare but not threatened species, which deserves wider recognition. It occurs in the North Coast District in dry sclerophyll forest where it grows in sandy soil. In cultivation it should be well suited to subtropical and temperate regions. Prefers well-drained acidic soil in a semi-shaded site. May be damaged by heavy frosts. It has potential for general use in gardens as well as in rockeries, perennial

borders and as a container plant. May not require regular pruning. Propagate from seed or from cuttings of firm young growth.

Olearia strigosa (Steetz) Benth.
(closely covered with pointed bristles)
WA   Bristly Daisy Bush
0.2–0.5 m × 0.3–1 m   Nov–June
Dwarf **shrub** with slightly sticky, hairy young growth; **branches** spreading to ascending; **branchlets** hairy; **leaves** 0.4–1 cm × about 0.1 cm, linear, alternate, sessile, spreading to ascending or sometimes recurved, scattered, green and hairy, margins strongly revolute, apex finely pointed; **flowerheads** daisy-like, about 1.5 cm across, solitary, terminal; **ray florets** about 8–12, white, lilac or blue; **disc florets** yellow; **achenes** silky-hairy.

This poorly known species is found in the Avon District where it grows in acidic sandy soils. Plants should adapt to semi-arid and temperate regions. They need well-drained soils with a semi-shaded aspect. Hardy to moderate frost. Because of its low growth habit it has potential for use in rock gardens and as a container plant, and for general planting. Propagate from seed or from cuttings of firm young growth.

The allied *O. muricata* has blunt leaves and smaller flowerheads.

Olearia stuartii (F. Muell.) F. Muell. ex Benth.
(after John McDouall Stuart, 19th-century Australian explorer)
Qld, SA, WA, NT
0.2–1 m × 0.2–1 m   June–Sept; also sporadic
Dwarf **shrub** with sticky, resinous, yellowish-green young growth; **branches** many, spreading to erect; **branchlets** sticky, resinous, becoming brown; **leaves** 1–2.8 cm × 0.2–0.8 cm, narrowly to broadly obovate, alternate, sessile, spreading to ascending, green and glandular-hairy above and below, flat, margins irregularly toothed, venation obscure, apex very blunt and toothed, strongly aromatic; **flowerheads** daisy-like, 1–3 cm across, on stalks to about 5 cm long, solitary or in loose terminal clusters, may be profuse and conspicuous; **ray florets** 20–65, blue to mauve or lilac; **disc florets** yellow; **achenes** silky-hairy.

An extremely widespread species which occurs in semi-arid and arid regions. It is found in woodland and shrubland on a variety of sandy and silty soils as well as in rocky sites. Rarely cultivated, this drought-tolerant species needs excellent drainage and plenty of heat to do well. Plants are hardy to moderate frosts. Pruning should promote bushy growth. Some people find the fragrance of the plant unpleasant but possibly the decorative flowerheads outweigh the foliage aroma. Propagate from seed or from cuttings of firm young growth.

There has been some confusion surrounding this species and a recent revision (1984) has separated a further 3 species. *O. gardneri* has very short-stalked, linear leaves with reticulate venation, while *O. xerophila* has prominently stalked, elliptical to broadly elliptical leaves and *O. humilis* has more or less glabrous, often bundled leaves with an obscure venation.

Olearia subspicata (Hook.) Benth.
(resembling a flower spike)
Qld, NSW, Vic, SA, WA, NT   Shrubby Daisy Bush
2–3 m × 1.5–2.5 m   July–Oct

Medium **shrub** with greyish-hairy young growth; **branches** ascending to erect; **branchlets** faintly hairy; **leaves** 0.5–2 cm × 0.1–0.3 cm, linear or obovate, alternate, more or less sessile, spreading to ascending, scattered, green and more or less glabrous above, woolly greyish-hairy below, margins entire and revolute, apex pointed; **flowerheads** daisy-like, to about 2.5 cm across, on stalks to about 1.2 cm long, axillary, forming leafy panicles, often profuse and conspicuous; **ray florets** 2–6, white; **disc florets** yellow; **achenes** silky-hairy.

This shrubby daisy bush is moderately ornamental. It is found in semi-arid and arid regions and is associated with mulga, shrubland, woodland and mallee. It inhabits sandy soils. Requires very good drainage and a warm to hot site. Hardy to moderate frosts. Pruning may encourage bushy growth on what can otherwise be an erect shrub. Propagate from seed or from cuttings of firm young growth.

Olearia suffruticosa D. A. Cooke
(like a shrub)
NSW, Vic, SA
0.4–1 m × 0.5–1 m   Jan–May

Dwarf **shrub** with a woody rootstock and glutinous young growth; **branches** spreading to ascending, short-lived; **branchlets** glutinous; **leaves** 0.3–2.4 cm × 0.5–1 cm, linear, folded, alternate, sessile, spreading to ascending, scattered or sometimes clustered, glandular and more or less glabrous above, glabrous below, flat, margins entire, apex pointed; **flowerheads** daisy-like, to about 1.8 cm across, on stalks to 5 cm long, in leafy clusters, may be profuse and conspicuous; **ray florets** 12–20, white to pink; **disc florets** pink; **achenes** hairy.

This small-flowered and moderately appealing daisy usually inhabits moist or swampy sites in poor quality soils. It is known from the Central Tablelands of NSW, south-western Vic and south-eastern SA. This species should adapt to a wide range of soils in temperate regions. An open or semi-shaded site should be suitable. Plants are hardy to moderate frosts. Removal of old stems and branches is beneficial. Propagate from seed or from cuttings of firm young growth.

O. glandulosa is closely allied but it has larger flowerheads and its stems do not die back.

Olearia tasmanica (Hook. f.) W. M. Curtis
(from Tasmania)
Tas   Alpine Daisy Bush
0.6–1.3 m × 0.6–1 m   Dec–Feb

Dwarf to small **shrub** with rusty-hairy young growth; **branches** spreading to ascending; **branchlets** rusty-hairy when young; **leaves** 1.2–2 cm × 0.5–0.8 cm, elliptical, alternate, short-stalked, spreading to ascending, scattered, leathery, deep green and shiny above, glabrous above, somewhat rusty-hairy below, margins entire and revolute, apex blunt; **flowerheads** daisy-like, about 2 cm across, solitary, on long slender stalks

in upper axils, forming leafy clusters, often profuse and very conspicuous; **ray florets** usually 5–6, white; **disc florets** yellow; **achenes** glabrous.

A charming species from near the western and south-western coast, as well as from above 1000 m altitude. It often grows as part of dense shrub vegetation. Uncommon in cultivation, it is best suited to temperate regions. Needs freely draining acidic soils with a semi-shaded or dapple-shaded aspect. Plants are hardy to most frosts and snowfalls. Should be better known in cultivation and has potential for general use and possibly dwarf hedging, and as a container plant. Propagate from seed or from cuttings of firm young growth.

O. alpina (Hook. f.) W. M. Curtis. is a synonym.

O. persoonioides is closely allied but it is much taller and has larger leaves.

Olearia tenuifolia (DC.) Benth.
(slender leaves)
NSW, Vic   Shiny Daisy Bush
1–2 m × 1–2 m   most of the year

Small **shrub** with glandular young growth; **branches** spreading to ascending; **branchlets** glandular; **leaves** 0.5–3.5 cm × 0.1–0.3 cm, linear, alternate, more or less sessile, spreading to ascending, scattered, glandular and glabrous above and below, margins entire or toothed and revolute, apex pointed; **flowerheads** daisy-like, to about 4 cm across, on stalks to about 3 cm long, terminal, solitary or in loose clusters, may be scattered to profuse but usually very conspicuous; **ray florets** 7–15, blue to mauve; **disc florets** yellow; **achenes** silky-hairy.

This ornamental species from southern NSW and eastern Victoria is found in shallow sandy or stony soils in mallee, woodland and dry sclerophyll forests. It is often a pioneer plant after soil disturbance. Plants need very good drainage and a sunny or semi-shaded site. They tolerate extended dry periods but can become a little unsightly. Pruning promotes bushy growth. Hardy to moderate frosts. Propagate from seed or from cuttings of firm young growth.

The allied *O. cordata* has scattered, hairy leaves with a broadened base.

Olearia teretifolia (Sond.) F. Muell. ex Benth.
(terete leaves)
Vic, SA   Cypress Daisy Bush
0.5–2 m × 0.6–1.5 m   Aug–Dec; also sporadic

Dwarf to small, conifer-like **shrub** with more or less glabrous, somewhat sticky young growth; **branches** many, spreading to ascending; **branchlets** short, many; **leaves** 0.2–0.5 cm × 0.1 cm, linear with broad base, alternate, sessile, ascending to erect, often appressed, deep green, glabrous, concave above, channelled below, apex blunt; **flowerheads** daisy-like, to about 1.5 cm across, sessile, solitary, terminal, forming leafy clusters, often very profuse and conspicuous; **ray florets** 4–8, white; **disc florets** yellow; **achenes** hairy to nearly glabrous.

A variable species which inhabits dry sclerophyll forest, woodland, mallee and scrubland on sandy or stony soils in western Vic and south-eastern SA. There

Olearia tasmanica — T. L. Blake

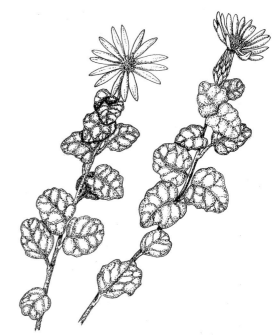

Olearia tomentosa × .6

are two distinct variants in cultivation. One is compact with crowded branches and dense foliage, and is sometimes marketed as 'Compacta' or Compact Form. The other is more upright and with a somewhat loose growth habit. They require good drainage and do best in acidic soils in sites which are semi-shaded or with dappled shade. Plants are hardy to moderate frosts and respond well to pruning. Propagate from seed or from cuttings of firm young growth.

Olearia tomentosa (H. L. Wendl.) DC.
(with short, soft hairs)
NSW   Toothed Daisy Bush
1–2.5 m × 1–2 m   Aug–May; sometimes sporadic
   Small to medium **shrub** with hairy young growth;

Olearia tenuifolia — D. L. Jones

branches spreading to erect; **branchlets** hairy; **leaves** 1.5–10 cm × 1–5 cm, ovate, alternate, stalked, spreading to ascending, scattered, dark green and becoming glabrous and shiny above, paler and somewhat felty-hairy below, more or less flat, margins toothed or shortly lobed, apex bluntly pointed; **flowerheads** daisy-like, 2.5–6 cm across, on stalks to 7 cm long, in loose terminal clusters, often profuse and very conspicuous; **ray florets** 13–29, white to blue; **disc florets** yellow; **achenes** glabrous or silky.

This handsome and variable species occurs in the Northern Central and Southern Coast Districts, as well as in the Central and Southern Tablelands. It is found in heath, shrubland and dry sclerophyll forest. An adaptable species in cultivation, it does best in a semi-shaded or dapple-shaded site with acidic soils which drain relatively well. Plants also tolerate plenty of sunshine and are suitable for semi-exposed coastal sites. Hardy to moderate frosts. There are two distinct selections commonly cultivated. One is a compact bushy selection which grows to about 1.5 m × 1.5 m and has pale mauve flowers. The other is taller and has a more open growth habit with larger white to pale blue-mauve flowerheads; it may be sold as 'Zane Grey's Bath'. All plants respond very well to pruning, which helps rejuvenate older plants. Sometimes plants may be subject to damage by powdery mildew, see Volume 1, page 173 for control. In 1793 plants of this species were introduced into England. Propagate from seed or from cuttings of firm young growth.
O. dentata Moench is a synonym.

Olearia tridens D. A. Cooke =
        Minuria tridens (D. A. Cooke) Lander

Olearia teretifolia, compact foliage variant   W. R. Elliot

Olearia tubiliflora (Sond.) Benth.
(tubular flowers)

Vic, SA	Rayless Daisy Bush
1–2 m × 0.6–1.5 m	Oct–Dec

Small **shrub** with cottony-hairy young growth; **branches** ascending to erect; **branchlets** hairy; **leaves** 0.5–1 cm × about 0.1 cm, linear, alternate, sessile, narrowing to base, spreading to ascending, green and somewhat hairy to glabrous above, densely whitish-hairy below, margins entire, revolute, apex blunt; **flowerheads** somewhat daisy-like, about 0.5 cm across, solitary, more or less sessile in axils, forming leafy racemes, often profuse and conspicuous; **ray florets** very small, 3–5, somewhat tubular, white; **disc florets** yellow; **achenes** hairy.

An inhabitant of rocky gullies and slopes in forest and woodland areas which have moderate rainfall. Lacks strong ornamental character and is rarely cultivated. Plants should adapt to a wide range of freely draining acidic soils in sites which are semi-shaded or dapple-shaded. Pruning promotes bushy growth. Plants are hardy to moderate frosts. Propagate from seed or from cuttings of firm young growth.

O. ramulosa has larger flowerheads to about 2 cm across and usually broader leaves.

Olearia viscidula (F. Muell.) Benth.
(slightly sticky)

NSW, Vic	Wallaby Weed; Viscid Daisy Bush
1.5–3.5 m × 1–2 m	July–Dec

Small to medium **shrub** with resinous young growth; **branches** spreading to ascending; **branchlets** hairy and often resinous; **leaves** 1.5–10 cm × 0.2–1.2 cm, narrowly elliptic to ovate, opposite or alternate, more or less sessile, spreading to ascending, scattered, dark green and resinous and often shiny above, greyish-hairy below, flat, margins entire, apex narrowly pointed; **flowerheads** daisy-like, to about 2 cm across, in axillary panicles, often profuse and very conspicuous; **ray florets** 8–21, white; **disc florets** cream to yellow; **achenes** silky-hairy.

A floriferous, far-ranging species from the North Coast, Northern Tablelands and North Western Slopes of NSW to eastern Victoria. It is found in woodland and dry sclerophyll forest where it grows in acidic to slightly alkaline soils. Rare in cultivation, plants should do best in a semi-shaded site with freely draining soils. Tolerates moderately heavy frosts and withstands short periods of dryness. Pruning can promote bushy growth. Propagate from seed or from cuttings of firm young growth.

Olearia viscosa (Labill.) Benth.
(sticky)

Vic, Tas	Viscid Daisy Bush
1–2.5 m × 1–2 m	Nov–Feb

Small to medium **shrub**, often with resinous young growth; **branches** ascending to erect; **branchlets** resinous; **leaves** 5–12.5 cm × 1–3.5 cm, lanceolate or elliptical, opposite, short-stalked, spreading to ascending, scattered, thin, shiny above, deep green and glabrous above, densely covered with silvery-white hairs below, margins entire, apex pointed, very strongly aromatic; **flowerheads** daisy-like, to about 1.5 cm across, on much-branched stalks in upper axils, forming crowded terminal clusters, very profuse and conspicuous; **ray florets** 4–6, with 1–2 well-developed, creamy-white or, rarely, mauve; **disc florets** yellow.

A floriferous species inhabiting coastal areas of East Gippsland, Vic, and widespread in Tas. It is usually found in wet sclerophyll forest. This fast-growing olearia is rarely cultivated. It prefers well-drained acidic soils which are moisture retentive, and a site which is semi-shaded or with dappled shade. Hardy to moderate frosts. Responds well to pruning. Propagate from seed or from cuttings of firm young growth.

Oligarrhena micrantha   B. W. Crafter

Olearia xerophila

Olearia viscosa × .5

Olearia xerophila (F. Muell.) F. Muell. ex Benth.
(dry-loving)
Qld, WA, NT
0.4–1 m × 0.6–1.5 m   July–Sept; also sporadic

Dwarf **shrub** with glandular hairy young growth; **branches** spreading to erect; **branchlets** hairy; **leaves** 2.5–7 cm × 0.7–2.2 cm, elliptic to spathulate, alternate, stalked, spreading to ascending, scattered, olive green, hairy above and below, flat, margins toothed, apex bluntly or finely pointed, strongly aromatic; **flower-heads** daisy-like, 2–4 cm across, on long stalks, forming loose terminal clusters, profuse and very conspicuous; **ray florets** 20–50, white, mauve, violet or purple; **disc florets** yellow; **achenes** silky-hairy.

This species occurs in seasonally dry subtropical and tropical regions where it is found in woodland or open shrubland. It inhabits gravelly or sandy soils on rocky ridges or cliff faces. Little is known of its reaction to cultivation. Probably best suited to frost-free regions. Plants will need excellent drainage and may do best in a warm, protected site. They are often straggly and open in nature, so they will probably need regular pruning to promote a bushy framework. Propagate from seed or from cuttings of firm young growth.

OLIGARRHENA R. Br.
(from the Greek *oligos*, few; *arrhen*, male; referring to the 2 stamens)
Epacridaceae

A monotypic genus endemic in south-western WA.

Oligarrhena micrantha R. Br.
(small flowers)
WA
0.2–0.6 m × 0.2–0.5 m   Sept–Nov; also sporadic

Dwarf **shrub**, with faintly hairy young growth; **branches** many, slender, ascending to erect; **leaves** 0.1–0.3 cm × up to 0.1 cm, lanceolate to ovate-lanceolate, sessile, closely appressed, overlapping, light green, concave above, margins hairy, apex bluntly pointed; **flowers** to about 0.25 cm long, bell-shaped, white to light yellow, erect, in axillary spikes forming a terminal raceme, often profuse and very conspicuous; **fruit** small, 1-celled drupe.

This very decorative species occurs close to the coast in the Eyre District. It inhabits sandy soils which may have a high lateritic gravel content and may be moist to wet for extended periods. This species is rarely cultivated, and regarded as very difficult to grow, but it deserves to be more widely known. Needs good drainage and a sunny or semi-shaded aspect. Hardy to light frosts. It has potential for floriculture, as well as for general garden and container planting. Responds well to pruning. Propagate from seed (although it is rarely available). Cuttings of fairly young growth which is barely firm are likely to give best results.

OMALANTHUS A. Juss.
(from the Greek *homalos*, smooth; and *anthos*, flower)
Euphorbiaceae

Shrubs or small **trees**; **sap** often milky; **stems** or **branchlets** often reddish; **leaves** simple, alternate, sometimes peltate, on long, slender petioles; **stipules** large, shedding quickly; **inflorescence** a slender terminal raceme; **flowers** unisexual; **male flowers** numerous, clustered; **female flowers** stalked, single at the base of the raceme; **fruit** a capsule; **seeds** arillate or not.

A genus of about 35 species distributed in Asia and Polynesia with 3 occurring in eastern Australia. One species is moderately common in cultivation. Propagation is from seed which has a short period of viability and is best sown fresh. Cuttings of firm young growth strike readily.

Previously known as *Homalanthus* (sometimes incorrectly spelt *Homolanthus*).

Omalanthus novoguineensis (Warb.) Lauterb. & K. Schum.
(from New Guinea)
Qld
6–10 m × 1–3 m   Oct–Mar

Tall **shrub** or small **tree** with an open habit; **leaves** 5–14 cm × 8–10 cm, broadly ovate, on slender petioles to 10 cm long, dark green above, pale green beneath; **racemes** 4–12 cm long, terminal; **flowers** green to pinkish; **capsules** 0.5–0.8 cm long, roughened; **seeds** blackish.

Plants are widespread in New Guinea. In Australia it occurs in the Torres Strait islands and northern parts of Cape York Peninsula, growing in or along the margins of rainforest. Not known to be in cultivation but could have potential for tropical regions. Requirements are probably similar to those of *O. nutans*. Propagate from fresh seed and possibly also from cuttings.

Omalanthus nutans (Forst.) Guill.
(nodding)
Qld, NSW   Native Poplar, Bleeding Heart
2–5 m × 1–2 m   Sept–Dec

Medium to tall bushy **shrub**, often with an open habit; young shoots enclosed by large stipules; young stems exuding whitish sap; **leaves** 5–15 cm × 6–12 cm, broadly ovate, on slender petioles to 9 cm long, green above, greyish-green beneath, old leaves turn red to crimson before falling; **racemes** 2–10 cm long, ter-minal; **flowers** yellowish-green to red; **capsules** 0.6–0.8 cm long, 2-lobed; **seeds** black with a fleshy yellow aril, mature Dec–Mar.

Widely distributed from central-eastern Qld to the south coast of NSW and also in New Caledonia, Vanuatu, Fiji and Tonga. Commonly grows along rain-forest margins and near streams in open forest but may also be seen along road verges and embankments. It is a fast-growing pioneer plant. The seedlings ger-minate on sites of disturbance. The species is popular in cultivation and is grown mainly for its foliage, espe-cially the colourful old leaves. It is also an excellent pioneer plant for use in reclamation projects within its natural range. It has the potential to become a weedy species in other areas. Plants can be grown successfully as far south as Melbourne. They adapt to different sites and soils providing drainage is unimpeded. Tolerates light to moderate frosts. Propagate from fresh seed, or from cuttings of firm young growth.

Previously well known as *O. populifolius* Graham.

Omalanthus populifolius Graham =
O. nutans (Forst.) Guill.

Omalanthus stillingifolius F. Muell.
(with leaves like the genus *Stillingia*)
Qld, NSW
1–2 m × 1–2 m   Aug–Dec

Small **shrub,** often straggly; **leaves** 2–6 cm × 1–3 cm, ovate to narrowly ovate, on petioles to 5 cm long, dark green above, paler and sparsely hairy to papillose beneath; **racemes** 4–10 cm long, terminal; **flowers** greenish to reddish; **capsules** 0.4–0.6 cm × 0.5 cm, globose; **seeds** with a small, hard knob.

Distributed from central-eastern Qld to south-eastern NSW growing on rainforest margins and in moist areas of open forest, sometimes among rocks. Easily grown in a partially sheltered situation but strongly inclined to become straggly. Benefits from regular tip pruning. Makes an interesting plant for containers. Requires well-drained soil and tolerates light to moderate frosts. Propagate from fresh seed or from cuttings of firm young growth.

OMPHACOMERIA (Endl.) A. DC.
(from the Greek *omphakos*, bitter and *meros*, a part; referring to the sour taste of the fruit)
Santalaceae
A monotypic genus endemic in Australia.

Omphacomeria acerba (R. Br.) A. DC.
(tart, sour, acidic)
NSW, Vic   Leafless Sourbush
0.3–1 m × 1–3 m   Aug–Dec
Wiry **shrub**, appearing leafless, suckering and often forming thickets; **branchlets** erect, green, rounded, grooved, finely striate, much-branched; **leaves** reduced to minute scales; **male flowers** about 0.3 cm

across, whitish, up to 7 in short axillary spikes; **female flowers** about 0.2 cm across, solitary; **drupe** 0.6–0.8 cm long, ellipsoid, purplish, sour tasting.

An interesting, wiry, broom-like shrub which grows in colonies and spreads by root suckers. It is a root parasite distributed from near Newcastle to eastern Vic in dry open forests. The fruit is edible but sour. It can be made into jams and preserves. Being a root parasite this species has proved difficult to cultivate and pro-vides a challenge for researchers. Propagation is from seed, which is slow and difficult to germinate, and pos-sibly also from root cuttings.

Leptomeria acerba R. Br. is a synonym.

OMPHALEA L.
(from the Greek *omphalos*, navel; also referring to flower carpels)
Euphorbiaceae
Robust **climbers**, rarely **trees; leaves** alternate, simple, entire, large, with 2 conspicuous glands at the top of the petiole; **inflorescence** cymes, simple or branched; **flowers unisexual**, small, lacking petals; **anther filaments** united at the base into a tube; **fruit** 2–3-lobed, leathery.

A genus of about 20 species widely distributed in tropical regions, particularly Malesia and Asia, with one endemic species in Australia. It is very rarely cul-tivated. Propagation is from seed which has a limited period of viability and is best sown fresh. Cuttings of hardened new growth can also be successful.

Omphalea queenslandiae F. M. Bailey
(from Queensland)
Qld
10–20 m tall   Aug–Dec
Robust **climber** forming a large canopy of foliage; **stems** to 20 m or more long and 15 cm in diameter; **younger stems** with decurrent ridges from the base of the petioles; **leaves** 10–18 cm × 15–20 cm, broadly oblong to ovate, dark green, leathery, 2 large glands at the junction with the petiole; **cymes** to 15 cm long, in the upper axils; **flowers** about 0.3 cm across, greenish to reddish, unisexual, several males arranged around a single female; **fruit** 6–10 cm across, globose, cream to yellow, leathery, containing 2–4 globular seeds, ripe Feb–May.

Occurs in north-eastern Qld where it is distributed from the coastal lowlands to moderate altitudes (about 800 m). It grows in rainforest. The stems reach the outer canopy and develop a large, heavy mass of foliage. Its excessive vigour precludes its widespread cultivation but it is grown to a limited extent as a host plant by butterfly enthusiasts. Must be pruned regu-larly to control growth and to keep the size of the plant manageable. Grows in a sunny position in well-drained soil. Frost tolerance is not high. Propagate from fresh seed, and from cuttings of firm new growth taken in summer.

ONAGRACEAE Juss.
A cosmopolitan family of dicotyledons consisting of about 640 species in 18 genera; 2 genera and about 20 species are native to Australia. In addition, a number of species of *Oenothera* are widely naturalised as weeds.

Omalanthus nutans   D. L. Jones

Members of this family are soft, herbaceous perennials, annuals or aquatics, with opposite or spirally arranged, sessile or petiolate leaves. The flowers have a distinctive basal tube formed by the fusion of the calyx, corolla and stamens. Australian genera: *Epilobium*, *Ludwigia*.

ONDINEA Hartog
(after Ondine, a class of water sprites)
Nymphaeaceae

A monotypic genus endemic in northern Western Australia.

Ondinea purpurea Hartog
(purple)
WA   Ondinea
Aquatic   Dec–Apr

Perennial **aquatic herb** with a tuberous rootstock; **tubers** 1–6 per plant, 1.5–2.5 cm × 1–2 cm, oblong, fleshy; **leaves** dimorphic, depending on whether submerged or emergent; **petioles** 10–40 cm long, spongy; **submerged leaves** 10–17 cm × 2–3 cm, yellowish-green above, purplish-brown beneath, shiny, thin-textured, margins intensely wavy to crisped, deeply cordate at the base, apex obtuse to emarginate; **emergent leaves** 5–7 cm × 1.5–2 cm, narrowly ovate, floating, leathery, bright green above, purplish beneath, margins entire or wavy, basal lobes often overlapping; **flowers** solitary, 2–5 cm across, purple to violet, on stalks to 3 m long; **sepals** 4, narrow; **petals** present or absent; **stamens** 15–34, in whorls; **berry** 1–2 cm × 0.8–1.5 cm, ovoid, green and purple-striped.

A remarkable aquatic which consists of 2 subspecies, both endemic in the Kimberley region of Western Australia. Plants grow actively during the wet season when water is plentiful and die back to the tuberous rootstock during the dry season. They form colonies along the margins of small ephemeral streams. The leaves of separate plants intermingle and flowers are held well above water level. Contractile roots, produced near the apex of the tuber, pull the tubers deep into the soil. With its incredible wavy leaves and interesting flowers *Ondinea* has tremendous prospects as an aquatic plant for aquaria and ponds in tropical regions. It has been introduced into cultivation with some success. Plants seem to require a minimum water temperature of about 15°C and benefit from an annual rest and repotting while dormant. They require bright light to grow actively and flower.

The 2 subspecies are easily distinguished. The flowers of ssp. *purpurea* lack petals whereas those of ssp. *petaloidea* Kenneally & Schneider have petals present.

Propagate both subspecies from seed which should be sown on the surface of moist to wet soil under glass in warm, humid conditions.

ONYCHOSEPALUM Steud.
(from the Greek *onyx*, *onychos*, claw; and Latin *sepalum*, sepal; or possibly from the Latin *onyx*, brown, referring to male flower colour)
Restionaceae

Dioecious **herbs**; with a branched rootstock; **stems** simple, usually a few, erect, slender, smooth, unbranched; **leaves** reduced to scales; **inflorescence** mainly terminal; **male spikelets** usually solitary; **female spikelets** few to several together; **perianth segments** 3; **stamens** 3; **fruit** a small nut.

A small genus with 2 species which is confined to south-western WA. Until recently (1996) it was thought to be a monotypic genus. Plants are rarely encountered in cultivation.

Onychosepalum laxiflorum Steud.
(loose flowers)
WA
0.1–0.3 m × 0.2–0.5 m   Sept–Oct

Perennial tufting, dioecious **herb**, with a branched rootstock; **stems** erect, slender, simple, a few short, overlapping, sheathing scales at base but none on upper part; **inflorescence** solitary terminal spikelets of several flowers; **male spikelets** to about 0.9 cm long, brownish, usually solitary, with overlapping often deciduous glumes, stamens 3, anthers exserted; **female spikelets** to about 0.9 cm long, purplish-red, style undivided; **perianth segments** 3, very thin; **fruit** a small nut.

This rare species is a dweller of swamps in the Irwin, north-western Darling and Eyre Districts. Evidently not cultivated, it needs to be introduced as a conservation measure. Plants should adapt to a range of moist-to-wet soils in temperate regions. They would appear to tolerate a sunny or semi-shaded site. Worth trying in bog gardens or as a container plant with a saucer to maintain readily available water. Propagate from seed, or by division of clumps which should not be too small.

The recently described (1996) *O. microcarpum* K. A. Meney & J. S. Pate is very closely allied but it differs in having spikelets to about 0.6 cm long, of which there are usually several on each female culm. It occurs in

Omalanthus stillingifolius   D. L. Jones

sandy soils of the northern Darling and southern Irwin Districts. Plants are short-lived.

Onychosepalum microcarpum K. A. Meney & J. S. Pate — see *O. laxiflorum* Steud.

OPERCULARIA Gaertner
(from the Latin *operculum*, cover; referring to the lid-like valves on the fruit)
Rubiaceae   Stinkweeds

Herbs, dwarf **shrubs** or rarely **twiners**; **leaves** opposite, strongly odorous when crushed; **stipules** forming sheath around leaf base; **inflorescence** flowers fused at base to form globular compound heads, or with a solitary head, terminal or in stem-forks, sometimes appearing axillary, on erect or recurved stems; **flowers** unisexual or bisexual, small, purplish; **calyx lobes** 3–5; **corolla** tubular, short; **stamens** inserted, with anthers exserted; **style** slender, usually 2-branched; **fruit** a 2-valved capsule.

An endemic genus of about 16 species. Roughly half the species occur in eastern Australia and the balance occur in the west. Plants have few ornamental features and are mainly of value for indigenous regeneration projects.

Propagate from seed, which does not require pre-sowing treatment, or from cuttings of firm young growth.

Opercularia aspera Gaertn.
(rough)
Qld, NSW, Vic   Common Stinkweed
0.3–1 m × 0.6–2 m   Sept–Jan

Dwarf perennial **herb**, with a woody rootstock; **stems** to about 1 m long, glabrous or rough-hairy; **leaves** 1–11.5 cm × 0.2–2 cm, lanceolate to ovate, opposite, stalked, glabrous or rough-hairy, mid green to deep-green, margins finely toothed or with hooked hairs, apex pointed; **flowers** small, in compound globular heads, with recurved stalks, purplish; **capsule** about 0.5–1 cm across, rough.

An extremely variable species with a widespread distribution in eastern Australia. It occurs from sea-level to the mountains and western slopes, and is found in heath, open forests and woodlands. Plants need fairly good drainage and a sunny or semi-shaded site is suitable. Hardy to most frosts. Introduced into England in 1790. Propagate from seed or from cuttings of firm young growth.

Opercularia hispida Spreng.
(covered with coarse, erect hairs)
Qld, NSW, Vic   Hairy Stinkweed
0.1–0.3 m × 0.2–0.5 m   Sept–Jan

Dwarf perennial **herb**, with a woody rootstock; **stems** to about 20 cm long, prostrate to erect, hairy; **leaves** 1.2–2.5 cm × 0.4–1 cm, lanceolate to ovate-lanceolate, opposite, stalked, hairy, dark green, margins entire or sometimes wavy, apex pointed; **flowers** small, purplish, in compound globular heads, with recurved stalks; **capsule** about 0.7 cm across, rough.

A widespread species which usually inhabits rocky sites and is often an understorey plant in eucalypt forests. Introduced into England during 1790. It has similar cultivation requirements to *O. aspera*. Propagate from seed or from cuttings of firm young growth.

Opercularia ovata Hook. f. — see *O. varia* Hook. f.

Opercularia varia Hook. f.
(variable)
NSW, Vic, Tas, SA   Variable Stinkweed
0.1–0.3 m × 0.3–0.6 m   June–Oct

Dwarf perennial **herb**; **stems** spreading to ascending, glabrous to slightly hairy; **leaves** 0.3–1.8 cm × 0.15–0.6 cm, linear, lanceolate or ovate, sessile, deep green, usually glabrous, apex pointed; **flowers** small, purplish, 2–7 in semi-globular head, often on twin-branched stems; **capsule** about 0.5 cm across.

A common and somewhat insignificant species which is found over a wide range of eastern Australia in heath, woodland and open forests. Needs fairly good drainage but will survive lengthy periods of wetness. Prefers semi-shade but tolerates some sunshine. Hardy to most frosts. Propagate from seed or from cuttings of firm young growth.

O. ovata Hook. f., from Vic, Tas and SA is very similar but differs because its branches arise from an underground rhizome and it has more flowers per head.

OPERCULINA Silva Manso
(from the Latin *operculum*, lid, cover; referring to the lid-like arrangement on the capsules)
Convolvulaceae

Annual or perennial **climbers** with twining **stems**; **leaves** simple, entire, alternate; **inflorescence** a cyme; **sepals** 5, often enlarged in fruit; **corolla** bell-shaped or funnel-shaped, glabrous or with hairy bands, short lasting; **fruit** a woody capsule, the upper layer separating as a lid at maturity.

A small genus of about 25 species; 4 occur in Australia. They have showy flowers but are rarely encountered in cultivation. Propagation is from seed. The woody seed coat must be sliced before sowing. Cuttings of firm young growth may be successful.

Operculina aequisepala (Domin) R. W. Johnson
(with equal sepals)
Qld, WA, NT
2–4 m long   July–Sept; also sporadic
Annual **climber** with twining stems; stems winged; young growth hairy; **leaves** 2–20 cm × 2–25 cm, broadly ovate to reniform, dark green, leathery, glabrous above, sparsely hairy beneath; **flowers** 4–5 cm across, white, funnel-shaped, 1–3 in the leaf axils; **sepals** glabrous, enlarging in fruit; **capsules** 1.5–2 cm across, globose.

Occurs in inland districts where it grows on clay soils in tussock grassland. An interesting climber for cultivation in arid and semi-arid zones. The species has had limited cultivation but seems to adapt readily. Requires well-drained soil and a sunny position. Tolerates moderate to heavy frosts. Propagate from seed.

Operculina brownii Ooststr.
(after Robert Brown, 18–19th-century British botanist)
Qld, WA, NT
3–5 m tall   June–Sept
Robust perennial **climber** with vigorous twining stems; young growth glabrous; **leaves** 6–12 cm × 6–8 cm, cordate to ovate-lanceolate, petioles winged, dark green and shiny above, paler beneath, accuminate; **flowers** 5–6 cm across, white, funnel-shaped, solitary in the upper axils; **capsules** 2–2.5 cm across, globular, brown, mature Feb–April.

A widespread species which extends to northern Australia where it grows in rock outcrops and monsoonal rainforests. Plants become deciduous and die back to the larger stems during the dry season. A vigorous climber which grows actively during the summer. Requires excellent drainage and a sunny position. Needs ample water when in active growth. Frost tolerance is very low. Propagate from seed and possibly also cuttings of firm young growth.

Operculina riedeliana (Oliver) Ooststr.
(after M. Riedely)
Qld
3–8 m tall   March–May
Vigorous perennial **climber** with strong twining stems; young growth glabrous; **leaves** 10–15 cm × 8–12 cm, cordate, bright green, with prominent veins; **cymes** short, densely flowered, in the upper axils; **flowers** 6–8 cm across, cream to pale yellow, trumpet shaped; **petals** with a dense, hairy exterior band; **capsules** 2–3 cm across, globose, rusty brown, held stiffly erect; **seeds** black, mature June–Nov.

Occurs in north-eastern Qld and extends to New Guinea and Indonesia. A component of coastal scrubs and rocky headlands, the plants grow actively during the summer season and become partially dormant

during the dry. With its large handsome leaves, colorful flowers and interesting fruit, the species has good prospects for cultivation in tropical and subtropical regions. It is especially well suited to coastal districts. Requires a sunny position in well-drained soil. Propagate from seed, and possibly also from cuttings of firm young growth.

Operculina turpethum (L.) Silva Manso
(from an Arabic name)
Qld, NT
5–8 m tall   Sept–Oct
Perennial **climber** with twining stems; old stems develop narrow, corky wings; young growth softly hairy; **leaves** 10–20 cm × 10–20 cm, broadly cordate, leathery, dark green above, paler beneath; **racemes** to 10 cm long, slender, axillary; **flowers** 5–6 cm across, white, funnel-shaped; **sepals** hairy; **capsules** 1.5–2 cm across, globular, brittle, mature Feb–March.

A very robust species which may blanket vegetation with its growth. It occurs in tropical regions and also extends overseas. Rarely cultivated because plants need plenty of room to spread. They will need regular pruning. Requires well-drained soil in a sunny location. Propagate from seed, and possibly also from cuttings of firm young growth.

OPHIOGLOSSACEAE Presl.
A small family of ferns consisting of about 70 species in 4 genera distributed in tropical and temperate regions. Most species are herbaceous perennials growing in the ground but a few are epiphytes. The frond is unusual in that it consists of a fertile section or lamina which piggy-backs on a sterile green lamina, these both being supported by a common stalk. The fertile lamina may be partly fused to the sterile lamina or supported on a distinct stalk. Australian genera: *Botrychium, Helminthostachys, Ophioglossum.*

OPHIOGLOSSUM L.
(from the Greek *ophis*, snake; *glossa*, tongue; referring to the shape of the fertile spike)
Ophioglossaceae
Terrestrial or less commonly **epiphytic ferns; rootstock** fleshy, tuberous or creeping; **roots** long, often wiry, often bearing vegetative buds; **fronds** one to many, tufted, complex, consisting of a fertile spike attached to a sterile lamina either directly or on a stalk, the pair being supported on a common stalk; **sporangia** in 2 rows immersed in the fertile spike; **spores** yellow.

A widely distributed genus of about 50 species; about 8 occur in Australia, very few of which are endemic. They are interesting ferns with the terrestrial species having a strict period of dormancy, usually to avoid dry conditions. They can be grown in containers of well-drained sandy mix fortified with crushed leaf mould. Propagation is by natural increase. Spore-raising is impossible as the developing plants have a stage which requires infection by a mycorrhizal fungus.

Ophioglossum coriaceum A. Cunn. = *O. lusitanicum* L.

Ophioglossum costatum R. Br.
(with veins or costae)
Qld, NT   Large Adder's Tongue
10–25 cm tall

Rootstock tuberous, more or less globose, with slender roots; **leaf bases** not persistent; **fronds** 10–25 cm long; **common stalk** 1–5 cm long; **sterile lamina** 3–9 cm × 1–2.5 cm, ovate to elliptic, pale green with a paler median band, apex obtuse to acute; **fertile spike** 1–6 cm long, on a stalk 5–20 cm long.

Occurs in tropical parts of Qld and NT and is also widely distributed from New Guinea to Asia and Africa. Often forms small colonies in moist to wet areas among boulders. Easily grown in a pot of well-drained mix. Requires warmth, bright light and plenty of water when in active growth. Must be kept dry while dormant in winter and spring. Propagate by natural increase.

Ophioglossum gramineum Willd.
(grass-like)
Qld, WA, NT   Narrow Adder's Tongue
5–15 cm tall

Rootstock tuberous, ovoid to globose, with slender roots; **leaf bases** not persistent; **fronds** 5–15 cm long; **common stalk** 1–5 cm long; **sterile lamina** 1–7 cm × 0.1–0.5 cm, linear to narrowly elliptic, dark green; **fertile spike** 0.5–5 cm long, on a stalk 3–8 cm long.

Widely distributed in tropical areas and forming large, loose colonies in woodland, rock outcrops and in open grassland. Plants grow actively during the wet season and become dormant in the dry. Easily grown in a pot of well-drained mix. Requires warmth, bright light and plenty of water when in active growth. Must be kept dry while dormant in winter and spring. Propagate by natural increase.

Ophioglossum intermedium Hook. f.
(intermediate)
5–30 cm tall

Rootstock tuberous, shortly creeping; **leaf bases** not persistent; **fronds** 1–6, erect; **common stalk** and sterile lamina united, 10–30 cm × 0.6–2 cm, dark green, fleshy, apex obtuse; **fertile spike** 3–5 cm long, on a stalk 3–8 cm long fused to the sterile lamina.

Occurs in northern parts of NT and also the Philippines and Indonesia. Grows as a terrestrial in rainforest. Experience in cultivation is limited but it may have similar requirements to *O. costatum*. Propagate by natural increase.

Ophioglossum lineare Schltr. & Brause
(linear)
Qld, NT
3–10 cm tall

Rootstock tuberous, ovoid to globose, with slender roots; **leaf bases** not persistent; **fronds** 3–8 cm tall; **common stalk** 1–5 cm long; **sterile lamina** absent; **fertile spike** 0.5–6 cm long, on a stalk 3–10 cm long.

Occurs in seasonally wet, sandy soil on rock outcrops. This species forms loose colonies. Plants grow actively during the wet season and become dormant during the dry. Not known in cultivation but probably has similar requirements to *O. gramineum*.

Ophioglossum lusitanicum L.
(from Portugal)
All states   Common Adder's Tongue
1–20 cm tall

Rootstock tuberous, erect, with widely spreading roots; **leaf bases** not persistent; **fronds** 1–20 cm long; **common stalk** 0.5–5 cm long; **sterile lamina** 1–6 cm × 0.3–1.5 cm, elliptic, ovate or obovate, dark green, fleshy, apex acute, tapered to the base; **fertile spike** 0.5–8 cm long, on a stalk 1–6 cm long.

Widespread across Australia but restricted to the south in WA and in the NT it is found only in Central Australia. Commonly seen in large, loose colonies on moist sheltered slopes in open forest and woodland, but also occurs on drier sites in grassland, among rocks and in heathland. An extremely variable species. Easily grown in a pot of well-drained mix, but watering must be curtailed when the plants are dormant over summer. Propagate by natural increase.

O. coriaceum A. Cunn. is a synonym.

Ophioglossum pendulum L.
(pendulous)
Qld, NSW   Ribbon Fern
0.5–2 m long

Epiphytic fern; **rootstock** creeping, white, fleshy; **fronds** 0.5–2 m × 1–2 cm, ribbon-like, sides parallel, dark green, often twisted; **fertile spike** fused to a sterile lamina by a stalk, 10–20 cm long, sausage-like, transversely ribbed.

A widespread species which extends to Australia where it is distributed from Cape York Peninsula to the Hunter River in northern NSW. It grows in rainforest, often hanging from clumps of large epiphytic ferns and orchids. An excellent basket fern which requires warm, humid conditions and free air movement. Plants must not be allowed to dry out. A tree fern basket or a wire basket full of elkhorn peat is a suitable medium. It has proved to be difficult to propagate from spore but clumps can be carefully divided.

Ophioglossum petiolatum Hook. = *O. reticulatum* L.

Ophioglossum polyphyllum A. Braun
(with many leaves)
Qld, NSW, Vic, SA, WA, NT   Inland Adder's
                                               Tongue
5–15 cm tall

Rootstock tuberous, erect, with spreading roots; **leaf bases** persistent; **fronds** 9–15 cm long; **common stalk** 1–5 cm long; **sterile lamina** 3–8 cm × 1–2 cm, ovate, lanceolate or elliptic, dark green to yellowish-green, very thick and leathery; **fertile spike** 1–3 cm long, on a stalk 2–6 cm long.

Occurs in inland regions, often growing on seasonally moist clay flats and depressions, but also in drier sites. It has also been recorded growing in coral rubble in southern islands near Gladstone, Qld. This is an interesting species which can be easily grown in a pot of well-drained mix. Requires regular watering when in active growth but must be kept dry when dormant. Propagate by natural increase.

Ophioglossum pendulum   D. L. Jones

Ophioglossum reticulatum L.
(with netted veins)
Qld, NSW, Vic, WA, NT   Adder's Tongue
5–35 cm tall

Rootstock tuberous, erect, with slender roots; **leaf bases** not persistent; **fronds** 5–30 cm long; **common stalk** 3–15 cm long; **sterile lamina** 1–10 cm × 1–4.5 cm, ovate to obovate, dark green, apex obtuse; **fertile spike** 1–6 cm long, on a stalk 5–15 cm long.

Grows among grass in open forest and woodland, often on moist slopes and among rocks. Forms colonies which reproduce vegetatively from the roots. Easily grown in a pot of well-drained mix. Needs water when in active growth but should be kept dry when dormant. Propagate by natural increase.

OPHIUROS Guertn.
(from the Greek *ophio*, snake; *oura*, tail; the racemes resemble snake tails)
Poaceae

Tall, clumping, perennial **grasses**; **rhizome** short creeping; **culms** stout, erect, branched; **leaves** rolled when young, becoming flat; **inflorescence** a panicle of clustered spikes; **spikelets** sessile, embedded in alternate cavities along the rachis.

A small genus of about 7 species, mainly from tropical Asia with 1 or 2 species occurring in Australia. They are rarely cultivated. Propagate from seed.

Ophiuros exaltatus (L.) O. Kuntze
(tall, lofty)
Qld, WA, NT
1.5–2.5 m tall   Dec–Mar

Tall clumping **grass**; **culms** 1.5–2.5 m tall, erect, crowded, branched, glabrous; **leaves** 30–60 cm × 1–1.5 cm, flat or rolled, margins slightly spiny; **panicles** compound, clustered, spreading racemes to 20 cm long; **spikelets** about 0.6 cm long, oblong, obtuse, crowded; **glumes** with longitudinal rows of pits.

Distribution is widespread in tropical Australia and extends to New Guinea. Grows in moist to wet depressions in clay soils. Useful for providing food and shelter for birds. Suitable for cultivation in tropical regions. Requires abundant water. Propagate from seed.

OPILIA Roxb.
(derivation unknown)
Opiliaceae

Shrubs or **climbers**; **leaves** simple, entire, alternate, leathery; **inflorescence** an axillary raceme; **flowers** small, bisexual; **petals**, **sepals** and **stamens** 4 or 5; **fruit** a drupe.

A small genus of 2 species; one in tropical Africa and the other widely distributed and extending to tropical Australia. Being root parasites they are rarely encountered in cultivation. Propagation is from seed, which probably has a short period of viability.

Opilia amentacea Roxb.
(flowering in catkins)
Qld, WA, NT
5–10 m tall   Aug–Oct

Bushy **climber** with robust stems to 20 cm across; **bark** grey, rough; young growth shiny, minutely hairy; **leaves** 5–15 cm × 2–5 cm, ovate to oblong, leathery, dull green, venation prominent; **racemes** to 3.5 cm long, clustered in the leaf axils; **flowers** about 0.3 cm across, green, sweetly scented; **drupes** 1.5–3 cm × 1.2–1.8 cm, oblong, pale yellow to orange-yellow, ripe Oct–Mar.

Widely distributed from Asia to New Guinea and occurring in Australia across the tropical north. The ripe fruit are edible and the bark and leaves can be used as a fish poison. This species lacks significant ornamental appeal and being a root parasite is probably difficult to cultivate. Propagate from seed, which probably has a short period of viability.

OPILIACEAE Valeton

A small family of dicotyledons consisting of 8 or 9 genera and about 30 species, distributed mainly in America, Africa and Asia. They are of interest because many species (perhaps all) are root parasites. They are trees, shrubs or climbers with alternate, simple leaves arranged distichously and drying yellowish-green. The small, bisexual or unisexual flowers are followed by fairly large yellow or orange fruit which have a thin fleshy layer and a large hard seed. Australian genera: *Cansjera*, *Opilia*.

OPISTHIOLEPIS L. S. Smith
(from the Greek *opisthen*, from the back; *lepis*, scale; referring to a scale-like nectary behind the ovary)
Proteaceae

A monotypic genus endemic in Australia.

Opilia amentacea   D. L. Jones

Opisthiolepis heterophylla L. S. Smith
(with variable leaves)
Qld   Blush Silky Oak
8–15 m × 5–8 m   Jan–March
Small to medium **tree** with a bushy habit; **young growth** covered with golden-yellow to brownish hairs; **juvenile leaves** pinnate, 30–50 cm × 10–15 cm, leaflets 5–18, 10–15 cm × 2.5–3.5 cm, lanceolate, dark green above, golden-hairy beneath, margins toothed; **mature leaves** usually simple, 10–15 cm × 7–10 cm, lanceolate, dark green above, golden-hairy beneath, margins entire; **racemes** 10–15 cm long in the upper axils, pendulous; **flowers** about 0.8 cm long, white, hairy; **follicles** 4–6 cm × 2.5–3.5 cm leathery or woody, mature July–Aug.
Occurs in north-eastern Qld where it grows in rainforests from the lowlands to the highlands. An extremely ornamental tree which has outstanding foliage. The pinnate juvenile leaves contrast markedly with the simple mature leaves. Plants have proved to be very amenable to cultivation and can be grown successfully at least as far south as Sydney. Best growth is achieved in well-structured loamy soils. Mulching, regular watering and light fertiliser applications are all beneficial. Tolerates light frosts. Propagate from seed, which has a limited period of viability and is best sown while fresh, or from cuttings of firm young growth.

OPLISMENUS P. Beauv.
(from the Greek *hoplismenus*, armed; referring to the awned glumes)
Poaceae
Annual or **perennial grasses** with a trailing or scrambling habit; **stems** smooth or grooved, branched, often rooting at the nodes; **ligule** hairy, with a tuft of hairs on each side; **leaves** hairy, margins entire or wavy; **panicles** short, with several short racemes; **spikelets** clustered or arranged on one side of the panicle; **glumes** awned.

A small genus of 7 species distributed in the tropics and warm temperate regions. Four species occur in Australia, mostly growing in cool shady forests in tropical and subtropical regions. Although they have some ornamental appeal these grasses are rarely grown. Propagation is from seed or by division of the clumps.

Oplismenus aemulus (R. Br.) Roem. & Schult.
(equalling, rivalling)
Qld, NSW, Vic   Australian Basket Grass;
                                       Wavy Beard Grass
0.1–0.5 m tall   sporadic all year
Weak scrambling or trailing **perennial grass**, sometimes forming mats; **culms** smooth, trailing, rooting at the nodes; **leaves** 2–5 cm × 0.4–1.8 cm, lanceolate, surface often wavy, glabrous, dark green, midrib whitish on underside, margins often wavy; **panicles** erect, sparsely branched, the branchlets to 5 cm long, bearing spikelets on one side only; **spikelets** crowded, 0.25–0.35 cm long; **awn** about 0.7 cm long, flexuose.
A widespread grass which is common in moist shady areas of open forest, moist gullies and on rainforest margins. It is commonly browsed by native animals and stock. This species has some ornamental qualities and can be readily established in a garden but it is inclined to become naturalised and weedy. A useful species for reclamation projects on shady, moist sites. Propagate from seed or by division.

Oplismenus imbecillis (R. Br.) Roem. & Schult.
(weak, feeble)
NT, Qld, NSW, Vic   Creeping Beard Grass
0.1–0.3 m tall   sporadic all year
Weak trailing **perennial grass** forming mats; **culms** trailing, much-branched, rooting at the nodes; **leaves** 3–6 cm × 0.5–0.7 cm, narrowly lanceolate, green, sparsely hairy, margins entire; **panicles** erect, the branches less than 0.15 cm long; **spikelets** 0.2–0.3 cm long, not compressed; **awn** about 0.7 cm long, slender, often wavy.
This species forms sparse to dense mats in shady moist areas of open forest, often on slopes near streams. Clumps are regularly browsed by native animals. This species has limited ornamental appeal and is rarely grown, although it may volunteer in gardens. May be a useful groundcover for rainforest gardens. Propagate from seed or by division.

Oplismenus undulatifolius (Ard.) P. Beauv. var. **mollis** Domin
(with wavy leaves; soft, pliant)
Qld, NSW   Soft Beard Grass
0.1–0.3 m tall   sporadic all year
Weak trailing or scrambling **perennial grass**; **culms** trailing, much branched, rooting at the nodes; **leaves** 4–8 cm × 0.4–1.8 cm, lanceolate, greyish-green, densely hairy, margins entire; **panicles** erect, to 1.5 cm long, the spikelets in clusters; **spikelets** 0.3–0.4 cm long, not compressed; **awns** about 1.5 cm long, straight.
Widely distributed in eastern Qld, northern NSW and New Guinea, these plants usually grow in shady, coastal forests and littoral rainforest. It has potential as

a groundcover in coastal reclamation projects after a canopy is formed. Rarely grown in gardens although it may appear as a volunteer. Propagate from seed and by division.

Orania appendiculata F. M. Bailey =
Oraniopsis apapendiculata (F. M. Bailey) J. Dransf.,
A. K. Irvine & N. W. Uhl

ORANIOPSIS (Becc.) J. Dransf., A. K. Irvine & N. Uhl (resembling the genus *Orania*)
Arecaceae
A monotypic genus endemic in Australia.

Oraniopsis appendiculata (F. M. Bailey) J. Dransf., A. K. Irvine & N. W. Uhl
(with an appendage on the petals)
Qld
5–12 m tall
   Trunk 20–40 cm thick, grey to greenish, prominently ringed; **fronds** pinnate, 2–4 m long, spreading in a graceful crown; **leaflets** 140–160 on each leaf, 50–75 cm × 3–4 cm, linear to lanceolate, dark green above, greyish-white to yellowish-brown beneath, midrib ridged, with 2 green lines; **panicles** 50–75 cm long, much-branched, arising among the leaves, the ends drooping; **flowers** about 0.8 cm across, white; **petals** with a prominent triangular appendage; **drupes** 3–3.5 cm across, globular, yellow.
   This attractive palm has a large crown of graceful fronds. It occurs in north-eastern Qld from south of Cooktown to near Innisfail, and grows in rainforests from the lowlands to about 1400 m altitude. Poorly known in cultivation because plants are slow-growing and need a sheltered position. They can be grown in the tropics (especially highland districts), as well as in the subtropical and warm temperate regions. Requires excellent drainage and responds to mulching, watering and light fertiliser applications. Propagate from seed, which germinates slowly and sporadically, sometimes over several years.
   Orania appendiculata F. M. Bailey is a synonym.

ORCHIDACEAE Juss.   Orchid Family
A huge family of monocotyledons consisting of more than 30,000 species in 600 genera. It is distributed in all the regions of the world, but is very strongly represented in the tropics, and rare in very cold zones. They are perennial herbs which grow as terrestrials, lithophytes, epiphytes or as leafless saprophytes. The epiphytes are found mainly in the tropical zones, whereas the terrestrial species mainly occur in temperate zones.
   Orchid flowers have two distinctive and recognisable floral features. The flowers have 3 sepals and 3 petals in 2 whorls, but the front petal (the lip or labellum) is greatly modified, differing from the others in shape, size and coloration. The labellum often has specialised glands and mimicry devices. The other modification involves the stamens and stigmas which are united into a single fleshy structure called the column. Both the labellum and the column exhibit a great variety of form and structure.
   Many species have complicated relationships with

insects to ensure fertilisation. The seeds of orchids lack any endosperm. They are winged in most species and when ripe are distributed by air currents. They are produced in huge numbers (often millions) from a single capsule.
   Australian orchids consist of more than 1200 species in about 100 genera. The epiphytic species occur mainly in tropical Qld and subtropical parts of Qld and NSW, reducing greatly in number in the temperate zones with only 2 species in Tas. By contrast the terrestrials have proliferated in southern temperate regions with very strong development in south-eastern and south-western Australia.

ORCHIDS
Orchids comprise the largest assemblage of flowering plants in the world and make up approximately ten per cent of the world's flora (see also Orchidaceae entry). Because their flowers exhibit a tremendous diversity in size, colour and shape, they have attracted considerable attention from horticulturists throughout the world. Some growers specialise to the extent of cultivating specific genera or species. This interest — often bordering on the fanatical — has resulted in the establishment of a huge number of specialist nurseries, hybridisation programmes, specialist books and some journals devoted entirely to their taxonomy and culture. It has also resulted in the wholesale stripping of plants from the wild for commercial gain and now many of the more exotic species are reduced to great rarity or are even extinct in the wild.
   Australian orchids, although lacking the flamboyance displayed by many of their large-flowered tropical relatives, have other endearing qualities such as massed flowering, colour variation, fragrance and charm. They have attracted their own band of enthusiastic disciples who study aspects of their biology, taxonomy and cultivation.
   For convenience, Australian native orchids can be considered in two groups based loosely on their growth habit.

Terrestrials
   These orchids grow in the ground. They are very well represented in Australia with about three-quarters of the native species growing this way. Terrestrial orchids abound in temperate Australia; in the tropics and subtropics they are less common; and they are absent from the arid and semi-arid regions of Central Australia. The majority of terrestrial orchids have a tuberous roots system which allows them to die back to avoid extremes of heat and dryness. Most species from temperate regions are dormant in summer. The tropical species grow actively in the wet season, and are dormant in the dry. A few genera of terrestrial orchids, such as *Calanthe*, *Cryptostylis* and *Phaius* are evergreen. Another specialised group lack leaves entirely and are known as saprophytes and they rely upon a symbiotic relationship with fungi for their survival.
   Tuberous terrestrial orchids grow in a range of habitats from exposed coastal headlands (*Caladenia latifolia*, *Pterostylis tasmanica*) to high mountain peaks (*Prasophyllum tadgellianum* is abundant near the top of Mt Kosciuszko). The majority of species are found in

sclerophyll forest and woodland. Other species grow in swamps (*Corybas fordhamii, Diuris drummondii*), and sometimes plants are wholly or partially submerged (*Microtis atrata, M. orbicularis*). Grassland is a particularly favoured environment and a number of species have suffered a drastic reduction in their populations since the widespread conversion of grassland to farmland. A few species extend into semi-arid regions (*Thelymitra sargentii, Pterostylis cobarensis*) and survive by specialised adaptations.

Many terrestrial orchids grow in colonies which may be loose and open or dense and spreading. Orchids with the latter type of growth often reproduce vegetatively — their tubers may increase in number by 2–5 times each year — and thus quickly colonise suitable habitats. *Caladenia flava, C. latifolia, Corybas incurvus, Pterostylis concinna* and *P. nutans* are examples of this orchid type. Some terrestrial orchids flower only after the stimulation of summer bushfires, for instance *Burnettia cuneata, Diuris purdiei* and *Pyrorchis nigricans*. A few terrestrial orchids have a climbing habit (*Erythrorchis cassythoides, Pseudovanilla foliata*).

Epiphytes and Lithophytes

These orchids grow either on trees (epiphytes) or rocks, boulders, cliff faces, and so on (lithophytes or epiliths). Some species adapt readily to both types of substrate. For convenience both types are considered here as epiphytes.

Australian epiphytic orchids are most abundant in north-eastern Qld. They also grow in the seasonally dry forests of the northern parts of the NT and WA, and in the colder and drier forests of temperate regions but in greatly reduced numbers. They are commonest in rainforests and in the moist humid forests along streams and surrounding swamps. Some species are also found in coastal mangroves and on high mountain peaks.

The larger epiphytes form clumps on tree trunks and the larger branches of trees. The smaller species are often twig epiphytes and they grow on the small to tiny branches and branchlets of the outer canopy. Epiphytic orchids always grow in positions which have free and unimpeded air movement. This is an extremely important criteria for their successful cultivation.

Terminology

Although orchids are basically similar to other plants, specific terms have been given to some organs. For definition of these terms, refer to the Glossary.

Cultivation of Terrestrials

The tuberous terrestrial species of orchids have strict cultivation requirements and are mainly grown by dedicated enthusiasts. Very few species (*Microtis parviflora* and *M. unifolia* are possible exceptions) adapt well to general garden culture and so the majority of these species are grown in specific soil mixes in pots. In rare instances, it is possible to duplicate natural conditions in a garden and to have some orchids volunteer from seed, but this is the exception rather than the rule.

Some species of native terrestrial orchids are difficult if not impossible to grow successfully (e.g. some *Caladenia* spp, *Calochilus* spp, and some *Prasophyllum* spp.); others can be grown with strict adherence to specialised requirements (most *Caladenia* spp., *Glossodia* spp.), and some are moderately easy (*Chiloglottis truncata, Corybas incurvus, Pterostylis curta, P. nutans*).

A suitable mix is be very freely drained and should include about one-third decayed leaf mould and some wood shavings. The pots are best maintained on benches in bush houses, shadehouses or cool glasshouses. They must have excellent ventilation. The plants need plenty of water while they are in active growth but when dormant, which is usually over summer, they should be kept dry. While in active growth the plants may respond to weak applications of diluted animal manures or liquid fertilisers low in phosphorus, but these materials should be used with great care.

Many species of terrestrial orchid increase their tuber number by vegetative reproduction and need to be repotted every 2–3 years. Repotting is carried out over the summer while the plants are dormant. The tubers are placed 2–3 cm deep in the new potting mix and are kept dry until the new shoot appears above ground. Some genera are very sensitive to disturbance (e.g. species of *Caladenia*) and should not be repotted.

Pests such as slugs and snails may cause major problems and consequently plants should be regularly protected by baits especially during rainy weather. Thrips and aphids can also be debilitating. Orchids may be affected by viral diseases which cause stunting and growth distortion. The effects of viruses are frequently promulgated by cultivation.

Epiphyte Cultivation

Native epiphytic orchids are very popular with orchid enthusiasts. Some groups, particularly *Dendrobium*, have been extensively hybridised in an attempt to improve on nature.

In the tropics, many native epiphytic orchids can be grown as garden plants either attached directly to suitable trees or grown in pots or on slabs. A few hardy species (e.g. *Dendrobium falcorostrum, D. speciosum, D. tarberi*), can be grown in a semi-protected situation as far south as Melbourne. Species from tropical areas, especially the lowlands, are cold-sensitive and if they are grown in areas with a cold winter climate they will need the warmth of a heated glasshouse to over-winter. By contrast, species from temperate regions do not generally grow and flower satisfactorily in the tropics.

Epiphytic orchids usually prefer fairly bright light with an abundance of air movement and humidity. Many of the larger, upright-growing species are planted in pots of a coarse mixture composed of materials such as softwood bark, charcoal and gravel. Some of the smaller species and those with a pendulous growth habit are grown on slabs of tree-fern fibre, cork or weathered hardwood. Repotting or remounting should be carried out when the pots become excessively full of growth or when the condition of the potting mix or the slab deteriorates significantly. Water should be applied in abundance during the summer months and the plants should be kept drier in winter and spring with only sufficient water to prevent dehydration. Most species respond favourably to regular light applications of liquid fertilisers throughout the growing season.

Some pests can cause significant damage. One of these, the Dendrobium Beetle, is a particular problem in tropical and subtropical regions. Other pests include slugs, snails, aphids and thrips. Viruses and fungus diseases can also be devastating.

Orchid Propagation

Terrestrial orchids are generally difficult to raise from seed except under specialised conditions. The seed is either sown aseptically in tubes or flasks of nutrient media or symbiotically in association with a suitable fungus. Some species can be successfully propagated by sprinkling seed around the base of the parent plant. A number of terrestrial orchids also increase naturally by vegetative means.

The larger epiphytic orchids can be propagated by dividing the clumps and some species also produce aerial growths which are easily removed and grown. They can also be propagated by sowing seed on sterilised media and from tissue culture.

OREOBOLUS R. Br.

(from the Greek *oros, oreos,* mountain; *bolos,* a lump; referring to the tufted growth habit in mountain sites)
Cyperaceae   Tuft-rushes

Dwarf grass-like **herbs**, forming cushion-like tufts; **leaves** narrow, inrolled and broadly sheathing at base; **inflorescence** a one-flowered spikelet, axillary or appearing terminal; **glumes** 3; **perianth** of 6 pointed scales in 2 whorls; **stamens** 3; **style** slender, not thickened at base; **fruit** a smooth ovoid nut.

A small genus of about 8 species distributed in the Pacific region, with 5 species found in Australia. They occur at high elevations. Plants have limited appeal but are suitable for growing amongst rock outcrops and in containers, including miniature gardens. Propagate from seed, which may benefit from stratification at below 4°C for 3–6 weeks, or by division.

Oreobolus acutifolius S. T. Blake —
see *O. oxycarpus* S. T. Blake

Oreobolus distichus F. Muell.
(in two opposite rows)
NSW, Vic, Tas   Fan Tuft-rush
0.03–0.15 m × 0.1–0.5 m   Nov–Feb

Dwarf perennial **herb**, forming dense tufts, cushions or mats from ascending rhizomes; **leaves** 2–6 cm × about 0.1 cm, linear, with broad basal sheath, in two opposite rows, channelled, stiff, pale-to-mid green, margins finely serrated, faintly 3-veined on undersurface, apex prickly pointed; **spikelets** 1-flowered, on more or less terete, erect stalks which elongate to as long as or longer than leaves; **glumes** 3, deciduous; **nut** about 0.2 cm × 0.1 cm, usually brownish, glabrous, bluntly pointed.

This tuft-forming herb is from alpine and subalpine regions where it occurs on the edges of sphagnum bogs and in moist-to-wet herbfields. Prefers well-drained moist soils and plenty of sunshine but may tolerate

some shade. Plants are frost hardy and withstand snowfalls. Propagate from seed or by division of clumps.

O. pumilio R. Br., the Alpine Tuft-rush, from NSW, Vic and Tas is similar but much smaller. Its leaves are up to 2.5 cm long, 5–6-veined on the undersurface and spirally arranged.

O. oligocephalus F. Muell. from the mountain plateaux of south-western Tas also develops as a dense mat of foliage. The leaves are spirally arranged and with 6–8 veins on both surfaces.

Oreobolus oligocephalus W. M. Curtis —
see *O. distichus* F. Muell.

Oreobolus oxycarpus S. T. Blake
(sharp fruit)
NSW, Vic, Tas
0.03–0.1 m × 0.1–0.3 m   Nov–Feb

Dwarf perennial **herb**, forming dense tufts or cushions from ascending rhizomes; **leaves** 0.25–0.7 cm × about 0.8 cm, linear with broader basal sheath, spirally arranged, nearly flat, pale green, single conspicuous vein on undersurface, apex pointed; **spikelets** 1-flowered, on stalks about as long as the leaves; **glumes** 3; **nut** to about 0.25 cm × 0.1 cm, narrow, whitish, grey or brownish, glabrous, sharply pointed.

This species occurs at high elevations in sphagnum bogs and in wet seepages. It has similar requirements and tolerances to *O. distichus.* Propagate from seed or by division of clumps.

O. acutifolius S. T. Blake has affinity to *O. oxycarpus* but it differs in having acutely pointed leaves and a nut which is broad and flattened. It is a small, rather inconspicuous species from the mountains of southern and western Tas.

Oreobolus pumilio R. Br. — see *O. distichus* F. Muell.

OREOCALLIS R. Br.

The Australian species formerly included in this genus are now placed in *Alloxylon* P. H. Weston & Crisp. *Oreocallis* is restricted to Peru and Ecuador in South America.

OREODENDRON C. T. White

(from the Greek *oros, oreos,* mountain; *dendron,* a tree)
Thymelaeaceae

A monotypic genus endemic in Qld. It is very similar to *Phaleria* and may in the future be included in that genus.

Oreodendron biflorum C. T. White
(with two flowers or flowers in pairs)
Qld   Nov–Jan
3–5 m × 2–4 m

Medium to tall **shrub** with an open habit; young growth glabrous; **leaves** 2.5–8 cm × 1.5–4 cm, ovate, dark green; **flowers** about 1.5 cm long, borne in pairs in the upper axils, pink with white lobes, tubular at the base, probably fragrant; **stamens** slightly exserted; **fruit** not seen.

A little-known species from north-eastern Qld where it grows in highland rainforests on the summit of

Thornton Peak. It is unknown in cultivation but as it closely resembles an ornamental species of *Phaleria* it probably has good prospects. It would be suited to subtropical and temperate regions. Propagation is prob-ably from fresh seed and perhaps also cuttings.

OREOMYRRHIS Endl.

(from the Greek *oros, oreos*, mountain; *myrrhis*, for the plant known as myrrh)
Apiaceae (alt. *Umbelliferae*)   Carraways

Tufted perennial **herbs; leaves** pinnate to compound, aromatic when crushed; **inflorescence** terminal, many-flowered umbels, on erect stalks, simple or branched; **flowers** small; **sepals** absent; **petals** 5, very small, white or reddish, with narrowed and incurved apex; **stamens** 5; **fruit** a schizocarp of 2 mericarps.

A genus of about 25 species which is mainly confined to the southern hemisphere. Most species occur in South America, New Zealand, and in Australia where there are 7 species. All Australian species are endemic.

Oreomyrrhis species are generally found at higher altitudes and reach their best development in the Australian Alps and Tasmania. They are not renowned as plants for cultivation but have potential for use in cool temperate regions. Many species have interesting, divided foliage which may be silvery (e.g. *O. argentea*), but they lack outstanding flowers. They are suitable for containers and possibly in rockeries, massed plantings or informal drifts such as in perennial borders. It could be worth trying some species as groundcovers incorporated into lawns, as is done with chamomille, because this would release the foliage fragrance. Plants may be useful for rehabilitation of eroded areas within their natural range. Their longevity in cultivation is not thoroughly documented but they may live for 2–3 years or longer.

Propagate from seed, which may benefit from stratification at 4°C for 4–6 weeks, or from cuttings of species with spreading stems.

Oreomyrrhis argentea (Hook. f.) Hook. f.
(silvery)
NSW, Vic, Tas   Silvery Carraway
prostrate × 0.1–0.3 m   Oct–Jan

Dwarf perennial **herb** with a short tap root, young growth silvery-hairy; **leaves** pinnate, 2.5–8 cm × 0.8–2 cm, with prominent, long, thick stalk, in a basal rosette, silvery-hairy, often appearing whitish, slightly aromatic when crushed; **leaflets** deeply pinnately lobed; **flowers** small, white, 15–25 per umbel, on a stout, erect, hairy stem to about 10 cm tall; **fruit** to about 0.5 cm long, becoming purplish-brown, glabrous or faintly hairy.

An attractive silvery-foliaged plant which occurs in montane grasslands often near sphagnum bogs. Prefers a moist but freely draining site and a fairly sunny aspect but also tolerates some shade. Hardy to most frosts and snowfalls. Recommended for perennial borders, rockeries and containers. Propagate from seed.

Oreomyrrhis brevipes E. M. Mathias & Constance
(short stalks)
NSW, Vic   Rock Carraway
0.1–0.4 m × 0.2–0.5 m   Oct–Jan

Oreomyrrhis argentea × .45

Dwarf perennial **herb,** with a short taproot; **stems** spreading to ascending, many, branched, covered in soft hairs; **leaves** 2–8 cm × 1.5–4 cm, pinnate, with stout stalk, sheathing at base, velvety-hairy, somewhat greyish-green, strongly aromatic; **leaflets** once or twice divided with narrow segments; **flowers** small, white to deep magenta, in umbels, on erect, hairy stems to about 5 cm tall; **fruit** to about 0.6 cm long, oblong, glabrous, ageing dark brown to nearly black.

An attractive plant with velvety foliage. This species is sparsely distributed in the Australian Alps, extending from the Kosciuszko region in NSW to Mt Skene in north-eastern Vic. It is usually found in rocky sites. Needs excellent drainage and a fairly sunny location. It has potential for use in rockeries and containers and

Oreomyrrhis pulvinifica   D. L. Jones

should succeed also in general plantings. Propagate from seed or stem cuttings.

O. sessiliflora Hook. f., is similar but lacks the spreading stems and has greenish leaves which are not velvety.

Oreomyrrhis ciliata Hook. f.
(fringed with fine hairs)

NSW, Vic, Tas   Bog Carraway
0.1–0.5 m × 0.1–0.3 m   Oct–Jan

Dwarf perennial **herb** with a slender taproot; **stems** very few or absent; **leaves** 1–9 cm × 0.5–3 cm, pinnate, in a basal rosette, stalks may be shorter or longer than the blade, with reddish-purple sheathing bases, glabrous except for ultimate leaflet, very strong, sweet fragrance when crushed; **leaflets** deeply lobed or pinnate, with very fine segments which have hairy margins; **flowers** small, white, in umbels on erect, densely hairy stems; **fruit** to about 0.5 cm long, ovoid to oblong.

This herb dwells in subalpine and alpine bogs and is distributed from Barrington Tops, NSW, to Tas. It has finely divided, attractive foliage but some people may find the sweetly fragrant foliage somewhat overpowering. Worth trying in bog gardens and other moist sites. Needs plenty of sunshine. May adapt to container cultivation if kept in a saucer of water. Propagate from seed.

Oreomyrrhis eriopoda (DC.) Hook. f.
(woolly stalks)

NSW, Vic, Tas, SA   Australian Carraway
0.05–0.5 m × 0.1–0.4 m   Sept–Jan

Dwarf perennial **herb** with a stout or slender taproot; **stems** absent or very few near base; **leaves** 2.5–20 cm × 1–8 cm, pinnate, hairy, usually in a basal rosette, strongly aromatic; **leaflets** deeply lobed or pinnate, with oblong to narrow segments; **flowers** small, white or tinged with pink, in umbels on stout hairy stalks, usually displayed well above the leaves; **fruit** to about 0.6 cm long, narrowly oblong, broadest near base, glabrous.

A wide-ranging species with its distribution extending from the north coast of NSW to the Lofty Range, SA (where it is very rare). It usually occurs on mountains and tablelands, growing in herbfields and heath. Plants require well-drained soils and will do well in sunny or semi-shaded sites. Hardy to most frosts and snowfalls. It has potential for use in borders, rockeries and containers. Roots are edible and the foliage has a faint carraway fragrance. Propagate from seed.

O. gunnii E. M. Mathias & Constance is from southern Tas where it has a scattered distribution. It is similar to *O. eriopoda* but differs in having hairy fruits which are broadest near the middle. Plants need to be introduced to cultivation as a conservation strategy.

Oreomyrrhis gunnii E. M. Mathias & Constance —
see *O. eriopoda* (DC.) Hook. f.

Oreomyrrhis pulvinifica F. Muell.
(cushion-like)

NSW, Vic   Cushion Carraway
0.02–0.1 m × 0.1–0.3 m   Oct–Jan

Dwarf perennial **herb**, mat-forming from a short-creeping rootstock; **stems** much-branched; **leaves** 0.5–2 cm × up to 1.5 cm, pinnate, glabrous except for margins of the sheathing bases and sometimes the ultimate segments; **leaflets** all deeply lobed or sometimes lower ones lobed and upper ones entire; **flowers** small, white, in umbels on erect, hairy stalks; **fruit** to about 0.4 cm × 0.15 cm, narrowly oblong.

This is the smallest species of the genus. It develops into a dense, low cushion-like plant in the Australian Alps where it is found mostly above the treeline. It has potential for use in miniature gardens, containers and rockeries. Plants require moist, well-drained soils and plenty of sunshine. They may adapt to bog garden conditions. Hardy to most frosts and snowfalls. Propagate from seed or from cuttings of firm young growth.

O. ciliata has similarities but differs in having glabrous leaf-sheaths and it is usually much taller.

Oreomyrrhis sessiliflora Hook. f.
(sessile flowers)

Tas
0.05–0.1 m × 0.1–0.2 m   Nov–Feb

Dwarf perennial **herb**, with a stout taproot; **stems** usually absent or few, erect and leafy; **leaves** 1–9 cm × 0.5–3 cm, pinnate, usually in a basal rosette, typical carrot fragrance; **leaflets** deeply pinnate or divided, dull green, glabrous or margins finely hairy; **flowers** small, white, in umbels on stout, simple or branched stalks, well above the foliage; **fruit** to 0.5 cm long, ovoid.

This species is from the north and western regions of the Central Plateau where it grows on mountain summits. Evidently not cultivated, it will need moist but freely draining soils and plenty of sunshine. It has fragrant foliage which gives it potential for use in gardens and containers and as a culinary herb. Propagate from seed.

ORITES R. Br.
(from the Greek *oros, oreos,* mountain; *ites,* belonging to; or *oreites,* a mountaineer; referring to the mountain habitats of species)
Proteaceae

Shrubs or **trees**; **leaves** alternate, small to large, entire, toothed or lobed; **inflorescence** axillary or terminal spikes; **flowers** bisexual, white or cream, paired, often heavily fragrant; **perianth** cylindrical in bud, 4 narrow-linear segments, free to base; **anthers** one near end of each segment; **style** slender, straight, with 4 nectar glands at base; **fruit** a leathery to woody, boat-shaped follicle which opens at maturity; **seeds** winged, 1 or 2 per follicle.

A small genus of about 10 species. All are endemic in Australia except for one species in the Andes Mountains in Chile. Within Australia they are in disjunct regions, with 4 species in Queensland or northeastern NSW rainforests, 1 species in the mountain ranges of south-eastern NSW and north-eastern Vic, and there are 4 species at higher elevations in Tas.

All have ornamental qualities with well-displayed spikes of cream flowers, most of which are sweetly scented. Some species have decorative leaves especially in their juvenile stage, and they have potential as indoor plants (eg. *O. excelsa* and *O. fragrans*). All species are cultivated to a limited degree but they often prove difficult to maintain for lengthy periods. They are usually slow-growing. Their response to long-term application of fertilisers is not well known and needs to be studied further because this may provide information on how to obtain better growth. They do not seem to be very susceptible to pest attack but some are susceptible to attack from *Phytophthora cinnamomi* (see Volume 1, page 168, for control methods). Plants need good drainage and most species prefer semi-shaded sites. Young plants, especially rainforest species, need protection from hot sun and strong wind.

Propagation is either from seed, which germinates readily without any pre-sowing treatment, or from cuttings of firm young growth, which may be slow to form roots. Application of rooting hormone can be beneficial.

Orites acicularis (R. Br.) Roem. & Schult.
(needle-like)

Tas	Yellow Bush
0.5–1.5 m × 0.5–2 m	Nov–Feb

Dwarf to small **shrub**; **branches** many, ascending to erect but sometimes spreading; **leaves** 1.5–3 cm × about 0.2 cm, linear-terete, ascending to erect, hard, shallowly grooved on upper surface, greenish-yellow, glabrous, apex pungent-pointed; **spikes** to about 1.5 cm × 1 cm, terminal or axillary, partially hidden by foliage; **flowers** about 0.5 cm across, creamy-white, sometimes with pinkish tips, sweetly-scented; **style** reddish; **follicles** to about 1.5 cm long, oblong, flattened, brown.

Yellow Bush is common in mountain vegetation over much of Tas where it grows in exposed and sheltered sites. Not highly ornamental, it is nevertheless valuable as a traffic control plant because of its highly prickly nature, and it can be used for background planting. It needs good drainage and may need supplementary watering during extended dry periods. A sunny or semi-shaded aspect is suitable. Plants tolerate strong winds and are hardy to most frosts and snowfalls. Propagate from seed or from cuttings of firm young growth.

Orites diversifolia R. Br.
(leaves of more than one kind)

Tas	
1–4 m × 0.6–1.5 m	Oct–Dec

Small to medium **shrub** with hairy young growth; **branches** few, ascending to erect or, rarely, spreading; **leaves** 3–12 cm × 0.3–1.8 cm, lanceolate, narrow-oblong or ovate, ascending to erect, short-stalked, tapered to base, rigid, flat, deep green and shining above, paler and glaucous or rusty below, margins recurved and entire or coarsely toothed, apex pointed; **spikes** to about 6 cm × 1.5 cm, in upper axils; **flowers** about 0.5 cm across, creamy-white to greenish-white, sweetly scented, usually profuse and very conspicuous; **follicles** 1.8–2.4 cm long, dark brown, more or less glabrous.

This outstanding if somewhat slow-growing and open-branched species is common in low altitude temperate rainforests and on mountains to about 1200 m altitude. It has had limited cultivation and deserves to be better known. Best suited to a cool temperate climate. Requires well-drained, moisture-retentive organic soils. A semi-shaded site is preferable. Plants may require supplementary watering during extended dry periods. They are suitable for general planting and worth trying in containers. Hardy to moderate frosts. Response to pruning is not recorded but tip pruning from an early stage should promote branching. Propagate from seed, which germinates readily, or from cuttings of firm young growth.

Orites excelsa R. Br.
(tall, noble)

Qld, NSW	Mountain Silky Oak
8–30 m × 4–8 m	June–Sept

Small to tall **tree**; young growth with reddish to rusty hairs; **bark** brown to grey, faintly scaly; **branches** spreading to ascending; **branchlets** rusty-hairy near tips; **juvenile leaves** simple or 3–5-lobed, often acutely toothed; **mature leaves** 9–20 cm × up to 8 cm, usually lanceolate, prominently stalked, dark green and shining above, whitish to grey below, margins usually entire or sometimes lobed or toothed, apex pointed; **spikes** to about 10 cm × 1.5 cm, axillary; **flowers** to about 0.6 cm across, white, sweetly scented, sometimes profuse and conspicuous; **follicles** 2–2.5 cm long, boat-shaped when open, ripe Feb–July.

This tallest species of the genus occurs in the rainforests of north-eastern NSW and south-eastern Qld, with a disjunct distribution near Cairns, Qld. It is a pioneer plant in disturbed sites and may colonise pasture. Plants have decorative foliage and although the floral display may be profuse during some seasons the flowers are not long-lasting. The tree itself may also be short-lived. *O. excelsa* has proved adaptable to a range of well-drained, acidic soils and is suitable for tropical and temperate regions. It does best in an open, sunny site, but needs protection from hot sun and strong winds when young. Moderate frosts can damage foliage. Plants may need supplementary watering during extended dry periods. Best suited to large gardens and parks but can be successful as an indoor plant while young. It produces fairly hard, pinkish wood which is well regarded for a number of uses including cabinet-making, interior lining and shingles. Propagate from fresh seed, which germinates readily, or from cuttings of firm young growth.

Orites fragrans F. M. Bailey
(sweet smelling)

Qld	Fragrant Silky Oak
3–10 m × 2–4 m	Aug–Sept

Medium to tall bushy **shrub** or small **tree**, with reddish young growth; **branches** spreading to ascending; **juvenile leaves** to 22 cm × 6 cm, prominently 3-lobed, glabrous, margins coarsely toothed; **mature leaves** 5–10 cm × 2.5–5 cm, ovate, prominently stalked, dark green and shining above, greyish below, glabrous, margins entire or shallowly lobed, apex bluntly pointed; **spikes** to about 10 cm × 2 cm, often clustered in upper axils, prominent bracts are crimson and deciduous; **flowers** to about 1 cm long, white, sweetly fragrant,

Orites excelsa

D. L. Jones

often profuse and very conspicuous; **follicles** to about 2.5 cm long, greyish.

An outstanding bushy species with ornamental foliage, flowers and fruits. It occurs in highland rainforests of the Bellenden Ker Range near Cairns. Uncommon in cultivation, it is suited to tropical and subtropical regions but may prove useful as an indoor plant, thus making it available also to people in temperate regions. Soils need to be well-drained and rich in organic matter. A semi-shaded site should give best results. Young plants need protection from hot sun and strong winds. May need supplementary watering during extended dry periods. Propagate from fresh seed, or possibly from cuttings of firm young growth.

Orites lancifolia F. Muell.
(lance-shaped leaves)
NSW, Vic   Alpine Orites
1–2 m × 1.5–2.5 m   Nov–Jan
Small **shrub** with pale to rusty-hairy young growth; **branches** many, spreading to ascending; **leaves** 1.5–3 cm × 0.4–1 cm, oblong-lanceolate to oblong-elliptic, spreading to ascending, short-stalked, thick, light green, becoming glabrous and shining above, prominent netted venation, margins usually entire or sometimes a few small teeth near the pointed apex; **spikes** 2–5 cm × about 1 cm, terminal or in upper axils; **flowers** to about 0.6 cm long, creamy-white, sweetly scented, often profuse and conspicuous; **follicles** about 2 cm long, boat-shaped, initially with grey to rusty hairs but becoming glabrous.

Confined to the Australian Alps and surrounding mountains where it grows as a bushy shrub with foliage often to ground level. Difficulty is often experienced in maintaining this species in cultivation. It needs freely draining soils and, although it tolerates some sunshine, a semi-shaded site is preferred. Hardy to most frosts and snowfalls. Needs low temperatures to flower well. May need supplementary watering during extended dry periods. Propagate from fresh seed, which may benefit from stratification at 4°C for 3–6 weeks. Cuttings of firm young growth can be slow to form roots.

Orites milliganii Meisn.
(after Dr Joseph Milligan, 19th-century surgeon and naturalist, Tas)
Tas
0.5–6 m × 0.5–3 m   Nov–Jan
Dwarf to small **shrub**, rarely a medium to tall **shrub**; **branches** ascending to erect, stiff; **leaves** 1.5–3 cm × 0.8–2 cm, ovate to obovate, spreading to ascending, short-stalked, thick, stiff, dark green above, paler below, glabrous, margins coarsely toothed with pungent points; **spikes** 2–6 cm × 1.5–2 cm, terminal, somewhat loosely arranged; **floral bracts** brown, densely hairy, deciduous; **flowers** to about 0.8 cm across, white, glabrous, sweetly perfumed, often profuse and very conspicuous; **follicles** to about 1.5 cm long, dark brown, glabrous.

An impressive species which is slow-growing and rarely likely to reach 3 m tall in cultivation. It occurs in scattered locations above 1200 m altitude in the southwestern region. Growers have found it difficult to establish in cultivation. Plants are best suited to a cool temperate climate with a semi-shaded aspect and require well-drained soils which are moist for most of the year. Hardy to most frosts and snowfalls. Very prickly and therefore may have application for foot traffic control. Propagate from seed or from cuttings of firm young growth.

Orites racemosa C. T. White =
Sphalmium racemosum (C. T. White) B. G. Briggs,
B. Hyland & L. A. S. Johnson

Orites lancifolia × .5

Orites milliganii T. L. Blake

Orites revoluta R. Br.
(margins rolled backwards)
Tas
0.5–2 m × 1–2.5 m Nov–Jan

Dwarf to small **shrub**, with rusty-hairy young growth; **branches** many, spreading to ascending; **branchlets** rusty-hairy; **leaves** 1–2 cm × 0.2–0.4 cm, linear to elliptical, spreading to ascending, more or less sessile, thick, leathery, dark green and becoming glabrous above, margins revolute nearly concealing the rusty-hairy undersurface, apex bluntly pointed; **spikes** about 0.5 cm across, creamy white, sweetly-scented, often profuse and conspicuous; **follicles** 1.2–2 cm long, brown, hairy.

A widespread and plentiful species which occurs on mountain plateaux from 600–1200 m altitude. The rusty new growth and flower-spikes are attractive. Cultivated to a limited degree, it needs a cool temperate climate, well-drained soils which are moist for most of the year, and sites which are semi-shaded or slightly sunny. Plants are hardy to most frosts and snowfalls. They should respond well to pruning. May need supplementary watering during extended dry periods. Hardy to most frosts and snowfalls. Useful for low informal hedging and general planting. Propagate from seed or from cuttings of firm young growth.

ORMOCARPUM Beauv.
(from the Greek *ormo*, cord or chain; *carpos*, fruit; referring to the use of seeds for necklaces)
Fabaceae (alt. Papilionaceae)

Shrubs or small **trees**, glabrous or hairy; **leaves** pinnate, often clustered on short shoots; **stipules** persistent, striate; **leaflets** often small, alternate; **inflorescence** an axillary raceme, rarely branched; **flowers**

pea-shaped, yellow or white, or sometimes streaked with purple; **fruit** indehiscent, breaking up into articles.

A small genus of about 20 species which is distributed from Africa and Madagascar to Asia and Polynesia. One widespread species extends to Australia. Propagation is from seed, which has a hard testa and requires pre-sowing treatment with boiling water. Cuttings may also be worth trying.

Ormocarpum cochinchinense (Lour.) Merr.
(from Cochin, China)
Qld
4–6 m × 2–3 m Oct–June

Medium to tall **shrub**; **leaves** pinnate; **leaflets** 9–15, 1–2 cm × 0.3–0.5 cm, broadly oblong, dark green, blunt; **racemes** 1–3 cm long, axillary, 1–3-flowered; **flowers** about 1 cm long, pea-shaped, yellow; **pods** 1.5–2.5 cm × 0.2–0.3 cm, linear, breaking up into small sections, each containing 1 or more seeds.

A little-known shrub which occurs in open forest on the Cape York Peninsula where it grows in well-drained sandy or gravelly soils. Apparently not in cultivation, but it has potential for use in the tropics and subtropics. Requires unimpeded drainage in a sunny position. Propagate from treated seed. Cuttings would also be worth trying.

ORMOSIA G. Jacks.
(from the Greek *ormo*, cord or a chain; referring to the use of seeds for necklaces)
Fabaceae (alt. Papilionaceae)

Trees; **leaves** alternate, pinnate, leathery; **leaflets** large; **inflorescence** usually terminal panicles; **flowers** pea-shaped, purple to pink; **stamens** 10, free, unequal; **pods** flat, usually thick, leathery; **seeds** circular, red or red and black.

A tropical genus of about 100 species distributed in eastern and southern America, eastern Asia and

Orites revoluta × .5

115

Australia where there is one species, endemic in north-eastern Qld.

Propagation is from hard-coated seeds which need pre-sowing treatment (see Volume 1, page 205).

Ormosia ormondii (F. Muell.) Merr.
(after Hon. Francis Ormond)
Qld — Yellow Bean
6–10 m × 4–6 m — Dec–Jan
Small bushy **tree**; **leaves** pinnate, to about 35 cm long; **leaflets** to 10 cm × 4 cm, ovate to elliptic, tapering to short stalk, dark green, leathery, apex pointed; **panicles** to about 25 cm long, loosely arranged; **flowers** pea-shaped, to about 3 cm long, pinkish to red, with prominent stamens, often profuse and conspicuous; **pods** to about 10 cm × 2 cm, woody; **seeds** about 1 cm long, red, ripe May–June.

This attractive tree from the Cook District has colourful flowers and ornamental foliage and pods. It has potential for use in tropical and subtropical areas and should do best in sheltered sites (especially when young), with moist, friable soils. Suitable for background planting, screening and providing shade. Mulching is beneficial and plants should respond well to slow-release fertilisers and supplementary watering during extended dry periods. Propagate from pre-treated seed.

Previously known as *Podopetalum ormondii* F. Muell.

OROBANCHACEAE Vent.

A small family of dicotyledons consisting of about 180 species in 13 genera. They are mainly found in temperate regions of the northern hemisphere. These plants are small root parasites which have reduced, scale-like leaves lacking chlorophyll. Their roots develop suckers when in contact with the roots of other plants, and the seeds of some species will only germinate when in contact with the root of a suitable host plant. The bisexual, 2-lipped flowers are arranged in erect spikes. Each fruit contains thousands of tiny seeds which are spread by wind. The family is represented in Australia by a single native species of *Orobanche* and another which is widely naturalised.

OROBANCHE L.

(from the Greek *orobos*, a vetch; *anche*, strangle; referring to the parasitic habit)
Orobanchaceae (alt. *Scrophulariaceae*)

Annual or perennial **herbs**, lacking chlorophyll and parasitic on the roots of other plants; **leaves** reduced, brownish, scale-like; **inflorescence** a terminal spike; **bracts** conspicuous; **sepals** 4 or 5, unequally united; **corolla** tubular, 2-lipped, the upper lip 2-lobed, lower lip 3-lobed; **stamens** 4; **fruit** a capsule; **seeds** tiny.

A genus of about 140 species distributed in temperate and tropical regions. A single species is native to Australia, and another, *O. minor* Smith, is widely naturalised. The seeds, which are tiny, must come into contact with the root of a suitable host plant before germination can occur.

Orobanche cernua Loefl.
var. **australiana** (F. Muell. ex Tate) J. M. Black ex Beck
NSW, Vic, Tas, SA, WA — Australian Broomrape
0.15–0.45 m tall — Aug–Nov

Herb with erect stems 15–45 cm tall, most parts hairy; **stems** undivided, sturdy; **leaves** 0.8–2 cm × 0.2–0.5 cm, broadly ovate to somewhat triangular, erect and partially stem-clasping; **spikes** about half the height of the stems; **floral bracts** often larger than the leaves; **sepals** 1–1.5 cm long, appressed to the base of the corolla; **corolla** 1.5–2 cm long, tube cylindrical, white, lobes purple, upper lip 2-lobed, lower lip 3-lobed, sparsely hairy, the hairs in clusters; **capsules** about 1 cm long, enclosed in the persistent corolla.

Widely distributed in inland regions, particularly in sandy soil in stream beds and other drainage lines. Occurs in small colonies around shrubby species of Asteraceae on which it is parasitic. This plant has interest value but is cultivated rarely if at all. May be introduced to a garden by sowing the seed within the root zone of a suitable plant such as a *Senecio* or *Ixiolaena* species.

ORTHOCERAS R. Br.

(from the Greek *orthos*, straight, upright and *ceras*, horn; referring to the horn-like lateral sepals)
Orchidaceae

Terrestrial **orchids** with a subterranean tuberous root system; **leaves** linear, in a basal tuft; **inflorescence** a raceme; **flowers** dull-coloured; **lateral sepals** slender, elongated and horn-like; **labellum** lobed; **fruit** a capsule.

A small genus of about 4 species occurring in New Caledonia, New Zealand and Australia. Two species are found in Australia — one is undescribed. Plants are generally difficult to grow and are rarely encountered in cultivation.

Orthoceras strictum R. Br.
(upright, erect)
Qld, NSW, Vic, Tas, SA
0.3–0.6 m tall — Oct–Feb
Leaves 2–5 in a basal tuft, 10–30 cm × 1–0.3 cm, linear, grass-like, erect, channelled, bright green; **inflorescence** to 60 cm tall, stiff, bearing 1–9 flowers in a terminal raceme; **flowers** about 1 cm across, yellowish-green, brownish or blackish; **dorsal sepal** about 1.2 cm long, broadly ovate, concave, closely hooding the column; **lateral sepals** to 0.4 cm × 0.1 cm, filiform, usually erect, sometimes spreading; **petals** about 0.8 cm long, hidden within the dorsal sepal; **labellum** 1–1.2 cm long, 3-lobed, with a conspicuous yellow central patch, recurved.

A widely distributed but variable species which grows in moist grassland, depressions and swamps, and less commonly in well-drained sites in open forest. The unusual flowers always arouse interest when noticed. Plants are grown mainly by orchid enthusiasts but unfortunately they are not easily maintained in cultivation. Usually plants lose vigour over 3 or 4 years and then they die. The successful raising of the species from seed may provide some answers to its cultivation.

ORTHOSIPHON Benth.

(with a straight tube)
Lamiaceae

Perennial **herbs** or soft **shrubs**; young growth hairy; **leaves** simple, decussate, entire or toothed; **flowers** showy, in whorls, in terminal, dense or leafy racemes; **corolla** tubular, straight or curved; **upper lobes** united into a broad 3–4-lobed lip; **stamens** 4, exserted; **fruiting calyx** enlarged, reflexed; **fruit** a nut.

A genus of about 30 species distributed from Africa and Madagascar to Asia and Polynesia. One species is native in Australia. It is highly ornamental and has become popular in cultivation mainly in tropical and subtropical regions. Propagation is from seed or more usually from cuttings.

Orthosiphon aristatus (Blume) Miq.
(awned)

Qld	Cat's Whiskers; Cat's Moustache
0.5–1 m × 0.5–1 m	sporadic all year

Dwarf to small, straggly to bushy shrub; **young growth** often purplish; **leaves** 3–5 cm × 2–3 cm, ovate to ovate-lanceolate, decussate, dark green, margins coarsely toothed; **racemes** mostly terminal, 4–10 cm long; **flowers** in whorls of 4–6, white or mauve, profuse and showy; **corolla tube** 0.8–1 cm long; **stamens** to 3 cm long, spreading; **fruiting** calyx about 1 cm long.

This shrub occurs in north-eastern Qld where it grows near streams. Plants are often found in alluvial soils which may be prone to flooding. A soft-looking shrub, it produces showy displays of flowers and has proved to be an excellent garden plant for tropical, subtropical and warm temperate regions. Very fast-growing and adaptable to a sunny position or shade. Requires regular pruning, especially after flowering, to prevent the plants from becoming straggly. Responds to light applications of fertiliser, and watering during dry periods. White and mauve-flowered variants are available. Propagate from cuttings of firm young growth, which strike readily.

Orthothylax glaberrimus (J. D. Hook.) Skottsb. =
Helmholtzia glaberrima (Hook. f.) Carvel

ORTHROSANTHUS Sweet
(from the Greek *orthros*, morning; *anthos*, flower; referring to the flowers which are at their best in the morning)

Iridaceae   Morning-flags; Morning Iris

Tufted perennial **herbs**, rarely with a short-creeping rhizome; **leaves** basal, linear, flat, erect to arching; **inflorescence** axillary panicle with spathes at base of branches or rarely single flowers, on erect, terete to flattened scape; **flowers** bisexual, with spreading petals and sepals, nearly sessile, short-lived, pale blue, blue or, rarely, white; **sepals** and **petals** 3, alike; **stamens** exserted, erect; **style** trifid; **fruit** a capsule, ovoid, 3-angled; **seeds** brown.

A small genus of 7 species; 4 are endemic in Australia and the others occur in South America. Three of the Australian species are confined to south-western WA while *O. multiflorus* is widely distributed from south-western Vic to southern WA. Plants are found mainly in coastal and adjacent regions and occur in a variety of freely draining soils which are usually wet to moist in winter and spring.

Morning Iris are highly regarded for their ornamental qualities and they are popular in cultivation.

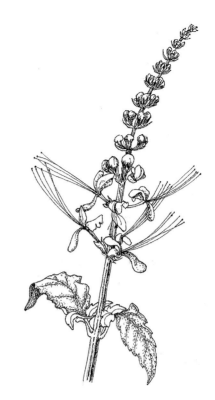

Orthosiphon aristatus × .6

Although individual flowers usually last for as little as one day, the long flowering period of most species makes them very appealing. Plants have tufting, grass-like foliage which provides an excellent alternative in terms of form and texture in plantings. This species is also an ideal plant around pools and boulders. They grow well in containers too. Old plants are rejuvenated by pruning to ground level or burning clumps.

Plants are rarely prone to pest and disease problems. Slugs, snails and some caterpillars may eat young leaves and some caterpillars may disfigure flower buds (see Volume 1 for control methods).

Propagation is from seed, which usually does not require any pre-sowing treatment and generally begins to germinate 25–60 days after sowing. Division of clumps is easily undertaken as they generally break into separate divisions readily.

Orthrosanthus laxus (Endl.) Benth.
(loose)

WA	Morning Iris
0.2–0.7 m × 0.2–0.5 m	Aug–Nov

Tufted perennial **herb**; **leaves** 10–45 cm × 0.3–0.5 cm, grass-like, flat, light-to-mid green, glabrous or faintly hairy, margins rough; **inflorescence** on scape to 70 cm tall, few-branched, 2–4 slender-stalked flowers per branch, with ovate spathes to about 1.8 cm long; **flowers** to about 4 cm across, pale to bright blue, rarely white, very conspicuous; **capsule** 1.2–1.5 cm long, 3-angled, apex pointed.

Orthrosanthus laxus   W. R. Elliot

Orthrosanthus multiflorus   W. R. Elliot

A charming, clumping herb which occurs over a wide range in south-western WA. It grows in a variety of soils which are often moist for extended periods. Adapts well to cultivation in a sunny or semi-shaded site although it prefers the latter. Soils can be sandy or clay loam. Suitable for general planting or rockeries and grows very well in containers. Plants are hardy to moderate frosts.

The var. *gramineus* (Endl.) Geerinck has narrow leaves to 0.2 cm wide and flower-scapes to about 40 cm tall. It has similar requirements to the typical variety. It is often marketed as a dwarf variant.

Propagate both varieties from seed, which generally begins to germinate 26–60 days after sowing. Division of clumps is also successful.

Orthrosanthus muelleri Benth.
(after Baron Sir Ferdinand von Mueller, first Government Botanist, Vic)
WA
0.2–0.3 m × 0.2–0.3 m   Sept–Oct

Tufted perennial **herb**; **leaves** 10–20 cm × up to 0.25 cm, grass-like, erect, flat, thin, green, margins densely hairy; **inflorescence** on scape to 30 cm tall, few-branched, with 4–6 flowers per branch, with ovate spathe to about 1–3 cm long, each enclosing 2 flowers; **flowers** to about 3 cm across, blue; **capsule** to about 1 cm long, included in spathe.

An attractive but poorly known species from the far western Eyre and south-western Roe Districts where it grows in sandy soils. Though rarely cultivated in Australia it has been grown to a small degree in western USA. Plants need good drainage and they prefer a semi-shaded site. They tolerate moderate frosts. This species has similar horticultural potential to *O. laxus*. Propagate from seed or by division of clumps.

Orthrosanthus multiflorus Sweet
(many-flowered)
Vic, SA, WA   Morning-flag
0.3–1 m × 0.3–1 m   Sept–Dec

Tufted perennial **herb**; **leaves** 20–60 cm × 0.2–0.8 cm, grass-like, erect to spreading, flat, green, margins glabrous and slightly rough; **inflorescence** on scape to 1 m tall, with a few spreading branches, usually 3–7 flowers per branch, usually 2 flowers per pair of ovate spathes to 2 cm long; **flowers** to about 4 cm across, pale blue, often profuse and conspicuous over a long period; **capsule** 1.2–2 cm long, 3-angled, apex pointed.

A very ornamental and widely distributed species which occurs in south-western Vic, on Kangaroo Island and the Eyre Peninsula in SA, and from Israelite Bay to the Stirling Range in WA. This species is found in coastal and adjacent areas in shrubby heathland. It inhabits sands, sandy loam and clay loam soils. Highly valued for its long flowering period, it is well proven in cultivation in temperate and subtropical regions. Adapts to a range of soils with good to reasonable drainage but prefers soils with moisture-retentive qualities. A semi-shaded site is best but it will tolerate plenty of sunshine. Hardy to moderate frosts. Excellent for general planting, in rockeries, or in containers. Old plants can be rejuvenated by cutting or burning the foliage to ground level, or by dividing and replanting clumps in late autumn or early spring. Propagate from seed, which usually begins to germinate 30–60 days after sowing, or by dividing clumps.

O. polystachyus differs in having inflorescences with erect branches and a single flower in each spathe.

Orthrosanthus polystachyus Benth.
(many flower-spikes)
WA
0.3–1 m × 0.3–1 m   Sept–Dec

Tufted perennial **herb**; **leaves** 0.2–1 m × 0.2–1 cm, grass-like, erect to spreading, flat, green, glabrous, margins slightly rough; **inflorescence** on scape to about 1 m tall, with many short-stemmed spikes of 3–5 flowers subtended by bracts and sometimes with 1–2 slender erect branches, each flower enclosed by a pair of spathes to 1 cm long; **flowers** to about 4 cm across, pale blue, often profuse and conspicuous; **capsule** 1.2–2 cm long, 3-angled, blunt apex.

A handsome species from south-western WA, in the Darling District between Pemberton and Busselton, where it usually grows in forest gullies and often along watercourses. Plants do well in cultivation and prefer a semi-shaded site in sandy or clay loam soils which are moist for most of the year. It has similar applications to *O. multiflorus*. This species is now becoming readily available and is grown to a limited degree overseas. Propagate from seed or by division of clumps.

O. multiflorus differs in having inflorescences with spreading branches and usually 2 flowers in each spathe.

Orthrosanthus polystachyus   W. R. Elliot

ORYZA L.
(from *oryza*; the name for rice plant in Greek and Latin)
Poaceae

Perennial, aquatic or semi-aquatic, tufted **grasses**; **culms** erect; **ligule** membraneous; **leaves** long, flat or folded, margins rough; **inflorescence** a loose, open panicle; **spikelets** stalked, laterally flattened, usually solitary; **glumes** minute; **lemmas** awned or not.

A small genus of about 20 species widely distributed in tropical regions. Two Australian species have been described but some authorities record up to 4 native species. The genus includes rice (*Oryza sativa* L.). This grain crop feeds more people than any other crop in the world. Rice is commonly grown in Australia and plants may naturalise for a short time. The native rice species have some ornamental value. They are a significant food source for wildlife and may also be an important genetic source for crop improvement. Propagation is from seed, which must be sown in shallow water.

Oryza australiensis Domin
(Australian)
Qld, WA, NT   Native Rice
1–1.5 m tall   Feb–April

Perennial **grass** forming sprawling clumps; **culms** 1–1.5 m tall, erect to decumbent, clustered, glabrous, smooth; **ligule** 0.3–0.6 cm long, becoming lacerated; **leaves** 20–65 cm × 1–1.7 cm, linear, spreading, dark green, acuminate; **panicles** much-branched, open, spreading; **spikelets** 0.6–0.7 cm long, whitish to yellowish, darkening with age; **glumes** membraneous; **lemmas** with an awn to 4.5 cm long.

Widespread in tropical Australia growing in swamps and seasonally flooded plains. The seed is a major source of food for many species of waterbirds. Native rice is ornamental and worth cultivating for its interest value. Best suited to the tropics. Requires moist to wet soil in a sunny position. Propagate from seed.

Oryza australiensis   D. L. Jones

Osbeckia australiana and leaf variations × .45

OSBECKIA L.
(after Pehr Osbeck, 18–19th century plant collector in China)
Melastomataceae

Herbs or **shrubs**, often hairy; **leaves** simple, entire, opposite, 3–5-nerved; **inflorescence** dense, terminal, flowers rarely solitary; **flowers** 4–5-merous; hypanthium hairy; **calyx** with triangular to ovate lobes, often falling early; **petals** flimsy, colourful; **stamens** 8 or 10; **fruit** a dry capsule, containing numerous small seeds.

A genus of about 60 species widely distributed in Asia. One in species occurs in Africa and 2 in Australia (one of which is endemic). Although they have showy, colourful flowers, these plants are not as commonly grown as they deserve to be. Propagation is from the small seed, which is best sown fresh and covered lightly with propagation medium, or from cuttings of semihardened new growth.

Osbeckia australiana Naudin
(from Australia)
Qld, WA, NT
1.5–2 m × 0.5–1.5 m   Feb–Sept
Slender to bushy **shrub**; **bark** reddish-brown on older stems; young growth light green, hairy; **leaves** 2–6 cm × 0.2–1.3 cm, narrowly linear to narrowly ovate, dark green, hairy; **inflorescence** a terminal, crowded panicle, bearing up to 5 flowers; **flowers** 2–4 cm across, short-lasting; **petals** 5, purple to mauve; **stamens** 10, yellow, prominent in a central cluster; **fruit** 0.8–1.2 cm across, nearly globular, brownish, fibrous and pulpy, ripe Feb–Sept.
Endemic and widely distributed in northern Australia where it grows in monsoonal rainforests, near streams and in swampy situations. Aborigines eat the young fruit. An excellent ornamental for tropical and warm subtropical regions. Plants flower over a long period especially if they are watered and fertilised regularly. They grow best in a protected position and will tolerate full shade. Very sensitive to frosts. Regular light pruning is recommended to maintain a bushy habit. This species is variable and offers scope for the selection of forms which have different growth habits, leaf shapes, flower colour and size. Propagate from fresh seed, and from cuttings of firm young growth, which strike readily.

O. perangusta F. Muell. and *O. koolpinyahensis* Sw. are synonyms.

Osbeckia chinensis L.
(from China)
Qld, NT
0.5–1 m × 0.3–1.5 m   July–Jan
Dwarf **shrub**, often sparse and straggly; young growth hairy; **leaves** 2.5–6 cm × 0.5–1 cm, narrowly ovate, often partly recurved, dark green above, greyish beneath, 3 veins prominent; **inflorescence** a dense terminal cluster bearing up to 6 flowers; **flowers** 2.5–4 cm across, short-lasting; **petals** 4, mauve to purple; **stamens** 8, yellow, in a loose central group; **fruit** 0.6–0.8 cm × 0.4–0.5 cm, cup-shaped, brown, glabrous.

Widely distributed from India to New Guinea, and occurring in Australia in north-eastern Qld and eastern Arnhem Land. It grows in open forest and woodland, and favours areas which are swampy or seasonally flooded. Although often sparse and straggly, this species has showy, colorful flowers and has potential as an ornamental in tropical regions. May respond well to regular watering and fertilisers. It may also be beneficial to cut plants back annually to the woody rootstock. Propagate from fresh seed and possibly also from cuttings.

Osbeckia koolpinyahensis S. W. = *O. australiana* Naudin

Osbeckia perangusta F. Muell. = *O. australiana* Naudin

OSBORNIA F. Muell.
(after John Walter Osborne, lithographer)
Myrtaceae
A monotypic genus which occurs in Australia, New Guinea, Indonesia and the Philippines.

Osbornia octodonta F. Muell.
(with eight teeth; referring to the calyx)
Qld, WA, NT   Myrtle Mangrove
2–4 m × 1–2 m   June–Dec
Densely bushy, small to medium **shrub**; **bark** grey to brown, fibrous, flaky; **leaves** opposite, 2–5 cm × 1–2 cm, obovate to spathulate, tapered to base, green often tinged with red, with numerous small, trans-lucent oil dots, margins minutely crenulate; **flowers** 0.7–1 cm across, pale greenish-cream, slightly tubular, fragrant, in groups of 1–3 in the upper axils or on the end of branchlets; **calyx** hairy, with 8 lobes; **petals** absent; **fruit** 0.7–1 cm × 0.4–0.6 cm, indehiscent, enclosed in the hairy calyx tube, containing 1 or 2 seeds, ripe Dec–Mar.

This mangrove is widely distributed along the coast and in the estuaries of northern Australia, extending to New Guinea, Indonesia and the Philippines. The leaves have a pleasant fragrance when crushed and have been used by the Aborigines of coastal districts to flavour cooking. The dark-colored wood is close-grained and very hard. An attractive, bushy tree which has limited prospects for cultivation, but it is one of a number of species which is of major significance for stabilisation of coastal areas. Propagate from seed.

OSCHATZIA Walp.
(after Dr Oschatz, the inventor of microtomy; cutting of thin specimen sections)
Apiaceae (alt. *Umbelliferae*)

Tufted perennial **herbs**; **leaves** radical, toothed or lobed; **inflorescence** irregular umbels of 2–9 flowers, or solitary flowers, on an erect, simple or branched leafless scape; **flowers** small; **petals** 5, white or tinged with pink; **stamens** 5, alternating with, and shorter than, the petals; **calyx lobes** triangular, short, less than half as long as petals; **fruit** consisting of two 5-ribbed mericarps.

An endemic genus of 2 species; one in the Australian Alps and the other in Tas. These somewhat insignificant plants are rarely cultivated but they can be dainty in flower and have potential for limited use, such as in containers, miniature gardens and rockeries.

Propagation is from seed, which may germinate better if stratified for 2–4 weeks at 2–4°C. Also worth trying from cuttings.

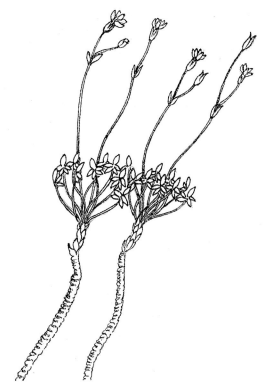

Oschatzia saxifraga × .8

Oschatzia cuneifolia (F. Muell.) Drude
(wedge-shaped leaves)

NSW, Vic	Wedge Oschatzia
0.1–0.3 m × 0.2–0.5 m	Dec–Feb

Perennial **herb**, with short-spreading rhizome; **leaves** 3–12 cm × 0.4–1 cm, narrowly wedge-shaped to somewhat spathulate, in a basal rosette, tapering to base with a long, often purple-tinted stalk, glabrous, apex lobed or deeply toothed; **flowers** about 0.5 cm across, white, 2–9 in a loosely arranged compound umbel.

This species is relatively rare in alpine and subalpine regions of the Australian Alps. It grows in wet grasslands or in bogs. Plants will require moist to wet soils which drain freely and an open sunny or slightly shaded site. May be best grown in containers which are placed in a saucer of water, or it may be worth trying a technique used for cushion plants (see Volume 3, page 135). Propagate from seed or from cuttings.

Oschatzia saxifraga (Hook. f.) Walp.
(rock-breaking)

Tas	
0.1–0.2 m × 0.5–1 m	Dec–Feb

Perennial **herb** with an erect, cylindrical rootstock and a rosette of leaves at apex; **leaves** to about 2.5 cm × 0.8 cm, leaf blade to about 1 cm long with 3–11 teeth or lobes, on a narrow stalk to about 1.5 cm long, glabrous; **flowers** about 0.5 cm across, white, often tinged with pink, 1–3 on slender deep red stalk to 15 cm tall; **fruit** about 0.4 cm long, ovoid, scarcely flattened.

A small delicate herb from peaty heaths in western and southern regions. It has had very limited cultivation. Plants are insignificant until in flower and although the flowers are not profuse they are still attractive. It needs similar conditions to *O. cuneifolia*. Propagate from seed or from cuttings.

OSMUNDACEAE R. Br.
A small family of ferns consisting of about 20 species in 3 genera, distributed in tropical and temperate regions. They are unusual because the sporangia are not grouped into sori and the spores are green and short-lived. Australian genera: *Leptopteris, Todea*.

OSTEOCARPUM F. Muell.
(from the Greek *osteon*, bone; *carpos*, fruit; referring to the hard fruiting perianth)
Chenopodiaceae

Perennial **herbs**, branching from base; **leaves** alternate, somewhat terete to club-shaped, glabrous, succulent; **flowers** bisexual, solitary, axillary, lobe margins ciliate; **perianth tube** somewhat fleshy; **stamens** 5; **fruiting perianth** glabrous, tube woody, with erect wing or wings or without wings.

A small endemic genus of 5 species, found in semi-arid and arid regions. Plants are rarely cultivated but they may be useful in saline soils, especially for revegetation projects. Some plants have greyish-green

Osteocarpum acropterum

Osbeckia australiana   D. L. Jones

foliage and some selections may adapt well to gardens
and containers. They have limited potential as forage
plants and, in fact, they are usually only eaten when no
other growth is available.

Propagation is from seed, which does not require
pre-sowing treatment. It may also be worth trying cuttings of firm wood.

Osteocarpum acropterum (F. Muell. & Tate) Volkens
(terminal wings)
Qld, NSW, SA, WA, NT   Water Weed; Babbagia
0.1–0.25 m × 0.3–0.5 m   Oct–March; also sporadic
Perennial **herb** of mounding habit; young growth
sometimes faintly hairy; **branches** many, spreading to
ascending; **leaves** 0.3–0.6 cm × 0.1–0.2 cm, club-
shaped, alternate, succulent, greyish-green, glabrous,
apex blunt; **flowers** solitary, axillary; **fruiting perianth**
about 0.5 cm long, lower half hollow, upper half somewhat globular, striate, with 2 unequal vertical wings or
sometimes a solitary wing.

A wide-ranging species which occurs in a variety of
soil types but usually uncommon on sandy soils. Often
colonises bare areas in nature. Needs maximum sunshine and should do best in loam or heavier soil types.

The var. *diminuta* J. M. Black, from western NSW,
western Vic and eastern SA, occurs on heavy soils
which are periodically waterlogged. It has a fruiting
perianth which lacks wings.

Propagate both varieties from seed and possibly
from cuttings.

O. dipterocarpum (F. Muell.) Volkens, from south-
eastern Qld, western NSW, north-eastern SA and
southern NT is allied, but it has a fruiting perianth
with 2–3 semi-circular erect wings.

O. pentapterum (F. Muell. & Tate) Volkens is allied
but it has a somewhat star-shaped fruiting perianth
with 5 wings. It occurs in south-western Qld, northern
and north-eastern SA and southern NT, where it

usually grows in heavy soils which are subject to periodic inundation.

O. scleropterum (F. Muell.) Volkens is closely allied
but it has fruiting perianths with 5 unequal wings, with
one more erect than the others. It occurs in south-
western Qld and north-western NSW.

Osteocarpum dipterocarpum (F. Muell.) Volkens —
see *O. acropterum* (F. Muell. & Tate) Volkens

Osteocarpum pentapterum (F. Muell. & Tate) Volkens —
see *O. acropterum* (F. Muell. & Tate) Volkens

Osteocarpum salsuginosum F. Muell.
(growing in or near salt marshes)
NSW, Vic, SA, WA, NT   Bonefruit
0.1–0.25 m × 0.2–0.5 m
Perennial **herb** of mounding habit; **branches** prostrate to ascending; **leaves** 0.4–0.6 cm × 0.1–0.2 cm,
semi-terete or club-shaped, alternate, succulent,
glabrous, apex pointed; **flowers** solitary, axillary; **fruiting perianth** about 0.15 cm tall, more or less globular,
with a small swelling on one side.

Plants are found in central and western NSW, western Vic, eastern SA, southern WA and southern NT. It
dwells on mainly clay soils, which are subject to
periodic inundation, or on margins of salt lakes.
Cultivation requirements are similar to *O. acropterum*,
to which it is closely allied, although it lacks winged
fruits. Propagate from seed and possibly from cuttings.

Osteocarpum pentapterum × .55

122

Ostrearia australiana × .4

Osteocarpum scleropterum (F. Muell.) Volkens —
　　　see *O. acropterum* (F. Muell. & Tate) Volkens

OSTREARIA Baill.

(from the Latin *ostrea*, oyster; referring to the fruit shape)
Hamamelidaceae
　　A monotypic genus endemic in north-eastern Qld.

Ostrearia australiana Baill.

(from Australia)
Qld	Hard Pink Alder
10–20 m × 5–8 m	Sept–Oct

　　Small to tall **tree** with a bushy habit; young growth stellate-hairy; **leaves** 5–15 cm × 2–6 cm, elliptic to lanceolate, dark green, thin but leathery, margins wavy, apex acuminate; **stipules** to 0.7 cm × 0.2 cm, stellate-hairy; **spikes** 2–2.5 cm across, short, dense, usually terminal; **flowers** about 1.8 cm across, pale green to yellow-green, densely crowded; **capsules** 1–1.7 cm × 1.2–2.4 cm, lobed, aggregated on a central axis, bearing 1 or 2 seeds; **seeds** 1–1.4 cm long, pale, shiny.
　　Endemic in north-eastern Qld where it grows near streams in rainforest, extending from the lowlands to about 1000 m. *O. australiana* is a dense bushy tree which is rarely grown but it has potential for planting in parks and large gardens. Needs well-drained, acidic soils, mulches and watering during dry periods. Light fertiliser applications are probably beneficial. The frost tolerance of provenances from higher altitudes is probably significant. Propagate from fresh seed.

OTANTHERA Blume

(from the Greek *our, otis,* an ear and *anthere* an anther; referring to ear-like appendages on the stamens)
Melastomataceae
　　Hairy **shrubs**; **leaves** simple, entire, opposite, petiolate; **inflorescence** a panicle or cyme, terminal or axillary, with numerous bracts; **flowers** 5-merous; hypanthium hairy or scaly; **calyx** with narrow lobes, falling early; **petals** flimsy; **stamens** 10, with a small bilobed appendage; **fruit** fleshy, containing numerous small seeds.
　　A small genus of about 15 species distributed from India to New Guinea. One species extends to Australia. Propagation is from cuttings or from seed.

Otanthera bracteata Korth.

(bearing bracts)
Qld	
1.5–2.5 m × 1–2 m	Aug–May

　　Small to medium **shrub** with an open to bushy habit; young growth covered with rusty hairs; **leaves** 6–12 cm × 2.5–6 cm, ovate to ovate-elliptical, dark green above, paler beneath, thin, roughened with bristly hairs; **panicles** crowded, terminal or in the upper axils, bearing up to 7 flowers; **flowers** 2.5–3 cm across, white, lasting a few days; **fruit** ovoid, to 0.7 cm × 0.6 cm, red.
　　Occurs in north-eastern Qld and extends to New Guinea and Indonesia. It inhabits lowland rainforest and is often prominent along road margins and in disturbed sites. An interesting shrub for tropical and subtropical regions. It grows readily in a range of positions from full sun to semi-shade and tolerates most well-drained soils. Frost tolerance is not high. Regular tip pruning is recommended and flowering can be promoted by watering during dry periods and regular light dressings of fertiliser. Propagate from seed, which germinates readily, or from cuttings of hardened new growth, which strike easily.
　　O. queenslandica Domin is a synonym.

Otoclinis macleayana F. Muell. =
　　　Callitris macleayana (F. Muell.) F. Muell.

OTTELIA Pers.

(from 'ottel-ambel', a native name for an Indian species)
Hydrocharitaceae
　　Annual or perennial **aquatic herbs**; roots fibrous; **leaves** dimorphic or not, submerged or floating; **petiole** spongy, with a sheathing base; **lamina** flattened; **inflorescence** a spathe containing unisexual or bisexual flowers; **sepals** 3; **petals** 3, large, **stamens** 6–15; **fruit** indehiscent, maturing underwater and disintegrating to release the seeds.
　　A genus of about 40 species widely distributed in the tropics. Two species occur in Australia but neither are endemic. They are prominent aquatics with short-lived, showy flowers produced in succession. Both species are worthy of cultivation. Established plants generally resent disturbance. Propagation is from seed, which germinates readily in warm shallow water over sand or mud.

Ottelia alismoides (L.) Pers.

(like the genus *Alisma*)
Qld, NT	
Aquatic	Sept–Mar

　　Tufted perennial **aquatic herb** forming dense clumps to 60 cm wide; **leaves** submerged or partly

123

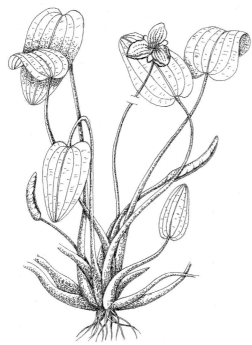

Ottelia ovalifolia × .45

emergent, up to 1 m long, on long spongy petioles; **leaf blade** 15–20 cm × 15–20 cm, broadly ovate to nearly orbicular, base strongly cordate, green to bronze, almost translucent, with prominent veins; **spathe** with 5–10, prominent, crisped wings, 2 especially prominent; **flowers** 5–6 cm across, white with darker veins, centre yellow; **stamens** 6–9; **capsule** winged; **seeds** glabrous.

Occurs in eastern NT and is widely distributed in eastern Qld extending south to near Ipswich. Also common in Asia and Africa. Cooked and eaten as a vegetable in Asia. An ornamental aquatic with outstanding leaves, it is highly suitable for tropical and subtropical regions. Plants can be grown in ponds, dams and also aquaria. Needs bright light and warm water. Propagate from seed, which germinates readily on mud or in warm shallow water.

Ottelia ovalifolia (R. Br.) Rich.
(with oval leaves)
Qld, NSW, Vic, SA, WA, NT   Swamp Lily
Aquatic   Nov–Apr

Tufted annual or perennial **aquatic herb**; **leaves** of 2 types, submerged and floating; **submerged leaves** strap-like; **floating leaves** with an elliptic blade, 5–16 cm × 3–6 cm, dark green to bronze, somewhat shiny, with 5–7 longitudinal veins; **flowers** of 2 types, non-opening and open-petalled; **open-petalled flowers** 5–6 cm across, white to cream with a reddish to purplish centre; **stamens** yellow, in 3 groups of 4; **fruit** bottle-shaped, ridged; **seeds** hairy.

A widely distributed species which grows in slowly flowing fresh water or in the still water of ponds, dams and lagoons. It can form extensive colonies especially

in nutrient-rich water. The flowers last about half a day but are produced regularly. This attractive aquatic is ideal for cultivation in ponds or dams but it has proved difficult to maintain in aquaria. Needs bright light to promote vigorous growth and flowering. Propagate from seed, which germinates readily on mud and in warm shallow water.

OTTOCHLOA Dandy
(after Dr Otto Stapf, 19–20th-century German botanist; and from the Greek *chloa*, grass)
Poaceae

Perennial **grasses** with long, slender scrambling or decumbent **culms**; **ligule** short, membraneous; **inflorescence** a panicle; **spikelets** glabrous, strongly compressed; **glumes** similar, shorter than the spikelet; **lemmas** more or less equal.

A small genus of 6 species; 2 are endemic in eastern Australia. They are rarely grown, if at all. Propagate from seed or by division of clumps.

Ottochloa gracillima C. E. Hubb.
(very slender)
Qld, NSW
0.1–0.3 m tall   Dec–Feb

Perennial **grass** forming sprawling clumps or mats; **culms** long, slender, rooting at the base; **leaves** 3–8 cm × 0.1–0.6 cm, linear-lanceolate, dark green, spreading; **panicles** to 13 cm long, sparsely branched (2–4 branches); **spikelets** about 0.2 cm long, in groups of 2–4; upper **lemma** wrinkled.

An interesting mat-forming grass found in moist shady sites. It grows in coastal forests in south-eastern Qld and northern and central NSW. Plants are grazed by native marsupials. It has potential for use around ponds and as a groundcover in shady situations. Propagate from seed and by division.

O. nodosa (Kunth) Dandy from south-eastern Qld is similar but taller-growing (to 0.5 m) with up to 12 branches in the inflorescence.

Ottochloa nodosa (Kunth) Dandy —
see *Ottochloa gracillima* C. E. Hubb.

OURISIA Comm. ex Juss.
(after Governor Ouris of the Falkland Islands, 18th-century)
Scrophulariaceae

Dwarf perennial **herbs**; **leaves** opposite, venation usually prominent and reticulate; **inflorescence** terminal verticillate panicles; **flowers** bisexual; **corolla** slightly zygomorphic; **calyx** 5-lobed; **petals** 5; **stamens** 4; not exserted; **fruit** a capsule.

A genus of about 25 species from the southern hemisphere. The majority of these occur in South America. One species occurs in New Zealand and there is one species in Australian which endemic in Tas. This genus is popular with rock garden enthusiasts but the Tasmanian species is cultivated only to a limited degree.

Propagation is from seed, which may benefit from stratification at 2–4°C for 3–6 weeks. Cuttings should strike readily as stems can self-layer.

Ourisia integrifolia R. Br.
(entire-leaved)
Tas
0.05–0.15 m × 0.1–0.3 m   Nov–Feb

Perennial glabrous **herb** with prostrate to spreading, self-layering stems; **leaves** 0.4–1.5 cm × 0.3–1 cm, ovate to nearly orbicular, opposite with long stalks, glabrous, glossy and deep green above, paler below, apex blunt or sometimes shallowly notched; **inflorescence** erect to about 15 cm tall, slender, reddish stems each with 1–2 flowers; **flowers** 1–2 cm across, tubular with spreading petals, white or pale bluish-purple; **calyx** about 0.6 cm long, deeply lobed; **capsule** about 0.6 cm long, ovoid-globular.

A delicate dwarf herb which occurs on mountains throughout Tas in constantly moist, shaded sites beside watercourses. It is often associated with rocks and in some cases grows on rocks. Although very limited in cultivation, it has potential for greater use especially in pots, miniature gardens and by rock garden enthusiasts. It needs constant moisture but simultaneously requires good drainage. Some enthusiasts cultivate plants in pots with a high percentage of peat moss and place them in a saucer of water in order to maintain high humidity. Further trials will establish the best procedure. Hardy to frost and light snowfalls. Propagate from seed or from cuttings.

OWENIA F. Muell.
(after Sir Richard Owen, 19th-century British biologist)
Meliaceae

Shrubs or **trees**, with reddish timber, sap usually milky, young growth sticky; **leaves** pinnate, with an even number of leaflets, alternate, somewhat pendulous; **inflorescence** axillary spike-like panicles, shorter than the leaves; **flowers** small, bisexual or unisexual, yellowish-white to brownish; **calyx** 5-lobed; **petals** 5, overlapping in bud; **stamens** 10, joined in a tube; **fruit** a globular drupe, reddish to purplish.

A small endemic genus of 6 species which occurs over a wide range of habitats and differing climatic conditions from desert to rainforest. Most species are cultivated but mainly by enthusiasts.

Propagation is from seed but germination results can be erratic. Further research is needed into the dormancy mechanism of seeds. Some species can be grown from root suckers.

Owenia acidula F. Muell.
(sour, tart)
Qld, NSW, SA, WA, NT   Gruie; Colane; Gooya;
                              Sour Apple; Emu Apple; Dillie Boolen
4–12 m × 4–8 m   sporadic through year

Bushy tall **shrub** or small **tree**, with sticky young growth, sometimes suckering from roots; **branches** many; **branchlets** often pendent; **leaves** pinnate, pendent, with a long slender stalk; **leaflets** 2–5 cm × 0.3–1 cm, 5–15 pairs, lanceolate to narrow-elliptic, glabrous, venation reticulate, apex blunt or pointed; **flowers** about 0.5 cm across, brownish-white, nearly sessile, in narrow panicles; **drupes** to about 4 cm across, globular, crimson when ripe, with edible outer flesh.

This species occurs in semi-arid regions where it grows in a range of heavy but relatively well-drained soils. Plants generally develop as shapely shrubs or trees with a short trunk and rounded canopy. This species has potential for much greater use as an ornamental or screening plant in semi-arid and temperate regions, but until seed propagation methods are unravelled it is not likely to be readily available. Tolerates a sunny or semi-shaded site and copes well with extended dry periods. Hardy to moderate frosts. Propagate from seed, which is extremely difficult to germinate. Some success has been achieved with root suckers.

A naturally occurring hybrid, *O.* × *reliqua* P. I. Forst., is known from the Burnett, Leichhardt and Port Curtis Districts in Qld. *O.* × *reliqua* has intermediate characteristics between *O. acidula* and *O. venosa*. It usually has 5–10 greyish-green leaflets which are linear-elliptic to ovate-elliptic and 3–5 cm × 0.75–1.4 cm. It may be in cultivation but this seems unlikely.

Owenia cepiodora F. Muell.
(onion-like odour)
Qld, NSW   Onion Cedar
6–30 m × 4–12 m   Nov–Dec

Small to tall **tree** with a dense canopy; young growth hairy, with onion fragrance; **bark** dark brown; **branchlets** thick, pale brown; **leaves** to about 50 cm long, pinnate, with stalk to about 30 cm long; **leaflets** to 15 cm × 6 cm, 12–20, broadly lanceolate, somewhat opposite, dark green, glabrous, glossy, not toothed, apex pointed; **flowers** about 0.8 cm across, white, in terminal panicles to about 23 cm long, sometimes profuse; **drupes** about 2 cm across, globular, red, ripe Jan–March.

Regarded as an endangered species, it occurs as a small scattered population in *Araucaria cunninghamii*

Owenia acidula × .5

forests. Rarely cultivated, it is a very attractive species which has potential for large gardens, parks and road-sides where it can be utilised for shade and screening. Needs friable loamy soils which are moist for most of the year and it appreciates a sunny aspect. Mulching is recommended and supplementary watering during extended dry periods should be beneficial. The red timber was highly valued for its similarity to red cedar, *Toona ciliata* (syn. *T. australis*) and was sometimes sold wrongly under that name. Propagate from seed, which is difficult to germinate.

Owenia × reliqua P. I. Forst. — see *O. acidula* F. Muell.

Owenia reticulata F. Muell.
(netted veins)
Qld, WA, NT   Desert Walnut
4–10 m × 3–8 m   Aug–Dec; also sporadic

Tall **shrub** to small **tree**, with sticky yellowish young growth; **bark** rough, dark grey; **branchlets** covered in sticky yellowish varnish; **leaves** to 30 cm or more long, pinnate; **leaflets** 6–20 cm × 3.5–10 cm, 1–6 pairs, oblong to elliptic, glabrous, pale-to-mid green, vena-tion prominently reticulate, apex blunt or notched; **flowers** to 0.5 cm across, white, dioecious, in loose widely branching panicles, often profuse; **drupes** to 4 cm across, pear-shaped to globular, red to purplish.

A handsome species, mainly from semi-arid and arid regions but also occurring in seasonally dry tropical areas. It usually grows in sandy soils of spinifex plains. Uncommon in cultivation, it is suited to arid, semi-arid and warm temperate regions where it will need excellent drainage and a sunny site. Plants are hardy to moderate frosts. Should respond well to supplementary watering during extended dry periods. The fruits, seeds and gum are edible and were utilised by Aborigines. Propagate from seed, which may be difficult to germinate.

O. vernicosa has much narrower leaflets.

Owenia venosa F. Muell.
(many veins)
Qld Rose Almond; Crow's Apple; Sour Plum
4–20 m × 3–12 m   Oct–Jan; also sporadic

Tall **shrub** or small to medium **tree**, with sticky yel-low or brownish young growth; **branchlets** thickish, sometimes reddish or purplish; **leaves** to about 20 cm long, pinnate; **leaflets** to 8 cm × 3 cm, 3–4 pairs, ellipt-ical to oblong, glabrous, dark green and glossy above, paler below, reticulate venation prominent below, apex blunt or notched; **flowers** about 0.5 cm across, white, in panicles about as long as the leaves; **drupes** to 4 cm across, globular, reddish, ripe Aug–Oct.

A very attractive tree with a lacy canopy. It occurs in dry rainforests from near Cairns to the Qld/NSW border region, and grows in a range of soils including clays. Should do well in tropical and subtropical regions and may succeed in temperate areas if given a semi-protected site. Needs good drainage and plenty of sunshine. Responds well to mulching and supplementary watering during extended dry periods. Young plants may need shaping. The reddish timber was used for cabinet-making, tool handles, flooring

Owenia vernicosa   T. L. Blake

and framing. Propagate from seed, which is difficult to germinate.

A naturally occurring hybrid between *O. acidula* and *O. venosa* is known (see *O. acidula* for further informa-tion).

Owenia vernicosa F. Muell.
(as if varnished)
Qld, WA, NT
4–12 m × 3–8 m   Oct–Nov

Tall **shrub** or small **tree**, with sticky, light green young growth; **bark** rough, grey to black; **branchlets** often sticky; **leaves** 30–35 cm long, pinnate, usually crowded towards ends of branches; **leaflets** 5–12 cm × 1.3–2.2 cm, 7–15 pairs, lanceolate to elliptic, usually paired and opposite, glabrous, dark green, venation reticul-ate, apex pointed; **flowers** to about 0.3 cm across, cream to creamish-green, male and female flowers on separate trees, often profuse, in terminal panicles to about 30 cm long; **drupes** 2–3.5 cm across, globular, on long stalks, reddish-brown, ripe April–Aug.

A tropical species from northern Australia where it is found in seasonally dry regions. It grows in sandy or gravelly soils on plateaux, gorges and escarpments, as part of open forest and woodland. Rarely cultivated due to propagation difficulties, it is well-suited to dry tropical regions and is worth trying also in subtropical areas. Needs excellent drainage and plenty of sun-shine. Young plants may require shaping. Aborigines utilised the sap as a general antiseptic. The leaves were used for treatment of headaches and sore eyes, and an infusion of bark for coughs. The bark and leaves were also used as a fish poison. Propagate from seed, which is very difficult to germinate.

O. reticulata differs in having much broader leaflets.

OXALIDACEAE R. Br.

A family of about 950 species of dicotyledons in 6 genera. It is mainly found in South Africa, and Central and South America. Plants are commonly perennial herbs with a bulbous root system, or rarely shrubs. The leaves, which often have terminal leaflets, contain oxalate crystals. The 5-merous flowers are frequently colourful and showy. In Australia, the family is represented by *Oxalis* which consists of about 5 native species with another 12–14 species naturalised as weeds, some of these being serious pests which can invade pasture and native bush.

OXALIS L.

(from the Greek *oxys* acid, sour, sharp; referring to the taste of the leaves)
Oxalidaceae

Annual or **perennial herbs**; **rootstock** a taproot, or fleshy and bulbous, often producing lateral bulblets; **leaves** in a rosette or cauline, mostly trifoliolate with 3 terminal leaflets; **leaflets** sessile or short-stalked, entire or lobed; **sepals** 5; **petals** 5; **stamens** 10; **fruit** a narrow capsule.

A large genus with about 800 species mostly from South Africa, and Central and South America. Six species are native to Australia. Many exotic species are grown as ornamentals and about 13 have become naturalised weeds. *O. pes-caprae* L. is the most serious weed because it invades established pasture, gardens, orchards and bushland. The native species of oxalis reproduce solely from seed and are non-invasive. Nevertheless they have limited ornamental appeal and are rarely grown. Propagation is from seed.

Oxalis exilis A. Cunn.
(thin, weak, slender)
NSW, Vic, Tas   Slender Wood Sorrel
Prostrate × 0.3 m   Oct–May

Annual or perennial mat-forming **herb** with fibrous roots; **stems** prostrate, slender, sparsely hairy; **leaves** trifoliolate, in tufts along the stem; **leaflets** 0.25–0.35 cm × 0.15–0.3 cm, cuneate, green, apex bilobed; **inflorescence** axillary, 1–2-flowered; **flowers** 1.5–1.8 cm across, yellow; **capsules** 0.5–0.6 cm × 0.2–0.25 cm, erect or reflexed, held above the leaves.

This species forms small mat-like clumps in moist-to-damp soil in coastal and near-coastal areas. It is not showy but may have some potential as a container plant. Propagate from seed.

Oxalis magellanica G. Forst.
(from the Straits of Magellan, South America)
Vic, Tas   White Wood Sorrel
0.05–0.1 × 0.2 –0.5 m   Sept–Feb

Perennial herb with branching rhizomes, often forming leafy tufts; **leaves** 3-foliolate; **leaflets** 0.4– 1 cm ×

0.3–0.8 cm, obcordate, green above, somewhat glaucous below; **flowers** 0.8 – 2 cm across, white, solitary, on slender stalks to about 4 cm long; **capsules** to about 0.5 cm across, ovoid to globular.

This species occurs in subalpine forests and alpine areas. Also in NZ and South America. Rarely cultivated in gardens, it requires moist, very well-drained soils in semi-shaded or shaded sites. Plants are attractive in flower and they are suitable for growing with ferns or in containers. Propagate from seed or division of rhizomes.

O lactea Hook. is a synonym.

Oxalis perennans Haw.
(perennial)
Qld, NSW, Vic, Tas, SA, WA   Scour Grass
0.1–0.2 m × 0.3 m   Aug–March

Perennial herb with a stout taproot; **stems** erect or spreading, to 25 cm long, slender, hairy; **leaves** trifoliolate; **leaflets** 0.4–1.5 cm × 0.2–0.8 cm, cuneate, bilobed, green; **inflorescence** axillary, 1–5-flowered; **flowers** 1.8–2.5 cm across, yellow; **capsules** 1–3 cm × 0.15–0.25 cm, cylindrical, hairy, red, held above the leaves.

Widespread in inland regions, it grows on a range of soil types and vegetation types. Generally forms loose, prostrate mats but the stems may also thread through shrubs. Suspected of causing oxalic acid poisoning of stock. Rarely cultivated but frequently adventive in gardens. Attractive when in flower and can add interest to plantings. Propagate from seed.

Frequently confused with the introduced, weedy *O. corniculata* L. which has smaller flowers and its capsules are not held above the leaves.

Oxalis magellanica   W. R. Elliot

Oxalis rubens Haw.
(bluish-red, reddish)
Qld, NSW, Vic, SA
0.1–0.2 m × 0.3 m   Feb–Nov

Biennial or perennial **herb** with fibrous roots; **stems** erect to ascending, to 35 cm long, reddish-brown, glabrous to sparsely hairy; **leaves** trifoliolate; **leaflets** 0.3–0.6 cm × 0.2–0.5 cm, cuneate, greenish-purple, bilobed; **inflorescence** axillary, 1–2-flowered; **flowers** about 1.5 cm across, yellow; **capsules** 1.5–1.8 cm × 0.3–0.4 cm, hairy, cylindrical, erect, held above the leaves.

Usually found growing in coastal scrub extending to stabilised dunes. Rarely cultivated but plants commonly naturalise in coastal gardens. It has limited ornamental appeal but is rarely weedy. Propagate from seed.

OXYLOBIUM Andrews
(from the Greek *oxys*, sharp; *lobos*, pod; referring to the sharp-pointed pods)
Fabaceae (alt. *Papilionaceae*)

Dwarf to tall **shrubs**; **leaves** usually opposite or whorled, sometimes scattered or rarely all alternate, simple, entire or lobed; **stipules** bristle-like, sometimes minute or absent; **inflorescence** terminal or axillary clusters, corymbs or racemes; **calyx** 2-lipped, 5-lobed more or less equal but upper lobes broader; **flowers** pea-shaped, yellow, orange, red, purplish-red, or in combinations; **stamens** free; **fruit** a 2-valved pod, inflated, hairy.

An endemic genus which has 6 species; all are restricted to eastern Australia. As a result of recent botanical revision many previously included species were transferred to other genera such as *Callistachys, Chorizema, Gastrolobium, Nemcia,* and *Podolobium,* and a couple of species are yet to be placed.

Oxylobium species occur in woodland, open forest and heathland. The distribution of the genus extends from south-eastern Qld to Tas and species are found from the coast to up in mountain ranges. Plants mainly occur in semi-shaded sites on soils that drain freely. Oxylobiums can be trailing groundcovers to tall shrubs. Their foliage is somewhat nondescript but young growth may be covered in rusty hairs. Generally plants are moderately long-lived. Many legumes have a breathtakingly bright and profuse display and oxylobium is no exception. Plants are not very common in cultivation and should be more widely appreciated for their beauty. The flowers of most species produce a sweet nectar which is gathered by insects such as butterflies.

Generally oxylobiums are grown as background shrubs but they warrant greater use as accent plants. They do best in well-drained acidic soils with some protection from full sunshine.

Cultivation does not pose many problems and plants usually grow well without supplementary fertilising. They may need watering during extended dry periods. Plants respond well to pruning and tall species could be utilised for hedging.

Pests and diseases are not a major problem except for Cinnamon Fungus, *Phytophthora cinnamomi,* which can be deadly for plants (see Volume 1, page 168 for further information). Caterpillar borers may do some damage (see Chewing Insects, Volume 1, page 143 for control methods). Scale can be present on plants and is usually readily controlled by applications of white oil.

Plants are propagated from seed, which needs pre-sowing treatment to soften or damage the seed coat to allow penetration of water (see Treatments and Techniques to Germinate Difficult Species, Volume 1, page 205). Cuttings of firm young growth usually root 6–12 weeks after they are prepared.

Oxylobium aciculiferum (F. Muell.) Benth. =
　　　　　　Podolobium aciculiferum F. Muell.

Oxylobium acutum (Lindl.) Benth. =
　　　　　　Nemcia acuta (Benth.) Crisp

Oxylobium alpestre F. Muell. =
Podolobium alpestre (F. Muell.) Crisp & P. H. Weston

Oxylobium arborescens R. Br.
(tree-like)
Qld, NSW, Vic, Tas   Tall Shaggy Pea
1.5–5 m × 1–3.5 m   Aug–Dec

Small to tall **shrub** with silky-hairy young growth; **branches** spreading to ascending, often becoming glabrous; **leaves** 2–6 cm × 0.3–0.8 cm, narrowly elliptical to narrowly ovate, opposite or in whorls of 3, glabrous and dark green above, silky-hairy below, midrib prominent, margins recurved, apex softly pointed; **stipules** absent; **flowers** pea-shaped to about

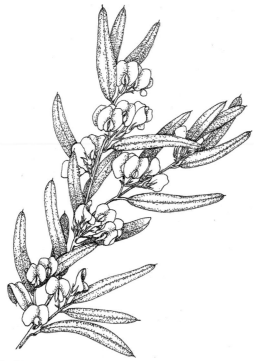

Oxylobium arborescens × .6

1 cm across, yellow to orange-yellow, in short, terminal or axillary racemes, sweetly fragrant, often profuse and very conspicuous; **calyx** hairy; **pods** to about 1 cm long, hairy.

A handsome plant which occurs in the open forests and woodlands on mountain ranges. It is distributed from south-eastern Qld, along eastern regions of NSW and Vic, and to northern and western Tas. Plants adapt well to a range of freely draining acidic soils and prefer a semi-shaded site. They are hardy to most frosts and tolerate lengthy dry periods. Plants may become open with age but they respond well to pruning during or after flowering. Excellent for general use and background planting as well as for hedging. Plant in a busy area so that passers-by can appreciate the floral perfume. Propagate from pre-treated seed, or from cuttings of firm young growth, which strike readily.

O. ellipticum var. *angustifolium* Benth. is a synonym.

Oxylobium atropurpureum Turcz. =
        Nemcia leakeana (Drumm.) Crisp

Oxylobium bennettsii C. A. Gardner =
        Gastrolobium racemosum (Turcz.) Crisp

Oxylobium capitatum Benth. =
        Nemcia capitata (Benth.) Domin

Oxylobium carinatum Meisn. =
        Chorizema carinatum (Meisn.) J. M. Taylor & Crisp

Oxylobium cordifolium Andrews
(heart-shaped leaves)
NSW   Heart-leaved Shaggy Pea
0.1–0.5 m × 1–2 m   Aug–Nov

Dwarf spreading **shrub** with densely hairy young growth; **branches** prostrate to slightly ascending; **branchlets** initially very hairy; **leaves** 0.1–1 cm × to about 0.6 cm, cordate, alternate or opposite or in whorls, dark green and faintly warty above, faintly hairy below, margins recurved, apex pointed and recurved; **stipules** absent; **flowers** pea-shaped, to about 1 cm across, orange-red, in short, terminal, clustered racemes, often profuse and very conspicuous; **pods** to about 1 cm long, ovoid, hairy.

An ornamental groundcover from the Central and South Coast, and Central and Southern Tablelands. *O. cordifolium* is distributed from the coast to the mountain ranges. It was introduced into England during 1807 but plants are rarely cultivated now and it deserves to be better known. Needs well-drained soils which are not prone to drying out for extended periods. A semi-shaded site should provide best results. Tolerates most frosts. This species may be useful in general planting as well as in rockeries and on embankments. A variant with small leaves and fruits from near Nowra is known. Propagate from pre-treated seed, or from cuttings of firm young growth, which should strike readily.

Oxylobium coriaceum (Sm.) C. A. Gardner =
        Nemcia coriacea (Sm.) Domin

Oxylobium ellipticum × .6

Oxylobium cuneatum Benth. =
        Nemcia dilatata (Benth.) Crisp

Oxylobium cuneatum Benth. var. **emarginatum** Benth. =
        Nemcia retusa (Lindl.) Crisp

Oxylobium drummondii Meisn. =
        Nemcia retusa (Lindl.) Crisp

Oxylobium ellipticum (Vent.) R. Br.
(elliptical)
NSW, Vic, Tas   Common Shaggy Pea
0.5–3.5 m × 1–2.5 m   Sept–Dec

Dwarf to medium **shrub** with densely hairy young growth; **branches** spreading to ascending; **branchlets** initially very hairy; **leaves** 0.5–3.5 cm × 0.3–1.2 cm, usually elliptical but may be ovate-lanceolate or cordate, usually in irregular whorls, deep green above, glabrous and faintly rough above, silky-hairy below, margins recurved, venation reticulate, apex ending in a somewhat pungent point; **stipules** present or absent; **flowers** pea-shaped to about 1 cm across; in short, crowded, terminal or axillary racemes, often very profuse and conspicuous; **standard** yellow to orange-yellow, often with reddish markings; **keel** red; **calyx** hairy; **pods** about 0.8 cm long, ovoid, densely hairy, apex pointed.

This floriferous and decorative species has many variants. It inhabits open forest and woodland and is often associated with shallow soils in rocky regions. Adapts well to a range of acidic soils which drain freely. A site with filtered sun or semi-shade should aid successful cultivation. Plants tolerate most frosts. It is not common in cultivation but a semi-erect, shrubby form is the most popular. Excellent for a wide range of general applications. *Oxylobium ellipticum* is very sensitive to *Phytophthora cinnamomi*. In 1805, plants were grown in England from collections made in Tasmania.

129

Oxylobium ellipticum   W. R. Elliot

Propagate from pre-treated seed, or from cuttings of firm young growth, which strike readily.

The var. *angustifolium* Benth. is referrable to *O. arborescens* R. Br. *O. robustum* differs in having longer leaves.

Oxylobium genistoides Meisn. =
           Chorizema genistoides (Meisn.) C. A. Gardner

Oxylobium graniticum S. Moore =
           Gastrolobium graniticum (S. Moore) Crisp

Oxylobium heterophyllum (Turcz.) Benth. =
           Gastrolobium heterophyllum (Turcz.) Crisp

Oxylobium ilicifolium (Andrews) Domin =
   Podolobium ilicifolium (Andrews) Crisp & P. H. Weston

Oxylobium lanceolatum (Vent.) Druce =
           Callistachys lanceolata Vent.

Oxylobium linearifolium (G. Don) Domin =
           O. lineare (Benth.) Benth.

Oxylobium lineare (Benth.) Benth.
This species is certain to be placed in another genus and a full description will be published in the Supplement at an appropriate time.

Oxylobium melinocaule E. Pritzel =
           Nemcia retusa (Lindl.) Crisp

Oxylobium microphyllum Benth.
This species will be transferred to an as yet undescribed new genus and will be included in the Supplement at an appropriate time.

Oxylobium obovatum Benth. =
           Nemcia dilatata (Benth.) Crisp

Oxylobium obtusifolium Sweet =
   Chorizema obtusifolium (Sweet) J. M. Taylor & Crisp

Oxylobium parviflorum Lindl. =
           Gastrolobium parviflorum (Benth.) Crisp

Oxylobium procumbens F. Muell. =
   Podolobium procumbens (F. Muell.) Crisp & P. H. Weston

Oxylobium pulteneae DC.
(after the genus *Pultenaea*)
NSW   Wiry Shaggy Pea
0.2–3 m × 1–2.5 m   Nov–Feb
   Dwarf to medium spreading **shrub** with a thick woody rootstock and hairy young growth; **branches** spreading to ascending, often intertwined; **branchlets** slender, hairy; **leaves** 0.5–1.5 cm × up to 0.3 cm, variable, linear, elliptic or triangular, alternate, opposite or in irregular whorls, deep green and faintly rough above, faintly hairy below, margins recurved, apex ending in a hooked point; **stipules** minute; **flowers** pea-shaped, about 1 cm across, orange-red, in terminal clusters, often profuse and conspicuous; **calyx** hairy; **pods** about 1 cm long, ovoid, faintly hairy, apex pointed.
   A densely foliaged, often wiry species from the North Coast, Central Coast and Central Western Slopes. It occurs in open forest and swamp forest. This species was first cultivated in 1824 in England but it is uncommon in cultivation today. Plants require semi-shaded sites with acidic soils which drain adequately although they tolerate extended moist periods. They

Oxylobium pulteneae × .75

130

tolerate moderate frosts and should respond well to regular pruning. Propagate from pre-treated seed, or from cuttings of firm young growth.

Oxylobium racemosum (Turcz.) C. A. Gardner =
Gastrolobium racemosum (Turcz.) Crisp

Oxylobium reticulatum Meisn. =
Nemcia reticulata (Meisn.) Domin

Oxylobium retusum R., Br. ex Lindl. =
Nemcia coriaceum (Sm.) Domin

Oxylobium rigidum C. A. Gardner =
Gastrolobium rigidum (C. A. Gardner) Crisp

Oxylobium robustum Joy Thomps.
(robust)
Qld, NSW
1.5–3.5 m × 1–2.5 m   Sept–Dec
Small to medium **shrub** with hairy young growth; **branches** spreading to ascending; **branchlets** hairy; **leaves** 2.5–8 cm × 0.3–1 cm, linear to narrowly ovate, often opposite or in whorls, usually not crowded, deep green, faintly rough and hairy to glabrous above, hairy below, margins recurved, apex pungent-pointed; **stipules** absent; **flowers** pea-shaped, about 1 cm across, yellow to orange often with reddish markings, in short axillary or terminal racemes, often profuse and very conspicuous; **calyx** rusty-hairy; **pods** about 1 cm long, oblong.

This floriferous species occurs in the Moreton and Wide Bay Districts in south-eastern Qld and in the North Coast District of NSW. It inhabits well-drained soils in open forests. It has been confused with *O. ellipticum* (see below) and may be marketed or cultivated under that name. Plants should adapt to a wide range of acidic soils. Well suited to filtered sunlight and semi-shaded sites. Useful for background and accent planting. Propagate from pre-treated seed, or from cuttings of firm young growth.

O. ellipticum is closely allied but it has shorter leaves.

Oxylobium scandens (Sm.) Benth. =
Podolobium scandens (Sm.) DC.

Oxylobium spathulatum Meisn. =
Chorizema spathulatum (Meisn.) J. M. Taylor & Crisp

Oxylobium spectabile Endl. =
Gastrolobium spectabile (Endl.) Crisp

Oxylobium tetragonophyllum E. Pritzel =
Gastrolobium tetragonophyllum (E. Pritzel) Crisp

Oxylobium tricuspidatum Meisn. =
Nemcia tricuspidata (Meisn.) Crisp

Oxylobium species (Gibraltar Range, NSW) =
Podolobium aestivum Crisp & P. H. Weston

OXYSTELMA R. Br. = *CYNANCHUM* L.

Oxylobium robustum   D. L. Jones

OZOTHAMNUS R. Br.
(from the Greek *ozo*, branch; *thamnus*, a shrub)
Asteraceae (alt. *Compositae*)

Dwarf to tall **shrubs** or rarely **herbs**; young growth usually woolly-hairy; **branches** usually woody; **leaves** alternate, entire except for rarely lobed base; **inflorescence** small flowerheads with a few female outer florets, in axillary or terminal corymbs, or panicles; **involucre** small, oblong ovoid or somewhat bell-shaped; **bracts** overlapping, white, creamy or straw-coloured; **achenes** glabrous, shortly villous.

Recently (1991) *Ozothamnus* was reinstated after having been regarded as a synonym of *Helichrysum*. It is still under study and the botanical status of some species is not finalised.

It is a sizeable genus which now contains about 56 species with 48 confined to Australia and a further 7 in NZ, and one species occurs in New Caledonia. It is represented in all states of Australia though the main concentration is in the eastern states. Generally *Ozothamnus* species are found from the coast to the slopes of low mountains but some such as *O. alpinus* and *O. hookeri* occur in subalpine and alpine regions where they are often subjected to low temperatures and lengthy periods of snow.

In nature, some species are pioneer plants of disturbed areas and may die after 5–10 years, (eg. *O. ferrugin-eus* and *O. rogersianus*).

Although not generally regarded as highly ornamental plants, many warrant cultivation because they are quick-growing and provide massed displays of flowerheads (eg. *O. argophyllus, O. blackallii, O. cordatus, O. diosmifolius, O. lepidophyllus, O. secundiflorus* and *O. thyrsoideus*). *Ozothamnus diosmifolius* is cultivated to a limited degree in gardens and is popular in the floriculture industry. A number of other species have potential for similar use including *O. occidentalis, O. obcordatus, O. reticulatus, O. retusus* and *O. stirlingii*. The flowers are a valuable source of nectar for butterflies .

In general, *Ozothamnus* species require neutral to

acidic soils with good drainage. Some tolerate open sunny sites but most do best in dappled shade or semi-shade. Many species are hardy to moderate or heavy frosts but they usually dislike long periods of high humidity combined with high temperatures. Little is known about their smog tolerance, but *O. ferrugineus* and *O. turbinatus* grow in areas known to have smog.

Fertilising of *Ozothamnus* species is usually not necessary but light applications of slow-release fertilisers can be beneficial for slow-growing plants.

Natural rainfall is usually all that is required for members of this genus but if plants are grown in soils which dry out readily, or in sheltered dry sites, then supplementary watering may be necessary.

Most *Ozothamnus* respond very well to tip pruning when young, which promotes bushy growth. Generally they tolerate hard pruning each year during or immediately after flowering but they usually do not respond very well to irregular hard pruning into old leafless wood.

Pests are not a major problem with *Ozothamnus* except for stem borers which may prove troublesome in some species. Borers are difficult to control. Borer activity is often signalled by sawdust on the ground around the trunk, or webs composed mainly of sawdust surrounding the trunk or branches. For control see Volume 1, page 143.

Propagation is from seed or from cuttings. The seed needs to be fresh and barely covered by the propagating medium. Some of the species from higher altitudes may benefit from stratification at 4°C for 2–3 weeks before sowing. See Volume 1, page 207 for further information. Cuttings of firm young growth, which is barely woody, usually give the best results — roots often form within 2–4 weeks. Soft growth wilts very readily and it may be difficult for the cuttings to return to their original turgidity. Cuttings need to have at least 4 nodes above the propagating medium because the nodes of some species are prone to die during the propagating period and although the cuttings do form roots they may not have any growth points.

Ozothamnus adnatus DC.
(attached to something)
NSW, Vic
1–4 m × 0.5–1.5 m   Dec–March

Small to medium **shrub** of heath-like appearance; **branches** erect; **branchlets** woolly-hairy, with lines of raised scars from leaf bases; **leaves** 0.6–0.8 cm × 0.1–0.2 cm, linear, dark green above, white woolly-hairy below, appressed against stem, with broad bases, blunt; **flowerheads** about 0.3 cm across, white and straw-coloured, in dense terminal clusters; **bracts** loose, blunt; **disc florets** white.

Restricted to the highlands of south-eastern NSW and eastern Vic. Grows in moist sites often close to streams. This species has interesting and unusual foliage. Plants adapt well to cultivation and succeed best in a partially protected position in well-drained soil. They respond well to mulches and watering during dry periods. Regular light pruning is recommended and plants should be cut back after flowering. Tolerates moderately heavy frosts. Propagate from seed, or from cuttings of firm woody growth, which strike readily.

This species is similar to many other *Ozothamnus* but readily distinguished by its appressed linear leaves and lines of raised leaf scars on stems.

Previously known as *Helichrysum adnatum* (DC.) Benth.

Ozothamnus alpinus (N. A. Wakef.) Anderb.
(alpine)
NSW, Vic   Alpine Everlasting
0.5–1.5 m × 0.5–1 m   Dec–March

Dwarf to small compact **shrub**; **branches** densely covered in white-woolly hairs; **leaves** 0.4–1 cm × 0.1–0.3 cm, oblong, dark green above, grey-green below, widely spreading, stiff, midrib prominent beneath, blunt; **flowerheads** about 0.3 cm across, white, shortly-stalked, in dense terminal clusters; **bracts** pink or rosy-red, prominent in bud; **disc florets** white, crowded.

Restricted to subalpine regions in south-eastern NSW and north-eastern Vic. Grows amongst low vegetation, often on slopes and near small streams. This attractive shrub is well suited to regions with a cold winter climate where best flowering is achieved. Plants will grow in full or partial sun and need abundant moisture in well-drained soil. An excellent container plant. Regular pruning is advisable. Tolerates moderately heavy frosts and snow. Propagate from seed, or from cuttings of firm young growth, which can be slow to strike.

Previously known as *Helichrysum alpinum* N. A. Wakef.

Ozothamnus antennaria Hook. f.
(feeler-like threads)
Tas
1–2.5 m × 0.6–1.5 m   Dec–Feb

Small to medium **shrub**; young growth sticky and hairy; **branches** usually erect; **branchlets** angled, short, slightly sticky, becoming glabrous; **leaves** about 1–3 cm × up to 1 cm, oblanceolate to obovate, short-stalked, aromatic, light to mid green, glabrous and slightly stickly above, paler and somewhat resinous below, margins recurved, apex rounded; **flowerheads** about 0.2 cm across, buff-cream, borne in terminal clusters about 2–3 cm across, profuse and conspicuous.

O. antennaria is a dweller of hills and slopes at 150–920 m altitude. This brightly-foliaged species is rarely encountered in cultivation. It has been found to tolerate freely draining, acidic or alkaline soils which have a high organic content. Prefers a semi-shaded site. Hardy to moderately heavy frosts but suffers from extended periods of frost and snow. Responds well to light pruning. Grown to a limited extent in England and Ireland. Propagate from seed, or from cuttings, which strike readily.

Previously known as *Helichrysum antennarium* (DC.) F. Muell. ex Benth.

Ozothamnus argophyllus (A. Cunn. ex DC.) Anderb.
(silver leaves)
NSW, Vic, Tas   Spicy Everlasting
1–3 m × 0.5–2 m   Oct–Dec

Small to medium **shrub** with pale green young growth; **branches** few, erect; **branchlets** silvery, hairy; **leaves** 3–9 cm × 0.5–1 cm, linear-lanceolate, dark green above, silvery-grey and hairy below, stiff; **flowerheads** about 0.4 cm across, white, fragrant, in dense flat clusters to about 5 cm across; **bracts** straw-coloured; **disc florets** white.

Distributed from central NSW through eastern Vic to Tas. Grows in open forest and is often prominent on hills close to the coast. Showy when in flower with a sweet spicy fragrance. Plants have ornamental foliage but can tend to become very straggly. Prune regularly especially after flowering. Best growth is achieved in partial or filtered sun, in well-drained soils. Tolerates light to moderate frosts. Propagate from seed or from cuttings of firm young growth.

Previously known as *Helichrysum argophyllum* (A. Cunn. ex. DC.) N. A. Wakef.

Ozothamnus backhousei Hook. f. —
see *O. rodwayi* Orchard

Ozothamnus bilobus (N. A. Wakef.) Anderb. =
O. retusus Sond. & F. Muell.

Ozothamnus blackallii N. T. Burb.
(after William E. Blackall, 19–20th-century medical practitioner, botanist and botanical artist, WA)
?SA, WA   Milky Everlasting
0.4–1 m × 0.5–1 m   Sept–Dec

Dwarf **shrub**; young growth hairy; **branches** erect; **branchlets** slender, ascending to erect, initially densely hairy; **leaves** to 1 cm × about 0.1 cm, linear, ascending, crowded, lower half attached to branchlet, green, faintly rough and glabrous above, densely hairy between recurved margins below, apex pointed; **flowerheads** about 0.3 cm across, oblong, milky-white, arranged in terminal clusters about 2.5 cm across, often profuse and conspicuous.

An interesting species with somewhat shell-like flowerheads. It is found in the Coolgardie and Eyre Districts of WA. Evidently not cultivated, it has potential for amenity planting and in cut-flower production. Suitable for semi-arid and temperate regions. Needs fairly well-drained soils. Plants will tolerate sunshine or semi-shade. Hardy to moderate frosts and lengthy dry periods. Response to pruning is not recorded. Propagate from seed or from cuttings.

Previously known as *Helichrysum blackallii* N.T. Burb.
O. cassiope is similar but has yellow-brown flowerheads and stem-clasping leaves.

Ozothamnus cassinioides (Benth.) Anderb.
(like the genus *Cassinia*)
Qld, NSW
1–3 m × 0.5–1.5 m   Aug–Dec

Small to medium **shrub**; **bark** papery; **stems** slender, erect, much-branched, with woolly hairs; **leaves** 1–3 cm × 0.1–0.25 cm, linear to narrowly elliptical, sessile to subsessile, dark green above, woolly-white below, margins recurved; **flowerheads** about 0.3 cm across, white and straw-coloured, in dense terminal corymbs; **bracts** straw-coloured; **disc florets** white; **pappus** about 0.4 cm long.

Restricted to southern Qld and northern NSW. It usually grows among rocks. Plants are never showy but have an appealing growth habit and would make an interesting addition to a garden. Very hardy to dryness once established. Regular light pruning is beneficial. Plants will grow in most sunny positions in well-drained soil. Tolerates light to moderate frosts. Propagate from seed or from cuttings of firm young wood.

Previously known as *Helichrysum cassinioides* Benth.

Ozothamnus cassiope (S. Moore) Anderb.
(probable reference to similarity to *Cassiope*)
WA
1–2 m × 0.6–1.5 m   Oct–Dec

Small **shrub**; young growth hairy; **branches** ascending to erect; **branchlets** slender, initially hairy, rigid; **leaves** 0.2–0.3 cm × about 0.1 cm, linear, sessile and stem-clasping at the lobed base, ascending to erect, green and glabrous above, hairy below, margins recurved, apex bluntly pointed; **flowerheads** about 0.2 cm across, conical, yellow-brown, arranged in somewhat loose terminal clusters of about 2.5 cm across, often profuse; **outer bracts** pale brown, shining; **achene** glabrous or slightly rough.

This poorly known species occurs in the Coolgardie District where it grows in sandy soils as part of the mallee or woodland. It has moderately ornamental features and should be best suited to semi-arid and warm temperate regions in well-drained warm sites.

Ozothamnus cassinioides × .5

133

Ozothamnus conditus

Ozothamnus diosmifolius   W. R. Elliot

Plants are hardy to moderate frosts. They may become leggy and pruning should begin at planting stage to promote bushy growth. May be suitable for floriculture. Propagate from seed, or from cuttings of firm young growth, which strike readily.

Previously known as *Helichrysum cassiope* S. Moore.

O. blackallii is allied but differs in having milky-white ovoid flowerheads and leaves which have their lower half attached to the branches.

Ozothamnus conditus (N. A. Wakef.) Anderb.
(stored)
NSW, Vic   Pepper Everlasting
1–3 m × 1–2 m   Nov–Jan
Small to medium, lightly branched **shrub** with woolly-white young growth; **branches** woolly-white, sticky; **leaves** 1–3.5 cm × 0.15–0.3 cm, linear to narrowly lanceolate, dark green above, white or brown-hairy below with a dark midrib; **flowerheads** 0.2–0.3 cm across; **bracts** with thick white tips; **pappus** with thickened tips.

An interesting shrub which is found in the Southern Tablelands of NSW and north-eastern Vic usually on dry stony slopes and ridges. The leaves have a spicy or peppery scent when crushed. Plants can be grown in filtered or partial sun, in well-drained soils. They tend to become straggly, so regular pruning is advisable. Tolerates moderately heavy frosts. Propagate from seed or from cuttings of firm young growth.

Previously known as *Helichrysum conditus* N. A. Wakef.

Ozothamnus costatifructus (R. V. Sm.) Anderb.
(ribbed fruits)
Tas
1–3 m × 0.6–2.5 m   Nov–Jan
Small to medium **shrub** with whitish-hairy young growth; **branches** ascending to erect; **branchlets** whitish-hairy when young; **leaves** 1.5–3 cm × to about 0.4 cm, narrowly linear, spreading to reflexed, numerous, green and somewhat glossy above, densely creamy-hairy below, margins recurved, apex pointed, aromatic; **flowerheads** to about 0.7 cm across, creamy white, with buff outer bracts, many in terminal clusters to 5 cm across, profuse and conspicuous.

This species is moderately ornamental. It has a very limited distribution in coastal forests of eastern and southern Tas. Although limited in cultivation, it is recorded as doing very well on the Scilly Isles of southwestern England. Requires good drainage and probably will do best with some shade. Hardy to moderately heavy frosts and dislikes extended dry periods. Tip pruning from an early stage is recommended for promoting a bushy framework, and regular pruning after flowering. It does well in containers. Propagate from seed or cuttings.

Previously known as *Helichrysum costatifructum* R. V. Sm.; *H. buftonii* N. T. Burb. is a synonym.

O. reticulatus (Labill.) DC. is allied but it has a spreading habit and longer, somewhat wrinkled leaves.

Ozothamnus cuneifolius (Benth.) Anderb.
(wedge-shaped leaves)
NSW, Vic   Wedge Everlasting
1–3 m × 0.5–2 m   Nov–Jan
Small to medium **shrub**, usually bushy; young growth silvery, densely woolly-hairy; **branchlets** densely woolly-hairy; **leaves** 1–3 cm × 0.4–0.8 cm, wedge-shaped, dark green and glabrous above, cream to green and woolly-hairy below, blunt, margins recurved, slightly irregular; **flowerheads** 0.3–0.5 cm across, white, in dense, rounded, terminal clusters; **bracts** white or pinkish; **pappus** about 0.5 cm long, white.

This species is known from eastern Vic and southeastern NSW. It grows in sheltered moist forests. Although not extremely showy in flower this species has potential because of its bushy habit and interesting foliage. May be suitable for floriculture. Plants grow well in a partially protected position and respond well to regular watering and mulches. Tip pruning is recommended from an early age to promote a bushy habit. Tolerates light to moderate frosts. Propagate from seed, or from cuttings of firm young growth, which strike readily.

Previously known as *Helichrysum cuneifolium* Benth.

Ozothamnus decurrens F. Muell.
(decurrent)
NSW, Vic, SA   Ridged Everlasting
0.2–1 m × 0.6–1.5 m   Nov–Feb
Dwarf **shrub**; young growth yellowish with glandular and woolly hairs; **branches** rigid, woolly-hairy with yellow glandular hairs; **leaves** 0.5–1 cm × about 0.1 cm, linear, with a broad decurrent base, green and glandular above, hairy below, apex broad or tapered with a

small recurved point; **flowerheads** about 0.3 cm across, in more or less crowded, terminal corymbs; **bracts** dull cream to straw-coloured; **florets** all bisexual.

Ridged Everlasting is usually found in sandy soils as an understorey shrub. It occurs in semi-arid open mallee shrubland and it should grow well in semi-arid and warm temperate regions. Plants have an appealing rounded growth habit. Good drainage is essential and a sunny or semi-shaded aspect will be suitable. Heavy frosts may damage plants. Response to pruning is not recorded but plants should respond well to light pruning. Propagate from seed or from cuttings of firm young growth.

Helichrysum catadromum N. A. Wakef. and *Ozothamnus catadromus* (N. A. Wakef.) Anderb. are synonyms.

O. retusus has similarities but is usually an upright shrub.

Ozothamnus diosmifolius (Vent.) DC.
(leaves like the genus *Diosma*)

Qld, NSW   Rice Flower; Sago Flower; Pill Flower; Ball Everlasting

2–5 m × 1–3 m   mainly Aug–Dec; sometimes Feb–April

Medium to tall **shrub** which is often sparse in the centre; **bark** papery; young growth light green; **leaves** 1–2.5 cm × 0.1–0.25 cm, linear, sessile, dark green and roughened above, woolly-hairy below, margins revolute, obtuse; **flowerheads** about 0.4 cm across, cream, white or pinkish, in broad, dense, domed to almost flat terminal heads; **bracts** white or pinkish; **disc florets** cream; **pappus** about 0.2 cm long.

Ozothamnus ericifolius   D. L. Jones

A common species extending from central Qld to south-eastern NSW and often prominent in open forests. The floral display is attractive and produced over a long period. Fresh or dried cut flowers are excellent for decoration.

Plants may become straggly and pruning during and after flowering is recommended. This species is adaptable and quick-growing but often short-lived. Suits a range of positions and soils. Best flowering is achieved in a sunny location. Tolerates light to moderately heavy frosts. Known to tolerate Cinnamon Fungus, *Phytophthora cinnamomi* as well as *P. nicotianae*.

This species shows variability in flower colour and some variants have been selected for cultivation. A soft-pink flowered variant from near Putty, NSW, performs well in south-eastern regions. A number of selections for cut-flower production have originated from selection and breeding trials in south-eastern Qld. 'Cook's Snow White' is a bushy selection which has round white buds in flat-topped heads, from 6–9 cm across, on slender stems. It is registered under the *Plant Breeder's Rights Act*. 'Cook's Tall Pink' has loose, flat-topped heads of oval to pointed pink buds which have a white base. It is also registered under the *Plant Breeder's Rights Act*. 'Cook's Salmon Pink' produces oval salmon buds, with white tips, in tight domed heads from 6.5–9.5 cm across. 'Redlands Sandra' is provisionally registered under the *Plant Breeder's Rights Act*. It is an upright selection which has creamy-white, pointed buds in heads from 6–8 cm across.

Propagate from seed or from cuttings of firm young wood. All selections need to be propagated vegetatively to retain their characteristics.

Previously known as *Helichrysum diosmifolium* (Vent.) Sweet

Ozothamnus diotophyllus (F. Muell.) Anderb.
(leaves like 2 ears)

Qld, NSW   Heath Everlasting

0.3–1 m × 0.3–1 m   Aug–Feb

Dwarf **shrub**, often straggly; **branches** slender, erect, covered with loose cottony wool; **leaves** 0.2–0.6 cm × 0.1 cm, narrowly lanceolate, scale-like, stem-clasping, with prominent ear-like auricles at the base; **flowerheads** about 0.3 cm across, yellow, in dense terminal panicles; **bracts** straw-coloured; **disc florets** yellow; **pappus** about 0.2 cm long.

An uncommon species from western regions, where it grows among mallee scrub in red sandy soils. Plants are generally inconspicuous but are very attractive when flowering. Well suited to cultivation as an ornamental in inland regions and also as an interesting container plant. Inclined to become straggly. Regular pruning from an early age is recommended. Best grown in an open sunny location in well-drained soils. Tolerates light to moderate frosts. Propagate from seed or from cuttings of firm young growth.

Previously known as *Helichrysum diotophyllum* F. Muell.

Ozothamnus ericeteus (W. M. Curtis) Anderb. =
O. ericifolius Hook. f.

Ozothamnus ericifolius Hook. f.
(leaves similar to those of the genus *Erica*)
Tas
1.5–3 m × 0.8–2 m   Dec–Feb

Small to medium **shrub**; young growth hairy, sticky; **branches** erect; **branchlets** short, ascending, with dense white hairs; **leaves** 0.4–0.8 cm × to 0.1 cm, oblong-linear, spreading or reflexed, sticky, sweetly spicy aromatic, hairy or becoming glabrous above densely hairy below, margins revolute, apex blunt; **flowerheads** about 0.5 cm across, in dense terminal clusters to about 2.5 cm across, profuse and most conspicuous; **ray florets** white; **bracts** buff-brown.

A profusely flowering species which usually develops as a column-like plant. It occurs in montane heaths and is common in the Great Lake region on the Central Plateau. It has potential for greater use in gardens and possibly for commercial cut-flower production. Needs freely draining soils. Plants are tolerant of open sites but need to have adequate soil moisture. They may do best in temperate regions where sites receive some shade. Hardy to moderately heavy frosts. Propagate from seed or from cuttings.

Previously known as *Helichrysum ericetum* W. M. Curtis; *H. ledifolium* ssp. *ericifolium* and *O. ericeteus* (W. M. Curtis) Anderb. are synonyms.

O. ledifolius is closely allied, but it grows as a spreading shrub and has brighter yellow-brown, yellow or reddish involucral bracts.

Ozothamnus expansifolius (Sieber ex P. Morris & J. H. Willis) Anderb.
(unfolded leaves)
Tas
1–2 m × 0.6–1.5 m   Dec–Feb

Small **shrub**; young growth hairy, sticky; **branches** erect, stout; **branchlets** densely covered in white-woolly hairs; **leaves** 0.2–0.6 cm × about 0.1 cm, more or less linear to narrowly triangular, broad base, spreading to ascending, crowded, greyish-green and sticky above, whitish-hairy below, margins nearly flat to revolute, apex pointed; **flowerheads** to about 0.6 cm × 0.3 cm, cream to white, or brown to reddish-brown, very short-stalked, 6–30 in compact terminal clusters of about 2 cm across; **outer bracts** straw-coloured to reddish.

Regarded as rare in the wild, this species inhabits mountain slopes and heaths at 900–1200 m altitude. It tolerates low temperatures and snowfalls. Plants need good drainage and will adapt to a semi-shaded, dapple-shaded or sunny site. They have potential for cut-flower production in cool temperate regions as well as general cultivation in gardens and containers. Propagate from seed or from cuttings of firm young growth.

Previously known as *Helichrysum expansifolium* (Sieber ex P. Morris & J. H. Willis) N. T. Burb. and *H. hookeri* (Sond.) Druce var. *expansifolium* Sieber ex P. Morris & J. H. Willis.

O. hookeri is allied but it has smaller leaves, thinner branches and more flowerheads per cluster.

Ozothamnus ferrugineus (Labill.) DC.
(rust-coloured)
NSW, Vic, Tas, SA   Tree Everlasting
2–5 m × 2–4 m   Nov–Feb

Small to tall **shrub** with an open habit; **bark** grey-brown, fibrous; young growth light green, hairy; **leaves** 2–7 cm × 0.2–0.4 cm, linear to linear-lanceolate, dark green and glabrous above, grey-green or yellow-green and woolly below, margins recurved or wavy; **flowerheads** about 0.4 cm across, white, in dense terminal corymbs, 5–12 cm across; **bracts** opaque; **pappus** plumed.

Widely distributed from north-eastern NSW to Tas and south-eastern SA. It grows in open forest and woodland in well-drained and moist to wet sites. Young plants are vigorous and bushy but the canopy becomes sparse with age. Flowering continues over a long period and displays can be profuse. Plants are very adaptable to a wide range of soils and positions. Hardy to most frosts. Regular pruning from an early age is strongly recommended. Propagate from seed or from cuttings of firm young growth.

Previously known as *Helichrysum dendroideum* N. A. Wakef.

Ozothamnus gunnii Hook. f.
(after Ronald G. Gunn, 19th-century botanical collector, Tas)
Tas
1–2 m × 0.6–1.5 m   Dec–March

Small **shrub** with white and woolly-hairy young growth; **branches** ascending to erect; **branchlets** woolly-hairy; **leaves** 2.5–4 cm × 0.2–0.3 cm, narrowly linear to linear-lanceolate, spreading, green and nearly glabrous or with scattered long white hairs above, dense long-tangled hairs below, margins recurved to revolute, apex pointed; **flowerheads** about 0.5 cm long, cream with pale brown, in terminal corymbose panicles, about 1 cm across, often profuse; **bracts** pale brown; **pappus** minutely barbed.

O. gunnii occurs on Flinders Island and along the coast of north-eastern Tas. It inhabits sandy soils. It is moderately attractive and is well-suited to cultivation in temperate regions in a sunny or semi-shaded, well-drained site. Plants may be useful for providing quick growth while slower-growing plants develop. They can become open and may need regular light pruning to promote bushy growth. Hardy to moderate frosts. Propagate from seed or from cuttings.

Previously known as *Helichrysum gunnii* (Hook. f.) F. Muell. ex Benth.

Ozothamnus hookeri Sond.
(after John D. Hooker, 19–20th-century English botanist)
NSW, ACT, Vic, Tas   Scaly Everlasting; Kerosene Bush
0.5–2 m × 1–2 m   Dec–Feb

Dwarf to small bushy **shrub**; **branches** erect; **branchlets** densely covered with white hairs; **leaves** 0.1–0.3 cm × 0.1–0.15 cm, narrowly triangular-ovate, closely appressed to stem, bright green to golden-green above, woolly-hairy beneath but usually hidden by closely revolute margins, sticky; **flowerheads** about 0.2 cm across, yellowish-green to cream, sessile, in small, dense, terminal clusters about 1.5 cm across; **pappus** about 0.3 cm long.

Restricted to montane and subalpine areas of south-eastern NSW, eastern Vic and Tas, but often locally common. Although not highly showy this species has an attractive dense growth habit and dark green or rarely golden-green scaly leaves which contrast with the woolly-white branches. Sometimes plants become straggly but they respond well to pruning. Suitable for use among rocks and makes an excellent container plant. Best suited to cool temperate regions. Plants will grow in full sun but need an abundance of water. Tolerates moderately heavy frosts and snowfalls. Propagate from seed or from cuttings of firm young growth.

Previously known as *Helichrysum hookeri* (Sond.) Druce.

O. expansifolius differs in having stouter stems, larger leaves and larger flower clusters.

Ozothamnus kempei (F. Muell.) Anderb.
(after Friedrich A. H. Kempe, co-founder of Hermannsburg Mission, NT)
WA, NT
0.5–1 m × 0.3–1 m   Aug–Oct; also sporadic
Dwarf **shrub**; **branches** and **branchlets** covered with woolly-white hairs; **leaves** 0.8–3.5 cm × 0.1–0.2 cm, linear, dark green and papillose above, woolly-hairy below, margins recurved to revolute; **flowerheads** about 0.5 cm across, cream to greenish, in dense terminal clusters about 2 cm across; **bracts** clear with a yellow central rib.

A little-known species restricted to Central Australia. It is not cultivated widely but has prospects for inland areas with a hot dry climate. Requires an open sunny position with free drainage and responds well to watering during long periods of dryness. Tolerates light to moderate frosts. Propagate from seed or from cuttings of firm young growth.

Previously known as *Helichrysum kempei* F. Muell.

Ozothamnus ledifolius (DC.) Hook. f.
(leaves similar to those of the genus *Ledum*)
Tas   Kerosene Bush
0.6–1.5 m × 0.8–2 m   mainly Nov–Feb
Dwarf to small **shrub**; young growth sticky, bright yellow; **branches** spreading to ascending, older ones with prominent leaf scars; **branchlets** spreading, sticky, densely hairy; **leaves** 0.7–1.5 cm × to about 0.2 cm, linear, spreading to reflexed, crowded, thick, sticky, hairy to nearly glabrous above, densely hairy below, margins revolute, apex pointed, sweetly spicy aromatic; **flowerheads** about 0.5 cm across, arranged in crowded terminal clusters to 3 cm across, profuse and most conspicuous; **ray florets** white; **bracts** yellow-brown, yellow or reddish.

The floriferous display and sweetly aromatic foliage (most noticeable in warm weather) make this a desirable species. It is found above an altitude of 750 m and is widespread in the mountains of Tas. Cultivated to a limited extent, it is valuable for temperate regions where it can be used for general planting and in rockeries. Needs freely draining soils. A fairly open or semi-shaded site is suitable. Tip pruning of young plants promotes bushy growth. Regular pruning after flowering may be required in warm temperate areas. Hardy to heavy frosts. It has potential as a pot plant, and lasts well as a cut flower. Plants are hardy on the southern

Ozothamnus kempei × .45, leaf × 1; flowerhead × 2

coast of England. Propagate from seed, which can be difficult to germinate, or from cuttings, which strike readily.

Previously known as *Helichrysum ledifolium* (DC.) Benth.

O. lycopodioides has smaller leaves and sessile flowerheads.

Ozothamnus lepidophyllus Steetz
(scaly leaves)
WA   Scaly-leaved Everlasting
0.4–0.8 m × 0.5–1 m   Oct–Jan
Dwarf **shrub**; young growth hairy; **branches** ascending to erect, old bark stringy or fibrous; **branchlets** short, slightly hairy; **leaves** to 0.1 cm long, minute, ovate to orbicular, crowded, strongly reflexed, green and glabrous above, hairy below, margins recurved to revolute, apex blunt; **flowerheads** to about 0.5 cm, oblong-cylindrical with milky-white bracts, in terminal clusters, often profuse and most conspicuous.

An interesting species with tiny crowded leaves and milky-white flowerheads. It occurs in the Avon, Eyre and Roe Districts where it is found inland as well as on or near the coast. It usually grows on sandy soils. While plants are evidently rarely cultivated, they have potential for greater use in semi-arid and temperate regions. Drainage needs to be excellent. A site with plenty of sunshine and some shade should be ideal. Hardy to light frosts. Withstands lengthy dry periods but supplementary watering would probably be beneficial. Reaction to pruning is not known. Suitable for gardens and containers and may be an appropriate plant for floriculture. Propagate from seed or cuttings.

Previously known as *Helichrysum lepidophyllum* (Steetz) Benth.

Ozothamnus hookeri W. R. Elliot

Ozothamnus ledifolius D. L. Jones

Ozothamnus lycopodioides Hook. f.
(similar to the clubmoss genus *Lycopodium*)
Tas
0.6–1 m × 0.6–1.5 m   July–Oct

Dwarf **shrub**; young growth sticky; **branches** ascending to erect, leaf scars prominent; **branchlets** spreading to ascending, sticky when young, glabrous; **leaves** 0.5–0.8 cm × to about 0.2 cm, linear-oblong, sessile, spreading, crowded, deep green, appearing glabrous and somewhat sticky above, margins recurved to revolute, apex blunt; **flowerheads** to about 0.5 cm across, white with buff to brown outer bracts, arranged in terminal clusters to about 1.5 cm across, profuse and conspicuous.

Enthusiasts praise this dwarf species, with its conifer-like foliage and often fulsome floral display, but it can flower poorly in some regions. It occurs on the eastern coast of Tas where it is frequently seen on rocky slopes. Usually adapts well to most soil types provided drainage is adequate, although some growers have experienced problems in maintaining plants. Suitable for sunny or semi-shaded sites. Withstands periods of wetness. Hardy to moderately heavy frosts. Responds well to pruning. Useful for general planting and does very well as a container plant. Propagate from seed, or from cuttings, which strike readily.

Previously known as *Helichrysum lycopodioides* (Hook. f.) Benth.

The allied *O. ledifolius* has yellowish new growth, much larger leaves and stalked flowerheads.

Ozothamnus obcordatus DC.
(inversely heart-shaped leaves)
Qld, NSW, Vic, Tas   Grey Everlasting
0.5–1.5 m × 0.3–1.5 m   Aug–Dec

Dwarf to small **shrub**, slender or spreading; **branches** erect, arising from near base, young growth somewhat shiny and often sticky; **leaves** 0.3–0.5 cm × 0.2–0.3 cm, obovate to obcordate, dark green and shiny above, woolly-grey below, tip reflexed, margins recurved; **flowerheads** about 0.5 cm across, golden-yellow and pale brownish below, profuse, in dense, almost flat, terminal panicles; **disc florets** 6–10; **pappus** slender.

A widely distributed, bushy species which consists of 2 subspecies. The typical subspecies is distributed in NSW, Vic and Tas, whereas ssp. *major* (Benth.) N. T. Burb. with larger leaves (up to 1.5 cm × 0.6 cm) occurs in south-eastern Qld, NSW and Vic. Both variants have ornamental appeal and are worth growing. Plants tend to become open and sparse with age so regular pruning from an early age is advisable. Pruning of old, sparse plants is usually unsuccessful. Suitable sites range from full sun to partial sun or even shade. Soils should be well drained. Tolerates light to moderately heavy frosts. It has potential as a fresh or dried cut flower. First introduced into England in 1829. Propagate from seed or from cuttings of firm young growth. Cuttings are sometimes difficult to strike and those taken from old, hard growth can be extremely difficult.

Previously known as *Helichrysum obcordatum* (DC.) F. Muell. ex Benth.

Ozothamnus obcordatus   W. R. Elliot

Ozothamnus occidentalis (N. T. Burb.) Anderb.

(western)

WA   Rough-leaved Everlasting
0.15–1 m × 0.2–0.75 m   July–Nov

Dwarf **shrub**; young growth hairy; **branches** many, spreading to ascending, ridged from node to node; **leaves** 0.3–0.6 cm × about 0.1 cm, linear, spreading, scattered, rough, margins revolute, apex blunt; **flowerheads** about 0.2 cm across, milky-white, in many-headed terminal clusters of about 2 cm diameter, often profuse; **ray florets** white to milky-yellow; **outer bracts** milky-white; **achenes** faintly warty.

This compact species occurs in the Roe and Coolgardie Districts where it usually is found in sandy soils amongst mallee scrub. Apparently not cultivated, it has potential for use in floriculture and in gardens. It is also worth trying as a container plant. It will need plenty of sunshine and good drainage. Plants should do best in semi-arid and warm temperate regions but may adapt to cool temperate areas if grown in a north-facing sunny site. Hardy to moderate frosts. Response to pruning is not known. Propagate from seed or from cuttings.

Previously known as *Helichrysum occidentale* N. T. Burb.

O. diosmifolius has some similarities but it lacks the very rough leaves and develops into a much taller shrub.

Ozothamnus purpurascens DC.

(becoming purple)

Tas
1–2.5 m × 0.6–1.5 m   Nov–Jan

Small to medium **shrub**, with columnar habit; young growth sticky; **branches** erect; **branchlets** short, hairy;

leaves 0.7–1 cm × up to 0.2 cm, narrowly-linear, spreading or reflexed, strongly aromatic, hairy, green, white-hairy below, margins recurved, apex blunt; **flowerheads** daisy-like, about 0.5 cm across, in crowded terminal clusters to 4 cm across, on branches and branchlets, usually profuse and most conspicuous; **ray florets** white; **disc florets** yellowish.

This aromatic species occurs from sea-level to about 250 m altitude and is a common plant of dry open hillsides in southern Tas. Rarely encountered in cultivation, it has potential for use as a coloniser of disturbed areas. May be useful for cut-flower production. Plants require well-drained soils and tolerate an open or semi-shaded site. Responds very well to regular pruning without which it becomes straggly. Hardy to most frosts. Propagate from seed, or from cuttings, which should root readily.

Previously known as *Helichrysum purpurascens* (DC.) W. M. Curtis. *H. rosmarinifolium* (Labill.) Benth. var. *ledifolium* (DC.) Tovey & P. Morris is also a synonym.

Ozothamnus reticulatus (Labill.) DC.

(netted veins)

Tas
1–3 m × 1–2.5 m   Sept–Dec

Small to medium **shrub**; young growth densely covered with white, woolly hairs; **branches** spreading to ascending, initially white woolly-hairy; **leaves** 3–6 cm × 0.2–0.5 cm, broadly linear, spreading, somewhat crowded, leathery, prominent sunken midrib and netted veins, dark green and glossy with scattered glandular hairs above, densely white woolly-hairy below, margins revolute, apex blunt; **flowerheads** to about 0.7 cm across, creamy-white and pale brown, in dense, terminal, corymbose panicles to about 7 cm across, often profuse and very conspicuous; **achenes** densely hairy.

A vigorous, much-branched species which is found on exposed coastal cliff-faces and rocks in southeastern and southern regions, extending from the Freycinet Peninsula to near Cox Bight. It is worthy of cultivation. Requires good drainage and prefers sunshine but also withstands semi-shaded sites. It has a bushy habit which makes it valuable for exposed coastal and windbreak planting, although it may not have a long lifespan. May be suitable for floriculture. Propagate from seed or cuttings.

Previously known as *Helichrysum reticulatum* (Labill.) Less.

O. costatifructus is closely allied but differs in having shorter leaves and nearly glabrous achenes.

Ozothamnus retusus Sond. & F. Muell.

(notched at apex)

Vic, SA   Rough Everlasting
0.5–1.5 m × 0.5–1 m   Sept–Jan

Dwarf to small **shrub**; young shoots with yellowish glandular hairs and roughened ridges; **leaves** 0.8–1.5 cm × 0.1–0.3 cm, linear to broadly linear-lanceolate, decurrent, dark green and very shiny above, yellow glandular-hairy below, base broad, apex rounded with a recurved mucro; **flowerheads** 0.2–0.3 cm across, cream to white, numerous, in dense terminal heads;

bracts white or straw-coloured; **inner florets** bisexual; **outer florets** female; **pappus** prominent, barbed.

An adaptable species, well suited to inland regions and cooler districts. Plants will grow in a hot sunny position. Soil drainage must be unimpeded and plants must be allowed room to develop as they may suffer from crowding. Regular light pruning is beneficial. Tolerates light to moderate frosts. Propagate from seed or from cuttings of firm young growth.

Previously known as *Helichrysum bilobum* N. A. Wakef. and *Ozothamnus bilobus* (N. A. Wakef.) Anderb.

Ozothamnus rodwayi Orchard
(after Leonard Rodway, 19–20th century Honorary Government Botanist, Tas)
Tas
0.6–1.3 m × 0.6–2 m   Dec–Feb

Dwarf to small **shrub**; young growth slightly sticky; **branches** many, usually spreading to ascending, sometimes procumbent; **branchlets** short, often slightly sticky; **leaves** about 0.7–1.5 cm × 0.4–0.6 cm, cuneate to broadly obovate, crowded, spreading to reflexed, aromatic, dark green and slightly hairy to glabrous above, cream and densely hairy below, apex rounded; **flowerheads** about 0.5 cm across, white with rusty outer bracts, borne in many-flowered, dense, terminal clusters to 3.5 cm across, profuse and conspicuous.

This small and appealing shrubby species provides a bright display in summer. It occurs in a wide range of habitats in Tas including subalpine, peaty heathlands and scrubland. It is relatively rare in cultivation. Plants were introduced into England in 1930 from collections by Harold Comber. They should adapt to a wide range of freely draining soils which may be moisture-retentive for most of the year or even prone to drying out. Plants withstand plenty of sunshine or an aspect with semi-shade or dappled shade. Hardy to most frosts, it has tolerated −15°C and snowfalls. Useful for general planting in rockeries and shrubberies as well as in containers. It has potential for low hedging.

There are two other varieties.

var. *kingii* (W. M. Curtis) P. S. Short has silvery-greyish foliage and the leaves have a hairy covering. This attractive variety should prove popular as it becomes better known.

var. *oreophilum* (W. M. Curtis) P. S. Short has narrow, very dark green leaves and less ornamental flowerheads.

Propagate all forms from seed, or from cuttings of firm young growth, which may be slow to form roots.

Previously known as *Helichrysum backhousei* (Hook. f.) F. Muell. ex Benth. and *Ozothamnus backhousei* Hook. f.

Ozothamnus rogersianus (J. H. Willis) Anderb.
(after Keith C. Rogers, 20th-century Victorian botanical collector)
Vic
1.5–3 m × 1–2 m   Oct–March

Small to medium **shrub**; young growth sticky, yellow-green; **branches** ascending to erect, stiff; **branchlets** short, ascending to erect, sticky, aromatic; **leaves** about 1–2 cm × up to 0.2 cm, linear, spreading to ascending, many, slightly rough, green and becoming nearly

glabrous above, cottony-hairy and often yellowish below, spicy-scented, margins revolute, apex pointed; **flowerheads** to 0.2 cm across, whitish, arranged in dense terminal clusters of up to 4 cm across, profuse and conspicuous.

This very quick-growing, erect species is rare in nature. It has 3 very restricted and disjunct occurrences in East Gippsland, Mt Wellington and the Otway Ranges. It is found in subalpine, mountainous and coastal heath woodland habitats. Evidently rarely cultivated, it needs to be grown for conservation purposes, but it also will appeal to some growers. The spicy fragrance of the foliage is noticeable on hot or rainy days. Needs a semi-shaded aspect in freely draining soils. Young plants respond well to pruning which helps promote bushy growth. It is short-lived in gardens but can be propagated readily from seed or from cuttings.

Previously known as *Helichrysum rogersianum* J. H. Willis.

Ozothamnus rosmarinifolius (Labill.) Sweet
(leaves similar to those of the genus *Rosmarinus*)
NSW, Vic, Tas   Rosemary Everlasting
1.5–3 m × 1–2 m   Sept–Feb

Small to medium **shrub**; **bark** woolly-white; young growth hairy; **branches** ascending to erect; **branchlets** usually ascending to erect, densely hairy; **leaves** 1.5–5 cm × up to 0.2 cm, linear, often crowded, shiny, dark green or rarely greyish, rough and channelled above, densely whitish-hairy below, margins revolute, apex pointed; **flowerheads** about 0.2 cm across, inner bracts small, white, spreading, outer bracts may be pink to reddish, arranged in dense terminal clusters to about 3 cm across, profuse and conspicuous.

Whitish bark contrasts with dark green leaves on this fast-growing and rather ornamental species. A pink-budded form is highly attractive, yet rarely seen in cultivation. *O. rosmarinifolius* is common throughout Tas, and extends to southern and eastern Vic and south-eastern NSW. It is often found in wet heath and beside creeks and rivers. Plants tolerate open, fairly sunny or semi-shaded sites. Soils need to drain readily even though plants withstand extended wet periods. Hardy to light or moderate frosts. Responds well to pruning, which is best begun at an early stage, otherwise plants can become leggy. The flowerheads are useful for floral decoration provided they are cut at an early stage, eg. in bud. A grey-foliaged selection marketed as 'Silver Jubilee' is cultivated in England. Propagate from seed, or from cuttings, which strike readily.

Previously known as *Helichrysum rosmarinifolium* (Labill.) Benth.

Ozothamnus scaber F. Muell.
(rough)
SA
0.7–2 m × 0.5–1.5 m   Sept–Dec

Dwarf to small **shrub**; young growth yellowish and glandular-hairy, rough; **branches** slightly woolly-hairy; **leaves** 0.8–2.5 cm × 0.1–0.3 cm, linear to narrowly oblanceolate, decurrent, dark green and glossy above, yellow glandular-hairy below, apex broad with a recurved point; **flowerheads** 0.2–0.3 cm across, cream to white, in crowded terminal heads; **bracts** subtending individual florets; **inner florets** bisexual; **outer florets** female.

O. scaber inhabits rocky hillsides in the Eastern and Flinders Ranges Districts. It is not strongly ornamental but will be useful in arid and semi-arid regions. Appreciates plenty of sunshine and needs excellent drainage. Plants are hardy to moderate frosts. Light pruning from an early stage will promote bushy growth. Propagate from seed or from cuttings.

Previously known as *Helichrysum bilobum* N. A. Wakef. var. *scabrum* (F. Muell.) N. T. Burb.

O. retusus is allied but differs in having leaves to 1.5 cm long and florets which lack subtending bracts.

Ozothamnus scutellifolius Hook. f.
(little shield-like leaves)
Tas
0.5–1.5 m × 0.5–1.5 m   Sept–Dec

Dwarf to small **shrub**; young growth hairy, sticky; **branches** ascending to erect; **branchlets** short, spreading to ascending, densely hairy; **leaves** to 0.2 cm long, orbicular, appressed base, reflexed, convex, fairly crowded, faintly sticky, dark green above, margins revolute; **flowerheads** very small, white to cream with brownish outer bracts, arranged in terminal clusters of 3–5 heads, profuse.

O. scutellifolius has a wide distribution throughout the state and is usually found on dry hillsides. It is cultivated to a very minor extent. Requires well-drained soils and tolerates a sunny or semi-shaded position. Hardy to moderate frosts and lengthy dry periods. Plants may need regular pruning to promote bushy growth. Suited to general planting, and flowering stems may be suitable for cutting and indoor decoration. Propagate from seed or from cuttings, which strike readily.

Previously known as *Helichrysum scutellifolium* (Hook. f.) Benth.

Ozothamnus secundiflorus (N. A. Wakef.) C. Jeffrey
(flowers arranged on one side)
NSW, Vic   Cascade Everlasting
1–2.5 m × 1.5–3 m   Nov–March

Small to medium **shrub**; young growth hairy; **branches** spreading to ascending, often arching; **branchlets** spreading to arching, striate, densely whitish-hairy; **leaves** 0.8–1.5 cm × to about 0.2 cm, narrowly oblong to narrowly cuneate, spreading, greyish due to dense covering of hairs above, whitish-hairy below, margins flat to slightly recurved, apex pointed; **flowerheads** about 0.2 cm across, white, arranged in dense clusters along and at the ends of short branchlets, profuse and most conspicuous; **outer bracts** usually brownish.

This floriferous, greyish-foliaged species is most common in the Australian Alps, but also occurs in the southern Blue Mountains of NSW and extends southwards to Wilson's Promontory in Vic. Usually found in open sites but also occurs as an undershrub in forests. Cultivated to a limited degree. Needs well-drained soils and does best with a sunny aspect. Hardy to most frosts and moderate snowfalls. Often develops cascading branches. Responds well to pruning. Useful for general planting and for low screening. Propagate from seed, or from cuttings, which strike readily.

Previously known as *Helichrysum secundiflorum* N. A. Wakef.

O. thyrsoideus is sometimes confused with this species, but it differs in having much longer, flattish leaves which are usually glabrous above.

Ozothamnus stirlingii (F. Muell.) Anderb.
(after James Stirling, early Government Geologist, Vic)
NSW, Vic   Ovens Everlasting
1–3 m × 1–2.5 m   Dec–March

Small to medium **shrub** with an open habit; **branchlets** tomentose; young growth somewhat sticky; **leaves** 4–10 cm × 1–2 cm, elliptical to lanceolate, dark green above, greenish-yellow to greenish-brown and hairy below; **flowerheads** 1–1.5 cm across, white, in dense, rounded terminal clusters 3–4 cm across; **bracts** brown and papery; **pappus** with thickened tips.

This species is prominent in montane and subalpine forests of south-eastern NSW and eastern Vic. It is worth growing for its attractive foliage and floral display. Crushed leaves are aromatic. Plants can be grown readily but succeed best if protected from long periods of hot sun. They respond well to mulching and watering during dry periods. May tend to become straggly so regular light pruning is advisable. Tolerant of moderately heavy frosts. Propagate from seed or from cuttings of firm young growth.

Previously known as *Helichrysum stirlingii* F. Muell.

Ozothamnus tesselatus (Maiden & R. T. Baker) Anderb.
(checkered in squares)
NSW
1–3 m × 0.6–2 m   Dec–March

Small to medium **shrub**; young growth woolly-hairy; **branches** ascending to erect; **branchlets** with lines of raised scars from leaf bases; **leaves** 1.2–2.5 cm × about 0.2 cm, linear, ascending to erect, lower third decurrent, dark green above, margins revolute, apex pointed; **flowerheads** about 0.3 cm across, straw-coloured, in crowded terminal clusters, often profuse and very conspicuous; **achenes** hairy.

An uncommon species which is regarded as vulnerable in the wild. It occurs on the Central Western Slopes where it grows in well-drained soils. Needs to be introduced into cultivation as part of a conservation strategy. The flowerheads are attractive and the species warrants being better known in cultivation. Plants probably require good drainage in a warm to hot site. They will tolerate moderate frosts. Pruning from an early age will promote bushy growth. Propagate from seed or from cuttings of firm young growth.

Previously known as *Helichrysum tesselatum* Maiden & R. Baker.

O. adnatus is allied but it has smaller leaves which have their lower half attached to the branches.

Ozothamnus thomsonii (F. Muell.) P. G. Wilson =
Cremnothamnus thomsonii (F. Muell.) Puttock

Ozothamnus thyrsoideus

Ozothamnus stirlingii   W. R. Elliot

Ozothamnus thyrsoideus DC.
(like a staff carried by Bacchus)
NSW, Vic, Tas   Sticky Everlasting
2–3 m × 2–3 m   Nov–Feb

Medium **shrub**; young growth sticky, hairy; **branches** spreading to ascending; **branchlets** spreading to arching, angular and striate, hairy; **leaves** 2–5 cm × up to 0.2 cm, narrowly linear, dark green and shiny above, hairy below, margins recurved to revolute, apex pointed to blunt; **flowerheads** about 0.2 cm across, white with brownish outer bracts, arranged in dense heads at ends of short branchlets, virtually covering the upper side of branches, profuse and most conspicuous.

Sticky Everlasting is amongst the most floriferous of the shrubby everlastings. It is widespread in south-eastern NSW, eastern Vic and Tas. Usually occurs on mountain slopes and foothills. Seldom cultivated, it deserves greater recognition. Adapts to most freely draining soils and does well in a sunny or semi-shaded site. Tolerant of fairly heavy frosts. Responds well to pruning which is best done from an early age if bushy growth is desired. Propagate from seed, or from cuttings, which strike readily.

Previously known as *Helichrysum thyrsoideum* (DC.) P. Morris & J. H. Willis.

O. secundiflorus differs in having shorter, densely hairy, flattish leaves.

Ozothamnus tuckeri (F. Muell. ex J. H. Willis) Anderb.
(after Gerard Tucker, 19–20th-century Victorian farmer)
NSW, ?Vic
0.6–2.5 m × 0.5–1.5 m   Oct–Jan

Dwarf to medium **shrub**; young growth woolly-hairy; **branches** ascending to erect, slender, stiff, slightly woolly-hairy; **leaves** 0.3–0.5 cm × about 0.1 cm, linear-oblong, ascending to erect, appressed, crowded, sessile with ear-shaped lobes at base, glabrous, apex blunt; **flowerheads** about 0.3 cm across, white or pink, somewhat globular, in terminal clusters of 1–5 cm across, often profuse and conspicuous; **achenes** densely hairy.

A decorative species from the Central Western Plains, North Western Plains and South Western Plains of NSW. There is a doubtful recorded occurrence in north-western Vic. It occurs in a range of well-drained sandy, loamy or calcareous soils. Plants have potential for cultivation in gardens and for floriculture. They need plenty of sunshine and good drainage. Probably best suited to semi-arid and warm temperate regions but may succeed in cooler conditions. Hardy to moderate frosts. Needs pruning from an early stage to produce a bushy framework. Propagate from seed or from cuttings of firm young growth.

Previously known as *Helichrysum tuckeri* F. Muell. ex. J. H. Willis.

O. diotophyllus is similar in habit but is distinguished mainly by its bright yellow flowerheads.

Ozothamnus turbinatus DC.
(shaped like a top)
NSW, Vic, Tas, SA   Coast Everlasting
1–3 m × 1–3 m   Feb–June

Small to medium bushy **shrub**; **branches** erect or decumbent; **branchlets** densely covered with woolly hairs; young growth grey or yellowish; **leaves** 1–2.5 cm × 0.1–0.2 cm, linear, sticky, spreading, crowded, green and glabrous above, woolly beneath, yellowish on the decurrent base, margins revolute; **flowerheads** about 0.4 cm across, cream to yellowish, crowded in elongated heads to about 3 cm across.

Widely distributed on coastal dunes and cliffs from south-eastern NSW to Victor Harbor, SA, and Tas. A very important sand-binding component of coastal vegetation in south-eastern Australia. This is a useful shrub for coastal gardens because it tolerates calcareous soils and considerable exposure to salt-laden winds. It may also be grown in heavier soils further inland but plants are somewhat susceptible to root-rotting fungi. Does best in an open sunny position. Light pruning after flowering is beneficial. Responds well to regular clipping. Tolerates light to moderate frosts. A grey-leaved selection from Robe, SA, is very ornamental. Propagate from seed or from cuttings of firm young growth.

Previously known as *Helichrysum paralium* (N. T. Burb.) W. M. Curtis.

Pileanthus peduncularis   B. .W. Crafter

P

PACHYCORNIA Hook. f.

(from the Greek *pachys*, thick, and the Latin *cornu*, horn; referring to the thick branches and spikes)
Chenopodiaceae

A monotypic genus endemic in Australia.

Pachycornia triandra (F. Muell.) J. M. Black
(three stamens)

NSW, Vic, SA, WA, NT — Desert Glasswort
0.3–0.8 m × 0.5–1 m — July–Feb

Glabrous dwarf **shrub**; **stems** succulent, cylindrical; **branches** many, succulent, bright green, internodes 1–2 cm long, with 2 prominent, short, horn-like lobes at their base, opposite, hairy grooves along length of internodes; **flowers** small, cream, clusters of 3 flowers in short spike-like arrangements; **fruit** fused and sunken in woody axis.

This plant dwells in saline clay soils in arid regions. It is important for soil conservation and wildlife habitat. The species has potential for use in reclamation of saline soils where overgrazing has depleted the vegetation. Propagate from seed or from cuttings. However, the cuttings are virtually impossible to strike because they usually become a soft gluggy mess.

Previously known as *Arthrocnemum triandrum* F. Muell.

PACHYGONE Miers

(from the Greek *pachys*, thick; *gone*, seed; referring to the thick seeds)
Menispermaceae

Climbers with twining stems; **leaves** simple, entire, alternate, petiolate, prominently veined; **inflorescence** axillary racemes or terminal panicles; **flowers** small, unisexual; **male flowers** in clusters; **female flowers** solitary; **sepals** 6 or 9; **petals** 6; **fruit** a horseshoe-shaped drupe.

A small genus of about 12 species distributed in China, South East Asia and the Pacific region. Three species occur in Australia and 2 of these are endemic. They are robust climbers with interesting, ornamental leaves. Birds feed on the fleshy fruit. In general, these plants are rare subjects in cultivation. Propagation is from seed, which has a short period of viability and should be sown while fresh. Cuttings of recently hardened new growth may be worth trying.

Pachygone hullsii F. Muell. =
P. ovata (Poir.) Hook. f. & Thomson

Pachygone longifolia F. M. Bailey
(with long leaves)

Qld
2–4 m tall — Oct–Dec

Vigorous **climber**; **stems** slender, striate; young growth pale green, glabrous; **juvenile leaves** to 30 cm × 12 cm; **mature leaves** 15–22 cm × 5–8 cm, oblong to lanceolate, dark green above, paler beneath, thin but leathery, base rounded, often peltate; **panicles** 15–30 cm long, much-branched, terminal or in the upper axils; **male flowers** about 0.2 cm across, greenish; **female flowers** about 0.3 cm across, greenish; **drupes** 0.2–0.3 cm × 0.2 cm, ovoid to pear-shaped, red, mature March.

This species occurs in the Cook and North Kennedy Districts. It grows in rainforests and is found from the coastal lowlands to the ranges and tablelands. Plants have interesting large leaves and colourful fruit. Most suitable for cultivation in tropical and subtropical regions. This climber must be given sufficient room to spread — excessive vigour may need to be curtailed by regular pruning. Requires filtered sun or partial sun, unimpeded drainage and water during dry periods. Tolerant of light frosts only. Propagate from fresh seed and perhaps also by cuttings of hardened new growth.

Pachygone ovata (Poir.) Hook. f. & Thompson
(ovate)

Qld, WA, NT
2–4 m tall — July–Sept

Vigorous **climber**; **stems** slender, wiry; young growth yellowish, short-hairy; **leaves** 6–10 cm × 5–7 cm, ovate, elliptical or cordate, 3 or 5 veins prominent, dark green above, hairy beneath, thin-textured but leathery; **racemes** 6–12 cm long, arising on the older stems; **male flowers** about 0.3 cm across, greenish; **female flowers** about 0.4 cm across, greenish; **drupes** 0.4–0.6 cm across, round, bluish, mature Oct.

A widely distributed species which extends to the tropical north of Australia. It grows in coastal scrubs, rainforest and moist areas in woodland, such as near streams. Not known to be in cultivation, it would be best suited to tropical and warm subtropical regions. The cultivation requirements are probably similar to those of *P. longifolia*. Propagate from fresh seed and perhaps by cuttings.

P. hullsii F. Muell. is a synonym.

145

Pachygone pubescens

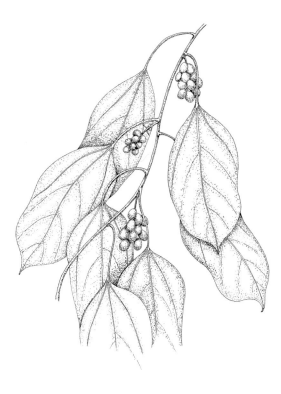

Pachygone ovata × .45

Pachygone pubescens Benth.
(short-hairy)
Qld
2–4 m tall   Sept–Nov
  Vigorous **climber**; **old stems** woody; young growth
short-hairy; **leaves** 7–11 cm × 5–8 cm, broadly ovate,
dark green and glossy, thin-textured but leathery, 3 or
5 veins prominent, short-hairy, apex obtuse to acuminate; **racemes** 6–12 cm long, often in groups, terminal
or in the upper axils; **male flowers** about 0.3 cm
across, greenish; **female flowers** about 0.4 cm across,
greenish-white; **drupes** about 0.5 cm across, ovoid,
red, mature Dec–March.
  A poorly known species which inhabits coastal rainforests in the Cook District in north-eastern Qld. It has
ornamental leaves which suggest that it has potential
for cultivation. Probably has similar cultural requirements to those of *P. longifolia*. Propagate from fresh
seed and possibly also by cuttings.

PACHYNEMA R. Br. ex DC.
(from the Greek *pachys*, thick; *nema*, thread; referring
to the thickened bases of the stamens)
Dilleniaceae
  Dwarf to medium **shrubs**, sometimes rhizomatous;
stems terete or flattened, often branching, foliage near
base while other parts are leafless; **leaves** simple, alternate or spirally arranged on base of stems, often deciduous; **flowers** bisexual, open-petalled, symmetrical,
solitary or in panicles, lateral or terminal; **petals** 5
(rarely 4); **sepals** 5; **stamens** 10; **fruiting carpels** 2.
  A small endemic genus of 7 species which is mainly
confined to the Top End of NT, although one species
extends to the Kimberley area in WA.
  Pachynema species are virtually untried in cultivation. Species such as *P. hooglandii* and *P. praestans* are
ornamental and deserving of wider recognition.
  This genus is most likely to succeed in subtropical
and seasonally dry tropical regions. Most species have
woody perennial rootstocks and are capable of reshooting after fire. This characteristic should enable
rejuvenation of cultivated plants by regular pruning.
Propagation is from seed, which probably will germinate best if it has been freshly collected. Cuttings of
firm young growth, especially from young branchlets,
should be possible.

Pachynema complanatum R. Br. ex DC.
(flattened, compressed)
NT
0.5–1.5 m × 0.5–1 m   March–Sept; also sporadic
  Dwarf to small **shrub** with a perennial rootstock;
stems lower stems much-branched, mainly flattened,
0.3–1.5 cm wide, upper stems less branched, terete
to flattened, all glabrous, pale-to-mid green; **leaves**
1.5–2.5 cm × 0.8–1.5 cm, elliptical, short-stalked,

Pachynema dilatatum   D. L. Jones

146

green, mainly glabrous, margins toothed in upper half; **flowers** open-petalled, 0.8–1.5 cm across, creamish, pinkish to reddish-white or pink, solitary or in few-flowered clusters on short stalks arising from stem notches; **petals** readily deciduous; **fruit** to 0.5 cm across, globular, enclosed in papery bracts.

This species occurs in a range of lateritic, gravelly and sandy soils in the Top End where it grows in open forest. Evidently not cultivated, it is worth trying in seasonally dry tropical and subtropical regions. Plants need good drainage in acidic soils. A sunny or semi-shaded site should be suitable. It has potential as a container plant. This species re-shoots from the rootstock after fire so it should also respond well to hard pruning to rejuvenate plants. Propagate from fresh seed or from stem cuttings.

Pachynema conspicuum (J. Drumm. ex Harv.) Benth. = *Hibbertia conspicua* J. Drumm. ex Harv.

Pachynema diffusum Craven & Dunlop — see *P. sphenandrum* F. Muell. & Tate

Pachynema dilatatum Benth.
(enlarged or widened)
NT
0.5–1.5 m × 0.5–1 m   Nov–Sept

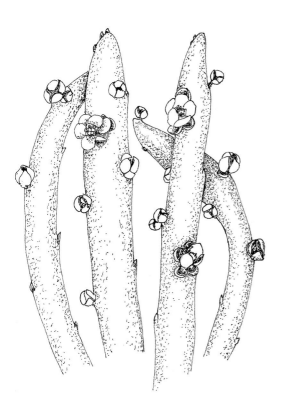

Pachynema dilatatum × .75

Pachynema praestans   D. L. Jones

Dwarf to small **shrub** with perennial rootstock; **stems** leafless, all flattened, much-branched; **branchlets** 0.5–3.5 cm wide, glabrous, pale-to-mid green, sometimes slightly glaucous; **flowers** open-petalled, 0.5–0.8 cm across, pale pink, bright pink, purplish-cream, cream or white with rose-red base, solitary, nearly sessile at notches; **petals** deciduous; **fruit** to about 0.4 cm across, globular, enclosed in papery bracts.

A dweller of shallow soils and gravels on rocky slopes and ridges in open woodland. It occurs in the Top End. Plants have interesting foliage and should be worth trying in cultivation. They must have excellent drainage and a warm, mainly sunny site. Suited to seasonally dry tropical regions. Propagate from seed or from cuttings of firm young branchlets.

Pachynema hooglandii Craven & Dunlop
(after Richard D. Hoogland, 20th-century Dutch botanist)
NT
0.6–1.5 m × 0.5–1 m   April–Aug; also sporadic

Dwarf to small **shrub** with a woody rootstock; **stems** leafless, glabrous, lower stems much-branched, flattened, to about 1 cm across, upper stems narrower, terete or flattened; **flowers** open-petalled, to about 2 cm across, white, pale pink or pink, on short stalks, solitary or few-flowered clusters, often profuse and very conspicuous; **petals** deciduous; **fruit** about 0.4 cm long.

This handsome, recently described (1992) species has the largest flowers in the genus. It occurs in woodland and grassland in the Top End where it grows in a range of sandy and clay soils which are often moist for extended periods. Worthy of a trial in subtropical and seasonally dry tropical regions and should adapt to a variety of acidic soils. A sunny or semi-shaded site is required. Plants can probably be rejuvenated by hard pruning. Propagate from seed or from cuttings of firm young branchlets.

Pachynema junceum Benth.
(rush-like)
NT
0.6–1.5 m × 0.3–0.6 m   Nov–Aug
Dwarf to small **shrub** with woody rootstock; **stems** leafless, dull green, glabrous, lower stems somewhat flattened to terete, much-branched, to about 0.5 cm wide, taller stems erect, terete, less branched, to about 0.2 cm wide; **flowers** open-petalled, to about 1 cm across, white, cream, pinkish-cream, or deep pink, on slender stalks to 0.8 cm long, moderately profuse and conspicuous; **petals** deciduous; **fruit** to about 0.5 cm across.

A rushlike species which inhabits a wide range of the Top End. It grows in shrubland, open forest and woodland on lateritic, sandy or rocky soils. Needs excellent drainage and a sunny or semi-shaded site. It reshoots after fire and older plants can probably be rejuvenated by hard pruning. This species is worth trying in subtropical and seasonally dry tropical regions. Propagate from seed and from cuttings of firm young regrowth.

P. sphenandrum is allied but differs in having more slender stems and a corona around the stamens.

Pachynema praestans Craven & Dunlop
(excellent, distinguished)
NT
1–2 m × 0.6–1.5 m   Jan–June
Small **shrub** with woody rootstock; **stems** solitary or several, erect, glabrous, terete, much-branched; **branchlets** pendent, angular or sometimes terete; **flowers** open-petalled, to about 0.7 cm across, dark red, on short thickened stalks, 1–2 per axil, often profuse and conspicuous; **petals** persistent; **fruit** to about 0.3 cm across.

This recently described (1992) species has pendent branchlets which is unique in the genus. It occurs in the Top End, where it grows on sandstone, and is often found in crevices and clefts or on scree. It warrants trials in subtropical and seasonally dry tropical regions. Although the flowers are small they are usually profuse and decorative. Response to pruning is not known. Propagate from seed or from cuttings of firm young branchlets.

Pachynema sphenandrum F. Muell. & Tate
(wedge-shaped stamens)
WA, NT
0.2–0.3 m × 0.2–0.4 m   Dec–June
Dwarf **shrub** with perennial rootstock; **stems** terete or nearly so, glabrous, much-branched; **branchlets** wiry; **leaves** 2–4 cm × 0.8–2.5 cm, broadly spathulate or obovate, on lower stems, often shed early, glabrous, upper margins toothed, apex blunt or rounded; **flowers** open-petalled, to about 0.8 cm across, cerise-red, solitary, near ends of branchlets, scattered to moderately profuse; **petals** persistent; **corona** encircling the stamens; **fruit** to about 0.4 cm across.

P. sphenandrum inhabits open forest and woodland where it grows in lateritic, granitic and sandy soils. It is distributed from the Kimberley, WA, to the Top End, NT. This moderately ornamental species should adapt to a range of freely draining acidic soils in sunny or semi-shaded sites, in subtropical or seasonally dry tropical regions. Plants reshoot from rootstock. Propagate from seed or from cuttings of firm young stem growth.

P. diffusum Craven & Dunlop, also from the Top End, is closely allied and has been confused with *P. sphenandrum.* It is a short-lived dwarf shrub, to about 0.3 m tall, with fine branches which are often sparsely arranged. It lacks the corona surrounding the stamens.

PACHYSTOMA Blume
(from the Greek *pachys,* thick; *stoma,* an opening or mouth)
Orchidaceae
Terrestrial **orchids** forming small clumps; leaves and inflorescences arising from a slender underground rhizome; **leaves** 1–few, narrow, pleated; inflorescence a raceme; **flowers** more or less tubular, dull coloured; **sepals** hairy outside; **labellum** 3-lobed, with a basal pouch.

A small genus of about 3 species distributed in Africa and from Asia to New Guinea with one species extending to Australia. This species is rarely encountered in cultivation. Propagation is from seed which must be sown aseptically together with a symbiotic fungus.

Pachystoma holtzei (F. Muell.) F. Muell. =
P. pubescens Blume

Pachystoma pubescens Blume
(softly hairy)
Qld, NT
0.3–0.6 m tall   Nov–Dec
Rhizome slender, branched; **leaf** erect, 20–45 cm × 1–1.2 cm, narrowly lanceolate, dark green, pleated, narrowed to a long basal stalk; **raceme** to 60 cm tall, with 5–10 bracts; **flowers** 4–10, dull pink with yellow tips, hairy, the segments not spreading widely; **petals** narrower than the sepals; **labellum** to 1.1 cm × 0.5 cm, 3-lobed, the apex ending abruptly.

Occurs in Indonesia and extends to northern Australia but plants are rarely collected. They grow in moist grassy areas and the leaves are difficult to discern among the grass. Plants are easily grown in a pot of well-drained gravelly mixture fortified with decaying leaf litter. They require warm sheltered conditions with free air movement and should not be watered while dormant in winter. Propagate from seed.

PAGETIA F. Muell. = *BOSISTOA* Benth.

PALAQUIM Blanco
(from a Philippine name)
Sapotaceae
Trees with milky sap; **leaves** simple, entire, alternate; **stipules** deciduous; **flowers** solitary or clustered in the axils; **calyx** of 4–8 overlapping segments; **corolla** tubular at the base, 5-lobed; **stamens** 5 or 10; **fruit** a fleshy berry or drupe.

A large genus of about 120 species widely distributed in South East Asia and the Pacific region; one species is endemic in Australia. Many exotic species are valuable timber trees. Propagation is from seed, which has a limited period of viability and must be sown fresh.

Palaquium galactoxylum (F. Muell.) Lam.
(milky wood; referring to the sap)
Qld   Red Silkwood; Cairns
      Pencil Cedar; Daintree Maple
10–25 m × 10–20 m   April–June

Medium **tree** with a deciduous habit; **bark** fissured, brown; **sap** milky; young growth bearing rusty hairs; **leaves** 7–15 cm × 2.5–4 cm, narrowly obovate, oblong or wedge-shaped, dark green, tapered to the base, apex blunt; **flowers** about 0.7 cm long, greenish-white, tubular at the base, with 5 pointed lobes; **stamens** 10; **drupes** 2.5–3.5 cm long, tapered to each end, blackish; **seeds** 1 or 2, brown, shiny.

Occurs in the coastal districts of north-eastern Qld. It grows near streams in lowland rainforest. Plants have a short deciduous period around Oct. It has red soft wood which has been used for joinery but is susceptible to attack by borers. This handsome tree is suitable for parks and large gardens in tropical and warm subtropical regions. Young plants require some protection from strong winds and hot sun. Needs well-drained soil and responds to mulches and light fertiliser applications. Frost tolerance is unlikely to be high. Propagate from fresh seed.

Bassia galactoxyla F. Muell. and *Lucuma galactoxyla* (F. Muell.) Benth. are synonyms.

PALMERIA F. Muell.
(after Sir James F. Palmer, 19th-century Victorian medical practitioner and politician)
Monimiaceae

Dioecious evergreen scrambling **shrubs** or woody **climbers**; young growth with rough stellate hairs; **leaves** opposite, mainly entire, rarely crenate; **flowers** unisexual, small, in axillary racemes; **male flowers** hemispherical, 4–7-lobed, with many stamens; **female flowers** globular or somewhat flask-shaped, lobes absent; **carpels** several; **fruiting perianth** more or less globular, fleshy, splitting at maturity to reveal the achenes.

A tropical genus of about 20 species which is mainly distributed in south-eastern Asia. There are 3 endemic species occurring in Australia, in tropical regions, and one species, *P. scandens*, extends to temperate areas. The Australian species are vigorous climbers which need plenty of space to reach their potential. They are not recommended for small gardens.

Propagation is from freshly collected seed. Cuttings are worth trying too — *P. scandens* produce roots fairly readily.

Palmeria coriacea C. T. White = *P. scandens* F. Muell.

Palmeria hypotephra Domin
(ash-coloured undersurface)
Qld
Climber   Aug–Sept

A very vigorous **climber**; **stems** to several metres long; **leaves** to about 9 cm × 4 cm, lanceolate, short-stalked, dark green above, brown to grey or glaucous and hairy below, margins entire or faintly toothed, apex pointed; **flowers** about 0.5 cm across, greenish, in axillary racemes to about 4 cm long; **fruit** about 1.5 cm across, globular, black, splitting on maturity during Jan–April to expose black seeds embedded in reddish pulp.

P. hypotephra occurs in the Cook, North Kennedy and South Kennedy Districts where it is regarded as rare but not threatened. It occurs in rainforest above 800 m altitude. Plants commonly reach the canopy of tall trees. It is rarely cultivated. Not highly ornamental but fruiting plants are interesting. Best suited to acidic soils in highland tropical and subtropical regions in protected sites which do not dry out very readily. May need to be pruned regularly to maintain plants at a desired size. Likely to suffer damage from heavy frosts. Propagate from fresh seed and possibly from cuttings of firm young growth.

Palmeria scandens F. Muell.
(climbing)
Qld, NSW   Anchor Vine; Pomegranate Vine
Climber   Aug–Sept

A very vigorous **climber** which may scramble when young; **stems** to several metres long, woody; **leaves** 5–12 cm × 2–5 cm, elliptical to oblong, stalked, deep green with scattered stellate hairs above, paler below, margins entire or faintly crenate, apex ending in a soft

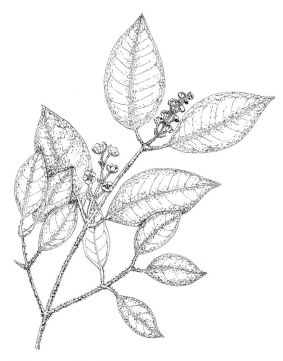

Palmeria scandens × .3

point; **flowers** to 0.3 cm across, whitish to pale green, in panicles on male plants and racemes on female plants; **fruit** to about 2 cm across, green to blackish, splitting into 4–6 lobes at maturity during Nov–Dec to reveal globular glossy brown seeds embedded in reddish pulp.

The distribution of Anchor Vine extends from north-eastern Qld to south-eastern NSW. It inhabits rainforest and moist areas of sclerophyll forest. This species is valuable for soil erosion control on large sloping sites because it can be pruned to develop a scrambling habit. It is also capable of climbing to the canopy of tall trees and can be grown on large strong structures. Plants do best in tropical and subtropical regions but adapt to cooler areas if frosts are not severe. A protected, well-drained but moist site with acidic soils is best for growth. Propagate from fresh seed, or from cuttings, which strike readily.

P. coriacea C. T. White is a synonym.

PANDANACEAE R. Br.   Pandan Family
A large family of monocotyledons consisting of about 900 species in 3 genera. It is particularly well developed in tropical parts of Africa, Asia and Polynesia. About 25 species in 2 genera are found in tropical and subtropical parts of Australia. They are climbers or shrubs and trees with erect stems supported by long (often spiny) prop roots. Long strap-like leaves are arranged in a distinct spiral up the stem. The leaves usually have spiny margins. Small unisexual flowers are borne in dense male or female heads at the ends of shoots and are subtended by often colourful bracts. Fruits are in compound heads and may be fleshy (*Freycinetia*). *Pandanus* fruits are fibrous to woody or fleshy and composed of numerous individual drupes, or fused into structures called phalanges. The drupes or phalanges fall separately as the fruit disintegrates. They contain edible seeds. Australian genera: *Freycinetia*, *Pandanus*.

PANDANUS Pierre
(a Latinised version of *pandan*, a Malayan name)
Pandanaceae   Screw Pines, Pandans
  Shrubs or **trees**, dioecious, often slender and sparsely branched; **trunk** 1 or a few, usually slender, fibrous; **prop roots** developing in some branch axils and growing to the ground; **leaves** long, sheathing at the base, usually spiny, spirally arranged, crowded at the ends of branches; **male flowers** tiny, crowded in spikes, the spikes arranged in a terminal inflorescence; **female flowers** tiny, crowded into an unbranched, often globose head, rarely several heads in a branched inflorescence; **fruiting head** large, fibrous, fleshy or woody, often colourful; **fruit** a drupe, either arranged singly or clustered into phalanges; **seeds** moderately large.

A large genus consisting of more than 600 species distributed throughout the world's tropics. Identification of the Australian species has been difficult and very confusing. At one stage more than 86 names were in use, and some species had been described from single plants with no allowance for variation. Much of this confusion has been solved by a recent study which recognises about 20 native species, most of which are endemic.

Typically for a tropical genus, the Australian species are concentrated in the northern part of the continent with significant diversity in Qld, NT and WA. One coastal species, *P. tectorius*, extends into northern NSW. In Australia, *Pandanus* are predominately a component of open forests and woodlands, but a number of species also occur in rainforest, and a couple grow in mangrove communities. Some species favour moist to wet soils and may be dominant along stream banks, whereas others grow in much drier situations.

The seeds of all species of *Pandanus* are edible and were an important source of nutrition for some Aboriginal tribes. They were eaten either raw or after roasting. The succulent fruit of some species is edible too, and the bases of the very young, developing leaves in the crown may also be eaten either raw or cooked. Leaves of *Pandanus* had a number of practical uses. Usually they were torn into strips and dried, then woven into utensils (mats, baskets and belts), or ornaments (arm bands and head bands), or made into twine, rope and even sails.

This genus has a significant contribution to make to horticulture, however, the spiny nature of plants is a major drawback to their much wider use. Sharp spines occur on the leaves of most species and some also have spines on their branches, trunks and in a few even the prop roots may be armed. The characteristic appearance of *Pandanus* imparts a tropical flavour to any landscape almost as strongly as a palm silhouette. Other ornamental features include large and often colourful fruit, and unusual (sometimes fragrant) flowers. Because plants are prickly they have limited use close to paths, or other points of human contact, although they can be used to direct foot traffic and in areas prone to vandalism. Young plants of some species are very tolerant of neglect and make excellent container subjects for indoor decoration.

Pests have not generally been a major problem in *Pandanus*, but recently an exotic mite has become established in Qld and this has resulted in the deaths of large numbers of *Pandanus* in some areas. Some species, particularly those from coastal districts or originating in drier climates, may be susceptible to root-rotting fungi.

Trials with *P. tectorius* show that large branches can be used successfully for propagation. The branches are dried for a couple of days before being placed upright with the basal metre or so of branch buried in sand. When solid roots are formed transplanting can occur. In coastal sand dunes some success has been achieved in propagating branches in situ. Some species produce stolons which may be useful for propagation and others produce numerous basal offsets. *P. gemmifer* produces numerous growths from branches which fall to the ground and take root.

Pandanus are mostly propagated from seed which has a limited period of viability and is best sown fresh. Germination may take 6–12 months or longer and seedlings may continue to appear sporadically for up to 3 years. Seed extraction can be difficult in those species which have very fibrous fruits. For particularly difficult species the whole segment, which contains many seeds, may be sown and the seedlings separated after germination.

Pandanus adscendens H. St John =
$\qquad$ *P. tectorius* Parkinson ex Z.

Pandanus angulatus H. St John = *P. whitei* Martellii

Pandanus aquaticus F. Muell.
(growing in or near water)
Qld, WA, NT $\qquad$ Water Pandan, River Pandan
5–7 m × 2–4 m $\qquad$ June–July

Tall **shrub** or small **tree**, growing singly or forming clumps; **prop roots** present or absent; **leaves** 100–140 cm × 5–6 cm, narrowly strap-shaped to narrowly triangular, dark green, erect to pendulous, midrib unarmed or spiny, margins with brown-tipped spines; **male inflorescence** branched, 20–30 cm long, with a whitish bract to 100 cm long subtending each branch; **male spikes** to 4.5 cm long, ellipsoid; **male flowers** cream to yellowish; **female inflorescence** unbranched, subtended by numerous whitish bracts to 100 cm long; **fruiting head** 10–15 cm across, globular, ripening yellowish; **drupes** separate, 3–4.4 cm × 0.5–0.9 cm, club-shaped, fleshy when ripe; **seeds** 0.6–0.7 cm long, ripe Dec–May.

This species is widely distributed in tropical Australia. It grows in shallow water and along the margins of watercourses. Often locally common and frequently dominating the vegetation which lines the banks of small or slow-flowing streams. Fruit disintegrates on ripening and the fleshy drupes are eaten by fish such as Sooty Grunter. Prop roots are used by the

Pandanus basedowii $\qquad$ D. L. Jones

Aborigines to make paint brushes. May be grown in tropical and warm subtropical climates. Requires a sunny location and an abundance of water. Very cold sensitive. Also makes an interesting container plant and can be useful for indoor decoration, conservatories and glasshouses. Propagate from fresh seed.

P. delestangii Martelli, *P. kimberleyanus* H. St John, *P. oblanceoloideus* H. St John and *P. spechtii* H. St John are synonyms.

Pandanus arhhemensis H. St John = *P. damannii* Warb.

Pandanus australiensis H. St John = *P. whitei* Martelli

Pandanus basedowii C. H. Wright
(after Herbert Basedow, 19–20th-century explorer and geologist, SA)
NT $\qquad$ Sandstone Screw Pine
4–7 m × 2–4 m $\qquad$ Nov–Jan

Medium **shrub** to small **tree** with a rounded crown; **branches** often horizontal; **prop roots** present; **leaves** 30–80 cm × 1–5 cm, narrowly strap-shaped, dull green, stiff, leathery, lower margins spiny, apex drawn out; **male inflorescence** branched, 10–20 cm long; **male spike** 2–3 cm long, subtended by whitish bracts; **female inflorescence** unbranched, with whitish bracts; **fruiting head** 15–20 cm across, more or less round, very lightweight, ripening cream-brown; **phalanges** about 10 per fruiting head, each containing 1 or 2 seeds; **seeds** about 2.5 cm long, ripe May–Oct.

Endemic in Arnhem Land where it is locally common on sandstone escarpments and crevices growing in pockets of sandy soil. Aborigines once ate the roasted seeds. It is one of the most ornamental of the Australian species of *Pandanus* because it develops a pleasant-looking bushy crown and has interesting fruit. Plants have potential for use in landscaping in

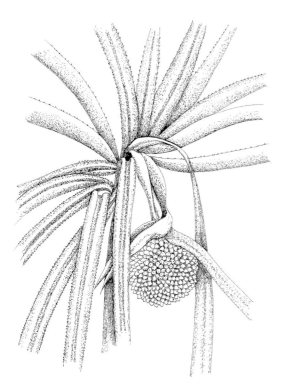

Pandanus aquaticus × .45

Pandanus basedowii × .45

tropical and warm subtropical regions. They require a sunny position, well-drained soil and respond to watering during the summer. Propagate from fresh seed.

Pandanus blakei H. St John =
　　　　　　　　　　　P. tectorius Parkinson ex Z.

Pandanus bowenensis H. St John =
　　　　　　　　　　　P. tectorius Parkinson ex Z.

Pandanus brownii H. St John =
　　　　　　　　　　　P. tectorius Parkinson ex Z.

Pandanus citraceus H. St John =
　　　　　　　　　　　P. solms-laubachii F. Muell.

Pandanus cochleatus H. St John = *P. conicus* H. St John

Pandanus conicus H. St John
(conical; referring to the fruit shape)
Qld　　　　　　　　　　　　　　　Mace Screw Pine
6–10 m × 3–5 m　　　　　　　　　　　　　Jan–Feb
Tall **shrub** or small **tree**, sparse to much-branched; **trunk** covered with short, squat spines; **prop roots** present; **leaves** 200–325 cm × 4–6 cm, narrowly strapshaped, glaucous to green above, leathery, margins spiny, apex long-tapered; **male inflorescence** branched, 30–40 cm long, semi-pendulous with white bracts; **male spikes** 8–10 cm long, white; **female inflorescence** unbranched; **fruiting heads** 15–20 cm × 10–15 cm,

pendulous, mace-shaped, with prominent protruding phalanges, ripening orange to red; **seeds** 1.2–1.4 cm long, ellipsoid.
Occurs on Cape York Peninsula where it grows in littoral rainforest, in mangroves on the edges of scrub, and in rainforest along streams up to about 400 m altitude. This interesting species has fruit which resembles a mace. It is rare in cultivation and has potential for wider planting. Tropical regions are most suitable but it may also succeed further south in the warm subtropics. Best in filtered sun or partial sun and needs plenty of water. Propagate from fresh seed and from stolons.
P. cochleatus H. St John and *P. sphaericus* H. St John are synonyms.

Pandanus convexus H. St John =
　　　　　　P. spiralis var. *convexus* (H. St John) Stone

Pandanus cookii Martelli
(after Cooktown)
Qld　　　　　　　　　　　　　　　　Screw Pine
6–8 m × 2–4 m　　　　　　　　　　　　Dec–March
Tall **shrub** or small **tree** with a spreading crown; **prop roots** absent; **leaves** 130–180 cm × 6–8 cm, strapshaped, M-shaped in cross-section, margins and midrib lacking spines; **male inflorescence** branched, about 30 cm long, with whitish bracts; **female inflorescence** unbranched; **fruiting heads** about 20 cm × 14 cm, ripening red; **phalanges** to 6.5 cm × 5 cm, with 9–12 seeds; **seeds** 2–2.2 cm long, ellipsoid, ripe March.
Occurs in coastal areas of north-eastern Qld between Cairns and Cooktown. Grows in open forest in sandy soils. This poorly known species is probably not in cultivation. Cultural requirements are likely to be similar to those of *P. tectorius*. Propagate from fresh seed.
P. pluriangulatus H. St John is a synonym.

Pandanus damannii Warb.
(after M. Damann)
Qld, NT
5–8 m × 3–5 m　　　　　　　　　　　　　Jan–May
Tall **shrub** or small **tree** with a sparse crown; **prop roots** present; **leaves** 1.3–1.9 m × 5–6 cm, swordshaped, M-shaped in cross-section, leathery, margins and midribs spiny, apex long-pointed; **male inflorescence** branched, pendulous, about 30 cm long, strongly fragrant; **male spikes** 5–7 cm long; **female inflorescence** unbranched; **fruiting heads** 20–25 cm × 17–20 cm, ovoid, ripening bright red; **phalanges** to 6.5 cm × 7.5 cm, containing 7–15 seeds; **seeds** 1.5–1.8 cm long, ellipsoid.
This species grows in open forest and along small streams from near sea level to the ranges and tablelands. It is widespread in northern Qld and Arnhem Land, NT. The male plants reproduce from stolons and frequently grow in colonies. A poorly known species, *P. damannii* has similar cultural requirements to *P. spiralis*. Propagate from seed and, at least in the male plants, from stolons.
P. arnhemensis H. St John, *P. medialinermis* H. St John, *P. orbicularis* H. St John and *P. stolonifer* H. St John are synonyms.

Pandanus darwinensis H. St John
(from the Darwin region)
NT, WA
4–6 m × 2–3 m   Dec–March

Tall **shrub** or small **tree**, sparsely branched; **prop roots** absent; **leaves** 90–140 cm × 5–6 cm, narrowly strap-shaped, glaucous, leathery, midrib unarmed or spiny, margins with brown-tipped spines, sometimes toothed; **male inflorescence** unknown; **female inflorescence** unbranched, subtended by whitish bracts; **fruiting head** 14–18 cm across, more or less globular in outline but irregular, ripening reddish; **drupes** aggregated into more or less rounded, woody phalanges to 8 cm × 7.5 cm, each containing 11–13 seeds; **seeds** 3.5–4 cm long, ripe April–Aug.

The typical variety occurs in the Top End, NT, and the Kimberley region of WA. It grows in moist areas of woodland and near swamps and streams. Plants are cultivated to a limited extent around Darwin. Requires a sunny position and an abundance of water. Very cold sensitive.

The var. *latifructus* (H. St John) Stone has smaller phalanges to 5 cm × 7 cm and up to 20 seeds in each phalange. The seeds are up to 2 cm long. It occurs near Darwin and in the Kimberley region of WA.

Propagate both variants by seed.

Pandanus delestangii Martellii = *P. aquaticus* F. Muell.

Pandanus endeavourensis H. St John =
P. whitei Martellii

Pandanus exarmatus H. St John = *P. whitei* Martellii

Pandanus extralittoralis H. St John =
P. tectorius Parkinson ex Z.

Pandanus ferrimontanus H. St John =
P. whitei Martellii

Pandanus forsteri C. Moore & F. Muell.
(after Johann R. Forster, 18th-century German botanist)
NSW (Lord Howe Island)   Screw Pine
8–12 m × 3–6 m   Nov–Jan

Small to medium **tree** with a spreading canopy; **prop roots** numerous, thick spiny; **leaves** 200–300 cm × 4–6 cm, strap-shaped, dark green, somewhat shiny, midrib and margins spiny; **male inflorescence** branched, 30–40 cm long; **male spikes** 3–4 cm long, subtended by white bracts; **female inflorescence** unbranched, subtended by papery bracts; **fruiting head** 20–30 cm × 10–15 cm, oblong to ovoid, ripening reddish; **phalanges** containing 4–7 seeds; **seeds** 2–3 cm long, ripe May–Aug.

Endemic on Lord Howe Island where it is common in the mountains and often grows in extensive stands. It has large prop roots which form an interesting framework in the vegetation. Plants have been grown successfully in temperate regions including large gardens in Sydney and Melbourne. An imposing species which must be given sufficient room to develop. Young plants require a sheltered situation, well-drained soil and an abundance of moisture. Tolerates

light frosts. Small plants make excellent container subjects for indoor decoration. Propagate from fresh seed.

Pandanus gemmifer H. St John
(produces vegetative reproductive buds)
Qld
6–10 m × 3–4 m   May–June

Tall **shrub** or small **tree** with a sparse crown; **prop roots** forming a prominent cone; **branches** clothed with clusters of shoots which fall to the ground and take root; **leaves** 1.5–4 m × 4–7 cm, strap-shaped, green to glaucous, margins spiny, midrib unarmed; **male inflorescence** branched, 20–30 cm long, with white bracts; **male spikes** 5–7 cm long, cream; **female inflorescence** unbranched, with white bracts; **fruiting heads** 20–25 cm × 10–18 cm, cylindrical, red; **phalanges** containing 13–17 seeds; **seeds** 1.3–1.5 cm long, fusiform, ripe Jan–Feb.

An interesting species from highland areas of northeastern Qld where it grows in rainforest on the edge of lakes and streams. Plants produce numerous vegetative shoots which take root after falling to the ground and by this means the species forms localised thickets. Interestingly, the spiral direction of the leaves becomes reversed after each flowering. *Pandanus gemmifer* has been introduced into cultivation but is grown on a very limited scale. Plants require a shady position with plenty of moisture. They should be suitable as far south as Melbourne. Propagate from seed or bulbils.

Pandanus hubbardii H. St John =
P. tectorius Parkinson ex Z.

Pandanus humifer H. St John = *P. whitei* Martellii

Pandanus kennedyensis H. St John = *P. whitei* Martellii

Pandanus kimberleyanus H. St John =
P. aquaticus F. Muell.

Pandanus kurandensis H. St John =
P. solms-laubachii F. Muell.

Pandanus latifructus H. St John =
P. darwinensis H. St John var. *latifructus* (H. St John) Stone

Pandanus lauterbachii K. Schum. & Warb.
(after C. A. G. Lauterbach, 19th–20th-century German explorer who collected in New Guinea)
Qld
4–6 m × 2–3 m   March–May

Tall sparsely branched **shrub**, often forming thickets; **prop roots** absent; **leaves** 100–150 cm × 4–6 cm, strap-shaped, dark green and shiny above, paler and dull beneath, margins and midrib spiny; **male inflorescence** 50–90 cm long, pendulous, with broad, white bracts; **male spikes** 10–15 cm long, with free stamens; **female inflorescence** 70–90 cm long, branched, consisting of 7–9 spikes, each subtended by a broad pale bract; **fruiting heads** 7–9 per infructescence, 10–17 cm × 8–12 cm oblong, red-brown when ripe, each segment tipped with a prickly point; **seeds** about 1.5 cm long.

Pandanus medialinermis

Pandanus conicus   D. L. Jones

Widespread in New Guinea but restricted in Australia to northern parts of Cape York Peninsula. It grows in rainforest which fringes mangroves and along large streams, where it often grows as fringing thickets. Sometimes these may extend into the water and form islands. Dense clusters of young plants are often prominent in these thickets. Apart from its growth habit this species is readily recognised by its branched fruiting head with the spiny fruit. Rare in cultivation, this plant has tremendous interest value and is suitable for planting on the margins of dams and lagoons in tropical regions. Propagate from fresh seed and from plantlets.

Pandanus medialinermis H. St John =
P. damannii Warb.

Pandanus monticola F. Muell.
(dwelling in the mountains)
Qld   Rainforest Screw Pine; Scrub Bread Fruit
3–5 m × 1–4 m   March
Slender medium to tall **shrub**, sometimes almost climbing; **stems** slender; **prop roots** absent; **leaves** 100–200 cm × 5–7 cm, strap-like, dark green, dull, midrib and margins spiny, tips often drooping; **male inflorescence** branched, 10–15 cm long; **male spikes** 5–7 cm long, subtended by cream or white bracts; **female inflorescence** unbranched, subtended by white bracts; **fruiting heads** 10–12 cm across, nearly globular, tips of carpels pungent-prickly, ripening orange-red; **seeds** about 2 cm long, ripe July–Sept.
This species occurs in the Cook and North Kennedy Districts. It is found in dense rainforest at moderate altitudes. Plants commonly grow in full shade often with their stems gaining support from neighbouring shrubs. Best fruiting is achieved on plants in well lit to sunny positions. This species has potential for wide use as a landscape plant. May be grown from the tropics to warm temperate regions. It will grow in positions ranging from full shade to full sun. Requires well-drained soils. Apply mulch and water during dry periods for best results. Propagate from fresh seed.
P. pluvisilvaticus H. St John is a synonym.

Pandanus mossmanicus H. St John =
P. solms-laubachii F. Muell.

Pandanus oblanceoloideus H. St John =
P. aquaticus F. Muell.

Pandanus oblatiapicalis H. St John =
P. tectorius Parkinson ex Z.

Pandanus oblatus H. St John
(oblate, depressed in shape)
Qld
10–14 m × 3–5 m   Feb
Small to medium **tree** with a sparse to rounded crown; **prop roots** present; **leaves** 100–160 cm × 8–9.5 cm, broadly sword-shaped, M-shaped in cross-section, dark green and glossy, thick and leathery, margins strongly toothed; **male inflorescence** branched, 30–40 cm long, pendulous, sweetly scented, with creamy-yellow bracts; **male spikes** about 5 cm long; **female inflorescence** unbranched, pendulous, with pale bracts; **fruiting heads** 15–22 cm × 10–13 cm, cylindric to oblong, pendulous, ripening bright yellow; **seeds** 2–2.5 cm long, ripe June.

Pandanus monticola × .45

A rare species from the Iron Range on Cape York Peninsula. Plants grow in monsoon rainforest thickets fringing large streams. It is not known to be in cultivation but probably has similar requirements to *P. zea*. Propagate from fresh seed.

P. somersetensis H. St John is a synonym.

Pandanus orbicularis H. St John = *P. damannii* Warb.

Pandanus papillosus H. St John =
P. solms-laubachii F. Muell.

Pandanus pedunculatus R. Br. =
P. tectorius Parkinson ex Z.

Pandanus pluriangulatus H. St John = *P. cookii* Martelli

Pandanus pluvisilvaticus H. St John =
P. monticola F. Muell.

Pandanus punctatus H. St John =
P. solms-laubachii F. Muell.

Pandanus radicifer H. St John =
P. solms-laubachii F. Muell.

Pandanus rheophilus Stone
(dwelling near streams)
WA
4–6 m × 2–3 m   Nov

Tall **shrub** or small **tree**, usually sparsely branched; **prop roots** absent; **leaves** 100–140 cm × 5–6.5 cm, narrowly strap-shaped, pale green to bluish green, erect to drooping, margins and midrib spiny, apex drawn out; **male inflorescence** not seen; **fruiting head** 15–18 cm across, globular, ripening orange; **drupes** 6–8 cm × 1.5–2.5 cm, wedge-shaped; **seeds** about 1 cm long, fusiform, ripe Dec.

Restricted to the King Leopold Ranges in the Kimberley region but apparently common along the margins of a single ephemeral stream. Not known to be in cultivation. Best suited to tropical regions in a sunny location with abundant water. Propagate from seed.

Pandanus rivularis H. St John =
P. solms-laubachii F. Muell.

Pandanus semiarmatus H. St John
(sparsely armed with spines)
WA, NT
4–6 m × 2–3 m   Dec

Medium to tall **shrub**, usually slender and sparsely branched; **prop roots** present or absent; **leaves** 100–150 cm × 5–6 cm, strap-shaped, dark green above, paler and glaucous beneath, thick and leathery, midrib unarmed, margins spiny; **male inflorescence** unknown; **female inflorescence** unbranched, with white bracts; **fruiting heads** 15–20 cm × 14–18 cm, broadly ellipsoid, red; **phalanges** to 6.5 cm × 6 cm, smooth, shiny, containing 8 or 9 seeds; **seeds** about 2 cm × 0.5 cm, ellipsoid, ripe June–Aug.

Occurs in the Top End of the NT and the Kimberleys in northern WA. Grows in moist areas of

Pandanus spiralis   D. L. Jones

open forest and woodland as well as near water. A poorly known species which is probably not cultivated. Cultural requirements would be similar to those of *P. spiralis*. Propagate from seed.

Pandanus sinuvadensis H. St John =
P. tectorius Parkinson ex Z.

Pandanus solms-laubachii F. Muell.
(after Hermann M. Solms-Laubach)
Qld   Screw Pine
8–20 m × 5–10 m   Dec–Jan

Small to medium **tree**, sparsely branched; **prop roots** present; **leaves** 100–200 cm × 6–8 cm, strap-shaped, glaucous, dull, erect to pendulous, midrib and margins spiny; **male inflorescence** branched, 30–40 cm long, pendulous; **male spikes** 6–10 cm long, white, subtended by white bracts to 130 cm long; **female inflorescence** unbranched, subtended by whitish bracts; **fruiting head** 20–26 cm × 15–20 cm, oblong, ripening red; **phalanges** numerous, containing 10–30 seeds; **seeds** 1–2.2 cm × 0.2 cm.

Occurs in north-eastern Qld where it is widely distributed on eastern Cape York Peninsula extending to south of Cairns. Ranges from coastal districts into the mountains and grows in open situations around permanent springs. Forms an interesting tree with a rounded, but often sparse crown. Not widely grown but well suited to planting in parks in tropical and warm subtropical regions. Requires a sunny location and plenty of water during dry periods. Propagate from fresh seed or from stolons.

P. yalna is similar but is distinguished by its glossy green leaves and yellow to orange fruit.

P. citraceus H. St John, *P. kurandensis* H. St John, *P. mossmanicus* H. St John, *P. papillosus* H. St John, *P. punctatus* H. St John, *P. radicifer* H. St John and *P. rivularis* H. St John are all synonyms.

155

Pandanus somersetensis

Pandanus somersetensis H. St John =
P. oblatus H. St John

Pandanus spechtii H. St John = P. aquaticus F. Muell.

Pandanus sphaericus H. St John = P. conicus H. St John

Pandanus spiralis R. Br.
(in a spiral)
WA, NT   Spiral Screw Pine
6–10 m × 2–4 m   Oct–Nov

Tall **shrub** or small **tree**, usually sparsely branched; **prop roots** absent; **leaves** 100–200 cm × 0.4–0.7 cm, narrowly strap-shaped, dark green to glaucous, midrib unarmed or spiny, margins spiny; **male inflorescence** branched, 20–30 cm long, **male spikes** 3–7 cm long, subtended by white bracts to 0.5 cm long; **female inflorescence** unbranched, subtended by white bracts; **fruiting head** 15–20 cm × 10–18 cm, ovoid to globose, ripening orange-red; **phalanges** to 7.5 cm × 7.5 cm, globular, containing 6–24 seeds; **seeds** 2–2.7 cm long, ripe May–Sept.

The typical variety is widespread in the Top End of NT and in the Kimberley region of WA. It grows on coastal headlands, stabilised dunes and near-coastal ranges, usually near water. A very important plant for Aborigines who once used preparations from the trunk pith to treat maladies such as diarrhoea, colds, sores, wounds and toothache. The leaves were woven into mats, bags or baskets or into twisted rope. Light trunks were suitable for making rafts, and the seeds were eaten either raw or roasted. This species often grows in clumps or colonies. Plants are extremely variable in growth habit ranging from spindly trees with few branches to stocky specimens with a more or less rounded crown. They are rarely grown but are most suitable for tropical and warm subtropical regions. May have some potential for coastal planting. Requires a sunny position with copious water during the summer.

Four further varieties have been described but only 2 are easily recognisable.

var. convexus (H. St John) Stone from the Kimberleys has broad phalanges to 6 cm × 7 cm containing 6–10 seeds. P. convexus H. St John is a synonym.

var. flammeus Stone is a very distinct variety with small, narrow almost spineless leaves and round fruiting heads about 13 cm across. It is restricted to the Edgar Range in the Kimberleys.

var. multimamillatus Stone is another distinctive variety which has phalanges to 6 cm × 6.5 cm containing 15–23 seeds, with each carpel having a distinct pyramidal point. It is known from the Mitchell Plateau in the Kimberleys.

var. thermalis (H. St John) Stone was described from plants growing around the margins of hot springs in the NT. It also grows in other moist sites and extends into northern WA. It is distinguished by its domed phalanges to 7.5 cm × 8.5 cm, each containing 9–11 seeds. P. thermalis H. St John is a synonym.

Propagate all variants from seed and possibly also by stolons.

Pandanus stolonifer H. St John = P. damannii Warb.

Pandanus stradbrookensis H. St John =
P. tectorius Parkinson ex Z.

Pandanus subinermis H. St John = P. whitei Martellii

Pandanus tectorius Parkinson ex Z.
(roof-like)
Qld, NSW   Coastal Screw Pine, Coastal
                                            Bread Fruit
8–12 m × 2–4 m   Oct–Feb

Tall **shrub** or small **tree**, sparsely branched to well rounded; **prop roots** present; **leaves** 80–160 cm × 0.5–1 cm, strap-shaped, dull green, M-shaped in cross-section, midrib and margins spiny; **male inflorescence** branched, 30–45 cm long; **male spikes** 4–7.5 cm long, each subtended by a papery white bract; **female inflorescence** unbranched, subtended by white bracts; **fruiting heads** 12–18 cm × 10–13 cm, globose to ellipsoid, ripening orange-red; **phalanges** numerous, containing 5–20 seeds; **seeds** about 3 cm long, ripe May–Dec.

A widely distributed littoral species which is common on the sea shores of many Pacific countries. It extends to Australia where it is distributed on various offshore islands and along the east coast between the tip of Cape York Peninsula, Qld, and Port Macquarie, NSW. Also occurs on Lord Howe Island. People often associate P. tectorius with the tropical shores of Australia. It is a prominent and familiar plant which is commonly seen growing on rocky coastal headlands and stabilised dunes — sometimes forming thickets. The plants have a very distinctive character with slender trunks supported by basal prop roots. They have sparse, slender, often curved branches, each ending in a dense cluster of narrow, curved, drooping leaves. The fruit is very knobby.

This species is a valuable component of maritime plant communities and is important for dune stabilisation. Plants also colonise the shores of coral islands. They have light, woody, seed-containing phalanges which are easily distributed by ocean currents. Birds and various animals feed on the seeds. The species is widely planted in coastal districts, and is often used in parks where its prickly nature prevents vandalism. It is also frequently planted to aid in dune stabilisation. Plants are intolerant of waterlogging and generally do not grow well on heavy or poorly drained soil. Tolerant of light frosts only. Propagate from fresh seed.

Pandanus tectorius is somewhat variable and a huge number of species have been described which are now regarded as synonyms. P. pedunculatus R. Br. is the best known of these. Some botanists recognise 5 or 6 varieties as occurring in Australia but these are generally difficult to identify and are not included here.

Pandanus terrireginae H. St John =
P. tectorius Parkinson ex Z.

Pandanus thermalis H. St John =
P. spiralis var. thermalis (H. St John) Stone

Pandanus truncatus H. St John = P. whitei Martelli

Pandanus viridinsularis H. St John =
P. tectorius Parkinson ex Z.

Pandanus whitei Martelli
(after C. T. White, 20th-century Qld botanist)

Qld   Screw Pine
6–8 m × 1–3 m   Feb

Tall **shrub** or small **tree**, usually sparsely branched; **prop roots** absent; **leaves** 130–160 cm × 5–7 cm, narrowly strap-shaped, M-shaped in cross-section, leathery, dark green, midrib unarmed, margins spiny, apex drawn out; **male inflorescence** branched, 40–50 cm long, with whitish bracts; **male spikes** 1.5–2.5 cm long, white; **female inflorescence** unbranched; **fruiting heads** 18–22 cm × 14–17 cm, ovoid, ripening reddish-brown to red; **phalanges** containing 9–15 seeds; **seeds** about 1.5 cm long, ellipsoid.

Widespread in north-eastern Qld from Cape York Peninsula to Rockhampton. Plants grow in open forest and woodland, and less commonly in rainforest. They are often very spindly with few branches and a sparse crown. Adapts well to garden cultivation in tropical and subtropical regions. Plants are relatively hardy to dryness once they are established. Tolerates full sunshine and needs good drainage. Withstands light frosts. Propagate from fresh seed.

P. angulatus H. St John, *P. australiensis* H. St John, *P. endeavourensis* H. St John, *P. exarmatus* H. St John, *P. ferrimontanus* H. St John, *P. humifer* H. St John, *P. kennedyensis* H. St John, *P. subinermis* H. St John and *P. truncatus* H. St John are all synonyms.

Pandanus yalna R. Tucker
(from an Aboriginal name)

Qld   Yalna
10–20 m × 5–10 m   Dec–Jan

Small to medium **tree**, sparsely branched; **trunk** to 8 m × 100 cm; **prop roots** present; **leaves** 2–3 m × 6–10 cm, strap-shaped, mid green, glossy, leathery, midrib and margins spiny; **male inflorescence** branched, 15–35 cm long, pendulous; **male spikes** 8–12 cm long, white, subtended by white bracts to 175 cm long; **female inflorescence** unbranched, subtended by white bracts; **fruiting head** 18–25 cm × 18–21 cm, ovoid to spherical, ripening yellow to orange-yellow; **phalanges** numerous, containing 10–30 seeds; **seeds** to 2.2 cm × 0.1 cm, ripe Dec–March.

Distribution is restricted to Cape York Peninsula in north-eastern Qld. *P. yalna* occurs from the coastal lowlands to the ranges. It grows in moist-to-wet sites and sometimes in water to 1.5 m deep. It also grows around the margins of mangrove communities. The young leaves, fruit and seeds are eaten by Aboriginal people. Plants may form leafy stolons, especially when growing in water, which take root and develop into separate plants. This species, which is cultivated on a very limited scale, has similar requirements to *P. solms-laubachii*. Propagate from fresh seed and from stolons.

Similar to *P. solms-laubachii* which has glaucous leaves and red fruit.

Pandanus yorkensis H. St John =
                                   P. tectorius Parkinson ex Z.

Pandanus zea H. St John
(like the genus *Zea*)

Qld   Corncob Screw Pine
5–7 m × 2–3 m   April–May

Tall **shrub** or small **tree** with an open crown; **prop roots** present; **leaves** 150–175 cm × 4–5, strap-shaped, M-shaped in cross-section, dark green above, glaucous beneath, midrib and margins spiny; **male inflorescence** branched, 25–35 cm long, with long spreading cream bracts; **male spikes** 10–15 cm long, greenish cream; **female inflorescence** unbranched, completely enclosed by large orange-yellow bracts and resembling an ear of corn in its husk; **fruiting heads** similar, the drupes exposed; **drupes** to 2 cm × 0.9 cm; **seeds** about 1.6 cm long.

Occurs in the Iron Range on Cape York Peninsula where it grows in rainforest fringes along streams and in gullies. One of the most interesting of the native *Pandanus* with its colourful fruit which bear a remarkable resemblance to an ear of corn. This species has been grown on a limited scale but is worthy of wider attention. Plants require a sheltered location in shade or filtered sun, well-drained soil and protection from strong winds. Frost tolerance is unlikely to be high. Propagate from fresh seed.

PANDOREA (Endl.) Spach.
(after *Pandora* the first mortal woman of Greek mythology)
Bignoniaceae

Evergreen woody **climbers** or rarely scrambling **shrubs**; **stems** usually twining, often very flexible; **leaves** pinnate, lacking tendrils; **inflorescence** terminal and axillary thyrses and/or racemes; **flowers** tubular with overlapping lobes, white, cream, yellow, brown, pink, pinkish-red, often with deeply coloured blotched throats; **stamens** 4, very rarely exserted; **fruit** an oblong, beaked capsule; **seeds** flat, with encircling membranous wing.

A small genus of about 8 species; 6 occur in Australia and 5 of these are endemic. *P. pandorana* is also distributed in Eastern Malesia and the south-western Pacific (e.g. New Caledonia). In Australia, *Pandorea* is distributed mainly along the coast, the coastal hinterland, and the Great Dividing Range from southern Qld to central-western Victoria. There is also the inland variant of *P. pandorana* which extends to the drier regions of Qld, NSW, SA and NT.

Pandoreas are usually vigorous climbers which occur in a range of habitats. Most species prefer areas of reasonably high rainfall and sites where protection is afforded by trees and shrubs. They are common in rainforest, closed forest, open forest and woodland where the soils are moist for most of the year. An exception is the inland variant of *P. pandorana*, mentioned above.

In nature, pandoreas rely on the structure of trees and shrubs on which to scramble and climb. Plants can often reach the canopy of tall trees. Most species and selections have glossy deep green leaves but it is the flowers which have brought popularity to the genus. The tubular flowers are usually produced in profusion and vary in colour and size depending on the species. *P. jasminoides* has the largest flowers of the genus and they range in colour from white to deep reddish-pink. *P. pandorana* has the greatest variety of flower colours including white, cream, pale yellow, deep yellow with brownish-gold tonings, burgundy, and various shades

Pandanus tectorius   D. L. Jones

of pink. This variation has resulted in a number of named cultivars being introduced for cultivation (see page 160).

Pandoreas are popular in cultivation and they are prominent in gardens and commercial landscapes. *P. pandorana* was introduced into England during 1793 and first flowered in 1805. *P. jasminoides* was introduced in the first half of the 19th century. Plants generally require a large space for optimal development and unless grown on well-established trees some form of sturdy artificial support is usually needed in order to display them at their best. As a generalisation, pandoreas need moderately well-drained, acidic soils which are moist for most of the year. Plants usually grow best if their root systems are not exposed to hot sunshine, however, foliage often does better with a sunny aspect.

Most pandoreas tolerate light to moderate frosts but apart from the selections of *P. pandorana* from the inland and moderately high altitudes many other selections suffer damage from heavy frosts. All species and selections seem to tolerate a wide range of humidity levels.

As *P. pandorana* and *P. jasminoides* grow well in inner urban areas it can be assumed that they are moderately to strongly smog-tolerant.

In most cases, plants grow extremely well without supplementary fertilisers but they do respond well to light doses of slow-release and water-soluble fertilisers. Pandoreas develop a strong and penetrating root system and are not recommended for planting in close proximity to underground pipelines. Once plants are well established it is rare for them to require supplementary watering unless they are being grown in hot areas with a low rainfall.

All species respond well to light or heavy pruning,

which may often be required to curb the spread of some plants. Some selections provide a dense foliage cover and are very useful for screening purposes if they are grown on a strong support. They can be clipped regularly but over-zealous pruning during winter and spring will reduce the floral display. Old plants can be rejuvenated by removal of mature stems but due to their entanglement this can be a time consuming task. Pruning of plants back to just above ground level will be easier, but it may take plants 1–2 years to develop foliage cover again. *P. pandorana* and possibly some other species are suitable for topiary if grown on frames.

Propagation is from seed or cuttings. Freshly collected seed usually germinates readily and does not require any pre-sowing treatment. The seed should be barely covered with the propagation medium. Cuttings of firm young growth (especially short side branches) strike readily. It is common practice not to retain a basal node when cuttings are prepared, which means that double the number of cuttings are obtained. Application of hormone rooting powder or liquid is often beneficial but not always necessary. *P. jasminoides* and *P. pandorana* may self-layer at the nodes when stems are lying on the ground. The rooted layers can be removed, potted, then kept in a warm, sheltered location until established and ready for planting.

Pandorea australis (R. Br.) Spach. =
P. *pandorana* (Andrews) Steenis

Pandorea austrocaledonica (Bureau) Seem.
(from southern Caledonia; referring to New Caledonia)
NSW (Lord Howe Island)   Boat Vine
Climber   Aug–Dec

Vigorous bushy **scrambler** or **climber**; young growth purplish; **leaves** pinnate, 2–12 cm long; **leaflets** 5–9, 2–7 cm × 1.5–3.5 cm, ovate to lanceolate, more or less sessile, bright-to-mid green, shiny, glabrous, prominent glands, margins bluntly toothed, apex softly pointed; **flowers** 1–2 cm long, cream with dark red to purplish blotches in throat, in many-flowered terminal sprays, sweetly fragrant; **capsules** to about 5 cm long.

This species occurs on Lord Howe Island as well as in New Caledonia and Vanuatu. It differs from most others because it has toothed leaflets and generally grows as a very bushy, scrambling climber. It adapts well to cultivation in subtropical and temperate regions but can suffer damage from moderately heavy frosts. It does best in light to medium, well-drained, acidic soils in sun, semi-shade or dappled shade. Plants respond very well to pruning or clipping. During warm weather the flowers emit a delightful, sweet vanilla-like fragrance. Propagate from seed, or from cuttings of firm young wood, which strike readily.

P. pandorana ssp. *austrocaledonica* (Bureau) P. S. is a synonym.

Pandorea baileyana (Maiden & Betche) Steenis
(after F. M. Bailey, 19–20th-century Queensland botanist)
Qld, NSW   Large-leaved Wonga Vine
Climber   Sept–March

158

Pandorea jasminoides 'Charisma'   W. R. Elliot

Vigorous **climber**; **stems** strongly twining, often blackish, with white lenticels; **branchlets** often pendent; **leaves** to 30 cm or more long, pinnate, opposite or in whorls of 3–4; **leaflets** 7–9, 5.5–14 cm × 2–6.5 cm, ovate, prominently stalked, base rounded, dark green, glabrous, leathery, veins prominent on undersurface, margins entire, apex pointed; **flowers** to about 1.5 cm long, tubular, curved, with short lobes, cream with a hairy pink interior, in terminal sprays to about 30 cm long, somewhat conspicuous; **capsules** to about 5 cm × 2 cm, inflated, somewhat papery.

P. baileyana is regarded as rare but not under threat in nature. It is confined to south-eastern Qld and north-eastern NSW. This species is not regarded as highly ornamental and is uncommon in cultivation but it warrants wider recognition. Plants need plenty of space. They can be kept well covered with foliage by regular pruning. Best grown in freely draining but moisture-retentive acidic soils in a semi-shaded site. Hardy to moderate frosts. Propagate from fresh seed or from cuttings of firm young growth.

Pandorea doratoxylon (J. M. Black) J. M. Black =
P. pandorana (Andrews) Steenis

Pandorea jasminoides (Lindl.) K. Schum.
(similar to the genus *Jasminum*)
Qld, NSW   Bower Climber; Bower Vine;
                                      Bower of Beauty
Climber or scrambler   Sept–April

Vigorous **climber** or **scrambler**; **stems** somewhat slender, lightly twining, glabrous; **leaves** pinnate, 12–20 cm long, opposite or in whorls of 3, prominently stalked; **leaflets** 4–7, 4.5–6 cm × 1.5–3 cm, lanceolate to ovate, short-stalked, base rounded, glabrous, deep green, margins entire, apex pointed; **flowers** 4–6 cm × to about 5 cm, white with cream to greenish or pink

throat, or pale to deep pink with deep pink to reddish-pink hairy throat, faintly hairy exterior, in terminal sprays to about 12 cm long, often profuse and very conspicuous; **capsules** to about 8 cm × 2 cm, oblong-ovoid, inflated, beaked, somewhat papery.

This somewhat variable but nevertheless very handsome species is confined to south-eastern Qld and the North Coast of NSW. It is regarded as one of Australia's best climbers. In nature it can reach the canopy of tall rainforest trees as well as scrambling over boulders and slopes. Plants adapt to tropical, subtropical and temperate regions. Young plants are very susceptible to frost damage while older plants may suffer some damage from heavy frosts but well-established plants usually reshoot. Acidic soils which are moisture retentive yet drain moderately well are suitable. A sunny or semi-shaded site is recommended with best flowering occurring where the foliage is exposed to plenty of sunshine. Plants respond very well to hard pruning which can help restrict them to a desired size. Plants may do well in hanging baskets.

Over recent years a number of selections have been marketed including the following:

'Alba' has white flowers with a pale creamish-green throat. It is very similar to 'Lady Di'.

'Charisma' has variegated foliage which is green with a cream margin. It has pale pink flowers which are not as profuse as most other selections.

'Deep Pink' produces bright pink flowers with a very deep pink throat. Similar if not the same as 'Pink Form'.

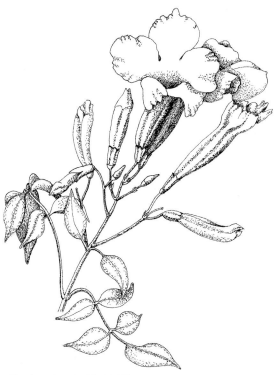

Pandorea jasminoides × .55

159

'Lady Di' was introduced during the 1970s. It is extremely popular and produces large white flowers with a yellowish throat.

'Lemon Bells' is a quick-growing cultivar. It produces large clusters of lemon flowers which may cover the foliage in late winter and spring.

'Pink Form' has flowers which are deep pink throughout. It flowers for most of the year.

'Southern Belle' is a very recently introduced selection which is registered under the *Plant Breeder's Rights Act*. It produces pale pink flowers which are larger in size during summer and early autumn.

Propagation is from fresh seed, or from cuttings, which do not need to have a basal node.

Pandorea nervosa Steenis
(prominent nerves or veins)
Qld
Climber   July–Nov; also sporadic
Moderately vigorous **climber** with glossy, purple-black young growth; **stems** slender, strongly twining; **leaves** pinnate, to about 25 cm long; **leaflets** 5–9, to about 8 cm × 3 cm, ovate, dark green, glossy, glabrous, venation prominent, apex pointed; **flowers** to about 5 cm × 2 cm, white with yellow interior, in crowded terminal sprays, usually with 1–2 flowers open at one time, rarely profuse but always conspicuous; **capsules** to about 12 cm × 2.5 cm, elliptical, brown, tapering to both ends.

A very ornamental, slow-growing climber from north-eastern Qld where it occurs in the highland rainforests. It deserves greater recognition by nurseries and gardeners. Plants require freely draining acidic soils which are moisture retentive and a site which is protected from drying winds and very strong sunshine. They suffer damage from moderate or stronger frosts. Responds well to pruning. Propagate from seed or from cuttings of hardened young growth.

Pandorea oxyleyi (DC.) Domin =
                    P. pandorana (Andrews) Steenis

Pandorea pandorana (Andrews) Steenis
(after *Pandora*, the first mortal woman of Greek mythology)
Qld, NSW, Vic, SA, WA, NT   Wonga Vine;
                              Wonga Wonga Vine
Climber or scrambler   June–Dec
Very variable, weak to vigorous **climber** or scrambling **shrub**; **stems** usually twining; **juvenile leaves** pinnate, 2–8 cm long; **leaflets** 8–17, small, margins usually coarsely toothed; **mature leaves** pinnate, 8–16 cm long, mainly opposite, prominently stalked; **leaflets** mostly 3–9, 1.5–8 cm × 0.3–4 cm, linear to ovate, base tapering to rounded, deep green, glabrous, margins usually entire, apex pointed; **flowers** 1–3 cm × 0.8–1.5 cm, tubular, more or less straight, variable in colour, white, cream, yellow, pink or brown, all with purplish blotches or stripes on the hairy interior, in terminal sprays to about 15 cm long, normally very profuse and conspicuous; **capsules** 3–8 cm × 1–2 cm, oblong, tapering to both ends, rough.

This species is one of Australia's most variable climbers and has an extremely diverse distribution.

Plants occur in rainforest, open forest, woodland and coastal scrub as well as in drier regions of the inland where the variant previously known as *P. doratoxylon* occurs. Generally most variants adapt well to cultivation. Acidic clay loam, loam or sandy soils are acceptable. The inland and southern variants are hardy to most frosts while those from northern areas may suffer frost damage. All respond well to pruning which may be needed to curb their spread.

A number of selections are available commercially. Further botanical revision could result in some of these selections being given different status.

'Alba' — see 'Snowbells'.

'Brown-flowered Variant' originated from north-eastern Qld. It has purplish new growth and purplish-brown stems. Flowers are mainly brownish, up to 3 cm long, and are well-displayed against large dark green leaflets.

'Burgundy-flowered Variant' is a lovely selection which is not commonly cultivated. It has flowers which are mainly burgundy but often with a touch of cream on the lobes, just near the top of the flower below the flower stalk.

'Fern Gully Form' has flowers about 2 cm long which are creamy white with reddish-purple to burgundy blotches on the interior of the throat. There is a degree of variation in the amount of colouration of the throat. This variant is common in the moist, closed forest, open forest, woodland and coastal scrub of southern NSW, Victoria and Tasmania.

'Fine-leaf Variant' has much narrower leaflets than many other selections and develops a light framework of foliage. Flowers are small and creamy white with deep purple blotches. Performs well as a basket plant if pruned regularly. It has been sold as *P. oxleyi* which is not a legitimate name.

'Golden Rain' — see 'Golden Showers'.

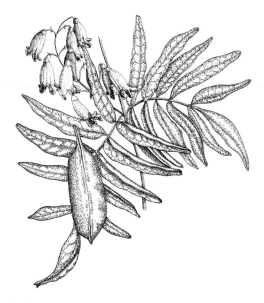

Pandorea pandorana × .4

'Golden Showers' has moderate sized leaflets. New growth may have bronze tonings. Flowers are yellow with light brownish tones and they are borne in pendulous sprays which are held on plants for much longer periods than most other selections.

'Herberton Variant' is not as vigorous as many selections. In nature it occurs in open forest in rocky soils of the Herberton region in Qld. It develops a strong rootstock and has lustrous, narrow, leathery leaflets. Flowers are yellow bells which are usually profuse and plants can also flower sporadically on new growth. Does well as a hanging basket plant.

'Inland Variant' often develops as an impressive large shrub with long switches of stems and foliage which wave in the breeze but will climb if they come in contact with suitable support. The leaflets are narrow and olive-green. Flowers are broad and cream with a hairy throat which is striped or blotched with purple. Often marketed as *P. doratoxylon*, which is a synonym of *P. pandorana*.

'Kingaroy Variant' grows quickly and covers large areas. It has small leaflets and profuse flowers. The flowers are about 6 cm long and may have a cream exterior with a purplish-red interior or they may be pinkish all over. This selection is highly recommended for subtropical and inland regions and may succeed in southern areas too.

'Pink-flowered Variant' (also known as 'Springbrook Pink') is a splendid selection from the highland areas of south-eastern Qld. It has shiny green leaves which enhance the display of reddish-pink flowers. May suffer frost damage. There also is a similar pink-flowered selection originated from the northern NSW coast.

'Rainforest Variant' is a vigorous climber with pale yellow flowers well-displayed against broad glossy green leaflets. New growth is pale yellowish-green.

'Snowbells' is an extremely vigorous form with mid green foliage. Massed pale creamy white flowers emit a sweet light fragrance especially during the early evening.

'White-flowered Variant' — see 'Snowbells'.

Propagation is from fresh seed, or from cuttings, which strike readily. Specific selections should be grown from cuttings or by other vegetative means to retain their distinctive characteristics.

Pandorea pandorana ssp. austrocaledonica (Bureau)
P. S. Green —
see *Pandorea austrocaledonica* (Bureau) Seem.

PANICUM L.
(from a Latin name for millet)
Poaceae

Annual or perennial **grasses** forming creeping mats, tussocks or aquatic clumps; base of the clump often thickened; **culms** erect to decumbent, glabrous or hairy; **ligules** membranous, hairy; **leaves** rolled when young, maturing flat, glabrous or hairy; **inflorescence** a slender to broad panicle, compact to sparse; **spikelets** solitary or in pairs, falling entire at maturity.

A large important genus of grasses consisting of about 500 species widely distributed in tropical, subtropical and temperate regions. About 25 species are native to Australia. They grow in diverse habitats including rainforests, swamps, stream margins, and sandhills and claypans in semi-arid regions. A number of exotic species are grown as pasture grasses especially in the tropics, and the grain of *P. miliaceum* L., commonly known as millet, is an important food source for stock, poultry and aviary birds. The seeds of many of the native species are eaten by seed-eating birds and small native rodents. Some of the perennial tufting species have the ability to withstand dry conditions and can produce new foliage, flowers and seeds quickly after rain. The foliage is palatable to stock and native animals but few species can withstand heavy persistent grazing.

Some of the native species of *Panicum* have ornamental qualities and are worth growing in containers or as garden plants. Untidy clumps can be rejuvenated by heavy cutting or burning. Propagation is by seed, transplant or division. Seed viability is frequently low.

Panicum australiense Domin
(Australian)

Qld, WA, NT	Bunch Panic Grass
0.05–0.25 m × 0.2–0.4 m	March–May

Annual grass forming dense, compact, green, reddish or purplish tufts; **culms** bent at the nodes, much-branched; **leaves** 3.5–6.5 cm × 0.3–0.5 cm, green, reddish or purplish, margins thickened, hairy; **panicles** 2–3 cm × 1 cm, with 2 or 3 branches, enclosed within the leaves; **spikelets** few, 0.3–0.4 cm long.

P. australiense is widespread in Central Australia where it grows on sand ridges and clayey sands, often among spinifex. An interesting annual grass with prospects for cultivation in arid areas. Plants form a dense, compact tuft and the leaves take on colourful hues as they age. Requires excellent drainage and a sunny aspect. Propagate from seed.

Panicum bisulcatum Thunb.
(with 2 grooves)

Qld, NSW, Vic	Black-seed Panic Grass
0.5–0.8 m × 1–2 m	Dec–April

Annual or **perennial aquatic grass** forming spreading clumps; **culms** erect, rooting at the nodes; **leaves** 10–15 cm × 0.5–1.2 cm, green to yellowish-green, flat to concave, acuminate; **panicles** to 40 cm × 35 cm, much-branched but open; **spikelets** 0.2–0.3 cm long, glabrous; **seeds** purplish-black.

Distributed from south-eastern Qld to Victoria, forming extensive colonies along stream banks and in swamps. An important species for stabilising streambanks, it also provides a significant habitat for aquatic wildlife and birds. It has been planted to reduce streambank erosion. Also useful for the banks of dams and ponds. Propagate from seed sown on mud and also by division.

Panicum decompositum R. Br.
(divided twice)

all states	Native Millet, Papa Grass
0.5–1 m × 0.3–0.8 m	Dec–April; also sporadic

Perennial grass forming stout tussocks; base of plant thickened, glabrous or sparsely hairy; **culms** hollow, entire or branched; **leaves** 15–30 cm × 1–1.2 cm, flat,

Pandorea pandorana 'Snowbells' and 'Golden Showers'
W. R. Elliot

green to bluish-green, midrib white, prominent, margins roughened; **panicles** to 40 cm × 40 cm, much-branched but open, the branchlets stiffly spreading, often wavy; **spikelets** 0.3–0.4 cm long, pale green to purplish; **seeds** dark brown, shiny.

A very widely distributed species which is a common component of grassland and woodland. It also colonises floodplains and alluvium. The seed heads break off at maturity and flow about on water, thus dispers-

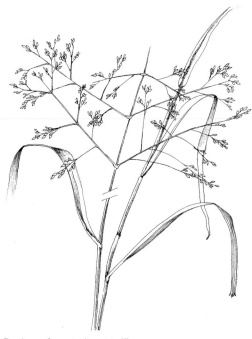

Panicum decompositum × .45

162

ing the seeds. Plants may become dormant during dry periods and then grow actively and flower soon after rain. The leaves provide good fodder for stock and native animals but plants rarely survive intensive grazing. The seeds are a very important food source for seed-eating birds and small native rodents. Plants have some ornamental appeal and can be planted among small shrubs and in rockeries. They adapt to different soil types but need a sunny location. Propagate from seed.

Panicum effusum R. Br.
(loosely spreading)
Qld, NSW, Vic, SA, WA, NT   Hairy Panic Grass
0.5–1 m × 0.1–0.3 m   Dec–April; also sporadic

Perennial grass forming tussocks; base of plant hairy; **culms** bent at the lower nodes, densely hairy; **leaves** 15–30 cm × 0.8–1.2 cm, flat, dark green, densely covered in long, white hairs, margins roughened; **panicles** to 30 cm × 30 cm, much-branched but open, stiff; **spikelets** 0.2–0.3 cm long, green to purplish, often borne in pairs; **seeds** dark brown, shiny.

Widely distributed in grassland, open forest and woodland. Often colonises disturbed sites. This interesting tussock grass has hairy leaves which are palatable and provide good fodder for stock and native animals. The seeds are an important food source for birds and native rodents. Requires well-drained soil in a sunny position. Propagate from seed.

Panicum lachnophyllum Benth.
(with woolly leaves)
Qld, NSW   Don't Panic
0.1–0.3 × 1–3 m   Dec–April

Perennial grass forming spreading patches; **culms** slender, branched, rooting at the nodes; **leaves** 1.5–10 cm × 0.3–0.9 cm, flat, dark green, sparsely hairy; **panicles** 2–7 cm × 1–3 cm, narrow, sparsely branched; **spikelets** 0.2–0.25 cm long, green, glabrous.

Distributed from central Qld to northern NSW. Plants grow in moist forests particularly in rainforest or along rainforest margins. The foliage is browsed by native marsupials. A useful groundcover grass for shady locations. Propagate from seed or by division.

Panicum paludosum Roxb.
(growing in swamps)
Qld, NSW, NT   Swamp Panic Grass
0.5–1.5 m × 1-2 m   Dec–April

Perennial aquatic grass forming floating clumps; **culms** erect, spongy, floating; **leaves** 15–30 cm × 0.8–2 cm, bright green, flat, glabrous; **panicles** 20–45 cm × 10–20 cm, sparsely branched, open; **spikelets** 0.35–0.45 cm long, glabrous, green.

A widely distributed species which grows in colonies on the margins of lagoons, dams, billabongs and sluggish streams. It is an important species for stabilising the banks of such sites and also provides a refuge and habitat for wildlife. It can be grown readily in water to 1 m deep. Propagate from seed sown on damp mud or by division.

Panicum prolutum F. Muell. =
     Homophalus proluta (F. Muell.) R. D. Webster

Pandorea pandorana 'Springbrook Pink'   D. L. Jones

Panicum pygmaeum R. Br.
(dwarf)

Qld, NSW	Pigmy Panic, Dwarf Panic
0.1–0.3 m × 0.5–1 m	Dec–April

Perennial grass forming mats or loose, spreading patches; **culms** slender, branched; **leaves** 1–6 cm × 0.3–0.7 cm, dark green, sparsely hairy; **panicles** 2–6 cm, with a few spreading branches; **spikelets** about 0.2 cm long, solitary or paired, green; **seeds** shiny, striate.

Occurs in moist, sheltered areas of dense eucalypt forest and along rainforest margins. The foliage is browsed by native marsupials. A useful groundcover grass for shady situations. Propagate from seed or by division.

Panicum queenslandicum Domin
(from Queensland)

Qld, NSW	Yabila Grass, Coolibah Grass
0.5–1 m × 0.3–0.5 m	Oct–Feb; also sporadic

Perennial grass forming dense tussocks; base of plant glabrous; **culms** erect, smooth, fibrous; **leaves** 15–30 cm × 0.2–0.3 cm, flat, dark green, smooth, glabrous; **panicles** to 40 cm × 40 cm, much-branched but open, the branchlets arising in whorls, erect to spreading; **spikelets** 0.3–0.7 cm long, glabrous, green to purplish, in pairs; **seeds** brown, shiny.

Distributed from central Qld to northern NSW. Plants grow on cracking clay soils in floodways. Seed heads break off at maturity and blow in the wind. The foliage is eaten by stock and native animals and the seeds are an important source of food for birds and small rodents. May be planted among shrubs or rocks in well-drained soil in a sunny position. Propagate from seed.

Panicum subxerophilum Domin
(somewhat dry-loving)

Qld, NSW	Gilgai Grass; Cane Panic Grass
0.5–0.8 m × 0.5–1 m	Dec–May

Perennial grass forming erect, slender, rigid to wiry, cone-like clumps; **culms** unbranched or sparsely branched, rigid; **leaves** 2–12 cm × 0.1–0.35 cm, flat or inrolled, bright green; **panicles** 5–30 cm × 5–20 cm, with widely spreading, wavy, branches; **spikelets** 0.2–0.3 cm long, singly or in pairs, green to dark purple.

Grows in floodways and in crabhole country in sparse woodland on clay soils — also occasionally in mallee shrubland. The foliage is palatable to stock and native animals and the seeds are eaten by birds and small native rodents. This is an interesting grass which has a fresh green appealing appearance when in active growth. Hardy to dryness but responds well to watering. Old tussocks can be rejuvenated by heavy cutting or burning. Requires a sunny position. Propagate from seed or by division.

PAPILLILABIUM Dockrill
(from the Latin, *papillatus*, papillae; *labium*, lip; referring to small bumps on the labellum)
Orchidaceae

A monotypic genus which is endemic in Australia.

Papillilabium beckleri (F. Muell. ex Benth.) Dockrill
(after Dr H. Beckler, original collector)

Qld, NSW	
epiphyte	Sept–Oct

Epiphytic orchid growing in very small clumps; **stems** to 4 cm long, unbranched; roots long, thin, wiry; **leaves** 2–6, 3–5 cm × 0.3–0.4 cm, linear-lanceolate, falcate, green, usually spotted with purple; **racemes** 2–4 cm long, slender; **flowers** 2–8, about 0.7 cm across, pale green with purplish or brownish markings, fragrant; **labellum** covered with masses of small bumps.

A tiny epiphytic orchid which may be locally common, but is often very difficult to find. It is distributed from south-eastern Qld to south-eastern NSW and grows on shrubs and trees in rainforest and along the margins of streams. It is grown by enthusiasts but can be difficult to maintain in cultivation. Plants require a protected shady situation, abundant humidity and free and unimpeded air movement. Propagate from seed sown under sterile conditions.

Previously known as *Sarcochilus beckleri* (F. Muell. ex Benth.) F. Muell.

PAPPOPHORUM Schreber —
see *ENNEOPOGON* Desv.

PARACALEANA Blaxell = *CALEANA* R. Br.

PARACETERACH Copel.
(similar to or near the genus *Ceterach*)
Hemionitidaceae

Terrestrial ferns often growing among rocks; **rootstock** creeping, usually scaly; **fronds** pinnate, with a terminal pinna; **pinnae** in opposite or subopposite pairs, thick, leathery, scaly; **sori** lacking indusia, in a broad marginal band.

Paraceterach muelleri

A small genus of 2 or 3 species endemic in Australia. They are drought-resistant ferns. The fronds are able to regreen from a dried state following substantial rain. Plants have an ornamental appearance but generally they have not proved to be amenable to cultivation in the ground, although some success has been achieved with pot culture. Propagation is from spores.

Paraceterach muelleri (Hook.) Copel.
(after Baron F. von Mueller)
Qld
0.1–0.4 m tall
Rootstock short-creeping, scaly; **fronds** pinnate, 10–40 cm long, erect to semi-pendent; **stipes** and rhachises brown to black, densely scaly; **pinnae** 1–3 cm × 1.3–1.8 cm, ovate, pale green, both surfaces densely scaly, thick; **sori** in a marginal band.

A highly ornamental fern from north-eastern Qld where it forms colonies in moist soil in open sites among rocks and along streams. They have proved to be difficult to establish in cultivation with the general response being a slow decline. Best in the tropics and subtropics where it may be successfully grown in a pot in warm, humid conditions with free air movement. Propagate from spores.

Paraceterach reynoldsii (F. Muell.) Tindale
(after J. Reynolds, 19th-century Australian politician)
Qld, SA, WA, NT
0.1–0.25 m tall
Rootstock medium-creeping, scaly; **fronds** pinnate, 10–25 cm long, erect; **stipes** and rhachises brown, shiny, scaly; **pinnae** 1–2 cm × 1–1.5 cm, ovate to circular, dark green, both surfaces scaly; **sori** in a marginal band.

Widely distributed in drier inland areas, usually in sheltered sites among rocks and in crevices. An attractive fern which has proved difficult to establish in the garden. It has been grown successfully as a pot subject in glasshouses. Propagate from spores.

PARAHEBE W. R. B. Oliv. — see *DERWENTIA* Raf. (in the supplementary volume of the Encylopaedia, plant entries section).

PARAMIGNYA Wight
(from the Greek *paramignumai*; to mix with)
Rutaceae
Evergreen **shrubs** or **climbers**; **stems** with axillary spines; **leaves** alternate, glandular-dotted, somewhat leathery; **flowers** solitary or in axillary clusters, panicles or racemes, white, fragrant; **petals** 3–5, overlapping in bud; **calyx** 3–5-lobed; **stamens** 6–10 or sometimes more; **ovary** 2–5-celled; **fruit** a berry, with thick resinous peel.

This tropical genus comprises about 20 species. The main representation is in Indonesia but it also occurs in India, China and the Philippines. There is one species in Australia.

Paramignya trimera (Oliv.) Burkill
(with three parts)
WA, NT
1.5–3 m × 1–3 m   mainly Aug–Nov

Small **shrub** or **climber** with glabrous young growth; **stems** glabrous, armed with spreading or recurved spines to about 3 cm long; **leaves** 2–10 cm × 1–6 cm, elliptic to obovate, short-stalked, glabrous, glossy green above, dull green below, margins entire, midrib and venation prominent, apex blunt or slightly notched; **flowers** to about 0.7 cm across, 3-petalled, white, in axillary panicles, sweetly fragrant; **stamens** 6; **fruit** about 1 cm across, globular, orange.

An inhabitant of vine thickets in close proximity to the coast. Plants grow on basalt in the Kimberley, WA, and in sandy soils in the Top End, NT. Unlikely to become popular in cultivation, it is valuable for wildlife habitat because of its spiny nature. Suitable for seasonally dry tropical regions in well-drained soils. Propagate from fresh seed and possibly from cuttings.

PARANTENNARIA Beauverd
(from the Greek *para*, near, beside; the genus *Antennaria*)
Asteraceae (alt. *Compositae*)
A monotypic endemic genus from the highest parts of the Australian Alps.

Parantennaria uniceps (F. Muell.) Beauverd
(one-head)
NSW, Vic
prostrate × 0.3 m   Dec–Feb
Dwarf, spreading, perennial, dioecious **herb** with faintly hairy young growth; **stems** spreading to ascending; **leaves** 0.7–1.5 cm × up to 0.2 cm, linear to linear-cuneate, alternate, stem-clasping, spreading to ascending, crowded, stiff, glossy and concave above, green with base often purplish-hairy, apex pungent-pointed; **flowerheads** to about 0.6 cm across, white with yellow, initially sessile with fertile heads becoming stalked; **ray florets** absent; **fruit** an achene.

An interesting groundcover which can form extensive patches in the Australian Alps. It is found in feldmark and herbfields as well as in wet gullies. Evidently rarely cultivated, it could have appeal for daisy family enthusiasts. Plants will need well-drained acidic soils in a sunny or semi-shaded site. It has potential for use in miniature gardens and in rockeries in cool temperate regions. Plants tolerate very cold conditions including heavy frost and snowfalls. Propagate from seed and possibly from cuttings.

PARARCHIDENDRON Nielsen
(from the Greek *para*, near; and the genus *Archidendron*)
Mimosaceae
A monotypic genus which extends to eastern Australia. Propagation is from seed which has a limited period of viability and should be sown fresh.

Pararchidendron pruinosum (Benth.) Nielsen
(as if covered by a powdery bloom)
Qld, NSW   Snow Wood
6–12 m × 2–5 m   Sept–Nov
Tall **shrub** or small **tree** with a bushy canopy; young growth light green; **leaves** bipinnate, 5–15 cm long; **pinnae** 1–4 pairs, with a gland between the upper pair; **leaflets** 5–11 per pinna, 2–8 cm × 1.5–2.5 cm,

Pararchidendron pruinosum × .55

lanceolate to oblanceolate, dark green and shiny above, paler and dull beneath; **flowerheads** about 4 cm across, opening greenish-white or yellowish, ageing to orange or brown, fragrant; **corolla** about 0.5 cm long; **stamens** to 1.3 cm long; **pods** 8–12 cm × 1–1.5 cm, twisted, yellow to orange outside, red inside; **seeds** black, glossy, mature Feb–June.

Widely distributed from north-eastern Qld to Nowra in southern NSW and also occurs in Java. It grows in rainforest often near streams. The leaves are eaten by the interesting larvae of the attractive Tailed Emperor Butterfly. It has become moderately popular in cultivation due to its fast growth, bushy habit and fragrant flowers. Plants will grow in a wide range of well-drained acidic soils and they respond favourably to mulches, light fertiliser applications and watering during dry periods. Tolerates moderate frosts. Propagate from fresh seed.

PARASARCOCHILUS Dockrill =
SARCOCHILUS R. Br.

PARASERIANTHES
(from the Greek *para*, near; and the genus *Serianthes*)
Mimosaceae

Shrubs and **trees** with a rounded often sparse crown; **bark** smooth or flaky; **branches** with prominent lenticels; **leaves** bipinnate, often large; **leaflets** opposite, usually small but numerous; **stipules** not spiny; **inflorescence** axillary spikes or racemes; **flowers** with prominent stamens; **sepals** and **petals** small; **stamens** numerous, long, colourful; **fruit** a pod, flat.

A genus of 4 species; 2 of these occur in Australia. *P. toona* is restricted to a small area of Qld. *P. lophantha* was considered indigenous but it is now regarded as introduced in south-western WA. It is weedy and has spread to many other areas of Australia.

Paraserianthes lophantha (Willd.) I. C. Nielsen.
This species is no longer regarded as native to Australia. It occurs in Indonesia and has been introduced into Australia and South Africa.

Previously known as *Albizia lophantha* (Willd.) Benth.

Paraserianthes toona (F. M. Bailey) I. C. Nielson
(an Indian name)
Qld Mackay Cedar; Red Siris
10–30 m × 5–15 m Aug–Oct

Small to medium **tree** with brownish-hairy young growth; **bark** grey to brown, often scaly; **branches** spreading to ascending, with yellow or white lenticels; **leaves** 15–30 cm long, bipinnate, alternate, deciduous during June–Sept; **leaflets** 10–20 pairs, 0.3–0.8 cm long, linear, dark green above, paler below; **flowers** about 0.5 cm across, cream, in oblong spikes with prominent stamens, arranged in terminal and upper axillary panicles, somewhat conspicuous, fragrant; **pods** 10–13 cm × 2–3 cm, flat, ripe during April–Aug; **seeds** 4–6.

This fast-growing species is restricted to central-eastern and north-eastern Qld where it grows in moist sclerophyll forest and rainforest of the lowland regions. It needs a sunny site in acidic soils which are moderately well drained. Best suited to tropical and subtropical regions. Plants are excellent as shade providers. Responds well to supplementary watering during extended dry periods. Not recommended for small gardens. Butterflies are attracted to the flowers. The timber is red with yellowish streaks and is highly regarded for decorative cabinet-making. It is also used to make long-lasting fence posts. Propagate from seed, which usually germinates without pre-sowing treatment.

Previously known as *Albizia toona* F. M. Bailey.

PARATEPHROSIA Domin
(from the Greek, *para*, near; and the genus *Tephrosia*)
Fabaceae

An endemic and monotypic genus restricted to Qld, WA and NT. It is allied to *Tephrosia*, which differs in having many axillary racemes and its pods can have 1–20 seeds.

Paratephrosia lanata (Benth.) Domin
(woolly)
Qld, WA, NT
1–2.5 m × 1–2.5 m March–Sept

Small to medium **shrub** with greyish-hairy or rarely rusty-hairy young growth; **branches** spreading to ascending, woolly-hairy; **stipules** to about 0.6 cm long, deciduous; **leaves** usually trifoliolate, sometimes unifoliolate, with stalk to 1 cm long; **leaflets** central one largest, 1.7–5 cm × 0.6–3.5 cm, others smaller, elliptic to obovate, covered in greyish hairs on both surfaces, margins entire, apex blunt with a small point; **flowers** pea-shaped, about 0.6 cm across, arranged in crowded often leafy somewhat terminal racemes to about 8 cm long, may be profuse but rarely highly conspicuous; **standard** orange front with yellow back; **wings** yellow with pink tips; **pods** to about 1 cm × 0.5 cm, ovoid, velvety-hairy, 1-seeded.

Paratrophis

Paraceterach reynoldsii   D. L. Jones

An interesting species from the arid regions where it grows in a range of well-drained acidic or alkaline soils which can be rocky or sandy. Rarely cultivated, it deserves greater attention for cultivation in arid, semi-arid and warm temperate regions. Good drainage and plenty of sunshine are crucial. Plants tolerate moderate frosts. Pruning of young plants should promote a bushy framework. Suitable as an accent plant in private and public gardens. Propagate from pre-treated seed (see Volume 1, page 205) or possibly from cuttings of firm young growth.

Lespedeza lanata Benth. is a synonym.

PARATROPHIS Bl. = *STREBLUS* Lour.

PARINARI Aubl.
(a native Guianan name)
Chrysobalanaceae
   Shrubs or **trees**; **stipules** present; **leaves** alternate, simple, entire; **inflorescence** much-branched panicle; **flowers** bisexual, somewhat bell-shaped, faintly swollen on one side, exterior and interior hairy; **sepals** 5, often unequal; **petals** 5; **stamens** 6–8, not exceeding sepals; **fruit** a drupe.
   A tropical and subtropical genus of about 50 species. One species occurs in Australia.
   Propagate from seed.

Parinari corymbosum (Blume) Benth. =
                              Maranthes corymbosa Blume

Parinari nonda Benth.
(an Aboriginal name)
Qld, WA, NT   Nonda
2–5 m × 1.5–5 m   Aug–Oct
   Medium to tall **shrub** with open growth habit, hairy young growth; **bark** fissured; **branches** spreading to ascending; **branchlets** densely hairy; **leaves** 3.5–8 cm × 2–4.5 cm, narrowly elliptic to broadly ovate, alternate, short-stalked, leathery, dark green and glabrous above, densely whitish-hairy below, midrib prominent, venation reticulate, apex blunt or pointed; **panicles** terminal and in upper axils, with hairy deciduous bracts; **flowers** to about 0.5 cm long, tubular with spreading, unequal, deciduous petals, whitish, cream, brown or yellow, exterior and interior hairy; **stamens** unequal in length; **drupe** to about 3.7 cm × 2.5 cm, ellipsoid, dry, initially green to reddish-brown, covered in small brownish scales or flakes.
   Widely distributed across northern Australia extending to New Guinea and the Solomon Islands. It is recorded as growing on sand dunes, on red sandy loam of the plains and in open *Melaleuca viridiflora–Eucalyptus* woodland. Evidently rarely cultivated because of its somewhat drab appearance. It is suitable for seasonally dry tropical regions. The fruit is edible but distasteful. Propagate from fresh seed and possibly from cuttings of semi-ripened wood.

PARKERIACEAE Hook.
A small family of tropical ferns consisting of 4 species in the genus *Ceratopteris*. They are interesting annual ferns which grow as aquatics either rooting in mud or free floating. They are excellent plants for aquaria and a couple are also eaten as vegetables. Two species of *Ceratopteris* occur in Australia.

PARSONSIA R. Br.
(after James Parsons, 18th-century English physician and naturalist)
Apocynaceae   Silkpods
   Woody or semi-woody **climbers** with twining **stems**; juvenile plants often have creepers with adventitious roots; **sap** clear and watery, or milky; **leaves** opposite, stalked, juvenile and mature leaves are often different; **inflorescence** in cymes or panicles; **flowers** mostly 5-merous, sometimes 4-merous, tubular with hairy throat; **ovary** 2-celled; **fruit** a capsule, pod-like, terete to fusiform or ovoid, 2-valved.
   This predominantly tropical and subtropical genus consists of about 130 species which occur in South-East Asia, New Guinea, New Caledonia, New Zealand and Australia. In Australia, there are 35 species with 33

Parinari nonda   D. L. Jones

Parsonia straminea   W. R. Elliot

endemic mainland species. One species is endemic on Lord Howe Island, and another species is widespread in adjacent tropical regions.

Sometimes *Lyonsia* has been classified as a separate genus but it is now accepted as a synonym.

Parsonsias are found from near sea-level to about 1200 m altitude. They inhabit rainforests, vine thickets, open forests and a few species extend into semi-arid regions.

Some species are extremely vigorous climbers with thick, woody, twining stems (eg. *P. latifolia*, *P. plaesiophylla* and *P. straminea*), while others are less vigorous (e.g. *P. lenticellata* and *P. leichhardtii*), and a limited number of species are weak climbers, for example *P. induplicata*.

Generally parsonsias are not regarded as ornamental plants but nevertheless some have decorative features. *P. plaesiophylla* and *P. straminea* are among the better known and they are cultivated, along with *P. brownii*, to a limited degree.

The main attributes which make parsonsias horticulturally desirable are quick growth, screening properties and the perfumed flowers. *P. densivestita*, *P. eucalyptophylla*, *P. lilacina*, *P. plaesiophylla*, *P. straminea* and *P. velutina* are examples of species with fragrant flowers.

Some species also have slender, pendent fruits which split open on ripening to release large quantities of silky-plumed seeds. These seeds are spread readily by the wind.

Parsonsias usually do best in protected sites with moist but well-drained acidic soils. Although appreciating constant moisture they do not tolerate extended waterlogging. Some species such as *P. eucalyptophylla* withstand drier conditions that last for extended periods. Frost tolerance is not well understood for most species but a number are hardy to moderate frosts — *P. straminea* and *P. eucalyptophylla* are possibly the most reliable. Most species react very well to extended periods of high humidity.

Parsonsias respond well to light applications of slow-release fertiliser which is best applied during early spring.

Supplementary watering may be required for some plants during extended dry periods. Organic mulches can be used to conserve moisture, and after they break down they can be mixed with the soil to provide excellent growing conditions.

Results from pruning parsonsias are inconclusive due to the limited number in cultivation but species such as *P. brownii* and *P. straminea* respond very well.

Evidently pests and diseases do not pose many problems. The leaves of some plants may be chewed by caterpillars (possibly the larvae of butterflies).

Propagation is from seed, or for some species cuttings are successful. Seed has a short viability and should be sown as soon as possible after it has been collected. Seedlings are best pricked out as soon as they can be safely handled otherwise they may become entangled and very difficult to transplant.

Propagation from cuttings has so far been very limit-ed and there needs to be further experimentation with this method. Cuttings of firm young growth generally give the best results. Species such as *P. eucalyptophylla* and *P. straminea* are readily propagated from sections of stems, which have adventitious roots.

Parsonsia alboflavescens (Dennst.) Mabb.
(white to pale yellow)
NT
Climber   March; possibly also at other times
Climber with glabrous young growth; **stems** twining; **sap** watery; **leaves** 8–15 cm × 3.4–8.2 cm, elliptic, opposite, base tapered or sometimes rounded, spreading to ascending, pale green, glabrous, soft, thin, venation conspicuous, apex pointed; **flowers** to about 1.4 cm across, yellowish or green, mainly glabrous, lobes spreading, in open axillary clusters which extend beyond leaves, moderately conspicuous; **capsules** 10–15 cm × less than 1 cm, tapered to both ends, glabrous.

A wide-ranging, attractive, tropical species which is common in South-East Asia, Malesia, Papua New Guinea and the Solomon Islands. In Australia, it occurs only in north-eastern Arnhem Land. This species is evidently not cultivated in Australia but it warrants greater attention. It is suitable for seasonally dry tropical regions and may adapt to subtropical areas too. Plants will need well-drained soils with a semi-shaded or sunny aspect. Propagate from fresh seed, or possibly from cuttings of firm young growth.

Parsonsia bartlensis J. B. Williams
(from Mt Bartle Frere)
Qld
Climber   June
Low to moderately vigorous **climber** with glabrous young growth; **stems** twining, glabrous; **sap** watery; **leaves** 3.8–5.2 cm × 1.4–2.3 cm, lanceolate, opposite, base rounded, not crowded, spreading to ascending, glabrous, margins recurved, venation prominent, apex pointed; **flowers** about 0.3 cm across, yellowish, with erect lobes, exterior glabrous or hairy, in short, compact, terminal or axillary panicles of 7–10 flowers; **fruit** not known.

167

Parsonsia blakeana

A recently described (1996) and apparently rare species which is known only from the upper regions of Mt Bartle Frere. It lacks strong ornamental properties. Plants should be suitable for subtropical and temperate regions. They need excellent drainage and regular watering. Propagate from seed, or from cuttings of firm young growth.

Parsonsia blakeana J. B. Williams
(after Stanley T. Blake, 20th-century Australian botanist)
Qld
Climber   Aug–Oct
Scrambling **climber** with glabrous young growth; **stems** twining, glabrous; **sap** watery; **leaves** 7–16.5 cm × 2.5–6.5 cm, oblong-ovate, opposite, base rounded, spreading to ascending, green above and below, glabrous, thin-textured, venation prominent, apex pointed; **flowers** to about 0.8 cm across, white to yellowish, lobes spreading, in mainly axillary panicles about as long as leaves, of 8–20 flowers; **capsules** 11–13 cm × 1.5–3 cm, narrowly ovoid, grooved.

This recently described (1996) species is restricted to one area of northern Cape York Peninsula where it grows in sandy soils of dry rainforest, open forest, vine thickets and exposed coastal heathland. Suited to seasonally dry tropical regions. Plants will need excellent drainage and a sunny or semi-shaded site. Propagate from seed, or possibly from cuttings of firm young growth.

P. bartlensis has smaller leaves and yellowish flowers with erect lobes.

Parsonsia brisbanensis J. B. Williams
(from the Brisbane region)
Qld
Climber   Oct–Jan
Moderately vigorous **climber** with faintly hairy young growth; **stems** initially with clinging roots, becoming twining and glabrous; **sap** watery; **juvenile leaves** 1.5–5 cm long, base cordate; **mature leaves** 7.5–23.5 cm × 2.4–7 cm, broadly lanceolate to ovate, opposite, rounded or cordate base, spreading to ascending, dull to deep green, with minute brownish hairs, stiff to softly pliable, venation prominent, apex tapering to a point; **flowers** to about 0.8 cm across, yellowish with revolute lobes, exterior faintly brownish-hairy, in axillary or terminal many-flowered panicles; **capsules** 6–10 cm × 1.8–2.6 cm, narrowly ovoid, faintly hairy.

This species is found in eucalypt forests and woodlands in the Brisbane–Kingaroy region where it is often prevalent in disturbed road verges. It was described in 1996. Plants should adapt to subtropical and temperate regions with well-drained soils. A sunny or semi-shaded site should be suitable. Propagate from seed, or possibly from cuttings of firm young growth.

Parsonsia brownii (Britten) Pichon
(after Robert Brown, 18–19th-century British botanist)
NSW, Vic, Tas   Mountain Silkpod; Twining Silkpod
Climber   Sept–Feb

Parsonsia brownii × .5, flower × 1.5

Vigorous **climber** with short-hairy young growth; **stems** twining, woody; **sap** watery; **leaves** 4–14.5 cm × 1–4 cm, narrowly-lanceolate to lanceolate, narrow when young becoming broader with age, opposite, not crowded, spreading to ascending, glabrous and glossy, deep green above, pale yellowish-green to fawn or greyish below, leathery to firm, venation visible, apex tapering to a fine point; **flowers** to about 0.7 cm across, yellow to greenish, faintly hairy, in loose axillary clusters with fawnish stems, rarely profuse; **capsules** 5–10 cm × about 1 cm, narrowly ovoid, hanging vertically, ripe Aug–Oct.

This vigorous climber inhabits acidic soils in rainforest, wet sclerophyll forest, fern gullies and moist sites in open forest. It is distributed on the mainland from north-eastern NSW to eastern and southern Vic, and is widespread in Tas. Adapts well to cultivation in subtropical and temperate areas and although it prefers moist, shaded sites it will tolerate limited expos-ure in soils which are dry for short periods. Plants can strangle or disfigure small shrubs and are best grown on wire or timber supports, or only with well-established trees. Regular pruning promotes bushy growth. Plants are hardy to moderate frosts. Propagate from seed or from cuttings of firm young growth.

P. brownii has been confused with *P. straminea* which differs in having stems that often produce adventitious roots. *P. straminea* also has broader yellow-green leaves, clear yellowish sap and capsules 10–20 cm long.

Parsonsia densivestita C. T. White
(densely hairy)
Qld   Silkpod
Climber   Oct–Dec

Vigorous **climber** with densely brownish-hairy young growth; **stems** woody, twining; **sap** watery; **leaves** 11–28 cm × 4–10 cm, ovate to broadly elliptic, oppos-ite, with prominent stalks, base rounded or cordate, spreading to ascending, not crowded, deep green with brownish hairs, thin-textured, flat, venation promin-ent below, apex long-tapering to a fine point; **flowers** 0.6–0.8 cm across, cream to greenish with spreading lobes, sweetly fragrant, in crowded axillary clusters with rusty-hairy stems, sometimes profuse and moder-ately conspicuous; **capsules** 13–23 cm × 1.8–2.5 cm, cylindrical, ripe Feb–March.

An extremely vigorous climber which often reaches the tree canopy. It is restricted to the Cook District where it grows in rainforest at an altitude of 200–400 m. Suitable for subtropical and highland tropical regions, it requires plenty of space to grow unless pruned regularly to contain its spread. Needs strong constructed supports or well-established trees on which to climb. Does best in well-drained acidic loams with a semi-shaded or dapple-shaded aspect. Plants tol-erate moderate frosts. Propagate from fresh seed. Cuttings of firm young growth are also worth trying.

Parsonsia diaphanophleba F. Muell.
(prominent veins)
WA
Climber   Jan–Feb; May–June;
                                  possibly sporadic

Weak to moderately vigorous **climber** with faintly hairy young growth; **stems** woody, twining; **leaves** 6–20 cm × 0.6–2 cm, linear to very narrowly ovate, oppos-ite, with petiole to about 1 cm long, spreading to ascending, dark green above, paler green to whitish below, becoming glabrous except sometimes hairy below, venation prominent, apex tapering to a fine point; **flowers** to about 0.6 cm across, cream with pale green or pinkish to purplish tonings, in terminal and axillary clusters of 8–30 flowers, rarely profuse; **cap-sules** to about 10 cm × 0.8 cm, tapered to both ends, hairy.

This rare and endangered species is known only from a small area of the coastal plain, south of Perth, where it grows on alluvial soil. Lacking strong orna-mental features, it nevertheless is cultivated to a very limited degree. Suited to temperate regions and may succeed in semi-arid areas. Plants grow in a sunny or semi-shaded site with acidic well-drained soils but they are frost sensitive. Pruning promotes bushiness. Propag-ate from seed, or from cuttings of firm young growth. Sometimes misspelt as *P. diaphanophlebia*.

Parsonsia dorrigoensis J. B. Williams
(from Dorrigo)
NSW   Milky Silkpod
Climber   Nov–Feb

Moderate **climber** with faintly hairy young growth; **stems** slender, twining; **sap** milky; **leaves** 4–12 cm × 1.5–5 cm, narrowly triangular to broadly ovate, oppos-ite, base often cordate, with petiole 1–2 cm long,

spreading to ascending, not crowded, glabrous, deep green above, green to purplish below, venation promin-ent, apex finely pointed; **flowers** to about 0.8 cm across, cream to yellowish with strongly recurved lobes, exterior glabrous, interior hairy at base, in axil-lary clusters; **capsules** 5–7 cm long, narrowly terete, glabrous.

A recently described (1996) rare species which is restricted mainly to the Dorrigo region inland from Coffs Harbour in north-eastern NSW. It grows in rain-forest and sclerophyll forest. Plants are deserving of cultivation and should do well in subtropical and tem-perate regions. They need a sheltered moist site with acidic soils. Evidently not as rampant as many other species. Frost hardiness is not known. Propagate from seed or from cuttings of firm young growth.

Parsonsia eucalyptophylla F. Muell.
(leaves like a eucalypt)
Qld, NSW   Gargalou
Climber   Oct–April

Vigorous **climber** with softly hairy young growth; **stems** initially soft, often with adventitious roots, becoming woody and tips twining; **sap** watery; **leaves** 8–30 cm × 0.5–2 cm, ovate and dark green on juvenile plants, otherwise linear to lanceolate, opposite, with petiole to 3 cm long, rarely crowded, yellowish-green and usually glabrous above, paler and hairy below, flat, thick-textured, venation prominent, apex tapering to a fine point; **flowers** to about 1.2 cm across, cream to pale yellow with revolute lobes, sweetly fragrant, in few-flowered axillary clusters; **capsules** 5–8 cm × 1–2 cm, broadest in lower half, faintly hairy.

This inland species occurs in woodland and scrub-land on the western slopes of Central Qld to the west-ern slopes and the far western plains of NSW. It is a valuable species which deserves to be better known in cultivation. The flowers are sweetly fragrant and bloom over a long period. It is well-suited to temperate and semi-arid regions and adapts to a sunny or semi-shaded site. Hardy to moderately heavy frosts and extended dry periods. Plants usually have very dense foliage. They respond well to pruning. Propagate from seed, or from cuttings of firm young growth, which strike readily.

The allied *P. blakeana* has glabrous stems and leaves, and creamy flowers with ascending lobes.

Parsonsia ferruginea J. B. Williams
(rust-coloured)
Qld
Climber   April–Dec

Vigorous **climber** with rusty-hairy young growth; **stems** long, woody, twining; **sap** watery; **leaves** 7–15 cm × 4–9 cm, broadly elliptic to ovate, opposite, densely hairy petiole to about 4 cm long, base often cordate, spreading to ascending, bright green, thin-textured, initially hairy over all, becoming somewhat glabrous above, venation prominent, apex shortly pointed; **flowers** about 0.5 cm across, cream to mauve or brownish, lobes spreading and exterior rusty-hairy, in axillary and terminal many-flowered panicles; **capsules** 5–12 cm × to about 2 cm, narrow, tapering to base, rusty-hairy.

Parsonsia fulva

Parsonsia eucalyptophylla   D. L. Jones

A recently described (1996) species which inhabits the tropical rainforests of Cape York. It is distinctive and worthy of cultivation in tropical and subtropical regions. Plants should do best in warm to hot, sunny sites in well-drained acidic soils. Pruning from an early stage will help promote bushy growth. Frost tolerance is not known. Propagate from fresh seed. Cuttings are also worth trying.

This species has been known as *P*. species affinity *velutina*, but *P. velutina* has hairier foliage and narrower capsules.

Parsonsia fulva S. T. Blake
(reddish-yellow)
Qld, NSW   Furry Silkpod
Climber   Oct–Jan
Vigorous **climber** with silky-brown young growth; **stems** long, woody, twining; **sap** watery; **leaves** 8–20 cm × 0.5–1.5 cm, elliptic to broadly ovate, opposite, with petiole to 3 cm long, base 2–3-lobed on juvenile leaves otherwise mainly blunt to faintly cordate, spreading to ascending, bright green and becoming more or less glabrous above, paler with brown hairs and small domatia below, thin-textured, often somewhat concave, venation prominent, apex pointed; **flowers** about 0.5 cm across, brownish, lobes spreading to recurved with hairy exterior, in somewhat large, loose, terminal or axillary panicles, rarely prominent; **capsules** 14–23 cm × 1 cm, cylindrical, faintly hairy.

This climber is often extremely vigorous and may cover large areas in its natural habitat of subtropical rainforest in south-eastern Qld and north-eastern NSW. Adapts well to cultivation in tropical, subtropical and temperate regions. Needs well-drained acidic loam soils with a semi-shaded or dapple-shaded aspect for best results. Plants may need regular pruning to contain their spread. Pruning also promotes the attractive rusty-hairy young growth. Hardy to light or moderate frost. This climber will need a sturdy constructed framework or well-established trees on which to grow. Propagate from seed, or cuttings of firm young growth, which usually strike readily but can suffer from lack of air movement.

Parsonsia grayana J. B. Williams
(after Bruce Gray, 20th-century botanical collector from Atherton, Qld)
Qld
Climber   throughout the year
Climber with glabrous young growth; **stems** woody, twining; **sap** watery; **leaves** 5–12 cm × 2–5 cm, ovate-lanceolate to ovate-elliptic, opposite, with petiole to 1.2 cm long, base rounded or slightly cordate, green above, green to brownish below, thin, glabrous, venation prominent, apex tapering to a fine point; **flowers** about 0.5 cm across, cream or green with reddish tips, lobes ascending to erect, in axillary or terminal somewhat open clusters, on sometimes hairy stalks; **capsules** 10–21 cm × up to 1.4 cm, very slender, glabrous.

This species from the Cape Tribulation–Atherton Tableland region was described in 1996. It occurs in rainforest and vine thickets at 200–1350 m altitude. Evidently rarely cultivated, it should be best suited to tropical and subtropical regions but may succeed in cooler areas if grown in warm protected sites. Plants require well-drained acidic soils. Propagate from fresh seed and possibly from cuttings.

Parsonsia howeana J. B. Williams
(from Lord Howe Island)
NSW (Lord Howe Island)
Climber   sporadic through year
Vigorous **climber** with faintly hairy young growth; **stems** woody, twining; **sap** yellowish; **leaves** 4–9 cm × 1.5–3 cm, elliptic, opposite, base blunt or tapered, with petiole to 1.5 cm long, spreading to ascending, glossy mid green above, paler below, glabrous, thick, venation prominent, apex pointed; **flowers** to about 0.7 cm across, orange to reddish-brown or sometimes yellowish, with honey fragrance, lobes spreading to recurved, exterior hairy, in many-flowered axillary or terminal clusters, moderately conspicuous; **capsules** 10–12 cm × 1–1.5 cm, more or less cylindrical.

Until 1994 this species was regarded as a variety of *P. straminea*. It is common on Lord Howe Island in lowland forests. Suitable for cultivation in subtropical and temperate regions and requires well-drained acidic soils. Does best with some shelter but also tolerates a sunny site. Withstands light frosts. Propagate from fresh seed, or cuttings of firm young growth, which strike readily.

Parsonsia induplicata F. Muell.
(infolded margins)
Qld, NSW   Thin-leaved Silkpod
Climber   Aug–Oct
Weak **climber** with glabrous young growth; **stems** slender, twining, glabrous; **sap** watery; **mature leaves** 3–15 cm × 1–4 cm, narrowly lanceolate to ovate, opposite, with petiole 0.3–1.5 cm long, spreading to ascending, not crowded, glabrous, deep green above, brown to purplish below, thin-textured, venation prominent, apex tapering to a fine point; **flowers** to about 0.4 cm

Parsonsia fulva   D. L. Jones

across, white or cream, not scented, in small, sparse, terminal panicles; **capsules** 6–10 cm × to about 0.8 cm, cylindrical, glabrous, pendent.

This species occurs in rainforest along the coast and on mountain ranges to an altitude of about 1000 m in south-eastern Qld and the North Coast of north-eastern NSW. It is a light climber which lacks strong appeal. Suitable for subtropical and temperate regions. Needs well-drained acidic loams and a site which is sheltered from hot sunshine. Plants are hardy to moderate frosts. They respond well to pruning. Propagate from seed, or from cuttings of firm young growth, which strike readily.

Parsonsia kimberleyensis J. B. Williams
(from the Kimberley region)
WA
Climber   May–June; possibly also sporadic
Moderate **climber** with hairy young growth; **stems** woody, twining; **sap** watery; **leaves** 6–11 cm × 3–6.5 cm, ovate, opposite, base rounded, with petiole to 1.4 cm long, spreading to ascending, not crowded, green, faintly hairy, thin-textured, venation prominent, apex pointed; **flowers** about 0.5 cm across, yellow to greenish, lobes more or less erect, in many-flowered, compact, axillary and terminal panicles; **capsules** not known.

A recently described (1996) species from the Dampier Peninsula of the Kimberley region where it grows in vine thickets. Evidently not in cultivation, it should be suited to tropical regions and may succeed in subtropical zones. Plants will need excellent drainage and plenty of sunshine. Propagate from seed, and cuttings would also be worth trying.

P. lanceolata is closely allied but it has smaller flowers and faint venation on the lower surface of leaves.

Parsonsia kroombitensis J. B. Williams
(from the Kroombit Tops region)
Qld
Climber   Sept–Dec
Moderate **climber** with faintly hairy young growth; **stems** woody, twining; **sap** watery; **leaves** 2–6 cm × 0.6–2 cm, linear to linear-lanceolate or somewhat tri-angular, opposite, with petiole to about 0.7 cm long, base rounded to cordate, green (brown to purplish when dry), glabrous or faintly hairy below, margins recurved, venation prominent, apex tapering to a point; **flowers** about 0.8 cm across, orange-yellow with red and white markings, lobes spreading to recurved, in few-flowered, open, mainly terminal clusters; **capsules** 4–10 cm × up to 1 cm, narrowly ovoid, glabrous or faintly hairy.

This rare species from central-eastern Qld was described in 1996. It occurs in woodland and shrubland on escarpments in acidic volcanic rock and soils. In the wild, it sometimes develops as a clumping plant. It has potential for planting in subtropical areas and may succeed in temperate regions. Needs good drainage and plenty of sunshine. Hardy to light frosts. Propagate from seed, and possibly from cuttings of firm young growth.

Parsonsia lanceolata R. Br.
(lanceolate)
Qld, NSW   Rough Silkpod
Climber   Dec–June; sometimes sporadic

Parsonsia howeana   W. R. Elliot

Vigorous **climber** with hairy or glabrous young growth; **stems** twining; **sap** watery; **leaves** 3–10 cm × 0.5–5 cm, variable, lanceolate to broadly elliptic, opposite, with petiole to about 1.2 cm long, spreading to ascending, not crowded, dark green and glabrous to hairy, leathery, margins usually recurved, apex blunt or finely pointed; **flowers** about 0.6 cm across, cream, exterior hairy, sweetly fragrant, on hairy stems in compact axillary or terminal clusters; **capsules** 6–12 cm × about 1.5 cm, tapered to both ends, apex blunt.

P. lanceolata occurs in a range of habitats including rainforest, river and creek banks, and rocky gorges. The distribution extends from south-eastern Qld to central western NSW. Plants adapt very well to cultivation in tropical and subtropical regions. It has attractive foliage and fragrant flowers. Needs well-drained acidic soils in a semi-shaded or dapple-shaded position. Responds well to supplementary watering over extended dry periods. Propagate from fresh seed, and cuttings of firm young growth are worth trying.

P. kimberleyensis is closely allied but differs in having larger, brownish flowers and prominently veined leaves.

Parsonsia langiana F. Muell.
(after Thomas Lang)
Qld
Climber   Feb–July; Nov

Vigorous **climber** with rusty-brown hairy young growth; **stems** woody, twining; **sap** watery; **leaves** 3–11 cm × 1–6 cm, ovate-lanceolate to ovate, opposite, base rounded to cordate, spreading to ascending, not crowded, dark green and glossy above, dull green and hairy below, venation prominent, apex tapering to a fine point; **flowers** to about 0.4 cm across, cream to pale yellow, on slender stalks, can be profuse, in open many-branched clusters; **capsules** to about 20 cm × 1 cm, cylindrical, pendent, faintly hairy.

This tropical rainforest species is restricted to the Cook and North Kennedy Districts of north-eastern Qld where it grows from 200–1200 m altitude. Mainly cultivated by rainforest enthusiasts, it adapts well to cultivation in tropical and subtropical regions in well-drained acidic soils with a semi-shaded to dapple-shaded aspect. Plants respond well to pruning which is often necessary to contain the rapid growth. Propagate from fresh seed, and cuttings of firm young growth are worth trying.

Parsonsia larcomensis J. B. Williams
(from Mt Larcom)
Qld
prostrate × 3–5 m   Jan–May

Creeping **shrub** with adventitious roots along the semi-woody **stems**; young growth glabrous; **sap** watery; **leaves** 1.1–4.5 cm × 1–2.2 cm, ovate to broadly elliptic, opposite, with petiole to about 0.7 cm long, base rounded or cordate, spreading to ascending, glabrous, green above, glaucous below, venation prominent, apex pointed; **flowers** to about 0.8 cm long, whitish with 5 interior red spots, lobes short and spreading to recurved, in few-flowered axillary or terminal panicles; **capsules** 7–11 cm × to about 1 cm, tapered to both ends, brown, faintly hairy.

This unusual, rare and interesting species was described in 1996. It occurs between Maryborough and Rockhampton where it grows in open heathland and shrubland on shallow, acidic soils of the cliffs and outcrops at 350–750 m altitude. Plants may be useful groundcovers in subtropical regions and possibly in cooler areas. They tolerate plenty of sunshine but frost tolerance is not fully understood. Propagate from fresh seed, cuttings or division of layered stems.

This species has been referred to as *P.* species (Mt Larcom) and *P.* species (Mt Perry).

P. straminea has larger capsules and shorter flowers.

Parsonsia largiflorens (Benth.) S. T. Blake
(large clusters of flowers)
Qld, ?NSW
Climber   Sept–Jan

Vigorous **climber** with glabrous or faintly hairy young growth; **stems** woody, twining; **sap** watery; **leaves** 6.5–16 cm × 4–8.5 cm, ovate to oblong-ovate, opposite, base rounded or slightly cordate, petiole 1.5–2.5 cm long, spreading to ascending, dark green, thin-textured, not pliable, apex abruptly pointed; **flowers** about 0.5 cm across, cream, hairy, in crowded axillary clusters, can be profuse; **capsules** to about 13 cm × 1.5 cm, long-tapering to apex.

This poorly known and rare species occurs in south-eastern Qld. It is thought to be extinct in north-eastern NSW. Plants inhabit rainforests and moist sclerophyll forests and they need to be cultivated as part of a conservation strategy. Should adapt to cultivation in tropical, subtropical and temperate regions. Plants will need well-drained acidic soils which do not dry out readily, in a semi-shaded or dapple-shaded site. Propagate from fresh seed, and cuttings of firm growth are worth trying.

Parsonsia latifolia (Benth.) S. T. Blake
(broad leaves)
Qld
Climber   Oct–Feb

Vigorous **climber** with glabrous young growth; **stems** woody, twining; **sap** somewhat milky; **juvenile leaves** large, ovate to oblong-ovate, 3-lobed; **mature leaves** 7–14 cm × 3.5–8 cm, broadly ovate, opposite, with petiole to 3.5 cm long, base rounded or cordate, spreading to recurved, glabrous, dark green, glossy, stiff, venation prominent, apex ending in a small point; **flowers** about 0.8 cm across, white or cream, lobes recurved, sweetly fragrant, in many-flowered, loose to crowded, mainly terminal panicles to about 12 cm across, moderately conspicuous; **capsules** 14–19 cm × 2–4 cm, cylindrical, somewhat woody, glabrous.

A floriferous species from high rainfall tropical rainforests of the coast and hinterland in north-eastern and central Qld. Needs well-drained acidic soils and tolerates a semi-shaded or somewhat open site. Well suited for screening purposes. Responds well to pruning and supplementary watering. Recommended for coastal regions. Propagate from fresh seed and from cuttings of firm young growth.

Plants from south-eastern Qld and NSW previously included in *P. latifolia* are now referrable to *P. longipetiolata*.

Parsonsia leichhardtii F. Muell.
(after Ludwig Leichhardt, 19th-century explorer)
Qld
Climber   July–Feb
Moderately vigorous **climber** with glabrous or faintly hairy young growth; **stems** slender, often wiry; **sap** watery; **juvenile leaves** often lobed; **mature leaves** 4–11 cm × 2–5 cm, ovate to oblong-ovate, opposite, base rounded or broadly wedge-shaped, with petiole to 2.5 cm long, spreading, dark green, glabrous, thin, venation pale green, apex tapering to a fine point; **flowers** about 0.4 cm across, cream to yellowish, tube inflated towards base, in small, crowded, axillary or terminal clusters, rarely profuse; **capsules** 5–8 cm × 2–3 cm, ovoid, faintly hairy.

P. leichhardtii is endemic to south-eastern regions where it occurs in dryish rainforest along the Great Dividing Range. It is often found growing in basalt-derived acidic soils. Poorly known in cultivation. Suitable for subtropical regions and may succeed in protected sites in temperate zones. A semi-shaded site should provide best results. Propagate from seed and possibly cuttings may be successful.

Parsonsia lenticellata C. T. White
(with small raised pores)
Qld
Climber   Nov–July
Moderately vigorous **climber** with glabrous young growth; **stems** slender, twining, long; **sap** watery; **leaves** 4.5–10 cm × 0.5–3 cm, lanceolate to narrowly elliptic, opposite, base cordate, with petiole to about 0.3 cm long, spreading to recurved, dark green above, paler below, thin, venation prominent, apex pointed; **flowers** to about 0.4 cm across, cream, lobes erect to incurved, in many-flowered, hairy-stemmed, compact, axillary and terminal panicles; **calyx lobes** recurved, faintly hairy; **capsules** 7.5–11 cm × to about 1 cm, cylindrical to narrowly ovoid, glabrous.

A rare species which grows in coastal or slightly inland regions of north-eastern Qld from near Port Douglas to the Mackay region. Found in depauperate rainforests, open forest and woodland. Rarely cultivated, it is suitable for tropical and subtropical regions. Propagate from fresh seed and possibly from cuttings.

Plants from south-eastern Qld were previously referred to as *P. lenticellata* but they are now known as *P. paulforsteri*, which differs in having thicker, narrowly lanceolate leaves with a rounded base and tapering gradually to a point.

Parsonsia lilacina F. Muell.
(lilac-coloured)
Qld, NSW   Crisped Silkpod
Climber   Aug–Nov
Light **climber** with faintly hairy young growth; **stems** 1–2 m long, slender, wiry or twining; **sap** watery; **leaves** 1.5–6 cm × 0.5–3 cm, oblong-lanceolate to oblong-ovate, opposite, base cordate, with petiole to 0.8 cm long, spreading to ascending, glossy and dark green above, paler below, thin, margins crisped or toothed, venation prominent, apex pointed; **flowers** about 0.4 cm across, cream with lilac, sweetly fragrant, lobes spreading, in few-flowered, loose, axillary clusters;

capsules 5–13 cm × 0.5–1 cm, slender, tapered to both ends, glabrous.

An attractive species from south-eastern Qld and north-eastern NSW, where it grows in dryish rainforests on basaltic soils at up to 700 m altitude. A moderately slow growing climber which is suitable for subtropical and temperate regions. Prefers a cool, semi-shaded site with moist acidic soils. Hardy to light frosts. Propagate from fresh seed or possibly from cuttings of firm young growth.

Parsonsia longiopetiolata J. B. Williams
(long petioles)
Qld, NSW   Green-leaved Silkpod
Climber   Dec–April
Vigorous **climber** with glabrous young growth; **stems** woody, twining; **sap** milky; **juvenile leaves** somewhat triangular and 3-lobed; **mature leaves** 3.5–8 cm × 2.5–5 cm, ovate to somewhat triangular, opposite, base cordate to very blunt, with petiole to 5 cm long, spreading, green, glabrous, soft, venation conspicuous, apex prominently tapering to a point; **flowers** to about 0.6 cm across, yellow to greenish, exterior faintly hairy, lobes spreading, in compact, many-flowered, terminal panicles; **capsules** 11–15 cm × 0.5–1.3 cm, narrowly terete, faintly hairy.

This recently described (1996) species has been confused with *P. latifolia* (see below). It occurs in eastern Qld and north-eastern NSW. The main distribution extends from near Gympie, Qld, to the Grafton region and disjunctly near Rockhampton and Atherton. Best suited to tropical and subtropical regions and needs a sheltered site with acidic soils. Propagate from fresh seed or possibly also from cuttings of firm young growth.

The allied *P. latifolia* has longer, broader capsules, white to cream flowers and ovate leaves.

Parsonsia nesophila F. M. Bailey = *P. velutina* R. Br.

Parsonsia paulforsteri J. B. Williams
(after Paul I. Forster, 20th-century Australian botanist)
Qld
Climber   most of the year
Slender **climber** with glabrous young growth; **stems** thin, twining, many small swellings on older growth; **sap** watery; **leaves** 4–10 cm × 0.2–2.5 cm, linear to broadly lanceolate, opposite, base rounded or faintly cordate, with petiole to about 0.7 cm long, spreading to ascending, dark green above, pale green to whitish below, glabrous, thick, venation prominent, apex tapering to a fine point; **flowers** to about 0.5 cm across, whitish, lobes erect to incurved, calyx lobes spreading, in short, crowded, axillary or terminal clusters; **capsules** 6–8 cm × up to 1 cm, tapered to both ends, glabrous.

A recently described (1996) species which inhabits dryish rainforest and vine thickets from the Rockhampton region to near Brisbane. It grows mainly on acidic clay soils. Probably best suited to tropical and subtropical regions but may succeed in cooler climates. Moderate drainage is required and plants tolerate a semi-shaded or sunny site. Frost tolerance is not known. Propagate from fresh seed or possibly from cuttings of firm young growth.

Parsonsia plaesiophylla

P. lenticellata is closely allied but it has thinner leaves with a cordate base and usually larger fruits.

Parsonsia plaesiophylla S. T. Blake
(with oblong leaves)
Qld
Climber   Oct–April

Vigorous **climber** with glabrous young growth; **stems** woody, twining; **sap** watery; **leaves** 4–12 cm × 1.5–5 cm, oblong to oblong-ovate, opposite, base rounded or tapered, with petiole to 3.5 cm long, spreading to ascending, bright green, glabrous, or sometimes faintly hairy below, thick, leathery, venation prominent, apex ending in a very small point; **flowers** to about 0.8 cm across, white or yellowish, fragrant, lobes strongly recurved, in many-flowered, spreading, terminal panicles to about 10 cm across, often moderately conspicuous; **capsules** 6.5–11 cm × about 1.5 cm, somewhat cylindrical, tapering near tip.

A very ornamental, fragrant-flowered species which occurs in dry rainforests and vine thickets from Cape York to near Maryborough. It is found near the coast and slightly inland in acidic and alkaline soils. Best suited to tropical and subtropical regions and may succeed in warm protected sites in temperate areas. Does well in a semi-shaded or sunny situation with freely draining soil. Plants withstand light frosts. Propagate from seed. Cuttings of firm young growth are also worth trying.

Parsonsia purpurascens J. B. Williams
(becoming purple)
NSW
Climber   Black Silkpod

Moderately weak **climber** with glabrous, often purplish young growth; **stems** slender, twining; **sap** watery; **leaves** 4–11 cm × 1.5–4 cm, lanceolate to narrowly ovate, opposite, base rounded or slightly cordate, with petiole to 1.5 cm long, spreading, green, purplish below when young, becoming purplish on shedding, glabrous, venation conspicuous, apex tapering to a soft point; **flowers** about 0.5 cm across, pale yellow to cream or brownish, lobes spreading and recurved, in loose terminal or axillary panicles; **capsules** 7–10 cm × 0.5–0.8 cm, slender, tapering to tip, glabrous.

A recently described species which was confused with *P. leichhardtii* (see below). It is an inhabitant of temperate rainforest and closed forest in coastal ranges and on the eastern side of the Great Dividing Range in northern and central regions. This species is not regarded as very ornamental. It is suited to subtropical and temperate regions. Needs a sheltered site with freely draining soils which are moist for most of the year. Withstands light frosts. Propagate from fresh seed and possibly from cuttings of firm young growth.

P. leichhardtii differs in having lobed juvenile leaves, mature leaves which are green when fresh or dried, and much broader capsules.

Parsonsia rotata Maiden & Betche
(wheel-shaped)
Qld, NSW   Corky Silkpod, Veinless Silkpod
Climber   Jan–May

Weak to vigorous **climber** with glabrous young growth; **stems** woody, twining, often with corky patches on older growth; **sap** watery; **leaves** 5–15 cm × 2–5 cm, lanceolate to oblong-elliptic, opposite, base usually rounded, with petiole to about 2 cm long, spreading, dark green and glossy above, paler below, venation usually faintly conspicuous, apex tapering to a fine soft point; **flowers** to about 1 cm across, cream to pale yellow, lobes pointed and spreading to ascending, stamens spirally twisted, in axillary or terminal clusters of 3 to about 20 flowers; **capsules** 15–25 cm × 2–5 cm, broadly cylindrical to narrowly ovoid, semi-woody, glabrous.

This coastal or near-coastal species is distributed from the Atherton Tableland in north Qld to the Hastings River near Port Macquarie. It occurs in tropical and subtropical rainforests where it grows mainly on clay loam soils. Plants are successful in subtropical and temperate regions and do best in semi-shaded sites with freely draining soils. Pruning from an early stage may help to promote better foliage production. Withstands light frosts. Propagate from fresh seed, or possibly from cuttings of firm young growth.

Parsonsia sankowskyana J. B. Williams
(after Gary Sankowsky, 20th-century Australian plant enthusiast and original collector)
Qld
Climber   Nov; possibly sporadic

Vigorous **climber** with hairy young growth; **stems** woody, twining; **sap** watery; **leaves** 11–22 cm × 3–11.5 cm, ovate to broadly ovate-elliptic, opposite, base rounded or slightly cordate, with petiole 2–7 cm long, spreading to ascending, dark green above, paler below, glabrous, venation prominent, apex pointed; **flowers** to about 0.9 cm across, pale yellow, lobes spreading to revolute, exterior hairy, in many-flowered, loose, axillary or terminal panicles; **capsules** 8–11 cm × 1.6–3.5 cm, ovoid, grooved, glabrous.

A recently described (1996) species which is confined to coastal subtropical rainforest in the Hervey Bay region of south-eastern Qld. Responds well to cultivation in tropical and subtropical regions in freely draining acidic soils. Tolerates a semi-shaded or slightly sunny location. Frost tolerance is unknown. It may be too vigorous for small gardens. Propagate from seed or from cuttings of firm young growth.

P. straminea has smaller leaves, narrower, faintly hairy capsules, and its stems often have adventitious roots. *P. eucalyptophylla* has narrower leaves and smaller flowers. *P. brisbanensis* also has smaller flowers.

Parsonsia straminea (R. Br.) F. Muell.
(straw-coloured)
Qld, NSW   Common Silkpod
Climber   most of the year

Vigorous **climber** with faintly hairy young growth; **stems** woody, twining, often with adventitious roots; **sap** yellowish; **juvenile leaves** 1–5 cm long, ovate, base cordate, green above, purplish below; **mature leaves** 4–24 cm × 1.5–8 cm, elliptic to oblong-ovate, opposite, base rounded or cordate, with petiole to 4.5 cm long, spreading to ascending, yellow-green to deep green above, usually glaucous below, glossy and glabrous or faintly hairy above, prominent thick venation, apex pointed; **flowers** to about 0.7 cm across, cream, yellowish or pink, sweetly fragrant, lobes spreading to recurved, exterior hairy, in many-flowered, axillary or terminal panicles, often moderately conspicuous; **capsules** 10–20 cm × about 1 cm, tapered to both ends, semi-woody, faintly hairy.

The distribution of this fragrant species extends from northern Cape York to northern NSW. It occurs in a wide variety of habitats including heathland, shrubland, open forest and rainforest and ranges from near sea-level to more than 1000 m altitude. Adapts very well to cultivation and does well in most soil types which are at least moderately well drained. Grows well in shade through to moderately exposed positions. Tolerates extended wet periods. It is capable of smothering nearby young plants but it is excellent for screening large areas and is suitable for climbing on well-established trees. Withstands most frosts although plants of coastal provenance may suffer some damage. Responds very well to pruning. Propagate from fresh seed, or from cuttings of firm young growth, which strike readily. Division of layered stems is successful too.

The var. *glabrata* Pichon is a synonym.

P. howeana differs in having smaller leaves and usually orange to reddish-brown flowers.

Parsonsia velutina

D. L. Jones

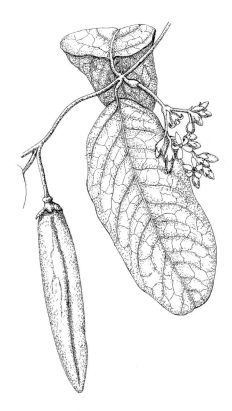

Parsonsia rotata × .5

Parsonsia tenuis S. T. Blake
(slender)
Qld, NSW   Slender Silkpod
Climber   June–Nov

Weak **climber** with hairy young growth; **stems** slender, twining, to about 2 m long; **sap** yellowish; **juvenile leaves** narrow; **mature leaves** 3.5–11 cm × 1–2.5 cm, oblong-lanceolate, opposite, base rounded, slightly cordate or blunt, with petiole to about 1.2 cm long, spreading to ascending, bright to dull green and glabrous to rough above, paler and softly hairy below, thin, soft, venation prominent, apex pointed; **flowers** about 0.7 cm across, cream, lobes slightly spreading, in loose axillary or terminal clusters to about 4.5 cm long; **capsules** 4–9 cm × up to 1 cm, slender, tapered to both ends, hairy, can become glabrous.

This species is restricted to the highland rainforests of the McPherson Ranges and Tweed Ranges of the Qld/NSW border region. It withstands light to moderate frosts and needs moist but well-drained acidic soils with a sheltered aspect. It is unlikely to grow so strongly that it smothers nearby shrubs. Propagate from fresh seed, or from cuttings of firm young growth, which strike readily.

Parsonsia velutina R. Br.
(velvety)
Qld, NSW, WA, NT
Climber   Sept–May; also
sporadic

Moderately vigorous **climber** with brownish, velvety young growth; **stems** woody, often twining; **sap** watery; **juvenile leaves** often 3-lobed; **mature leaves** 3–20 cm × 2–14 cm, ovate to oblong-ovate, opposite, base cordate,

175

Parsonsia velutina × .5

with petiole to about 4 cm long, spreading to ascending, deep green with rusty or fawn hairs to almost glabrous, soft, venation prominent below, apex often with small point; **flowers** to about 0.4 cm across, cream to brown, lobes erect, exterior hairy, in compact, many-flowered, axillary panicles; **capsules** 7–20 cm × 0.4–1.5 cm, slender, tapered to both ends, hairy or becoming glabrous.

This is the most widely distributed member of the genus. It is found mainly in coastal, subtropical and tropical areas from near Gosford, NSW, to the Kimberley region in WA. It also occurs in Papua New Guinea. *P. velutina* inhabits open forests, rainforests and vine thickets often growing along river and stream banks. Plants have attractive leaves and are moderately ornamental. They are useful for screening in subtropical and seasonally dry tropical regions. Needs plenty of sunshine and moderately good drainage. Generally plants from southern regions withstand light to moderate frosts. Propagate from fresh seed, or possibly also from cuttings of firm young growth.

P. nesophila F. M. Bailey is a synonym.

Parsonsia ventricosa F. Muell.
(swollen, inflated)
Qld, NSW   Acuminate Silkpod
Climber   Aug–May
Slender **climber** with glabrous or hairy young growth; **stems** slender, twining; **sap** watery; **leaves** 4–15 cm × 1.5–5 cm, oblong-lanceolate to ovate, opposite, base rounded or cordate, with petiole to about 1.2 cm long, spreading, green, soft, thin,

glabrous or sometimes softly hairy below, venation prominent, apex pointed; **flowers** about 0.6 cm across, cream, yellowish or brown, inflated base, lobes recurved, exter-ior glabrous, in axillary or terminal clusters; **capsules** 9–15 cm × 0.6–1.2 cm, tapered to both ends, faintly hairy.

This species is common in lowland rainforests of south-eastern Qld and north-eastern NSW. It grows in wet or dryish sites. Suitable for subtropical and seasonally dry tropical areas. Plants require acidic soils which drain freely and a semi-shaded aspect. Withstands light frosts. Usually adapts well to cultivation. Propagate from fresh seed. Cuttings of firm young growth are also worth trying.

Parsonsia wildensis J. B. Williams
(from the Wild River of north-eastern Qld)
Qld
Climber   Jan–May
Small **climber** with glabrous young growth; **stems** slender, twining; **sap** watery; **leaves** 2–6.5 cm × 0.4–1.7 cm, lanceolate to narrowly triangular, opposite, base rounded to cordate, with petiole to about 0.7 cm long, spreading to slightly recurved, green, thin, glabrous, venation conspicuous, apex finely pointed; **flowers** to about 0.8 cm across, cream, sometimes with pink spreading to recurved lobes, exterior faintly hairy, base of interior densely hairy, in compact mainly axillary clusters; **capsules** about 13 cm × 0.7 cm, slender, glabrous.

A recently described (1996) species from the Atherton Tableland where it grows in open forest of river gorges, often amongst granite rocks, at 600–1000 m altitude. Best suited to tropical and subtropical regions but may adapt to cooler areas too. Plants will need excellent drainage and a semi-shaded to sunny site. Frost tolerance is not known but plants will withstand light frosts. Propagate from fresh seed or possibly from cuttings of firm young growth.

Parsonsia wongabelensis J. B. Williams
(from the Wongabel State Forest)
Qld
Climber   Jan; possibly sporadic
Vigorous **climber** with brownish-hairy young growth; **stems** woody, twining; **sap** watery; **leaves** 6.5–14 cm × 4.5–8.5 cm, broadly ovate, opposite, base rounded, very blunt or slightly cordate, with petiole to about 4 cm long, soft, green with dense covering of dark brown hairs above, paler hairs plus conspicuous venation below, apex finely pointed; **flowers** about 0.8 cm across, cream, lobes recurved, exterior hairy, in axillary or terminal many-flowered clusters with brownish-hairy stalks; **capsules** 17–19 cm × 1.7–2 cm, ellipsoidal, widest near base, dark brown, hairy.

This recently described (1996) species is very rare. It is known from the Atherton Tableland where it occurs in tropical rainforest on acidic loam. Should do best in highland tropical and subtropical regions. Warrants cultivation as part of a conservation strategy. Plants will need a sheltered well-drained site. Hardiness to frost is not known. Propagate from fresh seed or possibly from cuttings of firm young growth.

PASPALIDIUM Stapf.
(after the genus *Paspalum*; and the diminutive *idium*; a smaller form of *Paspalum*)
Poaceae   Panic Grasses; Summer Grasses
   Annual or perennial **grasses**; **leaves** folded in bud, becoming flat; **culms** branched; **inflorescence** a panicle of 1-sided racemes; **spikelets** solitary or paired, on short branches, deciduous at maturity; **florets** 2, differ-ent; **glumes** lower one smallest; **lemma** acute.
   This genus of tufting grasses comprises about 40 species and 22 of these are native to Australia. They are represented in all mainland states in a range of habitats and soils with a preference for inland regions.
   Some species are ornamental and deserve cultivation for this aspect alone but many also are grazed for forage feed by stock. Species such as *P. jubiliflorum* are also valuable for stabilising clay and clay loam soils and the edges of waterways. Birds eat the seed of many species. Propagate from seed or by division of clumps. Species which form adventitious roots at nodes may possibly be grown from cut stems.

Paspalidium albovillosum S. T. Blake
(white hairs)
Qld, NSW   Panic Grass
0.3–0.6 m × 0.3–0.5 m   Dec–March
   Tufting perennial **grass**, forming erect tussocks; **stems** straight or slightly bent at lower nodes, loosely branched, glabrous, pale green to yellowish-green; **leaf-sheaths** with long hairs; **leaves** to about 15 cm × 0.5 cm, flat or inrolled, with soft white hairs on both surfaces, margins slightly rough; **panicles** 5–20 cm long, borne well above the leaves, with 6–10 erect heads to about 3 cm long; **spikelets** to about 0.2 cm long, pale green to purplish; **lemmas** ovate to elliptic.
   This species is widespread in Qld and occurs in north-eastern NSW. It is usually found in open woodland or mallee growing on poor loam, sandy loam or gravelly soils. Possibly useful to farmers for forage grazing of domestic stock and should be useful for soil conservation purposes. Propagate from seed.
   Previously known as *P. radiatum* var. *hirsutum* Vickery.

Paspalidium aversum Vickery
(bent back)
Qld, NSW   Panic Grass
0.4–0.9 m × 0.3–0.6 m   Dec–March
   Tufting perennial **grass**; **stems** erect to decumbent, sometimes rooting at lower nodes, branching, nodes glabrous; **leaf sheaths** glabrous or with hairy margins; **leaves** to about 25 cm × 1 cm, flat or loosely inrolled, glabrous or slightly rough; **panicles** 10–30 cm long, very narrow, with 2–10 heads to about 7 cm long; **spikelets** to about 0.3 cm long, may appear to bend backwards; **lemmas** acute.
   This summer- and autumn-growing grass is widespread in Qld and is often prominent in central and eastern NSW. Usually it is found on good quality loam which is moist for extended periods. In many areas it is valued as a foraging grass. Should be useful for soil conservation projects. Propagate from seed or by division.

Paspalidium breviflorum Vickery =
                    P. disjunctum S. T. Blake

Paspalidium caespitosum C. E. Hubb
(tufted)
Qld, NSW   Brigalow Grass
0.3–0.5 m × 0.3–0.5 m   Dec–March
   Tufted, more or less erect, perennial **grass**; **stems** simple or branched, slightly bent at lower nodes, nodes glabrous; **leaf sheaths** mainly glabrous; **leaves** usually 10–15 cm × up to 0.15 cm, inrolled, mainly glabrous, green; **panicles** to 30 cm long, erect, very narrow, with 7–10 heads, to about 2.5 cm long; **spikelets** about 0.2 cm long, pale green or sometimes with purple tonings; **lemmas** apex finely pointed.
   A widespread species which is most common in brigalow habitat. It has limited ornamental appeal but should be useful for soil conservation. Propagate from seed.

Paspalidium constrictum (Domin) C. E. Hubb.
(constricted, bound together)
Qld, NSW, Vic, SA, WA, NT   Knottybutt Grass;
                              Box Grass; Slender Panic
   Tufted perennial **grass** with a spreading, knotted rootstock; **stems** to about 0.8 m tall, many, somewhat cane-like, much-branched, covered in very small stiff hairs; **leaves** to about 12 cm × 0.1–0.4 cm, linear, flat or inrolled, dull green or greyish-green to greenish-blue; **panicles** to about 20 cm long, much-branched, somewhat slender beyond the leaves; **spikelets** about 0.25 cm long; **lemmas** elliptic, acute.
   A very widespread summer-growing species from grasslands, woodlands and open forests. It occurs in a range of clay loams, loams and sandy soils always near drainage channels and creeks in arid regions. Stock find this drought-tolerant grass very palatable and it recovers well after grazing. Useful for soil erosion control within its natural distribution. Propagate from seed and possibly by division of tussocks.
   P. gracile var. *rugosum* Hughes is a synonym.

Paspalidium criniforme S. T. Blake
(hair-like)
Qld, NSW
0.4–0.6 m × 0.2–0.5 m   Dec–March
   Tufted perennial **grass.** forming dense tussocks; **stems** ascending to erect; **leaf sheaths** sometimes hairy; **leaves** to about 50 cm × 0.1–0.25 cm, flat, often hairy, mid to deep green; **panicles** 2.5–25 cm long, with irregularly spaced spikelets; **spikelets** to about 0.3 cm long, pale green; **lemmas** acute.
   This species is from south-eastern Qld and eastern NSW. It can have moderately graceful foliage. Plant trials will establish its adaptability and performance as an ornamental grass in cultivation. Should do best in acidic well-drained soils in a sunny or semi-shaded site. Hardy to most frosts. Propagate from seed and possibly by division in winter.
   P. gracile var. *debile* Vickery is a synonym.

Paspalidium disjunctum S. T. Blake
(separate, disjunct)
Qld, NSW, NT
0.3–0.5 m × 0.3–0.5 m   Dec–March

177

Tufting perennial **grass**; **stems** branched, often rooting at lower nodes; **leaf sheaths** hairy on one margin; **leaves** to about 10 cm × 0.35 cm, linear, flat, glabrous except for rough nerves, green; **panicles** to about 30 cm long, with up to 10 racemes, narrow; **spikelets** about 0.2 cm long, pale green; **lemmas** broadly elliptic, apex prominently pointed.

A widespread species which usually inhabits better quality soils and often grows in sheltered sites. This grass should adapt well to tropical or subtropical regions. It has stems which root at the lower nodes so it may be useful for soil stabilisation. Propagate from seed or possibly by division.

P. breviflorum Vickery is considered a synonym although some botanists prefer to recognise it as a separate species.

Paspalidium distans (Trin.) Hughes
(spaced far apart)
Qld, NSW, NT
0.5–0.8 m × 0.3–0.6 m   Dec–March

Tufting perennial **grass**; **stems** ascending to erect, unbranched; **leaf sheaths** glabrous or with soft hairs; **leaves** to about 15 cm × 0.1–0.4 cm, linear, flat to loosely folded, glabrous, green; **panicles** to about 30 cm long, very narrow, with 3–10 racemes; **spikelets** about 0.2 cm long, pale green, sometimes with purple tonings; **lemmas** elliptic, acute.

This species inhabits scrub and woodland and usually grows in poor soils which are moist for extended periods. Should do well in exposed or shaded sites. Propagate from seed or possibly by division.

P. radiatum Vickery is a synonym.

Paspalidium globoideum (Domin) Hughes
(globe-shaped)
Qld, NSW   Shot Grass; Sago Grass
0.5–1 m × 0.3–0.7 m   Dec–March;
                                       also sporadic

Tufting perennial **grass** forming a dense tussock; **stems** unbranched, nodes glabrous; **leaf sheaths** mainly glabrous; **leaves** to about 20 cm × 0.6 cm, glabrous, green; **panicles** to about 45 cm long, with about 10 racemes, lower ones distant, upper ones often reduced to single spikelet; **spikelet** to about 0.5 cm long; **lemmas** 9–11-nerved.

An inhabitant of grasslands on heavy soils in south-eastern Qld and north-eastern NSW. Suitable as a stock fodder grass and the seed is eaten by birds. Propagate from seed or possibly by division of tussocks.

Paspalidium gracile (R. Br.) Hughes
(slender)
All states   Slender Panic; Graceful Panic Grass
0.5–0.8 m × 0.5–0.8 m   Dec–March

Tufting, somewhat wiry, perennial **grass** with a knotted rootstock; **stems** branched, nodes glabrous or faintly hairy; **leaf sheaths** glabrous to hairy; **leaves** to about 14 cm × 0.05–0.3 cm, narrowly linear, mainly convolute, often hairy, yellowish-green to deep green; **panicles** to about 25 cm long, with up to about 10 racemes to 2 cm long; **spikelets** to about 0.3 cm long; **lemmas** obovate to elliptic, acute.

This species has the widest distribution of the genus and occurs in a range of habitats. It is most common in rocky sites associated with heavy or skeletal soils. Grows in summer and is commonly grazed. May be useful for soil conservation projects. Propagate from seed or possibly by division of the creeping rootstock in winter.

The var. *debile* Vickery is a synonym of *P. criniforme*. The var. *rugosum* is now included in *P. constrictum*.

Paspalidium jubiflorum (Trin.) Hughes
(mane-like arrangement of flowers)
Qld, NSW, Vic, SA, WA, NT   Warrego Grass;
                    Yellow-flowered Panic Grass
0.3–1.3 m × 0.3–1 m   Dec–March; also sporadic

Tufting perennial **grass**; **stems** branched or unbranched, nodes glabrous; **leaf sheaths** glabrous; **leaves** to about 25 cm × 0.7 cm, linear, flat, somewhat stiff to lax, glabrous, green to blue green; **panicles** to about 50 cm long, with up to about 16 racemes to 4 cm long; **spikelets** to about 0.3 cm long, pale green to yellowish-brown; **lemmas** ovate to elliptic, acute.

A far-ranging species which inhabits the edges of creeks, rivers and drainage channels and is often associated with heavy black soils. It is a summer-growing grass which responds well to periodic inundation. Suitable as a stock fodder grass. Cultivated to a very limited degree. It has potential for soil erosion control in heavy soils. Propagate from seed or by division of tussocks.

Paspalidium radiatum Vickery var. **radiatum** =
P. distans (Trin.) Hughes

Paspalidium radiatum Vickery var. **hirsutum** Vickery =
P. albovillosum S. T. Blake

PASSIFLORA L.
(from the Latin *passio*, passion; *flos, floris*, flower; referring to the similarity of flower parts to the crown of thorns, nails and cross of Christ's Passion)
Passifloraceae

Herbaceous or woody **vines** with slender stems supported by axillary tendrils; **leaves** alternate, with or without lobes, margins entire or toothed, with prominent petiole; **flowers** usually 1–2 per axil, bisexual, often conspicuous; **sepals** 5; **petals** 5 or sometimes absent; **corona** of 1 to several free or united appendages, with flat, membranous, inner corona; **stamens** 5; **styles** 3; **fruit** a globular berry.

This is a large genus of about 370 species; 20 are from the Indo-Australian region and 4 of these are indigenous in Australia (3 of them are in cultivation). A further 7 are introduced species.

The Australian species generally adapt well to cultivation and some are moderately popular. They are quick-growing and therefore make excellent screening plants but they may need regular pruning to curtail their spread. Propagate from seed or from cuttings. The seed may be difficult to germinate and it is best sown fresh. Often better results are achieved by fermenting the seed for a few days before sowing. Cuttings of firm young growth taken over summer months usually provide best results. The use of a rooting hormone may be beneficial but is not essential.

178

Passiflora aurantia G. Forst.
(orange)
Qld, NSW   Red Passionflower;
                                   Blunt-leaved Passionfruit
Climber   July–Dec; also sporadic

Vigorous **climber** with glabrous young growth; **stems** very long, slender, wiry; **leaves** 2–10 cm × 2–8 cm, deep green, shallowly to deeply 3-lobed (rarely entire), lobes rounded, glabrous, stalk to 3 cm long, usually with 2 glands near top; **flowers** to about 11 cm across, usually salmon pink to bright red, sometimes white or pink, often profuse and very conspicuous; **fruit** 3–5 cm × 2–4.5 cm, more or less globular, pale green, becoming purplish with maturity.

A very showy, quick-growing, often short-lived climber. The typical variety occurs from north-eastern Qld to north-eastern NSW, and also extends to Malesia and the south-western Pacific region. Plants are found on rainforest margins and in coastal scrub (especially on disturbed sites). It grows in acidic loams and sands. Adapts very well to cultivation in tropical, subtropical and temperate regions. Does best in a semi-shaded and well-drained site but also grows well in filtered sunlight or where the foliage receives plenty of sunlight but the root region is protected. Although often initially forming dense foliage cover, plants can become sparse within a few years. Pruning from an early age promotes a bushy framework. Plants are hardy to moderate frosts. They respond well to slow-release fertilisers and mulching. May require supplementary watering during extended dry periods. In nature, *P. aurantia* is an important host plant for the larvae of the Glasswing Butterfly (which is also often known as Little Greasy).

The var. *pubescens* F. M. Bailey differs in having faintly hairy stems and leaves with glandless petioles. It is restricted to south-eastern Qld.

Propagate both varieties from seed or from cuttings of firm young growth, which strike readily.

Passiflora brachystephanea (F. Muell.) Benth. = *P. aurantia* G. Forst.

Passiflora cinnabarina Lindl.
(vermilion red)
NSW, Vic, Tas   Red Passionflower
Climber   Sept–Jan

Vigorous **climber** with bright green, glabrous young growth; **stems** very long, slender, wiry; **leaves** to about 10 cm × 10 cm, 3-lobed, moderately to deeply lobed with blunt to pointed apices, dark green, somewhat glossy, stalk 2–5 cm long, lacking glands; **flowers** to about 6.5 cm across, bright red, solitary in upper axils, scattered to profuse, very conspicuous; **fruit** to about 3.5 cm × 2.5 cm, ovoid to globular, green to greenish-grey, edible but distasteful.

A handsome and very vigorous climber/creeper from eastern NSW, eastern Vic and Tas. It is often associated with rocky sites on the coast and hinterland ranges especially where there has been disturbance of the soil. Relatively common in cultivation, it adapts to a wide range of acidic soils which have moderate to good drainage but do not dry out too readily. A sunny or semi-shaded site is suitable. Plants are

Passiflora aurantia   D. L. Jones

hardy to moderate frosts and respond well to pruning, which may be required to maintain luxuriant foliage. This species is cultivated in many other countries. Propagate from fresh seed, or from cuttings of firm young growth, which strike readily.

Passiflora herbertiana Ker Gawl.
(after Lady Carnarvon, neé Herbert)
Qld, NSW   Native Passionfruit
Climber   Aug–Nov

Vigorous **climber** with hairy young growth; **stems** very long, slender, faintly hairy, wiry; **leaves** 6–12 cm × up to 8 cm, 3–5-lobed (or rarely entire) with broad and shallow lobes, apices pointed, deep green, stalk to 7 cm long with 2 dark glands near tip; **flowers** to about 7 cm across, initially cream to greenish-yellow, maturing to pink or orange, solitary in upper axils, very conspicuous; **fruit** to about 5 cm × 4 cm, ovoid,

Passiflora cinnabarina   W. R. Elliot

pale green with paler spots, edible but not highly desirable.

This is a quick-growing climber but it is often short-lived. The typical variant occurs from north-eastern Qld to near Narooma in south-eastern NSW. It is found in wet sclerophyll forest or on rainforest margins. Plants may only last 4–6 years but in that time they are often excellent screen plants. Pruning from an early stage can help maintain foliage cover as well as prolong the life of plants. This species does well in a range of moist but well-drained acidic soils in tropical, subtropical and temperate regions. Prefers a semi-shaded situation. It has had limited cultivation overseas.

The subspecies *insulae-howei* P. S. Green is restricted to Lord Howe Island. It differs in being glabrous and having entire or 3-lobed leaves of 3–11 cm × 2.5–12 cm.

Propagate both variants from fresh seed, or from cuttings of firm young growth, which strike readily.

PASSIFLORACEAE Juss.   Passion Flowers

This large family of dicotyledons consists of about 600 species in 12 genera. More than 500 of these species are in the genus *Passiflora*, which is particularly well developed in South America. It is also represented in Australia by 4 native species and about 7 naturalised species. Many species of *Passiflora* are grown for their ornamental flowers or edible fruit. Most species in the family are climbers with axillary tendrils and with sexual organs arranged in a corona, which is frequently adorned with colourful sterile petals or sepals. Apart from *Passiflora*, the family is represented by two other species in Australia in the genus *Adenia*.

PATERSONIA R. Br.

(after William Paterson, 18th & 19th-century botanical collector and military commander in NSW and Tas)

Iridaceae   Purple Flags

Perennial herbs, mostly evergreen, rarely deciduous; **rootstock** woody, rarely a fleshy segmented rhizome; **leaves** linear or sword-shaped, mostly in fan-like groups; **inflorescence** terminal on a stem, composed of 2 tough spathes containing bracts and producing flowers at intervals; **flowers** lasting a few hours, mostly blue to violet, with a tubular base; **sepals** 3, large and showy; **petals** 3, small, erect; **fruit** a capsule; **seeds** with or without an aril.

A genus of about 20 species; about 18 are endemic in Australia and 1 or 2 species occur in Borneo and New Guinea. The distribution within Australia is concentrated in southern areas with a prominent group in south-western WA. Two of the WA species, *P. babianoides* and *P. graminea*, have a growth habit unusual for the genus. One species, *P. occidentalis*, extends across all of southern Australia although the western populations are much superior horticulturally to those from the eastern states. A single species, *P. macrantha*, is found in the northern part of the NT.

Most species of *Patersonia* have a tussock-like growth habit and many have flowers well displayed on long stems. The flowers themselves are flimsy and short-lasting but in most species they are large and colourful. They appear at regular intervals during the flowering period — sometimes several are produced together from a single spathe — and can make an impressive floriferous display. Patersonias are already popular in cultivation but they could be much more widely grown. They are attractive when planted among rocks, mingled with small shrubs or generally treated as a garden plant. Many species such as *P. glabrata*, *P. sericea* and *P. occidentalis* are at their best in a massed planting which creates drifts of colour during the flowering season. Patersonias also make very decorative plants for containers and miniature gardens.

Patersonias do not suffer greatly from pests. Aphids can be a problem on the young leaves and developing scapes but the damage is usually minor. Severe attacks on the root systems by curl grubs may cause partial death of plants. The grubs sometimes also damage the flowering spathes. Some species of *Patersonia* are very sensitive to root-rotting fungi especially *Phytophthora cinnamomi*. Symptoms include browning of leaves and partial or complete collapse of the whole plant.

Propagation of *Patersonia* is mainly from seed, which requires no pre-sowing treatment, although some species may be slow to germinate. Division of clumps is also successful for many species.

Patersonia argyrea D. Cooke

(silvery)

WA

0.2 -0.4 m × 0.2–0.4 m   Sept–Nov

Tufts dense; **leaves** 20–40 cm × 0.2–0.5 cm, linear to sword-shaped, biconvex in cross-section, grooved, densely covered with white hairs; **scapes** 20–35 cm long, densely hairy; **spathes** 3.5–5 cm long, dark brown, hairy; **flowers** violet; **capsules** and **seeds** unknown.

This poorly known but distinctive species from the Gairdner Range of the southern Irwin District grows in heath in grey sandy soils. It is apparently unknown in cultivation but its silvery-grey leaves suggest it has good ornamental qualities. Requires excellent drainage and a sunny location. Propagation is probably from seed and perhaps also by division.

Patersonia babianoides Benth.

(like the genus *Babiana*)

WA

0.05–0.15 m × 0.05–0.1 m   Sept–Nov

Tufts sparse; **rootstock** forms short, annual fleshy, rhizomatous segments, each segment bearing annual leaves and flowers; **leaves** 1- or 2-stalked at base, 7–18 cm × 0.5–1.4 cm, linear to narrowly elliptic, dull green, pleated or ribbed, prominently hairy; **scapes** 2–3 cm long, hairy; **spathes** 2–3.5 cm long, triangular, green, hairy; **floral tube** 1.5–2 cm long, glabrous; **flowers** 3–4 cm across, blue to purple; **sepals** 1.5–2 cm × 1.2–1.4 cm, obovate; **capsules** not seen.

Distributed between the Darling Range and Dunsborough in the south-western Darling District where it grows on laterite in jarrah forest. This species is unusual for its deciduous habit and stalked, ribbed leaves. There is some evidence that it may have a mycorrhizal relationship with soil fungi and its performance in cultivation is unknown. Propagation is from seed.

Patersonia diesingii Endl. = *P. occidentalis* R. Br.

Patersonia drummondii F. Muell.
(after James Drummond, botanical collector and first government naturalist in WA)
WA
0.1–0.3 m × 0.1–0.3 m   Aug–Oct
   Tufts moderately dense to dense; **leaves** 10–30 cm × 0.15–0.4 cm, narrowly linear, often twisted, dull green, grooved, margins and bases brown, hairy; **scapes** 15–27 cm long, striate, glabrous to sparsely hairy; **spathes** 3–6 cm long, green, glabrous, grooved; **floral tube** 3–4.5 cm long, glabrous to shortly hairy; **flowers** 3–5 cm across, pale blue; **sepals** 1.5–2.5 cm × 1.2–2 cm, obovate; **capsules** 1.5–3 cm long, ovoid; **seeds** black, with an aril.
   P. drummondii occurs in a narrow band from coastal areas near the Murchison River southwards and inland to near Southern Cross. Grows in coastal heath, sand-plain heath and mallee woodland in deep sands and laterite. Plants from the Murchison River area have larger, flatter leaves and larger flowers than those from around Southern Cross. This interesting species is poorly known in cultivation but it is grown in western USA by rockgarden enthusiasts. Requires excellent drainage in a sunny position. Frost tolerance is unlikely to be high. Propagate from seed or by division.

Patersonia flaccida Endl. = *P. occidentalis* R. Br.

Patersonia fragilis (Labill.) Asch. & Graebner
(fragile, delicate)
Qld, NSW, Vic, Tas, SA   Short Purple Flag
0.2–0.6 m × 0.1–0.3 m   Aug–Dec
   Tufts dense; **leaves** 20–60 cm × 0.1–0.6 cm, linear to terete, often grooved, glabrous, dull green to glaucous or pruinose, tip pointed, sometimes pungent; **scapes** 4–25 cm long, glabrous, with a reduced clasping leaf; **spathes** 2.5–4.5 cm long, green to light brown, margins often darker; **floral tube** 2.5–3.5 cm long, glabrous; **flowers** 2.5–4.5 cm across, pale blue to violet; **sepals** 1.2–2.3 cm × 1–1.5 cm, obovate; **capsules** 2.5–3 cm long, cylindrical; **seeds** black, with an aril.
   A variable species which is widely distributed from south-eastern Qld to south-eastern SA including Kangaroo Island. It is most prominent in coastal heaths and wallum but in some parts also extends into the tablelands and low mountains growing in forests and woodlands. Soil moisture ranges from well-drained to seasonally wet. Populations are variable in leaf shape (terete or flat), leaf width, leaf colour and glaucousness, scape length and flower colour. Plants from southern Vic and SA are notable for their strongly glaucous, terete leaves and very short scapes with flowers that are often nearly hidden in the base of the plants. *Patersonia fragilis* offers scope for the selection of horticultural forms. Some variants have proved to be difficult to establish in gardens probably because of sensitivity to root-rotting fungi.

Patersonia glabrata × .5

Others are extremely frost sensitive. Propagate from seed. Division of this species is often difficult.
   Patersonia glauca R. Br. is a synonym.

Patersonia glabrata R. Br.
(glabrous, hairless)
Qld, NSW, Vic   Leafy Purple Flag
0.3–0.8 m × 0.1–0.3 m   Aug–Jan
   Tufts sparse to dense; **stems** to 40 cm × 0.5 cm, woody, sparsely branched; **leaves** 10–40 cm × 0.2–0.5 cm, linear, widely spreading, flat, green to blue-green, glabrous, basal margins minutely hairy; **scapes** 10–30 cm long, glabrous; **spathes** 4–6 cm long, dark brown, sparsely hairy, spreading to 4 cm apart; **floral tube** 4–5 cm long, glabrous or sparsely hairy; **flowers** 4–6 cm across, pale blue to violet; **sepals** 2–3 cm × 1.5–2.5 cm, ovate to orbicular; **capsules** 2–4 cm long, cylindrical; **seeds** brown, with an aril.
   Widely distributed in coastal and near-coastal districts, including the ranges, extending from near Cardwell in north-eastern Qld to east of Wilsons Promontory, Vic. Occurs in coastal heaths growing in sandy soils and in clay loams in open forests. Plants are often locally common and may produce an impressive floral display. Adapts readily to cultivation in subtropical and temperate regions. One of the best for massed planting. Requires well-drained acid soil in semi-shade or full sun. Tolerates light to moderate frosts. Propagate from seed or by division.

Patersonia glauca R. Br. =
          P. fragilis (Labill.) Asch. & Graebner

Patersonia graminea

Patersonia babianoides   D. L. Jones

Patersonia graminea Benth.
(grass-like)
WA
0.05–0.2 m × 0.1–0.3 m   Sept–Oct
   Tufts dense; **stems** with swollen tuber-like growths at
soil level; **leaves** 5–20 cm × 0.1–0.3 cm, narrowly linear,
keeled on the lower side, withering annually, margins
hairy; **scapes** 20–33 cm long, deeply grooved, hairy at
base, remaining fleshy and green for more than a year
after flowering; **spathes** 1.5–2.5 cm long, lanceolate,
glabrous, green; **floral tube** 1.5 cm long, glabrous;
flowers 4–5 cm across, pale blue; **sepals** 2–2.5 cm ×
1.5–2 cm, obovate; **capsules** 1–1.4 cm long, ovoid;
seeds dark brown, with a tiny aril.
   Distributed between the Murchison River and
Watheroo, this unusual species grows in sand in
coastal heath, sandplain scrub and scrub on granite
outcrops. The rootstock has peculiar swollen knobby
growths and the leaves are short-lived. Inflorescences
remain green and fleshy long after flowering is
finished. The species is unknown in cultivation but
probably has similar requirements to *P. drummondii*.
Propagate from seed or by division.

Patersonia inaequalis Benth.
(unequal; referring to bracts)
WA
0.1–0.3 m × 0.1–0.3 m   Aug–Oct
   Tufts sparse, sometimes bushy; **stems** erect, to 6 cm
long, branched, covered by persistent leaf bases; **leaves**
10–32 cm × 0.1–0.25 cm, narrowly linear, often twisted,
deeply grooved, margins and bases shortly hairy;
scapes 18–25 cm long, glabrous; **spathes** 2–3 cm long,
one straight, one bent, green with brown margins;
floral tube 2–3 cm long, glabrous; **flowers** 2–3 cm
across, white; **sepals** 1–1.5 cm × 0.8–1.4 cm, rhomboid;
capsules 1.5–2 cm, cylindrical; **seeds** brown, with an aril.

This unusual species has an erect growth habit, distinctive spathes and small white flowers. It has a
restricted distribution between Stokes Inlet and Cape
Le Grand on the south coast. Plants grow in coastal
heath. They are not known to be in cultivation but
probably require excellent drainage in a sunny location. Frost tolerance is unlikely to be high. Propagate
from seed.

Patersonia juncea Lindley
(rush-like)
WA   Purple Flag
0.05–0.2 m × 0.05–0.15 m   Aug–Oct
   Tufts dense; **leaves** 7–22 cm × 0.06–0.14 cm, narrowly linear to terete, dark green, deeply grooved,
papillose, glabrous to sparsely hairy; **scapes** 5–25 cm
long, glabrous, finely striate; **spathes** 3–4.5 cm long,
brown, glabrous; **floral tube** 2.5–3.5 cm long, brown,
glabrous; **flowers** 3–5 cm across, pale blue to violet;
sepals 1.5–2.5 cm × 1.3–1.8 cm, obovate to orbicular;
capsules 3–4 cm long, cylindrical; **seeds** brown, with
an aril.
   P. juncea is widely distributed between Eneabba and
Israelite Bay. It grows in deep sand and laterite in a
wide range of habitats. Plants are adaptable and have
been grown successfully in rockery gardens, in among
low shrubs, and also in containers. Requires excellent
drainage and prefers sunshine. Tolerates light to moderate frosts. Propagate from seed or by division.

Patersonia lanata R. Br.
(woolly, with tangled hairs)
WA   Woolly Patersonia
0.15–0.4 m × 0.1–0.3 m   Aug–Oct
   Tufts dense; **leaves** 15–40 cm × 0.2–0.7 cm, sword-
shaped, flat, green to glaucous, glabrous except for
the basal margins; **scapes** 12–40 cm long, glabrous to
hairy; **spathes** 2.5–3 cm long, triangular, dark brown,
veined; **floral tube** 2–2.5 cm long, hairy near the base;
flowers 4–6 cm across, blue to violet; **sepals** 2–3 cm ×
1.8–2.7 cm, broadly elliptic; **capsules** about 2 cm long,
ovoid; **seeds** brown, without an aril.
   Restricted to the southern coastal plain between
Two Peoples Bay and Israelite Bay. This species grows
in deep sandy soil in heathland and coastal scrub on
stabilised dunes. It is unknown in cultivation and probably has similar requirements to *P. rudis*.
   P. lanata forma *calvata* D. Cooke has glabrous leaf
margins and glabrous scapes. It grows within the range
of the typical variant.
   Propagate both forms from seed or by division.

Patersonia limbata Endl.
(bordered with a different colour)
WA   Bordered Purple Flag
0.2–0.4 m × 0.1–0.2 m   Sept–Oct
   Tufts sparse to moderately dense; **stems** short,
branched; **leaves** 15–40 cm × 0.4–1 cm, ensiform, dull
green, glabrous, with prominently thickened brown
margins; **scapes** 25–40 cm long, glabrous, with 2 pale,
basal, sheathing leaves; **spathes** 4–5 cm long, dark
brown, glabrous, with paler margins; **floral tube**
3–3.5 cm long, hairy towards the base; **flowers** 4–6 cm

across, blue to violet; **sepals** 2.5–3 cm × 2–2.5 cm, ovate to orbicular; **capsules** 2–3 cm long, cylindrical; **seeds** dark brown, without an aril.

Distributed mainly between the Stirling Ranges and Albany and Cape Arid with small populations near Busselton. Grows in moist to wet sandy soils in heathland, banksia woodland and jarrah forest. Poorly known in cultivation but probably with similar requirements to the western provenances of *P. occidentalis*. Propagate from seed or by division.

Patersonia longifolia R. Br.
(with long leaves)

NSW, Vic	Purple Flag
0.2–0.3 m × 0.1–0.3 m	Sept–Dec

Tufts mostly sparse or sometimes dense; **stems** to 10 cm × 0.5 cm, branched; **leaves** 10–30 cm × 0.1–0.2 cm, linear, lax and usually trailing on the ground, glaucous, biconvex in cross-section, often twisted, margins with a single row of hairs incurved across the leaf blade; **scapes** 8–15 cm long; **spathes** 2–3.5 cm long, brown, silky-hairy; **floral tube** 1.5–3 cm long, short-hairy; **flowers** 4–6 cm across, dark purple; **sepals** 2–3 cm × 1.5–2.5 cm, broadly ovate; **capsules** 1.5–2.5 cm long, cylindrical; **seeds** brown, without an aril.

This species is distributed between the Hunter River and Genoa in eastern Vic. It grows in coastal heathland and the forests of near-coastal ranges. Plants form sparse clumps and are not particularly free-flowering. Requires a sunny location and excellent drainage. Very sensitive to root-rotting fungi and tolerant only of mild frosts. Propagate from seed and by division.

Similar to *P. sericea* which forms dense tufts and has broad, erect, flat leaves, long scapes and large spathes. *Patersonia sericea* var. *longifolia* (R. Br.) C. Moore & E. Betche is a synonym.

Patersonia longiscapa Sweet = *P. occidentalis* R. Br.

Patersonia macrantha Benth.
(with large flowers)

NT	Tropical Purple Flag
0.2–0.6 m × 0.2–0.4 m	Dec–March

Tufts dense; **leaves** 20–60 cm × 0.4–0.9 cm, sword-shaped, lax to erect, dull green to glaucous, glabrous or with hairy margins; **scapes** 20–50 cm long, hairy towards the apex; **spathes** 4–7 cm long, elliptic, brown, sparsely hairy; **floral tube** 2.5–3 cm long, hairy towards the base; **flowers** 6–8 cm across, blue to purple; **sepals** 3–4 cm × 2.5–3 cm, obovate; **capsules** 3–4 cm long, cylindrical, acute; **seeds** pale brown, without an aril.

A widely distributed species of the Darwin and Gulf Region of the Top End where it grows in open forest and woodland in well-drained sandy soils and laterite outcrops. It is often locally common and occurs in groups or loose colonies. The only species of *Patersonia* from the tropics, it has excellent potential for cultivation in tropical and subtropical regions. Plants are robust and produce impressive floral displays. They will grow in situations from filtered sun to full sun and need unimpeded drainage. Excellent for mingling with small shrubs. Propagate from seed or by division.

Patersonia macrantha   D. L. Jones

Patersonia maxwellii (F. Muell.) F. Muell. ex Benth.
(after George Maxwell, 19th-century botanical collector in WA)

WA	Purple Flag
0.1–0.25 m × 0.1–0.2 m	Sept–Nov

Tufts sparse; **leaves** 10–25 cm × 0.1–0.2 cm, linear, dull green, biconvex in cross-section, deeply grooved, papillose, margins with tiny, incurved brown hairs; **scapes** 5–20 cm long, glabrous; **spathes** 2.5–3.5 cm long, brown, glabrous; **floral tube** 2–2.5 cm long, glabrous; **flowers** 2.5–4 cm across, violet-blue; **sepals** 1.2–2 cm × 1–1.5 cm, obovate; **capsules** 2–2.5 cm long, cylindrical, sharply pointed; **seeds** dark brown, without an aril.

This species is mainly found between Albany and Israelite Bay. It grows in moist to wet, sandy soils in coastal heath. Also known from near Bunbury in jarrah forest. Virtually unknown in cultivation, this species probably has similar requirements to western provenances of *P. occidentalis*. May be well suited to container culture. Propagate from seed or by division.

Patersonia nana Endl. = *P. occidentalis* R. Br.

Patersonia occidentalis, white-flowered variant   W. R. Elliot

Patersonia occidentalis × .6

Patersonia occidentalis R. Br.
(western)

Vic, Tas, SA, WA	Long Purple Flag
0.1–0.8 m × 0.1–0.3 m	Sept–Dec

Tufts dense; **stems** much-branched; **leaves** 10–60 cm × 0.2–1 cm, linear to ensiform, flat, dull green, glabrous or with brown hairs on the margins; **scapes** 10–80 cm long, glabrous, with a basal clasping leaf; **spathes** 3–5 cm long, brown, glabrous or with short hairs on the keel; **floral tube** 2.5–4 cm long, with long hairs; **flowers** 4–7 cm across, pale blue, mauve or purple, rarely white; **sepals** 2–3.5 cm × 1.2–2.2 cm, ovate; **capsules** 2–2.5 cm long, cylindrical; **seeds** dark brown, without an aril.

A widespread species which occurs mainly in coastal and near-coastal localities — particularly in heathland but also in open forests and coastal scrub. It grows in sandy soils and clay loams which range from well-drained to poorly drained. An excellent ornamental which has become well entrenched in cultivation. Ideal for planting in rockeries, amongst dwarf shrubs or in drifts. Plants will grow in positions from semi-shade to full sun. They may suffer badly if the root system dries out. Mulches are beneficial. Old clumps, which often contain many dead leaves, can be rejuvenated by burning.

Plants from WA are much more robust and floriferous than those of the eastern states and in nature commonly produce massed floral displays. They also grow in much wetter sites, frequently in winter-wet swamps, with the bases of the plants partly submerged. Propagate from seed or by division.

Patersonia longiscapa Sweet, *P. sapphirina* Lindl., *P. diesingii* Endl., *P. flaccida* Endl., *P. nana* Endl., *P. turfosa* Endl. and *P. tenuispatha* Endl. are all synonyms.

Patersonia pygmaea Lindley
(short, dwarf)

WA	Small Purple Flag
0.1–0.2 m × 0.05–0.2 m	Aug–Nov

Tufts dense; **rootstock** compact, branched; **leaves** 5–20 cm × 0.15–0.3 cm, sword-shaped, flat, margins with long hairs; **scapes** 1–6 cm long, glabrous to hairy; **spathes** 3–4.5 cm long, dark brown, glabrous to short-hairy; **floral tube** 2.5–3.5 cm long, glabrous; **flowers** 3–5 cm across, blue to purple; **sepals** 1.5–2.5 cm × 1.2–1.8 cm, broadly ovate; **capsules** 2–2.5 cm long, ovoid; **seeds** brown, with an aril.

Distributed from near Perth to the Stirling Ranges and Albany. Grows in jarrah forest and heathlands in sands and laterite. This small species is well suited to growing in containers. Requires excellent drainage in filtered sun or semi-shade. Frost tolerance is not high. Propagate from seed and perhaps also by division.

Patersonia rudis Endl.
(rough)

WA	Hairy Flag
0.2–0.7 m × 0.1–0.3 m	Oct–Dec

Tufts dense; **stems** covered with brown resinous leaf bases; **leaves** 20–70 cm × 0.2–0.7 cm, linear to ensiform, flat, dull green, margins and base hairy; **scapes** 20–50 cm long, hairy; **spathes** 3–6 cm long, blackish, with silky hairs; **floral tube** 2–3 cm long, hairy; **flowers** 5–7 cm across, blue to violet; **sepals** 2.5–3.5 cm × 2–3 cm, obovate; **capsules** 2–3 cm long, ovoid; **seeds** black, without an aril.

The typical subspecies grows in sand and laterite in jarrah forest on the Darling Range. It is cultivated on a limited scale but has proved to be sensitive to root-rotting fungi. Requires soils with excellent drainage. Plants have been successfully grown in containers.

The ssp. *velutina* D. Cooke occurs in inland regions between Southern Cross and Coolgardie. It grows in semi-arid woodland. Not known to be in cultivation.

Propagate both subspecies from seed and by division.

Patersonia sapphirina Lindl. = *P. occidentalis* R. Br.

Patersonia sericea R. Br.
(with silky hairs)

Qld, NSW, Vic	Silky Purple Flag
0.2–0.5 m × 0.1–0.3 m	June–Jan

Tufts sparse to moderately dense; **stems** to 20 cm × 0.5 cm, branched; **leaves** 15–50 cm × 0.1–0.6 cm, linear to ensiform, erect, flat to terete, green, glabrous, basal margins hairy; **scapes** 5–50 cm long, upper part hairy, with a reduced clasping leaf; **spathes** 2–6 cm long, blackish, veined, silky-hairy; **floral tube** 1.5–3 cm long, short-hairy; **flowers** 4–6 cm across, dark purple; **sepals** 2–3 cm × 1.5–2.5 cm, broadly ovate; **capsules** 1.5–2.5 cm long, cylindrical; **seeds** dark brown, wrinkled, without an aril.

Patersonia sericea × .6

A widely distributed species which extends from south-eastern Qld to eastern Vic. It ranges from the coastal lowlands to subalpine woodland. Plants grow in a wide variety of habitats. A very floriferous species with established clumps producing impressive floral displays. Best flowering is achieved in a sunny location. Requires well-drained soil. Provenances from high altitudes are hardy to moderate frosts and flower later. Propagate from seed or by division.

P. longifolia is similar. It has narrow, prostrate, glaucous leaves which are biconvex in cross-section, short scapes and small spathes.

Patersonia sericea var. **longifolia** (R. Br.) C. Moore & E. Betche = *P. longifolia* R. Br.

Patersonia tenuispatha Endl. = *P. occidentalis* R. Br.

Patersonia turfosa Endl. = *P. occidentalis* R. Br.

Patersonia umbrosa Endl.
(growing in shade)
WA   Purple Flag
0.3–0.9 m × 0.1–0.5 m   Aug–Nov
   Tufts sparse to moderately dense; **stems** 2–5 cm long, sparsely branched; **leaves** 30–90 cm × 0.4–0.6 cm, flat, green, bases with brown margins, glabrous to short-hairy; **scapes** 30–80 cm long, glabrous, with a reduced clasping leaf; **spathes** 6–8.5 cm

long, green to brown, glabrous or with a few hairs on the keel; **floral tube** 4–5 cm long, silky-hairy; **flowers** 5–7 cm across, blue to violet; **sepals** 2.5–3.5 cm × 2–2.5 cm, ovate, with the midvein darker; **capsules** 2.5–3.5 cm long, cylindrical; **seeds** pale brown, with an aril.

Grows in winter-wet swamps in heathland and open forest of the lower south-west. An attractive tall species which merits wider cultivation. Best grown in shade to filtered sun in moisture-retentive soils. Plants may die if the root system dries out severely. Propagate from seed or by division.

Patersonia umbrosa var. **xanthina** (Oldfield & F. Muell. ex F. Muell.) Domin =
          P. xanthina Oldfield & F. Muell. ex F. Muell.

Patersonia xanthina Oldfield & F. Muell. ex F. Muell.
(yellow)
WA   Yellow Flag
0.5–1.2 m × 0.1–0.5 m   Aug–Oct
   Tufts moderately dense; **stems** 2–5 cm long, branched; **leaves** 50–100 cm × 0.4–0.6 cm, flat, green, bases with brown margins, glabrous to short-hairy; **scapes** 50–120 cm long, glabrous, stiffly erect; **spathes** 6–9 cm long, green, glabrous; **floral tube** 4–6 cm long, silky-hairy; **flowers** 5–8 cm across, yellow; **sepals** 3–4 cm × 2–3 cm, ovate; **capsules** 3–4 cm long, cylindrical; **seeds** brown, with an aril.

Distributed from Busselton to Cape Leeuwin growing in sands and laterite in jarrah and karri forests. An outstanding ornamental which is locally common and produces impressive floral displays. It has excellent prospects to become widely grown as a garden plant although it has proved to be sensitive to root-rotting fungi. Requires excellent drainage and does best when planted among low shrubs. This species will grow in a shady location. May be slow to flower in cultivation. Flowering can often be irregular (it may not occur at all in some years). Propagate from seed or by division.

Patersonia umbrosa var. *xanthina* (Oldfield & F. Muell. ex F. Muell.) Domin is a synonym.

PAVETTA L.
(from *Pawatta*, a Sinhalese name for *P. indica*)
Rubiaceae
   Shrubs or small **trees**; **leaves** opposite; **stipules** between petioles, sheathing, pointed; **inflorescences** terminal corymbs or panicles arising from fused bracts; **flowers** bisexual; **calyx tube** 4-lobed or 4-toothed; **corolla tube** cylindrical, lobes 4; **stamens** 4, with short filaments; **style** prominently exserted, slender; **fruit** a somewhat globular drupe.

A sizeable tropical genus comprising about 400 species. At least 10 species occur in Australia. This genus is closely allied to *Ixora*, which has sometimes been combined with *Pavetta*. *Ixora* differs in having free bracts at the base of the flowerheads.

These plants are not commonly cultivated but deserve to be better known because they produce sweetly-scented flowers in large, showy, terminal heads. Some species may do very well in large containers while others are suitable for background planting. Pruning from an early stage is very important as plants generally have a tendency to become leggy. They are

Patersonia xanthina   D. L. Jones

frost sensitive and should respond well to supplementary watering during extended dry periods. Butterflies are attracted to the flowers. Propagation is from freshly collected seed, which does not require any pre-sowing treatment. Trials of propagation from cuttings are required. Firm young growth should give best results and the application of rooting hormones could be beneficial.

Pavetta australiensis Bremek.
(from Australia)
Qld, NSW
2–5 m × 1.5–4 m   May–Nov
   Medium to tall **shrub**; young growth glabrous or nearly so; **bark** grey; **branches** spreading to ascending; **leaves** 5.5–15 cm × 2–6.2 cm, elliptic to lanceolate, extremely variable in shape and size, clustered near tips, thin-textured, deep green, somewhat glossy, margins entire, apex pointed; **flowerheads** to 10 cm × 12 cm, terminal clusters, sometimes loosely arranged; **flowers** white to cream, sweetly scented, often profuse and very conspicuous; **calyx** densely hairy; **corolla tube** 0.7–1.7 cm long, with flared lobes; **style** twice the length of floral tube; **drupes** to about 0.9 cm × 0.7 cm, somewhat globular, black, usually ripe May–Aug.
   The typical variety is a handsome shrub which is found mainly near the coast, from north-eastern Qld to north-eastern NSW, where it grows in forest and scrubland. Plants are highly desirable for cultivation in tropical and subtropical regions. They prefer moist, well-drained, acidic soils in a situation with semi-shade or partial sun. Plants respond well to supplementary watering during extended dry periods. They may develop an open habit with age. Pruning can help to promote bushy growth.
   The var. *pubigera* S. T. Reynolds occurs in north-eastern Qld, from Innisfail to the McIlwraith Range, where it grows in eucalyptus woodland. It differs in having leaves with a finely hairy undersurface, hairy flower stalks and flowers to 1–2 cm long. Cultivation requirements are similar.
   Propagate both varieties from seed or possibly from cuttings.

Pavetta brownii Bremek.
(after Robert Brown, 18–19th-century British botanist)
Qld, NT
1–5 m × 1–4 m   Dec–May
   Small to tall **shrub** with usually hairy young growth; **bark** grey, usually somewhat stringy; **branches** spreading to ascending; **branchlets** usually with fine hairs; **leaves** 7.5–21 cm × 3.5–9.2 cm, usually elliptic to obovate, with stalks to about 3 cm long, somewhat leathery, deep green, hairy above and below, midrib broad, margins entire, apex blunt or pointed; **flowerheads** to about 10 cm across, terminal, somewhat loose, flat-topped clusters; **flowers** white, sweetly scented, often profuse and conspicuous; **corolla tube** to about 1 cm long, with lobes to about 0.6 cm long,; **drupes** to about 0.8 cm × 0.75 cm, globular, black, glabrous.
   This is an extremely variable species. The typical variety occurs in coastal scrubs and on sand dunes in northern Qld, and extends to Arnhem Land, NT. Evidently rarely cultivated, it is best suited to subtropical and seasonally dry tropical regions. Plants should adapt to a range of freely draining soils. Tolerates extended dry periods.
   The var. *glabrata* S. T. Reynolds has glabrous branchlets, leaves and flowerheads. It is poorly known and has a similar distribution.
   Propagate both varieties from fresh seed; cuttings of firm young growth are worth trying.

Pavetta conferta S. T. Reynolds
(crowded)
NT
1.5–4 m × 1–3 m   Oct–Nov
   Small to medium **shrub** with hairy young growth; **branchlets** finely hairy; **leaves** 10.5–13 cm × 4–6 cm, elliptic to somewhat obovate, short-stalked, deep green and somewhat glossy, hairy above and below but

Pavetta australiensis × .4

186

Pavetta australiensis   D. L. Jones

sometimes becoming glabrous above, margins entire, apex blunt to rounded; **flowerheads** 5–15 cm across, compact, crowded, terminal; **flowers** white to cream, often profuse and conspicuous; **corolla tube** to about 1.3 cm long, with spreading lobes to about 0.6 cm long,; **fruit** not seen.

A recently described (1993) species which is thought to be restricted to a small area of the western Top End. It grows on sandy lateritic soil in low wood-

Pavetta brownii var. *brownii* × .5

Pavetta kimberleyana

land. Plants are worthy of trial in seasonally dry tropical regions and may succeed in subtropical areas. They will need good drainage. Propagate from fresh seed, if available, and possibly from cuttings.

Pavetta granitica Bremek.
(growing in soil derived from granite)
Qld
1–5 m × 1–4 m   Dec–Feb

Small to tall **shrub** with densely hairy young growth; **bark** light grey, flaky; **branches** spreading to ascending; **branchlets** pale grey to whitish; **leaves** 4–16.5 cm × 0.8–3.2 cm, narrowly elliptic to oblanceolate, tapering to both ends, short-stalked, dark green above, paler below, hairy (rarely glabrous) above and below, margins entire; **flowerheads** to about 7.5 cm across, terminal clusters shorter than leaves; **flowers** white, sweetly scented, often profuse and conspicuous; **corolla tube** to about 1.4 cm long, with lobes about 0.6 cm long; **drupes** to about 1 cm × 1 cm, globular to ovoid, black, in loose bunches.

This attractive species occurs in northern Qld. It grows in rocky outcrops on granite or sandstone hillsides. Plants are often found in open forest. In cultivation it grows well in tropical and subtropical regions in a range of well-drained, acidic soils with a sunny or semi-shaded aspect. Plants generally keep within the 1–3 m range. They tolerate extended dry periods and light frosts. Worth trying as a patio plant for large containers. Prune when young to promote bushy growth. Propagate from fresh seed or possibly from cuttings of firm young growth.

P. modesta Bremek. is a synonym. *P. muelleri* is closely allied but differs in having much broader leaves.

Pavetta kimberleyana S. T. Reynolds
(from the Kimberley Region)
WA
3–8 m × 2–6 m   Dec–March

Medium to tall **shrub** or small **tree** with long white-hairy young growth; **bark** stringy, silvery-grey to

Pavetta brownii var. *glabrata*   D. L. Jones

187

greyish-brown; **branches** spreading to ascending; **branchlets** lightly hairy; **leaves** 6–25 cm × 2.5–8.5 cm, broadly lanceolate to nearly obovate, base tapering to a long stalk, slightly glossy, faintly hairy above, becoming glabrous, prominent whitish or pale yellow midrib, margins entire, apex pointed or very blunt; **flowerheads** to about 14.5 cm across, terminal clusters, loosely arranged; **flowers** white to cream, sweetly scented, may be profuse and conspicuous; **corolla tube** to about 1.5 cm long, with flared lobes to about 0.7 cm long, hairy near throat, exterior glabrous; **drupes** to about 0.7 cm × 0.8 cm, somewhat globular, black, faintly hairy or glabrous.

This recently described (1993) species from the Kimberley Region was confused with *P. brownii*. It occurs mainly in vine thickets along the coastal region from the Mitchell Plateau to near Broome and on nearby islands. Plants will be suited to seasonally dry tropical and subtropical regions. Must have excellent drainage. Propagate from seed or possibly from cuttings.

P. muelleri differs in having short-stalked, hairier leaves.

Pavetta modesta Bremek. = *P. granitica* Bremek.

Pavetta muelleri Bremek.
(after Baron Sir Ferdinand von Mueller, 19th-century Government Botanist of Vic)
WA, NT
1.5–8 m × 1–5 m   Dec–March

Small **shrub** to small **tree** with hairy young growth; **bark** stringy or flaky, grey to blackish; **branches** spreading to ascending; **branchlets** hairy; **leaves** 5.2–16.5 cm × 2.5–7 cm, elliptic to elliptic-oblong or nearly obovate, tapering to both ends, deep green above, paler below, hairy (rarely glabrous) above and below, somewhat glossy, margins entire, apex pointed to somewhat blunt; **flowerheads** 6–11 cm across, terminal clusters, often somewhat loosely arranged; **flowers** white to cream, sweetly scented, may be profuse and conspicuous; **corolla tube** to about 1 cm long, with spreading lobes to 0.7 cm long; **drupes** to about 0.7 cm × 0.8 cm, more or less globular, black.

P. muelleri occurs in the north-eastern Kimberley region, WA, and in northern NT. It is usually found growing on sandstone plateaus, outcrops, escarpments, rocky hillsides and ridges. Plants should be best suited to subtropical and seasonally dry tropical regions. They must have excellent drainage but will tolerate a sunny or semi-shaded site. May need protection from cold when young. Propagate from seed and possibly from cuttings of firm young growth.

The allied *P. brownii* has broad, elliptic to oblong leaves. The allied *P. kimberleyana* has broader, less hairy leaves and longer stalks on the leaves, flowers and fruit.

Pavetta rupicola S. T. Reynolds
(of rocky areas)
Qld, NT
1.5–2.5 m × 1–2.5 m   Dec–March

Small to medium **shrub**; young growth densely hairy; **bark** grey, flaky; **branches** spreading to ascending;

branchlets densely hairy; **leaves** 5.5–13.5 cm × 1.5–5.5 cm, elliptic to nearly obovate, short-stalked, somewhat leathery, deep green, sparsely hairy above, densely white-hairy below, margins entire, apex pointed; **flowerheads** to about 6 cm across, terminal, loosely arranged; **flowers** white, often profuse and conspicuous; **corolla tube** to about 1.6 cm long, slender, with spreading lobes to 0.7 cm long; **calyx** densely covered in white hairs; **drupes** to about 0.7 cm × 0.8 cm, somewhat globular.

This species was described in 1993. It occurs in rocky gorges and hills of north-western Qld, and its distribution is thought to extend to the Barkly Tableland, NT. It has potential for cultivation in subtropical and seasonally dry tropical regions. Excellent drainage is extremely important. Propagate from fresh seed or from cuttings of firm young growth.

Pavetta speciosa S. T. Reynolds
(showy, beautiful)
NT
1.5–5 m × 1–4 m   Dec–March

Small to tall **shrub** with hairy young growth; **bark** pale greyish-brown; **branches** spreading to ascending; **branchlets** faintly hairy near tips; **leaves** 6–12 cm × 4–6 cm, elliptic to obovate, with stalks to about 2 cm long, deep green, glossy above and below, sometimes faintly hairy, margins entire, midrib and venation prominent, apex blunt to rounded and sometimes notched; **flowerheads** to about 12 cm across, terminal, crowded or loosely arranged; **flowers** white, often profuse and conspicuous; **corolla tube** to about 1.6 cm long, slender, exterior faintly hairy, with spreading lobes to about 0.7 cm long; **fruit** to 0.8 cm × 0.7 cm.

This showy and distinctive yet poorly known species was described in 1993. It occurs in a restricted area of the Top End where it grows in lateritic soils of open forests. Plants are deserving of cultivation and should do well in freely draining acidic soils in subtropical and seasonally dry tropical regions. A sunny or semi-shaded site should be suitable. Propagate from fresh seed and possibly from cuttings of firm young growth.

P. brownii is similar but it has glabrous flowers.

Pavetta tenella S. T. Reynolds
(somewhat dainty)
NT
3–8 m × 2–6 m   Nov–March

Medium to tall **shrub** or small **tree** with densely white-hairy young growth; **bark** light grey, somewhat smooth; **branches** spreading to ascending; **branchlets** covered in white hairs; **leaves** 12–25 cm × 4.7–10 cm, elliptic, tapering to both ends, thin, deep green, faintly hairy above and below, often becoming glabrous above, venation prominent, apex pointed; **flowerheads** to about 8 cm across, in loose, somewhat dainty, terminal clusters; **flowers** white, often moderately profuse and conspicuous; **corolla tube** to about 1.3 cm long, slender, with spreading lobes to about 0.6 cm long; **fruit** to about 0.7 cm × 0.7 cm, globular, black.

A recently described (1993) species from Melville Island and the northern coast of Arnhem Land. It is found along creeks and springs on the edges of rainforest and in vine thickets. Evidently not in cultivation, it deserves to be better known. Plants will need freely draining sites which are moist for extended periods. Best suited to tropical and subtropical regions. Propagate from fresh seed or possibly from cuttings.

Pavetta vaga S. T. Reynolds
(uncertain; referring to its status)
NT
1.5–4 m × 1–3 m   Oct–Feb

Small to medium **shrub** with silky-hairy young growth; **branchlets** hairy; **leaves** 6–15 cm × 4–7 cm, elliptic to nearly obovate, with stalks to 2.5 cm long, somewhat leathery, deep green and hairy above and below, often becoming glabrous above, margins entire, midrib and venation prominent, apex blunt to crowded; **flowerheads** to about 10 cm across, terminal, loosely arranged clusters; **flowers** white, often profuse and conspicuous; **corolla tube** to about 1.4 cm long, slender, broadest near tip, with spreading lobes to about 0.7 cm long, exterior hairy, on long hairy stalks; **calyx tube** with curved hairs; **drupes** about 0.5 cm × 0.5 cm.

This species was described in 1995 but there is some doubt about its relationship with other species. It occurs near Darwin where it grows in open forests on lateritic soils. Plants will need good drainage. They are best suited to seasonally dry tropical regions and sub-tropical areas. Propagate from fresh seed or possibly from cuttings of firm young growth.

PAVONIA Cav.
(after Jose Antonio Pavon, 18–19th-century Spanish botanist who travelled widely in South America)
Malvaceae

P. hastata is the only species of this genus to occur in Australia. It was regarded as native but further research has revealed that it was introduced from South America in the late 18th-century and has become naturalised in Qld, NSW and SA.

PEDALIACEAE R.Br.

A small family of dicotyledons consisting of about 50 species in 12 genera. It occurs mainly in Africa and Asia. Plants are mainly annual or perennial herbs, or rarely shrubs, with opposite leaves and glandular hairs. They have tubular, white, pink or purple flowers with 5 lobes, which are usually directed forwards. The family in Australia is represented by 3 species in the genus *Josephinia*.

PELARGONIUM
(from the Greek, *pelargos*, a stork; referring to the stork's head and bill form of the fruits)
Geraniaceae

Annual of perennial **herbs** or dwarf **shrubs**; **root-stock** often fleshy; **leaves** opposite or alternate, margins entire, toothed or lobed, hairy or nearly glabrous, often aromatic; **inflorescence** cymose umbels; **flowers** open-petalled; **sepals** 5, fused near base with rear one extended downwards as a fused spur; **petals** 5, white to deep pink or magenta, upper 2 usually larger; **stamens** 10; **fruit** composed of 5 dehiscent mericarps.

A genus of over 250 species which is distributed in temperate regions with its main concentration in southern Africa. In Australia there are about 10 species of which 6 are endemic and 4 are naturalised introductions.

The genus is represented in all states and mainly occurs in coastal or nearby inland regions. In general, flower production is not dramatic but *P. rodneyanum* is an exception with its brilliantly coloured and moderately-sized flowers.

Pelargoniums are very useful as fillers especially in new gardens because they grow quickly and provide cover for open spaces.

Cultivation requirements include good drainage and a sunny or semi-shaded site.

Some species such as *P. australe* are subject to thrips and whitefly attack (see Volume 1, page 153 for control methods) but rarely is the damage of much consequence.

Pelargoniums respond well to light applications of slow-release fertilisers but sometimes at the expense of flowering.

Propagation is from seed or from cuttings. Seed does not require pre-sowing treatment, however, results are best when fresh seed is used. Cuttings of firm growth usually produce roots readily. Rooting hormones can be used although cuttings have been successful without them.

Pelargonium australe Willd.
(southern)
Qld, NSW, Vic, Tas, SA, WA   Austral Storksbill;
                                                Native Storksbill
0.3–0.7 m × 0.5–1.5 m   Sept–April; also sporadic

Pelargonium australe × .6

189

Pelargonium drummondii

Pelargonium australe   W. R. Elliot

Dwarf perennial **herb** with a fleshy taproot and hairy young growth; **branches** spreading to ascending, hairy; **branchlets** hairy; **leaves** 2–10 cm × 2–9 cm, ovate to circular, opposite, shallowly 4 or 5–7-lobed, with hairy stalk to about 13 cm long, mid green, hairy above and below, margins wavy and broadly toothed, with a faint, sweet, spicy fragrance when crushed; **flowerheads** 4–12-flowered, with slender hairy stalks; **flowers** to about 1.5 cm across, pale to mid pink with purplish markings, often profuse and conspicuous; **fruits** to about 1.6 cm long.

This variable and widespread species occurs in a range of habitats but is usually found in sandy or rocky soils. It extends from coastal areas to far inland. *P. australe* has been cultivated for a long period and was introduced into England during 1792. It adapts to acidic or slightly alkaline soils which drain freely. Does well in a sunny or semi-shaded site. Hardy to most frosts and extended dry periods. Responds well to hard pruning which aids rejuvenation of old plants. Useful for border plantings, rockeries and containers. Plants can regenerate readily in gardens and may become slightly weedy. Often attacked by whitefly. Apparently Aborigines once used the root of this species as a food source. Propagate from seed, or from cuttings, which strike readily.

The naturalised *P. fragrans* Willd. has greyish, usually smaller leaves than *P. australe* but it has often been marketed as *P. australe*.

Pelargonium drummondii Turcz.
(after James Drummond, first Government Botanist, WA)
WA
0.2–0.5 m × 0.3–1 m   May–Oct
Dwarf **herb** with softly hairy young growth; **branches** semi-succulent, spreading to erect, softly hairy; **branchlets** hairy; **basal leaves** to about 4 cm × 3 cm, cordate to reniform, often alternate, long-stalked, spreading, mid green, hairy above and below, margins crenate,

faintly fragrant; **stem leaves** similar but much smaller and usually opposite, with shorter stalks; **flowerheads** usually 3–8-flowered, on hairy stems; **flowers** to about 2.5 cm across, pink, often profuse and conspicuous; **fruit** about 1.4 cm long.

This decorative species is confined to the southern regions where it is usually found amongst granite boulders. It adapts well to cultivation in temperate regions. *P. drummondii* requires well-drained acidic soils and will grow in a sunny or semi-shaded site. Plants may become straggly but if pruned regularly they maintain a fairly compact growth habit. They can be damaged by frosts. Suited to gardens and containers. Propagate from fresh seed, or from cuttings of firm growth, which strike readily.

P. drummondii has been confused with the naturalised *P. capitatum* which is strongly aromatic and does not have semi-succulent stems. *P. capitatum* occurs in sandy coastal regions.

Pelargonium havlasae Domin
(after Miss Havlas)
WA
0.05–0.15 m × 0.3–1 m   Sept–Nov
Dwarf **herb** with underground spreading tubers; **stems** erect, short, hairy; **leaves** 1–2 cm × 1.5–2.2 cm, ovate-cordate, shallowly 5–7-lobed, slender stalks to 8 cm long, mid green, faintly hairy to nearly glabrous, margins crenate; **flowerheads** usually 4-flowered but sometimes a solitary flower, on erect basal stalks to about 15 cm long; **flowers** about 2 cm across, white to pale pink with reddish veins, can be profuse and conspicuous; **fruit** to about 0.25 cm long.

This is a moderately poorly known species from southern regions of WA where it occurs in sandy clay and clay soils in the Avon, Eyre and Roe Districts. Evidently not cultivated, it should prove suitable for temperate regions and may not be too fussy about soil type. Needs plenty of sunshine. Hardy to moderate frosts. Worth trying in gardens and containers. Propagate from seed or division of tuberous stems.

P. rodneyanum is allied but much taller and has deep pink to rose-magenta flowers.

Pelargonium helmsii Carolin
(after Richard Helms, 19–20th-century naturalist)
NSW, Vic   Storksbill
0.15–0.3 m × 0.3–0.6 m   Nov–March; also sporadic
Dwarf perennial **herb** with a fleshy taproot and softly hairy young growth; **branches** spreading to ascending, hairy; **branchlets** hairy; **leaves** 1–4 cm × 1–5 cm, kidney-shaped to circular, opposite, hairy, often shallowly-lobed, margins crenate, aromatic; **flowerheads** 5–12-flowered, crowded, with long, slender, hairy stalks; **flowers** to about 1.5 cm across, deep pink, often with deeper markings, can be profuse and conspicuous; **calyx** lobes bluntly pointed; **fruit** to about 1.6 cm long.

This pelargonium occurs at higher altitudes of the Australian alps where it often grows in heathland. Evidently not cultivated, it has potential for wider use. Plants are hardy to frost and snowfalls. They require well-drained acidic soils in a sunny or semi-shaded site. May be suitable for borders and rockeries. Also worth trying as a container plant. Propagate from seed or from cuttings.

P. indorum differs in having sharply pointed and very hairy calyx lobes.

Pelargonium inodorum Willd.
(lacking scent)
Qld, NSW, Vic, Tas   Kopata
0.2–0.4 m × 0.3–0.6 m   Nov–March; also sporadic

Dwarf annual or short-lived perennial **herb** with a fleshy taproot and hairy young growth; **branches** spreading to ascending; **branchlets** hairy; **leaves** 1–4 cm × 1–5 cm, ovate to cordate, opposite, shallowly 5–7-lobed, with slender hairy stalks to about 5 cm long, hairy above and below or glabrous above, margins crenate, often lacking fragrance; **flowerheads** 3–14-flowered, on long, slender, hairy stalks; **flowers** to about 0.8 cm across, pinkish with darker striations, often very profuse and moderately conspicuous; **calyx lobes** sharply pointed; **fruit** to about 1–4 cm long, hairy.

An extremely widespread species which also occurs in New Zealand. Generally inhabits well-drained soils in open forest or woodland in mountainous areas. Adapts well to cultivation and although it produces small flowers they can be profuse. Needs freely draining acidic soils and a sunny or semi-shaded site for best results. Hardy to most frosts. Plants are usually bushy but they respond well to pruning if the habit is open. Does well in rockeries, containers and general planting. First introduced into cultivation in England during 1796. Propagate from fresh seed or from cuttings of firm growth.

The allied *P. helmsii* has blunt-pointed calyx lobes.

Pelargonium littorale Endl.
(coastal)
Vic, Tas, SA, WA   Storksbill; Kopata
0.2–0.5 m × 0.3–1 m   Aug–April; also sporadic

Dwarf perennial **herb** with a fleshy, non-tuberous taproot and hairy young growth; **branches** spreading to ascending, faintly hairy; **branchlets** faintly hairy; **basal leaves** 1.5–5.5 cm × 1.8–8 cm, ovate-cordate, in a loose rosette; **stem leaves** opposite, on stalks to about

Pelargonium rodneyanum   D. L. Jones

Pellaea falcata   T. L. Blake

7 cm long, shallowly 5–7-lobed, mid green, faintly hairy, margins faintly lobed; **flowerheads** 2–5-flowered, on long, slender, faintly hairy stalks; **flowers** to about 1.4 cm across, pink to deep pink, often conspicuous; **fruit** about 0.3 cm long.

This widely distributed species inhabits a variety of soils on the coast and hinterland. Plants adapt well to cultivation. They prefer freely draining soils in a sunny or semi-shaded site. *P. littorale* is hardy to most frosts and responds well to pruning. Excellent for containers, rockeries and general planting. May regenerate readily as seedlings in gardens. Plants were grown in England as early as 1837 and originated from near Perth, WA. Propagate from seed or from stem cuttings.

Pelargonium rodneyanum Lindl.
(after Carol S. Riddell (née Rodney) wife of the 19th-century Colonial Treasurer of NSW)
NSW, Vic, SA   Magenta Storksbill
0.2–0.5 m × 0.1–0.4 m   Nov–May; also sporadic

Dwarf perennial **herb** with tuberous roots and hairy young growth; **stems** hairy, usually ascending to erect; **basal leaves** 2–5 cm × 1.5–4 cm, narrowly ovate to ovate, shallowly 5–7-lobed, on stalks to 10 cm long, mid-to-deep green, veins faintly hairy, margins crenate; **flowerheads** 2–7-flowered, on long slender stalks; **flowers** to about 3 cm across, deep pink to rose-magenta, with deeper markings, sometimes profuse and always very conspicuous; **fruit** to about 2.2 cm long.

This delightful and often very showy storksbill occurs in a range of often rocky habitats. Frequently found in open forest, woodland and heathland. Adapts well to cultivation but is often not long-lived. Good drainage is important and usually acidic soils are best. A full sun or semi-shaded aspect is imperative. *P. rodneyanum* can be an excellent container plant and also grows well in rockeries and as a border plant. It is hardy to most frosts. Plants could be selected for variants with large flowers. Propagate from seed, which usually germinates readily.

191

PELLAEA Link
(from the Greek *pellaios*, dark; apparently referring to the dark stipes of these ferns)
Sinopteridaceae

Terrestrial ferns growing in clumps or spreading patches; **rootstock** creeping, slender, much-branched; **fronds** simple or 1–2-pinnate; **pinnae** spreading from the rachis, simple, entire, alternate; **sori** usually in a narrow marginal band.

A genus of about 40 species widely distributed in tropical and temperate regions of the world, with 2 species native in Australia and another naturalised. One of the Australian species is popular in cultivation. A few exotic species are grown as ornamental pot plants. Propagation is from spores or by division of the clumps.

Pellaea falcata (R. Br.) Fee
(sickle-shaped)
Qld, NSW, Vic, Tas   Sickle Fern
0.2–0.6 m tall

Terrestrial fern forming large spreading patches; **fronds** 20–60 cm tall, erect or arching; **pinnae** 27–89 per frond, 2.5–6 cm × 0.5–1.3 cm, oblong to linear-oblong, usually falcate, leathery, dark green above, paler beneath, margins entire or short-toothed; **sporangia** in a marginal band about 0.1 cm wide.

The typical variety, distributed from tropical areas of Qld to Tas, also extends overseas to New Caledonia, Norfolk Island, New Zealand and tropical Asia. It grows in moist sites in rainforest, wet sclerophyll forest, coastal scrubs and gullies in open forest. Plants often form extensive colonies several metres across. An excellent fern which is popular in cultivation. It grows readily in a range of situations but demands good drainage and moisture during dry periods. Tolerates moderate frosts. Plants from southern areas of Australia have large, dark green, shiny pinnae, whereas those from the coastal areas of northern NSW have smaller, duller pinnae.

The var. *nana* Hook. is distributed in Qld, NSW, Vic and Tas. It has shorter fronds (to 40 cm long) and pinnae to 2 cm long. It is usually found among rocks in rainforest and the plants generally form small patches of very dark green fronds. It is easily grown but is not as adaptable as the typical variety.

Propagate both variants by spore or division.

Pellaea paradoxa (R. Br.) Hook.
(unusual, contrary to type)
Qld, NSW
0.3–0.65 m tall

Terrestrial fern forming small patches; **rootstock** long-creeping, slender, much-branched; **young fronds** with a single heart-shaped to circular pinna; **mature fronds** 30–65 cm × 4–20 cm, erect; **pinnae** 9–23 per frond, narrowly ovate to ovate, 2–9.5 cm × 1–3 cm, base heart-shaped, dark green above, paler beneath, entire or crenulate, apex blunt to pointed; **sporangia** in a marginal band 0.2–0.4 cm wide.

Widely distributed from north-eastern Qld to the north coast of NSW and also in NZ. Plants often grow among rocks in rainforest, but they also extend to sheltered slopes and gullies in open forest. Distributed from coastal districts to the ranges and tablelands. Attractive when well grown but can be somewhat tricky to establish. Needs well-drained soil in shade or filtered sun. Best planted where it will receive minimum disturbance. Suited to planting among rocks and can also be successful in a container. Propagate from spores and by division.

PELTOPHORUM (Vogel) Benth.
(from the Greek *pelte*, a shield; *phoreo*, to bear; referring to form of the stigma)
Caesalpiniaceae

Hardwood trees; **leaves** bipinnate, with many leaflets; **inflorescence** terminal panicle; **flowers** open-petalled, yellow; **sepals** 5, united at base; **petals** 5, spreading, 2 lowest are largest; **stamens** 10, free; **fruit** a pod, thin, margins flat.

A tropical genus of about 9 species with 1 occurring in northern Australia. Highly desirable for horticultural uses. Propagation is from seed which requires treatment before sowing, see Volume 1, page 205.

Peltophorum ferrugineum (Decne.) Benth. =
          P. pterocarpum (DC.) Backer ex K. Heyne

Peltophorum pterocarpum (DC.) Backer ex K. Heyne
(winged fruit)
Qld, NT   Peltophorum; Yellow Flame Tree
6–15 m × 4–7 m   Aug–Dec; also sporadic

Small to medium **tree** with densely rusty-hairy young growth; **bark** grey to brown; **branches** spreading to ascending; **branchlets** densely rusty-hairy; **leaves** bipinnate, 10–45 cm long; **pinnae** 6–12 pairs, 3.5–12 cm long; **leaflets** 9–17 pairs, 0.6–2.1 cm × 0.3–1 cm, opposite, oblong, deep green above, paler below, becoming faintly hairy, apex blunt or notched; **panicles** to 60 cm long, terminal, very conspicuous; **flowers** about 5 cm across, bright yellow with central brown blotch, arranged in racemes; **pods** 5–10 cm × 1.4–3 cm, brown, flattish, with wing-like margin, somewhat woody.

This extremely showy species is native to northern Qld and the Top End of NT. It also extends to Malesia. In nature it occurs in coastal areas, on a range of soils, in vine thickets, mangroves and the margins of flood plains. Commonly cultivated, it is highly suitable for tropical and subtropical regions in gardens, parks and on roadsides. It is fast-growing and can be initially spindly but eventually it develops a rounded, bushy crown of foliage. Prefers well-drained soils which are moist for extended periods, and a sunny site. Plants are used for stabilisation of coastal sites. Subject to frost damage. Responds well to pruning. The bark has medicinal properties and is also used in the preparation of soga, a commonly used yellow-brown dye for batik. Propagate from treated seed and possibly from cuttings.
P. ferrugineum (Decne.) Benth. is a synonym.

PEMPHIS J. R. Forst. & G. Forst.
(from the Greek *pemphis*, a swelling; referring to the capsule)
Lythraceae

A monotypic genus which has a distribution ranging from the Pacific Islands to Africa.

Pemphis acidula J. R. Forst. & G. Forst
(acidic, sour, tart)
Qld, WA, NT   Digging Stick Tree
2–10 m × 1.5–6 m   April–Nov; possibly sporadic
Medium to tall **shrub** or rarely a small **tree** with grey-ish, silky-hairy young growth; **bark** brown, flaky; **branches** spreading to ascending, brittle; **branchlets** angular when young, silky-hairy; **leaves** 1.3–2.5 cm × 0.3–0.7 cm, elliptic to somewhat obovate, opposite, decussate, short-stalked, spreading to ascending, somewhat crowded, dull greyish-green, faintly silky-hairy above and below, thick and faintly succulent, apex blunt with glandular tip; **flowers** open petalled, 1–1.5 cm across, white, solitary, axillary, prominently stalked, usually scattered; **sepals** 6; **petals** 6; **stamens** 12, in whorls; **fruit** a capsule, to about 0.7 cm across, initially reddish, glabrous.

An inhabitant of beach edges and mangrove margins, in sand and rocky headlands of northern Qld, the Top End, NT, and the Kimberley Region, WA. Usually develops as a spreading medium-sized shrub and is somewhat *Leptospermum*-like in appearance. It is suitable for cultivation in tropical and subtropical regions. Good drainage is essential and maximum sunshine benefits good growth. It has potential for coastal soil stabilisation. Aborigines used wood for digging sticks and a mixture made from branchlets and roots was used for treatment of toothache. Propagate from seed, which does not require pre-sowing treatment, or possibly from cuttings of firm young growth.

PENNANTIA J. R. Forst. & G. Forst.
(after Thomas Pennant, 18th-century British zoologist)
Icacinaceae
Trees; **leaves** alternate, simple, entire or toothed, domatia present; **inflorescences** terminal or axillary corymbose panicles; **flowers** bisexual or unisexual; **sepals** ring-like or shallowly 5-lobed or toothed; **petals** 5, free; **stamens** 5; **stigma** 3-lobed; **fruit** a 1-seeded drupe.
A small genus of 3 species, represented in Norfolk Island, NZ and 1 species endemic in eastern Australia.

Pennantia cunninghamii Miers
(after Alan Cunningham, 19th-century botanist and explorer)
Qld, NSW   Brown Beech
8–20 m × 4–10 m   Aug–Feb; also sporadic
Small to medium **tree**, initially somewhat climber-like, with glossy, bright green, glabrous young growth; **bark** brown, scaly; **branches** spreading to ascending; **branchlets** glabrous, flexuose; **leaves** 10–16 cm × 5–8 cm, elliptic to ovate, spreading to ascending, stalked, leathery, glabrous, glossy green above and below, domatia prominent below, margins entire or sometimes toothed when young, apex pointed; **panicles** to about 12 cm long, mixture of male and bisexual flowers, terminal or in upper axils; **flowers** to about 0.5 cm across, white, often crowded and profuse; **fruit** to about 1.5 cm long, ovoid, black.
This ornamental species can reach sizeable tree proportions but usually grows to about 10–15 m tall. It occurs in warm temperate and tropical rainforests

Pennantia cunninghamii × .45

from north-eastern Qld to near Ulladulla, NSW, and is often in rocky sites near waterways. It is suited to cultivation in large gardens and parks, in temperate, sub-tropical and tropical regions. Needs freely draining but moisture-retentive acidic soils. Young plants require plenty of moisture and protection from hot sunshine but mature plants are more adaptable. Birds eat the ripe fruits. The timber is hard and resembles English beech. Propagate from fresh seed. Cuttings of firm young growth are also worth trying.

PENNISETUM Rich. ex Pers.
(from the Latin, *penna*, feather; *seta*, bristle)
Poaceae
Tufted or stoloniferous annual or perennial **herbs**; **culms** often branched; **leaves** rolled or folded in bud becoming flat, folded or convolute, **inflorescence** crowded spike-like panicle, rarely reduced and surrounded by sheaths; **spikelets** solitary or in groups of 2 – 4 surrounded by bristles; **florets** 2, upper bisexual, lower male or sterile; **glumes** unequal;
This cosmopolitan genus is comprised of over 130 species. A number of species occur in Australia but they are all introduced except for *P. alopecuroides*. In other countries, many of them are cultivated or utilised for grain and animal food. *P. clandestina* is the well-known Kikuyu that is commonly used for pasture and lawns. Some species are weedy and can infiltrate bushland. Propagation is from seed or division of clumps or rooted stems.

Pennisetum alopecuroides (L.) Spreng.
(resembling the genus *Alopecurus*)
Qld, NSW, Vic, Tas, SA, WA   Swamp Foxtail;
                                         Fountain Grass
0.6–1 m × 0.6–1 m   Nov–March; also sporadic
Tufted perennial **herb**; **leaves** 0.6–1 m × to about 0.6 cm, basal, flat to inrolled, pale green, rough on

Peltophorum pterocarpum   D. L. Jones

nerves, margins hairy. ligule densely hairy; **panicle** 7–20 cm × 2–3 cm, pale greyish-pink to purplish, spike-like, usually solitary on erect to arching stems to about 1 m long, often profuse and very conspicuous.

This very attractive species has been the centre of much discussion about whether it is native to Australia. Recent botanical studies show it is definitely not native to some parts of this continent having been spread after introduction in the early period of European settlement. It is offered for sale by many nurseries as an Australian plant and has become popular in cultivation over much of urban Australia. Plants adapt well to cutivation in a range of soils and climatic conditions, but can suffer damage from heavy frosts. This species has the capability of becoming weedy particularly in disturbed areas. It is recommended that it should not be cultivated within the vicinity of bushland. Propagate from seed or division of clumps.

P. alopecuroides is often confused with the closely allied weedy exotic *P. setaceum* which differs in having narrower leaves with rough but glabrous margins and shorter spikelets.

PENTACERAS Hook. f.
(from the Greek *pente*, five; *cera*, horn; referring to the carpel appendage)
Rutaceae
A monotypic genus which is endemic in Australia.

Pentaceras australis (F. Muell.) Hook. f. ex Benth.
(southern)
Qld, NSW   Bastard Crow's Ash; Black Teak;
                                   Penta Ash
6–15 m × 4–8 m   June–Oct
Small to medium **tree** with glabrous young growth; **bark** brown, smooth; **branches** spreading to ascending; **branchlets** glabrous, brown; **leaves** pinnate, to about 50 cm long; **leaflets** 7–15, 5–13 cm × up to 4 cm, lanceolate to ovate, obliquely angled at base, thin-textured, glabrous, dark green above, paler below, apex blunt or finely pointed; **flowers** about 0.6 cm across, 5-petalled, white, in large terminal clusters, with honey

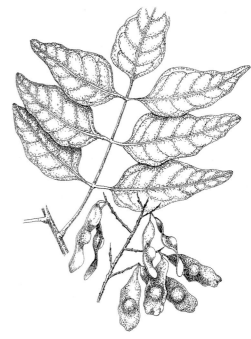

Pentaceras australis × .45

fragrance, very conspicuous; **fruit** 2–4 cm long, 1–5 carpels, surrounded by broad papery wing, ripe Nov–Jan.

A handsome tree when in full flower, it deserves to be better known in cultivation. It occurs in coastal rainforest from near Gympie in south-eastern Qld to the North Coast District of NSW. This species regenerates readily in disturbed areas. Plants require freely draining acidic soils and tolerate a shady or sunny site. They can be straggly while young but judicious pruning may promote bushier growth. Hardy to moderate frosts. Ideal for large gardens, parks and rural properties. Propagation from seed has proved difficult but success has been achieved with root cuttings.

Pentachondra ericifolia × .6

PENTACHONDRA R. Br.
(from the Greek *pente*, five; *chondros*, grain; fruits have 5 parts)
Epacridaceae

Dwarf **shrubs**; **branches** prostrate to ascending; **branchlets** glabrous or hairy; **leaves** alternate, flat or somewhat concave, crowded, more or less sessile, venation longitudinal; **flowers** solitary, terminal or axillary; **corolla tube** cylindrical, with spreading hairy lobes; **stamens** 5, not exserted; **fruit** a fleshy drupe.

A small genus of 5 species occurring in Australia and NZ. There are 4 species in Australia and 3 of these are endemic. In Australia they are confined to subalpine south-eastern regions of the mainland and Tas.

Pentachondras have ornamental characteristics that make them appealing for cultivation, however, they are not commonly grown mainly because propagation is difficult (see below). They need moisture-retentive acidic soils which have good drainage, and they prefer a sunny or semi-shaded site.

Propagation from seed has proved very difficult but, generally, cuttings of very young and somewhat soft growth will produce roots. Plants may be very slow or difficult to establish. The application of a rooting hormone is usually beneficial.

Pentachondra ericifolia Hook. f.
(leaves similar to the genus *Erica*)
Tas
0.15–0.7 m × 0.3–1.2 m   Nov–Feb

Dwarf, spreading to somewhat erect **shrub** with glabrous or hairy young growth; **branches** prostrate to ascending, often forming mats; **leaves** 0.3–0.6 cm × about 0.1 cm, narrowly linear, ascending to erect, crowded, more or less sessile, concave, hairy, especially on margins, undersurface 5-veined, apex pointed; **flowers** tubular, to about 0.5 cm long, white, with spreading and recurved hairy lobes, sessile in upper axils, can be profuse and conspicuous; **drupes** small, splitting into 5 parts when ripe.

A heath-like species with its main representation at high elevations on the Central Plateau, but also occurs near Hobart. Rarely encountered in cultivation, it will require exceptionally good drainage in acidic soils which do not dry out very readily. A sunny site will enhance flowering. Plants are hardy to frost and snowfalls. They have potential for use in rockeries and miniature gardens in temperate regions. Propagate from cuttings of very soft young growth.

Sometimes misspelt as *P. ericaefolia*.

Pentachondra involucrata R. Br.
(provided with covering)
Tas
0.1–2 m × 0.2–1 m   Dec–April

Dwarf to small **shrub** with densely hairy young growth; **branches** ascending to erect; **branchlets** hairy; **leaves** 0.8–2 cm × to about 0.2 cm, broadly lanceolate to broadly elliptical, ascending to erect, somewhat crowded on small branchlets, flat to slightly concave, mid to deep green, glabrous above, faintly hairy below, margins with long spreading hairs, prominent longitudinal veins, apex bluntly or finely pointed; **flowers** tubular, to about 1 cm long, pinkish in bud, usually

Pentachondra pumila   D. L. Jones

becoming white with maturity, 1–3 in upper axils, with narrow, spreading to recurved, hairy lobes, sweetly fragrant, often profuse and conspicuous; **drupes** small, splitting into 5 parts when ripe.

A variable but handsome species which may be prostrate or erect. It occurs in southern Tasmania where it grows at above 1100 m altitude, as well as at lower elevations. Some selections have flowers which have a pink corolla and white lobes, which is very attractive. It is cultivated to a limited degree. It needs freely draining acidic soils in a sunny or semi-shaded site which does not get extremely hot during summer. Plants

Pentachondra involucrata × .9, flower × 1.5

195

Pentachondra pumila

Pentachondra pumila × .7

respond very well to pruning and they are hardy to snowfalls and heavy frosts. An excellent plant for containers. Propagate from cuttings of soft new growth, which usually strike readily.

Pentachondra pumila (J. R. Forst. & G. Forst.) R. Br.
(dwarf, little)
NSW, Vic, Tas   Carpet Heath
0.05–0.15 m × 0.5–1.5 m   Dec–March
Mat-forming dwarf **shrub** with glabrous or faintly hairy young growth; **branches** long, wiry, often self-layering; **branchlets** rough; **leaves** 0.3–0.6 cm × up to 0.2 cm, oblong to elliptic, crowded, ascending to erect, concave, glossy, stiff, margins entire, apex pointed; **flowers** tubular, about 0.4 cm long, white, with short, spreading, hairy lobes, scattered to profuse; **drupes** to about 0.8 cm across, more or less globular, bright red, glabrous.
An eyecatching and common plant above the tree-line in the highest subalpine regions where it often grows on herbfields, grasslands and feldmark. It also extends to NZ. Slow growing in cultivation, it requires very well-drained acidic soils which rarely dry out. Full sunshine is best for flower production but plants will grow in semi-shaded sites too. Plants tolerate very low temperatures and snowfalls. Highly suitable for rockeries, miniature gardens and containers. Propagate from cuttings of very soft young growth, or by division of layered stems.

PENTALEPIS F. Muell.
(from the Greek *pente*, five; *lepis*, scale)
Asteraceae (alt. *Compositae*)
Perennial **herbs** or small **shrubs**; **leaves** opposite, entire or toothed, hairy; **flowerheads** cymose, on long peduncles, yellow, receptacle with chaffy bracts; **ray florets** 5, female; **disc florets** functionally male; **fruit** a narrow-linear cypsella.
Pentalepis is an endemic genus of 3 species. It is restricted to northern WA and NT. Species were previously included in *Chrysogonum* (see Volume 3,

page 36), which is now recognised as only occurring in North America.

Pentalepis ecliptoides F. Muell. —
see *P. trichodesmoides* F. Muell.

Pentalepis trichodesmoides F. Muell.
(similar to the genus *Trichodesma*)
WA, NT
0.3–1.5 m × 0.5–1 m   May–Sept; also sporadic
Dwarf to small **shrub**; **branches** ascending to erect, stiff, rough, hairy; **leaves** 2.5–8 cm × 0.6–2 cm, lanceolate, opposite, nearly sessile, 3–5-nerved, brittle, green, smooth, hairy to greyish-woolly, margins entire or faintly toothed, apex pointed; **flowerheads** to about 4 cm across, composed of 5–10 heads; **ray florets** 5, bright yellow.
This little-known species has interesting flowerheads. It occurs in northern WA, and the Victoria River Region of NT. Usually found on stony sand or loam of watercourses. It had limited cultivation and is best suited to seasonally dry tropical and subtropical regions, but it may succeed as a container plant in temperate regions. Needs freely draining acidic soils and plenty of sunshine. Likely to suffer frost damage. Propagate from seed, which germinates readily, or from cuttings.
Previously known as *Chrysogonum trichodesmoides* (F. Muell.) F. Muell.
P. ecliptoides F. Muell. from the Kimberley Region, WA, and adjacent areas in NT is an annual herb of up to 0.7 m tall, with rough linear to narrowly ovate leaves and loosely arranged flowerheads. Previously known as *Chrysogonum ecliptoides* (F. Muell.) F. Muell.
P. species (Kimberley) is allied to *P. ecliptoides* but differs in growing to about 1 m tall and it has slender racemes.

Pentalepis species (Kimberley) —
see *P. trichodesmoides* F. Muell.

PENTAPANAX Seemann —
see *POLYSCIAS* J. R. Forst. & G. Forst.

PENTAPELTIS (Endl.) Bunge
(from the Greek *pente*, five; *pelte*, small shield; referring to the peltately attached calyx lobes)
Apiaceae
Dwarf **herbs**; **leaves** kidney-shaped, orbicular or broadly triangular, coarsely toothed; **stipules** present; **inflorescence** compound, slender-stalked umbels, subtended by pointed involucral bracts; **flowers** bisexual, white; **calyx lobes** circular, peltate; **petals** acutely pointed, not inflexed; **fruit** 2 flattened mericarps.
A genus of 2 species endemic to south-western WA. Closely allied to *Xanthosia* which has inflexed petals and calyx lobes that are not peltately attached.

Pentapeltis peltigera (Hook.) Bunge
(shield or crescent-shaped)
WA
0.3–0.6 m × 0.5–1.5 m   Nov–March
Dwarf perennial **herb** with faintly hairy young growth; **stems** spreading to ascending, glabrous; **leaves**

1–2.6 cm × 2.5–7 cm, kidney shaped to orbicular-cordate or crescent-shaped, long-stalked, leathery, glabrous, margins thickened, shallowly toothed; **flowerheads** to about 3.5 cm across, with 3–5 slender rigid bracts, long-stalked; **flowers** small, white; **fruit** small, ribbed.

An interesting creeping species from the southern Darling District where it usually occurs in lateritic soils. Evidently rarely grown, it has potential for greater cultivation. Plants require freely draining acidic soils and will probably do best in a semi-shaded site. Should be suited to general planting, rockeries and as container plants. Plants respond well to pruning. They will suffer damage from heavy frosts and dislike extended dry periods. Propagate from fresh seed or from cuttings of firm non-flowering young growth.

Previously known as *Xanthosia peltigera* Benth.

P. silvatica (Diels) Domin is a closely allied, semi-prostrate herb, also from the southern Darling District. It differs in having semi-orbicular to broadly triangular leaves, and it has inconspicuous floral bracts. It may prove to be conspecific with *P. peltigera*.

Pentapeltis silvatica (Diels) Domin —
see *P. peltigera* (Hook.) Bunge

PENTAPETES L.
(from the Greek *pente*, five; *peta*, broad, spreading; referring to broad petals or sepals)
Sterculiaceae

A monotypic genus which is widespread in Indomalaysia and Malesia, and extends to the Northern Territory.

Pentapetes phoenicea L.
(purple-red, from the Phoenician dye)
NT

1–2 m × 0.6–1.5 m   March–May; also sporadic

Small annual **herb** with stellate-hairy young growth; **branches** many, ascending to erect; **branchlets** faintly stellate-hairy; **leaves** 7–15 cm × 1.5–3 cm, linear to somewhat triangular, alternate, stalked, spreading to ascending, green, glabrous above, stellate hairs on veins below, margins prominently toothed, apex pointed; **flowers** to about 2 cm across, 1–3 per axil; **petals** 5; **stamens** 15, in 5 groups, alternating with 5 staminodes which are as long as petals; **sepals** to 1 cm long, joined at base, hairy; **fruit** 1–1.5 cm × 0.6–1 cm, globular to oblong, 5-valved, hairy.

A widespread tropical species which is found throughout Asia and extends to the Darwin and Gulf Region in the NT. It inhabits swamps and low-lying areas. *P. phoenicea* has limited potential but may be useful for revegetation projects within its natural range. Suitable for tropical regions and requires moist to wet soils. Propagate from seed, which does not require pre-sowing treatment.

PENTAPOGON R. Br.
(from the Greek *pente*, five; *pogon*, beard; referring to the 5-awned lemma)
Poaceae

A monotypic genus endemic in south-eastern Australia.

Pentapogon quadrifidus (Labill.) Baill.
(with four parts)
NSW, Vic, Tas, SA   Five-awn Speargrass
0.5–1 m × 0.2–0.5 m   Sept–March

Tufting annual or short-lived perennial **grass**; **culms** erect or bent, nodes hairy; **leaves** to about 20 cm × 0.3 cm, flat to inrolled, moderately to densely hairy, rarely glabrous; **panicle** 3–20 cm long, moderately narrow; **spikelets** 1-flowered; **lemma** to about 0.7 cm long, 2-lobed, with twisted central awn and on each side 2 weaker awns.

A widespread, variable species which inhabits mainly woodland and lowland heaths where the soils are moist for extended periods. It is ornamental and has potential for cultivation in temperate regions. Also withstands grazing, which could make it useful for soil erosion control within its natural range.

The var. *parviflorus* (Benth.) D. I. Morris (previously *P. billardieri* R. Br. var. *parviflorus* Benth.) is endemic in Tas where it grows from sea-level to about 1200 m altitude. It differs markedly from the typical variety in having much smaller flowerheads.

Propagate from seed and possibly by division of clumps during early spring or late autumn.

PENTAPTILON E. Pritz.
(from the Greek *pente*, five; *ptilon*, wing; referring to the five-winged fruits)
Goodeniaceae

A monotypic genus endemic in the Austin and Irwin Districts of south-western WA.

Pentaptilon careyi (F. Muell.) E. Pritz.
(after H. Stuart Carey, the original collector)
WA

0.2–0.5 m × 0.1–0.2 m   Sept–Dec

Perennial **herb** with densely woolly-hairy young growth; **leaves** 3–8 cm × 1–3.5 cm, mostly basal, obovate to elliptic, crowded, tapering to both ends, densely hairy above and below; **inflorescence** loosely arranged terminal cluster on erect stalks to about 50 cm tall; **flowers** to about 1 cm across, 5-lobed, violet to purple or brownish throat and white to yellow wings, often profuse and conspicuous; **fruit** to about 0.8 cm × 0.8 cm, with prominent swollen wings.

An impressive herb of the Northampton-Murchison River Region. It grows in mallee and heath communities in sandy and gravelly soil. Regenerates well after bushfires. Rarely cultivated, it deserves to be better known because it has potential for massed or general planting, as well as for growing in rockeries and containers. Plants are best suited to semi-arid and temperate regions. They need excellent drainage, acidic soils and plenty of sunshine. Removal of spent flower stems should promote further flowering. Propagate from seed or possibly from leaf cuttings.

PENTATROPIS R. Br. ex Wight & Arn. —
see *RHYNCHARRHENA* F. Muell.

PEPEROMIA Ruiz & Pav.
(from the Greek *peperi*, pepper; *homios*, alike; referring to resemblance and relationship to the genus *Piper*)
Piperaceae (alt. *Peperomiaceae*)

Peperomia affinis

Succulent, epiphytic or terrestrial **herbs**; **rootstock** compact; **leaves** opposite or whorled, lower leaves often reduced, blades dotted and with palmate venation in Australian species; **inflorescence** a fleshy spike, axillary or terminal; **flowers** tiny, subtended by stalked, shield-like bract; **stamens** 2; **fruit** a tiny drupe, sticky.

A large tropical and subtropical genus of about 1000 species with 6 occurring in north-eastern Australia. Two of these are endemic and one, *P. pellucida*, is a naturalised introduction. They inhabit sheltered sites which are dry for extended periods or continually moist often with high humidity. Many exotic species are popular as potted and indoor plants because of their appealing foliage.

The Australian species are uncommon in cultivation but they deserve greater recognition from the horticultural industry because of their ornamental characteristics. Propagation is from fresh seed, or from leaf cuttings (which strike readily), or by division.

Peperomia affinis Domin =
P. *tetraphylla* (G. Forst.) Hook. & Arn.

Peperomia bellendenkerensis Domin
(from Bellenden Ker region)
Qld
0.1–0.2 m × 0.05–0.2 m   most of the year
Dwarf, succulent, lithophytic **herb** with faintly hairy young growth; **stems** spreading to ascending, often rooting at nodes; **leaves** to about 1.6 cm × 1.3 cm, ovate-elliptic to circular, alternate, tapering to base, spreading to ascending, thick, deep green, glabrous to faintly hairy, 5-nerved, with outer 2 somewhat indistinct, apex blunt to rounded; **spikes** 2–4 cm × up to 0.1 cm, terminal, solitary, slender, on short stalks; **flowers** very small; **drupes** very small.

Until 1994, this species was known only from the original collection but it has now been collected from a further five lowland sites in the Mt Bellenden Ker and Mt Bartle Frere region. It grows in humus on deeply shaded granite boulders in rainforest. Best suited to tropical and subtropical regions. Prefers containers with a coarse potting mixture or rock crevices in a sheltered garden situation. Propagate from fresh seed, cuttings, or divisions of rooted stems.

Peperomia blanda (Jacq.) Kunth.
                    var. **floribunda** (Miq.) H. Huber.
(pleasant, charming; many flowers)
Qld, NSW
0.02–0.04 m × 0.03–0.06 m   most of the year
Dwarf, succulent, lithophytic or terrestrial **herb** with hairy young growth; **stems** spreading to erect, often rooting at nodes; **leaves** to about 3 cm × 2.5 cm, ovate-elliptic to obovate, mainly opposite or rarely in whorls, spreading to ascending, base rounded or tapering to short stalk, thick, deep green, glabrous to hairy, 5-nerved with outer 2 somewhat indistinct, apex blunt to pointed; **spikes** to about 13 cm × 0.2 cm, terminal, slender, on short stalks; **flowers** very small; **drupes** to about 0.1 cm long, sticky.

A wide-ranging variety which is distributed in rainforest from the east coast of Cape York, Qld, to central NSW. Also occurs in Africa, India, Melanesia and

Peperomia blanda var. *floribunda*   W. R. Elliot

Malesia. Needs shaded and moist conditions. Plants should be grown in a pot of coarse acidic mixture, or in rock crevices where there is plenty of humus. Best suited to tropical and subtropical regions but can succeed in cooler regions if grown in protected sites, glasshouses or conservatories. Propagate from seed, cuttings, or by division of rooted stems.
P. *leptostachya* Hook. & Arn. is a synonym.

Peperomia enervis C. DC. & F. Muell.
(without nerves)
Qld
0.1–0.3 m × 0.1–0.2 m   most of the year
Dwarf, succulent, epiphytic or lithophytic **herbs** with glabrous young growth; **stems** spreading to erect, rooting at nodes, glabrous, reddish-brown; **leaves** to about 1.5 cm × 0.7 cm, obovate to cuneate, opposite or in whorls of 3, spreading to ascending, not crowded, thick, deep green, glossy, glabrous, nerves faint, apex blunt; **spikes** to about 6 cm × 0.2 cm, terminal, usually solitary, sometimes paired, slender, on short stalks; **flowers** very small, slightly sunken; **drupes** about 0.1 cm long, sticky.

An inhabitant of the wet tropics of northern Qld where it grows in locations which are permanently moist. Rarely cultivated, it deserves greater attention in tropical and subtropical areas. *P. enervis* is suited to growing with ferns and in terraria. Needs excellent drainage. Propagate from fresh seed, cuttings, or by divisions of rooting stems.
P. *johnsonii* C. DC. is a synonym.

Peperomia johnsonii C. DC. =
P. *enervis* C. DC. & F. Muell.

Peperomia leptostachya Hook. & Arn. =
P. *blanda* (Jacq.) Kunth. var. *floribunda* (Miq.) H. Huber

Peperomia urvilleana   D. L. Jones

Peperomia tetraphylla (G. Forst.) Hook. & Arn.
(four-leaved)
Qld, NSW
0.05–0.15 m × 0.1–0.2 m   most of the year

Dwarf, succulent, epiphytic **herb** with glabrous or faintly hairy young growth; **stems** spreading to ascending, rooting at nodes, glabrous or faintly hairy at nodes; **leaves** to about 1.4 cm × 0.9 cm, broadly ovate to nearly circular, in whorls of 4, thick, spreading to ascending, can be moderately crowded, deep green, glabrous above, initially faintly hairy below, midrib prominent, other venation obscure, apex blunt; **spikes** to about 4.5 cm × 0.3 cm, erect, terminal, solitary, slender, on short stalks; **flowers** very small, deeply sunken; **drupes** about 0.1 cm long, sticky.

Within Australia this species occurs from northeastern Qld to north-eastern NSW and Lord Howe Island. It is mainly an epiphyte in rainforest. Also occurs in Africa, Malesia and Melanesia. It is cultivated to a limited degree and requires well-drained humus-rich soils. Does well in a pot of coarse mix or in crevices of rocks where there is plenty of humus. Best suited to tropical and subtropical regions but could be grown in terraria or glasshouses in cooler regions. Propagate from fresh seed, cuttings, or by division of rooted stems.

P. affinis Domin is a synonym.

Peperomia urvilleana A. Rich.
(after Captain J. S. C. Dumont d'Urville, 18–19th-century naval explorer)
NSW (Lord Howe Island)   Two-leaved Peperomia
0.2–0.4 m × 0.1–0.2 m   most of the year

Dwarf, succulent, epiphytic or terrestrial **herb** with glabrous young growth; **stems** spreading to ascending, rooting at nodes, glabrous; **leaves** to about 1.5 cm × 1 cm, elliptic, alternate, spreading to ascending, rarely

crowded, mid-to-deep green, shiny, glabrous, thick, 3-nerved, apex blunt to rounded; **spikes** 2–6 cm long, usually terminal, solitary; **flowers** tiny; **drupes** to 0.1 cm long, ovoid, brownish.

This species occurs in shady sites of wet forests where it is found on tree trunks and old stumps as well as on mossy rocks. It also occurs on Norfolk Island and on a number of Pacific Islands. *P. urvilleana* is rarely cultivated and should do best in subtropical and tropical regions but may be successful in glasshouses or terraria in cooler climates. Needs good drainage but constant moisture. Appreciates plenty of humus. Propagate from seed, cuttings, or by divisions of rooted stems.

PEPEROMIACEAE (Miq.) Wettst.
A family of dicotyledons consisting of about 1000 species in 4 genera. There are about 6 species of *Peperomia* native to Australia. They are usually epiphytic herbs with succulent stems and opposite, alternate or whorled leaves. Flowers are tiny and arranged densely in a fleshy spike. Many species of *Peperomia* have decorative foliage and are grown for indoor decoration or in glasshouses and conservatories. This family is included in Piperaceae by some authors.

PEPLIDIUM Delile
(similar to the genus *Peplis*)
Scrophulariaceae

Perennial or annual, aquatic or semi-aquatic **herbs**; **stems** prostrate, rooting at nodes; **leaves** opposite, able to float on water, stalked, margins entire; **flowers** small, solitary or raceme-like, in axils of leaf-like bracts; **calyx** tubular, 5-toothed; **corolla** tubular, 2-lipped, upper lip 2-lobed, lower lip 3-lobed; **stamens** 2 or 4; **fruit** a capsule.

A genus of about 10 species with a distribution from northern Africa to Australia. There are 8 species in Australia and 7 of these are endemic. They inhabit eph-emeral pools, swamps, claypans or swale. They lack strong ornamental qualities but are important as colonisers of shallow aquatic areas and have potential for use on edges of pools and dams within their natural range.

Propagate from seed, cuttings, or by division of layered stems.

Peplidium humifusum Delile =
                    P. maritimum (L. f.) Asch.

Peplidium foecundum W. R. Barker
(fruitful or bountiful)
Qld, NSW, SA, WA, NT
prostrate × 0.3–1 m   Aug–Nov; also sporadic

Prostrate annual **herb**; **stems** creeping, rooting at nodes; **leaves** usually 0.3–1 cm × 0.2–0.5 cm, ovate to spathulate, opposite, short-stalked, glabrous, green, apex blunt; **calyx** to 0.3 cm long; **flowers** tubular, to about 0.4 cm long, pale blue to white; **stamens** 2; **capsule** to about 0.5 cm long.

This species is found in arid and semi-arid regions where it grows in or on the margins of ephemeral pools that occur in loam or clay soils. Needs similar

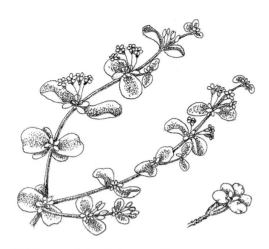

Peplidium muelleri × .6, flower × 2.5

conditions to those of its natural occurrence. Propagate from seed, cuttings, or by division of rooted stems.

Peplidium maritimum (L. f.) Asch.
(growing by the sea)
Qld, WA, NT
prostrate × 0.2–0.6 m   March–April;
                                        possibly sporadic
    Prostrate aquatic or semi-aquatic annual **herb**; **stems** creeping, rooting at nodes; **leaves** to about 1.5 cm × 0.9 cm, elliptic, glabrous, short-stalked, green, apex bluntly pointed; **calyx** tubular, about 0.2 cm long; **flowers** very small, white or pink, 1 per axil (in Australia); **stamens** 2; **capsule** to 0.25 cm × 0.25 cm, partially enclosed by calyx.
    This species is a dweller of saline mud flats and pools, and often grows in rocky areas. It is widely distributed along the northern coast and extends to Malesia, India and north-eastern Africa. Suitable for tropical and subtropical regions. Unlikely to become very popular but may have application for use in aquaria. Propagate from seed, cuttings, or by division of rooted stems.
    P. humifusum Delile is a synonym.

Peplidium muelleri Benth.
(after Baron Sir Ferdinand von Mueller, 19th-century Government Botanist of Victoria)
SA, WA, NT
prostrate × 0.3–1 m   Aug–Jan; also sporadic
    Prostrate perennial **herb** with a woody rootstock; young growth hairy; **stems** creeping, rooting at nodes; **leaves** to about 2 cm × 1.5 cm, obovate to circular, short-stalked, becoming glabrous, apex rounded to blunt; **calyx** to about 0.35 cm long; **flowers** tubular, barely longer than calyx, white with yellow, in racemes of axillary clusters, often profuse but rarely highly conspicuous; **capsule** enclosed by calyx.
    An inhabitant of arid and semi-arid regions where it grows in sandy clay or clay soils on margins of pools and waterways. It has limited application for cultivation but is worth trying in aquaria or clay-based

ornamental pools and on edges of dams within its natural range. Tolerates saline conditions. Propagate from seed, cuttings, or by division of rooted stems.

PERICALYMMA (Endl.) Endl.
(from the Greek *peri*, near, around; *calymma*, veil)
Myrtaceae
    A monotypic genus endemic in south-western WA.

Pericalymma crassipes Lehm. ex Schauer =
                                        P. ellipticum (Endl.) Schauer

Pericalymma ellipticum (Endl.) Schauer
(elliptical)
WA   Swamp Tea-tree
0.6–1.5 m × 0.5–1.5 m   Sept–Dec; also sporadic
    Dwarf to small **shrub** with glabrous young growth; **branches** flexuose, glabrous; **branchlets** slender, short; **leaves** 0.3–2 cm × up to 0.5 cm, narrowly obovate to obovate, alternate, short-stalked, concave to folded, green to grey-green, glabrous, margins entire, apex pointed and recurved; **flowers** 5-petalled, to about 1.2 cm across, white to pink with a pink to reddish centre, terminal, solitary or paired, scattered to profuse, very conspicuous; **calyx** to 0.5 cm × 0.35 cm, woody, 3-valved.
    A widespread species of the south-western region where it grows in a range of soils which are usually subject to periodic winter inundation. Inhabits coastal and near coastal areas. Responds well to cultivation in temperate regions and adapts to a range of acidic soils

Pericalymma ellipticum × .55, seeds × 2

which have moisture-holding capacity. A sunny site is best. Plants are hardy to moderately heavy frosts. They respond well to pruning and can be coppiced. Generally they do not require fertilising but applications of slow-release fertilisers can be beneficial in promoting quicker growth. Propagate from seed, which does not need pre-sowing treatment, and usually begins to germinate within 15–25 days after sowing. Cuttings of firm young growth form roots fairly readily. Application of rooting hormone is usually beneficial.

Leptospermum crassipes Lehm., *L. ellipticum* Endl, *L. floridum* (Schauer) Benth., *Pericalymma crassipes* Lehm. ex Schauer and *P. floridum* Schauer are all synonyms.

PERIPENTADENIA L. S. Sm.
(from the Greek *peri*, near, around; *pente*, five; *aden*, gland)
Elaeocarpaceae

Trees; **branchlets** mainly glabrous; **leaves** simple, alternate, elliptic to oblong, glabrous, margins toothed or crenate; **inflorescence** cluster of flowers or solitary flower, axillary or along old wood; **flowers** bisexual, 4- or 5-merous, large, showy; **calyx** white, cream or green; **sepals** free; **petals** free, lobed, white or cream; **stamens** 50–75; **fruit** globular, green or red, with 1 or rarely 2 seeds.

An endemic genus of 2 species confined to north-eastern Qld. Both species have qualities which make them desirable for cultivation.

Propagate from fresh untreated seed. Cuttings of firm young growth are slow to strike and application of rooting hormones is beneficial.

Peripentadenia mearsii (C. T. White) L. S. Sm.
(after J. E. Mears, original collector)
Qld   Buff Quandong; Boonjie
8–12 m × 5–6 m   Aug–Oct

Small **tree** with a bushy canopy; **trunks** sometimes multiple; **leaves** to about 20 cm × 6 cm, lanceolate to elliptical, short-stalked, spreading to ascending, glossy, deep green, glabrous, margins usually coarsely toothed, venation prominent, apex pointed; **flowers** about 1.8 cm across, white, ribbed in bud, solitary, axillary, on slender pendent stalks, scattered to profuse, conspicuous; **petals** 3-lobed, with lobes 2–3-toothed; **sepals** glabrous exterior, white, cream or green; **fruit** to about 3 cm × 2 cm, initially green, ripening brown; **seeds** red, edible.

A handsome bushy species from the Atherton Tableland where it grows in rainforest or on its margins. Uncommon in cultivation, it has potential for greater application in tropical, subtropical and temperate regions. Plants will need freely draining acidic soils and prefer to be sheltered from hot sunshine and winds. Should respond well to mulching and supplementary watering during extended dry periods. They are suitable for containers and are often attractive. The edible seed is reported as sweet and tasty and it may be worth selecting plants for seed production. Seeds are also eaten by possums and birds. Propagate from fresh seed, which is difficult to germinate. Cuttings of firm young growth are slow to strike.

Actephila mearsii C. T. White is a synonym.

Peripentadenia phelpsii B. Hyland & Coode
(after Roy Phelps, 20th-century Australian botanical collector)
Qld
8–25 m × 4–8 m   July–Sept

Small to medium **tree** with a bushy canopy; **trunk** usually solitary; **bark** greyish; **branchlets** slender, glabrous; **stipules** minute, deciduous; **juvenile leaves** some with 2–3 small lobes near base; **mature leaves** 6–15 cm × 2–3.2 cm, oblong to elliptic, with prominent stalk, spreading to ascending, deep green, glabrous, leathery, margins faintly toothed to wavy, venation prominent, apex bluntly pointed; **flowers** to about 1.8 cm across, white, with slender stalks to about 2 cm long, in axillary clusters of 2–5, can be profuse; **petals** finely fringed; **sepals** faintly hairy exterior, becoming reddish-brown; **fruit** to about 4.5 cm × 3.5 cm, greenish, somewhat globular; **seeds** reddish.

A recently described (1982) species which is restricted to the Mossman River catchment area of north-eastern Qld where it grows in rainforest at 40–400 m altitude. Rarely cultivated, it will need sheltered sites in acidic, freely draining soils. Plants will do best in tropical and subtropical regions because they suffer frost damage. Mulching and supplementary watering during extended dry periods should be beneficial. Propagate from fresh seed, which has a very low viability and sporadic germination, but can begin to germinate about 20 days after sowing. Cuttings of firm young growth are slow to strike.

PERIPTERYGIUM Hassk.
(from the Greek *peri*, around, near; and the genus *Pterygium*; both genera have similar winged fruit)
Cardiopteridaceae

Climbers with milky sap; **leaves** simple, entire, alternate, on long petioles; **inflorescence** a panicle; **flowers** small, 4 or 5-merous; **corolla** deciduous; **stamens** 4 or 5; **fruit** compressed, broadly winged.

A small genus of 3 species with 1 extending to Australia. Although interesting, the species is too vigorous for general cultivation. Propagate from seed, which has a limited period of viability and for best results should be sown fresh. Cuttings of firm young growth would be worth trying.

Peripterygium moluccanum (Blume) Sleumer
(from the Moluccas)
Qld
5–10 m tall   Sept–Nov

Robust **climber** with twining stems to 10 cm or more thick; **bark** grey-brown, corky; **sap** milky; **leaves** 8–15 cm × 1–12 cm, cordate, dark green with prominent lighter veins, on slender petioles to 10 cm long; **panicles** axillary, longer than the leaves, arching to pendulous; **flowers** about 0.5 cm across, greenish-white; **fruit** 2–2.5 cm × 1–1.5 cm, flat, conspicuously winged, ripe Nov.

Occurs in Indonesia, New Guinea and in north-eastern Qld where it grows in lowland rainforest in the Cook District. A vigorous climber with ornamental leaves and interesting fruit, but one which must have plenty of room to grow. Most suitable for parks and large gardens in tropical and subtropical regions.

Grows in sun or semi-shade in well-drained soil. Propagate from fresh seed and perhaps also by cuttings of firm young growth.

PERISTERANTHUS T. Hunt
(from the Greek, *peristera*, pigeon; *anthos*, flower; referring to an apparent resemblance of the flower to a pigeon)
Orchidaceae
A monotypic genus which is endemic in Australia.

Peristeranthus hillii (F. Muell.) T. Hunt
(after Walter Hill, original collector)
Qld, NSW
Epiphyte   Sept–Oct
Epiphytic orchid forming semi-pendulous clumps; **stems** 10–30 cm long, branched near the base, turning up near the apex; **leaves** 3–10, 15–25 cm × 3–4 cm, oblong, spreading to pendulous, light green, numerous veins prominent; **racemes** 15–25 cm long, pendulous; **flowers** about 0.7 cm across, numerous, downward-facing, pale green with brown markings and crimson spots, fragrant; **dorsal sepal** incurved; **lateral sepals** divergent; **petals** erect; **labellum** hinged, with a prominent, finger-like, central column.

A distinctive species distributed from north-eastern Qld to north-eastern NSW. Grows on trees in rainforest and is often prominent in littoral rainforests but also extends to the ranges and tablelands. Readily recognised by its drooping, prominently veined leaves. Plants are easily grown if given warm, humid conditions with plenty of air movement. They are best attached to a slab of cork, weathered hardwood or treefern, and must be hung away from direct sun. Propagate from seed sown under sterile conditions.
Saccolabium hillii F. Muell. is a synonym.

PERISTROPHE Nees
(from the Greek *peri*, near, around; *strophi*, turn, twist; referring to the twisted corolla tube)
Acanthaceae
Herbs; **leaves** opposite, sessile or stalked; **inflorescence** 2–3-flowered, on long slender stalks, flowers subtended by 2 slender often unequal bracts; **corolla** 2-lipped, upper lip entire, lower lip 3-lobed; **stamens** 2, exserted; **fruit** a capsule.

Peristrophe consists of 15–20 species which are mainly distributed in Africa, India, Asia and Malesia. Two species extend to Australia.

There has been discussion about whether the Australian member should be included in *Peristrophe*.

Peristrophe brassii R. M. Barker
(after Leonard J. Brass, 20th-century American botanist)
Qld
0.3–1 m × 0.2–0.6 m   June–Aug
Dwarf **herb**, sometimes rooting at lower nodes; **stems** more or less 6-angled, green, with white striations, more or less glabrous; **leaves** 3.3–7.5 cm × 0.2–1.5 cm, linear-lanceolate, opposite, jointed at base, more or less sessile, green, paler below, both surfaces have lime-like concretions; **corolla** tubular, about 1 cm long with spreading lobes to 1 cm long, white with

pink, purple or red spots in throat, exterior and interior hairy; **capsule** to about 1 cm long, densely hairy.

P. brassii is restricted to Cape York Peninsula where it grows in woodland and rainforest and often on sandy soils. Evidently not cultivated, it is suited to tropical and subtropical regions. May need protection from hot sun and winds. Drainage needs to be excellent. It has potential for massed planting as well as in containers. May be useful for quick cover while slower-growing plants develop. Propagate from seed or possibly from cuttings.

PERISTYLUS Blume
(from the Greek *peri*, around, *stylos*, column; referring to the overlapping dorsal sepal and petals)
Orchidaceae
Terrestrial orchids; **tubers** ovoid; **leaves** in a loose rosette at the base or near the centre of the stem; **inflorescence** a terminal raceme or spike; **flowers** white or green, often fragrant; **dorsal sepal** and **petals** overlapping to form a loose hood; **lateral sepals** often spreading widely; **labellum** 3-lobed nearly to the base, with basal spur, the lobes filiform or triangular; **fruit** a capsule.

A genus of about 65 species distributed across China and India through South-East Asia to New Guinea and Australia where there are 3 species (1 endemic). One species grows in rainforest and the others occur in moist grassy areas. The growth cycle is closely linked with the wet season. These orchids are difficult subjects to grow and they are mainly found in the collections of enthusiasts. They have very similar requirements to species of *Habenaria*, the cultivation of which is detailed in Volume 5, page 179. Propagation of *Peristylus* species is from seed sown symbiotically in flasks. Some success may be achieved by sprinkling seed around parent plants.

Peristylus banfieldii (F. M. Bailey) Lavarack
(after E. J. Banfield, original collector)
Qld
0.3–0.5 m tall   Jan–March
Terrestrial orchid; **leaves** 8–12.5 cm × 6–8 cm, ovate-lanceolate, 4–6 per plant scattered on basal half of stem, dark green, dull, margins wavy; **flower-stem** 30–50 cm tall, slender, bearing 15–50 flowers in a loose raceme; **flowers** 1–1.2 cm across, cream to pale yellow; **dorsal sepal** strongly hooded; **lateral sepals** spreading; **petals** spreading or incurved; **labellum** 3-lobed, lateral lobes longest, midlobe triangular; **spur** about 0.1 cm long.

Originally collected on Dunk Island and now also known from near Cardwell and Mareeba in north-eastern Qld. It grows in moist areas of open forest and near streams. A rare species which has been cultivated on a very limited scale by enthusiasts. It has similar requirements to *P. candidus*. Propagate from seed or by natural increase. Previously known as *Habenaria banfieldii* F. M. Bailey.

Peristylus candidus J. J. Sm.
(white)
Qld
0.1–0.35 m tall   March–April

Terrestrial orchid; **leaves** 6–8 cm × 2–2.5 cm, ovate, 2–3 per plant in a flat basal rosette, dull green to slightly bluish, margins entire; **flower-stem** 20–35 cm tall, slender, bearing 8–15 flowers in a loose raceme; **flowers** 0.6–0.7 cm across, white, tubular in the basal half, upper half of sepals and petals spreading; **dorsal sepal** strongly hooded; **petals** almost triangular; **labellum** 3-lobed, lateral lobes longest, nearly triangular; **spur** about 0.15 cm long.

A widespread species which extends to north-eastern Qld where it is distributed between Cooktown and Proserpine. It forms loose colonies in moist grassy areas in open forest and woodland. Plants are grown on a limited scale by enthusiasts and their requirements are as outlined in the genus introduction. Propagate from seed or by natural increase.

Peristylus papuanus (Kranzlin) J. J. Sm.
(from Papua New Guinea)

Qld	Green Habenaria
0.3–0.6 m tall	May–July

Terrestrial orchid; **leaves** 10–17 cm × 2–3 cm, narrowly ovate, dark green, somewhat fleshy, 4–7 per plant in a loose rosette near the centre of the stem; **flower-stem** 30–60 cm tall, very thin and wiry, bearing 6–36 flowers in a loose raceme; **flowers** about 1 cm across, green; **dorsal sepal** ovate, hooded; **lateral sepals** widely divergent; **petals** blunt or notched; **labellum** deeply 3-lobed, lateral lobes longest, often curled; **spur** about 0.8 cm long.

Plants occur in New Guinea and north-eastern Qld. They grow in lowland rainforest. It is a very difficult species to locate in the wild. Very rare in cultivation and can be difficult to maintain. Requires a terrestrial orchid mixture with plenty of decaying leaf mould added. Propagate from seed or by natural increase.

PERNETTYA Gaudich. = *GAULTHERIA* L.
Species previously included under *Pernettya* can be found in the Supplement to the Encyclopaedia.

PERROTTETIA Kunth
(after George Samuel Perrottet, 18–19th-century French botanical explorer)
Celastraceae

Shrubs or small **trees**; **leaves** alternate, entire or toothed; **stipules** present; **inflorescence** cymes in axillary thyrses; **flowers** mainly bisexual; **sepals** 4 or 5 (rarely 6–8); **petals** 5 (rarely 4); **stamens** 5 (rarely up to 8); **fruit** a berry with 1–4 seeds.

A widely represented genus of 25–30 species which occurs in Central America, Hawaiian Islands, Solomon Islands, Malesia and China. A solitary species occurs in Australia.

Perrottetia alpestris (Blume) Loes. =
P. arborescens (F. Muell.) Loes.

Perrottetia arborescens (F. Muell.) Loes.
(tree-like)

Qld	
3–5 m × 3–5 m	Aug–Feb

Medium to tall, tree-like **shrub** with faintly hairy young growth; **branches** spreading to ascending;

Persicaria barbata   W. R. Elliot

branchlets glabrous; **leaves** 4–12 cm × 2.5–5 cm, lanceolate to ovate, alternate, distichous, tapering to base, becoming glabrous, deep green, margins toothed or crenate, venation reticulate, apex softly pointed; **inflorescence** to about 3 cm long, in upper axils; **flowers** very small, cream, sometimes profuse, rarely very conspicuous; **petals** faintly hairy; **fruit** to about 1.2 cm diameter, red, usually 1-seeded.

P. arborescens is found in north-eastern Qld where it occurs in moist upland and mountain rainforest. This location indicates that it should adapt to subtropical and temperate regions. Plants need moist but freely draining acidic soils and appreciate protection from drying winds when young. The use of organic mulch should be beneficial and plants may need supplementary watering during extended dry periods. Propagate from fresh seed or possibly from cuttings.

PERSICARIA (L.) Mill.
(ancient name for a knotweed; referring to the leaves which appear similar to those of *Persica*, a peach)
Polygonaceae   Smartweeds; Knotweeds

Annual or perennial **herbs**; **stems** erect or decumbent, with tubular brown sheaths (ocrea); **leaves** simple, usually entire, stalked to nearly sessile; **inflorescence** spike-like, terminal or axillary, often panicle-like arrangement; **flowers** usually bisexual; **perianth segments** petal-like, 4–5, pink, white or sometimes green; **stamens** 4–8; **fruit** a nut.

A cosmopolitan genus of about 150 species; 15 species occur in Australia. The genus is represented in all states. They are moisture-lovers and most species inhabit moist depressions, swamps and lagoons or their margins, creek banks and other damp sites. They usually tolerate prolonged inundation and waterlogging.

In recent years, the renewed interest in conservation of wetlands — and recognition that wetlands provide wildlife habitat — has led to a better understanding of the importance of indigenous species. It is recommended that planting of *Persicaria* species should only occur within their natural range as some do have weedy properties.

Some species have attractive reddish to purple leaf coloration and profuse pinkish flowerheads, eg. *P. decipiens* and *P. orientalis*. These plants can add an extra dimension of interest in or beside water areas.

Some exotic species are cultivated for their food value with the introduced *P. odorata*, Vietnamese Mint, amongst the most popular. People who have sensitive skin may find that skin contact with species of *Persicaria* will cause dermatitis.

Propagation is from fresh seed, cuttings, or division of clumps or layered stems.

Persicaria attenuata (R. Br.) Sajak
(drawn out)
Qld, NSW, SA, WA, NT
0.6–1 m × 0.6–1.5 m   June–Dec; possibly sporadic

Aquatic or terrestrial perennial **herb**; **stems** spreading to erect, glabrous or with long hairs; **sheaths** reddish-brown, to 3 cm long; **leaves** 8–20 cm × 1.5–5 cm, narrowly ovate, somewhat falcate, stalked, hairy above and below, apex pointed; **spikes** to about 11 cm × 1 cm; **flowers** to about 0.4 cm long, white, moderately profuse; **nut** small, dark brown to blackish.

A widespread species which also occurs in New Guinea and Timor. It inhabits margins of swamps and waterways. Useful for planting around pools, dams and lake margins. Propagate from seed or from cuttings.

Previously known as *Polygonum attenuatum* R. Br.

Persicaria barbata (L.) H. Hara
(bearded)
Qld, WA, NT
0.5–1.2 m × 0.6–2 m   May–July

Aquatic or terrestrial perennial **herb**; **stems** erect to creeping, often floating, glabrous to hairy; **sheaths** to 2 cm long, pale, translucent; **leaves** 7–20 cm × 1.5–3 cm, narrowly elliptic to narrowly ovate, short-stalked, green, hairy lower midrib and margins, apex pointed; **panicle** to about 10 cm long, terminal; **flowers** to about 0.2 cm long, white, greenish-white or pink, moderately profuse; **nut** small, glossy black.

This species occurs from Africa to Asia and in the Solomon Islands. In Australia it is confined to northern areas where it is usually found growing in or beside waterways. Best suited to tropical and subtropical regions. Useful for wildlife habitat on edges of pools, dams and lakes. Propagate from seed, or by division of clumps with roots on lower nodes.

Previously known as *Polygonum barbatum* L.

Persicaria decipiens (R. Br.) K. L. Wilson
(misleading)
All states   Slender Knotweed
0.2–0.5 m × 0.3–1 m   May–July

Aquatic or terrestrial, annual or perennial **herb**; **stems** creeping to erect, rooting at lower nodes; **sheaths** to about 2 cm long, brown, translucent; **leaves** 5–12 cm × 0.5–1.5 cm, lanceolate to narrowly elliptic, short-stalked, green or often with reddish or pinkish tonings, usually with purplish blotch near middle, margins and veins hairy, apex pointed; **spikes** to about

6 cm × 0.4 cm, slender; **flowers** to about 3.5 cm long, pinkish, often profuse and conspicuous; **nut** small, dark brown to black.

A common, widespread species which occurs in a range of damp to wet situations. Very useful for wildlife habitat and may be decorative especially when plants are in flower or when they have reddish to purplish tinged leaves. Ideal for edges of pools, dams and lakes. Propagate from seed, cuttings, or division of stems with roots at nodes.

Previously known as *Polygonum decipiens* R. Br. and *Polygonum salicifolium* Brouss. ex Willd.

The following species are allied to *P. decipiens* but are doubtfully indigenous to Australia.

P. hydropiper (L.) Spach, Water Pepper, grows to about 1.2 m tall. It is found in Qld, NSW, Vic and WA. It is common in the northern hemisphere and is possibly not native to Australia. The leaves of 5–12 cm × 1–2.5 cm lack the purplish blotches of *P. decipiens* and the flower-spikes are loosely arranged. Previously known as *Polygonum hydropiper* L.

P. lapathifolia (L.) Gray, Pale Knotweed, can reach 2 m in height. It is possibly an introduction to Australia and is widespread in all states. Stems may be reddish and the leaves which are 6–21 cm × 1–4.5 cm do not have purple blotches. Creamy pink flowers are arranged in tight drooping heads. Previously known as *Polygonum lapathifolia* L.

Persicaria dichotoma (Blume) Masam. —
see *P. praetermissa* (Hook. f.) H. Hara

Persicaria hydropiper (L.) Spach —
see *P. decipiens* (R. Br.) K. L. Wilson

Persicaria lapathifolia (L.) Gray —
see *P. decipiens* (R. Br.) K. L. Wilson

Persicaria orientalis (L.) Spach
(eastern)
Qld, NSW, NT   Prince's Plume; Prince's Feathers
1–2 m × 1–2 m   May–Aug

Aquatic or terrestrial, annual or perennial **herb** with densely hairy young growth; **stems** erect, densely hairy; **sheaths** hairy; **leaves** 6–25 cm × 3.5–12 cm, ovate, prominently stalked, green, hairy above and below, apex pointed; **spikes** 2–8 cm long, drooping, in terminal panicle; **flowers** to 0.4 cm long, pink, rarely deep pink to reddish; **nut** small, brown to black.

A far-ranging species which extends to Asia. It inhabits moist depressions and swamp margins. Adapts very well to cultivation in most regions. Propagate from seed or from cuttings.

Persicaria praetermissa (Hook. f.) H. Hara
(neglected, overlooked)
NSW, Vic
0.8–1 m × 0.6–2 m   May–July; also sporadic

Aquatic or terrestrial perennial **herb** with hairy young growth; **stems** creeping to ascending, slender, hairy; **sheaths** hairy; **leaves** 2.5–8 cm × 0.6–2 cm, hastate, hairy on veins and margins, green, sometimes with purplish blotches, apex pointed; **inflorescence** with 1–6 small scattered clusters on 2–3 slender stems;

flowers to about 0.4 cm long, pink or white; **nut** small, mid-brown.

This species also extends to Asia. It is a moderately densely-foliaged plant which usually occurs on edges of swamps, lagoons and in creek beds. It is valuable for stabilising margins of pools, dams and lakes and is useful for wildlife habitat. Propagate from seed or cuttings.

Previously known as *Polygonum praetermissum* Hook. f.

The closely allied *P. strigosa* (R. Br.) H. Gross has broader leaves and the flowers are borne on up to 4 branches. It has dark reddish-brown nuts. It occurs in Qld, NSW, Malesia and Asia and grows in similar conditions. Previously known as *Polygonum strigosum* R. Br.

P. dichotoma (Blume) Masam. from north-eastern NSW and Qld has a small inflorescence and white flowers but differs markedly by having ovate leaves.

Persicaria strigosa (R. Br.) H. Gross —
    see *P. praetermissa* (Hook. f.) H. Hara

PERSOONIA Sm.
(after Christian Hendrik Persoon, 18–19th-century botanist who specialised in fungi)
Proteaceae   Geebungs, Snotty Gobbles

Dwarf to tall **shrubs** or small **trees**; **bark** smooth and hard to somewhat papery and soft; **leaves** mainly alternate, sometimes opposite or subopposite, rarely in whorls, simple, entire, glabrous to hairy; **inflorescence** usually somewhat leafy terminal or axillary racemes; **flowers** mainly yellow, tubular, with 4 spreading to reflexed tepals, exterior glabrous or hairy; **fruit** a 1- or 2-celled drupe, glabrous or hairy.

Persoonia comprises about 90 species and is an extremely important endemic member of the *Proteaceae* family. About one-tenth of the Australian members of *Proteaceae* are persoonias. Recent botanical studies have clarified the relationship between most species and has resulted in naming and describing many new species. There are still a small number of unnamed species to be described. *Persoonia* is distributed in all states with the main concentration in temperate regions of south-western WA, and eastern areas from south-eastern Qld to eastern Vic. Only two species, *P. falcata* and *P. tropica*, occur in tropical regions. Persoonias are distributed from near sea-level to the higher mountains and are generally more common in heathland and dry sclerophyll forests. They are also found in mallee shrubland and on the margins of rainforest. In most cases, species occur in freely draining acidic soils such as sand, sandy loam or well-drained clay loams which may overlie laterite.

Growth habit varies from prostrate plants with ground-hugging stems and foliage to erect-trunked tall shrubs or small trees, but the majority of species are bushy, spreading to erect, small to medium shrubs.

A number of species develop lignotubers from which many stems develop. In nature branching of the stems can occur beneath the soil level and it will be interesting to observe if this occurs in cultivation.

Generally persoonias have hard greyish to blackish bark but a highlight of a limited number of species is their loose and somewhat papery to flaky bark which has a blackish exterior and an often beautiful coppery-red inner area, eg. *P. levis, P. linearis, P. longifolia, P. moscalii* and *P. muelleri.*

The predominant feature of most persoonias is the foliage rather than the floral display. In fact, the genus offers a wide variation in foliage. Some have very small, pine-like leaves which can be glabrous or hairy while others are much broader and longer, eg. *P. levis.* Some of the broad-leaved species are extremely decorative when backlit by sunshine early or late in the day.

A number of species have attractive flushes of new growth with rusty to reddish tones, eg. *P. arborea* and *P. myrtilloides.* Some species such as *P. pinifolia* have pendent branchlets. Plants of *P. longifolia* produce long, virtually straight stems and these are harvested for the floricultural industry and marketed as Snotty Gobble, a name which refers to the succulent fruits. Cut foliage of most species can be a useful addition as a filler in floral arrangements. *P. linearis, P. pinifolia* and *P. oxycoccoides* respond very well to treatment with glycerine which preserves the foliage in a pleasing, blackish-brown, pliable form.

Variation of flower size and colour from species to species is not very marked but the manner in which they are displayed varies. Most species have yellow flowers and a minority produce cream flowers, eg. *P. gunnii*, or orange flowers, which are produced by a selection of *P. chamaepitys.* Often the flowers are somewhat scattered and arranged in leafy racemes. The more spectacular members of the genus such as *P. acerosa, P. pinifolia* and *P. tenuifolia* have very prominently displayed terminal and apparently leafless racemes. *P. pinifolia* is extremely long-flowering and often begins to flower around New Year's Day and may still be flowering at the end of August.

Olive-like fruits are an ornamental feature of some species. In many cases they are purely green, but a number of species produce fruits which are green with purplish stripes or streaks, eg. *P. media, P. rufa* and *P. terminalis.* Other species such as *P. pinifolia, P. rigida* and *P. virgata* produce fruits that usually gain strong purplish tonings as they reach maturity.

In nature it is not uncommon for hybridising to occur when there are a number of species in close proximity. Among those known for producing hybrid offspring in the wild are *P. acerosa, P. chamaepeuce, P. confertiflora, P. cornifolia, P. levis, P. myrtilloides* and *P. tenuifolia.* Some of these hybrids are cultivated and have ornamental characteristics, eg. *P. cornifolia* × *P. tenuifolia.*

Some *Persoonia* species such as *P. nutans* and *P. rudis* are regarded as rare or endangered in nature. *P. leucopogon* was until recently thought to be extinct but has since been relocated in the wild.

The raw fruit of tropical *P. falcata* was once a valuable food source for Aboriginals. Timber was used in the making of tools such as axe handles, boomerangs, spear-throwers and music sticks. There were a number of medicinal uses and these included infusions of bark and inner wood shavings for treatment of eye disease and other mixtures were used for treatment of chest congestion, colds, sore throats and diarrhoea.

Some species have bark which is highly sought after for using in bark pictures. *P. falcata, P. levis, P. linearis* and *P. silvatica* have flaky bark and the inner parts have marvellous tones of purple, red or brown (see Volume 2, page 232).

Cultivation

Persoonias are not common in cultivation yet and their potential has not been fully tapped. Mainly they are grown for amenity planting and only a very limited number of species are used. The small number of species on offer for cultivation is due to difficulties in propagation and that aspect is covered in more detail under Propagation below.

The first species introduced into cultivation, *P. lanceolata* from NSW, was propagated and grown in England in 1791. Shortly after in 1794, *P. linearis* was also cultivated. By the 1850s over 20 species had been introduced to English horticulture. During that period plants were mainly cultivated in a mixture of equal parts of loam, sand and peat.

In general, persoonias require well-drained acidic soils in a sunny or semi-shaded site. Good results have been achieved in sand, sandy loam, loam and clay loam. Most species dislike extended wet periods but they generally tolerate extended dry periods successfully.

Cold tolerance varies from species to species but those which occur naturally at higher altitudes, such as *P. oxycoccoides*, *P. gunnii* and *P. muelleri*, are capable of withstanding very heavy frosts and snowfalls. Most species are rarely damaged by moderate frosts (down to −4°C).

Many persoonias have a dislike for long periods of high humidity and some groundcovering species with crowded narrow leaves can sweat. This causes blackening and foliage drop.

Persoonias usually grow very well without supplementary fertilising. They can react poorly to excess nutrients and most species have a low resistance to excessive phosphorus. Phosphorus toxicity is usually very difficult to rectify. Light application of a low-phosphorus, slow-release fertiliser is frequently beneficial for nursery production and when planting in gardens. Symptoms of lack of iron and magnesium can be corrected with applications of iron chelates, iron sulphate or magnesium sulphate respectively. See Volume 1, page 95, for further details on plant nutrition. As with most plants, excessive nitrogen promotes foliage growth at the expense of root growth and plants may become top-heavy. Judicious pruning of foliage can often help to overcome instability of plants.

Persoonias rarely require supplementary water once they are well established unless plants are grown in semi-arid and arid regions or there is an extended dry period. Watering during periods of high temperatures can lead to the development of root fungal diseases (see Pests and Diseases below).

Pruning is a practice to which most persoonias respond very well. Most species have the capacity to reshoot from bare leafless wood. Species which have lignotubers, such as *P. longifolia*, can be coppiced regularly (ie. every 2–3 years). Some of the taller persoonias such as *P. linearis* and *P. pinifolia* are excellent as long-lived hedging plants which respond well to regular clipping.

Persoonia pinifolia, fruits   W. R. Elliot

Pests and Diseases

Predators are not usually a problem with persoonias. Now and again caterpillars such as the Grevillea Looper or the Double-headed Hawk Moth may cause minimal damage and leaf miners may be a minor problem. For control methods see Volume 1, page 145. The larvae of the Macadamia Twig Girdler may attack young shoots, see Volume 1, page 144. Galls can become apparent on stems and leaves of some species. They are usually difficult to control (see Volume 1, page 154). Fungal leaf spots are known to occur on some of the broader-leaved species (see Volume 1, page 173, for control methods). Some persoonias are susceptible to Cinnamon Fungus, *Phytophthora cinnamomi*, and its presence can be exacerbated by watering during summer. It is extremely difficult to control but recent research has shown that phosphonate can alleviate some of the symptoms although it does not kill the fungus. Also see Volume 1, page 168 for further information on fungal control.

Propagation

Most persoonias are difficult to propagate by conventional methods. It is rare to have success with germination of seed in nursery conditions, yet it can be common to have many seedlings emerge from beneath well-established plants of species such as *P. pinifolia*. Research into germination techniques for *Persoonia* is continuing. Sometimes identifying persoonia seedlings can be confusing because they have linear cotyledons similar to those of conifers and are usually 3–7 in number. If seed is sown it may not germinate for 4 or more years. Therefore it is best to sow seed in long-lasting containers.

Many of the persoonias offered in nurseries have been propagated from cuttings. The success rate can vary greatly. It has been shown that best results are gained from cuttings taken from young vigorous plants which are yet to produce flowers. As a generalisation, it is best to use the current season's soft young growth for cuttings. This growth can be promoted by hard pruning at the beginning or middle of spring. For most species it is best not to remove the soft tips of cuttings even if they are wilting. Removal of tips can be quickly followed by blackening of foliage and the cuttings rarely recover to produce vigorous plants.

Cutting trials have been undertaken using different strengths of rooting hormone in power and liquid form. These provided little evidence of any one mixture being significantly better than the others. Good results have been achieved using a combination powder mixture of 4000 ppm IBA/2000 ppm NAA, but differing results can be expected in different areas.

Persoonia acerosa Sieber ex Schult. & Schult. f.
(needle-shaped)
NSW
0.5–2 m × 0.5–1.5 m   Feb–May

Dwarf to small **shrub** with glabrous or faintly hairy young growth; **branches** spreading to erect; **branchlets** becoming glabrous; **leaves** 1.2–2.3 cm × about 0.05 cm, linear-terete, curved inwards, channelled above, ascending to erect, crowded, becoming glabrous, mid green, apex finely pointed; **flowers** to about 0.8 cm across, pale yellow, axillary, forming leafy spikes near ends of branchlets, often profuse and conspicuous; **drupes** to about 1 cm × 0.8 cm, green.

A handsome, spreading to erect species from the Central Coast and Central Tablelands. It is regarded as vulnerable. Plants grow in heath or open forest on sandstone. Cultivated to a limited degree, it requires well-drained acidic soils and a sunny or semi-shaded site. Tolerates heavy frosts and dry periods. Responds well to pruning and clipping. Warrants being better known in cultivation. Introduced into cultivation in England during 1824 as *P. pallida*. Propagate from seed or from cuttings of very young growth.

P. acerosa is known to hybridise in nature with *P. levis* and *P. myrtilloides*, and possibly with *P. linearis*. Some of these hybrids may be in cultivation.

Persoonia acicularis F. Muell.
(needle-like)
WA
0.2–1 m × 0.2–0.6 m   Aug–Jan

Dwarf stiff **shrub** with a lignotuber and stolon-like rhizomes; young growth hairy; **branches** ascending to erect; **branchlets** initially angular, becoming terete, hairy; **leaves** 1–2 cm × about 0.1 cm, linear-subulate, alternate, spreading to ascending, crowded, stiff, twisted, glaucous, initially hairy, becoming glabrous, 2-grooved, apex pungent-pointed; **flowers** to about 0.8 cm across, yellow, solitary, terminal or in upper axils, on slender glabrous stalks, scattered to profuse and conspicuous; **drupes** to about 1 cm × 0.5 cm, green, smooth.

An inhabitant of sandplains in the Irwin District. Evidently not cultivated, it should adapt to freely draining acidic, or slightly alkaline, sandy or loam soils. Plants will need plenty of sunshine. Being prickly, it

may not be very popular but it could help control foot traffic in selected areas. Propagate from seed or from cuttings of very young growth.

P. sulcata is sometimes confused with this species but it has shorter and broader, non-glaucous leaves.

Persoonia acuminata L. A. S. Johnson & P. H. Weston
(narrowing to a point)
NSW
0.3–1.7 m × 1–2.5 m   Dec–April

Dwarf to small, spreading **shrub** with hairy young growth; **branches** prostrate to ascending; **branchlets** hairy; **leaves** 0.8–2.2 cm × up to 0.1 cm, elliptic, ovate to narrowly oblong, alternate to nearly opposite, spreading to ascending, sessile, moderately crowded, green, generally hairy but less so with maturity, flat, apex pointed; **flowers** to about 0.8 cm across, yellow, glabrous, short-stalked, spreading to erect along ends of branchlets, moderately profuse and fairly conspicuous; **drupes** green.

This recently described (1991) species has a disjunct distribution. It occurs in the Barrington Tops in the north-east and in the Blue Mountains. Plants are found in montane heath and wet sclerophyll forest often in rocky sites. Should adapt to a range of freely draining acidic soils in temperate regions in sunny or semi-shaded sites. Plants tolerate most frosts. Propagate from seed or from cuttings of very young growth.

It has been raised to species status but was previously included in the *P. nutans* complex.

P. oxycoccoides var. *longifolia* Benth. is a synonym.

Persoonia adenantha Domin
(glandular flowers)
Qld, NSW
2.5–6 m × 2–4 m   most of the year

Medium to tall **shrub** with very hairy young growth; **branches** ascending to erect; **branchlets** hairy; **leaves** 3–14 cm × 0.6–3 cm, lanceolate to narrowly elliptic, spreading to ascending, green, initially hairy becoming glabrous, flat with recurved margins, apex pointed; **flowers** to about 0.7 cm across, yellow, hairy, usually solitary in upper axils, subtended by leaves or scale leaves, can be profuse and moderately conspicuous; **drupes** to about 1.5 cm × 1 cm, greenish-purple, becoming glabrous.

An interesting species from south-eastern Qld and the North Coast of NSW. Occurs in a range of habitats from heathland to wet sclerophyll forest. Rarely cultivated, it has potential for use in subtropical and temperate regions. Needs excellently drained acidic soils in sites with a sunny or semi-shaded aspect. Plants could be used as formal or informal hedging. Propagate from seed or from cuttings of very young growth.

Previously known as *P. cornifolia* ssp. B. *P. cornifolia* is closely allied but it differs in having broadly elliptic to ovate leaves and flowers which are mainly subtended by scale leaves.

Persoonia amaliae Domin
(after Amalia Dietrich, early botanical collector in Qld)
Qld
2–8 m × 1.5–5 m   Jan–July

Medium to tall **shrub** with moderately hairy young growth; **bark** grooved at base, otherwise smooth, grey; **branches** spreading to ascending; **branchlets** pendent, hairy; **leaves** 3–8 cm × 0.6–2 cm, lanceolate to narrowly elliptic or spathulate, spreading to ascending, sometimes incurved, green above and below, initially hairy becoming glabrous, margins recurved, apex softly pointed; **flowers** to about 1.3 cm long, yellow, exterior faintly to moderately hairy; **drupes** to about 1.8 cm × 1.2 cm, green, smooth.

This species occurs along the coast and on the hinterland of eastern Qld, in the Leichhardt, Moreton, Port Curtis, South Kennedy and Wide Bay Districts. It is often associated with granite and grows in well-drained soils. Plants are rare in cultivation and best suited to subtropical regions. May also succeed in temperate areas. Freely draining acidic soils and a sunny or semi-shaded site should be suitable. Plants are hardy to light frosts. Propagate from seed, or from barely firm young growth, which may be slow to form roots.

Persoonia angulata R. Br.
This name is no longer recognised. Botanical research has shown that it was attributed to the hybrids *P. acerosa* × *levis* and *P. acerosa* × *myrtilloides*.

Persoonia angustiflora Benth.
(narrow flowers)
WA
0.2–2 m × 0.3–1.5 m   Sept–Nov; also sporadic
Dwarf to small **shrub** with a lignotuber; young growth hairy; **branches** ascending to erect; **branchlets** initially angular, usually becoming terete, hairy; **leaves** 0.8–13 cm × about 0.1 cm, linear, alternate, flattish to nearly terete, ascending to erect, green, initially hairy, usually becoming glabrous, apex ending in a soft point; **flowers** to about 0.8 cm across, greenish-yellow, exterior silky hairy, on hairy slender stalks to 1.2 cm long, subtended by leaf scales or leaves, terminal or axillary, often profuse and conspicuous; **drupes** to about 1 cm × 0.35 cm, green, smooth.

This species occurs in the Avon and Irwin Districts and inhabits sandy or loam soils which often overlie laterite in heathland, mallee woodland, banksia woodland or tall woodland. Evidently rarely cultivated, it should be best suited to semi-arid and warm temperate regions. Needs freely draining acidic soils with a sunny or semi-shaded site. Hardy to moderate frosts. Plants were first cultivated in England in 1837 under the misapplied name of *P. fraseri*. Propagate from seed or from cuttings of very young growth.

The var. *burracoppinensis* D. A. Herb. is regarded as not worthy of varietal status, and the var. *pedicellaris* Benth. is a synonym of *P. hexagona* P. H. Weston.

Persoonia arborea F. Muell.
(tree-like)
Vic   Tree Geebung
4–9 m × 2–5 m   Nov–Feb
Tall **shrub** to small **tree** with reddish or whitish hairs on young growth; **branches** spreading to ascending; **branchlets** initially reddish-brown, hairy; **leaves** 3–10.5 cm × 1–3 cm, oblong-lanceolate, alternate, scattered, spreading to ascending, dark green above, paler below, hairy, flattish, apex blunt but ending in a small point; **flowers** to about 1 cm across, pale yellow, silky-hairy, on short stalks, solitary in upper axils, scattered to moderately profuse; **drupes** to about 1 cm × 0.8 cm, green to yellowish.

This is one of the tallest geebungs. It occurs in moist to wet forests in the Central Highlands, north-east of Melbourne. Cultivated to a limited extent, it requires moist but well-drained acidic loams or clay loams for best results. Tolerates sunshine but a semi-shaded site is best. Hardy to heavy frosts. Regular tip-pruning of young plants promotes bushier growth, otherwise it will remain fairly narrow with an open habit. Good for growing amongst taller trees and shrubs. Aborigines used the wood, which is very hard, for spear shafts and tools. Propagate from seed or from cuttings of reddish-brown young growth.

Persoonia articulata R. Br. = *P. longifolia* R. Br.

Persoonia asperula L. A. S. Johnson & P. H. Weston
(slightly rough with small bristles)
NSW, Vic
0.2–2 m × 0.6–2 m   Jan–Feb; also sporadic
Dwarf to small **shrub** with hairy young growth; **branches** prostrate to ascending; **branchlets** hairy; **leaves** 0.3–2.2 cm × 0.1–0.6 cm, narrowly oblong to ovate, alternate to somewhat opposite, crowded, slightly rough and hairy, green, sometimes paler below, flat with recurved margins, apex pointed; **flowers** to about 0.8 cm across, yellow, glabrous to hairy, on short, erect to spreading stalks, solitary, subtended by leaves or leaf-scales, moderately profuse and conspicuous; **drupes** to 1 cm × 0.7 cm, green with purple stripes.

A variable and attractive species from the Southern Tablelands of NSW and north-eastern Vic. Grows in heathland and wet sclerophyll forest on shallow, often stony soils. It is cultivated to a very limited extent and deserves to be better known. Plants should adapt to a range of freely draining acidic soils in sunny or semi-shaded sites. In temperate regions they are hardy to most frosts. Should respond well to pruning. Propagate from seed or from cuttings of very young growth.

The NSW populations have been known as *P. nutans* ssp. G, while the Victorian populations were sometimes placed in *P. oxycoccoides*.

Persoonia attenuata R. Br. = *P. media* R. Br.

Persoonia baeckeoides P. H. Weston
(similar to the genus *Baeckea*)
WA
0.5–1 m × 0.5–1.3 m   Nov–Dec
Dwarf **shrub** with greyish to pale brown hairs on young growth; **stems** many from the base; **branches** ascending to erect; **branchlets** hairy becoming glabrous, angular becoming terete; **leaves** 0.3–1.1 cm × 0.2–0.4 cm, spathulate, alternate, ascending to erect, crowded, twisted at base, usually faintly glaucous, glabrous, flat, apex blunt; **flowers** to about 0.8 cm across, greenish-yellow, glabrous exterior, on very short stalks,

usually spreading to erect, solitary, terminal or in upper axils, sometimes profuse; **drupes** to about 0.8 cm × 0.6 cm, smooth.

This poorly known species from the Roe District was described in 1994. It occurs in yellow sandy loam overlying laterite in heathland. Needs to be cultivated as part of a conservation strategy. Plants should do well in semi-arid and warm temperate regions. Well-drained acidic soils and a sunny site should be suitable. Plants are hardy to moderate frosts and extended dry periods. Propagate from seed or from very young growth.

Persoonia bargoensis P. H. Weston & L. A. S. Johnson
(from the Bargo region)
NSW
0.6–2.5 m × 0.5–2 m   Dec–Jan
Dwarf to medium **shrub** with hairy young growth; **branches** ascending to erect; **branchlets** hairy; **leaves** 0.8–2.5 cm × up to 0.3 cm, linear-lanceolate to lanceolate, alternate, spreading to ascending, moderately crowded, green, initially glabrous, or hairy becoming glabrous, smooth, flat or convex, margins recurved, apex finely pointed; **flowers** about 0.8 cm across, yellow, on spreading to recurved, short, glabrous stalks, terminal or in upper axils, often moderately profuse and conspicuous; **drupes** green.

A recently described (1991) species which is known from the Bargo area, north-east of Wollongong, in the Central Coast District. It grows on sandstone and shale in woodland and dry sclerophyll forest. Not well known in cultivation but it should adapt to most well-drained acidic soils in temperate regions. A semi-shaded site is preferred to a sunny one. Hardy to most frosts and extended dry periods. Responds well to pruning. Propagate from seed or from cuttings of very young growth.

Persoonia biglandulosa P. H. Weston
(with two glands)
WA
0.2–1.5 m × 0.6–2.5 m   Oct–Dec
Dwarf to small **shrub** with greyish to brown hairs on young growth; **stems** many, branching from base or underground; **branches** spreading to ascending; **branchlets** initially angular and densely hairy, becoming terete and glabrous; **leaves** 2–10 cm × about 0.1 cm, linear, nearly terete with groove below, alternate, ascending to erect, incurved, often crowded, flexible, green, becoming glabrous, apex pointed; **flowers** about 0.7 cm across, bright yellow, exterior hairy, on densely hairy stalks to about 1 cm long, in terminal racemes about 2–11 cm long, can be profuse and conspicuous; **drupes** to about 1.5 cm × 0.8 cm, smooth.

This species is regarded as rare in nature but not threatened. It was described in 1994 and can be found in low heathland of the central Irwin District where it grows in sandy soils that often overlie laterite. Plants should be successful in semi-arid and warm temperate areas and may adapt to cooler regions. Very well-drained acidic soils in a sunny site are appropriate. Plants withstand extended dry periods and tolerate moderate frosts. Low-growing selections may have potential as groundcovers. Propagate from seed or from cuttings of very young growth.

Previously known as *P. teretifolia* R. Br. var. *amblyanthera* Benth.

Persoonia bowgada P. H. Weston
(after Bowgada scrub, a plant community)
WA
1–3.5 m × 1–3 m   Oct–Nov
Small to medium **shrub** with greyish to pale brown hairs on young growth; **bark** smooth or rarely fissured, grey; **stems** many, branching from near base; **branches** spreading to ascending; **branchlets** faintly angular becoming terete, hairy becoming glabrous; **leaves** 2.5–11 cm × about 0.15 cm, linear, alternate, ascending to erect, often crowded at ends of branchlets, green, initially hairy becoming glabrous, apex pungent-pointed; **flowers** to about 0.8 cm across, yellow, exterior hairy, on short, densely hairy stalks, terminal and in upper axils, subtended by scale-leaves, reduced leaves or leaves, in racemes to about 4 cm long, sometimes profuse; **drupes** to about 1.5 cm × 0.7 cm, smooth.

A recently described (1994) species which inhabits scrub, mallee-heath and woodland in the Irwin and Carnarvon Districts, and occurs in sand and loam which can contain laterite gravel. Apparently not cultivated, it is best suited to semi-arid and warm temperate regions. Acidic soils with excellent drainage and plenty of sunshine are likely to be prerequisites for successful cultivation. Hardy to moderate frosts. Propagate from seed or from cuttings of very young growth.

The allied *P. hexagona* differs in having grooved leaves, and *P. angustiflora* has softly pointed leaves.

Persoonia brachystylis F. Muell.
(short style)
WA
1–2 m × 1–2 m   Nov–Jan
Small **shrub** with greyish to brownish hairs on young growth; **stems** branching from base or from underground; **branches** spreading to erect; **branchlets** terete, initially densely hairy, becoming glabrous; **leaves** 3.5–12 cm × 0.2–1 cm, narrowly spathulate, alternate, spreading to ascending, crowded near ends of branchlets, green, flat to convex, midrib prominent, margins recurved to revolute, apex pointed; **flowers** to about 0.8 cm across, bright yellow, exterior hairy, on hairy stalks to 1.5 cm long, mainly subtended by leaves, in racemes to 25 cm long, can be profuse and conspicuous; **drupes** to about 1.3 cm × 0.8 cm, smooth.

This interesting and ornamental species is considered to be vulnerable in nature. It has a known geographic range of less than 100 km in the northern Irwin District. Grows in yellow sand overlying laterite. Needs to be cultivated as part of a conservation strategy. Plants should do best in semi-arid and warm temperate regions. They will need excellently-drained acidic soils in a warm to hot, sunny or semi-shaded site. Propagate from seed or cuttings of very young growth.

P. biglandulosa, *P. comata* and *P. stricta* have similarities to *P. brachystylis* but they differ as flowers are usually not subtended by mature leaves.

Persoonia chamaepeuce   W. R. Elliot

Persoonia brevifolia (Benth.) L. A. S. Johnson & P. H. Weston
(short leaves)
NSW, Vic
0.8–1.7 m × 0.6–2 m   Dec–March

Dwarf to small **shrub** with hairy young growth; **branches** spreading to ascending; **branchlets** hairy, often initially reddish; **leaves** 1–2.5 cm × 0.3–1.2 cm, elliptic to obovate, alternate, spreading to ascending, moderately crowded, deep green, often paler below, becoming glabrous, flat, margins slightly recurved, apex pointed; **flowers** to about 0.8 cm across, yellow, exterior glabrous to hairy, on short, erect to recurved stalks, in upper axils, subtended by leaves, racemes to about 5 cm long, often moderately profuse and conspicuous; **drupes** to about 1 cm × 0.8 cm, green.

P. brevifolia is an attractive species which is known only from a small area in far south-eastern NSW and from eastern Vic. Plants grow in wet and dry sclerophyll forest, in well-drained soils derived from granite or sandstone. Rarely cultivated, it needs freely draining acidic soils in a semi-shaded site. Plants respond well to pruning and are hardy to moderate frosts. Propagate from seed or from cuttings of very young growth.

Previously known as *P. myrtilloides* var. *brevifolia* Benth.

Persoonia brevirhachis P. H. Weston
(short rhachis)
WA
0.3–2 m × 0.3–2 m   Aug–Oct

Dwarf to small **shrub** with hairy young growth; **stems** solitary or several to many, branching from near base; **branches** spreading to ascending; **branchlets** hairy becoming glabrous, angular becoming terete; **leaves**

0.8–5 cm × 0.25–0.6 cm, narrowly spathulate to oblanceolate, alternate, not twisted at base, ascending to erect, often crowded near ends of branchlets, becoming glabrous, often glaucous, stiff, margins recurved to revolute, apex ending in a soft point; **flowers** to about 0.8 cm across, yellow to greenish, exterior very hairy, on short hairy stalks, scattered; **drupes** to about 0.6 cm × 0.3 cm, smooth.

Occurs in the western Roe District in yellow sand overlying laterite in heathland or mallee-heathland. Evidently not cultivated, it lacks strong ornamental features. Plants will need very well-drained acidic soil in a sunny or semi-shaded aspect. They are hardy to moderate frosts. Propagate from seed or from very young growth.

This species has often been misidentified as *P. scabra*, which has narrowly oblong to sometimes narrowly spathulate leaves twisted at their base.

Persoonia caleyi R. Br. = *P. mollis* R. Br. ssp. *caleyi* (R. Br.) S. L. Krauss & L. A. S. Johnson

Persoonia chamaepeuce Lhotsky ex Meisn.
(dwarf pine)
NSW, Vic
0.1–0.2 m × 1–2 m   Dec–March

Dwarf spreading **shrub** with faintly hairy young growth; **branches** prostrate to ascending; **branchlets** faintly hairy to glabrous; **leaves** 0.8–2.5 cm × up to 0.2 cm, linear, alternate, spreading to ascending, fairly crowded, deep green, usually glabrous, concave above, apex pointed; **flowers** about 0.7 cm across, yellow, exterior glabrous or slightly hairy, on short stalks near ends of branchlets, sparse to rarely moderately profuse; **drupes** small, green.

A lovely groundcover which occurs in mountainous regions from New England to eastern Vic. Grows in a range of freely draining soils in woodland and sclerophyll forest. A reliable species in temperate regions which adapts to most well-drained acidic soils. A sunny or semi-shaded site is suitable. Sometimes it may be slow to become established. Plants respond very well to pruning and are hardy to most frosts. Suitable for general planting, rockeries and containers. Propagate from seed or from cuttings of very young growth.

P. chamaepitys is sometimes confused with this species, but it has very crowded, more or less terete leaves.

Persoonia chamaepitys A. Cunn.
(dwarf pine or fir)
NSW   Prostrate Geebung; Mountain Geebung
0.1–0.3 m × 1–3 m   mainly Sept–March

Dwarf **shrub** with hairy light green young growth; **branches** prostrate; **branchlets** hairy; **leaves** 0.7–2 cm × up to 0.1 cm, linear, often terete, alternate, spreading to ascending, crowded, mid green, becoming faintly hairy to glabrous, apex pointed; **flowers** to about 0.8 cm across, deep yellow to orange-yellow, exterior hairy, on short hairy stalks subtended by scale-leaves or leaves, in short clusters or racemes, strongly fragrant, often profuse and conspicuous; **drupes** to about 1.2 cm × 1 cm, green.

Persoonia chamaepitys   W. R. Elliot

A very appealing groundcover from the Central Coast, Central Tablelands and Central Western Slopes. It inhabits sandy soils in heathland or dry sclerophyll forest. Plants perform well in well-drained, acidic, light to medium soils in a sunny or semi-shaded site. They tolerate extended dry periods and are hardy to most frosts. Excessive rain and humid overcast weather may result in sweating and foliage blackening. First grown in England in 1824. A very attractive selection from Rylstone has orange-yellow flowers and very hairy branchlets, and it has proved very successful in cultivation. Propagate from seed or from cuttings of very young growth.

Persoonia chapmaniana P. H. Weston
(after Charles Chapman, 20th-century botanical collector, WA)
WA
1–2 m × 1–2 m   Sept–Nov
Small **shrub** with whitish to pale brown hairs on young growth; **branches** ascend from near ground level; **branchlets** terete, initially densely hairy, becoming glabrous; **leaves** 0.5–8 cm × to about 0.15 cm, inear, nearly terete, alternate, spreading to ascending, crowded, stiff, faintly 5-ribbed, green, becoming glabrous, smooth to faintly rough, apex pungent-pointed; **flowers** about 0.7 cm across, bright yellow, exterior glabrous, on very short hairy stalks, usually subtended by scale-leaves, in racemes to about 6 cm long, can be profuse and moderately conspicuous; **drupes** to about 1.3 cm × 0.6 cm, warty, green.

A recently described (1994) species from the Avon and Irwin Districts where it occurs in disjunct localities. Grows in sandy soils near salt lakes, in heath or woodland. Evidently not cultivated, it should be best suited to semi-arid and warm temperate areas but may adapt to cooler regions. Needs freely draining acidic soils and plenty of sunshine. Hardy to moderate frosts. Propagate from seed or from cuttings of very young growth.

Persoonia comata Meisn.
(tufted)
WA
0.2–1.5 m × 0.5–2 m   Sept–Feb
Dwarf to small **shrub** with a lignotuber and greyish-hairy young growth; **stems** 1 to many, branching from underground; **bark** smooth, grey; **branches** spreading to ascending; **branchlets** often initially angular becoming terete, hairy becoming more or less glabrous; **leaves** 2.5–15 cm × 0.25–1.75 cm, oblanceolate to narrow spathulate, alternate, ascending to erect, can be crowded, stiff to flexible, green to glaucous, more or less flat, margins recurved, apex softly pointed, blunt or notched; **flowers** to about 0.8 cm across, bright yellow, often with pink tonings, exterior hairy, prominently pouched, on hairy stalks to 2.5 cm long, subtended by scale-leaves or leaves, mainly in terminal racemes 2–45 cm long, sometimes profuse and very conspicuous; **drupes** to about 1.25 cm × 0.8 cm.

A handsome persoonia from the northern Darling and southern Irwin Districts where it grows in sandy soils overlying laterite, in dry sclerophyll forest, woodland, mallee-heaths or heathland. It is best suited to semi-arid and warm temperate regions. Acidic soils which drain freely and a sunny site are required. Plants are hardy to moderate frosts and extended dry periods. May have potential for cut-flower production as well as being suitable for general planting. Propagate from seed and from cuttings of very young growth.

The allied *P. saccata* has narrower leaves.

Persoonia confertiflora Benth.
(crowded flowers)
NSW, Vic   Cluster-flower Geebung
0.3–1.5 m × 0.5–1.5 m   Sept–Feb; also sporadic
Dwarf to small **shrub** with rusty-hairy young growth; **branches** spreading to ascending; **branchlets** hairy; **leaves** 3–9 cm × 1.3–3 cm, narrowly elliptic to ovate, short-stalked, alternate or opposite, spreading to ascending, green, initially hairy becoming more or less glabrous, smooth, flat, margins recurved, apex blunt to rounded; **flowers** about 0.7 cm across, yellow with rusty hairs, on very short, densely hairy, erect stalks, subtended by scale-leaves, clustered or in short axillary racemes, sometimes profuse and moderately conspicuous; **drupes** to about 1.5 cm × 1 cm, ovoid.

An admirable species with attractive rusty young growth. *P. confertiflora* occurs in the South Coast and Southern Tableland Districts of NSW and the Victorian Alps. It inhabits wet sclerophyll forest and woodland. Plants are cultivated to a limited extent and deserve to be better known. Should adapt to a range of well-drained acidic soils in temperate regions. Plants are hardy to most frosts and they withstand snowfalls. Tolerant of limited dry periods. Propagate from seed or from cuttings of very young growth.

P. confertiflora hybridises with *P. chamaepeuce* in

nature and at least one selection has been introduced to cultivation.

P. laurina is allied but has rough leaves which can be longer and broader than those of *P. confertiflora*.

Persoonia conjuncta L. A. S. Johnson & P. H. Weston
(joined together)
NSW
2–7 m × 1.5–4 m   Jan–Feb

Medium to tall **shrub** or small **tree**, young growth with greyish to tawny hairs; **bark** mainly smooth but finely fissured; **branches** spreading to ascending; **branchlets** initially hairy; **leaves** 6–14 cm × 1–2.6 cm, lanceolate to narrowly elliptic, alternate, spreading to erect, green, becoming glabrous, smooth, flat, margins recurved, apex pointed; **flowers** about 0.8 cm across, yellow, exterior hairy, on short hairy stalks, subtended by scale-leaves or leaves, in terminal or axillary racemes which sometimes reach 14 cm long, sometimes profuse, moderately conspicuous; **drupes** green.

A recently described (1991) species which occurs to the north of Port Macquarie over a limited range. Grows in wet to dry sclerophyll forest. Evidently not cultivated, it should succeed in temperate regions in well-drained, acidic, sandy or loam soils, in a semi-shaded or sunny site. May suffer damage from heavy frosts. Propagate from seed or from cuttings of very young growth.

It is known to hybridise with *P. linearis* in the wild. The allied *P. iogyna* differs in having smooth, non-fissured bark.

Persoonia cordifolia P. H. Weston
(heart-shaped leaves)
WA
1–2 m × 1–2 m   Dec–Jan

Small **shrub** with greyish-hairy young growth; **stems** many, branching from near base; **branches** spreading to ascending; **branchlets** terete, becoming glabrous; **leaves** 0.7–1.2 cm × 0.6–1.5 cm, broadly cordate, opposite, decussate, not twisted, scattered, spreading, stiff, green, flat or faintly concave, becoming glabrous, apex pointed; **flowers** about 0.7 cm across, bright yellow, exterior faintly hairy, erect to pendent on short faintly hairy stalks, in short terminal racemes; **drupes** not known.

This recently described (1994) and poorly known species occurs in a very restricted region of the Roe District. It inhabits heathland in sand or sandy loam. Worthy of cultivation as part of a conservation strategy, as well as for its distinctive foliage. Plants are suited to temperate and semi-arid regions and require excellently drained acidic soils in a sunny or semi-shaded site. Hardy to light frosts. Propagate from seed or from cuttings of very young growth.

Persoonia coriacea Audas & P. Morris
(leathery)
WA   Leathery-leaved Persoonia
0.5–2 m × 0.4–2 m   Nov–Feb

Dwarf to small **shrub** with a lignotuber; young growth greyish-hairy; **stems** branching from base or slightly underground; **branches** ascending to erect; **branchlets** initially angular to flattened, becoming

terete, hairy becoming glabrous; **leaves** 1–6.5 cm × 0.3–1.3 cm, spathulate to obovate or elliptical, alternate, twisted at base, spreading to erect, somewhat crowded, leathery, usually glaucous, becoming glabrous, prominent margin venation, apex pointed; **flowers** to about 0.7 cm across, bright yellow, exterior hairy, on ascending to pendent, glabrous to hairy stalks to about 1 cm long, in axillary racemes to about 7 cm long, moderately conspicuous; **drupes** to about 1.5 cm × 0.8 cm, smooth.

An erect to spreading species, once thought to be endangered but now regarded as not under threat in the wild. It occurs over a moderately wide range in the south-western area of WA. Grows on sand or loam often overlying laterite in mallee-heath and heath. Probably best suited to semi-arid and warm temperate regions but may succeed in cooler areas. Should adapt to a variety of well-drained acidic soils in a sunny or semi-shaded site. Hardy to moderate frosts. Can sucker if roots are damaged. Propagate from seed or from cuttings of very young growth.

Persoonia cornifolia A. Cunn. ex R. Br.
(leaves similar to those of the genus *Cornus*)
Qld, NSW   Broad-leaved Geebung
2–7 m × 1.5–4 m   sporadic throughout the year

Medium to tall **shrub** or small **tree**, young growth densely greyish- to rusty-hairy; **bark** smooth; **branches** spreading to ascending; **branchlets** hairy; **leaves** 2–10 cm × 1–5.5 cm, elliptic to ovate or rarely orbicular, alternate, spreading to ascending, green, initially faintly greyish-hairy, becoming glabrous, smooth to slightly rough, flat but margins somewhat recurved, apex bluntly pointed; **flowers** to about 0.8 cm across, yellow, exterior greyish-hairy, on very short, densely hairy, erect stalks, in very short, cluster-like racemes; **drupes** 1–1.5 cm × 0.7–1 cm, purplish, becoming glabrous.

Occurs in woodland and dry sclerophyll forest in sandy or gravelly soils and is usually associated with granite. It is distributed in south-eastern Qld, and in NSW it is found in the Northern Tablelands and North Western Slopes. Plants are cultivated to a limited extent. Best suited to very well-drained acidic soils in temperate and subtropical regions. A sunny or semi-shaded site is required. Hardy to moderate frosts. Propagate from seed or from cuttings of very young growth.

It has also been known as *P. cornifolia* ssp. D.

Botanical study has shown that some other closely allied taxa, which have been known as subspecies A, B and C, are now recognised at species level.
P. cornifolia ssp. A = *P. stradbrokensis* Domin
P. cornifolia ssp. B = *P. adenantha* Domin
P. cornifolia ssp. C = *P. rufa* L. A. S. Johnson & P. H. Weston

Persoonia cunninghamii R. Br. = *P. myrtilloides* ssp. *cunninghamii* (R. Br.) L. A. S. Johnson & P. H. Weston

Persoonia curvifolia R. Br.
(curved leaves)
NSW
0.3–2 m × 1–3 m   Sept–Feb

Dwarf to small **shrub** with hairy young growth; **branches** spreading to ascending; **branchlets** hairy; **leaves** 1.5–4.5 cm × to about 0.1 cm, linear to oblong or linear-spathulate, strongly incurved, terete to convex, alternate, ascending to erect, somewhat crowded, becoming more or less glabrous, apex softly pointed; **flowers** to about 0.8 cm across, yellow, exterior usually hairy, on short, spreading to erect, hairy stalks, usually subtended by leaves, in short racemes; **drupes** to about 1.3 cm × 0.9 cm.

P. curvifolia is found in the Central Tablelands, North Western Slopes, Central Western Slopes and South Western Plains. It occurs in sandy and gravelly soil in woodland and open forest. Although rarely cultivated, plants have interesting foliage. It should adapt to a range of freely draining soils in temperate regions. A sunny or semi-shaded site is suitable. Hardy to most frosts and tolerates extended dry periods. Propagate from seed or from cuttings of very young growth.

The allied *P. chamaepitys* has usually straight, terete and more crowded leaves, while *P. cuspidifera* has broader leaves.

Persoonia cuspidifera L. A. S. Johnson & P. H. Weston
(bearing a sharp point)
NSW
0.3–2 m × 0.6–2 m   Nov–March

Dwarf to small **shrub** with greyish to tawny hairs on young growth; **bark** smooth; **branches** ascending to erect; **branchlets** hairy; **leaves** 1–2 cm × up to 0.5 cm, spathulate, alternate, spreading to erect, incurved, convex, green, initially hairy becoming glabrous, rough, apex blunt with a small point; **flowers** about 0.8 cm across, yellow with maroon, exterior hairy, on short, erect, hairy stalks, subtended by leaves, in axillary or terminal racemes to about 7 cm long, can be profuse and conspicuous; **drupes** small, green with purple stripes.

Persoonia daphnoides × .6

This attractive species from the North Western Slopes was described in 1991. It grows on sandstone in dry sclerophyll forest. Plants are poorly known in cultivation but should adapt to a range of well-drained acidic soils with a sunny or semi-shaded aspect in temperate regions. They tolerate heavy frosts and extended dry periods. May be worth trying as low hedging. Propagate from seed or from cuttings of very young growth.

P. curvifolia has some similarities but differs in having much narrower leaves.

Persoonia cymbifolia P. H. Weston
(boat-shaped leaves)
WA
0.2–0.8 m × 0.3–1.5 m   Dec–Jan

Dwarf **shrub** with greyish to brown hairs on young growth; **bark** smooth, mottled grey; **branches** spreading to ascending; **branchlets** terete, becoming glabrous; **leaves** 0.4–4.5 cm × up to 0.3 cm, linear to narrowly oblong, alternate, twisted at base, ascending to erect, crowded, straight to curved upwards, green, flat to concave, becoming glabrous, apex sharply pointed; **flowers** about 0.7 cm across, yellow, exterior hairy, on short hairy stalks, subtended by scale leaves, solitary or up to 3 flowers per axillary or terminal cluster; **drupes** to about 1.2 cm × 0.7 cm, smooth.

A recently described (1994) and very distinct species from the Roe District. It has a limited distribution but is not regarded as rare. Occurs in sandy soils or rock crevices in heathland. Probably will do best in semi-arid and warm temperate regions in freely draining acidic soils with a sunny or warm semi-shaded site. Plants withstand moderately heavy frosts and extended dry periods. Propagate from seed or from cuttings of very young growth.

Persoonia daphnoides A. Cunn. ex R. Br.
(similar to the genus *Daphne*)
Qld, NSW
0.1–0.3 m × 0.7–2.5 m   Dec–Feb

Dwarf **shrub** with hairy light green young growth; **branches** prostrate; **branchlets** initially densely hairy, becoming less so; **leaves** 1.5–6 cm × 0.4–2.5 cm, obovate to spathulate, alternate, short-stalked, mid green, becoming glabrous, flat, margins slightly recurved, apex blunt with a very small point; **flowers** about 0.7 cm across, yellow, exterior moderately hairy, on short hairy stalks, subtended by scale-leaves or leaves, in short racemes, sometimes profuse and moderately conspicuous; **drupes** to about 1.5 cm × 1 cm, glabrous.

An appealing groundcovering species which occurs in south-eastern Qld and the Northern Tablelands of NSW. Usually associated with granite in woodland and dry sclerophyll forest. It has had limited cultivation but is proving adaptable in temperate regions. Needs well-drained acidic soils with a sunny or semi-shaded aspect. Plants withstand heavy frosts and extended dry periods. Excellent for underplanting and in rockeries. Also worth trying in large containers. Propagate from seed or from cuttings of very young growth, which strike moderately well.

This species has been known as *P. prostrata* ssp. B.

Persoonia diadena

Persoonia falcata   T. L. Blake

The closely allied *P. procumbens* has sparsely hairy, slightly shorter flowers.

Persoonia diadena F. Muell. = *P. saundersiana* Kippist

Persoonia dillwynioides Meisn.
(similar to the genus *Dillwynia*)

WA	Fitzgerald Persoonia
0.6–2 m × 0.6–2 m	Nov–Dec

Dwarf to small **shrub** with faintly greyish-hairy young growth; **branches** many, ascending to erect; **branchlets** initially angular and hairy, becoming terete and glabrous; **leaves** 0.5–2 cm × up to 0.15 cm, linear, alternate, twisted at base, ascending to erect, crowded, green, becoming glabrous, deeply concave, apex sharply pointed; **flowers** about 0.8 cm across, bright yellow, exterior glabrous, on short glabrous stalks, subtended by scale-leaves or leaves, in 1–4-flowered terminal or sometimes axillary clusters; **drupes** to about 1.2 cm × 0.6 cm, smooth.

This rare but not threatened species occurs in the Roe District, in the Ravensthorpe–Hopetoun region and is found in a range of sandy, loam or clay soils which can be winter-wet. Evidently not cultivated, it is best suited to temperate and semi-arid regions. Plants will need acid soils which drain moderately well in a site which is sunny or semi-shaded. They tolerate moderate frosts. Propagate from seed or from cuttings of very young growth.

Persoonia drummondii Lindl. = *P. longifolia* R. Br.

Persoonia elliptica R. Br.
(elliptical)

WA	Spreading Snotty Gobble
3–8 m × 2–5 m	Oct–Feb

Medium to tall **shrub** or small **tree** with a lignotuber and greyish-hairy young growth; **trunk** usually solitary; **bark** corky, thick, grey; **branches** spreading to ascending; **branchlets** initially angular and hairy, becoming

terete and glabrous; **leaves** 1.5–11 cm × 1–5 cm, oblanceolate to spathulate, alternate, usually twisted at base, spreading to ascending, crowded near tips, green, glabrous, flat or margins sometimes recurved, apex blunt or pointed; **flowers** about 0.9 cm across, pale greenish-yellow, exterior glabrous to faintly hairy, on hairy stalks to 0.7 cm long, subtended by scale-leaves or leaves, in terminal or axillary racemes to 12 cm long, sometimes profuse and moderately conspicuous; **drupes** to about 1.5 cm × 0.8 cm, glabrous.

A sizeable species from the northern Darling District where it extends from Perth to Albany within about 50 km of the coast. It grows in sands, clay loam or lateritic soils in woodland and dry sclerophyll forest. Not commonly cultivated, it is suitable for under-planting of tall trees as it tolerates shady sites. Best suited to temperate regions, it needs acidic soils with good drainage, and will also tolerate some sunshine. Hardy to moderate frosts and withstands moderate dry periods. Introduced to cultivation in England in 1840. Propagate from seed or from cuttings of very young growth.

Persoonia falcata R. Br.
(sickle-shaped)

Qld, WA, NT	Wild Pear; Gandala
2–9 m × 1–5 m	June–Nov

Medium to tall **shrub** or small **tree** with a lignotuber; young growth greyish-hairy and often pink to reddish; **trunk** usually solitary; **bark** deeply grooved, flaky exterior, dark grey to black, interior reddish-purple; **branches** spreading to ascending; **branchlets** terete, initially hairy becoming glabrous; **leaves** 3–35 cm × 0.2–7 cm, linear to oblanceolate, alternate, twisted at base, spreading to ascending, flexible, green and usually glaucous, glabrous, leathery, usually flat, apex bluntly pointed; **flowers** to about 0.8 cm across, cream yellow to bright yellow, exterior glabrous to hairy, on glabrous to hairy stalks to 1.5 cm long, subtended by scale-leaves or leaves, solitary or in terminal or axillary racemes 3–20 cm long, often profuse and conspicuous; **drupes** to about 2 cm × 1 cm, glabrous, green-yellow when ripe.

A wide-ranging and variable tropical species which occurs commonly on sandy soils and also on lateritic and stony hillsides, but is rarely found on clay. Grows in a wide range of habitats. Rarely cultivated, it is suited to subtropical and seasonally dry tropical regions. Should adapt to most freely draining soils in a sunny or semi-shaded site. This species is important to Aboriginal people as the wood is used for making many tools such as boomerangs and spear-throwers, and for music sticks. Infusions of wood, bark shavings and leaves have been used for treating various ailments. Inner bark used for bark pictures. Propagate from seed or possibly from cuttings of very young growth.

Persoonia fastigiata R. Br.
(with upright branches)

NSW	
0.2–1 m × 0.6–1 m	Oct–Jan

Dwarf **shrub** with hairy young growth; **branches** ascending to erect; **branchlets** initially hairy, becoming glabrous; **leaves** 1.5–4 cm × up to 0.2 cm, linear to oblong or linear-spathulate, spreading and incurved, green to yellow-green, initially hairy becoming glabrous, apex slightly pungent-pointed; **flowers** about 0.7 cm across, yellow, exterior hairy, on ascending to recurved, short, hairy stalks, subtended by scale-leaves and leaves, in short racemes; **drupes** to about 0.9 cm × 0.7 cm.

A poorly known but interesting species from the New England Tableland where it grows on granitic soils in woodland or dry sclerophyll forest. Worth growing because of its erect growth habit and unusual leaves. Plants will need freely draining acidic soils with a sunny or semi-shaded aspect. They are hardy to most frosts and tolerate dry periods. Propagate from seed or from cuttings of very young growth.

This species has been confused with *P. curvifolia* which has incurved leaves that do not spread from the branches at their bases.

Persoonia ferruginea Smith =
P. laurina Pers. ssp. *laurina*

Persoonia filiformis P. H. Weston
(thread-like)
WA
0.1–0.5 m × 1–2 m   Nov–Dec

Dwarf **shrub** with lignotuber; young growth whitish- to greyish-hairy; **stems** many, branching from below ground level; **branches** prostrate to ascending; **branchlets** initially angular and hairy, becoming terete and glabrous; **leaves** 0.5–2 cm × about 0.1 cm, linear, alternate, usually not twisted, ascending to erect, crowded, green to glaucous, stiff, becoming glabrous, smooth to somewhat rough, apex pungently pointed; **flowers** about 0.8 cm across, greenish-yellow, exterior glabrous, on short glabrous stalks, subtended by scale-leaves or leaves, in axillary or terminal racemes to about 3 cm long; **drupes** not known.

This poorly known species was described in 1994. It has a prostrate growth habit which gives it potential for horticultural applications. Occurs in heathland of the southern Irwin District in sand which often overlies lateritic gravel. Should adapt to freely draining acidic soils in temperate and semi-arid regions. Needs plenty of sunshine. Plants tolerate extended dry periods and moderate frosts. Propagate from seed or from cuttings of very young growth.

Persoonia flexifolia R. Br.
(with bent leaves)
WA
not known   Dec–Jan

Shrub with whitish- to greyish-hairy young growth; **branches** ascending to erect; **branchlets** initially angular and hairy, becoming terete and glabrous; **leaves** 0.4–2.5 cm × up to 0.3 cm, usually narrowly oblong, alternate, usually twisted at base, ascending to erect, often crowded, green, becoming glabrous, flat to concave or convex, apex sharply pointed; **flowers** about 0.8 cm across, colour not known (most likely yellow), exterior glabrous or glabrous to faintly hairy, on short stalks, subtended by scale-leaves or leaves, in terminal or axillary clusters; **drupes** not known.

This is an extremely poorly known species from the

Persoonia gunnii   T. L. Blake

Eyre District. It has been collected only twice, in 1802 and 1979, and is recorded as growing on lateritic or stony soils as part of heath vegetation. Warrants cultivation as part of a conservation strategy. Plants are suitable for semi-arid and temperate regions and will need good drainage in acidic soils with a sunny aspect. Frost hardiness is not known but it should tolerate light to moderate frosts. Propagate from seed or from cuttings of very young growth.

Persoonia helix   W. R. Elliot

Persoonia glaucescens

Persoonia glaucescens Sieber ex Spreng.
(somewhat glaucous)
NSW
1–2.5 m × 0.8–2.5 m   Dec–Feb; possibly sporadic
Small to medium **shrub** with hairy glaucous young growth; **branches** spreading to erect; **branchlets** hairy, glaucous when young; **leaves** 3–8 cm × 0.5–2 cm, oblanceolate to narrowly spathulate, alternate, ascending, especially glaucous when young, becoming glabrous, flat, smooth, apex pointed; **flowers** to about 0.8 cm across, yellow, exterior hairy, on very short hairy stalks, in leafy racemes, moderately profuse; **drupes** to about 1 cm × 0.8 cm, green, smooth.

This species is regarded as vulnerable in its natural habitat. It occurs in woodland and dry sclerophyll forest on the Central Coast and Central Tablelands. Grows on sandstone. It is cultivated to a very limited extent and should adapt to temperate and possibly subtropical regions. Plants require well-drained acidic soils and a sunny or semi-shaded site is appropriate. They tolerate moderate frosts. Propagate from seed or from cuttings of very young growth.

Persoonia graminea R. Br.
(grass-like)
WA
0.3–0.6 m × 0.6–1.5 m   Oct–Jan
Dwarf spreading **shrub** without a lignotuber; young growth greyish-hairy; **branches** few to many, branching near base, spreading to erect; **branchlets** initially angular and hairy, slowly becoming terete and glabrous; **leaves** 4–35 cm × up to 0.1 cm, linear, alternate to somewhat whorled, sometimes twisted, spreading to erect, in often crowded clusters, can be slightly incurved, green, becoming glabrous, prominent midrib and marginal veins, apex softly pointed; **flowers** about 0.5 cm across, bright yellow to greenish, fragrant, exterior hairy, pendent, on glabrous to hairy stalks to about 0.6 cm long, in terminal racemes to about 22 cm long, can be profuse and conspicuous; **drupes** to about 0.8 cm × 0.4 cm, smooth.

This species occurs in the southern Darling District where it is confined to coastal and hinterland regions from Margaret River to Albany. It has a somewhat herbaceous growth habit and grows in sandy or loam soils which are often swampy for most of the year. Evidently not cultivated, it should be suitable for temperate regions in friable acidic soils with a sunny or semi-shaded site. In nature it can be straggly but pruning from an early age should promote bushy growth. Propagate from seed or from cuttings of very young growth.

Persoonia gunnii Hook. f.
(after Ronald C. Gunn, 19th-century Tasmanian botanical collector)
Tas
1–4 m × 1–3 m   Jan–May
Small to medium **shrub** with silky-hairy young growth; **bark** scaly, dark brown on old branches; **branches** spreading to ascending; **branchlets** initially hairy, usually becoming glabrous; **leaves** 1.5–3.2 cm × 0.3–1 cm, obovate to spathulate, alternate, spreading to erect, usually bent near base and incurved, crowded,

yellow-green to green, becoming glabrous, flat, apex rounded with small point; **flowers** to about 1.5 cm across, cream-yellow, sometimes with pinkish tonings, exterior glabrous to faintly hairy, on short hairy stalks, subtended by leaves, in short terminal racemes or clusters, sometimes profuse and moderately conspicuous; **drupes** to about 1 cm × 0.9 cm, purple-black, smooth.

The typical variety is confined to the highest parts of western, central and southern regions of Tas. It grows in open forest, moist shrubland and moist heath. Cultivated to a limited extent, it is slow-growing and requires freely draining, light to medium, acidic soils in a semi-shaded or slightly sunny site. Plants tolerate snowfalls and are hardy to heavy frosts.

The var. *oblanceolata* Orchard occurs in the central highlands but it is rare. It differs in having oblanceolate leaves of up to 5 cm × about 0.7 cm. Flowers are white and known to bloom in March–April. Needs similar conditions to the var. *gunnii*.

Propagation has been successful from seed or from cuttings of very young growth, which can be very slow to form roots.

The var. *angustifolia* Benth. is a synonym of *P. muelleri* (P. Parm.) Orchard var. *angustifolia* (Benth.) Orchard. The var. *alpina* Hook. f. is a synonym of *P. muelleri* (P. Parm.) Orchard var. *muelleri*.

Persoonia hakeiformis Meisn.
(like the genus *Hakea*)
WA
0.5–2 m × 0.5–2 m   Nov–Jan
Dwarf to small **shrub** with greyish-hairy young growth; **bark** greyish, smooth, flaky near base; **stems** many, branching near base; **branches** spreading to ascending; **branchlets** slightly angular and hairy, becoming terete and glabrous; **leaves** 0.5–5 cm × up to 0.15 cm, linear, alternate, somewhat terete, grooved below, not twisted, often curved, spreading to ascending, crowded, green, becoming glabrous, stiff, apex pointed; **flowers** about 0.8 cm across, bright yellow, prominently pouched, exterior glabrous, on densely hairy stalks to 0.7 cm long, subtended by scale-leaves or leaves, in mainly terminal racemes to about 10 cm long, can be profuse and very conspicuous; **drupes** to about 1.5 cm × 0.8 cm, smooth.

This very distinctive species is regarded as vulnerable in nature. It occurs in the southern Avon and far western Roe Districts. It inhabits heath, mallee-heath and woodland in sandy loams overlying laterite. Warrants cultivating as part of a conservation strategy, as well as for its beauty. Plants need freely draining acidic soils in a sunny or warm semi-shaded site. They tolerate moderate frosts and extended dry periods. Selected spreading, lower-growing variants may have appeal. Propagate from seed or from cuttings of very young growth.

Persoonia helix P. H. Weston
(spiral; referring to leaves)
WA
0.4–2.8 m × 0.5–1.5 m   Nov–Feb
Dwarf to medium **shrub** with a lignotuber; young growth greyish-hairy; **stems** many, branching from

below or just above ground level; **branches** spreading to ascending; **branchlets** initially angular and hairy, becoming terete and glabrous; **leaves** 0.6–6 cm × up to 0.4 cm, mainly linear and usually spirally twisted, alternate, ascending to erect, crowded, becoming glabrous, glaucous-green, somewhat stiff, apex pointed; **flowers** about 0.8 cm wide, bright yellow, exterior glabrous to faintly hairy, on glabrous or hairy stalks to about 0.7 cm long, subtended by scale-leaves or leaves, terminal or axillary, solitary or in small clusters; **drupes** to about 1 cm × 0.8 cm, smooth.

A fascinating species with intriguing corkscrew-like foliage. *P. helix* was described in 1994. It occurs in the Coolgardie, southern Eyre and Roe Districts, and grows in sand or sandy loam soils which often overlie lateritic gravel. Plants are very uncommon in cultivation. They require excellently drained acidic soils and plenty of sunshine but will tolerate some shade. Hardy to moderately heavy frosts and able to withstand extended dry periods. They make excellent container plants. Propagate from seed or from cuttings of very young growth, which are usually slow to form roots.

This species has sometimes been misidentified as *P. tortifolia*, which is a synonym of *P. trinervis*. *P. trinervis* differs in having usually less-twisted leaves which have 3 grooves on the undersurface.

Persoonia hexagona P. H. Weston

(six-angled, six-sided)
WA
1–3.5 m × 0.6–2.5 m   Nov–Dec

Small to medium **shrub** with greyish to pale brown hairs on young growth; **stems** many, branching near base; **branches** spreading to ascending; **branchlets** terete, becoming glabrous; **leaves** 2.5–13 cm × about 0.1 cm, linear, somewhat terete, alternate, mainly ascending to erect, crowded, often incurved, green, with 6 ridges, apex pungent-pointed; **flowers** about 0.7 cm across, bright yellow, exterior hairy, on hairy stalks to about 1 cm long, subtended by scale-leaves or leaves, in terminal axillary racemes to about 4 cm long; **drupes** to about 1.5 cm × 0.7 cm, smooth.

A recently described (1994) species from the far northern Avon and northern Irwin Districts. It grows on sand, sandy loam and stony hillsides. It will be best suited to semi-arid and warm temperate regions and needs freely draining acidic soils in a sunny or warm semi-shaded site. Plants tolerate moderate frosts and are hardy to extended dry periods. Propagate from seed or from cuttings of very young growth.

Previously known as *P. angustiflora* Benth. var. *pedicellaris* Benth.

Persoonia hirsuta Pers.

(long hairs)
NSW   Hairy Geebung
0.3–1.3 m × 0.6–2 m   Nov–Jan

Dwarf to small **shrub** with densely rusty-hairy young growth; **branches** many, spreading to ascending; **branchlets** with long and stiff hairs; **leaves** 0.5–1.5 cm × up to 0.15 cm, linear to narrowly oblong, alternate, ascending to erect, crowded, green, initially densely hairy becoming more or less glabrous, margins revolute, apex pointed; **flowers** about 0.8 cm across,

golden-yellow, exterior rusty-hairy, on spreading to erect, densely hairy, short stalks, subtended by scale-leaves or leaves, in short leafy racemes; **drupes** to about 1.4 cm × 0.7 cm, purplish-green, hairy.

The typical subspecies occurs in the Central Coast District in heath, woodland and dry sclerophyll forest. Cultivated to a limited extent, it should do best in freely draining acidic soils in temperate and possibly subtropical regions. A sunny or semi-shaded site should be appropriate. Plants are hardy to moderate frosts. Groundcovering selections should become popular.

The ssp. *evoluta* L. A. S. Johnson & P. H. Weston differs in having narrowly elliptic to spathulate leaves (to about 0.5 cm wide) with recurved margins. It occurs in the Central Coast and Central Tablelands Districts and is more tolerant of low temperatures than the typical subspecies.

Propagate both subspecies from seed or from cuttings of very young growth.

Persoonia inconspicua P. H. Weston

(inconspicuous)
WA
0.5–2.5 m × 1–2.5 m   June–Sept

Dwarf to medium **shrub** with greyish to pale brown hairs on young growth; **stems** single to many, branching near base; **branches** spreading to ascending; **branchlets** terete, becoming glabrous; **leaves** 1–6.5 cm × about 0.1 cm, linear, somewhat terete, alternate, not twisted, ascending to erect, often crowded, green, flexible to stiff, grooved undersurface, apex pointed; **flowers** about 0.6 cm across, greenish-yellow, insignificant, exterior glabrous to hairy, on very short hairy stalks, subtended by scale-leaves, axillary, solitary or 2–3 together; **drupes** to about 1 cm × 0.6 cm, smooth.

A somewhat insignificant and poorly known species from the eastern Avon, western Coolgardie and northwestern Roe Districts. It inhabits sand or sandy loam in heath and mallee heath. Plants should adapt to arid, semi-arid and warm temperate regions. They are hardy to moderately heavy frosts. Low-growing groundcovering variants may be worthy of selection. Propagate from seed or from cuttings of very young growth.

This species has sometimes been misidentified as *P. microcarpa* which has convex leaves.

Persoonia iogyna P. H. Weston & L. A. S. Johnson

(referring to rusty-hairy ovary)
Qld
1.6–4 m × 1–3 m   Dec–Feb

Small to medium **shrub**; young growth rusty-hairy; **trunk** usually solitary; **bark** grey, smooth; **branches** spreading to ascending; **branchlets** initially hairy, becoming glabrous; **leaves** 1–11 cm × 0.4–2 cm, narrowly elliptical to oblanceolate, alternate, not twisted, spreading to ascending, somewhat crowded, deep green above, paler below, initially slightly hairy becoming more or less glabrous, flat, margins recurved, apex pointed; **flowers** to about 0.8 cm across, yellow, exterior hairy, on erect, short, hairy stalks, in terminal and axillary racemes to about 6 cm long, rarely profuse; **drupes** to about 1 cm × 0.5 cm, green, smooth.

A recently described (1994) species from the Moreton District of south-eastern Qld where it grows

Persoonia juniperina × .85

in dry or wet sclerophyll forests. It is suitable for subtropical and temperate regions and should adapt to a variety of well-drained acidic soils in sunny, semi-shaded or dapple-shaded sites. May be useful for informal or formal hedging. Frost tolerance is not known. Propagate from seed or from cuttings of very young growth.

This species has been referred to as a variant of *P. attenuata* which is now a synonym of *P. media*. That species differs from *P. iogyna* in having mainly glabrous branches and generally larger leaves. *P. oleoides* is allied, but is usually a spreading shrub to 1 m tall.

Persoonia isophylla L. A. S. Johnson & P. H. Weston
(equal leaves)
NSW
0.3–1.5 m × 0.6–2 m   Jan–July
    Dwarf to small **shrub** with greyish-hairy young growth; **bark** smooth, grey; **branches** many, spreading to ascending; **branchlets** faintly hairy, becoming glabrous; **leaves** 1.2–3 cm × less than 0.1 cm, linear, terete, alternate, spreading to recurved, crowded, bright green, becoming glabrous, apex softly pointed; **flowers** to about 0.7 cm across, yellow, exterior glabrous to faintly hairy, on very short stalks, subtended by leaves, in terminal racemes to about 10 cm long, often profuse and conspicuous; **drupes** green with purplish tonings, smooth.

    A handsome, long-flowering species which was described in 1991. It occurs north of Sydney where it inhabits heath and dry sclerophyll forest on Hawkesbury Sandstone on the coast and slightly inland. Sometimes cultivated mistakenly as *P. pinifolia*, which differs in being larger and in having flowers subtended by reduced leaves. It should adapt to a range of freely draining acidic soils in a sunny or semi-shaded site. Tolerates extended dry periods but its frost tolerance is not known. May be useful for low hedging. Propagate from seed or from cuttings of very young growth.

Persoonia juniperina Labill.
(similar to the genus *Juniperus*)
NSW, Vic, Tas, SA   Prickly Geebung
0.6–2 m × 0.5–2 m   Nov–March; also sporadic
    Dwarf to small **shrub** with hairy light green young growth; **branches** spreading to ascending; **branchlets** hairy; **leaves** 0.8–3 cm × up to 0.15 cm, linear, alternate, mainly spreading, sometimes incurved, crowded, deep green, becoming glabrous, concave, rigid, smooth, apex pungent-pointed; **flowers** about 0.8 cm across, yellow, exterior glabrous to hairy, on very short usually hairy stalks, mainly subtended by leaves, in leafy racemes, often profuse and moderately conspicuous; **drupes** to about 1 cm × 0.7 cm, smooth.

    A far-ranging species which occurs in a variety of acidic soils and habitats from the Southern Coast, NSW, to south-eastern SA. Moderately commonly cultivated, it adapts to most freely to moderately well-drained soils which can be winter-wet. A sunny, semi-shaded or dapple-shaded site is suitable. Hardy to most frosts and light snowfalls. Responds very well to light or hard pruning. Suitable for hedging. Excellent as a refuge or nesting plant for small birds. Introduced to cultivation in England in 1826. Propagate from seed, or cuttings of very young growth, which can be slow to form roots.

Persoonia karare P. H. Weston
(after Karara Pastoral Station)
WA
1–5 m × 1.5–5 m   Sept–Nov

Persoonia lanceolata × .7

Small to tall **shrub** with greyish-hairy young growth; **branchlets** terete, becoming glabrous; **leaves** 2–14 cm × up to 0.4 cm, linear, alternate, not twisted, spreading to ascending, often crowded, green, becoming glabrous, apex pointed; **flowers** about 0.7 cm across, yellow, exterior hairy, on short hairy stalks, in terminal racemes to about 1 cm long; **drupes** more or less globular, smooth.

A very poorly known species from sandplains in the south-western Austin District near the northern Avon District border. It was described in 1994. Worthy of cultivation as part of a conservation strategy. Plants will be best suited to arid, semi-arid and warm temperate regions. Should tolerate moderate frosts. Propagate from seed or from cuttings of very young growth.

This species has been misidentified as *P. saundersiana*, which has faintly hairy flowers and ribbed leaves that can be triangular to somewhat terete in cross-section. *P. stricta* is allied too, but has erect leaves.

Persoonia katerae P. H. Weston & L. A. S. Johnson
(after Penelope Kater, 20th-century volunteer botanical assistant, NSW)
NSW
2.5–9 m × 1.5–5 m   Jan–Feb
Medium to tall **shrub** or small **tree** with greyish-hairy young growth; **bark** base rough, other parts smooth, grey; **branches** spreading to ascending; **branchlets** hairy; **leaves** 6–17 cm × 0.8–2.2 cm, narrowly elliptical to oblanceolate, alternate, slightly twisted; spreading to ascending, green, becoming glabrous, flat, margins slightly recurved, apex pointed; **flowers** about 0.8 cm across, yellow, exterior hairy, on erect hairy stalks to about 3.5 cm long, in terminal leafy racemes to 16 cm long, sometimes profuse; **drupes** green, sometimes with purplish tonings.

This splendid species is one of the tallest persoonias but is not well known even though it is fairly common in the North Coast District, between Myall Lakes and Port Macquarie. It grows in coastal sand dunes, heath and dry sclerophyll forest. Cultivated to a limited extent. Plants need excellently drained acidic soils and a sunny or semi-shaded site is suitable. They should adapt to temperate or subtropical regions. It has potential for screening and windbreak purposes. Cold tolerance is not known. Propagate from seed or from cuttings of very young growth.

Persoonia laevis — see *P. levis* (Cav.) Domin

Persoonia lanceolata Andrews
(lanceolate)
NSW   Lance-leaved Geebung
1–2.5 m × 0.6–3 m   Dec–Feb; also sporadic
Small to medium **shrub**; young growth yellow-green with whitish hairs; **branches** spreading to ascending; **branchlets** whitish-hairy; **leaves** 3–10 cm × 0.5–3.2 cm, oblanceolate to narrowly spathulate, alternate, spreading to ascending, fairly crowded, yellow-green to pale green or rarely bluish-green, initially glabrous or hairy and becoming glabrous, flat, apex pointed; **flowers** to about 1 cm across, yellow, exterior hairy, on erect hairy stalks to 0.5 cm long, subtended by leaves, in leafy racemes, sometimes profuse, rarely highly conspicuous; **drupes** to about 1 cm × 0.8 cm, green, smooth.

An attractive species from the coastal regions and Central Tablelands District. Grows from sea-level to the mountain ranges on sand and sandstone in heath and dry sclerophyll forest. Adapts well to cultivation in temperate and subtropical regions in a variety of well-drained acidic soils with a sunny or semi-shaded site. Tolerates moderate frosts and withstands extended dry periods. Suitable for general planting and hedging. First cultivated in England in 1791. Propagate from seed or from cuttings of very young growth.

P. lanceolata ssp. B. is now known as *P. glaucescens* Sieber ex Spreng.

Persoonia laurina Pers.
(similar to the genus *Laurus*)
NSW   Laurel Geebung
0.6–1.5 m × 0.6–2 m   Oct–Dec
Dwarf to small **shrub** with rusty-hairy young growth; **branches** spreading to ascending; **branchlets** rusty-hairy; **leaves** 3.5–11 cm × 1–6 cm, ovate to oblong, mainly opposite and decussate, spreading to ascending, light green, becoming glabrous, flat, smooth, margins recurved, thick, leathery, apex rounded to blunt, ending in a hard point; **flowers** about 0.8 cm across, bright yellow with rusty hairs, on very short, erect, hairy stalks, in axillary clusters, often profuse and conspicuous; **drupes** about 1.3 cm × 1 cm, green becoming purplish.

A variable but very attractive species with the typical subspecies occurring in the North and Central Coast and Central Tablelands Districts. It usually grows on sandstone in heath or dry sclerophyll forest. Responds well to cultivation in temperate and subtropical regions. Needs freely draining acidic soils in a sunny or semi-shaded site and will tolerate dappled shade. Plants tolerate moderate frosts. Worthy of greater use in gardens and has potential for formal plantings.

The ssp. *intermedia* L. A. S. Johnson & P. H. Weston is from the Central Coast and Central Tableland Districts and it differs in having rough mature leaves.

The ssp. *leiogyna* L. A. S. Johnson & P. H. Weston occurs in the Central Coast, Central and Southern Tableland Districts. The leaves are rough and it differs in having a glabrous ovary.

Propagate all variants from seed or from cuttings of very young growth.

Persoonia laxa L. A. S. Johnson & P. H. Weston
(loose)
0.2–0.3 m × 1–2 m   Nov–Jan
Dwarf **shrub** with hairy young growth; **branches** prostrate to decumbent; **branchlets** faintly hairy; **leaves** 0.8–1.5 cm × up to 0.2 cm, linear, alternate, spreading to ascending, green, paler below, becoming glabrous, flat, margins recurved, apex pointed; **flowers** to about 0.8 cm across, yellow, exterior glabrous, on glabrous spreading to recurved stalks to about 0.8 cm long, solitary or in leafy clusters; **drupes** not known.

This species, which has affinities to *P. nutans*, has not been collected since 1908 and is presumed extinct. Collections were on the Central Coast between Manly

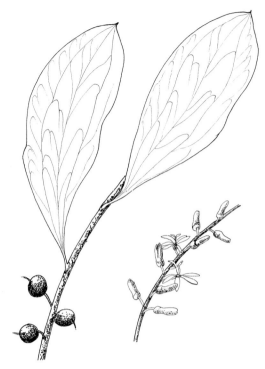

Persoonia levis × .55

and Broken Bay, where it is thought to have grown in heath, or dry sclerophyll forest or woodland. If it is ever collected again it should be cultivated as a conservation strategy. Plants will be suitable for temperate and subtropical regions and need freely draining acidic soils with a sunny or semi-shaded aspect. Its growth habit would make it a desirable addition to horticulture. Propagate from seed or from cuttings of very young growth.

It has been known as *P. nutans* ssp. B but *P. nutans* differs in a taller growth habit, longer leaves and longer flower-stalks.

Persoonia ledifolia A. Cunn. ex Meisn. =
P. mollis ssp. *ledifolia* (A. Cunn. ex Meisn.) S. L. Krauss
& L. A. S. Johnson

Persoonia leucopogon S. Moore
(white beard)
WA
0.3–0.8 m × 0.5–1 m   Nov–March
Dwarf **shrub** with greyish to rusty hairs on young growth; **branches** spreading to ascending; **branchlets** initially hairy, becoming glabrous; **leaves** 0.7–1.5 cm × to about 0.2 cm, narrowly elliptical to narrowly oblong, alternate, twisted, spreading to erect, crowded, greyish-green to glaucous, flat, stiff, apex sharply pointed; **flowers** about 0.7 cm across, yellow, exterior densely hairy, on short hairy stalks, in terminal or axillary clusters, often profuse; **drupes** to about 0.9 cm × 0.7 cm, more or less globular, smooth.

A poorly known attractive species which at one stage was presumed extinct until collected again in 1991. It is recorded as occurring on sand or loam in heath in the Coolgardie District. Worthy of cultivation as part of a conservation strategy. Best suited to semi-arid and warm temperate regions in excellently drained acidic soils with a sunny aspect. Plants should withstand moderately heavy frosts. Propagate from seed or from cuttings of very young growth.

Persoonia levis (Cav.) Domin
(smooth)
NSW, Vic   Broad-leaved Geebung
1.5–5 m × 1.5–3.5 m   Sept–Nov; also sporadic
Small to tall **shrub** with faintly hairy, often reddish young growth; **trunk** usually solitary; **bark** flaky, exterior black, undersurface brownish-red; **branches** spreading to ascending; **branchlets** initially glabrous or hairy becoming glabrous, often reddish; **leaves** 6–20 cm × 1.3–8 cm, oblanceolate to obovate or spathulate, alternate, spreading to ascending, light-to-mid green, becoming glabrous, flat, often curved, thick, smooth, apex round to blunt, ending in a small point; **flowers** about 1 cm across, golden-yellow, exterior glabrous to faintly hairy, on erect to spreading hairy stalks to about 0.8 cm long, in leafy racemes, can be profuse and conspicuous; **drupes** about 1 cm × 0.8 cm, somewhat globular, green, smooth.

A very handsome, variable-foliaged species from the Coast Districts, the Central and Southern Tablelands of NSW, and in eastern Vic. Grows in heath and dry sclerophyll forest. The coastal variant is usually a small to medium shrubby plant with broader leaves than the taller forest variant. Adapts well to cultivation but can be slow-growing. Does best in well-drained acidic soils

Persoonia linearis × .45

n semi-shaded site or dappled shade but tolerates full sunshine. Hardy to moderately heavy frosts and extended dry periods. Deserves to be better known but s rarely offered by nurseries. Inner bark is used for bark pictures. Propagate from seed or cuttings of very young growth.

P. levis is known to hybridise in the wild with *P. acerosa*, *P. linearis*, *P. mollis* ssp. *ledifolia*, *P. myrtilloides* and *P. oxycoccoides*. Some of these hybrids are known to be in cultivation.

P. levis × *linearis* is cultivated. Previously known as *P. lucida* R. Br., it is a medium to tall shrub of up to 5 m, with pendent, lanceolate leaves and bright golden-yellow flowers. Plants generally have characteristics intermediate between the two parents.

Persoonia linearis Andrews
(linear)
NSW, Vic Narrow-leaved Geebung
2.5–5 m × 1.5–3.5 m Nov–March
Medium to tall **shrub** with greyish-hairy young growth; **trunk** usually solitary; **bark** flaky, blackish exterior, brownish-red undersurface; **branches** spreading to ascending; **branchlets** whitish-hairy; **leaves** 2–9 cm × 0.1–0.6 cm, linear to linear-spathulate, alternate, spreading to ascending, crowded, dull greyish-green, becoming glabrous, flat, apex softly pointed; **flowers** to about 0.8 cm across, golden-yellow, exterior hairy, on spreading to erect, hairy, short stalks, in leafy racemes, can be profuse and conspicuous; **drupes** to about 1.3 cm × 1.3 cm, globular, green.

An ornamental species with appealing habit, bark, foliage and flower. It has a wide distribution in eastern NSW and eastern Vic. Occurs on a range of acidic soils and grows in woodland and open forest. Does well if cultivated in freely draining acidic soils in a semi-shaded site but also tolerates plenty of sunshine or dappled shade. Plants are hardy to most frosts and withstand extended dry periods. They respond well to pruning or clipping. May be used for screening and hedging. First introduced to cultivation in England in 1794. This plant was also grown as *P. pruinosa* in 1824. Propagate from seed, or from cuttings of very young growth, which are often slow to form roots.

Hybrids with *P. levis* occur and are cultivated — see *P. levis* (Cav.) Domin.

Persoonia longifolia R. Br.
(long leaves)
WA Snotty Gobble; Long-leaved Persoonia
1.5–6 m × 1–3.5 m Oct–Feb
Small to tall **shrub** with a lignotuber; young growth with rusty-brown hairs; **trunk** usually solitary and can have suckers at base; **bark** flaky, dark brown to grey exterior, undersurface reddish-purple; **bark** spreading to ascending; **branchlets** initially angular and hairy, becoming terete and glabrous, often pendent; **leaves** 2.5–20 cm × 0.2–1.6 cm, alternate, linear to oblanceolate, often slightly twisted, mainly flat, spreading to ascending, sometimes crowded, deep green, flexible, becoming glabrous, midrib prominent, apex pointed; **flowers** about 0.7 cm across, bright yellow, exterior glabrous, on pendent to erect hairy stalks to 1.2 cm long, in terminal or axillary leafy racemes to about

Persoonia marginata × .5

7 cm long, can be profuse; **drupes** to about 1.2 cm × 1 cm, purplish when ripe.

A very attractive persoonia from the Southern Darling District. Its distribution extends from near Perth to Albany. Occurs in woodland and dry sclerophyll forest where it grows in sand, sandy loam or laterite. Cultivated to a limited extent. It was introduced into England in 1850 and was cultivated as *P. drummondii* in 1857. Best suited to temperate regions. Plants need freely draining acidic soils in a site with semi-shade or dappled shade but also tolerate a fair amount of sunshine. They are hardy to moderate frosts and respond well to light or hard pruning. It is harvested for cut foliage which is renowned for its long-lasting qualities. Propagate from seed or from cuttings of very young growth.

Persoonia lucida R. Br. — see *P. levis* (Cav.) Domin

Persoonia macrostachya Lindl. = *P. saccata* R. Br.

Persoonia marginata A. Cunn. ex R. Br.
(bordered with a margin)
NSW
0.2–0.6 m × 1–2 m Oct–Feb
Dwarf **shrub** with hairy light green young growth; **branches** many, spreading to ascending; **branchlets** hairy or becoming glabrous; **leaves** 2–5 cm × 0.6–3 cm, elliptic to obovate, alternate, spreading to ascending, somewhat crowded, light-to-mid green, becoming glabrous, smooth to faintly rough, flat, margins thickened and often wavy, apex blunt with small point;

flowers to about 0.8 cm across, yellow, exterior faintly hairy, on erect to spreading hairy stalks to about 0.7 cm long, in leafy racemes, conspicuous but rarely profuse; **drupes** to about 1.2 cm long, moderately hairy.

An interesting, low-spreading species from the Central Tablelands where it is considered vulnerable. It occurs on sandstone in dry sclerophyll forest. Worthy of greater cultivation and should adapt to a range of well-drained acidic soils in temperate regions. May succeed in subtropical areas too. A sunny or semi-shaded site is appropriate. Hardy to most frosts. Well suited to growing among boulders and rocks, as well as for underplanting taller shrubs and trees. Propagate from seed, or from cuttings of very young growth, which may be slow to form roots.

Persoonia media R. Br.
(medium or inbetween)
Qld, NSW
3–8 m × 1.5–4 m   Oct–Feb; also sporadic

Medium to tall, sometimes lignotuberous **shrub**, or a small **tree**; young growth glabrous or faintly hairy; **trunks** solitary or sometimes multiple; **bark** rough at base; **branches** spreading to ascending; **branchlets** becoming glabrous; **leaves** 3–14 cm × 0.4–3.5 cm, linear-elliptic to obovate, alternate, ascending to erect, moderately crowded, green, becoming glabrous and smooth, flat with recurved margins, apex softly pointed; **flowers** to about 1 cm across, yellow, glabrous to faintly hairy, solitary in axils, forming a short leafy raceme, moderately conspicuous; **drupes** to about 1.5 cm × 0.9 cm, green with purple, smooth.

An attractive persoonia from south-eastern Qld, and the North Coast and Northern Tablelands of NSW. Generally occurs in dry and wet forest, or rainforest margins, in mountain ranges. Often develops as a multiple-trunked shrub in dryish sites. Cultivated to a minor degree, it warrants greater recognition in subtropical and temperate regions. Requires good drainage and a sunny or semi-shaded site with acidic soils. Hardy to moderate frosts. Useful for low screening and general planting. Propagate from seed or from cuttings of very young growth.

Previously often regarded as a synonym of *P. attenuata* R. Br. The allied *P. volcanica* has smooth bark, spreading hairs and generally smaller leaves.

Persoonia micranthera P. H. Weston
(small anthers)
WA
0.1–0.5 m × 0.8–2 m   Feb

Dwarf spreading **shrub** without a lignotuber; young growth with greyish to pale brown hairs; **branches** prostrate; **branchlets** initially angular and hairy, becoming terete and glabrous; **leaves** 2–8 cm × 0.4–3 cm, spathulate to oblanceolate, alternate or opposite, often twisted at base, spreading to erect, often in crowded clusters, soft, green, flat, margins faintly recurved, apex blunt with a soft point; **flowers** to about 0.8 cm across, yellow, exterior hairy, mainly upright, on hairy stalks to 0.8 cm long, can form leafy racemes 1–6 cm long; **drupes** to about 0.8 cm × 0.5 cm, smooth.

This recently described (1994) species is endangered and confined to a small area in the western Eyre District where it grows in sandy, stony soils. Warrants cultivation as part of a conservation strategy. Plants will need excellent drainage in acidic soils and a semi-shaded site should be suitable. Hardy to moderate frosts. It has potential for rockeries and as a groundcover for general planting. Propagate from seed or from cuttings of very young growth.

Persoonia microcarpa R. Br. = *Acidonia microcarpa* (R. Br.)
L. A. S. Johnson & B. G. Briggs

Persoonia microphylla R. Br.
(small leaves)
NSW
0.2–2 m × 1.5–3 m   Nov–Feb; also sporadic

Dwarf to small **shrub** with hairy young growth; **branches** prostrate to ascending; **branchlets** moderately hairy; **leaves** 0.3–1 cm × 0.2–0.5 cm, broadly elliptical to broadly ovate, alternate to somewhat opposite, spreading to ascending, crowded, deep green above, slightly paler below, initially hairy becoming less so, rough, margins recurved to revolute, apex bluntly pointed; **flowers** to about 1 cm across, yellow, glabrous to faintly hairy, on short, erect to spreading stalks, in leafy racemes 0.1–3 cm long, somewhat conspicuous; **drupes** green with purple stripes, smooth.

A moderately attractive species from the Central and Southern Tableland Districts where it grows in dry sclerophyll forest and montane heath. Poorly known in cultivation, it is usually slow-growing. Best suited to temperate regions. Plants require excellently drained acidic soils in a sunny or semi-shaded site. There is ample opportunity for clonal selection of different growth habits. Plants are hardy to moderately heavy frosts and extended dry periods. Propagate from seed or from cuttings of very young growth.

It has also been known as *P. nutans* ssp. F. and *P. oxycoccoides* var. *microphylla* (R. Br.) Domin.

Persoonia mitchellii Meisn. =
P. sericea A. Cunn. ex R. Br.

Persoonia mollis R. Br.
(softly hairy)
NSW   Soft Geebung
1.5–5 m × 1–4.5 m   Dec–April; also sporadic

Small to tall **shrub** with whitish- to coppery-hairy young growth; **bark** pale grey to brownish; **branches** spreading to ascending; **branchlets** densely hairy when young; **leaves** 4–10 cm × 0.6–1.5 cm, narrowly lanceolate, alternate or opposite or in whorls, spreading to ascending, moderately crowded, soft, green above with whitish to coppery hairs below, margins faintly recurved, midrib prominent, apex softly pointed; **flowers** about 1 cm across, yellow, hairy, solitary, on very short hairy stalks, in leafy racemes to about 15 cm long, often moderately conspicuous; **drupes** to about 0.8 cm × 0.7 cm, green to purplish brown.

The typical variant or this extremely variable species occurs in the lower and upper Blue Mountains. It inhabits sandy soils of wet and dry sclerophyll forest and is often described as the 'forest form'. All the sub-species are cultivated but some only to a very limited degree. They require good drainage and acidic soils. A semi-shaded site is best but plants do tolerate plenty of sunshine. They withstand moderately heavy frosts. The young growth of some subspecies is especially attractive.

There are a further 8 subspecies and these differ in such features as growth habit, leaf size, and the degree of hairiness.

P. mollis ssp. *budawangensis* S. L. Krauss & L. A. S. Johnson

Occurs in the dry and wet sclerophyll forests of the Budawang Range in south-eastern NSW. It grows 1–2.5 m × 1–2 m and has silky-hairy young growth. Leaves are 2–4 cm × 0.3–0.6 cm, linear-oblong to oblong-lanceolate and somewhat leathery.

P. mollis ssp. *caleyi* (R. Br.) S. L. Krauss & L. A. S. Johnson

This variant is from the South Coast District and is found near Jervis Bay and extending south to near Batemans Bay, NSW. It inhabits dry and wet sclerophyll forest and grow in sandy soils. Grows as a small to tall shrub of 1.5–4 m × 1–3 m and has softish lanceolate leaves, 3–6 cm × 0.2–0.6 cm, with a pointed apex.

P. mollis ssp. *ledifolia* (A. Cunn. ex Meisn.) S. L. Krauss & L. A. S. Johnson

This subspecies is from an area south of the Shoalhaven River in the Central Coast and Central Tablelands of NSW. Plants grow 1–2.5 m × 1–2.5 m and have linear-oblong to oblong-lanceolate green leaves, 2–4 cm × 0.3–0.6 cm, which are initially silky-hairy below. Margins are recurved to revolute and the midrib is not distinct.

Previously known as *P. ledifolia* A. Cunn. ex Meisn.

P. mollis ssp. *leptophylla* S. L. Krauss & L. A. S. Johnson

This variant is from the Central and Southern Coast and Southern Tableland Districts of NSW. It occurs in heath and dry sclerophyll forest and grows on sandstone. Plants are usually 0.5–1.5 m × 0.5–1.5 m with a compact habit. The bright green, stiff, linear leaves are 1.5–4 cm × up to 0.2 cm. Young growth is silky-hairy.

P. mollis ssp. *livens* S. L. Krauss & L. A. S. Johnson

An inhabitant of the Central and Southern Tablelands of NSW. Plants are found in dry sclerophyll forests on sandy or stony loams. It is a dwarf to compact shrub of 0.5–1.5 m × 0.5–1.5 m. It has soft, linear, green to grey-green leaves, 1.5–3 cm × up to 0.2 cm, which are usually densely hairy below.

P. mollis ssp. *maxima* S. L. Krauss & L. A. S. Johnson

This small to tall shrub of 2–5 m × 1.5–3.5 m occurs in the Central Coast District of NSW, where it is regarded as endangered. It is found in dry and wet sclerophyll forest and grows in shallow sandy soils. It is larger than other *P. mollis* subspecies in all its characteristics. It has coppery-coloured young growth and narrowly lanceolate leaves, 6–12 cm × 1–2 cm, which are initially hairy. Leaf margins are faintly recurved and the midrib is prominent.

P. mollis ssp. *nectens* S. L. Krauss & L. A. S. Johnson

A variant from the Central Coast of NSW which inhabits usually sandy moist soils in dry and wet sclerophyll forest. It is a small to medium shrub, 1.2–3 m × 1–2.5 m, with narrowly lanceolate leaves of 4–10 cm × up to 1.5 cm which are initially silky-hairy below. Flowers are prominently silky-hairy.

P. mollis ssp. *revoluta* (Sieber ex Schult. & Schult. f.) S. L. Krauss & L. A. S. Johnson

A prostrate shrub, 0.1–0.5 m × 1.5–4 m, from the Central Tablelands of NSW. It is found in dry sclerophyll forest where it grows in deep greyish-white sand. The shining green elliptical to oblong leaves are 2.5–4 cm × 0.4–1.5 cm, and are initially faintly hairy below. The margins are revolute and the midrib is faint to sometimes prominent. Previously known as *P. revoluta* Sieber ex Schult. & Schult. f.

All subspecies are propagated from seed or from cuttings of very young growth.

Persoonia moscalii Orchard
(after Tony Moscal, 20th-century Tasmanian botanist and original collector)
Tas
0.05–0.1 m × 0.8–1 m   Jan–March

Dwarf spreading **shrub** with hairy young growth; **bark** reddish-brown, flaky on older parts; **branches** prostrate with tips ascending, apparently not self-layering; **branchlets** hairy when young; **leaves** 0.6–1 cm × about 0.4 cm, obovate to oblanceolate, tapering to base, alternate, erect, crowded, somewhat fleshy, deep green, initially very hairy, becoming hairy only at base, veins indistinct, apex blunt to rounded; **flowers** about 1.3 cm across, yellow, exterior faintly hairy, solitary, nearly sessile, in upper axils; **drupes** to about 1 cm × 0.8 cm, deep reddish-purple, smooth.

This is a rare species from the south-western region near Bathurst Harbour, but it is not regarded as threatened. *P. moscalii* inhabits alpine heath shrubland in well-drained acidic soils at 640–760 m altitude. It has been cultivated to a very limited degree and has maintained its distinctive growth habit. May be slow to establish but should adapt to a range of freely draining soils with a sunny or semi-shaded aspect in cool temperate regions. Hardy to frost and snowfalls. Potential for rockeries, containers and general planting. Propagate from seed or from cuttings of very young growth.

Persoonia muelleri (P. Parm.) Orchard
(after Baron Sir Ferdinand von Mueller, 19th-century Government Botanist of Victoria)
Tas
1–5 m × 0.6–3 m   Dec–March

Small or rarely medium to tall **shrub** with hairy young growth; **bark** flaky, dark brown, on older parts; **branches** spreading to ascending; **branchlets** initially densely hairy; **leaves** 1.8–4 cm × 0.4–0.8 cm, oblanceolate, alternate, spreading to ascending, moderately crowded, thick, deep green, becoming faintly hairy, flat, veins indistinct, apex rounded or with small point; **flowers** to about 1.5 cm across, yellow, exterior hairy, solitary in upper axils, on faintly hairy stalks to about 0.5 cm long, in leafy racemes; **drupes** to about 1.2 cm × 0.9 cm, purplish, smooth.

The typical variant is confined to the Central Plateau mountains and the north-eastern mountain ranges. It occurs in open forest, in soils which can be moist for extended periods. Worthy of cultivation, it should adapt to temperate regions in a range of acidic soils with a sunny or semi-shaded aspect. Most likely slow-growing. Hardy to frost and snowfalls. Historically interesting because Mueller initially thought it was a species of *Drimys* (now *Tasmannia*) and the Frenchman, Parmentier, named it *Drimys muelleri*.

Previously known as *P. gunnii* Hook. f. var. *alpina* Hook. f. *P. muelleri* has been confused with *P. gunnii* which has much thinner leaves that become virtually glabrous with time.

There are a further 2 varieties.

P. muelleri var. *angustifolia* (Benth.) Orchard differs in having ascending linear to narrowly oblanceolate leaves of 4–5.5 cm × 0.3–0.7 cm. It occurs mainly on the mid-west coast at 300–350 m altitude. Previously known as *P. gunnii* Hook. f. var. *angustifolia* Benth.

P. muelleri var. *densifolia* Orchard usually occurs in exposed soils on the south-west to south coast region mainly near sea-level but it has been found up to 720 m. Plants have very crowded, spathulate leaves, 2–5 cm × 0.6–1 cm, which are hairy at maturity. Produces reddish-purple drupes of about 1 cm × 0.7 cm.

Propagate all varieties from seed or from cuttings of very young growth.

Persoonia myrtilloides Schult. & Schult. f.
(similar to *Vaccinium myrtillus*, Whortleberry)
NSW   Myrtle Geebung
0.5–2.5 m × 1–2.5 m   Oct–April; also sporadic

Dwarf to medium **shrubs** with hairy, sometimes reddish young growth; **branches** spreading to ascending; **branchlets** often initially reddish, becoming less hairy; **leaves** 1.2–5 cm × 0.4–3 cm, narrowly to broadly elliptic to ovate, alternate, spreading to ascending, rarely crowded, light-to-mid green, becoming glabrous, flat, margins faintly recurved, midrib distinct, apex pointed; **flowers** about 1 cm across, yellow, hairy, solitary on slender, spreading or recurved, hairy stalks to 1 cm long, forming leafy racemes to about 17 cm long, often moderately conspicuous; **drupes** green with purplish tonings, smooth.

The typical variant of this handsome species occurs in the upper Blue Mountains of the Central Tablelands where it grows in a range of soils derived from sandstone, in dry sclerophyll forest, woodland and heath. Adapts very well to cultivation in temperate regions and has also been successfully grown in subtropical areas. Requires well-drained acidic soils with a semi-shaded or dapple-shaded site. Hardy to most frosts and tolerates moderately lengthy dry periods. Responds well to pruning or clipping. Does well in large containers. Introduced to cultivation in England during 1837.

The typical variant has been known as *P. myrtilloides* ssp. A. It hybridises with *P. acerosa, P. levis* and *P. recedens* in nature, and some of these hybrids may be in cultivation.

The ssp. *cunninghamii* (R. Br.) L. A. S. Johnson & P. H. Weston is from the Central Tablelands north of the Blue Mountains, where it inhabits dry sclerophyll forest in *Callitris* woodland, on sandy or stony soils. It has lanceolate to broadly elliptic leaves of 1.2–3.8 cm × 0.6–3 cm, and tepal-tips which are prominently reflexed.

Previously known as *P. cunninghamii* R. Br. and *P. myrtilloides* ssp. B.

Propagate both subspecies from seed, or from cuttings of very young growth, which can give sporadic results.

P myrtilloides var. *brevifolia* Benth. is known as *P. brevifolia* L. A. S. Johnson & P. H. Weston.

Persoonia nutans R. Br.
(nodding)
NSW
0.5–1.5 m × 0.6–2 m   Nov–April; also sporadic

Dwarf to small **shrub** with faintly hairy young growth; **branches** spreading to erect; **branchlets** slender, often reddish when young, becoming glabrous; **leaves** 1–3.5 cm × 0.1–0.2 cm, linear, can be falcate, alternate, spreading to ascending, not crowded, deep green above, paler below, becoming glabrous, margins flat to faintly recurved, sometimes recurved near the softly pointed apex; **flowers** about 0.7 cm across, yellow, exterior glabrous, on slender pendent stalks to about 1.2 cm long, scattered but conspicuous; **drupes** green with purplish markings, smooth.

A dainty species from the Central Coast where it is regarded as endangered. It occurs in woodland and dry sclerophyll forest on a range of soils. Moderately popular in cultivation, it adapts well to subtropical and temperate regions. Does best in well-drained acidic soils in a semi-shaded site but will tolerate an open aspect. Hardy to moderately heavy frosts and extended dry periods. Responds very well to pruning or clipping and is suitable for use as dwarf hedging and in general planting, or in large containers. It was introduced into cultivation in England in 1824. Propagate from seed, or from cuttings of very young growth, which may be slow to form roots but generally a good percentage strike.

P. apiculata Meisn. and *P. nutans* R. Br. var. *apiculata* (Meisn.) Benth. are synonyms. Also has been known as *P. nutans* ssp. A.

Before recent botanical revision there were a number of closely allied taxa placed as various subspecies of *P. nutans*. They have since been given species ranking.

P. nutans ssp. B =
                    P. laxa L. A. S. Johnson & P. H. Weston
P. nutans ssp. C = *P. recedens* Gand.
P. nutans ssp. D =
                    P. terminalis L. A. S. Johnson & P. H. Weston
P. nutans ssp. E = *P. oxycoccoides* Sieber ex Spreng.
P. nutans ssp. F = *P. microphylla* R. Br.
P. nutans ssp. G =
                    P. asperula L. A. S. Johnson & P. H. Weston

Persoonia oblongata R. Br.
(oblong)
NSW
1–2 m × 0.8–1.6 m   Feb–July

Small **shrub** with hairy young growth; **branches** spreading to erect; **branchlets** initially hairy, often becoming nearly glabrous; **leaves** 1.5–6 cm × 0.4–2.5 cm, lanceolate to broadly elliptic, alternate, spreading to ascending, moderately crowded, flexible, bright green, becoming glabrous, flat, smooth, midrib often indistinct, apex ending in a small soft point; **flowers** about 0.9 cm across, yellow, exterior glabrous to faintly hairy, solitary on slender, pendent, glabrous stalks to 2.3 cm long, forming leafy racemes, moderately conspicuous; **drupes** to about 1.4 cm × 1.2 cm, green with purplish markings, becoming glabrous.

A bushy species from the Central Coast, Central Tablelands and Central Western Slopes. It inhabits dry sclerophyll forest and heath on soils derived from sandstone. It is uncommon in cultivation but should adapt to temperate regions and will possibly succeed in subtropical areas. Needs freely draining acidic soils in a sunny or semi-shaded site. Hardy to moderately heavy frosts and withstands extended dry periods. May be suitable for hedging. Propagate from seed or from cuttings of very young growth.

Persoonia oleoides L. A. S. Johnson & P. H. Weston
(similar to the olive genus, *Olea*)
NSW
0.2–1 m × 0.8–1.5 m   Jan–March
  Dwarf **shrub**, young growth with greyish to rusty hairs; **branches** many from near ground level, prostrate to ascending or erect; **branchlets** hairy; **leaves** 2–9 cm × 0.4–2.6 cm, mainly oblong to spathulate, alternate, spreading to ascending, moderately crowded, green, sometimes paler below, initially hairy, especially on margins, becoming glabrous, margins faintly recurved, apex pointed; **flowers** about 0.9 cm across, yellow, exterior hairy, solitary on short, erect, hairy stalks, in

Persoonia oxycoccoides

leafy racemes to about 13 cm long, usually moderately conspicuous; **drupes** green to reddish-purple or green with purple streaks.

An ornamental, recently described (1991) species from the North Coast and Northern Tablelands Districts, where it grows in clay loam, in dry and wet sclerophyll forest. Apparently not cultivated, it deserves to be much better known. There is scope to select plants for differing growth habits to suit rockeries, containers and general planting. Plants should be suitable for subtropical and temperate regions. They need well-drained acidic soils with a sunny or semi-shaded aspect. Should tolerate moderate frosts. Propagate from seed or from cuttings of very young growth.

Persoonia oxycoccoides Sieber ex Spreng.
(similar to the genus *Oxycoccus*)
NSW
0.5–1 m × 1–2.5 m   Nov–Feb
  Dwarf spreading **shrub** with hairy young growth; **branches** many, prostrate to ascending; **branchlets** often reddish, initially hairy, often becoming glabrous; **leaves** 0.4–1.2 cm × up to 0.6 cm, narrowly to broadly elliptical to ovate, alternate to somewhat opposite, spreading, dark green above, paler below, becoming more or less glabrous, more or less smooth, margins recurved, venation indistinct, apex softly pointed; **flowers** about 0.8 cm across, yellow, exterior glabrous, on short glabrous to hairy stalks, terminal or axillary, in leafy racemes to about 4 cm long, moderately conspicuous; **drupes** to about 1 cm × 0.6 cm, green, smooth.

An admirable species from the Central Tablelands where it grows in dry sclerophyll forest or heath in sandy soils. Moderately common in cultivation, it has proved adaptable in a range of acidic soils which drain very well to relatively well. Plants will grow in a sunny or semi-shaded site. They are hardy to heavy frosts and withstand extended dry or moist periods. They respond well to pruning or regular clipping. Excellent for underplanting of taller vegetation. Propagate from seed, or from cuttings of very young growth, which can be slow to form roots.

It has been marketed as *P. nutans*.

The var. *microphylla* (R. Br.) Domin is a synonym of *P. microphylla* R. Br. and var. *longifolia* (Benth.) is applicable to *P. acuminata* L. A. S. Johnson & P. H. Weston.

Persoonia papillosa P. H. Weston
(with many small projections)
WA
0.2–0.5 m × 0.2–0.6 m   Sept–Jan
  Dwarf **shrub** with brownish-hairy young growth; **branches** spreading to erect; **branchlets** initially angular and hairy, becoming terete and glabrous; **leaves** 0.6–3 cm × up to 0.15 cm, linear, alternate, not twisted, sometimes incurved, ascending to erect, crowded, rough, deep green, becoming less hairy to nearly glabrous, apex softly pointed; **flowers** about 0.8 cm across, yellow, exterior very hairy, solitary on hairy stalks to about 1.5 cm long, in terminal or axillary leafy racemes to 6 cm long; **drupes** not known.

Persoonia pinifolia   W. R. Elliot

A recently described (1994) and poorly known species. It occurs over a small range in the sandplains of the northern Irwin District where it grows in sandy soil. Evidently not cultivated, it warrants propagation as part of a conservation strategy. Plants will be best suited to warm temperate regions and require excellent drainage with plenty of sunshine, but may adapt to a semi-shaded site. Propagate from seed or possibly from cuttings of very young growth.

The allied *P. angustifolia* has smoother, grooved leaves and *P. trinervis* has broader leaves.

Persoonia pentasticha P. H. Weston
(five lines or rows)
WA
0.4–2 m × 0.6–2 m   Aug–Nov
Dwarf to small **shrub** with whitish to pale brown hairs on young growth; **branches** spreading to erect; **branchlets** terete, initially densely hairy, becoming glabrous; **leaves** 1–12 cm × about 0.1 cm, linear-terete, alternate, usually not twisted, 5-grooved, spreading to erect, often crowded, green, often becoming nearly glabrous, apex pungent-pointed; **flowers** about 0.8 cm across, yellow, exterior hairy, on short hairy stalks, in terminal or axillary leafy racemes to about 4 cm long; **drupes** not known.

This is a poorly known species from the northern Avon and south-western Austin Districts which was described in 1994. It inhabits heathland and grows in loam, gravels and sandy soils. Warrants cultivation as part of a conservation strategy. Plants will need excellently drained acidic soils and a sunny site. They are hardy to moderate frosts and tolerate extended dry periods. Best for warm temperate regions. Propagate from seed or from cuttings of very young growth.

The closely allied *P. chapmaniana* has broader leaves and generally longer racemes of flowers.

Persoonia pertinax P. H. Weston
(tenacious)
WA
1–2.5 m × 1–2.5 m   Jan–March
Small to medium **shrub** with greyish-hairy young growth; **bark** smooth, grey; **branches** spreading to erect; **branchlets** terete, initially densely hairy, becoming glabrous; **leaves** 0.7–5.5 cm × up to 0.25 cm, linear, alternate, usually twisted at base, spreading to ascending, sometimes crowded, flexible, green, becoming glabrous, apex softly pointed; **flowers** about 0.8 cm across, yellow, exterior hairy, on usually spreading to erect, short, hairy stalks, in terminal or axillary leafy racemes to about 6 cm long; **drupes** to about 1.3 cm × 0.8 cm, ovoid, smooth.

A recently described (1994) species which is locally common in woodland in the Coolgardie and Helms Districts, where it grows on sandy soil. Apparently not cultivated, it will be best suited to warm temperate and semi-arid regions. Plants will need excellent drainage and a warm to hot site for best results. They probably tolerate moderate frosts. Propagate from seed or possibly from cuttings of very young growth.

P. coriacea and *P. helix* are allied but they have

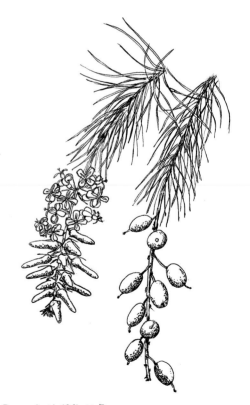

Persoonia pinifolia × .7

226

broader leaves. *P. pertinax* has been confused with *P. striata* which has glabrous flowers and grooved leaves.

Persoonia pinifolia R. Br.
(leaves similar to the genus *Pinus*)
NSW   Pine-leaved Geebung
3–5 m × 2.5–4 m   Jan–Aug
     Medium to tall **shrub** with hairy bright green young growth; **bark** grey, smooth; **branches** spreading to ascending; **branchlets** terete, initially whitish-hairy, often pendent; **leaves** 3–7 cm × about 0.05 cm, linear, alternate, terete or grooved below, spreading to ascending, crowded, often slightly curved, deep green, becoming glabrous, apex with soft hooked point; **flowers** about 0.8 cm across, golden yellow, fragrant, exterior faintly hairy, on short, spreading to erect, hairy stalks, subtended by reduced leaves, in terminal racemes to 30 cm or more long, often profuse and very conspicuous; **drupes** to about 1.5 cm × 1.2 cm, initially green, becoming reddish-purple, smooth.
     A highly decorative, often graceful, desirable species from the Central Coast. It grows on sandstone in dry sclerophyll forest and heath. Flowering can commence early January and continue through to August followed by olive-like fruits which are usually ornamental too. Very popular in cultivation and suitable for temperate and subtropical regions. Grows in a range of sandy or clay soils which have adequate drainage. Adapts to a sunny or semi-shaded site. Hardy to moderately heavy frosts. Responds very well to pruning and regular clipping. It is an excellent hedging plant. Also suitable as an accent plant or for screening and general planting. Sometimes grown in containers and can be used as a Christmas tree. Cut foliage and flowers are used for floral decoration. Propagate from seed, or from cuttings of very soft young growth, which can be slow to produce roots. It is best not to remove soft tip growth from cuttings even if it wilts because blackening of growth usually occurs and the cuttings often die.
     P. isophylla is similar but has leafy racemes of flowers.

Persoonia procumbens L. A. S. Johnson & P. H. Weston
(procumbent)
NSW
0.05–0.2 m × 0.8–2 m   Dec–Feb
     Dwarf **shrubs** often with rusty-hairy young growth; **branches** prostrate; **branchlets** with rusty hairs; **leaves** 1.5–3.6 cm × 0.4–2 cm, mainly obovate to spathulate, alternate, sometimes slightly twisted at base, spreading to ascending, scattered to moderately crowded, pale-to-mid green, becoming glabrous, flat, margins faintly recurved, venation usually indistinct, apex blunt with a small point; **flowers** about 0.8 cm across, yellow, exterior faintly hairy, on short, erect, hairy stalks, in axillary or terminal leafy racemes to about 2 cm long; **drupes** green, smooth.
     An appealing groundcovering species which was described in 1991. It occurs in dry sclerophyll woodland and forest in the Armidale region of the Northern Tablelands District. Grows in sandy to clayey soils associated with granite. Adapts well to cultivation in temperate regions and may succeed in subtropical areas. Requires light to medium acidic soils which

Persoonia rigida   W. R. Elliot

drain moderately well and tolerates a sunny or semi-shaded site. Responds well to pruning. Hardy to most frosts and withstands extended dry periods. Suitable for containers, rockeries and general planting. Propagate from seed, or from cuttings of very young growth, which may be slow to produce roots. It has been cited as *P. prostrata* ssp. C.
     The closely allied *P. daphnoides* has greyish to tawny hairs and has thinner leaves with distinct veins.

Persoonia prostrata R. Br.
(prostrate)
Qld
0.1–0.2 m × 1–2 m   Dec–Feb
     Dwarf spreading **shrub** with hairy young growth; **branches** prostrate; **branchlets** becoming glabrous; **leaves** 1.2–5.5 cm × 0.5–2.5 cm, elliptic to obovate, alternate, spreading to ascending, scattered to moderately crowded, pale-to-mid green, becoming glabrous, smooth, thick, margins flat, midrib distinct, apex blunt with point; **flowers** about 0.8 cm across, yellow, exterior hairy, on short hairy stalks, in leafy racemes; **drupes** to about 1.5 cm × 1 cm, smooth.
     This desirable species is confined to Fraser Island where it grows in sandy soils along the coast. Probably best suited to subtropical regions but may adapt to temperate areas. Soils must be freely draining and a sunny or semi-shaded site should be suitable. It has potential for planting in containers, rockeries and general areas. Propagate from seed or from cuttings of very young growth.
     P. stradbrokensis is allied but grows as a much taller shrub with larger leaves and larger flowers.

227

Persoonia pungens W. Fitz.
(ending in a hard sharp point)
WA
0.2–0.8 m × 0.8–1.5 m   Sept–Dec

Dwarf, spreading to erect **shrub** with a lignotuber; young growth greyish-hairy; **branches** from near base, spreading to ascending; **branchlets** short, terete, becoming glabrous; **leaves** 0.3–1.5 cm × up to 0.5 cm, elliptical to narrowly oblong, alternate, twisted at base, spreading to ascending, crowded, leathery, green, becoming glabrous, flat, midvein sometimes distinct, apex pungent-pointed; **flowers** to about 1 cm across, bright yellow, exterior glabrous, on short glabrous to hairy stalks, in short terminal or axillary leafy racemes, often profuse and conspicuous; **drupes** to about 1.2 cm × 0.8 cm, green with purple streaks.

An attractive inhabitant of heathland in the north-western Avon and southern Irwin Districts where it grows in sandy or loam soils. It is regarded as vulnerable in the wild and warrants cultivation as part of a conservation strategy. Suited to warm temperate and semi-arid regions and requires well-drained acidic soils in sunny or semi-shaded sites. Plants are hardy to moderate frosts and withstand extended dry periods. It has potential for garden planting and rockeries. Propagate from seed or from cuttings of very young growth.

P. leucopogon is allied but it has densely hairy flower stalks and exterior of flowers.

Persoonia quinquenervis Hook.
(five-nerved)
WA   Kanberry
0.3–2.5 m × 0.3–1.6 m   Oct–Dec; also sporadic

Dwarf to medium **shrub** with a lignotuber; young growth hairy; **branches** many, from near ground level, ascending to erect; **branchlets** short, initially angular and brownish-hairy, becoming terete and glabrous; **leaves** 0.5–7.5 cm × up to 1 cm, linear to narrowly spathulate or narrowly oblong, alternate, twisted at base and can often appear spiralled, sometimes incurved, ascending to erect, can be crowded, dark green and often glabrous, becoming glabrous, flat to nearly terete with 3–13 ridges, smooth to slightly rough, apex softly pointed; **flowers** about 0.8 cm across, bright yellow, exterior sometimes faintly hairy, on ascending to erect hairy stalks to about 1.7 cm long, in terminal or axillary leafy racemes to about 6 cm long; **drupes** to about 1.5 cm × 0.9 cm, smooth.

P. quinquenervis is a slow-growing species with interesting foliage. It has a wide distribution in the south-western region where it grows in a range of habitats and soil types but most commonly in sand or loam over laterite or gravel. Uncommon in cultivation, it should adapt well to temperate and semi-arid regions in a range of acidic soils with fair to excellent drainage. Plants tolerate a sunny or semi-shaded site. Hardy to moderately heavy frosts. Suitable for containers, rockeries and as an accent plant. Propagate from seed, or from cuttings of very young growth, which can be slow to form roots.

The allied *P. striata* has much narrower leaves with 3 prominent longitudinal ridges.

Persoonia recedens Gand.
(receding, away from another)
NSW
0.3–1 m × 0.8–2.5 m   Dec–Jan; also sporadic

Dwarf spreading **shrub** with hairy young growth; **branches** many, spreading to ascending; **branchlets** becoming glabrous; **leaves** 1–2 cm × up to 0.4 cm, narrowly oblong to narrowly elliptical, alternate to somewhat opposite, spreading to ascending, moderately crowded, mid green, sometimes paler below, becoming glabrous, margins flat to slightly recurved, apex softly pointed; **flowers** about 0.8 cm across, yellow, exterior glabrous, on short glabrous stalks, in short, leafy, axillary or terminal racemes; **drupes** small, green, smooth.

This species occurs in the upper Blue Mountains of the Central Tablelands where it grows on sandstone in dry sclerophyll forest. Extremely rare in cultivation, it deserves to be better known. Should do well in temperate regions and may succeed in subtropical areas. Needs freely draining acidic soils and a sunny or semi-shaded site. Hardy to moderately heavy frosts. It has potential for general planting and may be suitable for low hedging. Propagate from seed or from cuttings of very young growth. It has been cited as *P. nutans* ssp. C.

The closely allied *P. acuminata* has broader leaves which have prominently recurved margins.

Persoonia revoluta Sieber ex Schult. & Schult. f. = *P. mollis* ssp. *revoluta* (Sieber ex Schult. & Schult. f.) S. L. Krauss & L. A. S. Johnson

Persoonia rigida R. Br.
(rigid, stiff)
NSW, Vic   Hairy Geebung; Stiff Geebung
1–2.5 m × 0.7–2.5 m   Oct–Dec

Small to medium **shrub** with very hairy, often reddish to bronze new growth; **branches** many, spreading to erect; **branchlets** terete, hairy; **leaves** 1.5–5 cm × 0.4–2 cm, oblanceolate to spathulate, alternate, spreading to ascending, often incurved, moderately crowded, deep green, becoming glabrous or nearly so, flat to convex, margins recurved, apex blunt with a small point; **flowers** about 1 cm across, yellow, exterior hairy, on short, erect, hairy stalks, in short, leafy, terminal racemes, rarely profuse; **drupes** of about 1.2 cm × 1 cm, green to purplish green.

A sturdy often rounded species with a wide distribution from the Northern Tablelands of NSW to the Little Desert in western Vic. It occurs in dry sclerophyll forests and dry heath and is often associated with rocky soils. Rarely cultivated, it is deserving of greater attention. Plants are best suited to temperate regions and need excellently drained acidic soils. Prefers a sunny or semi-shaded site and also does well in dappled shade. It has potential for cut-foliage production and general planting. Propagate from seed, or from cuttings of very young growth, which can be slow to produce roots.

Persoonia rudis Meisn.
(rough)
WA
0.2–1 m × 0.6–2 m   Sept–Jan

Dwarf, erect to spreading **shrub** with long, brown to greyish hairs on the young growth; **stems** many, branching from below ground level; **branches** many, spreading to erect; **branchlets** terete and glabrous when mature; **leaves** 0.5–4.5 cm × up to 0.15 cm, linear, alternate, ascending to erect, crowded, stiff to flexible, deep green, initially densely hairy, becoming less hairy, often concave above, apex sharply pointed; **flowers** about 0.8 cm across, yellow, exterior moderately hairy, on hairy stalks to about 1 cm long, in terminal or axillary somewhat leafy racemes to about 10 cm long; **drupes** to about 1.3 cm × 0.9 cm, hairy.

This is an endangered species from the southern Irwin and northern Darling Districts. Occurs in dry sclerophyll forest or low heath and grows in sandy soil overlying lateritic gravel. Warrants cultivation as part of a conservation strategy. Probably will do best in warm temperate regions in freely draining acidic soils with a warm to hot aspect. Propagate from seed or from cuttings of very young growth.

P. filiformis is closely allied but it has shorter hairs and the flowers have distinctive slender anther appendages.

Persoonia rufa L. A. S. Johnson & P. H. Weston
(red to reddish-brown)
NSW
1–2.5 m × 1–3 m   Dec–Feb

Small to medium **shrub** with tawny to rusty-hairy young growth; **bark** smooth; **branches** spreading to ascending, often from near ground level; **branchlets** rusty-hairy; **leaves** 3–8 cm × 1–2.5 cm, elliptical to broadly elliptical, alternate, usually twisted at base, ascending to erect, sometimes incurved, green, paler below, initially hairy becoming less so with maturity, flat, margins recurved, apex softly pointed; **flowers** about 0.7 cm across, yellow, exterior with rusty hairs, on short, erect, hairy stalks, in short, leafy, terminal or axillary racemes, often profuse and conspicuous; **drupes** green or also with purplish streaks.

A handsome, recently described (1991) species which is confined to the Gibraltar Range in the Northern Tablelands. It is found on granitic soils in heath or dry sclerophyll forest. Evidently rarely cultivated, it should be grown more. Plants will be suitable for temperate regions and may succeed in subtropical areas. Requires well-drained acidic soils and tolerates a sunny or semi-shaded site. Hardy to most frosts. Useful for low screening and general planting. Propagate from seed or from cuttings of very young growth.

It has been known as *P. cornifolia* ssp. C. *P. cornifolia* differs in not having rusty hairs. The allied *P. oleoides* has narrow leaves and greyish to rusty hairs.

Persoonia rufiflora Meisn.
(reddish to reddish-brown flowers)
WA
0.5–2.5 m × 0.6–2 m   June–Oct

Dwarf to medium **shrub** with a lignotuber; young growth with greyish to rusty hairs; **stems** many, branching near base; **bark** mottled grey; **branches** spreading to ascending; **branchlets** terete, becoming less hairy; **leaves** 0.8–4.5 cm × up to 0.8 cm, linear to oblanceolate, alternate, ascending to erect, often incurved,

crowded, deep green, often slightly glaucous, usually stiff, becoming more or less glabrous, flat to convex, margins recurved to revolute, 3 prominent ridges above, apex softly pointed; **flowers** about 0.8 cm across, yellow, exterior rusty-hairy, sessile, solitary or paired, in middle to upper axils, moderately conspicuous; **drupes** to about 0.8 cm × 0.5 cm, kidney-shaped to obovoid, smooth.

An interesting species from the north-western Avon, northern Darling and Irwin Districts where it grows on sand or sandy loam which often overlies laterite. It also occurs on ironstone breakaways. Rare in cultivation, it should perform best in warm temperate and semi-arid regions. Needs freely draining acidic soils and a warm to hot aspect but will tolerate some shade. Hardy to moderate frosts. Propagate from seed or from cuttings of very young growth.

P. scabrella Meisn. is a synonym.

P. brevirhachis and *P. inconspicua* have similarities but both lack the 3 ridges on leaves and have flowers which are stalked.

Persoonia saccata R. Br.
(pouched)
WA   Snotty Gobble; Pouched Persoonia
0.2–2 m × 1–2 m   July–Jan

Dwarf to small **shrub** with a lignotuber; young growth with grey hairs; **stems** usually branching from underground; **bark** smooth, greyish, sometimes flaky at base; **branches** spreading to ascending; **branchlets** becoming terete and more or less glabrous; **leaves** 1–17 cm × to about 0.15 cm, linear to somewhat terete, alternate, spreading to erect, slightly incurved, crowded, deep green, often glaucous, becoming glabrous, margins recurved to revolute, grooved below, apex softly pointed; **flowers** to about 0.8 cm across, yellow to greenish-yellow, prominently pouched, exterior hairy, on whitish-hairy stalks to about 1.2 cm long, in terminal or axillary leafy racemes 2–45 cm long, often profuse and conspicuous; **drupes** to about 1.3 cm × 0.9 cm.

A very handsome species from the western Darling District where it inhabits forest and woodland and grows in sandy or gravelly soils. Needs to become better known in cultivation. Plants are best suited to temperate regions and require very well-drained acidic soils with a sunny or semi-shaded aspect. They are hardy to moderately heavy frosts and tolerate extended dry periods. Plants respond very well to pruning and flowering stems are excellent for indoor decoration. Propagate from seed or from cuttings of very young growth.

P. fraseri R. Br. and *P. macrostachya* Lindl. are synonyms and were cultivated under these names in England in 1837. The allied *P. comata* differs in having broader leaves.

Persoonia saundersiana Kippist
(after William Wilson Saunders, 19th-century English botanist)
WA
1–5 m × 1.5–5 m   Sept–Nov; also sporadic

Small to tall **shrub** with greyish-hairy young growth; **stems** many, branching from near base; **bark** mottled grey, smooth; **branches** spreading to ascending; **branchlets** initially angular and hairy, becoming terete and glabrous; **leaves** 1.5–21 cm × to about 0.3 cm, linear, alternate, sometimes twisted, spreading to erect, sometimes incurved, green to glaucous, becoming glabrous, flat, 2- to 4-grooved above and below, apex softly pointed; **flowers** to about 1 cm across, bright yellow, exterior more or less glabrous, on hairy stalks to 2 cm long, usually in terminal leafy racemes to about 10 cm long; **drupes** to about 1.3 cm × 1 cm, smooth.

This spreading species occurs in a large area of the central south-western region and is found growing in sand or loam overlying laterite. It is associated with She-oak or *Acacia* shrubland, heath or mallee heath. Should do best in temperate and semi-arid regions. Needs acidic soils which have good drainage and will tolerate a sunny or semi-shaded site. Plants are hardy to moderately heavy frosts. Propagate from seed or from cuttings of very young growth.

P. diadena F. Muell. is a synonym.

Persoonia scabra R. Br.
(rough)
WA
0.3–1 m × 0.5–1.5 m   Nov–Jan

Dwarf **shrub** with a lignotuber; young growth with grey to whitish hairs; **stems** usually branching from below ground level; **branches** spreading to ascending; **branchlets** terete, becoming glabrous; **leaves** 0.5–3.5 cm × to about 0.6 cm, narrowly oblong to narrowly spathulate, alternate, usually twisted at base, ascending to erect, can be incurved, flat, stiff, green, becoming glabrous, midvein distinct, apex sharply to pungent-pointed; **flowers** to about 0.8 cm across, yellow, exterior glabrous or hairy, on short, glabrous or hairy, erect stalks, terminal or in upper axils; **drupes** to about 1.2 cm × 0.8 cm, smooth.

Until recently *P. scabra* was regarded as rare in nature but subsequent collections have corrected this point of view. It occurs in the eastern Eyre and central Roe Districts where it grows on sand or sandy loam in shrub mallee communities. Evidently not cultivated in Australia but is recorded as growing in England in 1824. Plants are suited to temperate and semi-arid regions. They need excellently drained acidic soils and an open, moderately sunny, warm to hot site should be best. Hardy to moderate frosts. Propagate from seed or from cuttings of very young growth.

The closely allied *P. spathulata* has mixed glandular and non-glandular hairs and very rough leaves.

Persoonia scabrella Meisn. = *P. rufiflora* Meisn.

Persoonia sericea A. Cunn. ex R. Br.
(silky)
Qld, NSW
0.6–2 m × 0.8–2 m   Oct–April; also sporadic

Dwarf to small **shrub** with hairy young growth; **branches** spreading to erect; **branchlets** initially moderately to densely hairy, becoming less so; **leaves** 1–6 cm × 0.15–2.1 cm, mainly obovate to oblanceolate or spathulate, alternate, spreading to ascending, rarely crowded, deep green, initially hairy, becoming more or less glabrous, flat, margins sometimes recurved, midrib distinct, apex blunt to finely pointed; **flowers** about 1 cm across, yellow, exterior hairy, on spreading or recurved hairy stalks to 1.2 cm long, in short leafy racemes; **drupes** to about 1.4 cm × 0.8 cm, faintly hairy, green to purplish.

The distribution of this species is from south-eastern Qld to near Grenfell in NSW. Plants occur in sandy soils in woodland or dry sclerophyll forest. It is rarely cultivated and should do best in acidic soils in temperate and subtropical regions. Drainage must be excellent. A sunny or semi-shaded site is suitable. Plants are hardy to moderate frosts and can withstand extended dry periods. Propagate from seed or from cuttings of very young growth.

Hybrids between *P. sericea* and *P. cornifolia, P. tenuifolia* or *P. terminalis* occur in the wild and some may be cultivated.

Persoonia silvatica L. A. S. Johnson
(growing in woods)
NSW, Vic   Forest Geebung
3–6 m × 2–4.5 m

Medium to tall **shrub** with hairy young growth; **bark** smooth, hard; **branches** spreading to ascending; **branchlets** hairy and often reddish when young; **leaves** 3–12 cm × 0.6–2.5 cm, lanceolate to oblanceolate or narrowly spathulate, alternate, spreading to ascending, deep green above, paler below, becoming more or less glabrous, flat, margins recurved, smooth, venation distinct, apex pointed; **flowers** to about 1 cm across, yellow, exterior glabrous or hairy, on short erect stalks, in short leafy racemes or clusters; **drupes** to about 1.3 cm × 0.7 cm, hairy, becoming blackish.

Forest Geebung occurs on the South Coast and Southern Tablelands of NSW and in East Gippsland. It grows in woodland and wet sclerophyll forest. Very rare in cultivation, it should do best in cool temperate regions. Well-drained sandy or loamy acidic soils are required. A semi-shaded site is preferred but plants should tolerate greater sunshine. They appreciate organic mulches and may need supplementary watering during extended dry periods. Propagate from seed or possibly from cuttings of very young growth.

P. subvelutina is similar but has hairy mature leaves with strongly recurved to revolute margins.

Persoonia spathulata R. Br.
(spoon-shaped)
WA
0.2–0.6 m × 0.5–1.5 m   Dec–Jan

Dwarf **shrub** with hairy young growth; **stems** many, branching from below ground level; **branches** spreading to ascending; **branchlets** terete, initially with spreading to incurved glandular and non-glandular hairs, becoming more or less glabrous; **leaves** 0.7–3.8 cm × 0.5–1 cm, mainly narrow-spathulate to spathulate, alternate, usually twisted at base, ascending to erect, can be incurved, deep green, becoming glabrous, flat, venation distinct, apex sharply pointed; **flowers** about 0.8 cm across, yellow, exterior glandular-hairy, on erect hairy stalks to about 1 cm long, 1–2 in upper axils; **drupes** smooth.

This poorly known species was apparently not collected from the wild between 1802–1979. It occurs in

Persoonia stradbrokensis   W. R. Elliot

the eastern Eyre and eastern Roe Districts where it grows in sandheath. Warrants cultivation as part of a conservation strategy. Should grow best in warm temperate and semi-arid regions. Plants require freely draining acidic soils and a warm to hot aspect which can be sunny or semi-shaded. They are hardy to moderate frosts. May have potential for rockeries and containers. Its growth habit suggests that it has a lignotuber. Propagate from seed or from cuttings of very young growth.

P. scabra is allied but does not have glandular hairs.

Persoonia stradbrokensis Domin
(from the Stradbroke Island region)
Qld, NSW
4–8 m × 2–5 m   Dec–Feb

Tall **shrub** to small **tree** with hairy young growth; **bark** smooth, grey; **branches** spreading to erect; **branchlets** initially hairy, becoming more or less glabrous; **leaves** 3–11 cm × 1–4 cm, narrowly elliptic to broadly ovate or obovate, alternate, spreading to ascending, sometimes crowded, deep green, becoming glabrous, flat to convex, smooth, margins slightly recurved, apex softly pointed; **flowers** about 0.8 cm across, yellow, exterior hairy, on erect hairy stalks to about 0.4 cm long, in short, leafy, axillary or terminal racemes; **drupes** green with purple streaks.

A tall handsome species from south-eastern Qld and the North Coast District of NSW. It is found in coastal heath and dry sclerophyll forest. Plants are very limited in cultivation. Recommended for subtropical and cool temperate regions. Freely draining acidic soils and a semi-shaded site should be suitable. Worth trying as a screening or hedging plant. Should respond well to regular pruning and/or clipping. It has been cultivated as *P. cornifolia* ssp. A. Propagate from seed or from cuttings of very young growth.

P. prostrata is allied in flower and foliage but has distinct dwarf growth characteristics.

Persoonia striata R. Br.
(finely striped or grooved)
WA
0.2–0.7 m × 0.8–2 m   Oct–Dec

Dwarf **shrub** with a lignotuber; young growth with purplish to whitish hairs; **stems** often branching from near base or underground; **branches** spreading to erect; **branchlets** initially angular and hairy, becoming terete and glabrous; **leaves** 0.5–4.5 cm × up to 0.3 cm, linear-oblong to linear-spathulate, alternate, mainly spreading to ascending, often incurved, crowded, stiff, deep green, becoming glabrous, flat to convex with 3 elongated ridges above and below, apex softly pointed; **flowers** to about 1 cm across, bright yellow, exterior glabrous, on erect glabrous to faintly hairy stalks to about 1 cm long, solitary or in axillary clusters; **drupes** to about 1.3 cm × 0.8 cm, smooth.

P. striata occurs in the Eyre and Roe Districts. It grows in sand, loam or clay which usually overlies gravel or laterite in heath of mallee-heath. Evidently not cultivated, it has potential for containers, rockeries and general planting. Should do well in temperate and semi-arid areas. Acidic soils which drain freely and a sunny or semi-shaded site are recommended. Plants tolerate moderate frosts and withstand extended dry periods. Propagate from seed or from cuttings of very young growth.

The narrow-leaved forms of *P. quinquenervis* are similar to *P. striata* but have more grooves in the leaves.

Persoonia stricta C. A. Gardner ex P. H. Weston
(erect, upright)
WA
1–5 m × 0.8–4 m   Aug–Dec

Small to tall **shrub** with greyish-hairy young growth; **stems** many, branching from near base; **branches** ascending to erect; **branchlets** initially angular and hairy, becoming terete and glabrous; **leaves** 1.7–15 cm × to about 0.8 cm, linear-oblong to linear-spathulate, alternate, can be twisted at base, leathery, ascending to erect, often incurved, crowded, deep green and slightly glaucous, becoming glabrous, apex blunt or fine, softly pointed; **flowers** to about 1 cm across, bright yellow, exterior glabrous or faintly hairy, on spreading hairy stalks to about 1 cm long, in mainly terminal often leafy racemes to 10 cm long; **drupes** to about 1.3 cm × 1 cm, smooth.

A recently described (1994) species which inhabits the Irwin and north-western Avon Districts. Plants are found in sand or sandy loam, usually overlying laterite, in heath, shrubland or woodland. Apparently not cultivated, it should be best suited to warm temperate and semi-arid regions. Plants require very well-drained acidic soils and a warm to hot, sunny or semi-shaded site. They tolerate moderately heavy frosts and extended dry periods. May be useful for screening and hedging purposes. Propagate from seed or from cuttings of very young growth.

The allied *P. saundersiana* has grooved leaves while *P. comata* has creeping underground rhizomes.

231

Persoonia subtilis P. H. Weston & L. A. S. Johnson
(slender, fine)
Qld
0.2–1 m × 0.6–2 m   Nov–April

Dwarf **shrub** with a lignotuber; young growth with appressed greyish to tawny hairs; **stems** usually branching below ground level; **branches** prostrate to ascending; **branchlets** initially hairy; **leaves** 0.5–5 cm × up to 0.1 cm, linear, alternate, somewhat terete, spreading to erect, recurved to incurved, bright green, hairy or nearly glabrous on maturity, grooved below, apex softly pointed; **flowers** about 0.7 cm across, yellow, exterior hairy, on hairy stalks to about 0.8 cm long, in short, leafy, terminal or axillary racemes, moderately conspicuous; **drupes** to about 1.2 cm × 0.8 cm, green with purple streaks.

A recently described (1994) species which is moderately common in the south-eastern inland regions where it grows in sandy soils of dry sclerophyll forest and woodland. It is rarely cultivated and deserves to be better known. Plants should adapt to subtropical and temperate regions with very well-drained acidic soils. A sunny or semi-shaded site is suitable. Plants are hardy to moderately heavy frosts. It has potential for use in containers, rockeries and general planting. Propagate from seed and from cuttings of very young growth.

This species has been confused with *P. fastigiata* which is restricted to north-eastern NSW and has broader leaves and spreading hairs.

Persoonia subvelutina L. A. S. Johnson
(somewhat velvety)
NSW, Vic   Velvety Geebung
2–6 m × 1.5–3.5 m   Jan–March

Medium to tall **shrub** with velvety young growth; **bark** smooth, grey; **branches** spreading to ascending; **branchlets** velvety, hairy; **leaves** 3–7 cm × 0.6–1.5 cm, narrowly elliptic to obovate or spathulate, alternate, spreading to ascending, can be crowded, deep green above, much paler below, initially densely hairy becoming less so, flat, margins recurved to revolute, apex softly pointed; **flowers** to about 1.4 cm across, yellow, exterior hairy, on short erect stalks to 0.4 cm long, in upper axils forming leafy racemes; **drupes** to about 1.2 cm × 1.2 cm, hairy.

An inhabitant of subalpine woodland and wet sclerophyll forests in the Southern Tablelands of NSW and north-eastern Vic. Often grows on sheltered valley hillsides. Rarely encountered in cultivation and warrants greater recognition. Plants are best grown in cool temperate regions with freely draining acidic soils in sites which are shaded for most of the time. Should appreciate organic mulches and supplementary watering during extended dry periods. Propagate from seed or from cuttings of very young growth.

P. silvatica is allied but it has more-or-less glabrous mature leaves with slightly recurved margins.

Persoonia sulcata Meisn.
(longitudinally grooved)
WA
0.2–1 m × 0.8–2 m   Sept–Nov

Spreading to erect, dwarf **shrub** with greyish-hairy young growth; **stems** branching near base; **branches** spreading to ascending; **branchlets** initially angular and hairy, becoming terete and glabrous; **leaves** 0.5–5 cm × about 0.1 cm, linear, alternate, slightly twisted at base, spreading to ascending, often crowded, deep green, becoming glabrous, flattish to convex or concave, 6-ridged, apex pungent-pointed; **flowers** to about 0.7 cm across, yellow, exterior glabrous, on glabrous spreading to erect stalks to about 1.2 cm long, 1–3 flowers per axil; **drupes** to about 1 cm × 0.8 cm, smooth.

A very prickly-foliaged species which is regarded as vulnerable in the wild. It occurs in the northern Darling District where it grows in woodland on laterite or on granite slopes. Needs cultivating as part of a conservation strategy. Plants are best suited to warm temperate regions and will need well-drained acidic soils with a sunny or semi-shaded aspect. They tolerate moderate frosts and extended dry periods. May be useful for foot traffic control. Propagate from seed or from cuttings of very young growth.

The allied *P. striata* lacks pungent leaves while *P. acicularis* has branches with long hairs and somewhat glaucous leaves.

Persoonia tenuifolia R. Br.
(slender leaves)
Qld, NSW
0.2–1.3 m × 0.6–2.5 m   Sept–April

Dwarf to small **shrub** with hairy young growth; **branches** prostrate to ascending; **branchlets** initially hairy, becoming glabrous; **leaves** 0.6–2.5 cm × about 0.05 cm, linear, somewhat terete, grooved above, alternate, spreading to erect, incurved, crowded, green, becoming glabrous, apex softly pointed; **flowers** about 0.7 cm across, yellow, exterior glabrous, on short, glabrous or hairy stalks, in short, leafy, terminal racemes; **drupes** to about 1.1 cm × 0.8 cm, glabrous.

An attractive groundcover or dwarf shrub from south-eastern Qld and on the North Coast, Northern Tablelands and North Western Plains in NSW. The distribution extends from coastal heathland to dry sclerophyll forest of mountains. *P. tenuifolia* is rarely cultivated and needs to be better known. Plants should adapt to subtropical and temperate regions in well-drained acidic soils and will tolerate a sunny or semi-shaded site. Hardy to moderately heavy frosts. Suitable for general planting, rockeries and containers. It was introduced to cultivation in England in 1822. Propagate from seed or from cuttings of very young growth.

In nature, *P. tenuifolia* can hybridise with *P. cornifolia*, *P. sericea* and *P. stradbrokensis*. A hybrid *P. tenuifolia* × *cornifolia* is cultivated to a very limited degree.

Persoonia teretifolia R. Br.
(terete leaves)
WA
0.5–3 m × 0.6–2 m   Aug–Feb; possibly sporadic

Dwarf to medium **shrub** with greyish-hairy young growth; **stems** many, branching from base; **branches** spreading to erect; **branchlets** terete, becoming glabrous; **leaves** 0.5–7 cm × to about 0.15 cm, linear, terete, alternate, not twisted, mainly ascending to erect, usually recurved, often crowded, green, mainly glabrous, apex softly pointed; **flowers** to about 1 cm across, bright yellow, exterior hairy, on densely hairy short stalks, in terminal leafy racemes to about 10 cm long; **drupes** to about 1.5 cm × 0.7 cm, smooth.

A bushy species which is distributed between Israelite Bay and Albany, where it grows on the coast or some distance inland. It occurs in a range of sand and clay soils which often overlie laterite and is common in heath and mallee-heath. Evidently rarely cultivated, it is suitable for temperate and semi-arid regions. Needs well-drained acidic soils and plenty of sunshine. Plants tolerate moderate frosts. Propagate from seed or from cuttings of very young growth.

The var. *amblyanthera* Benth. is a synonym of *P. biglandulosa* P. H. Weston.

Persoonia terminalis L. A. S. Johnson & P. H. Weston
(terminal; at the apex)
NSW
0.7–1.5 m × 1–2.5 m   Dec–Jan; also sporadic
Dwarf to small **shrub** with faintly hairy young growth; **bark** smooth; **branches** spreading to erect; **branchlets** faintly hairy; **leaves** 0.35–1 cm × up to 0.2 cm, narrowly oblong, alternate, spreading to ascending and slightly recurved near apex, crowded, convex, green, becoming glabrous or faintly hairy, margins recurved, apex softly pointed; **flowers** about 0.8 cm across, yellow, exterior faintly hairy, on very short stalks, up to 5 per terminal cluster, conspicuous; **drupes** green with purple streaks, smooth.

The typical variant is restricted to the Torrington region in the Northern Tablelands where it grows on granitic soils in dry sclerophyll forest. It needs to be cultivated as part of a conservation strategy. Probably best suited to temperate regions but may succeed in subtropical areas. Requires very well-drained acidic soils with a sunny or semi-shaded aspect. They are hardy to heavy frosts and withstand extended dry periods.

The ssp. *recurva* L. A. S. Johnson & P. H. Weston is found in south-eastern Qld and also extends to the North Western Slopes of north-eastern NSW where it occurs on sandstone. It has leaves to about 0.75 cm long which are prominently recurved at the apex. It has been successful in cultivation.

Propagate both subspecies from seed or from cuttings of very young growth.

P. terminalis has been cited as *P. nutans* ssp. D, and also as a variant of *P. oxycoccoides*.

Persoonia tortifolia Meisn. = *P. trinervis* Meisn.

Persoonia trinervis Meisn.
(three-nerved)
WA
0.3–1.8 m × 0.5–1.5 m   Sept–Dec
Dwarf to small **shrub** with a lignotuber; young growth with greyish to rusty hairs; **stems** solitary to many, branching from below ground level; **branches** ascending to erect; **branchlets** terete, initially densely hairy, becoming glabrous; **leaves** 0.7–7 cm × 0.3–2 cm, linear-oblanceolate to spathulate, alternate, prominently spirally twisted, spreading to erect, often crowded, deep green, becoming glabrous, 3-ridged below, apex finely to bluntly pointed; **flowers** to about 0.7 cm

across, yellow, exterior densely hairy, on short, very hairy stalks, in terminal or axillary clusters; **drupes** to about 1.2 cm × 0.6 cm, smooth.

An intriguing and variable species from the Avon, Irwin and Roe Districts where it grows in sandy or loamy soils, which usually overlie laterite. Plants are found in heath, mallee-heath or mallee woodland. Rarely cultivated, it certainly has oddity value with its interesting foliage. Attempts at cultivation have usually failed probably because a mycorrhizal relationship is necessary. It is best suited to warm temperate and semi-arid regions but may also succeed in cooler areas. Soils must be acidic and drain freely. Plants need plenty of sunshine. They are hardy to moderately heavy frosts. May be planted in rockeries and containers and is likely to have penjing-bonsai potential. Propagate from seed or from cuttings of very young growth which can be slow to form roots.

P. tortifolia Meisn. is a synonym.

There are some allied species. *P. papillosa* has more flowers in longer racemes and *P. angustiflora* has very narrow, nearly terete leaves. *P. rufiflora* has leaves with 3 ridges on the upper surface and prominently recurved margins.

Persoonia tropica P. H. Weston & L. A. S. Johnson
(of the tropics)
Qld
2–3.5 m × 1.5–3.5 m   mainly Sept–Nov
Medium **shrub** with greyish to brownish hairs on young growth; **bark** greyish, smooth; **branches** spreading to erect; **branchlets** hairy, becoming glabrous; **leaves** 0.6–11 cm × 0.2–2.1 cm, narrowly elliptical to oblanceolate, alternate, not twisted at base, spreading to ascending, deep green above, slightly paler below, becoming glabrous except for basal hairs, flat, margins recurved, apex softly pointed; **flowers** about 0.8 cm across, yellow, exterior faintly hairy, on short, erect, hairy stalks, in short terminal racemes; **drupes** to about 1.4 cm × 0.9 cm, pale yellow to green.

This recently described (1994) species is found west of Innisfail in the Great Dividing Range where it grows in well-drained soils in dry or wet sclerophyll forests. Suitable for tropical and subtropical regions and should succeed in frost-free temperate regions. Acidic, freely draining soils and a semi-shaded site should be suitable. Plants may need supplementary watering during extended dry periods. Propagate from seed or from cuttings of very young growth.

Persoonia virgata R. Br.
(twiggy)
Qld, NSW   Small-leaved Geebung
2–6 m × 1.5–4 m   Aug–Feb
Medium to tall **shrub** with hairy young growth; **branches** ascending to erect, many, often slender; **branchlets** often reddish, initially hairy, becoming glabrous; **leaves** 2–6 cm × 0.1–0.5 cm, linear to linear-obovate, alternate, spreading to ascending, often somewhat crowded, bright green, becoming glabrous, flat, margins softly pointed; **flowers** to about 1 cm across, yellow, exterior glabrous, on spreading to erect, glabrous stalks to about 1 cm long, in terminal leafy racemes, often profuse and conspicuous; **drupes** to

Persoonia virgata × .5

about 1.2 cm × 0.9 cm, becoming purplish, smooth.

A handsome bushy species from south-eastern Qld and the North Coast District in NSW. It inhabits sand or sandy loam which sometimes overlies clay. Plants are found in moist soils of wallum scrub and dry

Petalostigma banksii　　　　　　　　　D. L. Jones

sclerophyll forest. Not commonly cultivated, it deserves to be better known. Should be suitable for temperate regions as well as for subtropical areas. Needs acidic soils which are moisture-retentive but not prone to waterlogging. It has potential for use in hedging, screening and general planting. Propagate from seed or from cuttings of very young growth.

Persoonia volcanica L. A. S. Johnson & P. H. Weston
(from volcanic soils)
Qld, NSW
1.8–6 m × 1.5–4.5 m　　　　　　Dec–Feb; also sporadic

Small to tall **shrub** with greyish to rusty-hairy young growth; **bark** greyish, smooth; **branches** spreading to erect; **branchlets** with greyish to rusty hairs; **leaves** 4–9 cm × 0.5–1.5 cm, linear to oblong, alternate, not twisted at base, spreading to ascending, green, becoming faintly hairy to glabrous at maturity, flat, margins recurved, apex softly pointed; **flowers** about 0.8 cm across, yellow, exterior hairy, on erect hairy stalks to 0.8 cm long, in short to long, leafy, axillary or terminal racemes; **drupes** 1.5 cm × 0.9 cm, green, glabrous.

An interesting, recently described (1991) species from a small area of the McPherson Range region of south-eastern Qld and north-eastern NSW. It inhabits sandy soils, or clay soils derived from granite, in dry sclerophyll woodland or forest. Warrants cultivation and plants should succeed if grown in acidic soils with a sunny or semi-shaded aspect in subtropical and temperate regions. Plants tolerate light frosts. May have application for screening and hedging as well as for general planting. Propagate from seed or from cuttings of very young growth.

P. media has been confused with this species. It differs in having rough lower bark, appressed hairs and generally larger leaves.

PETALOSTIGMA F. Muell.
(from the Greek *petalon*, petal; *stigma*, stigma; the stigmas are broad and petal-like)
Euphorbiaceae　　　　　　　　　　　　Bitter Barks

Shrubs or small **trees**; **bark** rough, fibrous or fissured; **leaves** simple, entire, alternate, glabrous or hairy; **flowers** small, unisexual; **male flowers** in axillary clusters, sometimes on leafless shoots, glabrous or hairy; **female flowers** solitary, glabrous or hairy; **tepals** 4; **fruit** a drupe, or schizocarp which separates into 3 or 4 segments.

A small genus of 7 species restricted to Australia with one species extending to New Guinea. They are mostly hardy shrubs and are often prominent in inland regions although one species grows in rainforest. Generally, they have limited ornamental appeal and are not much cultivated. Propagation is from seed but it can be very difficult. Seed from the rainforest species must be sown while fresh. Cuttings of firm young growth would be well worth trying.

Petalostigma banksii Britten & S. Moore
(after Sir Joseph Banks)
Qld, NT　　　　　　　　　　　　　　　Bitter Bark
2–3.5 m × 1–2 m　　　　　　　　　　　　Dec–Feb

Medium **shrub**, often straggly; **bark** dark grey, rough; young growth glabrous, shiny; **leaves** 2–3.5 cm ×

234

1.5–2.5 cm, broadly ovate to obovate or orbicular, dark green, shiny above, dull beneath; **male flowers** about 0.8 cm across, yellow in clusters, sometimes on leafless branches; **female flowers** about 0.8 cm across, yellow; **drupes** about 1.5 cm across, globular, orange-yellow, ripe May–July.

Occurs on rocky slopes and in woodland in northern areas. It sometimes dominates small communities. Plants have attractive foliage and colourful fruit. They are hardy shrubs for inland districts and tropical regions which have a seasonally dry climate. Needs a sunny position in well-drained soil. Propagate from seed and perhaps also by cuttings.

Petalostigma nummularium Airy Shaw
(button-like or coin-like)
Qld, WA, NT
1–2 m × 1–2 m   May–June
Dwarf bushy **shrub** with a dense habit; young growth densely grey-hairy; **leaves** 1–2.5 cm × 1–2 cm, broadly ovate to orbicular, thin-textured, somewhat papery, grey-green to brownish-green, hairy, apex blunt or shallowly notched; **male flowers** about 0.4 cm across, cream to yellowish, densely hairy, in dense axillary clusters; **female flowers** about 0.8 cm across, cream to yellowish, hairy; **drupes** about 1.5 cm across.

Scattered in open forest and woodland in tropical regions. Often grows in sand or laterite. Unknown in cultivation but has potential as a hardy shrub for semi-arid climates and seasonally dry tropical regions. Probably has similar requirements to *P. quadriloculare*. Propagate from seed and perhaps also by cuttings.

Petalostigma pachyphyllum Airy Shaw
(with thick leaves)
Qld
1–2 m × 1–2 m   July–Sept
Dwarf multi-stemmed **shrub** with a rounded habit; young growth short-hairy; **branches** brown; **leaves** 3–7 cm × 2–5 cm, elliptic to ovate, leathery, somewhat rigid, dark green, dull or shiny, apex blunt, ageing pinkish-brown; **male flowers** about 0.5 cm across, yellowish, hairy, in axillary clusters; **female flowers** about 0.8 cm across, yellowish, solitary or in pairs; **drupes** about 1.4 cm across, yellow to orange-red, ripe Sept.

Restricted to the Leichhardt district in Qld where it grows on sandstone slopes and outcrops in sandy or gravelly loams. Unknown in cultivation but it has potential for planting in subtropical and warm temperate regions. Attributes include a bushy habit, attractive foliage and colourful fruit. Would probably require a sunny location and well-drained soils. Propagate from seed and perhaps also by cuttings.

Petalostigma pubescens Domin
(shortly hairy)
Qld, NSW, WA, NT   Bitter Bark, Quinine Tree,
                                                   Native Quince
6–10 m × 3–4 m   Oct–Jan
Small to medium **tree** with a spreading crown; **bark** grey and black; young growth with grey silky hairs; **leaves** 2–6 cm × 1–2.5 cm, ovate to orbicular, dark green and glossy above, dull and hairy beneath; **male flowers** 0.4–0.8 cm across, creamy fawn, hairy, in

Petalostigma pubescens × .4

axillary clusters of 3 or 4; **female flowers** about 0.8 cm across, creamy-fawn, hairy; **drupes** 1.2–1.7 cm across, round, orange-yellow, hairy, ripe April–Nov.

A very widely distributed species which is a common component of open forest and woodlands especially along watercourses. It also grows in dry rainforest. Distribution extends north to New Guinea and south to the Clarence River, NSW. Attractive in form, this species has spreading to drooping branches, shiny foliage and interesting fruit. Adapts to conditions in near-coastal localities to inland towns. Requires well-drained soil in positions ranging from semi-shade to full sun. Mulches and watering during dry periods are beneficial. Tolerates moderate frosts. Propagate from fresh seed and perhaps also from cuttings of hardened new growth.

Petalostylis cassioides   T. L. Blake

Petalostigma quadriloculare F. Muell.
(with 4 locules or compartments)
Qld, WA, NT   Quinine Bush
0.5–1.5 m × 0.5–2 m   Feb–Nov
   Dwarf to small **shrub**, often growing annually from
a woody rootstock; young growth silky-hairy; **leaves**
1.5–6 cm × 0.8–3.5 cm, elliptic, dark green and shiny
above, dull and hairy beneath; **male flowers** 0.4–0.8 cm
across, yellow, in axillary clusters; **female flowers**
0.6–0.8 cm across, yellow, hairy; **drupes** 1.2–1.4 cm
across, globular, yellow, ripe June–Aug.
   Widespread and common in woodlands and forests
of northern Australia. In areas of frequent fire, the
plants form short multi-stemmed shrubs and sprout
quickly from the rootstock after a fire. An interesting,
hardy shrub for inland districts and tropical regions
which have a seasonally dry climate. Requires a sunny
position in well-drained soil. Responds well to pruning. Propagate from seed and perhaps also by cuttings.

Petalostigma quadriloculare F. Muell.
   var. **glabrescens** Benth. = *P. triloculare* Müll. Arg.

Petalostigma triloculare Müll. Arg.
(in 3 locules or compartments)
Qld, NSW   Long-leaved Bitter Bark
8–15 m × 3–5 m   Oct–Jan
   Small to medium **tree** with a straggly or rounded
crown; **bark** greyish-black; young growth with short
grey hairs; **leaves** 4–9 cm × 1–3.5 cm, narrowly elliptic,
dark green and shiny above, dull and hairy beneath;
male flowers 0.5–0.8 cm across, green to creamy-fawn,
lemon-scented, in axillary clusters of 3–5; **female
flowers** solitary, about 0.8 cm across, greenish; **drupes**
1–1.8 cm across, round, orange-brown, ripe March–
Oct.
   Occurs in dry rainforest and on rainforest margins
between Tewantin, south-eastern Qld, and Nymboida,
north-eastern NSW. The fruit explode when ripe scat-
tering the segments for up to 4 m. An interesting but
not highly ornamental species which is grown on a limited scale. Suitable for subtropical and temperate
regions. Young plants require a sheltered position in
well-drained soil. Tolerates light to moderate frosts.
Propagate from fresh seed, which takes 37–51 days to
germinate. Cuttings may be worth trying.
   P. quadriloculare F. Muell. var. *glabrescens* Benth. is a
synonym.

PETALOSTYLIS R. BR.
(from the Greek *petalon*, petal; *stylos*, style; referring to
the petal-like style)
Caesalpiniaceae
   Shrub lacking thorns or spines; **leaves** pinnate with
terminal leaf largest, hairy; **inflorescence** short axillary
racemes; **flowers** showy, sweetly scented; **sepals** 5;
petals 5, more or less equal; **fertile stamens** 3; **staminodes** 2; **style** petal-like; **fruit** a flat, erect pod.
   A small, endemic genus of 2–3 species widely distributed in arid and semi-arid regions. Plants can be
very ornamental when in flower and they are cultivated to a limited degree. They are best suited to arid,
semi-arid and warm temperate regions. In general,

Petalostylis cassioides × .65

they tolerate moderate frosts. Plants respond well to
regular pruning and can be used for hedging.
Butterflies are attracted to flowering plants. They
rarely suffer from pests and diseases. Propagate from
pre-treated seed which can be soaked in freshly boiled
water for 2–24 hours. For other seed treatments, see
Volume One, page 205. Cuttings of firm young growth
will produce roots. The application of rooting hormones may be beneficial.

Petalostylis cassioides (F. Muell.) Symon
(similar to the genus *Cassia*)
Qld, SA, WA, NT
0.7–2 m × 1–2.5 m   July–Oct; also sporadic
   Dwarf to small **shrub** with hairy young growth;
branches spreading to ascending, often maintained to
near ground level; **branchlets** hairy; **leaves** to about
15 cm long, pinnate; **leaflets** 11–80, 0.3–1.8 cm × up to
0.5 cm, obovate, green, faintly hairy, apex usually
slightly notched; **flowers** to about 3 cm across, yellow
to orange-yellow with reddish markings on top petal,
solitary or up to 3 in a short raceme, sweetly fragrant,
scattered to profuse, very conspicuous; **pods** about
2.5 cm × 0.8 cm, flattish, brown.
   A showy species from the semi-arid and arid regions
where it grows in a range of gravelly, sandy and sandy
clay soils. It is cultivated to a limited degree and
deserves to be better known. Plants need freely draining soils and can tolerate low alkalinity. A site with
plenty of sunshine is best but they will tolerate some
shade. Hardy to moderate frosts. Propagate from
treated seed or from cuttings of firm young growth.
   P. labicheoides R. Br. var. *cassioides* Benth. and var.
microphylla Ewart & Morrison are synonyms.

Petalostylis labicheoides R. Br.
(similar to the genus *Labichea*)
Qld, NSW, SA, WA   Butterfly Bush; Slender
                                              Petalostylis
1–3 m × 1.5–4 m   May–Dec; also sporadic
    Small to medium **shrub** with faintly hairy or glab-
rous and pruinose young growth; **branches** spreading
to ascending; **branchlets** sometimes pruinose; **leaves**
to about 8 cm long, pinnate; **leaflets** 3–19, lanceolate,
0.8–3 cm × up to 0.8 cm, green, flat to concave, apex
ending in a small point; **flowers** to about 4 cm across,
bright yellow with reddish markings, solitary or in
loose racemes of up to 5 flowers, sweetly fragrant, scattered to profuse, conspicuous; **pod** to about 2.5 cm ×
0.8 cm, flattish, brown.
    This very attractive species is widely distributed in
arid and semi-arid regions in similar conditions to
P. cassioides. It is sometimes encountered in cultivation
and is desirable because it has a long flowering period.
It adapts to a range of acidic and slightly alkaline soils
and does best if grown in a warm to hot site. Even
though plants tolerate extended dry periods they will
respond well to supplementary watering and may start
flowering again. Hardy to moderate frosts. Responds
well to pruning and can be used for hedging. Plants
may be partially deciduous during winter but once
spring comes new growth is produced quickly.
Propagate from treated seed or from cuttings of firm
young growth.

PETERMANNIA F. Muell.
(after Dr Augustus Petermann, 19th-century German
geographer)
Petermanniaceae (alt. *Philesiaceae, Smilacaceae*)
    A monotypic genus which is endemic in south-
eastern Qld and north-eastern NSW.

Petermannia cirrosa F. Muell.
(with tendrils)
Qld, NSW
Climber   Oct–Jan
    Vigorous **climber** or **scrambler**; **stems** to about 6 m
long, wiry, prickly, with leaf-opposed tendrils 5–15 cm
long; **leaves** 3.5–10 cm × 1–4 cm, lanceolate to ovate,
tapering to both ends, deep green, glossy, glabrous,
prominent longitudinal veins, apex pointed; **flowers**
to about 0.8 cm across, in few-flowered racemes; **tepals**
6, white, pink or rarely reddish, in 2 whorls; **stamens**
usually 6; **berry** to about 1.5 cm across, red, glabrous,
ripe Feb–May.
    An inhabitant of cool rainforests from near the
Qld/NSW border to north of the Hastings River, NSW.
Occurs from near sea-level to about 1,200 m altitude.
It lacks strong ornamental character but is useful for
cool sheltered sites in subtropical and temperate
regions. Needs moist but freely draining acidic soils
and responds well to application of organic mulches
and slow-release fertilisers. Tip pruning of young
plants promotes bushy growth. Withstands exposure
to moderate frosts. Propagate from fresh seed.

PETERMANNIACEAE Hutch.
A monogeneric family consisting of one endemic
species in the genus *Petermannia*. It is a perennial vine
with a fleshy rootstock. The climbing stems are wiry
and prickly with tendrils arising opposite the leaves.
Flowers are small and reddish with protruding stamens.

PETROPHILE R. Br. ex Knight
(from the Greek *petra*, rock; *phileo*, to love; referring to
the habitat of early collected species)
Proteaceae   Cone Bushes; Conesticks
    Dwarf to medium **shrubs**; **leaves** variable, simple
or 3-lobed or many-lobed, terete or flat, glabrous or
hairy; **inflorescence** cone-like, globular, ovoid or cylindrical, terminal or axillary, sessile or stalked, often
profuse and usually very conspicuous; **involucral
bracts** variable, sometimes deciduous; **cone scales** usually broad, somewhat woody, persistent; **flowers** white,
cream, yellow, pink, mauve or greyish-mauve, sessile;
tepals 4, partially united or split to base, exterior hairy
or glabrous; **stamens** 4; **style** slender, straight; **pollen
presenter** tapered to both ends or enlarged, apex
brush-like, hairy or glabrous; **fruiting cones** woody,
sessile or stalked, persistent; **fruit** a small usually compressed nut, usually hairy, sometimes winged.
    Petrophile is an outstanding member of the family
Proteaceae with many beautiful species which have
excellent potential for cultivation. It comprises 53
species which are all endemic in Australia. The greatest representation is in WA where 47 species occur —
the majority in the south-western region. *P. multisecta*
is confined to Kangaroo Island, SA, while the remaining 5 species are restricted to eastern Australia and are
distributed in south-eastern Qld and eastern NSW. For
many years this genus was known as *Petrophila* but the
name *Petrophile* was published first and has priority.
    Petrophile is very closely allied to *Isopogon* but the
latter has deciduous cone scales and nuts which are
not strongly flattened like those of *Petrophile*.
    Petrophiles occur from sea-level to moderate alti-
tudes with most species within 100 km of the coast.
Most are never subject to snowfalls with the exception
of *P. canescens, P. pedunculata, P. pulchella* and *P. sessilis*
which on very rare occasions may experience extremely short-term light falls. The majority of species
are found in areas where frosts occur but in most cases
these are only light to moderate in intensity.
    Petrophiles occur in a wide range of soil types which
are acidic in most cases. Plants inhabit predominantly
sandy or gravelly soil but the soil can also contain or
overly laterite and in some instances it can have a
moderately high clay or loam content or may contain
limestone. Usually soils are extremely well drained but
they may be subject to winter wetness and summer
dryness. *P. diversifolia* and *P. longifolia* can occur in
poorly drained winter-wet soils.
    The climatic range of *Petrophile* spans mainly tem-
perate and semi-arid zones with a very limited number
of species including *P. canescens, P. pulchella* and
P. shirleyae occurring in subtropical areas. Plants are
distributed in coastal and inland heath, shrubland,
mallee, woodland and sclerophyll forests usually in
open, often sparse vegetation where they receive
plenty of sunshine.
    In nature, some species are regarded as endangered
or rare. *P. aculeata, P. biloba, P. plumosa* and *P. trifurcata*
are among the most vulnerable.

Petrophile

Petrophile linearis   W. R. Elliot

Growth Habits and Features

Petrophiles have not achieved the popularity of their allied family members such as *Banksia, Grevillea* and *Hakea*, but nevertheless they have much to offer in cultivation. Floral displays are often profuse and plants show an interesting variation of foliage.

Some plants are dwarf and ground-hugging, eg. *P. longifolia*, or they may be bushy and spreading such as *P. serruriae*. Other species are erect with only a single or few stems, eg. *P. diversifolia*.

Foliage colour ranges from grey shades to deep green. Leaves can be simple or needle-like and often incurved. Some leaves are intricately much-divided with terete segments, eg. *P. divaricata*, or they may have flattened lobes, eg. *P. macrostachya, P. megalostegia* and *P. striata. P. glauca* (known wrongly as *P. trifida* for many years) has many-lobed, usually greyish-green leaves on a low spreading bush, and the result is highly alluring especially when combined with the cream to pale yellow flowerheads. The juvenile fern-like leaves of *P. diversifolia* are particularly attractive and can be maintained by judicious pruning although this may restrict flower production.

A limited number of species produce wonderful flushes of bronzish- to coppery-red young growth — *P. fastigiata, P. seminuda* and *P. striata* are excellent examples. *P. plumosa* has densely rusty-hairy young growth. *P. circinata* has flushes of deep red, hairy young growth which contrast vividly with the finely-lobed, glaucous, mature leaves.

Flower production can be profuse and the flowers are also very well displayed beyond the foliage. Most species produce terminal, globular or ovoid flower-heads. The majority of species have creamy yellow to yellow flowers and the main flowering occurs over late spring to early summer. Species renowned for flower production include *P. biloba*, greyish-mauve to pink flowers; *P. serruriae*, two distinct variants, one with creamy yellow flowers, the other with greyish-mauve to pink blooms; and *P. brevifolia, P. ericifolia* (and its close

allies), *P. media, P. seminuda* and *P. squamata* which all have particularly attractive yellow flowers.

P. linearis does not always produce profuse flower-heads but they are spectacular. It has deep pink flowers, mauve-tipped in bud and, on opening, the pollen presenter changes from yellow to deep orange. Most species are excellent cut flowers for indoor decoration as are stems with partially or fully developed fruiting cones.

Cultivation

Petrophiles are primarily used in urban landscape plantings especially in the home gardens of Australian plant enthusiasts. In the late 18th-century and up to the mid 19th-century there was considerable interest in petrophiles in England and Europe with a number of species cultivated in greenhouses for their ornamental value. *P. pulchella* (syn. *P. fucifolia*) was first cultivated in 1790 and *P. diversifolia* was introduced in 1803.

Within Australia it was once quite rare to encounter petrophiles in gardens because the closely allied genus *Isopogon* was more popular. Up to the 1950s, the two species most common in cultivation were *P. biloba* from WA and *P. sessilis* from NSW. In the 1960s and 1970s, *P. serruriae, P. squamata* and *P. teretifolia* became more readily available and in recent years *P. conifera, P. diversifolia, P. ericifolia* and *P. linearis* have been offered by nurseries.

Most petrophiles perform best in warm temperate and semi-arid regions. They require freely draining acidic soils. Excellent drainage is vital because many species are susceptible to fungal root diseases (see below). They usually do best in sites which receive plenty of sunshine and flower production is better in such sites. Heavy frost is damaging to most species. Light to moderate frosts may damage young growth but plants usually reshoot without any other ill effects. Most species will tolerate −2°C for short periods and some species such as *P. biloba, P. canescens, P. ericifolia, P. linearis, P. longifolia, P. pedunculata, P. pulchella, P. seminuda, P. sessilis, P. squamata* and *P. teretifolia* will tolerate −4°C with minimal or no damage.

Fertilising has an important bearing on successful cultivation of petrophiles. As with most other proteaceous plants they have a dislike for excess phosphorus. In most cases plants will grow adequately without supplementary feeding. If plant food is required it is recommended that a low-phosphorus fertiliser be used at about half the manufacturer's recommended rate. See Volume 1, pages 82, 95–99, 117 for further information on plant nutrition. Plants can sometimes indicate symptoms of iron chlorosis — young growth is yellow except for green midrib and veins — and iron chelates or sulphate of iron can be helpful in remedying this situation. Applications of magnesium sulphate (Epsom salts) may also be beneficial.

Quite a number of petrophiles develop as bushy, fairly dense plants but others have a limited number of branches and stems, eg. *P. diversifolia*. Although these plants are excellent for narrow sites, some species may need pruning to promote lateral growth. Tip pruning from an early stage should encourage bushy growth. The reaction of many petrophiles to pruning is poorly understood and further trials are needed. Therefore it is recommended that plants should not be pruned back to mature leafless wood. There are only a few species including *P. helicophylla, P. linearis* and *P. multisecta* which have lignotubers. They are capable of reshooting from very hard pruning and are suitable for coppicing.

Once petrophiles are established they will rarely require supplementary watering. Nearly all species tolerate extended dry periods. If watering is to be done it is best applied during cool spells of weather. *Phytophthora cinnamomi*, Cinnamon Fungus, can spread rapidly when conditions are moist and temperatures high. This fungus is often responsible for the death of many petrophiles. See Volume 1, page 168–70 for further information.

Petrophiles are sometimes susceptible to attack from other diseases, pests or ailments. Caterpillar larvae may chew on leaves but they are rarely a problem. Leaf miners can cause insignificant damage. Larvae may also be found in flowerheads and they will eat seed during the plant's development rendering the fruiting cones barren of seed. This is readily indicated by cutting through the cones just above the base. If there is no indicator of predator tunnels then seed is likely to be in good condition.

Propagation

Propagation of petrophiles is generally from seed or from stem cuttings.

Seed is not often readily available outside specialist Australian seed suppliers who may offer about one-third of the total species. Cones which are over 1 year old should contain ripe viable seed. On some occasions they may open to release seed. The cones can be collected and placed in a paper or cloth bag and kept in a warm place. As the cones dry out they will release the hairy nuts which are then ready for sowing.

Seed will germinate without pre-sowing treatment, although it may take 20–100 days before germination begins. Usually better results are obtained if the seed is placed in a wire strainer and passed over a light flame a few times to burn the hairs and singe the nuts. Recent experimentation with smoke treatment of seeds has shown that some species will respond with better germination. For further information on this treatment, see the Supplement to the Encyclopaedia, under P.

Petrophiles are sometimes propagated from cuttings. Cutting material taken from the current season's growth usually gives best results. After flowering there is a flush of new growth. When this reaches the semi-firm stage, cuttings 4–10 cm long are taken. It is advisable to retain at least 5 nodes with leaves because some nodes may die while in the propagation area before rooting occurs. The tips should be removed if they are soft and likely to wilt. Application of a hormone rooting compound may be beneficial in producing a strong root system. Roots are usually produced within 4–12 weeks, but may take longer to form.

For good growth of petrophiles in containers it is worth using a mix similar to the proteaceae potting mix in Volume 5, page 16.

Petrophile aspera   W. R. Elliot

Petrophile acicularis R. Br.
(needle-like)
WA
0.4–0.8 m × 0.5–1.2 m   Sept–Oct

Dwarf **shrub** with glabrous young growth; **branches** spreading to ascending; **branchlets** glabrous; **leaves** 5–18 cm × up to 0.2 cm, terete, straight to slightly curled, spreading to erect, green, glabrous, apex blunt or with curved tip; **flowerheads** about 2 cm across, ovoid, terminal, sessile, often appearing clustered, sometimes profuse and very conspicuous; **involucral bracts** very narrow, many, glabrous; **cone scales** ovate, grooved, glabrous; **flowers** about 1 cm long, cream, exterior densely hairy; **pollen presenter** to about 0.4 cm long, top-shaped below brush; **cones** to about 2 cm long, more or less globular.

A poorly known species from near Albany, and to the west of Albany, where it grows in sandy soils. Rarely cultivated, it was first grown in England during 1810 as *P. filifolia* which is now a synonym. Plants require freely draining acidic soils and a sunny or warm, semi-shaded. Hardy to light frosts. Worth trying in containers or rock gardens as well as in general planting. Propagate from seed or from cuttings of firm young growth, which may be slow to form roots.

This species has been confused with *P. longifolia* which is a prostrate shrub with cream to yellowish flowers.

Petrophile aculeata Foreman
(prickly)
WA
0.25–0.4 m × 0.3–0.5 m   Nov

Dwarf **shrub** with densely hairy young growth; **branches** spreading to erect; **branchlets** initially with long hairs, becoming glabrous; **leaves** 4–9 cm × 0.4–1.3 cm, narrowly obovate, scattered or somewhat clustered, spreading to ascending, deep green, initially hairy becoming glabrous, slightly rough, margins with irregular curved teeth on upper parts, venation prominent; **flowerheads** to about 2 cm across, terminal, solitary, often on short branchlets, somewhat hidden by foliage; **involucral bracts** narrowly ovate, grey, hairy; **cone scales** to broadly ovate, grey, hairy; **flowers** to about 1.1 cm long, colour unknown, exterior hairy near apex, base glabrous; **cones** about 1.2 cm across, globular.

A poorly known species which is restricted to a couple of locations in the southern Irwin District where it inhabits lateritic sand in low open heath. Needs to be cultivated as part of a conservation strategy. Plants will need freely draining acidic soils and plenty of sunshine. They will tolerate light frosts and extended dry periods. Propagate from seed, or from cuttings of firm young growth, which may be slow to produce roots.

Petrophile anceps R. Br.
(two-winged)
WA
0.3–0.6 m × 0.3–1 m   Sept–Oct

Dwarf **shrub** with mainly glabrous young growth; **branches** spreading to ascending; **branchlets** glabrous; **leaves** 3–11 cm × up to 0.3 cm, linear, straight or incurved, spreading to ascending, rarely crowded, deep green, glabrous, margins narrowly winged, apex blunt with pungent point; **flowerheads** to about 3 cm across, ovoid, terminal, solitary, can be moderately profuse and very conspicuous; **involucral bracts** very narrow, glabrous, becoming greyish; **cone scales** ovate, glabrous except for fringed margin; **flowers** about 1.5 cm long, yellow, exterior very hairy; **pollen presenter** to about 0.4 cm long, tapered below brush; **cones** to about 2.5 cm × 1.5 cm, ovoid.

P. anceps is confined to the Stirling Range where it grows in gravelly sand and loam in shrubland and mallee. Rarely encountered in cultivation, it must have excellently drained acidic soils and a sunny or semi-shaded site. Plants are hardy to moderately heavy frosts. It has potential for use in containers and rockeries. Propagate from seed or from cuttings of firm young growth.

P. linearis var. *anceps* (R. Br.) Benth. is a synonym.

Petrophile arcuata Foreman
(arching or curved)
WA
0.6–1.2 m × 0.6–1.5 m   Sept–Oct

Dwarf to small **shrub** with sparsely hairy young growth; **branches** many, spreading to erect; **branchlets** many, usually becoming glabrous; **leaves** 1–1.5 cm × about 0.1 cm, linear-terete, usually incurved, spreading to ascending, deep green, glabrous, apex pointed; **flowerheads** about 1.2 cm across, terminal, often profuse and conspicuous; **involucral bracts** ovate to oblong, glabrous, somewhat sticky; **cone scales** ovate, blunt to pointed, base densely hairy; **flowers** about 1.2 cm long, creamy yellow to yellow, densely hairy;

pollen presenter about 0.4 cm long, tapering below the brush; **cones** to about 1.5 cm across, globular.

An attractive, recently described (1995) species which occurs in the Avon, Coolgardie and Roe Districts. Grows in deep sand or sandy clay soils in heathland and mallee shrubland. Needs excellent drainage and plenty of sunshine and would be best suited to semi-arid and warm temperate regions. Hardy to moderate frosts. It has potential for cut-flower production as well as for general planting. Propagate from seed or from cuttings of firm young growth.

P. phylicoides is similar but it has glabrous flowers.

Petrophile aspera C. A. Gardner ex Foreman
(rough)
WA
0.2–1.3 m × 0.3–1 m   Aug–Nov

Dwarf to small **shrub** with more or less glabrous young growth; **branches** spreading to ascending; **branchlets** glabrous; **leaves** 15–30 cm × about 0.2 cm, terete, ascending to erect, usually incurved, with slight curling at apex, sometimes slightly twisted, deep green, glabrous, faintly rough, apex pointed; **flowerheads** about 5 cm across, globular, terminal, sessile, solitary, can be profuse and very conspicuous; **involucral bracts** linear-lanceolate; **cone scales** very broad, tapered to sometimes reflexed, apex glabrous; **flowers** about 2 cm long, white, creamy white, pale yellow, pale pink, exterior hairy, sweetly fragrant; **pollen presenter** about 0.5 cm long, top-shaped below brush; **cones** about 2.5 cm × 1.5 cm.

A recently described (1989) species from the southern Avon, far-western Eyre and south-western Roe Districts where it grows in sandy soils which often contain lateritic gravel. In nature, most plants are rarely more than 0.5 m tall. They require freely draining acidic soils and preferably a sunny site. Suited to temperate and semi-arid regions. Plants tolerate moderately heavy frosts and extended dry periods. Propagate from seed or from cuttings of firm young growth.

This species has previously been referred to as the shrubby variant of *P. longifolia*, which is a prostrate species. *P. teretifolia* differs in having smooth leaves which do not curl at the apex.

Petrophile biloba R. Br.
(two-lobed)
WA   Granite Petrophile
1–2.5 m × 0.5–1.5 m   June–Oct

Small to medium **shrub** with densely hairy young growth; **branches** ascending to erect, often few; **branchlets** densely hairy, sometimes becoming glabrous; **leaves** 2–4.5 cm × 1–4 cm, ovate-rhomboid in outline, tapered to base, lobes 3–4, 0.5–2 cm long, flat, grey-green to green, glabrous, apex pungent-pointed; **flowerheads** to about 3 cm across, axillary, sessile, forming leafy spikes, often profuse and very conspicuous; **involucral bracts** few, deciduous; **cone scales** silky-hairy except for glabrous apex; **flowers** to about 2.2 cm long, usually greyish to pink, exterior faintly hairy to glabrous; **pollen presenter** to about 0.4 cm long, somewhat club-shaped, white, cream, yellow, orange or mauve; **cones** to about 1.4 cm long, ovoid.

Petrophile brevifolia × .55

This outstanding member of the genus is regarded as vulnerable in nature. It occurs in the central and southern Darling District and inhabits a range of soils including granitic sand overlying laterite, and clay with laterite gravel. Moderately popular in cultivation, it is often difficult to maintain because of susceptibility to root-rotting fungi such as *Phytophthora cinnamomi.* Plants need excellently drained acidic soils with a sunny aspect. Frost can damage young growth. Pruning from an early stage promotes bushy growth. Flowering stems are ideal for indoor decoration. Introduced to England in 1850. Propagate from seed, which usually begins to germinate between 25–50 days after sowing, or from cuttings if firm young growth, which can be slow to produce roots.

Petrophile biternata Meisn.
(two lobes with three segments)
WA
0.6–1.5 m × 0.4–1.5 m   Aug–Oct
Dwarf to small **shrub** with faintly hairy, often bronzish young growth; **branches** spreading to erect, many; **branchlets** initially hairy, usually becoming glabrous; **leaves** to about 6.5 cm long, biternate or pinnate, channelled, stiff, thick, glabrous, greyish-green to green; segments 1–3.6 cm long, lanceolate to broadly elliptic or with further lobes, apex acute to blunt and pungent-tipped; **flowerheads** to about 2.5 cm across, ovoid to globular, terminal, solitary, sessile, can be profuse and very conspicuous; **involucral bracts** broad, glabrous, sticky; **cone scales** broad, more or less glabrous, sticky; **flowers** about 1 cm long, creamy-yellow to bright yellow, exterior glabrous, sticky; **pollen presenter** to about 0.5 cm long, tapered to both ends, yellow, hairy; **cones** about 2.5 cm long, ovoid.

This species has attractive foliage and flowers. It has a disjunct distribution in the Irwin and northern Darling Districts. Plants are cultivated to a limited degree and adapt well to warm temperate and semi-arid regions, with some success also in cool temperate areas. Needs acidic soils which drain freely and prefers plenty of sunshine but tolerates a semi-shaded site. Hardy to moderate frosts. Responds very well to pruning and flowering stems are excellent for decoration. Well suited to large containers and gardens. Propagate from seed, or from cuttings of firm young growth, which can be slow to produce roots.

Some enthusiasts and nurseries have confused this species with *P. fastigiata,* which has leaves with many narrow segments.

Petrophile brevifolia Lindl.
(short leaves)
WA
0.5–2 m × 0.5–1.5 m   July–Dec
Dwarf to small **shrub** with glabrous young growth; **branches** many, arising from base, spreading to erect; **branchlets** glabrous; **leaves** 2–9 cm × about 0.1 cm, linear, terete, spreading to erect, somewhat crowded, deep green, smooth, apex pungent-pointed; **flowerheads** to about 2.5 cm across, terminal, sessile, solitary, can be profuse and very conspicuous; **involucral bracts** linear, many, apex curled; **cone scales** narrowly ovate to ovate, glabrous; **flowers** to about 2 cm long, creamy-yellow, exterior densely hairy; **pollen presenter** with a glabrous tip; **cones** about 1 cm across.

A far-ranging species of southern WA. It is represented in the Avon, northern Darling, Irwin and western Roe Districts where it usually grows in sandy soils. Uncommon in cultivation, it requires excellent drainage and plenty of sunshine. Plants are hardy to moderate frosts and are drought tolerant. They respond well to moderate pruning. The pungent foliage makes plants useful for foot traffic control and wildlife habitat. Introduced to gardens in England in 1837. Propagate from seed or from cuttings of firm young growth.

P. media differs in having longer leaves and pollen presenters which are prominently hairy.

Petrophile canescens A. Cunn. ex. R. Br.
(hoary)
Qld, NSW   Conesticks
0.5–1.5 m × 0.5–1.2 m   Aug–Nov
Dwarf to small **shrub** with greyish-hairy young growth; **branches** ascending to erect; **branchlets** becoming glabrous; **leaves** 3–9 cm long, upper half with many terete ascending segments which are usually further divided, greyish-hairy, softly pointed; **flowerheads** to about 3 cm × 2 cm, terminal or axillary, sessile or short-stalked, solitary or with basal clusters, scattered to somewhat profuse; **involucral bracts** narrowly triangular, hairy; **cone scales** broadly ovate, rusty-hairy; **flowers** to about 1.5 cm long, white to pale cream, exterior silky-hairy; **cones** to about 3 cm long, oblong.

A dweller of coastal and inland sandy soils in dry and wet heath and dry sclerophyll forest. Distributed in south-eastern Qld and eastern NSW extending

Petrophile carduacea   W. R. Elliot

south to near Nerriga, NSW. Cultivated to a limited degree, it deserves to be better known for its greyish-hairy foliage. Plants require freely draining acidic soils in a sunny or semi-shaded site. They are hardy to moderately heavy frosts and withstand extended dry periods. Pruning young plants promotes bushier growth. Suitable for temperate and subtropical regions. Introduced into gardens of England in 1830. Propagate from seed or from cuttings of firm young growth.

The closely allied *P. sessilis* has stiff and spreading pungent-tipped leaf segments.

Petrophile carduacea Meisn.
(resembling a thistle)
WA
0.6–2 m × 0.5–1.5 m   Sept–Nov

Dwarf to medium **shrub** with densely hairy young growth; **branches** ascending to erect; **branchlets** hairy, can become glabrous; **leaves** 5–12 cm × 1–5 cm, oblong-lanceolate in outline, spreading to ascending, often recurved, deep green, initially hairy, becoming glabrous, flat, wavy margins, deeply toothed or lobed, with pungent tips, venation prominent; **flowerheads** to about 1.5 cm across, axillary, short-stalked, sometimes profuse and very conspicuous; **involucral bracts** triangular, few, glabrous; **cone scales** broad, brownish, glabrous, blunt; **flowers** about 0.8 cm long, creamy yellow to yellow, exterior silky-hairy; **pollen presenter** about 0.2 cm long, tapered to both ends, yellow; **cones** 2–3.5 cm long, more or less ovoid.

A moderately quick-growing species from the Stirling Range and adjacent regions. Usually occurs on gravelly soils or sand. Cultivated to a very limited degree. Well-grown plants produce attractive foliage.

It is best suited to temperate regions but can cope with semi-arid conditions. Prune early to promote bushy growth. Hardy to moderate frosts. Propagate from seed, which usually begins to germinate 28–60 days after sowing, or from cuttings of firm young growth, which may be slow to produce roots.

P. diversifolia differs in having leaves with many lobes.

Petrophile chrysantha Meisn.
(gold flowers)
WA
0.6–1.3 m × 0.5–1 m   June–Oct

Dwarf to small **shrub** with densely hairy young growth; **branches** ascending to erect, many; **branchlets** initially very hairy, often becoming glabrous; **leaves** 0.5–2 cm long, pinnate, deep green, crowded, spreading to ascending, often recurved, initially hairy usually becoming glabrous; segments 0.1–1 cm long, terete, faintly rough, apex pungent-pointed; **flowerheads** to about 1.5 cm across, ovoid, terminal, sessile, can be profuse and very conspicuous; **involucral bracts** broad, overlapping, margin fringed; **cone scales** broad, overlapping, inner ones narrower, margins fringed; **flowers** to about 1.4 cm long, cream to deep yellow, exterior densely hairy; **pollen presenter** about 0.3 cm long, tapered to both ends, yellow, nearly glabrous; **cones** about 1 cm across, globular.

A very attractive species from the southern Irwin and far-northern Darling Districts where it inhabits

Petrophile chrysantha × .95

242

sandy and gravelly soils in shrubland and woodland. Apparently not cultivated, it warrants being better known. Probably best suited to warm temperate or semi-arid regions with freely draining acidic soils and plenty of sunshine. Plants are drought tolerant and hardy to moderate frosts. Response to pruning is not known. Worth trying in gardens and containers. Propagate from seed, or from cuttings of firm young growth, which may be slow to produce roots.

Petrophile circinata Kippist. ex Meisn.
(coiled, rolled in a circle)
WA
0.25–1.2 m × 0.5–2 m   June–Nov
   Dwarf to small **shrub** with hairy, often reddish young growth; **branches** spreading to ascending, many; **branchlets** initially densely hairy, often becoming glabrous; **leaves** 5–20 cm long, 2–3-pinnate, long-stalked, spreading to ascending, strongly recurved, often crowded and entangled; segments 0.5–2.2 cm long, terete, many; **flowerheads** about 2.5 cm across, globular, terminal, sessile, rarely profuse, moderately conspicuous; **involucral bracts** broad, to about 2 cm long, many, exterior finely hairy; **cone scales** linear to narrowly ovate, brown, faintly hairy; **flowers** to about 4 cm long, white, cream or yellow, exterior hairy; **pollen presenter** about 0.3 cm long, tapered to each end, yellow, hairy; **cones** about 4 cm across, globular.
   A lovely foliage plant which occurs over a wide area of the south-western inland region. It inhabits sandy, gravelly and lateritic soils in a range of communities from heath to woodland. Extremely rare in cultivation, it has much to offer. Plants require freely draining acidic soils with a sunny or semi-shaded site and should perform best in warm temperate and semi-arid zones. They are hardy to moderate frosts and withstand extended dry periods. It has potential for use in containers, rockeries and general planting. Propagate from seed, which usually begins to germinate 26–60 days after sowing, or from cuttings of firm young growth, which may be slow to form roots.
   P. glauca has deeply-lobed leaves with flat, broader segments.

Petrophile colorata Meisn. = *P. squamata* R. Br.

Petrophile conifera Meisn.
(cone-bearing)
WA
0.3–1.5 m × 0.5–2 m   July–Oct
   Dwarf to small **shrub** with woolly-hairy young growth; **branches** spreading to ascending, many, stiff; **branchlets** initially hairy, can become glabrous; **leaves** 4–11 cm long, pinnate, spreading to ascending, green, glabrous, stiff; segments 0.4–3.5 cm, terete, spreading, often forked, apex pungent-tipped; **flowerheads** 2–3 cm × 1.5–2 cm, ovoid, terminal, solitary, sessile, can be moderately profuse, always conspicuous; **involucral bracts** lanceolate, deciduous, exterior hairy; **cone scales** broad, rounded, exterior hairy except for apex; **flowers** to about 1.5 cm long, cream to creamy-yellow, sometimes with greenish tonings, exterior hairy; **pollen presenter** about 0.4 cm long, tapered to both ends, yellow, faintly hairy; **cones** 1–3 cm long, ovoid.

A prickly, moderately attractive species from the northern Irwin District where it inhabits sandy soils of sandplains and heath. It is rarely encountered in cultivation and should do best in warm temperate and semi-arid regions. Must have excellently drained acidic soils and plenty of sunshine. Plants tolerate moderate frosts and extended dry periods. May be useful for foot traffic control. Propagate from seed, which usually begins to germinate 24–60 days after sowing, or from cuttings of firm young growth, which may be slow to produce roots.
   P. incurvata is closely allied but differs in having flat leaves.

Petrophile crispata R. Br.
(crinkled or crisped)
WA
0.2–0.8 m × 0.3–1.3 m   Sept–Nov
   Dwarf **shrub** with faintly hairy young growth; **branches** spreading to ascending, many; **branchlets** initially hairy, usually becoming glabrous; **leaves** 4–11 cm long, pinnate, spreading to ascending, dark green, glabrous, stiff; segments 0.6–4 cm long, terete, divided 2–3 times, apex pungent-tipped; **flowerheads** about 3.5 cm × 2 cm, ovoid, terminal, solitary, sessile, can be profuse, moderately conspicuous; **involucral bracts** ovate, exterior glabrous, slightly sticky, deciduous; **cone scales** lanceolate, long hairs at base; **flowers** about 1 cm long, yellow, exterior glabrous; **pollen presenter** about 0.3 cm long, tapered to both ends, ridged, yellow, hairs short and erect; **cones** 1.5–2.5 cm long, ovoid, greyish.
   This finely-foliaged species is poorly known in the wild. It occurs in the southern Avon, western Eyre and western Roe Districts where it grows in lateritic soils and sand overlying laterite as part of shrubland and open woodland. It is not popular in cultivation. Plants require relatively well-drained acidic soils and a site which receives plenty of sunshine. Best results are likely to be achieved in warm temperate and semi-arid

Petrophile divaricata   W. R. Elliot

regions. Plants are hardy to moderate frosts and extended dry periods. Propagate from seed, or from cuttings of firm young growth, which may be slow to produce roots.

P. seminuda differs in having glabrous branchlets, short, woolly hairs at the base of cone scales and a pollen presenter which is not ridged.

Petrophile cyathiforma Foreman
(cup-shaped)
WA

0.3–0.75 m × 0.4–1 m   Sept–Dec

Dwarf **shrub** with densely hairy young growth; **branches** many, spreading to erect; **branchlets** initially hairy, becoming glabrous; **leaves** 1–1.5 cm × about 0.1 cm, linear-terete, spreading to ascending, usually incurved, moderately crowded, deep green, becoming glabrous, apex pungent-pointed; **flowerheads** to about 1.5 cm across, terminal, very pointed in bud, can be profuse and very conspicuous; **involucral bracts** narrowly ovate, overlapping, sticky, persistent; **cone scales** narrowly ovate, exterior densely hairy; **flowers** to about 2 cm long, bright yellow, glabrous near tips; **pollen presenter** to 0.6 cm long, yellow, tapered to both ends, with short hairs; **cones** to about 2 cm across, somewhat hemispherical.

An ornamental, recently described (1995) species which occurs on sand and clayey sand in heath, shrubland and mallee communities in the Avon, Eyre and Roe Districts. Worthy of cultivation, it will need well-drained acidic soils in sunny or semi-shaded sites. Plants tolerate moderate frosts and extended dry periods. They have potential for use in gardens and containers. Propagate from seed or from cuttings of firm young growth.

P. imbricata differs in having faintly sticky involucral bracts and ovoid cones.

Petrophile divaricata R. Br.
(forked or widely spreading)
WA

1–2 m × 1–2 m   Aug–Dec

Small **shrub** with hairy young growth; **branches** spreading to ascending, thick, rigid; **branchlets** initially hairy, becoming glabrous; **leaves** 4–11 cm long, bipinnate, sparse to crowded, spreading to ascending, deep green, glabrous, stiff; segments 0.6–2 cm long, terete, grooved, apex pungent-tipped; **flowerheads** to about 3.5 cm across, oblong to ovoid, terminal or axillary, sessile, can be profuse and very conspicuous; **involucral bracts** broad with tapered apex, exterior glabrous, interior silky-hairy, deciduous; **cone scales** broad, hairy except for pointed apex; **flowers** 2–2.5 cm long, pale creamy yellow to bright yellow, exterior hairy, sweetly fragrant; **pollen presenter** to about 0.25 cm long, tapered to both ends, ridged, short reflexed hairs; **cones** to about 3 cm long, ovoid to cylindrical.

A handsome, prickly cone bush from the south-western region. It has a distribution extending from the southern Irwin District to the central Eyre District. Inhabits a range of soils which are often gravelly and sometimes clayey. Plants are best suited to temperate regions. They require freely draining acidic soils and plenty of sunshine. Hardy to moderate frosts and they respond well to pruning. Excellent for foot traffic control. Propagate from seed, which usually begins to germinate 30–75 days after sowing, or from cuttings of firm young growth, which can be slow to produce roots.

Petrophile diversifolia R. Br.
(more than one kind of leaf)
WA

1–3.5 m × 0.5–2 m   Sept–Dec

Small to medium **shrub** with densely hairy, often reddish, fern-like young growth; **branches** ascending to erect, often few; **branchlets** initially densely hairy, often becoming glabrous; **leaves** 3–11 cm long, pinnate, bipinnate or tripinnatifid, spreading to ascending, sometimes recurved, deep green, becoming glabrous; segments variable, toothed or lobed, soft; **flowerheads** about 2.5 cm long, ovoid, terminal or axillary, on stalks 1–2 cm long, can be profuse and moderately conspicuous; **involucral bracts** ovate, glabrous; **cone scales** hairy at base, tips glabrous, deciduous; **flowers** to about 1.2 cm long, white to creamy-white, sometimes pinkish, exterior densely hairy; **pollen presenter** about 0.3 cm long, tapered to both ends, yellow, more or less glabrous; **cones** to about 3 cm long, ovoid, with stalk to about 2 cm long.

A very distinct species which inhabits a range of sandy and gravelly soils that are sometimes poorly drained. It occurs in the southern Darling and western Eyre Districts. Cultivated to a limited degree, it can grow very quickly but it is often short-lived. Plants are suited to temperate regions. They do best in well-drained acidic soils which are moist for the greater part of the year. A sunny or semi-shaded site is suitable. Pruning of young and developing plants can promote bushy growth. Hardy to light frosts. It has potential for cut-flower and foliage production. First cultivated in England during 1803. Propagate from seed, which usually begins to germinate between 40–70 days after sowing, or from cuttings of firm young growth, which strike readily.

P. carduacea has simple leaves with deep triangular teeth.

Petrophile drummondii Meisn.
(after James Drummond, first Government Naturalist of WA)
WA

0.4–1.2 m × 0.3–1 m   Aug–Dec

Dwarf to small **shrub** with glabrous or hairy young growth; **branches** ascending to erect, stiff; **branchlets** usually glabrous; **leaves** 2–5 cm long, pinnate or much-divided, to 2 cm long, spreading to ascending or recurved, light-to-deep green, glabrous, stiff; segments to 2 cm long, terete, ternately divided, apex pungent-tipped; **flowerheads** 3–4 cm × 4–5 cm, globular, terminal, sessile, often profuse and very conspicuous; **involucral bracts** ovate-lanceolate, many, glabrous, sticky, deciduous; **cone scales** broad, base woolly-hairy, apex glabrous; **flowers** about 2 cm long, yellow, sticky, hairy, sweetly fragrant; **pollen presenter** about 0.5 cm long, tapered to both ends, orange, hairy; **cones** to about 3 cm long, ovoid.

An appealing, prickly species from the Irwin and Avon Districts where it grows in a range of sandy or sandy lateritic soils in heath and shrubland. Cultivated to a limited degree, it warrants greater attention. Suited to warm temperate and semi-arid regions and requires freely draining acidic soils and plenty of sunshine. Plants tolerate moderate frosts and extended dry periods. They respond well to pruning. Suitable for foot traffic control. Worth trying in containers. Propagate from seed, which usually begins to germinate between 20–70 days after sowing, or from cuttings of firm young growth, which strike readily.

Petrophile ericifolia R. Br.
(leaves similar to those of the genus *Erica*)
WA
0.6–1.5 m × 0.6–2 m   Aug–Nov
Dwarf to small **shrub** with glabrous or faintly hairy young growth; **branches** ascending to erect; **branchlets** glabrous to faintly hairy; **leaves** 0.4–1.2 cm × about 0.1 cm, linear-terete, sessile, spreading to erect, often crowded, becoming glabrous, deep green, apex blunt to pointed; **involucral bracts** obovate to ovate, many, often crowded, sticky; **cone scales** ovate, base densely hairy; **flowerheads** to about 2 cm across, globular, terminal, can be profuse and always conspicuous; **flowers** to about 2 cm long, yellow, hairy, somewhat sticky; **pollen presenter** about 0.4 cm long, tapered to both ends, hairy; **cones** to about 2.5 cm across, ovoid.

The typical subspecies occurs in the heathlands of the Avon, Eyre, Roe and Coolgardie Districts and is distributed from near the coast to some distance inland. It grows in deep sand which is often gravelly. Plants are uncommon in cultivation and require excellently drained acidic soils and plenty of sunshine. They are hardy to moderate frosts and extended dry periods. Best suited to semi-arid and warm temperate regions. Suitable for gardens and worth trying as container plants. Also has potential for cut-flower production.

The ssp. *subpubescens* (Domin) Foreman differs in having hairy branchlets, leaves to about 0.6 cm long which are usually hairy, and non-sticky flowers about 1.6 cm long. It occurs on the sandplains of the northern Avon District. Needs similar conditions to the typical subspecies. Previously known as *P. ericifolia* forma *subpubescens* Domin.

Propagate both subspecies from seed, or from cuttings of firm young growth, which may be slow to form roots.

P. ericifolia var. *glabriflora* Benth. is a synonym of *P. phylicoides* R. Br. and var. *scabriuscula* (Meisn.) Benth. is a synonym of *P. scabriuscula* Meisn.

Petrophile fastigiata R. Br.
(upright-branched)
WA
0.6–1.5 m × 0.4–1.2 m   Sept–Nov
Dwarf to small **shrub** with bronze-red, glabrous or hairy young growth; **branches** ascending to erect; **branchlets** glabrous or becoming so; **leaves** 5–11 cm long, 2–3 times ternately divided, spreading to ascending, mid-to-dark green, glabrous; **segments** terete, with yellowish softly pointed tips; **flowerheads** to about 4.5 cm long, ovoid, terminal, sessile, solitary or clustered, can be profuse and very conspicuous;

involucral bracts broadly triangular, many, crowded, exterior glabrous, deciduous; **cone scales** more or less circular, base woolly-hairy; **flowers** about 1 cm long, cream to yellow, glabrous; **pollen presenter** about 0.3 cm long, tapered to both ends, yellow, hairy; **cones** 2–4 cm × about 2 cm, ovoid.

An appealing cone-bush which occurs mainly in the central Eyre District and adjacent regions. It grows in shrubland and mallee on gravelly or sandy soils which often have clay content. Generally adapts well to cultivation in temperate and semi-arid regions. Does best in freely draining acidic soils and requires plenty of sunshine otherwise it can become very leggy. Responds well to pruning, which promotes bushy growth. Hardy to moderate frosts. Flowering stems are useful for indoor decoration. Propagate from seed, which usually begins to germinate 33–70 days after sowing, or from cuttings of firm young growth, which may be slow to produce roots.

P. biternata has channelled leaves with broad, flat segments.

Petrophile filifolia R. Br. = *P. acicularis* R. Br.

Petrophile fucifolia (Salisb.) Knight =
                    P. pulchella (Schrad.) R. Br.

Petrophile glabriflora Domin. = *P. phylicoides* R. Br.

Petrophile glanduligera Lindl. = *P. serruriae* R. Br.

Petrophile glauca Foreman
(glaucous)
WA
0.3–1 m × 0.6–1.5 m   Aug–Nov; also sporadic
Dwarf **shrub** with glabrous or faintly hairy young growth; **branches** prostrate to ascending; **branchlets** usually glabrous, often initially glaucous; **leaves** 7–20 cm long, pinnate, spreading to ascending, glaucous, glabrous, often incurved; **segments** to about 2.5 cm long, widely spaced, flat, apex pungent-tipped; **flowerheads** to about 2 cm across, more or less globular, terminal or axillary, sessile, can be profuse, sometimes partially hidden by foliage; **involucral bracts** triangular, crowded, dark brown, exterior glabrous; **cone scales** broad, exterior hairy except for pointed apex; **flowers** about 1.4 cm long, creamy white, creamy yellow or yellow, exterior silky-hairy; **pollen presenter** about 0.2 cm long, tapered to both ends, yellow; **cones** about 1.5 cm across, globular to ovoid.

A delightful species with attractive foliage and flowers. It occurs in the southern Avon, western Eyre and western Roe Districts where it grows in a range of sandy, gravelly sites which often contain lateritic gravel. This species has had limited cultivation. It is suited to temperate and semi-arid regions. Requires very well-drained acidic soils and does well in a sunny or semi-shaded site. Plants are hardy to moderate frosts and moderately long dry periods. Plants respond well to pruning. Suitable for containers, rockeries and general planting. Propagate from seed, or from cuttings of firm young growth, which strike readily.

This species has been grown and marketed as *P. trifida*, which is a synonym of *P. squamata*.

Petrophile fastigiata   W. R. Elliot

Petrophile helicophylla Foreman
(spirally twisted leaves)
WA
0.2–0.4 m × 0.3–1.6 m   Oct–Feb

Dwarf spreading **shrub** with a lignotuber; young growth glabrous; **branches** prostrate to ascending; **branchlets** glabrous; **leaves** 15–30 cm × about 0.2 cm, terete, spirally twisted, usually erect, crowded near ends of branchlets, deep green, glabrous, smooth, apex finely pointed; **flowerheads** to about 6 cm across, terminal, sessile, solitary, very conspicuous; **involucral bracts** very narrow, few, glabrous; **cone scales** very broad, glabrous; **flowers** to about 3.5 cm long, white, cream, or pale creamish-pink to pale pink, exterior hairy; **pollen presenter** about 0.8 cm long, bright yellow; **cones** to about 2.5 cm across, more or less globular.

An intriguing and highly decorative, recently described (1989) species from the western Eyre and south-western Roe Districts. It occurs in sandy or sandy-clay soils in heath and woodland. Although rarely cultivated, it is a valuable novelty plant because of its spiral leaves and growth habit. Plants should do best in temperate and semi-arid regions. Needs excellent drainage and plenty of sunshine. Hardy to moderate frosts and extended dry periods. It has potential for use in rockeries and large containers. Propagate from seed or from cuttings of firm young growth.

Petrophile heterophylla Lindl.
(leaves of more than one kind)
WA   Variable-leaved Cone Bush
1–2.5 m × 0.6–2 m   Aug–Oct

Small to medium **shrub** with more or less glabrous young growth; **branches** ascending to erect, many, stiff; **branchlets** glabrous; **leaves** 5–15 cm long,

linear to narrowly obovate, entire or with 2–3 lobes, 0.2–0.6 cm wide, spreading to ascending, sometimes incurved, mid-to-deep green, glabrous, venation conspicuous, apex pungent-pointed; **flowerheads** to about 1.5 cm long, ovoid, mainly axillary, sessile, can be profuse, moderately conspicuous; **involucral bracts** small, many, crowded, exterior glabrous, interior hairy, margins fringed; **cone scales** ovate, exterior hairy; **flowers** 1–1.5 cm long, cream, pale yellow or yellow, exterior silky-hairy, sweetly fragrant; **pollen presenter** about 0.3 cm long, somewhat club-shaped, 4-angled below brush; **cones** about 1.2 cm across, more or less ovoid.

P. heterophylla is widely distributed over the south-western corner of WA. It grows on sand, gravelly sand and lateritic sands as part of the heath, shrubland, open woodland and jarrah forests. Cultivated to a limited extent, it has potential for greater application. Plants are best suited to temperate regions and may succeed in semi-arid areas. They require well-drained acidic soils with a sunny or semi-shaded aspect. Tolerates light frost and limited dry periods. Responds well to pruning which can promote bushy growth. Worth trying as a hedging plant. Propagate from seed which usually begins to germinate 20–60 days after sowing, or from cuttings of firm young growth, which may be slow to produce roots.

Petrophile imbricata Foreman
(overlapping)
WA
1–2 m × 0.6–1.6 m   Aug–Sep

Small **shrub** with densely greyish-hairy young growth; **branches** ascending to erect; **branchlets** greyish-hairy; **leaves** 0.4–1.5 cm × about 0.1 cm, linear-terete, sessile, ascending to erect, often slightly incurved, crowded, deep green, initially covered with short hairs, becoming glabrous, apex blunt often with a small point; **flowerheads** to about 1.8 cm across, terminal, can be profuse and very conspicuous; **involucral bracts** narrowly ovate, crowded, faintly sticky, persistent; **cone scales** linear-lanceolate, base hairy; **flowers** to about 2 cm long, cream, densely

Petrophile glauca   D. L. Jones

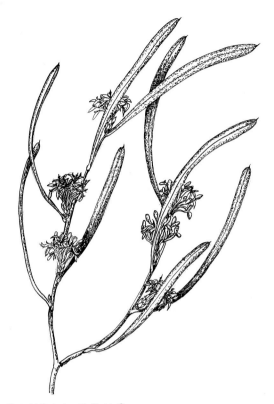

Petrophile heterophylla × .6

Petrophile linearis × .55

hairy; **pollen presenter** about 0.6 cm long, yellow swollen base, tapered to tip; **cones** to about 1.8 cm across, ovoid or somewhat cup-shaped.

A handsome, recently described (1995) species which occurs in heath, open woodland and forest, on lateritic sandy soils in the Avon District. Plants require very well-drained acidic soils and a sunny or semi-shaded aspect. Should do best in warm temperate and semi-arid regions. Hardy to moderate frosts. It has potential for cut-flower production and as a garden plant. Propagate from seed or from cuttings of firm young growth.

P. ericifolia differs in having sticky flowers and a shorter pollen presenter.

Petrophile inconspicua Meisn. =
> *Isopogon inconspicuus* (Meisn.) Foreman

Petrophile incurvata W. Fitzg.
(incurved)
WA
1–1.5 m × 1–1.5 m   Sept–Oct
Small **shrub** with hairy young growth; **branches** ascending to erect, many, stiff; **branchlets** becoming glabrous; **leaves** 7–12 cm × 0.5–2.5 cm, variable, lanceolate-falcate or with 2–5 short flat lobes, spreading to ascending, incurved, deep green, glabrous, apex pungent-pointed; **flowerheads** to about 3.5 cm

long, cylindrical to ovoid, terminal, solitary or 2–3 in clusters, can be profuse and moderately conspicuous; **involucral bracts** linear, deciduous; **cone scales** broad, overlapping, stiff, velvety; **flowers** 1–1.4 cm long, yellowish-white or cream, exterior silky-hairy; **pollen presenter** about 0.5 cm long, tapered to both ends, hairs short and erect; **cones** 2–3 cm × 1.2–2 cm, cylindrical to ovoid.

This species has a fairly limited distribution in shrubland. It grows on sand in the far-northern Avon District and adjacent areas of the Austin District. Apparently uncommon in cultivation. Plants are suited to semi-arid and warm temperate regions and require excellently drained acidic soils with maximum sunshine. They tolerate moderate frosts and extended dry periods. It has potential for garden and container use. Propagate from seed, or from cuttings of firm young growth, which may be slow to produce roots.

Petrophile juncifolia Lindl. = *P. media* R. Br.

Petrophile linearis R. Br.
(linear)
WA   Pixie Mops
0.4–0.8 m × 0.5–1 m   Aug–Nov
Dwarf **shrub** with a lignotuber; young growth glabrous; **branches** spreading to erect, few; **branchlets** glabrous, often glaucous when young; **leaves** 5–12 cm ×

247

0.15–1 cm, narrowly obovate, straight to strongly curved, spreading to ascending, rarely crowded, greyish-green to deep green, glabrous, thick, apex recurved with a short point; **flowerheads** to about 5 cm across, globular to ovoid, terminal, sessile, solitary, can be profuse, always very conspicuous; **involucral bracts** very narrow, many, glabrous; **cone scales** broadly obovate, base hairy, apex glabrous; **flowers** to about 3.5 cm long, commonly greyish-pink to mauve or rarely whitish, densely hairy; **pollen presenter** to about 0.5 cm long, narrowly top-shaped below brush, yellow to reddish-orange; **cones** about 2.5 cm long, ovoid.

An extremely showy member of the genus. It is found in open heath and woodland on the coastal plain and Darling Range from the southern Irwin District to the central Darling District. Grows in sand and lateritic soils. Highly sought after for cultivation, it needs excellently drained acidic soils and plenty of sunshine, although it does tolerate some shade. Plants are hardy to light frosts and withstand extended dry periods. Should do best in warm temperate and semi-arid regions. Potentially very useful for containers and rockeries as well as for general planting in gardens. Responds very well to light or hard pruning. Propagate from seed, which usually begins to germinate 21–70 days after sowing, or from cuttings of firm young growth, which strike readily.

The var. *microcephala* Domin is a synonym, while var. *anceps* (R. Br.) Benth. is referrable to *P. anceps* R. Br.

Petrophile longifolia R. Br.

(long leaves)

WA   Long-leaved Cone Bush
0.2–0.6 m × 0.5–1 m   July–March; mainly Sept–Nov

Dwarf **shrub** with glabrous, sometimes reddish-bronze young growth; **branches** prostrate to spreading, few; **branchlets** short, glabrous, brittle; **leaves** 15–30 cm × about 0.2 cm, terete, ascending to erect, often incurved, not twisted, crowded, mid green, glabrous, apex blunt or softly pointed; **flowerheads** to about 4.5 cm across, terminal, sessile, solitary, can be profuse and conspicuous, nestling among the leaves; **involucral bracts** linear-lanceolate, glabrous, stiff; **cone scales** broad, glabrous; **flowers** to about 2.5 cm long, cream to yellow, hairy, sweetly fragrant; **pollen presenter** about 0.4 cm long, top-shaped below the densely hairy brush; **cones** to about 3 cm long, ovoid.

A very attractive groundcovering species from the south-western Darling and western Eyre Districts. It grows in a variety of soils including sand, sandy loam, gravels and clay loam, all of which may be wet for extended periods. This species has been tried in cultivation but it has proved difficult to grow. Best for temperate regions but may succeed in semi-arid areas too. Plants need well-drained soils which are moisture-retentive. A sunny site is better than one which is semi-shaded. Plants tolerate light frosts. Well suited to rockeries, borders, containers and as an accent plant. Propagate from seed, which usually begins to germinate 20–60 days after sowing, or from cuttings of firm young growth, which usually strike readily.

The var. *tenuifolia* Benth. is a synonym.

P. aspera is the correct name for plants which have been marketed as a shrubby selection of *P. longifolia*.

Petrophile macrostachya R. Br.

(large flower spikes)

WA
0.6–1.5 m × 0.6–1.5 m   July–Nov

Dwarf to small **shrub** with hairy young growth; **branches** ascending to erect, stiff; **branchlets** initially greyish-hairy, usually becoming glabrous; **leaves** 3–8 cm long, pinnate or with 3 broad segments, spreading to ascending, stiff, mid-to-deep green, initially hairy, becoming glabrous and glossy; segments 2–4 cm × 0.25 cm, flat, sometimes lobed, pungent-tipped; **flowerheads** 2–6 cm long, cylindrical to oblong, terminal or axillary, sessile, sometimes clustered, often profuse and very conspicuous; **involucral bracts** ovate, crowded, deciduous, margins fringed; **cone scales** broad, base and margin hairy, apex glabrous; **flowers** about 0.9 cm long, cream to yellow, glabrous; **pollen presenter** about 0.3 cm long, tapered to both ends, yellow, sparsely hairy; **cones** to about 6.5 cm long, narrowly ovoid to nearly cylindrical.

An ornamental conebush from the Irwin and northern Darling Districts where it occurs in a range of sandy soils which sometimes overlie limestone. Plants are found in sandplain, heath, shrubland and woodland communities. Rarely cultivated and warrants greater attention. It requires well-drained soils which are acidic but it can also tolerate low alkalinity. Needs a sunny site for best results. Hardy to moderate frosts and extended dry periods. Usually has a fairly bushy habit from ground level. Propagate from seed, which usually begins to germinate 20–80 days after sowing, or from cuttings of firm young growth, which may be slow to produce roots.

Petrophile media R. Br.

(medium or intermediate)

WA
0.2–0.8 m × 0.5–1.2 m   Aug–Feb

Dwarf **shrub** with glabrous young growth; **branches** spreading to ascending, many; **branchlets** glabrous; **leaves** 5–30 cm × to about 0.2 cm, terete, ascending to erect, often incurved, scattered to somewhat crowded, mid-to-deep green, glabrous, smooth, apex often shortly curved and pungent-pointed; **flowerheads** to about 2.5 cm across, ovoid, terminal, solitary, sessile, often profuse, very conspicuous; **involucral bracts** linear to lanceolate, many; **cone scales** ovate-lanceolate, somewhat reflexed, glabrous; **flowers** 1.5–2 cm long, cream or yellow, densely hairy, sweetly fragrant, often with golden-brown tips in bud; **pollen presenter** about 0.4 cm long, top-shaped below the hairy brush; **cones** 1–2.5 cm long, mainly ovoid.

An attractive, floriferous species from the southern Darling and western Eyre Districts. Inhabits sandy, gravelly and clay soils in heath, shrubland, mallee and woodland. It has had limited cultivation. Best suited to warm temperate and semi-arid regions but will succeed in protected sites in cool temperate areas. Needs moderately well-drained acidic soils and plenty of sunshine. Plants tolerate moderate frosts and extended dry periods. They respond well to pruning. It has potential for use in containers and rockeries and well as in general planting. First cultivated in England in 1840 as var. *juncifolia* (Lindl.) Benth. which is now regarded as a synonym. Propagate from seed, which usually begins to germinate 26–70 days after sowing, or from cuttings of firm young growth, which strike readily.

Petrophile megalostegia F. Muell.
(large involucral bracts)
WA
0.3–1.3 m × 0.3–1.5 m   Aug–Oct

Dwarf to small **shrub** with glabrous young growth; **branches** spreading to ascending; **branchlets** glabrous; **leaves** 2.5–8.5 cm × about 0.2 cm, linear, terete or flattened, straight or curved to somewhat S-shaped, ascending to erect, deep green, glabrous, apex ending in a fine, elongated, pungent point; **flowerheads** to about 4 cm across, globular, terminal, solitary, sessile, often profuse, very conspicuous; **involucral bracts** ovate to elliptic, many, glabrous; **cone scales** elliptic to narrowly obovate, margin irregular, pale brown, glabrous; **flowers** to about 3 cm long, cream, creamy yellow or yellow, silky-hairy, sweetly fragrant; **pollen presenter** about 0.45 cm long, somewhat top-shaped below the orange brush, with spreading hairs; **cones** about 1.5 cm long, ovoid.

A handsome species from the north-western Avon and central and southern Irwin Districts where it grows in sand, clay loam and lateritic soils. Rarely encountered in cultivation, it deserves greater attention. Plants should do best in warm temperate and semi-arid zones. They require moderately well-drained acidic soils and plenty of sunshine although plants have been grown successfully in semi-shaded sites. Withstands moderate frosts and extended dry periods. Responds well to pruning. Worth growing in containers and rockeries as well as in general planting. Propagate from seed, which usually begins to germinate 30–70 days after sowing, or from cuttings of firm young growth, which may be slow to strike.

Petrophile merrallii Foreman
(after Edwin Merrall, 19–20th-century gold prospector and botanical collector)
WA
0.6–1.3 m × 0.5–1.3 m   Aug–Oct

Dwarf to small **shrub** with hairy young growth; **branches** ascending to erect; **branchlets** becoming glabrous; **leaves** 0.4–1.4 cm × about 0.1 cm, linear-terete, sessile, spreading to ascending, crowded, often slightly curved, deep green, initially with short curled hairs, becoming glabrous, slightly rough, apex pointed; **flowerheads** about 1.5 cm across, terminal, can be profuse and conspicuous; **involucral bracts** ovate, many, crowded, sticky; **cone scales** ovate, exterior of base densely hairy; **flowers** to about 1.5 cm long, yellow, densely hairy; **pollen presenter** about 0.4 cm long, tapered to both ends, hairy; **cones** to about 2 cm across, more or less globular.

A recently described (1995) species which is a member of the *P. ericifolia* complex. It occurs in the Avon, Coolgardie and Roe Districts where it inhabits mainly sandy soils of heath and mallee shrubland. Plants require very well-drained acidic soils and plenty of sunshine. They are best suited to semi-arid and warm temperate regions but may succeed in sheltered warm sites in cooler areas. Hardy to moderate frosts and extended dry periods. It has potential for cut-flower production as well as in general planting. Propagate from seed or from cuttings of firm young growth.

P. ericifolia has sticky flowers and larger cones.

Petrophile misturata Foreman
(mixture; referring to simple and divided leaves)
WA
0.6–1.2 m × 0.6–1.3 m   Sept–Nov

Dwarf to small **shrub** with densely hairy young growth; **branches** many, spreading to erect; **branchlets** hairy; **leaves** 0.8–1.6 cm long, simple to pinnate; segments terete, spreading to ascending, crowded, hairy, greyish-green, apex pungent-pointed; **flowerheads** about 1 cm across, globular, sessile, terminal, solitary, often profuse and very conspicuous; **involucral bracts** very narrow, few, deciduous; **cone scales** ovate, exterior hairy; **flowers** about 1 cm long, dull yellow, with long hairs; **pollen presenter** about 0.35 cm long, tapered to both ends, with few hairs; **cones** about 1.2 cm across, globular.

An interesting, recently described (1995) species which occurs mainly in the central-western Avon District where it grows in sandy soils. It has potential for use in warm temperate and semi-arid regions and may succeed in cooler areas if grown in a sheltered sunny site. Needs excellently drained acidic soils and plenty of sunshine. Plants withstand moderate frosts. Propagate from seed or from cuttings of firm young growth.

Petrophile multisecta F. Muell.
(many incisions)
SA
0.3–1 m × 0.3–1 m   Oct–Feb

Dwarf **shrub** with a lignotuber; young growth hairy, brownish-red; **branches** spreading to ascending, many; **branchlets** greyish-hairy; **leaves** 4–8 cm × to about 4 cm, much-divided with narrowly terete pungent-tipped segments to 0.2–1 × about 0.1 cm, mid green, becoming glabrous, stiff; **flowerheads** to about 2.5 cm × 2 cm, ovoid to globular, axillary, sessile, solitary, can be moderately profuse and conspicuous; **involucral bracts** lanceolate, exterior glabrous, interior silky-hairy; **cone scales** circular, silky-hairy except for glabrous apex; **flowers** 0.8–1.5 cm long, cream, densely hairy; **pollen presenter** to about 0.35 cm long, tapered to both ends, creamy-yellow; **cones** to about 4 cm long, ovoid.

This moderately compact species is confined to Kangaroo Island where it inhabits calcareous and lateritic sands. Cultivated mainly by enthusiasts, it has lovely brownish-red young growth and warrants greater attention. Suitable for temperate and semi-arid regions. Needs good drainage and tolerates some alkalinity. A sunny site gives best results but plants tolerate semi-shade. Hardy to moderate frosts. Responds well to pruning. Does well in containers and rockeries and suits general planting. Propagate from seed, or from cuttings of firm young growth, which may be slow to produce roots.

Petrophile pauciflora

Petrophile longifolia   W. R. Elliot

Petrophile pauciflora Foreman
(few-flowered)
WA
0.5–1.2 m × 0.5–1.2 m   Sept; possibly other times
   Dwarf to small **shrub** with hairy young growth;
branches ascending to erect; **branchlets** initially woolly-
hairy, becoming glabrous; **leaves** 1.5–3.5 cm long,
ascending to erect, scattered, usually with 3 to 5 terete
pungent-tipped segments, becoming glabrous and
mid green; **flowerheads** about 0.8 cm across, in upper
axils, solitary, few-flowered, prominently stalked,
apparently rarely very conspicuous; **involucral bracts**
ovate, exterior glabrous; **cone scales** ovate, exterior
woolly-hairy; **flowers** about 0.7 cm long, probably yellow, densely hairy; **pollen presenter** about 0.45 cm
long, straight, with short reflexed hairs; **cones**
0.6–0.8 cm long, ovoid to globular.
   A very poorly known species which has been found
in only two disjunct areas west of Meekatharra and
south-east of Paynes Find in the Austin District. Needs
to be cultivated as part of a conservation strategy. Best
suited to warm temperate and semi-arid regions. Soils
will need to drain freely and a very sunny site is a

Petrophile macrostachya   W. R. Elliot

prerequisite. Propagate from seed, or from cuttings of
firm young growth, which may be slow to produce
roots.

Petrophile pedunculata R. Br.
(with a stalk)
NSW
1.5–2.5 m × 1–2 m   Nov–Feb
   Small to medium **shrub** with faintly hairy or glabrous young growth; **branches** ascending to erect;
branchlets glabrous; **leaves** usually 8–16 cm long, with
many terete, spreading, softly pointed segments on
upper two-thirds, green, glabrous; **flowerheads** to
about 3 cm × 2 cm, axillary, on stalks 1–3 cm long, usually scattered, rarely profuse; **involucral bracts** narrow,
mainly deciduous; **cone scales** broad, hairy; **flowers** to
about 1 cm long, yellow, exterior glabrous; **cones**
to about 3 cm long.
   This species is found in the Central Coast, Southern
Coast and the Central Tableland Districts where it
grows on sand or skeletal soils in heath or dry sclerophyll forest. Cultivated mainly by enthusiasts, it does
well in temperate regions. Needs well-drained acidic
soils with a sunny or semi-shaded aspect. Hardy to
moderate frosts. Bushy growth is promoted by pruning from an early age. Grown in England as early as
1824. Propagate from seed or from cuttings of firm
young growth.

Petrophile phylicoides R. Br.
(like the genus *Phylica*)
WA
0.5–1.5 m × 0.5–2 m   Sept–Dec; sporadic to March
   Dwarf to small **shrub** with hairy young growth;
branches ascending to erect, many; **branchlets** initially
hairy, becoming glabrous; **leaves** 0.6–1.5 cm × up to
0.2 cm, terete, spreading to erect, incurved, crowded
especially at ends of branchlets, deep green, becoming glabrous, apex hard-pointed; **flowerheads** to
about 2 cm across, globular, terminal, solitary, sessile,
often profuse and very conspicuous; **involucral bracts**
narrow, faintly hairy, deciduous; **cone scales** broad,
base woolly-hairy, apex glabrous; **flowers** about 1.2 cm
long, pale to bright yellow, glabrous; **pollen presenter**
about 0.45 cm long, narrowly ovoid, orange, faintly
hairy; **cones** about 2.8 cm across, globular.
   A floriferous species from the Eyre and south-
western Roe Districts where it grows on sand, gravelly
sand and laterite in heath, shrubland and mallee communities. Cultivated mainly by botanical gardens and
enthusiasts, it warrants greater recognition. Plants
should do best in warm temperate and semi-arid
regions. They require freely draining acidic soils
and plenty of sunshine. Hardy to moderate frosts and
extended dry periods. Responds well to pruning.
Excellent for containers and rockeries. Propagate
from seed, which usually begins to germinate
30–50 days after sowing, or from cuttings of firm
young growth, which strike readily.
   P. ericifolia is closely allied but has densely hairy,
sticky flowers while *P. scabriuscula* differs in having
densely hairy, non-sticky flowers.
   P. ericifolia var. *glabriflora* Benth. and *P. glabriflora*
(Benth.) Domin are synonyms.

Petrophile pedunculata × .4

Petrophile plumosa Meisn.
(feathery, plumed)
WA
0.5–1.3 m × 0.5–1.3 m   July–Nov

Dwarf to small **shrub** with rusty-hairy young growth; **branches** spreading to erect, many; **branchlets** hairy; **leaves** 1.3–3.2 cm long, linear-spathulate and entire, or 2–3-lobed near apex, spreading to ascending, green to greyish-green, initially very hairy, becoming rough; segments 0.2–0.6 cm long, pungent-tipped; **flowerheads** to about 2–5 cm across, globular, terminal, solitary, often profuse and conspicuous; **involucral bracts** ovate-oblong, base hairy, apex glabrous, margins fringed; **cone scales** ovate-oblong, hairy to nearly glabrous; **flowers** to about 1.5 cm long, pale yellow, densely hairy; **pollen presenter** to about 0.35 cm long, tapered to both ends, yellow, with short reflexed hairs; **cones** about 2.5 cm long, ovoid.

This species is regarded as vulnerable in its natural habitat of sand, sandy gravel and gravelly soils of the northern Darling District. It occurs in shrubland communities. Very rare in cultivation. Needs very well-drained acidic soils. Prefers a sunny site but will tolerate some shade. Plants should do best in warm temperate and semi-arid areas. They tolerate extended dry periods and are hardy to moderate frosts. Worth trying as a container plant. Response to pruning is not known. Propagate from seed, or from cuttings of firm young growth, which may be slow to strike.

Petrophile propinqua R. Br. = *P. squamata* R. Br.

Petrophile pulchella (Schrad.) R. Br.
(beautiful)
Qld, NSW
1.6–3 m × 1–2 m   Aug–March; also sporadic

Small to medium **shrub** with more or less glabrous young growth; **branches** ascending to erect; **branchlets** glabrous; **leaves** mainly 4–10 cm long, with terete, further divided, softly pointed, ascending to erect segments in upper half, flexible, mid-to-dark green, glabrous; **flowerheads** to about 5 cm × 3 cm, spike-like, terminal or axillary, sessile or stalked, solitary or with basal clusters, rarely profuse but always conspicuous; **involucral bracts** lanceolate, few; **cone bracts** broad,

rusty-hairy, becoming glabrous; **flowers** to about 1.4 cm long, yellow, exterior hairy; **cones** to about 5 cm long, ovoid.

This species occurs in south-eastern Qld and it extends throughout the Central Tablelands, North, Central and South Coast Districts of NSW. It is usually found on shallow sandy or clay soils in heath and dry sclerophyll forest. This was the first species of the genus to be cultivated and was introduced into England in 1790. Plants adapt well in temperate regions. They require acidic soils with good drainage and a sunny or semi-shaded site. Hardy to moderate frosts and extended dry periods. Prune while young to promote lateral branching. Propagate from seed or from cuttings of firm young growth.

P. fucifolia (Salsb.) Knight is a synonym. *P. canescens* is similar but has greyish-hairy leaves with ascending segments. There are intermediate forms with both *P. canescens* and *P. sessilis* and some of these may be in cultivation.

Petrophile recurva Foreman
(recurved)
WA
1–2.3 m × 0.6–1.6 m   Aug–Oct

Small to medium **shrub** with densely hairy young growth; **branches** many, ascending to erect; **branchlets** becoming glabrous; **leaves** 0.7–1.3 cm × about 0.1 cm, linear-terete, sessile, ascending to erect, crowded, overlapping, with recurved tips, hairy becoming

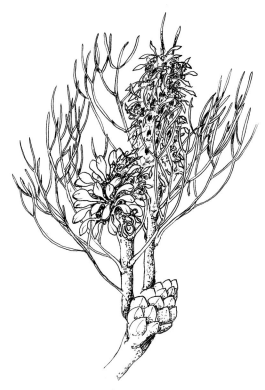

Petrophile pulchella × .75

glabrous, apex pungent-pointed; **flowerheads** to about 1.5 cm across, more or less globular, terminal, often profuse and very conspicuous; **involucral bracts** narrowly ovate, many, sticky; **cone scales** narrowly elliptic, hairy except for glabrous apex; **flowers** about 1 cm long, cream to pale yellow, densely hairy; **pollen presenter** about 0.3 cm long, yellow or orange-yellow, tapered to both ends, faintly hairy; **cones** 1–1.5 cm across, ovoid to globular.

Outstanding when in full flower, this recently described (1995) species is a distinctive member of the *P. ericifolia* complex. It inhabits the sandy and gravelly soils of heathland, shrubland and open woodland in the southern Irwin District. It has had limited cultivation and is likely to do best in warm temperate and semi-arid regions. Plants require excellent drainage, acidic soils and plenty of sunshine. Heavy frosts can cause damage. They respond well to pruning. Worth trying as a cut-flower as well as for use in gardens. Propagate from seed, or from cuttings of firm young growth, which can be slow to form roots.

Petrophile rigida R. Br.
(rigid, stiff)
WA
0.6–1.5 m × 0.6–1.5 m   Sept–Oct

Dwarf to small **shrub** with glabrous young growth; **branches** spreading to erect, many, often entangled; **branchlets** glabrous; **leaves** 3–8.5 cm long, much divided, spreading to ascending, crowded, mid-to-deep green, stiff; segments variable, 0.1–2.5 cm long, terete, grooved above, pungent-tipped; **flowerheads** to about 3 cm across, globular, terminal or axillary, solitary, sessile, often profuse and very conspicuous; **involucral bracts** lanceolate, exterior glabrous, interior silky-hairy, deciduous; **cone scales** broad, base silky-hairy, apex glabrous; **flowers** about 1.5 cm long, cream with yellowish tip which is golden-brown in bud, densely hairy; **pollen presenter** about 0.4 cm long, tapered to both ends, golden-yellow, faintly hairy; **cones** about 2 cm across, ovoid.

This species has a widespread but scattered distribution in the south-western region of Wa where it grows on sandy soils. It was first cultivated in England during 1833 but now is apparently rarely grown in Australia. Needs excellently drained acidic soil and plenty of sunshine. Withstands extended dry periods and is hardy to moderate frosts. Should do best in warm temperate and semi-arid regions. Plants are prickly and would probably be excellent for foot traffic control. Propagate from seed, which usually begins to germinate 28–70 days after sowing, or from cuttings of firm young growth, which may be slow to strike.

P. drummondii is allied but always has terminal flowerheads which are often elongated.

Petrophile scabriuscula Meisn.
(minutely scabrous)
WA
0.6–1.2 m × 0.5–1 m   May–Oct

Dwarf to small **shrub** with hairy young growth; **branches** ascending to erect, many; **branchlets** greyish-hairy, becoming glabrous; **leaves** 0.6–1.5 cm × about 0.2 cm, terete, ascending to erect, often appressed,

crowded, mid green, initially hairy, becoming glabrous and somewhat rough, apex pungent-pointed; **flowerheads** to about 2 cm across, ovoid, terminal, sessile, solitary, may be profuse and conspicuous; **involucral bracts** narrowly ovate, crowded, slightly sticky, hairy, margins fringed; **cone scales** normally ovate to broadly ovate, base hairy; **flowers** 0.8–1.4 cm long, cream-yellow to yellow, densely hairy; **pollen presenter** to about 0.35 cm long, yellow, tapered to both ends, with short erect hairs; **cones** to about 2 cm across, globular to ovoid.

P. scabriuscula is confined mainly to the Irwin and far-northern Darling Districts where it inhabits sandy soils in sandplain, heath and shrubland. Apparently not cultivated, it is best suited to warm temperate and semi-arid regions and requires very well-drained acidic soils and lots of sunshine. Plants should tolerate light frosts and extended dry periods. Reaction to pruning is not known. Propagate from seed or from cuttings of firm young growth, which may be slow to produce roots.

P. ericifolia var. *scabriuscula* (Meisn.) Benth. is a synonym. *P. phylicoides* is similar but has glabrous flowers and *P. ericifolia* has sticky involucral bracts and flowers.

Petrophile semifurcata Benth.
(shallowly-forked)
WA
0.3–1.5 m × 0.3–1.2 m   Sept–Nov

Dwarf to small **shrub** with hairy young growth; **branches** ascending to erect, many; **branchlets** hairy; **leaves** 7–14.5 cm long, terete, entire or 2–3-lobed at tips, rarely further divided, ascending to erect, dark green, becoming glabrous, stiff, pungent-tipped; **flowerheads** 2–4 cm × to about 3 cm, ovoid, terminal, solitary, stalked, can be profuse and conspicuous; **involucral bracts** narrowly triangular, deciduous; **cone scales** broad, woolly to velvety; **flowers** about 1.4 cm long, whitish, cream or lemon-yellow, silky-hairy; **pollen presenter** about 0.6 cm long, orange, faintly thickened, with short hairs; **cones** 2.5–4.5 cm × to about 3 cm, ovoid, stalk to about 2 cm long.

This species inhabits sandy soils of the northern Irwin District where it grows in heathland. Evidently rarely cultivated, it is best suited to warm temperate and semi-arid areas. Needs extremely well-drained acidic soils and maximum sunshine. Plants tolerate light frosts and extended dry periods. Propagate from seed, or from cuttings of firm young growth, which may be slow to produce roots.

Petrophile seminuda Lindl.
(partly naked)
WA
0.5–1.2 m × 0.5–1 m   Sept–Nov

Dwarf to small **shrub** with glabrous, reddish-bronze young growth; **branches** ascending to erect, many; **branchlets** glabrous, reddish-brown; **leaves** 3–12.8 cm long, much-divided, spreading to ascending, light-to-mid green, glabrous, stiff; segments 0.5–5 cm long, variable, 2–3 main segments usually further divided, pungent-tipped; **flowerheads** about 2 cm across, globular or ovoid, terminal, sessile, can be profuse and very conspicuous; **involucral bracts** ovate-lanceolate, exterior glabrous, interior densely hairy; **cone scales** broad, base woolly-hairy, apex glabrous, deciduous; **flowers** to about 1.2 cm long, yellow, glabrous; **pollen presenter** about 0.3 cm long, yellow, faintly hairy; **cones** 1–4 cm × to about 1.5 cm, ovoid to globular.

An appealing species with attractive foliage and flowers. It grows over a wide range of the south-western region and occurs in a variety of winter-wet sandy and gravelly soils. It has had limited cultivation. Adapts to most acidic soils which are relatively free draining. Likes plenty of sunshine but will tolerate some shade. Best suited to temperate and semi-arid regions. Withstands moderate frost and extended dry periods. Responds well to pruning. Propagate from seed, which usually begins to germinate 40–60 days after sowing, or from cuttings of firm young growth, which may be slow to form roots.

P. seminuda var. *indivisa* Benth. is a synonym.

Petrophile serruriae R. Br.
(like the genus *Serruria*)
WA
0.6–1.5 m × 0.8–2 m   July–Dec

Dwarf to small **shrub** with hairy, often pinkish young growth; **branches** many, spreading to ascending, often arching; **branchlets** densely silky-hairy, may become glabrous, often arching; **leaves** 1.5–3.5 cm long, bipinnate or tripinnate, mainly spreading, sometimes crowded, silky-hairy, may become glabrous; segments many, terete, grooved above, pungent-tipped; **flowerheads** to about 1 cm across, terminal or in upper axils, sessile or short-stalked, clusters to about 3 cm across, often profuse and very conspicuous; **involucral bracts** narrow, few, deciduous; **cone scales** narrowly ovate, base and exterior densely hairy, apex glabrous; **flowers** to about 1.6 cm long, yellow or pink to greyish-mauve, silky- or rusty-hairy; **pollen presenter** about 0.4 cm long, orange-yellow, tapered to both ends, hairy; **cones** to about 1.2 cm long, ovoid to globular.

A handsome species of the Irwin, Darling, western Avon and western Eyre Districts where it grows in coastal as well as inland regions. It occurs in heath, shrubland and woodland on a range of sandy and gravelly soils which often contain limestone and laterite. It has had limited cultivation. Does well in temperate and semi-arid regions. Needs well-drained soils which can be acidic or slightly alkaline and a sunny or semi-shaded site is appropriate. Tolerates moderate frosts but dislikes extended dry periods. A long dry period often leads to dieback of branches which initiates release of seeds from the cones. May become a weed in some areas. Plants respond well to hard pruning. Propagate from seed, which usually begins to germinate between 25–60 days after sowing, or from cuttings of firm young growth, which strike readily.

P. glanduligera Lindl. is a synonym.

Petrophile sessilis Sieber ex Schult. & Schult. f.
(sessile; without stalks)
NSW   Conesticks
1.5–3 m × 1–2 m   Oct–Jan; also sporadic

Small to medium **shrub** with bronze to brownish, hairy young growth; **branches** ascending to erect; **branchlets** becoming glabrous; **leaves** mainly 3–10 cm long, much-divided, spreading, mid-to-deep green, becoming glabrous, rigid; segments terete, pungent-pointed; **flowerheads** to about 4 cm × 2.5 cm, spike-like, solitary, terminal or axillary, rarely profuse, always conspicuous; **involucral bracts** broadly ovate, initially rusty-hairy, becoming glabrous; **cone scales** broadly ovate, initially hairy; **flowers** to about 1.4 cm long, yellow, exterior silky-hairy; **pollen presenter** about 0.25 cm long, yellow, tapered to both ends, hairy; **cones** to about 4 cm × 2 cm.

This species is found in heath and dry sclerophyll forest on sandstone. Occurs from Sydney and south-wards to near Jervis Bay in coastal regions and the Central and Southern Tablelands. Cultivated to a limited degree, it adapts well in temperate regions. Plants require freely draining acidic soils with a sunny or semi-shaded aspect. They are hardy to moderately heavy frosts and tolerate long periods of dryness. Bushiness is promoted by pruning young plants. Propagate from seed or from cuttings of firm young growth.

P. canescens is similar but it has flexible greyish-hairy leaves.

Petrophile shirleyae F. M. Bail.
(after Mrs J. Shirley)
Qld   Moreton Bay Conesticks
0.3–1.5 m × 0.3–1.5 m   Aug–Feb

Dwarf to small **shrub** with a lignotuber; young growth faintly hairy to glabrous, often reddish-bronze; **branches** ascending to erect; **branchlets** more or less glabrous; **leaves** 8–20 cm long; segments pinnate or bipinnate, terete, becoming glabrous, mid-to-deep green, rigid, pungent-pointed; **flowerheads** 2–8 cm × about 2 cm, spike-like, terminal, solitary or with 1–2 basal heads, on stalks to about 4 cm long, scattered to somewhat profuse, conspicuous; **involucral bracts** broadly ovate, few, becoming glabrous; **cone scales** broadly ovate, initially velvety, becoming glabrous; **flowers** to about 1 cm long, white to yellowish, exterior silky-hairy; **cones** to about 8 cm long.

An appealing inhabitant of coastal regions in the Moreton and Wide Bay Districts of the south-east. Grows in sandy or rocky soils which may be moist for extended periods. Plants may have lovely reddish-bronze new growth. Cultivated mainly by enthusiasts. Needs good drainage and does best in a sunny site but tolerates semi-shade. Plants are hardy to light frost and withstand limited coastal exposure. Well suited to rockeries and large containers, as well as for general planting. Very sensitive to sweating and root rot. Propagate from seed or from cuttings of firm young growth.

Sometimes marketed incorrectly as *P. shirliae*.

Petrophile shuttleworthiana Meisn.
(after Robert Shuttleworth, 19th-century English botanist)
WA
0.6–2.5 m × 0.5–2 m   Sept–Oct

Dwarf to medium **shrub** with glabrous young growth; **branches** spreading to erect; **branchlets** glabrous; **leaves** 3.5–8 cm long, pinnate or deeply

Petrophile serruriae, pink-flowered variant   W. R. Elliot

3-lobed in upper half, flattened, spreading to
ascending, light-to-dark green, glabrous; segments
2–4 cm × 0.25–1 cm, pungent-tipped; **flowerheads**
2.5–6.5 cm × about 2 cm, cylindrical to narrowly ovoid,
terminal or axillary, sessile or with stalk to about
1.5 cm long, often profuse and very conspicuous;
involucral bracts very narrow, deciduous; **cone scales**
small and hairy to broad and glabrous, deciduous;

Petrophile shirleyae   D. L. Jones

Petrophile shuttleworthiana   W. R. Elliot

flowers to about 1.2 cm long, white, creamy white,
cream or yellow, glabrous; **pollen presenter** to about
0.4 cm long, tapered to both ends, faintly hairy; **cones**
3.5–6 cm × about 1.5 cm, narrowly ovoid.

An interesting species mainly from the Irwin and
north-western Avon Districts where it inhabits sandy
clay, sandy loam and lateritic clay soils in heath, shrubland and mallee. Doubtful if in cultivation and warrants greater attention with its well-displayed flowers
and deeply-lobed leaves. Plants should adapt to a
range of acidic soils in temperate and semi-arid
regions. In nature plants tolerate moderate frosts and
extended dry periods. Reaction to pruning is not
known. Propagate from seed or from cuttings of firm
young growth, which may be slow to produce roots.

P. macrostachya is closely allied but has hairy young
growth and shorter flowers.

Petrophile squamata R. Br.
(with scales)
WA
1–1.8 m × 0.6–1.5 m   July–Dec

Small **shrub** with glabrous or faintly hairy young
growth; **branches** mainly ascending to erect; **branchlets** glabrous or becoming so; **leaves** 2–7 cm long,
once or twice ternately divided in upper half, or rarely
simple, spreading to erect, scattered to moderately
crowded, glabrous or faintly hairy, mid-to-deep green,
stiff; segments 0.3–3.5 cm long, variable, sometimes
lobed, pungent-tipped; **flowerheads** to about 2 cm ×
1.5 cm, ovoid, sessile, in upper axils forming leafy
spike to about 30 cm long, often profuse and very conspicuous; **flowerheads** small, deciduous; **cone scales**
ovate, hairy except for glabrous apex; **flowers** to about
1 cm long, creamy yellow to bright yellow, hairy;

pollen presenter about 0.4 cm long, yellow, thickened below the brush which is faintly hairy or glabrous; **cones** to about 1.6 cm × 1 cm, ovoid.

This is an attractive, widespread, variable species from southern WA where it grows in a wide range of acidic soils in a variety of plant communities. It has had limited cultivation but has adapted well in temperate and semi-arid regions. Needs moderately well-drained soils. Prefers sunny sites but will tolerate a semi-shaded aspect. Plants are hardy to moderately heavy frosts and extended dry periods. They respond well to pruning which promotes bushier growth. Often develops long, straight stems which are suitable for cut flowers. Propagate from seed, which usually begins to germinate 20–50 days after sowing, or from cuttings of firm young growth, which strike readily.

P. propinqua R. Br. and *P. trifida* R. Br. are synonyms. Plants of *P. glauca* have been marketed wrongly under the name of *P. trifida*.

Petrophile striata R. Br.
(finely striped or grooved)
WA
0.2–1 m × 0.3–1.5 m   Aug–Dec

Dwarf **shrub** with bronzish-red hairy young growth; **branches** spreading to ascending, many; **branchlets** initially hairy, usually becoming glabrous; **leaves** 3.5–8 cm long, pinnate or bipinnate, spreading to ascending often crowded, bright-to-mid green, hairy to glabrous, finely grooved; segments to about 1 cm × 0.1–0.3 cm, variable, entire or lobed, pungent-tipped; **flowerheads** to about 2 cm across, ovoid, axillary, sessile, solitary or clustered, often profuse and very conspicuous; **involucral bracts** ovate, rich brown, many, crowded; **cone scales** lanceolate to ovate, exterior mainly hairy; **flowers** to about 2 cm long, creamy white, cream, creamy-yellow or yellow, silky-hairy, small appendages at tip of segments; **pollen presenter** about 0.3 cm long, yellow, broadened below the faintly hairy to glabrous brush; **cones** to about 1.5 cm × 1.5 cm, ovoid.

A very pretty species from the southern Irwin, western Avon and northern and central Darling Districts where it occurs mainly in the Darling Range and adjacent coastal plain. It grows on sandy, gravelly, lateritic and clay soils. Not well known in cultivation but it has succeeded in freely draining acidic soils with a fairly sunny aspect. Plants are hardy to moderate frosts and withstand extended dry periods. Valuable for wildlife habitat and for foot traffic control. Propagate from seed, which usually begins to germinate 25–70 days after sowing, or from cuttings of firm young growth, which may be slow to form roots.

P. macrostegia is sometimes confused with this species. It differs in having hairy flowers which lack appendages.

Petrophile stricta C. A. Gardner ex Foreman
(erect, upright)
WA
0.6–1.6 m × 0.4–1.3 m   Oct–Dec

Dwarf to small **shrub** with more or less glabrous young growth; **branches** ascending to erect; **branchlets** glabrous; **leaves** 4.5–13.5 cm × about 0.2 cm,

Petrophile squamata   W. R. Elliot

terete, scattered to moderately crowded, ascending to erect, usually slightly incurved, deep green, glabrous, entire, apex ending in a pungent point; **flowerheads** to about 4 cm × 3 cm, somewhat ovoid, terminal, solitary, sessile, can be profuse and very conspicuous; **involucral bracts** linear, deciduous; **cone scales** broad, exterior velvety, finely pointed; **flowers** to about 1.2 cm long, cream to pink, exterior with long hairs; **pollen presenter** about 0.5 cm long, tapering to both ends, with short hairs; **cones** to about 4.7 cm long, woody, becoming glabrous.

This species was formally described in 1989 although it had been known for many years. It occurs mainly in the Avon, western Coolgardie and Roe Districts where it inhabits sandy and gravelly soils of the shrublands. Suited to growing in warm temperate and semi-arid regions. Needs very well-drained acidic soils and a very sunny aspect. Plants are drought tolerant and withstand moderately heavy frosts. They are prickly which makes them useful for wildlife habitat. Propagate from seed or from cuttings of firm young growth.

P. semifurcata is allied but sometimes can have lobed thicker leaves and a hairy style.

Petrophile teretifolia R. Br.
(terete leaves)
WA   Southern Pixie Mops
0.6–2 m × 0.6–2 m   Sept–Jan

Dwarf to small **shrub** with glabrous young growth; **branches** often few, spreading to erect; **branchlets** glabrous, brownish; **leaves** 4–22 cm × about 0.2 cm, terete, spreading to ascending, prominently incurved, glabrous, mid green, smooth, apex pointed or blunt; **flowerheads** to about 4 cm across, ovoid to globular,

255

terminal or axillary, sessile or on short stalks to about 1–5 cm long, often profuse and very conspicuous; **involucral bracts** few, outer ones narrow, inner ones broadest; **cone scales** more or less circular, pointed, base hairy and margins sometimes fringed, otherwise glabrous; **flowers** to about 2.5 cm long, pink to mauve, usually fading to whitish, densely hairy; **pollen presenter** about 0.45 cm long, yellow, top-shaped below the densely hairy brush; **cones** to about 2.5 cm long, elliptic to globular.

This attractive species is distributed throughout the Eyre District where it grows in sandy soils which are often associated with granite or laterite. It was originally cultivated in England during 1824 and it is grown to a limited extent in Australia. Plants require excellently drained acidic soils and plenty of sunshine for best results. They tolerate extended dry periods and moderate frosts, and they respond well to pruning. Cones are often retained for many years. Low-growing variants are suitable for rockeries and containers. Propagate from seed, which usually begins to germinate between 26–80 days after sowing, or from firm young growth, which may be slow to form roots.

Petrophile trifida R. Br. = *P. squamata* R. Br.
Plants marketed in past years as *P. trifida* are generally referrable to *P. glauca*.

Petrophile trifurcata Foreman
(three-forked)
WA
0.4–0.7 m × 0.4–1 m   Aug–Sept
Dwarf **shrub** with hairy young growth; **branches** many, ascending to erect; **branchlets** becoming glabrous; **leaves** 1–1.6 cm long, terete, usually with 3 short lobes or sometimes simple, spreading to ascending, somewhat crowded, usually becoming glabrous, apex pungent-pointed; **flowerheads** to about 1 cm across, globular, terminal, sessile; **involucral bracts** ovate, fringed, glabrous exterior; **cone scales** obovate, exterior densely hairy; **flowers** about 1 cm long, yellow, with long hairs; **pollen presenter** about 0.3 cm long, tapered to both ends, with very short hairs; **cones** about 1.2 cm across, globular.

A recently described (1995) species which occurs in the north-western Avon and far south-eastern Irwin Districts where it grows in sandy soils. Warrants cultivation as part of a conservation strategy. Plants will need very well-drained acidic soils with a sunny or warm semi-shaded site. They are hardy to moderate frosts and withstand extended dry periods. Probably best suited to warm temperate or semi-arid regions but may succeed in cooler areas especially if grown as a container plant. Propagate from seed or from cuttings of firm young growth.

P. misturata differs in having permanently hairy stems and leaves as well as an elongated pollen-presenter tip.

Petrophile wonganensis Foreman
(from the Wongan Hills region)
WA
0.5–1.5 m × 0.8–2 m   Aug–Jan

Dwarf to small **shrub** with whitish-hairy young growth; **branches** many, ascending to erect; **branchlets** becoming glabrous; **leaves** 0.5–1 cm × about 0.1 cm, linear-terete, sessile, ascending to erect, crowded, initially with whitish hairs, becoming glabrous, greyish-green, apex blunt; **flowerheads** to about 1.5 cm across, somewhat globular, terminal, often profuse and conspicuous; **involucral bracts** ovate to elliptic, faintly sticky; **cone scales** elliptic, slightly sticky, woolly-hairy; **flowers** about 1.5 cm long, yellow, with long hairs; **pollen presenter** about 0.4 cm long, covered with short hairs; **cones** to about 2 cm across, more or less globular.

A floriferous, recently described (1995) species from the Avon District where it is found mainly in the Wongan Hills region. It occurs in sand and sandy loam of heathlands and shrubland. Deserving of cultivation, it should do best in very well-drained acidic soils with a sunny aspect. Plants tolerate moderate frosts and extended dry periods and should do best in warm temperate and semi-arid regions. Worthy of trial for cut-flower production. Propagate from seed or from cuttings of firm young growth.

P. ericifolia ssp. *ericifolia* is allied but it is less hairy and it has larger flowers and cones.

PHACELLOTRIX F. Muell.
(from the Greek *phacellos*, cluster, bundle; *thrix* hair; referring to the bundles of pappus bristles)
Asteraceae
A monotypic genus endemic in northern and north-eastern Qld.

Phacellotrix cladochaeta (F. Muell.) F. Muell.
(branched bristles)
Qld, NT
0.1– 0.5 m × 0.2– 0.5 m   Feb–May
Annual **herb** with woolly-hairy young growth; **branches** woolly-hairy, spreading to erect; **leaves** 2.5–5 cm × 0.4– 0.7 cm, narrowly oblong to lanceolate, alternate, sessile, soft, spreading to ascending, faintly hairy above, densely woolly-hairy below, apex pointed; **flowerheads** daisy-like, to about 1 cm across, yellow, solitary, terminal; **achenes** brown, oblong.

A wide-ranging species from northern Qld which extends to the Darwin and Gulf Region, NT, and is also found in New Guinea. Evidently rarely cultivated, it is suitable for seasonally dry tropical regions. Probably does best in freely draining soils in a sunny or semi-shaded site. Worth trying in gardens or containers. Propagate from seed which does not require pre-sowing treatment.

PHAIUS Lour.
(from the Greek *phaios*, dark, dusky)
Orchidaceae
Terrestrial, clumping **orchids** with fleshy pseudobulbs or erect fleshy stems; **leaves** large, thin-textured, plicate, sheathing at the base; **inflorescence** an erect raceme; **flowers** large, often colourful and showy; **tepals** fleshy, incurved or spreading; **labellum** 3-lobed near the apex, base often tubular; **fruit** a capsule.

A genus of about 50 species ranging from Africa and Asia to Polynesia, New Guinea and Australia. In all 5 species are known to occur in Australia and 3 of them are endemic. Because of the large showy flowers of some species, the genus is popular in cultivation and natural populations have suffered badly from poaching. In general *Phaius* species are readily amenable to cultivation if given shade, warmth, high humidity and air movement. They are mostly grown as pot plants in a free-draining mix which includes some sandy loam and decaying humus. Watering is necessary throughout the year slackening off in winter when the plants are mostly quiescent. Propagation can be by division of clumps, from flower stems, and by seed. After flowering, sections of flower stems laid on moist sphagnum moss can develop small plantlets from some of the nodes. Seed from these orchids is mainly sown under sterile conditions but some success has been achieved by sprinkling seed on the surface of the potting mix around established plants.

Phaius amboinensis Blume
(from Ambon)
NT
1–2 m tall   June–Aug
  Terrestrial orchid forming small clumps; **stems** erect, 30–40 cm × 1.5–2.5 cm, fleshy, fluted; **leaves** 3–6 per stem, 50–80 cm × 8–12 cm, lanceolate, dark green to yellowish-green, pleated; **racemes** 1–2 m tall, bearing 10–20 flowers; **flowers** white, some yellow on the labellum; **labellum** about 3 cm × 2–6 cm, margins incurved around the column.
  Distributed in Indonesia, New Guinea and Arnhem Land, NT, growing on the margins of swamps. Very easily cultivated if given warm, humid conditions with plenty of air movement. Needs filtered light to flower well but must be protected from direct sun. Propagate by division or from seed.

Phaius australis F. Muell.
(southern)
Qld, NSW   Southern Swamp Orchid
1–2 m tall   Sept–Nov
  Terrestrial orchid forming large clumps; **pseudobulbs** 4–7 cm × 4–7 cm, ovoid, green, crowded, with coarse, thick roots; **leaves** 4–7 per pseudobulb, 0.8–1.3 m × 8–10 cm, lanceolate, dark green to yellowish-green, pleated, narrowed to the base; **racemes** 1–2 m tall, thick, bearing 4–16 flowers, in an impressive raceme; **flowers** about 10 cm across, exterior white, interior brick-red; **labellum** to 5 cm × 4.5 cm, lateral lobes strongly incurved over the column.
  Distributed from north-eastern Qld to north-eastern NSW, growing in wet sites, particularly swamps, but also in depressions and seepage areas. Once widespread and common but now greatly reduced and fragmented due to habitat clearing, drainage of swamp and collection by orchid enthusiasts. A shade-loving species which needs warm humid conditions with good air movement. Plants can be grown in a warm bush-house or cool glasshouse as far south as Melbourne. Best grown in a large pot with well-drained potting mixture which is rich in decaying humus. Watering must be regular throughout the year. Propagate from flower stems, division of clumps, or from seed.

Phaius bernaysii Rowland ex Rchb.f.
(after Dr Lewis Bernays, original collector)
Qld   Yellow Swamp Orchid
1–2 m tall   Sept–Nov
  Terrestrial orchid forming large clumps; **pseudobulbs** 4–7 cm × 4–7 cm, ovoid, green, crowded, with coarse, thick roots; **leaves** 4–7 per pseudobulb, 0.8–1.3 m × 8–10 cm, lanceolate, dark green to yellowish-green, pleated, narrowed to the base; **racemes** 1–2 m tall, thick, bearing 4–16 flowers in an impressive raceme; **flowers** about 10 cm across, exterior white, interior sulphur-yellow; **labellum** to 5 cm × 4.5 cm, lateral lobes incurved over the column.
  Occurs on some islands of Moreton Bay, particularly Stradbroke Island where it grows in swamps. It has become rare due to mineral mining, clearing and poaching. Cultivation and propagation are as for *P. australis*. To obtain the best results it is extremely important to water regularly.

Phaius pictus Hunt
(painted, colourful)
Qld
0.6–1 m tall   April–June
  Terrestrial orchid forming untidy clumps; **stems** erect, 4–6 cm × 1.5–2 cm, fleshy, angular; **leaves** 3–5 per stem, 50–70 cm × 8–10 cm, lanceolate, dark green, pleated; **racemes** 60–90 cm tall, bearing 4–20 flowers; **flowers** 4–5 cm across, exterior yellow, interior red with yellow stripes; **labellum** about 2 cm × 1.8 cm, tubular.
  Endemic in coastal and near-coastal areas of north-eastern Qld between the Bloomfield River and Cardwell. It grows in fairly dark situations in rainforest and sometimes grows in piles of litter. It is very rarely grown and orchid enthusiasts have proved that it can be either somewhat or very difficult to maintain in cultivation. Requires shade, warmth, high humidity and free and unimpeded air movement. The potting mixture must be moisture-retentive but very well-drained. Propagate from seed.

Phaius tankervilleae (Banks ex L'Herit) Blume
(after Lady Emma Tankerville)
Qld   Northern Swamp Orchid
1–2 m tall   Sept–Nov
  Terrestrial orchid forming large clumps; **pseudobulbs** 4–7 cm × 4–7 cm, ovoid, green, crowded, with coarse, thick roots; **leaves** 4–7 per pseudobulb, 0.8–1.3 m × 8–12 cm, lanceolate, dark green to yellowish-green, pleated, narrowed to the base; **racemes** 1–2 m tall, thick, bearing 4–16 flowers in an impressive raceme; **flowers** 10–15 cm across, exterior white, interior red; **labellum** to 5 cm × 4.5 cm, lateral lobes inrolled over the column to form a tube.
  A widespread species which extends to north-eastern Qld where it mainly grows along streams and in soakage areas and swamps in the higher parts of the ranges and tablelands. Popular with orchid growers because of its large, colourful flowers. Cultivation and propagation requirements are as for *P. australis* but the plants are more cold-sensitive and may require a heated glasshouse in temperate regions.

Phaleria chermsideana   D. L. Jones

PHALAENOPSIS Blume
(from the Greek *phalaina*, a moth, *opsis*, resemblance;
referring to the white flowers resembling large moths)
Orchidaceae   Moth Orchids

Epiphytic orchids forming small to large clumps;
stems branched or unbranched; **leaves** small to large,
often pendulous or nearly so, apex notched; **inflorescence** a raceme, often branched; **flowers** small to
large, white or colourful; **petals** much wider than the
sepals; **labellum** hinged to the apex of the column, 3-
lobed; **fruit** a capsule.

A small genus of about 30 species distributed from
India, China and South East Asia through the Pacific
to New Guinea and Australia. One species is endemic
to north-eastern Qld. This genus of orchids is very
popular with orchid growers. Many exotic species and
a huge number of hybrids are in cultivation as a result
of breeding programmes. The cut flowers are commonly used in corsages and wedding bouquets.
Propagation is from seed sown under sterile conditions. Occasionally growths develop on nodes of the
inflorescence and these can be potted as separate
plants when sufficiently developed.

Phalaenopsis rosenstromii F. M. Bailey
(after Gus Rosenstrom, original collector)
Qld   Daintree White Orchid, Moth Orchid
Epiphyte   Dec–April

Epiphytic orchid forming small to moderately large,
semi-pendulous clumps; **roots** coarse, flat, with red-
growing tips; **stems** short, branched; **leaves** 2–8 per
stem, 20–30 cm × 5–8 cm, oblong to obovate, dark
green, fleshy, pendulous, the tip unequally notched;
racemes 30–75 cm long, sparsely branched, arching,
each branch bearing up to 10 flowers; **flowers** 5–8 cm
across, white, nearly flat when viewed from the side;
labellum about 2 cm × 1.2 cm, white with yellow or
orange margins, the apex with 2 filiform curved arms.

Occurs in north-eastern Qld where it is distributed
between the Iron Range on Cape York Peninsula and

Mt Spec near Townsville. It grows on rainforest trees in
situations of high humidity combined with strong air
movement. A very handsome orchid which has
become rare in nature due to overcollecting. Plants
are easily grown in the tropics but can be very difficult
to maintain in temperate regions. They require
warmth, humidity and free air movement. They can be
grown on a slab of weathered hardwood or cork, but
are mostly grown in pots or slatted wood baskets in an
open very well-drained potting mixture. In temperate
regions they require the protection of a heated
glasshouse with a minimum temperature of 15°C. The
flowers last well after cutting. Propagation is mainly
from seed sown under sterile conditions. Occasional
plantlets develop from nodes on the racemes.

PHALERIA Jack
(from the Latin *phalero*, to adorn, display)
Thymelaeaceae

Shrubs or **trees**; **bark** 'bootlace type', can be removed
in long unbreakable strands; **leaves** opposite, often
decussate, stalked, glabrous; **inflorescence** lateral or
terminal, sessile or stalked clusters, subtended by 4
bracts; **flowers** bisexual; **calyx** tubular; **sepals** petal-
like, usually 4, sometimes 5 or rarely 6; **petals** absent;
stamens twice number of sepals; **fruit** a succulent
drupe.

This tropical and subtropical genus comprises about
20 species and is well represented in Malesia and
Pacific Islands. Four species occur in Australia where
they are found in Queensland and north-eastern NSW.

The Australian species are attractive shrubs and
trees which deserve to be better known in cultivation.
They have delightful fragrant flowers which contrast
with the glossy green leaves. Plants are best suited to
tropical and subtropical regions but are worth trying
in temperate regions.

Propagation is from fresh seed and possibly from
cuttings although trials are required.

Phaleria blumei Benth. var. latifolia Benth. =
                                        P. octandra (L.) Baill.

Phaleria chermsideana (F. M. Bailey) C. T. White
(after Lady Chermside, wife of early Governor of
Queensland)
Qld, NSW   Scrub Daphne
2–10 m × 1.5–6 m   Oct–Dec

Small to tall **shrub** or rarely a small **tree** with glabrous young growth; **branches** spreading to ascending;
branchlets glabrous; **leaves** 2–10 cm × 1–4 cm, elliptic,
opposite, short-stalked, spreading to ascending, dark
green, glossy, glabrous, margins entire, venation
prominent, apex pointed; **flowers** to about 1.7 cm
long, tubular with spreading lobes, white, creamish or
pink, hairy, in 3–7-flowered clusters, sweetly fragrant;
fruit about 1 cm long, elliptical, bright red, ripe
Jan–Feb.

A handsome, fragrant-flowered species from dry
rainforest on the Atherton Tableland, and in south-
eastern Qld, and north-eastern NSW. Deserves to be
better known in cultivation. Should prove adaptable in
subtropical and temperate regions. Plants need moist,

well-drained, acidic soils. They tolerate an open sunny site and also grow well in semi-shade. Supplementary watering in extended dry periods is required. Application of organic mulch is beneficial. Propagate from fresh seed, or cuttings of firm young growth, which strike readily.

Phaleria clerodendron Benth.
(with uncertain medical properties)
Qld, NT
3.5–10 m × 2.5–6 m   Oct–April
    Medium to tall **shrub** or small **tree** with glabrous young growth; **trunks** often multi-stemmed; **branches** spreading to ascending; **branchlets** glabrous; **leaves** 9–21 cm × 4–7.5 cm, elliptic, opposite, very short-stalked, glossy dark green above, paler below, glabrous, margins entire, apex pointed; **flowers** 2.8–4 cm long, tubular with spreading lobes, white, glabrous, in 5–7-flowered clusters, usually produced on older branches and trunks, sweetly fragrant, often profuse; **fruit** to about 2.5 cm long, ovoid, bright red, ripe March–May.
    This attractive species is reputed to have poisonous properties. It occurs in north-eastern Qld and north-eastern Arnhem Land, NT, where it grows in lowland and foothill rainforest. Plants usually grow quickly in cultivation and adapt to sunny or shaded sites in moist, well-drained, acidic soils in tropical and subtropical regions. Plants may succeed in temperate regions if grown in a warm to hot protected site. Application of organic mulch is beneficial. May require supplementary watering during dry periods. Propagate from fresh seed or possibly from cuttings.

Phaleria octandra (L.) Baill.
(eight anthers)
Qld, NT
1–3 m × 2–3 m   Nov–March
    Small to medium **shrub** with glabrous young growth; **branches** spreading to ascending; **branchlets** often pendent, glabrous; **leaves** 7.5–23 cm × 2–10 cm, elliptic, opposite, very short-stalked, bright green, glabrous, glossy, margins entire, venation prominent, apex pointed; **flowers** to about 1.7 cm long, tubular with spreading lobes, white, faintly hairy, usually in terminal clusters of 8–25 flowers, sweetly fragrant, often profuse; **stamens** 8; **fruit** to about 1.5 cm long, ovoid, red, ripe March–May.
    A very appealing species which occurs in northern Qld and NT, and also extends to New Guinea, Indonesia and India. Plants are recommended for tropical and subtropical regions. They require excellent drainage, acidic soils and do best in semi-shaded sites. Addition of organic material to soil is beneficial. Plants will need supplementary watering during dry periods. Judicious pruning should enhance the pendulous growth habit. Propagate from fresh seed, and cuttings of young growth are worth trying.

PHEBALIUM Vent.
(from the Greek *phibaleo*, a kind of fig; but was thought to be a Greek name for myrtle when this genus was described)
Rutaceae

Phebalium ovatifolium   D. L. Jones

    Dwarf to tall **shrubs** or small **trees**; **leaves** alternate, simple, terete to more or less flat, fleshy to somewhat papery, glandular and often warty, glabrous or with simple or stellate hairs, or with small scales; **inflorescence** umbels or cymes or solitary flowers, axillary or terminal; **flowers** 5-merous or rarely to 8-merous (eg. *P. nottii*); **calyx** entire and hemispherical or lobed or with sepals; **petals** usually free, white, cream, yellow,

Phebalium sqamulosum ssp. *squamulosum*   W. R. Elliot

259

pink, reddish or greenish, often inflexed at tip; **stamens** 10, free, spreading; **fruit** composed of up to 5 cocci.

Phebalium comprises about 50 species of which all but one are endemic in Australia. The non-endemic species, *P. nudum*, is from the North Island of N.Z. Within Australia *Phebalium* is represented in all states but not in the Northern Territory. Species are found in coastal and mountainous areas, on inland slopes and plains, and they extend to the margins of arid zones. Plants grow in a range of communities including marginal rainforest, wet and dry sclerophyll forests, woodland, shrubland and heathland. They are found in a variety of soil types such as deep sand, shallow sand, sandy loam, loam, clay loam and clay soils. Sometimes they are associated with lateritic gravels and they are often in rocky sites in eastern Australia, such as amongst granite outcrops. In general, they occur in freely draining soils.

Phebaliums can be dwarf spreading shrubs, eg. *P. montanum,* but the majority are shrubs in the small to medium size range of 1–4 m in height. Some species can reach 7–10 m tall including *P. squameum* and *P. wilsonii.*

Most species have aromatic foliage. *P. dentatum* and *P. frondosum* and some other species have a distinctive passionfruit fragrance. Other plants smell very strongly when crushed. The upper leaf surface is usually deep green but the undersurface is paler and often has very small, silvery or rusty scales. These scales are also apparent on the branchlets of many members, eg. *P. squamulosum* ssp. *squamulosum* and *P. whitei.*

The flowers are small and starry in all species except for *P. sympetalum* which has tubular flowers. Many species produce a profuse display of flowers. The flower buds are often decoratively covered with silvery, tawny, brownish or rusty scales which contrast with the white, cream, yellow or pink flowers. This can be a highlight of many phebaliums when buds and flowers are present at the same time on plants such as *P. nottii, P. squamulosum, P. tuberculosum* and *P. whitei.*

Relatively few cultivars have been selected from this genus and although there are many hybrids in nature, especially in Western Australia, virtually none have emerged from within the horticultural industry. A handsome variegated selection of *P. squameum* ssp. *squameum,* marketed as 'Illumination', has proved reliable in cultivation.

Phebaliums are mainly utilised for amenity planting in urban and rural gardens. Some have also had minimal use in public areas including *P. lamprophyllum, P. squameum* and *P. squamulosum.*

The timber and allied products are of very limited value although in the 19th-century *P. squameum* was regarded as having the characteristics of Red Cedar (*Toona ciliata* syn *T. australis*) but slightly lighter in colour and heavier. *P. squameum* was also used for telegraph poles and fence posts because of its long-lasting qualities. It is now marketed as Satin Box in Victoria.

Phebaliums are not commonly used as cut flowers but they do have great potential for this purpose. The buds of many species are decorative and the starry flowers of phebaliums do not usually shatter. A number of species are worth testing for cut material. These include *P. bilobum, P. bullatum, P. filifolium, P. frondosum, P. microphyllum, P. squameum, P. squamulosum, P. whitei* and *P. woombye.* The foliage of species such as *P. ambiens* and *P. frondosum* has potential for use as a floricultural filler. It responds very well to treatment with glycerine which makes the foliage useable as pliable dried material for an extended period.

Cultivation

Initial interest in phebaliums for cultivation was recorded in England during the early 19th century. It was in the 1820s when *P. dentatum, P. lachnaeoides, P. squameum, P. squamulosum* and others were first cultivated there in heated greenhouses. In ensuing years there was not much interest in the genus in Australia until the 1960s when specialist nurseries began offering a range of species including *P. bilobum, P. bullatum, P. dentatum, P. glandulosum, P. lamprophyllum, P. nottii, P. obcoradum, P. squameum, P. squamulosum* ssp. *ozothamnoides* (as *P. ozothamnoides*), *P. rotundifolium, P. stenophyllum* and *P. woombye.*

In general many phebaliums adapt well to cultivation and have a wide range of tolerances. The eastern Australian species are much better known in cultivation than those from Western Australia. Hopefully this state of affairs will change over the next decade. Most require acidic soils which drain freely and the majority tolerate a sunny or semi-shaded site. They are usually intolerant of waterlogged conditions. Light to moderate frosts are unlikely to damage most plants. *P. frondosum, P. ovatifolium, P. squamulosum* ssp. *alpinum, P. squamulosum* ssp. *argentum* and *P. wilsonii* tolerate heavy frosts and snowfalls without major damage.

Generally the eastern species from forests and shrubland are more tolerant of high humidity than those from western Victoria, SA and WA.

Like many other members of the Rutaceae family, most phebaliums are prone to root rot and collar rot, although some species such as *P. squameum* require moist, often shady conditions to reach their potential. Phebaliums respond well to light applications of fertilisers with a low phosphorus ratio, but they rarely need additional nutrients in cultivation. The majority can withstand extended dry periods and species such as *P. glandulosum, P. bullatum, P. canaliculatum, P. filifolium, P. microphyllum, P. stenophyllum* and *P. tuberculosum* are drought tolerant.

Pruning of phebaliums can be beneficial in promoting bushy growth as well as possibly extending their longevity. Pruning of flowering stems provides excellent material for indoor decoration. Most species withstand regular clipping and many including *P. dentatum, P. lamprophyllum, P. squameum* and *P. squamulosum* have potential for hedging.

Phebaliums may be attacked by certain pests and diseases — various scale insects can be a major problem. Citrus mealy bug is a problem which can be difficult to control in some regions, and common white fly can play havoc in the right conditions. See Volume 1, page 157–160 for control methods.

Phebaliums may be subject to attack by collar rot and other root fungal diseases. Organic mulches should be kept away from the trunk of most species in order to limit the possibility of collar rot.

Propagation

To date there has been very little success in the germination of phebalium seed and this area needs further research. Some species such as *P. squameum*, *P. squamulosum* ssp. *squamulosum* and *P. wilsonii* are known to regenerate well from seed after disturbance of bush areas. This seed, which has a hard coating, may have been lying in the soil for many years. See Treatments and Techniques to Germinate Difficult Species, Volume 1, page 205

Propagation is generally undertaken from cuttings. For best results it is recommended that propagation material be collected from young plants — preferably fresh growth which has just become firm. If the material is too young it can wilt readily and it rarely regains its turgidity. In practice it has been found that eastern species are easier to propagate from cuttings, but it may be that material from many western species has not been at optimum condition when prepared. The use of hormone rooting compounds can be beneficial in promoting quicker and stronger root production. Both powders and liquids have been used with success.

Tissue culture could be worth trying with phebaliums as success has been achieved with related *Eriostemon* species.

Phebalium alpinum (Benth.) Maiden & Betche =
   P. squamulosum ssp. *alpinum* (Benth.) P. G. Wilson

Phebalium ambiens (F. Muell.) Maiden & Betche
(around, surrounding)
Qld, NSW   Forest Phebalium
2–3.5 m × 2–4 m   Aug–Nov
Medium **shrub** with glabrous, often bronzish young growth; **branches** ascending to erect, many; **branchlets** terete or angled, glabrous; **leaves** 2.5–14 cm × 1–4.5 cm, oblong-elliptic to obovate, stem-clasping at base, with rounded lobes, spreading to ascending, bright-to-mid green, glabrous, glandular, margins finely toothed, midrib prominent, apex rounded or pointed, highly aromatic when crushed; **flowers** to about 1 cm across, white, in crowded terminal heads, usually profuse and very conspicuous; **cocci** about 0.3 cm high.

A handsome and very distinctive, quick-growing, floriferous species which inhabits rocky granitic soils in the most southern Darling Downs District, Qld, and the Northern Tablelands, NSW. This species deserves wider recognition. Adapts well to cultivation in subtropical and temperate regions. Needs freely draining acidic soils and does best in a semi-shaded site. Hardy to moderately heavy frosts. Suitable for general planting and screening. Responds very well to hard pruning or clipping. Foliage and flowers are excellent for indoor decoration. Propagate from cuttings of firm young growth, which strike readily.

A yellow-flowered plant from near Wallangarra, Qld, is possibly a hybrid of *P. ambiens* and *P. rotundifolium*.

Phebalium ambiguum C. A. Gardner
(doubtful)
WA
0.2–0.6 m × 0.2–0.5 m mainly Aug–Nov; also sporadic
Dwarf **shrub** with silvery to rusty young growth; **branches** many, spreading to ascending; **branchlets** glabrous with small, silvery or rusty scales; **leaves** 0.4–0.6 cm × to about 0.1 cm, linear-terete, ascending to erect, crowded, becoming glabrous above, sometimes slightly warty, stellate-scaly below, margins revolute, apex blunt; **flowers** to about 0.9 cm across, yellow, solitary, sessile, terminal on short branchlets, rarely profuse; **cocci** about 0.35 cm high.

P. ambiguum is the only phebalium with solitary sessile flowers. It occurs in the Avon and Roe Districts where it grows in yellow sand and sandy loam of the sandplain heath. Plants lack strong ornamental appeal. Evidently rarely cultivated, it should do best in semi-arid and warm temperate regions in a sunny or semi-shaded site with very well-drained acidic soils. Hardy to moderately heavy frosts. Worth trying in containers in cooler regions. Propagate from cuttings.

Microcybe pauciflora var. *uniflora* D. A. Herbert is a synonym.

Phebalium amblycarpum (F. Muell.) Benth. =
   P. rude ssp. *amblycarpum* (F. Muell.) P. G. Wilson

Phebalium anceps DC.
(two-edged, two-winged)
WA   Blister Bush
1.5–3.5 m × 1–2.5 m   May–Dec; also sporadic
Small to medium **shrub;** young growth with silvery scales; **branches** spreading to ascending; **branchlets** prominently angled, smooth; **leaves** 3–12 cm × 1–3.3 cm, elliptical, short-stalked, spreading to ascending, initially with small silvery scales, becoming glabrous, green, somewhat thin, margins faintly recurved, midrib prominent, apex pointed; **flowers** about 1 cm across, white, in terminal and axillary cymes to about 5 cm across, often profuse and very conspicuous; **cocci** to about 0.35 cm high.

A dweller of river banks and swampy soils in the Darling District where it occurs mainly in coastal regions. It is an attractive species which adapts to cultivation in subtropical and temperate regions. Needs moist but well-drained acidic soils and a semi-shaded site for best results. Plants tolerate light frosts. Pruning from an early stage promotes bushy growth. May be worth trying as hedging. Sap from plants can cause skin blisters on some people, hence the common name. Propagate from cuttings of firm young growth.

This species has been mistakenly known as *P. argenteum*.

Phebalium asteriscophorum F. Muell. =
   Asterolasia asteriscophora (F. Muell.) Druce

Phebalium aureum A. Cunn. = *P. squamulosum* Vent.

Phebalium beckleri F. Muell. =
   P. elatius ssp. *beckleri* (F. Muell.) P. G. Wilson

Phebalium billardieri A. Juss =
   P. squameum (Labill.) Engl.

Phebalium bilobum Lindl.
(bilobed, notched)
Vic, Tas   Notched Phebalium
0.6–4 m × 0.6–2 m   July–Nov

Phebalium ambiens   D. L. Jones

Dwarf to medium **shrub** with hairy young growth; **branches** many, spreading to erect; **branchlets** becoming terete, faintly to prominently stellate-hairy; **leaves** 0.8–5.5 cm × 0.4–2 cm, narrowly oblong to ovate-oblong, short-stalked, spreading to ascending, deep green, faintly stellate-hairy to glabrous, gland-dotted, margins flat or slightly recurved, faintly to prominently toothed, apex truncate or notched, spicy aroma when crushed; **flowers** to about 1 cm across, often red in bud, opening white and often ageing to red, in compact terminal clusters, often profuse and conspicuous; **cocci** about 0.5 cm high.

This appealing species has at least 3 distinct variants. The Grampians variant has short broad leaves with recurved faintly toothed margins and usually pink to reddish flower-buds. The Gippsland variant (originally known as *Eriostemon serrulatus*) has long flat leaves with prominently toothed margins. The Tasmanian variant (originally known as *Eriostemon truncatus*) has short, very blunt leaves and white buds. Plants have often proved difficult to maintain in cultivation. They require well-drained acidic soils which are somewhat moisture-retentive. A semi-shaded site is usually best. Plants are hardy to heavy frosts and respond well to pruning. They are grown successfully in containers. Excellent cut flower. Propagate from cuttings of firm young growth, which strike readily.

P. hillebrandii (F. Muell.) Engl. is a synonym, while *P. bilobum* Bartl. is a synonym of *P. rude* Bartl.

Phebalium brachycalyx P. G. Wilson
(short calyx)
WA
1–1.5 m × 1–1.5 m   July–Nov

Small **shrub** with glandular and scaly young growth; **branches** ascending to erect; **branchlets** slender, warty; **leaves** 1–1.5 cm × about 0.15 cm, oblong-cuneate, spreading to ascending, deep green, glabrous, warty and channelled above, silvery to rusty scales below, midrib prominent, margins crenulate and warty, apex rounded or slightly notched; **flowers** to about 0.9 cm across, white, usually in 3–6-flowered terminal clusters; **petals** with silvery or rusty scales on exterior; **cocci** about 0.3 cm high.

This poorly known phebalium occurs in the region between Southern Cross and Wongan Hills where it usually grows in lateritic soils. Rarely cultivated, it has proved difficult to maintain. Plants should do best in semi-arid and warm temperate regions. They require freely draining acidic soils and prefer plenty of sunshine although they will tolerate a semi-shaded site. May succeed as a container plant in cool temperate regions. Hardy to moderately heavy frosts and withstands extended dry periods. Propagate from cuttings, which can be slow to produce roots.

Hybrids with both *P. filifolium* and *P. tuberculosum* occur in nature and it is possible that they are cultivated.

P. bullatum has similar leaves but differs in its bright yellow flowers.

Phebalium brachyphyllum Benth.
(short leaves)
Vic, SA   Spreading Phebalium
0.5–1 m × 0.8–1.5 m   Sept–Nov

Dwarf **shrub** with glabrous to faintly hairy young growth; **branches** many, spreading to ascending; **branchlets** slender, terete, usually faintly hairy; **leaves** 0.3–0.5 cm × 0.2–0.4 cm, ovate to broadly ovate, short-stalked, somewhat cordate base, spreading to ascending, deep green, glabrous to faintly hairy, flat, apex blunt to rounded; **flowers** to about 0.8 cm across, pinkish buds, opening white, in few-flowered terminal clusters to about 2 cm across, sometimes profuse and very conspicuous; **cocci** about 0.4 cm high.

This moderately appealing species is regarded as rare but not threatened in nature. It inhabits mallee country in far western Vic and south-eastern SA, extending to the Eyre Peninsula. Plants are rarely encountered in cultivation. They require excellent drainage and are best suited to warm temperate and semi-arid regions. Hardy to moderately heavy frosts. Propagate from cuttings of firm young growth.

The allied *P. hillebrandii* has bilobed or very blunt leaves.

Phebalium bullatum J. M. Black
(blistered, bubble-like)
Vic, SA   Desert Phebalium, Silvery Phebalium
1–2 m × 0.6–1.5 m   Aug–Oct

Small **shrub** with rusty or silvery, scaly young growth; **branches** spreading to ascending; **branchlets** slender, somewhat warty, with silvery to rusty scales; **leaves** 0.6–1.2 cm × up to 0.2 cm, narrowly oblong, short-stalked, spreading to ascending, deep green and glabrous above, silvery-scaly below, margins warty, apex very blunt; **flowers** to about 0.8 cm across, bright yellow, in terminal clusters, rusty-scaly exterior to petals, often profuse and very conspicuous; **cocci** about 0.3 cm high.

An outstandingly decorative species of the Victorian and South Australian mallee communities where plants usually grow in deep sand. Often not easy to maintain in cultivation. Recommended for semi-arid and warm temperate regions, but may succeed as a container plant or in a sheltered warm to hot site in cooler regions. Plants are drought tolerant and hardy to moderately heavy frosts. Often develops as an open shrub. Responds well to pruning and flowering stems are useful for indoor decoration. Propagate from cuttings of firm young growth, which usually strike readily.

P. glandulosum (J. M. Black) A. B. Court is a synonym. *P. glandulosum* Hook. differs in having deeply channelled leaves and paler flowers.

Phebalium canaliculatum (F. Muell. & Tate) J. H. Willis
(longitudinally grooved or channelled)
WA
1.5–2 m × 1.5–3 m   Aug–Oct
Small **shrub** with silvery or rusty young growth; **branches** many, spreading to ascending; **branchlets** slender, glandular-warty with silvery or rusty scales; **leaves** 1.5–2.8 cm × about 0.1 cm, linear, terete to slightly flattened, straight to slightly curved, very short-stalked, spreading to ascending, moderately crowded, warty, covered in small silvery scales, channelled above, apex blunt; **flowers** to about 1 cm across, pale pinkish-mauve to deep pink, often fading to white, in terminal clusters, usually profuse and very conspicuous; **stamens** pale mauve, prominent; **anthers** yellow.

An extremely handsome species from the Austin, Avon and Coolgardie Districts where it grows in red or yellow sands or laterite of mallee and native pine communities. It is rarely cultivated and is most likely to succeed in semi-arid and warm temperate regions in extremely well-drained soils. Needs plenty of sunshine. It is drought tolerant and hardy to moderately heavy frosts. Responds well to pruning. Useful for low screening and general planting. Propagate from cuttings of firm young growth, which may be slow to form roots.

Phebalium bullatum   W. R. Elliot

Phebalium canaliculatum   W. R. Elliot

In the Coolgardie District there are plants which have intermediate characteristics between *P. canaliculatum* and *P. tuberculosum* and these may be in cultivation.

The allied *P. filifolium* has greyish-green leaves which are not channelled above.

Phebalium carruthersii (F. Muell.) Maiden & Betche
(after John Carruthers, 19th-century NSW parliamentarian)
NSW
0.6–1.3 m × 0.6–1.5 m   sporadic throughout the year
Dwarf to small **shrub** with hairy young growth; **branches** many, spreading to ascending; **branchlets** more or less terete, densely hairy; **leaves** 0.7–1.2 cm × 0.15–0.5 cm, ovate to lanceolate, short-stalked with often cordate base, spreading to ascending, somewhat crowded, grey-green to green, glandular and hairy, margins strongly recurved, apex blunt; **flowers** to about 2 cm long, in terminal pendent heads; **petals** yellowish-green, glandular; **stamens** dark red, about twice as long as petals; **cocci** to about 0.6 cm high.

This unusual and distinctive species is rare in nature but not threatened because it occurs in reserved areas in the South Coast District. It is rarely encountered in cultivation but should adapt to a variety of freely draining acidic soils in semi-shaded or shaded sites. Plants are hardy to moderately heavy frosts. Pruning from an early stage should promote bushy growth. May do well as a container plant. Propagate from cuttings of firm young growth, which usually strike readily.

Phebalium clavatum C. A. Gardner
(club-shaped)
WA
1–1.5 m × 1–1.5 m   Aug–Sept

263

Small **shrubs** with warty and silvery-scaly young growth; **branches** spreading to ascending; **branchlets** terete, warty with reddish glands and silvery scales; **leaves** to about 0.3 cm × 0.15 cm, club-shaped, more or less terete, short- and thick-stalked, ascending to erect, crowded on short branchlets, with silvery scales, apex becoming glabrous and very blunt to rounded; **flowers** to about 0.4 cm across, white, solitary, terminal, sessile; **petals** with silvery scaly exterior; **cocci** to about 0.3 cm high.

This species with its unusual foliage and flowers should be of interest to enthusiasts. It is restricted to the Coolgardie region where it grows in woodland on red sand. Will need excellent drainage and plenty of warmth. Probably best suited to semi-arid and warm temperate regions. Tolerates moderately heavy frosts and is drought-resistant. Propagate from cuttings of firm young growth.

Phebalium coxii (F. Muell.) Maiden & Betche
(after Dr J. C. Cox)
NSW
1–4 m × 0.6–2 m   Sept–Feb
Small to medium **shrub** with glabrous young growth; **branches** many, spreading to ascending; **branchlets** slender, often reddish-brown, glabrous, angular, slightly warty; **leaves** 3–7 cm × 1–1.5 cm, narrowly elliptic to oblanceolate, spreading to ascending, glabrous, deep green and glossy above, paler below, margins finely toothed, midrib prominent, apex pointed, passionfruit fragrance when crushed; **flowers** to about 1 cm across, white, in many-flowered terminal umbels, often profuse and very conspicuous; **cocci** about 0.5 cm high.

This species has a moderately restricted distribution in the Southern Tablelands and is found on creek banks and ridges in open forest. Does well in cultivation where it is usually quick-growing. It prefers a semi-shaded site and well-drained acidic soils which don't

Phebalium coxii × .5

dry out too readily. It has potential for greater use in gardens as a screening or hedging plant. Hardy to heavy frosts. Propagate from cuttings of firm young growth, which strike readily.

Phebalium daviesii Hook. f.
(after R. N. Davies, the original collector)
Tas
1–1.5 m × 0.6–1.5 m   Sept–Nov
Small **shrub** with smooth young growth; **branches** spreading to ascending; **branchlets** terete, smooth or faintly glandular warty; **leaves** 2–3 cm × 0.1–0.3 cm, linear-cuneate, short-stalked, glabrous and faintly glandular warty above, silvery below, midrib deeply grooved, apex notched or bilobed; **flowers** small, in terminal umbels, often profuse; **petals** white to deep cream, exterior with silvery or rusty scales; **cocci** unknown.

This species was considered extinct until it was recently found in the wild many years after the last collection in 1892. It is known only from the eastern coast in the north-eastern region. It is very new to cultivation, but has adapted well in freely draining acidic soils and does best in a semi-shaded or partially sunny site. Plants respond very well to pruning and regular clipping. They are well-suited for growing in containers. Hardy to moderately heavy frosts. Propagate from cuttings of firm young growth, which strike readily.

Phebalium dentatum Sm.
(toothed)
NSW   Toothed Phebalium
1.5–6 m × 1–3 m   Aug–Nov
Small to tall, tree-like **shrub** with stellate-hairy young growth; **branches** many, spreading to ascending; **branchlets** more or less terete, becoming glabrous; **leaves** 4–8 cm × 0.1–0.8 cm, linear to oblanceolate, short-stalked, spreading to erect, scattered, deep green, gland-dotted, glabrous to faintly hairy and glossy above, silvery to white stellate hairs below, margins recurved to revolute and sometimes faintly toothed, midrib prominent, apex very blunt or notched with 2 teeth; **flowers** to about 0.8 cm across, white to pale yellow, in axillary clusters, often profuse and very conspicuous; **cocci** to about 0.4 cm high.

A very floriferous and attractive species from dry sclerophyll forest where it usually grows on sandstone. Occurs in the North and Central Coast Districts, and Northern and Central Tablelands. Successfully cultivated in temperate and subtropical regions in freely draining acidic soils. Prefers a semi-shaded site but tolerates plenty of sunshine. Excellent as an understorey plant for light screening. May be a useful hedging plant. Hardy to most frosts. May need supplementary watering in extended dry periods. Foliage has a pleasant passionfruit fragrance. Introduced into cultivation in England in 1825 as *P. salicifolium* A. Juss., which is a synonym. Propagate from cuttings of firm young growth, which strike readily.

P. umbellatus (Turcz.) Turcz. is a synonym. *P. phylicifolium* is similar but has much shorter leaves and is rarely more than 1.5 m tall while *P. obtusifolium* has glabrous stems and leaves.

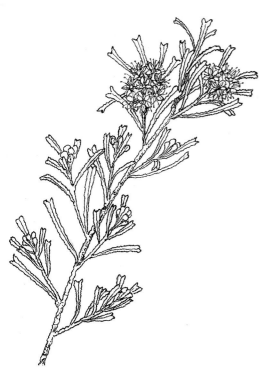

Phebalium daviesii × .8

Phebalium diosmeum A. Juss.
(after the genus *Diosma*)
NSW, Vic
1–2 m × 1–2 m   Aug–Oct

Small **shrub** with loosely hairy young growth; **branches** many, ascending to erect; **branchlets** hairy, initially angular; **leaves** 0.4–2 cm × 0.1–0.4 cm, linear and more or less terete or sometimes lanceolate to ovate, short-stalked, spreading to ascending, crowded, dull green and glabrous or faintly hairy above, simple or stellate-hairy below, margins recurved to revolute, apex blunt, pleasantly aromatic when crushed; **flowers** about 1 cm across, pale yellow, in crowded terminal clusters to about 2.5 cm across, profuse and very conspicuous; **cocci** about 0.4 cm high.

This highly desirable species inhabits sandy soils in heath and dry sclerophyll forest in coastal and adjacent areas. It is distributed from just south of Sydney to far eastern Vic. Plants adapt well to cultivation and require excellently drained acidic soils with a sunny or semi-shaded aspect. They are hardy to moderately heavy frosts. Responds well to pruning and can be used for low hedging. Propagate from cuttings of firm young growth, which produce roots readily.

P. lachnaeoides is closely allied and may be a hybrid of *P. diosmeum* and *P. dentatum*.

Phebalium drummondii Benth.
(after James Drummond, first Government Naturalist of WA)
WA
1–1.5 m × 1–1.5 m   Aug–Nov

Small **shrub** with smooth young growth; **branches** spreading to erect; **branchlets** smooth, with silvery scales; **leaves** 0.3–0.5 cm × up to 0.2 cm, narrowly to broadly elliptic to obovate, ascending to erect, flat to slightly concave, glossy green and glabrous above, silvery scales below, midrib obscure, apex blunt; **flowers** to about 1 cm across, bright yellow, in few-flowered terminal umbels to about 2 cm across, often profuse and very conspicuous.

A rare species with a geographic range of less than 100 km. It occurs in the Avon District where it grows in sandy and lateritic soils of mallee and heath communities. Uncommon in cultivation and deserves to be better known. Plants require freely draining acidic soils in a warm to hot site. They are hardy to moderately heavy frosts and tolerate extended dry periods. Probably best suited to warm temperate regions but worth trying in cooler areas. Propagate from cuttings of firm young growth.

P. filifolium and *P. microphyllum* have similar flowers to *P drummondii*, but very different leaves.

Phebalium elatius (F. Muell.) Benth.
(taller)
NSW   Tall Phebalium
1.5–2.5 m × 1.5–2 m   Sept–Nov

Small to medium **shrub** with glandular young growth; **branches** spreading to erect; **branchlets** angular to terete, glandular, warty; **leaves** 1.5–2 cm × 0.3–0.5 cm, narrowly obovate, tapering to base, ascending to erect, flat, glossy green and glabrous above, gland-dotted, midrib prominent below, faintly toothed near apex which is blunt to rounded or rarely faintly notched; **flowers** to about 1 cm across, white, in short terminal clusters to about 2 cm across, often profuse and very conspicuous; **cocci** to about 0.4 cm high.

The typical variant is known from the North Coast and Northern Tablelands Districts where it is uncommon. Often grows in gravelly soils. Rarely encountered in cultivation, it adapts to subtropical and temperate regions and needs well-drained acidic soils with a semi-shaded aspect. Hardy to moderate frosts.

The ssp. *beckleri* (F. Muell.) P. G. Wilson grows 2–5 m × 1.5–3.5 m and has aromatic obovate leaves of 2–3.5 cm × 0.6–1 cm with a rounded, usually notched apex. It occurs in far south-eastern Qld and extends to near Buladelah, NSW. Is found in warm temperate rainforest and wet sclerophyll forest. It is more common and reliable in cultivation and requires similar conditions to the typical variant. Responds very well to pruning and clipping and is very suitable for hedging.

P. beckleri F. Muell. is a synonym.

Propagate both variants from cuttings of firm young growth, which strike readily.

Phebalium ellipticum P. G. Wilson
(elliptical)
NSW
0.6–1.5 m × 0.6–1.5 m   Sept–Feb

Dwarf to small **shrub** with glandular-warty young growth; **branches** ascending to erect; **branchlets** terete, scaly, glandular-warty; **leaves** 2–3.5 cm × 0.8–1.5 cm, elliptical to ovate, short-stalked, spreading to ascending, moderately crowded, flat, deep green

Phebalium diosmeum   T. L. Blake

and glabrous above, silvery scales below, gland-dotted, apex rounded; **flowers** to about 1 cm across, white, in 2–5-flowered axillary clusters, often profuse and moderately conspicuous; **cocci** about 0.3 cm long, spreading.

This species is restricted to the mountain ranges south and east of Cooma in the Southern Tablelands where it occurs in shrubland and dry sclerophyll forest and is often found in association with rocky outcrops. Cultivated to a limited degree, it should be suitable for well-drained acidic soils in temperate regions. A semi-shaded site is preferred but plants will tolerate an open aspect. Hardy to frost and light snowfalls. It is not a highly ornamental species but it may be useful for low hedging. Propagate from cuttings of firm young growth.

The allied *P. ovatifolium* has rusty scales on branchlets, smaller leaves and 1–3 pink-budded flowers in axils.

Phebalium elatius ssp. *beckleri*   D. L. Jones

Phebalium equestre D. A. Cooke
(horsemen; referring to saddle-shaped leaves)
SA
0.1–0.4 m × 0.3–0.8 m   Aug–Oct
Dwarf spreading **shrub** with hairy young growth; **branches** many, somewhat tangled, spreading to ascending; **branchlets** slender, hairy, green to reddish; **leaves** 0.1–0.4 cm × 0.2–0.4 cm, saddle-shaped, very short-stalked, base cordate, spreading, scattered to somewhat crowded, glandular, rough above, glabrous below, margins recurved, apex blunt and decurved; **flowers** to about 0.8 cm across, pink buds opening to white, on reddish stalks, solitary or 2–3 in terminal clusters, rarely profuse; **anthers** yellow to pale pink; **cocci** about 0.3 cm long, usually bright pink.

A recently described (1987) species which is restricted to a small area on Kangaroo Island where it grows in deep sand in mallee communities. Mainly grown by enthusiasts probably because it has fascinating leaves. Plants need excellent drainage and a warm to hot site. Best suited to temperate regions. It could make an interesting container plant. Propagate from cuttings of firm young growth.

The allied *P. brachyphyllum* grows larger, lacks the entangled branching habit and produces more flower clusters.

Phebalium filifolium Turcz.
(thread-like leaves)
WA
1–1.5 m × 1–1.5 m   July–Oct
Small **shrub** with scaly young growth; **branches** many, spreading to ascending; **branchlets** slender, with rusty scales; **leaves** 1–1.8 cm × about 0.1 cm, linear, somewhat terete, spreading to ascending, somewhat crowded, grey-green to green, glossy and gland-dotted above, silvery to rusty scales and prominent midrib below, apex blunt; **flowers** to about 1 cm across, pale to bright yellow or rarely white, in 3–8-flowered terminal clusters, often profuse and very conspicuous; **cocci** to about 0.25 cm high.

Superb when in full flower, this species usually develops as a rounded shrub. It occurs in the Avon, Austin, Coolgardie and Roe Districts where it inhabits sand or sandy loam which can be overlying sandy clay. Recommended for semi-arid and temperate regions. Plants require very good drainage and plenty of sunshine. They tolerate moderately heavy frosts and extended dry periods. Responds well to pruning and flowering stems are useful for indoor decoration. Propagate from cuttings of firm young growth, which can be slow to produce roots.

P. filifolium hybridises in nature with *P. brachycalyx*, *P. maxwellii* and *P. tuberculosum* var. *megaphyllum.* Some of these hybrids may be in cultivation.

The allied *P. canaliculatum* has warty stems and pink flowers.

Phebalium frondosum N. G. Walsh & D. E. Albrecht
(many leaves, and frond-like appearance)
Vic
2–7 m × 1–3.5 m   Sept–Nov
Small to tall **shrub** or small **tree**; young growth pale; **branches** spreading to ascending, often horizontal;

Phebalium filifolium × .8, flower × 3

Phebalium glandulosum ssp. *glandulosum* W. R. Elliot

Phebalium gracile D. L. Jones

branchlets slender, angled, with rusty scales; **leaves** 0.8–2.5 cm × 0.6–1.7 cm, ovate, short-stalked, spreading usually horizontally, thin, flat to slightly concave, deep green, glabrous and gland-dotted above, silvery below, apex rounded or slightly notched, spicy passionfruit fragrance when crushed; **flowers** to about 1.3 cm across, white, 1 (rarely to 3) per axil on curved stalks, often moderately profuse and conspicuous; **cocci** to about 0.4 cm high.

A very desirable species which was described in 1988. Before then it was often wrongly named as *P. ovatifolium* or *P. squameum* ssp. *coriaceum*. It is rare in nature and occurs in a restricted area of East Gippsland at 830–940 m altitude, on loamy or skeletal soils. Does very well in cultivation in temperate regions and may adapt to subtropical areas. Plants need good drainage and perform best in semi-shaded sites. Hardy to heavy frosts and light snowfalls. Responds well to pruning and clipping. It has excellent potential for use as a hedge or general screening plant in sheltered sites. Foliage and flowering stems very good for indoor decoration. Propagate from cuttings of firm young growth, which form roots readily.

Phebalium glandulosum Hook.
(bearing many glands)
Qld, NSW, Vic, SA   Desert Phebalium
0.5–2.5 m × 0.5–2 m   Sept–Nov
Dwarf to medium **shrub** with faintly warty young growth; **branches** many, spreading to ascending; **branchlets** terete, glandular-warty with silvery or rusty

scales; **leaves** 0.5–3 cm × up to 0.3 cm, oblong-cuneate, spreading to ascending, moderately crowded, dull green, glandular-warty above, silvery below, margins recurved, apex blunt or notched; **flowers** to about 0.8 cm across, pale to bright yellow, in terminal umbels to about 2 cm across, often profuse and very conspicuous; **cocci** to about 0.4 cm high.

A splendid, floriferous and variable phebalium which is widely distributed from southern Qld to southern SA. The typical variant usually inhabits red sandy soils in mallee or on rocky slopes. Fairly popular in cultivation, it does well in temperate and semi-arid regions. Must have good drainage and a sunny or warm semi-shaded site is suitable. Hardy to moderately heavy frosts and withstands extended dry periods. Responds very well to pruning. Recommended for general planting. A dwarf selection does well as a container plant.

There are a further 2 subspecies.

The ssp. *angustifolium* P. G. Wilson has narrowly wedge-shaped leaves, 0.5–0.8 cm × 0.1 cm, and flowers to about 0.5 cm across. It is restricted to a small area of the Central Western Slopes of NSW where it grows along gullies in sclerophyll woodland. It is very uncommon in cultivation but should do well in subtropical and temperate regions in acidic soils.

The ssp. *eglandulosum* (Blakely) P. G. Wilson has narrowly oblong to wedge-shaped leaves which are faintly glandular-warty and about 0.5 cm × 0.1–0.2 cm with revolute margins. This subspecies occurs in south-eastern Qld and far north-eastern NSW. It is uncommon in cultivation but warrants greater trial in temperate and subtropical regions because of its decorative features.

Propagate all variants from cuttings of firm young growth, which strike readily.

The var. *bullatum* (J. M. Black) A. B. Court is now referrable to *P. bullatum* J. M. Black while var. *daviesii* (Hook. f.) Benth. is now known as *P. daviesii* Hook. f.

Phebalium gracile C. T. White
(slender)
Qld
0.6–1.5 m × 0.5–1 m   Feb–Oct

Dwarf to small **shrub** with hairy young growth; **branches** many, spreading to erect; **branches** slender, hairy; **leaves** 0.3–1 cm × 0.15–0.4 cm, oblong-elliptical to broadly ovate, very short-stalked, spreading to ascending, crowded, mid green, glabrous, gland-dotted, margins revolute, apex blunt to rounded, pleasant spicy fragrance when crushed; **flowers** to about 1 cm across, white, in terminal heads to about 2 cm across, often profuse and very conspicuous; **cocci** to about 0.4 cm long.

A bushy species from the extreme south-eastern area of Qld where it is known from only a few mountains and is often associated with rocky outcrops. It is very decorative and deserves to be better known in cultivation. Should do well in subtropical and temperate areas in freely draining acidic soils with a semi-shaded aspect. Plants are hardy to heavy frosts. They respond well to pruning and are excellent as low screening. Propagate from cuttings of firm young growth, which strike readily.

Phebalium hillebrandii (F. Muell.) J. H. Willis
(after Dr Wilhelm Hillebrand)
SA   Hillebrand's Phebalium
0.3–0.6 m × 0.3–0.6 m   Aug–Oct

Dwarf **shrub** with faintly hairy young growth; **branches** ascending to erect; **branchlets** slender, terete, faintly stellate-hairy; **leaves** 0.3–1.7 cm × 0.2–0.5 cm, ovate, very short-stalked, cordate base, spreading to ascending, not crowded, mainly glabrous, margins recurved, apex very blunt or notched; **flowers** to about 1 cm across, pink in bud opening to white, on slender stalks, in terminal 2–16-flowered loosely arranged clusters, scattered to sometimes profuse; **cocci** about 0.4 cm high.

A rare species from the Southern Mt Lofty Ranges where it grows in moist gullies of sclerophyll forests. Uncommon in cultivation and mainly of interest to enthusiasts. Plants need acidic soils which are moisture-retentive but at the same time freely draining. A semi-shaded aspect is probably best. Tolerates moderate frosts. May be an interesting container plant. Propagate from cuttings of firm young growth.

P. hillebrandii (F. Muell.) Engl. is a synonym of *P. bilobum* Lindl.

Phebalium lachnaeoides A. Cunn.
(similar to the genus *Lachnaea*)
NSW
2–5 m × 1.5–3.5 m   Aug–Oct

Medium to tall **shrub** with whitish-hairy young growth; **branches** spreading to ascending; **branchlets** initially angular and stellate-hairy, becoming terete and glabrous; **leaves** 0.5–1.5 cm × 0.1 cm, linear, terete, spreading to ascending, crowded, becoming glabrous, gland-dotted, margins strongly revolute, apex softly pointed; **flowers** about 1 cm across, creamy yellow, solitary in upper axils.

This rare species is known only from a few locations in the upper Blue Mountains, west of Sydney, where it grows in open rocky sites. Originally cultivated in England in 1824 but evidently currently not grown. It needs to be cultivated as a conservation strategy. Plants will need excellent drainage and a sunny or semi-shaded site. They withstand heavy frosts. The crowded foliage on plants may be useful for screening. Propagate from cuttings of firm young growth.

P. phylicifolium var. *lachnaeoides* (A. Cunn.) C. Moore is a synonym.

Some botanists theorise that *P. lachnaeoides* may be a hybrid of *P. diosmeum* and *P. dentatum*.

Phebalium lamprophyllum (F. Muell.) Benth.
(shining leaves)
NSW, Vic
1–2 m × 1–2 m   Aug–Nov

Small compact **shrub** with faintly hairy young growth; **branches** many, spreading to erect; **branchlets** terete to angled, prominently warty, initially hairy; **leaves** 0.3–1 cm × 0.2–0.4 cm, elliptic to nearly circular, very short-stalked, spreading to ascending, crowded, deep green and glossy above, paler below, margins entire or slightly irregular near the pointed or rounded apex; **flowers** to about 0.8 cm across, white to pale pink in bud, opening white with prominent stamens, solitary in upper axils, forming dense terminal clusters, usually profuse and very conspicuous; **cocci** about 0.3 cm high.

This variable species is highly adaptable and floriferous. It occurs from the northern region of the Central Tablelands and extends south through the Southern Tablelands of NSW to the mountains of eastern Vic. The most commonly cultivated variant has elliptic leaves. A very desirable shrub for subtropical and temperate regions. It grows in a range of heavy to light acidic soils and tolerates full sun, partial sun and shade. Hardy to most frosts and withstands extended dry periods. Responds well to pruning and clipping and is excellent for low hedging. This highly recommended phebalium also has a variant with pale pink buds which is very attractive. Propagate from cuttings of firm young growth, which strike readily.

Phebalium lepidotum (Turcz.) P. G. Wilson
(scaly)
WA
0.6–1.5 m × 0.5–1 m   June–Oct
Dwarf to small **shrub** with rusty-scaly young growth; **branches** spreading to ascending; **branchlets** slender, smooth with rusty scales; **leaves** 1–2.5 cm × about 0.2 cm, linear, very short-stalked, spreading to ascending, somewhat crowded, green above, glabrous and channelled above, silvery-scaly below, apex pointed to blunt; **flowers** about 1 cm across, white to cream, with rusty scales on exterior of petals, in 3–6-flowered terminal umbels about 1.5 cm across, often profuse and very conspicuous; **cocci** about 0.3 cm high, erect.

The typical variety occurs to the south and southeast of Coolgardie where it grows in sand and sandy loam which often overlies a clay subsoil. It is an attractive phebalium that is suitable for cultivation in semiarid and warm temperate regions. Soils must be very well drained and plants prefer plenty of sunshine. Plants are hardy to moderately heavy frosts and are drought tolerant. Previously known as *P. maxwellii* (F. Muell.) Engl.

The var. *obovatum* P. G. Wilson has thick, narrow-obovate leaves about 0.5 cm × 0.15 cm with a rounded apex. It occurs in the far-western region of the Great Australian Bight where it grows in sandy soils which often contain laterite.

Propagate both varieties from cuttings of firm young growth.

Phebalium lineare C. A. Gardner =
    P. rude ssp. *lineare* (C. A. Gardner) P. G. Wilson

Phebalium longifolium S. T. Blake =
P. squamulosum ssp. *longifolium* (S. T. Blake) P. G. Wilson

Phebalium lowanense J. H. Willis
(after Lowan, a western Victorian region)
Vic, SA   Lowan Phebalium
0.4–0.7 m × 0.4–1 m   Aug–Sept
Dwarf **shrub** with rusty- to silvery-scaly young growth; **branches** spreading to ascending; **branchlets** densely covered with rusty or silvery scales; **leaves** 0.4–1.2 cm × 0.1 cm, linear, somewhat terete, more or less sessile, spreading, moderately crowded, green to

greyish and smooth or slightly rough above, scaly below, margins revolute, apex blunt; **flowers** to about 1 cm across, bright yellow, with silvery or rusty scales on exterior of petals, on thick stalks in terminal heads to about 2 cm across, often profuse and very conspicuous; **cocci** to about 0.3 cm high, erect.

A handsome species which is regarded as vulnerable in nature. It is from far-western Vic and south-eastern SA, where it grows as an undershrub in deep sand of the mallee region. Rarely encountered in cultivation, it requires extremely well-drained soils and a warm to hot site for best growth. Tolerates moderately heavy frosts and extended dry periods. Should do best in semi-arid and warm temperate regions. Worth trying in rockeries and as a container plant. Propagate from cuttings of firm young growth.

The closely allied *P. stenophyllum* has slender flower stalks and usually slightly smaller flowers.

Phebalium maxwellii (F. Muell.) Engl. =
                    P. lepidotum (Turcz.) P. G. Wilson

Phebalium microphyllum Turcz.
(small leaves)
1–1.5 m × 1–1.5 m   July–Oct
Small **shrub** with scaly young growth; **branches** spreading to ascending; **branchlets** slender, smooth, scaly; **leaves** 0.3–0.4 cm × about 0.15 cm, oblong, very short-stalked, spreading to ascending, somewhat crowded, mid-to-dark green, smooth and glabrous above, silvery-scaly below, margins strongly recurved, apex blunt to rounded; **flowers** to about 0.8 cm across, creamy yellow to yellow, undersurface of petals rusty-scaly, in 3–6-flowered terminal umbels to about 1.5 cm across, often profuse and very conspicuous; **cocci** about 0.3 cm high, erect.

This phebalium is very floriferous. It occurs in the Avon, Eyre and Roe Districts where it grows in granitic sands and sandy loam which can overlie clay or laterite. Suitable for cultivation in semi-arid and warm temperate regions. Plants require very good drainage and plenty of sunshine. They are drought tolerant and hardy to moderately heavy frosts. Plants respond well to pruning. Propagate from cuttings of firm young growth, which may be slow to form roots.

The allied *P. filifolium* has narrower leaves. *P. tuberculosum* is also allied but it has warty stems and leaves.

Phebalium montanum Hook.
(of the mountains)
Tas
0.1–0.3 m × 0.3–1 m   Oct–Feb
Dwarf spreading **shrub** with faintly hairy young growth; **branches** prostrate to decumbent; **branchlets** deep reddish, faintly hairy; **leaves** 0.5–1.2 cm × about 0.2 cm, terete, with short reddish stalks, spreading to ascending, often slightly curved, crowded, somewhat fleshy, deep green, glabrous, gland-dotted and grooved above, apex bluntly pointed; **flowers** to about 0.7 cm across, pinkish-red buds opening to white or pale pink, solitary in upper axils and forming leafy spikes or clusters, sometimes profuse and moderately conspicuous; **anthers** initially reddish, becoming yellow; **cocci** about 0.4 cm high.

Phebalium lamprophyllum   T. L. Blake

This rare Tasmanian endemic species inhabits mountains at 1050–1370 m altitude in the north-east and central regions where it often grows in rocky sites. Very uncommon in cultivation, it is best suited to cool temperate regions. Plants can be difficult to maintain. Needs good drainage and a sunny or semi-shaded site is suitable. Most likely to succeed if grown as a container plant. Hardy to most frosts and snowfalls. Responds well to pruning. Propagate from cuttings of firm young growth, which may be slow to form roots.

P. oldfieldii differs in having flat leaves which often have recurved margins.

Phebalium nottii (F. Muell.) Maiden & Betche
(after Dr Nott, supporter of a search expedition for Leichhardt)
Qld, NSW   Pink Phebalium
1–3 m × 0.7–2 m   Sept–Nov
Small to medium **shrub** with rusty- to silvery-scaly young growth; **branches** spreading to erect; **branchlets** terete, densely covered in rusty to silvery scales; **leaves** 2–5 cm × 0.4–1.3 cm, elliptic to oblong-elliptic, short-stalked, spreading to ascending, moderately scattered, deep green, glabrous and channelled above, silvery scales below, margins flat to recurved, apex rounded, strongly aromatic when crushed; **flowers** to about 1.4 cm across, with 5–8 petals, pale to dark pink on the same plant or, rarely, white, solitary or in terminal clusters, often profuse and very conspicuous; **cocci** to about 0.5 cm high.

A beautiful member of the genus though it is sometimes difficult to maintain in cultivation. It has a wide distribution extending from the North Kennedy District to south-eastern Qld, and in NSW it occurs in the North Coast, North Western Plains and Central Western Slopes Districts. Often found on shallow and rocky soils in dry sclerophyll forest and on sand in native pine woodland. For successful cultivation, well-drained acidic soils are essential. Plants like a warm to

hot site and tolerate sun or semi-shade. They are hardy to moderately heavy frosts and withstand extended dry periods. Plants respond well to pruning. Propagate from cuttings of firm young growth, which usually form roots readily.

In nature, intermediate forms between *P. nottii* and *P. woombye* occur and some of these are in cultivation, often under the name of *P.* species aff. *nottii*. One attractive form has lanceolate leaves which have a rusty-scaly undersurface.

Phebalium obcordatum Benth.
(heart-shaped and broadest above middle)
NSW, Vic   Club-leaved Phebalium; Dainty Phebalium
0.7–1.5 m × 0.7–1.5 m   Aug–Oct
Dwarf to small **shrub** with rusty young growth; **branches** spreading to ascending; **branchlets** smooth or faintly warty, with rusty scales; **leaves** 0.2–0.4 cm × up to 0.25 cm, broadly obovate to obcordate, very short-stalked, spreading to ascending, scattered to somewhat crowded, deep green and warty or gland-dotted above, silky-scaly below, flat and grooved to somewhat convex, apex rounded or notched; **flowers** to about 0.6 cm across, white to pale yellow, in small sessile terminal clusters, often profuse and moderately conspicuous; **cocci** to about 0.4 cm high, warty.

This species is rare in nature. It occurs in south-central NSW and western Vic and grows in very well-drained soils. Adapts well to cultivation in temperate regions in soils which drain freely. A warm to hot site which is sunny or semi-shaded is best. Plants tolerate moderately heavy frosts and extended dry periods. Worth growing as a container plant. Often develops as a compact rounded shrub but also responds very well to pruning. Propagate from cuttings of firm young growth, which usually form roots readily.

Phebalium obtusifolium P. G. Wilson
(blunt leaves)
Qld
0.6–1.2 m × 0.5–1 m   Sept–Nov
Dwarf to small **shrub** with often reddish young growth; **branches** spreading to ascending; **branchlets**

Phebalium nottii   D. L. Jones

prominently angular, often reddish, faintly warty, shining; **leaves** 1.5–5 cm × 0.3–0.6 cm, narrowly elliptic to narrowly obovate, sessile, spreading to ascending, not crowded, shiny deep green and glabrous above, smooth, margins often faintly toothed near the blunt apex; **flowers** to about 0.8 cm across, creamy-white, in 10–20-flowered terminal clusters, often profuse and very conspicuous.

This elegant species has a very restricted distribution on sandstone hills in the Moreton District and is regarded as vulnerable in nature. It is cultivated to a limited degree and has potential for greater use. Should adapt to a range of freely draining acidic soils in subtropical and temperate regions. A somewhat sunny or semi-shaded site should be suitable. Plants tolerate light frosts. Worth trying in gardens and large containers. Propagate from cuttings of firm young growth.

The allied *P. dentatum* has leaves with stellate hairs on the undersurface. *P. elatius* is also allied but it has terete stems.

Phebalium oldfieldii (F. Muell.) Benth.
(after Augustus Oldfield, 19th-century English botanist and zoologist)
Tas
0.5–1.3 m × 0.5–1.5 m   Nov–Jan
Dwarf to small **shrub** with stellate-hairy young growth; **branches** spreading to erect; **branchlets** become terete, stellate-hairy; **leaves** 0.7–1.5 cm × 0.2–0.6 cm, oblong to oblong-cuneate, short-stalked, spreading to ascending, crowded, leathery, shiny, mid green and glabrous above, gland-dotted below, margins sometimes faintly toothed near rounded or notched apex, highly aromatic; **flowers** to about 0.7 cm across, pinkish in bud, opening with white to cream interior, in few-flowered terminal clusters, rarely profuse; **cocci** to about 0.3 cm high, slightly spreading.

The foliage of this species emits a pleasing fragrance on warm days. It is rare in nature and has a disjunct distribution in the mountains of the central and southern regions where it often grows in exposed conditions with other dwarf shrubs. Rarely cultivated, it is

Phebalium ralstonii   D. L. Jones

Phebalium oldfieldii × .5

best suited to cool temperate regions in sunny sites which have freely draining acidic soils. Plants are subjected to heavy frosts and snowfalls. Suitable for rockeries and containers. Propagate from cuttings of firm young growth, which may be slow to form roots.

The allied *P. montanum* has leaves with a grooved upper surface.

Phebalium ovatifolium F. Muell.
NSW
(ovate leaves)
0.5–1.5 m × 0.8–2 m   Nov–Feb
Dwarf to small **shrub** with pale rusty young growth; **branches** many, spreading to ascending; **branchlets** angular to terete, faintly glandular-warty with pale rusty scales; **leaves** 0.9–1.5 cm × 0.5–1 cm, broadly elliptical to broadly ovate, short-stalked, leathery, spreading to ascending, somewhat crowded, glabrous above, dark green to olive-green and gland-dotted above, densely silvery-scaly below, margins flat, apex blunt to rounded; **flowers** about 1 cm across, buds often pink, opening white, 1–3 per upper axils often forming leafy sprays, can be profuse and very conspicuous; **cocci** about 0.3 cm high.

An appealing shrub from the Kosciuszko region where it is most common in snowgum woodland. Often found in peaty soil amongst granite rocks. Needs to be more widely cultivated. It has potential for general planting, rockeries and hedging, and should perform very well when espaliered. Plants will grow best in cool temperate regions but it is worth trying them in warmer zones. They require moisture-retentive but freely draining acidic soils and a sunny or semi-shaded site is suitable. Hardy to frost and snow.

Phebalium ovatifolium × .85

May tend to die suddenly. Propagate from cuttings of firm young growth, which form roots readily.

P. ellipticum is similar but it has large glandular-warty leaves with a notched apex.

Phebalium ozothamnoides F. Muell. =
P. squamulosum ssp. *ozothamnoides* (F. Muell.) P. G. Wilson

Phebalium phylicifolium F. Muell.
(leaves similar to the genus *Phylica*)
NSW, Vic   Mountain Phebalium; Alpine Phebalium
0.6–1.5 m × 0.6–2 m   Oct–Dec
    Dwarf to small **shrub** with faintly hairy young growth; **branches** many, spreading to erect; **branchlets** angular becoming terete, faintly hairy; **leaves** 0.8–1.7 cm × up to 0.25 cm, linear to narrowly elliptic, short-stalked, spreading to ascending, somewhat crowded, deep green and glabrous above, usually stellate-hairy below, margins recurved to revolute, apex pointed to blunt; **flowers** to about 0.8 cm across, pale yellow, 1–3 in upper axils forming leafy sprays, often profuse and very conspicuous; **cocci** about 0.3 cm high.
    A common species of subalpine tracts in south-eastern NSW and eastern Vic. In nature it can provide alluring floral displays. Plants adapt well to cultivation in temperate regions and better flowering occurs in cool areas. Well-drained acidic soils and a sunny or semi-shaded aspect are essential for good results. Plants are hardy to frost and snow. They respond well

to pruning and can be used for low hedging. Propagate from cuttings of firm young growth, which form roots readily.
    The var. *lachnaeoides* (A. Cunn.) C. Moore is a synonym of *P. lachnaeoides*.

Phebalium podocarpoides F. Muell. =
    P. squamulosum ssp. *alpinum* (Benth.) P. G. Wilson

Phebalium pungens (Lindl.) Benth. =
                                  Eriostemon pungens Lindl.

Phebalium ralstonii (F. Muell.) Benth.
(after A. J. Ralston)
NSW
0.4–1 m × 0.5–1 m   mainly June–Sept; also sporadic
    Dwarf **shrub**; young growth glabrous or faintly hairy; **branches** spreading to erect; **branchlets** very angular, smooth, glabrous; **leaves** 2.5–5 cm × 0.5–0.8 cm, oblanceolate, tapering to base, ascending to erect, moderately crowded, deep green and mainly glabrous above, paler with prominent midrib below, margins recurved, apex shortly notched; **flowers** initially bell-like, to about 1.8 cm long, yellow-green or yellow-green and pink, stamens prominently extended, in pendent terminal clusters of about 1.5 cm across; **cocci** to about 0.5 cm high.
    This unusual member of the genus is regarded as vulnerable in its natural habitat of creek banks and ridges in open forest of the South Coast District. Plants have succeeded in cultivation and require well-drained acidic soils. Should do best in a semi-shaded area. Plant in sites which help to show off the interesting flowers. May do well as a container plant. Hardy to moderately heavy frosts. Propagate from cuttings of firm young growth.
    The allied *P. sympetalum* is a much taller shrub with stellate-hairy stems and the stamens are barely exserted from the tubular corolla. *P. viridiflorum* is also similar but it has terete stems and erect, narrowly-oblong leaves.

Phebalium retusum Hook. =
                    P. squameum ssp. *retusum* (Hook.) P. G. Wilson

Phebalium rhytidophyllum D. E. Albrecht & N. G. Walsh
(wrinkled leaves)
NSW
1.5–3 m × 1–2 m   Sept–Nov
    Small to medium **shrub** with glandular and scaly young growth; **branches** many, spreading to erect; **branchlets** angular, with brownish scales, glandular-warty; **leaves** 0.3–1.2 cm × 0.25–1 cm, obovate, stalked, spreading to ascending, somewhat crowded, glossy deep green and glabrous above, silvery-scaly below, margins flat to recurved, apex rounded with prominent notch; **flowers** about 1 cm across, white, solitary or rarely up to 3 in upper axils, forming leafy sprays, can be profuse and moderately conspicuous; **cocci** about 0.3 cm high.
    A recently described (1988) vulnerable species which is restricted to an isolated region in far south-eastern NSW where it grows at about 1000 m altitude

in shrubland and open forest, often amongst granite outcrops. Cultivated only to a very limited degree at present but it has potential for greater use in temperate regions. Plants need freely draining acidic soils and although they tolerate sunshine, they should do best in a semi-shaded site. Hardy to moderately heavy frosts. Worth trying for screening and/or hedging because it has a bushy habit. Propagate from cuttings of firm young growth, which strike readily.

P. frondosum is similar but does not have notched leaves. *P. squamulosum* ssp. *coriaceum* and ssp. *retusum* are also allied but have longer, ovate to oblong leaves.

Phebalium rotundifolium (A. Cunn. ex Endl.) Benth.
(round leaves)
Qld, NSW   Round-leaved Phebalium
1–2.5 m × 1–2.5 m   Sept–Nov

Small to medium, compact **shrub** with faintly hairy young growth; **branches** many, spreading to erect; **branchlets** terete, smooth, faintly stellate-hairy; **leaves** 0.6–1 cm × 0.4–0.6 cm, broadly ovate to circular, short-stalked, ascending to erect, crowded, deep green and glabrous above, paler below, margins flat and slightly irregular near the rounded apex, spicy fragrance when crushed; **flowers** to about 1.2 cm across, white to pale yellow, usually solitary in upper axils and forming terminal globular clusters of about 1.5 cm across, often profuse and very conspicuous; **cocci** to about 0.6 cm high.

A distinctive species which is very ornamental when in full flower. It occurs in sandy and often peaty soils associated with granite outcrops in south-eastern Qld and far north-eastern NSW. Not extremely well known in cultivation, it is deserving of greater attention in subtropical and temperate regions. Plants are usually very bushy. They need moisture-retentive but well-drained acidic soils and will do well in sunny or semi-shaded sites. Hardy to moderately heavy frosts. Propagate from cuttings of firm young growth.

Phebalium rude Bartl.
(rough)
WA
0.5–1 m × 0.5–1.5 m   June–Dec

Dwarf **shrub**; young growth silvery-scaly; **branches** spreading to ascending; **branchlets** initially prominently angular and silvery-scaly; **leaves** 0.7–3.5 cm × 0.5–2 cm, obcordate to nearly orbicular, short-stalked, base rounded to tapering, smooth, deep green, becoming glabrous, initially silvery-scaly, apex rounded or notched to bilobed; **flowers** to about 1.2 cm across, white, solitary, axillary, often scattered, rarely profuse; **cocci** about 0.5 cm high.

The typical subspecies occurs along the coast and nearby islands or slightly inland from near Cape Arid to west of Albany. It grows on a range of soil types including sand dunes, sandy loam, lateritic loam, limestone, granite and clay. Evidently rarely cultivated, it is probably only of interest to enthusiasts. Plants should adapt to a wide range of soils in temperate and semi-arid regions.

There are a further 2 subspecies.

ssp. *amblycarpum* (F. Muell.) P. G. Wilson differs from

ssp. *rude* in having cuneate-obcordate to broadly lanceolate leaves of up to 1.7 cm × 0.9 cm. It occurs in the southern Avon, weston Eyre and south-western Roe Districts on sandy soils which can contain lateritic gravel or limestone. Previously *P. amblycarpum* F. Muell.

ssp. *lineare* (C. A. Gardner) P. G. Wilson grows to about 0.5 m × 0.5 m and has linear leaves to about 2 cm × 0.15 cm. It occurs in the eastern Eyre District where it grows in quartzite sand. Previously *P. lineare* C. A. Gardner.

Propagate all subspecies from cuttings of firm young growth.

Phebalium salicifolium A. Juss. = *P. dentatum* Sm.

Phebalium squameum (Labill.)Engl.
(scaly)
Qld, NSW, Vic, Tas   Satinwood; Satin Box; Tallow-
                                  wood; Cheesewood; Bobie-bobie
3–12 m × 2–4.5 m   Sept–Nov

Medium to tall **shrub** or small **tree** with silvery- to rusty-scaly young growth; **branches** many, spreading to erect; **branchlets** angular, with silvery or rusty scales; **leaves** 5–12 cm × 1–2.2 cm, narrowly to broadly elliptic, short-stalked, spreading to ascending, not usually crowded, flat, thinnish, glossy mid-to-deep green and glabrous above, silvery-scaly and midrib prominent below, apex bluntly to finely pointed, pleasant to strongly fragrant when crushed; **flowers** to about 1.5 cm across, white, usually in many-flowered, loose to compact, axillary clusters shorter than the leaves, often profuse and very conspicuous; **cocci** about 0.3 cm high.

A very handsome and variable species. The typical subspecies has a wide distribution which extends from south-eastern Qld, along or near the coast of NSW, in the Otway Ranges, Vic, and on King Island and mainland Tas. It is confined mainly to moist gullies of dry sclerophyll forest and rainforest. Plants adapt very well to cultivation in a range of acidic soils. They do best in a semi-shaded site but also tolerate plenty of sunshine. Plants are hardy to moderately heavy frosts. This species was first cultivated in England in 1822 as *P. billardieri*, which is now a synonym, and has subsequently been grown in many other regions. It makes an excellent hedge or screening plant and has potential for use in topiary and for espaliering. The wood is hard and long-lasting and was used for telegraph poles and fencing posts. It has potential for wider use in woodturning.

P. squameum 'Illumination' is an excellent variegated cultivar which has vigour. It grows to about 3–4.5 m × 2–3 m. The leaf margins are deep cream and contrast with the deep green centre. Flowers are the same as those of the typical species. It has excellent potential as a hedging plant.

There are 2 further subspecies.

The ssp. *coriaceum* P. G. Wilson has smooth branches with leathery, broadly ovate to broadly elliptic leaves to about 2 cm × 0.8 cm with a rounded apex, and few-flowered cymes. It is known only from the mountains near the head of the Macallister River. It has been confused with *P. ovatifolium* which has warty branchlets.

The ssp. *retusum* (Hook.) P. G. Wilson grows 2–4 m ×

Phebalium squamulosum

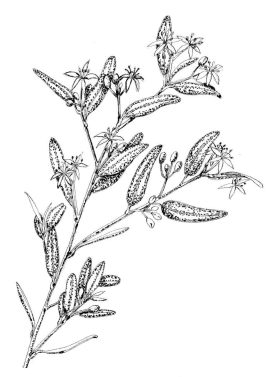

Phebalium squameum ssp. *retusum* × .8

1.5–3 m and has warty branchlets. The leaves are quite thinnish, strongly aromatic and oblong-ovate to about 0.3 cm × 0.8 cm with a blunt or notched apex. Flowers are white and arranged in attractive and profuse few-flowered cymes. Occurs in north-eastern Tas where it is usually associated with rocky sites.

Propagate all subspecies and cultivars from firm young cuttings, which strike readily.

A taxon from the Victorian Central Highlands which has been referred to as *P.* species aff. *squameum* is now known as *P. wilsonii*.

Phebalium squameum ssp. *squameum*   W. R. Elliot

Phebalium squamulosum Vent.
(very scaly)
Qld, NSW, Vic   Forest Phebalium; Scaly Phebalium
1–8 m × 0.6–4 m   Aug–Nov

Small to tall **shrub** or rarely a small **tree** with rusty scaly young growth; **branches** many, spreading to erect; **branchlets** becoming smooth, with rusty scales; **leaves** 1.5–5.5 cm × 0.15–1 cm, oblong to elliptic, short-stalked, spreading to ascending, scattered to somewhat crowded, mid-to-deep green and mainly smooth above, mainly rusty-scaly below, margins flat to recurved, apex pointed or blunt or sometimes notched; **flowers** to about 0.8 cm across, deep cream to yellow in terminal clusters, often profuse and very conspicuous; **cocci** to about 0.4 cm high.

This is an extremely variable species which has many handsome and highly desirable variants.

The typical subspecies has an extended distribution from south-eastern Qld, along eastern NSW and into eastern Vic to near Healesville, north-east of Melbourne. It usually develops as a bushy small to medium shrub but in some areas such as far-eastern Vic it can reach about 7 m in height. Plants are suitable for subtropical and temperate regions and grow best in freely-draining acicid soils in semi-shaded or somewhat sunny sites. They are hardy to moderately heavy frost and withstand extended dry periods.

This species was first cultivated in 1823 as *P. aureum* and in 1824 as *P. squamulosum*.

At least 2 distinct variants are cultivated. One is a shrub of 1–2 m tall with densely rusty-scaly stems and blunt deep-green leaves with a densely rusty-scaly undersurface. The flowers are bright yellow. Another has longer, lighter green, narrow leaves with fewer

Phebalium squamulosum ssp. *alpinum* × .65

rusty scales and the flowers are creamy yellow. A variant which matches this description is marketed as 'Healesville Form' and it can be difficult to maintain in cultivation.

All variants respond very well to pruning or clipping. There are a further 9 subspecies. These differ mainly in features such as leaf shape and the degree of wartiness or scaliness.

P. squamulosum ssp. *alpinum* (Benth.) P. G. Wilson.

A bushy shrub of 1–1.5 m × 1–2 m. It has oblong to oblong-elliptic leaves of 0.7–1.7 cm × 0.5–0.25 cm becoming smooth and glossy above, rusty or silvery-scaly below and with a blunt or rounded apex. Usually about 5 creamy-white to yellow flowers per cluster. Responds well to cultivation in temperate regions. The common name of 'Yellow Heather' describes this subspecies well. It is restricted to north-eastern and eastern Vic where it grows often in rocky sites. This subspecies can intergrade with ssp. *ozothamnoides.*

P. squamulosum ssp. *argenteum* P. G. Wilson

This attractive, compact, small shrub is 1–2 m × 1.5–2.5 m. The leaves are oblong to obovate and 1–3.5 cm × 0.3–1 cm. Stellate hairs give the leaves a greyish-green appearance when young. The leaf undersurface has silvery scales. Flowers are cream to pale yellow and often very profuse. It is an excellent plant for low screening and general planting. Distribution extends from the southern North Coast District of NSW to extreme eastern Vic. Usually grows near the coast or slightly inland and can withstand slightly alkaline soils.

P. squamulosum ssp. *coriaceum* P. G. Wilson

A shrub of about 1–2 m × 1–2 m. It lacks warty stems. Leaves are leathery, elliptic to broadly oblong and about 2.5 cm × 0.5 cm. They are densely silvery-scaly below. Bright yellow flowers are profuse, in many-flowered clusters. It occurs in woodland on rocky slopes mainly in the Warrumbungles region of NSW.

P. squamulosum ssp. *gracile* P. G. Wilson

This shrub of about 1–2 m × 1–2 m has narrowly elliptic to oblong leaves of 0.7–2 cm × about 0.2 cm. It has profuse and very conspicuous yellow flowers in spring. This subspecies is cultivated mainly by enthusiasts and deserves to be better known. It occurs in the Darling Downs, South Leichhardt and Maranoa Districts of Qld and its distribution extends to the Upper Hunter Valley and Pilliga Scrub of NSW.

P. squamulosum ssp. *lineare* P. G. Wilson

An small erect shrub with smooth branchlets and smooth linear leaves of 1–2.5 cm × 0.1 cm. The leaves are strongly channelled above and have a blunt apex. Flowers are pale yellow and displayed on slender stalks in profuse clusters. It occurs in the Central Coast and Central Western Slopes where it usually grows in sandstone soils derived from sandstone conglomerate.

P. squamulosum ssp. *longifolium* (S. T. Blake) P. G. Wilson

This is a shrub of 2–3 m × 1.5–2.5 m. It has warty stems with smooth, glabrous, narrowly elliptic leaves, 2–7 cm × 0.4–1 cm, which have rusty scales below. The flowers have white petals with rusty scales on their exterior. It is restricted to the Atherton Tableland of north-eastern Qld, and is cultivated to a limited degree. Suitable for tropical, subtropical and temperate regions.

Phebalium stenophyllum   W. R. Elliot

P. squamulosum ssp. *ozothamnoides* (F. Muell.) P. G. Wilson

This lovely variant develops into a bushy shrub of 1–2 m × 1–2 m. It has hairy young growth and dull green, obovate to nearly circular, pleasantly aromatic leaves of 0.7–1.1 cm × 0.4–0.7 cm. The leaves become glabrous above and have stellate hairs with scales below. Flowers are pale yellow and are usually profuse. It occurs in mountainous country and is distributed from near Glen Innes in the Northern Tablelands of NSW to north-eastern Vic. It adapts very well to cultivation in temperate regions. Previously known as *P. ozothamnoides* F. Muell.

P. squamulosum ssp. *parvifolium* P. G. Wilson

A subspecies of about 0.6–1.3 m × 0.6–1.5 m with glandular-warty stems and leaves. The narrowly-oblong leaves of 0.6–1 cm × 0.1 cm are deeply channelled above and often appear terete. It occurs in the Central and Southern Western Slopes Districts of NSW where it grows in sandy soils of mallee and rocky ridges.

P. squamulosum ssp. *verrucosum* P. G. Wilson

A shrub of about 1–2 m × 1–2 m with prominently glandular-warty branches and oblong-elliptic leaves to 1.5–3 cm × 0.25–0.6 cm. The leaves are hairy while young. Yellow flowers are profuse. It occurs in north-eastern NSW, in the North Coast District and North Tablelands. Plants usually grow on dry ridges in woodland and rainforest.

All subspecies are propagated from cuttings of firm young growth, which usually form roots readily.

P. squamulosum var. *grandiflorum* C. T. White =
P. whitei P. G. Wilson

P. squamulosum var. *stenophyllum* Benth. =
P. stenophyllum (Benth.) Maiden & Betche

Phebalium stenophyllum (Benth.) Maiden & Betche
(narrow leaves)
NSW, Vic, SA   Narrow-leaved Phebalium
1–1.5 m × 1–1.5 m   Aug–Nov

Small **shrub** with brownish-scaly young growth; **branches** spreading to erect; **branchlets** terete, with brownish scales; **leaves** 0.4–2 cm × up to 0.2 cm, linear to narrowly oblong, more or less terete, often incurved, spreading to ascending, initially greyish, usually becoming greener and glabrous, margins revolute, often obscuring the scaly undersurface, apex blunt; **flowers** to about 1 cm across, bright yellow, in sessile terminal clusters to about 1.5 cm across, often profuse and very conspicuous; **cocci** to about 0.4 cm high.

This elegant species has a disjunct distribution. It occurs in the North Western Slopes District in NSW, in the Wimmera and Mallee areas of Vic and across the border in SA. It is cultivated mainly by enthusiasts but deserves to be better known. Suitable in semi-arid and warm temperate regions. Plants need very well-drained soils and plenty of sunshine to reach their potential. They are hardy to moderately heavy frosts and are drought tolerant. They respond well to pruning and do well as container plants. A grey-leaved selection is highly ornamental. Propagate from cuttings of firm young growth, which strike readily.

P. squamulosum var. *stenophyllum* Benth. is a synonym. Plants marketed as *P.* species aff. *stenophyllum* are referrable to *P. stenophyllum*.

Phebalium sympetalum P. G. Wilson
(united petals)
NSW   Rylstone Bell
2–3 m × 1–2 m   June–Oct

Medium **shrub** with stellate-hairy young growth; **branches** spreading to erect; **branchlets** angular, becoming glabrous; **leaves** 1.5–3.5 cm × 0.4–0.8 cm, elliptic to obcuneate, very short-stalked, spreading to ascending, scattered to somewhat crowded, green, smooth and becoming glabrous, midrib prominent below, margins faintly toothed near the blunt or notched apex; **flowers** tubular, to about 1.5 cm long, greenish-yellow with slightly exserted stamens, terminal, solitary or up to 3 per cluster, scattered, rarely profuse; **cocci** about 0.4 cm high.

A unique species with *Correa*-like flowers which is regarded as vulnerable in nature. It occurs in the Rylstone region in the Central Tablelands where it grows among rocky outcrops in dry sclerophyll forest. Rarely cultivated, it will add interest to plantings. Plants need excellently drained acidic soils in a sunny or semi-shaded site. They tolerate fairly heavy frosts and extended dry periods. Nectar-feeding birds should visit the flowers. Propagate from cuttings of firm young growth.

P. ralstonii and *P. viridiflorum* are allied but both have flowers with free petals and exserted stamens.

Phebalium tuberculosum (F. Muell.) Benth.
(with small wart-like swellings)
WA
1–2 m × 1–2 m   July–Dec

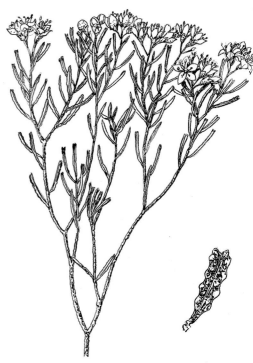

Phebalium tuberculosum ssp. *megaphyllum* × .75, leaf × 3

Small **shrub** with glandular-warty young growth; **branches** many, spreading to ascending; **branchlets** terete, prominently warty; **leaves** 0.4–0.7 cm × up to 0.15 cm, oblong and more or less terete, with revolute margins, short-stalked, spreading to ascending, rarely crowded, deep green and glossy with glandular warts above, apex blunt; **flowers** to about 0.9 cm across, white to pale yellow, usually in 3–4-flowered terminal clusters, often profuse and very conspicuous; **cocci** to about 0.4 cm high.

This is a variable species with the typical subspecies occurring over much of south-western WA where it grows in a range of soil types including sand, lateritic sand, loam and clay. It is rarely cultivated and deserving of greater attention. Plants are best suited to warm temperate and semi-arid regions but must have good drainage and plenty of sunshine. They tolerate drought and are hardy to moderately heavy frosts. Plants usually have fairly dense foliage and are useful for screening.

There are a further 2 subspecies.

The ssp. *brachyphyllum* P. G. Wilson is a small shrub, about 1 m × 1 m, with sessile leaves to about 0.25 cm × 0.15 cm. It has a very restricted distribution in sandy soils north-east of Kalgoorlie.

The ssp. *megaphyllum* (Ewart) P. G. Wilson grows to about 1.5 m × 1.5 m. It has slightly recurved, oblong-cuneate leaves of 0.7–1.5 cm × 0.2–0.4 cm and it can have up to 6 white or rarely yellow flowers per cluster. It occurs in the central-western Avon and western Coolgardie Districts where it grows in sandy and lateritic soils.

Propagate all subspecies from cuttings of firm young growth, which may be slow to produce roots.

Phebalium viridiflorum P. G. Wilson
(green flowers)

NSW — Green Phebalium
0.7–2 m × 0.5–1.5 m — June–Sept

Dwarf to small **shrub** with stellate-hairy young growth; **branches** spreading to erect; **branchlets** terete, stellate-hairy; **leaves** 2–4 cm × 0.4–0.8 cm, oblong-elliptic, tapering to short stalk, spreading to ascending, rarely crowded, deep green, gland-dotted and becoming more or less glabrous above, stellate-hairy below, margins recurved, apex with 2 very short lobes; **flowers** to about 2 cm long, 6–12 in pendent clusters, rarely profuse but always conspicuous; **petals** to about 1 cm long, greenish; **stamens** twice as long as petals, pale greenish-yellow to yellow; **cocci** to about 0.6 cm high.

A rare and unusual species from the Northern Tableland and North Western Slopes. It grows in heathland and woodland often in rocky crevices. Plants are very uncommon in cultivation. They require excellently drained acidic soils and a sunny or semi-shaded site. Frost rarely troubles plants and they tolerate extended dry periods. Probably best suited to temperate regions. Worthy trying as a container plant and in gardens. Propagate from cuttings of firm young growth.

P. ralstonii is allied but differs in having angular branchlets and glabrous stems and leaves.

Phebalium whitei P. G. Wilson
(after C. T. White, 20th-century Queensland Government Botanist)

Qld
0.7–1.6 m × 0.5–1.3 m — Aug–Nov

Dwarf to small **shrub**; young growth brownish-scaly; **branches** ascending to erect; **branchlets** terete, with many brownish scales; **leaves** 1–6 cm × 0.4–1.2 cm, narrowly oblong, tapering to short stalk, spreading to ascending, scattered to moderately crowded, deep green and glabrous above, silvery-scaly below, midrib prominent, margin flat to recurved and often irregular near the rounded apex; **flowers** to about 1.7 cm across, rusty-brown in bud, opening brilliant yellow, terminal, sessile, solitary or up to 6 flowers per cluster, often profuse and very conspicuous.

An outstanding species from south-eastern Qld where it occurs in the border ranges growing amongst granite outcrops. Popular in cultivation but frequently has limited longevity possibly due to pampering of plants. Plants are best suited to temperate regions and require well-drained acidic soils and a sunny or semi-shaded site. They are hardy to most frosts and tolerate extended dry periods. They respond very well to pruning and flowering stems are excellent for indoor decoration. If leggy plants become unstable, removal of excessive top growth is recommended. An excellent container plant. Propagate from cuttings of firm young growth, which usually strike readily.

P. squamulosum var. *grandiflorum* C. T. White is a synonym.

Phebalium wilsonii N. G. Walsh & D. E. Albrecht
(after Paul G. Wilson, 20th-century Western Australian botanist)

Vic
3–10 m × 1.5–4 m — Aug–Oct

Medium to tall **shrub** or small **tree** with warty young growth; **branches** spreading to ascending; **branchlets** terete, silvery-scaly, glandular-warty; **leaves** 3–8 cm × 0.5–1.5 cm, lanceolate to narrowly elliptic, stalked, spreading to ascending, scattered, deep green, mainly smooth and glossy above, densely silvery-scaly below, midrib prominent, margins flat to recurved, apex pointed; **flowers** to about 1 cm across, white, axillary, forming leafy panicles, often profuse and conspicuous; **cocci** about 0.4 cm high.

A recently described (1988) species which was often referred to as *P.* species aff. *squameum*. It occurs in a very restricted area of the Victorian Central Highlands where it grows in open forest and cool temperate rainforest. Plants are uncommon in cultivation. They require well-drained, moisture-retentive, acidic soils with a semi-shaded or shaded site. They are hardy to most frosts and light snowfalls. Plants may need supplementary water during extended dry periods and the use of mulches is beneficial. Should be useful for underplanting of tall trees. Propagate from cuttings of firm young growth, which usually strike readily.

The closely allied *P. squameum* ssp. *squameum* has smooth branchlets.

Phebalium woombye (F. M. Bailey) Domin
(from the Woombye region)

Qld — Woombye
0.3–3 m × 0.6–2 m — July–Nov

Phebalium woombye × .45, flower × 1

Phebalium whitei

W. R. Elliot

Dwarf to medium **shrub** with rusty-scaly young growth; **branches** many, spreading to erect; **branchlets** terete, rusty-scaly; **leaves** 1–6 cm × 0.4–1.2 cm, nar-rowly elliptic to broadly elliptic, short-stalked, spreading to ascending, rarely crowded, deep green, smooth and glabrous above, silvery-scaly below, midrib prominent, margins usually flat, apex blunt to rounded; **flowers** to about 1.2 cm across, rusty buds, opening white to pale pink, in terminal many-flowered clusters, often profuse and very conspicuous; **cocci** about 0.4 cm high.

A handsome species from coastal regions of south-eastern Qld where it grows in the drier parts of the Wallum and heathland. Somewhat popular in cultivation, it does well in subtropical and temperate regions. Plants prefer freely draining acidic soils with a sunny or semi-shaded aspect. They are hardy to moderate frosts. Sometimes plants may become slightly unkempt but they respond well to moderately hard pruning and regular clipping. A number of selections have been introduced to cultivation including dwarf variants with pink or white flowers. Propagate from cuttings of firm young growth, which strike readily.

This species is known to intergrade and possibly hybridise with *P. nottii* and some selections are some-times cultivated. Plants can have leaf characteristics of one species and floral characteristics of the other. All variants are attractive and should be popular in cultivation.

PHEROSPHAERA W. Archer =
  MICROSTROBUS J. Garden & L. A. S. Johnson

PHILESIACEAE Dum.
A small family of monocotyledons consisting of about 12 species in 6 genera. One species of *Eustrephus* and one *Geitonoplesium* species are found in Australia. The family is prominent in South America with *Lapageria rosea* being a prized horticultural subject. Most species in the family are strong climbers with wiry, twining stems, and a few are shrubs.

PHILOTHECA Rudge
(from the Greek *philos*, loving; *thece*, covering, box)
Rutaceae

Dwarf to small **shrubs** with simple hairs; **leaves** alternate, simple; **inflorescence** terminal cymes or solitary flowers; **flowers** bisexual, open-petalled; **sepals** 5, free; **petals** 5, overlapping in bud; **stamens** 10; **fruit** of 1–5 cocci.

A small endemic genus of 6 species; 4 of these occur in south-western WA.

P. salsolifolia is the only species commonly cultivated. It is extremely attractive when in full flower whereas most of the other species lack dramatic characteristics and are evidently rarely cultivated.

Seed may be very difficult to germinate, see Treatments and Techniques to Germinate Difficult Species, Volume 1, page 205. Usually best results are gained from cuttings of barely firm growth.

Philotheca australis Rudge = *P. salsolifolia* (Sm.) Druce

Philotheca basistyla Mollemans
(base of the style; referring to the expanded style base)
WA
0.5–1 m × 0.4–0.8 m   Aug–Oct

Dwarf **shrub** with glabrous young growth; **branches** spreading to ascending; **branchlets** glandular; **leaves** 0.7–1.4 cm × up to 0.15 cm, narrowly club-shaped, ascending to erect, crowded, glabrous, glandular, apex rounded; **flowers** to about 1.5 cm across, white, solitary, terminal, nearly sessile; **cocci** not known.

A very poorly known and apparently rare, recently described (1993) species from east of Perth, on the boundary of the Avon and Coolgardie Districts where it grows in deep yellow sand. Cultivation is advised as part of a conservation management programme. Plants should do best in very well-drained, light, acidic soils in a sunny or lightly shaded site. They should withstand moderate frosts and extended dry periods. Propagation from cuttings of barely firm young growth may provide best results.

P. langei is allied but differs in having pointed leaves and smaller flowers.

Philotheca calida F. Muell. =
  Drummondita calida (F. Muell.) P. G. Wilson

Philotheca ciliata Hook. —
  see *P. salsolifolia* (Sm.) Druce

Philotheca citrina P. G. Wilson
(lemon-coloured)
WA
0.6–1.3 m × 0.5–1.5 m   Aug–Sept

Dwarf to small **shrub** with resinous young growth; **branches** many, spreading to ascending, greyish; **branchlets** pale green, resinous, brittle; **leaves** about 1 cm × up to 0.15 cm, narrowly clavate, alternate, sessile and decurrent, spreading to ascending, often crowded on branchlets, bright green, more or less glabrous, glandular-warty, flat above, convex below, apex blunt with a small point; **flowers** about 2 cm across, pale greenish-yellow, terminal, solitary, on short stalks, very conspicuous.

A recently described (1992) species which is known only from a small area in the far north-western Austin District where it grows in crevices on a low red granite outcrop. It needs to be cultivated as part of a conservation strategy. Should do best in warm temperate and semi-arid regions. Plants will need excellently drained acidic soils and plenty of sunshine. Other information is scant. Propagate possibly from seed, or from cuttings of firm young growth, which may be slow to form roots.

Philotheca ericoides (Harv.) F. Muell. =
Drummondita ericoides Harv.

Philotheca hassellii F. Muell. =
Drummondita hassellii (F. Muell.) P. G. Wilson

Philotheca langei Mollemans
(after Dr Robert T. Lange, 20th-century Australian ecologist and palaeobotanist)
WA
0.8–1.2 m × 0.4–0.7 m   Aug–Oct
Dwarf to small **shrub** with glabrous young growth; **branches** ascending to erect; **branchlets** glandular; **leaves** 0.4–0.6 cm × about 0.1 cm, club-shaped, spreading to erect, crowded, grooved, glabrous, glandular, apex pointed; **flowers** to about 1 cm across, white, 1–3 at ends of branchlets, on short stalks.

A recently described (1993) species which is known from a very restricted area north-west of Southern Cross on the boundary of the Avon and Coolgardie Districts. It occurs in granite country on shallow sand over clay. Warrants cultivation as part of a conservation strategy. Needs a warm to hot, sunny or semi-shaded site in freely draining acidic soils. Hardy to moderate frosts and extended dry periods. Propagate from cuttings of barely firm young growth.

Allied to *P. basistyla* which has very blunt leaves and slightly larger flowers.

Philotheca miniata C. A. Gardner =
Drummondita miniata (C. A. Gardner) P. G. Wilson

Philotheca salsolifolia (Sm.) Druce
(leaves similar to the genus *Salsola*)
Qld, NSW
0.5–2 m × 0.4–1.5 m   Sept–Dec
Dwarf to small **shrub** with glabrous or faintly hairy young growth; **branches** ascending to erect; **branchlets** slightly warty, usually glabrous; **leaves** 0.5–2.2 cm × about 0.15 cm, linear to nearly terete, spreading to ascending, crowded, glandular, green, glabrous, apex pointed; **flowers** 1.4–2.7 cm across, pink to purplish, terminal, solitary or 2–3 together, can be profuse and prominent; **cocci** to about 0.6 cm long, pointed.

A highly decorative species from south-eastern Qld and over a wide range in eastern NSW. It occurs in coastal heath as well as in often rocky sites in open forest and woodland. It is cultivated to a limited degree mainly by enthusiasts, and has proved moderately reliable. Plants require very good drainage and acidic soils. A sunny or semi-shaded site is suitable. Plants are hardy to moderate frosts and extended dry periods. They can become top-heavy and pruning from an

Philotheca salsolifolia × 1

early age is beneficial. Does well in containers. Introduced to cultivation in England in 1822. Propagate from cuttings of barely firm young growth, which can be slow to form roots.

P. australis Rudge is a synonym.

P. ciliata Hook. is closely allied to *P. salsolifolia* but differs in having fringed leaves, which are crowded, and smaller flowers 0.8–1 cm across. It is from the Leichhardt, Maranoa, Mitchell and Warrego Districts of Qld where it often grows on rocky outcrops.

A closely allied but as yet undescribed species from south-eastern Qld and the Northern and Central Western Slopes and North Western Plains of NSW differs in having narrower leaves to about 1 cm long and white to pink flowers of about 1 cm across. Evidently not cultivated, it will need similar conditions to *P. salsolifolia*.

Philotheca tubiflora A. S. George
(tubular flowers)
WA
0.3–0.6 m × 0.3–0.6 m   June–Sept
Dwarf **shrub** with faintly hairy young growth; **branches** many; **branchlets** becoming glabrous; **leaves** 0.25–0.4 cm × less than 0.1 cm, club-shaped, spreading to ascending, crowded, glabrous, glandular, apex pointed, ending in brown gland; **flowers** about 0.5 cm across, white to pale pink, tubular base with reflexed hairy lobes, solitary and usually terminal, can be profuse and very conspicuous; **cocci** blunt, one-seeded.

279

Philydraceae

This appealing species has an isolated distribution on the western edge of the Great Victoria Desert. It is found on rocky rises or eroded granitic breakaways and grows in brown clay. Not highly ornamental. Suited to arid, semi-arid and warm temperate regions. Needs plenty of sunshine and good drainage. Probably tolerates moderate frosts. Propagate from barely firm young growth.

PHILYDRACEAE Link   Frogsmouth Family

A small family of monocotyledons consisting of 3 genera and about 6 species restricted to Australia and South-East Asia. They are annual or perennial herbs with a fleshy rootstock and narrow leaves in a fan-like tuft. The flowers are colourful, each consisting of two large flimsy outer segments, two small inner segments and a single stamen. Australian genera: *Helmholtzia, Philydrella, Philydrum.*

PHILYDRELLA Caruel

(from the Greek *philos*, loving; *hydro*, water; and *ella*, diminutive suffix)

Philydraceae   Butterfly Flowers

Terrestrial and aquatic, glabrous perennial **herbs**; **scape** solitary, not branched; **leaves** broad and cauline, few, fleshy; **inflorescence** 1–10-flowered spike; **flowers** with 4 petal-like segments, outer 2 much larger than inner 2, many-veined, with sheathing bracteoles, glabrous; **fruit** a narrowly oblong capsule.

A small genus of 2 species restricted to south-western WA. Plants are rarely encountered in cultivation but have potential for use in and on the edges of pools. Propagate from seed.

Philydrella drummondii L. G. Adams

(after James Drummond, first Government Botanist, WA)

WA   Greater Butterfly Flower
0.15–0.25 m × 0.1–0.2 m   mainly Sept–Oct

Dwarf perennial **herb**; **leaves** to about 10 cm long, subulate, few, basal and cauline; **scape** to about 0.25 m tall; **flowers** to about 2 cm long, lemon-yellow, in 3–9-flowered spikes to about 6 cm long; **fruit** to about 0.7 cm long.

A dweller of freshwater swamps and seepage lines in the southern Darling District. It has potential for use in pools and bog gardens in temperate regions. May suffer frost damage. Propagate from seed.

Philydrella pygmaea (R. Br.) Caruel

(dwarf)

WA   Lesser Butterfly Flower
0.05–0.2 m × 0.4 –1 m   mainly Sept–Oct

Dwarf perennial **herb**; **leaves** 2; **basal leaf** to 10 cm long; **cauline leaf** to about 2 cm long; **scape** to about 0.2 m tall, green to reddish; **flowers** to about 1.5 cm long, lemon-yellow, in 2–8-flowered spikes to about 4 cm long; **fruit** to about 0.6 cm long.

The typical variant of this aquatic herb occurs over a wide range in south-western WA, extending from near Geraldton to east of Esperance. Worth trying in pools and bog gardens. Should withstand light to moderate frosts. The ssp. *minima* L. G. Adams is a very poorly known variant. It has shorter leaves on scapes to about 10 cm long. Propagate both variants from seed, which can begin to germinate 40–60 days after sowing.

PHILYDRUM Banks & Sol. ex Gaertn.

(from the Greek *philos*, loving; *hydro*, water)
Philydraceae

A monotypic genus which has a wide distribution in Australia and also extends through south-eastern and eastern Asia to southern Japan.

Philydrum lanuginosum Gaertn.

(long cottony hairs)

Qld, NSW, Vic, WA, NT   Frogsmouth
0.3–1.25 m × 0.1–0.3 m   Oct–March

Emergent, perennial, aquatic **herb** with short-creeping rhizome, often tufting; **leaves** 0.3–0.6 m × 1–2 cm, linear, mostly basal, spreading to ascending, somewhat succulent, dull green; **scapes** to about 1.2 m tall; **inflorescence** solitary spike or sometimes branched, with 10–25 flowers; **flowers** about 1 cm across, yellow, sessile, subtended by grey woolly-hairy bracts, usually short-lived; **fruit** to about 1 cm long, hairy, 3-valved.

This widespread species inhabits swamps, margins of pools, dams, streams and moist soils in temperate and tropical regions. Often colonises disturbed sites in the wild. Adapts well to cultivation but can be short-lived. Although the flowers are usually short-lived plants are a pleasing adjunct to pools or bog gardens. They usually tolerate light frosts and often reshoot if foliage is damaged. It has the potential to become weedy. Propagate from seed, which does not require pre-sowing treatment, or from divisions of clumps, which re-establish readily.

PHLEBOCARYA R. Br.

(from the Greek *phlebos*, vein; *caryon*, nut; referring to the prominent veins on the fruit)
Haemodoraceae

Tufting perennial **herb** with short-creeping rhizomes; **leaves** basal, linear, terete or flattened, spreading to erect; **inflorescence** a panicle; **perianth** white or cream, glabrous; **sepals** 3, petal-like, free; **petals** 3, free; **stamens** 6; **style** longer than stamens; **fruit** 1-seeded.

A small endemic genus of 3 species restricted to south-western WA. It is allied to *Anigozanthos*. They are rarely encountered in cultivation with only *P. ciliata* sometimes offered by specialist nurseries. They are difficult to cultivate because of their reliance on mycorrhiza being present in the soil. However, a compact, tufting growth habit and dainty small flowers make them desirable for rockeries and containers as well as general garden planting.

Propagation is from seed, which may germinate more reliably with treatment to soften or break the outer coating. See Treatments and techniques to germinate difficult species, Volume 1, page 205. Division of established clumps is sometimes successful. This is best done in late autumn or early spring.

Phlebocarya ciliata R. Br.

(fringed with fine hairs)
WA
0.25–0.7 m × 0.2–0.8 m   mainly Sept–Nov

Phlebocarya ciliata × .45, leaf × 2

Dwarf, tufting, perennial **herb** with hairy young growth; **leaves** 25–70 cm × 0.15–4 cm, linear, flat, deep green, glabrous except for fringed margins; **sheaths** glabrous or slightly hairy; **inflorescence** shorter than leaves, much-branched, usually glabrous but sometimes hairy on lower part; **flowers** to about 0.7 cm across, white to cream, starry, often profuse and moderately conspicuous amongst the foliage.

P. ciliata is distributed from near Jurien Bay to Albany where it is found in heath and woodland, on sandy soils which can be swampy or well drained. Suitable for temperate and semi-arid regions. Needs freely draining acidic soils in a sunny or semi-shaded site. Appreciates growing amongst other plants which protect its root system from hot sunshine. Plants are hardy to light frosts. Propagate from seed or by division of clumps.

Phlebocarya filifolia (F. Muell.) Benth.
(thread-like leaves)
WA
0.25–0.4 m × 0.25–0.5 m   mainly Oct–Dec
Dwarf, tufting, perennial **herb** with hairy young growth; **leaves** 25–40 cm × up to 0.2 cm, linear, flattened to terete, faintly hairy, green; **sheath** hairy; **inflorescence** about as long as or longer than leaves, much-branched, on a glabrous scape; **flowers** about 1.2 cm across, whitish with bluish-grey exterior, profuse and conspicuous.

This attractive species is from the Southern Irwin and Northern Darling Districts where it is found in shrubland and woodland growing in sandy soils. Should be best suited to freely draining acidic soils with a sunny or semi-shaded site in temperate and semi-arid regions. Use in gardens, rockeries or containers. Propagate from seed or by division of clumps.

Phlebocarya pilosissima (F. Muell.) Benth.
(many long soft hairs)
WA
0.15–0.35 m × 0.2–0.5 m   mainly Aug–Oct
Dwarf, compact, tufting, perennial **herb** with hairy young growth; **leaves** 14–35 cm × up to 0.2 cm, linear, terete or flattened, usually hairy especially on margins, green; **sheaths** densely hairy; **inflorescence** shorter than leaves, on hairy scape; **flowers** to about 1 cm across, white to creamish, often profuse but somewhat hidden by foliage.

The typical subspecies occurs in the southern Irwin and far-northern Darling Districts where it grows in sandy soils or lateritic gravelly soils in shrubland. Plants should adapt well to cultivation in semi-arid and temperate regions if grown in a sunny or semi-shaded site with freely draining acidic soils. They tolerate light frosts and extended dry periods.

The ssp. *teretifolia* T. D. Macfarl. occurs in the southern Irwin District in sandy soils. It has loose tufts of terete leaves which are glabrous except near the apex, and taller flower scapes.

Both subspecies are best grown in rockeries and containers. Propagate from seed or by division of clumps.

PHOLIDOTA Lindl. ex Hook.f.
(from the Greek *pholidotos*, scaly; referring to the prominent, scale-like bracts on the inflorescence)
Orchidaceae
Clump-forming **epiphytic orchids**; **roots** fine and wiry; **pseudobulbs** fleshy, crowded; **leaves** 1 or 2, on the apex of a pseudobulb, plicate; **inflorescence** a wiry raceme, usually pendulous; **flowers** small, in 2 rows, each subtended by a large bract; **fruit** a capsule.

A genus of about 55 species distributed from Asia to New Guinea with a solitary species extending to Australia. Although easily grown it is rarely encountered in cultivation. Propagation is by division or from seed sown under sterile conditions.

Pholidota imbricata Hook. f.
(overlapping; referring to the bracts)
Qld   Rattlesnake Orchid
Epiphyte   March–May
Epiphytic orchid forming dense clumps; **pseudobulbs** crowded, 8–12 cm × 3–5 cm, conical to cylindrical, smooth or ridged, grey-green; **leaf** 1 per pseudobulb, 30–40 cm × 6–8 cm, lanceolate, dark green, thick-textured, pleated, tapered to a basal stalk; **racemes** 30–40 cm long, arching to pendulous; **flowers** 20–60, about 0.8 cm across, white to cream, crowded in 2 rows, each partly enclosed by a pinkish bract; **tepals** cupped; **labellum** 3-lobed.

Distributed in Indonesia, New Guinea and northeastern Qld from the Torres Strait Islands and the tip

Philydrum lanuginosum   W. R. Elliot

of Cape York Peninsula to Townsville. Grows mainly in rainforest but also extends to gullies and other moist sites in more open habitats. A bulky epiphytic orchid which is very easy to grow but is not particularly popular because of its small flowers. May be attached to suitable garden trees in tropical regions. In temperate regions the protection of a heated glasshouse is essential. Needs warm to hot humid conditions, bright light and plenty of air movement. Can be grown in a pot of coarse mix or attached to a slab of treefern, cork or weathered hardwood. Propagate by division or from seed sown under sterile conditions.

The name *P. pallida* Lindl. was applied to the Australian species but it is now known to be restricted to India and Thailand.

Pholidota pallida Lindl. —
see under *P. imbricata* Hook. f.

PHRAGMITES Adans.
(from the Greek *phragma*, hedge; referring to the dense, hedge-like growth habit)
Poaceae

Vigorous, semi-woody, bamboo-like **perennials** with creeping rhizomes; **culms** erect; **leaves** cauline, blade flat; **inflorescence** dense to loose, softly-hairy panicles; **spikelets** 2–10-flowered; **glumes** shorter than spikelet; **florets** articulate above glumes; **lemma** finely tapered; **stamens** 3.

A cosmopolitan genus of 3 species; 2 species extend to Australia. *P. australis* is the most common and is sometimes cultivated. It is often retained in rural areas to bind soil and also for wildlife habitat. Propagation is from seed or by division of clumps.

Phragmites australis (Cav.) Trin. ex Steud.
(southern)
All states   Common Reed
2–3 m × 2–5 m   mainly Nov–May

Vigorous bamboo-like **perennial** with vertical and horizontal rhizomes; **stems** to about 3 cm × 1 cm, not branched, semi-woody, stiff, erect, leafy; **leaves** to about 80 cm × 4 cm, sheathed at base, with flat or slightly inrolled blade, mid green, glabrous, under-surface smooth, spreading to ascending, often arching; **panicles** to about 40 cm long, initially green with purplish tonings, becoming creamy-white, tips often pendent, profuse and very conspicuous; **spikelets** 1–1.8 cm long.

A handsome, strong-growing, tall member of the grass family. Grows on the margins of ponds, pools and waterways, in water to about 2 m deep, and also in regularly inundated areas. It tolerates low salinity. Plant communities provide extremely valuable protection and breeding sites for many animals. A very adaptable species that does best in sunny sites in clay or clay loam soils but also subsists in sandy soils. Hardy to frosts and tolerates seasonal dryness. Not recommended for small properties or pools as it can dominate. Young shoots are edible. Sometimes treated as a weed by unknowing landowners. Propagate from seed, which can be difficult or slow to germinate. Division of clumps in autumn to late winter is usually successful but plants can be slow to establish again.

P. communis Trin. is a synonym.

The closely allied *P. karka* (Retz.) Steud. has narrower leaves which have a rough undersurface and end in a stiff, sometimes pungent point. It occurs in Qld, SA, WA, and NT. Plants are also highly valued for wildlife habitat.

Phragmites communis Trin. =
P. australis (Cav.) Trin. ex Steud.

Phragmites karka (Retz.) Trin. ex Steud. —
see *P. australis* (Cav.) Trin. ex Steud.

PHREATIA Lindley
(from the Greek *phreatia*, well; apparently referring to the close grouping of the tepals)
Orchidaceae

Small **epiphytic orchids**, solitary or clumping; **stems** short, branched or unbranched; **leaves** entire, thin or fleshy, sessile; **inflorescence** a slender raceme; **flowers** small, white, green or cream; **sepals** and **petals** similar; **labellum** nearly entire; **fruit** a capsule.

A genus of about 200 species distributed from Asia to Polynesia and New Guinea; 1 species is endemic in Qld. It has limited ornamental appeal and is grown mostly by orchid enthusiasts. Propagation is from seed, which must be sown under sterile conditions.

Phreatia baileyana Schltr. =
Octarrhena pusilla (F. M. Bailey) M. A. Clem. & D. L. Jones

Phreatia crassiuscula F. Muell. ex Nicholls
(thickened; referring to the leaves)
Qld
Epiphyte   Jan–April

Small **epiphytic orchid** growing singly or in small groups; **stems** short, unbranched, with a mass of thread-like roots; **leaves** 3–6, 4–6 cm × 0.7–1 cm, thick and fleshy, spreading like a fan, dark green, upper surface deeply channelled; **racemes** 2.5–3.5 cm long, fleshy, apex often decurved; **flowers** numerous, about 0.2 cm across, cream to greenish, crowded; **tepals** spreading widely; **labellum** nearly entire.

A common species of highland rainforests in north-eastern Qld between Cooktown and Townsville. It grows on the mossy branches and trunks of rainforest trees and sometimes also on boulders. Not very popular in cultivation but grown by orchid enthusiasts. Adapts readily to cultivation on a small slab of treefern. Requires cool, humid, shady conditions with free air movement. Propagate from seed sown under sterile conditions.

Phreatia pusilla (F. M. Bailey) Rolfe =
Octarrhena pusilla (F. M. Bailey) M. A. Clem. & D. L. Jones

Phreatia robusta R. S. Rogers =
Rhynchophreatia micrantha (A. Rich.) N. Halle

PHYLA Lour.
(from the Greek *phyle*, tribe, race; referring to many flowers in the calyx)
Verbenaceae

Phyla nodiflora (L.) Greene is popular in cultivation especially as a substitute for lawn grasses in warm temperate or semi-arid regions. There has been confusion as to whether or not it is an Australian indigenous species. Recent botanical research has shown that it was introduced after European settlement and therefore it has been excluded from this volume.

PHYLACIUM Benn.
(from the Greek *phylas*, guardian; referring to the large floral bracts)
Fabaceae (alt. *Papilionaceae*)

Climbing perennial **herbs**; **leaves** trifoliolate; **stipules** present; **inflorescence** more or less enclosed by large papery bracts; **flowers** pea-shaped; **stamens** 10, 9 joined in a tube with the other faintly joined; **fruit** a 1-seeded pod.

A tropical genus of 3 species which occurs in Asia with one species extending to north-eastern Qld.

Phylacium bracteosum Benn.
(conspicuous bracts)
Qld
Climber   Aug–Nov

A vigorous **climber** with hairy young growth; **stems** twining, thin, wiry; **leaves** trifoliolate; **leaflets** 2–10 cm × 1.2–4.5 cm, elliptical to ovate, deep green and glossy, often glaucous, sometimes hairy below, thin-textured, apex blunt, sometimes notched; **flowers** pea-shaped, about 0.8 cm long, white to pinkish, in few-flowered racemes about 5 cm long, partially enclosed by 1–3 shell-shaped floral bracts (of about 4.5 cm × 4 cm), bracts are white at flowering and mature to green at

fruiting; **standard** white; **wings** and **keel** white or sometimes with pinkish-blue markings; **pods** to about 1 cm × 0.7 cm, flattened, faintly hairy.

This species with fascinating floral bracts occurs in the Iron Range of north-eastern Qld, on the edges of rainforest and often on banks of waterways. It is rarely cultivated and deserves to be better known. Plants are suited to tropical regions and may adapt to subtropical areas. They do best in freely draining acidic soils in a sheltered site but tolerate some sunshine. Hardy to light frosts. Propagate from seed, which requires pre-sowing treatment to soften or crack the seed coating. See Treatments and techniques, Volume 1, page 205. Cuttings of firm young growth are also worth trying.

PHYLLACHNE J. R. Forst. & G. Forst.
(from the Greek *phyllon*, leaf; *achne*, chaff; referring to the grass-like appearance)
Stylidiaceae

Cushion-like perennial **herbs**; **leaves** small, thick, glabrous, densely crowded, undersurface convex; **flowers** solitary, sessile; **calyx lobes** 5–9; **corolla** a very short tube, with 4–9 lobes; **staminal column** erect; **fruit** a dehiscent or indehiscent capsule.

A small genus of 4 species with representation in South America, New Zealand and Australia, which has 1 species. They are fascinating cushion plants found at high altitudes.

Propagation is from seed or cuttings.

Phyllachne colensoi (Hook. f.) S. Berggren
(after John William Colenso, 19th-century missionary and botanical explorer in New Zealand)
Tas   Cushion Plant
0.05–0.1 m × 0.2–0.6 m   Nov–Feb

Tightly matted, cushion-like, perennial **herb**; **stems** short, many, few-branched, ascending to erect, self-layering; **leaves** to about 0.4 cm × 0.2 cm, somewhat wedge-shaped, broadest at base, dark green, glabrous, glossy, apex pointed, often brownish-yellow; **flowers** about 0.4 cm across, white, 5–6-lobed, solitary, barely exposed through leaves.

This intriguing cushion plant occurs at about 1200 m altitude on the exposed slopes of a limited number of mountains in Tas, and also in New Zealand. It is an extremely slow-growing species and has had very limited cultivation. Plants require constantly moist but freely draining acidic soils and plenty of sunshine. Best suited to cool temperate regions for growing in rockeries and miniature gardens. Plants must not be allowed to dry out and the addition of water-absorbing organic matter to the growing medium is usually beneficial. Other cushion plants such as *Abrotanella forsteroides* and *Donatia novae-zealandiae* have been grown successfully using the method explained under Cushion Plants, Volume 3, pages 135–6 and it may be worth trying this for *P. colensoi*. Plants can also be grown in pots and these are best placed in a vessel with a constant supply of shallow water.

Propagate from seed, or from cuttings, or by division of self-layered stems, which can be slow to become established.

Phyllanthus

PHYLLANTHUS L.
(from the Greek *phyllon*, leaf; *anthos*, flower; referring to some species having leaf-like flowering branches)
Euphorbiaceae

Mainly monoecious **herbs**, **shrubs** or **trees**; **leaves** alternate often spirally arranged, entire, sessile or short-stalked; **stipules** small; **flowers** often unisexual, solitary or in clusters, axillary; **calyx lobes** 4–6, petal-like, overlapping; **petals** absent; **male flowers** larger than female flowers, clustered, with 2–6 stamens; **female flowers** solitary or clustered; **styles** 3, usually divided at apex; **fruit** a capsule, separating into 3 mericarps.

A cosmopolitan genus of about 750 species mainly found in tropical and subtropical regions. There are about 51 species in Australia with representation in all states.

Most Australian species are generally lacking in strong ornamental qualities and are therefore rarely cultivated except by enthusiasts. Some species do have attractive foliage and have potential for background planting and for use as cut foliage for interior decoration.

Propagation is from fresh seed or from cuttings of firm young growth.

Phyllanthus albiflorus F. Muell. ex Müll. Arg. = *Sauropus albiflorus* (F. Muell. ex Müll. Arg.) Airy Shaw

Phyllanthus brassii C. T. White
(after Leonard J. Brass, 20th-century American botanist)
Qld
0.4–0.8 m × 0.4–1 m   Feb–April
Dwarf **shrub** with glabrous young growth; **branches** spreading to ascending, angular, glabrous; **leaves** 4–19 cm × 1.5–4.5 cm, lanceolate to elliptic-lanceolate, petiole short and thick, green, glabrous and glossy above and below, thick, leathery, distinct reticulate venation, apex pointed; **flowers** small, reddish, on pendent slender stalks to about 1 cm long, in axillary clusters; **capsules** about 0.5 cm across, globose.

An uncommon species from north-eastern Qld where it grows in rainforest. Cultivated to a limited degree, mainly by enthusiasts, it needs freely draining soils with a semi-shaded aspect. Should adapt to subtropical and temperate regions and it may naturalise in shady conditions. Propagate from seed or from cuttings of firm young growth.

Phyllanthus calycinus Labill.
(distinctive calyx)
SA, WA   False Boronia;Snowdrop Spurge
0.3–1.3 m × 0.3–1.5 m   June–Nov; also sporadic
Monoecious or dioecious, dwarf to small **shrub** with glabrous young growth; **branches** many, spreading to erect, greyish to greenish, glabrous; **leaves** 0.5–2.3 cm × 0.15–0.6 cm, narrowly elliptic to narrowly obovate, spreading to ascending, scattered to moderately crowded, glabrous, pale green to grey-green, more or less flat, margins entire, apex blunt; **stipules** very small, brown; **flowers** about 0.6 cm across, white to cream or sometimes pinkish, often pinkish buds, on slender sometimes pendent stalks, often profuse and

conspicuous, sweetly fragrant; **capsules** to about 0.6 cm across, globular.

An attractive, wide-ranging species which occurs in central southern SA and south-western WA. It is found in sandy or gravelly soils in woodland and open forest. Cultivated to a limited degree, it warrants greater attention. Does best in temperate and semi-arid regions in well-drained soils with a sunny or semi-shaded aspect. Suitable for gardens and containers. Plants tolerate moderate frosts and respond well to fairly hard pruning or clipping. They are susceptible to root rotting fungi and are prone to sweating during overcast and humid weather. Introduced to cultivation in England during 1822. Propagate from seed, or from cuttings of firm young growth, which strike readily.

Phyllanthus ciccoides Müll. Arg. —
see *P. reticulatus* Poir.

Phyllanthus cuscutiflorus S. Moore
(flowers similar to those of the genus *Cuscuta*)
Qld   Pink Phyllanthus
3–7 m × 2–4 m   Sept–Feb
Medium **shrub** to small **tree**; young growth glossy pink, glabrous; **branches** spreading to ascending, many, glabrous; **leaves** 2.5–7.5 cm × 1–3.5 cm, ovate, somewhat concave, spreading to ascending, somewhat crowded, green and glabrous above, paler and glabrous below, pink and pendent when young, venation distinct, apex pointed; **flowers** small, white or pink, on pendent slender stalks in axillary clusters, often profuse and moderately conspicuous; **capsules** 0.3–0.5 cm across, globose.

An appealing, fast-growing species which usually maintains its dense foliage cover to near ground level. It occurs in the northern Cook District where it grows in lowland rainforest and is often on banks or in close proximity to waterways. It is cultivated to a limited degree and is suitable for cultivation in tropical and subtropical regions. Plants prefer freely draining acidic soils, which are somewhat moisture-retentive. A sunny or semi-shaded site is suitable. Useful for screening and as an accent plant. Regular pruning may promote the decorative soft young growth. Propagate from seed or from cuttings of firm young growth.

Phyllanthus fuernrohrii F. Muell.
(after August E. Fürnrohr, 19th-century German botanist)
Qld, NSW, SA, WA, NT   Sand Spurge
0.2–0.5 m × 0.3–0.6 m   most of the year
Monoecious dwarf **shrub** with hairy young growth; **branches** often many, spreading to ascending; **branchlets** hairy; **leaves** 0.8–3 cm × 0.3–1 cm, elliptic to obovate, spreading to ascending, moderately crowded, covered with fine greyish hairs, margins flat, apex blunt with small point; **flowers** small, greenish, on pendent axillary stalks, can be profuse, rarely very conspicuous; **capsules** to about 0.5 cm across, hairy, greenish to reddish.

A wide-ranging species from warm temperate, semi-arid and arid regions where it usually occurs on sandy soils. It is rarely cultivated and possibly of interest mainly to enthusiasts. Needs excellent drainage and a warm to hot aspect for best growth. Plants are suspected of poisoning stock. Propagate from seed or from cuttings of firm young growth.

Phyllanthus gasstroemii Müll. Arg.
(after M. Gasstroem)
Qld, NSW   Blunt Spurge
1–2.5 m × 0.6–2 m   Sept–Jan
  Dioecious small to medium **shrub** with glabrous young growth; **branches** spreading to ascending; **branchlets** spreading, glabrous; **leaves** 0.5–2.2 cm × 0.4–1 cm, obovate-oblong to broadly obovate, in 2 distinct rows, glabrous, pale green to greyish-green, flat, apex blunt with small point; **flowers** small, greenish with whitish margins, on pendent axillary stalks, often profuse but rarely very conspicuous; **capsules** to about 0.4 cm across, more or less 3-lobed, greenish, glabrous.
  An interesting species which inhabits moist sites on margins of rainforest or wet forest in eastern Qld and eastern NSW. Cultivated mainly by enthusiasts, it is best suited to tropical and subtropical regions. Needs well-drained acidic soils and a semi-shaded site. Pruning from an early stage promotes bushier growth. This species is suspected of being poisonous to stock. Propagate from seed or from cuttings of firm young growth.

Phyllanthus gunnii Hook. f.
(after Ronald C. Gunn, 19th-century botanical collector, Tasmania)
NSW, Vic, Tas   Scrubby Spurge
1–3 m × 0.6–2 m   Sept–Jan
  Dioecious small to medium **shrub** with glabrous young growth; **branches** spreading to ascending, slender, usually angled; **branchlets** glabrous; **leaves** 0.8–2 cm × 0.8–1.2 cm, mainly broadly ovate to nearly circular, yellow-green to greyish-green, paler below, glabrous, margins flat, venation distinct, apex blunt and usually notched; **flowers** small, greenish-yellow to greenish, on pendent axillary stalks to 0.8 cm long, often profuse but rarely very conspicuous; **capsules** about 0.4 cm across, often maturing with reddish tonings.
  Scrubby Spurge inhabits open forest on rocky slopes often near waterways and in coastal areas. Plants require good drainage and acidic soils with a semi-shaded or moderately sunny site. Grown mainly for its attractive foliage. Plants respond well to pruning or clipping and are suitable for short-term hedging. Cut foliage may be used as decoration. Propagate from seed or from cuttings of firm young growth.
  The closely allied *P. saxosus* F. Muell. differs in having male and female flowers on different plants. It was known as *P. gunnii* var. *saxosus* (F. Muell.) Benth. and is still regarded as a variety by some botanists. It occurs in southern SA and near Esperance, WA.

Phyllanthus hirtellus F. Muell. ex Müll. Arg.
(shaggy)
Qld, NSW, Vic, ?SA   Thyme Spurge
0.3–0.8 m × 0.2–0.6 m   Sept–Feb
  Monoecious dwarf **shrub** with hairy young growth; **branches** ascending to erect, slender, many; **branchlets** green with short stiff hairs; **leaves** 0.4–1.2 cm × 0.2–0.4 cm, oblanceolate to cuneate, more or less sessile, spreading to ascending, green, hairy, margins flat to revolute, midrib prominent, apex blunt, often notched; **flowers** small, yellowish, on spreading to pendent axillary stalks, can be profuse but rarely very conspicuous; **capsules** about 0.4 cm across, hairy, green.
  A widespread and variable species which inhabits heath, woodland and open forest. Rarely cultivated but does best in well-drained acidic soils in a semi-shaded or lightly sunny site. Hardy to most frosts and extended dry periods. Responds well to pruning which rejuvenates plants. Introduced to cultivation in 1835 as *P. thymoides* Müll. Arg., which is a synonym. Propagate from seed or from cuttings of firm young growth.

Phyllanthus lamprophyllus Müll. Arg.
(shining leaves)
Qld
1–2.5 m × 1–2 m   May–July
  Small to medium **shrub** with glabrous young growth; **branches** spreading to ascending, many; **branchlets** spreading to ascending, many, slender, fern-like with many small leaves; **leaves** 0.5–1 cm × 0.3–0.5 cm, broadly elliptic, nearly sessile, spreading to ascending, crowded, more or less overlapping, dark green and shiny above, paler below, stiff, apex pointed; **flowers** small, cream, monoecious, solitary, sessile in upper axils, somewhat insignificant; **capsules** to about 0.6 cm long.
  This species has ornamental foliage somewhat like that of the maidenhair-fern. It occurs in Malesia and New Guinea, and extends to Queensland where it is found in the Burke, Cook, North Kennedy, South Kennedy and Port Curtis Districts. Often grows in rocky gorges on granite rocks. Quick-growing and suitable for seasonally dry tropical regions and subtropical areas. May succeed in cooler regions if grown in protected sites. Soils should be well drained with a sunny or semi-shaded aspect. May tolerate light frosts. Foliage is moderately dense and could be useful as low screening as well as for general planting. Worth trying as an indoor plant. Propagate from seed or from cuttings of firm young growth.

Phyllanthus pusillifolius S. Moore
(very small leaves)
Qld   Stick Bush
1–2 m × 0.5–1 m   May–July
  Dwarf to small **shrub** with glabrous young growth; **branches** erect; **leaves** 0.1–0.5 cm × 0.1–0.3 cm, elliptic to obovate, sessile, spirally arranged, clustered, more or less overlapping, apex blunt; **flowers** to about 0.2 cm across, white to creamy yellow, monoecious, solitary, sessile in upper axils, somewhat insignificant.
  This somewhat insignificant species occurs in north-eastern Qld where it is found on rainforest margins. It is cultivated to a limited degree and requires a shady, well-drained site with acidic soils for best results. Pruning may be helpful in promoting bushier growth. Propagate from seed or from cuttings of firm young growth.

Phyllanthus hirtellus   W. R. Elliot

Phyllanthus pusillifolius × .6

Phyllanthus lamprophyllus   W. R. Elliot

Phyllanthus reticulatus Poir.
(netted veins)
Qld, WA, NT
2–5 m × 1.5–3 m   May–Oct

Medium to tall **shrub** or sometimes a **climber**, usually with glabrous young growth; **branches** spreading to ascending; **branchlets** glabrous or faintly hairy; **leaves** 2.2–7 cm × 1.6–4 cm, elliptic to broadly elliptic, deep green and usually glabrous above, paler below, venation distinct, apex blunt; **stipules** to about 0.3 cm long, brown; **flowers** small, greenish, male and female flowers together in axillary clusters, on short stalks, can be profuse but rarely very conspicuous; **sepals** usually 5; **capsules** to 0.4 cm × 0.8 cm, initially succulent, red, purple or black when ripe; **seeds** reddish.

A tropical species which often is found along creeks and waterways. Rarely encountered in cultivation, it has limited appeal but is suitable for seasonally dry tropical areas and may succeed in subtropical regions. Needs moderately well-drained soils in a semi-shaded or sunny site. Pruning may be necessary to promote bushy growth. Propagate from seed or from cuttings of firm young growth.

P. reticulata is similar to and sometimes confused with *Flueggea virosa*, which has flowers with 4–7 sepals and brown seeds.

Phyllanthus saxosus F. Muell. — see *P. gunnii* Hook. f.

Phyllanthus subcrenulatus F. Muell.
(somewhat crenulate)
Qld, NSW
0.6–2.5 m × 0.5–1.5 m   Sept–Feb; also sporadic

Monoecious dwarf to medium **shrub** with glabrous young growth; **branches** spreading to erect; **branchlets** glabrous, angular; **leaves** 0.8–6 cm × 0.4–2 cm, lanceolate to ovate, more or less sessile, spreading to ascending, deep green above, paler below, glabrous, flat, margins faintly scalloped, apex pointed; **flowers** small, greenish with white borders, solitary or few together on stalks to 1 cm long, can be profuse but rarely very conspicuous; **capsules** about 0.6 cm across, globular, green, glabrous.

This species occurs in south-eastern Qld and north-eastern NSW where it inhabits rocky sites in dry forests and woodland. Evidently cultivated to a limited degree, it is best suited to subtropical regions but may adapt to temperate areas. Plants require well-drained acidic soils with a sunny or semi-shaded aspect. Hardy to light frosts. Propagate from seed or from cuttings of firm young growth.

Phyllanthus thymoides (Müll. Arg.) Müll. Arg. —
  see *P. hirtellus* F. Muell. ex Müll. Arg.

PHYLLOCLADUS A. & L. C. Rich. ex Mirb.
(from the Greek *phyllon*, leaf; *clados*, branch; referring to the leaf-like branchlets)
Phyllocladaceae (alt. *Podocarpaceae*)   Celery Pines
Monoecious or dioecious **shrubs** or **trees**; **bark** smooth; **branches** whorled; **branchlets** flattened; **leaves** spirally arranged, more or less deciduous; **male cones** cylindrical, stalked, in terminal clusters; **female cones** with spirally arranged scales which become pink to red as seeds develop; **seed** half-covered with white aril.

A small genus of about 6 species, represented in Philippines, Borneo, New Guinea, NZ and 1 endemic species in Tas.

Phyllocladus aspleniifolius (Labill.) Hook. f.
(leaves similar to the genus *Asplenium*)
Tas   Celery-top Pine
4–18 m × 1.5–5 m
Monoecious or dioecious, slow-growing, tall **shrub** or small to medium **tree**; lower **branches** whorled and horizontal; **cladodes** 1.5–8 cm long, flattened, leaf-like, opposite, alternate or whorled, green, thick, leathery, narrowing to base, cuneate, margins coarsely toothed or lobed; **leaves** to about 1 cm long on seedlings, reduced to small deciduous scales on mature plants; **male cones** to about 2 cm long, up to 10 clustered at ends of branches; **female cones** to about 1 cm long, borne on lower part of cladodes, often 3–4 together; **bract scales** become pink to red and partially envelop the seed.

A slow-growing and extremely long-lived species which is endemic in Tas where it grows from sea-level to about 900 m altitude. It is widely distributed and common in wet sclerophyll forests and temperate rainforest. Slow-growing in cultivation and is best suited to temperate regions. Plants need moisture-retentive but well-drained acidic soils in a partially sunny or shaded site. Application of organic mulches is beneficial. Plants may need supplementary watering during extended dry periods. They are hardy to frost and snowfalls. An excellent container plant while young, it is also used in bonsai/penjing. Bead-like nodules on the fine roots have an important symbiotic relationship with mycorrhiza, which are essential for

Phyllocladus aspleniifolius   D. L. Jones

good growth. The timber is highly prized because it shrinks only marginally on drying. It has been used for flooring, boat-building and wood-turning. Propagate from fresh seed, which can be difficult to germinate, or from cuttings of firm semi-woody growth. The cuttings should strike if mycorrhiza are present.

PHYLLOCLADACEAE (Pilger) Core
A small family of gymnosperms consisting of about 8 species in the genus *Phyllocladus*. They are distributed from the Philippines and Malaysia to New Zealand. There is 1 species in Australia. They are tall shrubs or trees with unusual, flattened leaf-like stems which are called phylloclades. The seed is subtended by a prominent, fleshy aril.

PHYLLODIUM Desv.
(from the Greek *phyllon*, leaf; *ium*, small; referring to the small leaves)
Fabaceae (alt. *Papilionaceae*)
Shrubs; **leaves** alternate, pinnate, with 3 leaflets; **stipules** free; **inflorescence** racemes or panicles, clustered in axils, with small leaf-like bracts; **flowers** pea-shaped, white or yellow, rarely purple; **calyx** 5-lobed; **stamens** 10, 9 joined, 1 free; **style** glabrous except for base; **fruit** a pod, with 1–7 joints.

A small tropical genus of about 6 species which has its distribution in southern Asia and northern Australia where there are 2 species. At one stage this genus was included in *Desmodium*. Propagation is from seed or cuttings. Seed requires pre-sowing treatment, see Treatments and Techniques for Difficult to Germinate Species, Volume 1, page 205. Cuttings of young but firm growth are worth trying.

Phyllodium pulchellum (L.) Desv.
(beautiful)
Qld, WA, NT
0.6–2.5 m × 0.6–2 m   Jan–March

Dwarf to small **shrub** with densely white to greyish-hairy young growth; **branches** many, spreading to ascending; **branchlets** densely hairy; **leaves** to about 18 cm long, pinnate; **leaflets** 3, 1–17.5 cm × 1–8 cm, terminal leaflet much longer than lateral ones, ovate, greyish-green, faintly hairy above and below, venation prominent, apex blunt; **stipules** to about 0.8 cm long, triangular, dryish; **racemes** to 50 cm long, with 3–6 flowers subtended by obovate to circular bracts; **flowers** pea-shaped; **standard** to about 0.7 cm × 0.5 cm, white to pale yellow; **wings** and **keel** white to pale yellow; **pods** to about 0.8 cm × 0.4 cm, oblong, often curved, with 1–3 constricted joints.

In Australia, this species is regarded as rare and is known from the Cook District, Qld, West Gardner District, WA, and the Top End, NT. It also occurs in PNG, Indonesia, India and China. It has interesting hairy young growth. Suited to seasonally dry tropical regions. Plants should adapt to a range of well-drained acidic soils. They can become open but pruning at an early stage should promote a bushy framework. Propagate from seed or possibly from cuttings of firm young growth.

Desmodium pulchellum (L.) Benth. is a synonym.

An apparently undescribed species from near Irvinebank in north-eastern Qld grows to about 1.5 m × 1.5 m. It has smaller leaves, clusters of up to about 15 flowers in racemes to about 20 cm long, and larger pods to about 3 cm × 0.6 cm with 3–4 constricted joints. It will need a shaded but warm site with moisture-retentive but well-drained soils. Plants should be hardy to moderate frosts.

PHYLLOGLOSSUM Kunze
(from the Greek *phyllon*, leaf; *glossa*, tongue; referring to the tongue-like leaves)
Lycopodiaceae

A monotypic genus which is endemic in Australia and New Zealand.

Phylloglossum drummondii Kunze
(after James Drummond, first government naturalist of WA)
NSW, Vic, Tas, SA, WA   Pigmy Clubmoss
0.02–0.06 m tall

Tiny terrestrial **fern ally**, growing in loose colonies; **rootstock** tuberous; **leaves** 5–10 in an erect tuft, 1–2 cm × 0.1 cm, linear, terete, fleshy, green to yellowish-green; **fruiting spike** 3–5 cm tall, ending in an ovoid, yellowish, cone-like structure about 0.6 cm long; **cone scales** overlapping; **sporangia** yellow.

Widely distributed in temperate regions. It grows in colonies, which may sometimes be extensive, on wet, peaty soil in heathland and around the margins of large rock sheets. Plants are deciduous with one or more replacement tubers being formed as the plant matures and dies back. It can be grown in a pot for a few years but tends to fade away. Plants have a mycorrhizal relationship and may need annual repotting into a potting mix similar to that needed for terrestrial orchids (see *Diuris* entry Volume 3, page 304 for example). Propagation is by natural increase.

PHYLLOTA (DC.) Benth.
(from the Greek *phyllon*, leaf; *ous, otus,* ear; referring to the leaf-like bracteoles of some species)
Fabaceae (alt. *Papilionaceae*)

Shrubs; **branches** terete; **leaves** alternate or irregular, simple, margins revolute, entire; **stipules** minute or absent; **flowers** pea-shaped, solitary or clustered in upper axils or terminal; **bracteoles** leaf-like, often persistent; **calyx** 5-lobed, upper 2 broadest and sometimes united to form a lip; **standard** circular; **wings** oblong; **stamens** 10, free; **fruit** an ovoid pod.

A small endemic genus of about 10 species which has representation in Qld, NSW, Vic, Tas and WA. Plants are found mainly in heathland, woodland and dry sclerophyll forests. Some species such as *P. phylicoides* can be very floriferous. The genus is closely allied to *Pultenaea* which has fused stipules and to *Aotus* which lacks bracteoles.

Phyllota species are grown mainly by enthusiasts and they are uncommon in cultivation.

Propagation is from seed or cuttings. The seed needs to be pre-treated before sowing to soften or crack the seed coating. See Treatments and Techniques to Germinate Difficult Species, Volume 1, page 205. Cuttings of firm current season's growth usually strike readily.

Phyllota aspera (DC.) Benth. =
P. phylicoides (Sieber ex DC.) Benth.

Phyllota barbata Benth.
(bearded)
WA
0.3–0.7 m × 0.3–1 m   possibly sporadic
                                                    throughout year

Dwarf **shrub** with very hairy young growth; **branches** many, spreading to ascending; **branchlets** hairy; **leaves** 0.3–1.2 cm × about 0.1 cm, linear, alternate, spreading to erect and recurved to incurved, deep green, glabrous or faintly hairy or scabrous, margins strongly revolute, apex blunt; **flowers** pea-shaped, to about 1 cm across, solitary, in upper axils, often forming leafy heads, can be profuse and very conspicuous; **standard** and **wings** yellow to orange; **keel** purplish-red; **pods** to about 0.6 cm long, ovoid, inflated.

A showy, dwarf species which deserves to be better known in cultivation. It occurs in the most south-eastern Darling and far-western Eyre Districts where it inhabits swampy, sandy soils. Best suited to temperate regions. Plants require acidic soils which are moisture-retentive but still moderately well drained. A semi-shaded site is preferred but plants tolerate plenty of sunshine. Hardy to light frosts. It has potential for use in gardens and containers. Propagate from seed or from cuttings.

Phyllota comosa (DC.) Benth. =
P. phylicoides (Sieber ex DC.) Benth.

Phyllota diffusa (Hook. f.) F. Muell.
(spreading)
Tas
0.1–0.4 m × 0.3–0.7 m   Oct–Nov

Spreading dwarf **shrub** with hairy young growth; **branches** many, spreading to ascending; **branchlets** hairy; **leaves** 0.5–0.8 cm × about 0.1 cm, linear, appearing terete because of strongly revolute margins, spreading to ascending, moderately crowded, deep green and glabrous above, paler and hairy below, apex ending in a small point; **flowers** pea-shaped, solitary or a few in upper axils, often profuse and very conspicuous; **standard** to about 0.8 cm across, bright yellow with reddish or rarely green markings; **wings** bright yellow; **keel** reddish; **pods** to about 0.5 cm long, ovoid, inflated.

A handsome, dwarf species from the heaths of the north-eastern and eastern coastal regions where it grows in sandy soils. Rarely cultivated, it needs excellently drained acidic soils and a sunny or semi-shaded aspect. Plants will be best suited to temperate regions. Pruning young plants will promote dense growth. Hardy to moderate frosts. Suitable for gardens and containers. Propagate from seed or from cuttings.

Phyllota gracilis Turcz.
(slender)
WA
unknown   unknown

Shrub with hairy young growth; **branches** slender, long; **branchlets** terete, hoary; **leaves** to about 0.3 cm long, linear, greyish-hairy, margins strongly revolute, apex blunt; **flowers** pea-shaped, dark red, solitary on slender axillary stalks to about 0.5 cm long; **standard** about 0.5 cm tall; **wings** shorter than standard; **keel** about 0.5 cm long; **calyx** densely hairy; **pods** about 0.4 cm long, ovoid, faintly hairy;

An extremely poorly known species from the south western region. Plants are likely to need very well-drained soils in a sunny or semi-shaded site. Propagate from seed or from cuttings.

Phyllota grandiflora Benth.
(large flowers)
NSW
0.6–1.3 m × 0.5–1 m   Nov–Feb

Dwarf to small **shrub** with very hairy young growth; **branches** many, spreading to erect; **branchlets** hairy; **leaves** 0.6–2 cm × 0.1–0.2 cm, linear, alternate, spreading to ascending or reflexed, can be crowded, dark green, becoming glabrous, margins revolute, apex blunt or pointed; **flowers** pea-shaped, in leafy spikes to about 3 cm across at ends of branches, often profuse and very conspicuous; **standard** to about 1.5 cm across, bright yellow with reddish markings; **wings** bright yellow; **keel** yellow with greenish markings; **pods** not known.

This floriferous *Phyllota* warrants attention. It occurs in the Central Coast District, between the Hawkesbury River and Sydney Harbour, where it grows in heath or dry sclerophyll forest on Hawkesbury sandstone. Apparently very rare in cultivation. Plants should adapt to well-drained acidic soils with a semi-shaded or sunny site in temperate and subtropical regions. Frost tolerance is not known. Worth trying as a general garden plant or in containers. Propagate from seed or from cuttings.

Phyllota humifusa Benth.
(spreading on the ground)
NSW   Dwarf Phyllota
0.1–0.2 m × 0.6–1.5 m   Oct–Feb

Spreading dwarf **shrub** with hairy young growth; **branches** many, prostrate to ascending; **branchlets** initially hairy, usually short; **leaves** 0.3–1 cm × up to 0.1 cm, linear, alternate, sometimes on one side of branchlets, spreading to ascending, moderately crowded, becoming glabrous, margins revolute, apex blunt or pointed, ending in a small point; **flowers** pea-shaped, in leafy spikes to about 2 cm across, at ends of branches, often profuse and conspicuous; **standard** to about 0.8 cm across, orange-yellow to reddish-brown; **wings** yellow; **keel** reddish-brown; **pods** about 0.5 cm long, with long hairs.

A very attractive, ground-hugging species which is cultivated to a limited degree. It inhabits sandy shale soils of dry sclerophyll forests in the southern Blue Mountains where it is regarded as endangered. Plants need freely draining acidic soils with a semi-shaded or partially sunny aspect. They tolerate heavy frosts and light snowfalls. Worth trying in gardens, rockeries and containers. Propagate from seed or from cuttings.

Phyllota luehmannii F. Muell.
(after Johann. G. W. Luehmann, early Victorian Government Botanist)
WA   Mop Pea
0.2–1 m × 0.3–1.3 m   July–Nov

Dwarf **shrub** with densely woolly-hairy young growth; **branches** many, spreading to erect; **branchlets** terete, densely hairy; **leaves** 0.3–1 cm × about 0.1 cm, linear, sessile, ascending to erect, crowdwd, deep green, with greyish to pale golden hairs, margins revolute, apex pointed; **flowers** pea-shaped, in leafy terminal heads to about 3 cm across, often profuse and very conspicuous; **standard** to about 1.5 cm across, golden-yellow to orange with deep red basal markings; **wings** golden-yellow to orange; **keel** deep reddish; **pods** to about 0.8 cm × 0.4 cm.

An interesting, highly desirable species from the drier regions of the south-west. It occurs in the Avon, southern Austin, Roe and Coolgardie and Helms Districts where it grows on sandy soils in mallee communities. Should do best in warm temperate and semi-arid regions. Soils will need to be extremely well-drained and have a sunny or semi-shaded aspect. Plants may succeed in containers in temperate regions. They tolerate moderate frosts. Propagate from seed or from cuttings.

This species may be transferred to another genus as a result of botanical revision.

Phyllota phylicoides (Sieber ex DC.) Benth.
(similar to the genus *Phylica*)
Qld, NSW   Heath Phyllota
0.6–1.5 m × 0.4–1.3 m   Aug–Nov; also sporadic

Dwarf to small **shrub** with hairy young growth; **branches** spreading to erect, can be slender; **branchlets** initially hairy; **leaves** 0.6–1.5 cm × to about 0.15 cm, narrowly linear, alternate, spreading to ascending, especially crowded near ends of branches,

Phyllota pleurandroides

Phyllota phylicoides × .8

mid to deep green, becoming glabrous, margins revolute, apex blunt to pointed; **flowers** pea-shaped, in terminal leafy spikes to about 3 cm across, often profuse and very conspicuous; **standard** to about 1 cm across, yellow to orange sometimes with reddish markings; **wings** and **keel** yellow to orange; **pods** to about 0.5 cm long, inflated, with long hairs.

A widely distributed, ornamental species which occurs from south-eastern Qld to south-eastern NSW, along the coast and eastern tablelands, where it grows in dry sclerophyll forest and heath. It is the most commonly grown species. Plants can be leggy but pruning when young, and also during or after flowering, promotes bushy growth. Plants need acidic soils and a semi-shaded or moderately sunny site is suitable. They are hardy to moderately heavy frosts. First cultivated in England during 1824. Bees visit the flowers for pollen. Propagate from seed or from cuttings of firm young growth, which strike readily.

P. aspera (DC.) Benth. and *P. comosa* (DC.) Benth. are synonyms.

Phyllota pleurandroides F. Muell.
(similar to the genus *Pleurandra*)
Vic, SA   Heathy Phyllota
0.4–1.5 m × 0.6–2 m   sporadic throughout the year
Dwarf to small, often suckering **shrub** with hairy young growth; **branches** spreading to ascending; **branchlets** hairy; **leaves** 0.6–1 cm × to about 0.1 cm, linear, alternate, ascending to erect, often clustered especially near ends of short branchlets, along branches, hairy, apex pointed and recurved; **flowers**

pea-shaped, usually solitary or paired in upper axils and can be partially hidden, rarely profuse; **standard** about 0.5 cm across, pale yellow to yellow; **wings** oblong, yellow; **keel** yellow with red; **pods** to about 0.7 cm long, brown, hairy.

Heathy Phyllota is moderately attractive. It occurs in sandy soils of south-western Vic and south-eastern SA. Rarely cultivated, it will do best in freely draining acidic soils with a sunny or semi-shaded aspect. Regular pruning promotes bushy growth and the hairy new foliage can be an attractive feature. Hardy to moderate frosts and extended dry periods. The suckering growth habit makes this species useful for erosion control in sandy soils. Propagate from seed, or from cuttings of firm young growth, which strike readily.

Phyllota remota J. H. Willis
(remote)
Vic, SA   Slender Phyllota
0.3–0.6 m × 0.3–0.6 m   mainly Oct–July
Dwarf **shrub** with hairy young growth; **branches** ascending to erect, slender; **branchlets** initially hairy; **leaves** 0.5–1 cm × about 0.1 cm, linear, scattered, ascending to erect, usually with hairs, somewhat glandular, margins recurved, apex straight-pointed; **flowers** pea-shaped, solitary, axillary, conspicuous or concealed by papery bracteoles; **standard** to about 0.6 cm across, reddish; **wings** and **keel** reddish; **pods** not known.

A somewhat insignificant, non-suckering phyllota from south-western Vic and extending through south-eastern SA to the Eyre Peninsula. Rarely cultivated, it will need very well-drained soils and a sunny or semi-shaded aspect. Plants are hardy to moderate frosts and extended dry periods. Propagate from seed or from cuttings.

Phyllota squarrosa (Sieber ex DC.) Benth.
(spreading widely and bent backwards)
NSW   Dense Phyllota
0.3–0.8 m × 0.5–1.3 m   Dec–March
Spreading, suckering, dwarf **shrub** with hairy young growth; **branches** many, prostrate to ascending; **branchlets** initially hairy, usually becoming glabrous; **leaves** 0.6–1.5 cm × about 0.1 cm, linear, alternate, spreading to ascending, often crowded, sometimes clustered, becoming glabrous, margins revolute, apex with a recurved yellow point; **flowers** pea-shaped, can form leafy spikes to about 2 cm across at ends of branchlets, rarely profuse but always conspicuous; **standard** to about 1 cm across, yellow-orange; **wings** yellow; **keel** orange to reddish; **pods** not seen.

This species was first cultivated in England in 1824 but evidently is rarely grown now. It occurs mainly in the Lower Blue Mountains where it grows on dry ridges in dry sclerophyll forest and heathland. Warrants greater attention by gardeners and designers. It has potential for a range of uses due to its suckering growth habit. Should adapt to a range of freely draining acidic soils in temperate regions. Hardy to moderately heavy frosts. Propagate from seed, or cuttings of sucker growth, which should give excellent results.

PHYMATOCARPUS F. Muell.
(from the Greek *phymatos*, swelling; *carpos*, seed; referring to the warty appearance of fruits of the type species)
Myrtaceae
   Shrubs; **leaves** alternate, small, spreading, aromatic
when crushed; **inflorescence** terminal, globular heads
with new growth emerging from apex during or after
flowering; **flowers** 5-merous, sessile, pinkish-mauve,
mauve, cream and mauve or rarely all cream, often
many are male; **stamens** many, much longer than
petals, united at base, in 5 bundles opposite petals;
anthers basifixed; **fruit** a woody capsule in globular
clusters.
   Phymatocarpus is endemic in south-western WA and it
comprises 2 species. Both are cultivated to a limited
degree by enthusiasts. They have proved adaptable
over a wide range of temperate regions.
   Propagation is from seed or from cuttings. Seed
does not require any pre-sowing treatment. It is best
barely covered and the Bog Method (see Volume 1,
page 204 for further information) is usually very successful. Cuttings of firm new season's growth usually
provide best results. Application of a hormone rooting
powder or liquid can be beneficial.
   Beaufortia and *Regelia* are closely allied genera.
Beaufortia is distinguished by all species (except *B.
sparsa*) having opposite leaves and staminal filaments
of unequal length. *Regelia* species have opposite leaves
and filaments of equal length.

Phymatocarpus maxwellii F. Muell.
(after George Maxwell, 19th-century botanical collector, WA)
WA
1.3–2.5 m × 1–2 m   Sept–Nov

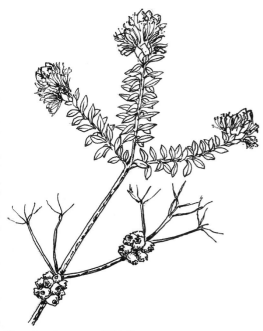

Phymatocarpus maxwellii × .9

Phymatocarpus porphyrocephalus   W. R. Elliot

   Small to medium **shrub** with faintly hairy young
growth; **branches** many, spreading to erect; **branchlets**
initially hairy; **leaves** 0.3–0.7 cm × to about 0.2 cm, narrowly lanceolate to elliptical, alternate, short-stalked,
spreading to ascending, usually recurved, rarely
crowded, green to greyish-green, apex blunt; **flowerheads** to about 1.2 cm across, globular, terminal, compact, often profuse and very conspicuous; **flowers**
about 0.4 cm across, pinkish-mauve to pinkish-purple;
stamens about 0.4 cm long in bundles of 3, **anthers**
yellow; **capsules** about 0.25 cm diameter, cylindrical,
clustered.
   A floriferous and attractive-foliaged species from the
Eyre and Roe Districts where it occurs in sandy soils of
the heathland. Does best in warm temperate regions
in freely to moderately draining acidic soils. Needs
plenty of sunshine to flower well but also tolerates a
semi-shaded site. Plants are hardy to moderate frosts
and they withstand extended dry periods. Responds
well to pruning or clipping. Butterflies visit the
flowers. Propagate from seed or from cuttings of firm
young growth which strike readily.
   Superficially this species is very similar to many of
the small 'pom-pom' flowered *Melaleuca* species, and
sometimes nurseries and growers are confused.
   P. sparsiflorus (W. Fitzg.) C. A. Gardner and *Regelia
sparsiflora* W. Fitzg. are synonyms.

Phymatocarpus porphyrocephalus F. Muell.
(purple heads)
WA
1–2 m × 1–2.5 m   Aug–Dec
   Small **shrub** with glabrous young growth; **branches**
many, spreading to ascending; **branchlets** glabrous,
greyish; **leaves** 0.2–0.6 cm × 0.2–0.5 cm, broadly ovate
to nearly circular, alternate or sometimes more or less
opposite, spreading to ascending, sometimes crowded
near ends of branchlets, glabrous, grey to grey-green,
flat to concave, apex blunt; **flowerheads** to about 2 cm
across, globular, terminal, compact, rarely profuse but

always conspicuous; **flowers** to about 1 cm across, pink, purple, yellow, cream or sometimes purple and cream; **stamens** to about 0.6 cm long, in bundles of 11–15; **anthers** yellow; **capsules** about 0.5 cm diameter, semi-globular, warty.

A spreading species from coastal and inland regions of the northern Darling and Irwin Districts where it grows in sandy soils. It is adaptable in cultivation and will grow and flower best in warm temperate regions. Needs a warm to hot, protected site in cooler zones. Suitable for clay loams and sandy soils which drain well. Plants can become open but pruning at an early stage, and during or after flowering, promotes bushy growth. Responds well to clipping. Hardy to moderate frosts and limited coastal exposure. Withstands extended dry periods. Propagate from seed, or from cuttings of firm young growth, which strike readily.

PHYMATOSORUS Pic. Serm.
(from the Greek *phymatos*, swelling, *sorus*, heap; referring to the fertile parts of the fern)
Polypodiaceae

Epiphytic, lithophytic or terrestrial **ferns** growing in clumps or forming spreading patches; **rootstock** creeping, scaly; **fronds** simple or pinnately lobed; **lobes** widely spreading to erect; **sori** lacking indusia, round to elliptical, large, usually in rows.

A genus of about 20 species distributed in tropical regions. Five species occur in Australia, some confined to the tropics and others extending to temperate regions. The Australian species are all decorative ferns which are worthy of cultivation. Propagation is from spores or by division of clumps.

Phymatosorus diversifolius (Willd.) Pic. Serm.
(with variable leaves)
Qld, NSW, Vic, Tas   Kangaroo Fern
5–50 cm tall

Epiphytic, lithophytic or rarely terrestrial **fern** forming large clumps; **rootstock** long-creeping, about 0.6 cm across, scaly, becoming strongly glaucous; **fronds** 5–50 cm × 4–30 cm, extremely variable in shape from simple and undivided to deeply lobed, dark green, leathery, shiny, margins entire or wavy, lobes widely spreading; **sori** round, brown to yellowish-brown, in a single row on each side of the indusia.

Widely distributed from south-eastern Qld to southern Tas and also in Norfolk Island and New Zealand. Forms extensive conspicuous colonies, clothing tree trunks and branches, tree ferns, rocks, boulders and cliff faces and sometimes in sandy soils. Often very prominent at high altitudes but also frequent in coastal districts where it may inhabit cliffs overlooking the sea. Plants in exposed situations commonly have stunted yellowish fronds. Frond shape is extremely variable in a colony but sometimes extensive patches may have mostly simple fronds. A very popular fern which adapts well to cultivation. Excellent in hanging containers and pots but can also be successful in the ground in well-drained soil. Tolerates moderate frosts. Propagate by division or from spores.
Microsorum diversifolium (Willd.) Copel. is a synonym.

Phymatosorus grossus (Langsd. & Fisch.) Brownlie
(thick, coarse)
Qld
10–60 cm tall

Terrestrial or lithophytic **fern** forming spreading colonies; **rootstock** long-creeping, about 0.8 cm across, scaly towards the apex; **fronds** 10–60 cm × 10–40 cm, erect, rarely simple, usually lobed, pale green, bluish-green or yellowish-green, thin-textured, lobes spreading widely, apical lobe often long and strap-like; **sori** round, brown, in 1 or 2 irregular rows on either side of the midrib.

Widespread but disjunct across northern Australia and extending to New Caledonia, Fiji and various Pacific islands. In Australia, it forms colonies in coastal scrubs and monsoon rainforest thickets usually in shady locations but sometimes exposed. An excellent groundcover fern for tropical and subtropical regions. Also makes an attractive container plant. Requires the protection of a heated glasshouse in cold climates. Propagate by division or from spores.
Microsorum grossum (Langsd. & Fisch.) S. B. Andrews is a synonym.

Phymatosorus membranifolius (R. Br.) Pic. Serm.
(with very thin-textured leaves)
Qld   Pimple Fern
0.4–1.2 m tall

Epiphytic, lithophytic or terrestrial **fern** growing in small clumps; **rootstock** creeping, about 1.2 cm across, green, fleshy, scaly towards the apex; **fronds** 40–120 cm × 20–60 cm, erect, variously lobed, rarely simple, greyish-green to dark green, thick and leathery, margins usually wavy, lobes with a long drawn out apex; **sori** round, brown, in one row on each side of the midvein, sunken on the lower surface, raised like a pimple on the upper surface.

Occurs in north-eastern Qld growing on trees or rocks in lowland rainforest, and occasionally plants are found growing on the ground in sandy soil. An interesting fern which grows readily in a pot or basket. Requires warm, humid conditions and needs the protection of a heated glasshouse in southern Australia. Propagate by division or from spores.
Microsorum membranifolium (R. Br.) Ching is a synonym.

Phymatosorus scandens (G. Forst.) Pic. Serm.
(climbing, creeping)
Qld, NSW, Vic   Fragrant Fern
10–50 cm tall

Epiphytic or lithophytic **fern** forming large spreading clumps; **rootstock** long-creeping, about 0.3 cm across, densely scaly; **fronds** 10–50 cm × 10–20 cm, variable in shape from simple to deeply lobed, thin-textured, dark green, fragrant, lobes falcate, margins often wavy; **sori** round, brown, in a single row close to the margin.

Distributed from north-eastern Qld to eastern Victoria and also in Norfolk Island and New Zealand. Forms extensive, dense colonies on tree trunks, tree-fern trunks and rocks, usually in cool, moist situations. The fronds, which are usually pendulous, have a distinct fragrance which is more obvious to some people than others. An excellent fern for hanging baskets and other containers. Best in shady, humid conditions with ample air movement. Plants may be established as a groundcover in a shady location. Tolerates light to moderate frosts. Propagate by division or from spores.

Microsorum scandens (G. Forst.) Tindale is a synonym.

Phymatosorus simplicissima (F. Muell.) Pic. Ser.
(very simple)
Qld
5–25 cm tall

Epiphytic or lithophytic **fern; rootstock** long-creeping, slender, covered with brownish scales; **sterile fronds** 3–8 cm × 1.5–2.5 cm, ovate-lanceolate to lanceolate, erect, on long stalks, dark green, with conspicuous venation, margins scalloped, apex drawn out; **fertile fronds** to 25 cm × 1.5 cm, linear to narrowly ovate; **sori** large, circular, brown, situated midway between the margin and the midrib.

Endemic in north-eastern Qld. growing on trees and rocks at high altitudes. Usually forms creeping strands or sparse clumps. An attractive fern for a treefern slab or in a small basket. It can also be grown in a pot of coarse, epiphytic mixture. Hardy in a bush house or unheated glasshouse in southern Australia. Propagate from spores or by division of the rhizomes.

Crypsinus simplicissimus (F. Muell.) S. B. Andrews is a synonym.

PHYSOPSIS Turcz.
(after the genus *Physa*, and from the Greek *opsis*; resemblance, appearance)
Chloanthaceae (alt. *Verbenaceae*)   Lamb's Tails

Woolly-hairy **shrubs; leaves** opposite, decussate, simple, sessile, entire, densely hairy; **inflorescence** elongated, woolly-hairy spikes; **corolla** tubular, with 4 lobes, glabrous, virtually enclosed by calyx; **calyx** tubular, 4-lobed, densely woolly-hairy; **fruit** usually single-celled, one-seeded, dry, indehiscent.

After recent botanical revision *Physopsis* now contains 5 species; 2 of these are very ornamental. The genus is endemic in south-western WA. Plants are uncommon in cultivation because of propagation difficulties but warrant greater attention for use in gardens and as cut flowers. Cut flowers have been harvested from the wild by the flower industry for many years. There is now a need for commercial plantations to be established. Propagation is from seed or cuttings. Seed germination results are not well documented. The seed will germinate without pre-sowing treatment but results are irregular. Further germination research would be very useful to ascertain which methods are likely to give best results. Cuttings of firm but fairly young growth can be struck but there are often problems encountered because of the densely hairy stems. Application of hormone rooting powders or liquids can be beneficial.

Mallophora is allied but has mainly cymose flower-heads and the floral tube is hairy.

Physopsis chrysophylla (C. A. Gardner) Rye
(golden leaves)
WA
2–5 m × 1.5–3 m   Oct–Jan

Medium to tall **shrub; branches** ascending to erect, densely covered in brownish-yellow woolly hairs; **leaves** 1.3–3.7 cm × 0.5–1.5 cm, obovate, decussate, ascending to erect, short-stalked, wedge-shaped base, olive green and somewhat rough above, densely covered with golden-yellow hairs below, margins slightly recurved and often wavy and toothed towards the blunt apex; **inflorescence** of 1 or more spikes 3–5.5 cm × 0.5–0.8 cm, golden-yellow, terminal, cylindrical, sometimes in panicles, most conspicuous; **flowers** to 0.4 cm long, bell-shaped, golden-yellow, bluntly lobed; **calyx** 5- or rarely 4-lobed, exterior densely golden-hairy; **fruit** to about 0.2 cm across, somewhat globular, slightly hairy.

This is the tallest species of the genus. It inhabits the northern Irwin District in the Murchison River and Shark Bay region where it grows in red or yellow sandy soil which can overlie limestone. Rarely cultivated, it has high potential for semi-arid and warm temperate regions. Must have excellent drainage and plenty of sunshine. Hardy to moderate frosts. Tip pruning from an early stage will promote bushy growth. Propagate from seed or from cuttings of firm young growth.

Previously known as *Newcastelia chrysophylla* C. A. Gardner.

P. spicata is allied but has leaves with a white-hairy undersurface. *P. chrysotricha* also differs in having lanceolate leaves.

Physopsis chrysotricha (F. Muell.) Rye
(golden hairs)
WA
0.5–1 m × 0.6–1.5 m   Aug–Oct

Dwarf **shrub; branches** usually decussate, densely covered with yellow-brown hairs; **leaves** 1–2.5 cm × 0.15–0.3 cm, narrowly linear, usually in whorls of 3, ascending to erect, sessile, green, glabrous, sticky-glandular above, golden-hairy below, margins recurved, apex pointed; **inflorescence** terminal branched spikes, crowded to somewhat interrupted, single spikes 5–14 cm × 0.8–1 cm, woolly-hairy, golden-yellow; **flowers** to 0.4 cm long, tubular, colour is not recorded but most likely white to golden-yellow, bluntly lobed; **calyx** 5-lobed or rarely 4-lobed, densely woolly-hairy; **fruit** not seen.

This poorly known species is recorded from the arid regions of the Coolgardie and Helms Districts, but apparently has not been collected since the late 19th-century. If found it needs to be introduced into cultivation as a conservation measure. Plants will need excellent drainage and a warm to hot aspect. They should be hardy to moderate frosts. This species has potential for use in floriculture. Propagate from seed or from cuttings of firm young growth.

Previously known as *Newcastelia chrysotricha* F. Muell.

P. viscida is allied but differs in its stems, leaves and flowers having whitish-grey hairs.

Physopsis lachnostachya C. A. Gardner
(softly woolly spikes)
WA   Lamb's Tail
0.6–1.5 m × 0.5–1 m   Sept–Nov

Dwarf to small **shrub** with densely whitish-woolly young growth; **branches** spreading to erect, often

Physopsis spicata

Physopsis lachnostachya   D. L. Jones

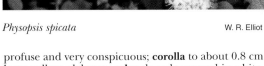

Physopsis spicata   W. R. Elliot

becoming more or less glabrous; **branchlets** terete, woolly-hairy; **leaves** 0.5–2 cm × 0.2–0.6 cm, narrowly ovate, opposite, sessile, spreading to ascending or sometimes erect, crowded on young branchlets, often overlapping, leathery, becoming deep green and more or less glabrous above, tawny-woolly hairs below, margins strongly revolute, apex pointed; **spikes** to about 9 cm × 1.5 cm, terminal, solitary or up to 3 together, compact, often profuse and very conspicuous; **corolla** about 0.5 cm long, yellow, glabrous; **calyx** densely covered with greyish- to white-woolly hairs.

A handsome shrub of the western Roe and southern Avon Districts where it grows in sandy soils which often overlie clay and contain lateritic gravel. Rarely encountered in cultivation and best suited to warm temperate regions. Needs moderately well-drained acidic soils and plenty of sunshine. Worth trying in large containers in a sheltered position in cooler areas. Plants can be straggly but tip pruning when young and cutting long-flowering flowering stems for indoor decoration helps promote a bushy framework. Hardy to moderate frosts and extended dry periods. Propagate from seed or from cuttings.

Physopsis spicata Turcz.
(with flower spikes)
WA   Hill River Lamb's Tail
0.4–0.8 m × 0.3–1 m   Oct–Feb

Dwarf **shrub** with densely woolly-hairy young growth; **branches** many, spreading to ascending; **branchlets** densely clothed with white-woolly hairs; **leaves** 0.8–3.3 cm × 0.2–1.3 cm, oblong to ovate-oblong, usually opposite, sometimes whorled, sessile, spreading to erect, sometimes crowded, faintly leathery, becoming glabrous to slightly rough above, densely white-woolly hairy below, margins recurved, apex blunt; **spikes** to about 4.5 cm × 1.5 cm, terminal, solitary or in clusters, compact or moderately so, often profuse and very conspicuous; **corolla** to about 0.8 cm long, yellow, glabrous; **calyx** densely covered in white-woolly hairs.

An eye-catching species from the western Avon, northern Darling and southern Irwin Districts. It occurs in sandy soils of heathland. This species has been a popular cut flower and needs to be cultivated for that purpose as well as for ornamental use. Plants require excellent drainage and maximum sunshine for best results. Recommended for warm temperate and semi-arid regions but may succeed in warm to hot protected sites or in containers in cooler areas. Plants are hardy to moderate frosts and withstand extended dry periods. They dislike extended wet periods. Propagate from seed or from cuttings.

Physopsis viscida (E. Pritzel) Rye
(sticky)
WA
0.5–2 m × 0.6–2.5 m   Sept–Nov; also sporadic

Dwarf to small **shrub**; **branches** densely covered in short grey hairs; **leaves** 1–3.5 cm × 0.1–0.4 cm, narrow-linear, in whorls of 3 or 4, ascending to erect, sessile, sticky and glabrous above, hairy, reticulate venation prominent below, margins recurved, apex pointed; **inflorescence** terminal, solitary spike 3–11.5 cm × 1–1.4 cm, or with pairs of short lateral spikes, whitish-grey and woolly-hairy; **flowers** to 0.4 cm long, bell-shaped, white to pale yellow, with 5 broad lobes; **stamens** and **style** not exserted; **calyx** 5-lobed, densely covered in whitish-grey hairs; **fruit** to about 0.25 cm long, obovoid, mainly glabrous.

P. viscida is found mainly in the region bounded by Comet Vale, Coolgardie and Kalgoorlie, where it grows in red sand, sandy clay soils, rocky sites and is often associated with lateritic gravels. For cultivation it will need very well-drained soils and plenty of sunshine. Hardy to moderate frosts. Worth trying in semi-arid and warm temperate regions. Propagate from seed or from cuttings of firm young growth.

Previously known as *Newcastelia viscida* E. Pritzel.

P. chrysotricha has some affinity but is readily distinguished by its yellow-brown hairs on young growth and its broader leaves.

PILEANTHUS Labill.

(from the Greek *pilos*, hat; *anthos*, flower; referring to the 2 bracteoles covering a bud)

Myrtaceae   Coppercups

Dwarf to small **shrubs**; **leaves** mainly opposite, simple, linear, entire, glabrous, aromatic when crushed; **inflorescence** terminal leafy corymbs, in upper axils, with 2 bracteoles covering each bud; **flowers** very showy; **petals** 5, spreading, fringed, the fringe exceeding the calyx; **calyx** 10-ribbed, 10 petal-like lobes, glabrous, entire; **stamens** 20, in a single row; **ovary** 1-celled; **fruit** a capsule.

A small genus of 3–5 highly attractive species. It is endemic in WA with its main distribution in northern parts of the south-western region. They are cultivated with varying degrees of success but they have excellent prospects because of their splendid floral displays. Plants are subject to attack by powdery mildew and grey mould especially in cool temperate regions (for control see Volume 1, page 173). They can also sweat badly in humid weather and are best grown in well-ventilated sites. Propagation is from seed or cuttings. Results from seed are often irregular. Reasearch into germination should enable some successful methods to be developed. Cuttings of firm young growth usu-ally strike readily. Application of hormone rooting compounds may be beneficial.

Pileanthus sp. aff. *filifolius* (Murchison)   B. W. Crafter

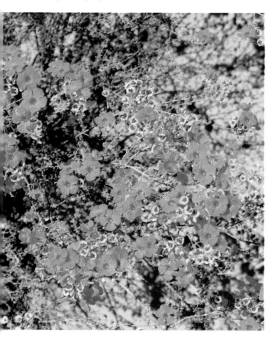

Pileanthus filifolius Meisn.

(thread-like leaves)

WA   Summer Coppercups

0.6–1.5 m × 0.6–1.5 m   Nov–Jan

Dwarf to small **shrub** with glabrous young growth; **branches** many, spreading to ascending; **branchlets** terete, slender, glabrous; **leaves** 0.4–1.3 cm × to about 0.2 cm, linear-terete or 3-angled, opposite, spreading to ascending, somewhat crowded especially near ends of branchlets, grey-green to green, apex pointed; **flowers** to about 2 cm across, deep red to scarlet, on very slender stalks 1.2–2.5 cm long, near ends of branchlets, often profuse and highly conspicuous.

In full flower, this species is very dramatic. It occurs in the Irwin and northern Darling Districts where it grows in sandy soils of heathland. To succeed it needs extremely well-drained acidic soils and plenty of sunshine. Best suited to warm temperate and semi-arid regions but can succeed in cooler areas if grown in a site which is sunny and has protection from cold winds. Highly suited to pot culture. Plants tolerate moderate frost and extended dry periods. Pruning when young promotes bushy growth. Propagate from seed, which can begin to germinate 20–70 days after sowing, or from cuttings of firm young growth, which usually strike readily.

There are a couple of apparently undescribed species closely allied to *P. filifolius* which have been cultivated at various times. One has bright orange-red flowers to about 2.5 cm across while the other has vibrant pink flowers. Both are highly desirable shrubs and need similar conditions to *P. filifolius*.

Pileanthus limacis Labill.

(snail, slug; referring to slug-like leaf shape)

WA   Coastal Coppercups

0.5–1.5 m × 0.6–2 m   Aug–Jan

Dwarf to small **shrub** with faintly hairy young growth; **branches** many, spreading to ascending; **branchlets** terete, glabrous; **leaves** 0.4–0.7 cm × to about 0.2 cm, linear, semi-terete, somewhat club-shaped, spreading to ascending, often incurved, moderately crowded, thick, smooth or glandular-warty, sometimes ciliate, apex very blunt; **flowers** to about 2 cm across, white to pale pink or sometimes salmon-red, on slender stalks to about 1 cm long, in upper axils, often profuse and highly conspicuous.

A lovely, floriferous species from coastal and slightly inland regions from Shark Bay to near Exmouth. Plants inhabit sand dunes. This species is uncommon in cultivation but it should do very well in warm temperate and semi-arid zones. Plants must have excellent drainage and maximum sunshine. Frost hardiness is not known but plants tolerate extended dry periods. Propagate from seed, which can begin to germinate 25–50 days after sowing, or from cuttings of firm young growth.

Pileanthus peduncularis Endl.

(with a stalk)

WA   Coppercups

0.4–1 m × 0.4–1.5 m   Aug–Dec

Dwarf **shrub** with glabrous young growth; **branches** many, spreading to ascending, twiggy; **branchlets** glabrous; **leaves** 0.2–0.5 cm × to about 0.2 cm, linear-terete or 3-angled, opposite, spreading to ascending, crowded, thick, glabrous, apex blunt; **flowers** about 1.5 cm across, orange, on slender stalks to about 2 cm long, often profuse and very conspicuous; **petals** prominently fringed.

A very charming species which has a wide distribution from near the coast to some distance inland. It occurs in the north-western Avon, northern Darling, Irwin and southern Carnarvon Districts. Plants are usually found on sandy soils which overlie laterite. Popular in cultivation but it has often proved short-lived in cool temperate regions. Does best in warm temperate and semi-arid regions. Needs good drainage and plenty of sunshine. Usually thrives in a fairly open site but appreciates other low plants in proximity. Hardy to moderate frosts and extended dry periods. Prone to attack by powdery mildew and grey mould. Propagate from seed, which can begin to germinate 25–80 days after sowing. Cuttings of firm young growth usually strike readily.

PILIDIOSTIGMA Burret

(from the Greek *pilidion*, a night-cap; *stigma*, stigma; referring to shape of the stigma)
Myrtaceae

Shrubs or small **trees; leaves** opposite, petiolate, venation pinnate; **stipules** present; **inflorescence** axillary, 1-flowered or raceme-like clusters; **flowers** 4- or 5-merous; **sepals** small, persistent; **petals** free; **stamens** many, in 2 or more rows; **ovary** 2–3-celled; **fruit** a succulent berry.

A small endemic genus comprised of 5 species. The main representation is in north-eastern Qld, with 2 species in south-eastern Qld and north-eastern NSW. All species have potential for cultivation but are uncommon in gardens and parks. Propagation is from fresh seed, which does not require pre-sowing treatment, or from cuttings of firm young growth, which

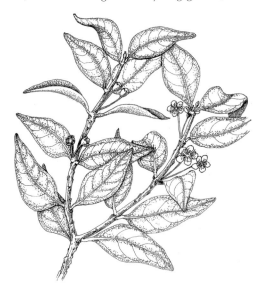

Pilidiostigma glabrum × .45

may be slow to produce roots. Application of hormone rooting compounds may be beneficial.

Pilidiostigma cuneatum Burrett =
    Archirhodomyrtus beckleri (F. Muell.) A. J. Scott

Pilidiostigma glabrum Burrett
(smooth, bare)
Qld, NSW   Plum Myrtle
3.5–5 m × 2.5–4 m   Oct–March

Medium to tall **shrub** with shiny glabrous young growth; **bark** wrinkled, brown, flaky on old branches; **branches** many, spreading to ascending; **branchlets** glabrous, initially purplish-brown; **leaves** 2–10 cm × 1–3.5 cm, elliptical to ovate, opposite, tapered to base, glabrous, dark green and shiny above, paler below, venation faint, apex pointed; **flowers** to about 1.5 cm across, white to pale pink, solitary or rarely paired, on slender stalks to about 2.5 cm long, rarely profuse; **stamens** prominent, white or pink, fluffy; **fruit** to about 1.7 cm × 1.5 cm, pear-shaped, purplish-black.

Plum Myrtle develops as a bushy shrub. It is confined to south-eastern Qld and north-eastern NSW. This species grows in rainforest and moist sclerophyll forest and is a common pioneering species in disturbed sites or neglected pasture. It adapts to cultivation in subtropical and temperate regions in a range of acidic soil types which are moist for most of the year. Tolerates sunshine as well as semi-shade. Appreciates organic mulches. Hardy to light frosts. Birds eat the fruits. Leaves can be subject to attack by psyllid insects which can be very difficult to control. Young plants can be used for indoor decoration. Propagate from seed, or from cuttings, which can be slow to strike.

Pilidiostigma parviflorum Burrett =
    Archirhodomyrtus beckleri (F. Muell.) A. J. Scott

Pilidiostigma recurvum (C. T. White) A. J. Scott
(bent backwards)
Qld
3–6 m × 2–4 m   Oct–March

Medium to tall **shrub** with glabrous young growth; **bark** pale grey; **branches** spreading to ascending; **branchlets** terete, glabrous; **leaves** to about 18 cm × 6.5 cm, lanceolate to broadly ovate, opposite, stalk about 1 cm long, tapering to base, thick and somewhat brittle, glabrous, dark green above, paler below, margins recurved, midrib prominent, venation faint, apex pointed; **flowers** to about 1.2 cm across, white, usually 2–6, in short axillary racemes; **stamens** many, white, fluffy; **fruit** to about 1 cm across, ovoid, black, ripe July.

A bushy species from Cape York Peninsula where it grows in the lowlands. Evidently not cultivated, it is worthy of greater attention. Plants are suited to tropical and subtropical regions and should adapt to a range of acidic soils in a sunny or semi-shaded site. Propagate from seed and possibly from cuttings.

Pilidiostigma rhytispermum (F. Muell.) Burrett
(wrinkled seed)
Qld
2–5 m × 2–3.5 m   March–Oct

Medium to tall **shrub** with hairy young growth; **branches** spreading to ascending; **branchlets** hairy, brownish; **leaves** 1.5–5 cm × 0.5–2 cm, elliptical to obovate, opposite, tapered to base, short-stalked, dark green above, paler below, glabrous except for hairy midrib, apex blunt to rounded; **flowers** to about 1.5 cm across, white, solitary on slender axillary stalks to 3 cm long, rarely profuse but always conspicuous; **stamens** many, white, fluffy; **fruit** to about 1 cm across, oblong, black, ripe March–April.

This attractive species is very compact in exposed sites but becomes open in semi-shade. It is from the Moreton and Wide Bay Districts in the south-east where it grows in moist but well-drained acidic soils. Well suited to cultivation in subtropical and temperate regions. Plants are hardy to light frosts. They respond well to pruning and are excellent for screening. May also be suitable for low hedging. Worth trying young plants indoors. Birds eat the fruits. Propagate from seed, or from cuttings of firm young growth, which can be slow to strike.

Pilidiostigma tetramerum L. S. Sm.
(four parts)
Qld
1.5–3 m × 2–3 m   Oct–Jan
Small to medium **shrub** with glabrous bronze to reddish young growth; **branches** many, spreading to ascending; **branchlets** terete to 4-angled, glabrous; **leaves** 1.5–9 cm × 0.6–2.5 cm, lanceolate to ovate-lanceolate, opposite, tapered to base, glabrous, thin-textured, dark green above, paler below, margins wavy, venation visible, apex pointed; **flowers** to about 1 cm across, creamy-white, 4–12 in axillary racemes, moderately profuse and conspicuous; **petals** 4; **stamens** white, many, fluffy; **calyx** 4-lobed; **fruit** 1–2 cm × up to 1.7 cm, ovoid, blackish, edible, ripe Oct–Jan.

A poorly known species which inhabits the lowland and highland rainforest in the Cook District. Evidently not cultivated, it has potential for use in gardens. It is best suited to subtropical and tropical regions and should adapt to a range of acidic soils. Plants are likely to need a sheltered moist site and application of organic mulches is recommended. May need supplementary watering during extended dry periods. Propagate from fresh seed or possibly from cuttings.

Pilidiostigma tropicum L. S. Sm.
(of the tropics)
Qld   Apricot Myrtle
4–12 m × 3–5.5 m   June–Nov
Medium to tall **shrub** or small **tree** with glabrous, apricot-coloured young growth; **bark** flaky; **branches** many, spreading to ascending; **branchlets** flattened to 4-angled, reddish; **leaves** 4–10 cm × 1–3 cm, elliptical to ovate, tapered to base, opposite, glabrous, deep green above, paler below, somewhat leathery, margins wavy, often reddish, apex pointed; **flowers** to about 1.3 cm across, white, cream or pink, 2–6 in axillary racemes, moderately profuse and conspicuous; **stamens** many, white to pink, fluffy; **fruit** to 2.2 cm × 1.4 cm, ovoid, purple to black, edible, fragrance of green peas when crushed, ripe Dec–Feb.

A handsome, bushy species from the Cook District where it grows in lowland and highland rainforest. Suitable for subtropical and tropical regions in a range of moist but freely draining acidic soils. Adapts to a sunny or semi-shaded site. Applications of mulch are very beneficial. Plants may need supplementary watering during extended dry periods. Valuable for windbreak and screening purposes as well as for general planting. Worth trying as a formal hedge. Fruit is eaten by birds and fruit bats. Propagate from fresh seed and cuttings are worth trying.

PILIOSTIGMA Hochst.
(from the Latin *pileus*, felt hat; *stigma*, stigma; apparently referring to the furry calyx)
Caesalpiniaceae
Shrubs or small **trees**, usually deciduous; **leaves** simple, entire, deeply notched to 2-lobed, petiolate; **inflorescence** a raceme; **flowers** unisexual or bisexual; **calyx** 2–5 lobed, deeply slit on one side; **floral tube** present; **petals** 5; **male flowers** with 10 free, fertile stamens; **female flowers** with staminodes; **fruit** an indehiscent woody pod.

A small genus of 6 species distributed in tropical areas of America, Africa and Asia. In Australia, there is 1 species and it is rarely encountered in cultivation. The genus has affinities with *Bauhinia*. Propagation is by seed which has a hard coat that must be broken or softened before sowing. See Treatments and Techniques to Germinate Difficult Species, Volume 1, page 205.

Piliostigma malabaricum (Roxb.) Benth.
(from Malabar in south-west India)
Qld, NT, WA
4–8 m × 2–4 m   Dec–April
Medium to tall deciduous **shrub** with a dense rounded canopy; **bark** grey to brown; **leaves** 4–10 cm × 6–12 cm, broadly elliptical, apex deeply notched, dark green above, greyish beneath, 7–9 veins prominent; **racemes** 4–6.5 cm long, axillary and terminal; **flowers** 3–4 cm across, pink to white; **calyx** split on one side, densely hairy, with 5 lobes; **petals** about 1.5 cm long, minutely hairy, blunt, tapered to base; **stamens** unequal; **pods** 10–30 cm × 1–1.2 cm, indehiscent, dark brown, pointed; **seeds** about 0.5 cm long, ripe May–Sept.

This species ranges from India to Australia where it is widely distributed in the tropical north, growing in vine thickets, woodland and along the banks of streams. It is a tough, hardy shrub with interesting foliage and attractive flowers. Plants are deciduous for a period during the dry season. It has excellent prospects as an ornamental shrub for semi-arid regions and tropical regions with a seasonally dry climate. Requires well-drained soil and a sunny location. Propagate from seed, which requires scarification or softening before sowing.

The var. *acidum* (Korth.) de Wit is regarded as synonymous.

PILULARIA L.
(from the Latin *pilula*, a little ball; referring to the fruiting bodies)
Marsileaceae

Pilidiostigma rhytispermum   D. L. Jones

Tiny **ferns** growing in clumps or mats; **rootstock** creeping, filiform; **fronds** narrowly linear, lacking any lamina, simple, entire; **sori** enclosed in a hard, pill-like, globular sporocarp, arising from the rhizome.

A small genus of 7 species occurring in Africa, Europe, North America, New Zealand and Australia, where there is 1 endemic species. Although singularly distinct, they are true ferns which are related to the genus *Marsilea*. Plants are rarely encountered in cultivation. Propagation is by division of the clumps.

Pilularia novae-hollandiae A. Braun
(from New Holland)
NSW, Vic, Tas, SA, WA   Austral Pillwort
0.02–0.07 m × 0.3–0.5 m

Tiny terrestrial **fern** forming mats; **rootstock** thread-like, much-branched, with fine roots; **fronds** 2–7 cm × 0.05 cm, narrowly linear, simple, coiled when young, dark green, scattered along the rootstock or 2–3 at the nodes; **sporocarps** 0.2–0.4 cm across, globular, hard, short-stalked.

Widespread in southern Australia but rarely observed. It grows in the mud of bogs, swamps and depressions, usually among grasses, sedges and rushes. Often the most obvious feature is the coiled young fronds. The small pill-like reproductive structures are usually buried in the mud. Although lacking ornamental virtue it has tremendous botanical interest. Austral Pillwort is very easily grown as a bog plant or in a pot standing in water. Prefers a sunny position and tolerates light to moderate frosts. Propagate by division.

PIMELEA Banks & Sol. ex Gaertn.
(from the Greek *pimele*, soft fat; referring to the oily seeds or possibly the fleshy cotyledons)
Thymelaeaceae   Rice-flower; Banjines

Perennial **herbs** to tall **shrubs**, mostly bisexual, sometimes unisexual; **bark** often very tough; **leaves** opposite and decussate or rarely alternate; **inflorescence** a condensed terminal raceme which rarely elongates, or axillary clusters, or solitary flowers; **floral bracts** many, few or absent, sometimes very conspicuous and ornamental; **flowers** tubular, with 4 petal-like sepals, exterior hairy or glabrous; **petals** absent; **stamens** 2 or very rarely 1; **fruit** usually a 1-celled nut.

Pimelea is a moderately large genus with about 108 species; about 90 of these are endemic in Australia including 1 species endemic on Lord Howe Island. The remainder occur in New Zealand. As the result of a recent botanical study, a number of new species have been named and described and some species were transferred to the genus *Thecanthes*.

The common name of Rice-flower possibly refers to the hard seed which is often enclosed in the husk-like spent flower, or because some species have flowers which look like clusters of rice. Banjine is an Aboriginal name that is commonly used for those species in WA that have large flowerheads.

Pimelea is distributed in all states of Australia. Plants are found in rainforest, wet sclerophyll forest, dry sclerophyll forest, woodland, mallee, desert, heathland, shrubland, exposed coastal sites, as well as on subalpine moors and slopes. The majority of species occur in well-drained acidic soils which can be sand, gravel, loam or clay loam. Some of these soils may be winter-wet and species such as *P. angustifolia*, *P. ligustrina* and *P. preissii* often inhabit such sites. On the other hand, there are species which occur in alkaline sand or soils that overlie limestone. In some instances, limestone is also present as outcrops. Species that grow in alkaline conditions include *P. calcicola*, *P. ferruginea*, *P. floribunda* and a coastal variant of *P. sylvestris*.

Most pimeleas are perennial woody shrubs but a few are herbaceous or semi-woody annuals or short-lived perennials, for example *P. elongata*, *P. simplex* and *P. trichostachya*.

Some dwarf ground-hugging species are very compact, such as *P. pygmaea*. Another species, *P. filiformis*, covers much larger areas because its stems are self-layering. *P. humilis* is usually a low spreading shrub which often suckers to form delightful small colonies. The majority of pimeleas are dwarf to small shrubs which are usually single-stemmed at ground level. They can also be much-branched like *P. ciliata*, *P. ferruginea*, *P. imbricata* and *P. spectabilis* or upright and sparse such as *P. axiflora* and *P. sulphurea*. A limited number are multi-stemmed with growth emanating from a thick elongated rootstock below ground level, eg. *P. suaveolens* and *P. sulphurea*, while *P. rara* may be single-stemmed or multi-stemmed. Taller examples are *P. lanata*, *P. neo-anglica* and *P. sylvestris*.

The foliage of some species can be quite dramatic when it is glaucous, eg. *P. decora* and *P. sulphurea*, or when young growth is covered in hairs, for instance *P. ammocharis*, *P. sericostachya* and *P. villifera*.

Flowering mainly occurs over spring and summer months and can be profuse. *P. imbricata* ssp. *imbricata* flowers in spring and if pruned hard immediately after flowering finishes another flowering may occur in late summer and/or early autumn. It may also be possible to similarly extend flowering on some other species.

Some species have very pleasantly perfumed flowers such as *P. aeruginosa* and *P. suaveolens* while others can have strong perfumes which some people find off-putting.

The individual flowers of pimeleas are very small but when they are arranged in terminal heads of about 4–6 cm diameter they make an impressive show in species such as *P. ciliata*, *P. cracens*, *P. decora*, *P. haemastostachya*, *P. ligustrina*, *P. spectabilis*, *P. suaveolens* and *P. sylvestris*. In some species, the floral bracts are very prominent in size and colouration. *P. physodes* has the largest bracts in the genus. They are cream to pale green with reddish markings or nearly all-red. It also has flowerheads which have similarities to Darwinia Mountain Bells. The yellowish bracts of *P. sulphurea* and *P. suaveolens* ssp. *flava* are an ornamental feature of these species as well as the floral parts.

Pimelea bark is generally very tough and it is quite difficult to break the branches of a number of species. *P. axiflora* is known as Bootlace Bush, a name which arose because early settlers stripped the bark and used it for bootlaces and tying material. In preparing cuttings it is often very easy to inadvertently strip the bark as the leaves are removed.

Pimeleas are mainly used in amenity planting and they are used to a limited degree for roadside plantings. A limited number of species are common in private and public gardens. There is potential for greater use — many of the horticulturally desirable species are not offered by nurseries yet. Australian plant enthusiasts have cultivated many species but due to propagation difficulties (see below) it is rare that these species are seen outside the gardens of enthusiasts. The most commonly cultivated pimeleas are: *P. brevifolia* ssp. *modesta*, *P. ciliata*, *P. ferruginea*, *P. filiformis*, *P. humilis*, *P. imbricata*, *P. linifolia*, *P. nivea*, *P. rosea*, *P. spectabilis*, *P. sylvestris* and *P. treyvaudii*.

Pimeleas are rarely exploited for cut-flower and cut-foliage production but some have long-lasting flowers and the foliage of some species could be utilised. Flowering stems of *P. physodes* were at one stage harvested from the wild. *P. nivea* is highly valued for its decorative foliage and is now becoming popular as a cut-foliage plant. As most pimeleas respond very well to regular moderately hard pruning they are worth testing for cut-flower and foliage production. Other species with potential include: *P. ammocharis*, *P. cracens*, *P. decora*, *P. flava*, *P. hispida*, *P. linifolia*, *P. macrostegia*, *P. rosea*, *P. sericea*, *P. spectabilis*, *P. suaveolens*, *P. sulphurea* *P. sylvestris* and *P. trichostachya*. Some species such as *P. ligustrina* may induce dermatitis and would need to be tested thoroughly to determine the offending parts.

Pimeleas are renowned for attracting insects such as butterflies to their flowers which often have a rich nectar supply.

Poisonous plants

The foliage and fruits of a number of pimeleas contain toxic chemicals and in some cases grazing animals such as cattle, horses and sheep have been affected. Plants known or suspected of having toxic properties include: *P. curviflora*, *P. decora*, *P. elongata*, *P. flava*, *P. haemastostachya*, *P. latifolia*, *P. linifolia*, *P. microcephala*, *P. pauciflora*, *P. simplex* and *P. trichostachya*.

Pimelea sericea   T. L. Blake

Cultivation

Pimeleas from eastern Australia were first cultivated in England during 1793 with the introduction of *P. linifolia*. Other species introduced at later dates include: *P. pauciflora* (1812), *P. drupacea* (1817), *P. glauca* and *P. humilis* (1824), and *P. nivea* (1833). In 1800, *P. rosea* from WA was cultivated and this was followed by other western species: *P. clavata* and *P. ferruginea* (1824), *P. hispida* and *P. sylvestris* (1830) and *P. lanata* (1834).

Best cultivation results are achieved for most species if they are grown in freely draining acidic soils, in sunny or semi-shaded sites. Plants prefer plenty of early morning and late afternoon sunshine rather than exposed sites which receive sunshine all day. Flowering of most species is better when plants receive at least several hours of sunshine but species such as *P. filiformis*, *P. ligustrina*, *P. nivea* and *P. sylvestris* flower moderately well in semi-shaded sites. Some of the species which occur in winter-wet sites in nature prefer moisture-retentive but well-drained soils when young and they may need supplementary watering during extended dry periods. Most species do not like waterlogging which may also provide ideal conditions for development of root rot (see page 300).

Most pimeleas are hardy to light or moderate frosts but some can suffer damage from heavy frosts. Those which are highly regarded for frost hardiness and unlikely to suffer damage at −7°C include: *P. alpina*, *P. axiflora*, *P. curviflora*, *P. drupacea*, *P. filiformis*, *P. humilis*, *P. ligustrina*, *P. nivea*, *P. pygmaea* and *P. sericea*. Many species from warm temperate and semi-arid regions dislike high temperatures combined with high humidity and it is common for excessive leaf drop to occur in such cases.

The majority of species originate from areas where soil nutrients are low and plants do not usually require supplementary feeding but they do respond well to light applications of slow-release complete fertilisers. Excess phosphorus can cause plant damage and death. Some plants can show symptoms of iron deficiency with yellowing of older leaves. Extra iron as iron sulphate or chelates of iron can usually correct the deficiency.

Once plants are well established they rarely require supplementary watering unless they are growing in exposed or steeply sloping sites. Overwatering can lead to root rot (see below). Generally most pimeleas have a tolerance of low moisture levels. Sporadic deep soaking of the root area can be beneficial during extended dry periods.

Tip pruning of pimeleas is recommended before or immediately after planting to promote a bushy framework. Most pimeleas respond very well to pruning or clipping and it can be an excellent method for rejuvenating old plants. It is best not to cut into leafless wood but species such as *P. ferruginea* and *P. imbricata* often produce new growth from leafless wood if plants are growing vigorously. Some moderately long-lived species such as *P. ligustrina* and *P. nivea* may be suitable for hedging. Further hedging trials are needed to ascertain which species can be recommended.

Pests are not a major problem. Sucking insects such as aphids, mealy bugs, passion vine hoppers, scale and white fly may cause minor problems. See Sucking Insects, Volume 1, page 153 for information on control methods. Light brown apple moth larvae eat foliage but rarely in large quantities. See Chewing Insects, Volume 1, page 143 for control methods.

Root rot caused by Cinnamon Fungus, *Phytophthora cinnamomi* and *Rhizoctania* can cause death of plants. This usually occurs when plants are grown in poorly drained soils or media in containers. For information on prevention and control methods see Volume 1, pages 168 and 172. Moulds may be a problem in poorly ventilated sites. *Botrytis cinerea*, Grey Mould, may affect hairy species during extended moist and still climatic conditions. See Volume 1, page 173 for control methods. Borers can attack pimeleas but are rarely of any major consequence. Sawdust at ground level can be an indicator of their presence. See Volume 1, page 143 for control methods. Powdery mildew may attack some of the dry region species and it can be difficult to detect on glaucous foliaged species such as *P. sulphurea*. For control methods see Volume 1, page 173.

Propagation

Pimeleas are usually propagated from seed or cuttings. There has also been limited success from tissue culture.

Good seed is often difficult to procure. Seeds have a hard coating which needs to be softened to achieve good germination results. See Treat-ments and Techniques to Germinate Difficult Species, Volume 1, page 205, for further information. Untreated seed can germinate but generally results are extremely disappointing with 1–5% germination. Often the seed takes 50–100 days or more to germinate so it pays not to discard seed trays until 2 years after sowing. Young seedlings are often very prone to damping-off, see Volume 1, page 172, for prevention and control methods.

Propagation from cuttings is the main method currently used and generally best results are achieved using new growth in late winter and early spring or immediately after flowering. For most species, cuttings should be prepared from young growth which is becoming firm but is not yet brown and woody. Cuttings taken from brown semi-hardwood material may form roots but generally it takes much longer for root initiation.

The bark strips easily on most species so care should be exercised when removing the lower leaves. Stripping occurs more readily in older wood. Cuttings can be from 2–8 cm long. It can be common for cuttings to drop most of their leaves during propagation but is rarely a cause for alarm as new leaves are usually produced once roots have initiated. Leaf drop is more common when older wood is used for cuttings.

With species such as *P. suaveolens* and *P. sulphurea* which produce stems from underground it may be worthwhile pruning plants heavily and using the new sucker growth as cuttings.

The use of hormone rooting powders or liquids can improve the percentage of cuttings that strike.

Grafting of difficult to cultivate species such as *P. physodes* shows promise for extending the cultivation life of this species. It is worth doing grafting trials with other species too.

Pimelea aeruginosa F. Muell.
(verdigris-coloured)
WA
0.2–1.5 m × 0.2–1 m   June–Oct; also sporadic

Dwarf to small **shrub** with glabrous young growth; **stems** usually solitary and erect; **branches** few, ascending to erect; **branchlets** yellowish to red-brown; **leaves** 0.7–2.2 cm × 0.25–0.8 cm, narrowly ovate to narrowly obovate, opposite, more or less sessile, spreading to ascending, bluish- to greyish-green above and below, glabrous, thick, apex pointed to blunt; **flowerheads** to about 3.5 cm across, pendent, many-flowered, very conspicuous; **floral bracts** to 2.5 cm × 1.7 cm, broadly elliptic to nearly circular; **outer bracts** green to bluish-green, glabrous; **inner bracts** yellowish, sometimes with hairy margins; **flowers** to about 1.5 cm long, yellow, bisexual, glabrous.

This is a wide-ranging species which occurs in the Avon, Eyre, Irwin and Roe Districts where it grows in sand or sandy clay that often has high gravel content and can overlie laterite. Plants will need plenty of sunshine and well-drained acidic soils. They should perform best in semi-arid and warm temperate regions but may succeed in cooler areas. Pruning from an early stage will promote bushy growth. Suitable for gardens, rockeries and containers. Propagate from seed or from cuttings of firm young growth.

P. sylvestris var. *aeruginosa* (F. Muell.) Benth. is a synonym.

Pimelea alpina (F. Muell.) Meisn.
(alpine)
NSW, Vic   Alpine Rice-flower
0.1–0.5 m × 0.5–1.5 m   July–March

Dwarf **shrub** with glabrous young growth; **branches** many, prostrate to ascending, often very crowded and

Pimelea alpina × .85

matted; **branchlets** often reddish when young; **leaves** 0.3–1.5 cm × 0.1–0.5 cm, narrowly elliptic, opposite, short-stalked, concave, spreading to ascending, often crowded, usually mid green above and below, glabrous, apex pointed; **flowerheads** to about 1.5 cm across, terminal, erect, often profuse; **floral bracts** to 1 cm long, elliptic to ovate, green; **flowers** to about 0.7 cm long, white to reddish-pink, bisexual or female.

A handsome inhabitant of subalpine regions in south-eastern NSW and eastern Vic. Occurs mainly at 1500–2000 m altitude where it grows in tussock grassland, heath and woodland. It is rarely cultivated and deserves greater recognition though it performs best at higher elevations. Requires very well-drained acidic soils in a sunny or semi-shaded site. Tolerates heavy frosts and snowfalls. Plants respond well to pruning which promotes compactness. Ideal for general planting, rockeries and containers. Propagate from seed or from cuttings of very young but firm growth.

Pimelea altior F. Muell. =
    P. latifolia ssp. *altior* (F. Muell.) Threlfall

Pimelea ammocharis F. Muell.
(lovely in the sand)
WA, NT
0.2–2 m × 0.2–1 m   March–Oct
Dwarf to small **shrub** with greyish-hairy young growth; **branches** few, ascending to erect; **branchlets** initially densely grey-hairy, becoming glabrous and brownish; **leaves** 0.4–1.7 cm × 0.1–0.5 cm, linear to narrowly elliptic, alternate, short-stalked, ascending to erect, usually overlapping, pale silvery-green above and below, hairy, apex pointed; **flowerheads** to about 2.2 cm across, pendent or erect, many-flowered; **floral bracts** 6–12, to about 1.4 cm × 0.4 cm, silvery-green to brownish, hairy; **flowers** to about 1.1 cm long, white to deep yellow, exterior hairy, bisexual or female.

An elegant species from arid and semi-arid regions in northern WA, extending to NT near the SA, WA and NT border region. It usually grows on red sand in dunes and rocky sites. Cultivated to a very limited extent, it requires exceedingly well-drained soils and plenty of sunshine. Difficulties are likely to arise when growing it in temperate regions. Plants are hardy to moderate frosts. Pruning may promote branching. Propagate from seed, which has been found to commence germination between 24–35 days after sowing, or from cuttings of firm young growth.

The allied *P. penicillaris* has broader leaves.

Pimelea angustifolia R. Br.
(narrow leaves)
?SA, WA   Narrow-leaved Pimelea
0.1–1.5 m × 0.1–0.8 m   Aug–Jan
Dwarf to small **shrub** with glabrous young growth; **branches** few, ascending to erect, slender; **branchlets** reddish-brown to greyish, glabrous; **leaves** 0.2–3 cm × 0.1–0.5 cm, linear to narrowly elliptic, opposite, short-stalked, ascending to erect, often incurved, mid green above and below, glabrous, apex pointed; **flowerheads** to about 2.5 cm across, erect to semi-pendent, terminal, many-flowered; **floral bracts** 4 or 6, to 1.5 cm × 0.8 cm, usually ovate, green to reddish; **flowers** to about 1.2 cm long, usually white to cream but also yellow or pale pink, exterior hairy, bisexual or rarely female.

An erect shrub of the south-western region. It is distributed from near Kalbarri to near Eucla, and it may occur over the border in SA. Inhabits sandy soils that often contain gravel or laterite, or sandy clay which can be inundated for short periods in winter. It is highly desirable for cultivation and adapts well to freely draining soils in warm temperate regions. Tolerates a sunny or semi-shaded site. Hardy to moderate frosts. Prune early to establish a bushy framework. Propagate from seed or from cuttings of very young growth.

P. tenuis M. Scott is a synonym.

P. floribunda has larger leaves and floral bracts. *P. linifolia* is also allied but can have pendent flowerheads.

Pimelea aquilonia Rye
(northern)
Qld   Northern Pimelea
0.7–3 m × 0.5–1.5 m   May–July
Dwarf to medium **shrub** with hairy young growth; **branches** ascending to erect; **branchlets** with crowded shiny hairs when young; **leaves** 0.8–3.5 cm × 0.2–0.7 cm, narrowly elliptic, opposite, short-stalked, spreading to ascending, often crowded near ends of branchlets, mid green above, paler below, glabrous or faintly hairy near the pointed apex; **flowers** about 1 cm long, white or cream, exterior hairy, in few-flowered terminal clusters, rarely profuse, male or female or sometimes bisexual.

Pimelea argentea

Pimelea angustifolia   W. R. Elliot

This recently described (1990) species is distributed on Cape York Peninsula from the tip to south of Cooktown. It is mainly found in exposed sandy sites. It has only moderately ornamental features and evidently is not in cultivation. Plants are suitable for seasonally dry tropical regions in very well-drained soils. Propagate from seed or possibly from cuttings of firm young growth.

Pimelea argentea R. Br.
(silvery)

WA   Silvery-leaved Pimelea
0.4–2 m × 0.3–1 m   July–Nov; also sporadic

Dwarf to small **shrub** with densely hairy young growth; **branches** ascending to erect; **branchlets** initially densely covered in whitish hairs; **leaves** 0.4–5 cm × 0.2–1 cm, linear or elliptic, alternate or opposite, usually short-stalked, spreading to ascending, moderately crowded, soft, pale green to silvery, very hairy, apex pointed; **flowers** to about 0.6 cm long, white, yellow or greenish, exterior faintly hairy, male or female in few- to many-flowered axillary clusters, often profuse but rarely very conspicuous.

P. argentea is a wide-ranging species with attractive foliage. It is from the south-western region extending from near Israelite Bay to the Murchison River and reaching inland to near Hyden. Plants occur in sands and sandy clay soils which are usually derived from granite. Cultivated to a limited degree. It has potential for harvesting of the foliage for indoor decoration. Requires plenty of sunshine but will tolerate a semi-shaded site. Needs soils which are moderately well drained. Plants are hardy to fairly heavy frosts. Propagate from seed, which may begin to germinate between 21–70 days after sowing. Cuttings of firm young growth are worth trying.

The var. *racemosa* F. Muell. is a synonym.

Pimelea avonensis Rye
(from the Avon Botanical District)

WA
0.4–1 m × 0.3–1 m   July–Oct

Dwarf **shrub** with glabrous young growth; **branches** ascending to erect; **branchlets** glabrous, greyish except reddish near flowerheads; **leaves** 0.5–2.5 cm × 0.5–3 cm, linear to narrowly elliptic, opposite, spreading to ascending, glabrous, deep green to bluish-green above, paler below, margins recurved to revolute, apex pointed; **flowerheads** to about 3.5 cm across, terminal, erect, many-flowered; **floral bracts** 4 or 6, to about 1.3 cm × 0.9 cm, narrowly to broadly ovate, green to yellowish-green, sometimes with reddish tones; **flowers** about 1 cm long, white with pinkish base, pink in bud, exterior very hairy.

This species occurs in the northern Avon District and grows in sandy soils in heathland or open woodland. It can be sparsely foliaged but is very attractive when in flower. Plants require excellent drainage and a warm to hot, sunny or semi-shaded site. Probably best suited to warm temperate and semi-arid regions. They tolerate moderately heavy frosts and extended dry periods. Prune from an early stage to promote bushy growth. Propagate from seed or from cuttings of barely firm young growth.

The closely allied *P. ciliata* has outer floral bracts with prominently hairy margins.

Pimelea axiflora F. Muell. ex Meisn.
(flowers in the axils)

NSW, Vic, Tas   Bootlace Bush
1–3 m × 0.6–1.5 m   June–Nov

Small to medium **shrub** with usually glabrous young growth; **stems** ascending to erect; **branches** spreading to ascending, slender; **branchlets** mainly glabrous, often slightly pendent; **leaves** 1–7.5 cm × 0.2–1.2 cm, linear to narrowly elliptic, opposite, rarely crowded, spreading to ascending, glabrous, mid green above,

Pimelea axiflora ssp. *axiflora* × .75

302

paler below, flat, apex pointed; **flowers** about 0.4 cm long, white to cream, exterior hairy, in few-flowered axillary clusters along the branches and branchlets; **floral bracts** 2–4, to about 0.7 cm long, elliptic, brown to green, exterior hairy.

A wide-ranging and variable species with the typical subspecies occurring from near Braidwood, NSW, through Vic to Tas. It is usually found in moist habitats from the coast to inland mountain ranges. Plants grow best in moisture-retentive but moderately well-drained acidic soils which are semi-shaded or in dappled shade. Well suited to growing with ferns. This species can be very spindly and somewhat graceful but hard pruning promotes bushy growth. Plants are hardy to most frosts and light snowfalls. The bark can be removed in thin long strands which are very tough and these were used by early settlers as bootlaces and for tying material. A hairy-stemmed variant is known from central Victoria.

There are a further 2 subspecies:

ssp. *alpina* (F. Muell. ex Benth.) Threlfall occurs in the Australian Alps of south-eastern NSW and north-eastern Vic, where it grows mainly at 1500–2000 m altitude. It differs from ssp. *axiflora* in having flowers to about 0.5 cm long, with a less-hairy exterior and only brown floral bracts.

ssp. *pubescens* Rye is from the Bungonia Gorge area of the Southern Tablelands, NSW. It has hairy stems and bases of leaves, and the leaves are larger, usually to 2.2 cm × 0.5 cm.

All subspecies are propagated from seed or more easily from cuttings of very young growth.

Pimelea biflora N. A. Wakef.
(two-flowered)
NSW, Vic   Matted Rice-flower
prostrate × 0.6–1.5 m   Nov–Jan

Dwarf **shrub** with hairy self-layering stems; young growth hairy; **stems** slender; **leaves** 0.2–1 cm × 0.1–0.5 cm, narrowly elliptic to elliptic, opposite, spreading to ascending, somewhat crowded, mid green to glabrous above, faintly hairy below, midrib prominent, apex pointed; **flowerheads** 2-flowered, terminal; **floral bracts** 2, to about 0.9 cm long; **flowers** to about 0.7 cm long, deep red, usually bisexual, scattered and somewhat inconspicuous.

A mat-forming species from higher elevations in south-eastern NSW and north-eastern Vic. It grows in subalpine forest and grassland. Evidently rarely cultivated. It has potential for use because it has a self-layering growth habit. Plants should adapt well to temperate regions in well-drained acidic soils with a sunny or semi-shaded aspect. They are hardy to most frosts and withstand snowfalls. Propagate from seed, cuttings, or by division of layered stems.

P. curviflora var. *alpina* F. Muell. ex Benth. is a synonym.

Pimelea 'Bonne Petite' = *P. ferruginea* 'Bonne Petite'

Pimelea brachyphylla Benth.
(short leaves)
WA
0.1–1 m × 0.1–0.6 m   July–Oct

Pimelea avonensis   W. R. Elliot

Pimelea brevifolia ssp. *modesta*   W. R. Elliot

Pimelea bracteata

Dwarf **shrub** with glabrous young growth; **branches** ascending to erect, slender; **branchlets** twiggy, glabrous, grey except for blackish to deep red-brown near flowerheads; **leaves** 0.2–1 cm × 0.1–0.3 cm, linear to elliptic-oblong, opposite and decussate, sessile or short-stalked, spreading to ascending, often incurved, crowded, glabrous, mid-to-dark green and often glaucous above, paler below, margins revolute, apex pointed; **flowerheads** to about 2.5 cm across, somewhat globular, terminal, erect, many-flowered, often profuse and very conspicuous; **floral bracts** 4 or 6, to about 0.8 cm × 0.7 cm, elliptic to ovate, green with reddish margins or mainly red; **flowers** to about 0.8 cm long, white, exterior hairy, bisexual or female, style exserted.

A pretty, floriferous species from a wide area of southern WA extending from near Israelite Bay to near Wagin. It is found in shrubland and mallee woodland and grows in a variety of soil types. Cultivated to a limited extent, it has potential for greater use in semi-arid and temperate regions. It should adapt to a range of acidic soils which are moderately well drained and will need a sunny or semi-shaded site. Hardy to moderate frosts. Useful in gardens and containers. Propagate from seed or from cuttings of firm young growth.

Pimelea bracteata Threlfall
(with bracts)
NSW
0.6–2 m × 0.4–1.5 m   Nov–Feb

Dwarf to small **shrub** with glabrous young growth; **branches** ascending to erect; **branchlets** glabrous, reddish-brown to green; **leaves** 0.5–2 cm × 0.2–0.6 cm, elliptic to narrowly obovate, opposite, short-stalked, ascending to erect, glabrous, green with reddish tones, paler below, margins faintly recurved, midrib prominent, apex pointed; **flowerheads** to about 2 cm across, compact, nodding, on short stalk, many-flowered, usually borne on branchlets; **floral bracts** 6 or 8, to about 1.8 cm × 1.5 cm, broadly elliptical to broadly ovate, yellow-green, often with red or purplish tonings, very conspicuous; **flowers** to about 1.7 cm long, glabrous, bisexual.

An admirable rice-flower from the Kiandra region in the Southern Tablelands. It grows at over 1000 m altitude in wet heath and beside creeks. Plants respond well to cultivation and deserve to be better known in temperate regions. They need moisture-retentive but freely draining acidic soils and a sunny or semi-shaded aspect. Hardy to most frosts and light snowfalls. Responds well to pruning which promotes bushy growth. It may be useful for low hedging. Propagate from seed or from cuttings of firm young growth.

P. ligustrina var. *glabra* Maiden & Betche is a synonym. *P. bracteata* has also been known as *P.* species B.

P. macrostegia is similar but the flowerheads have hairy stalks and are on main branches.

Pimelea brevifolia R. Br.
(short leaves)
WA
0.2–1 m × 0.2–1.3 m   July–Oct

Dwarf **shrub** with glabrous young growth; **branches** many, spreading to ascending; **branchlets** many, slender, glabrous; **leaves** 0.3–1.6 cm × 0.1–0.3 cm, mainly narrowly elliptic, opposite and decussate, sessile or short-stalked, spreading to ascending, somewhat crowded, glabrous, mid-to-deep green above and below, apex pointed; **flowerheads** to about 2 cm across, terminal, erect, often profuse and conspicuous; **floral bracts** 4, to about 1.2 cm × 0.9 cm, narrowly to broadly ovate, green to yellowish-green, sometimes with reddish tonings; **flowers** to about 0.8 cm long, white or rarely pale yellow, exterior hairy, bisexual or female.

The typical subspecies occurs mainly in the Eyre and Roe Districts and in the far south-eastern Darling District. It grows in sands or clay soils which are often associated with laterite or granite. This subspecies is not as common in cultivation as ssp. *modesta* but it adapts well to moderately well-drained soils in temperate regions. Prefers a sunny or semi-shaded site but will tolerate dappled shade. Plants are hardy to moderate frosts and tolerate limited dry periods. Pruning promotes bushy growth and rejuvenates old plants. Excellent for general planting and containers.

The ssp. *modesta* (Meisn.) Rye is a very decorative, compact, dwarf shrub of 0.7–1 m × 0.7–1 m. It occurs further inland than ssp. *brevifolia* and is found in the Avon, Coolgardie and Roe Districts. It has crowded leaves and elliptic floral bracts. Flowers are often pinkish in bud, opening white, and the flowerheads are borne on short branchlets and usually profuse. Flower-stems are excellent for posies.

Propagate both subspecies from seed, which can begin to germinate 21–45 days after sowing, or from cuttings of firm but young growth, which usually strike readily.

Pimelea brevistyla Rye
(short style)
WA
0.5–1.5 m × 0.3–1 m   Aug–Oct

Dwarf to small **shrub** with glabrous young growth; **branches** ascending to erect; **branchlets** greyish, glabrous; **leaves** 1.4–3 cm × 0.2–0.4 cm, narrowly ovate, opposite, short-stalked, spreading to ascending, rarely crowded, glabrous, mid green to deep bluish-green above and below, margins recurved, apex pointed; **flowerheads** to about 4 cm across, terminal, erect, many-flowered, often profuse and very conspicuous; **floral bracts** 2 or 4, to about 2 cm × 1.2 cm, ovate, green with reddish tonings; **flowers** to about 1.5 cm long, white, exterior hairy, bisexual.

A very desirable species with two variants. The typical subspecies occurs in the Darling Ranges near Perth where it grows in clay or gravelly sands derived from granite. Evidently rarely cultivated, it should adapt to a range of moderately well-drained acidic soils in temperate regions. A sunny or semi-shaded site is suitable. Plants tolerate light frosts and dry summers. It has potential for gardens and as a container plant.

The ssp. *minor* Rye occurs in the wheatbelt region of the Avon and Roe Districts where it is found in sandy clay soils overlying laterite and granite. It differs in growing 0.3–1 m high and has floral bracts to about

1 cm long. The flowers are about the same length, and flowerheads up to 3 cm across.

Propagate both subspecies from seed or from cuttings of very young firm growth.

Pimelea calcicola Rye
(growing on limestone)
WA
0.3–1 m × 0.2–0.6 m   Sept–Nov

Dwarf **shrub** with glabrous young growth; **branches** spreading to erect; **branchlets** mainly glabrous; **leaves** 1.3–2.7 cm × 0.3–0.7 cm, elliptic, opposite, short-stalked, spreading to ascending, sometimes incurved, glabrous, pale-to-mid green, flat to slightly concave, apex pointed; **flowerheads** to about 2.5 cm across, terminal, erect, many-flowered, often profuse; **floral bracts** 6 or 8, to about 2 cm × 0.8 cm, ovate, green, glabrous; **flowers** to about 1.6 cm long, pale-to-deep pink, exterior glabrous, bisexual.

An admirable species which occurs on calcareous coastal sand from near north of Perth to near Bunbury in the Darling District. It is cultivated to a limited extent and deserves to be better known. Plants need excellent drainage and should adapt to both alkaline and acidic soils. They need a sunny or semi-shaded site. May suffer damage from frosts but withstands extended dry periods. Suitable for gardens and may be an excellent container plant. Propagate from seed or from cuttings of very young but firm growth.

Pimelea cernua R. Br. = *P. linifolia* Sm. ssp. *linifolia*

Pimelea ciliata Rye
(fringed with fine hairs)
WA   White Banjine
0.5–1 m × 0.5–1.3 m   Aug–Dec

Dwarf **shrub**; young growth glabrous; **branches** many, ascending to erect; **branchlets** brownish when young, glabrous; **leaves** 0.5–2.5 cm × 0.15–0.8 cm, ovate to narrowly obovate, opposite, spreading to ascending, rarely crowded, glabrous, mid green to bluish-green above, paler below, margins recurved to revolute, apex pointed and recurved; **flowerheads** to about 3.5 cm across, terminal, erect, many-flowered, often profuse and very conspicuous; **floral bracts** 4 or 6, to about 1.3 cm × 1 cm, ovate to broadly ovate, green with pinkish base or sometimes mainly reddish, margins hairy; **flowers** to about 1.2 cm long, white or sometimes pink, exterior hairy, usually bisexual, style and stamens prominently exserted.

A stunning, floriferous and variable species with the typical variant distributed mainly in the southern Darling District from north of Perth to near Albany. It inhabits hillsides with heavy lateritic or granitic soils which are subject to limited winter-wet periods. This highly desirable pimelea has been cultivated and often incorrectly described for many years as the white-flowered variant or forest form of *P. rosea*. It grows successfully in a range of reasonably well-drained acidic soils with a sunny or semi-shaded aspect. Plants tolerate moderate frosts and extended dry periods. They can become top-heavy but respond well to hard pruning which can help to stabilise plants. The selection 'Snow Clouds' is moderately compact and has dark green leaves.

The ssp. *longituba* Rye occurs south of Busselton in heavy soils. It differs in having pale pink flowers to about 1.4 cm long, with scarcely exserted stamens and leaves of 0.8–1.5 cm × up to 0.5 cm, with a more or less erect apex.

Propagate both subspecies from seed, or cuttings of very young but firm growth, which strike readily.

P. rosea only occurs on sandy soils and has an open growth habit and deep pink flowers.

Pimelea ciliolaris (Threlfall) Rye
(fringed with small hairs)
NSW
0.2–0.5 m × 0.2–0.3 m   Oct–Dec

Dwarf **shrub** with white-hairy or glabrous young growth; **branches** spreading to ascending; **branchlets** with white hairs, often reddish beneath flowerheads; **leaves** 0.5–1.6 cm × 0.1–0.3 cm, linear to narrowly elliptic, alternate, spreading to ascending, usually crowded on branchlets, mid-to-deep green above, faintly hairy to glabrous, margins hairy, apex pointed; **flowerheads** to about 2.5 cm across, terminal, erect, many-flowered; **floral bracts** 9–15, to about 1.5 cm × 0.4 cm, narrowly ovate, green, glabrous to hairy, margins hairy; **flowers** to about 1 cm long, white, exterior densely hairy, bisexual.

A poorly known and rare species from the North Western Slopes where it grows mainly in rocky sites. Evidently not cultivated, it deserves to be introduced as part of a conservation strategy. Plants are best suited to temperate regions in well-drained acidic soils and should adapt to a sunny or semi-shaded site. Hardy to most frosts. Propagate from seed or possibly from cuttings of firm young growth.

P. octophylla R. Br. ssp. *ciliolaris* Threlfall is a synonym.

Pimelea cinerea R. Br.
(ash-grey)
Tas
1–2.5 m × 0.6–2 m   Nov–Jan

Small to medium **shrub** with whitish-hairy young growth; **branches** many, spreading to erect; **branchlets** whitish-hairy; **leaves** 0.5–2.5 cm × 0.2–1 cm, narrowly elliptic to elliptic, opposite, short-stalked, spreading to ascending, somewhat crowded on young branchlets, yellow-green to mid green and glabrous above, densely white-hairy below, margins slightly recurved, venation prominent, apex pointed; **flowerheads** about 1 cm across, terminal, few-flowered; **floral bracts** 4–6, smaller than leaves; **flowers** about 0.5 cm long, white to greenish-white, exterior hairy, bisexual.

This species has lovely foliage. It is fairly widespread but most common in southern and western areas growing in wet forests. Plants should adapt well to temperate regions with freely draining, moisture-retentive soils and a semi-shaded aspect. Hardy to most frosts and tolerates snowfalls. It has potential as a cut foliage crop and may do well as a hedging plant. Propagate from seed or from cuttings of very young growth.

P. nivea is similar but has many-flowered flowerheads and shorter leaves.

Pimelea ciliata   W. R. Elliot

Pimelea clavata Labill.
(club-shaped)
WA
1–6 m × 0.6–3 m   Sept–Feb; also sporadic

Small to tall **shrub** with hairy young growth; **branches** many, ascending to erect; **branchlets** initially hairy, greenish-brown to brown; **leaves** 0.6–4.5 cm × 0.15–1 cm, linear to narrowly elliptic, opposite, short-stalked, spreading to ascending, mid-to-dark green and becoming glabrous above, paler and hairy below, margins flat to faintly recurved, apex softly pointed; **flowerheads** to about 1.2 cm across, terminal, usually erect, compact, many-flowered, rarely profuse; **floral bracts** usually 2, to about 0.5 cm × 0.15 cm, leaf-like; **flowers** to about 0.4 cm long, white, cream or pale yellow, exterior hairy, usually separate male and female flowers.

P. clavata occurs on the mainland in coastal and adjacent areas of the southern tip of the south-western region, and also on islands from near Albany to the Archipelago of the Recherche. It is cultivated to a very limited extent and should grow well in temperate regions. Probably prefers moisture-retentive but freely draining acidic or alkaline soils with a sheltered site. Not highly ornamental in flower, but it has pleasant foliage and potential for hedging and screening. Hardy to moderately heavy frosts. It was introduced to cultivation in England in 1824. Propagate from seed or from cuttings of very young, barely firm growth.

Pimelea collina R. Br. =
P. *linifolia* Sm. ssp. *collina* (R. Br.) Threlfall.

Pimelea colorans A. Cunn. ex Meisn. = *P. stricta* Meisn.

Pimelea concreta F. Muell. =
Thecanthes concreta (F. Muell.) Rye

Pimelea congesta C. T. Moore & F. Muell.
(congested, crowded)
NSW (Lord Howe Island)
1–2 m × 0.6–1.6 m   July–Oct

Medium **shrub** with faintly hairy young growth; **bark** reddish, tough; **branches** ascending to erect; **branchlets** slender, reddish; **leaves** 0.8–2 cm × 0.2–0.6 cm, elliptical, opposite, short-stalked, spreading to ascending, often incurved, leathery, dull pale green, glabrous except for hairs on stalk, margins recurved, apex pointed; **flowerheads** about 2 cm across, terminal, erect, with about 9 flowers; **floral bracts** 4, leaflike; **flowers** to about 1.5 cm long, white, exterior hairy, bisexual.

Endemic on Lord Howe Island where it grows on exposed and often dryish ridges. Very rarely cultivated, it has moderate ornamental features. Plants should adapt to a range of freely draining acidic soils. A sunny or semi-shaded site should be satisfactory. Frost tolerance is not known. Withstands limited dry periods and strong winds. Propagate from seed or from cuttings of barely firm young growth.

Pimelea continua J. M. Black =
P. *simplex* (J. M. Black) Threlfall

Pimelea cornucopiae (Vahl) =
Thecanthes cornucopiae (Vahl) Wikström

Pimelea cracens Rye
(graceful, slender)
WA
0.4–1.5 m × 0.3–1 m   July–Dec

Dwarf to small **shrub** with glabrous young growth; **branches** ascending to erect; **branchlets** glabrous, reddish to grey-green; **leaves** 0.6–2.2 cm × 0.2–0.6 cm, narrowly elliptic to ovate, opposite, short-stalked, often somewhat crowded at ends of branchlets, spreading to ascending, sometimes reflexed, glabrous, yellow-green to deep bluish-green above and below, flat to concave, apex pointed; **flowerheads** to about 3 cm across, terminal, pendent, many-flowered, often profuse and very conspicuous; **floral bracts** 6, 8 or rarely 10, 1.2–3 cm × 0.6–1.5 cm, yellowish to pale green or bluish-green, many have reddish tonings; **flowers** to about 1.2 cm long, cream to pale yellow, sweetly scented, exterior hairy, bisexual or rarely female.

A superb ornamental species with 2 subspecies. The typical variant occurs in the far south-eastern tip of the Darling District, as well as in the Eyre and southern Roe Districts. Poorly known in cultivation, it may prove difficult to maintain for extended periods. Plants are most likely to succeed in well-drained acidic soils in a fairly sunny site. They tolerate moderately heavy frosts and extended dry periods. Probably best suited to warm temperate regions. Bushy growth is promoted by regular pruning during or after flowering.

The ssp. *glabra* Rye has more-or-less glabrous flowers and occurs in moist-to-wet soils of the southern Darling District. It may prove easier to cultivate than the typical subspecies.

Propagate both subspecies from seed or from cuttings of barely firm young growth.

Pimelea curviflora R. Br.
(curved flowers)
NSW　　　　　　　　Tough-barked Rice-flower
0.2–0.5 m × 0.1–0.3 m　　　Oct–Jan; also sporadic

Dwarf slender **shrub** with hairy young growth arising from a taproot; **branches** few, ascending to erect; **branchlets** hairy; **leaves** 0.15–1.1 cm × 0.1–0.3 cm, narrowly elliptic to obovate, opposite or alternate, spreading to ascending, deep green and becoming glabrous above, hairy below especially on margins and midrib, apex pointed; **flowerheads** about 0.8 cm across, terminal, few-flowered; **floral bracts** absent or insignificant; **flowers** to about 0.8 cm long, reddish to yellowish, curved.

This is an extremely variable species. The typical subspecies is confined to the Sydney region where it occurs on sandstone. It is not very ornamental and plants are rarely encountered in cultivation. They require good drainage and a semi-shaded site and will tolerate moderate frosts. The main likely application is for revegetation purposes.

There are a further 5 subspecies.

P. curviflora var. *acuta* Threlfall

A dwarf shrub of 0.2–0.5 m × 0.2–0.3 m which has yellowish-green leaves and yellowish-green flowers blooming mainly in Nov–Feb. It occurs in the Southern Tablelands District and its distribution extends from the Budawang Range to near Mt Kosciuszko. *P. curviflora* ssp. *gracilis* var. *acuta* Threlfall is a synonym.

P. curviflora var. *alpina* F. Muell. ex Benth. =
　　　　　　　　　　P. biflora N.A. Wakef.

P. curviflora var. *divergens* Threlfall

This somewhat tufting dwarf shrub of 0.1–0.5 m × 0.1–0.6 m has mid-to-deep green leaves which have sparse long hairs on the undersurface. It has its main distribution from south-eastern Qld to eastern NSW, north of Sydney. *P. curviflora* ssp. *gracilis* (R. Br.) Threlfall var. *divergens* Threlfall is a synonym.

P. curviflora var. *gracilis* (R. Br.) Threlfall

Develops as a dwarf to small plant of 0.2–1.5 m × 0.2–1 m. Leaves are hairy, pale green and have recurved or incurved margins. The greenish-yellow flowers bloom during July–Jan. This subspecies is distributed from south-eastern Qld to Tas, and westwards to the Yorke Peninsula of SA.

P. curviflora ssp. *micrantha* (F. Muell. ex Meisn.) Threlfall =
　　　　　　　P. micrantha F. Muell. ex Meisn.

P. curviflora var. *micrantha* (F. Muell. ex Meisn.) Benth. =
　　　　　　　P. micrantha F. Muell. ex Meisn.

P. curviflora var. *pedunculata* Benth. =
　　　　　　　P. strigosa Gand. (see below)

P. curviflora var. *sericea* Benth.

This dwarf shrub is 0.1–0.3 m × 0.1–0.5 m and has pale to dull green, densely hairy, elliptic to obovate leaves to 2 cm × 0.7 cm which have incurved margins. Flowering is mainly in Oct–Jan. It is widely distributed from south-eastern Qld to Tas and across to south-eastern SA. *P. curviflora* ssp. *gracilis* var. *sericea* Benth. is a synonym.

P. curviflora var. *subglabrata* Threlfall

Of similar dimensions to var. *sericea*, but the narrowly elliptic to narrowly obovate leaves are more-or-less glabrous and to about 1.7 cm × 0.35 cm, and can be paler below. *P. curviflora* ssp. *gracilis* var. *subglabrata* Threlfall is a synonym.

Propagate all subspecies from seed or from cuttings of barely firm young growth.

P. strigosa Gand. from south-eastern Qld and north-eastern NSW is allied to *P. curviflora* but differs because it has longer flowerhead stalks and lacks floral bracts. *P. curviflora* var. *pedunculata* Benth. is a synonym.

Pimelea decora Domin
(showy)
Qld　　　　　　　　　　Flinders Poppy
0.3–1.2 m × 0.3–1 m　　　　　　　　sporadic

Dwarf to small, single or multi-stemmed, annual or short-lived **shrub** with a woody rootstock and glaucous young growth; **branches** ascending to erect; **branchlets** pale green to glaucous; **leaves** 1.5–5.5 cm × 0.6–4 cm, elliptic to ovate, opposite or rarely alternate, sessile and somewhat stem-clasping, spreading to ascending, not crowded, glaucous and glabrous, soft, thickish and somewhat succulent, apex blunt to rounded; **flowerheads** to about 5 cm across and extending to 22 cm long at maturity, terminal, erect, many-flowered, very conspicuous; **floral bracts** 5–8, to 1.6 cm × 0.7 cm, similar to leaves except hairy; **flowers** to about 1.7 cm long, cream to creamy yellow near base and bright red tips, or rarely all creamy yellow, very strongly scented, older flowers fading and becoming deciduous.

A very handsome species from central Qld where it grows in heavy, often rocky soils which are subject to periodic inundation. Plants require an open, hot, sunny site in clay or loam soils and are best suited to subtropical or semi-arid regions. They have been difficult to maintain in temperate regions. Removal of old stems is usually beneficial in promoting bushier growth. Plants are hardy to moderately heavy frosts. Suitable as a strong accent plant for gardens, rockeries and containers. Leaves and stems have poisonous properties and although known to be toxic to livestock are rarely eaten by them. Propagate from seed or from cuttings of very young, barely firm growth.

The closely allied *P. haematostachya* has narrower leaves and shorter flowerheads.

Pimelea decussata R. Br. = *P. ferruginea* Labill.

Pimelea dioica C. T. White = *P. penicillaris* C. T. White

Pimelea drummondii (Turcz.) Rye
(after James Drummond, first Government Naturalist of WA)
WA
0.4–2 m × 0.2–1 m　　　　　　　　　June–Aug

Dwarf to small **shrub** with faintly hairy young growth; **branches** ascending to erect, glabrous; **branchlets** pale green or yellowish-green to reddish-brown; **leaves** 1–5 cm × 0.5–2 cm, elliptical, opposite, sessile, spreading to ascending or sometimes reflexed, mid green and glabrous above and below, midrib prominent, apex pointed; **flowerheads** about 4 cm across, terminal, pendent, many-flowered, compact, very conspicuous; **floral bracts** 6 or 8, 1–2 cm × 0.7–1.8 cm, ovate, outer bracts pale green to yellow-green and sometimes reddish, glabrous; **flowers** to about 2.7 cm

Pimelea drupacea

long, white or cream, exterior hairy, bisexual, stamens and style exserted.

An elegant coastal or near-coastal species from the Eyre District where it grows on sand dunes or in sandy soils which overlie granite. Cultivated to a limited degree, it requires very well-drained acidic soils and a sunny site. Probably best suited to warm temperate regions. Plants can become leggy and pruning usually promotes bushy growth. They tolerate light to moderate frosts. Suitable for gardens or containers. Propagate from seed or from cuttings of barely firm young growth.

Pimelea drupacea Labill.
(drupe-like fruit)
Vic, Tas   Cherry Rice-flower
1–3 m × 0.7–2.5 m   Sept–Jan; also sporadic
Small to medium **shrub** with hairy young growth; **branches** spreading to erect; **branchlets** terete, hairy; **leaves** 5–7 cm × 0.2–1.5 cm, narrowly elliptic to elliptic, opposite, short-stalked, spreading to ascending, not crowded, medium to deep green above and paler below, becoming glabrous, margins flat, venation prominent, apex softly pointed; **flowerheads** to about 2 cm across, terminal or in upper axils, loosely arranged, sometimes profuse, moderately conspicuous; **floral bracts** 2, leaf-like, small, hairy; **flowers** to about 0.6 cm long, white or pale cream, exterior hairy, bisexual; **fruit** to about 0.7 cm long, ovoid, succulent.

P. drupacea is widespread in Tas and the Bass Strait Islands and it is also distributed in Wilson's Promontory, Vic. Usually inhabits coastal shrubland, woodland and sclerophyll forest. May be found in low-lying areas. Best suited to temperate regions. Adapts well to cultivation but is rarely offered by nurseries. A sunny to somewhat shaded site with freely draining acidic soils is required. Plants are hardy to most frosts and they withstand extended dry periods. They respond well to

Pimelea drupacea × .6

pruning and are suitable for hedging. Introduced to cultivation in England in 1817. Propagate from seed or from cuttings of barely firm young growth.

Pimelea elata F. Muell. ex Meisn. =
      P. ligustrina Labill. ssp. *ligustrina*

Pimelea elongata Threlfall
(elongated)
Qld, NSW, SA
0.1–0.6 m × 0.1–0.5 m April–Oct; also sporadic
Dwarf, somewhat herbaceous **shrub** with hairy young growth; **stems** ascending to erect, woody at base, hairy; **leaves** 0.3–1.8 cm × up to 0.25 cm, narrowly elliptic, alternate, spreading to ascending, usually crowded at ends of stems, glabrous or hairy, yellow-green, apex pointed; **flowerheads** about 1 cm across, in terminal, elongating, often interrupted racemes to about 15 cm long; **floral bracts** absent; **flowers** to about 0.3 cm long, white, exterior densely covered with white hairs, bisexual.

This somewhat insignificant species has a scattered distribution in south-eastern Qld, the North Western Plains of northern NSW and in the northern Flinders Ranges, SA. Usually occurs in heavy soils with a sandy topsoil or in clayey sand. It has little to recommend it for cultivation except for indigenous revegetation purposes. Best suited to warm temperate and semi-arid regions. Should adapt to a range of soils. Hardy to most frosts. It is poisonous to sheep, cattle and horses. Propagate from seed or from cuttings of barely firm young growth. This species has also been known as *P. trichostachya* 'Form B'.

P. trichostachya Lindl. is closely allied and also has toxic properties but differs in having leaves which are alternate to somewhat opposite and it has long floral hairs. It occurs mainly in drier inland regions of mainland states. May be worth trying this species for cut-flower production.

Pimelea erecta Rye
(erect)
WA
0.3–2 m × 0.2–1.5 m   Oct–Jan; also sporadic
Dwarf to small **shrub**; young growth faintly hairy; **branches** ascending to erect; **branchlets** mainly glabrous except for leaf axils, pale brown to grey; **leaves** 0.5–2 cm × 0.15–0.5 cm, narrowly elliptic to ovate, opposite, spreading to ascending, rarely crowded, mid green and glabrous above and below, flat to concave, apex softly pointed; **flowerheads** to about 3 cm across, terminal, erect, many-flowered, often profuse and very conspicuous; **floral bracts** usually 8 or 10, to about 1 cm × 0.4 cm, elliptic to ovate, green with deep reddish base, straight, exterior glabrous, interior and margins hairy; **flowers** to about 0.8 cm long, white to pale pink, exterior hairy, bisexual.

This handsome, recently described (1988) species occurs in the Eyre and southern Roe Districts where it grows in sand, clay loam or clay soils, often in mallee woodland. Evidently rarely (if ever) cultivated, it deserves greater recognition. Plants should adapt readily to temperate and semi-arid regions. They will need relatively well-drained soils and plenty of sunshine. Light frosts should not damage plants and they can withstand extended dry periods. Worth trying in containers as well as in gardens. Propagate from seed or from cuttings of barely firm young growth.

P. villifera Meisn. differs in having hairy branches and 12–20 floral bracts which become reflexed after flowering.

Pimelea eyrei F. Muell. =
>*P. longiflora* R. Br. ssp. *eyrei* (F. Muell.) Rye

Pimelea ferruginea Labill.
(rust-coloured)

WA	Rice-flower
0.3–1.5 m × 0.7–2.5 m	Aug–Feb

Dwarf to small **shrub** with hairy young growth; **branches** many, spreading to ascending; **branchlets** deep reddish-brown to blackish, becoming glabrous; **leaves** 0.5–2 cm × 0.15–0.8 cm, narrowly elliptic to elliptic, opposite, short-stalked, mainly spreading, usually crowded near ends of branchlets, deep green and glossy glabrous above, much paler below, margins recurved, apex softly pointed; **flowerheads** to about 3 cm across, terminal, erect, many-flowered, often very profuse and extremely conspicuous; **floral bracts** 4, to about 2 cm × 1 cm, ovate, green with yellowish to reddish base, glabrous except for margins; **flowers** to about 1.3 cm long, pale pink or deep pink to reddish-purple, exterior hairy, bisexual.

This outstanding, well-known coastal species usually inhabits rocky headlands and sand dunes where plants grow in heath and shrubland. Popular in cultivation, it was introduced in England in 1824, as *P. decussata*. Over recent years, a number of cultivars have been introduced (see below). The most commonly cultivated selection has clear pink flowers. Plants need good drainage and a sunny or semi-shaded site is suitable. They are hardy to moderately heavy frosts and tolerate extended dry periods. Plants can lose their denseness with age but if pruned every 1–2 years bushy growth continues. Odd branches can die back and this is usually a symptom of Cinnamon Fungus, *Phytophthora cinnamomi*, which will eventually cause the plant to die. Diseased branches should be removed by pruning back to healthy growth because in some cases the disease has spread from where the fungus has entered the branch. This can happen from water splashing off infected soil. For information on *Phytophthora cinnamomi*, see Volume 1, page 168.

Some of the best known cultivars are:

'Bonne Petite' — sometimes also sold as 'Bon Petite'. It has lovely deep pink flowerheads.

'Magenta Mist' — produces deep pinkish-purple flowerheads.

'Pink Bouquet' — is a very compact selection with variegated foliage. It arose in Western Australia as a 'sport' of 'Bonne Petite'. The leaves are medium green with pale yellow margins. It is registered under the *Plant Breeders Rights Act.*

'Red' — differs in having reddish-purple flowers.

Some un-named selections with pastel pink or pale pink with white flowers have failed to be popular but can create a lovely effect when they are planted in mixed colour selections.

Propagate from seed, which can begin to germinate 15–70 days after sowing, or from cuttings of barely firm young growth. Cultivars need to be propagated vegetatively to retain their distinctive characteristics. This species has been successfully used as a grafting stock with *P. physodes*.

Pimelea filamentosa Rudge =
>*P. linifolia* Sm. ssp. *linifolia*

Pimelea filiformis Hook. f.
(thread-like)

Tas	
0.1–0.4 m × 0.6–2.5 m	Nov–Jan; also sporadic

Spreading, matting, dwarf **shrub** with glabrous young growth; **stems** prostrate to decumbent, very slender, many, often self-layering, glabrous; **leaves** 0.3–1.2 cm × 0.1–0.7 cm, elliptic, opposite, short-stalked, spreading to ascending, rarely crowded, glabrous, mid-to-deep green above, paler with prominent midrib below, apex bluntly pointed; **flowerheads** to about 2 cm across, terminal, often erect, many-flowered but usually loosely arranged, can be profuse; **floral bracts** absent; **flowers** about 0.8 cm long, white, pale pink, deep pink, or white with pink, exterior faintly hairy, glabrous, bisexual; **stamen** solitary.

A densely foliaged, mat-forming or semi-prostrate species from near Launceston in northern Tas. It grows in woodland and adapts extremely well to cultivation in temperate regions. Occurs in heavy or light acidic soils. Usually does best in a semi-shaded site but tolerates plenty of sunshine. Hardy to heavy frosts and extended dry periods. Responds well to pruning or shearing. Tolerates slightly alkaline soils. Excellent as a groundcover beneath shrubs and for cascading over rocks and down embankments. Propagate from cuttings of young growth, which root readily, or by division of layered stems.

Pimelea flava R. Br.
(yellow)

Vic, Tas, SA	Yellow Rice-flower
0.5–2.5 m × 0.3–1.5 m	Aug–Dec; also sporadic

Dwarf to medium **shrub**; young growth hairy; **branches** spreading to erect, slender; **branchlets** brown, becoming glabrous; **leaves** 0.2–2 cm × 0.2–1 cm, narrowly elliptic to nearly circular, opposite, short-stalked, spreading to ascending, often crowded, especially near ends of branchlets, yellow-green to mid green and glabrous above and below, apex blunt or pointed; **flowerheads** to about 1.5 cm across, terminal, erect, few to many-flowered, sometimes profuse and moderately conspicuous; **floral bracts** to about 1.2 cm × 1.1 cm, usually broadly elliptic, yellow to greenish-yellow; **flowers** about 0.6 cm long, bright yellow, exterior hairy, male and female.

The typical variant usually inhabits sandy soils. It is often present in coastal or near-coastal sites but is also found further inland. Not common in cultivation. Plants require at least moderately well-drained soils and usually prefer a semi-shaded site. They are hardy to most frosts. Suitable for gardens and containers (if pruned or clipped regularly).

The ssp. *dichotoma* (Schltdl.) Threlfall, Diosma Rice-

Pimelea flava ssp. *flava* × .55, flower × 3

Pimelea flava ssp. *dichotoma* × .8

flower, is from central-southern NSW, Vic and south-eastern SA. It occurs in sandy or heavy soils and is often associated with sclerophyll forest and mallee shrubland. Usually grows 0.3–1 m × 0.2–0.7 m and has white flowers with floral bracts to about 1 cm × 0.7 cm.

Both subspecies are known to have toxic properties and may cause vomiting if consumed. Propagate both subspecies from seed or from cuttings of barely firm young growth.

Pimelea ferruginea 'Magenta Mist'   D. L. Jones

Pimelea floribunda Meisn.
(abundant flowers)
WA
0.3–1 m × 0.3–1 m   July–Oct

Dwarf **shrub** with faintly hairy young growth; **branches** spreading to erect; **branchlets** yellowish to brown; **leaves** 0.7–4 cm × 0.2–1.5 cm, narrowly elliptic to ovate, opposite, spreading to erect, moderately crowded, mid green to bluish-green above, usually glaucous below, glabrous except for axillary hairs, apex softly pointed; **flowerheads** to about 5 cm across, terminal, pendent, many-flowered, can be profuse and very conspicuous; **floral bracts** 4 or rarely 6, 1–3 cm × 0.8–2 cm, ovate to broadly ovate, green with base red or pinkish, exterior glabrous, margins rarely hairy; **flowers** to about 1.7 cm long, white or cream, exterior finely hairy, bisexual or female.

An outstanding species from the northern Darling and Irwin Districts. It grows in sandy soils overlying limestone on the coast and further inland it is found on deep sand and lateritic breakaways. This species is grown to a limited degree and warrants being better known in cultivation. Plants should adapt to a range of freely draining soils in sites which receive plenty of sunshine. They are hardy to moderate frosts and withstand extended dry periods. Should make lovely container plants as well as valuable additions to garden planting. Propagate from seed or from cuttings of barely firm young growth.

Pimelea forrestiana F. Muell.
(after Sir John Forrest, 19–20th-century Western Australian explorer and parliamentarian)
WA
0.3–1.5 m × 0.2–1 m   June–Sept

Dwarf to small **shrub** with more or less glabrous young growth; **branches** many, ascending to erect; **branchlets** yellowish-green to brownish; **leaves** 0.6–4 cm × 0.15–0.6 cm, linear to narrowly elliptic, opposite, short-stalked, spreading to ascending, not crowded, mid-to-deep green and glabrous above and below, apex finely pointed; **flowerheads** spike-like, about 1 cm × 2.5 cm, initially compact becoming elongated, each terminal spike often has 2 axillary branches; **floral bracts** absent or 1 per spike; **flowers** to about 0.7 cm long, yellow, glabrous, male and female.

This pimelea occurs on rocky hillsides and amongst granite outcrops in the northern Avon, Ashburton, Austin and Fortescue Districts. It is not very ornamental but may be useful in semi-arid and arid regions. Plants must have excellent drainage and plenty of sunshine. They are hardy to moderately heavy frosts. Propagate from seed or from cuttings of barely firm young growth.

P. spiculigera F. Muell. is a closely allied, somewhat insignificant species which differs in having densely hairy pedicels. It has 2 subspecies. The typical subspecies occurs in the south-western Coolgardie and eastern Roe Districts of WA. It produces elongated and often interrupted flowerheads while ssp. *thesioides* (S. Moore) Rye has compact flowerheads and occurs over a wide range of drier parts of southern WA.

Pimelea gilgiana E. Pritz.
(after Ernst F. Gilg, 19–20th-century German botanist)
WA
0.4–1.2 m × 0.3–1 m   May–Sept

Dwarf to small, dioecious **shrub** with glabrous young growth; **branches** many, ascending to erect; **branchlets** glabrous, brown to greyish; **leaves** 0.3–2.5 cm × 0.15–0.7 cm, narrowly ovate to narrowly obovate, short-stalked, spreading to ascending, sometimes crowded on young branchlets, glabrous, mid green above and below, flat to concave, apex softly pointed; **flowerheads** about 2 cm across, terminal, erect to slightly pendent, many-flowered, can be profuse and conspicuous; **floral bracts** 0.5–1.5 cm × 0.7–1.5 cm, ovate to nearly circular, broadest and reddish-green to deep reddish-purple on female heads, green on male heads; **flowers** to about 1.3 cm long, male white or pink, female white, exterior hairy.

This handsome species is found on or along the coast in the Irwin District from Dirk Hartog Island to near Leeman. It grows in sand among limestone outcrops. It deserves to be better known in cultivation. Should adapt to freely draining acidic or alkaline soils in temperate and semi-arid regions. Plants need maximum sunshine. Frost tolerance is not known. It has potential for gardens and containers. Propagate from seed or from cuttings of barely firm young growth.

This species has been confused with *P. brevifolia* ssp. *modesta* and *P. hispida*. The former has much smaller leaves and white bisexual flowers, while *P. hispida* has

Pimelea filiformis   W. R. Elliot

broader leaves and pale pink to rose-pink bisexual flowers.

Pimelea glauca R. Br.
(glaucous)
Qld, NSW, Vic, Tas, SA   Smooth Rice-flower
0.1–1 m × 0.3–1.5 m   July–Feb; also sporadic

Dwarf spreading **shrub** arising from a taproot; young growth glabrous; **branches** few to many, prostrate to erect; **branchlets** glabrous; **leaves** 0.3–2 cm × 0.1–0.7 cm, normally narrowly elliptic to lanceolate, opposite, short-stalked, spreading to ascending, sometimes crowded, glabrous, bluish-green, midrib prominent, apex pointed; **flowerheads** to about 3 cm across, terminal, erect, compact, many-flowered, often profuse and very conspicuous; **floral bracts** 4, to 1.5 cm × 0.75 cm, lanceolate to ovate, exterior glabrous, margins hairy; **flowers** to about 1.5 cm long, creamy-white, exterior hairy, bisexual, sweetly scented.

An appealing, leafy species which has an extensive distribution from south-eastern Qld to south-eastern SA and northern and eastern Tas. It occurs in a range of soils and habitats in coastal to inland sites. Cultivated to a limited degree, it adapts well. Plants grow in moderately to very well-drained soils in sunny or semi-shaded sites. They tolerate heavy frosts and extended dry periods. Excellent for rockeries, containers and general planting. Responds well to hard pruning which rejuvenates mature plants. Propagate from seed or from cuttings of barely firm young growth.

Sometimes confused with *P. humilis* which is a suckering species with hairy stems.

Pimelea graciliflora Hook. = *P. sylvestris* R. Br.

Pimelea gracilis R. Br. =
          P. curviflora ssp. *gracilis* (R. Br.) Threlfall

Pimelea graniticola

Pimelea graniticola Rye
(growing in soil derived from granite)
WA
0.2–1 m × 0.2–1 m   Sept–Dec
    Dwarf **shrub** with glabrous young growth; **branches** many, spreading to erect; **branchlets** twiggy, yellow-green or reddish-brown to grey; **leaves** 0.4–1.7 cm × up to 0.1 cm, linear, alternate, short-stalked, ascending to erect, often incurved, crowded, glabrous, pale green to bluish-green, flat or concave, apex pointed or blunt; **flowerheads** about 2 cm across, terminal, erect, many-flowered, compact, often profuse and very conspicuous; **floral bracts** to about 40, about 0.8 cm × 0.2 cm, linear to narrowly triangular, pale green to bluish-green; **flowers** to about 1 cm long, cream or white, exterior prominently hairy, bisexual.
    This recently described (1988) species is poorly known and thought to be rare in nature. It inhabits granite outcrops in the central Avon District. Needs to be cultivated as a conservation strategy. Plants will need very well-drained acidic soils and plenty of sunshine. They should adapt to temperate or semi-arid regions. Hardy to moderate frosts. Propagate from seed or from cuttings of barely firm young growth.
    P. villifera is allied but differs in having hairy leaves and up to 20 floral bracts while *P. imbricata* has hairy, usually broader leaves.

Pimelea haematostachya F. Muell.
(blood-red spikes)
Qld   Pimelea Poppy
0.2–1 m × 0.2–0.8 m   June–Feb; also sporadic
    Dwarf semi-woody **herb** with a perennial rootstock; young growth glabrous; **stems** few, arising from base, greenish-yellow; **leaves** 1.8–8.5 cm × 0.15–2.8 cm, narrowly elliptic to narrowly ovate, opposite or alternate, spreading to ascending, not crowded, glabrous, green to glaucous, flat, midrib prominent below, apex bluntly pointed; **flowerheads** spike-like, about 5 cm across, elongating, terminal, many-flowered, very conspicuous; **floral bracts** 7–12, to 1.4 cm × 0.4 cm, narrowly ovate, hairy, deciduous; **flowers** to about 1.5 cm long, reddish tips with yellow base, exterior hairy, bisexual.
    This striking species inhabits grasslands on heavy soils in the Cook, North Kennedy, South Kennedy, Port Curtis, Leichhardt and Mitchell Districts. It is cultivated to a limited degree by enthusiasts. Plants are best suited to subtropical and seasonally dry tropical regions. Needs acidic soils with moderate drainage and prefers sun for most of the day. Plants are hardy to moderate frosts. Pruning of spent stems after flowering promotes vigorous young growth. Farm stock have possibly died from eating this species. Propagate from seed or when available from cuttings of barely firm young growth.
    The closely allied *P. decora* usually has elliptic to ovate leaves.

Pimelea halophila Rye
(salt-loving)
WA
prostrate–0.15 m × 0.3–0.8 m   Aug–Oct
    Dwarf, often cushion-like **shrub** with glabrous young growth; **branches** spreading to ascending; **branchlets** twiggy, yellow-green to purplish and grey; **leaves** to about 0.3 cm × 0.15 cm, elliptic to nearly circular, alternate, spreading to ascending, crowded, glabrous, medium green to bluish-green, apex bluntly pointed; **flowerheads** to about 1.3 cm across, terminal, compact, few- to many-flowered, rarely profuse; **floral bracts** 3–4, similar to leaves, exterior glabrous; **flowers** to about 0.3 cm long, tube pink, lobes white or cream, male and female.
    An unusual, recently described (1988) species which is known only from islands and the edges of salt lakes in the central Roe District. It is regarded as rare and endangered and should be cultivated as a conservation strategy. It is unknown whether plants will grow in non-saline soils. They will need moderately good drainage and maximum sunshine for any chance of success. Hardy to heavy frosts. If plants adapt to cultivation they have potential for gardens, rockeries and containers. Propagate from seed or possibly from cuttings of barely firm young growth.

Pimelea hewardiana Meisn.
(after Robert Heward, 19th-century English botanist)
Vic   Forked Rice-flower
0.3–1 m × 0.2–0.6 m   April–Nov; possibly sporadic
    Dwarf dioecious **shrub**; young growth hairy; **branches** spreading to erect; **branchlets** hairy, brownish, forking below flowers; **leaves** 0.3–1.2 cm × 0.1–0.4 cm, narrowly elliptic to obovate, opposite, short-stalked, spreading to erect, moderately crowded on branchlets, glabrous, mid green to dark green above, paler below, apex pointed or blunt; **flowerheads** to about 0.6 cm across, terminal or sometimes axillary, erect, few- to many-flowered, rarely profuse and highly conspicuous; **floral bracts** 4 or rarely 2, leaf-like; **flowers** to about 0.3 cm long, yellow, exterior hairy.
    A somewhat insignificant species from central and western Vic where it is most common in rocky sites such as gullies and in mallee scrubland. Evidently rarely cultivated. Needs excellent drainage and a warm to hot, sunny or semi-shaded site. Plants are hardy to most frosts and withstand extended dry periods. Propagate from seed or from cuttings of barely firm young growth.

Pimelea hirsuta Meisn. =
    P. latifolia ssp. *hirsuta* (Meisn.) Threlfall

Pimelea hispida R. Br.
(covered with coarse hairs)
WA   Bristly Pimelea
0.4–1.5 m × 0.3–1 m   Sept–Dec
    Dwarf to small **shrub** with glabrous young growth; **branches** spreading to erect; **branchlets** glabrous, reddish-brown to greyish-brown; **leaves** 0.9–3 cm × 0.2–1 cm, more or less elliptic, opposite, short-stalked, spreading to ascending, often incurved, rarely crowded, glabrous, mid green above, paler green below, flat or margins recurved, midrib prominent, apex bluntly pointed; **flowerheads** to about 3.5 cm across, terminal, erect, compact, many-flowered, often profuse and very conspicuous; **floral bracts** 4, 1–2 cm × 0.7–1.3 cm, broadly ovate, green with pink or

yellow markings, exterior glabrous; **flowers** to about 1.5 cm long, pale pink to rose-pink, exterior prominently hairy, bisexual or rarely female.

A splendid species from the coastal and nearby regions of the southern Darling and far-western Eyre Districts. It grows in sandy soils which may be subject to periodic waterlogging. Rarely encountered in cultivation, it is deserving of greater attention. Plants should adapt to a range of well-drained acidic soils in temperate regions and will need a sunny or semi-shaded site. Tolerates moderate frosts. Likely to be useful in gardens and containers. It was introduced to cultivation in England during 1830. Propagate from seed, which may take 60 days to begin to germinate after it is sown. Cuttings of barely firm young growth are worth trying.

The var. *lanata* (R. Br.) Diels & Pritz. is a synonym of *P. lanata* R. Br. which differs in its floral bracts having distinct yellow margins.

Pimelea holroydii F. Muell.
(after Arthur T. Holroyd, 19th-century NSW parliamentarian and patron of horticulture)
WA
0.3–1 m × 0.2–0.8 m   Jan–Feb; Aug–Oct
Dwarf **shrub** with glabrous young growth; **branches** spreading to ascending; **branchlets** reddish-brown to yellow-brown, glabrous; **leaves** 1.2–2.7 cm × 0.6–2 cm, ovate to broadly ovate, opposite or sometimes alternate, somewhat stem-clasping, spreading to ascending, not crowded, glabrous, mid green to bluish-green above and below, apex blunt; **flowerheads** to about 3.5 cm across, elongating with age, terminal, erect, compact, many-flowered, can be profuse and very conspicuous; **floral bracts** 4–7, 0.9–2 cm × 0.7–1.6 cm, green to bluish-green, exterior glabrous except for hairy apex; **flowers** to about 1.5 cm, white to cream, exterior prominently hairy, bisexual.

An attractive species from the Ashburton and Fortescue Districts where it inhabits red clay soils in the Pilbara region. It is cultivated to a very limited degree. Plants are best suited to warm temperate and semi-arid regions. They should adapt to a range of soil types and will need plenty of sunshine. Hardy to moderate frosts and extended dry periods. It may succeed as a container plant in cooler regions. Propagate from seed, which usually begins to germinate 15–30 days after sowing, or from cuttings of barely firm young growth.

P. decora and *P. haematostachya* are allied but differ markedly in having red and cream, or red and yellow flowers respectively.

Pimelea humilis R. Br.
(low, small)
NSW, Vic, Tas, SA   Common Rice-flower;
                                    Dwarf Rice-flower
0.2–0.5 m × 0.4–1.2 m   Sept–Jan
Spreading to erect, often suckering dwarf **shrub**; young growth hairy; **branches** few to many, prostrate to erect; **branchlets** brown, hairy; **leaves** 0.6–1.6 cm × 0.15–1 cm, mainly narrowly elliptic to lanceolate, opposite, short-stalked, spreading to ascending, often

Pimelea humilis × .85

somewhat crowded, glabrous, mid green to greyish-green above, paler below, midrib prominent, apex somewhat blunt; **flowerheads** to about 2.5 cm across, terminal, erect, moderately compact, many-flowered, often profuse and very conspicuous; **floral bracts** 4 or 6, to about 1.8 cm × 1.1 cm, ovate, green, exterior glabrous except for hairy apex; **flowers** to about 1.5 cm long, creamy-white, exterior hairy, bisexual or female, sweetly scented.

This species often develops as a pleasing, lightly suckering plant. It occurs over a wide range from south-eastern NSW to the Mt Lofty Ranges, SA, and to north-eastern Tas. It is found in a variety of soil types and adapts very well to cultivation in temperate regions. Soils need to drain moderately well and a sunny or semi-shaded site is suitable. Plants are hardy to most frosts and withstand extended dry periods. They respond well to pruning and mature plants can be sheared hard to rejuvenate them. May be used in general planting under shrubs, also in rockeries, borders and containers. Flowers attract nectar-seeking butterflies on sunny days. Introduced to cultivation in England during 1824. Propagate from seed, or from cuttings of barely firm young growth, which strike readily. Young sucker growth is ideal for cuttings.

This species is sometimes confused with *P. glauca*, which has glabrous branches.

Pimelea hypericina A. Cunn. =
       P. ligustrina ssp. *hypericina* (A. Cunn.) Threlfall

313

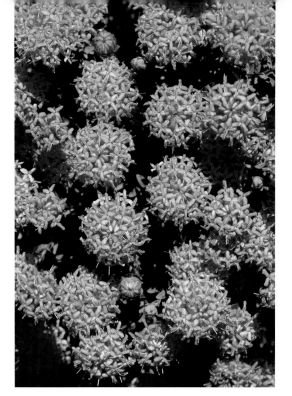

Pimelea imbricata ssp. *imbricata*   W. R. Elliot

Pimelea imbricata R. Br.
(overlapping)
WA
0.1–1 m × 0.4–1.5 m   Oct–Jan; also sporadic
Dwarf **shrub** with hairy young growth; **branches** many, spreading to ascending; **branchlets** hairy, brownish; **leaves** 0.1–1.6 cm × up to 0.5 cm, narrowly elliptic, alternate or sometimes opposite, spreading to ascending, crowded, mid green to greyish-green, usually glabrous except for hairy margins but sometimes sparsely to densely hairy, apex pointed; **flowerheads** to about 2.5 cm across, terminal, erect, compact, many-flowered, often profuse and very conspicuous; **floral bracts** usually 10–20, narrowly ovate to ovate, exterior hairy, margins entire; **flowers** to about 1 cm long, pale to deep pink, exterior hairy, bisexual.

A very attractive and variable species with 5 varieties. The typical variety occurs on or near the southern coast, from Albany to near Point D'Entrecasteaux in the west. It grows in sandy to gravelly soils of granitic areas which are subject to seasonal inundation. Adapts well to cultivation but may be short-lived. Suited to a range of moderately well-drained, acidic clay loam, loam or sandy soils with a sunny or semi-shaded aspect. Plants are hardy to light frosts but dislike extended dry periods. They respond very well to pruning or shearing which can promote further flowering in late summer or early autumn. Plants are excellent for general planting, rockeries and containers.

The further varieties have similar cultivation requirements.

P. imbricata var. *major* (Meisn.) Rye
This variant is restricted to the coastal plains of the Darling District, between Gin Gin and Serpentine, where it grows in sand overlying clay or sandy clay soils which are wet for extended periods in winter and spring. It has glabrous stems and the leaves are usually opposite. Flowers are white in heads to about 1.8 cm across. *P. imbricata* var. *gracillima* Meisn. and *P. microcephala* var. *major* Meisn. are synonyms.

P. imbricata var. *nana* (Graham) Ostenf. —
see var. *piligera* (Benth.) Diels & E. Pritz. below.
P. imbricata var. *petraea* (Meisn.) Rye
This densely hairy variety has alternate leaves. Plants occur in the Eyre and Yorke Peninsulas, and in the Flinders Ranges of SA. Usually found in hilly locations. The cream to white flowers are in heads to about 2.5 cm across. *P. petraea* Meisn. is a synonym.
P. imbricata var. *piligera* (Benth.) Diels & E. Pritz.
A handsome variant which has a widespread distribution extending from the northern Irwin District to east of Esperance in the Eyre District. It is found in coastal and inland regions growing in sandy or loam soils which may be seasonally waterlogged. Alternate leaves and stems are hairy. Flowers are white to cream in heads to about 2 cm across and usually profuse.
This variety has been referred to as var. *nana* (Graham) Ostenf. but this is a doubtful name.
P. imbricata var. *simulans* Rye
This poorly known variety occurs inland in the central Avon District where it is found in sandy soils which often contain lateritic gravel. It has more-or-less opposite leaves on glabrous branchlets and white flowers in heads to about 1.5 cm across.

All the varieties can be propagated from seed, which can begin to germinate 20–55 days after sowing, but cuttings of barely firm young growth usually strike readily and provide better results.

Pimelea incana R. Br. = *P. nivea* Labill.

Pimelea interioris Rye
(of the interior)
NT
0.7–1.5 m × 0.5–1.5 m   July–Nov
Dwarf to small **shrub** with hairy young growth; **branches** spreading to ascending; **branchlets** hairy when young; **leaves** 1.5–3 cm × 0.25–0.5 cm, narrowly elliptic, usually alternate, short-stalked, spreading to ascending, bright green, hairy but may become glabrous above, paler and hairy below, midrib prominent, apex softly pointed; **flowerheads** small in few-flowered axillary clusters; **floral bracts** absent; **flowers** to about 1 cm long, creamy-white to pale yellow, exterior very hairy, male and female.

A recently described (1990) species which is known only from a small region in southern NT where it grows in rocky gullies in often sheltered sites. Apparently not cultivated, it is not a very ornamental species but warrants cultivation as part of a conservation strategy. Plants are suited to semi-arid and arid regions and would need excellent drainage. May require a semi-shaded site for best growth. Frost hardiness is not known but plants should tolerate light frosts. Propagate from seed or from cuttings of barely firm young growth.

Pimelea lanata R. Br.
(woolly)
WA
1–4 m × 0.6–2.5 m   Dec–Feb; also sporadic

Small to medium **shrub** with glabrous young growth; **branches** many, slender, ascending to erect; **branchlets** reddish-brown to grey-brown; **leaves** 0.9–2.5 cm × 0.2–1 cm, mainly narrowly elliptic, opposite, usually short-stalked, ascending to erect, often incurved, mid green with reddish to yellowish margins, glabrous, flat or concave, apex softly pointed; **flowerheads** to about 2.5 cm across, terminal, erect, compact, many-flowered, often profuse and very conspicuous; **floral bracts** 4, 0.6–1.3 cm × 0.4–0.8 cm, green with cream, yellowish or pink-tinged margins, exterior glabrous; **flowers** to about 1.2 cm long, white or pale to deep pink, or base white with upper pink, exterior very hairy, bisexual.

This species can be striking when in full flower. It occurs in the coastal plains of the Southern Darling and far-western Eyre Districts where it usually grows in moist-to-wet sandy soils which are often overlying clay. Not commonly cultivated, it deserves wider recognition. Plants are suitable for temperate regions and will need moisture-retentive but moderately well-drained acidic soils and a site which is sunny or semi-shaded. Pruning can help to promote bushy growth. Hardy to light frosts. Best suited to garden planting but may do well as a container plant if pruned regularly. Propagate from seed or from cuttings of barely firm young growth.

P. hispida var. *lanata* (R. Br.) Diels & E. Pritz. is a synonym.

Pimelea latifolia R. Br.
(broad leaves)
Qld, NSW
0.7–3 m × 0.5–2 m   Aug–Oct; (sporadic April–July)

Dwarf to medium **shrub** with hairy young growth; **branches** ascending to erect; **branchlets** hairy, green to brownish; **leaves** 1–9.5 cm × 0.4–3 cm, narrowly elliptic to narrowly obovate, alternate or opposite, short-stalked, spreading to ascending, rarely crowded, hairy but may become glabrous above, venation visible, apex pointed; **flowerheads** to about 2 cm across, terminal, erect, few-flowered, rarely profuse and very conspicuous; **floral bracts** absent or similar to leaves; **flowers** to about 1 cm long, white, exterior hairy, bisexual or female.

The typical subspecies of this variable species is widely distributed. It extends down the eastern region from near Cairns to near Lismore, NSW. Plants usually grow on rainforest margins in good quality soils. Evidently rarely cultivated because of its poor ornamental qualities. Plants are best suited to subtropical and tropical regions and should adapt to a range of freely draining acidic loam and clay loam soils. Prefers semi-shaded or shaded sites.

P. latifolia ssp. *altior* (F. Muell.) Threlfall

This dwarf to small shrub is 0.3–2 m × 0.2–1.5 m. It has narrowly elliptic to nearly circular hairy leaves of up to 4.4 cm × 2 cm. The white flowers are in few-flowered heads to about 1 cm across. It occurs in south-eastern Qld and north-eastern NSW in or adjacent to rainforest margins. *P. altior* F. Muell. is a synonym.

P. latifolia ssp. *elliptifolia* Threlfall

Develops into a dwarf shrub of 0.1–0.5 m × 0.1–0.4 m. Leaves are mainly narrowly elliptic and 0.5–1.8 cm × 0.3–1 cm with hairy margins. The yellow to greenish-yellow flowers are in small heads. It is distributed in the North Coast, Central Tablelands and Central Western Slopes of NSW and occurs in rocky sites.

P. latifolia ssp. *hirsuta* (Meisn.) Threlfall

This subspecies grows to about 0.5 m × 1 m. It has slender hairy branches with more-or-less elliptical, hairy leaves of 0.2–1.5 cm × up to 0.6 cm. The yellow to greenish-yellow flowers are in small few-flowered heads. It is found mainly along the coast and hinterland between Newcastle and Nowra, NSW, where it usually grows in good quality soils in sclerophyll forest.

Propagate all subspecies from seed or from cuttings of barely firm young growth.

Pimelea lehmanniana Meisn.
(after J. G. C. Lehmann, 19th-century German botanist)
WA
0.3–1.2 m × 0.2–1 m   Aug–Nov

Dwarf to small **shrub** with glabrous young growth; **branches** ascending to erect; **branchlets** glabrous, deep reddish-brown to grey-brown; **leaves** 0.4–3.5 cm × 0.1–0.8 cm, opposite, narrowly ovate to narrowly obovate, short-stalked, spreading to ascending, somewhat crowded, glabrous, deep bluish-green above, yellowish-green to grey-green below, flat, apex softly pointed and sometimes recurved; **flowerheads** to about 4 cm across, terminal, erect or pendent, many-flowered, often profuse and very conspicuous; **floral bracts** 4 or 6, to about 2 cm × 1.4 cm, ovate to broadly ovate, pale yellow-green, can have reddish tonings; **flowers** to about 1.5 cm long, white to pale yellow, exterior hairy all over, bisexual or sometimes female.

Pimelea ligustrina ssp. *ciliata*   D. L. Jones

Pimelea leptospermoides

This is a highly ornamental species. The typical subspecies occurs in the southern Darling and western Eyre Districts where it is found in hilly country growing in sandy and gravelly soils. Uncommon in cultivation, it has potential for greater use in gardens and as a container plant in temperate regions. Plants need freely draining acidic soils and a sunny or semi-shaded site. They tolerate light frosts. Tip pruning of young plants promotes bushier growth.

The var. *nervosa* (Meisn.) Rye is from the southern Darling District. It is distributed from east of Perth to near Donnybrook and it usually occurs in lateritic soils. The white flowers often have pink tonings when in bud and the flowerheads are usually erect. The lowest part of the flower is glabrous.

Propagate both subspecies from seed, which can begin to germinate 35–60 days after sowing, or from cuttings of barely firm young growth.

P. lehmanniana var. *ligustrinoides* Benth. is a synonym of *P. rara* Rye.

Pimelea leptospermoides F. Muell.
(similar to the genus *Leptospermum*)
Qld
0.3–1 m × 0.2–0.8 m   May–Oct

Dwarf **shrub** with hairy young growth; **branches** ascending to erect; **branchlets** initially hairy, becoming glabrous; **leaves** 0.7–2.2 cm × 0.2–0.7 cm, elliptic to narrowly obovate, alternate, short-stalked, ascending to erect, often crowded, usually hairy, olive-green to deep bluish-green above, paler below, midrib prominent, apex pointed; **flowerheads** small, few-flowered axillary clusters, usually partially hidden by leaves; **floral bracts** absent; **flowers** to about 1 cm long, white, exterior densely hairy, bisexual or male.

This is a species with interesting foliage. It is known from the Port Curtis District where it is regarded as vulnerable. It occurs on sandy clay soils in sclerophyll forest and on stony hillsides. Should do best in subtropical and seasonally dry tropical regions. Needs freely draining soils in a sunny or semi-shaded site. May tolerate light frosts. Trials of this plant for cut-foliage production may be worthwhile. Propagate from seed or possibly from cuttings of barely firm young growth.

This species has been included in *P. umbratica*, which differs in being taller, having densely hairy branchlets and opposite leaves.

Pimelea leptostachya Benth. =
P. *sericostachya* F. Muell. ssp. *sericostachya*

Pimelea leucantha Diels
(white flowers)
WA
0.4–2 m × 0.3–1.5 m   Aug–Oct; also sporadic

Dwarf to small **shrub** with glabrous young growth; **branches** ascending to erect; **branchlets** glabrous, dark grey to brownish; **leaves** 0.8–2.7 cm × 0.15–0.7 cm, narrowly ovate-elliptic to ovate, opposite, short-stalked, ascending to erect, somewhat crowded near ends of branchlets, glabrous, mid green to bluish-green above, paler below, margins recurved to revolute, apex softly pointed; **flowerheads** to about 5 cm across, terminal, erect, compact, many-flowered, can be profuse and

very conspicuous; **floral bracts** 4 or rarely 6, to about 2.5 cm × 1.4 cm, ovate, green to yellow-green and often with pink tonings, exterior glabrous except for lower margins; **flowers** to about 1.8 cm long, cream to pale yellow, exterior hairy, bisexual.

A very handsome pimelea from the Irwin and northern Darling Districts where it occurs in sandy soils which can have sandstone or limestone outcrops. Cultivated to a limited degree, it warrants being better known in gardens. Plants should do best in warm temperate regions. They require excellently drained soils and a warm to hot, sunny or semi-shaded site. Pruning young plants promotes bushy growth. They tolerate light frosts and extended dry periods. Suitable as container plants. Propagate from seed, which may begin to germinate 26–60 days after sowing, or from cuttings of barely firm young growth.

P. spectabilis differs in having flat leaves, pink-budded flowers, slightly larger flowerheads and more-or-less glabrous floral bracts.

Pimelea ligustrina Labill.
(similar to the privet genus *Ligustrum*)
Qld, NSW, Vic, Tas, SA   Tall Rice-flower
0.5–2.5 m × 0.5–3 m   Oct–Dec; also sporadic

Dwarf to medium **shrub** with glabrous or faintly hairy young growth; **branches** many, spreading to ascending; **branchlets** mainly glabrous, brownish; **leaves** 1–8 cm × 0.2–2.5 cm, lanceolate to broadly elliptic, opposite, short-stalked, spreading to ascending, rarely crowded, becoming glabrous, mid green to deep-green above, paler below, flat or margins slightly recurved, venation prominent, apex pointed; **flowerheads** to about 4 cm across, semi-globular, terminal on more or less glabrous stalks, erect, very compact, many-flowered, often profuse and very conspicuous;

Pimelea ligustrina ssp. *ciliata* × .5

316

Pimelea linifolia ssp. *collina* × .95

Pimelea linifolia ssp. *linifolia* × .75

floral bracts usually 4, rarely 5–8, to about 1.6 cm × 1 cm, elliptical to ovate, green, often with reddish tonings, exterior glabrous except for occasionally hairy margins; **flowers** to about 1.7 cm long, whitish to cream, moderately fragrant, exterior hairy, bisexual and female.

This wide-ranging and variable species is regarded as one of the most ornamental of the eastern pimeleas. The typical subspecies has an extended distribution from the Moreton District in south-eastern Qld, through eastern NSW, eastern and southern Vic, far south-eastern SA and also in Tas. It is found from the coast to about 1400 m altitude, mainly in open forest and wet sclerophyll forest. It adapts well to cultivation in temperate and subtropical regions, growing in a range of freely to moderately well-drained acidic soils. Tolerates plenty of sunshine but prefers a semi-shaded site. Hardy to most frosts. Plants respond very well to pruning or clipping and they are suitable for hedging. Also useful for general planting. It was introduced into cultivation in England during 1823.

The ssp. *ciliata* Threlfall is from south-eastern NSW and eastern Vic. It occurs in subalpine areas at 1400–2000 m altitude. The leaves are to about 4.5 cm × 2 cm and it has prominent, attractive reddish to purplish floral bracts which have hairy margins. Plants have excellent potential for cultivation.

The ssp. *hypericina* (A. Cunn.) Threlfall grows to about 3 m tall and occurs in the North and Central Coast as well as the North and Central Tablelands. It grows on the margins of rainforest and wet sclerophyll forests. Floral bracts are silky-hoary on both surfaces and the flowerhead stalks usually have golden hairs. It was introduced to cultivation in England during 1834 as *P. hypericina* but is rarely grown in Australia.

The fruiting heads of this species can cause skin irritation for people with sensitive skin.

Propagate all variants from seed or from cuttings of barely firm young growth.

P. ligustrina var. *macrostegia* Benth. is a synonym of *P. macrostegia* (Benth.) J. M. Black.

Pimelea linifolia Sm.
(leaves similar to those of the genus *Linum*)
Qld, NSW, Vic, Tas, SA   Queen of the Bush;
        Slender Rice-flower; Flax-leaf Rice-flower
0.1–1.5 m × 0.3–1.5 m   July–Jan; also sporadic
Dwarf to small **shrub** with glabrous young growth; **branches** few to many, spreading to ascending, slender; **branchlets** glabrous, yellow-brown to reddish-brown; **leaves** 0.4–3 cm × 0.1–1 cm, mainly narrowly elliptic, opposite, short-stalked, spreading to ascending, sometimes crowded, glabrous, green, sometimes with reddish to purplish tonings, margins usually incurved, midrib prominent, apex pointed; **flowerheads** to about 4 cm across, terminal, erect or pendent, semi-compact, few- to many-flowered, often profuse and very conspicuous; **floral bracts** 4 or rarely 8, to

317

Pimelea longiflora

about 2 cm × 1 cm, ovate, green and usually with reddish toning, exterior glabrous; **flowers** to about 2 cm long, white, pink or rarely yellow, exterior finely hairy.

The typical subspecies is widely distributed in sometimes disjunct populations from near Cairns, Qld, to south-eastern SA. It occurs in a wide range of habitats from coastal to mountain regions. Moderately popular in cultivation, it adapts to a variety of soils which have excellent to moderate drainage. A semi-shaded site seems best but plants tolerate plenty of sunshine. They are hardy to most frosts. Plants may become leggy but respond well to hard pruning when they are growing well. This helps promote bushy growth. Selections such as 'Pygmy White' and 'Snowfall' from exposed coastal areas often retain their compact growth habit in cultivation. An elegant, slender-stemmed, pink-flowered selection was popular in the 1980s, but rarely could be maintained in cultivation for more than 5 years. Plants are suitable for container cultivation and for rockeries as well as for general planting. First introduced into cultivation in England in 1793 as *P. linifolia* and subsequently as *P. filamentosa* and *P. paludosa* in 1826, then as *P. cernua* in 1834.

This extremely variable species has 3 other subspecies.

P. linifolia ssp. *caesia* Threlfall

This variant occurs mainly at high altitudes in rocky sites on the Great Dividing Range. The distribution extends from the Northern Tablelands and North Western Slopes in NSW to near Mt Howitt in north-eastern Vic. It is a dwarf shrub to 0.5 m × 1 m with greyish-green leaves to 1.8 cm × 0.6 cm. The white to pink flowers appear during Oct–Feb. Deserves greater attention for cultivation.

P. linifolia ssp. *collina* (R. Br.) Threlfall

A dwarf to small shrub of 0.2–1.2 m × 0.3–1.m. It occurs in the Burnett, Darling and Moreton Districts in Qld, and extends to the Northern and Central Tablelands, North Western Slopes and Western Plains in NSW. Plants are found in damp heaths and sclerophyll forests. They have green leaves to 2.6 cm × 0.4 cm with prominent veins. Flowers are white, pink or yellow and bloom most of the year with a peak in Aug–Jan. *P. collina* R. Br. is a synonym and was introduced to cultivation in England in 1824 under that name.

P. linifolia ssp. *linoides* (A. Cunn.) Threlfall

A very ornamental variant with a disjunct distribution. It occurs in the Blue Mountains, NSW, the Grampians, Vic, on King Island and in western and southern Tas. Plants are found in sandstone ranges which receive a moderate to high rainfall. Rarely cultivated, it is deserving of greater attention. Plants grow to 0.3–2.5 m × 0.3–1.8 m and have linear to oblanceolate, dark bluish-green to deep green leaves of 0.4–4 cm × 0.1–0.7 cm, which are paler above than below. White flowerheads contrast with the green and reddish floral bracts, and bloom mainly in Aug–Feb.

P. linoides A. Cunn. is a synonym and was introduced to cultivation in England under that name in 1824.

Propagate all subspecies from seed, or from cuttings of barely firm young growth, which usually strike readily.

Pimelea longiflora R. Br.
(long flowers)
WA
0.3–1.5 m × 0.2–1 m   Aug–Feb

Dwarf to small **shrub** with hairy young growth; **branches** ascending to erect, slender; **branchlets** hairy, reddish-brown; **leaves** 0.4–1.8 cm × up to 0.3 cm, linear to narrowly elliptic, alternate to opposite, ascending to erect, rarely crowded, mid-to-deep green, becoming more or less glabrous above, usually hairy below, apex pointed; **flowerheads** to about 3 cm across, terminal, erect, moderately compact to somewhat loose, can be profuse and very conspicuous; **floral bracts** 4–6, to 1.2 cm × 0.3 cm, narrowly ovate to ovate, mid-to-deep green, exterior hairy, interior glabrous; **flowers** to about 1.2 cm long, white or cream, exterior hairy, bisexual or rarely female.

The typical subspecies is widespread in the southern Darling District from south of Bunbury extending to the Stirling Range. Usually found in seasonally waterlogged sand or sandy clay. It is cultivated to a limited degree now but was first introduced to cultivation in England during 1831. Plants are best suited to temperate regions and should adapt to most acidic soils which drain moderately well. A sunny or semi-shaded site is suitable. Hardy to light frosts. Bushy growth is promoted by pruning at an early stage.

The ssp. *eyrei* (F. Muell.) Rye differs in having densely hairy, narrowly elliptic leaves to about 1 cm × 0.25 cm. It flowers during Aug–Nov and has floral bracts which are hairy on both surfaces. Occurs in the Fitzgerald River National Park region. Previously known as *P. eyrei* F. Muell.

Propagate both subspecies from seed or from cuttings of barely firm young growth.

Pimelea macrostegia (Benth.) J. M. Black
(large floral bracts)
SA
0.7–1.5 m × 0.5–1 m   July–Feb

Dwarf to small **shrub** with glabrous young growth; **branches** spreading to erect; **branchlets** glabrous, yellow-brown to brown; **leaves** 0.7–3.3 cm × 0.2–1.3 cm, narrowly elliptic, opposite, short-stalked, spreading to ascending, moderately crowded, glabrous, pale bluish-green, paler below, leathery, prominent midrib below, apex pointed; **flowerheads** about 3 cm across, pendent, compact, many-flowered, often profuse and very conspicuous; **floral bracts** 4–6, to about 2.5 cm × 2 cm, ovate to broadly ovate, pale green, sometimes with purplish tonings, glabrous; **flowers** to about 1.7 cm long, pale yellow, exterior hairy, bisexual.

An appealing rice-flower which is confined to Kangaroo Island where it occurs in sandy soils in scrubland. It has had limited cultivation and is suitable for temperate regions. Needs freely draining acidic soils. Flowers best if grown in a sunny site but will tolerate a semi-shaded position. Hardy to moderate frosts. Suitable for gardens and also makes an ideal container plant if pruned immediately after flowering. Flowers attract many butterflies. Propagate from seed or from cuttings of barely firm young growth.

P. ligustrina var. *macrostegia* Benth. is a synonym.

Pimelea 'Magenta Mist' —
see *P. ferruginea* 'Magenta Mist'

Pimelea maxwellii F. Muell. ex Benth. =
P. brevifolia R. Br. ssp. *brevifolia*

Pimelea micrantha F. Muell. ex Meisn.
(small flowers)
NSW, Vic, SA, WA   Silky Rice-flower
0.1–0.5 m × 0.2–0.8 mz   Aug–Dec; also sporadic

Dwarf **shrub** with whitish-hairy young growth; **branches** many, spreading to ascending; **branchlets** whitish-hairy, slender; **leaves** 0.2–1.1 cm × 0.1–0.5 cm, narrowly elliptic to elliptic, alternate or opposite, ascending to erect, somewhat crowded, pale greyish-green, hairy above and below, apex pointed; **flowerheads** to about 0.5 cm across, terminal, few-flowered, can be profuse but rarely very conspicuous; **floral bracts** absent or leaf-like; **flowers** to about 0.4 cm long, whitish, exterior very hairy, bisexual; **stamens** not exserted.

A somewhat insignificant species which occurs in the drier regions of southern NSW, north-western Vic, southern SA and southern WA. Plants are found in a range of sandy, clay or rocky soils. Best suited to semi-arid and warm temperate regions. Prefers plenty of sunshine. Hardy to heavy frosts and withstands extended dry periods. Propagate from seed or from cuttings of barely firm young growth.

P. curviflora var. *micrantha* (F. Muell. ex Meisn.) Benth. and *P. curviflora* ssp. *micrantha* (F. Muell. ex Meisn.) Threlfall are synonyms.

Pimelea microcephala R. Br.
(small heads)
Qld, NSW, Vic, SA, WA, NT   Shrubby Rice-flower;
                               Mallee Rice-flower
1–4 m × 0.6–2.5 m   May–Oct; also sporadic

Small to medium **shrub** with more or less glabrous young growth; **branches** many, spreading to erect; **branchlets** mainly glabrous, reddish-brown to dark brown; **leaves** 0.6–6 cm × 0.1–0.7 cm, narrowly elliptic, opposite, short-stalked, spreading to ascending, not crowded, glabrous, mid green, flat or faintly concave, apex pointed; **flowerheads** to about 1 cm across, on hairy stalks, terminal, compact, few- to many-flowered, often profuse but somewhat inconspicuous; **floral bracts** 2–4, to 1.5 cm × 0.6 cm, leaf-like, glabrous; **flowers** to about 0.7 cm long, white, yellow or greenish, exterior hairy, male and female.

This species lacks strong ornamental features. It has a wide distribution over the drier regions of mainland Australia where it grows in a range of soils although it is more common in sand. Rarely encountered in cultivation, it is best suited to semi-arid and warm temperate regions with freely draining soils. Plants are hardy to moderately heavy frosts. May be useful for soil conservation purposes. It is recorded that birds feed on the seeds.

The ssp. *glabra* (F. Muell. & Tate ex J. M. Black) Threlfall has glaucous leaves to about 3.5 cm × 0.3 cm, and glabrous flowers. It is confined to north-western SA where it occurs in granitic soils.

Propagate both subspecies from seed, which may

Pimelea nivea   W. R. Elliot

begin to germinate 15–60 days after sowing, or from cuttings of barely firm young growth.

The somewhat insignificant but closely allied *P. pelinos* Rye flowers in June–July and differs in having flower clusters on short branchlets. It is a rare species confined to the margins of a number of salt lakes north of Esperance.

Pimelea milliganii Meisn.
(after Dr Joseph Milligan, 19th-century Tasmanian surgeon and naturalist)
Tas
0.3–1 m × 0.5–1.5 m   Dec–March

Dwarf **shrub** with densely hairy young growth; **branches** spreading to ascending; **branchlets** hairy; **leaves** 0.4–1.2 cm × 0.15–0.8 cm, elliptic, opposite, short-stalked, spreading to ascending, crowded, densely hairy, silvery-green above and below, apex bluntly pointed; **flowerheads** to about 1.5 cm across, terminal, compact, few-flowered; **floral bracts** usually 2, leaf-like; **flowers** to about 0.6 cm long, white, exterior densely hairy, usually bisexual.

A poorly known and apparently rare species from near the western coastal region where it occurs at 900–1250 m altitude. Evidently not cultivated and plants may not flower well at low altitude. Best results are likely to be achieved in cool temperate regions in well-drained acidic soils in semi-shaded or sunny sites. Hardy to most frosts and snowfalls. Propagate from seed or from cuttings of barely firm young growth.

Pimelea modesta Meisn. =
P. brevifolia ssp. *modesta* (Meisn.) Rye

Pimelea neo-anglica Threlfall
(from the New England region)
Qld, NSW   Poison Pimelea
1.5–3 m × 1–2 m   Sept–Oct; also sporadic

Small to medium **shrub**; young growth glabrous; **branches** ascending to erect; **branchlets** glabrous, slender, yellow-brown to brown; **leaves** 0.4–4.8 cm × 0.1–0.6 cm, narrow-elliptical, opposite, often falcate, spreading to ascending, rarely crowded, glabrous, mid green above and below, midrib prominent, apex pointed, strong odour; **flowerheads** to about 1 cm across, terminal few- to many-flowered, not very conspicuous; **floral bracts** 2 or rarely 4, to 2.3 cm × 0.3 cm, deciduous, narrowly elliptic to narrowly ovate, mid green; **flowers** to about 0.5 cm long, yellow to yellow-green, exterior glabrous or faintly hairy, male and female; **fruit** small, deep yellow to scarlet.

This species, which lacks significant ornamental qualities, occurs in south-eastern Qld and north-eastern NSW. Plants grow mainly in clay soils at 500–1000 m altitude. Suitable for subtropical and temperate regions. It is cultivated to a very limited degree and should adapt to a range of acidic soils in sunny or semi-shaded sites. Hardy to moderately heavy frosts. Propagate from seed or from cuttings of barely firm growth.

P. microcephala R. Br. is similar but it has flowers with a densely hairy exterior. *P. pauciflora* R. Br. is very closely allied and differs mainly in having glabrous flowers and flower stalks. It is distributed in south-eastern NSW and southern Vic where it grows mainly along waterways.

Pimelea nervosa Meisn. = *P. angustifolia* R. Br.

Pimelea nivea Labill.
(snow white)
Tas   Snowy Pimelea
0.6–3.5 m × 0.8–3 m   Sept–Feb; also sporadic

Dwarf to medium **shrub** with densely white-hairy young growth; **branches** many, spreading to erect; **branchlets** densely covered with white hairs; **leaves** 0.2–1.6 cm × 0.3–1.3 cm, elliptic to circular, opposite, very short-stalked, spreading to ascending, rarely crowded, glabrous to deep green above, dense white-woolly hairs below, flat, apex blunt; **flowerheads** to about 3 cm across, terminal, erect, moderately compact, many-flowered, often profuse and very conspicuous; **floral bracts** leaf-like; **flowers** to about 1 cm long, white to cream or sometimes pinkish, strongly fragrant, exterior densely hairy, bisexual or rarely female.

An exquisite species with attractive foliage and flowers. It has an extensive distribution in Tas where it occurs mainly in the eastern half from sea-level to about 750 m altitude. Plants grow in a variety of soils but they are often found in rocky soils which can be acidic or alkaline. Popular in cultivation for gardens and becoming so as a cut-flower/foliage crop. There is ample opportunity for selection of low-growing variants from coastal headlands. Plants do well in most soil types which do not dry out too readily. Grows in full sunshine or in a semi-shaded site. Hardy to most frosts. Responds very well to pruning or clipping and is useful for low hedging. It was first cultivated in England in

Pimelea nivea × .7

1824 as *P. incana* R. Br., which is now a synonym. Propagate from seed or from very young, barely firm cuttings, which may be slow to form roots.

P. sericea differs in having usually narrower and ascending leaves with recurved margins.

Pimelea octophylla R. Br.
(eight leaves)
Vic, SA   Woolly Rice-flower; Downy Rice-flower
0.4–1 m × 0.2–0.5 m   Aug–Feb; also sporadic

Dwarf **shrub** with woolly-hairy young growth; **branches** ascending to erect; **branchlets** initially covered in whitish-woolly hairs, becoming glabrous; **leaves** 0.4–2 cm × 0.1–0.6 cm, narrowly elliptic, alternate or nearly opposite, short-stalked, spreading to ascending, initially densely hairy, silvery to mid green or bluish-green, flat, convex or concave, apex pointed; **flowerheads** to about 3.5 cm across, terminal, usually pendent, many-flowered, often profuse and very conspicuous; **floral bracts** 6–12, to about 1.2 cm × 0.4 cm, narrowly ovate, densely hairy; **flowers** to about 1.7 cm long, cream to pale yellow, sweetly fragrant, exterior densely hairy, bisexual or female.

A very ornamental inhabitant of mainly sandy soils from south of Melbourne, Vic, to the Eyre Peninsula, SA. Cultivated to a limited degree but plants have been difficult to maintain. Must have excellent drainage and a sunny or warm to hot, semi-shaded site. Best results likely to be achieved in warm temperate regions. Tip pruning from planting can promote bushier growth. Hardy to moderately heavy frosts and tolerates extended dry periods. Propagate from seed or from cuttings of very young, barely firm growth.

Revision in 1988 resulted in name changes for subspecies.

The ssp. *ciliolaris* Threlfall is a synonym of *P. ciliolaris* (Threlfall) Rye. The ssp. *petraea* (Meisn.) Threlfall is a synonym of *P. imbricata* var. *petraea* (Meisn.) Rye and ssp. *subvillifera* Threlfall is a synonym of *P. subvillifera* (Threlfall) Rye.

Pimelea pagophila Rye
(lover of hills or mountains)
Vic
0.3–1.3 m × 0.5–1.5 m   Oct–Nov

Dwarf to small **shrub** with glabrous young growth; **branches** many, spreading to ascending; **branchlets** glabrous, greenish to purplish; **leaves** 0.7–2 cm × 0.2–0.5 cm, narrowly elliptic to elliptic, opposite, short-stalked, spreading to ascending, somewhat crowded, mid green above, paler below, midrib prominent, apex pointed; **flowerheads** to about 3 cm across, terminal, usually pendent, stalks hairy, moderately compact, many-flowered, often profuse and very conspicuous; **floral bracts** 4–8, to 1.7 cm × 1 cm, elliptic to ovate, yellow-green to green, usually with some or much purple, glabrous; **flowers** to about 1.5 cm long, white to cream, fragrant, exterior glabrous, bisexual.

A handsome, recently described (1990) species which is restricted to the mountains of the Grampians in western Vic. It is cultivated to a limited degree and

Pimelea pendens × .75

warrants greater attention. Plants are suited to temperate regions in well-drained acidic soils and they prefer a semi-shaded site. They are hardy to most frosts and tolerate extended dry periods. Pruning promotes bushy growth. Excellent for gardens and containers. Propagate from seed or from cuttings of barely firm young growth.

This species was confused with *P. linifolia* which has flowers with a hairy exterior while *P. bracteata* has glabrous flower stalks and much broader floral bracts.

Pimelea paludosa R. Br. = *P. linifolia* Sm. ssp. *linifolia*

Pimelea pauciflora R. Br. — see *P. neo-anglica* Threlfall

Pimelea pelinos Rye — see *P. microcephala* R. Br.

Pimelea pendens Rye
(hanging down, pendent)
WA
0.2–1 m × 0.1–0.6 m   May–Aug

Dwarf **shrub** with more or less glabrous young growth; **branches** ascending to erect; **branchlets** glabrous except for axillary hairs, grey-brown to reddish-brown; **leaves** 0.6–2 cm × 0.3–0.8 cm, narrowly elliptic to ovate, opposite, sessile, spreading to ascending, mid green to yellow-green, sometimes glaucous, glabrous, midrib prominent, apex softly pointed; **flowerheads** about 3 cm across, terminal, pendent, compact, many-flowered, sometimes profuse, always very conspicuous; **floral bracts** 2, 4 or 6, to 2.5 cm × 2 cm, broadly elliptic to broadly ovate, green to yellowish-green, often with reddish tonings, glabrous; **flowers** to about 1.5 cm long, pale green, exterior glabrous, bisexual or rarely female.

An outstanding pimelea from the granitic coastal region of the eastern Eyre District. It is found in sandy-clay soils on rocky slopes. Evidently rarely cultivated, it

Pimelea octophylla × .9

321

Pimelea penicillaris

Pimelea phylicoides   W. R. Elliot

Pimelea physodes   W. R. Ell

warrants greater attention and should adapt well to a range of acidic soils in temperate and semi-arid regions. A sunny site is preferable. Plants are hardy to moderate frosts. Tip pruning of young plants may promote bushy growth. Worth trying in containers as well as in gardens. Propagate from seed or from cuttings of barely firm young growth.

The allied *P. cracens* has hairy flowers and *P. aeruginosa* has yellow floral bracts.

Pimelea penicillaris F. Muell.
(like a paint brush)
Qld, NSW, SA, NT
0.6–2 m × 0.3–1.5 m   July–Oct; also sporadic

Dwarf to small, dioecious **shrub** with velvety-hairy young growth; **branches** many, ascending to erect; **branchlets** initially silvery-hairy, becoming brownish; **leaves** 0.4–2 cm × 0.25–1 cm, narrowly to broadly elliptic, alternate, short-stalked, spreading to ascending, sometimes recurved, silvery-silky hairs above and below, flat to convex, apex bluntly pointed; **flowerheads** about 2 cm across, terminal, erect, compact, many-flowered, can be profuse and very conspicuous; **floral bracts** 6–12, to 1.3 cm × 1.2 cm, broadly elliptic to ovate, often hairier than leaves; **flowers** to about 1 cm long, male yellow, female often pinkish, exterior densely hairy.

P. penicillaris has attractive foliage and flowers. It has a disjunct distribution on the Qld/NSW and SA/NT borders and also has an isolated occurrence in the Flinders Ranges. Usually occurs on sandy soils. Evidently not in cultivation, it warrants becoming better known. Suited to warm temperate and semi-arid regions in excellently drained soils with a warm to hot, sunny or semi-shaded aspect. Hardy to moderately heavy frosts. Propagate from seed or from cuttings of barely firm young growth.

Pimelea petraea Meisn. =
                  P. imbricata var. *petraea* (Meisn.) Ry

Pimelea petrophila F. Muell.
(rock-loving)
NSW, SA
0.3–1 m × 0.2–0.6 m   Aug–No

Dwarf **shrub** with hairy young growth; **branche** ascending to erect; **branchlets** slender, hairy, brow **leaves** 0.6–3.3 cm × 0.15–0.8 cm, narrowly elliptic elliptic, opposite, short-stalked, ascending to erec somewhat crowded, more or less glabrous, green bluish-green, flat to concave, apex pointed; **flowe heads** about 2 cm across, terminal, erect, compac many-flowered, often profuse and moderately co spicuous; **floral bracts** 2–4, to about 1.6 cm × 1 cn elliptical to ovate, glabrous, longer than flowerhead **flowers** to about 1 cm long, white, exterior hairy, ma and female.

An inhabitant of sandy loam soils in rocky sites semi-arid areas. It occurs in south-eastern SA and the Barrier Range of south-western NSW. Suited cultivation in drier regions. Needs good drainage an plenty of sunshine but may also do well in semi-shad Hardy to moderate frosts. Pruning may promo bushier growth. Propagate from seed or from cuttin of barely firm young growth.

Allied to *P. flava* which differs in having floral brac shorter than the flowerheads.

Pimelea phylicoides Meisn.
(similar to the genus *Phylica*)
Vic, Tas, SA   Heath Rice-flowe
0.2–0.6 m × 0.3–1 m   Aug–Jan; also sporad

Dwarf **shrub** with densely hairy young growt **branches** many, spreading to ascending; **branchle** hairy, brownish; **leaves** 0.2–1 cm × 0.1–0.3 cm, ellipt to narrowly ovate, alternate or opposite, short-stalked, spreading to erect, somewhat crowded, hairy to glabrous except for margins, mid green or rarely bluish-green, midrib prominent, apex bluntly pointed; **flowerheads** to about 2 cm across, terminal, erect, moderately compact, few-flowered, often profuse and conspicuous; **floral bracts** 3–6, to about 1 cm × 0.4 cm, narrowly ovate to ovate, exterior hairy; **flowers** to about 1.3 cm long, white, exterior hairy, bisexual.

Usually develops as an attractive, densely foliaged shrub. It occurs mainly in sandy soils and its distribution extends from the Eyre Peninsula, SA, to near Melbourne, Vic, and to Tas, including King Island. Plants require freely draining soils in a sunny or semi-shaded site. They should do best in temperate regions. Pruning from an early stage promotes bushy growth and better flowering. Hardy to moderate frosts. Propagate from seed or from cuttings of barely firm young growth.

Pimelea physodes Hook
(inflated, bladder-like)
WA   Qualup Bell
0.2–1.5 m × 0.1–0.8 m   July–Oct
   Dwarf to small, usually slender **shrub** with glabrous young growth; **branches** ascending to erect; **branchlets** yellowish or reddish near flowers; **leaves** 0.8–3.2 cm × 0.3–1.3 cm, narrowly elliptic to ovate, opposite, more or less sessile, mid green to greyish-green above and below, glabrous, prominent midrib, apex pointed or blunt; **flowerheads** to about 6 cm × 4.5 cm, bell-like, pendent, terminal, usually scattered, rarely profuse but extremely conspicuous; **floral bracts** 6 or 8 or rarely 4, enveloping the flowers, to about 6 cm × 4.5 cm, elliptic to broadly elliptic, pale green to cream with reddish or purplish colouration of varying degrees, glabrous; **flowers** to about 1.5 cm long, cream-green to green with reddish style, exterior hairy.

This is an outstanding pimelea and probably the most highly prized by enthusiasts. It is restricted to the Ravensthorpe–Fitzgerald National Park region where it grows in sandy or gravelly soils and is often found in rocky sites. The species has proved difficult to maintain in cultivation and is probably best grown as a container plant. Grafted plants are available on a limited scale but their longevity in gardens is as yet unknown. Plants need a warm to hot site with freely draining acidic soils. They are hardy to moderate frosts. Plants respond well to pruning from an early stage which can counteract legginess. Nectar-feeding birds visit the flowers. Propagate from seed or from cuttings of barely firm young growth. Some success has been achieved with tissue culture.

Pimelea preissii Meisn.
(after Ludwig Preiss, 19th-century German botanist)
WA
0.2–1 m × 0.1–0.8 m   Sept–Dec
   Dwarf **shrub** with glabrous young growth; **branches** spreading to ascending; **branchlets** glabrous, pale green to reddish when young; **leaves** 0.5–2 cm × 0.2–0.7 cm, narrowly elliptic, opposite, short-stalked, spreading to ascending, somewhat crowded, glabrous, mid green, flat to concave, apex blunt to pointed;

Pimelea rosea   W. R. Elliot

flowerheads to about 3 cm across, terminal, erect, compact, many-flowered, can be profuse and very conspicuous; **floral bracts** 4, to about 1.5 cm × 1 cm, ovate to broadly ovate, green, exterior glabrous; **flowers** to about 1.6 cm long, white, pink or very rarely red, exterior hairy, bisexual, styles sometimes only slightly exserted.

A pretty pimelea from the south-western corner of WA where it occurs on or near the coast or slightly inland and grows in sandy or clay soils which are usually subject to limited waterlogging over winter. Evidently rarely cultivated, it deserves wider attention. Plants are best suited to temperate regions and require moderately well-drained acidic soils with a sunny or semi-shaded aspect. Hardy to light frosts. Should respond well to pruning. Highly suited to containers as well as gardens. Propagate from seed, which may take up to 65 days to begin germination, or from cuttings of barely firm young growth.

Pimelea punicea R. Br. =
                       Thecanthes punicea (R. Br.) Rye

Pimelea pygmaea F. Muell. & C. Stuart ex Meisn.
(dwarf)
Tas
prostrate–0.1 m × 0.1–0.4 m   Sept–Dec
   Cushion-like dwarf **shrub** with hairy young growth; **branches** many, prostrate to ascending; **branchlets** covered by leaves; **leaves** 0.15–0.4 cm × 0.1–0.2 cm, elliptic, opposite, sessile, ascending to erect, crowded, overlapping, olive-green, initially hairy becoming more or less glabrous except for margins, apex pointed; **flowers** to about 0.8 cm long, white, solitary, terminal, exterior glabrous, can be profuse and conspicuous, male or female or bisexual.

323

Pimelea 'Pygmy White'

A very appealing, slow-growing, groundcovering species which is restricted to central Tas. Occurs in subalpine moorlands at about 1000 m altitude. Rarely cultivated, it deserves to be better known. It must have moist but well-drained acidic soils with a fairly open aspect to have any chance of success. Plants tolerate frost and heavy snowfalls. May not flower well at low elevations. It has potential for use in rockeries, miniature gardens and containers. Propagate from seed or from cuttings of barely firm young growth.

Pimelea 'Pygmy White' — see *P. linifolia* Sm.

Pimelea rara Rye
(rare)
WA
0.2–0.4 m × 0.2–0.5 m   Dec–Jan
Dwarf **shrub** with glabrous young growth; **branches** ascending to erect; **branchlets** glabrous, brownish; **leaves** 1.5–3 cm × about 0.8 cm, elliptic to obovate, opposite, very short-stalked, spreading to ascending, somewhat crowded at ends of branchlets, glabrous, dull bluish-green above and below, flat, prominent midrib, apex mainly bluntly pointed; **flowerheads** about 4 cm across, terminal, pendent or erect, compact, many-flowered, very conspicuous; **floral bracts** 4, to about 2 cm × 1 cm, broadly ovate, exterior glabrous; **flowers** to about 1 cm long, white, exterior mainly with long sparse hairs, bisexual.

An endangered species from the Darling Range, east of Perth, where it grows in lateritic soils. Rarely cultivated, it should adapt to a range of moderately well-drained acidic soils in temperate regions. Plants will need plenty of sunshine but will probably tolerate a semi-shaded site. Frost hardiness is unknown. Should make a lovely plant for rockeries or containers. Propagate from seed or from cuttings of barely firm young growth.

Pimelea rosea R. Br.
(rosy)
WA   Rose Banjine; Native Rose; Pink Banjine
0.3–1.3 m × 0.3–1.5 m   Aug–Jan
Dwarf to small **shrub** with glabrous young growth; **branches** many, spreading to ascending; **branchlets** glabrous, red-brown to greyish-brown; **leaves** 0.6–3 cm × 0.2–0.6 cm, narrowly elliptic, opposite, short-stalked, spreading to ascending, rarely crowded, glabrous, mid green to bluish-green above, pale green below, flat or with recurved margins, apex pointed; **flowerheads** to about 4 cm across, terminal, erect, moderately compact, many-flowered, often profuse and very conspicuous; **floral bracts** 4, to about 2 cm × 1 cm, ovate, green with yellowish to reddish base, glabrous except for hairy margins; **flowers** to about 1.5 cm long, pale to deep pink or reddish-purple, exterior hairy, bisexual.

This is certainly one of the outstanding pimeleas. It occurs mainly in coastal or nearby regions from near Perth to east of Albany. Often grows on coastal dunes and sandy clay soils but also found in granitic or limestone areas of the coast. Popular in cultivation but can be short-lived. Needs good drainage and plenty of sunshine to do well. Hardy to moderate frosts. Plants respond very well to pruning. They develop an excellent bushy framework if the pruning is started early. Does well in large containers and worth trying in sunny rockeries as well as for general planting. The species was first introduced in England during 1800 and was grown as *P. hendersonii* in 1837. Propagate from seed, which can begin to germinate 40–7? days after sowing. Cuttings of barely firm young growth strike readily.

Pimelea sericea F. Muell.
(silky)
Tas   Silky Pimele?
0.2–0.8 m × 0.3–1.3 m   Nov–Feb
Dwarf **shrub** with densely hairy young growth; **branches** many, ascending to erect; **branchlets** initiall? greyish-hairy, becoming glabrous and brownish; **leave?** 0.3–1.2 cm × 0.2–0.6 cm, elliptic to broadly elliptic, opposite, short-stalked, ascending to erect, crowded, deep green and glabrous above, whitish silky-hair? below, flat, apex pointed; **flowerheads** about 2 cm across, terminal, erect, compact, many-flowered, often profuse and very conspicuous; **floral bracts** leaf-like but broader; **flowers** to about 1 cm long, white, pink or pink tube with white lobes, exterior hairy, bisexual or sometimes female.

A very admirable rice-flower which grows a? 750–1400 m altitude on the moors. Rarely encoun? tered in cultivation, it is deserving of greater attention. Plants require moist but well-drained acidic soil which has plenty of organic matter. They will tolerate some sunshine but should do best in a semi-shaded site. Plants withstand heavy frosts and snowfalls. Pruning from an early stage promotes a bushy framework. Worth trying in containers, rockeries and for general planting. Propagate from seed, or from cuttings of barely firm young growth, which can be difficult to strike.

Pimelea sericostachya F. Muell.
(silky spikes)
Qld
0.3–0.6 m × 0.2–0.6 m   Feb–Sep
Dwarf **shrub** with densely hairy young growth; **branches** ascending to erect; **branchlets** hairy, initiall? greenish-yellow, becoming brown; **leaves** 0.4–3.2 cm × 0.1–0.8 cm, narrowly elliptic, alternate or somewha? opposite, short-stalked, ascending to erect, rarel? crowded, initially with shining hairs, becoming glab? rous or finely hairy above, shining hairs below, ape? pointed; **flowerheads** to about 1.3 cm across, elonga? ed to about 13 cm long, terminal, erect, crowded o? loose; **floral bracts** absent; **flowers** to about 0.8 cm long, pale yellow to yellow-green, exterior hairy, b? sexual or female.

An interesting rather than ornamental species. The typical subspecies is from north-eastern Qld, in the Cook, North Kennedy and eastern South Kenned? Districts. It occurs in open forests on sandy soils. Probably best suited to subtropical and seasonally dr? tropical regions. Needs good drainage and a warm? semi-shaded site. Pruning of young plants may pro? mote lateral growth.

The ssp. *amabilis* (Domin) Threlfall differs i? having leaves of 0.9–4.2 cm × 0.3–1.4 cm which hav? shining hairs on both surfaces. The flowerhead?

elongate to about 7.5 cm long. It is known from the Cook District where it also grows in sandy soils of open forests.

Propagate both subspecies from seed or from cuttings of barely firm young growth.

Pimelea serpyllifolia R. Br.
(thyme-like leaves)
NSW, Vic, Tas, SA, WA   Thyme Rice-flower
0.2–1.5 m × 0.3–1.2 m   most of the year

Dwarf to small **shrub** with glabrous young growth; **branches** many, spreading to erect; **branchlets** glabrous, can be faintly spiny; **leaves** 0.2–1.3 cm × 0.1–0.4 cm, narrowly elliptic to obovate, opposite, spreading to ascending, crowded, mid green to bluish-green, glabrous, apex pointed; **flowerheads** to about 1 cm across, terminal, compact, few-flowered, often profuse, rarely highly prominent; **floral bracts** 2 or 4, elliptic to obovate, to about 0.6 cm × 0.4 cm, glabrous, green; **flowers** to about 0.3 cm long, yellow, white, or rarely yellow-green, exterior and interior glabrous, male and female.

The typical subspecies of this variable species has an extended distribution from far south-western NSW to the Bass Strait Islands and to just west of the SA/WA border. It is most common on calcareous soils in woodland and scrubland. Although not very ornamental, it is valuable for revegetation in its natural range. Plants adapt to most soil types and grow in sunny or semi-shaded sites. Responds well to pruning. Tolerates most frosts and extended dry periods.

The ssp. *occidentalis* Rye is a variant from Cocklebiddy–Caiguna region of south-eastern WA where it grows on calcareous soils near the coast. It has hairs on stems below the flowerheads and the interior of the usually cream flowers is hairy.

Propagate both subspecies from seed or from cuttings of barely firm young growth.

P. spinescens Rye is a closely allied species which is somewhat nondescript. It is distinguished by spiny-tipped branchlets which are unique for the genus. The typical subspecies occurs in central Vic and has glabrous flowers, while the ssp. *pubiflora* Rye has hairy flowers and occurs in western-central Vic.

Pimelea sessilis Rye
(sessile, without stalks)
WA
0.2–0.5 m × 0.2–0.6 m   Aug–Oct

Dwarf **shrub** with glabrous young growth; **branches** spreading to erect; **branchlets** glabrous, yellowish to red-brown, becoming deep brown; **leaves** 0.2–1.5 cm × 0.1–0.9 cm, elliptic to nearly circular, opposite, sessile and stem-clasping, spreading to ascending, somewhat crowded, glabrous, mid green to bluish-green, sometimes paler below, margins recurved, apex blunt with a small point; **flowerheads** to about 3.5 cm across, terminal, erect, moderately compact, many-flowered, often profuse and very conspicuous; **floral bracts** 4, to about 1.5 cm × 1.4 cm, broadly elliptic to nearly circular, green with reddish or yellowish markings; **flowers** to about 1.2 cm long, white to pale yellow, exterior with long hairs, bisexual or rarely female.

A very handsome, recently described (1988) species

from the Irwin District where it grows in shrublands on sandy soils which can contain lateritic gravel. Plants will need excellently drained acidic soils with a sunny or semi-shaded aspect. They will probably do best in warm temperate or semi-arid regions. Tolerates moderate frosts and extended dry periods. This species has potential for container cultivation as well as for rockeries and general planting. Propagate from seed or from cuttings of barely firm young growth.

Pimelea simplex F. Muell.
(simple)
Qld, NSW, Vic, SA   Desert Rice-flower
0.1–0.6 m × 0.1–0.5 m   June–Oct

Dwarf, herbaceous or semi-woody **annual** with hairy young growth; **branches** ascending to erect; **branchlets** green to yellow-green and brown, becoming glabrous; **leaves** 0.3–1.6 cm × 0.1–0.3 cm, linear to narrowly elliptic, alternate, short-stalked, ascending to erect, moderately crowded, mid green to deep bluish-green, becoming glabrous but margins usually hairy, apex pointed; **flowerheads** to about 1.5 cm × 2 cm, terminal, erect, compact to elongated, many-flowered, can be profuse and conspicuous; **floral bracts** absent; **flowers** to about 0.75 cm long, white to greenish-yellow, exterior hairy, bisexual.

An annual species with the typical subspecies distributed in drier regions from far-southern Qld to north-western Vic and southern SA. It occurs on loam or clay soils or rarely in sand. It has limited appeal and is best suited to warm temperate and semi-arid regions in a sunny or semi-shaded site. Hardy to moderately heavy frosts. Young plants need pruning to promote lateral growth.

The ssp. *continua* (J. M. Black) Threlfall differs in having leaves to about 3.5 cm long and the flowerhead stalks have long hairs. It occurs in dry areas of central Qld, western NSW and mainly northern SA.

There is intergradation between the 2 subspecies and sometimes it can be difficult to separate them botanically. Both subspecies are known to have toxic properties.

Propagate from seed or possibly from cuttings.

Pimelea spectabilis Lindl.
(remarkably spectacular)
WA   Bunjong; Banjine; Bush Rose
0.5–2 m × 0.6–2 m   Aug–Dec

Dwarf to small **shrub** with glabrous young growth; **branches** spreading to ascending, slender; **branchlets** glabrous, yellowish-green to brown; **leaves** 1–4.5 cm × 0.2–0.7 cm, narrowly elliptic, opposite, short-stalked, spreading to ascending, usually uncrowded, glabrous, pale-to-mid green to bluish-green above, paler below, flat, apex softly pointed; **flowerheads** to about 7 cm across, terminal, erect or pendent, compact, many-flowered, often profuse and very conspicuous; **floral bracts** 4 or 6, to 3 cm × 2 cm, ovate, green with red or nearly all reddish, exterior glabrous except for margins which can be hairy; **flowers** to about 2 cm long, white, pale pink, white with pink, or pale yellow, exterior hairy, bisexual or very rarely female.

Undoubtedly one of the most florally spectacular members of the genus with its large well-displayed

Pimelea spicata

Pimelea serpyllifolia   W. R. Elliot

flowerheads. It occurs in 3 disjunct areas, mainly in the southern Darling District, but also extends to the western Eyre District. Commonly grows in hilly locations in jarrah forest and is found in gravelly and stony, sandy soils. Plants can be short-lived. Cinnamon Fungus, *Phytophthora cinnamomi*, is usually the main pathogen. Best suited to temperate regions. Needs excellently drained acidic soils in a sunny or semi-shaded site. Plants often perform best when grown at their own rate without supplementary fertilising and watering. They tolerate moderate frosts. Plants respond extremely well to pruning, and tip-pruning from an early stage is beneficial. Butterflies are attracted to flowering plants. Introduced to cultivation in England in 1840. Plants grown in 1851 as *P. verschaffeltii* were possibly a selection of *P. spectabilis*.

Propagate from seed, which can begin to germinate 15–60 days after sowing, or from cuttings of barely firm young growth, which usually strike readily. Worth trying grafting trials with *P. ferruginea* as rootstock.

Pimelea spicata R. Br.
(with flower-spikes)
NSW
0.1–0.5 m × 0.3–1 m   May–Jan
Dwarf **shrub** with yellowish glabrous young growth; **branches** prostrate to ascending; **branchlets** glabrous, yellowish to brown; **leaves** 0.5–2.1 cm × 0.2–1.1 cm, narrowly elliptic to elliptic, more or less opposite, spreading to ascending, not crowded, glabrous, mid green above, paler below, midrib prominent, thin, apex bluntly pointed; **flowerheads** to about 2 cm across, terminal, initially compact, becoming elongated and raceme-like, few-flowered, often profuse but rarely highly conspicuous; **floral bracts** absent; **flowers** to about 1 cm long, white usually with pink tones, exterior mainly glabrous, bisexual.

An interesting, somewhat dainty but not highly ornamental species which is regarded as endangered in its natural habitat. It occurs in the Central Coast District along the coast and inland to the foothills of the Blue Mountains. Plants grow on clay soils. Cultivated to a limited extent in temperate regions. Should adapt to a range of acidic soils which drain well and have a sunny or semi-shaded aspect. Pruning promotes bushy growth. Hardy to moderate frosts. Compact low-growing variants are suitable for rockeries and containers. Propagate from seed, or from cuttings of barely firm young growth, which strike readily.

Pimelea spiculigera F. Muell. —
               see *P. forrestiana* F. Muell.

Pimelea spinescens Rye — see *P. serpyllifolia* R. Br.

Pimelea stricta Meisn.
(erect, upright)
NSW, Vic, SA   Gaunt Rice-flower
0.3–0.8 m × 0.4–1 m   most of the year
Dwarf to small **shrub** with glabrous young growth; **branches** ascending to erect, wiry; **branchlets** brown, glabrous; **leaves** 0.5–3.5 cm × 0.1–0.5 cm, linear to narrowly elliptic, opposite, short-stalked, ascending to erect, often somewhat overlapping, yellow-green, mid green or blue-green, glabrous, midrib prominent, apex softly pointed; **flowerheads** to about 2.5 cm across, terminal, compact, few- to many-flowered, can be profuse and moderately conspicuous; **floral bracts** 4, to about 1.3 cm × 1 cm, ovate, exterior silky-hairy; **flowers** to about 1.5 cm long, white to creamy-white, sometimes with pink lobes, exterior silky-hairy, bisexual.

This rice-flower occurs over a wide range, in drier regions, where it grows mainly in sandy soils which are alkaline or acidic. It is often seen as a pioneer plant after fires, and has a life-span of about 3–5 years. Evidently rarely cultivated, plants should adapt to a range of soil types and would be best suited to warm temperate and semi-arid regions in a sunny or semi-shaded site. Plants are somewhat open in habit, but can be bushier if pruned. Hardy to moderately heavy frosts. Propagate from seed or possibly from cuttings of barely firm young growth.

P. colorans A. Cunn. ex Meisn. is a synonym.

Pimelea strigosa Gand. — see *P. curviflora* R. Br.

Pimelea suaveolens Meisn.
(sweetly scented)
WA   Scented Banjine
0.3–1.3 m × 0.3–1.5 m   June–Oct
Dwarf to small **shrub** with glabrous young growth; **stems** can arise from below ground; **branches** few, spreading to erect; **branchlets** yellowish- to reddish-brown; **leaves** 1–3.5 cm × 0.3–0.6 cm, elliptic to narrowly ovate, opposite, short-stalked, spreading to ascending and usually incurved, often somewhat crowded near ends of branchlets, glabrous, mid green above and below, apex softly pointed; **flowerheads** about 4 cm across, terminal, usually pendent, many-flowered, can be profuse and very conspicuous; **floral bracts** 8–14, to about 3 cm × 1.5 cm, elliptic to narrowly ovate, yellow-green, sometimes green or reddish,

Pimelea spectabilis   D. L. Jones

Pimelea suaveolens   W. R. Elliot

outer 2–4 with glabrous exterior; **flowers** to about 2 cm long, pale yellow, sweetly fragrant, exterior hairy, bisexual or rarely female.

P. suaveolens is a very attractive species. The typical subspecies is from the southern Irwin and Darling Districts and is distributed from near Jurien Bay to the Stirling Ranges. Usually occurs in open forest or woodland often in hilly sites. May be difficult to maintain in cultivation. Best grown in temperate regions. Needs excellently drained acidic soils and although appreciating plenty of sunshine it is often best suited to a semi-shaded site. Grows well in containers or can be used in general planting and in rockeries. Heavy frosts can damage plants.

The ssp. *flava* Rye occurs in drier habitats often amongst mallees in the south-western Avon, south-western Coolgardie and north-central Roe Districts. It is found in sand or sandy clay which often contains lateritic gravel. It has shorter, glaucous leaves, deep yellow floral bracts and bright yellow flowers. This subspecies is sometimes confused with *P. sulphurea* Meisn. which has narrowly elliptic to nearly circular leaves.

Propagate both subspecies from seed, which may begin to germinate 50–100 days after sowing, or from cuttings of barely firm young growth, which are often slow to produce roots.

P. suaveolens ssp. *tinctoria* (Meisn.) Benth. is a synonym of *P. tinctoria* Meisn.

Pimelea subvillifera (Threlfall) Rye
(somewhat shaggy-hairy)
SA, WA
0.2–0.8 m × 0.3–1 m   Sept–Nov

Dwarf **shrub** with hairy young growth; **branches** many, ascending to erect; **branchlets** yellow-brown to reddish-brown, densely hairy near flowerheads; **leaves** 0.2–0.8 cm × 0.1–0.3 cm, narrowly elliptic to elliptic, alternate, short-stalked, ascending to erect, crowded,

densely hairy, greyish-green to yellow-green, flat or concave, apex softly pointed; **flowerheads** to about 2 cm across, terminal, erect, compact, many-flowered, can be profuse and conspicuous; **floral bracts** 8–18, to about 1 cm × 0.2 cm, linear to narrowly ovate, densely hairy; **flowers** to about 1 cm long, white, exterior densely hairy, female or bisexual.

An ornamental species from the Flinders Ranges and Eyre Peninsula, SA, and in the Austin and Coolgardie Districts in WA. Evidently rarely cultivated, it is best suited to warm temperate and semi-arid regions. Plants need excellent drainage and plenty of sunshine but should tolerate semi-shade. Hardy to

Pimelea sulphurea   W. R. Elliot

moderate frosts. It has potential for use in containers and rockeries. Propagate from seed or from cuttings of barely firm growth.

P. octophylla ssp. *subvillifera* Threlfall is a synonym.

Pimelea sulphurea Meisn.
(sulphur-coloured)

WA   Yellow Banjine; Bread; Yellow-flowered Pimelea
0.2–0.6 m × 0.1–0.5 m   July–Nov

Dwarf **shrub** with glabrous young growth; **stems** arise from woody rootstock; **branches** ascending to erect; **branchlets** glabrous, yellowish becoming reddish-brown to grey; **leaves** 0.2–1.6 cm × 0.2–1 cm, narrowly elliptic to circular, opposite, more or less sessile, spreading to ascending, not crowded, glabrous, mid green to bluish-green, flat to concave, apex pointed or very blunt; **flowerheads** to about 3.5 cm across, terminal, pendent, compact, many-flowered, can be profuse and very conspicuous; **floral bracts** 6, 8 or 10, to about 1.5 cm × 1 cm, mid green to bluish-green or yellowish, exterior glabrous, not reflexed; **flowers** to about 1.8 cm long, yellow, exterior hairy, bisexual or rarely female.

This pimelea can have an open growth habit but pruning helps promote bushier growth. It occurs over a wide range to the south-west with its distribution within a triangle extending from Eneabba to near Southern Cross and to the Fitzgerald River National Park. Plants are usually difficult to maintain for extended periods in cultivation. Best results are gained in warm temperate or semi-arid regions. They need very well-drained acidic soils and a warm to hot aspect. Plants tolerate moderate frosts without damage and extended dry periods. Well suited as a container plant. Propagate from seed which may begin to germinate about 50–100 days after sowing. Cuttings of barely firm young growth can be slow to form roots.

Pimelea sylvestris R. Br.
(of the woods or forests)

WA
0.6–2.5 m × 0.5–2.5 m   Sept–Dec; also sporadic

Dwarf to medium **shrub** with glabrous young growth; **branches** many, spreading to ascending; **branchlets** glabrous, pale-to-mid green to brownish; **leaves** 1.2–4.5 cm × 0.3–2 cm, usually narrowly elliptic to elliptic, opposite, short-stalked, spreading to ascending, rarely crowded, glabrous, greyish-green to mid green above, paler below, flat to concave, apex softly pointed; **flowerheads** 2–4 cm across, terminal, erect, compact, many-flowered, often profuse and very conspicuous; **floral bracts** usually 6, can be 4 or 8, to about 2 cm × 1 cm, narrowly ovate to ovate, similar to leaf colour or purplish on exterior; **flowers** to about 1.5 cm long, white or sometimes with pink lobes or rarely all pink, strongly fragrant, exterior glabrous, bisexual.

A handsome but variable species which occurs mainly near the coast and slightly inland from south of Perth to West Mt Barren. Grows in granitic or lateritic sand. A compact coastal variant occurs in the Irwin District usually on sandy soils with limestone outcrops. A vigorous shrubby selection was popular during the 1970s but is rarely offered for sale now. Plants adapt well to a range of soil types in cultivation in temperate and semi-arid regions. They are successful in a sunny or semi-shaded site. Plants respond very well to pruning and clipping and are hardy to moderate frosts. This species deserves to be better known. It was first cultivated in England during 1830 where it was also grown as *P. graciliflora*. Propagate from seed, which may begin to germinate 30–100 days after sowing, or from cuttings of barely firm young growth, which usually readily produce roots.

P. sylvestris var. *aeruginosa* (F. Muell.) Benth. is a synonym of *P. aeruginosa* F. Muell.

Pimelea tenuis M. Scott = *P. angustifolia* R. Br.

Pimelea thesioides S. Moore =
  P. spiculigera F. Muell. ssp. *thesioides* (S. Moore) Rye

Pimelea tinctoria Meisn.
(used for dyeing)

WA
0.5–1 m × 0.3–1 m   Aug–Oct

Dwarf **shrub** with glabrous or faintly hairy young growth; **branches** ascending to erect; **branchlets** more or less glabrous, bluish-green or reddish below flowerheads, otherwise grey; **leaves** 1–2.1 cm × 0.4–1 cm, more or less elliptic, opposite, short-stalked, usually ascending and incurved, can be crowded at ends of branchlets, glabrous, yellowish-green to bluish-green, midrib prominent, apex pointed; **flowerheads** to about 4 cm across, terminal, pendent, compact, many-flowered, can be profuse and very conspicuous; **floral bracts** usually 8, 10, 12 or 14, to about 2.7 cm × 1.7 cm, narrowly elliptic to ovate-elliptic, yellow with green markings, not reflexed, margins hairy; **flowers** to about 2.2 cm long, yellow or yellow-green, exterior hairy, bisexual.

An outstanding pimelea which is mainly distributed from the Stirling Range to near Denmark plus a disjunct occurrence to the west. Usually found in sandy or gravelly soils in low shrubland or where soil has been disturbed. Plants require excellently drained acidic soils and although they tolerate plenty of sunshine they can also do well in a warm semi-shaded site. Pruning from an early stage promotes bushy growth. Hardy to light frosts. Worth trying as a container plant as well as in gardens. Propagate from seed or from cuttings of barely firm young growth.

Pimelea treyvaudii F. Muell. ex Ewart & B. Rees
(after Hector H. Treyvaud, 19–20th-century Victorian schoolmaster)

NSW, Vic   Grey Rice-flower
0.2–1 m × 0.3–1 m   May–Jan

Dwarf **shrub** with glabrous young growth; **branches** many, spreading to ascending; **branchlets** glabrous, yellow-brown to brown; **leaves** 0.6–3.7 cm × 0.2–1 cm, narrowly elliptic, opposite, short-stalked, spreading to ascending, not crowded, glabrous, deep green above, paler and with prominent midrib below, flat, concave or convex, apex softly pointed; **flowerheads** to about 3.5 cm across, terminal, erect, somewhat elongated, many-flowered, often profuse and very conspicuous; **floral bracts** 9–11, to 1.5 cm × 0.4 cm, narrowly ovate, cream base with green or green and purple, exterior glabrous, margins often hairy; **flowers** to about 2.2 cm long, white, exterior hairy, bisexual.

A desirable pimelea which occurs in south-eastern NSW and north-eastern Vic. It grows in rocky sites in dry sclerophyll forest. Sometimes encountered in cultivation, it deserves to be better known. Plants are suited to temperate regions and require well-drained acidic soils in a sunny or semi-shaded site. They are hardy to moderately heavy frosts and withstand periods of dryness. This species has potential for use in containers and gardens. Propagate from seed or from cuttings of barely firm young growth.

Pimelea trichostachya Lindl. —
see P. elongata Threlfall

Pimelea umbratica A. Cunn. ex Meisn.
(growing in shade)
Qld, NSW
0.6–1.5 m × 0.5–1 m April–July; also sporadic

Dwarf to small **shrub** with densely hairy young growth; **branches** many, spreading to ascending; **branchlets** brown-hairy, becoming glabrous; **leaves** 0.5–2.6 cm × 0.2–0.8 cm, narrowly elliptic, opposite, short-stalked, ascending to erect, somewhat crowded, glabrous and mid green above, paler and with brown hairs below, venation visible, margins flat or slightly recurved, apex softly pointed; **floral bracts** absent; **flowers** to about 1.4 cm long, white, solitary or few clustered in axils, often partially or fully hidden by leaves, unisexual or bisexual.

This relatively rare species has attractive foliage and has potential as a cut-foliage crop. It occurs in woodland and shrubland above rainforest in the eastern Qld/NSW border region, where it grows on stony soils. Evidently rarely cultivated, it should adapt to a range of well-drained acidic soils. Probably prefers a semi-shaded site but may prove more tolerant. Propagate from seed or from cuttings of barely firm young growth.

The closely allied *P. venosa* Threlfall is regarded as vulnerable in the wild. It differs in having stems and leaves covered in spreading whitish to cream hairs and usually has 3–6 flowers per axil. It is known only from a small region in the Northern Tablelands where it grows in sandy granitic soils. Should be cultivated as part of a conservation strategy.

Pimelea verschaffeltii Morren —
see P. spectabilis Lindl.

Pimelea villifera Meisn.
(bearing shaggy hairs)
WA
0.2–1 m × 0.2–1.3 m Oct–April

Dwarf **shrub** with hairy young growth; **branches** many, spreading to ascending; **branchlets** densely hairy; **leaves** 0.6–2.1 cm × 0.1–0.4 cm, linear to elliptic, opposite, short-stalked, ascending to erect, somewhat crowded, mid green to bluish-green, hairy to glabrous, flat or concave, apex softly pointed; **flowerheads** to about 2.5 cm across, terminal, erect, compact, many-flowered, often profuse and conspicuous; **floral bracts**

usually 6–10 pairs, to about 1.8 cm × 0.4 cm, narrowly oblong to almost ovate, mostly hairy; **flowers** to about 0.9 cm, white with dark base, exterior very hairy, bisexual.

An interesting and hairy species from the northern Darling, southern Irwin and central-western Avon Districts. Plants occur in a range of acidic and alkaline sands and loams. Apparently rarely cultivated, this species is best suited to warm temperate regions. It needs good drainage and a fairly sunny site. Plants are hardy to light frosts. It has potential for use in gardens and containers. Propagate from seed or from cuttings of barely firm young growth.

P. subvillifera is closely allied but differs in having alternate leaves.

Pimelea williamsonii J. M. Black
(after Herbert B. Williamson, 19–20th-century Victorian botanist, naturalist)
Vic, SA Silky Rice-flower; Williamson's Rice-flower
0.1–0.3 m × 0.3–0.6 m Oct–Nov

Dwarf **shrub**; young growth densely hairy; **branches** many, spreading to ascending; **branchlets** with whitish hairs, becoming less hairy; **leaves** 0.5–1.6 cm × 0.2–0.5 cm, more or less elliptic, alternate to nearly opposite, short-stalked, ascending to erect, somewhat crowded, greyish-green, faintly hairy above, densely whitish-hairy below, apex bluntly pointed; **flowerheads** to about 1 cm across, maturing to about 7 cm long, terminal, erect, can be profuse and conspicuous; **floral bracts** absent; **flowers** to about 0.5 cm long, covered in whitish and brownish hairs, bisexual.

A compact, densely foliaged, sand-loving species from north-western Vic and the South-eastern and Eyre Peninsula Districts of SA. Evidently rarely cultivated, it is most likely to interest enthusiasts. Best suited to warm temperate and semi-arid regions in very well-drained soils. Plants need plenty of sunshine. Hardy to moderate frosts. Worth trying in containers as well as gardens. Propagate from seed or possibly from cuttings of barely firm young growth.

PIMELODENDRON Hassk.
(from the Greek *pimele*, fat, oily; *dendron*, a tree)
Euphorbiaceae

Trees; **sap** milky, often ageing yellowish; **leaves** simple, entire, alternate, petiolate, petiole bearing glands; **inflorescence** small spikes or racemes; **flowers** small, unisexual; **fruit** a drupe containing soft white flesh and a single seed.

A small genus of about 8 species found mainly in Malaysia with 1 species occurring in Australia. It is virtually unknown in cultivation. Propagate from seed which has a limited period of viability and should be sown while fresh.

Pimelodendron amboinicum Hassk.
(from Ambon, Malay Archipelago)
Qld
10–20 m × 5–15 m Oct–Dec

Medium **tree** with a bushy canopy; **sap** milky, exuding from damaged parts; **bark** smooth, light brown; **petioles** with 2 prominent swollen glands near the apex; **leaves** 8–15 cm × 6–12 cm, elliptical to oblong,

Pimelea sylvestris, coastal variant   W. R. Elliot

crowded towards the end of the branchlets, dark green, apex acuminate; **racemes** 2–5 cm long, in axillary clusters or from the old wood; **flowers** about 0.3 cm across, yellow; **fruit** 1.5–2 cm across, globose, pink to red.

Occurs in Indonesia and New Guinea and extends to lowland rainforests on Cape York Peninsula. The sap sets like glue and is used in the Solomon Islands for making bracelets and ornaments. Not known to be in cultivation. It has limited ornamental appeal although the fruit are colourful. May have prospects for cultivation in tropical and warm subtropical regions. Plants probably require some shelter when young and unimpeded drainage. Propagate from fresh seed.

PIPER

(a Latin name, derived from the Greek *peperi*, itself derived from the Benghalese name for the pepper plant)
Piperaceae   Peppers

Shrubs, **trees** or more usually **climbers**; **climbing stems** adhering by numerous roots; **mature stems** branched, lacking roots; **leaves** simple, entire, alternate, often aromatic, often prominently veined; **inflorescence** a dense spike or raceme; **flowers** tiny, unisexual or bisexual; **fruit** a separate drupe or immersed in the swollen, fleshy rachis.

A large genus of about 2000 species widely distributed in tropical regions of the world. In Australia, there are about 9 species but the exact number is uncertain — botanical revision of this genus would be very helpful. Black pepper is obtained from the fruit of *Piper nigrum*, an Indian species. The Australian species are poorly known in cultivation. Most have attractive foliage but tend to be very robust and occupy considerable space in a garden. They also need a very strong structure to support their growth. All of the Australian species grow well in containers where their vigour can be controlled by pruning and restricting nutrients. They are potentially useful as decorative indoor plants. Propagation is from seed, which has a limited period

of viability and is best sown while fresh. Cuttings of juvenile shoots strike readily but cuttings from mature shoots can be difficult.

Piper banksii Miq. = *P. caninum* Blume.

Piper caninum Blume
(common)
Qld   Common Pepper Vine
10–20 m tall   May–July
Robust **climber** forming bushy clumps; **young stems** clinging by numerous roots; **mature stems** much-branched, spreading or pendulous; **leaves** 6–9 cm × 4–6 cm, ovate to broadly ovate, dark green and glossy above, paler and finely hairy beneath, with a prominent midrib; **spikes** 5–12 cm long, cylindrical; **flowers** tiny, greenish, bisexual; **drupes** 0.6–0.8 cm long, ovoid, brownish-red, on stalks about 0.8 cm long, ripe Oct–Dec.

A bushy climber with ornamental foliage. Common in the rainforests of north-eastern Qld where it is distributed from the coastal lowlands to the ranges and tablelands. May be too vigorous for general garden planting but could be useful as an indoor plant or perhaps in hanging containers. Outdoor plants require a protected position in moist, well-drained soil. Mulches and watering during dry periods are beneficial. Indoor plants require humid conditions with good light and air movement. Propagate from fresh seed and from cuttings of juvenile growth.
Piper banksii Miq. is a synonym.

Piper harveyanum Domin = *P. mestonii* F. M. Bailey

Piper interruptum Opiz
(interrupted; referring to the arrangement of the flowers)
Qld
10–15 m tall   Aug–Oct
Slender to bushy **climber**; **stems** to 5 cm across, branching mainly in the upper parts; **leaves** 7–12 cm × 6–8 cm, ovate, bright green, thin-textured, 5–7 veins prominent, acute; **male spikes** 10–20 cm long, very slender; **female spikes** 5–12 cm long; **flowers** tiny, unisexual, whitish; **drupes** 0.3–0.5 cm long, ovoid, yellow, ripe March–May.

Occurs in New Guinea and north-eastern Qld, and grows in rainforests. A little-known bushy climber which has some ornamental appeal. Plants probably have similar cultivation requirements to *P. mestonii*. Propagate from fresh seed and from cuttings.
Piper triandrum F. Muell. is a synonym.

Piper mestonii F. M. Bailey
(after Archie Meston, original collector)
Qld   Long Pepper
10–20 m tall   Jan–March
Robust **climber** forming bushy clumps; **main stems** upright, with clinging roots; **lateral branches** pendulous, without roots; **leaves** 15–25 cm × 15–20 cm, broadly ovate, dark green, glossy, with 3 veins prominent, acuminate; **spikes** 2–4 cm long; **flowers** tiny, cream; **fruit** 3–5 cm × 1–1.5 cm, cylindrical, soft, red, with numerous tiny seeds, ripe Aug–Oct.

Occurs in New Guinea and in lowland rainforests near the Russell River, north-eastern Qld. A very attractive climber which has excellent prospects for cultivation in tropical and warm temperate regions. Ideal for growing on an existing tree or strong supporting trellis. Young plants require shelter, moist but well-drained soil, and mulching and watering during dry periods. They are very frost sensitive. Container plants should be given a trial as indoor decoration. Propagate from fresh seed or from cuttings of juvenile growth.

Piper harveyanum Domin is a synonym.

Piper novae-hollandiae Miq.
(from New Holland)
Qld, NSW, NT   Giant Pepper Vine
10–20 m tall   April–Aug
Robust **climber** forming bushy clumps; **young stems** slender, clinging by numerous roots; **mature stems** woody, spreading; **bark** grey, rough; **juvenile leaves** to 8 cm × 8 cm, ovate, thin-textured, base cordate; **mature leaves** 12 cm × 9 cm, broadly ovate, thick, leathery, base rounded, dark green, veins prominent; **male spikes** 1–2 cm long, cylindrical; **female spikes** 0.8–1 cm ovoid; **flowers** unisexual, tiny, cream; **drupes** 0.5 cm long, ovoid, red, on slender stalks 1.5–2 cm long, ripe Nov–Jan.

Distributed between north-eastern Qld and south-eastern NSW with a disjunct occurrence in the Top End, NT. It grows in rainforest from the coast to the ranges and tablelands. A large, bushy climber with attractive leaves and interesting fruit which are eaten by birds. Much too large for the average home garden but an interesting plant for parks and botanic gardens. Young plants need shade and a large host or structure to support their growth. The younger plants also make excellent container plants for indoor decoration. They withstand relatively dark conditions and their vigour can be controlled by pruning and restricting fertiliser use. Propagate from fresh seed and from cuttings of juvenile shoots.

Piper rothianum F. M. Bailey
(after Dr Walter E. Roth, noted Australian anthropologist)
Qld
10–15 cm tall   Oct–March
Robust, straggly or bushy **climber**; **main stems** erect, with numerous roots; **lateral branches** pendulous, without roots; **leaves** 10–16 cm × 5–9 cm, broadly ovate, dark green, dull to shiny, 5 veins prominent, with a curved apical point; **spikes** 8–12 cm long; **flowers** tiny, cream; **fruit** 8–12 cm × 0.6–0.8 cm, cylindrical, fleshy, dull red, with numerous tiny seeds, ripe Aug–Oct.

Occurs in New Guinea and in north-eastern Qld where it is widely distributed in coastal rainforests. The ripe fruit are edible and have a hot taste. Suitable for parks and large gardens in tropical and warm temperate regions. It has similar cultivation requirements to *P. mestonii*. Worth trying as a container plant for indoor decoration. Propagate from fresh seed and from cuttings of juvenile growth.

Piper novae-hollandiae × .4

Piper triandrum F. Muell. = *P. interruptum* Opiz

Piper umbellatum L.
(flowers in umbels)
Qld   Shrubby Pepper
1–2 m × 1–2 m   April–June
Small **shrub** with a rounded to spreading habit; young growth short-hairy; **stems** fleshy; **leaves** 15–30 cm × 15–30 cm, heart-shaped to kidney-shaped, on slender petioles 10–20 cm long, dark green, thin-textured, apex pointed; **spikes** 3–10 cm long, very slender, 2–7 in a

Piptocalyx moorei   W. R. Elliot

loose umbel; **flowers** tiny, bisexual, whitish, densely packed; **drupes** 0.3–0.4 cm × 0.2 cm, obovoid, with 3 ridges, reddish, ripe Dec–March.

An interesting shrub with large handsome leaves. Occurs in New Guinea and at moderate altitudes in the ranges and tablelands of north-eastern Qld. It is found in rainforest and has been grown to a limited extent as an ornamental in the subtropics. Requires a sheltered location in moist well-drained soil. It has potential for use as a decorative container plant. Frost tolerance is unlikely to be high. Propagate from seed and from cuttings.

Piper species affinity **abbreviatum**
Qld
8–15 m tall   Aug–Oct
Slender to bushy **climber**; **stems** to 2 cm across, branching mainly in the upper parts; **leaves** 5–8 cm × 3–3.5 cm, ovate, dark green, dull, thin-textured, 3–5 veins prominent, acuminate; **spikes** about 1 cm × 0.5 cm; **flowers** tiny, cream; **fruit** 1–1.5 cm × 1 cm, cylindrical, red, with numerous seeds, ripe Jan–March.

An undescribed species distributed in coastal districts of Cape York Peninsula where it grows in rainforest. Plants are usually slender but may sometimes become bushy towards the top. A very ornamental species which has potential for use in hanging containers and indoor decoration. Requirements are similar to those of *P. mestonii*. Propagate from seed or from cuttings of juvenile growth.

Piper species affinity **caninum**
Qld
10–20 m tall   Aug–Oct
Robust **climber** forming bushy clumps; **stems** to 6 cm across, much-branched; **leaves** 5–9 cm × 4–6 cm, broadly ovate, glossy green, with a prominent pale midrib; **spikes** 3–7 cm long; **flowers** tiny, cream; **drupes** about 0.6 cm long, bright red, in cylindrical clusters about 7 cm × 3 cm, ripe Dec–Feb.

An undescribed species from the Cook and North Kennedy districts of Qld where it grows in rainforest. Plants have attractive lustrous leaves and interesting fruit. They are grown on a limited scale by enthusiasts. Suitable for tropical and warm subtropical regions with requirements similar to those of *P. mestonii*. It has potential for use as indoor decoration. Propagate from fresh seed and from cuttings of juvenile growth.

Differs from *P. caninum* which does not have glabrous leaves or short, crowded fruit clusters.

PIPERACEAE C. A. Agardh.   Pepper Family
A large family of dicotyledons consisting of about 2100 species in 5 genera. It is particularly well represented in tropical Asia, Africa and South America. Plants are climbers, shrubs or small trees with thickened nodes and alternate or whorled simple leaves, which may have a spicy or aromatic odour. The tiny flowers are densely crowded in a spike. The family includes many spices such as black pepper, betel pepper and *Piper methysticum*, used in the preparation of the drink called kava. The family in Australia is represented by about 9 species of *Piper* and 2 species of *Macropiper*.

PIPTOCALYX Oliv. ex Benth.
(from the Greek *pipto*, to fall; *calyx*, calyx; referring to the deciduous calyx)
Trimeniaceae
Climbers or scandent **shrubs**; **leaves** opposite, simple, oil-dotted; **inflorescence** racemes, mainly axillary, sometimes terminal; **flowers** bisexual or male, small, with 2–11 perianth segments; **stamens** usually many; **fruit** a berry.

A small genus of 2 species with one endemic in eastern Australia and the other occurring in New Guinea.

Further botanical studies may result in this genus being placed in *Trimenia*.

Piptocalyx moorei Benth.
(after Charles Moore, late 19th-century Superintendent of Royal Botanic Gardens, Sydney)
Qld, NSW   Bitter Vine
Climber   Nov–Jan
Vigorous **climber** or scrambling **trailer** with long woody stems and densely rusty-hairy young growth; **leaves** 5–10 cm × 1.5–4 cm, lanceolate to ovate, opposite, stalked, olive-green, with small oil dots, more or less glabrous, undersurface woolly-hairy, flat, margins entire or crenate, apex pointed, aromatic when crushed; **flowers** to about 0.5 cm across, whitish to greenish, rusty-hairy in bud, terminal, usually bisexual, lower ones usually male, in terminal or axillary racemes, can be profuse but never conspicuous; **berry** to about 0.8 × 0.6 cm, black, ripe Feb–March.

A vigorous species from south-eastern Qld and north-eastern NSW. It grows in temperate rainforest as well as eucalypt forest on the coastal ranges. Cultivated to a very limited degree. Plants are not highly ornamental but they have excellent screening properties and when grown as a groundcover they are useful for soil erosion control. Adapts well to temperate and subtropical regions in a range of acidic soils which drain well. Plants tolerate a sunny site but do best in semishade. They are hardy to moderate frost. Pruning may be required to keep plants within bounds. Propagate from fresh seed, which does not require pre-sowing treatment. Cuttings of semi-hard growth strike readily.

PIPTURUS Wedd.
(from the Greek *pipto*, to fall; *oura*, tail; referring to the long-stalked leaves)
Urticaceae
Dioecious **shrubs** or small **trees**, sometimes scandent; **leaves** alternate, 3–5-veined, undersurface usually woolly-hairy; **stipules** early deciduous; **inflorescence** globular clusters in axillary spikes; **male flowers** usually 4- or 5-merous; **female perianth** ovoid, apex toothed, stigma exserted; **fruit** an achene, in globular clusters at nodes.

A tropical genus of 40–50 species distributed in the Mascarene Islands (east of Madagascar), Malesia, Polynesia and Australia which has 1 species.

Pipturus argenteus (G. Forst.) Wedd.
(silvery)
Qld, NSW   White Nettle; Koomeroo-Koomeroo;
                                           Native Mulberry
4–8 m × 4–8 m   Nov–Feb; also sporadic

Pipturus argenteus × .4

Tall **shrub** to small **tree** with silvery-hairy young growth; **bark** grey to greyish-brown, smooth with vertical rows of small swellings; **branches** spreading to ascending; **branchlets** terete, covered in whitish hairs; **leaves** 7–26 cm × 2–6.5 cm, more or less ovate, alternate, prominently stalked, spreading to ascending, thin, mid-to-deep green and glabrous above, densely covered with short silvery-white hairs below, venation prominent, margins finely toothed, apex pointed; **flowers** tiny, white, in small globular clusters forming interrupted spikes to about 6 cm long, on older wood, insignificant; **fruit** to about 0.6 cm diameter, white, succulent, mulberry-like, edible, sweet, juicy, ripe May–July.

A quick-growing, often rangy species which is a non-stinging member of the nettle family. It is somewhat ornamental during windy weather which exposes the silvery undersurface of the leaves. In Australia, it occurs on the north-eastern coast from north Qld to Lismore, NSW. It is also in Polynesia and Malesia. Not commonly cultivated, it grows well as far south as Melbourne but can suffer damage from frost. Adapts to a range of acidic soils which drain well but are moist for most of the year. Grows in a shady or sunny site. Responds very well to regular hard pruning. Leaves may be eaten by larvae of butterflies. Propagate from fresh seed, or from firm semi-hardwood cuttings, which strike readily.

Australian Aborigines extracted a rich brown dye from the bark. Pacific Islanders used the bark as a source of fibre for making fishing nets.

PISONIA L.

(after Willem Piso, 17th-century Dutch doctor who wrote a book about Brazilian medicinal plants)
Nyctaginaceae

Shrubs or **trees**, unarmed or thorny **climbers**; **leaves** simple, entire, opposite, alternate or crowded in pseudo-whorls; **inflorescence** an axillary or terminal cyme or umbel; **flowers** small, bisexual or unisexual; **corolla** 5–10-merous; **fruit** smooth or ribbed, sticky, with or without sticky glandular hairs.

A small tropical genus of 35 species found particularly in America and South-East Asia. Three species occur in Australia although none is endemic. The trees have ornamental foliage but are rarely cultivated. Sticky fruits are dispersed by birds but a heavy load of the fruit can gum the birds' feathers and cause death. The bird mortality rate in years of heavy fruiting can be very high. Propagation is from seed, which has a limited period of viability and for best results should be sown fresh.

Pisonia aculeata L.
(with prickles or thorns)
Qld, NSW, WA, NT   Thorny Pisonia
10–20 m tall   Dec–March

Robust **climber** with long woody stems armed with recurved hooks; young growth short-hairy; **leaves** 6–10 cm × 5–7 cm, elliptic, ovate or broadly ovate, thin-textured, dark green; **cymes** axillary, dense, expanding in fruit; **flowers** unisexual, greenish to creamy yellow, fragrant; **male flowers** about 0.4 cm across, funnel-shaped; **female flowers** about 0.3 cm across, bell-shaped; **fruit** 1–1.2 cm × 0.2 cm, cylindrical or club-shaped, ribbed, with glandular hairs, ripe March–May.

A pantropical species which is common in northern Australia and extends into north-eastern NSW. It is often abundant in coastal districts and grows along streams and rainforest margins, sometimes forming dense, thorny thickets. This species has very limited ornamental appeal but may be useful for regenerating degraded sites. The thorny thickets also provide a useful wildlife habitat. Propagate from fresh seed.

Pisonia brunoniana Endl. —
see under *P. umbellifera* (J. R. Forst. & G. Forst.) Seemann

Pisonia grandis R. Br.
(great, large, tall)
Qld, NT   Giant Pisonia; Bird-killer Tree
15–25 m × 5–10 m   Jan–June

Medium to tall **tree** with a spreading canopy; **bark** pale grey; **wood** soft; **branches** brittle; young growth finely hairy; **leaves** 20–30 cm × 10–18 cm, elliptic to broadly ovate, light yellowish-green, thin-textured, short-hairy, becoming glabrous; **cymes** terminal, much-branched; **flowers** bisexual or unisexual, about 0.4 cm long, funnel-shaped, greenish-white; **fruit** 0.8–1.2 cm × 0.2 cm, cylindrical to club-shaped, ribbed, with rows of sticky glandular hairs, ripe July–Oct.

A littoral species which is widely distributed on islands in the Indian and Pacific Oceans. It often dominates the vegetation on small atolls and coral cays. The trees form an extensive light canopy of foliage, but the branches are very brittle and are readily torn off during storms. The young leaves of this tree are cooked as a green vegetable. Sea birds use the trees extensively for roosting and nesting, however, the

Pisonia umbellifera × .33, close-up × .65

masses of sticky seeds can adhere tenaciously to the birds' plumage restricting their movement and resulting in a high death toll.

In Australia, the species is confined to coral cays off the Qld coast and a few small islands in the Gulf of Carpentaria. It is easily grown but rarely cultivated. Best suited to sandy soils in coastal districts of the tropics. Propagate from fresh seed.

Pisonia umbellifera (J. R. Forst. & G. Forst.) Seemann (umbrella-shaped)

Qld, NSW   Bird Lime Tree
15–25 m × 5–10 m   July–Nov

Medium to tall **tree**, with a rounded or spreading crown; young growth glabrous or short-hairy; **leaves** 20–35 cm × 10–15 cm, ovate to elliptic or oblong, crowded towards the end of branchlets, dark green, shiny, often drooping; **umbels** axillary or terminal; **flowers** bisexual or unisexual, funnel-shaped, 0.3–0.7 cm long, white, pink or yellow, scented; **fruit** 2–4 cm × 0.6 cm, spindle-shaped, with 5 sticky ribs.

A widely distributed species which extends to Australia where it occurs from Cape York Peninsula, Qld, to the Illawarra regions of south-eastern NSW, and also on Lord Howe Island. It commonly grows in rainforest but may also extend to sheltered gullies in more open habitats. The viscid fruits adhere to the plumage of roosting birds and in severe cases may cause death by restricting movement. These trees are

fast-growing and have large ornamental leaves. They are useful shelter plants and provide protection to slow-growing species and those which need shade. Requires well-drained soil in filtered sun or semi-shade. Responds well to mulches, light fertiliser applications and watering during dry periods. Propagate from fresh seed.

There is some conjecture that the southern populations of this species may be distinct from the northern ones. If so, the applicable name for southern populations is *P. brunoniana* Endl.

PITHECELLOBIUM Mart. =
                              ARCHIDENDRON F. Muell.

PITHOCARPA Lindl.
(from the Greek *pithos*, wide-mouthed, large jar; *carpos*, fruit; referring to shape of fruit)
Asteraceae

Semi-woody, annual or short-lived perennial **herbs; branches** many, ascending to erect; **leaves** alternate, mainly on lower parts, simple, entire; **inflorescence** solitary, terminal flowerhead or in corymbs of 2–9 heads; **receptacle** flat, bracts absent; **fruit** an angular achene, glabrous or with small swellings.

Pithocarpa is a genus of 4 species which is endemic in south-western WA. They are attractive plants but are rarely encountered in cultivation. Propagation is from seed, which does not require pre-sowing treatment. Cuttings of non-flowering firm young growth from young plants strike readily, even without application of hormone rooting compounds.

Pithocarpa achilleoides P. Lewis & Summerh. —
                              see *P. pulchella* Lindl.

Pithocarpa corymbulosa Lindl.
(flowers borne in small corymbs)
WA
0.5–1 m × 0.3–0.6 m   Jan–Feb

Dwarf, semi-woody, annual **herb** with whitish-hairy young growth; **branchlets** few, hairy; **leaves** 0.5–2.5 cm × 0.1–0.5 cm, linear to narrowly elliptic, alternate, scattered, few, lower ones broadest, sessile, glabrous and deep green above, white-hairy below, apex pointed; **flowerheads** to about 0.5 cm across, with up to about 9 per corymb, can be profuse and very conspicuous; **involucral bracts** white; **florets** yellow, bisexual; **achenes** to about 0.1 cm long, glabrous.

This interesting herb is restricted to the Darling Range where it usually grows in lateritic soils overlying granite. It is cultivated to a very limited degree and is best suited to temperate regions. Plants need plenty of sunshine to flower profusely but they will tolerate a semi-shaded site. Soils need to be acidic and well-drained. Pruning of very young plants promotes bushy growth. Hardy to light frosts. Propagate from fresh seed, which can be slow to germinate, or from cuttings of young plants.

Pithocarpa major Steetz = *P. pulchella* Lindl.

Pithocarpa melanostigma P. Lewis & Summerh.
(black stigma)
WA
0.2–0.4 m × 0.3–0.6 m   Nov–April
Dwarf, semi-woody, annual **herb** with hairy young growth; **branches** many, spreading to erect; **branchlets** slender, hairy; **leaves** linear, alternate, scattered; **flowerheads** about 0.6 cm across, terminal, solitary or a few per corymb, on slender stems, can be profuse and conspicuous; **involucral bracts** dark brown to blackish; **florets** tubular, yellow, bisexual; **achene** small, hairy.

Distributed in the most southern region of the Darling District where it grows in open forest and woodland in sandy soils which are moist for extended periods. Evidently not cultivated. It has similar requirements to *P. corymbulosa*.

Pithocarpa pulchella Lindl.
(beautiful)
WA
0.2–0.4 m × 0.1–0.3 m   June–Sept
Dwarf, semi-woody, annual or short-lived perennial **herb** with white-hairy young growth; **branches** ascending to erect; **branchlets** white to greyish-hairy, slender, rigid; **leaves** 0.5–2.5 cm × 0.1–0.4 cm, linear to narrowly elliptic, alternate, scattered, few, sessile, glabrous and deep green above, white woolly-hairy below, apex pointed; **flowerheads** to about 1.5 cm across, terminal, usually solitary, can be profuse and very conspicuous; **involucral bracts** bright white; **florets** tubular, yellow, bisexual; **achenes** about 0.1 cm long, densely hairy.

A charming species from the Avon, northern Darling and southern Irwin Districts. It grows in woodland in sandy soils which can overlie limestone or laterite. Cultivated to a limited degree, it should perform best in temperate regions in a sunny or semi-shaded site with very well-drained soils. Pruning of young plants should provide more flowering stems. Propagate from fresh seed or from cuttings of young plants.

P. achilleoides is closely allied but differs in having flowerheads to about 0.8 cm across. Botanical studies may prove it to be conspecific with *P. pulchella*.

PITTOSPORACEAE R.Br.

A family of dicotyledons consisting of about 225 species in 9 genera. The family is particularly diverse in Australia, Africa and the Pacific region. All 9 genera are represented in Australia and there are about 75 species. They are shrubs, subshrubs, trees or climbers with simple, entire, alternate or spirally arranged leaves. The bisexual flowers, which have 5 sepals and 5 petals, may produce colourful displays. The fruits may be dry or fleshy and are sometimes brightly coloured too. Many native species of this family have become popular horticultural subjects. Australian genera: *Billardiera, Bursaria, Cheiranthera, Citriobatus, Hymenosporum, Marianthus, Pronaya, Rhytidosporum, Sollya.*

PITTOSPORUM Banks ex Gaertn.
(from the Greek *pitta*, pitch, resin; *spora*, seed; referring to the sticky, resinous seeds)
Pittosporaceae

Dioecious **trees** or **shrubs**; **branchlets** sometimes spiny; **leaves** alternate or appearing in whorls at ends of branches, simple, margins entire or sometimes toothed; **inflorescence** solitary or few- to many-flowered terminal racemes, panicles or clusters; **flowers** with mainly free petals, sometimes cohering in the throat and then tubular with spreading lobes, often fragrant; **sepals** 5, free; **petals** 5; **stamens** free; **fruit** a capsule, ovoid to globular, sometimes compressed, initially often fleshy and resinous, orange-red; **seeds** very sticky in Australian species.

A widespread genus of about 150 species which is represented in Africa, southern Asia, western Pacific, Hawaiian Islands and in Australia where there are about 20 species. The main representation is in the forests of coastal or near coastal regions of the eastern states. The one exception is *P. angustifolium* which occurs in all mainland states and is common in semi-arid and arid regions. In nature, pittosporums grow in a range of soils from sand to clay and they are also found in gravelly and rocky sites.

A recent revision of *Pittosporum* will bring about some changes. For example, the genus *Citriobatus* will be included in *Pittosporum* because it is not significantly different. Another new genus, *Auranticarpa* L. Cayzer, Crisp & I. Telford, will be erected to encompass *P. melanospermum*, *P. resinosum* (a poorly known species from the Kimberley, WA) and *P. rhombifolium*. The correct name for the familiar species previously well known as *P. phylliraeoides* is now *P. angustifolium*.

A number of exotic species and cultivars, particularly from New Zealand, have become popular garden plants and a limited number of native species are also valued for planting. The delicious floral fragrance of *P. undulatum* has made it popular, but because it can regenerate vigorously when grown outside its natural range it is now known as a weedy species (see entry below). Most species have sticky seeds which are eaten

Pittosporum angustifolium   W. R. Elliot

by birds so it will be important to monitor the regeneration capacity of any cultivated *Pittosporum* to ascertain whether, like *P. undulatum*, they pose a threat to native bushland. *P. angustifolium*, *P. revolutum*, *P. rhombifolium* and *P. bicolor* are relatively common in cultivation, while *P. venulosum*, which tolerates drier conditions, is becoming better known.

Most species have fragrant flowers which can be a highlight on still, warm spring or early summer nights when the perfume permeates the air. Male plants tend to have more flowers than female plants and consequently produce a more concentrated aroma.

The fruit of some species is extremely decorative eg. *P. revolutum*, *P. rhombifolium* and *P. venulosum*. Colour varies from pale yellow to deep orange. An added decorative feature is provided by the red or reddish-brown seeds which are displayed against the often colourful flesh of the opened fruits — *P. revolutum* and *P. rubiginosum* are excellent examples.

Most pittosporums are long-lived in cultivation and can be very useful for screening as well as for informal or formal hedging. Generally they respond very well to pruning and regular clipping. *P. undulatum* has been in cultivation for many years. It is an excellent general-purpose plant and has also been used to protect fruit crops from harsh winds.

By and large pittosporums do best in acidic soils which are at least moderately well drained. *P. angustifolium* is the most adaptable as it tolerates moderate alkalinity. Tolerance of sunshine varies. Most species from north-eastern regions prefer a semi-shaded site instead of full sunshine. In full sun, plants are liable to attack by scale insects.

Once established most species withstand periods of dryness. *P. angustifolium* and *P. undulatum* are very tolerant of extended periods of dryness.

Pests and diseases are not a major cause for concern. *P. undulatum* is prone to attack from aphids, passion vine hoppers, scale and pittosporum leaf miner, and leaf distortion thrips can attack *P. revolutum* and *P. undulatum*. See Volume 1, or *Pests, Diseases and Ailments of Australian Plants* for control methods.

Propagation

Propagation is from seed or cuttings. Seed has proved to be generally difficult to raise and the results obtained can be very sporadic — consistent germination is a rarity. Often only a very few seedlings will appear in the first year and stragglers may continue to germinate over several years. Seed is ready for sowing once it starts to drop from the fruits. The seeds are covered with a sticky substance which should be removed by washing in warm water. After the seed is dry again it should be barely covered with seed-raising medium. Germination should occur within 15–60 days; 15–20°C is the optimum temperature range for best results. Germination may be improved by soaking the seeds in 1 part NappySan to 10 parts water after lightly scratching the outer seed coat.

Cuttings may be slow to produce roots and in some species the root systems may be poor. Generally best results are achieved with semi-hardwood cuttings from current season's growth. Application of hormone rooting compounds is usually beneficial. Some propagators

obtain good results by taking cuttings early to mid autumn and propagating without the use of bottom heat.

Pittosporum angustifolium Lodd.
(with narrow leaves)
Qld, NSW, Vic, SA, WA, NT   Weeping Pittosporum;
       Butterbush; Berrigan
3–15 m × 1.5–6 m   June–Nov

Medium to tall **shrub** or small **tree** with glabrous young growth; **bark** pale grey, often flaky or grooved; **branches** spreading to ascending or pendent; **branchlets** pendent, glabrous; **leaves** 4–12.5 cm × 0.4–1.2 cm, narrowly elliptic to oblong, alternate, short-stalked, pendent, olive green, glabrous, margins flat, apex usually ending in a short curved point; **flowers** to about 1.2 cm across, cream to yellow, the females solitary and terminal on short shoots, the males in few-flowered terminal or axillary clusters, lightly fragrant, scattered to profuse, rarely highly conspicuous; **capsules** to about 1.8 cm long, ovoid to globular, usually compressed near stalk, yellowish, scattered to profuse.

A highly desirable species with an appealing weeping growth habit, fragrant flowers and ornamental fruits. It occurs over a wide range of semi-arid and arid areas where it grows in woodland, particularly mallee woodland, and to a lesser extent in shrubland. It is popular in cultivation having been grown for many years in Australia as well as overseas in areas such as south-western USA and the Middle East region. Plants are suitable for temperate, semi-arid and arid areas and they succeed in acidic or moderately alkaline soils which are not prone to waterlogging. Moderately fast-growing. Does best in a site which receives maximum sunshine but will tolerate intermittent shade. Plants are hardy to most frosts and grow well without supplementary watering. Although grown primarily for their pendent foliage, plants can be clipped and used for hedging. Fruit production is usually variable from year to year. Two or more plants in close proximity may enhance fruiting. Foliage is valuable as fodder for stock during drought periods. Flowers attract butterflies. The very hard, close-grained timber is used for wood-turning.

Propagate from seed, or from cuttings of firm young growth, which may be slow to form roots.

Previously well known as *P. phylliraeoides* DC. but that species is now known to be confined to coastal areas of WA.

P. phylliraeoides var. *microcarpa* S. Moore is a synonym of *P. angustifolium*.

The southern WA variant has smaller leaves, non-pendulous stems and a rounded canopy. It will be described as a separate species in the ongoing revision of this genus.

Pittosporum bicolor Hook.
(two-coloured)
NSW, Vic, Tas   Banyalla
2.5–12 m × 1.5–5 m   Sept–Nov

Medium to tall **shrub** or small **tree** with hairy young growth; **branches** spreading to ascending; **branchlets** terete, initially hairy; **leaves** 3–8 cm × 0.5–1.8 cm, lanceolate to narrowly oblong, alternate, short-stalked

Pittosporum bicolor × .45

spreading to ascending, sometimes crowded, deep green above, rusty to silvery hairs below, margins flat to recurved, leathery, apex bluntly to finely pointed; **flowers** 1–1.5 cm long, pale yellow with purplish-maroon markings, mainly terminal, solitary (female) or in small clusters (male), usually pendent on hairy stalks, with light sweet fragrance, rarely profuse and conspicuous; **petals** strongly reflexed; **capsules** to about 1 cm across, globular, greyish, hairy; **seeds** bright orange-red to deep red.

An inhabitant of moist forests and humid gullies where it may initially grow as an epiphyte. Its distribution extends from the South Coast and Southern Tablelands of NSW, through eastern Vic to the Otway Ranges, and it is common throughout Tas. Plants are somewhat slow-growing and usually bushy. They adapt well to cultivation in moist but well-drained acidic soils with a semi-shaded or shaded aspect. Hardy to most frosts and snowfalls. Useful for general planting and is suitable for hedging and topiary as it responds well to clipping and pruning. Dislikes drying out. The timber was used by Tasmanian Aborigines for making clubs. It has also been used for axe handles and billiard cues and is highly regarded for wood-turning. Propagate from seed, or from cuttings of firm young growth, which can be slow to form roots.

P. bicolor hybridises with native *P. undulatum* in NSW and Victoria. In Tas, hybridising also occurs with *P. undulatum* which has become naturalised there. The offspring in Tasmania have been known as *P.* × *emmettii* W. M. Curtis. This hybrid has wider, glabrescent leaves (about 4–9 cm × 1.5–3 cm), and larger, strongly fragrant, creamy yellow, prominently unisexual flowers. The fruits are larger too and similar to *P. bicolor* fruits but often bright yellow inside with orange rather than red, seeds.

Pittosporum × emmettii W. M. Curtis —
        see under *P. bicolor* and *P. undulatum*

Pittosporum erioloma C. Moore & F. Muell.
(woolly border)
NSW (Lord Howe Island)
3–8 m × 2–5 m   Aug–Oct; also sporadic

Medium to tall **shrub** or small **tree** with woolly-hairy young growth; **branches** spreading to ascending; **branchlets** initially woolly-hairy; **leaves** 2.5–5 cm × 0.7–1.5 cm, oblanceolate, alternate, spreading to ascending, often crowded near ends of branchlets, pale green, becoming glabrous, margins recurved, apex blunt with small point; **flowers** to about 1.4 cm × 1 cm, reddish-lilac at base, grading to cream or white tips, 2–3 per terminal cluster, lightly fragrant, rarely profuse or conspicuous; **petals** reflexed; **calyx lobes** margins prominently fringed; **capsules** to about 2 cm across, globular, 3-valved, green, hairy; **seeds** red-brown.

Endemic on Lord Howe Island where it grows mainly in the higher regions, often in rocky sites. More closely related to the New Zealand species than to Australian ones. It is rarely encountered in cultivation but adapts well in subtropical and temperate regions. Needs excellently drained acidic soils and while it tolerates sunshine it is likely to do best in a semi-shaded site. It has potential for hedging and general planting. Propagate from seed, or from cuttings of firm young growth, which may be slow to form roots.

Pittosporum ferrugineum Dryander
(rust-coloured)
Qld, NT   Rusty Pittosporum
2–6 m × 1.5–3 m   May–June

Medium to tall **shrub**, often sparse, with golden-brown hairy young growth; **branches** spreading to ascending; **branchlets** rusty-hairy; **leaves** to about

Pittosporum erioloma × .33

337

Pittosporum melanospermum

25 cm × 8 cm, lanceolate to obovate, alternate, tapering to base, spreading to ascending, dark green and becoming glabrous above, rusty-hairy below, margins entire, apex softly pointed; **flowers** to about 0.8 cm × 0.8 cm, white to creamy-yellow, in terminal corymbs, sweetly fragrant, sometimes profuse and conspicuous; **capsules** to about 1.5 cm across, ovoid to globular, yellow, rough; **seeds** orange-red, few per fruit.

A widespread tropical species which occurs in northern regions as well as overseas. Rarely cultivated now, but was grown in England as early as 1810. Plants are best suited to seasonally dry tropical or subtropical regions. Soils need to be at least moderately well drained and a semi-shaded or shaded site is suitable. Plants can become open but pruning should promote bushier development. Propagate from seed, or from cuttings of firm young growth, which can be slow to produce roots.

Pittosporum melanospermum F. Muell.
(black seed)
Qld, WA, NT   Black-seed Pittosporum
3–10 m × 2–5 m   Jan–April

Medium to tall **shrub** or small **tree** with glabrous young growth; **bark** rough, grey, grooved, somewhat flaky; **branches** spreading to ascending; **branchlets** glabrous; **leaves** 6.5–14 cm × 2.5–5 cm, lanceolate to obovate, alternate, prominently stalked, spreading to ascending, pale green above, lighter below, glabrous, venation prominent, margins entire, flat to wavy, apex bluntly pointed; **flowers** to about 1 cm × 1 cm, white to creamy white, in crowded terminal corymbs to about 10 cm across, sweetly fragrant, often profuse and very conspicuous; **petals** spreading from base; **capsules** to about 1.5 cm across, somewhat pear-shaped, yellow to orange; **seeds** black, not sticky.

A highly attractive species from northern regions of Australia where it grows in sandy or rocky sites in open forest and woodland. Generally has a shrubby habit in coastal scrubs and is taller and tree-like in grassland. Uncommon in cultivation, it deserves to be better known because of its highly fragrant flowers. Plants will do best in subtropical and seasonally dry tropical regions. Needs good drainage and a sunny or semi-shaded site is suitable. Propagate from seed, or from cuttings of firm young growth, which may be slow to form roots.

Because of recent botanical studies, this species will be transferred to a new genus (*Auranticarpa*).

Pittosporum moluccanum (Lam.) Miq.
(from the moluccas)
WA, NT
2.5–6 m × 1.5–4 m   Feb–March

Medium to tall **shrub** with glabrous young growth; **bark** pale grey; **branches** many, spreading to ascending; **branchlets** glabrous; **leaves** 7–14.5 cm × 2–5.5 cm, narrowly elliptic, spirally arranged, prominently stalked, spreading to ascending , pale green, glabrous, glossy, margins recurved and wavy, apex bluntly pointed; **flowers** to about 0.8 cm × 1 cm, white to greenish-cream, in somewhat loose, terminal clusters, sweetly fragrant, can be profuse and conspicuous; **capsules** to about 2.5 cm × 2.5 cm, ovoid to globular, opening very

Pittosporum moluccanum × .45

widely, orange, somewhat woody; **seeds** orange-red.

An attractive inhabitant of coastal tropical regions in northern areas of Australia and south-eastern Asia. It grows on dunes and rocky sites often amongst vine thickets. This species has potential for wider cultivation in subtropical and seasonally dry tropical regions and is suited to coastal sites. Needs good drainage and a sunny or semi-shaded aspect. Plants are densely foliaged and worth trying as hedging. Propagate from seed, or from cuttings of firm young growth, which may be slow to produce roots. Seed germinates relatively easily after about 3 months of mist irrigation.

Pittosporum oreillyanum C. T. White
(after Bernard O'Reilly)
Qld, NSW   Thorny Pittosporum
1.5–4 m × 1–4 m   Sept–Nov

Small to medium, open **shrub** with hairy, coppery young growth; **branches** spreading to ascending; **branchlets** becoming glabrous, always spiny; **leaves** 0.2–2 cm × 0.2–0.5 cm, ovate to circular, short-stalked, spreading to ascending, can be crowded, dark green and glabrous above, hairy below, margins entire, wavy or toothed, apex softly pointed; **flowers** about 1 cm × 1 cm, white to pink, axillary, solitary on short stalks, never profuse or very conspicuous; **capsules** with thin valves, to about 1 cm across, globular, yellowish to yellow-brown, smooth; **seeds** bright red.

A very distinctive, spiny species from highland rainforests of the Macpherson and Tweed Ranges in south-eastern Qld and north-eastern NSW respectively. Cultivated mainly by enthusiasts, it is best suited to subtropical regions but will grow in temperate zones. Plants are often straggly. Best grown in moist, shady sites and does well amongst ferns. Soils should be acidic. Makes an interesting plant for containers and should be an unusual bonsai subject. Propagate from seed, or from cuttings of firm young growth, which may be slow to form roots.

Pittosporum oreillyanum × .5, leaf and flower × 2

Pittosporum rhombifolium   W. R. Elliot

Pittosporum phylliraeoides DC.
(resembling the genus *Phillyrea*)
WA
6–10 m × 3–5 m   Aug–Dec
Tall **shrub** or small **tree** with densely grey-hairy young growth; **branches** spreading to erect; **leaves** 2.4–4 cm × 0.6–1.7 cm, elliptic, lower surface with dense silver-grey hairs, margins thick, apex with a short point; **flowers** about 1 cm across, yellow, solitary or

Pittosporum revolutum   D. L. Jones

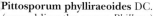

paired, terminal or axillary, hairy; **capsules** to about 1.3 cm long, dusky-orange, hairy; **seeds** red-brown.

Grows in isolated patches on the coastal plain between Kalbarri and the North West Cape, commonly on sand over limestone. Poorly known in cultivation and its requirements are largely unknown. Previously confused with *P. angustifolium*, a much more widely distributed species which is popular in cultivation. Propagate from seed or from cuttings of firm young growth.

Pittosporum revolutum Dryander
(margins rolled backwards)
Qld, NSW, Vic   Yellow Pittosporum; Mock Orange;
                                    Rough-fruit Pittosporum
1.5–5 m × 1–3 m   Oct–Dec
Small to tall **shrub** with densely rusty-hairy young growth; **branches** spreading to ascending; **branchlets** initially rusty-hairy; **leaves** 5–15 cm × 1.5–6 cm, elliptic to obovate, alternate or whorled, stalked, spreading to ascending, dark green, becoming glossy and glabrous above, usually faintly to densely rusty-hairy below, margins slightly recurved, entire, venation prominent, apex bluntly pointed; **flowers** to about 1.5 cm across, yellow, few (female) to several (male) in terminal clusters, sweetly fragrant, rarely profuse and conspicuous; **calyx lobes** reddish-brown; **capsules** to about 2.5 cm long, green, eventually reddish-brown to yellowish-orange, rough to warty; **seeds** red to reddish-brown, copious.

A handsome, shrubby, rhizomatous species which inhabits coastal rainforest, wet sclerophyll forest and dry sclerophyll forest. It grows on good quality soils in eastern Qld, eastern NSW and far East Gippsland, Vic. This adaptable species grows well as far south as Melbourne. Suitable for acidic soils with a semi-shaded or sunny aspect. Dislikes drying out and is sensitive to iron deficiency. Hardy to moderate frosts. Plants respond well to pruning which promotes decorative flushes of new growth. Makes an attractive container plant and can be placed indoors for limited periods. Introduced into English gardens in 1795. A quantity of fresh leaves can be rubbed together with water to form a pleasant lather which can be used as a substitute for bath soap. Propagate from seed, or from cuttings of firm young growth, which can be slow to produce roots.

Pittosporum rhombifolium Hook.
(rhomboid-shaped leaves)

Qld, NSW	Hollywood; Diamond Laurel; White Holly
4–15 m × 2–6 m	Nov–Jan

Tall **shrub** or small to medium **tree** with glabrous, pale green young growth; **bark** pale grey; **branches** spreading to erect; **branchlets** glabrous; **leaves** 6–11 cm × 3–5 cm, ovate to rhomboidal, alternate or clustered, stalked, spreading to ascending, glossy, glabrous, dark green above, paler below, margins irregularly toothed, apex softly pointed, parsley-like fragrance when crushed; **flowers** about 1 cm across, white, in crowded terminal clusters, sweetly fragrant, often profuse and conspicuous; **petals** spreading from the base; **capsules** about 1 cm long, somewhat globular to ovoid, often compressed, in dense clusters, yellow to orange; **seeds** black, few, not sticky.

An outstanding species from north-eastern Qld to north-eastern NSW. It occurs in freely draining soils from the coast to slightly inland and is found in rainforest and woodland. Adapts well to cultivation and deserves to be better known. Needs well-drained acidic soils and will grow in a sunny or semi-shaded site. Hardy to moderate frosts and withstands strong winds. Excellent for gardens in inland towns. The massed colourful fruits can remain on plants for many months over autumn and winter providing an impressive display. May be slow-growing. Responds well to pruning. Suited to container cultivation and can be used indoors. The close-grained white timber is suitable for wood-turning. Propagate from seed, or from cuttings of firm young growth, which are slow to produce roots.

Because of recent botanical studies, this species will be transferred to a new genus (*Auranticarpa*).

Pittosporum rubiginosum A. Cunn.
(rusty)

Qld	Red Pittosporum
1–3 m × 0.6–1.5 m	Aug–Nov

Small to medium, often single-stemmed **shrub** with rusty-hairy young growth; **branches** ascending to erect, few; **branchlets** initially rusty-hairy; **leaves** 6–15 cm × 2–4 cm, oblong-lanceolate, alternate or clustered, tapering to base, nearly sessile, spreading to ascending, dark green, whitish hairs below, margins entire or faintly toothed and wavy; **flowers** bisexual, about 1.5 cm across, white to pale yellow, in somewhat open,

terminal clusters, sweetly fragrant, rarely profuse but always conspicuous; **capsules** to about 2 cm long, thin-valved, yellow to orange, faintly hairy; **seeds** reddish.

An interesting and appealing species from north-eastern Qld where it is often prominent in highland areas. It usually develops as a single-stemmed plant but may become more branched with maturity. New leaves are often purple to mauve on the undersurface. Pruning at an early age may promote lateral growth. Plants inhabit rainforest or rainforest margins in acidic soils which rarely dry out. Not commonly cultivated, it requires good drainage and a semi-shaded aspect to do well. Applications of mulch, fertiliser and supplementary watering can promote good growth. It is worth trying in gardens and containers in tropical and subtropical regions. Propagate from seed, or from cuttings of firm young growth, which may be slow to form roots.

The ssp. *wingii* (F. Muell.) R. C. Cooper is a synonym of *P. wingii* F. Muell.

Pittosporum undulatum Vent.
(wavy)

Qld, NSW, Vic	Sweet Pittosporum
5–12 m × 4–7 m	Aug–Dec

Tall **shrub** or small **tree** with glabrous or faintly hairy, pale green young growth; **branches** spreading to ascending; **branchlets** glabrous; **leaves** mainly 6–15 cm × 1.5–4 cm, elliptic to oblanceolate or ovate, alternate or clustered, stalked, spreading to ascending, glabrous, glossy, dark green, margins entire and wavy, apex softly pointed; **flowers** about 1.5 cm across, creamy white, in few- to many-flowered terminal clusters, sweetly fragrant, often profuse but rarely highly conspicuous; **capsules** to about 1.4 cm × 1 cm,

Pittosporum undulatum × .65

340

globular to obovoid, yellowish-brown to pale orange, smooth; **seeds** orange.

A delightful and popular species for cultivation which has become entrenched as an environmental weed in parts of south-eastern Australia. Within its natural range from south-eastern Qld and eastern NSW to eastern Vic, it occurs in rainforest, wet and dry sclerophyll forest and woodland. It often grows in moist protected sites. This species is highly desirable for cultivation because it has a beautiful floral perfume which can pervade large areas. It was grown in England in 1789 and has been popular in Mediterranean climatic regions such as California, USA, where it is known as Victorian Box or Mock Orange. However, *P. undulatum* should definitely not be grown in close vicinity to bushland outside of its natural distribution. Plants adapt to most acidic soils in sunny or semi-shaded sites. They are hardy to moderate frosts and withstand extended dry periods. Responds well to pruning or clipping and has been used for hedging for many years. The resinous sap from fresh cuts is very sticky. Well developed plants provide dense shade and are commonly used by birds as nesting sites. The close-grained white to whitish-brown timber has been used to a limited degree for wood-turning. Propagate from seed, or from cuttings of firm young growth, which may be slow to produce roots.

The cultivar 'Variegatum' is a bushy variant which has cream and green variegated leaves. It is cultivated to a limited degree and maintains its variegated habit. It is propagated by cuttings, aerial layers or grafting.

Hybrids of *P. undulatum* and *P. bicolor* occur in nature on the Australian mainland, and similar hybrids have occurred in Tas where *P. undulatum* is naturalised. The Tasmanian progeny are sometimes known as *P.* × *emmettii* W. M. Curtis.

Pittosporum venulosum F. Muell.
(closely veined)
Qld   Rusty Pittosporum
3–8 m × 2–4 m   Sept–Dec

Medium to tall **shrub** or small, narrow **tree** with rusty-hairy young growth; **branches** spreading to ascending; **branchlets** rusty-hairy; **leaves** 4–10 cm × 2–3.5 cm, lanceolate, alternate or clustered, stalked, spreading to ascending, bright green, becoming glabrous above, whitish to rusty-hairy below, margins faintly recurved, venation prominent, apex bluntly pointed; **flowers** about 1.5 cm across, creamy-white, hairy, in somewhat loose, terminal clusters, sweetly fragrant, can be profuse and conspicuous; **capsules** to about 2 cm across, yellow-orange; **seeds** orange-red to reddish-brown.

A handsome, bushy species from the Cook, North and South Kennedy and Port Curtis Districts of eastern Qld where it occurs in a range of soils and habitats. Adapts very well to cultivation in subtropical and tropical regions and is worth trying in protected positions in temperate areas. Needs acidic soils which can be light to moderately heavy. A sunny or semi-shaded site is suitable. Useful for screening and hedging. Propagate from seed, or from cuttings of firm young growth, which may be slow to form roots.

Pittosporum wingii F. Muell.
(after W. Wing)
Qld
3–6 m × 0.5–1 m   Oct–Dec

Medium to tall, sparsely branched **shrub** mostly covered in dense golden-brown hairs; young growth purple-red, hairy; **branches** erect, few; **branchlets** hairy; **leaves** 8–18 cm × 2–4 cm, oblong-lanceolate to obovate, whorled, nearly sessile, spreading to ascending, dark green, purplish hairs below, margins wavy; **flowers** about 1.5 cm across, cream to pale yellow, in dense terminal clusters, sweetly fragrant, conspicuous; **capsules** to about 2 cm long, yellow to orange, faintly hairy; **seeds** reddish.

Occurs in north-eastern Qld from Cooktown to near Townsville, and is found growing in rainforest. It is an interesting small species which has potential for gardens in tropical and subtropical regions. Best grown in a shady position in well-drained soil. Mulches, regular light pruning and fertilisers are beneficial. Frost tolerance is not high. Propagate from seed, or from cuttings of firm young growth, which may be slow to produce roots.

P. rubiginosum ssp. *wingii* (F. Muell.) R. C. Cooper is a synonym.

PITYRODIA R. Br.
(from the Greek *pityron*, bran, husk, scale; *odes*, like; referring to the leaves)
Verbenaceae (alt. *Chloanthaceae*)   Native Foxgloves

Evergreen, dwarf to medium **shrubs**; **stems** usually erect, woody, terete or 4-angled; **branches** hairy; **leaves** simple, decussate or in whorls of 3 or scattered, non-decurrent, venation reticulate; **stipules** absent; **inflorescences** cymes, clusters or solitary flowers, in terminal or axillary leafy spikes, racemes or panicles; **flowers** tubular, bisexual, usually 2-lipped, upper lip 2-lobed, lower lip 3-lobed, lobes usually spreading; **calyx** 5-lobed, persistent; **stamens** 4, sometimes exserted beyond tube; **anthers** 2-lobed; **fruit** a 4-celled, non-fleshy drupe which may split into 2 parts on ripening.

Pityrodia is an endemic genus comprised of 41 species There are 27 species in WA, 16 species occurring in NT and 1 species in central eastern Qld. Only 3 species, *P. loricata*, *P. loxicarpa* and *P. ternifolia*, occur in more than one state and they are found in both WA and NT.

Within the genus there are some highly desirable ornamental species which should be better known in cultivation, while other species are of interest mainly to enthusiasts.

Pityrodia is distributed from the coast to the harsh desert regions. Generally plants occur in sandy or gravelly soils usually in sandy heath, eg. in southern WA, or on sandstone escarpments which occur in northern WA and the Top End of NT.

Many pityrodias are dwarf to small, somewhat erect, woody shrubs with densely hairy stems and leaves which can have a sticky texture because of glandular hairs. The foliage is often one of the highlights of members of this genus although brilliant flowers and foliage are combined in species such as *P. axillaris*, *P. dilatata*, *P. oldfieldii* and *P. terminalis*.

The common name, Native Foxglove, was soon

accepted because the flowers of many pityrodias reminded early settlers of the unrelated European foxgloves, which were already cultivated in European gardens.

To date pityrodias have only been cultivated to a limited degree because economic methods of propagation have not been developed yet. Recent experimentation with smoke treatment of seeds may lead to plants becoming more readily available.

Cultivation requirements for most species include freely draining acidic soils, and a site which receives full sunshine or sunshine for the greater part of the day. Most species tolerate extended dry periods and are rarely damaged by light frosts. Success with southern species is more likely to be achieved in semi-arid and warm temperate regions than in cool temperate areas. The tropical species have a dislike for low temperatures. Regular air movement is critical for successful cultivation.

Response to applications of fertilisers is not well understood but nursery plants respond well to light application of slow-release fertilisers which have low levels of phosphorus.

Some species can have an open, often leggy growth habit, so it is usually beneficial to tip prune young plants to promote side-branching. Pruning of semi-mature and mature plants is not well understood to date as there have not been enough trials to recommend one procedure over another. However, *P. dilatata*, *P. oldfieldii* and *P. terminalis* respond well to light or moderate pruning.

Pests and diseases may cause some problems. Chewing insects may disfigure leaves but are rarely of great consequence (see Volume 1, page 143, for control methods). Plants are sensitive to root-rotting fungi such as *Phytophthora cinnamomi* (see Volume 1, page 168 for further information). Sweating during extended periods of high humidity and rain is usually detrimental to plant growth.

Grey Mould, *Botrytis cinerea*, is likely to cause major damage in areas which have extended overcast and wet weather during flowering. The old flowers are a perfect host for Grey Mould which will quickly envelop healthy growth especially on plants with hairy foliage (see Volume 1, page 173 for control methods).

Propagation of pityrodias is from seed or from cuttings. Seed is often difficult to procure. Germination results are not always satisfactory. Seedlings can be common after bushfires have passed through an area and with the recent successful seed germination of a range of genera after smoke treatment, it may be worthwhile treating seed of *Pityrodia* species in a similar way (see the Propagation entry in the Supplement for further information).

Another seed treatment involves placing the seed in a wire strainer and passing it over a naked flame to thoroughly scorch the seed, which should then be sown immediately. Placing the seeds in warm water (about 60°C) and leaving them to soak for 24 hours before sowing may also be beneficial. Untreated seed of species such as *P. atriplicina*, *P. axillaris*, *P. oldfieldii*, *P. paniculata*, *P. teckiana*, *P. terminalis*, *P. uncinata* and *P. verbascina* have germinated after varying periods of time, some beginning as early as 20 days after sowing,

while other species have taken up to 115 days to germinate. Damping off may be a problem with young seedlings during hot humid weather conditions. See Volume 1, page 172 for further information.

Pityrodia cuttings are very prone to severe damage from Grey Mould, *Botrytis cinerea*, and other fungal problems because of their hairy and sometimes sticky stems and leaves. It is advisable to dip propagation material in a dilution of 1 part bleach to 19 parts water for 5–10 seconds before preparing cuttings. The material is then washed thoroughly in fresh acidic water. Cuttings usually benefit from the application of rooting hormones. Roots may take 4–30 weeks or more to appear. Cuttings of firm young growth seem to produce the best results.

Pityrodia angustisepala Munir
(narrow sepals)
NT
1–2 m × 0.6–1.5 m   Feb–April

Small **shrub** with densely hairy young growth; **branches** ascending to erect, somewhat spindly, 4-angled, densely hairy; **leaves** 3–9.5 cm × 0.6–2.5 cm, elliptic to elliptic-lanceolate, opposite, base tapered to long petiole, flat, glandular and faintly hairy above and below, margins crenulate, apex softly pointed; **flowers** to about 2.5 cm long, pale yellow, exterior glandular and hairy, solitary or clusters of 3 in upper axils, moderately conspicuous; **calyx lobes** 1–1.8 cm × 0.1–0.3 cm; **drupes** to about 0.5 cm × 0.4 cm.

This species is from the Top End where it occurs mainly in Arnhem Land on sandstone gorges and

Pityrodia angustisepala × .55

plateaux. Evidently not cultivated, it should be best suited to seasonally dry tropical regions. Needs freely draining acidic soils and plenty of sunshine. Pruning from an early stage may promote bushy growth. Propagate from fresh seed or possibly from cuttings of firm young growth.

P. quadrangulata differs in having more-or-less sessile leaves.

Pityrodia atriplicina (F. Muell.) Benth.
(similar to the genus *Atriplex*)

WA   Saltbush Foxglove
1–2.5 m × 0.6–1.8 m   Aug–Dec

Small to medium **shrub** with yellowish-hairy young growth; **branches** ascending to erect, densely greyish-hairy; **leaves** 1–3.5 cm × 0.5–3 cm, broadly elliptic to nearly orbicular, opposite, more or less sessile, becoming greyish to whitish-hairy, margins entire, apex blunt; **flowers** to about 2.5 cm long, pink with purple-spotted throat, exterior hairy, in leafy panicles in upper axils, often profuse and very conspicuous; **calyx** velvety-hairy; **drupes** enclosed within the calyx.

A handsome member from the northern Irwin and nearby Austin Districts where it grows in sandheath on the coast and slightly inland. Warrants greater cultivation and should grow to its potential in semi-arid and warm temperate regions. Requires excellently drained soils and plenty of sunshine. Plants withstand light frosts. Pruning at an early age promotes bushy growth. Propagate from seed, or from cuttings of firm young growth, which can be difficult to strike and are prone to leaf drop in excessively humid conditions.

P. paniculata differs in having oblong to obovate leaves and a more open arrangement of flowers.

Pityrodia augustensis Munir
(from the Mt Augustus region)

WA   Mt Augustus Foxglove
0.8–1.5 m × 0.6–1.3 m   Aug–Oct

Dwarf to small **shrub** with densely hairy young growth; **branches** spreading to erect, covered in woolly greenish-white hairs; **leaves** 3–6 cm × 0.6–1.5 cm, narrowly elliptic to elliptic-lanceolate, opposite, sessile, base tapered, thick, covered in woolly greenish-white hairs, margins entire, apex bluntly pointed; **flowers** 1.8–2.5 cm long, deep lilac, exterior faintly hairy and glandular, usually in 3–5-flowered clusters or sometimes solitary, often profuse and very conspicuous; **calyx** to about 1.2 cm long, deep purple-lilac, glandular and densely woolly-hairy; **drupes** to about 0.35 cm long, enclosed within calyx.

A very showy species from the Mt Augustus region in the Western Ashburton District. It occurs in sandy, gravelly soils of the mountain slopes. Recommended for semi-arid regions and may also succeed in warm temperate zones where it would be worth trying as a container plant. Good drainage and plenty of sunshine are essential. Plants are hardy to light frosts. Propagate from seed and possibly from cuttings of firm young growth.

P. axillaris and *P. terminalis* have similarities but both have leaves which are broadest towards the apex.

Pityrodia axillaris   W. R. Elliot

Pityrodia axillaris (Endl.) Druce
(axillary)

WA   Native Foxglove; Woolly Foxglove
0.2–0.4 m × 0.3–0.6 m   Sept–Oct

Spreading dwarf **shrub**; young growth densely woolly-hairy; **branches** spreading to ascending, densely covered with woolly-white hairs; **leaves** 1.5–5 cm × 0.5–1.8 cm, obovate to oblong-obovate, opposite, sessile, base tapered, soft, densely covered with whitish hairs, margins entire, apex very blunt; **flowers** 2–3 cm long, orange-red to deep red or burgundy, sometimes with violet streaks, exterior nearly glabrous, in clusters of 3–5 flowers or solitary, often profuse and very conspicuous; **calyx** to about 1.8 cm long, hairy, persistent; **drupes** densely hairy, enclosed in calyx.

This outstanding member of the genus has a limited distribution being confined to sandheath of the far north-western Avon District. Cultivated to a limited degree, it must have a warm to hot site with very well-drained acidic soils. Best for semi-arid and warm temperate regions. Withstands moderate frosts. May be propagated from seed, or cuttings of firm young growth, which may strike but hairiness causes problems.

P. spectabilis C. A. Gardner is a synonym. *P. terminalis* has been confused with *P. axillaris* within the nursery industry but *P. terminalis* is a taller shrub with flowers in leafy spikes. The allied *P. augustensis* has narrowly elliptic leaves and deep lilac flowers.

Pityrodia bartlingii (Lehm.) Benth.
(after Frederich G. Bartling, 19th-century German botanist)

WA   Woolly Dragon; Woolly Foxglove
0.3–1.5 m × 0.2–1 m   Aug–Jan

Dwarf to small **shrub**; young growth greyish- to rusty-hairy; **branches** ascending to erect, densely hairy; **leaves** 1–5.5 cm × 0.2–0.7 cm, linear to lanceolate, or

Pityrodia byrnesii

appearing terete, scattered or in whorls of 3, or sometimes opposite and decussate, sessile, base tapered, dull green, warty and hairy above, densely woolly-hairy below, margins revolute, apex bluntly pointed; **flowers** 1.2–2.3 cm long, light mauve to purplish-pink with brownish spots in throat, exterior hairy, solitary or in clusters of 3 per axil, forming leafy racemes to 35 cm long, often profuse and very conspicuous; **calyx** to about 1.3 cm long with narrow lobes, covered in grey- to white-woolly hairs, persistent; **drupes** to about 0.4 cm long, enclosed in calyx.

This distinctive species is widely distributed in the south-west. The main concentration is in the Irwin and Darling Districts with disjunct occurrences in the western Coolgardie and western Roe Districts. It inhabits sandy or gravelly soils in sandheath. Rarely encountered in cultivation, it must have extremely well-drained acidic soils and a sunny aspect. Plants are hardy to moderate frosts. Pruning may help to promote branching although response to pruning has not been documented. Propagate from seed, or from cuttings of firm young growth, which can be difficult to strike.

P. halganiacea has similarities but is distinguished by its purplish-blue flowers in somewhat open panicles.

Pityrodia byrnesii Munir
(after Norman Byrnes, contemporary Australian botanist)
NT
0.7–1.3 m × 0.5–1 m   Aug–Sept

Dwarf to small **shrub** with sticky young growth; **branches** spreading to ascending, glandular with stellate hairs, often rusty-orange when old; **leaves** 1.2–3 cm × 0.4–1 cm, oblong to narrowly elliptic, in whorls of 3, sessile, green, sparsely hairy, glandular, sticky, flat, margins entire, apex pointed; **flowers** to about 1.3 cm long, creamy white with purplish stripes on upper lip, exterior of lobes hairy, solitary in upper axils, sometimes profuse, rarely very conspicuous; **drupes** to about 0.4 cm long, obovoid, glandular and hairy.

This species occurs in Arnhem Land and would mainly be of interest to enthusiasts. Suitable for seasonally dry tropical regions and may survive in subtropical zones. Requires very good drainage in acidic soils and a warm to hot aspect. Propagate from seed and possibly from cuttings.

Pityrodia caerulea (F. Muell. & Tate) E. Pritz. =
P. halganiacea (F. Muell.) E. Pritz.

Pityrodia canaliculata A. S. George
(longitudinally grooved or channelled)
WA
1–2.5 m × 0.6–2 m   Aug–Sept

Small to medium **shrub** with hairy young growth; **branches** ascending to erect, many, hairy; **leaves** 1–4.5 cm × 0.2–0.4 cm, linear, opposite, more or less sessile, dark green, becoming glabrous and channelled above, small fawn scales and prominent midrib below, apex pointed; **flowers** to about 1 cm long, white with red to reddish-brown spotted throat, exterior scaly, solitary or in clusters of 3 in upper axils, can be profuse and moderately conspicuous; **calyx** to about 0.5 cm long, bell-shaped, scaly, persistent; **drupes** to about 0.4 cm long, flat-topped, glandular, enclosed by calyx.

A moderately ornamental yet poorly known species which is apparently restricted to a small area of the central Austin District where it grows on sandy soils. Suited to arid and semi-arid regions and may adapt to warm temperate zones if given a very sunny site. Drainage must be excellent. Plants withstand moderately heavy frosts. Propagate from seed and possibly from cuttings.

The closely allied *P. lepidota* is a smaller shrub which has much shorter, blunt leaves.

Pityrodia chorisepala Munir
(separate sepals)
NT
0.6–1 m × 0.5–0.8 m   July–Sept

Dwarf **shrub** with densely greyish-hairy young growth; **branches** ascending to erect, covered in greyish hairs; **leaves** 0.5–1.5 cm × to about 0.6 cm, ovate, opposite, sessile, base narrowed, spreading to ascending, densely covered in yellowish-grey woolly hairs, flat, margins entire, apex bluntly pointed; **flowers** to about 0.8 cm long, white, exterior glabrous, solitary or clusters of 3 in upper axils, forming leafy spikes, can be profuse and moderately conspicuous; **calyx** to 0.7 cm long, deeply lobed with narrow segments; **drupes** about 0.3 cm long, depressed at top, corrugated.

This species is known only from the western Central Australian Region where it occurs in a few localities near the WA border. Evidently not cultivated, it is likely to be of interest to enthusiasts. Suitable for arid and semi-arid regions. Hardy to moderate frosts. Propagate from seed and possibly from cuttings.

Pityrodia chrysocalyx (F. Muell.) C. A. Gardner
(golden calyx)
WA
0.3–0.8 m × 0.2–0.5 m   Sept–Nov

Dwarf **shrub** with rusty-hairy young growth; **branches** ascending to erect, slender, covered with yellow-brown scales; **leaves** 0.2–0.8 cm × 0.1–0.4 cm, ovate to broadly ovate, scattered or in 3s, nearly sessile, spreading to mainly reflexed, smooth and convex and sticky above, with yellow-brown scales below, apex pointed; **flowers** to 1.2 cm long, whitish, exterior of lobes hairy, solitary in upper leaf axils, sometimes profuse, rarely very conspicuous; **calyx** with yellowish-brown scales, persistent; **drupes** to about 0.5 cm long, depressed at top, hairy, enclosed within calyx.

A poorly known species, mainly from the Norseman–Esperance region where it occurs in sandy loam. Apparently not cultivated, it lacks strong ornamental appeal. Suitable for semi-arid and warm temperate regions. Needs good drainage and plenty of sunshine. Withstands moderately heavy frosts and extended dry periods. Propagate from seed and possibly from cuttings of firm young growth.

P. depremesnilia (F. Muell.) E. Pritz. is a synonym.

Pityrodia coerulea (F. Muell & Tate) E. Pritz. =
P. halganiacea (F. Muell.) E. Pritz.

Pityrodia cuneata × .5

Pityrodia cuneata (Gaudich.) Benth.
(wedge-shaped)
WA
0.6–3.5 m × 0.5–2.5 m   Aug–Oct
Dwarf to medium **shrub** with whitish- to brownish-hairy young growth; **branches** spreading to erect, many, with white- or brownish-woolly hairs; **leaves** 0.6–2.5 cm × 0.3–1 cm, obovate to cuneate, opposite, tapered to base, sessile, woolly hairs covering rough surface, flat, margins entire or finely toothed near apex, venation conspicuous below; **flowers** 0.8–1.4 cm long, blue fading to white with purple spots inside tube, exterior hairy, solitary or in 3s in upper axils, forming terminal leafy spikes, can be profuse and conspicuous; **calyx** to 0.7 cm long, with deep broad lobes, densely hairy, persistent; **drupes** about 0.2 cm long, globular, 2-celled, enclosed in calyx.
P. cuneata is restricted to the Shark Bay region of the northern Irwin District where it grows in sandy soils. Warrants cultivation and should be best suited to semi-arid and warm temperate regions. Plants withstand extended dry periods and are hardy to light frosts. They need exceedingly well-drained acidic soils and plenty of sunshine. Propagate from seed and cuttings of firm young growth are worth trying.
P. paniculata is allied but differs in having leafy racemes of flowers.

Pityrodia depremesnilia (F. Muell.) E. Pritz. =
            P. chrysocalyx (F. Muell.) C. A. Gardner

Pityrodia dilatata (F. Muell.) Benth.
(enlarged or widened)
WA
0.3–0.6 m × 0.5–1 m   Sept–Nov
Dwarf **shrub** with white woolly-hairy young growth; **branches** spreading to ascending, many, densely covered

in white-woolly hairs; **leaves** 1–3.8 cm × 0.4–1.5 cm, obovate to oblong-spathulate, sessile, tapered below middle to stem-clasping base, spreading to ascending, initially densely woolly-hairy, becoming glabrous above to reveal wrinkled surface, mainly flat, margins initially entire, becoming crenate, apex somewhat rounded; **flowers** to about 2.5 cm long, orange-red, exterior densely hairy, solitary or clustered in upper axils, forming leafy spikes, usually profuse and very conspicuous; **calyx** to about 1.2 cm long, densely hairy, long narrow lobes persistent; **drupes** to about 0.3 cm long, nearly globular, densely hairy, enclosed in calyx.
An extremely ornamental species which is cultivated to a limited degree. It occurs in the north-western Avon, northern Darling and South-eastern Irwin Districts where it grows in sandy soils which can contain lateritic gravel. Requires well-drained acidic soils and maximum sunshine. Performs best in semi-arid and warm temperate regions but can succeed in cooler areas. Excellent for rockeries and containers. Responds very well to hard pruning and withstands moderately heavy frosts although foliage can be damaged. Propagate from seed, or from cuttings of firm young growth, which strike readily.
P. quadrangulata differs in having leaves which are not stem-clasping and the flowers are pale yellow.

Pityrodia drummondii Turcz. =
            P. loxocarpa (F. Muell.) Druce

Pityrodia exserta (Benth.) Munir
(protruding)
WA
0.15–0.5 m × 0.3–1 m   July–Nov
Dwarf **shrub** with woolly-hairy young growth; **branches** spreading to ascending, initially woolly-hairy, becoming nearly glabrous; **leaves** 1–4.5 cm × 0.1–0.5 cm, linear to linear-lanceolate, decussate, scattered or in whorls of 3, sessile, crowded, deep green and wrinkled above, initially densely hairy below, margins recurved to revolute, apex bluntly pointed; **flowers** 1.5–3 cm, deep pink to dark red, exterior hairy, solitary or in 3s in upper axils, forming leafy spikes, usually profuse and very conspicuous; **stamens** and **style** exserted; **calyx** to about 1.2 cm long, deeply divided, hairy, persistent; **drupes** to about 0.5 cm long, nearly globular, hairy, enclosed in calyx.
The typical variant is a very ornamental dwarf plant which mainly inhabits the sandy and gravelly soils of the coastal and adjacent areas of the Fitzgerald National Park in the Eyre District. Rarely cultivated, it deserves greater recognition because it has potential for use in gardens and containers in semi-arid and temperate regions. Plants require well-drained acidic soils and maximum sunshine. They withstand light frosts and extended dry periods. It was previously known as *P. uncinata* var. *exserta* Benth.
The var. *lanata* Munir differs in having upper stems and leaves covered with whitish-woolly hairs and flowers which have the stamens and style barely exserted. It occurs further inland than the typical variant and has similar requirements but may struggle in cool temperate regions.

Pityrodia exsuccosa

Pityrodia dilatata   W. R. Elliot

Propagate both variants from seed or possibly from cuttings of firm young growth.

Pityrodia exsuccosa F. Muell. =
  Dicrastylis exsuccosa (F. Muell.) Druce

Pityrodia flexuosa Price =
  Dicrastylis flexuosa (Price) C. A. Gardner

Pityrodia gilruthiana Munir
(from the Mt Gilruth region)
NT
1–1.5 m × 1–1.5 m   Feb–April
Small **shrub** with hairy and sticky young growth; **branches** spreading to ascending, many, hairy, sticky; **leaves** 0.6–3 cm × up to 0.4 cm, linear to narrowly lanceolate, in whorls of 3 or sometimes scattered, sessile, spreading to ascending, moderately crowded near ends of branches, glabrous but very sticky, deep green above, paler below, margins flat or slightly recurved, apex pointed; **flowers** to about 1.3 cm long, white with purple streaks in upper lip, sessile and solitary in upper axils, forming leafy spikes, rarely profuse and very conspicuous; **calyx** to about 1 cm long, deeply lobed, with longitudinal ribs, sticky, persistent; **drupes** to about 0.4 cm long, obovoid, hairy, enclosed within calyx.

An uncommon species from Arnhem Land where it grows on sandstone ridges. Probably mainly of interest to enthusiasts. Suited to seasonally dry tropical and subtropical regions. Plants will need excellent drainage and plenty of sunshine. Propagate from seed and possibly from cuttings of firm young growth.

The closely allied *P. pungens* differs in its pungent-tipped leaves and less sticky stems and leaves.

Pityrodia glabra Munir
(hairless)
WA
0.5–1.3 m × 0.6–1.5 m   Aug–Sept
Dwarf to small **shrub** with glabrous, slightly sticky young growth; **branches** ascending to erect, many, glabrous, yellowish-brown; **leaves** 0.8–2 cm × 0.3–1 cm, oblong to narrowly obovate, opposite, nearly sessile, base tapered, spreading to ascending, deep green, glabrous, somewhat sticky, margins entire except for toothed apex, venation prominent below; **flowers** 0.9–1.4 cm long, white, exterior glabrous, solitary in upper axils, forming leafy spikes; **calyx** to 0.8 cm long, glabrous, sticky, persistent; **drupes** about 0.4 cm long, obovoid, sparsely hairy.

A poorly known and rare species from the Shark Bay region of the northern Irwin District where it grows in sandy soils. Plants should do best in semi-arid and warm temperate regions. Drainage needs to be excellent and plants require maximum sunshine. Frost tolerance is not known. Propagate from seed or possibly from cuttings of firm young growth.

This species has been confused with *P. teckiana*, which has hairy growth and stem-clasping leaves.

Pityrodia glutinosa Munir
(covered with a sticky secretion)
WA
1–1.5 m × 1–1.5 m   Aug–Nov
Dwarf to small **shrub** with sticky, glabrous young growth; **branches** spreading to erect, many, glabrous, sticky; **leaves** 0.5–1.5 cm × 0.2–0.5 cm, oblong, decussate, sessile, base somewhat rounded, spreading to ascending, green, glabrous, sticky, margins mainly flat except recurved near apex, venation visible, apex bluntly pointed; **flowers** 1–1.2 cm long, white, exterior glabrous, solitary in upper axils, rarely profuse and very conspicuous; **calyx** to about 0.7 cm long, sticky, persistent; **drupes** about 0.5 cm long, obovoid, faintly hairy.

This species is restricted to the far-northern Irwin District and occurs in deep sandy soils. Evidently not common in cultivation. For best results it will need excellent drainage and plenty of sunshine. Suitable for semi-arid and warm temperate regions. Tolerates light frosts. Propagate from seed or possibly from cuttings.

Pityrodia halganiacea (F. Muell.) E. Pritz.
(similar to the genus *Halgania*)
WA
0.3–1 m × 0.2–0.7 m   Sept–Nov
Dwarf **shrub** with greyish- to whitish-hairy young growth; **branches** ascending to erect, whitish- to greyish-hairy; **leaves** 0.7–3.5 cm × 0.2–0.4 cm, linear-lanceolate to nearly terete, in whorls of 3 or 4, sessile, spreading to ascending, often incurved, somewhat crowded on branchlets, initially woolly-hairy but becoming glabrous above, woolly-hairy below, margins revolute, apex pointed; **flowers** 1–1.5 cm long and up to 2 cm across, purplish-blue, exterior of lobes hairy, in few-flowered cymes in upper axils, forming loose panicles to about 13 cm long, often profuse and very conspicuous; **calyx** to about 0.6 cm long, pinkish, hairy, persistent; **drupes** to about 0.25 cm across, globular, 4-lobed, enclosed in calyx.

An extremely handsome species from the Avon, Coolgardie and northern Darling Districts where it grows in sandy soils which often contain lateritic gravel.

Pityrodia hemigenioides   W. R. Elliot

Pityrodia lanceolata   D. L. Jones

It deserves wider recognition in cultivation. Should succeed in semi-arid and warm temperate regions. Needs freely draining acidic soils and lots of sunshine. Tolerates moderate frosts and extended dry periods. Response to pruning is not known. Propagate from seed and cuttings of firm young growth are worth trying.

P. coerulea (F. Muell. & Tate) E. Pritz. is a synonym.

Pityrodia hemigenioides (F. Muell.) Benth.
(similar to the genus *Hemigenia*)
WA
0.2–1 m × 0.4–1.5 m   Aug–Sept

Dwarf spreading **shrub** with whitish- to greyish-hairy young growth; **branches** spreading to ascending, many, whitish- to greyish-hairy, often entangled; **leaves** 0.5–2 cm × 0.2–1.1 cm, linear-oblong to elliptic, opposite, sessile, base tapered, ascending to erect, somewhat crowded on branchlets, covered with whitish to greyish hairs, margins mainly recurved, apex blunt; **flowers** 0.9–1.2 cm long, white, interior with purple dots, exterior becoming glabrous, solitary in upper axils, forming leafy spikes, can be profuse and moderately conspicuous; **calyx** to about 1 cm long, greyish-hairy, persistent; **drupes** to 0.5 cm long, somewhat globular, enclosed within calyx.

An attractive species which inhabits mainly coastal and inland regions of the northern Irwin District where it grows in rocky or sandy soils. It may be found as an undershrub in *Banksia* woodland and heathland. Rarely cultivated, it requires excellent drainage and will grow in sunny or semi-shaded sites. Plants tolerate light frosts and respond well to pruning. Best suited to semi-arid and warm temperate regions but can succeed in cooler areas if grown in protected warm to hot sites. Propagate from seed, or from cuttings of firm young growth, which strike readily.

P. spenceri Munir has been confused with *P. hemigenioides*. *P. spenceri* is an extremely rare species known from only one area in the Top End, NT. It has stalked leaves and the flowers have a hairy exterior.

Pityrodia jamesii Specht
(after Stewart James)
NT
0.5–2 m × 0.3–1.5 m   Oct–March

Dwarf to small **shrub** with hairy and sticky young growth; **branches** terete, ascending to erect, hairy, yellowish-brown; **leaves** 0.5–7.5 cm × 0.2–1.5 cm, ovate to ovate-lanceolate, opposite, sessile, somewhat stem-clasping, ascending to erect, aromatic, crowded near ends of branches, hairy, sticky, margins toothed and recurved, venation conspicuous, apex pointed; **flowers** 0.8–1.3 cm long, creamy-white, exterior of lobes hairy, solitary in upper axils, rarely profuse; **calyx** to about 0.8 cm long, ribbed lobes, hairy, persistent; **drupes** to 0.4 cm long, obovoid, hairy, enclosed in calyx.

This species is restricted to Arnhem Land where it grows on the sandy soils and stony slopes of the sandstone escarpment. Rarely cultivated, it is of interest mainly to enthusiasts. Suitable for seasonally dry tropical regions and may succeed in subtropical areas. Requires excellent drainage and plenty of sunshine. Aborigines have used infusions of this pityrodia for colds, headaches and treating wounds. *P. jamesii* is also a host for the spectacular Leichhardt grasshopper. Propagate from seed or cuttings of firm young growth.

Pityrodia lanceolata Munir
(lanceolate)
NT
1–2 m × 0.6–1.5 m   June–Nov

Small **shrub** with greyish-hairy young growth; **branches** ascending to erect, more or less 4-angled,

347

greyish-hairy; **leaves** 3–13 cm × 0.5–3.5 cm, lanceolate, opposite, base rounded or tapered with petiole to 1.3 cm long, spreading to ascending, dark green and rough above, greyish hairs and conspicuous venation below, flat, margins entire, apex pointed; **flowers** 2–2.7 cm long, red, exterior glandular and woolly-hairy, solitary or in 3s in upper axils, forming leafy panicles, moderately profuse and conspicuous; **stamens** exserted; **calyx** 1–1.8 cm long, greyish-hairy, lobes ribbed, persistent; **drupes** to 0.6 cm long, oblong, woolly-hairy in upper half, glabrous below, ribbed.

A handsome, interesting and rare species from Arnhem Land east of Darwin where it occurs in rocky crevices. It has potential for cultivation in seasonally dry tropical regions and may succeed in subtropical areas too. Plants will need excellently drained acidic soils and plenty of sunshine. They are likely to suffer frost damage. Propagate from seed and cuttings of firm young growth are worth trying.

P. megalophylla has some affinity but differs in having ovate leaves.

Pityrodia lanuginosa Munir
(with long, woolly or cottony hairs)
NT
0.3–1 m × 0.3–1 m   Jan–June
Dwarf **shrub** with whitish-hairy young growth; **branches** spreading to ascending, terete, densely woolly-hairy; **leaves** 0.8–2.5 cm × 0.2–0.8 cm, narrowly ovate-lanceolate, decussate or in whorls of 3, sessile, base tapered, spreading to ascending, pale greyish-green, hairy, aromatic when crushed, margins flat to slightly recurved, venation conspicuous, apex pointed; **flowers** to about 1 cm long, white to pale pink, upper lobes and tube with deep reddish-purple stripes, sessile, solitary in upper axils, forming leafy few-flowered spikes, moderately conspicuous, rarely profuse; **calyx** to 0.7 cm long, deeply lobed, green, exterior faintly hairy, persistent; **stamens** exserted; **drupes** to about 0.4 cm long, obovoid, hairy.

P. lanuginosa inhabits the sandstone escarpments of Arnhem Land where it often grows in deep sand as a component of the shrubland and open forest. Evidently not cultivated. Best suited to seasonally dry tropical regions. Plants will need well-drained acidic soils with a sunny aspect. Propagate from seed and possibly from cuttings of firm young growth.

Pityrodia lepidota (F. Muell.) E. Pritz.
(scaly)
WA
0.2–1 m × 0.3–1.5 m   Aug–Nov
Dwarf **shrub** with yellow-rusty, scaly young growth; **branches** many, spreading to ascending, densely covered in brownish scales; **leaves** 0.3–1.5 cm × 0.2–0.4 cm, oblong to narrowly obovate, in whorls of 3 or sometimes opposite or scattered, sessile, spreading to ascending, somewhat crowded near ends of branches, greyish or dull green with brownish scales, flat, margins entire, apex blunt; **flowers** 0.7–1 cm long, whitish to pale pink or pale lilac, exterior of lobes scaly and hairy, solitary or in 3s in upper axils, forming leafy spikes to about 8 cm long, sometimes profuse and

moderately conspicuous; **calyx** about 0.6 cm long, densely covered with brownish scales, persistent; **drupes** to about 0.4 cm long, somewhat pear-shaped, hairy, enclosed within calyx.

A moderately attractive species which is widely distributed in the south-western region in the Avon, Coolgardie, Eyre and Irwin Districts where it grows on sandy soils which often contain lateritic gravel. Should do best in semi-arid and warm temperate regions. Needs freely draining acidic soils and plenty of sunshine. Plants tolerate moderate frosts. Response to pruning is not known. Propagate from seed and cuttings of firm young growth are worth trying.

The var. *verticillata* E. Pritz. is a synonym.

P. loricata differs in having silvery stems, oblong-lanceolate leaves and calyces.

Pityrodia lewellinii F. Muell. =
                    Dicrastylis lewellinii (F. Muell.) F. Muell.

Pityrodia loricata (F. Muell.) E. Pritz.
(clothed in scales)
WA, NT
0.3–0.6 m × 0.5–1 m   Aug–Sept
Dwarf **shrub** with silver-scaly young growth; **branches** spreading to ascending, many, silvery-scaly, without hairs; **leaves** 0.5–2.5 cm × 0.2–0.5 cm, oblong-lanceolate to narrowly elliptic-lanceolate, opposite, sessile, base tapered, ascending to erect, somewhat crowded, glabrous with silvery scales above and below, flat, margins entire, apex pointed; **flowers** 0.6–1 cm long, white to pale pink, exterior glabrous, usually in 3s in upper axils, forming leafy spikes; **stamens** exserted; **calyx** to 0.7 cm long, exterior densely covered in silvery scales, persistent; **drupes** to about 0.5 cm long, pear-shaped, hairy, enclosed within calyx.

A silvery-foliaged species from the Carnegie, Coolgardie and Eucla Districts of WA and is also recorded from near Mt Sonder, NT. Plants have potential for cultivation in arid and semi-arid regions and may also be worth trying in warm temperate zones. They require excellent drainage and maximum sunshine and they tolerate moderate frosts. May be suitable as a container plant too. Propagate from seed and possibly from cuttings of firm young growth.

P. lepidota is closely allied but differs in having brownish scales.

Pityrodia loxocarpa (F. Muell.) Druce
(oblique or slanting fruits)
WA, NT
0.5–1.5 m × 0.5–2.5 m   Aug–Oct
Dwarf to small, often sparse **shrub** with hairy to glabrous young growth; **branches** ascending to erect, often slender and can be flexuose, striate, hairy or glabrous; **leaves** 1–5 cm × 0.5–4 cm, elliptic-oblong to orbicular, decussate or in whorls of 3, base tapered to petiole, spreading to ascending, rarely crowded, initially woolly-hairy all over, usually becoming glabrous and dark green above, flat, margins usually toothed, venation conspicuous, apex blunt; **flowers** 0.8–2.4 cm long, white to pale pink with purplish spotted throat, exterior hairy, in loose, sometimes leafy, terminal panicles or interrupted spikes, often profuse but usually

moderately conspicuous; **stamens** not exserted; **calyx** to 0.6 cm long, with purplish hairs, persistent; **drupes** to 0.3 cm long, obovoid, oblique, hairy, enclosed in calyx.

A wide-ranging species which has its main representation on or near the coast from Hill River in the Irwin District to near Roeburne in the Fortescue District. Other plants have been recorded in central WA and a lone record exists from just across the NT border. It occurs in deep sand as well as sand overlying limestone. Best suited to semi-arid and warm temperate regions. Must have excellent drainage and lots of sunshine. Withstands light frosts. Plants can be leggy and pruning at an early stage may promote bushier growth. Propagate from seed or possibly from cuttings of firm young growth.

P. drummondii Turcz. and *P. petiolaris* E. Pritz. are synonyms.

P. paniculata is allied but differs in having glaucous, hairy stems and pale mauve to pale lilac flowers.

Pityrodia maculata C. A. Gardner =
P. teckiana (F. Muell.) E. Pritz.

Pityrodia megalophylla Munir
(large leaves)
NT
1–2 m × 1–2 m   April–June
Small **shrub** with densely hairy young growth; **branches** ascending to erect, 4-angled, hairy; **leaves** 4–14 cm × 2–6 cm, ovate, opposite, base more or less cordate, petiole to 2 cm long, spreading to ascending, somewhat crowded near ends of branches, densely greyish-hairy, flat, margins entire, venation prominent, apex pointed; **flowers** 2.2–2.7 cm long, reddish, exterior hairy, solitary or mainly 3–7 per axillary cluster, forming leafy panicles, moderately conspicuous; **stamens** exserted; **calyx** 1.5–2.4 cm long, hairy, persistent, prominently ribbed; **drupes** to 0.6 cm long, oblong, 4-angled in upper half, glabrous in lower half.

This *Pityrodia* has the largest leaves of any species in this genus. It is restricted to a small area of Arnhem Land where it grows in rocky outcrops. Evidently rarely cultivated, it has potential for use in seasonally dry tropical regions and may succeed in subtropical areas. Requires freely draining acidic soils and plenty of sunshine. Propagate from seed or possibly from cuttings of firm young growth.

P. lanceolata differs in having smaller, lanceolate leaves with a rounded or narrowed base. *P. obliqua* has smaller leaves, narrowly lobed calyces and globular drupes.

Pityrodia muelleriana E. Pritz. = *P. viscida* W. Fitzg.

Pityrodia myriantha F. Muell. =
Dicrastylis fulva J. Drumm. ex Harvey

Pityrodia obliqua W. Fitzg.
(oblique)
WA
0.6–2 m × 0.5–1.5 m   May–July
Dwarf to small **shrub**; young growth yellowish-hairy; **branches** ascending to erect, terete, with greenish-grey

hairs; **leaves** 5–9.5 cm × 2–4.2 cm, narrowly ovate to ovate, opposite, base cordate or rounded, spreading to ascending, somewhat crowded near ends of branches, dense covering of greenish-grey hairs, flat, venation conspicuous below, apex bluntly pointed; **flowers** 0.8–1.4 cm long, pink with purple-streaked throat, in axillary clusters of 3–7, forming leafy panicles, can be profuse and conspicuous; **stamens** partially exserted; **calyx** to about 0.7 cm long, deeply lobed with narrow hairy segments, persistent; **drupes** about 0.3 cm long, more or less globular, hairy.

This species is endemic in the Kimberley region where it grows on mountain rock faces. It is worthy of cultivation and is recommended for seasonally dry tropical regions. Plants will require excellently drained acidic soils and plenty of sunshine. Propagate from seed or possibly from cuttings of firm young growth.

P. megalophylla has some similarities but it is distinguished by its much larger leaves.

Pityrodia oldfieldii (F. Muell.) Benth.
(after Augustus Oldfield, 19th-century English botanist)
WA   Oldfield's Foxglove
0.5–1.7 m × 0.3–1.5 m   Aug–Dec
Dwarf to small **shrub** with densely woolly-hairy, reddish-brown or dark brown young growth; **branches** ascending to erect, terete, with yellowish-brown or pale brown hairs; **leaves** 1.5–5 cm × 1–2.5 cm, broadly ovate to somewhat rhomboidal, opposite, base strongly tapered, spreading to ascending, densely woolly-hairy, greyish-green above, yellowish-green below, margins

Pityrodia oldfieldii × .8

349

faintly wavy or crenate, apex rounded or very blunt; **flowers** 1.5–2.3 cm long, pale pink to pinkish-mauve, with purple-spotted throat, solitary or up to 15 in upper axils, forming loose leafy spikes, can be profuse and very conspicuous; **stamens** not exserted; **calyx** 0.8–1.3 cm long, with short rounded lobes, greenish-yellow to brownish hairs, persistent; **drupes** about 0.4 cm long, globular, hairy, enclosed in calyx.

A very handsome species of the western Austin and northern Irwin Districts where it grows in sandy soils which often contain lateritic gravels. Plants require freely draining soils with a very sunny aspect and are best suited to semi-arid and warm temperate regions. Pruning from an early age may promote bushy growth. They withstand moderate frosts and extended dry periods. Worth trying in large containers. Propagate from seed, which can begin to germinate between 35–50 days after sowing, or from cuttings of firm young growth, which are prone to rotting.

The closely allied *P. verbascina* has longer, elliptic to oblong leaves and pointed calyx lobes.

Pityrodia ovata Munir
(ovate, egg shaped)
WA
0.4–1.2 m × 0.4–1.5 m   June–Aug
Dwarf to small **shrub** with sticky, yellowish-grey, hairy young growth; **branches** ascending to erect, terete, with yellowish-grey hairs; **leaves** 0.5–1.5 cm × 0.3–0.6 cm, narrowly ovate to ovate, opposite, sessile, spreading to ascending, somewhat sticky, densely covered in glandular hairs, smooth above, margins entire, venation distinct below, apex pointed; **flowers** 0.5–0.7 cm long, white, exterior glabrous, solitary or in 3s, in upper axils, rarely profuse; **stamens** exserted; **calyx** to about 0.5 cm long, glandular-hairy, deeply lobed, persistent; **drupes** not seen.

A poorly known species from the Dampier District and the far north of the Canning District. Needs to be cultivated as a conservation strategy. Suitable for seasonally dry tropical regions and semi-arid areas. Likely to need extremely well-drained soils and maximum sunshine. Frost tolerance is not known. Propagate from seed or possibly from cuttings of firm young growth.

Pityrodia paniculata (F. Muell.) Benth.
(flowers in panicles)
WA
0.6–1.5 m × 0.6–2 m   July–Oct
Dwarf to small **shrub** with greyish-hairy young growth; **branches** spreading to ascending, many, greyish-hairy; **leaves** 1–5 cm × 0.5–2 cm, oblong to obovate, opposite, sessile, base tapered, spreading to ascending, moderately crowded near ends of branches, densely covered with greyish hairs, margins usually entire, sometimes toothed in upper half, venation distinct, apex blunt; **flowers** 1.3–1.8 cm long, pale mauve to pale lilac, lower lobe much larger than others, exterior hairy, sometimes solitary but usually 3–15 in axillary clusters, forming loose terminal panicles, often profuse and very conspicuous; **stamens** barely exserted; **calyx** to about 0.9 cm long, deeply divided, lobes rounded, hairy, persistent; **drupes** about 0.3 cm long, almost globular, hairy, enclosed in calyx.

Pityrodia paniculata × .5

A desirable species from the western Austin, Carnarvon and Fortescue Districts. It occurs mainly near the coast and slightly inland, and grows on deep sands, including sand dunes. Best results likely to be achieved in semi-arid and warm temperate regions. Excellent drainage and a hot sunny location are essential. Frost tolerance is unknown. Propagate from seed, which can begin to germinate between 25–50 days after sowing. Cuttings of firm young growth are also worth trying.

The allied *P. atriplicina* has much broader leaves and a much tighter floral arrangement.

Pityrodia petiolaris E. Pritz. =
                    P. loxocarpa (F. Muell.) Druce

Pityrodia puberula Munir
(faint covering of downy hairs)
NT
0.4–0.6 m × 0.5–0.8 m   May–June
Dwarf **shrub** with sticky and hairy young growth; **branches** spreading to ascending, terete, slender, sticky, hairy; **leaves** 0.8–4 cm × 0.2–0.8 cm, linear to narrowly lanceolate, decussate, sessile, spreading to ascending, sticky, faintly hairy, green, margins entire and recurved, apex pointed; **flowers** 1–1.3 cm long, creamy-white, upper lip and tube with deep purple stripes, exterior of lobes hairy, solitary in upper axils, forming leafy spikes, can be moderately profuse; **stamens** slightly exserted; **calyx** about 0.8 cm long,

Pityrodia terminalis   W. R. Elliot

with very narrow lobes, hairy, persistent; **drupes** to 0.4 cm long, obovoid, very hairy.

P. puberula is restricted to Arnhem Land where it grows in shrubby woodland on soils derived from sandstone. It lacks strong ornamental appeal and may only be of interest to enthusiasts. Suitable for seasonally dry tropical regions and may succeed in subtropical areas. Needs excellent drainage and plenty of sunshine. Propagate from seed or possibly from cuttings of firm young growth.

The allied *P. gilruthiana* has deep green, very sticky leaves which are usually in whorls of 3.

Pityrodia pungens Munir
(ending in a hard, sharp point)
NT
0.4–0.75 m × 0.3–0.6 m   Jan–June
Dwarf **shrub** with sticky and hairy young growth; **branches** ascending to erect, terete, sticky, hairy; **leaves** 0.8–6 cm × 0.2–0.6 cm, linear to narrowly lanceolate, usually in whorls of 3 or sometimes opposite, spreading to ascending, somewhat crowded near ends of branches, pale-to-mid green, somewhat sticky, becoming glabrous above, midrib prominent below, mainly flat, margins entire, apex ending in a sharp point; **flowers** to about 0.7 cm long, white, upper lobe with purple stripes, exterior of lobes hairy, solitary in upper axils, forming leafy spikes, can be moderately profuse; **stamens** slightly exserted; **calyx** to about 0.8 cm long, deeply divided, lobes pointed, hairy, persistent; **drupes** to about 0.3 cm long, obovoid, hairy.

This species is from the Top End. It is distributed from near Katherine to western Arnhem Land where it grows on the sandstone escarpments. Evidently rarely cultivated, it will interest mainly enthusiasts. It is

suitable for seasonally dry tropical regions. Plants require excellent drainage in acidic soils and plenty of sunshine. Propagate from seed and from cuttings of firm young growth.

P. ternifolia is similar except it has toothed ovate leaves.

Pityrodia quadrangulata Munir
(four-angled)
NT
0.8–2 m × 0.6–1.5 m   Jan–July
Dwarf to small **shrub** with white-woolly young growth; **branches** ascending to erect, 4-angled, covered with white-woolly hairs; **leaves** 3–10 cm × 1–2.5 cm, elliptic-oblong to obovate, opposite, base tapered, spreading to ascending, densely woolly-hairy, thick, flat, margins entire or finely toothed, apex blunt; **flowers** 1.4–2.5 cm long, cream to pale yellow, with small lobes, exterior faintly hairy, 3–12 in upper axillary clusters, forming moderately tight panicles, often profuse and very conspicuous; **stamens** sometimes slightly exserted; **calyx** to 1.5 cm long, deeply divided, lobes narrow, hairy, persistent; **drupes** to 0.6 cm long, obovoid, 4-angled, 4-ribbed, enclosed within calyx.

A highly desirable pityrodia from the central Top End where it grows on sandy soils in open woodland. It has potential for cultivation in seasonally dry tropical areas as well as subtropical regions. Plants require freely draining acidic soils and plenty of sunshine. Pruning of young plants may help promote bushy growth although response to pruning is not known. Propagate from seed or possibly from cuttings of firm young growth.

P. dilatata is allied but differs in having smaller leaves and orange-red flowers with larger lobes.

Pityrodia racemosa (Turcz.) Benth. =
                    P. terminalis (Endl.) A. S. George

Pityrodia uncinata   T. L. Blake

351

Pityrodia salviifolia R. Br.
(similar to the genus *Salvia*)
Qld
1.5–2.5 m × 1.5–3 m   Aug–Sept; also sporadic
Small to medium **shrub** with scaly young growth; **branches** spreading to erect, terete, covered in pale brown scales; **leaves** 5–13 cm × 0.5–3.3 cm, lanceolate, opposite, base tapered to petiole which is to 1.2 cm long, spreading to ascending and often drooping, covered in scales, dull green and somewhat rough above, rusty with midrib prominent below, flattish, venation distinct, apex pointed; **flowers** to about 0.7 cm long, white, exterior nearly glabrous, 5–9 in axillary clusters, forming leafy spikes, can be profuse, rarely highly conspicuous; **stamens** sometimes slightly exserted; **calyx** to about as long as flower, hairy, persistent; **drupes** to 0.5 cm long, somewhat top-shaped, hairy, loosely enclosed within calyx.

P. salviifolia occurs in tropical and subtropical areas of eastern Qld. The main concentration is on or near the coast and hinterland with further scattered representation inland. Not highly ornamental, it is mainly of interest to enthusiasts. Plants need excellent drainage and plenty of sunshine. Pruning at an early age may promote bushy growth. Propagate from seed or possibly from cuttings of firm young growth.

Pityrodia scabra A. S. George
(rough)
WA
0.6–1.3 m × 0.6–1.3 m   Aug–Sept
Dwarf to small **shrub** with sticky golden-rusty young growth; **branches** many, spreading to erect, terete, covered in sticky golden-rusty hairs; **leaves** 0.5–1.5 cm × 0.1–0.3 cm, linear to nearly terete, in whorls of 3, sessile, spreading to ascending, moderately crowded, sticky, becoming rough above, rusty-hairy below, margins shallowly lobed and revolute, apex bluntly pointed; **flowers** to about 1 cm long, white, exterior glabrous, clusters of flowers in upper axils, forming leafy spikes to about 8 cm long, rarely highly conspicuous; **stamens** exserted; **calyx** about 0.5 cm long, shortly lobed, hairy, persistent; **drupes** about 0.3 cm long, obovoid, hairy, enclosed in calyx.

This species is regarded as vulnerable in its natural habitat on sandy loam soils in the central Avon District. Needs to be cultivated as part of a conservation strategy. Mainly of interest to enthusiasts. Best suited to semi-arid and warm temperate regions. Requires good drainage and plenty of sunshine. Tolerates moderate frosts. Propagate from seed or possibly from cuttings of firm young growth.

Pityrodia serrata Munir
(saw-toothed)
NT
0.6–1 m × 0.5–1 m   May–June
Dwarf **shrub** with mainly glabrous young growth; **branches** ascending to erect, terete, becoming rusty-orange with age; **leaves** 0.5–1.3 cm × 0.2–0.4 cm, ovate to oblong-ovate, in whorls of 3, sessile, spreading to ascending, crowded and often overlapping, glabrous, smooth, margins prominently toothed, apex ending in pungent point; **flowers** to about 0.8 cm long, white,

upper lip purple-striped, exterior of lobes hairy, solitary in upper axils, forming leafy spikes, rarely profuse or very conspicuous; **stamens** slightly exserted; **calyx** to about 0.8 cm long, lobes narrow, glabrous, persistent; **drupes** about 0.3 cm long, obovoid, hairy.

P. serrata is rare and known only from one area in Arnhem Land where it grows in soils derived from the sandstone escarpments. Likely to be of interest to enthusiasts. Suitable for seasonally dry tropical regions and may also succeed in subtropical areas. Requires excellently drained acidic soils and a very sunny aspect. Reaction to pruning is not known. Propagate from seed or possibly from cuttings of firm young growth.

P. ternifolia is similar but it has less-crowded, cordate, glandular-hairy leaves.

Pityrodia spectabilis C. A. Gardner =
                    P. axillaris (Endl.) Druce

Pityrodia spenceri Munir —
          see *P. hemigenioides* (F. Muell.) Benth.

Pityrodia teckiana (F. Muell.) E. Pritz.
(after Francis P. L. Alexander, Duke of Teck, 19th-century President, Royal Horticultural Society, England)
WA
0.3–1.5 m × 0.5–2 m   Aug–Nov
Dwarf to small **shrub** with sticky and hairy young growth; **branches** spreading to erect, terete, very sticky, covered in glandular hairs; **leaves** 0.5–3.5 cm × 0.3–1.2 cm, ovate-oblong, opposite or sometimes in whorls of 3, sessile, stem-clasping, ascending to erect, green, glandular-hairy to glabrous, brittle, venation conspicuous, margins toothed, apex pointed; **flowers** 1.5–2.5 cm long, pale pink, blue-mauve or violet-lilac, often with brown-spotted throat, exterior faintly hairy, usually solitary, sometimes clusters of 3 in upper axils, forming leafy spikes, can be profuse and moderately conspicuous; **stamens** not exserted; **calyx** to 0.8 cm long, deeply divided, glandular-hairy, persistent; **drupes** about 0.3 cm long, somewhat globular with apical depression, hairy.

P. teckiana is commonly associated with granite outcrops in the Avon and Coolgardie Districts. It is cultivated to a limited degree and often is short-lived. Best suited to semi-arid and warm temperate regions. Plants require freely draining acidic soils and a warm to hot, sunny aspect. They withstand moderate frosts and extended dry periods. Leggy growth is curtailed by pruning from an early stage. Propagate from seed which usually begins to germinate between 30–60 days after sowing. Cuttings of firm young growth usually strike readily.

P. maculata C. A. Gardner is a synonym.

The allied *P. glabra* differs in having smooth, non-sticky stems and non stem-clasping leaves.

Pityrodia terminalis (Endl.) A. S. George
(terminal, at the apex)
WA   Native Foxglove; Woolly Foxglove
0.5–1.5 m × 0.3–1 m   Sept–Nov
Dwarf to small **shrub** with white-woolly young growth; **branches** ascending to erect, terete, densely

covered with white-woolly hairs; **leaves** 1.5–7.5 cm × 0.5–2 cm, oblong to narrowly elliptic-oblong, opposite, sessile and sometimes stem-clasping, recurved to ascending, green with greyish hairs, thick, soft, midrib prominent, margins entire, apex blunt; **flowers** 1.8–2.7 cm, pale pink, deep purplish-pink to claret-red, or rarely white, exterior faintly hairy, solitary or 3–5 in clusters in upper axils, forming terminal leafy racemes, often profuse and extremely conspicuous; **stamens** not exserted; **calyx** to 1.5 cm long, deeply divided, lobes narrow, covered in greyish hairs, persistent; **drupes** about 0.5 cm long, broadly pear-shaped, hairy, enclosed within calyx.

This outstanding native foxglove is widely distributed over the inland region of south-western WA in the sand heath of the Avon, Coolgardie, Eyre and Irwin Districts. Plants in the southern region of its distribution often have pale pink to whitish flowers. It requires excellently drained acidic soils and maximum sunshine. Best suited to warm temperate and semi-arid regions. Plants are hardy to moderate frosts and extended dry periods. They respond well to pruning from an early stage as well as during or immediately after flowering. Worth trying as a container plant. Propagate from seed, which usually begins to germinate between 60–90 days after sowing. Cuttings of firm young growth can produce roots but they are very prone to dying if propagation medium is kept too wet.

P. racemosa (Turcz.) Benth. is a synonym of *P. axillaris* which is closely allied and has been confused with *P. terminalis*. *P. axillaris* differs in having oblong-ovate leaves, obovate calyx lobes and deep red to orange-red flowers which are nearly glabrous on the exterior.

Pityrodia ternifolia (F. Muell.) Munir
(leaves in threes)
WA, NT
0.6–2 m × 0.5–1.5 m   Feb–Oct

Dwarf to small **shrub** with sticky and hairy young growth; **branches** ascending to erect, terete, sticky, hairy; **leaves** 1.2–7 cm × 0.5–3.5 cm, ovate or narrowly ovate, usually in whorls of 3 or rarely opposite to scattered, more or less sessile, spreading to ascending, green with glandular hairs, strongly aromatic, venation distinct below, margins prickly toothed, apex pointed; **flowers** 1–1.5 cm long, cream, pale mauve to pink-red, throat and upper lobes with purplish stripes, exterior of lobes greyish-hairy otherwise glabrous, solitary in upper axils, forming leafy spikes, rarely profuse but usually moderately conspicuous; **stamens** slightly exserted; **calyx** to about 1 cm long, 10-ribbed, lobes narrow, hairy, persistent; **drupes** to about 0.5 cm long, obovoid, hairy.

A very prickly species from northern WA and western Arnhem Land, NT. It inhabits woodland and shrubland of sandstone ridges and plateaus where it usually grows in sandy soils. Mainly of interest to enthusiasts. Suitable for seasonally dry tropical regions and may succeed in subtropical areas. Requires extremely well-drained acidic soils with a sunny or semi-shaded aspect. The extremely colourful Leichhardt grasshopper eats the leaves of *P. ternifolia*. Propagate from seed or possibly from cuttings of firm young growth.

Dennisonia ternifolia F. Muell. is a synonym.
P. serrata is similar but it has sessile, glabrous leaves and its calyx is about as long as the flower, while *P. pungens* has entire leaves.

Pityrodia uncinata (Turcz.) Benth.
(hooked)
WA
0.15–0.6 m × 0.25–1 m   July–Nov

Dwarf spreading **shrub** with woolly-hairy young growth; **branches** spreading to ascending, densely woolly-hairy; **leaves** 1–5 cm × 0.2–0.5 cm, linear to linear-lanceolate, in whorls of 3, scattered or decussate, sessile, crowded, dark green and wrinkled with woolly hairs above, usually woolly-hairy below, margins recurved to revolute, apex ending in a small hook; **flowers** 1–1.7 cm long, deep pink to red or rarely yellow, exterior hairy, solitary or in 3s in upper axils, forming leafy spikes, often profuse and very conspicuous; **stamens** not exserted; **calyx** to about 0.9 cm long, deeply divided, lobes narrow, hairy, persistent; **drupes** to about 0.3 cm long, somewhat globular, hairy, reticulate, enclosed within the calyx.

This stunning species has a disjunct distribution over a wide area of south-western WA. It occurs in the Avon, Darling, Eyre and Irwin Districts where it grows in sandy soils which may contain lateritic gravel or have a dry subsoil. Plants deserve greater recognition but they have proved difficult to cultivate with regular success. Suited to warm temperate and semi-arid regions. It may adapt to cooler areas if grown in sheltered sunny sites. Needs acidic soils which have relatively good drainage and lots of sunshine. Hardy to moderate frosts. Response to pruning is not recorded. Propagate from seed, which usually begins to germinate between 50–70 days after sowing. Cuttings of firm young growth can be slow to form roots and are very prone to rotting.

The var. *exserta* Benth. was given species status and is now known as *P. exserta* (Benth.) Munir. *P. exserta* var. *exserta* has rougher leaves and exserted stamens while *P. exserta* var. *lanata* has broader flowers in which the stamens are only slightly exserted.

Pityrodia verbascina (F. Muell.) Benth.
(like the genus *Verbascum*)
WA   Golden Bush
0.5–2.5 m × 0.5–1.5 m   Sept–Dec

Dwarf to medium **shrub** with brownish-red to brownish-yellow hairy young growth; **branches** ascending to erect, terete, with brownish-red or brownish-yellow to golden-yellow woolly hairs; **leaves** 2–10 cm × 0.5–4 cm, variable, mainly elliptic to elliptic-oblong, opposite, mainly sessile, reflexed to ascending, not crowded, densely covered with greyish hairs, thick, soft, margins entire, apex blunt or pointed; **flowers** 1.2–1.8 cm long, pinkish-white with pink-spotted throat, exterior hairy, 3–12 in stalked clusters in upper axils, forming leafy spikes, often profuse and very conspicuous; **stamens** sometimes slightly exserted; **calyx** to about 1.3 cm long, yellow, deeply divided, densely hairy, persistent; **drupes** about 0.3 cm long, somewhat globular, hairy, splitting in two, enclosed within calyx.

A striking species found in the Avon and Irwin

353

Pityrodia verbascina   B. W. Crafter

Districts where it grows in sandy and gravelly soils of the sandheath. It has had limited cultivation but deserves to be better known. Plants need very well-drained acidic soils and maximum sunshine. Best suited to semi-arid and warm temperate regions. It will tolerate light to moderate frosts as well as extended dry periods. Tip pruning of young plants promotes bushy growth. Response to pruning of mature plants is not well known but if they are growing vigorously they do respond well. Propagate from seed, which usually begins to germinate between 30 and 120 days after sowing. Cuttings of firm young growth can produce roots but often they die because excessive moisture causes rotting of stems.

The *forma aurea* E. Pritz. and *forma leucocalyx* E. Pritz. are not considered worthy of separate rank and are now regarded as synonyms.

The allied *P. oldfieldii* has broadly ovate to nearly rhomboidal leaves which taper strongly to their base, and broad, rounded calyx lobes.

Pityrodia viscida W. Fitzg.
(sticky)
WA
0.3–0.6 m × 0.3–0.6 m   Sept–Oct
Dwarf **shrub** with sticky and hairy young growth; **branches** ascending to erect, terete, covered in short, creamy to yellowish, sticky hairs; **leaves** 0.5–1.6 cm × 0.3–0.5 cm, oblong-obovate or narrowly elliptic, opposite or sometimes in whorls of 3, sessile, tapered to base, moderately crowded near ends of branches, dark green, glabrous and sticky above, creamish hairs below, margins entire and recurved, apex bluntly pointed; **flowers** 0.9–1.2 cm long, white, exterior glabrous, solitary in upper axils, forming short leafy spikes,

rarely profuse or very conspicuous; **stamens** exserted; **calyx** to about 0.9 cm long, ribbed, sticky-hairy, lobe narrow, persistent; **drupes** to about 0.5 cm long, obovoid, faintly glandular-hairy.

A poorly known species from the central Irwin District where it grows in sandheath. Lacks ornamental features but is possibly worth growing as part of a conservation strategy. Plants are likely to require excellently drained acidic soils with a sunny aspect. They should withstand moderate frosts and extended dry periods. Propagate from seed and possibly from cuttings of firm young growth.

P. glutinosa has affinities but differs in having glabrous stems and faintly toothed leaves. *P. hemigenioides* differs in having cottony-white hairs on stems, leaves and calyx.

PLACOSPERMUM C. T. White & W. D. Francis
(from the Greek *placos*, a flat body and *sperma*, seed, referring to the flat winged seeds)
Proteaceae
A monotypic genus endemic in Australia.

Placospermum coriaceum C. T. White & W. D. Francis
(thick, leathery)
Qld   Rose Silky Oak
6–20 m × 3–12 m   Oct–Dec
Tall **shrub** or small to tall, bushy **tree**; young growth usually pinkish, becoming bright green; **juvenile leaves** to 80 cm × 30 cm, usually lobed or entire; **mature leaves** to 25 cm × 8 cm, spathulate, usually entire, dark green and glossy above, leathery; **panicle** 2.5–11 cm long, terminal and in the upper axils; **flowers** 1.2–2 cm long, pink, not opening widely; **follicles** 3–4 cm × 3–3.8 cm, leathery, containing up to 20 seeds, ripe March–April.

Occurs in north-eastern Qld growing in rainforest from relatively low altitudes up to peaks on the ranges and tableland. An outstanding ornamental which has become very popular in cultivation. Young plants have a slender habit with impressive, large-lobed leaves whereas older plants develop a bushy habit. Plants are fast-growing and have proved to be adaptable and successful in tropical, subtropical and temperate regions. They require well-drained soil in positions ranging from full sun to partial sun. Plants need moisture when young otherwise they can wilt quickly resulting in leaf shedding or disfiguration. Tolerates moderate frosts. Propagate from seed, which has a limited period of viability and is best sown while fresh.

PLAGIOCARPUS Benth.
(from the Greek *plagio*, oblique, sideways; *carpos*, fruit)
Fabaceae (alt. *Papilionaceae*)
A monotypic genus endemic in tropical regions of WA and NT.

Plagiocarpus axillaris Benth.
(axillary)
WA, NT
0.4–2 m × 0.3–1.5 m   March–July
Dwarf to small **shrub** with densely hairy young growth; **branches** ascending to erect; **branchlets** densely woolly-hairy; **leaves** 3-foliolate or rarely simple

354

Planchonella australis × .45

alternate or arranged spirally, sessile, erect; **leaflets** 1–2.8 cm × up to 0.9 cm, mainly narrowly elliptic or oblong-elliptic, green, especially hairy above and on margins and midrib below, venation pinnate, apex blunt or pointed; **flowers** pea-shaped, about 0.8 cm across, pale yellow, solitary, axillary, nearly sessile or with very short stalks, rarely very conspicuous; **calyx** with 5 equal lobes, hairy; **pods** to about 1.3 cm × 0.8 cm, more or less sessile, inflated, oblique, glabrous.

A somewhat ornamentally insignificant species from tropical woodlands in the Kimberley, WA, and the Top End, NT. Suitable for tropical and possibly subtropical regions. Should adapt to a range of well-drained acidic soils with a sunny or semi-shaded aspect. Damage from frosts is likely. Propagate from seed, which needs pre-sowing treatment to damage or soften seed coat. See Treatments and Techniques for Difficult to Germinate Species, Volume 1, page 205.

PLANCHONELLA Pierre
(after Louis D. Planchon, 19–20th-century professor of pharmacy at University of Montpellier, France)
Sapotaceae

Shrubs or **trees**; **sap** milky; **leaves** alternate, simple, entire, shortly petiolate, papery to leathery; **flowers** unisexual or bisexual, in axillary clusters, often on long stalks; **sepals** 4–6-merous, forming a basal tube; **petals** glabrous; **stamens** inside the tube; **fruit** a berry; **seeds** large, with a prominent scar.

A large genus of about 100 species distributed mainly in Asia and the Pacific region with 1 species in the Seychelles and 2 in South America. The fruit of many species is edible. About 18 species occur in Australia but very few of these are grown to any extent. Species of this genus are included in *Pouteria* by some

authorities. Propagation is from seed, which has a limited period of viability and for best results should be sown while fresh. Some species can be propagated successfully from cuttings of firm young growth, but in many species the branchlets are too thick for this purpose.

Planchonella arnhemica (F. Muell.) P. Royen
(from Arnhem Land)
WA, NT   Nutwood
4–10 m × 3–5 m   Dec–May
Medium to tall **shrub** or small **tree**; young growth yellowish, hairy; **leaves** 7–18 cm × 2.5–8 cm, obovate, crowded towards the ends of branchlets, thin-textured, dark green, dull, hairy on both surfaces, margins wavy, apex blunt or notched; **flowers** about 0.4 cm long, greenish, clustered in the leaf axils; **berries** 2–3 cm × 2–2.7 cm, ovoid to globose, brownish, minutely hairy; **seeds** 4 or 5, to 1.3 cm × 0.8 cm, brown, shiny, ripe May–Sept.

Grows on sandstone outcrops and in woodland and vine thickets in the Kimberley Region, WA, and also the Top End of the NT. A hardy species able to withstand long dry periods during which the plants may become deciduous. Cultivated on a very limited scale, if at all. Requires well-drained soil in a sunny location. Propagate from fresh seed.

Planchonella australis (R. Br.) Pierre
(southern)
Qld, NSW   Black Apple; Wild Plum;
                                                    Yellow Buttonwood
10–25 m × 10–15 m   Sept–March
Medium to tall **tree** with an erect to spreading canopy; **bark** rough, fissured; young growth bright green, hairy; **leaves** 8–16 cm × 2–6 cm, obovate, ovate or elliptic, thick and leathery, dark green and shiny above, paler beneath, tapered to the base; **flowers** about 0.5 cm long, greenish-cream, solitary or clustered, axillary or on old wood; **berries** 2–5 cm × 2–5 cm, ovoid to ellipsoid, purplish-black to black; **seeds** 3–5, to 5 cm long, brown, shiny, ripe Dec–March.

Widely distributed between Tully, Qld, and the Illawarra region of south-eastern NSW. Common in coastal rainforests and also extending to the coastal ranges. The timber is suitable for light work, engraving and turning. The fruit is edible and has a pleasant flavour. It is produced in large quantities and can be used to make preserves. The decorative shiny seeds are very ornamental and have been used for necklaces.

Black Apple is a handsome tree which is grown to a limited extent, but deserves wider recognition. When grown in an open position plants develop a spreading habit. They are generally fast-growing and have proved to be very adaptable. An excellent species for revegetating degraded slopes and gullies. Requires well-drained moist soil and responds well to fertilisers. Tolerates moderate frosts. Propagate from fresh seed and from cuttings of firm young growth.

Planchonella brownlessiana (F. Muell.) P. Royen
(after Dr A. C. Brownless)
Qld   Northern Coondoo
8–20 m × 3–5 m   Feb–March

Planchonella chartacea

Small to medium **tree**; young growth with grey or yellowish hairs; **leaves** 5–15 cm × 1.5–5.5 cm, obovate or oblanceolate, thin and papery, greenish-brown and shiny above, greyish and hairy beneath, margins wavy, apex blunt; **flowers** about 0.8 cm long, greenish-cream, solitary or in clusters of 2–3; **berries** about 3.2 cm × 1 cm, oblong to ellipsoid, black; **seeds** 2, about 2 cm × 0.8 cm, dark brown, shiny, ripe Oct–Dec.

Endemic in north-eastern Qld growing in rainforest at moderate altitudes in the ranges and tablelands. A dense bushy tree with prospects for cultivation in large gardens and parks. It will probably succeed in temperate regions. Cultivation requirements are probably as for *P. australis*. Propagate from fresh seed.

Planchonella chartacea (F. Muell.) H. J. Lam
(papery)
Qld, NSW — Thin-leaved Coondoo
6–20 m × 4–10 m — March–May

Tall **shrub** or small to medium **tree**; young growth glabrous; **leaves** 4–20 cm × 2–6 cm, obovate to spathulate, thin and papery, glabrous, dark green, shiny above, dull beneath, midvein sometimes yellowish; **flowers** about 0.5 cm long, cream-green, in axillary clusters of 4–7; **berries** 1.5–2.2 cm × 0.5–1.5 cm, black; **seeds** 2–5, to 2 cm × 1 cm, ellipsoid, ripe May–Oct.

Widely distributed from Cape York Peninsula, Qld, to Byron Bay, northern NSW, and also in New Guinea. Grows in lowland rainforests and is often prominent in coastal districts. Grown on a limited scale by enthusiasts. Plants develop a spreading habit in open situations and are useful for planting in coastal areas from the tropics to warm temperate regions. Plants withstand sun when quite small but may benefit from some shelter. Soil drainage must be unimpeded. Propagate from fresh seed and perhaps also by cuttings.

Planchonella cotinifolia (A. DC.) Dubard
(with leaves like the genus *Cotinus*)
Qld, NSW — Yellow Lemon
4–10 m × 2–4 m — Oct–March

Medium **shrub** to small bushy **tree**; young growth hairy; **leaves** 0.5–5 cm × 0.3–4 cm, obovate, elliptic or orbicular, yellowish-green, shiny, stiff and papery, narrowed to base, apex blunt; **flowers** 0.6–0.8 cm long, greenish to yellowish, solitary or in clusters; **berries** 0.6–1.5 cm × 0.5–0.7 cm, ellipsoid to ovoid, green to black; **seed** 1, shiny.

The typical variety is distributed between Rockhampton, Qld, and the Richmond river, northern NSW. It grows in dry rainforest and vine thickets on ranges inland from the coast. Suitable for subtropical and temperate regions. A relatively hardy species which will withstand periods of dryness once established. Plants are bushy from an early age. Best in filtered sun or partial shade in well-drained soil. Tolerates moderate frosts.

The var. *pubescens* P. Royen occurs in the North Western Plains District of NSW and the Burnett District of Qld. It is hairy on most of its parts and the fruits have up to 4 seeds.

Propagate both varieties from fresh seed and possibly also from cuttings.

Planchonella crocodiliensis P. Royen
(from Crocodile Island, NT)
NT
4–8 m × 3–5 m — Jan–Feb

Medium to tall bushy **shrub**; young growth with greyish or brown hairs; **leaves** 2–4 cm × 0.2–1 cm, spathulate to obovate, crowded towards the apex of branchlets, greyish-green, short-hairy, thin-textured, papery; **flowers** about 1 cm long, cream, in axillary clusters; **berries** about 2.5 cm × 2 cm, obovoid, blackish; **seeds** 5, about 1.5 cm × 0.7 cm, brown, shiny, ripe Dec.

A poorly known species which has apparently been rarely collected. It may have potential for cultivation in tropical coastal districts. Cultivation and propagation as for *P. arnhemica*.

Planchonella eerwah (F. M. Bailey) P. Royen
(after Mt Eerwah, near Eumundi, Qld)
Qld — Black Plum
4–10 m × 3–6 m — Aug–Nov

Medium to tall **shrub** or small **tree**; **bark** scaly; young growth with grey hairs; **leaves** 4–14 cm × 1.5–6.5 cm, obovate to spathulate, leathery, glabrous, dark green and shiny above, duller beneath, tapered to the base; **flowers** about 0.4 cm long, cream-green, hairy outside; **berries** 3–6 cm × 2–4 cm, obovoid to globose, dark reddish-purple; **seeds** 1–5, 2.5–3 cm long, ripe March–April.

Occurs in the Moreton and Wide Bay Districts in southern Qld growing on rocky slopes and ridges in vine thickets and rainforest dominated by hoop pine (*Araucaria cunninghamii*). A rare species now reduced to a few small and widely disjunct populations. Two plants have been cultivated at the Melbourne Botanic Gardens since the original collection in the late 1800s and recently seed from existing Qld populations has been distributed widely to aid in the

Planchonella eerwah, all parts × .5

conservation of the species. Requires well-drained soil and responds strongly to mulches, watering during dry periods and light fertiliser applications. Tolerates light to moderate frosts. Propagate from fresh seed and from cuttings of firm young growth.

Planchonella euphlebia (F. Muell.) W. D. Francis
(with beautiful veins)

Qld	Candlewood
10–18 m × 5–12 m	March–April

Medium to tall **tree** with a rounded to spreading crown; **bark** dark brown, wrinkled; young growth brown-hairy; **leaves** 5–14 cm × 2.5–4.5 cm, obovate to oblanceolate, crowded towards the end of branchlets, leathery, dark green and shiny above, dull and hairy beneath; **flowers** about 0.5 cm long, greenish-cream, solitary or clustered; **berries** 2.5–3.5 cm × 2–2.5 cm, globose to ovoid, red; **seeds** 1–5, glossy, ripe Oct–Dec.

Occurs in north-eastern Qld between Cooktown and Townsville in montane rainforest. The bright yellowish wood is hard and has been used for flooring and scantling. Green wood burns freely. The fruit is edible with a pleasant flavour. A handsome tree for tropical, subtropical and warm temperate regions. Probably grows too large for the average home garden but excellent for parks and acreage planting. Requires well-drained soil and responds to watering, mulching and light fertiliser applications. Propagate from fresh seed and possibly also from cuttings.

Planchonella laurifolia (A. Rich.) Pierre
(leaves like the laurel genus *Laurus*)

Qld, NSW	Blush Coondoo
15–25 m × 10–15 m	April–July

Medium to tall **tree**; **bark** brown, shed in large pieces leaving a pattern on the trunk; young growth hairy; **leaves** 9–20 cm × 3–7.5 cm, elliptic to obovate, crowded towards the end of branchlets, leathery, dark green, dull above, glossy beneath; **flowers** about 0.4 cm long, greenish, hairy, in axillary clusters; **berries** 1–2.2 cm × 0.5–1 cm, obovoid, black; **seed** 1, 1–1.5 cm long, brown, shiny, ripe Aug–Dec.

Distributed mainly in coastal districts from north-eastern Qld to north-eastern NSW. Plants occur in rainforests and particularly littoral rainforests. The species is grown on a limited scale but has potential for wider use. Plants develop a spreading habit when planted in the open. They need well-drained soil and respond to mulching and fertiliser application. Tolerates light to moderate frosts. The timber is suitable for boxes, light work and turning. Propagate from fresh seed and from cuttings of firm young growth.

Planchonella macrocarpa P. Royen
(with large fruit)

Qld	Big-leaf Coondoo
12–20 m × 10–15 m	July–Sept

Medium to tall **tree**; young growth hairy; **leaves** 10–22 cm × 5–7 cm, narrowly elliptical, crowded towards the end of branchlets, thick and leathery, dark green and shiny above, dull beneath, narrow at the base; **flowers** not recorded; **berries** 5–6 cm × 3.5–4.5 cm, dark green to greenish-black, ellipsoid; **seeds** about 3.5 cm × 1.8 cm, brown, shiny, ripe Jan.

A poorly known species which grows in rainforest on the Evelyn Tableland, north-eastern Qld. A large tree suitable for parks and acreage planting. Rarely cultivated if at all. Probably has similar requirements to those of *P. australis*. Propagate from fresh seed.

Planchonella myrsinoides (A. Cunn. ex Benth.) S. T. Blake ex W. D. Francis
(resembling the genus *Myrsine*)

Qld, NSW Blunt-leaved Coondoo; Yellow Plumwood	
4–10 m × 2–5 m	Sept–Dec; also sporadic

Medium **shrub** to small bushy **tree**; young growth yellowish, hairy; **leaves** 2–10 cm × 1–5 cm, ovate to oblong or obovate, green, leathery, hairy when young, glabrous when mature, tapered to the base; **flowers** 0.5–1 cm long, greenish to cream, hairy outside; **berries** 1–3 cm × 0.5–2.5 cm, fusiform to ovoid, purple-black; **seeds** 1–3, 1.5–2.2 cm long, shiny.

Distributed between Bundaberg, Qld, and Forster, NSW; also occurs on Lord Howe Island. Found mainly in coastal districts in littoral rainforest and dry rainforest. A bushy species which develops a spreading habit in open situations. Grown to a limited extent. Suitable for subtropical and temperate regions especially in coastal localities. Requires excellent drainage in filtered sun or partial sun. Propagate by fresh seed and perhaps also by cuttings.

Pouteria myrsinoides A. Cunn. ex Benth. is a synonym.

Planchonella obovata (R. Br.) Pierre
(reversely ovate-obovate)

Qld	Northern Yellow Boxwood; Black Ash; Yellow Teak
10–20 m × 8–12 m	Aug–Oct; also sporadic

Small to medium **tree** with a bushy crown; young growth silky-hairy; **leaves** 6–24 cm × 1.5–15 cm, ovate, obovate, or lanceolate, thin-textured or papery, glabrous and shiny above, hairy beneath; **flowers** about 0.5 cm long, greenish-white, in dense axillary clusters, some flowers female only; **berries** 1–1.5 cm × 1–1.5 cm, globose, often lobed, red or bluish; **seeds** 1–5, 0.8–1.2 cm × 0.3 cm, yellow, shiny, ripe.

A coastal species which is widely distributed from South-east Asia to New Guinea and Australia. Grows in beach scrubs and littoral rainforest. A useful species for coastal planting and shelter in tropical and subtropical regions. Frost tolerance is unlikely to be high. Requires well-drained soils and a sunny location. The light, attractively marked wood is used for carving, turning and cabinet-making. Propagate from fresh seed.

Planchonella obovoidea H. J. Lam
(reversely ovoid, obovoid)

Qld	Northern Yellow Boxwood
15–25 m × 10–15 m	Jan–March

Medium to tall **tree**; young growth short-hairy; **leaves** 8–20 cm × 3–8 cm, ovate to obovate or elliptic, thin and papery, green and shiny above, greyish beneath, margins often wavy, apex blunt; **flowers** about 0.3 cm long, white, cream or yellow, in clusters of 3–6; **berries** 2–3 cm × 1–1.5 cm, obovoid, dark red to blackish-red; **seed** 1, 1.5–2.5 cm × 0.8–1.2 cm, brown or black, shiny, ripe Sept–Oct.

Planchonella papyracea

Planchonella laurifolia   D. L. Jones

Distributed from Indonesia to Qld where it occurs in the Cook, North Kennedy and South Kennedy Districts. Mainly found growing in lowland rainforest close to the coast. A large tree which has potential for planting in coastal districts of the tropics and subtropics. Its cultural requirements are probably similar to those of *P. obovata*. Propagate from fresh seed.

Planchonella papyracea P. Royen
(papery)
Qld   Pink Boxwood
8–15 m × 3–6 m   Jan–Feb
Small to medium **tree**; young growth densely rusty-hairy; **leaves** 8–18 cm × 3–7.5 cm, obovate to elliptic, leathery, dark green and shiny above, hairy beneath, basal margins inrolled, apex blunt; **flowers** about 0.5 cm long, cream, rusty-hairy, crowded in axillary clusters; **berries** about 2 cm × 1.5 cm, obovoid, greenish-black; **seeds** 2, about 1.8 cm × 1 cm, brown, ripe Nov–Feb.
Endemic in north-eastern Qld growing in rainforest on the ranges and tablelands. This species is not known to be in cultivation but should be suitable for tropical, subtropical and warm temperate regions. It probably has similar requirements to those of *P. laurifolia*. Propagate from fresh seed.

Planchonella pohlmaniana (F. Muell.) Pierre ex Dubard
(after R. W. Pohlman)
Qld, NSW, NT   Yellow Boxwood; Engraver's Wood
10–20 m × 10–15 m   Aug–Oct; also sporadic
Medium to tall **tree**; young growth with brownish or blackish hairs; **leaves** 7–20 cm × 1.5–6 cm, oblong to obovate, often clustered towards the end of branchlets, thin and papery, shiny above, dull or shiny beneath, margins yellowish, tapered to the base, apex blunt; **flowers** about 0.45 cm long, greenish-cream, in axillary clusters of 4–12; **berries** 1–3 cm across, globose to ovoid, green to black; **seeds** 3–5, 1–1.2 cm × 0.5–0.9 cm, brown, shiny, ripe April–July.
Endemic in Australia where it is widely distributed from the Atherton Tableland, Qld, to the Richmond River, northern NSW and also in the Top End of the NT. Grows in drier types of rainforest. Cultivated on a limited scale by enthusiasts and also useful for rainforest reclamation projects. It has similar cultural requirements to *P. laurifolia*. The timber is hard, yellow and close-grained which makes it excellent wood for carving and engraving. Propagate from fresh seed and possibly also from cuttings.

Planchonella xerocarpa (F. Muell. ex Benth.) H. J. Lam. (with dry fruit)
Qld, NT   Blush Coondoo; Northern Coondoo
6–15 m × 5–10 m   July–Sept
Tall **shrub** or small to medium **tree**; young growth with dark brown hairs; **leaves** 9–15 × 2.5–5 cm, elliptic to lanceolate or obovate, stiff and papery to leathery, shiny above, dull and sparsely hairy beneath, tapered to the base; **flowers** about 0.6 cm long, hairy, clustered in leaf axils and on leafless growth; **berries** 1.5–2 cm × 1–1.3 cm, obovoid, blackish, white-hairy in an apical sunken area; **seeds** 2–5, about 1.5 × 0.5 cm, brown, shiny, ripe Nov.
Endemic in north-eastern Qld growing in rainforest at moderate to high altitudes in the ranges and tablelands. A bushy species which can be grown as far south as Melbourne. Plants are generally slow-growing and young plants need some shelter. Soils must be well-drained and mulching is beneficial. Tolerates light to moderate frosts. Propagate from fresh seed.

PLANCHONIA Blume
(after Jules E. Planchon, 19th-century French botanist)
Lecythidaceae
Shrubs or **trees**; **leaves** simple, entire, often crowded towards the end of branchlets; **inflorescence** a terminal

Planchonella pohlmaniana × .4

Planchonia careya   D. L. Jones

or axillary raceme; **flowers** bisexual, large, showy, short-lived, with prominent stamens; **calyx lobes** 4; **petals** 4; **stamens** numerous, long; **fruit** a berry with a persistent calyx; **seeds** many.

A small genus of about 14 species distributed from the Andaman Islands to the Philippines and New Guinea. One species occurs in northern Australia. It is widespread and common but rarely cultivated. Propagation is from seed which is best sown while fresh.

Planchonia careya (F. Muell.) Knuth
(after Dr William Carey)
Qld, WA, NT   Cocky Apple; Billygoat Plum
6–15 m × 2–5 m   July–Dec

Tall **shrub** or small to medium **tree**, often straggly; **bark** grey, corky; **leaves** 2.5–10 cm × 3–6 cm, broadly ovate, obovate, or spathulate, dark green, ageing red, somewhat shiny above, dull beneath; **racemes** 4–8 cm long, axillary or terminal; **flowers** about 8 cm across, pink and white, showy, short-lived; **calyx tube** to 1.5 cm long, hairy; **petals** to 4.5 cm × 1.7 cm, white and pink; **stamens** to 4.5 cm long, numerous, white; **berries** 7–9 cm × 3–3.7 cm, pear-shaped, green, with a persistent calyx, ripe Nov–March.

Widespread and common in tropical Australia and also in New Guinea. Grows in open forest and woodland in a wide range of situations including plains, stony hills, gullies and sand dunes. Plants are deciduous for part of the dry season. The fruits are edible and are eaten by Aborigines either raw or roasted. Other parts have a wide range of uses. For instance, the fibrous bark is used for strings, bags and belts. The leaves are used to poison fish, and preparations from the bark and leaves are used to treat a wide range of sores and maladies. The flowers are large and showy but they open at night and collapse by midday. They are produced at intervals rather than in profusion.

Cocky Apple is a hardy, adaptable species suitable for many climates including warm inland districts, warm coastal districts and seasonally dry tropical and subtropical regions. It demands excellent drainage but is generally not fussy about other soil conditions. It grows best in a sunny location. Propagate from fresh seed.

PLANOCARPA C. M. Weiller
(from the Greek *planos*, flat; and *carpos*, fruit; referring to the depressed spherical fruits)
Epacridaceae

Shrubs to about 1 m high; **branches** many; **leaves** alternate, simple, spreading to erect, undersurface somewhat glaucous; **inflorescence** in upper axils, 1–3-flowered spikes; **flowers** female or bisexual, cream, tubular, small, subtended by 1 bract and 2 bracteoles; **sepals** 5; **petals** 5, joined to form corolla tube which is equal to or longer than calyx; **lobes** spreading; **stamens** 5; **fruit** a fleshy, depressed spherical drupe, red or reddish-black.

A recently described (1996) genus with 3 species. It is endemic in Tas where plants occur above 950 m altitude. They were previously included in *Cyathodes* which differs in having longer solitary, terminal or axillary flowers and spherical fruits.

The *Planocarpa* species have ornamental qualities but are rarely encountered in cultivation. The most striking attributes of this genus are the flushes of new growth and small, colourful, fleshy fruits.

Plants are slow-growing and need freely draining acidic soils with a sunny to semi-shaded aspect. They will be best suited to cultivation in temperate regions.

Propagation is from seed or cuttings, and aerial layering may be successful. Seed may be very slow to germinate as the closely allied *Cyathodes* species can take 1–3 years to germinate. Cuttings are most likely to be successful if very soft material is used (without any brown wood on the base of the cutting). Application of hormone rooting mixture may be beneficial in promoting root production.

Planocarpa nitida (S. J. Jarman) C. M. Weiller
(shining)
Tas
0.2–0.5 m × 0.5–1 m   Sept–Oct

Dwarf **shrub** with faintly hairy young growth; **branches** spreading to erect, many; **branchlets** reddish-brown to brown, faintly ribbed, faintly hairy; **leaves** 0.5–0.8 cm × up to 0.2 cm, ovate, ascending to erect, deep green and shiny above, slightly glaucous and striate below, flat to convex, margins recurved, apex tapering to slender pungent tip; **flowers** to about 0.5 cm long, cream, bisexual, solitary in upper axils near branchlet tips; **drupes** to about 0.4 cm × 0.7 cm, reddish-black, shiny.

This species has a very restricted distribution in the Central Plateau region at over 1100 m altitude. It grows in shallow stony soils. Plants will need freely draining acidic soils which are moist for most of the year and a semi-shaded site, but they probably also tolerate a sunny aspect at high altitude. Propagate from seed, or from cuttings of very young growth, which is still moderately soft.

Cyathodes nitida S. J. Jarman is a synonym.

Planocarpa petiolaris (DC.) C. M. Weiller
(with petioles)
Tas
0.2–1 m × 0.5–1 m   Oct–Feb
Dwarf **shrub** with faintly hairy young growth;
branches spreading to erect, many; **branchlets** brown,
faintly ribbed; **leaves** 0.6–0.9 cm × to about 0.35 cm,
oblong to ovate, erect, slightly crowded, thick, dark
green with hairy base above, glaucous and striate
below, flat, apex blunt to pointed, hard; **flowers** to
about 0.5 cm long, cream to pinkish, bisexual or
female, solitary or in 2–3-flowered spikes in upper axils
near branchlet tips; **drupes** to about 0.4 cm × 0.7 cm,
red to reddish-black, becoming glabrous.

An abundant species of the central region. It usually
grows above 1200 m altitude in soils derived from
dolerite. Rarely cultivated, it warrants being better
known. Needs freely draining but moisture-retentive
acidic soils with a sunny or semi-shaded site. Worth trying in rockeries and as a container plant. Propagate
from seed, cuttings of very young growth, or from aerial layers.

Cyathodes petiolaris (DC.) Druce is a synonym.

Planocarpa sulcata (Mihaich) C. M. Weiller
(longitudinally grooved)
Tas
0.2–0.4 m × 0.5–1 m   Oct–Nov
Dwarf **shrub** with faintly hairy young growth;
branches spreading to ascending, many; **branchlets**
reddish-brown, faintly hairy; **leaves** 0.8–1.1 cm ×
0.2–0.3 cm, oblong-ovate, spreading to ascending,
green and more or less shiny above with faintly hairy
base, glaucous and striate below, margins faintly
recurved, apex pointed to blunt; **flowers** to about
0.45 cm long, cream, bisexual and female, solitary or
rarely 2 together in upper axils near ends of branchlets; **drupes** to 0.3 cm × 0.7 cm, red, smooth, glossy.

This species is restricted to the western mountains
above 950 m altitude. It occurs in subalpine scrub and
heathland. Evidently rarely cultivated, it should adapt
to freely draining but moisture-retentive acidic soils
with a semi-shaded aspect. It may have potential for
use in containers and rockeries. Propagate from seed,
and from cuttings of very young growth. Aerial layering is worth trying.

Cyathodes sulcata Mihaich is a synonym.

PLANTAGINACEAE Juss.   Plantain Family
A cosmopolitan family of dicotyledons consisting of
about 260 species in 3 genera. About 21 species
of *Plantago* are native to Australia. Members of this
family are annual or perennial herbs with a rosette of
leaves and tiny flowers aggregated into erect spikes.

PLANTAGO L.
(from the Latin name Plantain)
Plantaginaceae   Plantains; Wild Sago
Annual or perennial **herbs**; **leaves** usually in a basal
rosette, veins 3–9, axils hairy; **inflorescence** spikes or
heads, rarely solitary flowers; **flowers** bisexual, small,
in axils of sepal-like bracts; **corolla** tubular, 4-lobed;
sepals 4; **stamens** 4; **fruit** a capsule.

A cosmopolitan genus of about 165 species; 33
species occur in Australia and of these 8 are introduced and often weedy. Of the Australian species, 5
are endemic in Tas and one is endemic on Lord Howe
Island.

They occur from sea-level to the highest altitudes
where there is the greatest representation. Australian
plantains have not been fully explored for their cultivation value to date. The smaller species, eg. *P. alpestris*,
P. glacialis, *P. gunnii*, *P. muelleri* and *P. tasmanica* have
appealing foliage, form and growth patterns. Those
which have hairy foliage can be very decorative after
rain, heavy dews or frost, when their hairiness is high-
lighted. Some have potential for use in containers,
miniature gardens and rock gardens while others are
more likely to be used in regeneration programmes.
Plants prefer to be grown in soils which are moist to
wet for some part of the year and they have a preference for good drainage. Apparently, early settlers collected seeds of an inland NSW species which they
soaked in boiling water and then mixed with sugar to
make a dessert similar to sago-pudding. Aborigines are
known to make a porridge-like mixture from *P. debilis*.
Other plantains have been used in fresh salads or
cooked as a vegetable.

In most cases, propagation is from seed, which does
not require pre-sowing treatment, but high altitude
species may respond well to stratification for 3–6 weeks
at 4°C before sowing. Species with adventitious roots
such as *P. alpestris*, *P. euryphylla*, *P. glacialis*, *P. muelleri*
and *P. palustris* can be propagated by division of the
mat-forming clumps.

Plantago alpestris B. G. Briggs, Carolin & Pulley
(of the highlands)
NSW, Vic
0.06–0.1 m × 0.1–0.2 m   Nov–March
Dwarf perennial **herb** with adventitious roots and
hairy young growth; **leaves** 6–8 cm × 1–2 cm, elliptic
to obovate, in a basal rosette, spreading to ascending,
mid-to-deep green, hairy to almost glabrous, somewhat fleshy, 3-veined, margins entire; **flowerheads**
1.5–3 cm long, ovoid to cylindrical, on erect stems
8–16 cm long; **flowers** cream; **capsules** to about 0.3 cm
long.

An inhabitant of alpine and subalpine tracts which
can form small colonies in moist-to-wet soils of creeks,
stream banks and margins of bogs. Should adapt
to a range of acidic soils in cool temperate areas.
Appreciates plenty of sunshine. May be used in rock
gardens and containers. Propagate from seed or division.

P. antarctica Decne. differs in always having hairy
leaves with more-or-less toothed margins and 3–5 veins.
It is found in the Australian Alps at 1100–1500 m
altitude, as well as in Tas.

P. cladarophylla B. G. Briggs, Carolin & Palley is a
perennial herb and is similar to *P. alpestris* in having
adventitious roots. It has thin-textured, linear-elliptic
to elliptic leaves, 7–26 cm × 0.8–2.2 cm, with 3–5 soft
veins. Leaf margins are entire or with a few small teeth.
It occurs in the Northern Tablelands of NSW where it
grows in swamps, beside waterways or in herbfields.

P. debilis R. Br. is an annual or perennial herb with

hairy leaves, 3–15 cm × 0.8–4 cm, that have toothed or entire margins. The flowerheads are 4–10 cm long but elongate to about 30 cm long when fruiting. It is widespread in coastal and inland areas of all states.

P. europhylla, B. G. Briggs, Carolin & Pulley is closely allied but has 5-veined leaves. It occurs above 1500 m altitude in the Australian Alps where it grows in subalpine grassland and alpine herbfields.

Plantago antarctica Decne. —
see under *P. alpestris* B. G. Briggs, Carolin & Pulley

Plantago bellidoides Decne. —
see under *P. varia* R. Br.

Plantago cladarophylla B. G. Briggs, Carolin & Pulley —
see under *P. alpestris* B. G. Briggs, Carolin & Pulley

Plantago daltonii Decne. —
see under *P. tasmanica* Hook. f.

Plantago debilis R.Br. —
see under *P. alpestris* B. G. Briggs, Carolin & Pulley

Plantago glacialis B. G. Briggs, Carolin & Pulley
(of the ice)
NSW, Vic
0.02–0.05 m × 0.3–0.5 m   Jan–March
Dwarf perennial **herbs** with adventitious roots and glabrous or faintly hairy young growth; **leaves** 2–5 cm × 0.8–2 cm, linear-elliptic, in a basal rosette, spreading to ascending, mid green, mainly glabrous, glossy, thick, concave, 1-veined, margins entire or with a few minute teeth; **flowerheads** small, few-flowered, on stalks about as long as leaves; **flowers** brownish to reddish-brown; **capsules** to about 0.4 cm long, blunt.

This attractive mat-forming perennial is restricted to above 2000 m altitude in the Australian Alps where it grows in wet, often very cold sites. Grown to a very limited degree. It is suited to cool temperate regions but may prove difficult to maintain. Needs well-drained but moisture-retentive, acidic, silty loam. Worth trying to grow it in similar conditions to those recommended for Cushion Plants in Volume 3, page 136. Propagate from seed or division.

P. gunnii Hook. f., from Tas, is an interesting and highly specialised species which is usually found growing in cushion plant mounds at high altitude. It forms a basal rosette of hairy leaves to about 2 cm across. May be almost impossible to cultivate but it is worth trying in conditions similar to those described in Volume 3, page 136.

Plantago gunnii Hook. f. —
see under *P. glacialis* B. G. Briggs, Carolin & Pulley

Plantago hedleyi Maiden
(after Charles Hedley, 19–20th-century Australian naturalist)
NSW (Lord Howe Island)
0.1–0.2 m × 0.15–0.3 m   Nov–Feb
Dwarf perennial **herb** with glabrous or faintly hairy young growth; **leaves** 7–20 cm × 1.5–4 cm, narrowly

oblanceolate to narrowly elliptic, prominently tapered to base, in a basal rosette, spreading to ascending, mid green, mainly glabrous except for tuft of hairs at base, 3–5-veined, margins mainly entire or with few minute teeth; **flowerheads** 2–10 cm long, cylindrical, on hairy stalk to 25 cm tall; **flowers** cream; **capsules** to about 0.3 cm long.

This plantain occurs in rocky sites on Mt Gower and Mt Lidgbird. It has leaves in interesting basal rosettes. Plants have potential for use in miniature gardens and rock gardens in subtropical and temperate regions. Plants need moist but well-drained soils with a semi-shaded or filtered sun aspect. Propagate from seed.

Plantago hispida R. Br. — see under *P. varia* R. Br.

Plantago muelleri Pilger
(after Baron Sir Ferdinand von Mueller, 19th-century Government Botanist of Victoria)
NSW, Vic   Star Plantain
0.05–0.15 m × 0.2–0.5 m   Dec–March
Dwarf perennial **herb** with adventitious roots and glabrous young growth; **leaves** 0.25–12 cm × 0.7–2.5 cm, lanceolate to elliptic, in a compact basal rosette, spreading to ascending, dark green, glabrous, glossy, thick, 3-veined, margins entire or with few minute teeth; **flowerheads** small, initially hidden amongst leaves, the rusty-hairy stalk elongates to about 10 cm long as flowers mature; **flowers** brownish; **capsules** to about 0.4 cm long, nearly globular.

A tightly rosetted species which is restricted to alpine and subalpine regions where it grows in moist-to-wet sites in bogs, beside waterways and in herbfields. It is cultivated to a limited degree. Needs moist but well-drained acidic soils with a sunny or semi-shaded aspect. May do best in cushion plant conditions as described in Volume 3, page 136. Propagate from seed or by division.

P. palustris L R. Fraser & Vickery is an allied mat-forming marsh-dweller which has duller, more-or-less erect leaves. It is confined to the Barrington Tops in the Northern Tablelands of NSW and is cultivated to a very limited degree.

Plantago palustris L R. Fraser & Vickery —
see under *P. muelleri* Pilger

Plantago paradoxa Hook. f.
(strange, unusual)
Tas
0.02–0.05 m × 0.05–0.1 m   Nov–Jan
Dwarf perennial **herb** with hairy young growth; **leaves** 1.5–5 cm × 0.3–0.8 cm, lanceolate to oblanceolate, in a basal rosette, tapered to base, spreading, mid-to-deep green, hairs in tufts near small teeth or in bands across surface, 1-veined; **flowerheads** small, initially partially hidden amongst leaves, the hairy stalks elongate erectly to about as long as leaves; **flowers** cream, small; **capsules** very small.

This Tasmanian endemic species has interesting foliage and is widespread in mountainous regions where it is common in grassland. Plants require moist but well-drained acidic soils and a sunny or semi-

Plantago tasmanica

Plantago muelleri   D. L. Jones

shaded aspect. They may do best in conditions recommended for growing Cushion Plants as described in Volume 3, page 136. Propagate from seed.

Plantago tasmanica Hook. f.
(from Tasmania)
Tas
0.03–0.05 m × 0.05–0.1 m   Dec–March
    Dwarf perennial **herb** with hairy young growth; **leaves** 0.8–6 cm × 0.6 cm, broadly lanceolate to oblanceolate, in a flat or slightly erect basal rosette, deep green, thick, long silky hairs at base otherwise faintly hairy, margins entire or faintly to coarsely toothed, apex blunt; **flowerheads** in short spikes on hairy stalks longer than leaves; **flowers** cream; **capsules** small.
    The typical variant forms a tight rosette. It has a wide distribution in the high mountains of Tas where it occurs in a range of habitats that include alpine bogs, open shrubland and exposed areas. Plants may

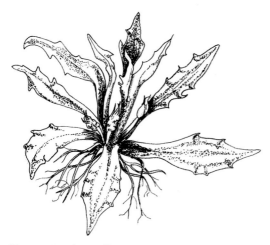

Plantago paradoxa × .8

do best if grown in the conditions recommended for Cushion Plants in Volume 3, page 136.
    The var. *archeri* W. M. Curtis has obovate leaves which are densely hairy on both surfaces. It has a similar distribution and cultivation requirements.
    P. daltonii Decne. is a closely allied species but it differs in having glabrous or faintly hairy, somewhat leathery, linear-lanceolate to elliptical leaves of 3–5 cm long. It is endemic in Tas where it occurs in alpine and subalpine sites which are subject to regular flooding.

Plantago turrifera B. G. Briggs, Carolin & Pulley —
                    see under *P. varia* R. Br.

Plantago varia R. Br.
(variable)
Qld, NSW, Vic, Tas, SA   Variable Plantain
0.1–2 m × 0.15–0.3 m   Aug–Feb; also sporadic
    Dwarf perennial **herb** with variable young growth; **leaves** mainly 6–20 cm × 1–2 cm, narrowly elliptic to oblong-oblanceolate, basal, spreading to ascending, deep green, hairy, 3–5-veined, margins usually with prominent often recurved teeth, apex pointed; **flowerheads** 3–11 cm long, somewhat loosely arranged, terminal, solitary on hairy stalks to about 20 cm long; **flowers** cream; **capsules** to about 0.4 cm long.
    A wide-ranging species which occurs in a variety of habitats such as grassland, woodland and forest, where soils are often dryish for part of the year. Although lacking strong ornamental qualities it is cultivated to a limited degree. Adapts to a range of acidic soils with a sunny or semi-shaded aspect. Propagate from seed.
    P. bellidioides Decne. from Tas is allied. It has broadly elliptical to ovate leaves 3–8 cm long which are in a spreading basal rosette. Plants are found in low altitude herbfields and grassland.
    P. hispida R. Br., has affinity but it differs in having compact flowerheads on stalks to 30 cm tall. The leaves are 1–5-veined and 4–9 cm × 0.4–1.2 cm. It occurs in NSW, Vic, Tas, SA and WA where it can be found in rocky sites and shallow soils on or near the coast as well as inland.

PLATYCERIUM Desv.
(from the Greek *platys*, broad; *ceras*, horn; referring to the horn-like fronds)
Polypodiaceae   Elkhorn Ferns
    Epiphytic or lithophytic **ferns**, often forming large clumps; whole plant covered with stellate hairs; **rhizomes** short, branched or unbranched; **fronds** of 2 types; **sterile fronds** (nest fronds) overlapping, closely appressed at the base, upper part erect, entire or more usually lobed; **fertile fronds** jointed to the rhizome, 6–several, leathery, forked; **sporangia** in dense patches on various parts of the lower surface of fertile fronds.
    A genus of about 18 species distributed in Africa, South America, tropical Asia, New Guinea and Australia. There are 4 species in Australia and 3 are endemic. They are large epiphytes or lithophytes which form bulky clumps. Three of the Australian species form multiple clumps by virtue of adventitious buds developing from the rhizome, whereas *P. superbum* entirely lacks this capacity and maintains a single

Plantago varia   D. L. Jones

growing point throughout its life. The sterile fronds, or nest fronds, trap litter from the forest canopy and curl inwards as they age — the whole mass eventually rots to form a compost into which the fern roots grow. The true fronds, which are produced at intervals, bear spores and fall off when they are old.

Elkhorn ferns, especially the native species, are very popular in cultivation and in general are adaptable and fairly easy to grow. Propagation is from spores, and the clumping species are propagated by division.

Platycerium bifurcatum (Cav.) C. Chr.
(twice-forked)
Qld, NSW   Elkhorn Fern
Epiphyte

Epiphytic or lithophytic **fern** forming large clumps; **sterile fronds** 15–30 cm across, initially pale green, becoming brown, upper margins deeply lobed, the lobes forked 1–3 times, blunt; **fertile fronds** 0.3–1 m long, erect to pendulous, dark green above, paler beneath, both surfaces stellate-hairy, the upper half of the fronds 1–5-times forked, the lobes to 2.5 cm wide, strap-shaped; **sporangia** brown, covering the lower surface of the apical part of the lobes.

A very widely spread, common, prominent epiphyte which grows on trees and rocks in rainforest, wet sclerophyll forest and moist gullies and slopes in more open habitats. It occurs from north-eastern Qld to south-eastern NSW and extends to Lord Howe Island and New Guinea. Plants develop into huge bulky clumps with the rhizomes branching to form new growths within the clump. Other epiphytic plants commonly grow in the large clumps.

Elkhorn ferns are very popular subjects for cultivation in tropical, subtropical and temperate regions. They can be attached to large slabs of materials such as hardwood or tree fern, tied directly onto the trunks of living trees (those which do not shed bark) or placed on boulders or large rocks. They can also be very successfully started in hanging baskets. Plants do best in a position that has filtered sun or partial sun and plenty of air movement. Once established they can withstand

dry periods, but regular watering will produce the best appearance in plants. Light dressings of well-rotted animal manure or blood and bone are very beneficial.

P. bifurcatum is variable. Plants from southern NSW often have semi-erect, thick, leathery, greyish-green, fertile fronds, whereas those from highland areas of north Qld have deep green, thin, fertile fronds which are strongly pendulous. This latter variant is sometimes treated as *P. willinckii* T. Moore.

Propagate all variants from spores or by division.

Platycerium bifurcatum var. **hillii** (T. Moore) Domin = *P. hillii* T. Moore

Platycerium grande (Fee) J. Sm. —
see under *P. superbum* de Jonch. & E. Hennipman

Platycerium hillii T. Moore
(after Walter Hill)
Qld   Northern Elkhorn Fern
Epiphyte

Epiphytic or lithophytic **fern** forming small to large clumps; **sterile fronds** 10–25 cm across, initially pale green, becoming brown and smooth, upper margins entire or shallowly lobed, often wavy; **fertile fronds** 0.5–1.2 m × 30–60 cm, dark green, nearly glabrous, the apical third is forked several times, lobes to 6 cm wide, strap-shaped; **sporangia** brown, covering the lower surface of the apical part of the lobes.

Endemic in north-eastern Qld where it is distributed from the top of Cape York Peninsula to near Ingham. It grows on trees and rocks in humid, lowland rainforest. Not as popular in cultivation as *P. bifurcatum* but nevertheless a very interesting species which is easily grown in tropical and warm subtropical regions. Needs warm, humid conditions with unimpeded air movement. Propagate from spores or by division.

Platycerium bifurcatum var. *hilli* (T. Moore) Domin is a synonym.

Platycerium bifurcatum   D. L. Jones

Platycerium superbum de Jonch. & E. Hennipman
(superb, magnificent)
Qld, NSW   Staghorn Fern
Epiphyte

Epiphytic or lithophytic **fern**; **sterile fronds** 30–60 cm across, pale green, densely stellate-hairy, upper margins irregularly divided into forked lobes; **fertile fronds** 30–200 cm long, often borne in pairs, pendulous, broadly wedge-shaped at the base then forked up to 5 times; **sporangia** brown, in a single patch on the lower surface of a broad area between the branches of the first fork.

Distributed from north-eastern Qld to north-eastern NSW growing mainly on trees in rainforest, but also extending to moist protected gullies in more open habitats. Also grows on large boulders and rock faces. A very impressive fern which remains as a solitary clump, increasing in size each year, with odd plants attaining very impressive dimensions. One of the most popular native ferns, the staghorn is grown in tropical to temperate regions. In warm relatively and frost-free zones plants may be tied direct onto garden trees, however, they are also commonly attached to large slabs of weathered hardwood or tree fern. In temperate regions plants should be sited in a position which offers some protection from the cold, such as on a north-facing wall under eaves, and kept dry over winter. Healthy plants which are watered regularly can withstand considerable exposure to sun. They respond well in summer to regular light applications of rotted animal manure, blood and bone or liquid fertilisers. Propagate from spores. Sporelings grow quickly until about 5 cm across, then tend to become difficult and are sensitive to rotting. At this stage they should be tied to paperbark poles and hung in a warm sheltered position with free and unimpeded air movement.

For many years this species was confused with *P. grande* (Fee) J. Sm., a Philippine species with 2 spore patches on each fertile frond.

Platycerium veitchii (Underw.) C. Chr.
(after John James Veitch, English nurseryman)
Qld   Silver Elkhorn Fern
Epiphyte

Lithophytic **fern** forming small to large clumps; **sterile fronds** 10–20 cm across, initially silvery-white, becoming brownish, densely stellate-hairy when young, upper margins deeply lobed, lobes entire or forked once, to 20 cm × 1.5 cm, blunt to pointed; **fertile fronds** 20–80 cm long, erect to semi-pendulous, initially silvery-white with dense stellate hairs, some hairs shedding with age, forked near the apex into 2–8 narrow lobes to 20 cm × 2.5 cm; **sporangia** brown, covering the lower surface of the apical part of the lobes.

Distributed disjunctly from northern Qld to southern parts of central Qld. Occurs on the western slopes of the Great Dividing Range growing on boulders, cliff faces and gorges in open forest and woodland, often in exposed situations. Commonly forms extensive colonies. Flourishing plants are an impressive spectacle with all parts densely clothed with silvery-white hairs. Plants adapt well to cultivation and can be grown successfully on a tree or slab as far south as Melbourne. Resents shade and is best placed in a position exposed to considerable sun. Once established, plants are very tolerant of dryness. Propagate from spores or by division.

Platycerium willinckii T. Moore —
                    see *P. bifurcatum* (Cav.) C. Chr.

PLATYLOBIUM Sm.
(from the Greek *platys*, broad, flat; *lobos*, pod; referring to the seed pods)
Fabaceae (alt. *Papilionaceae*)   Flat Peas

Prostrate to semi-climbing **shrubs** with woody rootstocks; **leaves** opposite or rarely alternate, simple, sessile or stalked; **stipules** ovate to narrowly ovate, persistent; **flowers** pea-shaped, axillary, 1 to several, sessile or stalked, subtended by papery scales and bracts; **bracteoles** 2, scale-like; **calyx** densely hairy, 5-lobed, upper 2 longest, rounded and much larger than 3 acutely pointed lower ones; **pods** flat, several-seeded, winged on upper side.

An endemic genus of 4 species which is confined to south-eastern Qld, eastern NSW, Vic, Tas and south-eastern SA.

Platylobium is found in coastal habitats and extends inland to mountainous regions. Plants occur in a range of acidic soils which drain freely. They can be prostrate or clumping and sometimes they grow as climbers if support is available. The woody rootstock enables them to survive fire and many other natural catastrophes. New stems will emerge after damage to growth. Flower production varies from species to species. They all have appeal but *P. formosum* is the most ornamental. It often produces a profuse display of yellow and reddish flowers in spring. Platylobiums are very adaptable to cultivation and usually do not pose any major problems. They grow well without supplementary feeding and watering is rarely needed once they become well-established, except during extended dry periods. They are rarely susceptible to major pests or diseases.

Propagation is from seed or from cuttings. Seed has a long viability but requires pre-sowing treatment to soften the hard seed-coat. See Treatments and Techniques for Difficult to Germinate Species, Volume 1, page 205. Plants can be grown from cuttings of barely firm young growth. *P. formosum* is the easiest to propagate by this method. Plants may also have self-layering stems which can provide excellent results when treated as cuttings.

Platylobium alternifolium F. Muell.
(alternate leaves)
Vic
prostrate × 1–1.5 m   Sept–Nov

Dwarf **shrub** with woody rootstock; young growth faintly hairy; **stems** prostrate, one to several, becoming less hairy to nearly glabrous; **leaves** 1–4.5 cm × 0.6–2.5 cm, cordate to ovate or triangular-ovate to almost circular, prominently stalked, spreading to ascending, not crowded, deep green, becoming more or less glabrous above and below, flat, venation prominent, apex pointed to rounded; **flowers** pea-shaped, to about 1.5 cm across, solitary, axillary, on glabrous to faintly hairy, scaly stalks to about 2 cm long; **standard** yellow with reddish-purple basal blotching and deep red exterior; **wings** yellow with central dark red or purplish markings; **keel** deep purple in upper half; **pods** to about 2 cm × 2 cm, oblong, flat, glabrous except for margins.

An attractive species which is confined to western Victoria with its main representation in the Grampians. Rarely cultivated, it should adapt to a range of acidic soils in temperate regions. A sunny or semi-shaded site is suitable. Plants tolerate heavy frosts and extended dry periods. Worth trying in rockeries and large containers. Pruning of young stems may promote denser growth. Propagate from seed or from cuttings.

Platylobium formosum Sm.
(beautiful)
Qld, NSW, Vic, Tas   Handsome Flat Pea
0.2–2.5 m × 1–3 m   Sept–Jan
Dwarf to medium **shrub** with a woody rootstock; young growth glabrous to hairy; **branches** prostrate to erect, sometimes climbing if support is available; **branchlets** glabrous to hairy, fairly rigid; **leaves** 2.5–9 cm × 0.7–4 cm, narrowly ovate to broadly ovate, mainly opposite, nearly sessile to prominently stalked, spreading to ascending, deep green, glabrous or somewhat rough above, paler below, prominent reticulate venation, apex pointed; **flowers** pea-shaped, to about 2 cm across, solitary on hairy stalks to 3 cm long, 1–4

Platylobium obtusangulum × .7

per axil, often profuse and very conspicuous; **standard** yellow to orange-yellow with red basal blotch; **wings** yellow, sometimes with reddish base; **keel** red to purplish; **pods** to 5.5 cm × 2 cm, oblong, dark brown to blackish, stalked, glabrous or hairy.

An extremely variable species in growth habit and leaf shape. It has a distribution extending from southeastern Qld, along the NSW coast and inland to the Central Tablelands and Central Western Slopes, to eastern and southern Vic, and eastern Tas. It is found in a range of habitats from rainforest to heathland. Adapts very well to cultivation in acidic soils which drain at least moderately well. Plants will cope with full sunshine but do better in a semi-shaded or shaded site. They are hardy to most frosts and can withstand moderately long dry periods. Useful for general planting as well as on embankments. Butterflies are frequent visitors to flowering plants. Plants with narrowly ovate leaves are sometimes classified as ssp. *parviflorum* (Sm.) A. Lee. Propagate from seed, or from cuttings of young growth, or layered stems.

Platylobium obtusangulum Hook.
(obtuse angle)
Vic, Tas, SA   Common Flat Pea
0.6–1 m × 0.6–1.8 m   Sept–Dec
Dwarf, often clumping **shrub** with a woody rootstock; young growth glabrous or hairy; **stems** many, slender, spreading to erect, often arching, sometimes trailing; **branchlets** slender, brown; **leaves** 1–3 cm × 0.6–3.8 cm, broadly triangular or arrowhead-shaped, opposite, more or less sessile or short-stalked, spreading to ascending, rarely crowded, deep green becoming glabrous or faintly rough above, glabrous to hairy below, flat, venation prominent, apices rounded or prickly pointed; **flowers** pea-shaped, to about 2 cm across, solitary on very short stalks which are hidden by

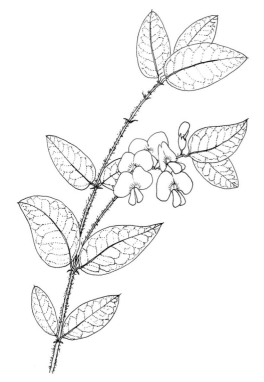

Platylobium formosum × .55

365

bracts and scales, 1–3 per axil, scattered to moderately profuse, always conspicuous; **standard** orange-yellow with reddish basal blotch, exterior brownish to pinkish-red; **wings** orange-yellow with pinkish-red base; **keel** reddish with whitish base; **pods** to about 1.5 cm × 1.5 cm, oblong, dark brown, nearly sessile, hairy.

A wide-ranging and moderately handsome species from Kangaroo Island, south-eastern SA, southern Vic and northern and eastern Tas. It is found in freely draining acidic sandy or loam soils in heathland, woodland and open forest. Adapts well to cultivation in temperate regions and does best in semi-shaded sites. Plants are hardy to most frosts and extended dry periods. Old plants are rejuvenated by hard pruning. Does well in containers and in general planting. Suited to planting beneath established tall trees. Propagate from seed or from cuttings of barely firm new growth emerging from woody rootstock.

P. triangulare differs in having stalked flowers and nearly glabrous pods.

Platylobium triangulare R. Br.
(triangular)

Vic, Tas · Ivy Flat Pea

0.3–0.6 m × 1–1.5 m · Sept–Nov

Dwarf **shrub** with a woody rootstock; young growth glabrous or hairy; **stems** many, slender, prostrate to ascending, glabrous to hairy; **leaves** 1–3.2 cm × 0.6–2.6 cm, broadly triangular, with cordate or hastate base, more or less sessile, spreading to ascending, rarely crowded, deep green, becoming glabrous or faintly hairy above, paler and faintly hairy to glabrous below, flat, venation prominent, apices usually prickly pointed; **flowers** to about 2 cm across, usually solitary on slender hairy stalks to about 2 cm long, 1–3 per axil, scattered to moderately profuse; **standard** orange-yellow with reddish-purple basal blotch, exterior pinkish-red to purplish-brown; **wings** orange-yellow with reddish base; **keel** deep purplish to reddish with whitish base; **pods** to about 3.3 cm × 1.8 cm, oblong, on hairy stalk, glabrous except for hairy margins.

An interesting species from south-western Vic and northern and eastern Tas. It grows in well-drained acidic soils. Rarely cultivated because of its similarity to *P. obtusangulum*. It has similar cultivation and propagation requirements to that species.

P. obtusangulum differs in having more-or-less sessile flowers and densely hairy pods.

PLATYSACE Bunge
(from the Greek *platys,* flat, broad; *sacos,* sack, bag; referring to the broad fruit)

Apiaceae (alt. *Umbelliferae*)

Shrubs or semi-woody perennial **herbs; stems** branched, terete, angular or flattened; **leaves** entire, toothed or lobed, narrowly linear to somewhat circular, hard; **inflorescence** usually terminal or axillary umbels; **flowers** small, bisexual; **sepals** minute or absent; **petals** small, white, cream, yellow or greenish, initially inflexed, becoming spreading; **fruit** of 2 mericarps, flattened laterally.

This endemic genus comprises about 25 species with the greatest representation (about 16 species) in southern WA. Two species, *P. dissecta* and *P. eatoniae*, are now presumed extinct. It needs botanical study in order to clarify the status of many species. *Platysace* is not generally regarded as having many species which are highly ornamental but within the genus there are some very attractive species, such as the variable *P. lanceolata*, which have proved reliable and desirable for cultivation. Some of the 'leafless' or near leafless species including *P. juncea* and *P. xerophila* warrant trials in cultivation.

The flowers, although small, can be produced in profusion to provide a lovely lace-like display of white and sometimes pinkish flowers over a long and often extended period, eg. *P. ericoides* and *P. lanceolata*.

For successful cultivation most species need well drained and preferably acidic soils. They prefer semi-shaded sites but will tolerate a fair amount of sunshine. Cold tolerance is not well known but most species withstand light to moderate frosts.

Plants respond well to light applications of low phosphorus, slow-release fertilisers. Most species tolerate quite dry conditions but they may need supplementary water during extended dry periods.

There have been very few pruning or clipping trials but *P. lanceolata* responds well if plants are in moderate to vigorous growth. Little is known regarding the tolerance of *Platysace* species to pests and diseases. Scale can sometimes be a problem but can be readily controlled by conventional methods (see Sucking Insects, Volume 1, page 153). Borers and caterpillars may be a minor problem and can be controlled as described under Chewing Insects, Volume 1, page 143.

Propagation is from seed or from cuttings. The viability and longevity of seed is poorly known so it is recommended that fresh seed be used. Seed does not require any pre-sowing treatment. Some species are regularly propagated from cuttings, eg. *P. lanceolata*. Cuttings of young growth which has just started to become firm are likely to give best results. Application of rooting hormones can help to promote a strong root system.

Platysace anceps (DC.) Norman —
see *P. compressa* (Labill.) Norman

Platysace arnhemica Specht
(from Arnhem Land)

NT

1.5–3 m × 0.6–1.5 m · Dec–April

Small to medium, slender **shrub; branches** ascending to erect, twiggy; **branchlets** with prominent leaf scars; **leaves** 6–15 cm × 0.1–0.2 cm, linear, sessile, in whorl-like clusters at ends of branchlets, green, glabrous, apex finely pointed; **flowers** to about 0.3 cm across, cream, in slender-stemmed, terminal, compound umbels, 8–15 cm across, sometimes profuse and moderately to very conspicuous; **fruit** about 0.2 cm × 0.2 cm, compressed with prominent ribs.

An interesting and decorative species from Arnhem Land where it grows often in crevices of the sandstone escarpment. It is cultivated to a limited degree and is best suited to seasonally dry tropical regions. Needs excellently drained acidic soils in a sunny or semi-shaded site. Should rarely require supplementary watering during extended dry periods. Propagate from seed or from cuttings of firm young growth.

Trachymene hemicarpa (F. Muell.) Benth. is a synonym.

Platysace cirrosa Bunge
(with tendrils)
WA   Karna; Kanna; Native Potato
Climber   Feb–April
A tuber-bearing, herbaceous **climber** with a perennial rootstock and glabrous young growth; **branches** wiry, to about 50 cm long, twining; **leaves** 2–3 cm × to about 0.3 cm, linear, alternate, sessile, green, glabrous, present on plant from about June to Nov; **flowers** to about 0.2 cm across, cream to yellow, in terminal umbels on leafless stems which originate from previous season's rootstock, can be moderately profuse; **fruit** small, flattened.

This species is dormant in summer in its natural habitat then new stems with foliage are produced from the rootstock during autumn. It occurs in stony soils in open forests of the Avon and Irwin Districts Although it lacks strong ornamental qualities, the tuberous roots were an important food source for local Aborigines. Probably best suited to cultivation in warm temperate regions. Plants require very well-drained soils in a sunny or semi-shaded site. They are hardy to moderate frosts. Worth trying as a vegetable crop. Propagate from seeds or cuttings and possibly by division of tubers.

Platysace clelandii (Maiden & Betche) L. A. S. Johnson
(after Dr J. B. Cleland, Australian botanical collector)
NSW
0.3–0.6 m × 0.5–1 m   Aug–Feb
Dwarf **shrub** with hairy young growth; **branches** scrambling, densely hairy; **leaves** 0.6–0.7 cm × to about 1 cm, fan-shaped to nearly circular, tapered to small stalk, spreading to ascending, green, faintly hairy, apex broad and usually with 3–5 shallow lobes; **flowers** small, white, in terminal compound umbels to about 3 cm across, can be profuse; **fruit** about 0.15 cm × 0.2 cm, with short bristles.

P. clelandii is a rare species with unusual foliage. It occurs in the Central Coast and Central Tablelands where it grows in dry sclerophyll forest in sandy soils derived from the surrounding sandstone rocks. Evidently not cultivated, it should adapt to subtropical and temperate regions if grown in well-drained acidic soils with a semi-shaded or sunny aspect. Plants are hardy to moderate frosts. They can develop unwieldy branches but pruning should encourage more bushy growth. Propagate from seed and probably from cuttings of firm young growth.

Platysace commutata (Turcz.) Norman
(changed)
WA
0.15–0.5 m × 0.2–0.5 m   Nov–Feb
Dwarf **shrub** with glabrous young growth; **branches** many, slender, ascending to erect; **leaves** 0.4–1 cm × 0.1–0.2 cm, linear, alternate, spreading to ascending, crowded, thick, deep green, glabrous, apex blunt or pointed; **flowers** small, white to cream, in short,

Platysace lanceolata   W. R. Elliot

compact, terminal umbels, often profuse and moderately conspicuous; **fruit** about 0.25 cm × 0.25 cm, 8-shaped in cross-section.

A poorly known species of the Avon, south-eastern Darling, Eyre and Roe Districts where it grows in heath and mallee. Apparently not cultivated, it lacks strong ornamental qualities. Suited to semi-arid and temperate regions. Plants will need excellent drainage and a warm to hot site. Tolerates moderate frosts. Propagate from seed or from cuttings of firm young growth.

Platysace compressa (Labill.) Norman
(flattened)
WA   Tapeworm Plant
0.4 m × 0.75 m × 0.5–1.5 m   Jan–May; also sporadic
Dwarf **shrub** with perennial rootstock; **branches** few to many, sometimes zig-zagged, often much-branched, with wings of varying widths, light green; **basal leaves** much-divided, withering early, light green; **stem leaves** 0.1–0.3 cm long, few, entire or 3-lobed or reduced to minute scales; **flowers** about 0.2 cm across, buds pink, opening white to cream, in loosely arranged umbels on ends of small branchlets, often profuse and moderately conspicuous; **fruit** about 0.25 cm across, granular or smooth.

This wide-ranging species is found in the southern Darling and Eyre Districts where it inhabits sandy soils of coastal or inland regions. Cultivated to a limited degree because of its interesting winged stems and growth habit. May appeal most to enthusiasts. Best for temperate regions in sites which are well drained, with a sunny or semi-shaded aspect. May suffer frost damage. Propagate from seed or from cuttings of firm young growth.

P. anceps is a many-stemmed herbaceous subshrub. It has narrower stem wings and smaller umbels than *P. compressa*. Flowers mainly bloom during April–Oct.

Platysace deflexa

Occurs in the southern Darling and western Eyre Districts.

The allied *P. filiformis* (Bunge) Norman is a somewhat insignificant species with slightly flattened stems and pendent branchlets. It occurs on or near the coast in the southern Darling District.

P. haplosciada (Benth.) Norman has a perennial rootstock and terete or angular stems. Flowers are small and white to pinkish-white. They are arranged in simple umbels at the end of slender stems. It is endemic in the eastern Eyre District.

Platysace deflexa (Turcz.) Norman
(bent or turned downwards)
WA
0.2–0.5 m × 0.3–0.5 m   Oct–March
Dwarf **shrub** with tuberous rootstock; **branches** spreading to ascending, many, slender; **leaves** 0.2–0.4 cm × 0.1–0.2 cm, linear to broadly lanceolate or sometimes nearly orbicular near branch tips, reflexed, crowded, glabrous, apex blunt; **flowers** about 0.2 cm across, white, in loosely arranged terminal umbels on long slender stems, displayed well beyond foliage; **fruits** to about 0.3 cm across.

This species is regarded as rare in its natural habitat of sandy soils in the Avon, Eyre, Roe and southern Darling Districts. Plants need good to excellent drainage and a sunny or semi-shaded site. They tolerate moderate frosts and should adapt to temperate regions. The egg-shaped root tubers were a food source for Aborigines. Propagate from seed or from cuttings of firm young growth.

Platysace effusa (Turcz.) Norman
(loosely spreading)
WA
0.15–0.5 m × 0.3–0.6 m   Feb–May; also sporadic
Dwarf **shrub**; young growth glabrous; **branches** many, spreading to ascending, slender; **leaves** 0.4–0.8 cm × 0.1–0.3 cm, linear to spathulate, tapered to base, ascending to erect, crowded near ends of branchlets, glabrous, apex blunt or pointed; **flowers** about 0.2 cm across, white, in loosely arranged terminal umbels to about 5 cm across, on slender stems beyond the foliage, can be profuse and moderately conspicuous; **fruit** about 0.25 cm across, slightly rough.

This species often develops as a somewhat open shrub. It occurs over a wide area in the south-western region and is found in the Avon, Coolgardie, Eyre, Irwin and Roe Districts where it grows in sandy and clay soils. Evidently not cultivated, it is worthy of trial in temperate and semi-arid regions. Should adapt to a range of acidic soils and will tolerate moderate frosts. Propagate from seed or from cuttings of firm young growth.

Platysace ericoides (Sieber & Spreng.) Norman
(resembles the genus *Erica*)
Qld, NSW, Vic   Heath Platysace
0.2–0.6 m × 0.3–1 m   Aug–April
Dwarf wispy **shrub** with glabrous or faintly hairy young growth; **branches** many, widely spreading to ascending, slightly rough or hairy; **leaves** 0.3–2 cm × up to 0.2 cm, linear to oblong, spreading to ascending

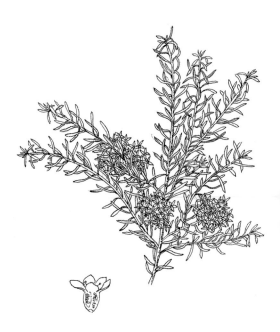

Platysace ericoides × .6, flower × 2.5

or sometimes reflexed, green, glabrous to faintly hairy, apex blunt or pointed; **flowers** about 0.2 cm across, white to cream, in terminal, few-flowered, compound umbels to about 2 cm across; **fruit** to about 0.2 cm wide.

An extremely variable species which usually inhabits moist-to-wettish sandy soils in heath and dry sclerophyll forest. It is found in coastal and slightly inland regions from south-eastern Qld to eastern Vic. This species has moderate ornamental appeal and is cultivated to a limited degree. Plants do well in temperate regions. They require well-drained acidic soils and a semi-shaded or moderately sunny site is suitable. Hardy to moderate frosts. Propagate from seed or from cuttings of firm young growth.

Platysace filiformis (Bunge) Norman —
see *P. compressa* (Labill.) Norman

Platysace haplosciadia (Benth.) Norman —
see *P. compressa* (Labill.) Norman

Platysace heterophylla (Benth.) Norman
(bearing leaves of more than one kind)
Vic, SA   Slender Platyscae; Corn Parsley
0.2–0.5 m × 0.3–0.6 m   Dec–Feb
Dwarf perennial **herb** or **subshrub** with a woody rootstock; **stems** spreading to ascending, slender, glabrous, to about 0.5 m long; **juvenile leaves** often trifid; **mature leaves** 0.5–1.6 cm × about 0.1 cm, narrowly-linear or filiform, ascending to erect, green, glabrous; **flowers** to about 0.2 cm across, white, in small, terminal, few-flowered, compound umbels, often profuse but rarely highly conspicuous; **fruit** about 0.25 cm across, faintly rough.

The typical variant is more-or-less confined to coastal areas. It is found in southern Vic and south-eastern SA where it usually grows in sandy soils. It is renowned as a pioneer species after bushfires. Plants have a lacy appearance when flowering and are moderately ornamental.. They are suited to temperate regions and require freely draining acidic soils. A semi-shaded site is probably best although they tolerate full sunshine and are hardy to light frosts. The longevity of plants in cultivation is not known.

The var. *tepperi* (J. M. Black) Eichler has cuneate leaves to about 1 cm long with a 3-lobed apex. It is endemic on Kangaroo Island, SA.

Propagate both variants from seed or from cuttings of firm young growth.

The closely allied *P. linearifolia* (Cav.) Norman differs in having glabrous stems and very narrow leaves 1–2.5 cm long. It occurs in south-eastern Qld, Central and South Coast and Central Tablelands of NSW and has similar cultivation requirements. This species was first introduced into England in 1824.

Platysace juncea (Bunge) Norman
(rush-like)
WA
0.3–0.6 m × 0.3–0.6 m   Nov–Feb

Dwarf rush-like **shrub** with woody rootstock; **stems** erect, mainly straight, angular, glabrous; **basal leaves** much-divided, withering early; **stem leaves** 0.5–2 cm, linear, scattered, deciduous early; **flowers** about 0.3 cm across, white to creamy yellow, in terminal, irregular, compound umbels 2–5 cm across, on long stems, often profuse; **fruit** to about 0.3 cm across, slightly warty.

A wide-ranging species in south-western WA where it grows in sandy soils of heath and woodland and is often associated with granite rocks. Evidently not cultivated but because interest in tufting plants is growing it has potential. Suited to temperate and semi-arid regions. Needs good drainage and tolerates a sunny or semi-shaded site. Hardy to moderate frosts. Propagate from seed or from cuttings of firm young growth.

The allied *P. pendula* (Benth.) Norman differs in having pendent heads on short stems. It occurs in the southern Darling and western Eyre Districts where it grows in sandy, well-drained soils. Should respond well to similar cultivation conditions as those recommended for *P. juncea*.

P. ramosissima (Benth.) Norman is almost a dwarf variant of *P. juncea* but it has flexuose stems which branch in the upper half. It occurs in the Darling District and will need cultivation conditions similar to those of *P. juncea*. See also *P. xerophila*.

Platysace lanceolata (Labill.) Druce
(lanceolate)
Qld, NSW, Vic   Shrubby Platysace; Native Parsnip
0.6–2 m × 0.6–2 m   Sept–March; also sporadic

Dwarf to small **shrub** with faintly hairy young growth; **branches** many, spreading to ascending, finely hairy; **leaves** mainly 1–5 cm × 0.4–1.5 cm, narrowly to broadly elliptic or sometimes nearly circular, spreading to ascending, moderately crowded, deep green, glabrous, apex blunt or pointed; **flowers** about 0.2 cm across, white or cream to pinkish, in terminal compound umbels, 1.5–5 cm across, often profuse and conspicuous; **fruit** about 0.2 cm across, warty.

A variable species which is often quite showy when in bud and flower. It occurs from south-eastern Qld to western Vic in a variety of soil types and vegetation. *P. lanceolata* has proved adaptable and an oval-leaved selection is one of the most popular in cultivation. Plants require moderately well-drained acidic soils and prefer a semi-shaded aspect but will tolerate a fairly sunny site. They withstand some dryness but may need supplementary watering during extended dry periods. Plants are hardy to moderately heavy frosts and they respond well to pruning. Flowering stems can be used for indoor decoration but need to be picked as first buds open. It was first cultivated in England during 1829 as *Trachymene lanceolata*. Propagate from seed or from cuttings of firm young growth.

P. lanceolata is known to hybridise with *P. stephensonii* in nature and some hybrids may be in cultivation.

Platysace linearifolia (Cav.) Norman —
              see *P. heterophylla* (Benth.) Norman

Platysace maxwellii (Cav.) Norman
(after George Maxwell, 19th-century botanical collector)
WA
0.5–1.2 m × 0.4–1.2 m   June–Aug; Jan–Feb

Dwarf to small **shrub** with creeping rhizomes; **stems** ascending to erect, slender; **leaves** 1.5–3 cm × 0.1–0.2 cm, narrowly linear, ascending to erect, crowded, green, glabrous, rigid, apex pointed; **flowers** about 0.3 cm across, cream to yellowish, in terminal compound umbels with involucral bracts about 2 cm long, can be profuse but rarely highly conspicuous; **fruit** about 0.2 cm across.

An inhabitant of sandy soils in the Avon, Irwin and Roe Districts. It lacks strong ornamental qualities and evidently is not cultivated. Requires excellent drainage and a warm to hot site. Probably best suited to warm temperate or semi-arid regions. Plants tolerate moderate frosts and extended dry periods. Propagate from seed or from cuttings of firm young growth.

Platysace pendula (Benth.) Norman —
              see *P. juncea* (Bunge) Norman

Platysace ramosissima (Benth.) Norman —
              see *P. juncea* (Bunge) Norman and
              P. xerophila (E. Pritz.) L. A. S. Johnson

Platysace stephensonii (Turcz.) Norman
(after Lawrence Stephenson, 19th-century Australian botanist)
NSW   Stephenson's Platysace
0.2–0.6 m × 0.3–0.6 m   Dec–Jan; also sporadic

Dwarf **shrub** with hairy young growth; **branches** spreading to ascending, many, hairy; **juvenile leaves** with 3 spreading, linear to ovate, pointed lobes; **mature leaves** 1.2–2 cm × 0.5–0.7 cm, ovate to obovate or 2-lobed, spreading to ascending, crowded, deep green, apex pointed; **flowers** about 0.2 cm across, white, in terminal compound umbels 1–3 cm across, some-

Platysace valida

times partially hidden by foliage; **fruit** to about 0.2 cm across, warty.

This species is regarded as rare in nature. It inhabits well-drained soils in heath and dry sclerophyll forest of the Central and South Coast Districts where it grows on the coast and the hinterland. Cultivated to a very limited degree, it should succeed in temperate and subtropical regions. Needs acidic soils and a semi-shaded or moderately sunny site will be suitable. Plants tolerate light frosts. Propagate from seed or from cuttings of firm young growth.

Hybrids with *P. stephensonii* and *P. lanceolata* as parents occur in nature and some of these may be in cultivation.

Platysace valida (F. Muell.) F. Muell.
(strong)
Qld
2–4 m × 1–2 m   Oct–Jan

Medium **shrub** with glabrous young growth; **stems** terete, brownish, smooth; **branches** spreading, somewhat slender, terete, greenish; **leaves** 2.5–5 cm × 0.3–1 cm, linear-oblong to lanceolate, mainly sessile, glabrous, glossy above and below, flat, leathery, 3-nerved, apex blunt; **flowers** small, white, in terminal compound umbels, to about 3.5 cm across, often profuse; **fruit** to about 0.5 cm × 0.7 cm, flattish.

This distinctive species occurs in the Cook, North and South Kennedy, Mitchell and Leichhardt Districts where it grows in sandy soils and rocky sites in open forest. It is cultivated to a limited degree and should grow best in subtropical regions in well-drained acidic soils with a semi-shaded aspect. Frost tolerance is not known. Propagate from seed or from cuttings of firm young growth.

Platysace xerophila (E. Pritz.) L. A. S. Johnson
(dry-loving)
WA
0.1–0.4 m × 0.2–0.6 m   Sept–Jan

Perennial dwarf **herb** with woody rootstock; **stems** many, prostrate to erect, stiff, somewhat ribbed to angular, zig-zagged, usually leafless; **basal leaves** deeply dissected, usually a few present; **stem leaves** small, 3-lobed or linear; **flowers** to about 0.3 cm across, white to cream, in terminal compound umbels about 3 cm across, can be profuse; **fruit** to about 0.25 cm across, usually warty.

An interesting, prostrate to erect species from the Avon, Irwin and northern Darling Districts where it grows in sandy soils. Apparently not cultivated, it has potential for cultivation because of its unusual foliage and low growth habit. Plants should do best in warm temperate and semi-arid regions. Drainage needs to be excellent and sunny or semi-shaded sites are recommended. Plants withstand light to moderate frost. Worth trying in rockeries, borders and containers. Propagate from seed or from cuttings of firm young growth.

A variant from the northern Darling District, with many slightly flexuose branches, was sometimes placed in *P. ramosissima* but is now regarded as referrable to *P. xerophila*.

Platytheca galioides   D. L. Jones

PLATYTHECA Steetz
(from the Greek *platys*, broad, flat; *theke* box; referring to the anthers)
Tremandraceae

Small **shrubs** with simple hairs; **leaves** linear to terete, sessile, whorled, 8–10 per whorl; **flowers** 5-merous, solitary on slender axillary stalks; **petals** purple to bluish-purple; **stamens** 10, arranged in 2 unequal whorls; **style** slender; **fruit** a capsule.

A small genus of 2 stunning species endemic in south-western WA. Propagation is usually from cuttings of firm young growth which generally strike readily.

Platytheca crassifolia Steetz — see *P. galioides* Steetz.

Platytheca galioides Steetz
(similar to the genus *Galium*)
WA   Platytheca
0.3–0.7 m × 0.2–0.5 m   June–Nov; also sporadic

Dwarf **shrub** with hairy young growth; **branches** few to many, slender, spreading to ascending, brittle; **branchlets** short, often purplish-green; **leaves** 0.8 cm × up to 0.1 cm, linear to terete, usually 8 per whorl, spreading to ascending, light-to-deep green, glabrous to hairy, margins strongly revolute, apex blunt; **flowers** to about 2 cm across, bluish-purple, pendent with spreading petals, solitary, on slender axillary stalks to about 2.5 cm long, scattered to moderately profuse, always very conspicuous.

A very striking species with iridescent, well-displayed small flowers. It is widespread in the Darling District and extends to the Eyre District. Usually grows in moist but well-drained acidic soils in heathland and woodland. There are 2 distinct variants in cultivation.

The typical one is a lax-stemmed dwarf shrub while the other variant, which is also sometimes referred to as *P. crassifolia* Steetz, has stiffer, shorter, often slightly warty leaves and occurs from east of the Stirling Range to near Bremer Bay. It is often difficult to maintain plants alive in cultivation for more than 1–2 years due to susceptibility to fungal root diseases or lack of mycorrhiza. Plants need excellent drainage and usually last longer if nutrients and supplementary watering are kept to a minimum. A sunny site is required but plants dislike full sunshine all day. They are brittle and easily broken especially if they become top-heavy. Pruning from an early stage is beneficial. Plants are hardy to moderate frosts but dislike periods of overcast weather which may promote serious attacks of Grey Mould, *Botrytis cinerea*. See Moulds, Volume 1, page 173 for prevention and control methods. An excellent container and rockery plant. Propagate from seed, which usually begins to germinate 40–60 days after sowing. Cuttings of firm young growth strike readily.

P. verticillata (Walp.) Baill. is a synonym.

P. juniperina which occurs on the higher parts of the Stirling Range and Barren Ranges is very similar. It mainly differs in having prickly pointed leaves on lax to fairly rigid stems. *P. juniperina* has similar cultivation requirements to those of *P. galioides* but is not commonly seen in gardens.

Platytheca juniperina Domin — see *P. galioides* Steetz

Platytheca verticillata (Walp.) Baill. = *P. galioides* Steetz

Platyzoma microphyllum   W. R. Elliot

PLATYZOMA R. Br.
(from the Greek *platys*, flat, broad; *zoma*, girdle, belt; referring to the spore case arranged in a broad band)
Platyzomataceae
   A monotypic genus endemic in Australia.

Platyzoma microphyllum R. Br.
(small leaves)
Qld, NSW, WA, NT   Braid Fern
0.4–1 m × 0.5–1.5 m
   Terrestrial **fern** forming tussocks or small colonies; **rootstock** short-creeping, covered in golden to golden-brown hairs; **fronds** of 2 types, simple or pinnate, 0.2–1 m × 0.3–0.5 cm, erect, many, crowded, young growth pale yellow-green, becoming bluish-green, leathery; **pinnae** to about 0.3 cm long, round to ovoid or nearly globular, somewhat pouch-like, margins strongly revolute; **sori** of 2–4 sporangia, initially enclosed by membrane.
   An extremely handsome fern from north-eastern Qld, Top End, NT, and northern WA, with an isolated occurrence in the North Western Plain of NSW. Plants resent disturbance and attempts to establish the species in cultivation have failed due to lack of correct mycorrhiza. Research is required to ascertain best methods of propagation. In nature it occurs in woodland in deep sandy soils where it withstands extended periods of dryness after being waterlogged during the wet season.

PLATYZOMATACEAE Nakai
A monotypic fern family consisting of a single species of *Platyzoma* which is endemic in Australia. This fern is sometimes included in Gleicheniaceae but it has a number of unique features which make it distinctive.

PLECTORRHIZA Dockrill
(from the Greek *plectos*, twisted; *rhizos*, a root; referring to the tangled roots)
Orchidaceae
   Small to moderate-sized **epiphytes**, often forming tangled clumps; **roots** numerous, wiry, long and tangled; **stems** slender, sparsely branched; **leaves** alternate, simple, spreading widely from the stems, sheathing at the base; **inflorescence** a raceme; **flowers** small, dull-coloured, often fragrant; **labellum** hinged to the column base.
   A small genus of 3 species (all are endemic in Australia). Two species are fairly commonly grown by orchid enthusiasts. Propagation is from seed which must be sown under sterile conditions.

Plectorrhiza brevilabris (F. Muell.) Dockrill
(with a short lip or labellum)
Qld
Epiphyte   Nov–Feb
   Epiphytic orchid forming small to moderately large, often straggly clumps; **roots** coarse, wiry often extensive; **stems** 10–50 cm long, erect to pendulous, flattened, sparsely branched; **leaves** 3–7, 5–8 cm × 1.2–1.5 cm, narrowly ovate-lanceolate, dark green, spreading, apex unequally notched; **racemes** 10–18 cm

long, pendulous, bearing 3–20 flowers; **flowers** about 1 cm across, green with red markings and a white patch on the labellum, fragrant.

Widely distributed from central Cape York Peninsula to the Noosa River. Grows on shrubs and trees in rainforest and is often found on slopes and near streams. Easily grown on suitable garden plants in tropical and subtropical regions and hardy in a bush house or cool glasshouse in temperate regions. Best attached to a slab or cork or treefern. Requires shade, warmth, humidity and air movement. Propagate from seed sown under sterile conditions.

P. tridentata is similar but the plants often hang free from the host and have a shorter, broader labellum.

Plectorrhiza erecta (Fitzg.) Dockrill
(upright, erect)
NSW (Lord Howe Island)
Epiphyte   Oct–Dec
Epiphyte forming straggly clumps of erect stems; **roots** numerous, coarse, wiry; **stems** 20–45 cm long, erect, branching from the base; **leaves** numerous, 2–4.5 cm × 1.2–2 cm, oblong to elliptical, bright green, fleshy, apex thickened and deflexed; **racemes** 2–4 cm long, thick, stiffly spreading, bearing 2–5 flowers; **flowers** about 0.8 cm across, greenish-yellow exterior, reddish-brown interior, segments remaining cupped.

P. erecta is endemic on Lord Howe Island and is found growing at the bases of low shrubs in coastal scrub. Plants are easy to grow but are rarely encountered in cultivation. Can be grown in a pot of coarse mixture or on a slab of weathered hardwood or cork. Requires bright filtered light, warmth, humidity and plenty of air movement. Propagate from seed sown under sterile conditions.

Plectorrhiza tridentata (Lindl.) Dockrill
(with three teeth)
Qld, NSW, Vic   Tangle Orchid; Tangle Root
Epiphyte   Sept–Jan
Epiphytic orchid forming small to large, straggly clumps; **roots** coarse, wiry, often contorted; **stems** 5–30 cm long, pendulous, often suspended by one to a few roots; **leaves** 5–20, 6–10 cm × 1.2–1.5 cm, narrowly ovate-lanceolate, dark green, spreading, sharply pointed; **racemes** 6–12 cm long, slender, bearing 3–15 flowers; **flowers** about 0.8 cm across, green to brown with a prominent white labellum, fragrant.

This orchid is widely distributed from north-eastern Qld to eastern Victoria. It grows in moist sheltered habitats. Clumps frequently dangle in the air held to the host by one or a few roots, or alternatively, the clumps form into a tangled mass. Very easily grown if provided with shade, humidity, warmth and air movement. Best on a slender slab of treefern, cork or on a piece of branch. Can also be tied to suitable garden plants. Propagate from seed sown under sterile conditions.

P. brevilabris is similar but the plants are firmly attached to the host and have a longer, narrow labellum.

PLECTRACHNE Henrard = *TRIODIA* R. Br.

PLECTRANTHUS L'Her.
(from the Greek *plectron*, cock's spur; *anthos*, flower; referring to the spur on the corolla of the first-named species)
Lamiaceae   Plectranthus; Cockspurs
Herbs and semi-woody, often somewhat succulent **shrubs; branches** spreading to ascending, sometimes rooting at nodes; **leaves** opposite, margins usually toothed or shallowly lobed, often aromtic, often fleshy; **inflorescence** terminal or upper axillary racemes or cymes; **flowers** tubular; **corolla** 2-lipped, straight to curved, longer than calyx, swollen or spurred at base, upper lip entire, lower lip 3- or 4-lobed; **calyx** 2-lipped, lower lip 4-lobed; **fruit** a small nut.

A wide-ranging, mainly tropical genus of about 250 species. They occur in Africa, Asia, Pacific Islands and Australia where there are about 35 natives and a few naturalised species.

The main concentration of this genus is in eastern Qld and all but 4 species are found there. A number of species occur in NSW and one in Vic. *P. intraterraneus* is the only species from arid and semi-arid regions and it occurs in SA, WA and NT where it grows in moist sites of gorges and near streams.

Australian *Plectranthus* occur from near sea-level to about 1200 m altitude. They are mainly found in or near rainforest although some species grow in open forest and woodland. It is uncommon for them to occur in heathland or desert areas except for *P. intraterraneus*. Generally they inhabit rocky sites such as outcrops and on cliffs. Plants grow in crevices and depressions often in accumulated well-decomposed organic matter which is ideal for successful development. Semi-shaded or shaded sites are preferred by plants, but some species such as *P. pulchellus* grow in exposed, often sunny sites.

Most species are semi-woody shrubs with the upper parts of stems somewhat succulent. A limited number of species have a tuberous base which the stems may die back to after flowering is finished, eg. *P. arenicolus*, *P. glabriflorus* and *P. parviflorus*.

Branches and stems are spreading to erect and sometimes can become straggly but most species respond well to hard pruning which promotes new growth very quickly.

The leaves are soft and often fleshy. They are also hairy to varying degrees and some species such as *P. argentatus*, *P. minutus*, *P. nitidus*, *P. omissus* and *P. torrenticolus* appear silvery from a distance. These species are very decorative in massed plantings. Generally the leaves have a green or silvery undersurface but certain species are purplish underneath eg. *P. gratus*, *P. nitidus* and *P. spectabilis*. The young stems mainly are green to silvery but *P. nitidus* may have deep reddish stems and *P. argentatus* can be pinkish.

Many species have aromatic foliage with *P. foetidus* and *P. graveolens* having among the most strongly scented leaves. Some species such as *P. alloplectus*, *P. graniticola* and *P. pulchellus* do not have aromatic foliage.

The flowers are small and delicate but are often produced in profusion on stems to 50 cm or more in length. Flower colour ranges from white to purple including lilac, mauve or pale blue flowers. In some cases, the calyx is purple (eg. *P. foetidus*) which adds more colour to the flowering stems. The flowers are arranged in opposite, half-wheel-like clusters on each side in succession up the stem, with some clusters sparsely arranged while others are very crowded (eg. *P. spectabilis*).

In Australia, not many of the native species are popular in cultivation, mainly because they are rarely offered by nurseries, but this situation should improve as gardeners and designers become better acquainted with the native species. They have potential for use in rockeries, borders and containers, and will be valuable in general planting. They have been used successfully in cultivation on dry rock walls which is a very logical application for them because in nature they frequently occur on rocks and cliff-faces.

There is little evidence of *Plectranthus* being used in Australia for medicinal purposes but it is recorded that crushed leaves and branches of *P. congestus* were soaked in water and the fluid was drunk for the treatment of internal complaints.

Plectranthus species are insect-pollinated and many insects visit flowering plants. They may be important as hosts to native animals is apparently this is not known. Butterflies will visit flowers of some species.

The basic requirement for cultivation is well-drained acidic soil. The soil should contain a reasonable quantity of well-composted organic matter to help retain moisture for most of the year. A sunny or semi-shaded site is suitable. Usually plants develop better growth and colour in full sun. Although many species will tolerate full shade most of them have a tendency to become open and leggy which can make them unattractive (see pruning, below).

The cold tolerance of many species is as yet unknown but a number are hardy to light-to-moderate frosts including *P. argentatus*, *P. graveolens*, *P. nitidus*, *P. parviflorus* and *P. suaveolens*.

Most species respond well to light applications of slow-release fertiliser. During extended dry periods it may be necessary to provide supplementary watering as plants may wilt if their roots do not have access to moisture.

Pruning is an important aspect of cultivating *Plectranthus*. Young plants should be pruned hard at planting in order to promote a bushy framework of branches. Tip pruning as plants develop is also beneficial. Species such as *P. argentatus*, *P. parviflorus* and *P. suaveolens* respond very well to hard pruning in late autumn or early spring. This will rejuvenate plants and promote an excellent foliage and flower display over summer months.

Pests do not cause serious problems to most species in this genus. Slugs and snails often attack young plants but rarely are troublesome to mature plants. Caterpillars may chew leaves but are rarely an important pest.

Information on diseases which attack *Plectranthus* species is scant. Sometimes an odd branch may wilt and never recover its vigour. This may be caused by an undetermined fungus. It is best to remove the affected stem as soon as possible by cutting above a node below the junction of the affected branch and healthy branch.

Propagation is from seed, or cuttings, or by division of layered stems. Seed is rarely offered commercially. Ripe seed is shed very quickly and plants must be observed regularly if seed is to be collected. The seed is ripe once the fruits become barely brown. Little is known about how long the seed remains viable and best results are usually achieved with fresh seed. The seed does not require pre-sowing treatment. Germination generally begins between 20–90 days after sowing.

The most commonly used method of propagation is from stem cuttings because they produce roots very quickly. Discarded prunings often produce roots if left on the soil surface. Best results are usually obtained from semi-hardwood. Tip cuttings of soft growth will strike readily if high humidity levels are maintained. Application of rooting hormones is not usually beneficial in promoting quicker rooting. Propagation using layered stems may produce larger plants more quickly than using conventional stem cuttings. It is common for the short broken branches of some species (eg. *P. parviflorus* and *P. suaveolens*) to initiate roots in situ when they drop to the ground during autumn and winter, if the weather is cool and soil conditions are moist.

Plectranthus actites P. I. Forst.
(a watcher; referring to it growing high above the surrounding area)
Qld
0.2–0.6 m × 0.8–2 m   Aug–Sept; also sporadic
Dwarf, aromatic and semi-woody **shrub** with hairy young growth; **branches** rounded, usually spreading and decumbent; **branchlets** rounded, succulent; **leaves** 1.8–5 cm × 1–3.2 cm, broadly ovate, mainly spreading, succulent, green and hairy above, silvery-green and hairy below, venation prominent, margins toothed, strongly aromatic when crushed, apex blunt; **flowers** to about 1.1 cm long, purple, in erect, solitary, terminal, spike-like cymes to about 10 cm long, conspicuous.

A recently described (1994) unusual species which is restricted to an isolated peak in the Leichhardt District where it grows as a lithophyte on the cliffs. It warrants cultivation because of its decumbent growth habit and very succulent foliage. Probably best suited to subtropical regions but may adapt to temperate areas. Plants require excellently drained soils with a moisture-retentive mulch in a sunny or semi-shaded aspect. They tolerate light frosts. May be worth trying in hanging baskets. Propagate from seed or from cuttings of firm young growth.

P. graveolens has similarities but is erect with 4-sided stems and widely spaced leaves.

Plectranthus alloplectus S. T. Blake
(different spur)
Qld, NSW
0.3–0.8 m × 0.5–1.5 m   Jan–Sept; also sporadic
Dwarf, non-aromatic, semi-woody **shrub** with whitish-hairy young growth; **branches** few, spreading to erect; **branchlets** hairy; **leaves** 3–9.5 cm × 1–3.5 cm, narrowly ovate, spreading to ascending, tapering to base, sage-green and hairy above, densely greyish-hairy below, margins crenulate, apex pointed or

Plectranthus amoenus   D. L. Jones

blunt; **flowers** to 1.2 cm long, violet-blue to purplish, in 1–3-branched racemes to about 25 cm long, can be profuse, moderately conspicuous.

An attractive species which occurs in the southern Moreton District in Qld and the North Coast District, NSW. It grows on cliffs and ledges. Rarely cultivated, it should adapt to subtropical and temperate areas. Needs excellent drainage and a semi-shaded site. Organic mulches are beneficial. Propagate from seed or from cuttings of firm young growth.

Plectranthus amicorum S. T. Blake
(of the friends; in recognition of those who helped the author)
Qld
1–1.5 m × 1–2 m   Dec–Aug; also sporadic
Small, sweetly aromatic, semi-woody **shrub** with hoary young growth; **branches** few, pale brown; **branchlets** purplish when young, hairy; **leaves** 2.5–10 cm × 1.5–5 cm, narrowly ovate to broadly ovate, tapering to base with long petioles, greyish-green to sage-green above, whitish-hairy below, margins bluntly toothed, apex blunt; **flowers** to about 1.2 cm long, blue to blue-purple with paler tube, usually in solitary racemes to about 12 cm long; **calyx** often maroon; **corolla lobes** faintly fringed.

P. amicorum has sweetly fragrant foliage. It occurs in gravelly soils in the Cook and North Kennedy Districts where it grows at about 660 m altitude. Plants are suitable for subtropical and temperate regions. Should adapt to a range of freely draining soils which do not dry out too readily. A semi-shaded site is best. Pruning of young plants promotes bushier growth and pruning moderately hard every couple of years may be required. Hardy to light frosts. Propagate from seed or from cuttings of firm young growth.

Plectranthus amoenus P. I. Forst.
(pleasant, delightful)
Qld   Plectranthus
0.3–0.6 m × 0.5–1.2 m   throughout the year
Dwarf, sweetly aromatic, semi-woody **shrub** with hairy young growth; **branches** mainly erect, 4-sided, rarely crowded, greenish; **leaves** 1.2–12 cm × 1–6 cm, ovate, base broad with prominent petiole, slightly reflexed to ascending, green and densely hairy above, silvery-green and densely hairy below, margins with rounded teeth, venation prominent, apex blunt to pointed; **flowers** to about 1.5 cm long, lilac-blue, decurved, in 1–3-branched racemes to about 20 cm long, often profuse and moderately conspicuous.

This recently described (1997) species has sweetly scented foliage. It has a restricted distribution near Herberton in the Cook District. Plants are found growing on rocky sites. This species is cultivated to a very limited degree and grows best in well-drained acidic soils with a semi-shaded aspect. Plants are hardy to light frosts and respond well to regular pruning which promotes more compact growth. Suitable for general planting, borders, rockeries and for large containers. Propagation is from seed or cuttings.

P. graveolens is allied but differs in having strongly scented leaves with pointed teeth and smaller flowers.

Plectranthus apreptus S. T. Blake
(undistinguished)
Qld, NSW
0.3–1.5 m × 0.6–2 m   Feb–May
Dwarf to small, non-aromatic, semi-woody **shrub** with glandular, hairy young growth; **branches** ascending to erect, few; **branchlets** sparsely hairy; **leaves** 4–13 cm × 2–7 cm, ovate to nearly circular, base broad with long petiole, somewhat glossy, green and faintly hairy above, paler and hairy below, margins toothed, apex blunt; **flowers** to 1.3 cm long, blue to mauve, in 1–3-branched racemes to about 25 cm long, can be profuse and moderately conspicuous.

This species has a disjunct distribution in the Cook District, Qld, and in the North Coast District, NSW. It is often found on the banks of waterways and edges of rainforest. Cultivated to a very limited degree, it needs a sheltered site and is probably best suited to tropical and subtropical areas. Tip prune plants from an early age to promote bushy growth. Propagate from seed or from cuttings of firm young growth.

P. apreptus is known to hybridise with *P. foetidus* in the wild.

P. dumicola has similarities but it is a woody shrub of about 1 m tall with aromatic foliage. It is a somewhat nondescript species which occurs in the semi-deciduous vine forests of the Cape York Peninsula.

Plectranthus apricus P. I. Forst.
(sunny; referring to natural habitat)
Qld
0.8–1.5 m × 1–2 m   Dec–May; also sporadic
Dwarf to small, faintly sweet-scented, semi-woody **shrub** with hairy young growth; **branches** many, spreading to erect, 4-sided; **branchlets** faintly hairy; **leaves** 1.5–7 cm × 0.8–3.5 cm, ovate to broadly ovate, base broad with prominent petiole, spreading to

ascending, green and hairy above, silvery-green and hairy below, margins short-toothed, venation prominent, apex pointed; **flowers** to about 1.2 cm long, pale lilac with cream base, in somewhat open 1–3-branched racemes to about 20 cm long, rarely highly conspicuous.

This species was described in 1994 and is known only from a small area in the southern Cape York Peninsula where it grows in open sunny sites in shallow soils on sandstone. Suitable for seasonally dry tropical and subtropical regions. Should adapt to a range of freely draining acidic soils with a sunny or semi-haded aspect. Propagate from seed or from cuttings of firm young growth.

P. apricus has been confused with *P. parviflorus*, which is a much smaller species. *P. gratus* has similarities but is a small species with strongly scented leaves.

Plectranthus arenicolus P. I. Forst.
(inhabiting a sandy place)
Qld   Plectranthus
0.2–0.3 m × 0.3–0.6 m   May–Aug; also sporadic

Dwarf, semi-woody, faintly aromatic **shrub** with densely hairy young growth; **stems** ascending to erect, with tuberous base; **branches** mainly erect, hairy; **leaves** 2.5–4.5 cm × 1.8–3 cm, narrowly ovate to ovate, base broad, tapering to prominent petiole, spreading to ascending, dull green and hairy above, paler below, often with purplish toning, margins with short broad

Plectranthus argentatus   W. R. Elliot

teeth, venation moderately prominent, apex bluntly pointed; **flowers** to about 1.2 cm long, deep blue, in somewhat open 1– 3-branched racemes to about 25 cm long, moderately conspicuous.

P. arenicolus is a distinctive species which was described in 1991. It is considered rare in nature and is known only from a limited area of Cape York Peninsula where it grows on sandstone cliffs. It has appealing deep blue flowers. Plants need excellent drainage and a warm to hot site. Best suited to seasonally dry tropical or subtropical regions. A light layer of organic mulch is likely to be beneficial. Worth trying as a container plant if pruned regularly. Propagate from seed or from cuttings of firm young growth.

P. gratus is similar but has smaller leaves and lacks a tuberous base.

Plectranthus cremnus   D. L. Jones

Plectranthus apricus × .5, flower × 1.5

Plectranthus argentatus S. T. Blake
(silvered)
Qld, NSW   Silver Plectranthus
0.6–1.5 m × 1–2 m   Dec–May; also sporadic

Dwarf to small, virtually non-aromatic, semi-woody **shrub** with silvery-hairy young growth; **stems** spreading to erect, 4-sided; **branchlets** silvery-hairy; **leaves** 3.5–12 cm × 2–6 cm, ovate to broadly ovate, base broad with long petiole, reflexed to ascending, silvery green and velvety above and below, margins toothed, venation prominently raised below, apex blunt; **flowers** to about 1.1 cm long, whitish, nearly straight, in 1- to 3-branched racemes to about 30 cm long, often profuse and moderately conspicuous.

A very handsome species from the border mountains in south-eastern Qld and north-eastern NSW. It grows in well-drained soils which often have a covering layer of humus. Plants adapt well to cultivation in subtropical and temperate areas. Best grown in a semi-shaded position but will tolerate plenty of sunshine. Prefers moist soils but withstands dry periods. Plants respond well to hard pruning which is often needed every 2–3 years to rejuvenate plants. A selection 'Green Silver' was marketed in the 1970s. Propagate from seed or from cuttings of firm young growth.

P. argentatus hybridises readily and a number of hybrids have originated in the wild and in gardens. Some of these are cultivated. The other parents include *P. graveolens, P. parviflorus* and *P. spectabilis.* Some plants marketed by nurseries in south-eastern Australia as *P. argentatus* are referable to *P. nitidus* which differs in having white-and-blue flowers.

Plectranthus australis R. Br. var **graveolens** (R. Br.) Domin =
P. graveolens R. Br.

Plectranthus blakei P. I. Forst.
(after Stanley T. Blake, 20th-century Australian botanist)
Qld   Plectranthus
0.3–0.5 m × 0.5–1 m   May–Dec; also sporadic

Dwarf, semi-woody, sweetly aromatic **shrub** with greyish-hairy young growth; **stems** ascending to erect; **branchlets** silvery-hairy; **leaves** 2.5–9 cm × 1.2–4 cm, ovate to somewhat triangular, base broad, tapering to long petiole, reflexed to ascending, dull green and hairy above, paler and hairy below, margins toothed, venation prominent, apex bluntly pointed; **flowers** to 1.6 cm long, lilac-blue, in 1–3-branched racemes to about 30 cm long, can be profuse and very conspicuous.

A very appealing, recently described (1994) species which is confined to the Blackdown Tableland in the Leichhardt District. It grows in humus and peaty sandy loam among or on rocks at 800–900 m altitude. Cultivated to a limited degree, it warrants wider recognition in subtropical and temperate regions. Soils need to drain freely and a semi-shaded site is preferred. The addition of organic mulch is recommended. May tolerate moderate frosts. Worth trying in containers and as a bedding plant. Propagate from seed or from cuttings of firm young growth.

P. apreptus is similar but has leaves with a glossy upper surface. *P. gratus* has many more flowers per whorl, while *P. arenicola* has shorter flowers.

Plectranthus congestus R. Br.
(congested, crowded)
Qld, WA   Kai-kai
0.5–2.5 m × 0.6–1.5 m   March–July; also sporadic

Dwarf to medium, lightly aromatic, perennial **herb** or semi-woody **shrub** with hairy young growth; **stems** ascending to erect, often purplish; **branches** hairy; **leaves** 3–12.5 cm × 1.3–6 cm, narrowly ovate to broadly ovate or somewhat triangular, base blunt, tapered to long petiole, reflexed to ascending, pale green and hairy above, grey-green and hairy below, margins crenately toothed, venation prominent, apex pointed or blunt; **flowers** to about 1.1 cm long, bluish-violet, often with white markings, in racemes to about 50 cm long, often forming leafy panicles, usually profuse and conspicuous.

A wide-ranging tropical species which is found in Timor, New Guinea and extends to northern WA and north-eastern Qld, where it occurs in the Cook and North Kennedy Districts. Cultivated to a limited degree, it should do best in seasonally dry tropical and subtropical regions. It is a vigorous, attractive species and will need regular pruning to maintain a bushy habit. Plants need good drainage with a semi-shaded site. Propagate from seed or from cuttings of firm young growth. In Queensland, Aborigines made a mixture from soaking crushed leaves and branches in water for the treatment of internal problems.

P. parviflorus is similar but less vigorous, not as floriferous and its leaves have less teeth.

Plectranthus cremnus B. J. Conn
(on coastal rocky cliffs)
NSW
prostrate–2 m × 1–1.5 m   throughout the year

Dwarf, aromatic, perennial **herb** with hairy young growth; **branches** prostrate to decumbent, covered in long hairs, base terete, upper parts 4-sided, sometimes self-layering; **leaves** 5–6.5 cm × 4.5–5 cm, broadly ovate to nearly semi-circular, with long distinct petiole, moderately crowded, reflexed to ascending, green, hairy, margins broadly toothed, venation prominent, apex blunt to rounded, pleasant geranium-like fragrance; **flowers** to about 0.6 cm long, white tube and blue-purple lobes, in spike-like racemes, often profuse and conspicuous.

This species inhabits rocky headlands of the North Coast District where it grows in crevices with shallow sandy soils. It has been cultivated as *P. graveolens* 'Headland Form'. This vigorous groundcovering plectranthus deserves to be better known. It requires moderately well-drained soils with a sunny or semi-shaded aspect. Responds well to pruning. Has potential for borders, rockeries, large pots and hanging baskets. Propagate from seed or from cuttings of firm young growth.

P. graveolens has some similarities but has strongly fragrant leaves with more teeth.

Plectranthus cyanophyllus P. I. Forst.
(bluish leaves)
Qld
0.3–0.6 m × 0.6–1.5 m   Sept–Dec; also sporadic

Dwarf, faintly sweetly aromatic, semi-woody **shrub** with hairy young growth; **branches** spreading to erect, 4-sided, hairy; **leaves** 1.3–5 cm × 0.8–3.2 cm, broadly ovate, broad base tapering to distinct petiole, reflexed to ascending, silvery-hairy, fleshy, venation prominent, margins wavy, apex blunt to rounded; **flowers** to about 1 cm long, pale purple, in racemes to about 10 cm long, 1–3-branched, can be profuse and conspicuous.

A very handsome and distinctive species. It was recently described (1994) and is known only from one mountain in the North Kennedy District where it grows on rocky outcrops in open woodland. It has had very limited cultivation but with lovely silvery-blue foliage it deserves to be more common. Needs good drainage and a semi-shaded site. Tolerates limited periods of dryness. Best suited to subtropical and warm temperate regions. Frost tolerance is not known. Pruning should promote dense foliage growth. Propagate from seed or from cuttings of firm young growth.

Plectranthus diversus S. T. Blake
(turned different ways)

Qld Plectranthus
0.5–1.5 m × 1–1.5 m throughout the year

Dwarf to small, aromatic, semi-woody **shrub** with hairs which spread in different directions; **branches** spreading to erect, thickish, with short internodes, densely glandular-hairy; **leaves** 3.8–11 cm × 1.7–6.5 cm, ovate to broadly ovate or somewhat triangular, with prominent petiole, reflexed to ascending, greyish-hairy above, greyish- to rusty-hairy below, margins with rounded teeth, venation prominent, apex blunt, with strong sweet fragrance; **flowers** to about 1.5 cm long, purple to violet, usually with paler lower lip, in 1–5-branched racemes to about 30 cm long, can be profuse and moderately conspicuous.

This species has sweetly fragrant silvery foliage and colourful flowers which make it worthy of wider cultiva-tion. It occurs in the Cook, North Kennedy and South Kennedy Districts. It often grows in exposed rocky sites which contain basalt, granite, limestone or sandstone. Adapts well to cultivation in subtropical regions and is worth trying in temperate areas. Plants have grown well on dry rock walls. Regular pruning promotes compactness. Frost tolerance is unknown. Propagate from seed or from cuttings of firm young growth.

P. excelsus has affinity but grows to about 3 m tall and has widely spaced, strongly fragrant leaves.

Plectranthus dumicola P. I. Forst. —
see *P. apreptus* S. T. Blake

Plectranthus excelsus P. I. Forst.
(tall, noble)

Qld Plectranthus
1–3 m × 1–3 m June–July

Small to medium, strongly aromatic, semi-woody **shrub** with hairy young growth; **branches** spreading to erect, 4-sided, densely hairy; **leaves** 2.5–17.5 cm × 2–12 cm, ovate to broadly ovate, base usually cordate, with prominent petiole, reflexed to ascending, fleshy, green and hairy above, paler green below, margins with many rounded teeth, venation prominent, apex

pointed; **flowers** to about 1 cm long, pale purple, bent at right angles near base, in 1- to 7-branched racemes to about 22 cm long, can be profuse and conspicuous.

A recently described (1994) species which is regarded as the tallest Australian member of the genus. It is known only from west of the Iron Range on Cape York Peninsula where it grows in granite rockpiles. It should be best suited to seasonally dry tropical and subtropical regions but may adapt to warm protected sites in other areas. Plants require well-drained acidic soils with a sunny or semi-shaded aspect. Pruning should help to promote compact growth. Propagate from seed or from cuttings of firm young growth.

P. diversus is allied but differs in being much smaller and having closer-spaced, sweetly fragrant leaves.

Plectranthus foetidus Benth.
(bad smelling)

Qld Plectranthus
0.5–1.5 m × 1–3 m May–Oct; also sporadic

Dwarf to small, moderately to strongly aromatic, semi-woody **shrub** with densely hairy young growth; **branches** spreading to ascending, thick, 4-sided, fleshy when young, densely hairy; **leaves** 2.5–15 cm × 1.5–10 cm, broadly to very broadly ovate, sessile or with short petiole, reflexed to ascending, thick, green and hairy above, whitish-hairy below, margins with many short teeth, often tinged purple, venation prominent, apex blunt to pointed; **flowers** to about 1.5 cm long, deep violet to blue, prominently bent near middle, in 1–3-branched (or more) densely arranged racemes, often profuse and moderately conspicuous.

A strong-growing species from the Cook District where it grows on cliff faces and steep rocky slopes. Rarely cultivated, it warrants greater attention for use in subtropical and seasonally dry tropical regions. Also worth trying in warm sheltered sites in temperate areas. Suitable for dry rock walls and general planting. Pruning should promote dense growth. Frost tolerance is not known. Propagate from seed or from cuttings of firm young growth.

P. graveolens is similar but it has longer petioles and more loosely arranged racemes.

Plectranthus glabriflorus P. I. Forst.
(glabrous or smooth flowers)

Qld Plectranthus
0.1–0.3 m × 0.5–1 m throughout the year

Dwarf, non-aromatic perennial **herb** with tuberous base; young growth faintly hairy; **branches** spreading to ascending, not woody, 4-sided; **leaves** 1.5–5 cm × 1–2.6 cm, broadly ovate, base often cordate, with prominent petiole, reflexed to ascending, dark green and glossy above, greenish-purple below with raised veins and scattered hairs, margins usually with 6–10 rounded teeth, apex moderately blunt; **flowers** to about 1 cm long, pale purple, glabrous, more or less straight, in 1–3-branched racemes to about 20 cm long, often profuse and moderately conspicuous.

This interesting and rare species was described in 1994. It is found in the Atherton Tableland of the Cook District where it grows on granite outcrops in open forest. Plants need good drainage and prefer a semi-shaded site. Should grow well in subtropical and temperate areas. Plants tolerate light frosts. Worth trying on dry rock walls as well as in rockeries, gardens and containers. Propagate from seed or from cuttings of firm young growth.

Plectranthus graniticola P. I. Forst.
(growing in soil derived from granite)
Qld
0.3–0.6 m × 0.6–1.5 m   March–May;
                                      possibly sporadic

Dwarf, non-aromatic, semi-woody **shrub** with silvery-hairy young growth; **branches** spreading to erect, with silvery hairs; **leaves** 11.5–11.5 cm × 0.6–5 cm, lanceolate-ovate, with short to long petiole, green with silvery hairs above and below, somewhat fleshy, margins toothed, venation prominent, apex pointed; **flowers** to about 1.3 cm long, dark purple, in 1–3-branched racemes to about 20 cm long, often profuse and moderately conspicuous.

Described in 1992, this species inhabits sunny open sites in granite rock outcrops in a small area of the South Kennedy District in Central Queensland. It warrants cultivation and should succeed in subtropical regions. It is also worth trying in warm protected sites in temperate areas. Plants require well-drained acidic soils with a sunny or semi-shaded aspect. They should tolerate light frosts. Suitable for general planting, rockeries, borders, dry rock walls and containers.

P. argentatus is closely related but has white to very pale bluish-white flowers, while *P. omissus* has lilac-blue flowers and *P. torrenticolus* has lilac flowers.

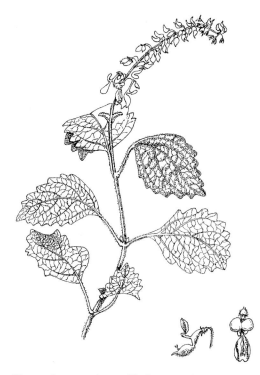

Plectranthus graveolens × .55, flowers × 1

Plectranthus gratus S. T. Blake
(agreeable or pleasant)
Qld   Plectranthus
0.6–1 m × 1–1.5 m   throughout the year

Dwarf, strongly aromatic, semi-woody **shrub** with hairy young growth; **branches** spreading to erect, 4-sided, hairy; **leaves** 3–6.5 cm × 1.5–3.5 cm, ovate to very broadly ovate, base broad, with prominent petiole, reflexed to spreading, olive-green and hairy above, purplish-green and hairy below, margins with short and often rounded teeth, venation prominent, apex bluntly pointed; **flowers** to about 1 cm long, lilac to blue, bent at base, 1–few-branched racemes with 10–20 flowers per whorl, often profuse and moderately conspicuous.

A very attractive and moderately compact species from the Cook District where it grows on mountain slopes in a small area. It is suitable for subtropical regions and may succeed further south if grown in a warm to hot, sheltered site. Requires well-drained acidic soils with a sunny or semi-shaded aspect. Frost tolerance is not known. Worth trying in general planting and on dry rock walls. Propagate from seed or from cuttings of firm young growth.

P. apreptus generally is taller with faintly fragrant leaves and less flowers per whorl.

Plectranthus graveolens R. Br.
(strongly smelling)
Qld, NSW   Plectranthus
0.3–1.3 m × 0.8–2 m   throughout the year

Dwarf to small, strongly aromatic, semi-woody **shrub** with hairy young growth; **branches** spreading to erect, 4-sided, rarely crowded, can be purplish when young; **leaves** 5.5–16.5 cm × 3.5–7.5 cm, elliptic to broadly ovate or nearly circular, base broad with prominent petiole, reflexed to ascending, green to dark green and densely hairy above, paler and densely hairy below, margins toothed, venation prominent, apex blunt to pointed; **flowers** to about 1 cm long, usually violet-blue, decurved, in racemes to about 25 cm long, often profuse and moderately conspicuous.

This species has strongly scented foliage and a wide-ranging distribution. It occurs in the Cook District and extends through eastern Qld to eastern NSW, as far south as Nowra in the South Coast District. Usually found in rocky sites. It is cultivated to a limited degree and grows best in well-drained acidic soils with a semi-shaded aspect. Plants are hardy to light frosts and respond well to regular pruning which promotes more compact growth. Suitable for general planting, borders, rockeries and for large containers. A groundcovering cultivar 'Benelong Frosty Carpet' has been selected from variants and is available in some nurseries. Propagation is from seed or cuttings. Cultivars are grown from cuttings to retain their distinctive characteristics.

It is fairly common in nature for *P. graveolens* to intergrade with *P. parviflorus* and *P. suaveolens*, which makes it difficult to distinguish each species. It is evident that some of these variants are cultivated.

P. australis R. Br. var. *graveolens* (R. Br.) Domin is a synonym.

Plectranthus spectabilis   D. L. Jones

Plectranthus habrophyllus P. I. Forst.
(shaggy leaves)

Qld   Plectranthus
0.3–0.5 m × 0.6–1.5 m   throughout the year

Dwarf, sweetly-scented, semi-woody **shrub** with hairy young growth; **branches** spreading to erect, 4-sided, densely hairy; **leaves** 2.6–6.5 cm × 1.4–3.5 cm, broadly ovate, base broad with prominent petiole, reflexed to spreading, fleshy, green with dense long hairs above, pale green with raised veins and hairy below, margins bluntly toothed, apex pointed; **flowers** to about 0.9 cm long, light purple, slightly curved near base, openly arranged in 1–3-branched racemes to about 16 cm long, can be moderately profuse, rarely highly conspicuous.

A recently described (1994) species with sweetly fragrant foliage. It is found in the Moreton District where it inhabits rock outcrops in the vicinity of vine forest in sites which are often shaded. This endangered species is uncommon in cultivation. It warrants further attention and has potential for use in rockeries, borders and containers. Should adapt to subtropical and temperate regions if grown in freely draining acidic soils with a semi-shaded or shaded aspect. Response to frost is not known. Propagate from seed or from cuttings of firm young growth.

The allied and silvery-foliaged *P. suaveolens* lacks long hairs and has a longer corolla.

Plectranthus intraterraneus S. T. Blake
(inland)

SA, WA, NT   Cockspur Flower
0.4–1 m × 0.5–1.5 m   July–Dec

Dwarf, sweetly fragrant **herb** to semi-woody **shrub** with densely hairy young growth; **branches** spreading to erect, few to many, 4-sided, densely hairy; **leaves** 1.8–5.5 cm × 1–4 cm, ovate to nearly triangular, base broad with short petiole, mainly spreading to ascending, light green and loosely hairy above, paler and hairy below, margins somewhat coarsely toothed, apex bluntly pointed; **flowers** to about 1.2 cm long, usually pale-violet, prominently bent below middle, in branched racemes to about 20 cm long, rarely profuse and very conspicuous.

This is the widest ranging plectranthus. It occurs in mainly inland areas of low rainfall and is disjunct from all other species. Plants are found in rocky, often open sites and usually along watercourses. It is cultivated to a limited degree and is suited to semi-arid and temperate regions. Plants need excellent drainage and a warm to hot aspect for best results. They are hardy to moderately heavy frosts. Pruning promotes compact growth. Propagate from seed or from cuttings of firm young growth.

The allied *P. parviflorus* is smaller and has strongly scented leaves which are very hairy above and less so below, while *P. suaveolens* which has sweetly scented leaves differs with its spreading growth habit and dense hairs.

Plectranthus leiperi P. I. Forst.
(after Glen Leiper, original collector in 1993)

Qld
0.3–0.5 m × 0.8–1.5 m   Feb–July; possibly sporadic

Dwarf, strongly and sweetly aromatic **herb** or semi-woody **shrub** with hairy young growth; **branches** spreading to erect, 4-sided, sparsely hairy; **leaves** 2.3–8 cm × 1.5–5.5 cm, ovate, base broad and flattish with prominent petiole, reflexed to spreading, fleshy, silvery-green and hairy above, pale green and densely hairy below, venation prominent, margins with rounded teeth, apex blunt; **flowers** to about 1.3 cm long, lilac with cream base, curved or 1–3-branched racemes to about 18 cm long, can be profuse and moderately conspicuous.

This species was described in 1994. It occurs in a small area of the Moreton District where it is considered vulnerable due to proposed development. Plants are found in rock flats and outcrops. Warrants cultivation as part of a conservation strategy. Should be suited to subtropical and temperate regions. Needs well-drained acidic soils and a sunny or semi-shaded site. It has potential for use in borders, rockeries, containers and general planting. Frost tolerance is not known. Propagate from seed or from cuttings of firm young growth.

P. suaveolens has affinity but it has silvery foliage and blue to violet flowers.

Plectranthus megadontus P. I. Forst.
(large teeth)

Qld   Plectranthus
0.1–0.3 m × 0.4–1 m   May–Aug; possibly sporadic

Dwarf, faintly aromatic, semi-woody **shrub** with hairy young growth; **branches** spreading to erect, 4-sided, hairy; **leaves** 1.4–8.5 cm × 0.9–3.5 cm, narrowly ovate to somewhat triangular, base broad and flattish, with slender petiole, reflexed to spreading, fleshy, green and sparsely hairy above, pale green and hairy below, margins with prominent deep teeth, venation prominent, apex pointed; **flowers** about 1 cm long, light purple, sharply curved near base, in 1–3-branched racemes to about 20 cm long, can be profuse and moderately conspicuous.

A recently described (1994) species which is known only from the McIlwraith Range on Cape York Peninsula where it grows on granite rock on the margins of rainforest. Best suited to subtropical and seasonally dry tropical regions. Soils will need to drain freely and the site should be semi-shaded. Plants are likely to be affected by frost. It has potential for use in general planting, rockeries, borders and in containers such as patio pots and hanging baskets. Propagate from seed or from cuttings of firm young growth.

P. gratus has similarities but it is a larger shrub with succulent leaves that have broader petioles and smaller teeth.

Plectranthus minutus P. I. Forst.
(minute, very small)
Qld
0.3–0.5 m × 0.8–1.3 m   Feb–May; possibly sporadic
Dwarf, semi-woody **shrub** with densely hairy young growth; **branches** spreading to erect, reddish when young, with silvery hairs; **leaves** 1–2.5 cm × 0.4–1 cm, narrowly ovate to obovate, tapering to a short slender petiole, reflexed to spreading, fleshy, green and shiny with velvety silvery hairs above, similar below, margins toothed in upper half, venation mainly obscured by hairs, apex blunt to pointed; **flowers** to about 1.1 cm long, deep mauve to blue, abruptly curved near base, in racemes with reddish axes to about 20 cm long, very conspicuous.

This distinctive, small-leaved, very attractive yet poorly known species was described in 1994. It is known from only one mountain on the Atherton Tableland where it grows on sandstone. Warrants cultivation because of its beauty and for conservation purposes. Plants will require freely draining acidic soils with a semi-shaded aspect. It should adapt to cultivation in subtropical and temperate areas and may possibly tolerate light frosts. It has potential for use in rockeries, borders, containers and in general planting. Pruning will promote compact growth. Propagate from seed or from cuttings of firm young growth.

Plectranthus mirus S. T. Blake
(quaint, extraordinary)
Qld   Plectranthus
0.6–1.5 m × 1–2 m   May–Nov; also sporadic
Dwarf to small, faintly sweetly aromatic, semi-woody **shrub** with glandular-hairy young growth; **branches** spreading to erect, densely hairy; **leaves** 3.5–11 cm × 3–6.5 cm, ovate to very broadly ovate, base broad with thickish petiole, fleshy, green and densely hairy above, paler and densely hairy below, margins with rounded teeth, venation prominent, apex blunt; **flowers** to about 1.4 cm long, blue to purple, sharply bent at middle, in 1–3-branched racemes to about 20 cm long, often profuse and very conspicuous.

A handsome species from the Cook District where it is found west and north of Cairns in rainforest or growing on granite outcrops. It is cultivated to a limited degree and warrants greater attention. Should adapt to subtropical and seasonally dry tropical regions and may succeed in temperate areas if grown in warm sheltered sites. Plants require excellently drained acidic soils with a semi-shaded aspect. Frost tolerance is not known. Pruning should promote compact growth. It has potential for use in borders and containers and in general planting. Propagate from seed or from cuttings of firm young growth.

Plectranthus nitidus P. I. Forst.
(shining)
Qld, NSW   Silver Plectranthus
0.3–2 m × 0.6 –3 m   Jan–June; also sporadic
Dwarf, non-aromatic, perennial **herb** with silver-hairy young growth; **branches** spreading to erect, fleshy, reddish to purple when young; **leaves** 3.5–8.5 cm × 1.5–3.5 cm, lanceolate-ovate, tapering to slender petiole, reflexed to spreading, fleshy, grey-green and hairy above, purplish and hairy below, margins toothed, venation prominent, apex pointed; **flowers** to about 1 cm long, very pale lilac to lilac sometimes with white, curved near base, in 1–5-branched racemes, can be profuse and conspicuous.

This ornamental species was described in 1992. It is found in the border ranges of south-eastern Qld and north-eastern NSW where it grows in rainforest on rock outcrops. It is becoming popular and warrants cultivation in subtropical and temperate regions. Plants require freely draining acidic soils with a shaded or semi-shaded aspect. Pruning at an early stage should promote compact growth and large plants respond very well to regular hard pruning. Tolerates light to moderate frosts. Excellent for growing in borders, rockeries or containers and may be worth trying indoors. A selection originating from the Numinbah Valley in south-eastern Qld has inadvertently been marketed as *P. argentatus*. It develops into a very ornamental shrub of 1–2 m × 1.5–3 m and has leaves of 5–16.5 cm × 3–8 cm. It grows best in semi-shaded or shaded sites but will tolerate plenty of sunshine if the site is not too dry.

Propagate from seed or from cuttings of firm young growth.

P. apreptus has some similarities but it is generally larger and has glandular hairs, while *P. argentatus* has white flowers.

Plectranthus omissus P. I. Forst.
(neglected)
Qld
0.6–1 m × 0.9–1.5 m   Jan–March; possibly sporadic
Dwarf to small, faintly fragrant, semi-woody **shrub** with silvery-hairy young growth; **branches** spreading to erect, 4-sided, densely covered in short hairs; **leaves** 4.5–12 cm × 3–6 cm, broadly ovate, base broad with long petiole, reflexed to spreading, somewhat fleshy, green with silvery hairs above and below, margins toothed, venation prominent, apex blunt; **flowers** to about 0.9 cm long, pale lilac to lilac, bent at base, in

1–5-branched racemes to about 35 cm long, can be profuse and conspicuous.

A recently described (1992) species which is known only from one location in the Wide Bay District where it grows on south-facing rock outcrops on the margins of vineforest. It deserves to be cultivated as part of a conservation strategy and for its ornamental qualities. Plants should adapt to subtropical and temperate regions in well-drained acidic soils with a shaded or semi-shaded aspect. Frost tolerance is not known. Pruning from an early stage should promote a bushy framework. Propagate from seed or from cuttings of firm young growth.

P. argentatus is allied but differs in having white to very pale bluish-white flowers.

Plectranthus parviflorus Willd.
(small flowers)
Qld, NSW, Vic   Cockspur Flower
0.1–0.8 m × 0.5–1.5 m   throughout the year
Dwarf perennial **herb** or semi-woody **shrub** with a tuberous base and hairy young growth; **branches** spreading to erect, terete, often pinkish when young, with reflexed hairs, sometimes rooting at nodes; **leaves** 2.5–11.5 cm × 2–4 cm, oblong-ovate to somewhat circular, broad base tapering to petiole, reflexed to ascending, fleshy, green and hairy above, paler and less hairy below, venation prominent, margins with rounded teeth, apex blunt to rounded; **flowers** to about 1.1 cm long, pale blue to violet-blue, bent near base, in 1-5-branched racemes to about 25 cm long, often profuse, rarely highly conspicuous.

This variable, wide-ranging species is distributed from north-eastern Qld to Vic. It is generally found in rocky areas such as the gorges of creeks and rivers in coastal forest and margins of rainforest, as well as on exposed mountain peaks at about 1200 m altitude. In some habitats, it receives over 2500 mm of rain while in other areas it receives about 600 mm. Moderately popular in cultivation, it adapts to a range of fairly well-drained acidic soils in sunny or semi-shaded sites. Hardy to moderate frosts and responds well to pruning. Plants are rejuvenated by removing old branches at the tuberous base. It seeds prolifically and often regenerates readily. May be a potential weed. Suited to general planting, borders, rockeries and containers. Flowers of some selections are very profuse, but small and pale in colour, while others are larger, in brighter blues and lilacs, but more sparse. Propagate from seed or from cuttings of firm young growth.

It can be difficult to distinguish this species from *P. graveolens* and *P. suaveolens* because they hybridise with each other readily.

Plectranthus pulchellus P. I. Forst.
(beautiful)
Qld   Plectranthus
0.2–0.4 m × 0.5–1 m   July; possibly at other times
Dwarf, non-aromatic, semi-woody **shrub** with hairy young growth; **branches** spreading to erect, 4-sided, hairy; **leaves** 1.2–10 cm × 0.6–5 cm, narrowly ovate to triangular, tapering to slender petiole, reflexed to ascending, fleshy, green and faintly hairy above, pale green and faintly hairy below, venation prominent, margins with broad teeth, apex pointed; **flowers** to about 1 cm long, light purple, abruptly curved near base, on purple stem, in 1–7-branched racemes, often profuse and conspicuous.

A highly desirable, recently described (1994) species which is known from only one area in the far north Cape York Peninsula where it grows on sandstone cliffs. This robust species warrants cultivation and should be best suited to subtropical and seasonally dry tropical regions. Plants require freely draining acidic soils and plenty of sunshine. It has potential for use in containers, rockeries, borders and in general planting. Propagate from seed or from cuttings of firm young growth.

P. arenicolus is closely allied. It has less flowers per cluster and is not as robust.

Plectranthus spectabilis S. T. Blake
(remarkably spectacular)
Qld
1–2 m × 1–2 m   May–Sept; also sporadic
Small, very faintly fragrant, semi-woody **shrub** with densely hairy young growth; **branches** spreading to erect, 4-sided, densely hairy; **leaves** 4–14 cm × 3.2–7 cm, ovate to somewhat circular, tapering to petiole which is about as long as the blade, reflexed to spreading, fleshy, sage-green and densely hairy above, paler to often purplish and densely hairy below, venation prominent, margins short-toothed, apex blunt to rounded; **flowers** to about 1 cm long, deep blue, usually with paler tube, curved at middle, often crowded, in 1–11-branched racemes to about 20 cm long or more, often profuse and very conspicuous.

This species is rarely cultivated although it has long been regarded as a striking plant. It occurs in the south-eastern Cook District where it grows in rocky sites on margins of rainforest. Should adapt to subtropical and seasonally dry tropical regions and may succeed in cooler areas if grown in protected warm sites. Needs freely draining acidic soils with a semi-shaded aspect. Frost tolerance is not known. Pruning may promote compact growth. It has potential for growing in borders and general planting. Propagate from seed or from cuttings of firm young growth.

P. foetidus is similar but differs in having leaves with more teeth and usually 1–5-branched racemes.

Plectranthus suaveolens S. T. Blake
(sweet-scented)
Qld, NSW
0.4–1 m × 0.6–2 m   throughout the year
Dwarf, sweetly fragrant, semi-woody **shrub** with densely hairy young growth; **branches** spreading to erect, often decumbent, 4-sided, hairy, sometimes rooting at nodes; **leaves** 3.5–12 cm × 2–6 cm, ovate to broadly ovate, base tapering to prominent petiole, reflexed to ascending, somewhat fleshy, green and hairy above, paler and hairy below, venation prominent, margins prominently toothed, apex pointed; **flowers** to 1.2 cm long, blue to violet, often with paler tube, abruptly bent at base, usually in 1–3-branched racemes to about 35 cm long, often profuse and conspicuous.

An attractive species which is cultivated to a limited

degree but deserves to be better known. It occurs in south-eastern Qld and in the North Coast and Northern Tablelands of north-eastern NSW. It often grows on granite outcrops. Adapts well to cultivation in subtropical and temperate regions but may suffer damage from frosts. Grows best in well-drained acidic soils with a semi-shaded aspect but will tolerate a fair amount of sunshine. Plants respond well to pruning and this is best done each year in late winter or early spring to promote new flowering growth. Suitable for general planting, borders and large containers. Propagate from seed or from cuttings of firm young growth.

P. suaveolens is sometimes difficult to distinguish from *P. parviflorus*, which usually is less hairy.

Plectranthus torrenticola P. I. Forst.
(from near rushing streams)
Qld   Plectranthus
0.2–0.4 m × 0.6–1.5 m   throughout the year
Dwarf, faintly fragrant, semi-woody **shrub** with silvery-hairy young growth; **branches** spreading to erect, 4-sided, with silvery hairs; **leaves** 3–8 cm × 1.5–3 cm, narrowly ovate, base tapering to prominent petiole, reflexed to spreading, fleshy, green with silvery hairs above, pale green with silvery hairs below, venation prominent, margins toothed, apex pointed; **flowers** to about 1.2 cm long, light purple, abruptly curved at base, in 1–5-branched racemes to about 20 cm long, often profuse and conspicuous.

The silvery appearance of this recently described (1992) species makes it desirable for cultivation. It is found in rocky outcrops, often near waterways, in the northern Moreton District. Should prove reliable in relatively frost-free subtropical and temperate regions. Needs excellently drained acidic soils and a semi-shaded or slightly sunny site. Pruning should promote compact growth. Suitable for rockeries, borders, containers and general planting. Propagate from seed or from cuttings of firm young growth.

P. argentatus is allied but differs in being larger and having white to very pale bluish-white flowers.

PLECTRONIA L. = *CANTHIUM* Lam.

PLEIOCOCCA F. Muell. =
            ACRONYCHIA J. R. & G. Forst.

PLEIOGYNE Miers
(from the Greek *pleion*, more; *gyne*, female, woman; referring to numerous stigmas of the flowers)
Menispermaceae
A monotypic genus endemic on the coast of eastern Qld.

Pleiogyne australis Benth.
(southern)
Qld
Climber   Aug–Dec
Twining, tall, deciduous **climber** with hairy young growth; **stems** woody, can be pendent; **juvenile leaves** lobed or toothed; **mature leaves** 3–11 cm × 2–5.5 cm, lanceolate to ovate, alternate, spreading to ascending, pale-to-mid green, initially very hairy above and below, becoming nearly glabrous and shiny above, reticulate

venation prominent, apex pointed; **flowers** unisexual, about 0.5 cm across, greenish, hairy, in racemes to about 8 cm long; **sepals** 6; **petals** 6; **fruit** about 1 cm across, rounded to kidney-shaped, hairy, reddish becoming black when ripe in Aug–Nov.

A poorly known species which occurs in somewhat open sites in coastal rainforests where it often grows among rocks. Suited to tropical and subtropical regions in well-drained acidic soils. Plants appreciate sunshine but will also grow in shaded areas. They need good drainage.

Propagate from fresh seed and cuttings are worth trying.

PLEIOGYNIUM Engl.
(from the Greek *pleion*, more; *gyne* female; referring to the large number of carpels in the ovary)
Anacardiaceae
Dioecious **tree**; **leaves** alternate, pinnate, with a terminal leaflet; **leaflets** entire, opposite or nearly so; **inflorescences** axillary, the male inflorescence a panicle, the female inflorescence a raceme; **flowers** 4–6-merous; **fruit** a drupe with 5–12 seeds.

A small genus of 1 or 2 species distributed from the Philippines to New Guinea and Australia. It also occurs on some Pacific islands. The species that extends to Australia is cultivated as an ornamental and for its fruit. Male and female trees are necessary for fruit production. Propagation is from seed which has a limited period of viability and should be sown while fresh.

Pleiogynium cerasiferum (F. Muell.) R. Parker =
           P. timorense (DC.) Leenh.

Pleiogynium solandri (Benth.) Engl. =
           P. timorense (DC.) Leenh.

Pleiogynium timorense (DC.) Leenh.
(from Timor)
Qld   Burdekin Plum
10–20 m × 8–15 m   Aug–Oct
Small to medium **tree** with a widely spreading crown; young growth usually hairy; **leaves** 10–18 cm long, pinnate; **leaflets** 5–11, 4–10 cm × 2–6 cm, lanceolate, elliptic or ovate, often unequal, dark green, glabrous to hairy; **male inflorescences** are panicles to 15 cm long; **female inflorescences** are racemes to 5 cm long; **flowers** about 0.8 cm across, whitish; **drupes** 2.5–3.5 cm across, round and flattened, dark purple, ripe Feb–May.

Occurs in Qld where it is distributed from Cape York to Maryborough; also occurs in the Philippines, Indonesia and New Guinea. Grows in open forest, coastal scrubs and drier rainforests. Plants have a deciduous period during the dry season. The fruit has become a popular item of bush tucker. It has a thin layer of flesh which is edible when soft. In most plants, the flesh is purple but sometimes it is white. The leaves were used by the Aborigines as a fish poison. This tough tree is grown on a limited scale. Suitable for tropical, subtropical and warm temperate regions. Requires well-drained soil in a sunny position. Tolerates light frosts. Propagate from fresh seed.

Pleiogynium timorense   D. L. Jones

Pleiogynium solandri (Benth.) Engl. and *P. cerasiferum* (F. Muell.) R. Parker are synonyms.

PLEOMELE Salisb.
(from the Greek *pleos*, full; *meli*, honey; referring to an abundance of nectar produced by one species)
Dracaenaceae

Woody monocotyledonous **shrubs**; **leaves** simple, elongate, appearing whorled, sheathing at the base; **inflorescence** a raceme or panicle; **flowers** bisexual; **sepals** and **petals** united at the base into a tube; **stamens** 6; **fruit** a berry.

A large genus of about 140 species with a single species endemic in northern Australia. Previously included in the genus *Dracaena* Vand. ex L. Many exotic species, hybrids and cultivars are grown in tropical and subtropical gardens for their colourful leaves. They also make handsome pot plants and are excellent for decoration indoors or in glasshouses. Propagation is from basal suckers, stem cuttings or seed. The seed has a limited period of viability and is best sown fresh.

Pleomele angustifolia (Medik.) N. E. Br.
(with narrow leaves)
Qld, NT   Native Dracaena
3–8 m × 1–2 m   Sept–March

Tall, slender, suckering **shrub**, often sprawling or climbing; **stems** slender, sparsely branched, pithy; **leaves** 12–40 cm × 2–3 cm, linear, spreading, dark green, sometimes with yellowish bands near the margins, long-pointed, crowded near the apex of a stem; **panicles** 15–40 cm long, terminal, much-branched; **flowers** 1.5–3 cm long, white, fragrant, tubular at the base, lobes linear, spreading or recurved; **berry** 1–1.5 cm across, brownish to red, pulpy, containing 1–3 large seeds, ripe Aug–Nov.

Endemic in Australia where it is distributed in the Top End, NT, Torres Strait Islands and in Qld from Cape York to near Cairns. It grows in sheltered rainforests often near the sea. The stems tend to be weak and thread their way through surrounding shrubs and trees which act as a support. Plants are easily grown

but are not commonly cultivated. They make an excellent indoor plant especially if placed in a brightly lit position. They are very sensitive to cold, especially frosts, and can be grown as a glasshouse plant in temperate regions. An interesting addition to gardens in the tropics and subtropics. Requires a shady position in well-drained soil and responds to mulches, watering and fertilisers.

The cultivar, 'Honoriae', differs in having cream to ivory-yellow marginal bands on the leaflets. It was collected on Stephens Island, Torres Strait, but does not appear to have become established in cultivation.

Propagate from fresh seed, or from cuttings, which strike readily.

Dracaena angustifolia (Modikus) Roxb. is a synonym.

PLESIONEURON (Holttum) Holttum
(from the Greek *plesios*, near; *neuron*, nerve or small vein; referring to the sori being near veins)
Thelypteridaceae

Terrestrial **ferns**; **rootstock** erect, rarely creeping, apex scaly; **fronds** pinnate, lowest pinnae not reduced in size; **pinnae** deeply lobed, narrowed at the base; **sori** with or without indusia.

A genus of about 39 species distributed from Malaysia to the Philippines, Polynesia and New Guinea with 1 species in Australia. It is a handsome fern mainly encountered in the collections of enthusiasts. Propagation is from spores, which germinate readily when fresh.

Plesioneuron tuberculatum (Ces.) Holttum
(with small tubercles)
Qld
1–1.8 m tall

Terrestrial **fern** forming small to medium-sized clumps; **rootstock** erect, short but thick, scaly; **fronds** 1–1.8 m × 35–50 cm, erect in a tussock, with a slender basal stalk to 75 cm long, pinnate to pinnatifid, dark green and glossy above, paler beneath; **pinnae** to 25 cm long, widely spreading, with drawn-out tips; **sori** round, black, with a small indusium which is soon shed.

Occurs in New Guinea, the Moluccas and in Qld where it is rare and restricted to rainforests of the Atherton Tableland. A very handsome fern, excellent for tropical, subtropical and warm temperate regions. May be grown in a large container or a sheltered position in the ground. Requires well-drained but moist soil, shade or filtered sun and humidity. Responds to watering, mulches and light fertiliser applications. Propagate from spores, which germinate readily when fresh.

PLEURANDROPSIS Baill. = *ASTEROLASIA* F. Muell.

PLEUROCARPAEA Benth.
(from the Greek *pleura*, rib; *carpos*, fruit; referring to the ribbed achenes)
Asteraceae (alt. *Compositae*)

Dwarf to small **shrubs** or perennial **herbs** with annual stems; **leaves** alternate, sessile or short-stalked, entire or toothed; **inflorescence** terminal, solitary or in loose corymbs; **involucral bracts** few, in 2 rows;

Pleurocarpaea denticulata

Pleurocarpaea denticulata × .8

florets bisexual, tubular, white, pink or purple, 5-lobed; **achenes** prominently ribbed.

An endemic tropical genus of 2 species distributed in far-northern Australia. These species are evidently untried in cultivation.

Pleurocarpaea denticulata Benth.
(minutely toothed)
Qld, WA, NT
0.3–0.8 m × 0.4–1 m   Oct–Feb

Dwarf perennial **herb** with woody rootstock; young growth sometimes hairy; **stems** semi-woody, annual, spreading to erect, angular or grooved, glabrous or becoming glabrous; **leaves** 2–10 cm × 0.5–5 cm, narrowly elliptic to ovate, more or less sessile, spreading to ascending, becoming glabrous above, pale green, margins irregularly toothed or entire, apex pointed to rounded; **flowerheads** 1.5–2.5 cm across, terminal, solitary, usually scattered; **flowers** to about 1 cm long, tubular with spreading lobes, white pink or purplish; **achenes** about 0.4 cm long, more or less 10-ribbed.

An interesting species which reshoots from its woody rootstock during Sept–Dec as the wet season approaches. Then it flowers and fruits before the end of the wet season in March–April. Plants are found in a range of soils which are subject to waterlogging. Suitable for subtropical and seasonally dry tropical regions and are worthy of attention for cultivation. Needs plenty of sunshine. Propagate from seed and possibly from cuttings of young emergent shoots.

P. fasciculata Dunlop is a recently described (1991) shrubby species of about 1 m tall. It occurs in Arnhem Land where it is regarded as rare. Flowerheads are white and about 0.8 cm across. They bloom during Jan–June. It requires similar conditions to *P. denticulata*.

Pleurocarpaea fasciculata Dunlop —
                              see *P. denticulata* Benth.

PLEUROMANES (C. Presl) C. Presl
(from the Greek *pleura*, rib; *manes*, soft; referring to the thickened hairy bands in the fronds)
Hymenophyllaceae

Tiny lithophytic or epiphytic **ferns** forming sparse clumps or strands; **rootstock** creeping, slender and thread-like; **fronds** 2–3 times pinnate, membranous, pendulous, grey to glaucous, with thickened hairy bands; **indusium** cup-shaped.

A small genus of 3 species distributed from Sri Lanka to Polynesia and New Guinea with 1 species in Queensland. It is very rarely seen in cultivation because it has very specialised requirements (see *P. pallidum*) and is generally difficult to grow. Propagation is possibly from spores or by division.

Pleuromanes pallidum (Blume) C. Presl
(pallid, pale)
Qld   Floury Filmy Fern
5–25 cm long

Tiny epiphytic or lithophytic **fern** often growing in strands; **rootstock** long-creeping, thread-like; **fronds** 5–25 cm long, 2–3-pinnate, whitish to glaucous, with thickened hairy areas; **sori** terminal, cup-shaped.

Widely distributed in Polynesia and extending to north-eastern Qld where it grows on trees and rocks in highland rainforest. Plants have proved to be difficult to maintain in cultivation. Culture in a bottle, terrarium or aquarium may be appropriate, but air movement is necessary. See entry for Filmy Ferns, Volume 4, page 294. Propagate by division.

PLEUROPAPPUS F. Muell.
(from the Greek *pleura*, side; *pappus*, down; referring to the asymmetrical attachment of the pappus to the achene)
Asteraceae (alt. *Compositae*)

A monotypic genus endemic in SA.

Pleuropappus phyllocalymmeus F. Muell.
(leaf covering)
SA
0.05–0.2 m × 0.1–0.3 m   Sept–Dec

Dwarf annual **herb** with greyish-hairy young growth; **stems** terete, slender, spreading to erect from the base, simple or lightly branched, hairy to becoming glabrous; **leaves** 0.7–1.5 cm × about 0.1 cm, linear, initially opposite, becoming alternate, sessile, spreading to ascending, covered in greyish-woolly hairs, apex pointed; **flowerheads** to about 2 cm × 0.5 cm, golden-yellow, terminal, usually solitary, very conspicuous; **outer bracts** leaf-like; **inner bracts** shorter; **florets** 2 per head, tubular, yellow.

This interesting annual occurs on the Eyre and Yorke Peninsulas where it grows mainly on the clay margins of saline depressions which are wet for extended periods. Evidently not cultivated, it will be best suited to warm temperate regions. Needs an open site for best results. Propagate from seed, which should not require pre-sowing treatment.

Angianthus phyllocalymmeus (F. Muell.) J. M. Black is a synonym.

PLEUROSORUS Feé
(from the Greek *pleura*, rib; *sorus*, heap; referring to the elongated sori along the veins)
Aspleniaceae

Small terrestrial **ferns**; **rootstock** short, creeping or erect, apex scaly; **fronds** crowded, pinnate, densely hairy, able to regreen from dry fronds; **sori** without indusia, elongate along the veins, coalescing with age.

A small genus of 4 species; 2 species occur in Australia. They are very attractive small ferns which have somewhat specialised growing requirements and are consequently cultivated mainly by enthusiasts. Propagation is from spores. Sporelings can rot readily if overwatered or overcrowded.

Pleurosorus rutifolius (R. Br.) Feé
(with leaves like the genus *Ruta*)
All states   Blanket Fern
4–18 cm × 10 cm

Small terrestrial **fern**; **rootstock** short-creeping, with dark brown scales; **fronds** 4–18 cm × 1.5–4 cm, pinnate, dark green, thin-textured but leathery, densely clothed with spreading brownish hairs; **pinnae** fan-shaped, often deeply notched or lobed; **sori** 0.2–0.5 cm long, linear, borne along the veins, spreading with age.

A widespread species which also occurs in New Zealand. It typically forms clumps in rock crevices and in the lee of boulders. Plants are very tolerant of dryness. The fronds have the ability to revive and regreen after becoming brown and shrivelled. An interesting pot subject but plants must be given good light and plenty of air movement to maintain them in cultivation. They also need excellent drainage and should not be overwatered. Propagate from spores which can damp off readily.

P. subglandulosus (Hook. & Grev.) Tindale differs by having the hairs tipped with a prominent, glistening gland. It has a similar distribution to *P. rutifolius* and requires similar cultural conditions.

Pleurosorus sublandulosus (Hook. & Grev.) Tindale —
see under *P. rutifolius* (R. Br.) Feé.

PLEUROSTYLIA Wight & Arn.
(from the Greek *pleura*, rib; *stylos*, style; referring to the ribbed style on the side of the fruit)
Celastraceae

Shrubs or **trees**, bisexual; **leaves** opposite, decussate; **stipules** minute, deciduous; **inflorescence** axillary cymes; **flowers** small, tubular; **sepals** 5; **petals** 5 or rarely 4; **stamens** 5; **fruit** a nut.

A tropical genus of about 6 species distributed in Africa, Madagascar, Mascarene Islands, South-East Asia, Sri Lanka, New Caledonia and in Australia where there is 1 species.

Pleurostylia opposita (Wall.) Alston
(opposite)
Qld
4–15 m × 2–8 m   Oct–Jan

Tall **shrub** to medium **tree** with glabrous young growth; **branches** spreading to ascending; **branchlets** glabrous; **leaves** 2.5–6 cm × 1–4 cm, ovate to obovate, opposite, decussate, tapered to base, spreading to ascending, thin-textured, deep green above, paler below, glabrous, margins entire, venation prominent, apex blunt to somewhat rounded; **flowers** to about 0.3 cm across, white to pale yellow or greenish, in axillary cymes; **petals** to about 1.5 cm long, elliptic to broadly ovate; **fruit** to about 0.7 cm long, obovoid, white to pale yellow, fleshy.

In Australia, this species occurs mainly in the seasonally dry rainforests of eastern Qld. It is found in two disjunct areas (one in the south-east and the other on Cape York) from sea-level to about 200 m altitude. Not highly ornamental, it is best suited to subtropical and tropical regions and requires well-drained acidic soils. Probably mainly of interest to rainforest enthusiasts. Propagate from fresh seed or possibly from cuttings.

PLUCHEA Cass.
(after Noel Pluche Antoine, 17–18th-century French abbot and author)
Asteraceae

Perennial **herbs**, **subshrubs** or **shrubs**; **stems** often woody; **leaves** simple, alternate, sessile, toothed, lobed or entire, glabrous or glandular; **flowerheads** terminal, solitary or in cymes or panicles; **outer disc florets** female; **inner disc florets** bisexual; **fruit** an achene, glabrous or hairy.

A genus of about 50 species distributed in temperate and tropical regions of America, Africa and Asia. Eleven species occur in Australia, some undescribed, of which about 5 are endemic. Most species grow in inland habitats and favour low-lying sites which are subject to periodic inundation. Although the flowerheads are colourful, few, if any, of the Australian species are grown. Propagation is from seed and possibly also from cuttings of recently hardened growth.

Pluchea baccharoides (F. Muell.) F. Muell. ex Benth.
(resembling the genus *Baccharus*)
Qld, NSW   Narrow-leaved Plains Bush
0.5–1.5 m × 0.5–1 m   Aug–Nov

Erect, much-branched low **shrub**; **stems** woody; young growth bright green and sticky; **leaves** 1–2.5 cm × 0.1–0.2 cm, narrowly linear, dark green, thick, entire, pointed; **flowerheads** about 0.4 cm across, mauve to purplish, clustered at the ends of the branches; **florets** crowded; achenes about 1 cm long.

Distributed in inland regions growing in sandy alluvial soils along the courses of ephemeral streams. This species is moderately ornamental but is probably not cultivated. It has similar cultivation requirements to *P. dentex*. Propagate from seed and possibly also from cuttings.

Pluchea dentex R. Br. ex Benth.
(with teeth)
Qld, NSW, SA, WA, NT   Bowl Daisy
0.3–0.6 m × 0.3–0.5 m   Oct–Jan; also sporadic
    Perennial **subshrub**, usually with an erect habit;
stems woody at the base, sticky, with short hairs; **leaves**
0.5–4 cm × 0.1–1 cm, linear to narrowly oblanceolate,
sparsely toothed or lobed, upper leaves entire, aromatic when crushed; **flowerheads** to 1.5 cm across, pink
to purplish, clustered at the end of stems; **florets**
crowded; **achenes** to 1.6 cm long, sparsely hairy.
    This species is widely distributed in semi-arid
regions. It often grows in low-lying areas which are subject to periodic inundation, especially those areas
where water lies for long periods. Flowering plants are
colourful and interesting but tend to become coarse
and woody. They are moderately attractive in appearance. This species may be useful in inland areas with a
hot dry climate, but it requires plenty of water. Tolerates poor drainage and moderate to heavy frosts.
Plants probably benefit from periodic heavy pruning.
Propagate from seed and possibly also from cuttings.

Pluchea eyrea F. Muell. =
                        P. rubelliflora (F. Muell.) B. L. Rob.

Pluchea rubelliflora (F. Muell.) B. L. Rob.
(with reddish flowers)
Qld, NSW, SA, WA, NT   Plains Bush
0.2–0.6 m × 0.3–0.8 m   sporadic all year
    Perennial dwarf **shrub** or **subshrub** with an erect
habit; **stems** terete, arising from a carrot-like rootstock; **leaves** 1–3 cm × 0.2–0.4 cm, narrowly lanceolate,
the base forming a decurrent wing along the stem,
margins coarsely toothed; **flowerheads** to 0.6 cm across,
pink, clustered at the end of the branches; **florets**
crowded; **achenes** abortive.
    An unusual species which forms clonal colonies
reproducing by suckers. Grows in wet claypans, swamps
and seasonal watercourses. An interesting species
which may be useful in inland regions. It probably has
similar cultivation requirements to *P. dentex*. Propagate
from cuttings and suckers.
    Pluchea eyrea F. Muell. is a synonym.

Pluchea tetranthera F. Muell.
(with 4 anthers)
Qld, NSW, SA, WA, NT   Pink Plains Bush
0.3–1.5 m × 0.5–1 m   Aug–Nov
    Rigid, much-branched, low **shrub**; young growth
glabrous or slightly hairy, often glandular; **leaves**
1–3 cm × 0.5–1.5 cm, obovate to cuneate, entire to
irregularly toothed, upper leaves linear, dark green;
flowerheads to 0.5 cm across and 1 cm long, pink,
clustered at the ends of branches; **florets** crowded;
achenes about 0.8 cm long, sparsely hairy.
    This widespread species from inland regions often
grows in deep, red sandy soils. It frequently occurs in
low-lying sites. Plants may be short-lived and last for
only 2–5 years. This species is fast-growing and may be
useful in the gardens of inland towns. Requires a
sunny location and an abundance of water. Propagate
from seed and possibly also from cuttings.

PLUMBAGINACEAE Juss.
A cosmopolitan family of dicotyledons consisting of
about 560 species in 10 genera. It is poorly represented in Australia by 3 genera and 5 species. Members
of this family are herbs with leaves in a basal rosette or
shrubs and climbers with alternate leaves. The regular-
shaped, often colourful flowers are 5-merous and
the bases of the petals may be fused into a long tube.
A number of exotic species are commonly grown as
garden plants. Australian genera: *Aegialitis*, *Limonium*,
Plumbago.

PLUMBAGO L.
(from the Latin *plumbum*, lead; European Leadwort,
P. europaea, was used for treatment of lead poisoning)
Plumbaginaceae
    Perennial **herbs** or **shrubs**, often sprawling or climb-
ing; **leaves** simple, entire; **inflorescence** few to many-
flowered, often spike-like racemes, terminal; **flowers**
tubular, with 5 spreading lobes; **calyx** 5-lobed, hairy;
style with 5 short branches; **fruit** a capsule enclosed in
calyx.
    A wide-ranging genus of about 12 species which are
found in temperate and tropical regions of the world.
One species occurs in Australia.

Plumbago zeylanica L.
(from Ceylon, now known as Sri Lanka)
Qld, NSW, WA, NT
0.3–1.3 m × 0.6–2 m   mainly April–Aug
    Dwarf to small **shrub** with glabrous young growth;
stems slender, scrambling, spreading to erect, ribbed,
sometimes reddish; **leaves** 3.5–11 cm × 1.5–4.5 cm,
ovate to elliptic, alternate, with sometimes winged and
often stem-clasping stalks to about 2.5 cm long,
spreading to ascending, glabrous, green above, paler
below, flat, margins entire, apex blunt to pointed;
flowers to 2.5 cm × about 1.5 cm, cream, white or pale
blue, with glandular-hairy calyx, in many-flowered
spike-like racemes, can be profuse and very conspicuous; **capsule** slender, to about 0.6 cm long, glandular-
hairy.
    A moderately ornamental species which is valuable
because of its often scrambling habit. In Australia, it
occurs over a wide range in frost-free areas and
extends to the Pacific Islands, Asia, Africa and India.
Uncommon in cultivation here, it deserves wider
recognition in tropical and subtropical regions. Plants
adapt to a variety of acidic and slightly alkaline soils.
They grow well in semi-shaded or shaded sites.
Responds well to pruning or clipping and is worth trying as a hedge. Suitable for sheltered embankments
and general underplanting. Propagate from seed,
which does not require pre-sowing treatment, or from
cuttings of firm young growth.

PNEUMATOPTERIS Nakai
(from the Greek *pneumatikos*, of the wind; *pteris*, fern)
Thelypteridaceae
    Terrestrial **ferns**; **rootstock** erect, often forming a
short fibrous trunk; **fronds** pinnate, large; **pinnae**
numerous, lobed, ending in a tail-like section, often
with coloured glands; **sori** round, with or without an
indusium.

A genus of about 76 species widely distributed in Africa, South-East Asia, Polynesia, Indonesia and New Guinea. Three species occur in Australia but none is endemic. The Australian species are large handsome ferns deserving of wider recognition for their ornamental qualities. Propagation is from spores, which germinate readily if fresh.

Pneumatopteris costata (Brack.) Holttum
(fluted, ribbed)
Qld
0.5–1 m tall

Robust terrestrial **fern**; **rootstock** erect, forming a short trunk; **fronds** erect in a tussock, 50–100 cm × 30–40 cm, dark green above, paler beneath, thin-textured; **pinnae** 15–20 cm × 1.5–2 cm, spreading at right angles, lower pinnae gradually reduced in size, deeply lobed, lower surface with yellow glands; **sori** round, usually without an indusium or sometimes a small indusium present.

Extends from the Philippines to Indonesia, New Caledonia, New Guinea and north-eastern Qld. Grows along stream banks in rainforest usually in very sheltered localities. A handsome fern which grows well in tropical and subtropical regions. Requires shade, well-drained but moist soil and an abundance of water. Frost tolerance is not high. Plants are fast-growing when planted in suitable conditions. Propagate from spores.

Pneumatopteris pennigera (G. Forst .) Holttum
(like a feather)
Qld, NSW, Vic, Tas   Lime Fern
0.4–1.1 m tall

Robust terrestrial **fern** forming small to large clumps; **rootstock** erect, forming a short trunk; **fronds** erect in a tussock, 40–110 cm × 12–30 cm, mid-to-light green, sometimes yellowish, thin-textured; **pinnae** 6–15 cm × 1.5–2 cm, spreading at right angles or the upper pinnae at a steeper angle, the lower 4 pairs reduced in size and more widely spaced; **sori** round, lacking indusia.

Widely distributed from south-eastern Qld to Tasmania and also in New Zealand. In southern Australia, it commonly grows on limestone whereas in Qld and New Zealand it is found in acid soils. A handsome fern which can be grown successfully with the addition of lime. Needs filtered sun or partial sun and an abundance of water. Can be difficult to maintain in acid soils unless dressed annually with lime. Tolerates most frosts. Propagate by division and from spores.

Pneumatopteris sogerensis (A. Gepp) Holttum
(from the area of Soger, New Guinea)
Qld, NSW   Giant Creek Fern
0.5–2 m tall

Robust terrestrial **fern** forming large clumps; **rootstock** erect, forming a short, slender trunk; **fronds** 50–200 cm × 40–60 cm, dark green, somewhat fleshy; **pinnae** 20–30 cm × 2.5–3 cm, spreading at right angles, the lowest 5 or 6 pairs reduced to small lobes; **sori** round, with indusia.

Occurs in New Guinea, Solomon Islands, the Moluccas, and in Australia where it is distributed from north-eastern Qld to north-eastern NSW. Grows along streams and sheltered slopes in rainforest often at moderate to high altitude. A handsome fern which is moderately popular in cultivation. Can be successfully grown in temperate regions. Best planted in a sheltered situation in moisture-retentive, well-drained soil. Responds to mulching, watering and fertiliser applications. Propagate from spores.

POA L.
(from the Greek *poa*, grass)
Poaceae (alt. *Graminae*)   Tussock Grasses

Tufted annual or perennial **herbs**, often stoloniferous or rhizomatous; **leaves** flat or inrolled with membranous ligules; **inflorescence** a panicle, open or somewhat condensed; **spikelets** solitary, flattened, 2–15-flowered; **glumes** 3, 1–3-nerved; **lemma** similar to glumes, keeled, mainly 5-nerved, lacking awns.

Poa is a cosmopolitan genus of about 300 species. It occurs in temperate regions. About 50 species are found in Australia of which 7 are introduced naturalised species. They occur from the coasts to the highest peak (Mt Kosciuszko, in NSW). In recent years, poas have become increasingly popular. Plants show very attractive variations in foliage texture and colour and they have an appealing tussocking habit. People also recognise many of the species for their value in soil conservation and wildlife habitat. Many small animals such as lizards use *Poa* species for protection from predators. Larvae of butterflies and moths feed on many of the poas. Small seed-eating birds are also regular visitors to plants as seed ripens in summer and autumn.

Some poas are small and relatively delicate, eg. *P. tenera*, while other species have a rhizomatous root system which can be vigorous and can cover large areas, eg. *P. ensiformis*.

Poas adapt very well to cultivation in temperate regions and some are also suitable for subtropical areas. It is recommended that indigenous *Poa* species be selected for broadscale planting projects and that

Poa fawcettiae   W. R. Elliot

wherever any non-indigenous poas are being cultivated they should be carefully monitored to ensure that a weed problem does not arise.

As most species occur in soils of low fertility they usually grow very well without supplementary fertilising. Most species tolerate extended dry periods except for some species such as *P. tenera* which requires moisture-retentive soils. Plants respond well to supplementary watering.

Tussock grasses can be rejuvenated very readily by shearing or pruning clumps to just above ground level. Burning clumps in autumn or early spring is also a very successful method of rejuvenation. Sometimes tussock grasses are pruned to half of their height, but the result can be unsightly and clumps may take a very long time to recover or they may never regain the typical appearance of a tussock. The removal of dead leaves by using a hand or rake as a comb is both easy and usually very successful.

Propagation is from seed or by division of clumps. Seed propagation is straightforward and the seed does not require pre-sowing treatment. Germination usually occurs about 2–3 weeks after it is sown. Some nurseries hand-broadcast seed over the top of growing medium in tube or plug trays. If the medium is kept moist, this practice produces excellent results and saves the need for 'pricking-out' seedlings.

Division of clumps is an alternative method of propagation which is used to a limited degree. It is most successful when plants are just beginning active growth. Late winter to early spring is usually a suitable time. It is best not to use very small divisions and it is important to retain as much of the root system as possible.

Poa australis R. Br.
For many years this name was used for a large number of different taxa particularly those often referred to as snow-grasses. As a result of botanical revision in 1970, a number of new species were described and some old names were resurrected. The name *P. australis* is no longer a legitimate name yet grasses are still marketed as *P. australis* by some nurseries.

Poa clivicola Vickery
(growing on slopes)
NSW, Vic   Fine-leaved Snow-grass
0.3–0.6 m × 0.3–1 m   Nov–Feb
Tufted perennial **grass**; **leaves** about 15 cm × 0.05 cm, inrolled, green to faintly glaucous, somewhat rough, sheath pale; **culms** to about 60 cm tall, erect; **panicles** to about 10 cm × 7 cm; **spikelets** to 0.6 cm long, 2–7-flowered, green to purplish; **glumes** to 0.4 cm long, 3-nerved, more or less equal.

A pleasing grass which occurs in moist sites in the subalpine and alpine grassland of south-eastern NSW and north-eastern Vic. Suitable for temperate regions in moisture-retentive acidic soils in sunny or semi-shaded sites. Hardy to frost and snow. Propagate from seed or by division of clumps.

P. clivicola is sometimes confused with *P. meionectes* Vickery, which differs in having finer, softer leaves with often purplish sheaths and smaller glumes. It occurs in south-eastern NSW, eastern Vic and south-eastern SA.

Poa costiniana Vickery
(after Dr Alec Costin, 20th-century Australian alpine ecologist)
NSW, Vic, Tas   Bog Snow-grass
0.4–0.8 m × 0.3–1 m   Nov–March
Tufted perennial **grass**; **leaves** to about 40 cm × 0.15 cm, closely folded, terete or faintly angular, green or rarely glaucous, smooth, glabrous, somewhat glossy, rigid, pungent-pointed, sheath pale or sometimes purplish; **culms** to about 80 cm tall, mainly erect; **panicles** to 20 cm × 10 cm, somewhat pyramidal; **spikelets** to 0.6 cm long, 2–5-flowered, mainly purplish; **glumes** to 0.4 cm long, 3-nerved, upper longest.

A fairly vigorous grass which is found mainly in moist-to-wet soils in low lying areas of alpine and subalpine regions. Adapts well to cultivation in acidic soils in sunny or semi-shaded sites. Plants are prickly and therefore unsuitable for using in areas where people are likely to sit or lie down. Propagate from seed or by division of clumps.

Poa crassicaudex Vickery =
          P. sieberiana Spreng. var. *hirtella* Vickery

Poa ensiformis Vickery
(sword-shaped)
NSW, Vic   Purple-sheathed Tussock-grass
0.8–1.2 m × 0.5–1.5 m   Oct–March
Tufted, rhizomatous, perennial **grass**; **leaves** to about 50 cm × 0.5–1 cm, flat to slightly folded, deep green, smooth to rough above, often smooth below,

Poa ensiformis × .5, close-up × .2

sheath deep purple; **culms** to about 1.2 cm tall, green, erect; **panicles** to about 30 cm × 18 cm, somewhat loosely arranged; **spikelets** to 1 cm long, 3–8-flowered, green or purplish; **glumes** to 0.4 cm long, 3-nerved, more or less equal.

A very vigorous, wide-ranging grass which is found mainly in hill and mountain forests where soils are usually moist for most of the year. Moderately popular in cultivation but can become somewhat invasive under suitable conditions. Best for loam or clay loam soils in shaded or semi-shaded sites but will tolerate plenty of sunshine. Hardy to frost and snow. Clumps are rejuvenated by shearing to just above ground level in late winter. Propagate from seed or by division of clumps.

Poa exilis Vickery =
    P. meionectes Vickery (see also *P. clivicola* Vickery)

Poa fawcettiae Vickery
(after Stella G. M. Carr, née Fawcett, 20th-century Australian botanist)
NSW, Vic, Tas   Horny Snow-grass
0.4–0.6 m × 0.5–1.5 m   Nov–Feb
Tufted perennial **grass** often with short-creeping rhizomes or stolons; **leaves** 10–30 cm × 0.1 cm, inrolled-terete, glaucous, smooth, glabrous, pungent-pointed, sheath purplish; **culms** to about 60 cm tall, often purplish, erect; **panicles** to about 15 cm × 10 cm, somewhat pyramidal; **spikelets** to 0.6 cm long, 3–5-flowered, often purplish and glossy; **glumes** to about 0.35 cm long, 3-nerved, more or less equal.

A very attractive grass from subalpine and alpine areas which are mainly above 1500 m altitude. Cultivated to a limited degree, it requires well-drained acidic soils and a sunny or semi-shaded site. Plants grown in semi-shade are softer and less purplish. Hardy to frost and snow and tolerates extended dry periods. Propagate from seed or by division of clumps.

The closely allied and attractive *P. phillipsiana* Vickery, Blue Snow-grass, differs in having narrower leaves with rough surfaces and flowers during Jan–March. It occurs in dryish sites in alpine and subalpine regions of south-eastern NSW and north-eastern Vic. It is cultivated to a limited degree and adapts well in temperate regions.

Poa gunnii Vickery
(after Ronald C. Gunn, 19th-century Tasmanian botanical collector)
Vic, Tas
0.1–0.7 m × 0.1–0.5 m   Dec–Feb
Tufted perennial **grass**; **leaves** to about 30 cm × 0.1 cm, unrolled or folded or rarely flat, green, smooth to faintly rough, rarely hairy on outer surface; **sheaths** glossy, green or sometimes purplish; **culms** to about 70 cm tall, slender, terete, erect; **panicles** to about 8 cm long, moderately open; **spikelets** to about 0.7 cm long, 2–6-flowered, green to purplish; **glumes** to 0.3 cm long, more or less equal, green to purplish; **anthers** purplish.

A handsome grass which often creates a purplish haze across a field when it is in flower. It is widespread and common at 800–1400 m altitude and is also found at lower elevations. Uncommon in cultivation, it should

Poa gunnii × .4

adapt well in temperate regions with freely draining acidic soils. Tolerates open or semi-shaded sites. The purplish anthers are very attractive. Propagate from seed or by division of clumps.

An attractive, bluish-leaved variant from eastern Vic differs in having purplish leaf sheaths and culms which are branched at their base. Further botanical study may prove this to be an undescribed endemic species.

Poa halmaturina J. M. Black —
          see *P. poiformis* (Labill.) Druce

Poa helmsii Vickery
(after Richard Helms, 19–20th-century Australian naturalist)
NSW, Vic
1–1.5 m × 0.5–1.5 m   Nov–March
Vigorous, tufted, perennial **grass**; **leaves** 15–80 cm × 0.3–0.8 cm, flat to slightly folded, pale green, rough, venation distinct, not pungent; **culms** to about 1.5 m tall, erect, flattened below panicle; **panicles** to about 35 cm × 15 cm, somewhat pyramidal and condensed; **spikelets** to about 0.8 cm long, 3–8-flowered, green or sometimes purplish; **glumes** to about 0.35 cm long, 3-nerved, more or less equal.

This impressive species inhabits the wet soils of creeks and swamps mainly above 1000 m altitude where plants are often subject to snow and heavy frosts. It occurs in south-eastern NSW and eastern Vic. Cultivated to a limited degree, plants need acidic soils which are moist for most of the year and a semi-shaded or somewhat sunny site. Propagate from seed or by division of clumps.

Poa hiemata

Poa hiemata × .5

Poa hiemata Vickery
(of winter)
NSW, Vic, Tas   Soft Snow-grass
0.3–0.6 m × 0.3–1 m   Nov–Feb

Tufted, or sometimes stoloniferous, perennial **grass**; **leaves** 15–50 cm × about 0.5 cm, inrolled, terete or slightly angular, bright green or sometimes glaucous, smooth, softly pointed, sheath pale green or rarely purplish; **culms** to about 60 cm tall, more or less erect; **panicles** to about 12 cm × 8 cm, somewhat open or condensed; **spikelets** to about 0.5 cm long, 3–5-flowered, green or purplish; **glumes** to 0.3 cm long, upper longest.

A lovely soft-foliaged grass which is often dominant in alpine areas where there are moist soils. Adapts well to cultivation in temperate regions. Prefers moisture-retentive acidic soils in sunny or semi-shaded sites. Hardy to frosts and snow. Propagate from seed or by division of clumps.

Poa hothamensis Vickery
(from Mt Hotham region)
Vic   Ledge-grass
0.5–1 m × 0.5–1.5 m   Nov–Feb

Tufted perennial **grass** often with rhizomatous roots; **leaves** to about 30 cm × 0.5 cm, flat or closely folded, somewhat stiff, both surfaces usually faintly to moderately hairy, sheath purplish; **culms** to about 1 m tall, erect, often slightly flattened; **panicles** usually about 12 cm × 6 cm, somewhat contracted; **spikelets** to 0.8 cm long, 2–6-flowered, often purplish; **glumes** to 0.3 cm long, 3-nerved, upper slightly longer.

The typical variant is restricted to the Victorian Alps where it grows in alpine and subalpine shrubland. Cultivated to a limited degree, it requires well-drained

acidic soils with a sunny or semi-shaded aspect. Plants are hardy to snow and frost.

The var. *parviflora* N. G. Walsh differs in having aerial stolons, lax and somewhat velvety leaves and smaller spikelets to about 0.35 cm long. It has a more restricted distribution in far eastern Vic. Evidently not in cultivation.

Propagate both varieties from seed or by division of clumps.

Poa induta Vickery
(covered, clothed)
NSW
0.6–1.1 m × 0.5–1 m   Nov–Feb

Loosely tufted perennial **grass**; **leaves** to about 30 cm long, loosely inrolled, greyish-green, densely hairy, somewhat lax; **sheath** densely hairy; **culms** to about 1.1 m tall, erect; **panicles** to about 22 cm long, narrow, often one-sided; **spikelets** to about 0.35 cm long, 2–7-flowered; **glumes** about 0.25 cm long, upper 3-nerved, lower 1–3-nerved.

This attractive species is restricted to the Central and Southern Tablelands Districts. It is usually found in woodland at high elevations. Uncommon in cultivation, it should adapt to a range of well-drained acidic soils in a semi-shaded or sunny site. Hardy to frosts and light snowfalls. Propagate from seed or by division of clumps.

The similar *P. petrophila* Vickery differs in having culms to about 0.7 m high, somewhat stiff leaves and shorter panicles. It occurs in the Australian Alps of NSW and Vic, at about 1300 m altitude, growing in shallow soils amongst granite boulders and rocks.

Poa labillardieri Steud.
(after Jacques J. H. de la Billardiére, 18–19th-century French naturalist)
Qld, NSW, Vic, Tas, SA   Common Tussock-grass
0.5–1.3 m × 0.5–1.5 m   Oct–Feb; also sporadic

Tufted perennial **grass**; **leaves** to about 80 cm × 0.35 cm, flat or channelled or sometimes inrolled, usually green, rarely greyish-green, rough, usually softly pointed, sheath pale green or rarely purplish; **culms** to about 1.3 m tall, more or less erect, sometimes flattened below the panicle; **panicles** to about 25 cm × 15 cm, pyramidal, somewhat open; **spikelets** to about 0.8 cm long, mostly 3–5-flowered, green or purplish; **glumes** mostly about 0.3 cm long, upper 3-nerved, lower sometimes 1-nerved.

The typical variant of this appealing and variable species has a wide distribution. It usually occurs in alluvial soils which are moist for extended periods. Adapts very well to cultivation in semi-shade or full sunshine and is one of the most common grasses used in a range of applications. Looks and grows well in massed plantings. Needs acidic soils for best results. Clumps are rejuvenated by shearing to just above ground level during late winter–early spring. Selections with bluish foliage are becoming popular.

The attractive var. *acris* Vickery has inrolled, sharply pointed, often bluish leaves to about 30 cm long, with purple sheaths. Culms are rarely above 80 cm tall. It is restricted to alpine and subalpine regions in eastern Vic and the Central Highlands of Tas.

Propagate both varieties from seed or by division of clumps.

Poa laevis R. Br. = *P. poiformis* (Labill.) Druce

Poa meionectes Vickery — see *P. clivicola* Vickery

Poa mollis Vickery — see *P. rodwayi* Vickery

Poa morrisii Vickery
(after Dennis I. Morris, 20th-century Tasmanian botanist)
Vic, SA   Velvet Tussock-grass
0.4–1 m × 0.3–0.6 m   Sept–Nov; also sporadic
Tufted perennial **grass** sometimes with rhizomes; **leaves** to about 30 cm × 2 cm, flat, folded or faintly inrolled, greyish-green, hairy, softly pointed; **sheaths** densely hairy; **culms** to about 90 cm tall, ascending to erect, faintly hairy; **panicles** to about 25 cm × 15 cm, pyramidal, moderately green; **spikelets** to 0.6 cm long, 3–8-flowered, green or purplish; **glumes** about 0.2 cm long, 3-nerved, hairy.
A highly desirable, soft-foliaged grass which inhabits a range of soils throughout Vic and south-eastern SA. Adapts well to cultivation in a variety of acidic soils in sunny or semi-shaded sites. Although tolerating dry periods it does best in soils which are moist for most of the year. Hardy to frosts. Clumps are rejuvenated by shearing to just above ground level in late autumn or late winter to early spring. Propagate from seed or by division of clumps.

Poa poiformis   D. L. Jones

Poa petrophila Vickery — see *P. induta* Vickery

Poa phillipsiana Vickery — see *P. fawcettiae* Vickery

Poa plebeia R. Br. = *P. poiformis* (Labill.) Druce

Poa poiformis (Labill.) Druce
(in the form of a *Poa*)
NSW, Vic, Tas, SA, WA   Coast Tussock-grass;
     Blue Tussock-grass
0.6–1.2 m × 0.5–1.5 m   Sept–Jan
Tufted perennial **grass**; **leaves** to about 1.2 m × 0.2 m, inrolled or folded, mainly erect, green to bluish-green, smooth to faintly hairy, sometimes roughish, pungent-pointed; **sheaths** pale; **culms** to about same length as leaves; **panicles** to about 30 cm long, narrow and condensed; **spikelets** to 0.8 cm long, 2–7-flowered, green to straw-coloured; **glumes** to 0.5 cm long, usually 3-nerved, more or less equal.
The typical variant of this prominent coastal species occurs on cliff tops and sand dunes. It is important in soil erosion control and is commonly planted for this purpose. Adapts to acidic or alkaline soils and needs plenty of sunshine to grow well but also tolerates semi-shaded sites. Clumps are rejuvenated by shearing to just above ground level or by burning.
The var. *ramifer* D. I. Morris has rhizomatous or stoloniferous roots and finer, very glaucous leaves which have purplish sheaths. It is widespread in Tas and Bass Strait Islands, and there are scattered populations on the Vic coast.
Propagate both varieties from seed or by division of clumps.
P. laevis R. Br. and *P. plebeia* R. Br. are synonyms.

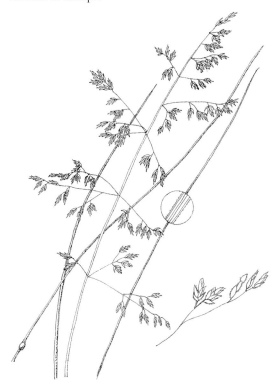

Poa labillardieri × .5, close-up × .2

Poa queenslandica

The allied *P. halmaturina* J. M. Black looks like a dwarf variant of *P. poiformis* but differs in its scaly rhizomes, culms to about 15 cm tall and leaves up to 10 cm long. It occurs on or near the coast in far south-western Vic and south-eastern SA. Plants have similar cultivation requirements to those of *P. poiformis*.

Poa queenslandica C. E. Hubb.
(from Queensland)
Qld, NSW   Queensland Poa
0.8–1.8 m × 0.8–2 m   most of the year

Loosely tufted perennial **grass**; **leaves** to about 40 cm × 0.4–1.5 cm, linear, flat, thin-textured, green, glabrous, softly pointed, sheaths striate, glabrous, becoming loose; **culms** to about 1.8 m tall; **panicles** to about 40 cm long, loosely arranged, with spreading branches; **spikelets** to about 0.6 cm long, 2–4-flowered, green; **glumes** about 0.2 cm long, upper 3-nerved, lower 1–3-nerved, more or less equal.

This tall grass occurs on margins of rainforest in south-eastern Qld and eastern NSW. It is cultivated to a limited degree and warrants further attention in subtropical and temperate regions. Plants require moisture-retentive, acidic soils with a semi-shaded aspect. Propagate from seed or by division of clumps.

Poa rodwayi Vickery
(after Leonard Rodway, 19–20th-century Honorary Government Botanist, Tas)
Vic, Tas
0.3–0.75 cm × 0.5–1 m   Nov–Feb

Tufted perennial **grass**; **leaves** to about 30 cm × 0.1 cm, inrolled-terete, somewhat stiff, greyish-green to green, with dense covering of short hairs, upper sheaths green and hairy; **culms** to about 75 cm tall, smooth; **panicles** to about 15 cm × 8 cm, pyramidal, loosely arranged; **spikelets** 3–5-flowered, green or purplish; **glumes** to about 0.3 cm long, 3-nerved.

An interesting species which is found in western Vic and is fairly widespread in Tas. Occurs in grassland and woodland. Rarely encountered in cultivation, it should adapt readily to a range of acidic soils in sunny or semi-shaded sites. Plants are hardy to frosts. Propagate from seed or by division of clumps.

P. mollis Vickery is allied to *P. rodwayi* and is of similar dimensions but differs in having hairier leaves with purple sheaths. It is endemic in Tas where it inhabits dryish sites on cliffs and hills in the north-western region.

Poa saxicola Vickery
(growing amongst rocks)
NSW, Vic, Tas   Rock Poa
0.3–0.75 m × 0.5–1 m   Nov–Feb

Tufted perennial **grass**, sometimes with rhizomatous roots; **leaves** to about 25 cm × 0.5 cm, flat to slightly folded, green, glabrous, smooth to slightly rough, blunt-tipped, upper sheaths green and striate; **culms** to about 75 cm tall, terete or slightly compressed; **panicles** 5–15 cm × 1–2 cm, arching to pendent at maturity, few-branched, compact; **spikelets** to about 0.7 cm long, 2–4-flowered, usually purplish; **glumes** to 0.4 cm long, upper 3-nerved, lower 1–2-nerved, more or less equal.

A distinctive species with arching panicles. It is confined to rocky grassland and shrubland in north-eastern NSW, north-eastern Vic and is widespread at 800–1400 m altitude in Tas. Rarely encountered in cultivation, it is suitable for cold areas. Needs acidic soils which drain freely but also retain moisture. A sunny or semi-shaded site is suitable. Propagate from seed or by division of clumps.

Poa sieberiana Spreng.
(after Franz W. Sieber, 19th-century Austrian botanist)
Qld, NSW, Vic, Tas   Tussock-grass
0.5–1 m × 0.3–1 m   Oct–March; also sporadic

Tufted perennial **grass**; **leaves** 10–60 cm × up to 0.7 cm, inrolled, terete or angular, green to greyish-green, rough or rarely smooth, often curled, sheaths pale green or rarely purplish; **culms** to about 1 m tall, mainly erect; **panicles** to about 25 cm × 12 cm, pyramidal, loosely arranged; **spikelets** to about 0.5 cm long, 2–7-flowered, green or purplish; **glumes** to 0.25 cm long, upper 3-nerved, lower 1–3-nerved, more or less equal.

This extremely variable species is one of the most common tussock-grasses over its range. The typical variety is widespread from sea-level to the highest mountains. It occurs in a range of soils. Adapts very well to cultivation in a variety of freely to moderately well-drained soils with a sunny or semi-shaded aspect. Excellent for underplanting established shrubs and trees. Hardy to frost and withstands limited periods of dryness. Clumps are rejuvenated by shearing to just above ground level or by burning during late autumn or early spring.

There are 2 other varieties.

var. *cyanophylla* Vickery is very attractive and highly desirable for cultivation. It has very fine bluish leaves and often a shortly stoloniferous root system. It occurs in south-eastern NSW and north-eastern Vic. Plants inhabit higher altitudes.

var. *hirtella* Vickery has faintly to moderately hairy leaves and is virtually coextensive with the typical variant (except for Tas), but not as common. *P. crassicaudex* Vickery is a synonym.

Propagate all varieties from seed or by division of clumps.

Poa tenera F. Muell. ex Hook. f.
(tender, soft)
NSW, Vic, Tas, SA   Slender Tussock-grass;
                              Soft Tussock-grass
0.1–0.5 m × 0.5–2 m   Oct–Jan; also sporadic

Tufting perennial **grass** with branched, trailing or aerial stolons; **leaves** 5–25 cm × about 0.15 cm, folded or inrolled, bright green, soft, glabrous; sheaths pale green to red or purplish; **culms** to about 40 cm long, weak, slender; **panicles** to about 8 cm × 6 cm, loosely arranged; **spikelets** to about 0.5 cm long, 2–4-flowered, greenish; **glumes** to 0.25 cm long, more or less equal.

A lovely soft grass which inhabits mainly moist shaded sites of forests and banks of waterways. In nature it often acts as a climber if support is provided by other plants. Well suited to cultivation in moisture-retentive acidic soils in shaded or semi-shaded sites. Hardy to frosts. Ideal for growing with ferns and other moisture-loving plants. Propagate from seed or by division of clumps and stolons.

POACEAE Nash (alt. Gramineae)   Grass Family
A very large cosmopolitan family of monocotyledons consisting of about 10,000 species in 670 genera. In Australia, the family is represented by about 925 species in 150 genera.

Grasses are annual or perennial herbs with a tufted or creeping habit. They are very rarely shrub-like or tree-like (eg. the bamboos). The stems, known as culms, are hollow or pithy except at the nodes which are solid. The leaves have a basal sheath and a blade with a membranous or hairy structure at the junction known as the ligule. Grass flowers are greatly modified and the parts can be difficult to determine. Known as spikelets they consist of 2 opposite rows of bracts. The outer bracts (glumes) are sterile and the inner bracts (lemmas) enclose the sex organs. The fruit is 1-seeded.

Grasses are not only diverse but are also abundant. Some authorities estimate that they constitute up to twenty per cent of the world's vegetation cover. They are particularly prominent in open habitats such as grasslands, tundra, pampas, sparse woodlands and so on, but are much less prominent in dense habitats such as rainforests and heathland.

Main Australian genera: *Agrostis, Amphipogon, Aristida, Brachiaria, Chloris, Danthonia, Deyeuxia, Dichelachne, Digitaria, Echinochloa, Echinopogon, Enneapogon, Eragrostis, Eriachne, Leptochloa, Micraira, Panicum, Paspalidum, Plectrachne, Poa, Setaria, Stipa, Themeda, Triodia.*

PODOCARPACEAE Endl.
A family of gymnosperms consisting of 6 genera and about 125 species distributed mainly in the southern hemisphere with elements in Japan, Central America and the West Indies. The family is represented in Australia by 5 genera and about 15 species. They are monoecious or dioecious trees or shrubs with small, often scale-like leaves. The male and female cones are small and the fruit often have a swollen, fleshy receptacle which may be colourful. Australian genera: *Lagarostrobus, Microcachrys, Microstrobus, Podocarpus, Prumnopitys, Sundarcarpus.*

PODOCARPUS L'Herit. ex Pers.
(from the Greek *pous, podos,* foot; *carpos,* fruit; referring to the fruit's fleshy stalk)
Podocarpaceae

Dioecious, or rarely monoecious, **trees** or **shrubs; leaves** mainly alternate, sometimes spirally arranged or apparently opposite, flat, oblong, scale-like or needle-like; **male cones** cylindrical, solitary, or clustered in upper axils, often appearing terminal; **female cones** terminal or axillary, usually solitary, of 1–45 scales; **fruiting cone** usually with 1 ovule attached to enlarged fleshy stalk.

A widespread genus with representation in the southern hemisphere and southern Asia. It comprises about 95 species but detailed studies will transfer many species to new genera. Within Australia there are currently 8 described species but revision will extend this to about 15 species. Most species are endemic and they occur in tropical and temperate regions. *Podocarpus* are found from sea-level to about 2000 m altitude. *P. drouynianus* is the only western species. Some species have a limited distribution, for example *P. dispermus,* which occurs only in north-eastern Qld.

Some species are scrambling shrubs, including *P. spinulosus* and *P. lawrencei* (can also be an upright shrub or small tree), while others such as *P. grayae* can be stately trees. *P. drouynianus* has underground creeping stems. It responds very well to coppicing and the foliage is sold in the florist trade as Emu Grass.

Cultivation of Australian species of *Podocarpus* was once quite popular and *P. elatus* was planted in many parks and large gardens. *P. lawrencei* is the most popular species today and selection of cultivars has occurred here and overseas.

Podocarpus require moderately to very well-drained soils. Plants respond well to light applications of slow-release, low phosphorus fertilisers. Once plants are established they can withstand extended dry periods.

P. elatus, P. lawrencei and *P. spinulosus* respond very well to regular pruning and clipping and the other species are likely to react similarly. Pests and diseases are rarely a major concern. Scale is likely to be the main problem but it can be readily controlled (see Sucking Insects, Volume 1, page 153, for control methods).

Propagation is from seed or cuttings. Fresh seed provides the best results and it does not require any pre-sowing treatment. Cuttings can provide varied results. Current season's growth which is still green but firm usually produces roots readily. Application of a hormone rooting compound can be beneficial.

Podocarpus alpinus Hook. f. —
                                        see *P. lawrencei* Hook. f.

Podocarpus amarus Blume =
        Sundarcarpus amarus (Blume) C. N. Page

Podocarpus dispermus C. T. White
(two seeds)
Qld
5–15 m × 3–8 m   March–July

Tall **shrub** or small to medium **tree** with glabrous young growth; **bark** slightly flaky, light brownish-grey; **branches** spreading to ascending; **branchlets** ribbed; **leaves** 10–20 cm × 2–3.5 cm, linear to oblong or lanceolate, alternate, short-stalked, glabrous, dark green, glossy, leathery, midrib prominent, apex softly pointed; **male cones** to about 5 cm × 0.3 cm, narrowly cylindrical; **female cones** axillary, on enlarged, red, fleshy stem to 3 cm × 3 cm.

This rainforest understorey species is confined to north-eastern Qld where it occurs on the Atherton Tableland and in the foothills of Mt Bartle Frere and the Bellenden Ker Range. Rarely cultivated, it should be best suited to tropical and subtropical regions but may succeed in cool temperate areas if grown in a protected site. Needs well-drained acidic soils. Propagate from seed, or from cuttings of firm young growth, which may be slow to form roots.

P. elatus is allied but differs in having much smaller leaves and purplish to bluish-black fruit stems.

393

Podocarpus drouynianus F. Muell.
(after Edouard Drouyn de Lhuys, 19th-century French politician)
WA   Wild Plum; Emu Berry; Kula
0.6–3 m × 0.5–2.5 m   Aug–April

Dwarf to medium **shrub** with creeping underground stems; young growth often pale green and glaucous; **stems** erect, usually glaucous, often many; **leaves** 4–11 cm × up to 0.5 cm, linear, alternate, crowded, spreading to ascending, often recurved, becoming deep green above, paler with prominent midrib below, glabrous, leathery, margins recurved, apex softly pointed; **male cones** 0.5–1.2 cm × up to 0.5 cm, usually solitary; **female cones** about 1.5 cm long, solitary, seed on enlarged, glaucous, dark blue stem to 2 cm long; **seeds** to 2.5 cm long, solitary.

This is the only species from the western region of Australia. It is a common plant in jarrah forests in the southern Darling District where it grows on sandy soils which are often rich in lateritic gravel. Although slow-growing in cultivation, it has potential for greater use. Suited to temperate regions and needs freely draining acidic soils in a semi-shaded or partially sunny site. Hardy to moderate frosts and tolerates extended dry periods. Responds well to heavy pruning. Foliage stems are harvested from the wild for the florist trade and marketed as Emu Grass. These cut stems are extremely long-lasting. Well established plants can be coppiced but may be slow to produce new growth. Makes an attractive container plant and can be grown indoors. Propagate from seed, or from cuttings of firm young growth, which can be slow to form roots.

Podocarpus elatus × .5

Podocarpus elatus Endl.
(tall)
Qld, NSW   Illawarra Plum; Plum Pine; Brown Pine;
                    She Pine; Yellow Pine
5–15 m × 3.5–8 m   March–July

Tall **shrub** or small to medium **tree** with yellowish-green young growth; **bark** greyish-brown to dark brown, fibrous, often grooved; **branches** spreading to ascending; **branchlets** ribbed; **leaves** 6–18 cm × 0.5–1.4 cm, linear to oblong, alternate or whorled, spreading to ascending, moderately crowded, deep green above, paler below, glabrous, leathery, midrib prominent, apex softly pointed; **male cones** to about 3 cm long, narrowly cylindrical, in more or less sessile axillary clusters; **female cones** axillary, solitary, 1 greenish fruit on enlarged, fleshy, pruinose, purplish-to bluish-black stem.

P. elatus dwells in rainforest and along the margins of rainforest. It occurs from the Cook District of north-eastern Qld to near Nowra, NSW. Often found in coastal regions and along the Great Dividing Range. It is often associated with waterways. Mainly grown in public parks and gardens but also by enthusiasts. Adapts well to a range of well-drained acidic soils in semi-shaded or sunny sites. Plants may be slow to become established but respond well to slow-release fertilisers. They are suitable for screening and hedging as they respond well to regular pruning or clipping. Sometimes used for bonsai (or penjing). Plants may not produce fruit every year although in some years they can have a very heavy crop. Birds eat the fruits and stems. The edible succulent fruit receptacle has a resinous flavour and a high Vitamin C content. The fruit is used in jams and jellies and is a special ingredient of delicious tarts and cakes. The highly regarded timber is used in wood-turning and for furniture and other wood products. Propagate from seed, or from cuttings of firm young growth, which may be slow to produce roots.

Podocarpus grayae de Laub.
(after Netta E. Gray, 20th-century American botanist)
Qld
4–5 m × 4–5 m   Oct

Tall **shrub** with glabrous young growth; **bark** greyish-brown, fibrous; **branches** spreading to ascending; **leaves** 10–25 cm × 0.8–2 cm, linear to falcate, alternate, tapering to base, pendent, deep green, glabrous, midrib prominent, leathery, apex finely pointed; **male cones** 2–6 cm × 0.3–0.5 cm, more or less sessile, 1–4 per axil; **female cones** axillary, solitary, 1 seed which can be green, pink, red, blue or blackish, on enlarged, red, fleshy stalk to about 1.8 cm × 1.7 cm.

A rainforest species from Cape York Peninsula and north-eastern regions where it grows in a range of soils from sea-level to about 1000 m altitude. An attractive species with pendent foliage. Suited to tropical and subtropical regions and should adapt to a variety of freely draining acidic soils. Worth trying in gardens and may succeed in large containers or trained as bonsai (penjing). Propagate from seed, or from cuttings of firm young growth, which may be slow to produce roots.

Sometimes incorrectly spelt as *P. grayii*.

Podocarpus lawrencei, an upright bushy selection from north-eastern Vic
W. R. Elliot

Podocarpus lawrencei Hook. f.
(after Robert B. Lawrence, 19th-century Tasmanian botanist/naturalist)

NSW, Vic, Tas	Mountain Plum Pine
0.4–4 m × 1–3 m	Dec–March

Dwarf to tall **shrub** with glabrous, pale green to glaucous young growth; **branches** prostrate to ascending; **branchlets** ribbed; **leaves** 0.5–1.5 cm × 0.15–0.4 cm, oblong-linear, alternate, crowded, spreading to ascending, deep green to glaucous above, paler below, glabrous, midrib prominent on undersurface, somewhat rigid, apex usually blunt; **male cones** to about 0.4 cm long, cylindrical, pinkish to purplish, solitary or clustered; **female cones** axillary, sessile, 1 greenish seed on enlarged, pink-to-bright red, fleshy stalk to about 0.5 cm long.

This extremely variable and handsome species is found in subalpine regions and at lower altitudes in south-eastern NSW and eastern Vic. It is also widespread in Tas. Recent botanical studies will result in the description of about 7 new species. The variant forms of this species are listed below. Plants adapt very well to cultivation in temperate and subtropical regions. They grow best in freely draining acidic soils and appreciate plenty of sunshine, but they also tolerate shady sites. Hardy to heavy frosts and snowfalls. They also withstand extreme winds. Plants respond very well to pruning or clipping. Suitable for many applications in gardens and containers. The low-spreading variants are excellent for growing in combination with boulders.

The typical form is from low altitudes in northern Tas. It is a shrub which grows to about 2 m and has green leaves and a spreading growth habit.

The following are variants which will be given species status in the future.

East Gippsland (Goonmirk Range), Vic — this is a long-lived, green-foliaged, bushy, tall shrub to medium tree. In nature, it can reach about 14 m tall but in cultivation it is more likely to grow about 3–6 m.

Montane south-eastern NSW and north-eastern Vic — a small, bushy shrub of about 2 m × 1.5–2.5 m with green leaves.

Pine Lakes region, Tas — this variant has decumbent stems with erect branchlets and green leaves.

Mt Mawson region, Tas — a small, more-or-less erect, bushy shrub of 1–2 m tall with blue-green leaves.

Mt Wellington region, Tas — a variant of about 1 m tall with bluish leaves.

Subalpine south-eastern NSW — this is a prostrate variant with sprawling stems and thick glaucous leaves.

Some dwarf spreading selections are available and variants with very glaucous foliage are becoming popular in cultivation. Cultivars such as 'Alpine Lass', 'Blue Gem' (purple-tinged, creamish young growth) and 'Bluey' are grown overseas and these are likely to be referable to the Mt Wellington and/or subalpine variants.

Propagate from seed, or from cuttings of firm young growth, which usually produce roots readily. Plants with distinctive foliage or growth characteristics should be propagated from cuttings.

P. alpinus Hook. f. is a synonym

Podocarpus neriifolius D. Don
(leaves like the genus *Nerium*, Oleander)

Qld	
4–5 m × 4–5 m	Oct

Tall **shrub** with glabrous young growth; **bark** light-brown, deciduous (in jig-saw shaped flakes); **branches** whorled, spreading to ascending; **leaves** 8–12 cm × about 1 cm, linear to falcate, alternate, tapering to

Podocarpus neriifolius
D. L. Jones

base, pendent, deep green, glabrous, midrib prominent, leathery, apex finely pointed; **male cones** 2–6 cm × 0.3–0.5 cm, more or less sessile, 1–4 per axil; **female cones** axillary, solitary, 1 seed which can be green, pink, red, blue or blackish, on enlarged, red, fleshy stalk to about 1.8 cm × 1.7 cm.

A rainforest species from the north-eastern region where it grows in a range of soils from sea-level to about 1000 m altitude. This species has been confused with *P. elatus* and *P. grayae* and further botanical study will clarify its status. It is suited to tropical and subtropical regions and should adapt to a variety of freely draining acidic soils. Plants are worth trying in gardens and they should succeed in large containers. May also be trained as bonsai (penjing). Propagate from seed, or from cuttings of firm young growth, which may be slow to produce roots.

Podocarpus smithii de Laub.
(after Lindsay S. Smith, 20th-century Queensland botanist)
Qld
6–20 m × 3–10 m   Nov–Dec
Small to medium **tree** with glaucous young growth; **bark** fibrous; **branches** spreading to ascending; **leaves** 6–16 cm × 1–2 cm, linear, alternate, short-stalked, spreading to ascending, glaucous, glabrous, midrib prominent, apex tapering to a fine point; **male cones** 2–4.5 cm × about 0.5 cm, on stalks to about 1 cm long; **female cones** axillary, 1 green or orange to reddish fruit on enlarged, firm, orange-red stalk.

An endemic species confined to north-eastern Qld where it grows in rainforest and is common on granitic soils at 50–1000 m altitude. Evidently rarely cultivated, it is suited to tropical and subtropical regions. Freely draining acidic soils are required and a semi-shaded or shaded site should be best, but plants may adapt to a sunnier situation. Propagate from seed, or from cuttings of firm young growth, which may be slow to produce roots.

Podocarpus spinulosus (Sm.) R. Br. ex Mirb.
(bearing small spines)
Qld, NSW
0.7–3 m × 1.5–4 m   March–Oct
Dwarf to medium **shrub**; young growth glabrous; **branches** prostrate to erect, often long and slender, sometimes rooting at nodes; **leaves** 1.5–8 cm × 0.2–0.4 cm, narrowly linear to linear-elliptic, alternate, spreading to ascending, moderately crowded, deep green above, paler below, apex pungent-pointed; **male cones** 0.4–0.8 cm long, axillary, clustered; **female cones** usually solitary, 1 seed or rarely 2, on end of glaucous, bluish-black, fleshy stems.

This spreading, often scrambling species occurs in sheltered sites among rocks or in sandy soils of the Leichhardt, Moreton and Wide Bay Districts of Qld. It also extends along the North, Central and South Coast Districts of NSW where it is more common south of the Gosford region. It is cultivated to a limited degree and has proved reliable in subtropical and temperate regions. Plants prefer well-drained acidic soils which are moist for most of the year. Does best in semi-shade but will tolerate greater sunshine. Plants are hardy to

moderately heavy frosts and are excellent in coastal locations. They respond well to pruning and clipping. Suitable for use in rockeries and general planting. Grows successfully as a basket plant. Propagate from seed, or from cuttings of firm young growth, which usually produce roots readily.

Podochilus australiensis (F. M. Bailey) Schltr. =
Appendicula australiensis (F. M. Bailey) M. A. Clem. &
D. L. Jones

PODOLEPIS Labill.
(from the Greek *podos*, foot; *lepis*, scale; referring to the stalked inner involucral bracts)
Asteraceae   Copper-wire Daisie
Annual or perennial **herbs** with woolly-hairy young growth, often glandular or becoming glabrous; **leaves** in a basal rosette and usually withering early, also on stems and alternate, entire; **inflorescence** daisy-like or somewhat hemispherical terminal heads, solitary or in cymes, often with scale-leaves merging with involucral bracts; **outer bracts** sessile; **intermediate bracts** with linear stalks; **ray florets** yellow or rarely white or purple, in 1 row, female; **disc florets** yellow or rarely white, bisexual, tubular; **achenes** terete, with pappus of many bristles.

Podolepis is endemic in Australia and comprises about 20 species. They occur mainly in eastern, southern and western regions with a limited number in Central Australia. Plants are found mainly at low elevations with the exception of *P. monticola* and *P. robusta* which occur in mountainous regions.

The annual species may have a single stem and erect growth habit, or they may have many branching stems. The perennial species have a woody rootstock from which they produce new growth in spring each year. Most species have spreading basal leaves which in many cases wither early while the stem leaves are persistent.

The flowers can be prolific on some species (eg. *P. davisiana, P. gracilis, P. jaceoides, P. lessonii, P. neglecta, P. robusta* and *P. rugata*).

To date *Podolepis* has been very much underrated in horticulture. They are not commonly cultivated but they have potential for use especially in massed plantings and in containers including window boxes and pots of varying sizes. In England in the early 1800s species such as *P. canescens, P. gracilis, P. jaceoides* and *P. rugata* were grown in greenhouses for floral displays. Annual species perform well if they are sown in autumn and are fertilised and watered regularly until flowering commences. Then watering and fertilising should be discontinued in order to promote a floriferous display. Breeding programs within the genus could result in some excellent progeny such as hybrids with compact growth habit and plentiful flowers. The cut-flower stems are also attractive for indoor decoration.

Flowering plants attract butterflies and other insects which drink the nectar.

Most species require freely draining soils and plenty of sunshine to succeed in cultivation. Soils should be acidic to neutral but some species, eg. *P. canescens*, will lerate low alkalinity. *P. monticola* prefers sites which re sunny for only part of the day.

Cold tolerance varies from species to species. *P. robusta* is hardy to snowfalls and most frosts while other species tolerate light to moderate frosts. This aspect needs further study for the majority of species.

Very little is known about the reaction of *Podolepis* to fertilisers but *P. jaceoides*, *P. lessonii*, *P. neglecta* and *P. rugata* var. *rugata* respond very well to low doses of slow-release fertilisers. Further research is required to ascertain the best dosage rates. In many cases plants will also develop well without additional fertilising.

Podolepis require moisture during late winter and spring to promote new growth which will bear flowers. Annual species tolerate wet soils during winter and early spring but need to be drier during late spring and early summer to produce their best flowering. Little is understood yet about their reaction to pruning. The long-stemmed, perennial species, eg. *P. rugata* var. *rugata*, may be coaxed to produce bushy growth by tip-pruning the initial new growth in early spring, otherwise long, unbranched stems can cause plants to become somewhat leggy. Removal of spent flower stems may help to promote further flowering on the perennial species such as *P. jaceoides*, *P. monticola* and *P. rugata*.

Pests and diseases do not seem to pose major problems but caterpillars may chew foliage. See Chewing Insects, Volume 1, page 143 for control methods.

Podolepis are usually propagated from seed. Occasionally division of clumps has been successful. Seed does not require any pre-sowing treatment and usually begins to germinate between 10–30 days after sowing if kept between 15–30°C. Root clumps of perennial species, eg. *P. jaceoides* and *P. monticola*, have been successfully established by dividing them and treating them as cuttings. Further research and testing is necessary before this becomes a popular method of propagation.

It may be possible to propagate some perennial species from stem cuttings. It is suggested that initial growth from rootstock be used in early to mid-spring. Propagation from tissue culture may also be successful for some of the perennial species.

Podolepis acuminata R. Br. = *P. jaceoides* (Sims) Voss

Podolepis acuminata var. **robusta** (Maiden & Betche) J. H. Willis = *P. robusta* (Maiden & Betche) J. H. Willis

Podolepis arachnoidea (Hook.) Druce
(covered with fine cobwebby hairs)
Qld, NSW, Vic, SA   Clustered Copper-wire Daisy;
                                             Cottony Podolepis
0.5–0.8 m × 0.5–1 m   June–Dec; also sporadic
Dwarf perennial **herb** with woolly-hairy young growth; **stems** slender, wiry, few, initially woolly-hairy, becoming glabrous and reddish-brown; **basal leaves** to about 13 cm × 1.8 cm, oblanceolate, many, initially woolly-hairy, often becoming glabrous above, woolly-hairy below; **stem leaves** 2–11 cm × up to 1.6 cm, narrowly lanceolate, few, stem-clasping, becoming glabrous and deep green above, woolly-hairy below, margins revolute, apex softly pointed; **involucral**

bracts reddish-brown; **flowerheads** daisy-like, to about 0.8 cm across and 1 cm long, in compact cymes on slender stems, moderately profuse and conspicuous; **ray florets** usually 5–7, yellow; **disk florets** yellow; **achenes** about 0.2 cm long.

Although widely distributed, this species is rarely abundant in nature. It is mainly represented in central and northern NSW and southern Qld. Plants are found on shallow stony and deep sandy soils of mallee and open woodland. Of moderate ornamental appeal, it is best suited to semi-arid and warm temperate regions. Must have excellent drainage and plenty of sunshine. Hardy to most frosts. Propagate from seed.

P. rhytidochlamys F. Muell. is a synonym.

Podolepis aristata Benth. = *P. canescens* DC.

Podolepis auriculata DC.
(ear-shaped appendages)
WA   Wrinkled Podolepis
0.15–0.3 m × 0.1–0.3 m   Aug–Nov
Dwarf annual **herb** with woolly-hairy young growth; **stems** many, slender, woolly-hairy; **basal leaves** oblanceolate, soon withering; **stem leaves** to about 7 cm × 1 cm, broadly linear to lanceolate, sessile, decurrent, deep green, apex pointed; **involucral bracts** acute, prominently wrinkled; **flowerheads** daisy-like, to about 3.5 cm across, solitary, on slender stems, often profuse and very conspicuous; **ray florets** golden-yellow to yellow-orange; **disc florets** yellow-orange; **achenes** about 0.2 cm long.

A handsome annual of the Avon, Coolgardie and northern Irwin Districts. It grows in red sandy soils and may be associated with saline flats. Cultivated to a limited degree, it has excellent potential as a container or bedding plant. Plants are often stunning when grown in large drifts. They require excellent drainage and maximum sunshine. Hardy to moderate frosts. Should adapt to semi-arid and temperate regions. Propagate from seed.

P. pallida Turcz. is a synonym.

Podolepis canescens DC.
(hoary)
Qld, NSW, Vic, SA, WA, NT   Large Copper-wire Daisy;
                                  Bright Podolepis; Grey Podolepis
0.3–0.9 m × 0.4–1 m   Aug–Dec
Dwarf annual **herb** with woolly-hairy young growth; **stems** slender, spreading to erect, wiry, initially woolly-hairy, becoming glabrous and reddish-brown; **basal leaves** to 12 cm × about 1.5 cm, oblanceolate, becoming deep green and somewhat glabrous above, woolly-hairy below; **stem leaves** to about 10 cm × 1.6 cm, lanceolate to narrowly obovate, stem-clasping, deep green and faintly hairy above, woolly-hairy below, margins recurved, apex softly pointed; **involucral bracts** shiny, pale yellow to golden, or reddish-brown; **flowerheads** daisy-like, 1.5–3 cm across, in loose cymes or rarely solitary, often profuse and very conspicuous; **ray florets** pale to bright yellow; **disc florets** yellow; **achenes** to about 0.2 cm long.

This lovely annual extends over much of the southern and central regions of Australia where it grows in a range of soil types which are often wet for extended

Podolepis capillaris

Podolepis canescens   W. R. Elliot

periods. It warrants wider cultivation and is worth trying as a bedding or container plant. Plants should adapt to most semi-arid, temperate and subtropical regions. They require freely draining soils and a sunny or semi-shaded aspect. Introduced to cultivation in England during the early 1800s when it was grown as *P. aristata* and *P. chrysantha*. Plants are hardy to most frosts. Propagate from seed, which usually begins to germinate 10–35 days after sowing.

Podolepis gardneri is similar but it is smaller and has shorter leaves.

Podolepis capillaris (Steetz) Diels —
see *P. microcephala* Benth.

Podolepis chrysantha Endl. = *P. canescens* DC.

Podolepis davisiana D. A. Cooke
(after Gwenda L. R. Davis, 20th-century Australian botanist)
SA
0.1–0.3 m × 0.1–0.3 m   Aug–Oct
Dwarf annual **herb** with woolly-hairy young growth; **stems** erect, slender, wiry, reddish-brown, becoming glabrous except for nodes; **basal leaves** 1–4 cm long, oblanceolate to elliptic, few, greyish-green, withering; **stem leaves** 2–5.5 cm × up to 0.7 cm, lanceolate, stem-clasping, mainly erect, somewhat woolly above, densely woolly-hairy below, apex softly pointed; **involucral bracts** straw-coloured with bluish-green toning; **flowerheads** to about 1 cm across, hemispherical, on wiry stems to about 4 cm long, in loose cymes, moderately profuse and conspicuous; **florets** tubular, yellow; **achenes** to about 0.15 cm long.

A pretty but poorly known annual from the gibber plains and stony slopes of central SA. It is worthy of cultivation and should do well in semi-arid and warm temperate regions and may succeed in cooler areas i grown in warm sheltered locations. Suitable as a bed ding or container plant. Needs excellent drainage and lots of sunshine. Hardy to moderately heavy frosts Propagate from seed.

Podolepis gardneri G. L. R. Davis
(after Charles Gardner, 20th-century botanist, WA)
WA
0.1–0.3 m × 0.2–0.5 m   July–No
Dwarf annual **herb** with hairy young growth; **stem** many, ascending to erect, much-branched, becom ing glabrous, reddish-brown; **basal leaves** to abou 7 cm × 0.5 cm, tapered to base; **stem leaves** to abou 7 cm × 0.3 cm, broadly linear, deep green, short-hairy apex blunt; **flowerheads** daisy-like, about 2.5 cm across, solitary, on slender wiry stems, often profuse and conspicuous; **ray florets** pale yellow; **disc floret** golden-yellow.

This inhabitant of sandy soils in the Carnarvon Austin and northern Irwin Districts has potential fo cultivation as a bedding or container plant in semi arid and warm temperate regions. It may also have some chance of succeeding in cool temperate areas Excellent drainage and plenty of sunshine are import ant prerequisites for success. Plants should tolerate light to moderate frosts. Propagate from seed.

P. canescens is similar but it is taller and has longe: leaves.

Podolepis gracilis (Lehm.) R. A. Graham
(slender)
WA   Wiry Podolepis; Slender Podolepi
0.2–0.5 m × 0.2–0.4 m   Aug–Jar
Dwarf annual **herb** with hairy young growth; **stem** few, wiry, erect, faintly hairy with age; **basal leave** usually absent; **stem leaves** 5–10 cm × 1–2 cm, broadl linear to obovate, sessile, slightly stem-clasping, dar green, slightly hairy, margins revolute, apex softl pointed; **involucral bracts** straw-coloured to reddish brown; **flowerheads** daisy-like, 1–3 cm across, solitary or in loose many-headed cymes, on slender stalks rarely profuse; **ray florets** white or pale pink to purplish-pink; **disk florets** yellow; **achenes** to abou 1.5 cm long.

P. gracilis is a dainty species when in flower. It occur in the Avon, Darling and Irwin Districts in gravelly and sandy soils which can contain limestone. It is commor in woodland. Uncommon in cultivation, it should be better known because of its ornamental qualities Plants are suited to semi-arid and temperate regions Should prove valuable as a bedding or container plant Needs very good drainage and plenty of sunshine Tolerates moderate frosts. Introduced to gardens in England in 1826. Propagate from seed, which usually begins to germinate 7–25 days after sowing.

The closely allied and decorative *P. nutans* AC. ha varnished involucral bracts and yellowish-white ray florets which may have purplish tips. It is poorly knowr and occurs in the Darling and western Eyre Districts Plants have similar cultivation requirements to *P. gra cilis*. Seed is sometimes offered by commercial seed companies.

Podolepis hieracioides F. Muell.
(similar to the genus *Hieracium*)
NSW, Vic　　　　　　　　　Long Podolepis
0.4–0.8 m × 0.2–0.7 m　　　　　　Oct–Feb

Dwarf perennial **herb** with sticky, hairy young growth; **stems** solitary or few, erect, hairy to becoming glabrous; **basal leaves** to about 16 cm × 3 cm, elliptic, stem-clasping, coarsely hairy; **stem leaves** to 13 cm × 1 cm, linear to narrowly lanceolate, stem-clasping, coarsely hairy, apex pointed; **involucral bracts** ovate, smooth, straw-coloured; **flowerheads** daisy-like to 2 cm or more across, usually 2–6 heads per loosely arranged cyme, can be profuse and conspicuous; **ray florets** golden-yellow; **disk florets** yellow; **achenes** about 0.3 cm long.

A decorative annual from the Tablelands of NSW and eastern and north-eastern Vic. It is found in deep soils in low-lying areas and on shallow often stony soils on hillsides. This species has rarely been cultivated with much success, but it warrants greater attention. Needs acidic soils with good drainage and a sunny or semi-shaded site. Suitable as a bedding or container plant in temperate regions and may succeed in sub-tropical areas. Hardy to moderately heavy frosts. Propagate from seed.

Podolepis jaceoides (Sims) Voss
(similar to Jacea, Spanish name for knapweed)
Qld, NSW, Vic, Tas, SA　　　Showy Copper-wire Daisy;
　　　　　　　　　　　　　　Showy Podolepis
0.3–0.8 m × 0.3–0.7 m　　　　　　Oct–Feb

Herbaceous **subshrub** with a perennial rootstock and hairy young growth; **stems** arise from rootstock annually, few to many, erect, usually becoming glabrous and deep reddish-brown; **basal leaves** 10–20 cm × 1–2 cm, oblanceolate, many, persistent; **stem leaves** 1–5 cm × 0.2–1 cm, narrowly lanceolate, few, faintly stem-clasping, becoming rough and deep green above, woolly-hairy below, margins revolute, apex pointed; **involucral bracts** ovate, straw-coloured, smooth; **flowerheads** daisy-like to about 3 cm across, solitary to a few heads per cyme, on slender stems, often profuse and very conspicuous; **ray florets** bright yellow; **disc florets** yellow; **achenes** about 0.3 cm long.

An extremely attractive species which is widely distributed over eastern and south-eastern regions where

Podolepis jaceoides　　　　　　　　　W. R. Elliot

Podolepis lessonii　　　　　　　　　W. R. Elliot

it is found in heavy clay, clay loam and sandy soils in woodland, mallee and grasslands. The roots are edible and were utilised by Aborigines and early European settlers. It is often found in moderately rich soils. Cultivated to a limited degree, but deserves to be better known because it has excellent potential as a plant for borders, containers and floral bedding displays. It is suitable for semi-arid, temperate and sub-tropical regions. Plants are hardy to most frosts and although tolerating extended dry periods they respond very well to supplementary watering. Removal of spent flowering stems promotes flowering. Introduced into England in 1803 as *P. acuminata*. Propagate from seed. Division of rootstock is also successful.

Podolepis kendallii (F. Muell.) F. Muell.
(after Franklin Kendall, 19–20th-century amateur botanical collector)
WA
0.1–0.4 m × 0.1–0.5 m　　　　　　Aug–Oct

Dwarf annual **herb** with faintly woolly-hairy young growth; **stems** solitary or a few, erect, becoming glabrous; **basal leaves** absent; **stem leaves** to about 5 cm × 0.2 cm, linear to filiform, stem-clasping, crowded, deep green, becoming glabrous; **involucral bracts** wrinkled, margins thin and dry; **flowerheads** to about 2.5 cm across, more or less hemispherical, solitary, often profuse and very conspicuous; **florets** all tubular, bright yellow; **achenes** about 0.2 cm long.

An attractive annual from the Austin, Fortescue and Irwin Districts. It grows in sandy and clay soils of open mallee woodland. Plants have potential for use in semi-arid and warm temperate regions and may succeed in cooler zones. They need moderately good drainage and maximum sunshine. They are hardy to moderate frosts. Suitable for massed planting and in containers. Propagate from seed, which usually begins to germinate 6–25 days after sowing.

Podolepis lessonii (Cass.) Benth.
(after René P. Lesson, 19th-century French pharmacist, ornithologist and botanist)
WA
0.1–0.4 m × 0.1–0.4 m   Aug–Jan

Dwarf annual **herb**; young growth woolly-hairy; **stems** usually many, slender, ascending to erect, sparsely woolly-hairy or becoming glabrous; **basal leaves** to about 4.5 cm × 1.2 cm, woolly-hairy; **stem leaves** to about 7 cm × 1.5 cm, narrowly obovate to ovate, stem-clasping, deep green and sparsely hairy above, woolly-hairy below, apex softly pointed; **involucral bracts** pale, semi-transparent, faintly wrinkled; **flowerheads** to about 1.2 cm across, semi-hemispherical, solitary, terminal, on long, slender, reddish-brown stems, often profuse and conspicuous; **florets** all tubular, bright yellow; **achenes** 0.1 cm long, terete.

A floriferous dwarf annual which grows in a range of well-drained soils in open woodland and mallee of the Austin, Avon, Coolgardie, Darling, Eyre and Irwin Districts. Plants adapt well to cultivation in temperate regions and should succeed in semi-arid zones. A sunny site with freely draining sandy or clay loam soils is suitable. Ideal for mass planting of borders or informal drifts as well as in containers. Propagate from seed, which usually begins to germinate 5–30 days after sowing.

P. muelleri is similar but it has smaller bright green leaves.

P. tepperi AC, Delicate Everlasting, is also similar but it differs in having flower-buds which are greater in length than breadth. It is widely distributed and occurs in north-western Vic, southern SA and in the Avon, Coolgardie and Roe Districts of southern WA. It has similar cultivation requirements to those of *P. lessonii*.

Podolepis longipedata DC.
(long stalks or stems)
Qld, NSW, SA   Tall Copper-wire Daisy
0.4–1 m × 0.3–0.8 m   Oct–Dec; also sporadic

Dwarf annual **herb** with woolly-hairy young growth; **stems** few, erect, branched, wiry, brownish, hairy or becoming glabrous; **basal leaves** 3–10 cm × 0.4–1.2 cm, oblanceolate, few, tapered to base, slightly woolly-hairy, becoming withered; **stem leaves** 1–10 cm × 0.1–0.5 cm, linear to narrowly lanceolate, somewhat stem-clasping, ascending to erect, slightly woolly-hairy to nearly glabrous, apex softly pointed; **flowerheads** daisy-like, to about 3 cm across, in loose terminal cymes, often profuse and very conspicuous; **involucral bracts** straw-coloured with thin margins; **ray florets** bright yellow; **disc florets** yellow; **achenes** about 0.2 cm long, with many bristles.

A wide-ranging species which has its main distribution in south-eastern Qld and north-eastern NSW, with a disjunct occurrence in the Gawler Range region of SA. It grows in sandy soils and is often found on low sand dunes. Warrants attention for cultivation in subtropical, temperate and semi-arid regions. Plants will need excellent drainage and a very sunny site. Hardy to light frosts. Worth trying in mass plantings and as a

container plant. Aborigines once ate the thickened roots. Propagate from seed.

The var. *robusta* Maiden & Betche is a synonym of *P. robusta* (Maiden & Betche) J. H. Willis.

Podolepis microcephala Benth.
(small heads)
WA   Small-headed Podolepis
0.1–0.3 m × 0.1–0.4 m   Sept–Nov

Dwarf annual **herb** with glabrous young growth; **stems** many, ascending to erect, slender, glabrous, reddish-purple and glaucous; **basal leaves** absent; **stem leaves** 0.5–1.2 cm × about 0.15 cm, narrowly linear, scattered, glabrous, apex blunt; **flowerheads** about 1 cm across, somewhat hemispherical, solitary, terminal, on slender stems, often profuse and conspicuous; **involucral bracts** obtuse, glandular; **ray florets** yellow, about as long as the tubular **disc florets**; **achenes** to about 0.1 cm long.

A dainty but poorly known annual from the Carnarvon and Irwin Districts where it grows in shell dunes and sandy soils which are often adjacent to saline flats. It has potential for use in semi-arid and warm temperate regions and may be successful in cooler zones. Freely draining soils are essential and a sunny site is important. Suitable as a bedding or container plant. Propagate from seed.

The Wiry Podolepis, *P. capillaris* (Steetz) Diels, differs in having smaller elongated flowerheads which can be yellow or white. The achenes are without hairs. It occurs in Qld, NSW, Vic, SA, WA and NT and has similar cultivation requirements to *P. microcephala*. Some botanists believe this species should be placed in the genus *Siemssenia*.

Podolepis monticola R. J. F. Hend.
(mountain dweller)
Qld
0.3–0.6 m × 0.3–0.6 m   Sept–Feb

Dwarf perennial **herb** with woolly-hairy young growth; **stems** ascending to erect, woolly-hairy; **basal leaves** 10–21 cm × 3.5–6.5 cm, ovate to obovate, tapered to long petiole, stem-clasping, rough above, woolly-hairy below; **stem leaves** similar to basal leaves but narrower and smaller, reducing in size ascending the stem, apex pointed; **flowerheads** daisy-like, to about 3 cm across, up to 10 in terminal or axillary cymes, with woolly-hairy stems, often profuse and very conspicuous; **involucral bracts** straw-coloured, ovate; **ray florets** yellow; **disc florets** yellow; **achenes** to about 0.3 cm long.

A very handsome species from the McPherson Ranges of south-eastern Qld where it grows in rock crevices on mountain hillsides at about 1100 m altitude. Worthy of greater cultivation. Suited to subtropical and temperate regions. Plants require freely draining acidic soils and should adapt to a sunny or semi-shaded site. They are hardy to moderate frosts. Worth trying in general planting and as a bedding plant. Should also do well in containers. Propagate from seed, or from divisions of rootstock, which are treated as cuttings.

P. jaceoides and *P. longipedata* are similar but they have smaller basal leaves.

Podolepis neglecta × .5

Podolepis muelleri (Sond.) G. L. R. Davis
(after Baron Sir Ferdinand von Mueller, 19th-century
Government Botanist of Victoria)
NSW, SA   Small Copper-wire Daisy
0.1–0.3 m × 0.1–0.5 m   Aug–Oct
    Dwarf annual **herb** with woolly-hairy young growth;
stems few, ascending to erect, slender, wiry, reddish-
brown, faintly cobwebby to becoming glabrous; **basal
leaves** 1–4 cm long, elliptical to oblanceolate, few,
becoming withered; **stem leaves** 1–4 cm × 0.3–0.7 cm,
lanceolate, scattered, stem-clasping, hairy, margins
revolute, apex softly pointed; **flowerheads** to about
1 cm across, somewhat cup-shaped, terminal, solitary,
on thread-like stems, can be profuse and conspicuous;
involucral bracts straw-coloured, broadly ovate; **florets**
tubular, yellow; **achenes** about 0.15 cm long, with few
bristles.
    This interesting annual is found in western NSW
and central southern SA on heavy soils and stony sites
ranging from the coast to some distance inland. Often
associated with grassland, saltbush shrubland and
woodland. It has potential for massed planting and as
a container plant. Plants require fairly good drainage
and plenty of sunshine. They tolerate moderately
heavy frosts. Propagate from seed.
    P. lessonii differs in having larger stem leaves which
are deep green.

Podolepis neglecta G. L. R. Davis
(neglected)
Qld, NSW
0.3–1 m × 0.3–0.8 m   Sept–Feb; also sporadic

Dwarf perennial **herb** with glabrous or faintly hairy
young growth; **stems** few, ascending to erect, glabrous
or faintly hairy; **basal leaves** withering early; **stem
leaves** 6–12 cm × 1–2 cm, narrowly lanceolate to
oblong, sessile, glabrous, hairy or faintly woolly,
margins flat, apex softly pointed; **flowerheads** daisy-
like, to about 3 cm across, up to about 16 heads in
terminal clusters, often profuse and very conspicuous;
involucral bracts smooth, shining, rigid; **ray florets**
yellow; **disc florets** tubular, yellow; **achenes** to about
0.3 cm long.
    A floriferous and ornamental species from north-
eastern NSW and south-eastern Qld. Plants occur in
well-drained acidic soils in woodland. Evidently rarely
cultivated, it should grow in subtropical and temperate
regions. Freely draining soils and a sunny site will
provide optimum conditions. Plants tolerate light to
moderate frosts. Propagate from seed.

Podolepis nutans Steetz —
                see *P. gracilis* (Lehm.) R. A. Graham

Podolepis pallida Turcz. = *P. auriculata* DC.

Podolepis robusta (Maiden & Betche) J. H. Willis
(robust)
NSW, Vic   Alpine Podolepis; Mountain Lettuce
0.4–0.8 m × 0.2–0.5 m   Dec–March
    Dwarf perennial **herb** with a woody rootstock and
faintly to densely hairy young growth; **stems** solitary to
a few, thick, erect, faintly to densely cottony-hairy;

Podolepis robusta × .5

Podolepis rugata

basal leaves to about 25 cm × 2–5 cm, oblong to narrowly spathulate, stem-clasping, faintly hairy or becoming glabrous, bright green, somewhat fleshy, margins smooth or crinkled; **stem leaves** to about 15 cm long, reducing in size up the stem, narrowly elliptic to oblong, bright green, more or less glabrous; **flowerheads** daisy-like, to about 3.5 cm across, 3–11 per contracted cyme, often profuse and very conspicuous; **involucral bracts** straw-coloured, shining, with thin margins; **ray florets** deep yellow to orange; **disc florets** tubular, golden-yellow to orange; **achenes** to about 0.4 cm long.

An extremely showy member of the genus which is a relatively common plant of subalpine regions especially in the grasslands and woodlands. It is cultivated with varying degrees of success. Plants require moist but well-drained acidic soils with a moderately sunny aspect. May prove useful in winter-cold regions for rock gardens, bedding displays and containers such as patio pots. Plants require winter chilling to produce strong growth and to flower well. They are hardy to frost and snow. Propagate from seed which may benefit from stratification at 4°C for 3–6 weeks before sowing. Seedlings are very sensitive to 'damping-off' fungi.

Podolepis rugata Labill.
(wrinkled)
Vic, SA, WA   Pleated Podolepis
0.1–0.7 m × 0.3–1 m   Sept–Jan

Dwarf perennial **herb** with a woody rootstock; young growth glabrous to cobwebby; **stems** several, spreading to erect, usually unbranched, glabrous to faintly hairy; **basal leaves** to about 8 cm long, soon withering; **stem leaves** 4–10 cm × 0.8–1.5 cm, linear to elliptical, somewhat stem-clasping, ascending to erect, deep green, glabrous to faintly cobwebby, margins flat, apex softly pointed; **flowerheads** to about 4 cm across, solitary or a few in loose terminal cymes, on stalks to about 10 cm

Podolepis monticola   D. L. Jones

Popolepis rugata   W. R. Elliot

long, often profuse and very conspicuous; **involucral bracts** ovate, straw-coloured to reddish-brown; **ray florets** bright yellow; **disc florets** yellow; **achenes** to about 0.3 cm long.

The typical variant of this ornamental podolepis occurs from western Victoria, through southern SA and across southern WA. It is found in woodland and mallee and can be common on sand dunes. Cultivated mainly by enthusiasts, it warrants greater attention. It is suitable for temperate and semi-arid regions. Adapts to sandy or clay loam soils which are relatively well drained. Maximum sunshine promotes bushier growth in comparison with semi-shaded sites. Hardy to moderately heavy frosts. Tip pruning of very young growth in early spring may promote branching of stems. Introduced to gardens in England in 1803.

The var. *littoralis* G. L. R. Davis is a coastal variant which inhabits cliffs and sand dunes. It has crowded, somewhat succulent leaves on short stems. Flowerheads are solitary and on distinct stalks to about 15 cm long.

Propagate both variants from seed which usually begins to germinate between 10–25 days after sowing.

Podolepis tepperi (F. Muell.) D. A. Cooke —
     see *P. lessonii* (Cass.) Benth.

PODOLOBIUM R. Br.
(from the Greek *podos*, a foot; *lobos*, a pod; referring to the stalked pods)
Fabaceae (alt. *Papilionaceae*)   Shaggy Peas

Shrubs; **leaves** opposite or sometimes whorled, entire or lobed; **stipules** more or less rigid, spreading to recurved, persistent; **inflorescence** solitary flowers or terminal axillary racemes; **calyx** 2-lipped, 5-lobed, more or less equal, upper lobes broader; **flowers** pea-shaped, yellow to orange; **stamens** free; **fruit** a pod, 2-valved, inflated, hairy.

An endemic genus of 6 species distributed in eastern Australia extending from south-eastern Qld to eastern Vic. Until recently plants were included in *Oxylobium* but are now placed in *Podolobium*, which was first described in 1811.

Podolobium species occur in woodland, open forest and heathland and are found from near sea-level to above the tree line in subalpine regions. They are renowned for their profuse floral displays which are generally seen for a 4–6 week period in spring and summer. The flowers of some species are sweetly fragrant and the nectar is often sought by insects.

They are cultivated to a limited degree but warrant greater attention. Plants do best in temperate or subtropical regions. They need well-drained soils with a sunny or semi-shaded aspect. Fertilising is rarely needed and once plants are established supplementary watering is required mainly during extended dry periods.

They are susceptible to Cinnamon Fungus, *Phytophthora cinnamomi*, see Volume 1, page 168 for treatment and control methods.

Plants are propagated from seed or cuttings. Seed requires pre-sowing treatment to soften or damage the seed coat and allow penetration of water. See Treatments and Techniques to Germinate Difficult Species, Volume 1, page 205. Cuttings of firm young growth usually produce roots in 4–12 weeks.

Podolobium aciculiferum F. Muell.
(bearing needles)

Qld, NSW	Needle Shaggy Pea
1.5–2.5 m × 1–2 m	Nov–March; also sporadic

Small to medium **shrub** with hairy young growth; **branches** hairy, often arching; **leaves** 1–3 cm × 0.3–1 cm, narrowly ovate to broadly ovate to nearly rhomboid, base rounded or very blunt, spreading to ascending, leathery, glabrous and glossy above, paler and sometimes faintly hairy below, margins entire, venation reticulate, apex pungent-pointed; **stipules** to 0.7 cm long needle-like; **flowers** pea-shaped, to about 0.8 cm across, yellow, solitary or in racemes in upper axils, scattered, rarely profuse; **calyx** slightly hairy; **pods** 1–1.5 cm long, curved.

An extremely prickly species which usually occurs on stony soils. Its distribution extends from the Port Curtis District, Qld, to the South Coast in NSW. Plants are found mainly on the coast and adjacent areas in open forest. Rarely cultivated, it should be an excellent foot traffic control plant for subtropical and temperate regions in sunny or semi-shaded sites with good drainage. Hardy to light frosts. Propagate from seed or from cuttings of firm young growth.

Oxylobium aciculiferum (F. Muell.) Benth. is a synonym.

Podolobium aestivum Crisp & P. H. Weston —
see *P. ilicifolium* (Andrews) Crisp & P. H. Weston

Podolobium alpestre (F. Muell.) Crisp & P. H. Weston
(of the highlands)

NSW, Vic	Alpine Shaggy Pea;
	Mountain Shaggy Pea; Alpine Oxylobium
0.5–2 m × 1.5–4 m	Oct–Dec

Podolobium procumbens   W. R. Elliot

Dwarf to small **shrub** with silky-hairy young growth; **branches** spreading to ascending, sometimes wiry; **leaves** 1–4 cm × 0.3–1.3 cm, lanceolate to oblong-elliptic, opposite or in whorls of 3, short-stalked, leathery, dark green and reticulate above, paler and hairy to glabrous below, margins recurved, apex blunt or pointed; **stipules** to 0.3 cm long, recurved; **flowers** pea-shaped, 1–1.2 cm across, terminal or axillary clusters, often profuse and very conspicuous, light sweet

Podolobium alpestre × .55

Podolobium ilicifolium

Podolobium ilicifolium × .8

fragrance; **standard** orange-yellow with reddish basal blotch; **wings** and **keel** usually reddish; **pods** 1–1.5 cm long, hairy, dark grey.

Plants are very common in subalpine forests in south-eastern NSW and eastern Vic and they also occur above the tree line. *P. alpestre* is often the predominant species and it can form dense impenetrable colonies. Cultivated to a limited degree, it has proved adaptable to a wide range of acidic soils in temperate regions. Hardy to frosts and snow and tolerates extended dry periods. Prefers a semi-shaded site but withstands plenty of sunshine. Responds well to pruning and is useful for informal hedging. Excellent for soil erosion control in mountainous regions. Propagate from seed or from cuttings of firm young growth.

Oxylobium alpestre F. Muell. is a synonym.

Podolobium ilicifolium (Andrews) Crisp & P. H. Weston
(leaves similar to the genus *Ilex*)
Qld, NSW, Vic   Prickly Shaggy Pea
1–3 m × 0.6–2 m   Sept–Dec; also sporadic
Small to medium **shrub** with rusty-hairy to nearly glabrous young growth; **branches** ascending to erect; **branchlets** glabrous to faintly hairy; **leaves** 2–10 cm × 1–3 cm, narrowly ovate to ovate in outline, opposite, spreading to ascending, sometimes recurved, deep green, glabrous and glossy above, prominent reticulate venation, irregularly and deeply lobed with pungent points, apex pungent-pointed; **flowers** pea-shaped, to

about 1 cm across, yellow to orange-yellow with reddish markings, in axillary racemes often longer than leaves, often profuse and very conspicuous; **pods** about 1 cm × 0.3 cm, ovoid to oblong, more or less hairy.

A very distinctive and ornamental species which is found in south-eastern Qld, eastern NSW and far-eastern Gippsland in Vic. It occurs in dry and wet sclerophyll forest and grows in sandy or clay soils. Plants are often associated with rocky sites. *P. ilicifolium* is a prickly plant which makes it useful for foot traffic control. Suitable for subtropical and temperate regions. Does best in moderately well-drained acidic soils with a sunny or semi-shaded aspect. Hardy to moderately heavy frosts. Responds well to pruning or clipping. Introduced to cultivation in England during 1791 as *P. trilobatum* and in 1822 as *P. staurophyllum*. Propagate from seed or from cuttings of firm young growth which may be slow to produce roots.

Oxylobium ilicifolium (Andrews) Domin is a synonym.

The recently described (1995) *P. aestivum* Crisp & P. H. Weston is closely allied but it has hairy, shallowly and more-or-less regularly lobed, oblong to ovate leaves. The flowers are orange and bloom in late spring and summer with the main display in December. It occurs in dry sclerophyll forest of north-eastern NSW often in rocky soils associated with basalt. Cultivation requirements are similar to those of *P. ilicifolium*.

Podolobium procumbens (F. Muell.) Crisp & P. H. Weston
(procumbent)
NSW, Vic   Trailing Shaggy Pea
0.1–0.3 m × 0.3–1.3 m   Oct–Jan
Dwarf spreading **shrub** with a woody rootstock; young growth faintly hairy; **branches** prostrate to ascending; **branchlets** becoming glabrous; **leaves** 1–2.5 cm × 0.6–1.2 cm, ovate, opposite or whorled, spreading to ascending, somewhat crowded at ends of branchlets, deep green, glabrous, prominent reticulate venation, apex pungent-pointed; **flowers** pea-shaped, to about 1.5 cm across, orange or rarely cream or yellow, in terminal and axillary racemes, often profuse and very conspicuous; **pods** to about 1.5 cm long, hairy, straight or curved.

An attractive species from south-eastern NSW to central Vic. It inhabits dry sclerophyll forest and woodland and is often found in rocky and shaley clay soils. Cultivated to a limited degree, it is best suited to temperate regions. Does well in semi-shade but also tolerates plenty of sunshine. Needs moderately well-drained acidic soils. Hardy to heavy frost and withstands light snowfalls. Old plants can be rejuvenated by hard pruning which promotes new growth from the woody rootstock. Does well in gardens or containers. Propagate from seed or from cuttings of firm young growth.

Oxylobium procumbens F. Muell. is a synonym.

Podolobium scandens (Sm.) DC.
(climbing)
Qld, NSW   Climbing Shaggy Pea; Netted Shaggy Pea
prostrate–1 m × 0.8–1.5 m   Sept–Nov

Dwarf trailing **shrub** with hairy young growth and a woody rootstock; **branches** slender, prostrate to ascending; **branchlets** initially hairy; **leaves** 1–7 cm × 0.5–2.5 cm, usually opposite, obovate or ovate to elliptic, not crowded, dark green and glossy above, faintly hairy below, margins more or less crenulate, venation reticulate, apex pointed; **stipules** bristle-like; **flowers** pea-shaped, about 1.5 cm across, in axillary or terminal racemes, on slender stalks beyond the leaves, very conspicuous; **standard** pale orange to yellow, usually with reddish-brown markings; **keel** and **wings** yellow; **calyx** rusty-hairy; **pods** to about 1.5 cm long, ovoid to oblong, hairy.

An appealing, trailing species which occurs in south-eastern Qld and extends to near Bodalla in the South Coast District, NSW. Usually found in gravelly clay soils of the coast and adjacent ranges. Adapts well to cultivation but is rarely encountered. Plants need moderately well-drained acidic soils in a semi-shaded site for best results. They will scramble through other plants or they can be used as a groundcover in open sites. Tolerant of short dry periods and hardy to moderate frosts. Responds well to hard pruning. Sensitive to *Phytophthora cinnamomi*. Introduced into England during 1824. Propagate from pre-treated seed or from cuttings of firm young growth.

Previously known as *Oxylobium scandens* (Sm.) Benth.

PODOPETALUM F. Muell. =*ORMOSIA* Jacks.

PODOSPERMA Labill. = *PODOTHECA* Cass.

PODOSTEMACEAE Rich. ex C. A. Agardh.
A family of dicotyledons consisting of about 130 species in 45 genera. It is strongly represented in the American tropics but there are only 2 genera and 2 species which occur in Australia. They are remarkable aquatic herbs, usually annuals, found growing on rocks in fast-flowing water, even in cataracts and rapids. The tiny growths occur in mats and often resemble mosses, with the leaves and stems being greatly modified. The flowers are often inconspicuous. Flowering occurs in the wet season and the seeds are shed as waters subside during the dry season, germinating when submerged by rising waters at the onset of the wet season. Australian genera: *Torrenticola, Tristicha*.

PODOTHECA Cass.
(from the Greek *podos*, foot; *thece*, box; referring to the stalked achenes)
Asteraceae (alt. *Compositae*)   Longheads
Spreading to erect annual **herbs** with sticky young growth; **stems** simple or branched; **leaves** mainly alternate, sessile, entire; **inflorescence** solitary, terminal, elongated flowerheads with many overlapping involucral bracts; **florets** all tubular, bisexual, yellow; **achenes** more or less cylindrical, short-stalked, hairy.

A small endemic genus of 6 species which is represented in all states except Qld and NT. The main distribution occurs in south-western WA where all 6 species are found. They are not common in cultivation. Some species such as *P. chrysantha* and *P. wilsonii*

are desirable for cultivation. All species are best suited for use as bedding plants or for growing in containers such as window boxes and patio pots. They require good drainage and plenty of sunshine. Pruning or shearing of young plants between 3–6 weeks old should promote lateral growth and therefore the production of more flowerheads. Response to regular applications of liquid fertiliser is not known.

Propagation is from seed which does not require pre-sowing treatment. Germination usually begins between 7–30 days after sowing. Trials of direct sowing of seed into garden beds or plantings should be undertaken. For further information on methods, see Annuals, Volume 2, page 202.

Podotheca angustifolia (Labill.) Less.
(narrow leaves)
NSW, Vic, Tas, SA, WA   Sticky Heads; Sticky
                                                         Longheads
0.05–0.3 m × 0.05–0.3 m   Sept–Nov
Dwarf annual **herb**; young growth glandular-hairy; **stems** prostrate to ascending, sometimes cottony, with glandular hairs; **leaves** 0.6–9 cm × 0.1–0.7 cm, linear to oblanceolate, alternate except for lower pairs, sessile, ascending to erect, green, glandular-hairy, thick, often somewhat succulent; **flowerheads** 1.2–4.5 cm × up to 1.2 cm, solitary, terminal, can be profuse; **bracts** ovate to narrowly triangular, green to purplish, hairy; **florets** tubular, yellow, sometimes with purplish tinges; **achenes** to about 0.4 cm long, with 5 bristles.

A widespread species ranging from far south-western NSW, mainly western Vic, Bass Strait Islands, southern SA and south-western WA. It occurs in varied coastal and inland habitats. *P. angustifolia* has minimal ornamental appeal but it may be useful for regeneration projects. Should adapt to a wide range of soil types and needs a moderate amount of sunshine. Hardy to moderately heavy frosts. First cultivated in England in 1835 where it was described as 'an annual of no great beauty'. Propagate from seed which usually begins to germinate 7–20 days after sowing.

Podotheca chrysantha (Steetz) Benth.
(golden flowers)
WA   Golden Longheads; Yellow Podotheca
0.1–0.35 m × 0.1–0.4 m   Aug–Dec
Dwarf annual **herb** with glandular-hairy young growth; **stems** spreading to erect, brownish, glandular-hairy; **leaves** 0.5–8.5 cm × 0.1–0.65 cm, linear to lanceolate, sessile, ascending to erect, green, glandular-hairy, apex often incurved; **flowerheads** 1–2.5 cm × 0.4–3 cm, terminal, solitary, often profuse and very conspicuous; **bracts** narrowly elliptic to narrowly triangular, hairy, green; **florets** tubular, golden yellow; **achenes** to about 0.3 cm long with 9–11 bristles.

This charming annual occurs in the Austin, Avon, Coolgardie and Darling Districts. It is found in woodland and heath growing in sandy soils which may overlie limestone. Plants need excellent drainage and plenty of sunshine. They tolerate moderately heavy frosts. It has potential for use as a formal and informal bedding plant and is well suited to cultivation in containers such as window boxes and patio pots. May be

Podotheca fuscescens

Podotheca chrysantha × .45

grown in temperate and semi-arid regions. Propagate from seed, which usually begins to germinate 7–20 days after sowing.

Podotheca fuscescens (Turcz.) Benth. =
Rhodanthe fuscescens (Turcz.) Paul G. Wilson

Podotheca gnaphalioides Graham
(similar to the genus *Gnaphalium*)
WA   Golden Longheads; Dwarf Longheads
0.1–0.6 m × 0.2–0.6 m   Aug–Nov
    Dwarf annual **herb**; young growth glandular-hairy; **stems** prostrate to erect, with glandular and non-glandular hairs or rarely glabrous, green to purple;

Podotheca angustifolia   D. L. Jones

leaves 0.1–10 cm × 0.1–0.5 cm, linear to narrowly elliptic, stem-clasping, ascending to erect, green, hairy, apex pointed; **flowerheads** 2–5 cm × 0.3–1.5 cm, terminal, solitary, often profuse and conspicuous; **bracts** ovate to narrowly triangular, dark green to purple-green, hairy; **florets** tubular, yellow to yellow-orange; **achenes** to about 0.3 cm long, with 5 bristles.
    An interesting annual which is cultivated to a very limited degree. It has potential for use in temperate and semi-arid regions in miniature gardens, as a bedding plant and for cultivation in containers. It occurs over a wide area of southern south-western WA extending from near Dirk Hartog Island to near the south coast. In nature, it favours sandy soils or clay loams and it is sometimes found in saline depressions. Plants are often associated with woodland, mallee and heath vegetation. Needs good drainage and plenty of sunshine. Tolerates moderately heavy frosts. First cultivated in England in 1841. Propagate from seed which usually begins to germinate 6–25 days after sowing.
    P. pygmaea A. Gray is a synonym. *P. pritzelii* is allied but has smaller flowerheads and succulent leaves and bracts.

Podotheca pollackii (F. Muell.) Diels =
    Rhodanthe pollackii (F. Muell.) Paul G. Wilson

Podotheca pritzelii P. S. Short
(after Ernst G. Pritzel, 19–20th-century German botanist)
WA
0.05–0.25 m × 0.1–0.3 m   Sept–Nov
    Dwarf annual **herb** with hairy young growth; **stems** ascending to erect, hairy; **leaves** 1–3.5 cm × 0.1–0.25 cm, linear to lanceolate, sessile, ascending to erect, pale green to purplish, hairy, succulent; **flowerheads** 2–2.6 cm × to 0.8 cm, terminal, sessile, can be profuse and conspicuous; **bracts** ovate to narrowly triangular, pale green or purplish, succulent; **florets** tubular, yellow-orange; **achenes** about 0.15 cm long, with 5 bristles.
    An apparently rare species that was described in 1989. It is known from only one location near Wongan Hills where it grows in saline, sandy soils with samphire and melaleuca. Warrants cultivation as part of a conservation strategy. Needs plenty of sunshine and seasonally wet soils. Hardy to moderately heavy frost. Propagate from seed.

Podotheca pygmaea A. Gray = *P. gnaphalioides* Graham

Podotheca uniseta P. S. Short
(one bristle)
WA
0.05–0.25 m × 0.1–0.3 m   Aug–Sept
    Dwarf annual **herb** with hairy young growth; **stems** ascending to erect, hairy; **leaves** 1–4.5 cm × 0.1–0.4 cm, linear to lanceolate, sessile, ascending to erect, succulent, green, red or purple; **flowerheads** 2–3 cm × up to 1 cm, terminal, solitary, profuse and conspicuous; **bracts** ovate to narrowly triangular, somewhat fleshy, green, sometimes with purple tinge; **florets** tubular, yellow and usually with purplish tips; **achenes** to about 0.2 cm long, with one bristle.

A recently described (1989) species which inhabits samphire zones surrounding salt lakes in the northern Avon District. It is moderately attractive and has potential for cultivation in semi-arid and temperate regions. Plants need plenty of sunshine and good drainage. They tolerate moderate frosts. Worth trying as a bedding and container plant. Propagate from seed.

P. pritzelii is closely allied but its bracts are bright green and more succulent. It also has achenes with one bristle.

Podotheca wilsonii P. S. Short
(after Paul G. Wilson, 20th-century botanist, WA)
WA
0.07–0.45 m × 0.15–0.6 m   Sept–Dec

Dwarf annual **herb** with glandular-hairy young growth; **stems** ascending to erect, purplish, glandular-hairy; **leaves** 0.5–13.5 cm × to about 1 cm, linear to lanceolate, sessile, ascending to erect, green to purplish, glandular-hairy, apex often incurved; **flowerheads** 1.5–3.5 cm × 0.7–3 cm, terminal, solitary, possibly profuse and very conspicuous; **bracts** linear to ovate or triangular, glandular-hairy; **florets** tubular, yellow; **achenes** to about 0.2 cm long, with 5 bristles.

This species from the Austin, Avon and Coolgardie Districts was described in 1989. It is found in saline soils which are usually sandy. Often grows in association with samphire, *Atriplex, Frankenia* and *Melaleuca* species. It is attractive during flowering and has potential for use as a bedding or container plant. Plants will be best suited to semi-arid and warm temperate regions. They need plenty of sunshine and moderately good drainage and they tolerate fairly heavy frosts. Propagate from seed.

P. chrysantha is allied but it is usually somewhat erect and it has achenes with 9–11 bristles.

POGONOLEPIS Steetz
(from the Greek *pogon*, beard; *lepis*, a scale; referring to the bearded involucral bracts)
Asteraceae (alt. *Compositae*)

Annual **herbs**; **stems** simple or branched; **leaves** alternate or sometimes opposite, sessile, entire, glabrous to faintly hairy; **inflorescence** compound head of 5–40 heads, more or less broadly obovoid, woolly-hairy; **involucral bracts** in many rows, with outer bracts leaf-like, inner bracts hairy at apex; **florets** 5-merous, 1 per head, bisexual, yellow; **stamens** 5; **achene** obovoid; **pappus** absent.

An endemic genus of 2 species found in southern Australia. Previously included in *Angianthus* and *Skirrhophorus*.

Pogonolepis lanigera (Ewart & Jean White) P. S. Short =
P. stricta Steetz

Pogonolepis muelleriana (Sond.) P. S. Short
(after Baron Sir Ferdinand von Mueller, 19th-century Government Botanist of Vic.)
NSW, Vic, SA, WA   Stiff Cup-flower
0.05–0.1 m × 0.1–0.5 m   Aug–Nov

Dwarf annual **herb** with more or less glabrous to hairy young growth; **stems** prostrate to ascending, slender, wiry, brown to reddish-brown, woolly-hairy near flowerheads; **leaves** 0.2–1.5 cm × about 0.1 cm, linear, alternate, spreading to ascending, usually recurved, dark green to purplish-green, greyish-hairy near base, stiff, apex pointed; **flowerheads** to about 0.4 cm across, whitish, woolly-hairy, terminal, bracts slightly longer than flowerheads, often profuse and conspicuous.

Although not highly ornamental, this interesting species occurs over a wide range of south-western NSW, western Vic, southern SA and south-western WA. Usually associated with mallee communities and is also found on margins of saline depressions. It has very limited potential for cultivation in warm temperate and semi-arid regions as an annual bedding plant. Also worth trying in containers. Propagate from seed or possibly from cuttings of non-flowering stems.

Pogonolepis stricta Steetz
(erect, upright)
WA   Stiff Angianthus
0.05–0.15 m × 0.2–0.5 m   Aug–Jan

Dwarf annual **herb** with glabrous to hairy young growth; **stems** prostrate to erect, often reddish, rigid; **leaves** 0.4–2.5 cm × up to 0.15 cm, linear to narrowly lanceolate, ascending to erect, can be moderately crowded, deep green, glabrous to densely hairy, stiff, apex pointed; **flowerheads** to about 0.4 cm across, whitish woolly-hairy, terminal, bracts slightly longer than flowerheads, often profuse and conspicuous.

This species is confined to southern WA where it grows mainly in sandy soils on the edges of saline depressions. It is closely allied to *P. muelleriana* but differs in having a more rigid growth habit and longer leaves. It has similar cultivation requirements. Propagate from seed and possibly from cuttings of non-flowering stems.

POGOSTEMON Desf.
(from the Greek *pogon*, beard; *stema*, stamen; referring to the bearded stamens)
Lamiaceae

Herbs or dwarf **shrubs**; **branches** usually hairy; **leaves** opposite or whorled, sessile or stalked, maybe strongly aromatic; **inflorescence** spike-like, densely hairy; **bracts** densely hairy; **calyx** tubular, 5-lobed, hairy; **corolla** tubular, short, 4-lobed; **stamens** 4, free, with hairy filaments; **style** with 2-branched stigma; **fruit** enclosed in calyx.

A tropical genus of about 70 species. It extends from Japan, China, India, South Eastern Asia and Indonesia to Australia where 1 species occurs in the northern regions. Sometimes misspelt as *Pogonostemon*.

Pogostemon stellatus (Lour.) Kuntze
(starry)
Qld, WA, NT
0.3–0.6 m × 0.5–1.5 m   May–Sept

Dwarf, aquatic or semi-aquatic, perennial **herb**; **stems** glabrous or faintly hairy, often reddish, rooting at nodes; **leaves** 3–9 cm × 0.2–0.6 cm, linear, mainly in whorls of 4–10, sometimes opposite, sessile or nearly so, soft, green, glabrous, margins entire or toothed, apex pointed; **flowerheads** spike-like, to about 2.5 cm ×

Pogostemon stellatus × .75, flower × 4

0.5 cm, terminal, conspicuous; **flowers** to about 0.2 cm long, pink or pinkish-purple.

A dweller of riverbeds, billabongs, swamps and the muddy margins of these water bodies. It is able to withstand fast-flowing water. Plants occur in the northern regions of each state. Mainly of interest to enthusiasts but could be useful for planting in water and permanently wet soils on margins of swamps and waterways. Best suited to tropical areas but may succeed in subtropical regions. Propagate from seed, cuttings of firm young growth, or by division of layered stems.

Dysophylla stellata (Lour.) Benth. is a synonym.

POLLIA Thunb.
(derivation uncertain)
Commelinaceae

Perennial **herbs**; **stems** prostrate to erect, often decumbent; **leaves** alternate, stem-clasping, lanceolate to oblong, parallel venation, glabrous; **inflorescence** terminal clusters or cymes; **flowers** bisexual, 3-merous; **tepals** 6, petal-like, outer tepals mainly green, inner tepals white or blue; **stamens** 6; **fruit** a 3-celled nut.

A cosmopolitan tropical genus of about 26 species; Two species occur in north-eastern Australia and one of these is endemic. Only *P. crispata* is usually cultivated.

Propagation is from seed, by division of layered stems, or from cuttings, which strike very quickly. Cuttings strike even when placed in water.

Pollia crispata (R. Br.) Benth.
(crinkled, crisped)
Qld, NSW   Pollia
0.3–1 m × 0.5–2.5 m   Sept–April; also sporadic

Dwarf perennial **herb** with faintly hairy young growth; **stems** prostrate to ascending, self-layering at nodes, somewhat fleshy, green; **leaves** 10–25 cm × 2–4.5 cm, narrowly ovate to ovate, spreading, often reflexed, crinkled margins on leaf sheath, mid-to-deep green and glossy above, pale green below, midrib prominent, apex finely pointed; **flowers** to about 1.5 cm across, white or rarely blue, in mainly erect terminal clusters to about 5 cm long, rarely profuse or very conspicuous; **fruit** about 0.6 cm long, ovoid, bluish.

This leafy species occurs in south-eastern Qld and in the Central Tablelands, North Coast and Central Coast of NSW. It grows in rainforest or its margins and is sometimes found along waterways and in other wet sites. Plants adapt very well to cultivation in subtropical and tropical regions. It is a valuable plant for rainforest revegetation. Grows as far south as Melbourne although it may suffer frost damage. It does well in most soils but needs a semi-shaded or shaded site in soils which are moist for most of the time. Appreciates organic mulches and will self-layer readily in this medium. Provides good foliage cover in areas where it

Pollia crispata × .55

can be difficult to maintain undershrubs and other groundcovers. Responds well to hard pruning. Propagates readily from cuttings or division of layered stems. Cuttings will root in water.

P. macrophylla is closely allied but differs in having entire leaf-sheaths, and flowers which are in distinct clusters or whorls. It occurs in eastern Qld and extends to New Guinea. This species has similar cultivation requirements to *P. crispata* but is more frost-sensitive.

Pollia macrophylla (R. Br.) Benth. —
see *P. crispata* (R. Br.) Benth.

POLYALTHIA Blume
(from the Greek *poly*, many; *altheis, althos*, healing; the bark is used to cure many ailments)
Annonaceae

Shrubs or **trees**; **leaves** simple, entire, alternate, stalked; **flowers** solitary or clustered on woody out-growths of older branches; **sepals** 3 or 4; **petals** 6–8, in 2 whorls, thick to succulent; **stamens** numerous; **carpels** numerous; **fruiting carpels** free, stalked, succulent; **seeds** 2–5.

A genus of about 120 species distributed from tropical Africa to Asia and the Pacific region. Four or five species occur in Australia. They are all bushy trees with colourful fruit which is eaten by birds. Propagation is from seed which has a limited period of viability and must be sown while fresh.

Polyalthia australis (Benth.) Jessup
(southern)
WA, NT
8–16 m × 5–10 m   March–Oct
Small to medium **tree** with a bushy habit; young growth glabrous; **leaves** 7–20 cm × 2.5–9 cm, ovate to elliptic, somewhat leathery, bright green and shiny above, dull beneath, base rounded, margins wavy, apex blunt; **flowers** about 4 cm across, green to yellowish-green, fragrant, solitary in the leaf axils; **sepals** to 0.25 cm long, minutely hairy; **petals** 2.8–3.5 cm long, narrow, glabrous or minutely hairy; **fruiting carpels** in pendulous clusters of up to 25, 1.5–2.5 cm × 1–1.5 cm, cylindrical to ellipsoid, stalked, ripening through yellow and orange to red and purple-black; mature Nov–Feb.

A common species which grows along streams in lowland rainforest and vine thickets, in scrub on stabilised dunes, and on cliffs and bluffs overlooking the sea. Aborigines use the wood for woomeras and an infusion from the leaves is used for treating ear ache. It is planted on a limited scale around Darwin and is deserving of wider use. Clusters of fruit are decorative over many weeks. In fertile soils, plants have proved to be fast-growing and they have a lush, pleasant appearance. They need plenty of water during dry periods and respond well to mulching and fertilisers. Propagate from fresh seed, which germinates readily.

Popowia australis Benth. and *Polyalthia holtzeana* F. Muell. are synonyms.

Polyalthia holtzeana F. Muell. =
P. australis (Benth.) Jessup

Polyalthia australis × .45

Polyalthia michaelii C. T. White
(after Canon N. Michael, amateur plant collector)
Qld   China Pine
8–12 m × 5–8 m   Oct–Jan
Small **tree** with a bushy habit; young growth shiny, glabrous; **leaves** 7.5–20 cm × 4.5–7.5 cm, ovate to oblong, somewhat leathery, dark green and shiny above, paler beneath; **flowers** about 2.5 cm across, greenish to yellow-green or purplish, fragrant, in axillary clusters on older leafless wood; **petals** to 1.4 cm × 0.4 cm, inner petals larger than outer petals, short-hairy interior; **fruiting carpels** 3–5, to 4 cm × 1 cm, ovoid to globose, red.

A rare species which occurs in north-eastern Qld. It grows in rainforest from the lowlands to moderate altitudes. Poorly known in cultivation but has ornamental qualities. Probably has similar cultivation requirements to *P. australis*. Propagate from fresh seed.

Polyalthia nitidissima (Dunal) Benth.
(very shiny)
Qld, NSW, NT   Canary Beech
8–15 m × 5–10 m   Nov–Feb
Small to medium **tree** with a bushy habit; **bark** grey to blackish, flaky; **branchlets** wrinkled; young growth hairy; **leaves** 5–11 cm × 3–5 cm, elliptic to oblong, bright green, very shiny above, paler and dull beneath; **flowers** about 1.8 cm across, cream or yellowish, in axillary clusters; **sepals** about 0.2 cm long; **petals** about 0.6 cm long, thick; **fruiting carpels** in clusters of 3–8, 0.8–1 cm long, yellow, orange or bright red, glossy, on stalks to 0.6 cm long, mature Nov–March.

Occurs in the Top End, NT, and from Cape York Peninsula to north-eastern NSW. Plants are also found in New Caledonia. This widespread and common species is often prominent in coastal rainforests. Birds

Polyalthia nitidissima × .5

2.5–7.5 cm × 1–2.5 cm, narrowly ovate to ovate or elliptic, thin-textured and papery, green to yellowish-green, shiny above, dull beneath, sparsely hairy along veins on the underside, acuminate; **flowers** 2–3 cm across, greenish, cream or yellow, fragrant, solitary in the axils; **sepals** to 0.3 cm long, hairy; **petals** 6, to 1.5 cm long, hairy exterior, inner petals shorter than outer petals; **stamens** numerous; **carpels** hairy; **fruiting carpels** in clusters of 2–6, 1.5–3 cm × 0.8–1 cm, cylindrical, short-stalked, orange-red, succulent, ripe Sept–Jan.

In Australia, this species is distributed in the tropical north. It grows in rainforest and vine thickets sometimes on rocky sites. The fruit is edible and has a pleasant taste. Grown on a limited scale in Darwin, it has potential for greater use in tropical and subtropical regions. Requires a sunny location in well-drained soil. Responds to mulching, watering and regular light fertiliser applications. Frost tolerance is not high. Propagate from fresh seed and possibly also from cuttings of firm young growth.

feed on the fruit. It is an attractive, moderately slow-growing species which is suitable for rainforest planting, larger gardens and parks. Young plants may require protection. Prefers moist well-drained soils. Tolerates light to moderate frosts. Propagate from fresh seed.

POLYAULAX Backer
(from the Greek *poly*, many; *aulax*, furrow; referring to the furrowed seeds)
Annonaceae
A monotypic genus which occurs in Australia, Java and New Guinea.

Polyaulax cylindrocarpa (Burck) Backer
(with cylindrical fruit)
Qld, WA, NT
4–8 m × 3–5 m   April–July
Medium to tall **shrub** or small **tree**; crown bushy but open; young growth with brown hairs; **leaves**

Polyaulax cylindrocarpa   D. L. Jones

POLYCALYMMA F. Muell. & Sond.
(from the Greek *poly*, many; *calymma*, veil)
Asteraceae (Alt. *Compositae*)
A monotypic genus endemic in semi-arid and arid regions of Australia.

Polycalymma stuartii F. Muell & Sond.
(after John McDouall Stuart, 19th-century Australian explorer)
Qld, NSW, Vic, SA, NT   Poached-egg Daisy;
                                           Ham-and-eggs Daisy
0.1–0.5 m × 0.1–0.3 m   May–Oct
Annual **herb**, young growth with dense, often sticky, woolly-white hairs; **stems** woolly-hairy, erect, often semi-woody at base, sometimes unbranched; **leaves** 2–7 cm × 0.1–0.3 cm, linear to linear-lanceolate, alternate, spreading to erect, grey-green, densely hairy to nearly glabrous above, hairy below, margins entire, apex softly pointed; **flowerheads** daisy-like, 2–4 cm across, somewhat hemispherical; **floral bracts** white, papery; **disc florets** deep yellow; **achenes** to about 0.3 cm long, silky-hairy.

A common component of inland sandy habitats where it can be one of the very prominent daisies which grow soon after soaking rains. A particularly attractive species which is rarely encountered in cultivation. It has the potential to be a highly desirable bedding plant or short-term garden plant. Needs maximum sunshine and excellent drainage. Hardy to moderately heavy frosts. Pruning young stems can produce bushy growth. For further information on cultivation see Annuals, Volume 2, page 202. Propagate from seed, which does not require pre-sowing treatment, or from cuttings.

Previously known as *Myriocephalus stuartii* (F. Muell. & Sond.) Benth.

POLYCARPAEA Lam.
(from the Greek *poly*, many; *carpos*, fruit; referring to the abundance of seeds)
Caryophyllaceae

Annual or perennial **herbs** which may develop a tap-root in some species; **stems** erect or spreading; **leaves** simple, entire, opposite or whorled, usually narrow; **stipules** small, papery; **inflorescence** of terminal cymes, with conspicuous, colourful papery bracts; **sepals** 5, papery; **petals** 5, entire or toothed; **stamens** 5; **fruit** a 3-valved capsule.

A genus of about 50 species; 12 of these occur in Australia. Although only moderately ornamental they are of interest because of the colourful papery bracts which subtend the flowers. Some species may grow in large masses in favourable seasons. Propagation is from seed which germinates well if fresh. Seedlings are very susceptible to damping off.

Polycarpaea arida Pedley
(dry, arid)
Qld, NSW, SA, WA, NT
0.1–0.2 m × 0.1 m   March; Aug–Nov

Dwarf perennial **herb** forming sparse to dense clumps; **stems** procumbent to erect, with white hairs, arising from a carrot-like rootstock; **stipules** shorter than leaves; **leaves** 0.4–1 cm × 0.1 cm, linear, dark green, pointed; **corymbs** short, dense, leafy; **bracts** about 0.35 cm long, membranous; **sepals** about 0.35 cm long, white or yellowish; **petals** short; **capsules** about 0.2 cm long.

Widely distributed in arid and semi-arid regions of inland districts growing in well-drained sandy soils. An interesting species that has potential for use in gardens in areas with a hot, dry climate. May also be a useful container plant. Requires a sunny position and free drainage. Propagate from seed.

Polycarpaea corymbosa (L.) Lam.
(in corymbs)
Qld, WA
0.1–0.2 m × 0.1–0.2 m   Dec–June

Annual or perennial **herb**; **stems** erect; **stipules** shorter than leaves; **leaves** 1.5–2.5 cm × 0.2–0.4 cm, linear, dark green to grey-green, glabrous or sparsely hairy; **corymbs** short, sparse to dense; **bracts** 0.2–0.45 cm long, 2-lobed, margins often fringed; **sepals** about 0.35 cm long, white, pink or reddish; **petals** short; **capsules** about 0.2 cm long.

The typical variety is a pantropical species. In Australia, it occurs in coastal and near-coastal districts of tropical Qld and WA. It grows in a variety of well-drained soils in open forest, woodland and on rocky outcrops. Plants have an interesting habit but are rarely showy.

The var. *minor* Pedley has smaller flowers (about 0.2 cm long), smaller capsules and the sepals often have a thickened, reddish, triangular area at the base. It is found in coastal and inland districts of Qld, north-eastern NSW and NT. The distribution also extends to New Guinea. Plants of this variety are sparse annuals.

The var. *torrensis* Pedley has leaves to 3.5 cm long, distinct midveins in the sepals and hairy capsules. It occurs in northern Cape York Peninsula, Qld, and islands of the Torres Strait.

All variants have limited potential for use in gardens or containers. Propagate from seed.

Polycarpaea longiflora   W. R. Elliot

Polycarpaea longiflora F. Muell.
(with long flowers)
Qld, WA, NT
0.3–0.8 m × 0.2–0.5 m   April–Aug

Shrubby perennial **herb** with woody base; **stems** many, erect, with curly hairs; **stipules** about 0.8 cm long, white; **leaves** 0.6–3.5 cm × 0.1–0.3 cm, linear, sessile, glabrous or sparsely hairy, margins recurved; **cymes** terminal, narrow, densely flowered; **bracts** 0.6–0.8 cm long, white, pointed; **flowers** 0.6–1 cm long, red, pink, purple or white; **petals** to 0.6 cm long; **capsule** about 0.3 cm long, ovoid.

Distributed in tropical regions growing in well-drained soil in woodland and open forest. A showy species with excellent prospects for cultivation in tropical and subtropical regions as well as inland districts. Requires a sunny location and unimpeded drainage. Frost tolerance is unlikely to be high. Propagate from seed and possibly also from cuttings.

Polycarpaea microphylla Pedley
(with small leaves)
Qld, WA, NT
0.03–0.1 m × 0.1–0.3 m   May–July

Perennial **herb** forming cushion-like clumps; **stems** spreading, much-branched, densely covered with white curly hairs; **stipules** about 0.15 cm long, white; **leaves** 0.2–0.5 cm × 0.05–0.1 cm, narrowly elliptic, thick, glabrous, spine-tipped; **flowers** in axillary or terminal clusters; **bracts** to 0.2 cm long, white; **sepals** to 0.2 cm long, white; **capsules** to 0.15 cm long, ovoid.

Widely distributed in tropical areas growing in woodland and among rocks in sandstone outcrops. Not known to be in cultivation but has potential for use in containers and in miniature gardens. Requires a sunny position and free drainage. Propagate from seed.

411

Polycarpaea spirostylis F. Muell.
(with twisted, screw-like style)
Qld, NSW, SA, WA, NT   Copper Plant
0.1–0.3 m × 0.1–0.3 m   April–Aug
   Annual or short-lived perennial **herb**; **stems** erect, branched, glabrous; **stipules** to 0.3 cm long, triangular, white; **leaves** 0.5–2 cm × 0.1 cm, linear, often clustered, lower leaves often spathulate, glabrous; **cymes** terminal, much-branched; **bracts** to 0.4 cm long, ovate; **sepals** to 0.5 cm long, white or pink; **petals** about 0.4 cm long; **capsules** to 0.35 cm long, ovoid.
   The typical subspecies is restricted to south-eastern parts of Cape York Peninsula, Qld. It occurs in open forest and on rock outcrops. This plant frequently grows on heavily mineralised soils and it has been used as an indicator of copper deposits. However, it also grows on non-mineralised soils. Three other subspecies have been described.
   ssp. *compacta* Pedley has flowers crowded in the cymes, sepals with red-brown midribs and shallowly notched petals. It extends from the Gulf of Carpentaria to central Qld.
   ssp. *densiflora* (Benth.) Pedley has very compact cymes, sepals with purple midribs and entire or slightly notched petals. It occurs on Cape York Peninsula, Qld and Arnhem Land, NT.
   ssp. *glabra* (C. T. White & W. D. Francis) Pedley has open, sparse cymes, sepals with red-brown midribs and deeply notched corolla lobes. It is widespread across western Qld, northern SA, northern WA and the NT. It has been noted that this subspecies often grows on soils containing silver-lead deposits.
   All variants of this species are attractive and merit cultivation in gardens or containers. They are best suited to freely draining soils and may struggle in cool temperate regions. Plants can also be useful for the reclamation of mine dumps since they have the ability to grow on mineralised soils without absorbing high levels of heavy metals.
   Propagate all variants from seed.

POLYGALA L.
(from the Greek *polys*, much; *gala*, milk; plants were used to aid in the production of milk)
Polygalaceae   Milkworts
   Annual or perennial **herbs**, **shrubs**, **trees** or rarely **climbers**; **leaves** simple, entire, alternate or occasionally opposite or whorled; **inflorescence** a raceme or spike, rarely panicles; **sepals** 5, the 2 inner sepals large and wing-like; **petals** 3, 2 fused to the stamens; **keel** entire or with a crest; **stamens** 8, in a tube; **fruit** a capsule, often winged.
   A large cosmopolitan genus of about 600 species. About 12 species are native to Australia and 3 other exotic species are widely naturalised as weeds. The Australian species are mostly small herbs with limited ornamental appeal. The flowers however are often colourful with an intricate structure, especially those species which have a prominent crest on the keel. These species have potential for cultivation in containers and miniature gardens. Propagation is from seed, which germinates readily if fresh. Perennial species can be propagated from cuttings of firm young growth.

Polygala japonica Houtt.
(from Japan)
Qld, NSW, Vic   Dwarf Milkwort
0.1–0.2 m × 0.1–0.2 m   Oct–Dec; also sporadic
   Perennial **herb** or **subshrub** with a woody taproot; **stems** erect or prostrate; young growth with curly hairs; **leaves** 0.5–2.2 cm × 0.3–0.8 cm, ovate to elliptic, dark green, with prominent netted venation; **racemes** 1–3 cm long, axillary; **flowers** about 0.6 cm across, mauve to purple; **keel** crested; **capsules** about 0.4 cm across, orbicular, winged.
   Distributed from south-eastern Qld to eastern Vic. Plants grow among grass in open forest and woodland. Never showy but with interesting flowers. May be grown as a garden plant but it is better suited to containers or miniature gardens. Requires good drainage and tolerates moderate frosts. Benefits from regular light pruning. Propagate from seed or from cuttings of firm young growth.
   Polygala veronicea F. Muell. is a synonym.

Polygala linariifolia Willd.
(with leaves like the genus *Linarium*)
Qld, NSW, WA
0.1–0.2 m × 0.1–0.15 m   Oct–Dec; also sporadic
   Annual or perennial **herb** with a woody taproot; **branches** erect, with curly or straight hairs; **leaves** 0.5–4.5 cm × 0.1–1 cm, variable in shape from linear to obovate, glabrous or hairy, veins inconspicuous, blunt; **racemes** 1–9 cm long, axillary; **flowers** 0.3–0.5 cm long, yellow or purple; **outer sepals** wing-like; **keel** with 2 small forked appendages; **capsules** about 0.4 cm long, oblong.
   Grows in open forest often in hilly districts well inland from the coast. This moderately ornamental species may have potential for use in containers and miniature gardens. Probably has similar cultural conditions to *P. japonica*. Propagate from seed and possibly also from cuttings.

Polygala orbicularis Benth.
(round, orbicular)
NT
0.1–0.2 m × 0.1 m   Jan–May
   Small annual **herb**, sparsely branched; **leaves** 2–2.5 cm × 2–3.5 cm, broadly obovate or orbicular, dark green; **racemes** 2–3 cm long; **flowers** about 0.6 cm across, brilliant blue; **inner sepals** about 0.5 cm long, ovate; **corolla** about 0.5 cm long, the side petals enlarged, the crest deeply fringed; **capsules** about 0.5 cm across, orbicular.
   Occurs in the Top End of the NT growing in woodland and among rocks. Although this species is small and annual, it has colourful flowers with an exquisite structure. It would make an ideal plant for containers or miniature gardens. Requires warm conditions, bright light and moisture. Propagate from seed.

Polygala veronicea F. Muell. = *P. japonica* Houtt.

POLYGALACEAE Juss.   Milkwort Family
A family of dicotyledons consisting of about 900 species in 17 genera. It is widely distributed in subtropical and temperate regions, but absent from New

Zealand. In Australia, the family is represented by about 40 species in 3 genera. They are herbs, shrubs or climbers with alternate and often small leaves. The flowers, which resemble those of the family Fabaceae, often have a prominent crest or brush on the front petal. Australian genera: *Comesperma, Polygala, Salomonia.*

POLYGONACEAE Juss.   Dock Family

A family of dicotyledons consisting of about 750 species in 30 genera. It is widely distributed on many continents but it is most prolific in temperate regions of the northern hemisphere. About 41 species in 5 genera are native to Australia and many more species have become naturalised as weeds. They are herbs, shrubs, climbers or, rarely, trees. The leaves are alternate and have a specialised, stipule-like growth (called an ochrea) which sheaths the stem above the leaf base. The small flowers lack petals and are borne in racemes or panicles. Australian genera: *Muehlenbeckia, Persicaria, Polygonum, Reynoutria, Rumex.*

POLYGONUM L.

(from the Greek *polys,* many; *gony,* knee; referring to stem-nodes; also *Polygonon arrhen;* name given for *Polygonum aviculare* by Dioscorides)
Polygonaceae   Wireweeds

Perennial **herbs**, usually glabrous; **stems** spreading or rarely erect; **leaves** simple, usually entire, with glabrous, lacerated, silvery or white sheaths (ocreae); **flowers** bisexual, solitary or in small axillary clusters; **perianth segments** petal-like, usually 5, without enlarged wings or keels, pink or white; **stamens** 5–8; **fruit** a nut enclosed in perianth.

A cosmopolitan genus of about 50 species; 5 of these occur within Australia but only *P. plebeium* R. Br. is indigenous. It is a somewhat nondescript, prostrate, mat-forming species which occurs in dry areas of Qld, NSW, Vic, SA and NT.

As a result of botanical revision in 1988 many species formerly in *Polygonum* were transferred to *Persicaria.* Species described in this volume are listed below.

Persicaria differs in having species which are either annual or perennial herbs. They inhabit moist or wet sites and often root at nodes along the stems. The flowers are in spike-like arrangements.

Polygonum attenuatum R. Br. =
    Persicaria attenuata (R. Br.) Sojak

Polygonum barbatum L. =
    Persicaria barbata (L.) H. Hara

Polygonum decipiens R. Br. =
    Persicaria decipiens (R. Br.) K. L. Wilson

Polygonum dichotomum Blume =
    P. dichotomum (Blume) Masam.

Polygonum hydropiper L. =
    Persicaria hydropiper (L.) Spach

Polygonum lanigerum R. Br. =
    Persicaria lapathifolia (L.) Gray

Polygonum lapathifolium L. =
    Peersicaria lapathifolia (L.) Gray

Polygonum strigosum R. Br. =
    Persicaria strigosa (R. Br.) H. Gross.

POLYMERIA R. Br.

(from the Greek *poly*, many; *merous*, part; referring to the stigmatic lobes)
Convolvulaceae

Annual or perennial **herbs**; **stems** trailing or rarely twining, often rooting at nodes; **leaves** simple, alternate, entire; **inflorescence** axillary, solitary or up to 3-flowered cymes, subtended by small bracts; **corolla** broadly bell-shaped to funnel-shaped, shallowly lobed; **stamens** 5, more or less equal; **sepals** 5; **fruit** a globular capsule, 1-celled.

A small tropical and subtropical genus of about 8 species; 6–7 of these are endemic in Australia. The genus is also represented in Timor and New Caledonia. Plants occur near the coast as well as some distance inland — often in grassland and woodlands. They are found growing in sandy loam or clay soils which are often subject to short periods of inundation. The flowers may only last one day but they are quickly replaced by newly opened flowers each day. Often the flowering is prolific if growing conditions are suitable.

Polymerias are not common in cultivation and deserve to be better known as they have potential as bedding plants (upright species) as well as for cultivation in containers.

Propagate plants from seed, cuttings or by division of layered stems. Seeds do not require any pre-sowing treatment and should begin to germinate between 15–60 days after sowing. Cuttings of firm young growth strike readily. Division of young stems which have rooted at the nodes can also be successful.

Polymeria ambigua R. Br.
(doubtful)
Qld, WA, NT
prostrate × 1.5–4 m   March–Sept

Perennial **herb** with hairy young growth; **stems** trailing, to about 2 m long, rooting at nodes, slender, hairy; **leaves** 1.5–7.5 cm × 0.5–2.5 cm, variable, ovate to broadly ovate or oblong to circular, often 2-lobed at base, with prominent stalk, dark green above, paler below, hairy, venation prominent, apex ending in a small sharp point; **flowerheads** to about 1.5 cm across, white, pink or mauve-pink to purple, on slender axillary stalks, can be profuse, very conspicuous; **style** usually 4–6 branches; **capsules** to about 0.6 cm across, ovoid to globular, glabrous.

A pretty, trailing species from northern regions where it occurs in a range of soils in woodland, shrubland and grassland. Cultivated to a limited degree, it has potential for greater use in seasonally dry tropical and subtropical regions. Should adapt to most acidic soils. Needs moderately good drainage and a sunny or semi-shaded site is suitable. Worth trying on embankments as well as in general planting. Frost hardiness is not known. Responds well to regular pruning. Propagate from seed, which can begin to germinate

Polygala orbicularis   D. L. Jones

15–30 days after sowing, or from cuttings of firm young growth.

An apparently undescribed species from the eastern Kimberley has strong affinity to *P. ambigua*. It differs in being more-or-less glabrous and it has much longer leaf-stalks.

P. distigma also has affinity but its style is 2-branched.

Polymeria angusta F. Muell.
(narrow)
WA
prostrate–0.3 m × 0.6–1.5 m   April–Sept

Perennial **herb** with silky-hairy young growth; **stems** mainly trailing, silky-hairy, slender, rooting at nodes; **leaves** 2–6 cm × 0.2–0.8 cm, linear to lanceolate, base cordate with distinct stalk, deep green above, paler below, silky-hairy above and below, venation prominent; **flowers** to about 1.5 cm across, pink, solitary, on axillary stalks, conspicuous; **style** with 6–8 branches; **capsules** about 0.6 cm across, glabrous.

A poorly known species from northern WA. Suitable for seasonally dry tropical areas and may succeed in subtropical regions. It is closely allied to *P. longifolia* and will probably have similar cultivation requirements. Propagate from seed or from cuttings of firm young growth.

Polymeria calycina R. Br.
(distinctive calyx)
Qld, NSW, WA, NT   Bindweed
prostrate–1 m × 1–3 m   Sept–April; also sporadic

Annual or perennial **herb** with hairy young growth; **stems** trailing, slender, hairy, often with adventitious roots; **leaves** 1.5–10 cm × 0.3–2 cm, linear to ovate-oblong, base cordate to auriculate, with prominent stalk, deep green above, paler below, faintly hairy to glabrous, venation prominent, apex bluntly pointed to somewhat rounded; **flowers** 0.7–2 cm across, usually pink, sometimes mauve to purplish-violet, often with pale yellow-green throat, 1–3 flowers on long slender stems, rarely profuse; **style** up to 9 branches; **capsules** to about 0.6 cm across, globular.

A far-ranging species with a distribution extending from near Batemans Bay, NSW, and northwards, then across Australia and as far south as the Pilbara in WA. Mainly grown by enthusiasts, this species deserves to be better known. It adapts to most moderately well-drained acidic soils and does well in a sunny or semi-shaded site. Southern populations tolerate light frosts. Plants can spread rapidly and they respond very well to pruning. Suitable for general planting, containers and hanging baskets. Propagate from seed or from cuttings of firm young growth.

Polymeria distigma Benth.
(double stigma)
WA
prostrate × 1.5–3.5 m   April–June

Perennial **herb** with hairy young growth; **stems** trailing, slender, greyish-hairy; **leaves** 2–6 cm × 0.3–3.3 cm, variable, narrowly ovate to broadly ovate, base cordate, prominently stalked, deep green above, paler below, greyish-hairy, apex bluntly pointed; **flowers** to about 0.8 cm across, pale pink, solitary or 2–3, on slender stems, somewhat profuse, rarely highly conspicuous; **style** 2-branched; **capsules** to about 0.5 cm across, globular, glabrous.

This moderately appealing species is endemic in the Dampier District of the Northern Botanical Province where it occurs in sandy soils near the coast and further inland. It is best suited to subtropical and seasonally dry tropical regions. *P. distigma* should be useful as a groundcover. Needs excellent drainage and plenty of sunshine. Frost tolerance is not known. Propagate from seed or from cuttings of firm young growth.

P. ambigua is allied but the style has 4–8 branches.

Polymeria longifolia Lindl.
(long leaves)
Qld, NSW, NT, WA   Polymeria; Peak Downs Curse
0.2–0.3 m × 0.2–0.3 m   most of the year

Perennial **herb** with a woody rootstock and silky-hairy young growth; **branches** ascending to erect, silky-hairy; **leaves** 2–7 cm × 0.2–1 cm, linear to narrowly ovate, base cordate to hastate, sessile to short-stalked, deep green, glabrous to silky-hairy, apex pointed; **flowers** to about 3 cm across, pale pink to pink, solitary, on stems to about 3.5 cm long, can be profuse and very conspicuous; **style** with 6–8 branches; **capsules** about 0.6 cm across, globular, glabrous.

An interesting herb which can have a creeping rootstock from which plants reshoot each year. It is wide-ranging in tropical and subtropical regions where it often grows in heavy alluvial soils which can be subject to cracking as they dry out. Deserves to be better known in cultivation. Needs plenty of sunshine. Frost tolerance is not known. Tip pruning of young regrowth may promote compact growth and more flowers. Suitable for gardens, perennial borders and containers. Propagate from seed or from cuttings of firm young growth.

Polymeria marginata Benth.
(bordered with a margin)
Qld
0.2–0.3 m × 0.3–0.5 m   Sept–May

Dwarf perennial **herb** with a creeping rhizome and hairy young growth; **branches** ascending to erect, hairy; **leaves** 2–12 cm × 0.8–1.8 cm, linear to oblong, base more or less cordate, prominently stalked, glabrous and deep green above, paler with a few hairs below, venation prominent, apex pointed or blunt; **flowers** to about 3 cm across, pink to pinkish-purple, solitary, on stems 2–8 cm long, can be profuse and conspicuous; **style** 6–8-branched; **capsules** about 0.5 cm across, globular.

This species is distributed from northern central to southern central regions where it grows in a range of soils. Plants need plenty of sunshine and moderately good drainage. They withstand extended dry periods but should respond well to supplementary watering. Evidently rarely cultivated, it is suitable for subtropical and semi-arid regions and may succeed in temperate zones if grown in warm to hot protected sites. Propagate from seed or from cuttings of firm young growth.

Polymeria pusilla R. Br.
(very small)
Qld, NSW, NT
prostrate × 1.5–2.5 m   Dec–May; also sporadic

Annual or perennial **herb** with glabrous or faintly hairy young growth; **stems** trailing, slender, rooting at nodes; **leaves** 1.2–3.5 cm × 0.7–2 cm, oblong to obovate-oblong, base truncate to cordate, prominently stalked, deep green above, paler below, glabrous to faintly hairy, venation prominent, apex blunt or sometimes notched; **flowers** about 1 cm across, pink with pale yellow throat, solitary, on slender stems, can be profuse and moderately conspicuous; **style** 4–8-branched; **capsules** to 0.6 cm across, more or less globular.

A wide-ranging species which inhabits clay or clay loam soils which are often subject to waterlogging for short periods. Suitable for subtropical, or seasonally dry tropical regions and may succeed in temperate areas. Needs plenty of sunshine and should adapt to a variety of soil types. Worth trying for general planting or on lightly sloping embankments and in hanging baskets. Propagate from seed or from cuttings of firm young growth.

POLYOSMA Blume
(from the Greek *poly*, many; *osme*, scent, smell; referring to the highly fragrant flowers of some species)
Grossulariaceae

Shrubs or **trees**; **leaves** simple, entire or margins toothed, opposite (or nearly so) or in whorls; **inflorescences** terminal racemes; **flowers** 4-merous; **calyx** 4-toothed; **petals** 4, narrow, united in a tube, apical lobes free; **stamens** 4; **fruit** a 1-seeded berry.

A genus of about 60 species distributed from the Himalayas to Australia. There are about 8 endemic species in Australia including 2 or 3 which are undescribed. The Australian species grow in rainforest and are generally uncommon in cultivation. All of these species possess useful horticultural qualities and should become more widely grown. Propagation is from seed, which has a limited period of viability and for best results should be sown while fresh. Cuttings are also worth trying.

Polyosma alangiacea F. Muell.
(like *Alangium*)
Qld
6–10 m × 3–5 m   June–July

Tall **shrub** or small **tree**; young growth glabrous, shiny; **leaves** 8–12 cm × 3–5 cm, on petioles to 5 cm long, oblong, thin-textured but leathery, dark green and shiny above, dull beneath, margins entire, apex acuminate; **racemes** 10–12 cm long, terminal; **flowers** about 1 cm long, crowded, greenish; **berry** 1–1.2 cm long, ovoid, dark blue to black, ripe Aug–Oct.

Occurs in north-eastern Qld and grows in rainforest. This handsome bushy species with attractive foliage has prospects for cultivation in tropical and subtropical regions. Requires moist well-drained soil. Young plants may require some protection for the first few years. Tolerates light frosts only. Propagate from fresh seed and possibly from cuttings.

Polyosma cunninghamii Bennett
(after Allan Cunningham, 19th-century botanical collector in Australia)
Qld, NSW   Featherwood
8–12 m × 3–6 m   March–Nov

Small to medium bushy **tree**; **bark** brown, ridged and wrinkled; young growth grey, hairy; **leaves** 5–10 cm × 3–5 cm, lanceolate, thin-textured, dark green, shiny, margins coarsely toothed; **racemes** 4–6 cm long, in the upper axils; **flowers** to 1 cm long, tubular, white or greenish, strongly fragrant; **berry** 1.5–2 cm × 1 cm, ovoid, black, fluted, ripe March–Aug.

This moderately fast-growing species is distributed from south-eastern Qld to southern NSW. It grows in rainforest. Plants have attractive foliage, racemes of inconspicuous but pleasantly fragrant flowers and interesting fruit which attracts birds. Useful for screening purposes. May be grown successfully in temperate regions. Requires well-drained, moist soil in partial sun or filtered sun. Mulches and light fertiliser applications are beneficial. Tolerates light to moderate frosts. Propagate from fresh seed, and from cuttings of firm young growth which strike fairly readily.

Polyosma hirsuta C. T. White
(hairy)
Qld
6–12 m × 2–4 m   April–June

Tall **shrub** or small bushy **tree**; young growth densely rusty-hairy; **leaves** 6–10 cm × 2–4 cm, ovate-lanceolate, hairy when young, dark green, veins on the undersurface densely hairy, margins irregularly notched and with a few teeth, apex sharply acuminate; **racemes** 4–7 cm long, terminal, densely hairy; **flowers** about 0.7 cm long, yellowish, hairy; **berry** about 1 cm across, ovoid, pointed, black, ripe June.

Occurs in lowland rainforest of north-eastern Qld. An interesting, moderate-sized species which has hairs on most of its parts. It is not known to be in cultivation

Polyosma reducta

Polyosma rhytophloia × .5

but probably has similar requirements to those of *P. cunninghamii*. Propagate from fresh seed and possibly also from cuttings.

Polyosma reducta F. Muell.
(reduced, made small)
Qld
4–8 m × 2–4 m   Dec–March
Medium to tall **shrub** or small **tree**; young growth with tiny hairs; **leaves** 3–7 cm × 1.5–2.5 cm, lanceolate, thin-textured and papery, pale green and dull above, paler beneath, brittle, suddenly tapered at the base, apex drawn out; **racemes** 3–4 cm long, terminal, densely flowered; **berry** 0.4–0.6 cm long, ovoid, black, on long stalks.

A poorly known species from coastal rainforests in north-eastern Qld. It is not known to be in cultivation but probably has similar requirements to *P. alangiacea*. Propagate from fresh seed and perhaps also from cuttings.

Polyosma rhytophloia C. T. White & W. D. Francis
(with wrinkled bark)
Qld
6–12 m × 2–4 m   June–Sept
Tall **shrub** or small bushy **tree**; **bark** greenish-grey with conspicuous wrinkles; young growth short-hairy; **leaves** 9–14 cm × 3–4 cm, ovate to elliptic, thin-textured, dark green, veins prominent especially on the underside, margins with peg-like teeth; **racemes** 5–10 cm long, terminal; **flowers** about 1.5 cm long, tubular, greenish-white, base pinkish; **petals** 4; **berry**

about 0.8 cm across, globose to ovoid, dark blue to black, ripe Jan–April.

Occurs in the Eungella Range to the west of Mackay and grows in rainforest. A small bushy species which has potential for gardens in subtropical zones. It probably has similar cultivation requirements to *P. cunninghamii*. Propagate from fresh seed and perhaps also from cuttings.

Polyosma rigidiuscula F. Muell. & F. M. Bailey
(somewhat rigid)
Qld
2–6 m × 1.5–3 m   Aug–Nov
Small to tall **shrub** with a dense habit; young growth bright green, glabrous; **leaves** 5–10 cm × 2–3 cm, lanceolate to ovate, rigid, thick and leathery, dark green and shiny above, dull beneath, margins with minute teeth; **racemes** 2–5 cm long, terminal, densely flowered; **flowers** about 1 cm long, tubular, dull purple; **berry** 0.8–1 cm long, ovoid to ellipsoid, black, slightly hairy, ripe Dec–March.

Restricted to stunted rainforests on the summits of mountains in north-eastern Qld. A bushy habit, dark foliage and interesting flowers provide this species with excellent prospects for cultivation in tropical and temperate regions. Probably can be grown successfully as far south as Melbourne. Also has excellent prospects as a bonsai subject. Requires good drainage. Propagate from seed and from cuttings of firm young growth.

POLYPHLEBIUM Copel.
(from the Greek *poly*, many; *phlebos*, vein; referring to the veins in the frond)
Hymenophyllaceae
A monotypic genus which occurs in Australia and New Zealand.

Polyphlebium venosum (R. Br.) Copel.
(veined)
Qld, NSW, Vic, Tas   Veined Bristle Fern
5–15 cm long
Small epiphytic **fern** forming dense masses; **rootstock** creeping, thread-like, much-branched; **fronds** 5–15 cm long, 1–2 times pinnate, translucent dark green, divisions somewhat irregular, margins wavy; **sori** terminal, bell-shaped, with a prominent protruding receptacle.

Widely distributed in eastern Australia and also found in New Zealand and some adjacent islands. Common in moist-to-wet shady habitats especially fern gullies where it may grow in dense masses on the fibrous trunks of tree ferns. Easily grown if given conditions of constant high humidity and regular daily misting. Requires shade and protection from drying winds. Culture in a bottle, terrarium or aquarium can also be successful. See entry for Filmy Ferns, Volume 4, page 294. Propagate by division.

POLYPODIACEAE S. F. Gray
A large family of ferns consisting of about 50 genera and about 1000 species. It is widely distributed but particularly diverse in the tropics. Many species are

epiphytes and a number are also lithophytes and terrestrials. As a group they have much to offer horticulture and many species are commonly grown. The family is represented in Australia by 9 genera and about 30 species — many are popular horticultural subjects. Australian genera: *Belvisia, Colysis, Crypsinus, Dictymia, Drynaria, Microsorum, Platycerium, Pyrrosia, Schellolepis.*

POLYPODIUM L.

This genus of ferns became a botanical dumping ground for a large number of distantly related species including a number from Australia. Botanical revision has shown that *Polypodium* is largely restricted to Europe and America. All Australian ferns previously placed in *Polypodium* are now placed in other genera (see Polypodiaceae which lists some Australian fern genera).

POLYPOMPHOLYX Lehm. = *UTRICULARIA* L.

POLYSCIAS J. R. Forst. & G. Forst.

(from the Greek *poly,* many; *scias,* canopy, umbel; referring to the often large and many-rayed umbels) *Araliaceae*

Tall **shrubs** or **trees**; **trunks** without buttresses; **branchlets** thick, often short, with prominent leaf scars; **leaves** compound, alternate, pinnate to tripinnate, mainly at ends of branches; **leaflets** opposite, ovate; **inflorescence** terminal, mainly compound umbels or panicles or rarely corymbs; **flowers** bisexual or unisexual, solitary or in umbels, stalked or rarely sessile; **calyx** white, cream, green or purple; **petals** 4–5 or sometimes to 8 or more; **stamens** usually 4–5 or sometimes to 8 or more; **fruit** a succulent drupe, globular or slightly compressed.

A genus of about 100 species which is represented in tropical and temperate regions of Australia, Africa, Asia, Malesia and the Pacific Islands. About 14 species occur in Australia and about 10 of these are endemic. Over the years, Australian species have previously been included in *Kissodendron* Seem., *Panax* L., *Pentapanax* Seem and *Tieghemopanax* R. Vig.

The majority of species are found in Qld and some also occur in NT and NSW while *P. sambucifolia* is found in Qld, NSW, Vic and Tas.

In Australia, *Polyscias* occurs mainly at low to moderate altitude. *P. sambucifolius* grows in areas to just below the snowline while *P. wilmottii* is found at up to 1500 m altitude.

Some species have very ornamental pinnate leaves which are usually spirally arranged near the ends of the branches. The leaves can be to about 1.3 m long. Flowers are in racemes which are often prominent and sizeable with many branches.

Polyscias are uncommon in cultivation but some species such as *P. elegans* and *P. murrayi* are very fast-growing, straight-trunked species and deserve to be better known in tropical and subtropical regions. Certain species are also used for timber products on a small scale. The wood of *P. elegans* has the appearance of European maple, however, it is susceptible to splitting and borer attack.

In cultivation, most species prefer well-drained acidic soils in sites which are sunny to semi-shaded. Species such as *P. australiana, P. bellendenkerensis* and *P. sambucifolia* are capable of withstanding fairly dense shade. Soils need to be moist for most of the year, and the application of mulches is beneficial.

Most species are not suitable for areas with low temperatures but *P. sambucifolia* will tolerate –7°C or less without suffering foliage damage.

Plants do not usually need supplementary feeding once they are established although light applications of low phosphorus, slow-release fertiliser can be beneficial in establishing healthy young plants. During extended dry periods plants welcome supplementary watering.

Borers may be a problem especially for plants which are not growing well (see Volume 1, page 143 for control methods).

Gummosis is sometimes apparent on *P. sambucifolia.* It is usually more common in fine-foliaged plants. This condition is often a result of borer activity or from attack by fungal diseases. It is difficult to control and removal of damaged branches and stems is essential. If the fungus begins to take over, plants should be removed, including as many roots as possible, and destroyed.

Propagation is from seed, cuttings, or by division of root suckers. Seed is often very difficult to obtain and particularly so for rainforest species. The seed does not require any pre-sowing treatment although records show that *P. elegans* seed found in currawong droppings has provided good germination results. Seed is best sown while fresh because viability longevity is not well understood. Cuttings of firm young growth are successful for species such as *P. sambucifolia* and if select foliage characteristics are to be retained then vegetative methods must be used. Application of hormone compounds can be beneficial aids in root production. Some species produce root suckers, eg. *P. sambucifolia,* and these can be used with best results likely during late winter and early spring for this species.

Polyscias australiana (F. Muell.) Philipson
(from Australia)

Qld, NT   Ivory Basswood
3–8 m × 2–5 m   Dec–Jan

Medium to tall **shrub** or small **tree** with rusty-hairy young growth; **bark** greyish; **branches** ascending to erect; **branchlets** rusty-hairy when young; **leaves** 30–80 cm long, pinnate; **leaflets** 8–20, 10–18 cm × up to 2 cm, ovate, opposite, tapered to base, glossy and dark green above, paler below, pendent, margins entire, apex pointed; **flowers** very small, greenish, in umbels forming large spreading panicles, often profuse; **fruit** to about 0.9 cm across, globular, black, ripe May–Aug.

A moderately fast-growing species from near sea-level to about 1000 m altitude. Found in wet sclerophyll forest and margins of rainforest in the Top End of NT, on Cape York Peninsula and in north-eastern Qld. Suitable for tropical and subtropical gardens. Plants require well-drained acidic soils and tolerate a sunny, semi-shaded or shaded aspect. Plants from higher altitudes tolerate moderate frosts. Propagate from fresh seed.

Polyscias australiana   D. L. Jones

Polyscias bellendenkerensis (F. M. Bailey) Philipson
(from Mt Bellenden Ker).
Qld
4–8 m × 3–6 m   Oct

Tall **shrub** to small tree with rusty-hairy young growth; **bark** greyish; **branches** ascending to erect; **branchlets** initially rusty-hairy; **sap** clear, aromatic; **leaves** to about 10 cm long, bipinnate, crowded near ends of branches; **leaflets** to about 4 cm long, ovate; **flowers** small, white or cream, in umbels arranged in panicles, often profuse; **fruit** to about 0.8 cm × 0.7 cm, obovoid, black.

This species is restricted to the Bellenden Ker Range in north-eastern Qld, at above 1500 m altitude, where it grows in mountain rainforest which is frequently enveloped in cloud. In nature, it often flowers and fruits at an early stage. Plants are suitable for tropical and subtropical regions and may succeed in temperate areas if grown in protected moist sites. Plants need very well-drained acidic soils. A semi-shaded or shaded aspect is recommended. Application of organic mulches and slow-release fertilisers are beneficial. Young plants may need supplementary watering during extended dry periods. Propagate from fresh seed.

Polyscias caroli Harms
(unknown)
Qld
0.6–1 m × 0.4–0.8 m   not known

Dwarf **shrub** with glabrous young growth; **branches** spreading to ascending; **branchlets** dull grey, glabrous; **leaves** mainly 15–25 cm long, pinnate; **leaflets** 6–9 cm × 3.5–5 cm, 3–4 pairs, oblong-lanceolate to obovate, green, papery, apex blunt to pointed; **flowers** small, in few-flowered umbels with rusty-hairy to nearly glabrous peduncles.

An uncommon species that has its main representation in New Guinea but it also extends to the Cook District. It occurs in rainforest where it grows in freely draining soils. Evidently not cultivated in Australia, it is suited to seasonally dry tropical regions and may succeed in subtropical areas. Propagate from seed or from cuttings of firm young growth.

Polyscias cissodendron (C. Moore & F. Muell.) Harms
(ivy-like tree)
NSW (Lord Howe Island)   Island Pine
5–15 m × 2.5–6 m   Sept–Nov

Dioecious tall **shrub** to medium **tree**; **trunk** greyish; **branches** ascending to erect; **branchlets** grey, smooth; **leaves** 10–35 cm long, pinnate; **leaflets** 7–13, 3–10 cm × 1.5–4.5 cm, narrowly lanceolate to ovate, tapered to base, mid green, glabrous, margins toothed, apex pointed; **flowers** small, yellow, in umbels arranged in panicles to about 30 cm long, often profuse and conspicuous; **fruit** to about 0.5 cm long, somewhat globular, slightly flattened, purplish-maroon.

This common and appealing forest tree is from the lowland regions of Lord Howe Island. It also occurs in Vanuatu and New Caledonia. Suitable for cultivation in subtropical and tropical regions and it succeeds in temperate areas if grown in warm protected sites. Plants must have good drainage and protection from strong drying winds. Propagate from fresh seed.

Tieghemopanax cissodendron (F. Muell.) R. Vig. is a synonym.

Polyscias elegans (C. Moore & F. Muell.) Harms
(elegant, neat)
Qld, NSW   Celery Wood; Mowbulan Whitewood;
                   Silver Basswood; Black Pencil Cedar
6–20 m × 3–5 m   Feb–Aug; also sporadic

Small to medium **tree** with umbrella-shaped canopy; young growth slightly hairy; **trunk** usually straight, often solitary; **bark** greyish; **branches** spreading to ascending; **branchlets** thick, grey; **leaves** to about 1 m or more, pinnate to bipinnate; **leaflets** many, to about 13 cm × 3–6.5 cm, ovate to elliptic, glossy and dark green above, paler below, margins entire, aromatic when crushed with a celery fragrance; **flowers** about 1 cm across, purple, bisexual or unisexual, in racemes

Polyscias cissodendron   D. L. Jones

418

Polyscias elegans, leaves × .1, raceme × .25

forming panicles, often profuse; **fruit** about 0.5 cm across, depressed globular, purplish to black.

This handsome, extremely fast-growing species is an excellent pioneer plant in rainforest areas which have been damaged. It is distributed from New Guinea and Thursday Island to near Jervis Bay, NSW, and usually occurs in volcanic or sedimentary soils. Adapts well to cultivation in tropical and subtropical regions and will succeed in cooler areas if grown with protection from cold winds. Plants require moist but well-drained acidic soils and do best in soils which have plenty of organic content. May need supplementary watering during extended dry periods. It is a valuable plant for regeneration programmes within its own range. Highly recommended for general planting and young plants are suitable for indoor pots. Fruits are eaten by many native birds. The soft white wood has the aroma of celery when freshly cut. Propagate from fresh seed. It is sometimes recommended that seed found in currawong droppings provides best germination.

Tieghemopanax elegans (C. T. Moore & F. Muell.) R. Vig. is a synonym.

Polyscias fruticosa (L.) Harms
(shrubby, bushy)
Qld
3–5 m × 2–3 m   April
Medium to tall **shrub** with glabrous young growth; **branches** spreading to ascending, with leaves clustered near tips; **leaves** tripinnate, to about 75 cm long, sheathing base; **pinnae** to about 25 cm long; **leaflets** 1–12 cm × 0.5–4 cm, variable, linear-lanceolate to oblong, short-stalked, often sparse, deep green, glabrous, aromatic, deeply lobed or toothed; **flowers** about 0.5 cm across, in a spreading panicle to about

60 cm long, often profuse; **fruit** about 0.5 cm across, fleshy, somewhat globular.

A wide-ranging tropical species of the Indo-Pacific and Malesian region. It occurs in the Cook and North Kennedy Districts of Qld. Rarely cultivated in Australia, it has potential for ornamental use in tropical and subtropical areas. Plants should adapt to a variety of well-drained acidic soils in sunny or semi-shaded sites. The foliage is used for culinary purposes in Asia. Propagate from fresh seed.

Polyscias guilfoylei (Cogn. & Marche) L. H. Bailey
(after W. R. Guilfoyle, 19th-century Director, Royal Botanic Gardens, Melbourne)
Qld
2–3 m × 1–2 m   not known
Small **shrub** with glabrous young growth; **branches** spreading to ascending, few; **leaves** to about 60 cm long, pinnate, glabrous, petiole to about 18 cm long, base sheathed; **leaflets** about 10–14 cm × 6–7 cm, oblong, rhomboidal or broadly ovate, green, glabrous, thin, limp, often with whitish or yellowish blotches near the toothed margins, apex pointed or blunt; **flowers** to about 0.5 cm across, in spreading panicles with branches to about 50 cm long, branchlets to about 8 cm long, highly conspicuous.

In Australia, this distinctive species is known only from the Cook District but it is more common in overseas tropical regions where it is widely cultivated as a hedge plant in Malesia and the tropical Pacific. Best suited to tropical areas and adapts to a range of well-drained soils. Regular trimming prevents flowering. Should benefit from organic mulches and supplementary watering during extended dry periods. Propagate from seed.

Polyscias macgillivrayi (Seem.) Harms
(after Dr J. MacGillivray)
Qld
4–7 m × 2–4 m   not known
Tall **shrub** to small **tree** with glabrous young growth; **trunk** usually solitary; **bark** greyish; **branches** spreading to ascending, few; **leaves** pinnate, to 1 m or more long; **leaflets** 20–25 cm × 8–10 cm, oblong, often broadest near base, green, glabrous, margins entire or

Polyscias sambucifolia   D. L. Jones

419

faintly toothed, revolute, apex rounded, ending in small point; **flowers** about 0.5 cm across, in few-flowered umbels, forming large loose panicles, often profuse and conspicuous; **fruit** to about 0.8 cm across, flattened.

This small tree occurs in rainforest in the Cook District of north-eastern Qld and also in New Guinea. Suitable for tropical and subtropical planting, it requires freely draining acidic soils. Should respond well to organic mulches and supplementary watering during dry periods. May prove suitable for indoors in other regions. Propagate from fresh seed.

Tieghemopanax macgillivrayi (Seem.) R. Vig. is a synonym.

Polyscias mollis (Benth.) Harms
(softly hairy)
Qld
4–10 m × 2–4 m   Dec–Jan

Tall **shrub** to small **tree** with softly hairy young growth; **trunk** usually solitary and erect; **bark** greyish; **branches** few, spiny; **branchlets** woolly-hairy, with sharp prickles; **leaves** to about 1 m long, pinnate or bipinnate; **leaflets** usually 13–17, to about 25 cm × 6 cm, lanceolate to ovate, deep green, glabrous above, softly hairy below, margins toothed, apex pointed; **flowers** small, blue or purple or reddish-purple, bisexual, 5-merous, in large terminal umbels or panicles, often profuse; **fruit** to about 0.6 cm across, somewhat globular but flattened, blue to purple or reddish-purple, fleshy, ripe March–April.

This highly ornamental species is endemic in the Cook District where it grows from sea-level to about 1100 m altitude. Cultivated mainly by enthusiasts, it deserves greater recognition in horticulture. Suited to tropical and subtropical regions in freely draining, but moist, acidic soils with a semi-shaded or shaded aspect. Addition of organic mulch is beneficial. May need supplementary watering during extended dry periods. Does well as an indoor pot plant. Propagate from fresh seed.

Tieghemopanax mollis (Benth.) R. Vig. is a synonym.

Polyscias murrayi (F. Muell.) Harms
(after Patrick Murray, 19th-century Australian patron of botany)
Qld, NSW, Vic`   Pencil Cedar; Umbrella Tree;
                                      White Basswood
8–15 m × 3–5 m   Jan–April

Small to medium **tree** with a palm-like crown; **trunk** usually solitary and branchless up to 6 m; **bark** brownish-grey, smooth except for corky pimples; **leaves** to more than 1 m long, pinnate; **leaflets** 8–30, 6–25 cm × 3–9 cm, lanceolate-ovate, opposite, dark green, margins entire or toothed, apex pointed; **flowers** to about 0.4 cm long, cream to greenish, bisexual, in umbels arranged in large panicles, often profuse; **fruit** to about 0.6 cm across, flattened, lobed, pale blue, brown or blackish.

A superb quick-growing pioneer species which inhabits roadsides and rainforest margins. In Australia, it is distributed along the eastern coastal regions from north-eastern Qld to just across the Victorian border in East Gippsland. It also occurs in New Guinea. Plants are cultivated to a limited degree. They require well-drained acidic soils and are suitable for tropical, subtropical and temperate regions. Plants need shelter from hot drying winds during initial establishment. They benefit from the addition of organic mulches and supplementary water during extended dry periods. Useful as an indoor plant. Birds visit fruiting trees. Propagate from seed, which is difficult to collect.

Tieghemopanax murrayi (F. Muell.) R. Vig. is a synonym.

Polyscias purpureus C. T. White
(purple)
Qld
2–3 m × 2–3 m   Feb–April

Medium **shrub** with glabrous young growth; **branches** spreading to ascending, few; **leaves** pinnate, to about 40 cm long; **leaflets** usually 7–9, to about 23 cm × 8 cm, lanceolate, somewhat fleshy, pale green; **flowers** to about 0.4 cm across, purplish, in panicles about 25 cm across, on purplish stems, often profuse and conspicuous; **fruit** about 0.3 cm across, compressed.

An attractive species which inhabits rainforest in the South Kennedy District. Evidently rarely cultivated, it is suitable for tropical and subtropical regions with acidic, moderately well-drained soils. Needs protection while young. Application of organic mulches is beneficial. May need supplementary watering during extended dry periods. May be useful as foot traffic control because of its prickly branches. Propagate from fresh seed.

Polyscias sambucifolia (Sieber ex DC.) Harms
(leaves similar to those of *Sambucus*)
Qld, NSW, Vic, Tas   Elderberry Panax;
                                      Ornamental Ash
4–10 m × 3–6 m   Nov–Feb

Tall **shrub** to small **tree**, often with suckering roots, able to form extensive clonal colonies; **bark** brownish-black with pimples in lines or ridges; **branches** spreading to ascending; **branchlets** can be slightly pendent on fine-leaved variants; **leaves** 4–35 cm × 4–15 cm, variable, pinnate, bipinnate or rarely tripinnate; **leaflets** usually 5–21, 4–20 cm × 2–6 cm, lanceolate or deeply lobed to pinnate, spreading to ascending, deep green and often glaucous, glabrous; **flowers** to about 0.4 cm long, yellow-green, in umbels forming open terminal panicles to about 12 cm long, often profuse but rarely conspicuous; **fruit** to about 0.6 cm long, ribbed, translucent bluish to mauve.

A fast-growing, extremely variable species which is distributed from south-eastern Qld to the Otway Ranges in western Vic. It is found in rainforest, open forest and coastal woodland. Current research recognises 3 subspecies. The typical ssp. *sambucifolia* has broad leaflets and is found from south-eastern Qld to east of Melbourne, Vic. A subspecies with narrow leaflets occurs from northern NSW to the Otway Ranges, Vic, and a subspecies with narrowly lobed, ferny leaves is found from northern NSW to eastern Vic. The most popular subspecies is the ferny-leaved variant but all subspecies are cultivated to a limited

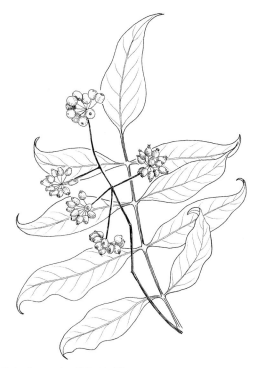

Polyscias sambucifolia × .45

degree. Plants adapt to a wide range of acidic soils including soils which are wet for extended periods, however, they do best in freely draining soils. A shaded or semi-shaded site is best. Plants can sucker and form small colonies. They respond well to pruning and old suckering growth is usually best removed at ground level. The coarser-leaved variants are valuable quick-growing screening plants if pruned regularly but the ferny-leaved subspecies is better over the longer term. Some plants may be subject to a wood rot disease which is identified by a gummy secretion on trunk and branches. Diseased plants are best destroyed including as much of the root as possible. Excellent for shaded, moist areas. Suitable for use as an indoor potted plant. Propagate from seed or from cuttings of firm young growth.

Tieghemopanax sambucifolius (Sieber ex DC.) R. Vig. is a synonym. *Tieghemopanax multifidus* N. A. Wakef. was a name given to the ferny-leaved variant with many narrow segments but it is now regarded as a synonym.

Polyscias scutellaria (Burm. f.) Fosberg
(like a dish or saucer)
Qld
3–6 m × 1.5–3 m   May
   Medium to tall **shrub** with glabrous young growth; **branches** ascending to erect, glabrous, with spirally arranged leaves near apex; **leaves** simple to trifoliolate, lamina 8–25 cm across, mainly oblong to circular or reniform, petiole 6–28 cm long, with sheathing base, bright green and shiny above, yellow-green below, midrib brown, glabrous, margin coarsely toothed or

faintly lobed near the rounded apex, sweetly aromatic; **flowers** to about 0.5 cm across, white to yellowish, in spreading panicles with branches 10–30 cm long, conspicuous; **fruit** about 0.7 cm across, somewhat globular, black to purplish-black, fleshy.

   A widespread tropical species which extends to Australia where it occurs in the Cook and North Kennedy Districts. It grows in well-drained soils in rainforest. Commonly cultivated in Malesia and the tropical Pacific where it is used as a hedging plant. Variegated foliage selections are popular. Should adapt well to cultivation in subtropical and seasonally dry tropical regions. May need supplementary watering during extended dry periods. In the tropics outside Australia, foliage is used for treatment of cancer, baldness and as a diuretic. Propagate from seed or from cuttings of firm young growth.

Polyscias willmottii (F. Muell.) Philipson
(after Dr J. Willmott)
Qld
5–10 m × 2–4 m   Dec–Jan
   Tall **shrub** to small **tree**; **branchlets** with leaves crowded near apex; **leaves** to about 25 cm long, pinnate; **leaflets** usually 3–5, to about 10 cm × 3 cm, lanceolate to narrowly ovate, opposite, glossy, dark green above, paler below, margin flat or wavy, apex pointed; **flowers** small, greenish to yellow, bisexual, in umbels arranged in spreading panicles, rarely profuse; **fruit** to about 0.9 cm long, more or less hemispherical, purple, fleshy.

   This rare endemic species has attractive, lacy foliage. It occurs in the Cook District where it inhabits the higher mountains at 1000–1500 m altitude. It grows as a tropical tree although it often starts life as an epiphyte or lithophyte. Worthy of greater cultivation in tropical and subtropical regions. Needs freely draining acidic soils and a warm site which is protected from drying winds. Organic mulches are beneficial as is supplementary watering during extended dry periods. Worth trying as an indoor plant and may be excellent for bonsai (penjing). Propagate from seed or from cuttings of firm young growth.

Pentapanax willmottii F. Muell. is a synonym.

POLYSTICHUM Roth
(from the Greek *poly*, many; *stichos*, rows; referring to the sori in rows)
Dryopteridaceae   Shield Ferns
   Terrestrial, clumping **ferns**; **rootstock** erect, sometimes forming a short trunk, densely scaly; **scales** large, papery; **fronds** clustered, 2–3 times pinnate, often harsh and leathery, proliferous bulbils sometimes present (they may develop into plantlets); **sori** circular, with a large peltate indusium.

   A large cosmopolitan genus of about 180 species; 4 of these are endemic in eastern Australia. A number of exotic species are commonly offered for sale in Australian nurseries. The native species are very adaptable ferns which are also popular in cultivation. Propagation is from spores and, in some species, by plantlets formed near the frond apex. Plantlets should not be removed until they are well developed — preferably with some small roots present.

Polystichum australiense

Polystichum fallax   D. L. Jones

Polystichum australiense Tindale
(Australian)
NSW   Shield Fern
0.3–0.9 cm tall

    Rootstock erect, thick, scaly; **fronds** 2–3-pinnate, oblong in outline, 30–90 cm × 10–14 cm, in a sparse to dense tussock, dark green above, paler beneath, harsh textured, with proliferous bulbils often present near the apex; **pinnules** ovate, with marginal spines and an apical tooth; **sori** round; **indusia** round, sometimes with a dark centre.

This species is widely spread from northern to southern NSW. It grows in open forest and along rainforest margins often on rock outcrops or in stony soils. A tough, hardy fern which is easily grown in a range of positions. For best appearance plants need limited exposure to direct hot sun. Tolerates moderate frosts. Propagate from spores and by bulbils.

Polystichum fallax Tindale
(spurious, not genuine, false)
Qld, NSW   Rock Shield Fern
0.3–0.8 m tall

    Rootstock erect, densely scaly; **fronds** 30–80 cm × 8–22 cm, 2–3-pinnate, narrowly ovate in outline, dark green, leathery, harsh-textured; **pinnules** ovate, the margins with sharp teeth, the apex with a short point; **sori** round; **indusium** round, brown, often with a dark centre.

P. fallax has a restricted distribution in south-eastern Qld and north-eastern NSW. It grows on rock outcrops in rainforest and open forest at moderate altitudes. A tough fern which grows readily in a shady or semi-shady position. Once established plants will withstand some dryness, but for best appearance they need

supplementary watering. Ideal for planting among rocks. Propagate from spores.

Polystichum formosum Tindale
(beautiful)
Qld, NSW, Vic   Broad Shield Fern
0.3–0.7 m tall

    Rootstock erect, thick, densely scaly; **fronds** 2–3-pinnate, 30–70 cm × 10–25 cm, ovate in outline, erect in a tussock, bright green above, paler beneath, leathery; **pinnules** ovate, margins with sharp teeth, apex with a short point; **sori** circular, in 2 rows; **indusia** round, uniformly pale coloured.

Broad Shield Fern is distributed from south-eastern Qld to eastern Vic. It grows on rock faces in sheltered forests often near streams. A handsome fern which adapts readily to cultivation. Ideal in a moist sheltered position and looks appealing when planted among rocks. May also be suitable in large containers. Propagate from spores.

Polystichum proliferum (R. Br.) Presl
(proliferous, bearing live plants)
NSW, Vic, Tas   Mother Shield Fern
0.3–1.5 m tall

    Rootstock erect, forming a short fibrous trunk with age, densely scaly; young fronds with sharply recurved tips; **fronds** 2-pinnate, 25–150 cm × 15–30 cm, oblong

Polystichum proliferum, frond base and scales (A) × 1, pinnule (B) × 4, plantlet (C) × 1

in outline, in an erect to spreading tussock, dark green above, paler beneath, somewhat harsh, with a proliferous bulbil often present near the apex; **pinnules** asymmetric, the tips pointed to shortly toothed; **sori** round; **indusia** round, light brown with a dark centre.

Widely distributed from near the Qld border in northern NSW to southern Tas. A common fern which often grows in extensive colonies. It is frequently prominent in fern gullies and in montane forests it may form a dominant groundstorey under tall gums. Very easily grown and adaptable to a wide range of situations, often tolerating moderate exposure to sun and even withstanding dry periods once established. Excellent for planting under established trees and is also a good container plant. Tolerates heavy frosts and snowfalls. Propagate from spores or by plantlets formed on the frond tips.

POMADERRIS Labill.

(from the Greek *poma*, lid; *derris*, skin or leather; referring to the membrane-like cover of capsules)
Rhamnaceae

Shrubs or small **trees** with hairy young growth; **leaves** simple, alternate, undersurface with stellate hairs or mixture of stellate and longer simple hairs; **stipules** brown, shed early or rarely persistent; **inflorescence** small cymes, mainly in terminal panicles or corymbs, sometimes in tight clusters; **flowers** small, whitish to yellow; **sepals** often reflexed; **petals** 5 or often absent, often shed early; **fruit** a capsule, separating into 3 mericarps.

A moderately sized genus of about 70 species which is restricted to Australia and New Zealand. Within Australia there are about 65 species and only 3 of these are not endemic. The best representation is in the eastern states with the majority of species in NSW and Vic. There are 7 species in WA of which 5 are endemic to that region. *Pomaderris* occurs in all states except NT and is found mainly in temperate regions. They grow in heathland, shrubland, woodland and open forest at altitudes ranging from near sea-level to about 2200 m.

Pomaderris species are mostly spreading to upright, small to medium shrubs but some may grow to tree size, eg. *P. aspera, P. brogoensis, P. cinerea, P. racemosa* and *P. virgata*. Foliage is one of the major ornamental features of *Pomaderris* species with flushes of new growth often covered in colourful rusty to reddish simple and/or starry hairs. Some species such as *P. argyrophylla* and *P. tropica* have silvery foliage, especially their new growth, and this is most noticeable during windy days when the silvery undersurface is exposed.

Individual flowers are usually cream to various shades of yellow. They are small but may be produced in sizeable terminal panicles which can nearly obscure the foliage. *P. aurea, P. elliptica, P. lanigera, P. ligustrina, P. obcordata* and *P. subplicata* are some of the species which provide an outstanding floral display.

Pomaderris do not produce lignotubers except for a variant of *P. lanigera* from far-eastern Vic which has a semi-lignotuberous root system.

Although many *Pomaderris* have excellent prospects for cultivation they are not well known and only a limited number offered by nurseries. *P. aspera* (often used for revegetation projects), *P. aurea* (which includes

Pomaderris aspera   W. R. Elliot

P. humilis), *P. lanigera, P. obcordata* and *P. pilifera* are among the most common species available.

In the early to mid 19th-century a limited number of species were cultivated in England. *P. apetala* was introduced in 1803, *P. elliptica* in 1805 and *P. lanigera* in 1806. In 1804, *P. discolor* was cultivated in France.

Pomaderris are very useful for planting in semi-shaded sites especially beneath established trees. *P. elliptica, P. paniculosa* ssp. *paralia* and *P. pilifera* are among those which tolerate plenty of sunshine. Most species respond well to pruning or clipping from an early age and can be used for short- to medium-term hedging.

P. paniculosa ssp. *paralia* copes very well with coastal exposure and *P. halmaturina* ssp. *halmaturina* is useful in coastal regions too. They both have sand-binding qualities.

Many different insects visit the flowers to obtain nectar and pollen, and the larvae of butterflies feed on the foliage of some species. Jewell butterflies lay their eggs on *P. apetala* and *P. aspera*.

Most pomaderris prefer well-drained acidic soils. Many of them need moisture for most of the year although they will tolerate extended dry periods once they are well established. In general, most species tolerate light frosts while some are hardy to heavy frosts, eg. *P. apetala, P. aspera, P. elliptica, P. ferruginea* and *P. pilifera*. Little is known about how well pomaderris will tolerate humidity or smog.

Pomaderris respond well to light applications of slow-release fertilisers which have a low phosphorus content. However, over-fertilising may cause excessive foliage growth at the expense of root growth and this often leads to instability of young plants. In such cases, the removal of some foliage is often beneficial in the short term.

Pomaderris adnata

Pests and diseases rarely cause major problems. Borers may be troublesome (see Chewing Insects, Volume 1, page 143 for control methods). Caterpillars may disfigure leaves but are of little consequence.

Propagation is from seed or from cuttings. Best results are achieved with fresh seed which is treated by heating for 10 minutes at 150°C. Seed placed in freshly boiled water for 15 seconds has also produced successful germination. Scarification or abrasion of seed coating can also be beneficial. See Volume 1, page 206 for further information.

Propagation from cuttings is a regular practice. Firm young growth taken from non-flowering tip or side growth generally strikes readily. If growth is too soft it wilts very readily and is unlikely to regain its turgidity. It is usually best to completely replace the cuttings if further material is available in order to achieve successful rooting.

Pomaderris adnata N. G. Walsh & F. Coates
(joined together)
NSW
1–2 m × 1–2 m   Sept

Small **shrub** with greyish starry hairs on young growth; **branches** spreading to erect; **branchlets** greyish-hairy; **leaves** 1.5–3 cm × 0.3–0.8 cm, narrowly ovate to obovate, spreading to ascending, deep green and glabrous above, greyish-hairy below, margins recurved, venation distinct, apex blunt; **stipules** to 0.2 cm, deciduous; **flowers** small, pale hairs on buds, opening cream, in terminal heads 1–3 cm across, can be profuse; **petals** shed early.

A recently described (1997) species which is known only from the Wollongong region where it grows on sandy loam in heath woodland and open forest. It is regarded as vulnerable in nature and needs to be cultivated as a conservation strategy. Plants require good drainage with a semi-shaded aspect. Frost tolerance is not known. Propagate from seed or from cuttings of firm young growth.

P. phylicifolia has similarly shaped leaves but they are hairy on the upper surface.

Pomaderris affinis N. A. Wakef. = *P. intermedia* Sieber

Pomaderris albicans Steud. =
Trymalium albicans (Steud.) Reissek

Pomaderris andromedifolia A. Cunn.
(leaves similar to those of the Andromeda or *Pieris*)
Qld, NSW, Vic   Pomaderris
1–3 m × 1–3 m   Sept–Nov

Small to medium **shrub** with rusty-hairy young growth; **branches** many, spreading to ascending; **branchlets** densely covered with straight rusty to tan hairs; **leaves** 0.8–9 cm × 0.4–3.5 cm, narrowly ovate to elliptic, base tapered or rounded, spreading to ascending, dull green to greyish-green and becoming glabrous above, rusty to whitish with silky simple hairs below, margin flat to recurved, apex blunt to pointed; **flowers** very small, yellow to golden-yellow, in terminal panicles 1–8 cm across, often profuse and very conspicuous; **capsules** with whitish hairs.

The typical variant occurs from south-eastern Qld

and along the Great Dividing Range to far-eastern Vic. It is found in shallow soils in heathland, woodland and open forest from 30–1150 m altitude. Plants adapt well to cultivation in temperate and subtropical regions. They require well-drained acidic soils with a semi-shaded aspect. Plants are hardy to moderate frosts and lengthy dry periods. They respond well to light or moderate pruning. Introduced to cultivation in England in 1824.

The ssp. *confusa* N. G. Walsh & F. Coates has ovate to obovate leaves, 1.5–4 cm × 0.8–1.5 cm, with curved to flexuose hairs on branchlets. It is restricted to south-eastern NSW where it is often found growing on shallow soils in rocky sites beside creeks and rivers. Needs similar conditions to the typical variant.

Propagate both variants from fresh seed or from cuttings of firm young growth.

P. delicata N. G. Walsh & F. Coates is very closely allied to the typical subspecies. It differs in lacking simple hairs and it also has very short stipules which deciduate very quickly. Known only from near Goulburn, NSW, where it grows on sandstone derived soils in dry open forest.

Pomaderris angustifolia N. A. Wakef.
(narrow leaves)
NSW, Vic   Pomaderris
1–3 m × 1–2.5 m   Oct–Nov

Small to medium **shrub** with hairy young growth; **branches** many, spreading to ascending; **branchlets** greyish-hairy; **leaves** 0.7–2.7 cm × 0.1–0.6 cm, linear to narrowly obovate, spreading to ascending, moderately crowded, dark green, rough and hairy above, densely greyish-hairy below, margins recurved to revolute, apex blunt; **flowers** small, cream to yellow, in axillary or terminal 2–20-flowered clusters, often profuse and conspicuous; **capsules** small, blackish.

A wide-ranging species with disjunct populations from north-eastern NSW and along the Great Dividing Range to near Maffra, Vic. It occurs at 250–2200 m altitude, often near creeks and rivers, and is found in shallow, often rocky soils. Plants lack very strong ornamental qualities but they are cultivated to a limited degree. A semi-shaded site with moderately well-drained acidic soils is suitable. Plants are hardy to moderately heavy frosts. Propagate from seed or from cuttings of firm young growth.

Pomaderris apetala Labill.
(without petals)
Vic, Tas   Hazel Pomaderris
3–15 m × 2–6 m   Oct–Dec

Medium to tall **shrub** or small to medium **tree** with greyish- or rarely rusty-hairy young growth; **trunks** smooth, dark grey; **branches** spreading to ascending; **branchlets** with greyish or rarely rusty hairs; **leaves** 3–12 cm × 1.5–4 cm, narrowly ovate to ovate, recurved to ascending, dark green and becoming glabrous above, dense covering of whitish to greyish starry hairs below, margins faintly toothed, venation prominent, apex pointed; **flowers** small, greyish to cream, in more or less pyramidal panicles to 25 cm long, often profuse, rarely highly conspicuous.

The typical subspecies is found mainly in wet forests

and gullies throughout Tas with a disjunct occurrence in the Grampians, Vic. Established plants often have many lichens on trunks. Plants do best in moist, well-drained, acidic soils with a semi-shaded or shaded aspect. They are hardy to heavy frosts. Planting this species in copses creates a marvellous effect. First introduced into England in 1803. Jewell butterflies use this species as a larval host plant which often results in noticeably eaten leaves.

The ssp. *maritima* N. G. Walsh & F. Coates is a small to medium coastal shrub of 1–3 m × 1–3 m. Leaves are ovate and 2.5–6 cm × 1.5–3 cm with a blunt apex. The inflorescence is up to 15 cm long. It inhabits coastal dunes and cliffs in the Wilson's Promontory region, Bass Strait Islands and the north coast of Tas.

Propagate both variants from fresh seed or from cuttings of firm young growth.

The allied *P. aspera* differs in having a light covering of mainly rusty hairs on the undersurface of the leaf.

Pomaderris argyrophylla N. A. Wakef.
(silvery leaves)

Qld, NSW	Silver Pomaderris
1.6–5 m × 1–3.5 m	June–Nov

Small to tall **shrub** with silvery-hairy young growth; **branches** spreading to ascending; **branchlets** silvery-hairy; **leaves** 5–12 cm × 1.5–4 cm, narrowly ovate, spreading to ascending, dark green and glabrous above, silvery-hairy below, margins flat, venation prominent, tapering to acute apex; **flowers** small, buds brownish, opening to white or cream, in loose panicles to about 10 cm across, often profuse; **fruit** silvery-hairy.

The typical variant of this handsome species occurs from near Herberton, Qld, to the Barrington area of NSW, often in disjunct populations. It is found at 80–1600 m altitude in rainforest margins, open forest and shrubland where it grows in clay loams and stony soils associated with waterways, rocky outcrops and mountain ridges. Evidently rarely cultivated, it has potential for use in subtropical and temperate regions. Good drainage, acidic soils and semi-shaded or open sites are prerequisites. Plants are hardy to moderate frosts. Worth trying as a hedging plant.

The ssp. *graniticola* N. G. Walsh & F. Coates has silvery to golden branchlets and blunt, ovate to elliptic leaves of 1.2–6.5 cm × 0.8–3 cm. The cream to yellow flowers are borne in compact pyramidal panicles of 2–8 cm across during Sept–Oct. It occurs in the granite soils of south-eastern Qld and north-eastern NSW. Cultivation requirements are probably similar to those for the typical variant.

Propagate both subspecies from fresh seed or from cuttings of firm young growth.

Pomaderris aspera Sieber ex DC.
(rough)

Qld, NSW, Vic, Tas, SA	Hazel Pomaderris
1–12 m × 1.5–5 m	Oct–Nov

Medium to tall **shrub** or small **tree** with rusty-hairy young growth; **branches** spreading to ascending; **branchlets** rusty-hairy; **leaves** 4–14 cm × 2–6 cm, ovate, recurved to ascending, dark green and usually becoming glabrous above, pale greenish with rusty starry hairs below, margins faintly toothed, venation

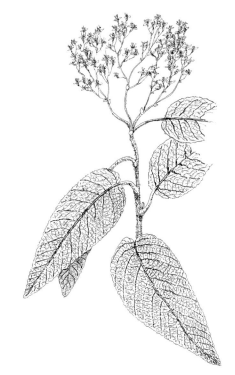

Pomaderris aspera × .5

prominent, apex tapering to pointed or rarely blunt; **flowers** small, cream to green, sometimes with crimson tonings, arranged in slightly open panicles 5–35 cm long, often profuse.

A quick-growing species which inhabits open forest and rainforest from south-eastern Qld and along the Great Dividing Range through NSW to Vic. Further populations are found in north-eastern Tas and south-eastern SA. Trunks often provide habitat for lichens. Excellent for copses and for planting beneath established trees. Adapts to a range of acidic soils which have moderate to excellent drainage. Does best in sheltered or semi-sheltered sites. Hardy to most frosts and light snowfalls. First cultivated in England during 1825. This species is a larval host for Jewell butterflies. Propagate from fresh seed or from cuttings of firm young growth.

The allied *P. apetala* differs in having a very dense covering of whitish to greyish hairs on the undersurface of the leaf.

Pomaderris aurea N. A. Wakef.
(gold)

?NSW, Vic	Golden Pomaderris
0.5–3 m × 0.5–2.5 m	Sept–Oct

Dwarf to medium **shrub** with greyish- to rusty-hairy young growth; **branches** spreading to ascending; **branchlets** with greyish to rusty hairs; **leaves** 1.5–6 cm × 0.8–2.3 cm, elliptic to ovate, spreading to ascending, somewhat crowded, deep green and hairy above, much paler below with dense greyish hairs and sparse

Pomaderris betulina

Pomaderris aurea   W. R. Elliot

rusty hairs, flat, venation distinct, apex pointed or blunt; **flowers** small, golden-yellow, in somewhat crowded, rounded to pyramidal clusters 3–8 cm across, often profuse and very conspicuous; **fruit** greenish, densely hairy.

This highly desirable species is restricted to eastern and north-eastern Vic (and possibly adjacent NSW), where it is found in open forest, woodland and shrubland at 150–900 m altitude. It occurs in sandy or loam soils. Cultivated to a limited degree, it is suitable for temperate regions and may adapt to subtropical zones. Needs well-drained acidic soils and a semi-shaded site is preferred although it tolerates plenty of sunshine. Hardy to most frosts. Responds well to pruning or clipping. Recommended for planting beneath established trees. May require supplementary watering during extended dry periods. Propagate from seed, or from cuttings of firm young growth, which strike readily.

P. humilis N. G. Walsh which has been known and marketed as *P.* species (Holey Plains) was recently reduced to a synonym.

P. lanigera is also allied but has much larger leaves and larger, paler yellow flowerheads.

Pomaderris betulina A. Cunn. ex Hook.
(similar to the birch genus, *Betula*)
NSW, Vic   Birch Pomaderris
1–4 m × 1–3 m   Oct–Nov

Small to medium **shrub** with rusty-hairy young growth; **branches** spreading to ascending, few to many; **branchlets** rusty-hairy; **leaves** 1.3–6 cm × 0.5–2 cm, ovate to obovate, spreading to ascending, moderately crowded, dark green and becoming glabrous above except for the recurved margins, pale starry hairs below, venation prominent, apex blunt or pointed; **flowers** small, cream, in small crowded clusters, can be profuse but rarely very conspicuous; **petals** 5; **fruit** with white and rusty hairs.

The typical variant is found at 80–520 m altitude in

eastern NSW and south from the New England Tableland. It also extends to eastern Vic. Plants occur in open forests and woodland on slopes and along rocky waterways. This species lacks strong ornamental qualities but was first cultivated in England in 1823. Rarely encountered now, it is suitable for temperate regions and requires excellently drained acidic soils with a semi-shaded aspect but will tolerate greater exposure.

The ssp. *actensis* N. G. Walsh & F. Coates differs mainly in having flat leaves of 1.5–4 cm × 0.8–2.2 cm which have faint venation. It is confined to the ACT region.

Propagate both subspecies from fresh seed or from cuttings of firm young growth.

Pomaderris bilocularis A. S. George
(two locules or ovaries)
WA   Tutanning Pomaderris
0.3–1 m × 0.3–0.6 m   Oct–Nov

Dwarf **shrub**; young growth with greyish to rusty hairs; **branches** few, spreading to ascending; **branchlets** faintly greyish- to rusty-hairy; **leaves** 0.6–1 cm × 0.25–0.7 cm, ovate to obovate, spreading to ascending, not crowded, greyish-green and very faintly hairy above, densely grey and rusty hairs below, margins flat or recurved, venation faint, apex blunt to notched; **flowers** small, cream to yellow, in clusters to about 1 cm across, often profuse but rarely very conspicuous.

A very restricted species which is endemic in the Avon District where it grows in shallow lateritic soils of woodland and mallee. Worthy of cultivation as a

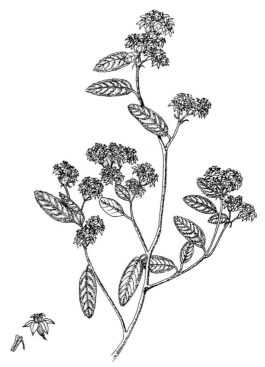

Pomaderris betulina × .55, flower × 2, anther × 5

426

conservation measure. Plants will need well-drained acidic soils and a warm semi-shaded or sunny site. Hardy to light frosts. Propagate from seed or from cuttings of firm young growth.

Pomaderris bodalla N. G. Walsh & F. Coates — see *P. brunnea* N. A. Wakef.

Pomaderris brevifolia N. G. Walsh
(short leaves)
WA   Pomaderris
0.5–1.5 m × 0.5–1.5 m   July–Sept
   Dwarf to small **shrub** with whitish-hairy young growth; **branches** spreading to ascending, few to many; **branchlets** whitish-hairy; **leaves** 0.3–0.7 cm × 0.2–0.4 cm, obovate to obcordate, spreading to ascending, moderately crowded, deep green and glabrous or faintly hairy above, densely hairy below, margins revolute, venation faint, apex blunt to notched; **flowers** small, cream to pale pink, in 10–20-flowered clusters to about 1.5 cm across, often profuse and moderately conspicuous; **petals** about 0.1 cm long.
   A recently described (1994) species which occurs from near Israelite Bay to the Stirling Range. It is found in heath and mallee communities growing in sandy and lateritic clay loam soils. Worthy of cultivation, it is suited to temperate and semi-arid regions. Needs a warm to hot site with moderate to good drainage. Plants may need pruning at an early stage to promote bushy growth. Propagate from seed or from cuttings of firm young growth.

Pomaderris brogoensis N. G. Walsh
(from the Brogo region)
NSW   Pomaderris
2–9 m × 1.5–4.5 m   Nov–Dec
   Medium to tall **shrub** or small **tree** with greyish-hairy young growth; **bark** mottled grey on mature plants; **branches** spreading to ascending; **branchlets** densely greyish-hairy; **leaves** 0.8–2.5 cm × 0.7–2 cm, broadly obovate, spreading to ascending, moderately crowded, green to greyish-green and hairy above, densely hairy below, margins flat or rarely wavy, apex blunt to very blunt; **stipules** to 0.6 cm long, persistent or deciduous; **flowers** small, cream, in pyramidal panicles to about 3.5 cm long, often profuse; **petals** absent.
   This species is confined to the foothills of the South Coast and the Southern Tablelands Districts where it grows at 40–470 m altitude in shallow soils beside waterways and on dry rocky ridges. It is cultivated to an extremely limited degree. Needs excellently drained acidic soils with a semi-shaded aspect. Hardy to moderate frosts. Propagate from seed or from cuttings of firm young growth.
   P. cinerea is similar but lacks simple hairs and has smaller yellow flowers and longer leaves.

Pomaderris brunnea N. A. Wakef.
(deep brown)
NSW, Vic   Pomaderris
2–4 m × 1.5–3 m   Sept–Oct
   Medium **shrub** with whitish- to greyish-hairy young growth; **branches** spreading to ascending, many; **branchlets** greyish-hairy; **leaves** 1.5–4 cm × 0.8–1.5 cm,

elliptic to obovate, recurved to spreading, moderately crowded, deep green and glabrous above, with greyish and rusty hairs below, margins recurved, apex blunt; **flowers** small, greyish to brownish in bud, opening cream, in clusters forming pyramidal panicles 3–5 cm long, often profuse and moderately conspicuous; **petals** absent; **fruit** rusty-hairy.
   P. brunnea occurs at low altitude in shallow sandy soils of gullies and slopes near waterways. It is mainly distributed near Sydney with further disjunct populations south of Armidale, NSW, and in far-eastern Vic. It lacks strong ornamental features and is only cultivated to a limited degree. Should prove adaptable in temperate regions and may succeed in subtropical areas. Good drainage, acidic soils and a semi-shaded site are required. Plants withstand moderate frosts. Propagate from seed or from cuttings of firm young growth.
   The allied *P. bodalla* N. G. Walsh & F. Coates has broader, flat leaves with less prominent venation. It is endemic in the foothills of the Great Dividing Range between Bodalla and Bega in NSW where it grows in open forest and dry rainforest. It has similar cultivation requirements to those of *P. brunnea*.

Pomaderris calvertiana F. Muell. = *P. ledifolia* A. Cunn.

Pomaderris canescens (Benth.) N. A. Wakef.
(hoary)
Qld
1–3 m × 1–3 m   May–June; also sporadic
   Small to medium **shrub** with greyish-hairy young growth; **branches** often multi-stemmed, spreading to ascending; **branchlets** greyish-hairy; **leaves** 2–15 cm × 1.2–5 cm, ovate to elliptic, prominently stalked, spreading to ascending, deep green and becoming glabrous above, dense covering of whitish and rusty hairs below, margins slightly recurved, apex pointed; **flowers** small, whitish in bud, opening cream, in somewhat hemispherical clusters 2–8 cm across, often profuse and conspicuous; **petals** 5.
   A pleasing species which has attractive foliage. It occurs in the Moreton, North Kennedy, Port Curtis and Wide Bay Districts where it grows in woodland and open forests of the coastal hinterland, on ridges and slopes beside waterways. Worthy of wider recognition. Suitable for subtropical and temperate regions. Should do well in freely draining acidic soils with a semi-shaded aspect. Frost tolerance is not known. Propagate from seed or from cuttings of firm young growth.
   P. ferruginea Sieber ex Fenzl var. *canescens* Benth. is a synonym.

Pomaderris cinerea Benth.
(ash-grey)
NSW
1.5–10 m × 1–5 m   Dec–Jan
   Small to tall **shrub** or small **tree** with densely hairy young growth; **branches** spreading to ascending; **branchlets** with fawn hairs; **leaves** 1.5–6 cm × 0.7–2.8 cm, ovate to elliptic, spreading to ascending, grey-green to greyish and somewhat velvety above, whitish-hairy below, flat, apex rounded to bluntly pointed; **stipules**

to about 0.5 cm long, usually deciduous; **flowers** small, pale yellow to yellow, in somewhat pyramidal panicles of 2–6 cm × 1.5–3.5 cm, often profuse and conspicuous; **petals** absent.

An ornamental species from the south-eastern region where it grows in sandy loam or loam soils of dryish to wettish sites in open forest and rainforest margins. Cultivated to a limited degree, it deserves to be better known. Plants require moderately well-drained acidic soils and a semi-shaded aspect but will also tolerate denser shade. Withstands moderate frosts. It has potential for use as formal or informal hedging. Propagate from seed or from cuttings of firm young growth.

Pomaderris clivicola E. M. Ross
(growing on slopes)

Qld	Pomaderris
3–6 m × 2–4 m	Dec–Jan

Medium to tall **shrub** with greyish-hairy young growth; **stems** often many; **branches** spreading to ascending; **branchlets** greyish-hairy, slender; **leaves** 1.5–3.5 cm × 0.6–1.5 cm, ovate, recurved to ascending, grey-green to greyish and hairy above, paler hairs below, flat, venation faint, apex blunt to pointed; **flowers** small, greyish-hairy in bud, opening cream to yellow, in pyramidal panicles to about 2 cm long, often profuse and conspicuous; **petals** absent.

A recently described (1990) species which occurs in only a limited region of the Burdett District where it grows in red loam soils. Plants are moderately ornamental. Suitable for subtropical and temperate regions. Needs freely draining acidic soils with a semi-shaded aspect. Frost tolerance unknown. Propagate from seed or from cuttings of firm young growth.

The closely allied *P. coomingalensis* N. G. Walsh & F. Coates is a shrub of 3–5 m high. It has smaller leaves which are glabrous on the upper surface. This species is known only from the northern Burdett District where it occurs on red loam in open eucalypt forest. Should have similar cultivation requirements to those of *P. clivicola*.

P. cinerea has similarities but differs in having elliptic leaves.

Pomaderris cocoparrana N. G. Walsh
(after the Cocoparra Range)

NSW	Pomaderris
1–3 m × 1–3 m	Sept–Oct

Small to medium **shrub** with rusty-hairy young growth; **branches** spreading to ascending; **branchlets** densely rusty-hairy; **leaves** 0.8–3.5 cm × 0.8–2 cm, broadly ovate to nearly circular, recurved to ascending, dark green and slightly velvety above, with greyish and rusty hairs below, flat, venation distinct, apex blunt to faintly notched; **flowers** small, yellow, in crowded pyramidal panicles to about 5 cm long, often profuse and very conspicuous; **petals** absent.

An attractive, recently described (1990) species which is known only from near Griffith where it grows in shallow soils or gravels derived from sandstone. It is worthy of cultivation and should prove reliable in temperate regions in well-drained acidic soil. A semi-shaded site is preferable but plants may tolerate more

sunshine. Withstands moderately heavy frosts. Propagate from seed or from cuttings of firm young growth.

Pomaderris coomingalensis N. G. Walsh & F. Coates —
see *P. clivicola* E. M. Ross

Pomaderris costata N. A. Wakef.
(prominent ribs)

NSW, Vic	Pomaderris
1–4 m × 1–3 m	Oct–Nov

Small to medium **shrub**; young growth with fawn to golden hairs; **branches** spreading to ascending; **branchlets** densely rusty-hairy; **leaves** 2.5–5 cm × 1.5–3.5 cm, ovate to elliptic, spreading to ascending, deep green and becoming glabrous above, with rusty and golden hairs below, margins flat to slightly wavy, apex blunt; **flowers** small, cream, in somewhat pyramidal panicles, 3–5 cm × about 4 cm, often profuse and conspicuous; **petals** absent.

This rare and endangered species has disjunct populations along the Great Dividing Range from the Upper Hunter Valley, NSW, to far-eastern Vic. It is often found in rocky sites in dry forest and shrubland. It warrants cultivation as part of a conservation strategy. Plants prefer very well-drained acidic soils with a semi-shaded aspect but they may adapt to sunnier sites. They tolerate extended dry periods and are hardy to moderate frosts. Propagate from seed or from cuttings of firm young growth.

Pomaderris cotoneaster N. A. Wakef. —
see *P. prunifolia* A. Cunn. ex Fenzl

Pomaderris crassifolia N. G. Walsh & F. Coates —
see *P. vellea* N. A. Wakef.

Pomaderris delicata N. G. Walsh & F. Coates —
see *P. andromedifolia* A. Cunn. ssp. *andromedifolia*.

Pomaderris discolor (Vent.) Poir.
(two distinct colours)

NSW, Vic	Pomaderris
2–5 m × 1.5–4 m	Sept–Oct

Medium to tall **shrub** with whitish to greyish young growth; **branches** spreading to ascending, many; **branchlets** greyish-hairy; **leaves** 3–9.5 cm × 1.5–4 cm, ovate to elliptic, spreading to ascending, prominently stalked, deep green and glabrous above, greyish-hairy below, thin, margins recurved, venation prominent, apex pointed; **stipules** to 0.6 cm long, deciduous; **flowers** small, pale cream to pale yellow, in pyramidal panicles to about 6 cm long, often profuse and conspicuous; **petals** often absent.

A wide-ranging and often common species from the North and Central Coast, Central Tablelands and South Western Slopes Districts of NSW. It also extends to eastern Vic. Usually found at 20–100 m altitude in moist open forest and rainforest margins. Cultivated to a limited degree it is best suited to freely draining but moisture-retentive acidic soils with a semi-shaded aspect. Plants tolerate light frosts. First cultivated in France in 1804 and subsequently in England in 1814. Propagate from seed or from cuttings of firm young growth.

Pomaderris elacophylla F. Muell.
(small or short leaves)
NSW, Vic, Tas   Lacy Pomaderris; Small-leaf
                                    Pomaderris
1–3.5 m × 0.8–3 m   Oct–Dec

Small to medium **shrub**; young growth with fawn to rusty hairs; **branches** spreading to ascending, can be semi-weeping; **branchlets** rusty-hairy; **leaves** 0.2–1.1 cm × 0.15–0.8 cm, ovate to obovate, spreading to ascending, moderately crowded, dark green and hairy to glabrous above, brownish-hairy below, margins flat or recurved, apex blunt or faintly notched; **stipules** to 0.5 cm long, very narrow, deciduous; **flowers** small, cream, solitary or in small axillary clusters, sometimes profuse but rarely highly conspicuous; **petals** absent.

This species occurs in south-eastern NSW, eastern Vic and western Vic. Plants are found in mountains at 100–1200 m altitude where soils are often moist for extended periods. Rarely encountered in cultivation but has proved adaptable in freely draining acidic soils with a semi-shaded or shaded aspect. Hardy to moderately heavy frosts. Responds well to pruning and may prove to be a useful hedging plant. Propagate from seed or from cuttings of firm young growth.

P. vacciniifolia is closely allied but differs in the flowers having petals.

Pomaderris elliptica Labill.
(elliptical)
NSW, Vic, Tas   Pomaderris
1.5–4 m × 1–3 m   Sept–Nov

Small to medium **shrub** with densely starry-hairy young growth; **branches** spreading to ascending; **branchlets** with cream to whitish hairs; **leaves** 3–9 cm × 1.5–4.5 cm, narrowly elliptic to ovate, mainly spreading to ascending, mid green and glabrous above, faintly whitish-hairy below, margins flat to wavy, venation prominent, apex blunt to pointed; **stipules** to 0.5 cm long, broad, deciduous; **flowers** small, pale yellow, in hemispherical, or pyramidal, terminal panicles to about 12 cm across, often profuse and very conspicuous; **petals** to 0.2 cm long.

The typical subspecies is widespread with its distribution extending from near Taree, NSW, to eastern Vic and south-eastern Tas. It grows on the coast and hinterland in moist forests. Adapts well to cultivation in temperate regions and prefers freely draining but moisture-retentive acidic soils with a semi-shaded or sunny aspect. Hardy to light frosts and responds well to pruning. Cultivated in England from as early as 1805.

P. multiflora Sieber ex Fenzl is a synonym.

The ssp. *diemenica* N. G. Walsh & F. Coates has simple and stellate hairs on its foliage. The stipules are up to 0.7 cm long and petals to 0.25 cm long. It is endemic in eastern Tas where it grows mainly in moist shrubland and open forest but is also found in exposed rocky sites.

Propagate both subspecies from seed or from cuttings of firm young growth.

P. pilifera is allied and has been confused with *P. elliptica*. *P. pilifera* has smaller leaves and long rusty hairs on leaf undersurface veins.

Pomaderris ericifolia Hook. =
                    P. phylicifolia Lodd. ex Link

Pomaderris eriocephala N. A. Wakef.
(woolly heads)
NSW, Vic   Pomaderris
1–3 m × 1.5–3.5 m   Sept–Oct

Small to medium **shrub** with rusty-hairy young growth; **branches** spreading to ascending; **branchlets** rusty-hairy; **leaves** 0.8–4.5 cm × 0.7–3 cm, ovate to obovate, spreading to ascending, dark green with greyish hairs above, whitish and rusty hairs below, margins more or less entire with tufts of hairs at ends of prominent veins, apex blunt or faintly notched; **stipules** to 1 cm long, narrowly triangular; **flowers** small, rusty-hairy in bud, opening to cream, in tight axillary globular clusters to about 1.3 cm across, often profuse and conspicuous; **petals** 5 or absent.

This decorative species inhabits open forest and woodland where it grows in rocky, shallow soils and is often associated with waterways. It is distributed from north-eastern NSW to eastern Vic and occurs at 300–870 m altitude. Cultivated to a limited degree, it adapts well to well-drained acidic soils with a semi-shaded aspect. Hardy to moderately heavy frosts and extended dry periods. Propagate from seed or from cuttings of firm young growth.

Pomaderris ferruginea Sieber ex Fenzl
(rust-coloured)
Qld, NSW, Vic   Rusty Pomaderris
1.5–4 m × 1–3 m   Aug–Oct

Small to medium **shrub** with densely rusty-hairy young growth; **branches** spreading to ascending; **branchlets** rusty-hairy; **leaves** 3–9 cm × 1.5–4 cm, narrowly ovate to ovate, recurved to ascending, prominently stalked, deep green and glabrous above, dense covering of cream and rusty hairs below, margins flat or slightly recurved, venation prominent, apex blunt to pointed; **stipules** to 0.6 cm long, ovate, deciduous; **flowers** small, greyish to golden-hairy in bud, opening cream or rarely yellow, in hemispherical to pyramidal terminal panicles to about 10 cm across, often profuse and very conspicuous; **petals** 5 or rarely absent.

A handsome wide-ranging species which has its distribution from south-eastern Qld to just west of Melbourne, Vic, mainly on the coastal side of the Great Dividing Range. It is found in woodland and open forest, usually associated with watercourses or rocky outcrops. Adapts well to cultivation in temperate and subtropical regions. Prefers well-drained acidic soils with a semi-shaded aspect but will tolerate more sunshine. Hardy to moderate frosts. Pruning from an early stage promotes bushy growth. Useful for planting beneath established trees. Propagate from seed or from cuttings of firm young growth.

The var. *canescens* Benth. is a synonym of *P. canescens* (Benth.) N. A. Wakef. and var. *pubescens* Benth. is a synonym of *P. lanigera* (Andrews) Sims.

P. ligustrina is allied but differs in having flowers without and usually smaller leaves. *P. lanigera* has been confused with *P. ferruginea* in the nursery trade. *P. lanigera* has leaves which are mid green above and paler below with pale rusty hairs on the veins.

Hybrids with *P. ferruginea* and *P. andromedifolia* as parents occur in the Blue Mountains of NSW and some of these may be in cultivation.

Pomaderris elliptica   W. R. Elliot

Pomaderris flabellaris (F. Muell. ex Reissek) J. M. Black
(fan-shaped)
SA   Fan Pomaderris
1–2.5 m × 0.6–2 m   Sept–Oct
   Small to medium **shrub** with rusty-hairy young
growth; **branches** spreading to ascending, slender;
branchlets rusty-hairy; **leaves** 0.4–1.2 cm × 0.5–1.3 cm,
fan-shaped, tapered to base, somewhat crowded,
spreading to ascending, greyish-hairy above and
below, flat to folded inwards, margins often crenate or
toothed in upper parts; **flowers** small, rusty-hairy in
bud, opening greenish to yellow, in few-flowered
racemes or panicles to about 3 cm long, sometimes
profuse but rarely very conspicuous; **petals** absent.
   P. flabellaris has interesting fan-shaped leaves. It occurs
in heathland, mallee and woodland in the central and
southern Eyre Peninsula where it grows on sandy loam
which often overlies ironstone gravel. Evidently rarely
cultivated, it is suitable for semi-arid and temperate
regions. Should adapt to acidic well-drained soils and
will tolerate sunny or semi-shaded sites. Worth trying
as a hedging plant. Hardy to moderately heavy frosts.
Propagate from seed or from cuttings of firm young
growth.

Pomaderris forrestiana F. Muell. —
                                see *P. myrtilloides* Fenzl

Pomaderris gilmourii N. G. Walsh
(after Phil Gilmour, contemporary botanical collector,
NSW)
NSW   Pomaderris
2–4 m × 1.5–3 m   Nov–Dec

Medium **shrub** with greyish-hairy young growth;
branches spreading to ascending; **branchlets** greyish-
hairy, shining; **leaves** 0.8–4 cm × 0.4–1.3 cm, elliptic to
obovate, spreading to ascending, deep green and glabrous above, shiny with greyish hairs below, margins
recurved, venation distinct, apex pointed to blunt;
flowers small, silvery- to greyish-hairy in bud, opening
cream, in globular to pyramidal panicles to about 6 cm
long, often profuse and conspicuous; **petals** absent.
   The typical variant of this ornamental species occurs
at 450–980 m altitude in south-eastern NSW where it
grows on a few peaks and rocky slopes. Cultivated to a
very limited degree, it has potential for use in temperate regions. Needs very well-drained acidic soils with a
semi-shaded aspect. Plants are hardy to moderately
heavy frosts and they withstand extended dry periods.
   The var. *cana* N. G. Walsh differs in having flat leaves
which have stellate hairs but lack shiny hairs. It is
restricted to one location in south-eastern NSW. Flowering plants are attractive. Should have similar cultivation requirements to the typical variant.
   Propagate both variants from seed or from cuttings
of firm young growth.

Pomaderris grandis F. Muell.
(tall or large)
WA   Large Pomaderris
1–3 m × 1–2.5 m   Sept
   Small to medium **shrub** with silky golden-hairy
young growth; **branches** spreading to ascending;
branchlets with silky golden hairs; **leaves** 3–10 cm ×
1.2–2.5 cm, narrowly elliptic to obovate, spreading to
ascending, deep green and glabrous above, densely
white-hairy below, margins flat or recurved, apex
pointed; **stipules** to 0.9 cm long, narrow, deciduous;
flowers small, whitish- to silvery-hairy in bud, opening
white to cream, in panicles to 6 cm × 6 cm, can be profuse and moderately conspicuous; **petals** absent.
   An interesting species from the Mt Manypeaks
region where it grows in moist sites in gravelly soils
overlying granite. Rarely cultivated, it has potential for
wider use in gardens. Plants need good drainage in
acidic soils and they should adapt to a sunny or semi-
shaded site in temperate regions. Frost tolerance is not
known. Propagate from seed or from cuttings of firm
young growth.

Pomaderris halmaturina J. M. Black
(pertaining to Kangaroo Island)
SA   Kangaroo Island Pomaderris
2–3 m × 1–2 m   Oct–Nov
   Medium **shrub** with rusty-hairy young growth;
branches spreading to erect; **branchlets** rusty-hairy;
leaves 2–5.5 cm × 1.2–2.5 cm, elliptic to obovate,
spreading to ascending, deep green and wrinkled with
stiff hairs above, greyish and rusty hairs below, margins
lightly toothed, flat, apex pointed; **stipules** to about
0.7 cm long, linear, deciduous; **flowers** small, yellow-
green, sometimes with deep red tonings, in short often
open panicles to about as long as leaves, sometimes
profuse but rarely very conspicuous; **petals** absent.
   The typical subspecies is regarded as under threat in
nature. It occurs mainly in low-lying coastal regions on

430

Kangaroo Island with disjunct populations in the south-eastern region of mainland SA. Plants are cultivated to a very limited degree. They require good drainage, acidic soils and a semi-shaded or moderately sunny site. Frost tolerance is not known.

The ssp. *continentis* N. G. Walsh is a shrub of 1.5–5 m × 1–3.5 m. It has ovate leaves of 1.5–10 cm × 0.7–3 cm which are rarely glabrous above. The leaf margins are shallowly lobed. It occurs in south-eastern SA and far south-western Vic with a disjunct occurrence near Torquay, Vic. Plants are found on old dune systems beside waterways or in open forest.

Propagate both subspecies from seed or from cuttings of firm young growth.

P. oblongifolia N. G. Walsh has affinity with *P. halmaturina* but it differs in having oblong to narrowly ovate leaves with a very fine covering of hairs. It is a slender shrub of 1–2.5 m × 0.6–1.6 m with greenish to maroon flowers which bloom during Nov–Jan. It is endemic in the gravelly sands and rocks of the Snowy River, north of Buchan, Vic. Evidently not cultivated but is suited to temperate regions. It would require excellent drainage.

Pomaderris helianthemifolia (Reissek) N. A. Wakef.
(leaves similar to those of *Helianthemum*)

Vic	Pomaderris
1–2 m × 1–2 m	Oct–Nov

Small **shrub** with greyish- to rusty-hairy young growth; **branches** spreading to ascending; **branchlets** with greyish or rusty hairs; **leaves** 1.5–4 cm × 0.4–1 cm, oblong to narrowly obovate, spreading to ascending, deep green and glabrous above, densely greyish-hairy below, margins recurved, apex blunt; **stipules** to 0.25 cm long, narrow, deciduous; **flowers** small, greyish-hairy in bud, opening cream to yellow, in terminal, or upper axillary, elongated panicles to about 3.5 cm across, often profuse and moderately conspicuous; **petals** absent.

The typical subspecies is endemic in a small region of Gippsland where it grows in rocky sites besides waterways at 80–250 m altitude. Cultivated to a very limited degree, it should adapt to well-drained acidic soils with a semi-shaded aspect in temperate regions. Plants are hardy to moderate frosts.

There are two other subspecies.

The ssp. *hispida* N. G. Walsh & F. Coates is of similar dimensions. It has oblong to narrowly obovate leaves of 1–4.5 cm × 0.2–1 cm which are hairy above. It is found in NSW and eastern Vic at 40–600 m altitude. Plants also grow in rocky sites besides waterways.

The ssp. *minor* N. G. Walsh & F. Coates has narrowly obovate leaves of 0.5–1.3 cm × 0.2–0.4 cm which are also hairy above. It is endemic in north-central and north-eastern Vic at 400–600 m altitude and inhabits stream margins.

All subspecies are propagated from seed or from cuttings of firm young growth.

Pomaderris humilis N. G. Walsh = *P. aurea* N. A. Wakef.

Pomaderris intangenda F. Muell. ex F. Muell. =
    Granitites intangendus (F. Muell. ex F. Muell.) Rye

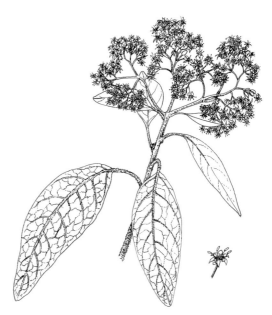

Pomaderris intermedia × .5, flower × 1.5

Pomaderris intermedia Sieber ex DC.
(intermediate)

NSW, Vic, Tas	Pomaderris
1–3.5 m × 1–2.5 m	Sept–Oct

Small to medium **shrub** with pale greyish-hairy young growth; **branches** spreading to ascending; **branchlets** densely covered with pale greyish hairs; **leaves** 3–10 cm × 1–5 cm, narrowly ovate to broadly elliptic, spreading to ascending, deep green and glabrous above, undersurface with short, pale, greyish hairs and longer brown hairs, margins flat, venation distinct, apex blunt to pointed; **stipules** to 0.8 cm long, ovate, deciduous; **flowers** small, greyish-hairy in bud, opening bright yellow, in hemispherical, or pyramidal, terminal panicles to 15 cm × about 12 cm, often profuse and highly conspicuous; **petals** 5.

An outstanding species which is distributed from near Glen Innes, NSW, and along mainly the eastern and southern slopes of the Great Dividing Range to southern-central Vic. It also extends to northern Tas. Plants are found in heathland, woodland and open forest. *P. intermedia* adapts well to cultivation in temperate regions and requires freely draining acidic soils in a semi-shaded site. Hardy to moderate frosts and withstands extended dry periods. Responds well to pruning which will promote bushy growth if it is done from an early stage. Excellent for planting beneath established trees provided it receives sunshine. First grown in England in 1825. Propagate from seed or from cuttings of firm young growth.

P. affinis N. A. Wakef. and *P. sieberiana* N. A. Wakef. are synonyms.

P. elliptica is closely allied. It usually lacks simple hairs on the leaf undersurface. Hybrids of *P. elliptica* and *P. intermedia* occur in the wild and may be in cultivation. *P. lanigera* is similar but it has reddish-brown young branches. A hybrid reported to have *P. intermedia* and *P. lanigera* as parents is cultivated. *P. ferruginea* is also similar but it has cream or rarely pale yellow flowers.

Pomaderris lanigera (Andrews) Sims
(woolly)
Qld, NSW, Vic   Woolly Pomaderris
0.5–3.5 m × 0.5–3 m   Aug–Oct

Dwarf to medium **shrub** with rusty to reddish-hairy young growth; **branches** spreading to erect; **branchlets** covered with rusty to reddish hairs; **leaves** 4–13 cm × 1.5–5 cm, narrowly or broadly ovate to elliptic, recurved to ascending, mid green and hairy (often appearing glabrous to naked eye) above, velvety below with fawn and rusty hairs, margins flat, venation distinct, apex pointed or blunt and sometimes faintly notched; **stipules** to 0.6 cm long, ovate, deciduous; **flowers** small, pale-to rusty-hairy in bud, opening yellow, in somewhat hemispherical terminal panicles to about 12 cm across, often profuse and highly conspicuous; **petals** 5.

A very ornamental species with a widespread distribution. Plants are found from near Carnarvon, Qld, to eastern regions of NSW. It also extends to near Melbourne, Vic. Usually inhabits dryish sites in heathland, woodland and open forest. Relatively common in cultivation, it adapts to temperate and subtropical regions. Must have good drainage and acidic soils. Although plants tolerate plenty of sunshine they are usually healthier in a semi-shaded site. They are hardy to moderately heavy frosts and withstand extended dry

periods. Responds well to pruning which helps promote bushy growth. Grown in England as early as 1806. Propagate from seed or from cuttings of firm young growth.

P. ferruginea var. *pubescens* Benth. is a synonym.

P. ferruginea has been confused with *P. lanigera* in the nursery trade. *P. ferruginea* has dark green leaves and cream or rarely pale yellow flowers. *P. aurea* differs in having smaller leaves and yellowish-grey hairs.

Pomaderris ledifolia A. Cunn.
(leaves similar to the genus *Ledum*)
Qld, NSW, Vic   Pomaderris
0.6–2 m × 0.6–2 m   Sept–Nov

Dwarf to small **shrub** with greyish- to rusty-hairy young growth; **branches** spreading to erect; **branchlets** with whitish to rusty hairs; **leaves** 0.6–2.5 cm × 0.2–0.6 cm, narrowly elliptic, recurved to ascending, scattered, light-to-mid green and glabrous above, densely greyish-hairy below except for rusty hairs on midrib, margins flat or recurved, venation indistinct, apex pointed; **stipules** very small, deciduous; **flowers** small, greyish-hairy in bud, opening yellow, in terminal clusters to about 2.5 cm across, often profuse and conspicuous; **petals** 5.

An interesting pomaderris which has disjunct populations from the Qld/NSW border to far eastern Vic. It grows in rocky sites in shrubland, woodland and open forest on mountain slopes and ridges at 240–1100 m altitude. Cultivated to a limited degree, it is worthy of wider recognition. Suitable for subtropical and temperate regions. Requires well-drained acidic soils in a semi-shaded or shaded site. Withstands moderately heavy frosts and extended dry periods. Pruning from an early stage should promote bushy growth. Introduced into cultivation in England in 1824. Propagate from seed or from cuttings of firm young growth.

P. calvertiana F. Muell. is a synonym.

P. ledifolia is sometimes confused with small-leaved variants of *P. andromedifolia*. Generally, *P. andromedifolia* has broader leaves with distinct venation on the undersurface.

Pomaderris ligustrina Sieber ex DC.
(similar to the genus *Ligustrum*)
Qld, NSW, Vic   Privet Pomaderrris
1.5–5 m × 1–4 m   Sept–Oct

Small to tall **shrub** with softly hairy young growth; **branches** spreading to erect; **branchlets** initially hairy but soon becoming glabrous; **leaves** 2.5–6 cm × 0.8–2 cm, narrowly elliptic to narrowly ovate, spreading to ascending, deep green and glabrous above, with dense pale to rusty hairs below, margins recurved, venation distinct, apex pointed; **stipules** to 0.6 cm long, deciduous; **flowers** small, greyish-hairy in bud, opening cream to yellow, in pyramidal or somewhat globular panicles to about 5 cm long, often profuse and conspicuous; **petals** absent.

A wide-ranging species which extends from southeastern Qld to eastern Vic. It occurs at 20–600 m altitude in dryish open forest and rainforest margins, often in rocky sites. Cultivated to a limited degree, it adapts well in temperate regions if grown in freely draining acidic soils with a semi-shaded aspect. Hardy

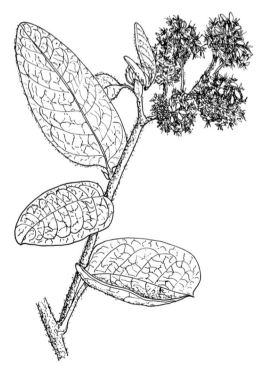

Pomaderris lanigera × .75

432

Pomaderris ledifolia × .8, flower × 3

to moderate frosts and tolerates dry periods. It has potential for use as a formal or informal hedge.

The ssp. *latifolia* N. G. Walsh & F. Coates grows 0.5–4.5 m tall and has broadly ovate to nearly circular leaves of 1.7–3 cm × 1–2 cm with a blunt or slightly pointed apex. It occurs from near Stanthorpe, Qld, to the Glen Innes region in NSW at 800–900 m altitude. Should have similar cultivation requirements to the typical subspecies.

Propagate from seed or from cuttings of firm young growth.

P. sericea N. A. Wakef. has similarities. It is a shrub of up to 2 m × 2 m with silky branchlets that have golden and greyish hairs. The leaves are narrowly elliptic, 0.6–3 cm × 0.5–1 cm, and are glabrous above and silky-hairy below. Yellow flowerheads of about 1.2 cm across form short leafy racemes mainly in Oct. This species occurs in two disjunct areas; near Bowral, NSW; and in far-eastern Vic. Plants are found on shallow gravelly soils overlying sandstone. Evidently not cultivated, it should have similar requirements to *P. ligustrina*.

Pomaderris mediora N. G. Walsh & F. Coates
(middle coast; referring to Central Coast District of NSW)

NSW   Pomaderris
0.5–3 m × 1.5–3 m   Sept–Oct

Dwarf to medium **shrub** with greyish to rusty-hairy young growth; **branches** procumbent to erect; **branchlets** with greyish and rusty hairs; **leaves** 1–1.8 cm × 0.15–0.5 cm, narrowly elliptic to narrowly obovate, spreading to ascending, deep green and glabrous

above, whitish or greyish-to-rusty hairs below, margins recurved to revolute, apex blunt, often recurved; **stipules** to 0.4 cm long, narrowly triangular, deciduous; **flowers** small, whitish-hairy in bud, opening cream, in pyramidal panicles to about 7 cm long, often profuse and conspicuous; **petals** absent.

P. mediora is a recently described (1997) species which is restricted to the Central Coast District where it grows in soils derived from sandstone and shale. It should adapt to temperate and subtropical regions. Plants will need freely draining acidic soils and should tolerate an open or semi-shaded site. There is scope to select low spreading variants for cultivation. Propagate from seed or from cuttings of firm young growth.

P. phylicifolia ssp. *phylicifolia* is allied but differs in having leaves with a hairy upper surface and persistent stipules, while the ssp. *ericifolia* has smaller, narrower leaves.

Pomaderris multiflora Sieber ex Fenzl =
                                          P. elliptica Labill.

Pomaderris myrtilloides Fenzl
(similar to the Whortleberry, *Vaccinium myrtillus*)

WA   Pomaderris
0.3–2 m × 0.5–2 m   April–Sept

Dwarf to small **shrub** with whitish-hairy young growth; **branches** spreading to ascending; **branchlets** densely covered with whitish hairs; **leaves** 1–2.6 cm × 0.7–1.5 cm, elliptic to obovate, spreading to ascending, deep green and faintly hairy to glabrous above, whitish hairs below, margins flat to revolute, apex blunt to

Pomaderris myrtilloides × 1, flower × 2

433

rounded, sometimes notched; **stipules** to 0.4 cm long, persistent; **flowers** small, whitish-hairy in bud, opening cream or pale pink and fading to white, in terminal clusters to about 3.5 cm across, often profuse and conspicuous; **petals** present.

This is a mainly coastal species which occurs from near Eucla to Bremer Bay in the Eyre District. Plants are found on alkaline sands as well as on laterite or acidic sand. *P. myrtilloides* inhabits heath and mallee communities. Cultivated to a limited degree, it deserves to be grown more widely. It is suitable for semi-arid and temperate regions. Needs excellent drainage and tolerates a sunny or semi-shaded site. Hardy to light frosts. Propagate from seed or from cuttings of firm young growth.

P. forrestiana is allied but has greyish- to rusty-hairy young growth and obovate leaves of 0.5–1.2 cm × 0.4–0.8 cm. It has yellow flowers in heads to about 2 cm across which are produced during Aug and Sept. It occurs in south-eastern WA and in the far southwestern region of SA. Plants are found in sandy soils overlying limestone. *P. forrestiana* hybridises with *P. myrtilloides* in the wild.

Pomaderris nitidula (Benth.) N. A. Wakef.
(shining)
Qld, NSW   Shining Pomaderris
1–4 m × 1–3 m   Sept–Oct
Small to medium **shrub** with silvery- to coppery-hairy young growth; **branches** spreading to erect; **branchlets** silvery-hairy; **leaves** 2–6 cm × 1–2.2 cm, elliptic, tapering to base, spreading to ascending, deep green and glabrous above, densely silvery-hairy below, margins flat to slightly recurved, apex pointed; **stipules** to 0.8 cm long, ovate, deciduous; **flowers** small, silky-hairy in bud, opening cream, in terminal clusters to about 5 cm across, often profuse and conspicuous; **petals** present.

An attractive species which is confined to southeastern Qld and north-eastern NSW. It grows in shallow soils on rocky hilltops at 500–920 m altitude. Plants are found in shrubland, open forest and rainforest margins. Evidently not cultivated, it should grow well in freely draining acidic soils with a semi-shaded aspect. Hardy to moderate frosts. Propagate from seed or from cuttings of firm young growth.

P. phillyreoides var. *nitidula* Benth. is a synonym.

Pomaderris notata S. T. Blake
(marked with lines or spots)
Qld   Pomaderris
1–4 m × 1–3 m   Oct–Dec
Small to medium **shrub** with greyish-hairy young growth; **bark** often whitish; **branches** spreading to erect; **branchlets** greyish-hairy; **leaves** 1.8–5 cm × 0.9–1.8 cm, elliptic to obovate, tapering to base, spreading to ascending, deep green and more or less glabrous above, greyish hairs and rusty hairs below, margins flat or faintly recurved, venation distinct, apex pointed; **stipules** to 0.5 cm long, narrow, deciduous; **flowers** small, greyish-hairy in bud, opening cream, in more or less pyramidal terminal panicles to about 6 cm long, often profuse and conspicuous; **petals** absent.

P. notata is regarded as rare in its natural habitat in far south-eastern Qld and far north-eastern NSW. It occurs in shallow soils in shrubland and rainforest margins at 800–900 m altitude. Worthy of cultivation as part of a conservation strategy. Plants should adapt to subtropical and temperate regions. They will need well-drained acidic soils with a semi-shaded aspect. Frost tolerance is not known. Propagate from seed or from cuttings of firm young growth.

Pomaderris obcordata Fenzl
(heart-shaped and broadest above the middle)
Vic, SA   Wedge-leaved Pomaderris
0.3–3 m × 0.5–2 m   July–Sept; also sporadic
Dwarf to medium **shrub** with densely greyish-hairy young growth; **branches** spreading to erect; **branchlets** greyish-hairy; **leaves** 0.6–2.6 cm × 0.3–2.3 cm, obovate to obcordate or wedge-shaped, spreading to ascending, deep green and glabrous, often shining above, greyish-hairy below, margins flat to revolute, apex bilobed to trilobed; **stipules** to 0.3 cm long, ovate, persistent; **flowers** small, white to pink in bud, opening white to pale pink, in tight terminal clusters to about 3 cm across, often profuse and very conspicuous; **petals** absent.

A very ornamental species which mainly occurs in south-eastern SA with a limited distribution in far-western Vic. Generally, plants inhabit heath, shrubland and mallee communities on sandy soils with associated limestone. Adaptable to cultivation in temperate and semi-arid regions. Needs excellent drainage and grows successfully in semi-shade or full

Pomaderris obcordata × .55, flower × 2.5

Pomaderris oraria ssp. *oraria* × .7, flower × 2

Pomaderris phylicifolia　　　　　　　　　D. L. Jones

sunshine. Hardy to moderately heavy frosts and extended dry periods. Responds very well to pruning. Both low-spreading and taller selections are cultivated. Propagate from seed or from cuttings of firm young growth.

Pomaderris oblongifolia N. G. Walsh —
　　　　　　　　　see *P. halmaturina* J. M. Black

Pomaderris oraria F. Muell. ex Reissek
(from the coast)
Vic, Tas　　　　　　　　　　　　Coast Pomaderris
0.5–1.3 m × 0.6–2 m　　　　　　　　　　Oct–Nov
　Dwarf to small **shrub** with greyish to rusty-hairy young growth; **branches** spreading to erect, many; **branchlets** usually rusty-hairy; **leaves** 1–3 cm × 0.8–2.3 cm, broadly elliptic, recurved to ascending, often moderately crowded, deep green and hairy above, whitish to grey hairs below, margins flat, apex blunt to rounded, often faintly notched or toothed; **stipules** to 0.7 cm long, linear, deciduous; **flowers** small, greenish often with cream or deep reddish tonings, in few- to many-flowered clusters forming a leafy panicle, can be profuse and moderately conspicuous; **petals** absent.
　The typical subspecies is confined to coastal areas. It occurs on Wilson's Promontory and the western end of the Ninety Mile Beach in Vic, as well as on the Bass Strait Islands and in north-eastern Tas. Plants are found on sand dunes and in shallow sandy soils overlying granite and shale. They adapt well to cultivation

in freely draining soils with a sunny or semi-shaded aspect. Hardy to moderate frosts and extended dry periods. Responds well to pruning and can be used for low hedging.
　The ssp. *calcicola* N. G. Walsh is often a somewhat open shrub of 1–2.3 m × 1.5–2.5 m. It has narrowly elliptic to ovate leaves of 2–7 cm × 1–2.5 cm which are deep green and somewhat velvety above. It occurs in eastern Vic near the coast and further inland on soils derived from limestone or marble.
　Propagate both subspecies from seed or from cuttings of firm young growth.
　Plants from coastal western Vic, Tas, SA and southern WA, previously referred to as *P. oraria*, are now known as *P. paniculosa* ssp. *paralia* N. G. Walsh.

Pomaderris pallida N. A. Wakef.
(pale)
NSW　　　　　　　　　　　　Pale Pomaderris
0.7–2 m × 1–2.5 m　　　　　　　　　　Oct–Nov
　Dwarf to small **shrub** with greyish-hairy young growth; **branches** spreading to ascending; **branchlets** densely greyish-hairy; **leaves** 0.7–2.2 cm × 0.3–0.7 cm, elliptic, recurved to ascending, deep green and somewhat velvety above, densely greyish-hairy below, margins faintly recurved, venation distinct, apex blunt; **stipules** to 0.6 cm long, triangular; **flowers** small, yellow, in few-flowered terminal and axillary clusters, forming leafy panicles to about 10 cm long, can be profuse and conspicuous; **petals** absent.
　This species is from the Canberra and Cooma region where it grows in shallow soils, often on steep slopes, beside waterways. It is found in dry open forest and woodland. Cultivated to a limited degree, it should prove adaptable in temperate regions. Plants need good drainage and acidic soils with a semi-shaded or somewhat sunny aspect. Hardy to heavy frosts and light snowfalls. Propagate from seed or from cuttings of firm young growth.

Pomaderris paniculosa

Pomaderris paniculosa F. Muell. ex Reissek
(in panicles)
NSW, Vic, SA, WA   Pomaderris
0.6–2.5 m × 1–2.5 m   Sept–Oct

Dwarf to medium **shrub** with greyish- to rusty-hairy young growth; **branches** spreading to erect; **branchlets** with grey and rusty hairs; **leaves** 0.8–1.5 cm × 0.6–1.2 cm, elliptic to obovate, recurved to ascending, deep green and more or less glabrous above, paler below with greyish hairs and rusty hairs, margins entire or rarely very faintly toothed, venation distinct, apex blunt or faintly notched; **stipules** to 0.4 cm long, linear; **flowers** small, greyish to rusty-hairy in bud, opening cream to greenish, in axillary clusters to 7 cm × 2 cm, forming leafy panicles, can be profuse and conspicuous; **petals** absent.

The typical subspecies is wide-ranging and occurs in disjunct, semi-arid locations in south-western NSW, central and western Vic, southern SA and south-eastern WA, mainly on soils derived from marine sediments. Usually associated with heath, mallee or woodland. Suited to cultivation in alkaline or acidic soils which drain freely. Plants tolerate plenty of sunshine but adapt better in semi-shade. They are hardy to moderately heavy frosts. May be useful as a low hedging plant.

The ssp. *paralia* N. G. Walsh is a bushy shrub of 1–2 m × 1–2 m. It has ovate to elliptic leaves of 1.5–5 cm × 1–2.5 cm which are glabrous above. The flowers are in narrow, axillary and terminal clusters to about 5 cm long. This subspecies occurs in coastal areas in Vic, northern Tas, south-eastern SA and very rarely in WA. It has similar cultivation requirements to the typical subspecies.

Propagate both subspecies from seed or from cuttings of firm young growth.

P. oraria ssp. *oraria* which often occurs with *P. paniculosa* ssp. *paralia* differs in having leaves with a hairy upper surface. *P. racemosa* is also similar but it is a tall shrub or small tree usually associated with waterways.

Pomaderris parrisiae N. G. Walsh
(after Margaret Parris, 20th-century botanical collector, NSW)
NSW   Pomaderris
1.5–9 m × 1–5 m   Oct–Nov

Small to tall **shrub** or small **tree** with silvery-hairy young growth; **branches** spreading to erect; **branchlets** silvery-hairy; **leaves** 2–8 cm × 0.6–2.5 cm, lanceolate to elliptic, spreading to ascending, deep green and glabrous above, silvery-hairy below, margins flat or faintly recurved, venation indistinct, apex pointed; **stipules** to 1 cm long, narrowly ovate, deciduous; **flowers** small, whitish-hairy in bud, opening cream to yellow, in spreading panicles to about 8 cm across, often profuse and very conspicuous; **petals** present.

A handsome species from the southern Tablelands where it occurs mainly in open forest at 150–900 m altitude. Cultivated to a limited degree but it deserves greater recognition. Probably best suited to temperate regions but may adapt to subtropical conditions. Needs well-drained acidic soils with a semi-shaded site. Hardy to moderately heavy frosts. It has potential for use as a screening or hedging plant in sheltered locations. Propagate from seed or from cuttings of firm young growth.

Pomaderris pauciflora N. A. Wakef. —
see *P. prunifolia* A. Cunn. ex Fenzl

Pomaderris phylicifolia Lodd. ex Link
(leaves similar to the genus *Phylica*)
NSW, Vic, Tas   Narrow-leaf Pomaderris
1–2.5 m × 0.7–2 m   Oct–Nov

Small to medium **shrub** with greyish-hairy young growth; **branches** spreading to erect; **branchlets** with greyish hairs; **leaves** 0.6–1.5 cm × 0.1–0.6 cm, narrowly elliptic to narrowly obovate, spreading to ascending, deep green and hairy above, greyish-hairy below, margins recurved to revolute, venation distinct, apex blunt; **stipules** to 0.6 cm long, narrow, persistent; **flowers** small, greyish-hairy in bud, opening cream to yellow, in small axillary clusters near ends of branchlets, moderately profuse and conspicuous; **petals** absent.

The typical variant occurs in south-eastern NSW, eastern Vic and eastern Tas. It grows at 200–1000 m altitude often in rocky sites beside waterways. Rarely cultivated, it is suitable for temperate regions with very freely draining acidic soils in sunny or semi-shaded sites. Plants withstand moderately heavy frosts. Introduced to cultivation in England in 1819.

The var. *ericoides* Maiden & Betche has broader leaves of 0.3–0.8 cm × about 0.1 cm, with revolute margins and a recurved apex. This variant occurs in south-eastern NSW and eastern Vic. It grows at 250–1400 m altitude in woodland and shrubland, and inhabits streams and swamp margins.

Propagate both varieties from seed or from cuttings of firm young growth.

P. adnata may be confused with *P. phylicifolia* var. *phylicifolia* because the leaf shape is similar, but the leaves of *P. adnata* have a glabrous upper surface.

Pomaderris pilifera N. A. Wakef.
(with slender hairs)
Vic, Tas   Pomaderris
2–4 m × 1.5–3 m   Sept–Oct

Medium **shrub** with greyish-hairy young growth; **branches** spreading to ascending; **branchlets** greyish-hairy; **leaves** 2–5 cm × 1–3 cm, ovate to obovate, spreading to ascending, deep green and glabrous above, paler with greyish hairs below, margins flat, venation distinct, apex blunt; **stipules** to 0.6 cm long, ovate, deciduous; **flowers** small, greyish-hairy in bud, lemon-yellow, in somewhat pyramidal panicles to 10 cm long, often profuse and very highly conspicuous; **petals** present.

A highly desirable species from eastern Vic and eastern Tas. Plants inhabit areas to about 700 m altitude often in shallow soils near creeks in woodland and open forest. They adapt very well to cultivation and will grow best in moderately well-drained acidic soils with a semi-shaded or somewhat sunny aspect. Hardy to moderately heavy frosts. Responds well to pruning and is suitable for hedging. Propagate from seed, or from cuttings of firm young growth, which strike readily.

Pomaderris pilifera × .55, flower × 2

P. pilifera has been marketed wrongly as *P. elliptica*, which differs in having larger leaves with wavy or faintly toothed margins.

Pomaderris precaria N. G. Walsh & F. Coates
(precarious; referring to its natural habitat)
NSW   Pomaderris
1.5–3 m × 1–2 m   Sept–Oct

Small to medium **shrub** with rusty-hairy young growth; **branches** spreading to ascending; **branchlets** rusty-hairy; **leaves** 1–4.5 cm × 0.8–2.5 cm, elliptic to obovate, reflexed to ascending, not crowded, deep green and slightly velvety above, with greyish and rusty hairs below, somewhat stiff, margins flat, apex bluntly pointed; **stipules** to 0.5 cm long, deciduous; **flowers** small, greyish-hairy in bud, opening golden-yellow, in more or less hemispherical panicles to about 8 cm across, can be moderately profuse and conspicuous; **petals** present.

This slender species was named in 1997. It is regarded as rare in the Central Tablelands–Central Western Slopes area where it grows in shrubland and woodland in gravelly soils derived from sandstone. Warrants cultivation as a conservation strategy. Requires excellently drained acidic soils with a semi-shaded aspect. Hardy to moderately heavy frosts and extended dry periods. This species has sometimes been referred to in publications as *Pomaderris* 'species D'. Propagate from seed or from cuttings of firm young growth.

Pomaderris prunifolia A. Cunn. ex Fenzl
(leaves similar to the genus *Prunus*)
Qld, NSW, Vic   Plum-leaf Pomaderris
1–4 m × 1–4 m   Oct–Nov

Small to medium **shrub** with rusty-hairy young growth; **branches** spreading to ascending; **branchlets** rusty-hairy; **leaves** 1.2–8 cm × 0.7–3.5 cm, ovate to obovate, recurved to ascending, deep green and faintly hairy to glabrous above, creamish- and rusty-hairy below, margins flat to faintly recurved, venation distinct, apex blunt or pointed; **stipules** to 0.5 cm long; **flowers** small, fawn- to rusty-hairy in bud, opening cream to yellow, in somewhat globular axillary and terminal clusters of about 1 cm across, forming leafy panicles to about 10 cm long, often profuse and moderately conspicuous; **petals** absent.

A quick-growing species which is distributed from south-eastern Qld and eastern NSW to southern-central Vic. Generally occurs in shallow soils in rocky sites. Inhabits shrubland, woodland and open forest at 50–1350 m altitude. Cultivated to a limited dgree, it is best suited to temperate areas but may succeed in subtropical regions. Well-drained acidic soils and a semi-shaded site are prerequisites. May become somewhat open but responds well to judicious pruning. Hardy to moderately heavy frosts. Propagate from seed or from cuttings of firm young growth.

P. cotoneaster N. A. Wakef. is allied. It is a shrub of 1–3 m × 1–3 m with leaves of 2–5 cm × 1.5–3 cm. Cream flowers are produced in panicles to 10 cm long during Oct–Nov. This poorly known species is from south-eastern NSW and far-eastern Vic. It inhabits rocky sites at 150–600 m altitude and has similar cultivation requirements to those of *P. prunifolia*. Propagate from seed or from cuttings of firm young growth.

P. pauciflora N. A. Wakef. is sometimes confused with small-leaved forms of *P. prunifolia*. It is a shrub of 1–3 m × 1–3 m with narrowly elliptic to obovate leaves of 0.5–3 cm × 0.2–1.5 cm. It has cream flowers which are produced during Oct–Nov. *P. pauciflora* is found at 250–850 m altitude in south-eastern NSW and far-eastern Vic. It inhabits rocky and sandy sites on edges of waterways. Cultivation and propagation requirements are similar to those of *P. prunifolia*.

Pomaderris queenslandica C. T. White
(from Queensland)
Qld, NSW   Pomaderris
0.5–3.5 m × 0.5–3 m   Aug–Oct

Dwarf to medium **shrub** with greyish- and rusty-hairy young growth; **branches** spreading to ascending; **branchlets** with greyish and rusty hairs; **leaves** 2–7 cm × 1–3.3 cm, elliptic to ovate, spreading to ascending, deep green and glabrous above, with grey hairs and rusty hairs below, margins flat, venation distinct, apex blunt or pointed; **stipules** to 1.2 cm long, broadly ovate, deciduous; **flowers** small, cream, in globular clusters 1–4 cm across, forming panicles, often profuse and very conspicuous; **petals** usually absent.

An attractive species which has disjunct populations from the Blackdown Tableland in Qld to near Gilgandra, NSW. It occurs in woodland, dry forest and rainforest margins and is often found on the slopes of gullies. Plants are found mainly in sandy soils derived from granite. It deserves to be better known in cultivation and should respond well in subtropical and temperate regions. Needs very well-drained acidic soils with a semi-shaded aspect. Tolerates moderate frosts.

Pomaderris racemosa

Pomaderris prunifolia   W. R. Elliot

Propagate from seed or from cuttings of firm young growth.

The allied *P. ferruginea* differs in having larger and usually looser panicles of flowers which have petals.

Pomaderris racemosa Hook.
(in racemes)
Vic, Tas, SA   Slender Pomaderris
2–8 m × 1.5–4 m   Oct–Dec

Medium to tall **shrub** or small **tree** with greyish- or rusty-hairy young growth; **branches** spreading to erect; **branchlets** densely covered in greyish to rusty, stellate hairs; **leaves** 1–3 cm × 0.5–1.5 cm, elliptic to broadly ovate, recurved to ascending, deep green and glabrous or faintly hairy above, pale greyish hairs and rusty hairs below, margins faintly wavy to shortly toothed, venation distinct, apex blunt; **stipules** to 0.4 cm long, slender, deciduous; **flowers** small, greyish- or rusty-hairy in bud, opening to cream, in terminal and axillary panicles to about 4 cm long, often profuse and moderately conspicuous; **petals** absent.

This species occurs in southern Vic, central and eastern Tas, and far south-eastern SA. It is found in moist forest and shrubland, often near watercourses, at 30–850 m altitude. Not common in cultivation. Plants require well-drained acidic sandy loam or loam soils with a semi-shaded aspect for best results. Hardy to moderately heavy frosts. Responds well to pruning. Propagate from seed or from cuttings of firm young growth.

P. elachophylla is similar to small-leaved variants of *P. racemosa*, but it has much smaller panicles of flowers.

Pomaderris reperta N. G. Walsh & F. Coates
(rediscovered)
NSW   Pomaderris
1–3 m × 1–3 m   Sept–Oct

Small to medium **shrub** with rusty-hairy young growth; **bark** silvery-grey; **branches** spreading to ascending, becoming somewhat glaucous; **branchlets** initially rusty-hairy; **leaves** 1–3.5 cm × 0.8–2 cm, elliptic to nearly circular, spreading to ascending, not crowded, dark green above, whitish-hairy with rusty hairs on veins below, margins flat, apex blunt or notched; **stipules** to 0.5 cm long, deciduous; **flowers** small, silvery- or rusty-hairy in bud, opening to cream or pale yellow, in globular clusters, forming loose panicles to about 4 cm × 4 cm, often profuse and conspicuous; **petals** usually absent.

A pleasing, recently described (1997) species which is regarded as vulnerable in the wild because of its restricted distribution. It is known only from a very small area in the Central Western Slopes District. Occurs on sandy loam in woodland. Needs to be cultivated as part of a conservation strategy. Requires well-drained acidic soils with a semi-shaded aspect and should tolerate moderate frosts. Propagate from seed or from cuttings of firm young growth.

Pomaderris rotundifolia (F. Muell.) Rye
(round leaves)
WA   Pomaderris
1–2 m × 1–2 m   Aug–Sept

Small, more or less lignotuberous **shrub** with greyish-hairy young growth; **branches** spreading to ascending; **branchlets** greyish-hairy; **leaves** 0.5–1.5 cm × 0.3–1.3 cm, broadly ovate to nearly circular, reflexed to ascending, rarely crowded, dark green and glabrous above, greyish-hairy below, margins recurved, apex notched and recurved; **stipules** to about 1 cm long, narrow, persistent; **flowers** small, greyish-hairy in bud, opening cream or pinkish, in tight globular heads of 1–2 cm across, often profuse and conspicuous; **petals** present, hooded.

This species occurs in the eastern Eyre and eastern Roe Districts where it grows in shrubland, mallee heath and mallee on sandy soils derived from limestone. Evidently rarely cultivated, it has potential for use in warm temperate and semi-arid regions. Should tolerate acidic or slightly alkaline soils which drain freely. Plants will perform best in a warm to hot site. They withstand light frosts. This species can reshoot from the lignotuberous-like base. Propagate from seed or from cuttings of firm young growth.

Cryptandra rotundifolia (F. Muell.) F. Muell. and *Spyridium rotundifolium* F. Muell. are synonyms.

Pomaderris sericea N. A. Wakef. —
                    see *P. ligustrina* Sieber ex DC.

Pomaderris sieberiana N. A. Wakef. =
                              P. intermedia Sieber

Pomaderris subcapitata N. A. Wakef.
(in head-like clusters)
NSW, Vic   Pomaderris
1.5–4 m × 1.5–3 m   Sept–Nov

Small to medium **shrub** with long rusty hairs; **branches** spreading to ascending; **branchlets** rusty-hairy; **leaves** 0.8–4.5 cm × 0.7–2.5 cm, broadly ovate to obovate, spreading to ascending, faintly velvety above, dense pale greyish hairs and sometimes long rusty hairs below, margins flat, venation distinct, apex blunt or faintly notched; **stipules** to 0.7 cm long, narrow, deciduous; **flowers** small, whitish- to coppery-hairy in bud, opening cream to yellow, in somewhat globular compact heads 2–5 cm across, often profuse and conspicuous; **petals** usually present.

Although this pomaderris is distributed from far north-eastern NSW to eastern Vic, it is not common. It occurs at disjunct locations in heath, woodland and open forest, usually in rocky soils, at 200–1000 m altitude. Cultivated to a limited degree, it is probably best suited to temperate regions with very well-drained acidic soils in a moderately sunny to semi-shaded site. Plants are hardy to most frosts. Propagate from seed or from cuttings of firm young growth.

P. subcapitata is sometimes confused with *P. velutina* which has stalked flowers in loosely arranged panicles.

Pomaderris subplicata N. G. Walsh
(somewhat pleated)
Vic   Pomaderris
1–2 m × 1–2 m   Oct

Small **shrub** with greyish- and coppery-hairy young growth; **branches** spreading to ascending, many; **branchlets** densely hairy; **leaves** 0.3–1 cm × 0.2–0.6 cm, ovate to obovate, mainly spreading to ascending, somewhat crowded, flat to V-shaped in cross-section, deep green and faintly hairy above, whitish-hairy below, venation indistinct, apex blunt; **stipules** to about 0.2 cm long, narrow, deciduous; **flowers** small, opening pale yellow to yellow, in small axillary heads to about 1 cm across, near ends of branchlets, forming leafy panicles, often profuse and moderately conspicuous; **petals** early deciduous.

This recently described (1992) species is regarded as vulnerable in its natural woodland habitat of north-eastern Vic. Plants need to be cultivated as a conservation strategy. They require well-drained acidic soils with a semi-shaded aspect. Hardy to most frosts and withstands extended dry periods. Propagate from seed or from cuttings of firm young growth.

Pomaderris tropica N. A. Wakef.
(of the tropics)
Qld   Tropical Pomaderris
2–3 m × 1.5–3 m   July–Nov

Medium **shrub** with silvery-hairy young growth; **branches** spreading to ascending; **branchlets** silvery-hairy; **leaves** 2.4–8.4 cm × 1.4–3.3 cm, ovate, mainly spreading to ascending, mid green, faintly hairy above, silvery-hairy below, flat, venation distinct, apex blunt; **stipules** to 0.6 cm long, deciduous; **flowers** small, whitish-hairy in bud, opening white to cream, in semi-globular terminal heads to about 7 cm across, often profuse and conspicuous; **petals** absent.

P. tropica is one of the northernmost-occurring species where it is found in rocky sites in the Bellenden Ker Range, west of Cairns. It is regarded as rare in nature and is worthy of cultivation. Plants need extremely well-drained acidic soils with a shaded or semi-shaded aspect. This species is suitable for tropical and subtropical regions. May require supplementary water during extended dry periods. Propagate from seed or from cuttings of firm young growth.

Pomaderris tropica   W. R. Elliot

Pomaderris vacciniifolia F. Muell. ex Reisseck
(leaves similar to the Blueberry genus, *Vaccinium*)
Vic   Round-leaf Pomaderris
1–4 m × 0.6–3 m   Sept–Nov

Small to medium **shrub** with greyish to pale rusty hairs on young growth; **branches** spreading to ascending, often many; **branchlets** with greyish to pale rusty hairs; **leaves** 0.8–2.2 cm × 0.6–1.3 cm, elliptic to broadly elliptic, recurved to ascending, deep green and glabrous above, paler and faintly hairy below, margins flat, venation distinct, apex blunt; **stipules** to 0.2 cm long, linear, deciduous; **flowers** small, greyish-hairy in bud, opening creamy-white, in axillary clusters of about 1–2 cm across, forming leafy panicles, can be profuse and conspicuous; **petals** 5.

The Round-leaf Pomaderris from north and north-east of Melbourne often has a somewhat open habit. It is found in moist shrubland and forest. Plants adapt well to cultivation in well-drained acidic loams with a semi-shaded or shaded aspect. Hardy to moderately heavy frosts. Responds well to pruning which is best begun at an early age. Propagate from seed or from cuttings of firm young growth.

P. elacophylla is similar but differs in having flowers without petals.

Pomaderris vellea N. A. Wakef.
(woolly)
NSW   Woolly Pomaderris
0.6–2 m × 0.6–2 m   Oct

Dwarf to small **shrub** with woolly young growth; **branches** spreading to ascending; **branchlets** densely woolly with rusty curly hairs; **leaves** 2–8 cm × 1.2–3.5 cm, more or less elliptic to oblong, recurved to ascending, densely hairy above, often greyish to

Pomaderris velutina

bluish, densely woolly hairs below, thick-textured, margins flat to slightly recurved, venation distinct, apex blunt to slightly notched; **stipules** to 0.5 cm long, narrowly triangular, deciduous; **flowers** small, densely hairy in bud, opening to cream, in crowded pyramidal or semi-globular racemes to about 3 cm long, can be profuse and conspicuous; **petals** 5.

This distinctive, woolly-foliaged species is endemic in eastern NSW. It has a distribution which extends north from near Sydney to the New England region. Plants grow at 100–1000 m altitude mainly on granite-derived soils in shrubland, woodland and dry open forest. Worthy of cultivation, it should adapt well to very well-drained acidic soils with a semi-shaded aspect. Plants tolerate moderate frosts. Propagate from seed or from cuttings of firm young growth.

P. crassifolia N. G. Walsh & F. Coates is a closely allied, recently described (1997) species, from south-eastern Qld and north-eastern NSW. It grows 1–2 m × 1–2 m and has elliptic to ovate leaves of 2–6 cm × 1–2.8 cm. The leaves are similar to those of *P. vellea* except that they have a glabrous upper surface. Flowers are cream and they lack petals. They bloom during Aug–Sept. This species has similar cultivation requirements to *P. vellea*.

Pomaderris velutina J. H. Willis
(velvety)
NSW, Vic   Velvety Pomaderris
1–2.5 m × 1–2.5 m   Sept–Nov

Small to medium **shrub** with greyish hairs and longer rusty hairs on young growth; **branches** spreading to erect; **branchlets** densely hairy, with long rusty hairs; **leaves** 0.8–4.5 cm × 0.5–2 cm, ovate to obovate, recurved to ascending, mid green and faintly velvety above, densely greyish-hairy below with longer rusty hairs, margins flat, venation distinct, apex blunt or faintly notched; **stipules** to 1 cm long, narrowly ovate, semi-persistent; **flowers** small, greyish- and rusty-hairy in bud, opening yellow, in clusters of about 1.5 cm across, forming pyramidal panicles to about 8 cm long, often profuse and very conspicuous; **petals** present.

Velvety Pomaderris occurs in south-eastern NSW and eastern Vic. Generally it is found at 240–1000 m altitude in rocky sites beside waterways. Plants occur in shrubland, woodland and open forest. Cultivated to a limited degree, it has grown well in temperate regions in well-drained acidic soils. Best suited to a semi-shaded site. Plants tolerate moderately heavy frosts and respond well to pruning. Propagate from seed or from cuttings of firm young growth.

P. velutina is sometimes confused with *P. pallida*, which differs in having larger elliptic leaves and flowers which lack petals.

Pomaderris virgata N. G. Walsh
(twiggy)
NSW   Pomaderris
2–8 m × 1–4 m   Nov–Dec

Medium to tall **shrub** or small **tree**, with coppery-hairy young growth; **bark** tessellated at base, otherwise smooth; **branches** mainly ascending; **branchlets** with pale-to-deep coppery hairs; **leaves** 2–9 cm × 0.7–3 cm, lanceolate to oblong, spreading to ascending, deep

Pomaderris velutina × .55, flower × 2

green and glabrous above, dense covering of cream to coppery hairs below, margins recurved, venation distinct, apex pointed; **stipules** to about 0.4 cm long, lanceolate, deciduous; **flowers** small, silky-hairy in bud, opening yellow to golden yellow, in terminal panicles to about 7 cm × 6 cm, often profuse and very conspicuous; **petals** absent.

A very attractive species which is regarded as rare in its natural habitat of south-eastern NSW and far-eastern Vic. It grows at 160–850 m altitude on shallow often rocky soils in shrubland and open forest. Warrants cultivation for its beauty and because it is rare. Plants should adapt to temperate regions with very well-drained acidic soils in a semi-shaded site. They tolerate moderate frosts and extended dry periods. Propagate from seed or from cuttings of firm young growth.

P. costata is closely allied but differs in lacking coppery young growth and it has broader leaves with wavy margins.

POMATOCALPA Breda, Kühlew. & Hasselt
(from the Greek *pomatos*, flask or cup; *calpe*, pitcher; referring to the flask-shaped or urn-shaped labellum)
Orchidaceae

Epiphytes or **lithophytes**; **roots** coarse, long; **stems** short, sparsely branched; **leaves** simple, alternate, spreading widely, thin-textured but tough, apex notched; **inflorescence** a raceme or panicle; **flowers** small, generally dull-coloured; **labellum** pouched.

A genus of about 60 species distributed from Asia to the Philippines, Indonesia, Polynesia and New Guinea. Two species occur in Australia and one of these is endemic. Both are easy to grow if given suitable conditions and are moderately popular with orchid growers. Propagation is from seed which must be sown under sterile conditions.

Pomatocalpa macphersonii (F. Muell.) T. E. Hunt
(after J. Alexander Macpherson, early Victorian parliamentary)
Qld
Epiphyte   July–Oct
Epiphytic orchid forming small clumps; **roots** cord-like; **stems** short, sparsely branched; **leaves** 2–8, 15–25 cm × 2.5–3.5 cm, oblong, crowded, dark green, thin-textured but stiff, main vein ridged; **racemes** 6–10 cm long, stiff, fleshy, bearing 3–30 flowers; **flowers** about 1 cm across, yellow with prominent red blotches and markings.
Endemic in Queensland where it is distributed from the top of Cape York Peninsula to near Rockhampton. It grows in rainforest from the lowlands to moderate altitudes. Easily grown if given warm conditions, humidity, ample air movement and filtered bright light. Best tied on a slab of treefern, weathered hardwood or cork. In temperate regions, plants require the protection of a heated glasshouse. Propagate from seed sown under sterile conditions.

Pomatocalpa marsupiale (Kranzlin) J. J. Smith
(pouched)
Qld
Epiphyte   Nov–May
Epiphyte forming robust, upright clumps; **roots** long, thick, cord-like; **stems** thick, branched near the base; **leaves** 20–30 cm × 4–5 cm, oblong to strap-shaped, leathery, yellowish-green, deeply channelled, apex unequally notched; **inflorescences** 30–45 cm long, stiffly erect, branched near the end, each branch bearing a dense head of crowded, upward-facing flowers; **flowers** 1.2–1.5 cm across, green to brown with a cream or yellowish labellum.
Occurs in New Guinea and also extends to the ranges of central Cape York Peninsula. Mainly grows on the upper branches of large trees in sparse rainforest and occasionally grows on rocks. The flowers are produced over a long period. A tough epiphytic orchid which is easily grown if given bright light, warmth, humidity and an abundance of air movement. Requires a heated glasshouse in temperate regions. Best in a hanging pot or basket. Requires a coarse, freely draining mix. Propagate from seed sown under sterile conditions.

PONGAMIA Adanson
(from *pongam;* the Tamil name for this tree)
Fabaceae
Shrubs or **trees**; **leaves** alternate, pinnate, with a terminal leaflet; **leaflets** entire, opposite; **inflorescences** axillary racemes; **flowers** pea-shaped, hairy; **fruit** an indehiscent, woody pod.
A small genus of 2 or 3 species which occurs in tropical Asia and the Pacific region. There are 2 species in Australia but 1 is undescribed. The other native species is widely cultivated in the tropics and subtropics as a shade tree. Propagation is from seed which is best sown fresh.

Pongamia pinnata (L.) Pierre
(with pinnate leaves)
Qld, NT   Pongamia
6–20 m × 5–15 m   Sept–Nov
Tall **shrub** to medium **tree** with a widely spreading, bushy canopy; **bark** greyish, smooth; young growth shiny; **leaves** pinnate, 6–15 cm long; **leaflets** 5–7, 4.5–15 cm × 4–9 cm, oblong, apical leaflet largest, dark green, glossy, thin-textured; **racemes** 8–12 cm long, axillary; **flowers** about 1 cm long, pea-shaped, creamy-pink to bluish or bluish-pink, fragrant; **pods** 4–6 cm × 1.5–3 cm, oblong, woody, pale brown; **seeds** 1–2, round, red-brown, ripe June–Oct.
Widely distributed from tropical Asia to the Pacific region, New Guinea and northern Australia. This species is a prominent component of beach scrubs (particularly scrubs on stabilised dunes) but it also extends to lowland rainforests and vine thickets especially near streams. All parts of the plant are toxic and induce nausea and vomiting if eaten. The flowers are fragrant at night. Pongamia is a fast-growing, adaptable tree that is widely planted in tropical and sub-tropical regions. It is a familiar sight in parks and is also excellent in coastal locations. An ideal shade tree which also provides screening and shelter. May be trained as a large hedge. Plants may become deciduous for a short period in dry periods. Requires excellent drainage and a sunny location. Propagate from fresh seed. Production of pods is low in some years and heavy in others.

PONTEDERIACEAE Kunth.
A small family of monocotyledons consisting of about 35 species in 9 genera. It is very well represented in North and South America. They are annual or perennial aquatics which may be submerged, emergent or floating. The petioles of some species are swollen and spongy to provide buoyancy. Flowers are large and colourful and often provide an attractive display. Water Hyacinth, *Eichornia crassipes*, is member of Pontederiaceae and is a serious introduced weed of waterways in Qld, NT and NSW. The family is represented in Australia by 2 species of *Monochoria*.

Popowia australis Benth. =
                          Polyalthia australis (Benth.) Jessup

PORANA Burm. f.
(derivation uncertain)
Convolvulaceae
Perennial **climbers** with twining stems; **leaves** alternate, simple, entire; **inflorescence** solitary axillary flower with subtending bracts; **sepals** 5, free, becoming much-enlarged, papery; **corolla** bell-shaped, sometimes with very short lobes; **stamens** 5, more or less equal; **style** 1; **fruit** a capsule, 2-valved or indehiscent.
A mainly tropical genus of about 20 species. It is distributed in Asia and Malesia, and there are 2 species endemic in Australia. They have attractive delicate flowers and are worthy of cultivation. Propagation is from seed, which can begin to germinate 10–50 days after sowing. Cuttings of firm young growth provide best results if taken during summer.

441

Porana commixta Staples
(mixed, intermingled)
Qld, NSW, WA
Climber   July–Oct
Perennial **climber** with silky-hairy young growth; **stems** to about 4 m long, twining, initially covered in silky-grey hairs, becoming woody; **leaves** 1.6–6 cm × 0.15–0.5 cm, linear to narrowly oblong, alternate, spreading to ascending, scattered, green and sparsely hairy above, densely silky-hairy below, flat, apex pointed or sometimes blunt or rounded; **flowers** to about 1.5 cm × 1.5 cm, white, purple, purple-blue or blue, solitary, in axils, on a hairy stalk, rarely profuse but always conspicuous; **sepals** narrowly elliptic, initially to 0.5 cm × 0.3 cm, then growing to 2 cm × 1 cm and enclosing a capsule; **bracts** to about 0.5 cm long; **capsules** about 0.5 cm long.
This species has interesting foliage and dainty flowers. It occurs in semi-arid regions of southern-central Qld, northern-central NSW and western-central WA. Cultivated to a limited degree, it is best suited to warm temperate and semi-arid regions. Plants enjoy sunshine and often intermingle with other shrubs. *P. commixta* will adapt to a range of soils. Hardy to moderate frosts and extended dry periods once established. Propagate from seed or from cuttings. It has been confused with *P. sericea*, which occurs only in WA and differs in having yellowish hairs. *P. sericea* also has broader sepals which enlarge to become near-circular during fruiting.

Porana sericea (Gaudich.) F. Muell.
(silky)
WA
Climber   July–Oct
Perennial **climber** with yellowish-hairy young growth; **stems** to about 3 cm long, twining, initially covered with yellowish hairs, becoming woody; **leaves** 2–4 cm × 0.2–0.3 cm, linear-lanceolate, alternate, spreading to ascending, scattered, green and sparsely hairy above, golden to tawny hairs below, flat, apex pointed; **flowers** to about 1.5 cm across, pale blue or mauve to deep violet, with white centre, solitary, in axils, on slender stalk to about 1.5 cm long, rarely profuse but always conspicuous; **sepals** broadly elliptic to broadly ovate, becoming nearly circular at fruiting; **bracts** to 0.5 cm long; **capsules** about 0.5 cm long.
A delicately-flowered twiner which is found in the Avon, Carnarvon, Irwin and Darling Districts where it grows near the coast, and some distance inland, in a range of soil types. It is cultivated to a limited degree and should adapt to temperate and semi-arid regions. Plants prefer sites with filtered sunshine or semi-shade. They are hardy to light frosts. Propagate from seed or from cuttings of firm young growth.

PORANTHERA Rudge
(from the Greek *poros*, passage, pore; *anthera* anther; referring to the terminal pores of the anthers)
Euphorbiaceae
Monoecious annual **herbs** or dwarf **shrubs**; **leaves** opposite or alternate, entire; **stipules** present; **inflorescence** more or less head-like raceme, solitary or in a leafy terminal corymb; **sepals** 5; **petals** 5; **stamens** 5; **fruit** a 6-valved capsule.
A genus of about 10 species; 8 of these are found in Australia. *Poranthera* is represented in all states. Plants are found in heath, shrubland, woodland or in dry and wet sclerophyll forest. They often grow on stony or gravelly soils of poor quality. Most species have low ornamental appeal but they can add interest to plantings. The annual *P. microphylla* is an excellent plant for filling small open spaces.
Propagation is from seed which is likely to give best results if fresh. Cuttings of firm young growth should strike readily during late spring to autumn.

Poranthera corymbosa Brongn.
(flowers borne in a corymb)
Qld, NSW, Vic   Clustered Poranthera
0.4–0.8 m × 0.2–0.3 m   Sept–April
Dwarf **shrub** with glabrous young growth; **stems** 1–3, erect, often reddish; **branches** few, at ends of stems, ascending to erect; **leaves** 3–5 cm × 0.15–0.4 cm, linear to narrowly ovate, spreading to ascending, crowded, mid green or sometimes reddish above, paler below, margins recurved to revolute, apex pointed; **stipules** entire; **flowers** to about 0.3 cm across, white, in many-flowered terminal corymbs 3–15 cm across, very conspicuous; **capsules** about 0.3 cm across, green.
P. corymbosa occurs in south-eastern Qld, eastern NSW and eastern Vic. It inhabits open forest and is usually found in gravelly or sandy soils. Rarely cultivated, it should do best in freely draining acidic soils in a moderately sunny or semi-shaded site. Hardy to moderately heavy frosts. Pruning of very young plants may promote branching and this would improve the floral display. Propagate from seed or from cuttings of firm young growth.

Poranthera ericifolia Rudge
(leaves similar to the genus *Erica*)
NSW   Heath-leaved Poranthera
0.15–0.3 m × 0.1–0.2 m   Sept–Jan
Dwarf **shrub** with glabrous to faintly hairy young growth; **stems** few, erect, sometimes without branches; **branches** few, near ends of stems; **leaves** 0.8–1.2 cm × 0.1–0.2 cm, linear to narrowly ovate, sessile, spreading to ascending, moderately crowded, mid green, margins

Poranthera microphylla   W. R. Elliot

Portulaca bicolor D. L. Jones

usually revolute, apex pointed; **stipules** jagged; **flowers** to about 0.3 cm across, white or with pinkish tonings, in crowded terminal corymbs, 3–8 cm across, often profuse and conspicuous; **capsules** about 0.3 cm across.

This attractive species is restricted to the coast and nearby ranges where it grows in heath and open forests on sandy soils. Evidently rarely cultivated, it requires good drainage in acidic soils and a sunny or semi-shaded aspect. It was first cultivated in England in 1824 where it was grown in pots of sand and peat. Tip pruning of young plants may promote bushy growth. The blackish dried foliage is ornamental and trials with glycerine treatment to preserve the foliage may be worthwhile. Propagate from seed or from cuttings of firm young growth.

P. ericoides Klotsch has some similarities but it differs in being a dense dwarf shrub of 0.1–0.2 cm tall and it has entire stipules. It occurs in the Avon, Darling, Eyre and Roe Districts of WA and is common on the southern Lofty Ranges and Kangaroo Island in SA. Plants have similar cultivation requirements to *P. ericifolia*.

Poranthera ericoides Klotsch — see *P. ericifolia* Rudge.

Poranthera huegelii Klotsch
(after Karl A. von Hugel, 19th-century Austrian botanist)
WA
0.15–0.6 m × 0.3–0.6 m   Oct–Dec

Dwarf **shrub** with glabrous young growth; **stems** glabrous; **branches** on upper part of stems; **leaves** 1–2.5 cm × 0.1–0.2 cm, linear, sessile, spreading, moderately crowded, deep green above, paler below, glabrous, margins recurved, apex pointed; **stipules** white, often toothed at base; **flowers** about 0.3 cm across, white or sometimes pinkish, in loose panicles, often profuse and conspicuous; **capsules** about 0.3 cm across, venation reticulate.

P. huegelii inhabits sandy and lateritic soils in heathland and woodland in the Darling and western Eyre Districts. Apparently rarely cultivated, it warrants greater attention in temperate regions. A sunny or semi-shaded site in freely draining acidic soils should be suitable. Tolerates light frosts. Tip pruning of

young plants may promote bushy growth. Propagate from seed or from cuttings of firm young growth.

Poranthera microphylla Brongn.
(small leaves)
All states   Small Poranthera
prostrate–0.15 m × 0.1–0.3 m   mainly Oct–May
Annual **herb**; young growth glabrous; **branches** spreading, slender, soft; **leaves** 0.4–0.15 cm × 0.15–0.3 cm, obovate to spathulate, recurved to spreading, crowded near ends of branchlets, green to grey-green above, paler below, glabrous, margins recurved, apex pointed; **flowers** about 0.2 cm across, white to pinkish, in tight terminal corymbs surrounded by leaves, often profuse and moderately conspicuous; **style** 6–branched; **fruit** to about 0.2 cm across.

A somewhat insignificant, wide-ranging annual which is valuable for filling spaces in gardens especially on slopes such as embankments. It occurs mainly in woodland and forest. Frequently plants seem to appear from out of the blue and they are often mistaken for a weed particularly when young. This species can regenerate readily in disturbed areas. Plants grow in a wide range of acidic soils which have at least moderate drainage. They prefer a semi-shaded site and are hardy to most frosts. As flowers mature they often gain pinkish tonings. Propagate from seed.

P. triandra J. M. Black is closely allied. It differs in being smaller in all aspects and has 3 styles per flower. It occurs on sandy soils in SA and WA.

Poranthera triandra J. M. Black —
see *P. microphylla* Brongn.

PORTULACA L.
(from the Latin name *purslane*, used for *P. oleracea*)
Portulacaceae

Annual or perennial **herbs**, usually succulent and fleshy; **stem** prostrate to erect, smooth or warty, much-branched; **stipules** modified to axillary hairs; **leaves**

Portulaca digyna D. L.. Jones

opposite or alternate, sessile or stalked, often fleshy; **inflorescence** an axillary or terminal cyme or head; **flowers** colourful, yellow, pink, red or purple, often conspicuous and showy; **sepals** united at the base into a tube; **petals** 4–6; **stamens** 4–numerous; **fruit** a capsule.

A cosmopolitan genus of about 120 species distributed in tropical and temperate regions. About 20 species are native to Australia. They are distributed in all states although they are particularly prominent in inland and tropical regions. One South American species, *P. grandiflora*, is commonly grown as a garden annual and it may sometimes naturalise locally. The Australian species (which are mostly annuals) include some which have large, showy, colourful flowers. These species are very worthy of cultivation as garden annuals and there is plenty of scope to select superior strains and begin hybridisation programmes. Most species have a compact growth habit and are well-suited to cultivation in containers and miniature gardens. Propagation is from seed which germinates readily in warm, moist conditions. Some species can be propagated readily from stem cuttings.

Portulaca australis Endl.
(southern)
Qld
prostrate–5 cm × 1–3 cm   Nov–April
Prostrate to shortly erect annual **herb**; **stems** few, fleshy; **leaves** 0.4–1.2 cm × 0.1–0.45 cm, oblong to elliptic, green, thick and fleshy, blunt, axillary hairs absent; **flowers** 1–1.2 cm across, yellow, often in pairs, in the upper axils; **sepals** about 0.3 cm long; **petals** about 0.6 cm long, joined at the base; **stamens** 25; **capsules** round, splitting in the middle; **seeds** black, shiny.

Distributed along the coast where it grows in flat rocky areas close to tidal influence, and also in drier inland sites. *P. australis* is often sparsely branched but the flowers are showy. It has excellent potential for use in containers and miniature gardens. Requires excellent drainage and a sunny location. Propagate from seed.

Portulaca bicolor F. Muell.
(in 2 colours)
Qld, NSW, WA, NT
prostrate × 1–10 cm   Nov–June
Prostrate annual **herb**; **stems** fleshy; **leaves** 0.3–0.8 cm × 0.3–0.8 cm, broadly ovate to circular, green, bronze or reddish, fleshy, blunt, axillary hairs absent; **flowers** 0.5–0.6 cm across, yellow, pink or purple, opening in the afternoon, solitary in the upper axils or terminal; **sepals** about 0.2 cm long; **petals** 0.2–0.3 cm long; **stamens** 6; **capsules** about 0.3 cm long; **seeds** black.

Widely distributed in inland districts, it often grows in shallow soil on boulders and in rocky situations. The flowers open for a short period in the afternoon. An interesting plant which could be useful in miniature gardens. Requires excellent drainage and a sunny location. Propagate from seed.

Portulaca clavigera Geesink
(clubbed)
WA
prostrate × 10–20 cm   Feb–June

Prostrate annual **herb**; **stems** spreading, fleshy, red, appearing jointed; **leaves** 0.5–0.8 cm × 0.3–0.5 cm, ovate to elliptic, margins revolute, acute, axillary hairs conspicuous; **flowers** 1–1.6 cm across, bright pink to red, showy, in 2–5-flowered heads; **sepals** 0.3–0.45 cm long, awned; **petals** 0.5–0.8 cm long, notched; **stamens** 12–15; **capsules** 0.3–0.4 cm long, ovoid; **seeds** black, shiny.

Endemic in the Kimberleys where it grows in shallow soils on rocky outcrops. A fast-growing species which is often very short-lived. Showy when in flower. It has good prospects as a garden annual or for containers in tropical and subtropical regions. Requires excellent drainage and a sunny location. Propagate from seed.

Portulaca digyna F. Muell.
(with 2 styles)
Qld, WA, NT
prostrate × 10–20 cm   April–Aug; also sporadic
Prostrate annual **herb**; **stems** spreading, fleshy, warty; **leaves** 0.3–0.6 cm × 0.3–0.5 cm, broadly ovate, elliptic or circular, green to bronze, thick and fleshy, blunt, axillary hairs short; **flowers** 0.5–0.7 cm across, pink to purple, in many-flowered cymes; **sepals** to 0.2 cm long; **petals** to 0.4 cm long; **stamens** 4; **capsules** about 0.5 cm long, obovoid; **seeds** black, shiny.

Widely distributed in open forest and woodland, particularly in inland districts. It has potential as a container plant in a sunny situation. Requires excellent drainage. Propagate from seed.

Portulaca filifolia F. Muell.
(with thread-like leaves)
Qld, NSW, SA, WA, NT
5–30 cm × 10–25 cm   Oct–April
Annual **herb** with erect fleshy stems; **leaves** 1–3 cm × 0.1–0.4 cm, linear to filiform, terete, fleshy, dark green or reddish, with prominent tufts of hairs to 1.5 cm long in the axils; **bracts** woolly; **flowers** 1.5–4 cm across, yellow, sessile, showy, in dense heads of up to 10 flowers; **sepals** to 0.45 cm long, ovate; **petals** 0.8–2 cm long, obovate; **stamens** 12–30; **capsules** 0.3–0.5 cm long, ovoid; **seeds** grey or black.

Widely distributed in woodland and grassland. This interesting species frequently colonises disturbed sites on the margins of roads and tracks. The flowers are colourful and showy. It has potential for use in massed planting in a hot, sunny site and would also be an attractive container plant. Propagate from seed.

Portulaca intraterranea J. M. Black
(inland)
Qld, NSW, SA, WA, NT   Buttercup Pigface
prostrate × 20–40 cm   Aug–Dec
Prostrate annual or biennial **herb**; **stems** fleshy; **leaves** 1–2 cm × 0.1–0.3 cm, oblanceolate to obovate, green to bronze, fleshy, axillary hairs inconspicuous; **flowers** 2.5–3.5 cm across, yellow, in 3–4-flowered heads, showy; **sepals** 0.5–0.8 cm long; **petals** 1–1.7 cm long; **stamens** more than 20; **capsules** to 0.5 cm long, conical; **seeds** greyish-black.

Widely distributed in inland regions growing on sandhills and clay pans. In good seasons, plants form

444

dense mats and flower prolifically. An excellent annual for inland districts with a hot, dry climate. Requires well-drained soil in a sunny location. Propagate from seed or cuttings.

Portulaca oleracea L.
(edible)

all states   Pigweed; Purslane; Munyeroo
prostrate × 30–50 cm   Aug–March

Prostrate annual or biennial **herb**; **stems** spreading, fleshy, green or bronze; **leaves** 0.5–2.5 cm × 0.3–1 cm, oblong to obovate, green to bronze, fleshy, blunt, axillary hairs inconspicuous; **flowers** 0.8–1.5 cm across, yellow, solitary or in many-flowered heads; **sepals** 0.3–0.6 cm long; **petals** 0.4–0.7 cm long; **stamens** 8–15; **capsules** 0.4–0.5 cm long, ovoid; **seeds** black, shiny.

A cosmopolitan species which occurs as weedy variants in Australia. Some variants are probably exotic in origin and others are indigenous. Widely distributed, especially in inland regions where it is often prominent in floodplains. All parts are edible, cooked or raw, and were widely eaten by Aborigines (especially the seeds). The species has limited appeal for cultivation but may be grown for culinary purposes. Propagate from seed or cuttings.

Portulaca oligosperma F. Muell.
(with few seeds)

Qld, WA, NT
prostrate × 10–20 cm   Feb–May

Prostrate annual **herb**; **stems** fleshy, smooth or warty, rooting at the nodes; **leaves** 0.4–0.8 cm × 0.2–0.3 cm, narrowly ovate to elliptic, acute; **flowers** 0.4–0.5 cm across, pink, in few-flowered cymes; **sepals** 0.15–0.25 mm long; **petals** 0.2–0.25 mm long; **stamens** 6; **capsules** about 0.3 cm long, ovoid; **seeds** grey.

Widespread in tropical regions of Australia where it grows in open forest and woodland. It has potential as an annual in tropical regions and may be cultivated in containers and miniature gardens. Requires a sunny position in well-drained soil. Propagate from seed.

Portulaca pilosa L.
(with long hairs)

Qld, NSW, NT, WA
prostrate × 30–50 cm   Oct–March

Prostrate annual **herb**; **stems** spreading, fleshy; **leaves** 1–2.5 cm × 0.1–0.4 cm, linear, narrowly ovate to narrowly obovate, green, acute, axillary hairs conspicuous; **flowers** 0.6–1.5 cm across, pink, crimson or purple, in heads of 1–6 flowers, showy; **sepals** 0.3–0.9 cm long, narrow; **petals** 0.3–0.8 cm long; **stamens** to 25; **capsules** to 0.7 cm long, ovoid; **seeds** black.

A complex species which is in need of further study. Regarded as a naturalised weed in Qld and NSW, but apparently native in northern Australia. Perhaps botanical studies will show these populations are distinct species. It has showy colourful flowers. Requires a sunny position in well-drained soil. Propagate from seed.

PORTULACACEAE Juss.   Parakelia Family
A cosmopolitan family of dicotyledons consisting of about 580 species in 10 genera. It is particularly diverse in North and South America. In Australia, the family is represented by 8 genera and about 55 species. They are mostly annual herbs with fleshy, swollen leaves arranged alternately or in opposite pairs. The flowers have 2 sepals, 5 or 6 petals (sometimes more) and a variable number of stamens. The fleshy roots, seeds and leaves of some species are eaten by the Aborigines. Many Australian members of the family have small flowers but some *Calandrinia* and *Portulaca* species have large, colourful and showy flowers. Australian genera: *Anacampseros, Calandrinia, Montia, Neopaxia, Portulaca, Rumicastrum, Sedopsis, Talinum.*

POSIDONIACEAE (Kunth) Lotsy
A small family of monocotyledons consisting of 3 species in the genus *Posidonia*; 2 of these species are endemic to Australia. They are marine aquatic plants which form colonies in shallow coastal waters to a depth of about 10 m. They have strap-like leaves which are often washed ashore after storms and the waves may roll these into fibre balls. A small industry developed in South Australia based on the collection of *Posidonia* leaves which were then dried and used for packing materials or blended with other fibres in textile manufacture. Seed dispersal in these plants is interesting. The fruit is fleshy and buoyant and it may be seen floating in huge numbers before it splits and releases the seeds which then sink.

POTAMOGETON
(from the Greek *potamos*, river; *geiton*, neighbour; referring to habitat)

Potamogetonaceae   Pondweeds

Annual or perennial aquatic **herbs** with slender creeping rhizomes; **stems** brittle; **leaves** alternate except usually opposite beneath flower stalk, floating or submerged; **sheath** at each leaf axil, membranous, more or less encircling the stem; **inflorescence** simple, cylindrical spike on axillary stalk; **flowers** bisexual, small; **perianth segments** 4; **stamens** 4; **anthers** 4, 2-celled; **carpels** 4, 1-celled; **fruiting carpels** indehiscent, shortly beaked.

Potamogeton is a cosmopolitan genus of about 100 species. About 8 species are native to Australia where they are widely distributed. There are 1–2 introduced species.

Potamogeton species are an important source of food and shelter habitat for freshwater wildlife. They are also very good indicators of pollution and are commonly used to monitor water quality. Pollination mechanisms vary. The flowers on most species are pollinated on emerged stems but some other species are pollinated beneath water.

Plants prefer fresh water but a few species, eg. *P. pectinatus* and *P. perfoliatus*, tolerate low salinity levels. Some species such as *P. crispus* grow very quickly when water is 15–20°C. They can easily cover the surface of ponds, dams and channels which may need to be cleared to provide an open surface of water.

Waterbirds aid in the distribution of plants as they eat the ripe fruits and deposit them in new sites.

Propagation is from seed, cuttings or by division of rhizomes. Fresh seed gives best results and does not require pre-sowing treatment. The Bog Method should provide good germination results (see Volume 1, page 204, for further information). Cuttings of stems are best placed in containers of coarse propagation sand or bundled together with a suitable weight and then submerged in water. Division of rhizomes usually provides excellent results.

Potamogeton australiensis A. Benn.
(from Australia)
NSW, Vic, Tas, SA   Thin Pondweed
2–6 m across   Dec–Feb

Emergent, perennial, aquatic **herb**; **stems** flattened, to about 2.7 m long; **leaves** thin, somewhat translucent; **submerged leaves** 5–23 cm × 0.5–9 cm, ovate to broadly ovate, sessile or nearly so, base rounded to tapered, green to reddish-brown, margins entire but often crinkled, prominent longitudinal veins, apex blunt; **emergent leaves** 1.5–9.5 cm × 0.9–6.5 cm, floating, ovate to nearly circular, stalked, green, often glossy, prominent longitudinal veins, apex blunt; **spikes** to about 4.5 cm long, cylindrical, crowded, emergent, on stalk to about 20 cm long.

A coastal or near coastal species which occurs in still or slowly flowing fresh water. It grows on a muddy base. Rarely cultivated, it is suitable for temperate regions in pools and dams with a sunny or semi-shaded aspect. Propagate from seed, cuttings or by division.

Potamogeton crispus L.
(crinkled)
Qld, NSW, Vic, SA, WA, NT   Curly Pondweed
2–8 m across   Nov–May

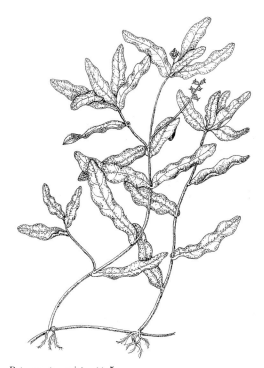

Potamogeton crispus × .5

Submerged, annual or perennial, aquatic **herb**; **stems** flattened, to about 4 m long; **leaves** 2.5–10 cm × 0.5–1.5 cm, narrowly oblong to narrowly ovate, sessile, base blunt to rounded, deep green, thin, translucent, margin finely toothed and wavy to crinkled, 3–5-nerved, apex blunt; **spikes** 0.8–2 cm long, cylindrical, emergent, on slender stalks, often profuse.

An attractive, widely distributed species which also occurs in Europe, Asia and Africa. It is introduced in North America and New Zealand. Found in ponds, dams and slowly flowing waterways. Tolerates very low salinity but best grown in fresh water. Can quickly cover large areas but usually dies back to a smaller area over a few months. Grows best in water more than 50 cm deep. Excess growth may need to be removed from small pools or dams. Very valuable for freshwater wildlife as habitat and as a food source. Readily transferred to other areas by visiting birds. Tolerates cold water. Propagate from seed, cuttings or by divisions.

Potamogeton drummondii Benth.
(after James Drummond, first Government Naturalist, WA)
WA
1.5–4 m across   Oct–Feb

Emergent, perennial, aquatic **herb**; **stems** to about 2 m long; **submerged leaves** 4–12 cm × 0.5–2 cm, linear to elliptic or ovate, sessile or nearly so, very thin, several-nerved; **emergent leaves** 2–4 cm × 1.2–2.5 cm, elliptic to broadly elliptic, long-stalked, 10–15-nerved, opaque; **spikes** 1–2 cm × about 0.5 cm, cylindrical, emergent, on stalk to about 6.5 cm long.

P. drummondii occurs in still or slow flowing water on the coastal plain and hinterland from Cape Arid to Jurien Bay. Evidently rarely cultivated, it is suitable for pools and dams in temperate regions. Propagate from seed, cuttings or by division.

Potamogeton javanicus Hassk.
(Javanese; of Java)
Qld, NSW, WA, NT
1–3 m across   Oct–March

Submerged to emergent, perennial, aquatic **herb** with creeping rhizomes and a somewhat turf-like appearance; **stems** short; **submerged leaves** many, 5–10 cm × 0.2–0.3 cm, linear, tapered to base and apex, translucent; **emergent leaves** few, 2–4 cm × up to 1 cm, narrowly elliptic, bright green, 5–7-nerved, apex pointed; **spikes** to about 2 cm long, cylindrical, crowded, emergent, on slender stalk.

A wide-ranging species from warm to hot, coastal or near coastal areas where it is found in shallow water of creeks and lakes. Needs warm water to grow well. It can form a dense mat of foliage. Suitable for shallow pools. Propagate from seed, cuttings or by division of rhizomes.

Potamogeton ochreatus Raoul
(with a sheath)
Qld, NSW, Vic, Tas, WA, NT   Blunt Pondweed
2–8 m across   Aug–April

Submerged, annual or perennial, aquatic **herb** with creeping rhizomes; **stems** to about 4 m long, flattened to terete; **leaves** 2.5–13 cm × 0.2–0.6 cm, linear, sessile, base tapered, translucent, flat, margin entire, apex blunt to rounded; **leaf sheath** to about 2 cm long, free; **spikes** 1.5–3 cm long, cylindrical, crowded, emergent.

A widespread species which is usually found in deep water in swamps, dams, lakes, channels, creeks and rivers. Adapts well to cultivation in still or flowing water. The water needs to be over 1 m deep for best results. Usually grows quickly in late spring and early summer and then commonly recedes because it breaks down or is eaten by aquatic wildlife. Tolerates low salinity levels. Propagate from seed, cuttings or by division of rhizomes.

Potamogeton pectinatus L.
(comb-like)
Qld, NSW, Vic, Tas, SA, WA   Fennel Pondweed;
Sago Pondweed
2–6 m across   Nov–May

Submerged, perennial, aquatic **herb** with long-creeping, tuber-producing rhizomes; **stems** to about 3 m long, terete, many, much-branched, crowded; **leaves** 1.5–15 cm × 0.2 cm, linear, base joined to sheath, translucent, margins flat and entire, apex pointed; **spikes** 1.5–6 cm long, cylindrical, submerged or slightly emergent.

A very dense-growing species with a wide distribution. Also occurs in Europe, Africa, Asia, North America and New Zealand. Usually found growing in moderately deep, still water. Rarely cultivated because of its dense submerged growth which is often a nuisance in pools. The fruits and leaves are eaten by waterfowl. Propagate from seed, cuttings or by division of rhizomes.

Potamogeton perfoliatus L.
(leaf surrounding the stem)
Qld, NSW, Vic, Tas   Clasped Pondweed;
Perfoliate Pondweed
2–6 m across   Dec–May

Submerged, perennial, aquatic **herb** with creeping rhizomes; **stems** to about 3 m long, terete, moderately branched; **leaves** submerged, 1–7 cm × 1–4 cm, ovate to circular, sessile, base cordate and stem-clasping, brown to greenish, thin, translucent, margins wavy,

Potamogeton javanicus   D. L. Jones

Potamogeton tricarinatus   W. R. Elliot

apex blunt; **spikes** 1–2 cm long, cylindrical, on thickish stalk, emergent.

Clasped Pondweed has ornamental foliage. It is distributed over its range in fresh water creeks and rivers, as well as in still water. Tolerates low salinity levels. Suitable for cultivation in pools and dams. Provides plentiful food and shelter for aquatic wildlife and waterbirds.

The following varieties of *P. perfoliata* are now recognised as synonyms: var. *minor* F. M. Bailey; var. *muelleri* A. Benn; and var. *rotundifolius* Wallr.

Potamogeton sulcatus A. Benn. =
     P. tricarinatus F. Muell. & A. Benn. ex A. Benn.

Potamogeton tricarinatus F. Muell. & A. Benn. ex A. Benn.
(three-ribbed)
Qld, NSW, Vic, Tas, SA, WA   Floating Pondweed;
Furrowed Pondweed
2–8 m across   Sept–April

Emergent, perennial, aquatic **herb** with creeping rhizome; **stems** to about 4 m long, flattened, much-branched; **submerged leaves** 4.5–23 cm × 0.5–9 cm, narrowly ovate to broadly ovate, sessile or nearly so, base rounded to tapered, thin, translucent, margins flat or faintly wavy, apex blunt or pointed; **emergent leaves** 1.5–10 cm × 1–7 cm, ovate to nearly circular, stalked, base rounded to cordate, dull green, thickish, margins flat, prominent longitudinal veins, apex blunt to pointed; **spikes** 1.5–5 cm long, cylindrical, emergent.

This is an extremely variable species with 4 distinct variants based mainly on leaf size and the number of longitudinal veins.

447

Potamogetonaceae

Variant I — vigorous, with trailing stems. It has broad leaves with up to about 27 longitudinal veins. It occurs in the Murray-Darling River system where it can cover large areas. The synonym, *P. sulcatus* A. Benn, is included in this variant.

Variant II — medium vigour with erect stems and cordate-based leaves with about 17–20 veins. Occurs in NSW, Vic, Tas, SA, WA and NT. It is found in dams and ponds.

Variant III — small plants with erect stems and elliptic to oval leaves with about 14 veins. Occurs in Qld, NSW, Vic, TAS, SA and WA as well as in New Caledonia.

Variant IV — vigorous plants. Leaves are narrowly ovate to ovate and fruits are reddish-brown. It occurs in tropical and subtropical areas of Qld, WA and NT.

All variants are suitable for ponds or dams. Propagate from seed, cuttings or by division of rhizomes.

POTAMOGETONACEAE Dum.   Pondweed Family

A small cosmopolitan family of monocotyledons which is comprised of 100 species in 2 genera. There are 10 species of *Potamogeton* which are native to Australia. They are perennial aquatic herbs which grow in either fresh or brackish water. Some species have submerged and floating leaves of different shape. The flowers are relatively small and insignificant. Some species overwinter by forming specialised fleshy storage organs called turions. These plants are important to aquatic life and are severely affected by pollution.

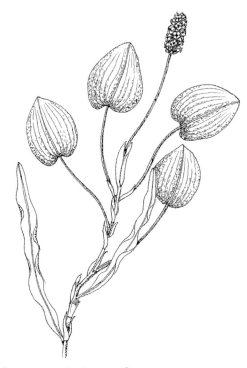

Potamogeton tricarinatus × .6

POTAMOPHILA R. Br.

(from the Greek *potamos*, river; *philos*, loving; referring to its habitat)

Poaceae

A monotypic genus endemic in north-eastern NSW.

Potamophila parviflora R. Br.

(small flowers)

NSW

1–1.5 m × 0.3–0.5 m   Dec–Feb

Tufting, reed-like, aquatic, perennial **grass** with short-creeping rhizome; **culms** to about 1.5 m × 0.7 cm; **leaves** to about 50 cm × 0.4–0.6 cm, erect, rolled in bud, becoming slightly inrolled or flat, green, slightly rough; **panicles** to about 45 cm × 2–5 cm; **spikelets** to about 0.5 cm long, glabrous, brownish with cream.

A tightly clumping species which occurs in the North Coast and North Tablelands Districts and inhabits river banks or river beds which are exposed for about half the year. Probably best grown within its natural range. It will need soils which are moist to wet for most of the year and plenty of sunshine. Worth trying on the edges of pools or dams. Plants have also been used in bank stabilisation and for soil erosion control with some success. Provides excellent wildlife habitat. Propagate from seed and possibly by division of rhizomes.

POTHOS L.

(from the Sinhalese name *potha*, used for a species in Sri Lanka)

Araceae

Slender to much branched **climbers**; **stems** with numerous adventitious roots; **leaves** simple, distichous, the petiole sheathing or flattened and photosynthetic; **flowering shoots** horizontal; **inflorescence** terminal, solitary or clustered, sometimes on scaly sidegrowth; **spathe** dull-coloured, inconspicuous; **spadix** sessile or stalked, cylindrical; **flowers** tiny, bisexual; **fruit** a fleshy berry.

A genus of about 50 species widely distributed from Madagascar to Polynesia and New Guinea. Two species occur in Australia and they are both endemic. One of these is cultivated on a limited scale. Propagation is from seed, which has a limited period of viability and is best sown while fresh. Cuttings of firm young growth are generally successful.

Pothos brassii B. L. Burtt

(after Leonard Brass, 20th-century American collector in Australia and New Guinea)

Qld   Northern Pothos

Climber   July–Sept

Slender **climber**; stems wiry, much-branched; **leaves** to 20 cm × 6 cm, ovate to elliptical, dull green, thick, fleshy, apex pointed; **petiole** 4–6 cm long, base sheathing; **inflorescence** terminal, on short growth; **spathe** to 3 cm × 1 cm, white or cream; **spadix** 1–2 cm long, cylindrical; **flowers** white or cream, tiny, crowded; **fruit** to 1 cm × 0.4 cm, ovoid, red or orange, ripe Nov–Feb.

Occurs in north-eastern Qld growing in wet coastal rainforests from the lowlands to about 600 m altitude. Suitable for gardens in tropical and subtropical

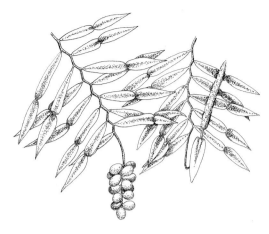

Pothos longipes × .45

regions and may have potential as a glasshouse plant. Requires a shady aspect and warm, humid conditions. Propagate from fresh seed and from cuttings of firm young growth.

Pothos brownii Domin = *P. longipes* Schott

Pothos longipes Schott
(with long stalks)
Qld, NSW   Native Pothos
Climber   Oct–April

Scrambling root **climber** forming a complex network of stems and leaves; **leaves** 1.5–5 cm × 0.5–1.5 cm, ovate to lanceolate, dark green, glossy, glabrous, apex acuminate; **petiole** broadly winged, leaf-like, usually much longer than the leaves, abruptly constricted where joined with the leaf; **inflorescence** solitary, on horizontal shoots; **peduncle** to 5 cm long; **spathe** to 2.5 cm × 0.5 cm, greenish to bronze or dark purplish, often reflexed; **spadix** to 6 cm × 0.7 cm, cylindrical, yellowing, becoming black; **flowers** minute, white; **fruit** to 1.3 cm × 0.6 cm, red, mature Jan–March.

Widely distributed from northern Qld to northern NSW. Grows on rocks and trees in rainforest. An interesting climber which commonly forms a dense covering consisting of complex interlaced growths. Clusters of red fruit are showy and are eaten by birds. The flesh of these fruit is edible but rather tasteless. Plants of this species resent disturbance and can be slow to establish. Once growing however they are easy to maintain. Can be grown from tropical to warm temperate regions. Best in a shady position and they need humidity and regular moisture. Propagate from fresh seed and from cuttings of young growth.
P. brownii Domin is a synonym.

POUTERIA Aubl.
(from the Guianan words 'pourama pouteri' for the type species of the genus)
Sapotaceae
Shrubs or **trees**; **sap** milky; **leaves** alternate, simple, entire, shortly petiolate, papery to leathery; **flowers** bisexual, solitary or in axillary clusters; **sepals**

5–6-merous, hairy exterior; **corolla** 4–8-merous, tubular or bell-shaped; **stamens** 4–8; **fruit** a berry; **seeds** large, with a prominent scar.

A large genus of about 150 species widely distributed in tropical regions of the world. The fruit of many species is edible. Most have ornamental foliage and a bushy habit but very few species are encountered in cultivation. Four species have been recorded in Australia, but of these only 2 have become at all well known. Some authorities consider that all or most species of *Planchonella* are better accommodated in *Pouteria*. Propagation is from seed, which has a limited period of viability and should be sown while fresh. Cuttings may be worth trying.

Pouteria castanosperma (C. T. White) Baehni
(with seeds like chestnuts)
Qld   Saffron Boxwood; Yellow Plum; Poison Plum;
                             Milky Plum
10–15 m × 5–8 m   Jan–April

Small to medium **tree**; young growth with rusty hairs; **leaves** 2.5–12 cm × 0.5–5 cm, lanceolate, thin and papery, green, shiny on both surfaces, glabrous, narrowed to the base, apex drawn out, blunt; **flowers** about 1 cm long, cream, on stalks 1–2 cm long, in axillary clusters of 2 or 3; **berries** 3–4 cm × 1.5–4.5 cm, ovoid to globose, purple to blackish; **seeds** 1–2, to 4 cm × 3 cm, brown, ripe Aug–Nov.

This species is endemic in north-eastern Qld. It grows in rainforest at low to moderate elevations. The fruit is reportedly edible but one of its common names suggests otherwise. Grown on a limited scale. A bushy small tree which has potential for cultivation in tropical and subtropical regions. Requires well-drained soil and responds to mulches and light fertiliser application. Propagate from fresh seed.

Pouteria sericea (Aiton) Baehni
(silky-hairy)
Qld, WA, NT   Mongo
6–10 m × 3–5 m   Nov–April

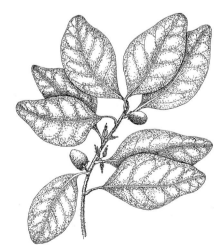

Pouteria sericea × .4

Tall **shrub** or small **tree** with a spreading canopy; **bark** brown to black, rough; young growth with rusty, silky hairs; **leaves** 5–14 cm × 2–6.5 cm, ovate to elliptic, sometimes nearly round, dark green and somewhat shiny above, silky-hairy beneath, apex blunt to notched; **flowers** about 0.6 cm long, cream to yellow, in axillary clusters; **berries** 1.5–2.5 cm × 1–1.5 cm, ellipsoid, purple; **seed** 1, about 1.8 cm × 1 cm, yellowish, speckled, shiny, ripe Nov–July.

P. sericea is widely distributed in tropical Australia. It grows in monsoon rainforest and vine thickets usually near streams. Also often prominent on coastal headlands and stabilised dunes. Ranges in habit from straggly to bushy. The fruit is edible and the attractively marked seeds can be used for necklaces. The wood burns well and is used for carving and cabinet-making. It was once used by Aborigines to make axe handles and woomeras. This species has ornamental features and it deserves to be given a trial in cultivation. Plants are best suited to tropical and subtropical regions. Especially useful in coastal districts. Requires excellent drainage and probably responds to mulches and light fertiliser applications. Propagate from fresh seed.

Botanical Regions of Australia

District

Barkly Tableland

RITORY

Cook

Burke

North Kennedy

South Kennedy

Gregory North

Mitchell

QUEENSLAND

Port
Curtis

Leichhardt

STRALIA

Lake Eyre Basin

airdner-Torrens Basin

Gregory South

Warrego

Maranoa

Burnett

Wide
Bay

Moreton

Darling Downs

Flinders
Ranges

North Far
Western Plains

North Western Plains

North
Tablelands

Northern
Tablelands

North Coast

Eastern

Western
Slopes

Eyre Peninsula

NEW   SOUTH   WALES

Northern
Lofty

Yorke
Peninsula

Murray

South Far
Western Plains

South Western
Plains

Central
Western
Slopes

Central
Tablelands

Central
Coast

Southern
Lofty

Kangaroo Island

Mallee

South
Western
Slopes

Southern
Tablelands

South Coast

South
Eastern

Wimmera

Northern District

North Eastern

North Central

VICTORIA

East
Gippsland

Western District

Central

Gippsland

TASMANIA

Glossary of Technical Terms

abaxial On the side of a lateral organ away from the axis; the lower side of a leaf or petiole.

aberrant Unusual or atypical; differing from the normal form.

abrupt Changing suddenly rather than gradually.

abscise To shed or throw off.

abscission Shedding of plant parts, eg. leaves. This may be natural resulting from old age or premature as a result of stress.

acaulescent Without a trunk.

accessory buds Lateral buds associated with a main bud such as in a leaf axil. They usually develop only if the main bud is damaged.

accessory root Lateral roots developed from the base of the trunk (as in palms) as opposed to those arising from the seed root system.

acerose Very slender or needle-shaped.

achene A small dry one-seeded fruit which does not split at maturity, eg. *Clematis, Senecio, Rhodanthe.*

acicular Needle-shaped.

actinomorphic Symmetrical and regular; usually applied to flowers, eg. *Wahlenbergia.*

aculeate Bearing short sharp prickles.

acuminate Tapering into a long drawn out point,

acute Bearing a short sharp point.

adaxial On the side of a lateral organ next to the axis; the upper side of a leaf or petiole.

adnate Fused tightly together so that separation without damage is impossible.

adventitious Arising in irregular position, eg. adventitious roots, adventitious buds.

aerial roots Adventitious roots arising on stems and growing in the air.

aff., affinity A botanical reference used to denote an undescribed species closely related to an already described species.

after-ripening The changes that occur in a dormant seed and render it capable of germinating.

aggregate fruit A fruit formed by the coherence of ovules that were distinct while in the flower, eg. *Rubus.*

albumen An old term used for the endosperm of seeds.

alternate Borne at different levels in a straight line or in a spiral.

amino acid An organic compound which is a structural unit of protein.

amplexicaul Stem-clasping, eg. the base of a sessile leaf.

androecium Collectively the male parts of a flower, ie. the stamens.

angiosperm A major group of plants which bear seeds within an ovary.

annual A plant completing its life cycle within 12 months.

annular Prominent ring scars left on the trunk of certain palms after leaf fall, eg. *Archontophoenix.*

annulate Bearing annular rings on the trunk.

anomalous An abnormal or freak form.

anther The pollen-bearing part of a stamen.

anthesis The process of flowering.

apetalous Without petals.

apical dominance The dominance of the apical growing shoot which produces hormones and prevents lateral buds developing while it is still growing actively.

apiculate With a short pointed tip.

apomixis Seed development without the benefit of sexual fusion.

appendage A small growth attached to an organ.

appressed Pressed flat against something.

aquatic A plant growing wholly or partially submerged in water.

arborescent With a tree-like growth habit.

arboretum A collection of planted trees.

arcuate Arched.

aril A fleshy or papery appendage produced as an outgrowth from the outer coat of a seed.

aristate Bearing a small bristle.

armed Bearing spines or prickles.

articles A part which separates readily; section of branchlet in Casuarinaceae.

articulate Jointed or having swollen nodes, eg. the culms of grasses.

asexual reproduction Reproduction by vegetative means without the fusion of sexual cells.

attenuated Drawn out.

auricle An ear-like appendage, eg. surrounding the bases of the fronds of *Angiopteris.*

auriculate Bearing auricles.

auxin A growth-regulating compound controlling many growth processes such as bud-break, root development, seed germination etc.

awn A bristle-like appendage, eg. on the seeds of many grasses.

axil Angle formed between adjacent organs in contact; commonly applied to the angle between a leaf and the stem.

axillary Borne within the axil.

axis The main stem of a plant or part of a plant.

Glossary of Technical Terms

barbed Bearing sharp backward-sloping hooks as on *Calamus* or *Rubus*.

basifixed Attached at or by the base; often referring to anthers or leaves.

basionym The published scientific name on which a change in botanical status is based.

bearded Bearing a tuft of hairs.

berry A simple, fleshy, many-seeded fruit with 2 or more compartments which do not split open when ripe.

biconvex Convex on both sides.

biennial A plant completing its life cycle in 2 years, usually growing vegetatively in the first year and flowering in the second year.

bifid Deeply notched for more than half its length.

bifoliolate With 2 leaflets to a leaf.

bifurcate Forked into 2 parts.

bilobed Two-lobed.

bilocular With 2 cavities.

bipinnate Twice pinnately divided.

bisexual Both male and female sexes present.

biternate With 3 divisions, each of which is divided into 3.

blade The expanded part of a leaf.

bole The trunk of a tree.

bottom heat A propagation term used to denote the application of artificial heat in the basal region of the cutting.

bract Leaf-like structure which subtends a flower-stem or inflorescence.

bracteole A small leaf-like structure found on a flower-stem.

branchlet A small slender branch.

bristle A short stiff hair.

bulb An enlarged thickened stem containing a bud surrounded by thickened leaf scales.

bulbil A small bulb produced in a leaf axil; a specialised bud produced at the junction of main veins on the fronds of certain ferns.

bulbous Bulb-shaped.

burr A prickly fruit.

bush A low, thick shrub, usually without a distinct trunk.

caducous Falling off early.

caespitose Growing in a tuft or tussock; also applied to clumping palms.

calcareous An excess of lime, as in soil.

callus Growth of undifferentiated cells; in orchids an organelle developed on the labellum.

calyx All of the sepals.

cambium The growing tissue lying just beneath the bark.

campanulate Bell-shaped.

cane A reed-like plant stem.

canopy The cover of foliage of a tree or community.

capitate Enlarged and head-like.

capitulate An inflorescence consisting of sessile flowers in a head, as in the family Asteraceae.

capsule A dehiscent dry fruit containing many seeds.

carinate Bearing a keel.

carpel Female reproductive organ.

catkin A pendent spike-like inflorescence composed of unisexual apetalous flowers; often used loosely for any rod-like inflorescence.

caudex A trunk-like axis in monocotyledons and some ferns.

cauliflory The production of flowers from the trunk and larger branches.

cauline Attached to the stem.

chaff Sterile packing around seeds found in some members of the family Myrtaceae, eg. *Eucalyptus*.

chartaceous Papery texture.

chlorophyll The green pigment of leaves and other organs, important as a light-absorbing agent in photosynthesis.

ciliate With a fringe of hairs.

cirrus A whip-like organ bearing recurved hooks used as an aid for climbing; it arises as an extension of the leaf rhachis and is present in some species of *Calamus*.

cladode A stem modified to serve as a leaf.

clavate Club-shaped.

claw The narrowed base of a sepal or petal; the staminal bundles eg. the flowers of the genera *Calothamnus* and *Melaleuca*.

cleistogamous A term applied to self pollinating flowers which do not open, eg. *Viola*.

clone A group of vegetatively propagated plants with a common ancestry.

coccus A single unit of a multiple fruit which splits at maturity.

column A fleshy growth in the flowers of orchids formed by the union of the stigmas and stamens.

combinatio nova A new combination of plant names.

compound leaf A leaf with two or more separate leaflets.

compressed Flattened laterally.

concolorous The same colour on both sides; usually refers to leaves.

cone A woody fruit in gymnosperms formed by sporophylls arranged spirally on an axis; other woody fruits such as those of the genus *Banksia* and *Casuarina* are also called cones.

confluent Running together; said of compound leaves where the leaflets remain united and do not separate.

congested Crowded closely together.

conifer A cone-bearing tree with needle-shaped or scale-like leaves; a gymnosperm.

connate Fused or joined together.

connective Tissue between uppermost part of filament and the 2 anther lobes.

contorted Twisted.

contracted Narrowed.

contractile root A specialised root developed by bulbs to maintain the bulb at a suitable level in the soil.

convoluted Rolled around and overlapping as in the leaves of a young shoot.

coppice shoot A shoot developing from a dormant bud in the trunk or larger branches of a tree; a very common feature of eucalypts.

cordate Heart-shaped.

coriaceous Leathery in texture.

corolla All of the petals.

corymb An inflorescence where the branches start at different points but reach about the same height to give a flat-topped effect.

456

costa The rib of a costapalmate leaf in palms; in ferns a main vein.

costapalmate Palmate leaves with a well developed rib which is an extension of the petiole into the blade and is the equivalent of the rhachis, eg. *Corypha elata.*

costate Ribbed.

cotyledon The seed leaf of a plant.

crenate The margin cut regularly into rounded teeth.

crenulate The margin cut into fine, rounded teeth.

crisped The margins very wavy or crumpled.

cross Offspring or hybrid.

cross-fertilisation Fertilisation by pollen from another flower.

cross-pollination Transfer of pollen from flower to flower.

crown That part of a shrub or tree above the first branch on the trunk.

crownshaft A series of tightly packed specialised tubular leaf-bases which terminate the trunk of some pinnate-leaved palms.

crozier Young coiled fern frond.

cryptogamist A person who studies non-flowering plants.

culm Flowering stem of grasses or sedges.

cultivar A horticultural variety of a plant or crop.

cuneate Wedge-shaped.

cupular Cup-shaped.

cupule A bowl or cup-shaped calyx developed at the base of some fruit.

cymbiform Boat-shaped.

cyme An inflorescence where the branches are opposite and the flowers open sequentially downwards starting from the terminal of each branch.

cymose Divided like a cyme.

cypsela A 1-seeded fruit developed from an inferior ovary.

damping off A condition in which young seedlings are attacked and killed by soil-borne fungi.

deciduous Falling or shedding of any plant part.

decorticate Regular shedding of bark.

decumbent Reclining on the ground but with the tips ascending; often referring to branches.

decurrent Running downwards beyond the point of junction; often referring to leaves, phyllodes, leaflets, lobes etc.

decussate Opposite leaves in 4 rows along the stem.

deflexed Abruptly turning downwards.

dehiscent Splitting or opening when mature.

deltate Triangular.

dentate Toothed.

denticulate Finely toothed.

depauperate A weak plant or one imperfectly developed.

depressed Flattened at the end.

determinate With a definite cessation of growth in the main axis.

dichasium A cyme with branches in regular opposite pairs.

dichotomous Regularly forking into equal branches.

dicotyledons A section of the Angiosperms with members usually bearing 2 seed leaves in the seedling stage.

diffuse Widely spreading and much-branched; or open growth.

digitate Spreading like the fingers of a hand from one point.

dimorphic Existing in 2 different forms.

dioecious Bearing male and female flowers on separate plants.

diploid Having 2 sets of chromosomes.

disc In orchids the callus adornment on the labellum; in eucalypts the tissue in the bud or fruit between the point where stamens arise and the top of the ovary.

disc floret The tubular flowers in the centre of heads of the family Asteraceae.

discolorous When surfaces differ in colour; usually refers to leaves.

dissected Deeply divided into segments.

distichous Alternate leaves arranged along the axis in 2 opposite rows.

divaricate Widely spreading and straggling.

divided Separated to the base.

domatia Small structures in the leaves of rainforest plants; they occur in the axils of the midrib and main lateral veins and may be either little tufts of hair or small, sunken, hooded enclosures (see *foveole*).

dormancy A physical or physiological condition that prevents growth or germination even though external factors are favourable.

dorsal sepal Sepal subtending the column of orchid flowers.

drupaceous Drupe-like.

drupe A fleshy indehiscent fruit with seed(s) enclosed in a stony endocarp.

drupelet One drupe of an aggregate fruit made up of drupes, eg. *Rubus.*

dune A mound formed from wind-blown sand.

ebracteate Without bracts.

ecology Study of the interaction of plants and animals within their natural environment.

effuse Very open and loosely spreading.

ellipsoid Elliptic in outline and solid.

elliptical Oval and flat and narrowed to each end which is rounded.

elongate Drawn out in length.

emarginate Having a notch at the apex.

embryo Dormant plant contained within a seed.

endemic Restricted to a particular country, region or area.

endocarp A woody layer surrounding a seed in a fleshy fruit.

endosperm Tissue rich in nutrients surrounding the embryo in seeds.

ensiform Sword-shaped.

entire Whole; not toothed or divided in any way.

enzyme A specialised protein capable of promoting a chemical reaction.

ephemeral A plant completing its life cycle within a very short period, ie. 3 – 6 months.

epicalyx Extra bract-like segments attached to the calyx; prominent in members of the Malvaceae, eg. *Hibiscus.*

epicarp The outermost layer of fruit.

Glossary of Technical Terms

epidermis The outer layer of cells which protects against drying and injury.

epigeal A term used for roots which grow above-ground.

epilith A plant growing on rock.

epiphyte A plant growing on or attached to another plant but not parasitic.

equable A term used to describe the endosperm of seed when it is smooth and uniform.

equitant Folded lengthwise with margins adhering to each other except for at and near the stem-clasping base.

erect Upright.

evergreen Remaining green and retaining leaves throughout the year.

exocarp Outermost layer of the fruit wall.

exotic A plant introduced from overseas.

exserted Protruding beyond the surrounding parts.

exstipulate Without stipules.

falcate Sickle-shaped.

family A taxonomic group of related genera.

farinaceous Containing starch; appearing as if covered by flour.

fasciculate Arranged in clusters.

ferruginous Rusty brown colour.

fertile bract Bract on an inflorescence which subtends the flowering branches (cf. *sterile bract*).

fertilisation The act of union of the pollen gametes and egg cells in the ovule.

fetid Having an offensive odour.

fibrillose Bearing fine fibres or threads.

fibrose Containing fibres.

fibrous Of non-decorticating bark which has long or short, usually dense fibres.

filament The stalk of the stamen supporting the anther.

filiform Long and very slender; thread-like.

fimbria The fringe; often applied to the fine hair-like fringes of a scale.

fimbriate Fringed with fine hairs.

flabellate Fan-shaped.

flabellum A term sometimes applied to the united pair of terminal leaflets of a pinnate leaf, eg. *Hydriastele*.

flaccid Soft, limp, lax.

flagellum A whip-like organ that bears curved hooks and is used as an aid to climbing; it is a modified inflorescence and arises in a leaf axil, eg. *Calamus*.

flared With a spreading rim.

flexuose Having a zigzag form.

floccose Having tufts of woolly hairs.

flora The plant population of a given region; also a book detailing the plant species of an area.

floral leaf A specialised leaf subtending flowers or an inflorescence and differing from normal foliage leaves.

floret The smallest unit of a compound flower.

floriferous Bearing numerous flowers.

flowerhead An arrangement where all the flowers are sessile.

foliaceous Leaf-like.

foliolate Bearing leaflets.

follicle A dry fruit formed from a single carpel and which splits along one line when ripe.

forest A plant community dominated by trees.

forked Divided into nearly equal parts.

form A botanical division below a species.

forma Latin for form (see above)

foveole A sunken, often hooded structure found on the leaves of some rainforest plants (see *domatia*).

free Not joined to any other part.

frond Leaf of a fern or palm.

fruit The seed-bearing organ developed after fertilisation.

fruitlet Small fruits forming part of an aggregate fruit, eg. *Rubus*.

fugacious Falling or withering away very early.

fungicide A chemical used to control fungus diseases.

fused Joined or growing together.

fusiform Spindle-shaped; narrowed to both ends from a swollen middle.

galea A hood of helmet-shaped structure formed by fusion of petals and sepals, eg. *Pterostylis*.

gamete One of the sex cells, either male or female.

gemma A vegetative bud by which a plant propagates.

gene A hereditary factor located in linear order on a chromosome.

geniculate Bent like a knee.

genus A taxonomic group of closely related species.

germination The active growth of an embryo resulting in the development of a young plant.

gibbous Humped.

glabrous Without hairs; smooth.

gland A fluid-secreting organ.

glandular Bearing glands.

glaucous Covered with a bloom giving a bluish lustre.

globoid Globe-like, globular, spherical.

globose Globular; almost spherical.

glume The bract subtending the spikelets of grasses and sedges.

glutinous Covered with a sticky exudation.

granular Covered with small grains.

growth regulator A synthetic compound which can control growth and flowering responses in plants and seeds.

growth split A vertical crack or split that develops in the trunk of a fast growing tree.

gymnosperm A major group of plants which bear seeds not enclosed within an ovary.

gynoecium Collectively the female parts of a flower.

habit The general appearance of a plant.

habitat The environment in which a plant grows.

halophyte A plant which grows in saline soils.

hapaxanthic A term describing clumping palms, individual stems of which die after flowering (cf. *monocarpic, pleonanthic*).

haploid Having a single set of chromosomes.

hastate Shaped like an arrow-head and with spreading basal lobes.

haustorial Said of a parasite which is able to absorb water and nutrients directly from a host by a specialised attachment.

head A composite cluster of flowers, as in the family Asteraceae.

458

herb A plant which produces a fleshy rather than woody stem.

herbaceous A perennial plant which dies down each year after flowering.

herbicide A chemical used to control weeds.

hermaphrodite Bearing both male and female sex organs in the same flower.

heterogamous With male, female, bisexual or neuter flowers in any combination in an inflorescence.

hilum The scar left on the seed at its point of detachment from the seed stalk.

hirsute Covered with long, spreading coarse hairs.

hispid Covered with stiff bristles or hairs.

hispidulous Minutely hispid.

hoary Covered with short white hairs giving the surface a greyish appearance.

hormone A chemical substance produced in one part of a plant and inducing a growth response when transferred to another part.

hybrid Progeny resulting from the cross-fertilisation of differing genera, species or varieties.

hybrid swarm Variable population resulting from complex crossing such as between the hybrids themselves or between the hybrids and the parents.

hybrid vigour The increase in vigour of hybrids over their inbred parents.

hypanthium In eucalypts the enlarged receptacle which encloses the ovary.

hypocotyl Part of the embryo between the cotyledons and primary root.

imbricate Overlapping.

imparipinnate Pinnate leaves bearing a single terminal leaflet which extends from the end of the rhachis.

incised Cut sharply and deeply.

incurved Curved inwards.

indehiscent Not splitting open at maturity.

indeterminate Growing on without termination.

indigenous Native to a country, region or area.

indumentum The hairy covering on plant parts.

induplicate Leaflets folded longitudinally with the Vs (leaf margins) opened upwards.

indusium A membrane covering a fern sorus; a cup-shaped structure protecting the stigma in Goodeniaceae.

inferior Below some other part; often used in reference to an ovary when held below other layers of the perianth.

inflorescence The flowering arrangement of a plant.

infructescence A term used to describe a fruiting inflorescence

inhibitor A chemical substance which prevents a growth process.

insecticide A chemical used to control insect pests.

internode The part of a stem between 2 nodes

involucre A cluster of overlapping bracts surrounding the base of the flower-heads.

involute Rolled inwards.

jointed Bearing joints or nodes.

juvenile The young stage of growth before the plant is capable of flowering.

keel A ridge like the base of a boat; in pea-shaped flowers the basal part formed by the union of 2 petals.

kino A resin-like exudate which is especially common in eucalypts.

labellum A lip; in orchids the petal in front of the column.

lacerate Irregularly cut or torn into narrow segments.

laciniate Cut into narrow slender segments.

lamina The expanded part of a leaf.

lanceolate Lance-shaped; narrow and tapering at each end, especially the apex.

lateral Arising from the main axis; arising at the side of.

lateritic Describes soils formed by weathering of rocks; composed chiefly of iron and aluminium hydroxides.

lax Open and loose.

leaf-base Specialised expanded and sheathing part of the petiole where it joins the trunk, as in palms.

leaflet A segment of a compound leaf.

legume A dry fruit formed from one carpel and splitting along 2 lines.

lemma The lower bract enclosing the flower of grasses.

lepidote Dotted with persistent, small, scurfy, peltate scales, as on some palm leaves.

liana, liane A large woody climber.

lignotuber woody swelling containing dormant buds at the base of a trunk, eg. mallee eucalypts.

ligulate Strap-shaped.

ligule A strap-shaped organ; in grasses a growth at the junction of leaf-sheath and blade.

linear Long and narrow with parallel sides.

lithophyte A plant growing on rocks, cliff faces, boulders etc.

littoral Growing in communities near the sea.

lobe A segment of an organ as the result of a division.

loculus A compartment within an ovary.

loricate Covered with overlapping scales, eg. *Calamus* fruits.

Malesia An area comprising New Guinea, the islands of Indonesia, Malaysia and the Philippines.

mallee A shrub or tree with many stems arising from at or below ground level.

mangrove A specialised plant growing in salt water and gathering oxygen through specialised roots (pneumatophores).

marcescent Short-lived and withering while still attached to the plant.

marginal Attached to or near the edge.

maritime Belonging to or growing near the sea.

marlock A shrub or tree with many stems arising from a point on the trunk above ground level.

marsh A swamp.

mealy Covered with flour-like powder.

membranous Thin-textured.

mericarp One segment of a fruit that breaks into individual carpels at maturity.

meristem A growing point or an area of active cell division.

merous The number of parts of a flower that makes up any whorl, eg. petals, stamens; usually written 4-merous etc.

459

Glossary of Technical Terms

mesocarp Middle layer of a fruit wall.

midrib The principal vein that runs the full length.

miticide (also *acaracide*) A chemical used to control mites.

moniliform Constricted at regular intervals and appearing bead-like.

monocarpic A term describing plants which flower once, then die (cf. *hapaxanthic, pleonanthic*).

monocotyledons A section of the Angiosperms bearing a single seed leaf at the seedling stage.

monoecious Bearing male and female flowers on the same plant.

monopodial A stem with a single main axis which grows forward at the tip.

monotypic A genus with a single species.

morphology The form and structure of a plant.

mucronate With a short sharp point.

mucronulate With a very small point.

mutation A change in the genetic constitution of a plant or part of a plant.

mycorrhiza A beneficial relationship between the roots of a plant and fungi or bacteria resulting in a nutrient exchange system. Some plants cannot grow without such a relationship.

nectar A sweet fluid secreted from a nectary.

nectary A specialised gland which secretes nectar.

nematicide A chemical used to control nematodes.

nematode A minute worm-like animal, some species of which attack plants.

nerves The fine veins which traverse the leaf-blade.

node A point on the stem where leaves or bracts arise.

nodule A small swollen lump on roots; in legumes the nodules contain symbiotic bacteria of the genus *Rhizobium.*

nut A dry indehiscent one-seeded fruit.

nutlet A small nut enclosing a single seed.

obcordate Cordate with the broadest part above the middle.

oblanceolate Lanceolate with the broadest part above the middle.

oblate Nearly spherical, but noticeably broader than long; flattened from above.

oblique Slanting; unequal-sided.

obovate Ovate with the broadest part above the middle.

obtuse Blunt or rounded at the apex.

offset A growth arising from the base of another plant.

olivaceous Dark olive green.

operculum A structure formed from the fusion of petals and sepals and protecting the stamens and style when in bud, eg. *Eucalyptus.*

opposite Arising on opposite sides but at the same level.

orbicular Nearly circular.

order A taxonomic group of related families.

organelle A small plant organ.

osmosis Diffusion of water through a membrane caused by different concentrations of salts on either side of the membrane.

oval Rounded but longer than wide.

ovary The part of the gynoecium which encloses the ovules.

ovate Egg-shaped in longitudinal section.

ovoid Egg-shaped.

ovule The structure within the ovary which becomes seed after fertilisation.

palea The upper bract enclosing the flower of grasses.

paleaceous Clothed with papery scales.

palmate Divided like a hand.

palmatifid Lobed like a hand.

palmet A dwarf-growing palm, eg. *Linospadix* spp.

panicle A much-branched racemose inflorescence.

paniculate Arranged in a panicle.

papillose Rough with small protuberances.

pappus A tuft of feathery bristles representing a modified calyx, on the seeds of Asteraceae.

parasite A plant growing or living on or in another plant, eg. mistletoe.

paripinnate Compound pinnate leaves lacking a terminal leaflet.

parthenocarpy Development of fruit without fertilisation and seed formation.

patent Spreading out.

pectinate Shaped like a comb.

pedicel The stalk of a flower in a compound inflorescence.

pedicellate Growing on a pedicel.

peduncle The main axis of a compound inflorescence or the stalk of a solitary flower.

pedunculate Growing on a peduncle.

peltate Circular with a stalk attached in the middle on the undersurface.

pendant, pendent Hanging downwards.

penninerved The veins branching pinnately.

perennate To survive from season to season, generally with a period of inactivity between each season.

perennial A plant living for more than 2 years.

perfoliate United around the stem, as in leaves.

perianth A collective term for all of the petals and sepals of a flower.

pericarp The hardened ovary wall that surrounds a seed.

persistent Remaining attached until mature; not falling prematurely.

pesticide A chemical used to control pests. In this case, pests is defined as a range of plant enemies including fungi, bacteria, nematodes, insects, mites etc.

petal A segment of the inner perianth whorl or corolla.

petaloid Petal-like; resembling a petal.

petiolate Bearing a petiole.

petiole The stem or stalk of a leaf.

phalange Segment of a compound fruit composed of fused carpels; eg. *Pandanus* fruits.

phloem Part of the vascular system of plants concerned with the movement and storage of nutrients and hormones.

photosynthesis The conversion of carbon dioxide from the atmosphere to sugars within green parts of the plant, using chlorophyll and energy from the sun's rays.

phylloclade A stem acting in the capacity of a leaf (cladode).

phyllode A modified petiole acting as a leaf.

pilose With scattered long simple hairs.

pinna A primary segment of a divided leaf.

pinnate Once-divided with the divisions extending to the midrib.

pinnatifid Once-divided with the divisions not extending to the midrib.

pinnatisect Pinnately divided nearly to the midrib without forming leaflets.

pinnule The segment of a compound leaf divided more than once.

pistil The ovule and seed-bearing organ of the gynoecium.

pistillate flowers Female flowers.

pistillode A sterile pistil, often found in male flowers.

pleonanthic Plants which flower regularly each year after reaching maturity (cf. *hapaxanthic, monocarpic*).

plicate Folded longitudinally.

plumose Feather-like from fine feathery hairs.

pneumatophore Specialised roots of mangroves carrying oxygen to the plant.

pod A dry non-fleshy fruit that splits when ripe to release its seeds.

pollen Haploid male cells produced by the anthers.

pollination The transference of the pollen from the anther to the stigma of a flower.

pollinium An aggregated mass of pollen grains found in the Orchidaceae and Asclepiadaceae.

polyembryony The condition where a seed produces more than one embryo.

polygamous Having mixed unisexual and bisexual flowers together.

polymorphic Consisting of many forms; a variable species, eg. *Grevillea rosmarinifolia*.

praemorse As though bitten off, as in the leaflet tips of *Ptychosperma*.

prickle A small spine borne irregularly in the bark or epidermis.

procumbent Spreading on the ground without rooting; often referring to branches.

proliferous Bearing offshoots and other processes of vegetative propagation.

prophyll The first (or outer) sheathing bract of a palm inflorescence.

prostrate Lying flat on the ground.

proteoid roots A specialised root development found in species of the family Proteaceae. Numerous short roots develop in a compact mop-like clump.

protuberance A swelling or bump.

provenance A particular geographic site of origin.

pruinose Covered with a waxy, greyish bloom.

pseudanthia False flowers.

pseudobulb Thickened bulb-like stems of orchids bearing nodes.

pubescent Covered with short soft downy hairs.

pulvinus A cushion-like growth of inflated cells on the leaf or leaflet-stalk at its junction with the blade, or on spines or inflorescence branches, usually in pairs.

punctate Marked with spots or glands.

pungent Very sharply pointed; also smelling strongly.

pyrene The stony seed(s) found in a succulent fruit.

pyriform Pear-shaped.

raceme A simple unbranched inflorescence with stalked flowers.

racemose In the form of a raceme

radical Arranged in a basal rosette.

radicle The undeveloped root of the embryo.

ray floret The outermost, flattened florets of the inflorescence in the family Asteraceae.

recurved Curved backwards.

reduplicate Leaflets folded with the Vs (leaf-margins) opened downwards.

reflexed Bent backwards and downwards.

regular Symmetrical, especially of flowers.

reniform Kidney-shaped.

repent With a creeping growth habit.

reticulate A network; eg. veins.

retuse With a slight notch at the apex.

revolute With the margins rolled backwards.

rhachilla A small rhachis; the secondary and lesser ones of a compound inflorescence.

rhachis The main axis of a compound leaf or an inflorescence.

rhizome An underground stem.

rhomboidal Diamond-shaped.

rib The section of the petiole of a costapalmate leaf that extends into the blade.

rootstock A term for a rhizome.

rosette A group of leaves radiating from a centre.

rostrate With a beak.

rotate The lobes of the corolla spreading horizontally like the spokes of a wheel.

rugose Wrinkled.

rugulose Finely wrinkled.

ruminate Folded like a stomach-lining; used to describe the folds in the endosperm of some seeds.

runner A slender trailing shoot forming roots at the nodes.

saccate Pouch or sac-like.

sagittate Shaped like an arrow-head, with the basal lobes pointing downwards.

samara An indehiscent winged seed.

samphire A plant community which occurs on saline soils and comprises members of the family Salicorniae, eg. glassworts being dominant.

saprophyte A plant which derives its food from dead or decaying organic matter.

scabrous Rough to the touch.

scale A dry, flattened, papery body; sometimes also used as a term for rudimentary leaf.

scandent Climbing.

scape Leafless peduncle arising near the ground; it may bear scales or bracts and the foliage leaves are radical.

scarious Thin, dry and membranous.

schizocarp A dry fruit which splits at maturity into carpels which are usually 1-seeded.

sclerophyll A plant (or forest) with hard stiff leaves.

scribbles The irregular markings in bark of eucalypts; caused by insect larvae.

scrub Strictly a plant community dominated by shrubs; often used loosely for rainforests.

scurfy Bearing small, flattened, papery scales.

secondary thickening The increase in trunk diameter as the result of cambial growth.

461

Glossary of Technical Terms

section A taxonomic subgroup of a genus containing closely related species.

secund With all parts directed to one side.

seed A mature ovule consisting of an embryo, endosperm and protective coat.

seed-coat The protective covering of a seed; also called testa.

seedling A young plant raised from seed.

segment A subdivision or part of an organ, eg. sepal is a segment of the calyx.

self pollination Transfer of pollen from stamen to stigma of same flower.

sepal A segment of the calyx or outer whorl of the perianth.

sepaloid Sepal-like.

septal With partitions.

septate Divided by partitions.

serrate Toothed with sharp, forward-pointing teeth.

serrulate Finely serrate.

sessile Without a stalk, pedicel or petiole.

seta A bristle.

setaceous Shaped like a bristle.

setose Bristly.

shrub A woody plant that remains low (less than 6 m) and usually with many stems or trunks.

silicula A dry fruit comprising 2 carpels with a separating partition, less than twice as long as its width.

siliqua A dry fruit comprising 2 carpels with a separating partition, more than 3 times as long as its width.

simple Undivided; of one piece.

sinuate With a wavy margin.

sinus A junction; a specialised term for flowers of the orchid genus *Pterostylis* describing the point of junction of the lateral sepals.

slip A cutting.

soboliferous Bearing creeping, rooting stems; sometimes interpreted as bearing suckers, and applied to clumping plants.

sorus A cluster of sporangia on the fronds of ferns.

spadix An inflorescence which is a fleshy spike with flowers more or less sunken in the axis, and usually enclosed by a spathe.

spathe A large sheathing bract which encloses a spadix.

spathulate Spatula-shaped; with a broad top and tapering base.

spear-leaf The erect, unopened young leaf of a palm.

species A taxonomic group of closely related plants, all possessing a common set of characters which set them apart from another species.

spicate Arranged like or resembling a spike.

spike A simple unbranched inflorescence with sessile flowers.

spikelet A small spike bearing one or more flowers; in grasses a small unit composed of glumes and florets.

spine A sharp rigid structure.

spinescent Bearing spines or ending in spines.

spinule A weak spine.

spinulose With small spines.

sporangium A case that bears spore.

spore A reproductive unit which does not contain an embryo.

sporeling A young fern plant.

spur A tubular sac-like projection of a flower, often containing nectar.

stamen The male part of a flower producing pollen, consisting of an anther and a filament.

staminate flowers Male flowers.

staminode A sterile stamen; often of different form, eg. petaloid.

standard The dorsal petal, usually confined to flowers of the family Fabaceae.

stellate Star-shaped or of star-like form.

stem-clasping Enfolding a stem.

sterile Unable to reproduce.

sterile bracts Bracts on an inflorescence which do not subtend a flower or flowering branch (cf. *fertile bracts*).

stigma The usually enlarged area of the style receptive to pollen.

stipe A stalk or leaf-stalk.

stipel A secondary stipule at the base of leaflets.

stipitate Stalked.

stipule Small bract-like appendage borne in pairs at the base of the petiole.

stocking A short section of persistent bark on the lower trunk of a smooth-barked tree.

stolon A basal stem growing just below the ground surface and rooting at intervals.

stoloniferous Bearing stolons; spreading by stolons.

strain An improved selection within a variety; also cultivar.

strand plant A plant growing near the sea.

stratification The technique of burying seed in coarse sand so as to expose it to periods of cold temperatures or to soften the seed-coat.

striate Marked with narrow lines or ridges.

strobilus A cone.

strophiolate Bearing a strophiole.

strophiole An appendage arising from the seed-coat near the hilum.

style Part of the gynoecium connecting the stigma with the ovary.

sub Beneath, nearly, approximately.

subclimber Shrubby plant that can produce long shoots which need support of surrounding plants.

subcordate Nearly cordate.

subfamily A taxonomic group of closely related genera within a family.

subspecies A taxonomic subgroup within a species used to differentiate geographically isolated variants.

subternate Nearly divided or arranged in threes.

subulate Narrow and drawn out to a fine point.

succulent Fleshy or juicy.

sucker A shoot arising from the roots or the trunk below ground level.

sulcate Grooved or furrowed.

superior Above some other part; often used in reference to an ovary when held above other layers of the perianth.

symbiosis A beneficial association of different organisms.

sympodial Lacking a single, persistent growing apex; lateral growth below the apex is dominant, resulting in change of growth direction.

taproot The perpendicular main root of a plant.

taxon A term used to describe any taxonomic group, eg. genus, species, variety.

taxonomy The classification of plants or animals.

tendril A plant organ modified to support stems used in climbing.

tepal A term used for the perianth segments when the sepals and petals are alike, eg. Liliaceae.

teratology The study of abnormal or aberrant forms.

terete Slender and cylindrical.

terminal The apex or end.

ternate Divided or arranged in threes.

terrestrial Growing in the ground.

testa The outer covering of the seed; the seed-coat.

tetragonous With 4 angles.

tetrahedral With 4 sides.

thorn A reduced branch ending in a hard sharp point.

thyrse A densely branched inflorescence in which the main axis is racemose and the secondary axes are cymose.

tomentose Densely covered with short, matted, soft hairs.

tomentum A covering of matted soft hairs.

tortuous Twisted; with irregular bending.

transpiration The loss of water vapour to the atmosphere through openings in the leaves of plants.

tree A woody plant that produces a single trunk and a distinct elevated head.

triad A group of 3.

tribe A taxonomic group of related genera within a family or subfamily.

trichomes A term used to describe outgrowths of the epidermis such as scales or hairs.

trifid Divided into 3 to about the middle.

trifoliolate A compound leaf with 3 leaflets.

trigonous With 3 angles.

trilobed, trilobate With 3 lobes.

tripartite Divided into 3 parts nearly to the base.

tripinnate Divided 3 times.

tripinnatifid Divided 3 times with divisions not extending to the midrib.

triquetrous With 3 angles or ridges.

truncate Ending abruptly as if cut off.

trunk The main stem of a tree.

tuber The swollen end of an underground stem or stolon.

tubercle Small tubers produced on aerial stems, eg. *Dioscorea*.

tuberculate With knobby or warty projections.

tuberoid The swollen end of an underground root.

tuberous Swollen and fleshy; resembling a tuber.

tufted Growing in small, erect clumps.

turbinate Shaped like a spinning top.

turgid Swollen or bloated.

turion A scaly shoot or sucker; a starch-rich attachment to a rhizome.

twiner Climbing plants which ascend by twining of their stems or rhachises.

type form The form of a variable species from which the species was originally described.

umbel An inflorescence in which all the stems arise at the same point and the flowers lie at the same level.

umbellate Like an umbel.

umbonate Having a central boss or knob.

undulate Wavy.

unequal Of different sizes.

unilateral One-sided.

unilocular With one cavity.

unisexual Of one sex only; staminate or pistillate.

united Joined together, wholly or partially.

urceolate Urn-shaped.

valvate Opening by valves; with perianth segments overlapping in the bud.

valve A segment of a woody fruit.

variegated Where the basic colour of a leaf or petal is broken by areas of another colour, usually white, pale green or yellow.

variety A taxonomic subgroup within a species used to differentiate variable populations.

vascular bundle The internal conducting system of plants.

vascular plant A plant bearing water conducting tissue in its organs.

vegetation The whole plant communities of an area.

vegetative Asexual development or propagation.

vein The conducting tissue of leaves.

veinlet A small or slender vein.

velamen A veil, spongy, epidermis of epiphytic roots.

venation The pattern formed by veins.

vernalisation The promotion of flowering.

verrucose Rough and warty.

verticillate Arranged in whorls.

vesicle A thin-walled globular swelling.

viable Alive and able to germinate, as of seeds.

villous Covered with long, soft, shaggy hairs.

virgate Twiggy.

viscid Coated with a sticky or glutinous substance.

viviparous Germinating while still attached to the parent plant, eg. mangroves.

wallum A vegetation community which occurs on coastal sands of south-eastern Qld and north-eastern NSW; often waterlogged in the wet season.

watershoot A strong rapid growing shoot arising from the trunk or main stem.

whorl Three or more segments (leaves, flowers etc.) in a circle at a node.

wing A thin, dry, membranous expansion of an organ; the side petals of flowers of the family Fabaceae.

woolly Bearing long, soft, matted hairs.

xeromorph A plant with drought resistant features.

xerophyte A drought resistant plant.

xerophytic Able to withstand drought or dry conditions.

zygomorphic Asymmetric and irregular; usually applied to flowers, eg. *Anigozanthos*.

zygote A fertilised egg.

Further Reading

Adams, G. M. (1980) *Birdscaping Your Garden*, Rigby, Adelaide.

Allaby, M. (ed.) (1985) *The Oxford Dictionary of Natural History*, Oxford University Press, Oxford, England.

Allen, H. H. (1961) *Flora of New Zealand*, Volume 1, Government Printer, Wellington, NZ.

Anderson, R. H. (1967) *The Trees of New South Wales*, Government Printer, NSW.

Armitage, I. (1978) *Acacias of New South Wales*, New South Wales Region, Society for Growing Australian Plants, Sydney.

Arthur, T. E. & Martin, S. D. (eds.) (1981) *Street Tree Directory*, Royal Australian Institute of Parks & Recreation, Victorian Region, Tullamarine.

Aston, H. I. (1973) *Aquatic Plants of Australia*, Melbourne University Press.

Aston, H. I. (1986) *An Elementary Index to Australian Herbarium Journals from their Inception to 1985 Inclusive*, H. I. Aston, Melbourne.

Audas, J. W. (193?) *Native Trees of Australia*, Whitcombe & Tombs Ltd., Melbourne.

Australian Daisy Study Group (1987) *Australian Daisies for Gardens and Floral Art*, LothianPublishing Co., Melbourne.

Australian Plant Study Group (1980) *Grow What Where*, Thomas Nelson (Australia) Ltd., Melbourne.

Australian Plant Study Group (1982) *Grow What Wet*, Thomas Nelson (Australia) Ltd., Melbourne.

Australian Systematic Botany Society (1981) *Flora of Central Australia* (ed. J. Jessop) A. H. & A. W. Reed, Sydney.

Austrobaileya (1977–) Periodic Journal of Queensland Herbarium, Department of Primary Industries, Brisbane, Government Printer, Qld.

Bailey, F. M. (1899–1902) *The Queensland Flora*, Queensland Government, H. J. Diddams, Qld.

Bailey, F. M. (1909) *A Comprehensive Catalogue of Queensland Plants*, Government Printer, Qld.

Bailey, L. H., Hortorium, Staff of the (1976) *Hortus Third*, Macmillan, New York.

Baines, J. A. (1981) *Australian Plant Genera*, Society for Growing Australian Plants, Sydney.

Baker, K. F. (ed.) (1957) *The U. C. System for Producing Healthy Container-Grown Plants*, University of California.

Barr, A., Chapman, J., Smith, N. & Beveridge, M. (1988) *Traditional Bush Medicines*, Greenhouse Publications, Richmond, Victoria.

Beadle, N. C. W. (1971–87) *Students Flora of North-Eastern New South Wales*, Parts 1–6, University of New England, Armidale.

Beadle, N. C. W., Evans, O. D. & Carolin, R. C. (1972) *Flora of the Sydney Region*, A. H. & A. W. Reed, Sydney.

Bean, T. (1983) *Eucalypts of the Sunshine Coast and the Coast from Bundaberg to Coffs Harbour*, T. Bean, Nambour, Qld.

Beard, J. S. (ed.) (1965) *Descriptive Catalogue of West Australian Plants*, Society for Growing Australian Plants, Sydney.

Beauglehole, A. C. (1980) *Victorian Vascular Plant Checklist*, Western Victorian Field Naturalists Clubs Association, Portland, Vic.

Bennett, E. M., (1993, 2nd edn) *Common and Aboriginal Names of Western Australian Plant Species*, Wildflower Society of Western Australia, Eastern Hills Branch, Glen Forrest, WA.

Bentham, G. & Mueller, F. (1863–78) *Flora Australiensis*, Parts 1–7, Lovell Reeve & Co., London.

Biological Survey of Eastern Goldfields of Western Australia, Parts 1 & 2 (1984) Western Australian Museum, Perth.

Black, J. M. (1943–57) *Flora of South Australia*, Parts 1–4, Government Printer, Adelaide.

Black, J. M. (1978) *Flora of South Australia*, Part 1 (3rd edn), Government Printer, SA.

Blackall, W. E. (1954) *How to Know Western Australian Wildflowers*, Part 1, University of Western Australia Press, Perth.

Blackall, W. E. & Grieve, B. J. (1956–80) *How to Know Western Australian Wildflowers*, Parts 2–4, University of Western Australia Press, Perth.

Blake, S. T. Roff, C. (1972) *The Honey Flora of Queensland*, Government Printer, Qld.

Blake, T. L. (1981) *A Guide to Darwinia and Homoranthus*, Society for Growing Australian Plants, Maroondah Group, Ringwood Vic.

Blakely, W. F. (1965) *A Key to the Eucalypts*, 3rd edn, Government Printer, Canberra.

Blombery, A. M. (1967) *A Guide to Native Australian Plants*, Angus & Robertson, Sydney.

Blombery, A. M. (1972) *What Wildflower is That?* Paul Hamlyn, Sydney.

Blombery, A. & Rodd, A. N. (1982) *Palms*, Angus & Robertson, Sydney.

Boland, D. J., Brooker, M. I. H., Kleinig, D. A. & Turnbull, J. W. (1980) *Eucalyptus Seed*, CSIRO, Melbourne.

Further Reading

Boland, D. J. et al (1984) *Forest Trees of Australia*, 4th edn, Thomas Nelson (Australia) Ltd., Melbourne & CSIRO, Melbourne.

Bonney, N. B. (1977) *An Introduction to the Identification of Native Flora in the Lower South-East of South Australia*, South-East Community College, Mt Gambier.

Boomsma, C. D. (1981) *Native Trees of South Australia*, 2nd edn, Woods & Forests Department, SA.

Boomsma, C. D. & Lewis, N. B. (1984) *The Native Forest & Woodland Vegetation of South Australia*, Woods & Forests Department, SA.

Borror, D. J. (1960) *Dictionary of Word Roots and Combining Forms*, Mayfield Publishing Co, Palo Alto, California.

Breckwoldt, R. (1983) *Wildlife in the Home Paddock*, Angus & Robertson, Sydney.

Bridgeman, P. H. (1976) *Tree Surgery*, David & Charles Ltd., London.

Briggs, J. D. & Leigh, J. H. (1996) *Rare or Threatened Australian Plants*, revised edn, CSIRO, Melbourne.

Brock, J. (1988) *Top End Native Plants*, John Brock, Darwin.

Brooks, A. E. (1973) *Australian Native Plants for Home Gardens*, 6th edn, Lothian Publishing Co., Melbourne.

Brown, A. & Hall, N. (1968) *Growing Trees on Australian Farms*, Forestry & Timber Bureau, Canberra.

Brown, M. J., Kirkpatrick, J. B. & Moscal, A. (1983) *An Atlas of Tasmania's Endemic Flora*, Tasmanian Conservation Trust Inc., Hobart.

Brummitt, R. K. & Powell, C. E. (1992) *Authors of Plant Names*, Royal Botanic Gardens, Kew, England.

Brunonia (1978–87) Periodical Journal of the Herbarium Australiense, CSIRO, Melbourne.

Buchanan, A. M. (ed.) (1995) *A Census of the Vascular Plants of Tasmania*, Tasmanian Herbarium, Hobart.

Burbidge, N. T. (1963) *Dictionary of Australian Plant Genera*, Angus & Robertson, Sydney.

Burbidge, N. T. & Gray, M. (1970) *Flora of the Australian Capital Territory*, Australian National University Press, Canberra.

Bureau of Flora and Fauna (1981–)*Flora of Australia* (ed. A. S. George et al), Australian Government Publishing Service, Canberra.

Burgman, M. A. & Hopper, S. D. (1982) *The Western Australian Wildflower Industry*, Department of Fisheries & Wildlife, Perth.

Canberra Botanic Gardens (1971–) *Growing Native Plants* (series) Australian Government Publishing Service, Canberra.

Carman, J. K. (1978) *Dyemaking with Eucalypts*, Rigby, Adelaide.

Chapman, A. D. (1991) *Australian Plant Name Index*, Australian Flora & Fauna Series No. 14, Australian Government Publishing Service, Canberra.

Chippendale, G. M. (ed.) (1968) *Eucalyptus Buds and Fruits*, Forestry & Timber Bureau, Canberra.

Chippendale, G. M. (1973) *Eucalypts of the Western Australian Goldfields*, Australian Government Publishing Service, Canberra.

Chippendale, G. M. (1983, unpublished) *Eucalyptus Nomenclature*, Division of Forest Research, CSIRO, Canberra.

Christophel, D. C. & Hyland, B. P. M. (1993) *Leaf Atlas of Australian Tropical Rain Forest Trees*, CSIRO, Melbourne.

Clements, M. A. (1982) *Preliminary Checklist of Australian Orchidaceae*, National Botanic Gardens, Canberra.

Clemson, A. (1985) *Honey and Pollen Flora*, Department of Agriculture, NSW & Inkata Press, Melbourne.

Clifford, H. T. & Constantine, J. (1980) *Ferns, Fern Allies and Conifers of Australia*, University of Queensland Press, St. Lucia, Qld.

Cochrane, R. G., Fuhrer, B. A., Rotherham, E. R., Simmons, J. & M. & Willis, J. H. (1980) *Flowers and Plants of Victoria and Tasmania*, A. H. & A. W. Reed, Sydney.

Conabere, B & Garnet, J. Ros (1974) *Wildflowers of South-Eastern Australia*, Volumes 1–2, Thomas Nelson (Australia) Ltd., Melbourne.

Contributions from the Queensland Herbarium (1968–1975) Department of Primary Industries, Brisbane, Government Printer, Qld.

Costermans, L. (1981) *Native Trees and Shrubs of South-Eastern Australia*, Rigby, Adelaide.

Costin, A. B., Gray, M., Totterdell, C. J. & Wimbush, D. J. (1979) *Kosciusko Alpine Flora*, CSIRO, Melbourne and William Collins, Sydney.

Court, A. B. (ed.) (1980) *Catalogue of the Living Plants Supported by Herbarium Vouchers*, Australian National Botanic Gardens, Canberra.

Cribb, A. B. & Cribb, J. W. (1974) *Wild Food in Australia*, William Collins, Sydney.

Cribb, A. B. & Cribb, J. W. (1981) *Useful Wild Plants in Australia*, William Collins, Sydney.

Cribb, A. B. & Cribb, J. W. (1981) *Wild Medicine in Australia*, William Collins, Sydney.

Cunningham, G. M., Mulham, W. E., Milthorpe, P. L. & Leigh, J. H. (1981) *Plants of Western New South Wales*, Soil Conservation Service of New South Wales.

Curtis, W. M. (1956–79) *The Students Flora of Tasmania*, Parts 1–4a, Government Printer, Tas.

Curtis, W. M. & Morris, D. I. (1975) *The Students Flora of Tasmania*, Part 1, 2nd edn, Government Printer, Tas.

Curtis, W. M. & Morris, D. I. (1994) *The Students Flora of Tasmania*, Part 4a, St David's Park Publishing, Tas.

Dockrill, A. W. (1969) *Australian Indigenous Orchids*, Volume 1, Society for Growing Australian Plants, Sydney.

Duncan, B. C. & Isaac, G. (1986) *Ferns and Allied Plants of Victoria, Tasmania and South Australia*, Melbourne University Press, Carlton, Vic.

Dunlop, C. R. (ed.) (1987) *Check list of Vascular Plants of the Northern Territory*, Conservation Commission of the Northern Territory, Darwin.

Eichler, H. (1965) *Supplement to J. M. Black's Flora of South Australia*, 2nd edn, Government Printer, Adelaide.

Elliot, G. (1984) *Colour Your Garden with Australian Plants*, Hyland House, Melbourne.

Elliot, G. M. (1981) *Fun With Australian Plants*, Hyland House, Melbourne.

Elliot, G. M. (1985) *The Gardener's Guide to Australian Plants*, Hyland House, Melbourne.

Elliot, G. M. (1988) *The New Australian Plants for Small Gardens and Containers*, Hyland House, Melbourne.

Elliot, G. M. (1996) *Gwen Elliot's Australian Garden*, Hyland House, Melbourne.

Elliot, G. & R. & Kuranga Nursery, (1996) *The Kuranga Handbook of Australian Plants*, Lothian Books, Port Melbourne.

Elliot, R., (1975) *An Introduction to the Grampians Flora*, Algona Guides, Melbourne.

Erickson, R. (1958) *Triggerplants*, Paterson Brokensha Pty. Ltd., Perth.

Erickson, R. (1968) *Plants of Prey in Australia*, Lamb Publications, Osborne Park, WA

Erickson, R., George, A. S., Marchant, N. G. & Morcombe, M. K. (1973) *Flowers and Plants of Western Australia*, A. H. & A. W. Reed, Sydney.

Everist, S. K. (1974) *Poisonous Plants of Australia*, Angus & Robertson, Sydney.

Ewart, A. J., (1930) *Flora of Victoria*, Government Printer, Vic.

Fairhall, A. R. (1970) *West Australian Native Plants in Cultivation*, Pergamon Press, Sydney.

Floyd, A. G. (1960–) *NSW Rainforest Trees* (series) Research Notes, Forestry Commission of New South Wales, Sydney.

Flora of Australia Series (1981–) Australian Government Publishing Service (1981–94), Australian Biological Resources Study with CSIRO (1995–), Melbourne

Foreman, D. B. & Walsh, N. G. (eds) (1993) *Flora of Victoria*, Volume 1, Inkata Press, Melbourne.

Forest Tree Series (1970–74) Forestry & Timber Bureau, Canberra.

Francis, W. D. (1970) *Australian Rainforest Trees*, 3rd edn, Forestry & Timber Bureau, Canberra.

Fuller, L. & Badans, R. (1980) *Wollongong's Native Trees*, Fuller, Wollongong.

Galbraith, J. (1977) *Collins Field Guide to the Wild Flowers of South-East Australia*, William Collins, Sydney.

Gardner, C. A. (1975) *Wildflowers of Western Australia*, West Australian Newspapers Ltd., Perth.

Gardner, C. A. (1979) *Eucalypts of Western Australia*, Western Australian Department of Agriculture, Perth.

Gardner, C. A. & Bennetts, H. W. (1956) *The Toxic Plants of Western Australia*, West Australian Newspapers Ltd., Perth.

Gemmell, N. (1980) *Trees and Places; a South Australian Study*, Investigator Press Pty. Ltd., SA.

George, A. S. (1984) *An Introduction to the Proteaceae of Western Australia*, Kangaroo Press, Kenthurst, NSW.

George, A. S. (1984) *The Banksia Book*, Kangaroo Press, Kenthurst, NSW.

Gill, A. M., Groves, R. H. & Noble, I. R. (eds) (1981) *Fire and the Australian Biota*, Australian Academy of Science, Canberra.

Goddard, C. & Kalotas, A. (1988) *Punu, Yankanytjatjara Plant Use*, Angus & Robertson, Sydney.

Goodman, R. D. (1973) *Honey Flora of Victoria*, Department of Agriculture, Victoria.

Green, J. W. (1981) *Census of the Vascular Plants of Western Australia*, Western Australian Herbarium, Department of Agriculture, Perth.

Groves, R. H. (1981) *Australian Vegetation*, Cambridge University Press, Cambridge, England.

Guilfoyle, W. R. (about 1909) *Australian Plants Suitable for Gardens, Parks, Timber Reserves etc.*Whitcombe & Tombs Ltd., Melbourne.

Hacker, J. B. (1990) *A Guide to Herbaceous and Shrub Legumes of Queensland*, University of Queensland Press, St Lucia, Qld.

Hadlington, P. W. & Johnston, J. A. (1977) *A Guide to the Care and Cure of Australian Trees*, New South Wales University Press Ltd., Sydney.

Hall, N. (1972) *The Use of Trees & Shrubs in the Dry Country of Australia*, Forestry & Timber Bureau, Canberra.

Hall, N., Johnston, R. D. & Chippendale, G. M. (1970) *Forest Trees of Australia*, Australian Government Publishing Service, Canberra.

Handreck, K. A. & Black, N. D. (1994) *Growing Media for Ornamental Plants and Turf*, revised edn, New South Wales University Press, Sydney.

Handreck, K. A. (1993) *Gardening Down Under*, CSIRO, Melbourne

Handweavers & Spinners Guild of Victoria (1974) *Dyemaking with Australian Flora*, Rigby, Adelaide.

Harden, G. J. (ed.) (1990 – 93) *Flora of New South Wales*, New South Wale University Press, Sydney.

Harmer, J. (1975) *North Australian Plants*, Part 1, Society for Growing Australian Plants, Sydney.

Harris, T. Y. (1977) *Gardening with Australian Plants— Shrubs*, Thomas Nelson (Australia) Ltd., Melbourne.

Harris, T. Y. (1979) *Gardening with Australian Plants— Small Plants and Climbers*, Thomas Nelson (Australia) Ltd., Melbourne.

Harris, T. Y. (1980) *Gardening with Australian Plants— Trees*, Thomas Nelson (Australia) Ltd., Melbourne.

Hartley, W. (1979) *A Checklist of Economic Plants in Australia*, CSIRO, Melbourne.

Hartmann, H. T. & Kester, D. E. (1975) *Plant Propagation Principles and Practices*, 3rd edn, Prentice Hall, USA

Hauser, P. J. (1992) *Fragments of Green; an Identification Field Guide for Rainforest Plants of the Greater Brisbane Region*, Rainforest Conservation Society Inc., Bardon, Qld.

Hearne, D. A. (1975) *Trees for Darwin & Northern Australia*, Australian Government Publishing Service, Canberra.

Heywood, V. H. (1978) *Flowering Plants of the World*, Mayflower Books Inc., New York, USA.

Hillis, W. E. & Brown, A. G. (eds) (1978) *Eucalypts for Wood Production*, CSIRO, Melbourne.

Further Reading

Hockings, F. D. (1980) *Friends and Foes of Australian Gardens*, A. H. & A. W. Reed, Sydney

Holliday, I. (1989) *A Field Guide to Melaleucas*, Hamlyn Australia, Melbourne.

Holliday, I. & Watton, G. (1975) *A Field Guide to Banksias*, Rigby, Adelaide.

Holliday, I. & Watton, G. (1980) *A Gardener's Guide to Eucalypts*, Rigby, Adelaide.

Hyland, B. P. M. (1982) *A Revised Card Key to the Rainforest Trees of North Queensland*, CSIRO, Melbourne.

Hyland, B. P. M. & Whiffin, T. (1993) *Australian Tropical Rain Forest Trees — An Interactive Identification System*, CSIRO, Melbourne.

Isaacs, J. (1987) *Bush Food, Aboriginal Food and Herbal Medicine*, Weldons, Sydney.

Jacobs, M. R. (1979) *Eucalypts for Planting*, Food & Agriculture Organisation of the United Nations, Rome.

Jacobs, S. W. L. & Pickard, J. (1980) *Plants of New South Wales*, Government Printer, Sydney.

Jessop, J. P. (ed.) (1984) *A List of the Vascular Plants of South Australia*, 2nd edn, Adelaide Botanic Gardens & State Herbarium & the Department of Environment & Planning, Adelaide.

Jessop, J. P. & Toelken, H. R. (eds) (1986) *Flora of South Australia*, Parts 1–4, South Australian Government Printer, Adelaide.

Jones, D. L. (1993) *Cycads of the World*, Reed Books, Sydney.

Jones, D. L. (1987) *Encyclopaedia of Ferns*, Lothian Publishing Co., Melbourne.

Jones, D. L. (1988) *Native Orchids of Australia*, Reed Books, Sydney.

Jones, D. L. (1986) *Ornamental Rainforest Plants in Australia*, Reed Books, Sydney.

Jones, D. L. (1996) *Palms in Australia*, 3rd edn, Reed Books Pty. Ltd., Frenchs Forest, NSW.

Jones, D. L. & Clemesha, S. C. (1976) *Australian Ferns and Fern Allies*, A. H. & A. W. Reed, Sydney.

Jones, D. L. & Elliot, W. R. (1986) *Pests Diseases and Ailments of Australian Plants*, Lothian Publishing Co., Melbourne.

Jones, D. L. & Gray, B. (1988) *Climbing Plants in Australia*, Reed Books, Sydney.

Journal of the Australian Botanic Gardens (1976–) Adelaide Botanic Gardens, SA.

Keighery, G. (1979, unpublished) *Notes on the Biology and Phytogeography of Western Australian Plants*, Kings Park & Botanic Gardens, West Perth.

Kelly, M. (1989) *An Introduction to the Wildflowers of "The Millewa"*, Margaret Kelly, Millewa

Kelly, S. (1983) *Eucalypts*, Volume 1, revised edn, Thomas Nelson (Australia) Ltd., Melbourne.

Kelly, S. (1983) *Eucalypts*, Volume 2, revised edn, Thomas Nelson (Australia) Ltd., Melbourne.

Kenneally, K. F., Edinger, D. C. & Willing, T. (1996) *Broome and Beyond*, Department of Conservation & Land Management, Como, WA.

Kings Park Research Notes (1973–) Kings Park Board, Kings Park & Botanic Gardens, West Perth.

Kleinschmidt, H. E. & Johnson, R. W. (1979) *Weeds of Queensland*, Queensland Department of Primary Industries, Brisbane.

Landscape Australia (1979–) Periodical Journal of the Australian Institute of Landscape Architects, Landscape Publications, Melbourne.

Langkamp, P. J. (ed.) (1987) *Germination of Australian Native Plant Seed*, Inkata Press, Melbourne.

Lassak, E. V. & McCarthy, T. (1983) *Australian Medicinal Plants*, Methuen Australia Pty. Ltd., North Ryde, NSW.

Launceston Field Naturalists Club (1981) *Guide to Flowers and Plants of Tasmania*, M. Cameron (ed.) A. H. & A. W. Reed, Sydney.

Lazarides, M. (1970) *Grasses of Central Australia*, Australian National University Press, Canberra.

Lear, R. & Turner, T. (1977) *Mangroves of Australia*, University of Queensland Press, St. Lucia, Qld.

Leeper, G. W. (ed.) (1970) *The Australian Environment*, CSIRO & Melbourne University Press.

Leigh, J. & Boden, R. (1979) *Australian Flora in the Endangered Species Convention* – CITES, Australian National Parks & Wildlife Service, Canberra.

Leigh, J., Boden, R. & Briggs, J. (1984) *Extinct and Endangered Plants of Australia*, Macmillan Co. Aust. Pty. Ltd., Melbourne.

Levitt, D. (1981) *Plants and People, Aboriginal Uses of Plants on Groote Eylandt*, Australian Institute of Aboriginal Studies, Canberra.

Lord, E. E. & Willis, J. H. (1982) *Shrubs and Trees for Australian Gardens*, 5th edn, Lothian Publishing Co., Melbourne.

Loudon, W. (ed. Mrs. Loudon) (1855) *Loudon's Encyclopaedia of Plants*, Longman, Brown, Green & Longmans, London.

McCubbin, C. (1971) *Australian Butterflies*, Thomas Nelson (Australia) Ltd., Melbourne.

Macdonald, B. (1986) *Practical Woody Plant Propagation for Nursery Growers*, Timber Press, Portland Oregon, USA.

McLuckie, J. & McKee, H. S. (1962) *Australian & New Zealand Botany*, Horwitz-Graeme, Sydney.

McMillan Browse, P. D. A. (1979) *Hardy, Woody Plants from Seed*, Grower Books, London.

Maiden, J. H. (1889) *The Useful Native Plants of Australia*, Facsimile edn, Alexander Bros Pty. Ltd., Melbourne.

Marchant, N. G. et al (1987) *Flora of the Perth Region*, 2 Vols, Western Australian Herbarium, Department of Agriculture, Perth.

Mathias, M. E. (1976) *Color for the Landscape*, Brooke House Publishers, California, USA.

Mitchell, A. A. & Wilcox, D. G. (1988) *Plants of the Arid Shrublands of Western Australia*, University of Western Australia Press, Nedlands, WA.

Molyneux, B (1980) *Grow Native*, Anne O'Donovan, Melbourne.

Molyneux, B. & Forrester, S. (1988) *The Austraflora Book of Australian Plants*, Viking O'Neil, Melbourne.

Molyneux, B. & Macdonald, R. (1983) *Native Gardens, How to Create an Australian Landscape*, Thomas Nelson (Australia) Ltd., Melbourne.

Moore, L. B. & Edgar, E. (1980) *Flora of New Zealand*, Volume 2, Government Printer, Wellington, NZ.

Muelleria (1965–) Periodic Journal of the National Herbarium of Victoria, Royal Botanic Gardens, Melbourne.

Newbey, K. (1968–72) *West Australian Plants for Horticulture*, Parts 1–2, Society for Growing Australian Plants, Sydney.

Newbey, K. (1982) *Growing Trees on Western Australian Farms*, Farm Management Foundation of Australia (Inc.) Mosman Park, WA.

Nicholls, W. A. (1964) *Orchids of Australia*, Complete edn (ed. D. L. Jones & T. B. Muir), Thomas Nelson (Australia) Ltd., Melbourne.

Nicholson, H. & N. (1985–96) *Australian Rainforest Plants*, Parts 1–4, Hugh & Nan Nicholson, Terrania Rainforest Nursery, The Channon, NSW

Nuytsia (1970–) Bulletin of the Western Australian Herbarium, Department of Conservation & Land Management, Western Australia, Government Printer, WA.

Olde, P. & Marriott, N. (1994–95) *The Grevillea Book*, Volumes 1–3, Kangaroo Press, Kenthurst, NSW.

Pate, J. S. & Beard, J. S. (eds) (1984) *Kwongan, Plant Life of the Sandplain*, University of Western Australia Press, Nedlands, WA.

Pate, J. S. & Dixon, K. W. (1982) *Tuberous, Cormous and Bulbous Plants*, University of Western Australia Press, Nedlands, WA.

Pate, J. S. & McComb, A. J. (1981) *The Biology of Australian Plants*, University of Western Australia Press, Nedlands, WA.

Patrick S. J. & Hopper, S. D. (1982) *A Guide to the Gazetted Rare Flora of Western Australia*, Supplement 1, Department of Fisheries & Wildlife, Perth.

Paxton, J. (1868) *Paxton's Botanical Dictionary,* revised and corrected by S. Hereman, Bradbury, Agnew & Co., London.

Pearson, S. & A. (n.d.) *Plants of Central Queensland,* The Society for Growing Australian Plants, NSW Ltd., Sydney NSW.

Penfold, A. R. & Willis, J. L. (1961) *The Eucalypts,* Leonard Hill, London.

Plumridge, J. (1976) *How to Propagate Plants,* Lothian Publishing Co., Melbourne.

Prescott, A. (1988) *It's Blue with Five Petals,* Ann Prescott, Prospect, SA.

Pryor, L. D. (1976) *The Biology of Eucalypts,* Edward Arnold, London.

Pryor, L. D. (1981) *Australian Endangered Species; Eucalypts,* Australian National Parks & Wildlife Service, Canberra.

Pryor. L. D. & Johnson, L. A. S. (1971) *Classification of the Eucalypts,* Australian National University, Canberra.

Queensland Herbarium, Staff of the (1993) *Queensland Vascular Plants, Names and Distribution,* Queensland Department of Environment & Heritage, Brisbane.

Radke P. & A., & Sankowsky G. & N., (1993) *Growing Australian Tropical Plants,* Frith & Frith Books, Malanda, Qld.

Ralph M. (1997) *Growing Australian Native Plants from Seed for Revegetation, Tree Planting and Direct Seeding,* Bushland Horticulture, Fitzroy, Victoria.

Ratcliffe, D. & P. (1987) *Australian Native Plants for Indoors,* Little Hills Press, Crows Nest, NSW.

Raven, P. H., Evert, R. F. & Curtis, H. (1982) *Biology of Plants,* 3rd edn, Worth Publishers Inc. New York USA.

Rayment, T. (circa 1930) *Profitable Honey Plants of Australia,* Whitcombe & Tombs Ltd., Melbourne.

Rogers, F. J. C. (1971) *Growing Australian Native Plants,* Thomas Nelson (Australia) Ltd., Melbourne.

Rogers, F. J. C. (1975) *Growing More Australian Native Plants,* Thomas Nelson (Australia) Ltd., Melbourne.

Rogers, F. J. C. (1978) *A Field Guide to Victorian Wattles,* revised edn, F. J. C. Rogers.

Romanowski, N. (1992) *Water and Wetland Plants for Southern Australia,* Lothian, Melbourne.

Ross, J. H. (1996) *A Census of the Vascular Plants of Victoria,* 5th edn, National Herbarium of Victoria, Royal Botanic Gardens, Melbourne.

Rotherham, E. R., Briggs, B. C., Blaxell, D. F. & Carolin, R. C. (1975) *Flowers & Plants of New South Wales and Southern Queensland,* A. H. & A. W. Reed, Sydney.

Royal Horticultural Society, (1956) *Dictionary of Gardening,* 4 Volumes plus Supplement, 2nd edn with corrections, Oxford University Press, England.

Rye, B. L. & Hopper, S. D. (1981) *A guide to the Gazetted Rare Flora of Western Australia,* plus Supplement 1, Department of Fisheries & Wildlife, Perth.

Rye, B. L., Hopper, S. D. & Watson, L. E. (1980) *Commercially Exploited Vascular Plants Native in Western Australia: Census, Atlas and Preliminary Assessment of Conservation Status,* Department of Fisheries & Wildlife, Perth.

Sainsbury, R. M. (1987) *A Field Guide to Isopogons and Petrophiles,* University of Western Australia Press, Perth.

Salter, B. (1977) *Australian Native Gardens and Birds,* Ure Smith, Sydney.

Sanders, W. T. (1949) *Sanders' Encyclopaedia of Gardening,* revised by Macself, A. J., W. H. & L. Collingridge Ltd., London.

Sharr, F. A. (1978) *Western Australian Plant Names and their Meanings,* University of Western Australia Press, WA.

Simon, B. K. (1978) *A Preliminary Check-list of Australian Grasses,* Botany Branch, Department of Primary Industry, Brisbane.

Simon, B. K. (1990) *A Key to Australian Grasses,* Queensland Department of Primary Industry, Brisbane.

Further Reading

Simpfendorfer, K. J. (1975) *An Introduction to Trees for South-Eastern Australia*, Inkata Press, Victoria.

Smith, A. W. & Stearn, W. T. (1972) *A Gardener's Dictionary of Plant Names,* Enlarged edn, Cassell & Company Ltd., London.

Snape, D. (1992) *Australian Native Gardens, Putting Visions into Practice,* Lothian Books, Port Melbourne.

Society for Growing Australian Plants, *Australian Plants, Quarterly,* Society for Growing Australian Plants, Sydney.

Society for Growing Australian Plants, Canberra Region (1976) *Australian Plants for Canberra Gardens,* 2nd edn, Society for Growing Australian Plants, Canberra.

Society for Growing Australian Plants, Keilor Plains Group, (1995) *Plants of Melbourne's Western Plains, A gardener's guide to the original flora,* Society for Growing Australian Plants, Keilor Plains Group, Niddrie, Victoria.

Society for Growing Australian Plants, Maroondah Group Inc., Vic. Miscellaneous publications, Society for Growing Australian Plants, Maroondah Group Inc., Ringwood Vic.

Society for Growing Australian Plants, Maroondah Group Inc., Vic. (1993) *Flora of Melbourne, A Guide to the Indigenous Plants of the Greater Melbourne area,* revised edn, Hyland House, South Melbourne.

Society for Growing Australian Plants, Western Suburbs Branch, Queensland (1979) *Perfumed and Aromatic Australian Plants,* Society for Growing Australian Plants, Western Suburbs Branch, Indooroopilly, Qld.

Sparnon, N. (1967) *The Beauty of Australia's Wildflowers,* Ure Smith, Sydney.

Spencer, R. (1995) *Horticultural Flora of South-eastern Australia,* University of New South Wales Press, Sydney.

Specht, R. L. (1972) *Vegetation of South Australia,* 2nd edn, Government Printer, Adelaide.

Spooner, P. (ed.) (1973) *Practical Guide to Home Landscaping,* Readers Digest, Sydney.

Stanley, T. D. F. & Ross, E. M. (1983–86) *Flora of South-eastern Queensland,* Volumes 1–2, Qld Department of Primary Industries, Brisbane.

Stead Memorial Wildlife Research Foundation, Miscellaneous publications, Stead Memorial Wildlife Research Foundation, Sydney.

Stearn, W. T. (1973) *Botanical Latin,* 2nd edn, David & Charles, Newton Abbot, England.

Stearn, W. T. (1992) *Stearn's Dictionary of Plant Names for Gardeners,* revised edn, Cassell. London.

Stones, M. & Curtis, W. (1967–78) *The Endemic Flora of Tasmania,* Parts 1–6, The Ariel Press, London.

Telopea (1975–) Contributions from the National Herbarium of New South Wales, Government Printer, NSW.

Thomas, M. B. & McDonald, W. J. F. (1987) *Rare and Threatened Plants of Queensland,* Department of Primary Industries, Brisbane.

Tothill, J. C. & Hacker, J. B. (1973) *The Grasses of Southeast Queensland,* University of Queensland Press, St. Lucia, Qld.

Thompson, R. P. (1991) *Water in Your Garden,* Lothian Books, Port Melbourne.

Van der Sommen, F. J., Boardman, R. & Squires, V. (Eds) (1983) *Trees in the Rural Environment,* Natural Resources Faculty, Roseworthy Agricultural College & Institute of Foresters of Australia, SA Division (Adelaide Branch).

Walsh, N. G. & Entwisle, T. J. (eds) (1994, 96) *Flora of Victoria,* Volumes 2 & 3, Inkata Press, Melbourne.

Wells, J. S. (1955) *Plant Propagation Practices,* Macmillan, New York, USA.

Western Australian Herbarium Research Notes (1978–85) Western Australian Department of Agriculture, Perth.

Wheeler, D. J. B., Jacobs, S. W. L. & Norton, B. E. (1982) *Grasses of New South Wales,* The University of New England, Armidale, NSW.

Wheeler, J. R. (ed.) (1972) *Flora of the Kimberley Region,* Department of Conservation & Land Management, Perth.

Whibley, D. J. E. (1980) *Acacias of South Australia,* Government Printer, Adelaide.

Williams, K. A. W. (1979–87) *Native Plants Queensland,* Volumes 1–3, K. A. W. Williams, North Ipswich, Qld.

Willis, J. C. (1973) *Dictionary of the Flowering Plants and Ferns,* 8th edn, revised by Airy-Shaw, H. K., Cambridge University Press, London.

Willis, J. H. (1962–72) *A Handbook to Plants in Victoria,* Volumes 1 & 2, Melbourne University Press.

Wilson, G. (1975) *Landscaping with Australian Plants,* Thomas Nelson (Australia) Ltd., Melbourne.

Wilson, G. (1980) *Amenity Planting in Arid Zones,* Canberra College of Advanced Education, Canberra.

Womersley, J. S. (ed.) (1978) *Handbook of the Flora of Papua New Guinea,* Volume 1, Melbourne University Press.

Woolcock, C. & D. (1984) *Australian Terrestrial Orchids,* Thomas Nelson, Melbourne.

Wrigley, J. W. & Fagg, M. (1988) *Australian Native Plants,* 3rd edn, William Collins, Sydney.

Wrigley, J. W. & Fagg, M. (1989) *Banksias, Waratahs and Grevilleas,* William Collins, Sydney.

Wrigley, J. W. & Fagg, M. (1993) *Bottlebrushes, Paperbarks & Tea Trees,* Angus & Robertson, Sydney.

Zacharin, R. F. (1978) *Emigrant Eucalypts, Gum Trees as Exotics,* Melbourne University Press, Melbourne.

Zimmer, G. F. (1956) *A Popular Dictionary of Botanical Names and Terms,* Routledge, Kegan & Paul Ltd., London.

Common Names Index

Common Names Index

Common Names Index

Common Names Index

I heard my name ⟨...⟩ ⟨...⟩ ⟨...⟩ ⟨...⟩ ⟨...⟩ later Uncle Zavier appeared in the bathroom doorway. By now I had slipped under the mountain of foam like they do in the soap ads. I glared at him.

'Do carry on,' he said, 'it's not every day I get the chance to watch a beautiful guest at her ablutions.'

I cut him short: 'It's nice to be admired, but I wonder if you might be developing into a dirty old man, Uncle.'

I began soaping myself, and he watched with a gleam in his eyes as I slipped the soap cake round my breasts and rubbed my nipples.

'I'm working on it,' he replied . . .

LAURE-ANNE

The story of Laure-Anne D. as told to
Nicholas Courtin

A STAR BOOK

published by
the Paperback Division of
W.H. Allen & Co PLC

A Star Book
Published in 1986
by the Paperback Division of
W.H. Allen & Co. plc
A Howard and Wyndham Company
44 Hill Street, London W1X 8LB

Copyright © Nicholas Courtin 1986

Printed and bound in Great Britain by
Anchor Brendon Ltd, Tiptree, Essex

Typeset in Plantin by Fleet Graphics, Enfield, Middlesex

ISBN 0 352 31735 3

Chapter One

The night train to Evian was rocking me gently in my couch-ette. In the normal way of events I would have liked it, the imperceptible quivering through my body as I lay passively in the sack-like sheets. But this time I was slowly going mad.

Running away from home into the arms of a gambler banned from every casino in France and Monaco, was the most exciting decision of my whole 20 years. It was a brave thing to do, it had flair, it gave me pride of achievement such as I had never experienced.

It was also scaring me out of my wits.

I was dreading waking up on arrival on the French shore of Lake Geneva, scared that this could turn out to be the crum-miest of all the fiascos I had encountered in a life that had never been anything but deadly dull.

But I would not awake, for the simple reason that I could not sleep. Between the moment we left the Gare de Lyon at 23.05 that Friday, and midnight, I had lain glumly listening to herds of tough guys and girls trading jibes and clumping about with skis and hold-alls. Seeking relief from this, I had tried to take an interest in the magazine I bought at the bookstall ('Widow-hood Without Shame', 'Across Greenland By Sled', 'The Truth About the G-Spot' . . .) and had abandoned the attempt.

Now I lay weak, trembling and perspiring. Once again at 3 a.m. I rehearsed the dialogue I would have with Uncle Xavier.

He was not a real uncle, but a remote relation of my

mother's by marriage. Nevertheless my future was in his hands. He might decide to smack my bottom (oh, he'd love that!) and promptly phone home, returning me to Evian station like a parcel delivered in error. On the other hand only 18 months earlier he had been my host, and that might conceivably colour his view. Our parents had packed my sister Sylvie and I off to Uncle Xavier and Aunt Ginette for a holiday. I had never been to the Club des Ducs near Evian, and was astonished at the size of the place. That they had money I knew, but was I tremendously impressed all the same. I naively assumed that they had actually paid for the *fin de siècle* hotel with its large annexe containing conference facilities, saunas, keep fit rooms, disco, piano bar and so on, the whole affair set in vast woody grounds for the delectation of the international glitterati. Uncle Xavier had flirted with crime in some way, and I knew about the casino ban from family talk. I assumed he was the owner of the Club des Ducs, instead of which he was only the manager with a small share in the group that owned it.

Sylvie and I were allowed the run of the premises and the park. We used the keep fit facilities, danced in the disco, and in my case sat in on a marketing conference run by a computer peripherals company. Most of the time I was as quiet as a mouse, and spent a while strolling around the golf course and watching people play tennis. It was a glimpse into the world of wealth and big business, and I was saddened to return to the toy boys, as the girls called the male students at the faculty.

This was really what caused me to flip, the contrast of it all. I was dissatisfied with my law studies, and could see myself settling down in a solicitor's offices, eventually marrying some reliable middle class fellow, having brats and contributing some of the next generation of lawn-mowing people. The fatality of it appalled me. The Law itself impressed me not one bit; I saw legal practice as a huge racket, a way of making money out of confused legislation as it hit ordinary folk and struggling businesses. The profession – incidentally third on the list for tax fraud – disgusted me before I even started. I refused to be cynical and consequently pursued my work under protest. School had been boring and university was as bad. I was close to deciding that education as we know it is downright immoral; at least if examiners set the question 'Should education be banned?', it would make the candidates think, even though the examiners might be past it!

All was not gloom and darkness, however. You cannot have a truly sick mind in a healthy body. The vigour of youth produced its own moments of delight: walking in the Luxembourg Gardens on a bright fresh day, sitting on café *terrasses*, gazing in shop windows on a rainy evening, running for the 6.35 train to the suburbs with giggling friends and just catching it as the station staff scolded us, the fragrance of flowers and fruit, the rich smell in the pork butcher shops.

Now I had fled, quit home in the first big, instinctive decision I had come to. I had simply woken up two days ago no longer able to stand my dependence on the family, resolved that I must fly from the nest. I had cash enough for a trip to Evian, and put a few things in a travel bag.

A train rushed by the other way, turning my stomach over with its suddenness. I was rumbling inside anyway, and I would soon need to go to the toilet. This would involve the awful business of struggling back into the skirt I had thought best to wear for Uncle Xavier's benefit, then taking all my things with me for fear they would be stolen, then quietly climbing down the ladder thing hoping the boy underneath wouldn't grope me, then easing the door open, and so on and so on until I was back in the couchette again. To make matters worse, I was in a compartment over the wheels, having got the couchette at the last moment.

I was torn between misery and anger, convinced that once again some mysterious powers were stopping me getting the most out of life. I gritted my teeth, if the worst came to the worst I would be sent back and there would be a hell of a row but I would have brought the festering sore to a head. At times during the journey it even occurred to me that the family might be glad to see the back of me. It was hard to tell.

7

My curt note had read:

Dear Parents,

After lengthy reflection, I have decided to leave home and abandon my studies. I know this will cause you pain and I am sorry to do it without warning, but I cannot face a long fruitless argument about it. Do not worry, I have money and my savings book. I have no intention of doing away with myself or taking drugs. Do not wring your hands over me, I will make out.

Love

Laure-Anne

Love? I wasted little of it on my parents. Love had done little to make my world go round; it was hate and smouldering anger that had finally spurred me into action.

Eventually I made the expedition along the corridor and back, by which time there were still more than three hours until 07.43 when the *Société Nationale des Chemins de Fer Français* would deliver me on time, the best-run train service in the world and the only public service in France that works properly, except for the fire service.

At Evian I slouched out of the station, griping with hunger, thirsty and unwashed, feeling as if I had jumped from adolescence to menopause overnight. Several *croissants* and a litre of coffee later, I glanced out of the café to see a minibus pulling into the station concourse. Discreetly painted on the door was the name *Club des Ducs*. I was out there in a flash.

'You going straight back now?' I yelled.

'Collecting newspapers, then I'm off,' the man said.

'I'll get in now.'

Which I did, sitting on the front bench as the driver went off for his packages. There were two others on board already, cleaning women I supposed. The seat was cold and I wriggled about dusting off my clothes so that I looked less of a waif. Nobody spoke.

Returning, the driver said: 'Come by the 7.43?'

'Yes.'

'You must have missed my last trip.'

'I had a bite to eat. Do you know if Monsieur Simmonnier is at the hotel this morning?'

'Should be, hardly ever away Saturdays.'

For all I cared Uncle Xavier could be an ex-killer, I liked him. In his own way he was rather romantic, built on short broad Mediterranean lines, which made him a kind of Aristotle Onassis in my eyes. His wife Ginette was a gentle soul who did what he told her, not quite the Jacqueline Kennedy type.

At the reception desk I declared: 'I would like to see Monsieur Simmonier please.'

'Have you an appointment?'

'It's a surprise, my name's Laure-Anne, he'll be glad to see me.'

A minute in which to study the tastefully-appointed foyer, and he walked smartly out of a wall and agreed with me: 'Well, this is a surprise, I didn't know you were coming to these parts, my little L'Anne.'

We embraced fondly: 'Nor did I until yesterday, it was a snap decision. Perhaps we could go into your office and I'll tell you all about it.'

We transferred from the hall to his cubby-hole, where he briefly dismissed a waiter he had been talking with; I later learned he was the *maitre d'hôtel*.

'Right, I'm all ears.'

'I've reached rock bottom, uncle, and I've left home and university,' I announced. His expression did not change. 'That doesn't surprise you?'

'No, plenty do it.' His eyes ran me up and down, terminating at my bosom. I forced my own blinkers open to stop myself blushing. I clamped my back teeth, it was my move next, and I launched into a confused list of reasons for my flight, ending with the statement: 'So I'm starting out on a new life, even if it kills me.'

A slit appeared in his face: 'So what am I supposed to do about it?'

'I want you to hide me here for a bit.'

9

'That's possible, but why should I? Your mother will never forgive me, I have to reassure her, it's my duty.'

'You have no choice. If you don't take me in, I'll tell her you touched me up last year, that you perverted me.'

The slit burst open and he roared with laughter. It went on for a full 20 seconds and he brought a handkerchief to his tear-filled eyes: 'My God, I haven't had a good laugh like that for years! What did I do?'

'You put your hand up my skirt, you squeezed my thing.'

Another bout of heaving and snorting, and then: 'So you want a bolt-hole. How many weeks is "a bit", and what do you intend to do after that?'

'I don't know, but I'll think of something a couple of days from now. Oh come on, Uncle, be decent, let me stay for a while. Please. You can have another feel, I won't give you away.'

Again he exploded into mirth: '*You* won't give *me* away! You've certainly got a cheek, I like your pluck.'

'I've got nothing to lose. It's this hotel or robbing a bank.'

His eyes popped: 'Don't make jokes like that, it makes me nervous.'

I rose from my chair and went to his side, ruffling his hair: 'Come on, for old times' sake, say yes.'

He calmed down: 'All right, L'Anne, we'll give it a try. Is that all the luggage you have? I'll get you fixed up.'

He stabbed at an intercom button and rasped: 'Madame Colin, I have a special friend, Mademoiselle D., who's turned up unexpectedly. She'll want a room on the house, what have we available?' The transaction was swiftly completed, and then he gave orders for me to have free meals, buzzing through to the *maitre d'hotel*.

'You'll eat with us at our table this evening, we go in early. Meanwhile I'll let you settle in. When did you last get any sleep?'

'About last Tuesday, by the feel of it.'

Shortly before noon next day, I returned from a long walk and flung off my clothes for a bath. I had slept for much of that Saturday, and again that night. Now I felt on top of the world.

Mine must have been one of their best rooms, snugly heated and impeccably furnished in an unusual wine hue that was exactly right for the dove grey walls. The bathroom was in grey, black and primrose, and had everything including a jar of bubble bath powder which I added in generous quantity to the over-hot water I ran from the luxurious mixer tap. I swooshed the water about me and it was starting to foam magnificently. The walk in the cold had stirred my blood and given me an appetite. I had even gone sentimental about a church bell in the distance. I was now a rosy, warm Laure-Anne, revelling in the scented soapsuds and looking forward to a good laze in the bath.

I stopped agitating the water and heard my name being called in the room. I stiffened, and a second later Uncle Xavier appeared in the bathroom doorway.

'Uncle Xavier! I locked the door, how did you get in?'

He looked to heaven: 'L'Anne, that's not worthy of you. Every hotel manager has pass keys.' I frowned and he excused his entry: 'I caught sight of you when you came back and knew you were in your room. I knocked but there was no reply, so I let myself in.'

'Just like that.'

'You might have fainted or something.'

By now I had slipped under the mountain of foam like they do in the soap ads. I glared at him.

He glanced politely at the wall and told me: 'Laure-Anne, I would like a little private talk with you. We have to sort something out.'

I jerked into the sitting position: 'You haven't told them I'm here!'

'Of course not, I promised remember?'

I beamed at him: 'Thank you, Uncle Xavier, I knew I could count on you.'

A pause and then he said: 'Do carry on, it's not every day I get the chance to watch a beautiful guest at her ablutions. Not so long ago I saw you like that . . . '

Covering my breasts, I cut him short: 'It's nice to be admired, but I wonder if you might be developing into a dirty old man, Uncle.'

'I'm working on it.' Not to lose face, he assumed a serious expression and went on: 'No, what I want to discuss is your future. To be quite open about it, what do you propose to live on? Money – if you see what I mean?' He sat down on the stool.

I began soaping myself, and he watched with a gleam in his eyes as I slipped the soap cake round my breasts and then rubbed my nipples. My breasts were firm, undoubtedly one of my better attributes. I was puzzled as to how I could do the rest of myself elegantly if he persisted in going on for any length of time about finances.

With this in mind, and because I was about to pop the winning question anyway, I twittered: 'It's so difficult thinking about money in a hot bath. I feel wonderfully warm and carefree. But I have actually been thinking about it this morning. How would it be if you found me a little job here? You employ dozens of people, and one more wouldn't make much difference, would it?'

He gaped and answered quickly: 'That's not quite the way it works. If I were really the owner here, I'd jump – I mean there'd be no trouble, but the company has ordered no recruiting without their direct consent. The crisis, you know. We all have to cut back on spending.'

I looked at him as innocently as I could: 'Then keep me until someone leaves, and then you can give me her job and no-one will know any different.' To help things along I rose from the now-creamy water to soap my bottom and between my legs with the washglove. 'I congratulate you on your colour scheme, it's superb; even the washglove is a lovely primrose, and so delightfully soft.'

He looked at the washglove, which was exactly what I intended, and he coughed: 'It's not that easy, the company likes to send down people of its own for jobs, girls it has trained up.'

'Oh don't give me that, you can fix it, Xavier.' I left out the 'Uncle' deliberately, but hoped I was not being too much of a tart. Overdoing it would be fatal.

'Sorry, *ma petite cherie*, there's no scope . . .'

I was now busy with the shower, as elegantly as I could: 'You're not sorry at all, you just find me a nuisance here. I embarrass you with what you did to me and what I know about the casinos and everything.'

I was shooting in the dark, moving into deep water. I pulled the bathplug and replaced the shower thing.

'What do you know about the casinos?'

'More than you think,' I lied.

I stared at him, hoping he would not take me up on it, and then changed tactics: 'Listen, Xavier, I've always held you in deep affection. After all, it was you who initiated me sensually, in a manner of speaking. I can't forget that at an intimate moment like this.' I stepped out of the bath, keeping my front towards him. 'Let's do a deal.'

'A deal?'

'Yes, you can dry me with the towel if you give me a job, and then we can keep our secret, stay real good friends.'

He rose from the stool and paced back and forth in the restricted area next to the bath, adjusting his trousers once or twice, pulling at his belt and turning away. I stood quite still; this would have to be his decision.

'Dry me, I'd like that,' I whispered, reaching over for the huge primrose bathtowel and handing it to him. His jacket was off in an instant and he was rubbing me awkwardly down the back and sides. I rotated and he dealt with my rear, then my breasts. I went, 'Mmmm'. This was the moment of truth, and as he patted my midriff I opened my legs a little. Would he respond?

His sallow face took on a light crimson hue and I said: 'Why, you're flustered, I'm just a girl like any other, you must have known lots. Oh, you're adorable, I do like you so much.' I carefully draped my arms round his neck and placed my head against his cheek, the towel covering my whole torso. His arms were round me at once and he croaked: 'Not like any other.' I was overjoyed to feel his hard member pressing against my belly through the material.

And then his mouth was on mine, and I was returning his kiss. There was no going back for him, he had committed himself, and I gave him a second kiss during which I shifted my lower body into a position that spoke for itself.

He made his decision and within seconds we were flinging back the bed cover and the sheet and the blankets. With astonishing alacrity he slicked his trousers off. Not bothering with his tie, he helped me onto the bed and stood for a moment staring at me darkly. I turned my head demurely to see his crumpled black suit on the floor.

'The door,' I whispered. I was pleased to see that his shirt had a straight edge; I could not have watched him scurry over with flapping shirt-tails to lock the door.

'It's fine,' the old fox said gruffly, fingering my neck, 'you have a splendid body.'

His jutting dark blue underpants were in some thin man-made fibre and I could see his rod outlined within. He whipped the garment off and his thick organ stood up firmly.

I caught my breath. I was not playing any more, not trapping him. I was in earnest and I felt my jaw drop in wonder at the experience I was about to undergo.

'Xavier, wonderful wonderful Xavier, take me now right away. Quickly. I know I am ready!' Hardly conscious of the further preliminaries, I felt him at my entrance. With a single push he forced through my barrier. I closed my eyes, wondering how far he would go inside, what it would be like.

For this was my first time, the real thing at last. I was coming already! The few thrusts he gave took me into another world; a slight stab and then this rigid flesh like gristle slipping in slowly and out again. I climaxed with a corner of the pillow stifling my cry. He followed within seconds, but all I could feel was a hotness.

14

Later he thanked me with gallant sincerity, and I told him it was the first time. He gaped in disbelief, and slowly separated from me. I persisted, saying that no-one could have done it more expertly, and that I was glad it was him. It was the most heavenly experience a girl could wish for. This convinced him and filled him with awe.

He was silent for a while and then declared: 'It's a deal, L'Anne, I'll find you a job.'

'You won't regret it,' I said with a final kiss. And meant it.

Chapter Two

My sexual experience had been pretty much of a shambles from the moment I entered university, which I took to like a duck to the Sahara desert.

In the early seventies we were still recovering from the May '68 national upheaval, the period when students by their tens of thousands battled nightly in the Latin Quarter with the police, and the nation's workforce crossed its arms. I had not been involved, however.

A year after the turmoil the De Gaulle Government consolidated its position, but one of his referendums produced a *non* vote and *Mongeneral* swept out of the Elysée Palace, never to return. Georges Pompidou took over the Presidency from 'the most illustrious of Frenchmen' and everybody settled for sullen acceptance. It lasted until 1973-74 when oil rates soared, and the West had to deal with a new balance of power. *Pompom* died amid great suffering and Valery Giscard d'Estaing became Head of State for seven years until 1981.

All this took a good deal of digesting and the political groups in the university jostled for position. The Liberal Right was firmly in the driving seat and the students were wooed conscientiously by the Left and the Far Right as they sought to undermine our complacency. Personally, I was worried and fought to believe that the economy was not out of control, for I was an avid reader of economic and business news. Even so I kept within my shell, endeavouring to come to grips with my studies. I merely scanned and threw away most of the political

16

literature we were handed, as it lacked any pretensions to a credible programme. But the *Parti Communiste Français* was in a category of its own, clearly the only force strong enough to bring about genuine change. Groups of communists regularly sold *L'Humanité* outside the faculty, and I bought it quite often. I also took out a subscription to the theoretical monthly *Cahiers du Communisme*, which would arrive at the apartment, triggering constant attacks from my catholic family, against whom my equally firm argument was that I sought to learn 'the other view' too.

The upshot of which was that I found myself at a communist party dance, where I met a promising young man called Clovis, a name I never knew existed outside the history books. He was the first boy who made me go weak at the knees, appearing at the door with a bunch of people and scrutinizing us all with piercing pale blue eyes, pushing back his luxurious dark hair from a pale forehead with the palm of his hand. He wore a grey suit and mustard tie and looked like he would pull a gun if he got any lip. He was a leading man sort of fellow. My partner of the moment was tangling us up in an intricate movement and I had to concentrate. He also had his hand well down my back and I knew my skirt was hitched up. I had had enough of him and excused myself on grounds of fatigue.

It was some time before I got to meet Clovis. We did a cross between a tango and a ballet dance, making a joke about it before leaving the floor to begin a complicated parley about what he did in the Party. By eleven o'clock I was sweaty and dizzy. I informed Clovis that I needed air and he collected some people for a walk to the Place de la République where there was a hot snacks place. We were three couples and we linked arms two by two. I was being accepted, and a youngster is always glad of that. As we stood at the snack bar, the others necked and Clovis put his hand on my shoulder. We were chatting away and I suddenly realized that the others had gone.

'We'll find them back at the dance,' said my leading man. 'Let's go for a stroll around.'

I could think of nothing more delightful, and we linked arms

17

round the waist. He was no longer the tough guy but simply a competent considerate youth any mother would have approved of, especially as he 'worked at a bank' although he failed to say at what.

Romance was in the air and some of the stars overhead had just been born, there were wings on my heels, and I knew that the crisp night was bringing a flush to my cheeks. Even the traffic lights took on new significance. He was such a handsome boy, probably about 22, and I was 18, a girl like any other walking along with her chap. I was mellow with joy and knew that the first kiss would be devastating.

We entered a minor street lined with shops and I kept stopping to look at shoes and clothes. Clovis guided me into a largish doorway with a twin way-in round a pillar thing. This was to be the great moment for the little *bourgeoise* from the suburbs. He pushed me back against the inner side of the pillar. Neither of us spoke. His hands cupped my head and I put up my lips to be kissed. He gave me a peck that lasted all of two seconds and I draped my arms round him with the message: 'You're not going to get away with that.' I was Anouk Aimée in *Un Homme Et Une Femme* with my coat collar prettily cradling my hair. I had said it so suavely! But it was I who kissed him properly, putting every ounce of tenderness I possessed into this first real embrace with a boy. I was so busy being glamorous that I did not even notice his reaction, knew only that I was parting my lips and advancing the tip of my tongue. He seemed reserved, which delighted me, and I increased the pressure and withdrew, doing it several times. He was unresponsive, but he was unbuttoning my coat. Of course, it would be much more of a clinch that way! Then his hands were on my bottom, then on my breasts and finally playing with my skirt hem. I was so amazed at this fast turn of events that I instinctively clamped my knees together, and was about to bring my arms down when he jammed a foot between my shoes and forced my legs apart. He was really quite rough about it, but I told myself he was the masterful type, knew what he was doing, and so much the better for that.

18

'Don't move, I won't hurt you,' he stated. This terrified me: who said anything about hurting? I had my arms down in microseconds and pushed at his waist to curb his ardour. Already his zip was undone and he was hoiking his shaft out above his underpants.

'Take it easy,' he growled, 'I only want a quick feel. You know you want it as much as I do.' Well, not like that I didn't. 'Let's stay like this for a while,' he added.

'I'm not on the pill, so you can't do it,' I said in a brittle high-class voice. True, if I made a fuss I would never live it down and it would be all over the university. On the other hand I did not want a baby. Humouring him I said: 'You take it easy too, we can work it out from now on.'

His hands flashed to my panties, however, rising directly to the target and he pressed my quim. Deftly he slipped some fingers under the elastic and touched me. I was now sure he would try to poke me and I wanted to get away. He withdrew his fingers, then put them in his mouth and returned them wet with spittle. He rubbed my clitoris, and only then did I begin to experience arousal. I was in a quandary; in the dim light I saw his penis, the length of it frightening me; on the other hand I could not take my eyes off it, the first full-size one I had ever seen. I stared at it in wonder and he guided my hand to it, fastening my fingers round, and I found it warm. His expert thumb was slithering about inside me, and I yielded myself up to the pleasure because I wanted him to get it off for me. To encourage him I moved my hand up and down his cock, and bestowed little bites around his mouth. Oh the years I had wasted compared with all those girls back in the dancehall! This was the most exciting fun imaginable!

'When was your last period?' he demanded.

'No,' I yelped, 'it's too risky, please don't put it in, please Clovis, please!'

He was hissing through his teeth and I went on: 'I like you Clovis, I think you're terrific, but we can do it properly next time.'

Much good did it do me, he was going to get what he could

right now! He whipped my briefs down to my knees, grabbed his penis and groaned. My skirt fell down into place and his gunk spurted over the outside of it.

For a moment we were both immobile, and then I grabbed my panties up. He got a handkerchief out and started dabbing at the fluid.

'Don't you dare touch me,' I said thickly. 'Give me the handkerchief.'

'Sorry,' he said, zipping himself up.

My eyes smarted: 'It's too late, you've ruined my skirt.'

'You wanted it.'

'You spoiled everything, why couldn't you wait?'

Still turgid, I sobbed in humiliation. He had stolen my first big thrill, and it ended with me scrabbling about in my bag for my own hankie.

'Take me back to the dance, please,' I snapped, vowing that never again would I engage in this kind of heavy petting. The clamminess between my legs as we hastened along added to my misery.

There was this much to thank him for: I slept badly, but had nice erotic dreams too. From then on I looked at men in a different way, thinking of their pricks as they ran for the Metro or stood close to me in the packed morning train.

At university there was a fair amount of horseplay. I bought a rust-coloured boiler suit which I was rather proud of, with a big zip right down the front, which was duly undone by the jokers once a day on average. My posterior was pinched as frequently, and I suppose on the whole I deserved it, for I was trying out my powers of seduction, as girls do. I was taken to see films and was pawed, squeezed and fingered, but none of the boys got any further, for I was scared of the adverse reports about the pill and avoided using it. More than once I was snarled at as a cock-teaser. Worse luck, I never grew fond of anyone, but I must have been a dreadful turn-off. The lads were only trying their luck.

The result of all this was a permanently pent-up existence

for me, a state of continuous semi-arousal with no relief except in the privacy of my own bed. Instinctively I was looking for the man who would deflower me, and meanwhile I was an utter prig. Even today I grow red at the thought of my own conceit.

Thus I was ripe for Uncle Xavier. But before I continue with my adult experiences, a brief word about the family.

We lived in a large modern flat in the West Paris suburb of Marnes La Coquette, with a lounge, the parents' room, a room for my other brothers Yves and Jean-Marie, another for myself and Sylvie who is 18 months older than me. My father, Charles D., was and indeed at the time of writing still is, an engineer with the *Compagnie Générale de Génie* and frequently went abroad to install equipment. Before marrying him, Adrienne had been a secretary of some kind in the same Group. He earned good money and lived much of the time on the firm, so that none of us wanted for anything. We ate well and lived amid costly furniture and furnishings. In turn we children went to a nearby private school, as we were practising catholics.

Experts are prone to refer to one's early formative years, but in my case I simply drifted along. None of us was actually killed, robbed, set fire to, crashed into, thrown out of a job or tortured. So we were a happy family in that sense. But dreary; one of those comfortable, humble, quietly-toiling households with nothing to be ashamed of, or proud of either. If we had had a lawn we would have mown it every Saturday from 10 to 11 a.m.

I grew to loathe this semi-affluent Lawnmower Society fairly soon. Meanwhile I was a keen reader, as we had hundreds of books in the flat. Possibly because of my father's profession, I liked poring over the boys' how-things-work volumes, fascinated by the illustrations. Encyclopaedias were another abundant source of joy with their pictures of cement works, cableships, combine harvesters and God knows what else. I was a perfect pest with my questions, and was often summarily dismissed when people were busy. I learned to read fast, I had a motive, and years later scored 138 in an IQ test. At school I was told I could be brilliant if I tried; I preferred to scour the magazines at home, my favourites being the weekly news magazine *Le Point* and the business periodical *L'Expansion*. These contained a wealth of facts and figures on production, exports and imports, wages and strikes, new technologies, marketing, takeovers and what-not, which hardly anyone cared about at the time. They were to wake up when the crisis got under way, but I was ahead of them, greedily devouring this news while the others went for strip cartoons and TV mags. I learned out of school, and it is not entirely surprising that my daily paper today is – wait for it! – the *Wall Street Journal*.

Chapter Three

I was no longer a virgin, but that was only the first of the change in my fortunes during the first two weeks at Evian.

The next day around 10 a.m., when the establishment was ticking over to the clink of tableware and the whine of a vacuum-cleaner, I slunk through the main hall intending to go for another of my walks. I pushed against the revolving door and felt someone shoving it round. It was Uncle Xavier.

'Good morning L'Anne. If you're going for a stroll I'd like to join you for a couple of minutes. I have news for you.'

I looked at him as if he was the angel Gabriel. He was quite an ugly runt of a man in reality, a craggy version of the singer Charles Aznavour in shape and style. But I was still in a post-orgasmic haze and simply saw him as the person who had given me the sole moment of utter bliss so far.

'You look as if you slept well,' he began.

'Ten hours.' He was right to notice. Only five minutes earlier the mirror had told me I bloomed, radiated a glow. I was a woman at last!

'I slept six hours, don't need much. I – er – have decided there is only one job I can find for you: a kind of extra hostess. I won't go into how I can fix it, that doesn't concern you, but you will be on the payroll like everyone else, with social security cover and everything else official.'

'But that's absolutely marvellous,' I cried, grabbing his arm.

He calmed me down, flapping his hands: 'However, I shall have to take you on as a learner, an apprentice if you prefer to put it that way. It's too complicated to explain, but that is the best way of avoiding trouble with the company. Later on, when the first of the other three hostesses hands in her notice, you will take her place. Assuming you like the work, of course.'

'I will, I know I will. Thanks so much Uncle Xavier, it's just the break I need, and I love you for it. I'll never forget this, believe me I won't.'

He remained stern-faced: 'But, Laure-Anne, you must promise me no funny business. What I mean is that this is going to be an official arrangement, you will have to work like everyone else, you'll call me Monsieur Simmonier, and Aunt Ginette Madame Simmonier. Another thing, as an apprentice, you won't even get the official minimum wage, you understand. Later it will be different. You will naturally obtain your share of the pooled tips, we put everything in a kitty, although I take nothing. It averages out at one-fifth extra. You understand, you will be merely another member of the staff with no special privileges. I shall be reprimanding you from time to time, that's certain, and it will do you no harm in any case.'

We walked up and down outside the hotel on the crunchy gravel in the sun. It was mild for March, a pleasant morning with no hint of a breeze, and I looked around me at the old hotel building and the cleverly spliced-in modern annexe. Softy-eyed, I scanned the grounds and the fields beyond, hugging myself with contentment. It was all panning out so perfectly, I could not have dreamed for better.

'I haven't finished yet,' my companion said. 'How much money have you got? Be honest, how much?'

'Two hundred cash and eight hundred in my savings book, which I have with me.'

He rapidly worked out with me what my net income would be per month including tips and a small clothing allowance.

Then: 'Keep your savings, and there's no reason why you should not slowly add to them. Meanwhile you will need clothes and things. No, don't answer. There is no question of you contacting your parents, at least until you are properly settled in, when we can break the news gently. I want you to take this envelope. Put it straight in your coat pocket, it's 3,000 francs, half of it a present from Ginette and me, the rest you will reimburse from your pay over the next two months.'

I made to interrupt but he continued: 'Now, L'Anne, I want you to look pretty and dishy, smarten yourself up. Dump your student stuff and dress like the real woman you're going to be from now on: dresses, new hair style, decent stockings – the lot. Go into Evian, look at the shops, beauty parlours, have a manicure, new shoes, everything. I want to see you transformed when you put on that uniform. That's an order.'

'Yes boss.'

'Cut that out. No familiarity from now on.'

I assumed a serious expression. It was not difficult. His instructions made me feel a wee bit like a kept woman. Was he setting me up as his private call-girl? Had he effectively kidnapped me, and was threatening me with immediate despatch to Marnes La Coquette unless I gave him sex? It occurred to me to bring the conversation round to our session of the previous evening, but feared this would wreck the new arrangement.

He must have read my thoughts: 'By the way, about yesterday, I don't want you to feel . . . that is, after all . . . I – er – went crazy and it's embarrassing for both of us, and I promise you it will not happen again.'

I laid a hand on his: 'It's too late, I am ruined for life. But I'm not a bit sorry, I wanted you and it was super. Let's remember it for ever and say no more.'

He was as relieved to drop the subject as I was, and we chatted a little more about the job. He took me inside and I did a formal walkabout, meeting the staff. I was introduced as Mademoiselle D. who would be starting as a trainee hostess two weeks from now.

That same afternoon I took the minibus into Evian and was measured for my uniform. It was then that I realized what was happening to me. I window-shopped with the 3,000 francs

25

hovering before my eyes, a larger sum than I had ever had in my possession, though most of it was back in the hotel safe at that instant. I got myself fixed up with some contraceptive pills; there was no point in resisting this now, and it was a fortunate thing that Xavier had bedded me at the right moment in the month, or so I hoped, at the same time glad I was not going to play Russian roulette. It was a snap decision, part of the 'two weeks folly' as I later named it. Who could say what nights of love lay in store? And I had no intention of missing out on them!

That night I slept even more soundly, waking around nine, which made it too late for breakfast. I promised myself I would cadge something from the kitchen, and meanwhile drank a glass of water and showered quickly. The morning was already far advanced and it struck me that there was no reason to get up with a rush; I would simply do my shopping faster that afternoon. Besides, I had days and days ahead of me.

Breathing deeply, I opened the curtains and then sat on the edge of the bed, then lay back and pulled the sheet and blankets over me. Voluptuously I writhed on the firm mattress, twisting my pelvis and flaying my legs about. Only two more nights and I would move to a smaller room in the zone occupied by some others of the staff who did not live locally. I had seen the room and did not care for it much: porridge walls, beige bedspread, nondescript furniture. I resolved to make it as tasteful as I could, although for the moment I had no idea how, apart from filling it with flowers.

After my burst of energy, my acts of boldness, I was a little girl again, revelling in the warmth of the covers. My left hand was already at my mound and I explored further, remembering dear Xavier who had been there 36 hours earlier. I was growing moist and forced myself to stop, pulling my hand away. There were more important things to work out and I frowned in concentration.

If I was fool enough to waste this immense opportunity at this juncture, the best thing I could do was to jump under a train.

I let my mind wander over the new image I was to acquire. I must get it right from the outset. What exactly was a hostess? Answer: a slinky doll kneeling at the foot of the male – or at least that was how it would be here. No women's lib nonsense, but a form of prostitution in a way. Oh, it wasn't as crude, but were we not high-class tarts, we hostesses? After all, what did the public want from a hostess on an airliner, at an exhibition, in a hotel foyer? They wanted a smile, a trim figure and serenity.

Frankly, I had seen enough of authoritative, hard-voiced women to know that that was not what men wanted, certainly not what Laure-Anne D. intended to be. A red-knuckled feminist was the last thing I meant to finish up as. No, I was going to be a real woman, not competing on men's territory. I began mentally ticking off the adjectives: I would be feminine, alert and bright-eyed, submissive, subtle, yielding, sexy without being vulgar. I would be the pleasant sort of creamy girl every man wanted to go to bed with. Wasn't I already feeling creamy inside at the thought of what people would say if they knew what was going through my mind at this moment? It would be terrific being able to say yes or no to any man as the mood took me. I could pick and choose my stallions, there was no limit to the fun an eager 20-year-old girl could have if she set her mind to it!

On an impulse I leaped out of bed and flung myself into an orgy of narcissism. I glided into the bathroom where there was a long mirror and twirled around in front of it, swaying my hips modestly. Yes, I had a good body: just the right slope and curve to the shoulders, adequate breasts that would develop, discreet browny nipples, slender waist, firm almost boyish bottom, thighs that were in need of a wee bit more flesh but they would pass. Hm, the knees were a little bony but they would develop too. I was desirable, there was no denying it. I did a belly dance, snaking my hands about. I would be a sensation in sheer stockings, primrose suspenders, lace-trimmed bra and boxer knickers – all in silk, or perhaps satin, or that new shiny material. That was another thing; my clothes

27

would be feminine: soft materials not too thin of course, no divided skirts, no tight ones either or jeans, no tight-fitting jumpers but preferably rather loose tops that would hint at but not conceal my alluring bust. Modest necklaces that looped perhaps, earrings to match. I caught sight of my broad flat feet on the bathroom tiles; from now on it would be high heels, better still mid-high ones that would shape my calves and make me look vulnerable without actually tottering. A whole host of magazine hints and tips flooded into my mind. I would learn to walk elegantly with the legs swinging freely from the hips, and pubis slightly advanced, head high and with a suspicion of a forward tilt that would help disguise my short neck; I would keep my feet parallel, the inside of the knees just avoiding contact as I walked. I sauntered in and out of the bathroom a bit on my toes, and was sure I would get the hang of it in a matter of days.

Then I saw my head! A disaster. My neck really was rather chunky, cheeks pudgy and pink, big mouth with puckered lips, thick eyebrows, characterless mousy hair, red ears, a sulky expression. Nose OK. I tried to smile, tried several smiles, and was delighted to see the change, especially in my eyes. Oh those grey-green eyes, thank goodness for them, they were my only quality facial feature; they were large, or at any rate fairly large, dreamy and kind. How was it possible for them to be so kind after what I'd been through in two decades? But there they were, nice friendly eyes giving the come-on. A mystery, my secret weapon!

It was getting chilly and I was hungry, but I was so impatient that I ran back to the bed, grabbing a magnifying mirror on the way, to read three expensive women's mags I bought in Evian. I flicked on the strip lamp over the bed and scrutinized the magazines, looking for faces and heads the same shape as mine, noting how the cosmetic ads made these difficult shapes and features into beautiful expressions. All I had ever used so far was lipstick and very little of that. I realized you could do anything with make-up.

I was falling in love with myself, and a little voice was telling me it would do no harm just once more to make love to myself. I was hungry for that too. Heroically, I resisted. Meanwhile, running round inside my head was a list of what I would buy: dresses, shoes, belts, handbag, lingerie, make-up, perfume. My brain was frantic with all these details, I would never have enough time.

In a final surge of exuberance I took hold of the magnifying mirror and pulled faces at myself, smiled, laughed, fluttered my eyelashes, looked disdainful, winked, blew a kiss, ran my tongue over my lips and finally chucked the mirror on the floor.

Downstairs at the reception desk Madame Colin called me over, handing me an envelope.

I read:

> Laure-Anne, *cherie* – Had to leave quickly on business. Your father phoned telling me you'd left home, asked me to let him know if you turned up here. Not to worry, I told him nothing. Destroy this note.
>
> X.

After begging a morsel from the kitchen ahead of lunch, I wrote a note to the family:

> Dear Parents,
> This is to tell you I am all right. Have landed on my feet, more news later. Will keep in touch.
>
> Laure-Anne

Writing even those few words was difficult. I stamped the envelope intending to have some hotel guest post it a long way away. Then I shrugged off the family.

They had a tape recorder in the main conference room and I asked if I could use it. About a year earlier at a party I had heard my voice on tape. It shocked me. The diction was not too bad, but what emerged was a horrible undulating sound, a sort of wheedling up-and-down voice the kind I hated in other females. Speech delivery was to be a vital component in my new image; I remembered a fascinating starlet interviewed on television once who ruined the whole effect she was making

29

when she spoke. It was Friday afternoon at the hotel with no conferences going, and I worked at my problem for two hours non-stop:

– Good morning sir, I hope you had a pleasant journey. If you'd just like to sign the register I will have your luggage attended to.

– Oh yes sir, we have everything you could wish for in the way of sporting facilities: tennis, golf, roller skating, deep-sea diving, hunt-the-hostess.

– The first session ends at noon, and then the hotel provides a free drink before lunch. Don't worry sir, we make sure our guests are sober for the afternoon session.

– Excuse me, are you Monsieur Dupont? Ah, your wife is on the phone sir from Paris, sounds like she's preparing the guillotine, if you'll pardon me saying so.

– I'm sorry, it's very kind of you, but I never drink while I'm working. And I don't think you'd better do that, the Marketing Director is looking at us. My room number's 14, first floor on the right as you leave the lift.

– No sir, we don't have live porn shows. May I suggest a glimpse through the latest *Playboy*? Unless you prefer *Gay News*.

– Our resident button sewer is off duty, but I'd be glad to do it for you. I beg your pardon? But of course you must keep your trousers on.

And so on, varying the pitch, adding and subtracting huskiness. Until I believed I had it about right. The basic ploy, I found, was to give urgency to the message, and to have gaps between sentences. Madame Colin rushed in and cried: 'I'm sure the whole place would love to know your room number, but you had better keep it to yourself. It's going over the entire sound system, you must have pushed the wrong knob.'

We put that right, and I went on with a page or two of the operating instructions and other less controversial material.

Finally the big day arrived, and I donned my uniform. Everyone called it a uniform but in fact it was merely a knee-length

30

skirt in heather mixture wool, a cream blouse with a diagonal bottle green and orange stripe pattern, a mid-brown waistcoat and sensible brown shoes with a shiny buckle. No hat or beret. The waistcoat had a Club des Ducs badge in gold lettering.

I paraded in front of the long mirror in my new room, making the skirt twirl and flipping the waistcoat open to see the effect. I then stood quite still and smiled at myself. I had done a wonderful make-up job.

As to my hair, I had chosen to give it ash-blonde highlights; the *coiffeur* and I had tried a variety of styles but in the end we decided to leave it as a straightish schoolgirl bob, the ears half hidden, half-neck length with a parting. We added a discreet slide on one side. Frankly there was little more we could do with it. On the whole I looked quite glamorous, and the gap in my front teeth of about three millimeters added a bit of interest.

Thus I became number four hostess and went down to nurse the conferees and so on. We revamped the working hours so that Jocelyne and Carole were on duty from 8 a.m. to 3 p.m. and Huguette and myself from 3 p.m. to 10 p.m. Working weeks were complex, as sometimes we had to do Saturdays and Sundays and sometimes there were no conferences at all, or just short ones, and we had days off then. The basic principle was that, working in pairs, we were 'in charge' of the conference people while they worked and fed.

That misty Monday morning, about two dozen men aged 25 to 60 arrived by coach and Huguette and I opened the coach door, climbing up inside.

'Good morning, gentlemen,' Huguette bawled. 'I hope you didn't have too tough a night on the train. My name is Huguette and this is Laure-Anne, and we shall be looking after you for the next five days.' (Murmurs from the passengers.) 'You should be in possession of your room numbers, but the porters will look after your luggage. There is a cloakroom for your coats, coffee is waiting for you in the restaurant before you start the first session. If you have any problems, just ask Laure-Anne or me.' We climbed down, followed by the conference leader who chatted with Huguette while I guided the men through the main entrance, whose revolving doors were folded back.

So far, so good, I hadn't opened my mouth except to say 'follow me, please'. Then one or two of the gents chatted me up about this and that, mostly to do with where things were and whether this was an ordinary hotel as well, which it was. I twisted and turned and kept my voice low and sultry, as per the tape recordings. My new shoes were hurting me but I smiled, smiled, smiled, gaining confidence minute by minute. A cheerful fatty asked if the badge was real gold, another said he hadn't expected to contend with a *nana* (a dolly bird), a third said he could do with a whisky and hadn't slept a wink and I said he'd feel better when he'd had a siesta. And so the first day got off to a satisfactory start.

It was fun being a woman when you were surrounded by eager bucks in the prime of their lives, all in expensive suits, for this conference had to do with insurance, and all keen to flirt while away from home and eyeing the handful of presentable females available. It was like being on a de luxe liner with a dozen men for each girl. Having just emerged from my chrysalis I was flattered by the constant glances at my legs, bosom, waist and face. I realized the power of the strip-tease dancer before an audience, marvelled at the power of the flesh, was puzzled why a centimetre too much or too little in this or that part of the female anatomy could make all the difference between indifference and desire on the part of the male. I positively blossomed and had difficulty keeping my voice subdued in this stimulating, ruttish atmosphere of ambition and financial greed.

It was on the Tuesday evening that the first attempt at seduction occurred. Not by me, of course, I would not have dared. Huguette and I sat in on the evening meal, as we were still 'in charge' of our flock. They were more carefree now, eating and drinking with gusto and trading jokes. In this bawdy-tinged ambiance I made sure I drank twice as much water as wine, and indeed was starting to find the bonhomie a tiny bit wearing. Ten o'clock came, and Huguette said we were going off duty now, whereupon a toast was drunk to 'our guardian angels' and the pair of us nodded our heads daintily to left and right, then skedaddled. In the foyer, Huguette said she would just check the lights and equipment in the conference hall, so I wished her goodnight and mounted the carpeted stairs.

I reached the first landing and a man's voice declared: 'You have the most enticing dimples behind your knees, Laure-Anne.'

I spun round and saw it was one of the seminar people. He came to a halt about six steps down from me.

'Oh,' I responded pleasantly, 'kind of you to notice, sir. Can I help you in any way?' I assured myself that if I stuck to my proper role, it would turn out well – whatever his intentions.

He remained where he was and grinned: 'I'd like to sell you some insurance.'

'Perhaps it's re-assurance I need.'

'I can give you good cover.'

'With profits, I hope.'

His eyes glaring at my hem, he climbed as far as the first landing: 'Perhaps we can work out a policy in some quiet spot, my room for example, it's along here. I have a whisky miniature for exactly this occasion.' His fists enclosed one of my hands. 'You have wonderful eyes.'

So this was to be my very first non-family man. He was one of those cavalry officer types with thick stubby fair hair and moustache, the kind every heroine went for in the novels but none would dream of trusting. The cad! I was certainly interested.

'Thankyou kindly, it's very observant of you.'

'Come on then.'

'It is most irregular, understand, the management would never approve.' I had not withdrawn my hand and he now nudged me into movement along the passage with a fist in the small of my back. Be passive, I told myself.

'It's after hours, so your time is your own.'

33

'Just a small whisky, then, and I must be off to bed.'

'You certainly must.'

The room was as well-appointed as the one I used to have. He locked the door quietly and I pretended not to notice. There were two low chairs and without further ado he pushed them together. I sat on the resultant settee while he found a couple of glasses and turned all the lights out but one – over the bed. He joined me on my right and promptly slid an arm over my shoulders. I eyed him in mock-anguish, we clinked glasses, grinned and sipped.

'As I said, you have wonderful eyes . . . ' I lowered them modestly.

There was some back and forth about what colour they were (penetrating glare from sir), about my *Mystère* perfume (big sniff), the attractive hue of my uniform (hand on thigh), the sheerness of my stockings (hand on knee).

'There's no need to check, you know, we're inspected every day before we go on duty.' I moved the knees slightly.

Inevitably: 'But I am more interested in off-duty inspections.'

'Your attentions are highly flattering, sir, but as an insurance man you may find that my yield may not be very high.'

'I'll take the risk, your surrender value is enormous, I am sure.'

His hand hovering at my knees he said urgently: 'Come on, Laure-Anne, just a kiss from those pouting lips, they've been fascinating me for two days.'

'Hmmm, just one, then I must be going,' I said, as guileless as they come.

Some kiss! I thought I would faint, it went on so long. It was simply scrumptuous with exactly the right pressure, the tiniest tremor and the stimulus of the blond moustache. I came out of it breathing hard, to find his hand lifting my breasts.

Before I could feign protest, he said: 'Your breath is so sweet.'

His was wine-laden with a hint of cigar, but I quipped: 'I use Brand X, it has that tingling feeling.'

This he seized upon: 'Are you tingling now, Laure-Anne?'

'Yes,' I exhaled, my nipples as hard as nuts and sending signals straight to my entrails. 'Kiss me again, I liked it. Then I really must go, I have to get up early.'

'That's not true, I checked your hours.' I sank before his arms and I had a blurred vision of his mouth the instant before I closed my eyes. This time his lips were wider apart and so were mine. I forgot about the moustache, and gave myself up to the embrace.

On the bed, in Position 1, we prepared to consolidate our good relations. But after a while I began to lose impetus, my lubrication was faulty and he began to hurt me. Whether it was the handicap of wearing all my underclothes plus shoes at his request, I don't know, or the excitement of the past two weeks, or the importance of the occasion, the absolute need to make it good. The fact was that my self-confidence waned, I sensed it might not go according to plan, and that the pefect moment we had been working up to was likelier to be a perfect anti-climax. Clovis sneaked into my brain, followed by Sister Thérèse and even my brothers. The cavalry man slowed down and I felt the panic rising. I would be frigid for weeks if it flopped.

'Keep going,' he ordered swiftly, then whipped off my knickers and buried his mouth in my genital zone. He pushed up my legs, I let myself go, and the tension eased. I told myself it would work out after all, and I concentrated: I was a man doing it to Laure-Anne, he was a bull, I was doubled up with my gaping sex oggled by an audience of men – I tried a dozen scenes, as he administered to my exposed part.

Which all goes to prove that sexual intercourse is never a foregone conclusion, and is the most intricate and unplannable of all relationships. Any small thing can upset it, and I think my difficulty then was that I was playing three roles at once: the innocent hostess being taken advantage of, the fancy whore, and Laure-Anne the fun girl.

He entered me again and I sensed the beginnings of desire, but now it was his turn to worry.

'I can't hold on much longer,' he growled.

'Oh no, not yet! Count sheep or something, wait for me, do wait for me!'

'For goodness sake, tell me a joke, a horror story, anything!'

He pinched the base of his penis and breathed hard.

I said: 'I'll tell you about my night with Auntie Charlotte. She lived in a village near Nantes with a husband called Valentin, I was 12 at the time. The village had only one ugly house, and Auntie Charlotte found it – a single-level place with a small garden for the dog. It was the only road without sewage and they had a cesspit. It was also on a dangerous bend, and the front wall kept getting knocked down. They had a cellar but never used it.'

'Why?'

'It was knee-deep in water. The roof sagged too. There was no front door and you went through a miniscule front garden and down a side passage to about half way. It was the only door to the place.'

'Laure-Anne, you gorgeous thing, why don't you straddle me? You have delightful hips.'

I mounted him and said: 'Oh, that's heaven, your thing's so firm, keep still and make it last, let me do it. Why didn't we think of this before? Oh it's lovely, I can really feel you go all the way. Isn't sex absolutely . . . ' I let out an almighty shriek as my climax overtook me and I clawed at his shoulders. I yelled again and saw his face contorting too. I collapsed onto him.

A man's voice came from the adjoining room, accompanied by three bangs on the wall: 'Are you all right in there?'

'Yes, I was having a nightmare,' the cavalry man said.

I gave him a tender kiss and we lay there for a while.

At length he said: 'What about the dog?'

'It died of rheumatic fever about five years ago.'

I was still astride my cavalry man and moved about hoping his organ would stiffen anew. 'Can you do it again?' I pleaded. But he said no and I didn't care to argue about it. He let me out of his room around midnight. I never did learn his name.

36

On the subsequent Friday, immediately after lunch, Madame Colin beckoned me over to the reception desk: 'Call for you.'

'Hello,' I crooned into the phone.

'Ah, is that you Laure-Anne? This is Marcel Giraud.'

'Yessir.' It was the insurance boss, the top guest. He had a voice like a trombone, stood a head higher than most people, had a few wisps of hair and wore gold spectacles. The personification of the capitalist moneybags.

'I'm missing a report, but I know where it is. It's in the conference room somewhere near the tape recorder and all that tackle. The title is "Towards a new methodology for computerized rate adjustment" and I'd be so grateful if you could bring it up. My room number's 103.'

'Yessir, right away, sir.'

I checked my appearance in the corridor mirror and tapped twice on the door.

'Come in,' the trombone brayed.

I entered and found him standing in the middle of the room in his shirt and tie, trouserless. 'Oh, excuse me, sir.'

'Don't worry, just getting ready for a quick nap, I'll pull the curtains. So you're Laure-Anne, nice name. You're a sort of public relations girl, I suppose one might say. Tell me, do you ever have anything to do with private relations. Ho-ho-ho.'

'I'm sorry sir, I don't understand.' Oh, the little innocent!

'Look, my girl, I'm absolutely dying for a little – er – affection, if you see what I mean. I've got this dreadful summing up to deliver at 3 o'clock and I need the compassion of a cool feminine hand. Don't you feel amorous, after a good meal?'

I lowered my eyes and fiddled with the report: 'Sometimes, I must confess, sir.'

He barked: 'Nothing to be ashamed of. I'm sure your methodology is infinitely soothing. You have the most wonderful eyes, has anyone ever told you?'

'Thankyou sir, very kind of you, sir.'

'D'you like your work?'

'I started on Monday, it's interesting, I'm a trainee.'

He twigged at once that this made me vulnerable, and it seemed to excite him, for he hopped on one leg: 'You know you're expected to be nice to guests, and I'm a very important one. Do put that report on the chair over there and relieve me of my burden. Put your soft hand on my brow, lovely hands you have.'

'If you wish sir.' I walked timidly up to him and did as he asked. This was clearly going to be a slave role for me.

'Listen, young lady,' he said, guiding my other hand under his shirt where he was bare, 'I'll make it worth your while. Do it there, caress me, I really need it.'

With every appearance of shock, I withdrew my hand. He added: 'Do you ever go without panties?'

'Really, sir!'

'Do fondle me. You realize I can make it difficult for you, report you, you might lose your job.'

'Oh no, that would be awful!' I wailed, my eyes huge. He could still not be sure I was falling in with his act, but I delicately took his shaft in my palm and stayed motionless, feeling it grow bigger. He told me to wait while he put the do-not-disturb sign outside and locked us in. We resumed our pose, and he knew he had me. At the same time I had to pretend unwillingness to give it ambiguity and spice.

He ordered: 'Undo your blouse and take your dress thing off. Keep your shoes on. Now, onto your knees.'

'I really ought not to, I shall get into trouble,' I acted, taking his organ into my mouth. He was still only semi-erect, perhaps due to his age, and I began running my lips along his joystick, tightening my mouth at the base and pulling gently.

'Oh, that's just right, you're so soft, let me hold you.' He clasped my head in his big meaty hands and disarranged my hair, lifting it up repeatedly and letting it fall again. He knelt down in turn and his prestigious member plopped out, wet and shiny. I felt his hands go into my panties and he pushed me to my feet, lowering my knickers slowly, then kneading my thighs at the suspenders.

'Do you like that, Laure-Anne?' he asked, stroking my legs.

38

'It makes me feel funny, sir.'

'Oh I must stop, but I can't. Get down on all fours, splay your bottom. Ah, so sweet.'

'I won't allow it there,' I declared firmly, thus setting a limit which I knew that, as a gentleman, he would respect.

His palms ran over my rump, pinching and patting the flesh so that it must have quivered. A masochistic thrill came over me, I was the mare waiting to be covered. Suddenly he was upon me, forcing my back down so that my posterior rose and he had direct access to my sex. He struggled for a while as he sought to penetrate me, I tried to help but it was no good. My new stallion, a thoroughbred this time, recovered from this *contretemps*, and in this he won my esteem, for he seized me ape-like round in the middle and half threw me onto the edge of the bed, so that my legs dangled. I made them part and he muttered: 'Oh those stockings!' He was against me, feeling the hosiery again, I closed my eyes tight and waited for his assault.

'Does it hurt?' It felt like a lethal weapon.

'Yes, a little,' I said. 'Please don't make it hurt, please, sir!'

He knew that a little meant quite a lot, and he thrust his missile in again and again, and I whimpered once or twice, and he knew I was not pretending, that I was dry and in pain. But he knew too that I had no choice but to submit. Because he was who he was, because I could do nothing to counter his power over me, because I was his object and he could do what he wished with me, because his lust for my body was an immense force that neither of us could resist.

Finally he came with a throaty roar, and I sobbed with pain. My vagina seemed to be on fire when he briskly withdrew.

That evening Huguette and I stood with Xavier waving goodbye to the coach, and we went back into the foyer. I was still feeling sore from the afternoon, but curiously satisfied even though I had had pain rather than pleasure. Was I a masochist, then?

Xavier retired to his office, and we girls headed for the conference room to clear up.

Madame Colin waylaid me with two envelopes this time. I pretended to be in a hurry and stuffed them unopened into my waistcoat pocket, for I suspected what they contained. When my hostess colleague was busy with the electronic system, I ripped them open.

One contained 400 francs, the other 600 francs. There was nothing else inside, and I never knew whose envelope was whose.

I was so proud of myself.

Chapter Four

A score of beauty aids, all my own, lay scattered on the glass top of the dresser in my room. It was a dull morning, and I was trying out some new facial effects culled from the magazines. Some 30 minutes had gone by and I was enjoying myself dipping into pots, squeezing tubes, massaging, finger-patting, plucking, outlining, eyelining, applying and wiping. New Laure-Anne faces smirked at me from the mirror one after another, then changed to a quizzical expression. My only raiment was a bathtowel over my shoulders and bust. Later, I reflected, I would have half an hour with the tape recorder to check on my low-profile husky voice.

The decision as to the future 'real me' was within my grasp, when I heard a heavy tread outside. Instinctively I braced myself; I was really a little too exposed for comfort. The foot-falls hesitated, turned back, the owner knocked on another door rather slowly, and finally on mine.

'Who is it?'

'Monsieur Simmonier.'

'Come straight in.'

Xavier in his manager's suit stood just inside the door: 'Hm, smells like candy floss in here. Nice though.'

'Open the window if you like, but not for too long.'

'Oh no, I like all these stinks. I'll watch, you're like a cat licking itself, must cost a fortune all this stuff.'

I turned my head towards him, running the tip of my tongue along the sheen I had just applied: 'We plain girls have to work hard at it.' I grinned in amusement at him. Yes, I thought, you are in two minds whether to make a pass at me after all, you are thinking that no-one need know, the passage is empty, your niece would be afraid to complain, and you would just love me to push my chair back so that you can see more. I fidgeted a little but kept my legs under the dresser, at the same time drawing the bathtowel closer round me. Maybe I was a cock-teaser after all, maybe I was obtaining my revenge for Giraud's missile attack, which still irked me somehow. Anyhow I carried on teasing prettily.

'Plain? What rubbish. And those sulky lips are just waiting to be kissed. But I didn't come up for that. I want to discuss your parents. I've been talking it over with Ginette.'

'Have they phoned again?'

'No, but you've been here a while now and I think we should let them know.'

'Depends if you're going to keep me, sulky or otherwise.'

'Of course I am, if you want to continue. Sincerely, you're an asset. Madame Colin says you keep yourself to yourself and don't say much, which is just what the customers like. You are also efficient.'

We agreed there was no purpose in holding out to my parents any longer, now that I had made a clean break and was self-supporting. I said I'd like to stay, and thanked him again, even squeezing his hand to confirm the bond between us. It was decided that he would phone home to say I had turned up unexpectedly, he would pass me the receiver and I would say that I had asked for a job and uncle had very kindly offered me one. We then sketched out two short letters we would send off, mine stressing that they need not worry, his affirming that he believed it would be best in the circumstances for Laure-Anne to be under his watchful eyes.

These eyes were at that instant watching the bathtowel slipping away from my forestructure millimetre by millimetre, but he said he supposed he ought to be doing some work.

'Laure-Anne,' he said, 'take pity on me, let me hold you for a minute. I promise I won't start anything.'

'All right, only it's not out of pity, it's genuine affection. Doesn't Aunt Ginette let you do anything? She's nice.'

'It's different,' was all he said. He had his thrill, storing up fantasies, and left five minutes later.

Since the insurance seminar, nobody had tried to make me, and I was worried. Hence the make-up session. It was good for me to be unsure of my appeal, because it told me I had to strive hard to get the sex I was increasingly hungry for. I needed to be pleasant, feline and above all a good listener. I was finding the rapt gaze was quite a weapon.

Mealtimes were my best opportunity to flirt. I was no Miss France and my only hope was to attend to my guests entirely, as they put the country to rights. I made them feel like gods.

I heard opinions across the entire political spectrum, except for the Far Left, as we had no such persons at Evian.

From the pseudo-socialists I heard our handsome President Valery Giscard d'Estaing lambasted for not being a genuine member of the *haute-bourgeoisie* because his daddy or someone had bought the family name. And yet he was something to do with Louis XV. I never did understand this. He was a 'boy scout' in his approach to the burden that lay upon his shoulders. As a former girl guide this interested me, and I used to wise-crack about that. VGE had completed misjudged the crisis with his assertion: 'This is a monetary crisis, not a crisis of the capitalist system.' I felt sorry for VGE because he had got it so wrong after all those years in the business. One guest informed me: 'He was guilty of the worst strategic error of the post-war era, when he launched the Western summit meetings to sabotage the budding North-South dialogue, which is our only chance of achieving the new world economic order that must come.' His family, it appeared, virtually owned the Centrafrican Republic, where he got all his diamonds. Decidedly, he was a President worth having, it seemed to me.

Then there were the liberals. Their big pitch was free trade, and if only market forces were allowed to work, the crisis would resolve itself. Demurely I would suggest that the world could not be expanding exponentially for ever, and asked why it was that the rich were always on the side of free trade. Naturally, I was given to understand, it needed 'really good people' to run the system, and was assured that we were now moving into 'organized free trade'. I would abandon the argument gracefully at that stage, leaving my interlocutor to promise me that a new prosperity would result from the new technology. The crisis would be over as soon as the computers could fix it, there would be no repeat of the 1929 crash. Unemployment was meanwhile 'simply a structural problem'. I would nod wisely at that.

The Far Right were the most inebriating, or inebriated, I could never decide which. Their great hope was the national saviour who would unite us all, eradicating the Marxists, the unemployed, the 'remnant of the world money power', and the health scheme. As to the sick and handicapped I discovered that they were the victims of Judeo-Christian decadence and that they would be the first to gain from the 'new order' (another one!). The blacks and the Arabs were occupying half the hospital beds anyway, and filling the queues at the job centres; they should be free to go home to their own countries, and we would keep ours. I was warned that I could be raped by these foreigners in the Metro and get the pox, and I assured our guests that I would make sure all my assailants were Europeans. I was toasted for that.

It was very perplexing, but on one thing everyone was agreed: the Russians were heading for the chateau de Vincennes – I presumed via the Autoroute de l'Est. Thank goodness we were in Evian! As to the Chinese, if we mowed them down at a rate of one million a day, there would be just as many left the next day. One of my informants revealed that, in Paris, you never saw any Asian funerals and their deaths were never recorded officially; the reason was of course that they recycled their old people through the Chinese restaurants, which explained why they could serve meals so cheap and undercut the French.

On the whole I tended to fall in with Uncle Xavier's pragmatic view of politics.

'It's funny,' I observed one day as the three of us ate together, 'this place seems to bring out all the political opinions. I never heard so much politics before I came here, it must be something we put in the salad.'

'Huh, politics,' he scoffed. 'I was born in September 1929 and within weeks Wall Steet financiers were jumping off sky-scrapers. Nobody could have foreseen that! Politics is the simplest thing in the world, it's economics. If you're rich you're Right Wing, if you're middle class you're Centrist, if you're neither you're Left. That's all there is to it. Except that aound my age you tend to forget about it completely unless you make a mistake and switch the news on.'

With so many experts knowing so many answers, I felt my best contribution would be to stay happy, particularly through the orgasm, which as far as I was concerned stretched before me as the path to salvation. If there is a heaven, I reckoned it must consist of a sort of perpetual arousal, a theory I understood to be shared by Islam, which probably explains why it is the fastest-growing religion in the 20th century.

I took time off from these serious matters to engage in another pursuit, painting. I worked hard at that too, for on my days off I would set up my portable easel in some woody corner of the estate, sometimes outside it, and forget every-thing. Lost to the world, I struggled to convey on canvas some of the awesome effects of light and shade. Taking up from my teen-age daubings, I bought a series of easy guides and jabbed away at the rectangle before me, the creative act affording me immense tranquillity of soul.

Rarely did anyone disturb me, but on a fine day in June when the ground mist seemed reluctant to waft away and the wind at 1,000 feet was pushing bulky white clouds across the horizon like floats in a carnival, all different but with a common theme, I had a visitor.

'Exciting shadows, aren't they?' he said, making me jump. I turned round and smiled politely, and he added in a sing-song:

'Yellow-greens, olive, Van Dyck brown, mauve, white and brick. I wish I could join you.' He was a nondescript man, neither handsome nor ugly, but with nothing special about him. I put his age at 30 or so.

'So you paint, too,' I said flatly, rubbing out with my cloth. I hoped he was not going to be a joker, for I was getting into rather a mess.

'Yes, I use a cloth a lot too.'

'It helps with the shadows. Get those right and the rest looks after itself, don't you think?'

'Of course. You don't mind me watching?'

'Not a bit. I'm only a beginner, so I've got everything to be ashamed of. Go ahead and criticize, I'll listen to any advice as long as it's free and I agree with it.'

'I wouldn't dream of criticizing, I'm only a learner myself. I like your sky.'

His accent was unusual, and I ventured: 'Excuse me asking, but where do you come from?'

'Nottingham.'

'Where's that?'

'In the middle of England.'

'Oh, so you're English, that's wonderful, I can practise my English!'

'If you like,' he said in his mother tongue.

I continued in English: 'Some marvellous paintings your people did, the 19th century – er – landscapes, is that right?'

'That's right. If you ever come over to England I'd be glad to show them to you. Are you a guest too?'

'No, a hostess, it's my day off. I did not mean to do painting, but they are hanging on the curtains in my room.'

He laughed: 'In my country we hang up curtains.'

'That's funny, we hang them down in France. But with the English they do everything upside down. Oh dear now I have spotted my robe, I mean dress. I should wear my own pantaloons for this.'

'Trousers,' he corrected.

46

'I will never learn. What is your name? Mine is Laure-Anne.'

'Terry.'

'You are here on your own, yes? We have no seminars at present.'

He explained he was the French representative for a North of England chemical company. He had been doing business in Lyons, and was taking a couple of days off at the Club des Ducs. No, he had no family over here and, no, he was not married.

'And you, are you married? I see you haven't a ring.'

'I am too young to think of that. I have a good time first. Babies I do not like, too. I must find a rich daddy-sugar. We can sit together at lunch, OK?'

'But I'm not rich.'

'I always know a man by his clothing. You have a good skirt.'

'Shirt. I say, may I make a small suggestion? Try a darkish mauve for that shadow, black or even grey kills it dead. May I?' He took another brush and improved the effect.

'We go to lunch now,' I announced. 'You arrange your hairs.'

He ran his hand through his hair: 'Actually, I'm trying to make the comb obsolete. Are you leaving all your tackle?'

'Of course, everyone here is a gentleman, nobody will touch it. Let's come back and do it afterwards.'

'Do it? Hah-ha, yes I see. I'd be delighted.'

Terry, I decided over lunch, was a *very* nice young man. I liked cheerful company, and we had laughed together several times already, he was quick to zip back a reply and my inadequate English made for plenty of confusion. As every woman knows, it is through incidents like this that we progress.

'I don't know who's waiting at this table, but I think it's us,' he cracked. We chuckled when my version of shellfish came out as selfish and I could not for the life of me say it properly as when I also said he should try the excellent wiped cream.

Uncle Xavier noted that we were getting along fine.

Intrigued or jealous, he came over and politely interrogated Terry, sitting with us to drink an Armagnac.

Which meant that we rose from the table some time after 2.30. I borrowed an apron from the kitchen for Terry and we returned to the easel. Frustrated of late, I was feeling quite mellow now, but thought it best not to rush things with my *Britannique*; it was well known they took a long time to warm up. I hoped he would not be turned off by my old student gear and the apron. As we stood looking at the now-unveiled canvas, he lightly placed a hand on my shoulder, initiating physical contact. So he was mellow too, at least I hoped so, for I had never had an encounter with a foreigner.

He said: 'Let me recite a poem for you, nice and slowly so that you can practise your English:

> While Titian was mixing rosemadder,
> His model ascended the ladder,
> Her position, to Titian,
> Suggested coition,
> So he nipped up the ladder and 'ad 'er!'

I laughed and said: 'I must start again immediately, otherwise I lose courage.'

'Hm,' he said after a moment. 'Funny, you're not even fiancéed then?'

'Is this a proposition? You want me to go up the ladder?'

We worked for an hour, making a joint operation of it, then I went indoors and rustled up a jug of lemon tea. I said I was fed up with this painting, and why didn't he do one to show me how?

He thought for a moment and then said: 'Do you know what I'd like to do? Paint your portrait. You have a certain quiet grace in your movements that is most attractive, and you have a charming way of holding your head.'

'Oh Terry, that is a lovely saying for a girl to hear, so you really think so?'

'I'd like to try catching it on canvas. Here, where's that pencil? Get over there and lean back against the tree. I'll use this board. That's right, tuck your legs to the side. Oh well, perhaps not, you'll soon be tired, just sit how you like, and take your bearings so that you can get up occasionally and then go back to the same position.'

Actually I was dying to get into position under him or on top of him, but all I could do was make my eyes smoulder. After a while I suggested: 'If the light fails, you can carry on tomorrow.' Did he want to make love or not? How much longer was he going to take? How could I signal to him?'

'Sorry old girl, have to leave at lunchtime. My sister and her husband are due in Paris tomorrow and I've got to look after them all weekend. Keep the same expression please, I'll be as quick as I can. Now let's have these tubes set out just right, I'll have this brush and this one. Shit!!!'

The palette slipped from his hand and smeared the bottom of one trouser leg. We fussed about, wiping it off, and I said: 'Take them off, nobody will see. It's not very promiscuous round here.' (In French the word means crowded.)

'It will be in a minute when I'm *sans culotte*!'

My chance at last: 'I am already *sans culotte*, I am so hot.' I had indeed divested myself of the said garment when I went in for the tea, and was now thoroughly randy.

Hesitating a second or two, he took off his trousers and I resumed my position. But it was the wrong one and he came over to adjust my head, squatting indecently before me in his trunks. I judged it was now or never! I pulled him forward onto his knees and drew his lips to mine. At last he got the message, and I accidentally touched his crotch with my knee, keeping it there. A sting of concupiscence triggered a muscular twitch somewhere inside me.

Effortlessly we moved into the act, murmuring in our respective languages. I helped him all I could; it was I for example who removed his trunks, which were in a lustrous black material. But first I softly played with him, while praising his choice. He said: 'I'm your knight in shining armour.' And I said that was very poetic of him. I wanted to go on playing when the trunks were off, but he said: 'In my country, the man is the boss; stand up against that tree.' I did

49

as I was told, whereupon he raised my skirt, spread my legs and administered to me from below. After that, he led me to one of the few areas bearing grass. I reclined but it was prickly, and we snatched off our aprons and made a blanket of them, clean side up on my insistence. He pushed my knees back and from the moment he nudged at my outer lips he was silent, carefully penetrating a little at a time. 'I hope I won't have a multicoloured bottom,' I said. 'Keep talking, it helps me,' he growled. So I helped him: 'Oh it's divine, Terry Terry Terry, sweaty male, give it to me, you're thick and you find me irresistible and you're going to make it last, because you want me, do anything, oh hurry I'm nearly there. AAAhhh! Yes hold back, it'll go on for ever, yes another one please. AAAhhh! I can't believe it, yes I'm insatiable, I can't get enough, I've never had a multiple and you still haven't come. How long can you wait? Oh please wait. I can feel you starting to throb. You can't understand it in French, I don't care. You're throbbing against my vagina. Your turn, *mon amour*, you are spurting yes I have it, inebriating and refreshing too, I'm going to faint with pleasure.' Terry hugged me fiercely and the ground prickled and dug into my bottom, so that I had a third orgasm as he collapsed on me.

We stayed like that for ages and ages. I wanted the world to stop.

'My God you're sexy,' he said at last. 'You were frantic.'

'I've never had three, or even two.'

Next day in the empty conference room we kissed goodbye. He seemed strangely off-hand in view of the momentous event. He said he would call me on Monday. I told him to come back as soon as he could.

Monday came and went, and I spent a restless night. I was sure he had said Monday. I had been busy with a new seminar, but beamed all the time at everyone. Even the normally taciturn Jocelyne said I looked as if I'd started a baby. Six o'clock and no call, but I realized he was waiting to get home from the office to make his call. By eight o'clock I was worried and at nine I asked Madame Colin's replacement if there were any messages for me. I hung about after dinner, chatting to guests in the foyer, sure that I had found the man of my life and that there was a simple explanation. Perhaps he was ill. At eleven I crept upstairs miserably. He must have said Tuesday.

Tuesday was hell, and at 6.30 p.m. it was I who rang him, no answer, and again at seven. We had a kiosk in the hall and I used that.

'Allo,' he stated. At last!

'Hello Terry, it's Laure-Anne,' I said brightly in English. 'I said I would telephone you. How is my knight tonight? Did you have a nice weekend?'

'Oh fine, fine,' he said airily. 'Everything went off perfectly.'

He carried on about what he had done, he and his sister and her husband. In fact he seemed likely to go on for about an hour on the wretched subject, and I decided it was time to force the issue: 'Doing any painting?'

'Ah no, far too busy. I can't really finish the portrait without seeing you. I haven't even got a photo of you.'

At last we were on the right lines. I said: 'If you think you can escape with that, you make a mistake. You come down again in a few days.'

Silence at the other end, and I had to ask if he was still there, which is always a nuisance. He said: 'It won't be for some time, I'm afraid.'

'Oh, how desolating.'

'Very busy at the moment, sudden rush of work, might even have to go to Brussels.'

I plodded on: 'How about our trip to London? I'm longing to see the National Gallery, and the one near Hereford (is that right?) with all the 19th century works?'

Another silence and then: 'Look, Laure-Anne, I'm afraid it'll have to be called off for a while. I – er – I'm getting married in June, we've just fixed the date, last week in June.'

I was shattered. I lurched against the side of the kiosk. 'Oh I see.' Blackness closed in on me and all I could see was the

51

phone. He said he would be in touch and went on for a while about it.

I breathed hard and when he stopped I said: 'Congratulations to you, best of luck. So you will contact me. Goodbye, Terry – goodbye.'

It was a full minute before I dared to move.

I excused myself from dinner, and went up to my room.

Writing slowly, I produced a long letter to my parents, explaining for the first time how I needed to find my feet, to sort things out. I told them I was sorry, I was in perfect health and happy, I would come and see them soon.

Chapter Five

What followed was far worse than an agonizing reappraisal. In my despair days went by. From the depths of my childhood came remorse, and the awful suspicion that I had been punished for my sins – those sins of the flesh I had cheerfully refused to acknowledge. Suppose there was a God after all, and I had been wicked all this time. Suppose even touching oneself was a mortal sin!

I refused to believe it was true. Such innocent fun could not be wrong. Or else everyone was damned, and nobody could convince me they could put it right by babbling in the confessional and saying a few prayers.

From the moment I first saw my brother Jean-Marie's little worm floating between his legs in the bath, until well beyond the age of 7 which was termed the Age of Reason, religion and I had circled each other warily like boxers in a ring. Until I finally gave it a knock-out blow around the age of 10. This unfortunately was not helping me in my present woes. What I sought to know was why my first real chance to fall in love had ended with me distraught in the phone booth. What had I done to deserve that, where was the sin?

My underlying difficulty was that I had never possessed the innate moral sense, which the Church told us was a natural gift. When religious instruction began at school, the early glowing world full of love, beauty and kindness became tangled with warnings and interdictions. There was a catch in it, there were things we must not do, like chattering idly and telling untruths, quarelling, stealing, thinking too much of ourselves and disobeying parents and teachers. None of which seemed to affect me at the time. The list included thinking bad thoughts and self-abuse, but nobody actually explained them and I lost interest.

Not surprisingly I took confession in my stride, almost certainly because I could never get the hang of it, going into the box in the chapel every Friday and answering the same questions over and over again. Even so, I was shaken from my lethargy on one visit to the sinkhole of venery.

It was a different priest this time, and when I knelt down in the gloom I could see through the grill that he had a fat red face and black glasses.

'Bless me, Father, for I have sinned,' I recited.

'Bless you my child, now what have you got to tell me?'

'I beg your pardon?'

'Tell me your sins first. You've been instructed how to do it, haven't you?'

'Oh yes, Father.'

'So what have you done?'

'I haven't done anything.'

'Don't be silly, you must have done something.' (Silence from my side of the grill.) 'Have you been telling untruths?'

'No Father.'

'Well, you mustn't keep things from God, you know. You are talking to God at the moment. Have you said your prayers night and morning?'

'Yes, Father.' A downright fib, of course, but I amended it to: 'Well, most times.'

'That's a good girl, God knows everything. How about self-abuse?'

'I don't understand.'

The priest said nothing, and then: 'God Almighty has given you a lovely body and it's for him, and you have to look after it. You must not abuse it.'

That word again! 'Abuse it, Father?'

'Yes – hrmph – play with yourself. Do you play with yourself?'

'Oh yes, always, everyone else is too busy.'

'No, my child, I mean touch yourself, your private parts.'

'Private parts, Father?'

'Hrmph, between your legs for example. Do you do that?'

'Oh yes, Father.'

'How many times?'

'Lots of times.' I was itching to do it then, and decided I would when I got out.

'How many times a day?' he snapped. I wondered what that had to do with the subject. Still, they did say that religion was a mystery.

'Not in the day, only the evenings in bed, and when I have my bath.' That was not strictly true either.

'Don't you think that's bad?'

'Oh no, I like it, Father.'

'It's sinful, my child, God doesn't like it.'

'Does he do it too?' I whispered.

'No, he doesn't like *you* doing it. You must stop it or you won't go to heaven, he'll punish you. Do you play with anyone else, boys especially?'

'No, I play by myself.'

'Anything else you'd like to tell me?'

'I can't think of anything.'

'I'm sure you can, what about your thoughts, do you think impure thoughts?'

'Impure, Father?'

'Nasty thoughts.'

'Like what, Father?'

Presumably he didn't know what he meant either, because he went on: 'You must not imagine naughty things, my child.'

'How do I know when they are naughty?'

'God will guide you.'

'Yes, Father.'

'You must be very careful to please God, he can punish you if you are very naughty and you may not join him in heaven.

Pray to be happy with him for ever in heaven. Now for your penance . . . '

The germ of a new idea had been planted. The idea of doing *it* with someone else, especially boys, opened up new vistas, and I dwelt on this as I lay with the lights out in bed. I imagined Jean-Marie's zizi, and Yves', and my fingers would creep to my own zizi-less cleft. Then suddenly I was afraid at what the priest had said; I would be punished. This seemed to excite me and I again saw the face of the confessor and his smell came back, like the liquor mummy put in the cakes. I wondered if the priest had a zizi and I seemed to see it, but only for a moment. Perhaps this was a naughty thought, and I realized my hand had wandered again. I snatched it away and soon fell asleep.

I appeared to be nearer understanding the difference between being good and being naughty, and resolved to be good. Sometimes I would be good for a long time, but in the end I would set about bringing on the nice marzipan feeling again, then be sorry afterwards.

Eventually it didn't seem to matter and I took my pleasures in my stride thereafter. Nobody noticed I was a budding hedonist, one of those people who believe that pleasure is the Chief Good and the Proper Aim. I won't go on about it, except to say that I long ago noticed that indulgence in sensual delights that harm nobody tends to be castigated chiefly by people who are prepared to ignore other evils: tricking others in business, milking their companies even if it means running down the business, speculating on the markets and otherwise exploiting simple souls. The Mafia, I notice, are very strict on sexual matters but don't mind sending people to the river bed in a concrete coffin. And what is one to think of the arrogant Western leaders who have spent a decade at their summits throwing stones at an old tin can? I suppose I simply have a different notion of sin.

Well, that's enough of that! My moral dilemma dissolved under the influence of an external factor which I shall now recount.

Chapter Six

Twenty is a good age for a girl, and I have never met one with this number of summers who sulked for long.

My new-found exuberance took over and I returned to the preoccupation that was uppermost in my mind. I was fascinated by my development into a mature woman. I suspected that I was filling out across the abdomen and had more flesh on my thighs and bosom. It was natural that I wanted to use my attributes, and my sensual proclivity seemed to augment day by day. I was a future Marilyn Monroe and seriously considered having my hair dyed blonde.

However, the chemistry of sex was one thing, what I wanted was the mechanics. I could have entertained a whole regiment of paratroops, such were my nubile propensities at that time. Oh, how I worked on those male guests, but to no avail! The luck of the draw was against me.

Then the insurance magnate Marcel Giraud came to my rescue, though in a surprising way. He turned up one midweek afternoon in early September, this time as a private guest.

He suddenly appeared in one of the smaller committee rooms and found me folding some promotional material. He declared in subdued tones: 'Ah, Mademoiselle D., Laure-Anne, nice to see you again.' We shook hands. 'I've been settling in and I wanted to see you.' He pushed the door to.

'I was most satisfied,' he went on quietly, 'with our last little encounter and would like to thank you again.'

'Thankyou too, sir,' I murmured, trying hard to blush and

lowering my eyes. 'It was somewhat rushed, I fear, but you were most generous.'

He placed a friendly hand on my arm: 'I really hope I didn't hurt you. I was quite taken away. You've only yourself to blame for being so pretty.' I tilted my head with a tiny shake to tell him there were no hard feelings. He rushed on: 'I'm staying just tonight, and I've actually come to see *you*. Do you possibly think you could oblige me with your services once more, but in a completely different kind of way that will prove far more pleasant for you? How can I put it? I have a strong leaning towards receiving corporal punishment, if you see what I mean, and I would be extremely grateful if you would oblige.'

He paused to see the effect, and I said, 'We-e-ll, I've never done anything like that before, but I'm sure it can be arranged. How nice of you to ask me, and in such a charming way – I mean that, Monsieur Giraud, the world seems so full of mean pigs these days – and in any case you've come all this way, I couldn't possibly say no. But – er – it's rather crowded and doesn't this sort of thing make a good deal of noise? There might be trouble, I feel.'

He smiled broadly: 'And you're so well-behaved, if I may return the compliment. You've no cause to worry, I've thought of everything, even brought you a nice little outfit. All you need are some boots, because I didn't know your size.'

'Oh dear, I haven't got any, just some ankle boots.'

'What colour are they?'

'Dark brown.'

'They'll have to do, it's a detail. Now listen, we can go to some woods I have located a few miles away and we'll be safe from prying eyes and ears. Say you agree, please. I promise you that if this works out a great future beckons you.'

We continued for a while to discuss details, and I arranged to sneak out just after he left by car the next morning, and meet him a short distance from the estate.

So it was that we turned off into a cart track that forenoon, and parked the car discreetly. We stood in our ordinary clothes listening for a while to a tractor far in the distance, and then he lugged a travel bag out of the back and we sought out a small clearing in the thick wood that could hardly have felt a human foot for years. I wondered if I was crazy to have come with this virtual stranger, but told myself he was known to the hotel, and was too important a personality to dare or need to turn rough. Men in his position simply do not murder women, and nobody with his obvious interest in the call of the flesh and the money to satisfy it can be dangerous. And if he did turn the plot round and decide to give me a whipping against my will, one word from me and he would be finished; Uncle Xavier would see to that. Thus I told myself I was safe.

'This is perfect,' he stated. He wore a dark blue business suit with thin stripes, I remember. I had on a simple wine-coloured skirt and cream jumper. 'You remember your lines, how it goes?'

'I think so, you can count on me.' We smiled our connivance, and then he gave me a look that would blow-dry your hair from 50 yards. 'Right, take the bag and get changed out of sight.'

I emerged from hiding a few minutes later in black shiny tights, my brown bootees (!), mid-thigh leather skirt and thin leather jerkin both in glistening black. Underneath, I wore a red G-string but no bra. My hair was pulled back and tied with a red ribbon, no easy task as it was really too short for this. I carried a leather whip a foot or so in length.

'Undo a couple of buttons,' he said. 'Right, now start.'

I gulped and stood with my legs slightly apart, the whip dangled down, and swayed a little. A last look round at the sun beaming down through the trees, a deep breath of the crisp earthy smell, and I began.

'Giraud Marcel,' I said in a clear voice. 'Stand to attention when I am speaking and look at me. I am given to understand that you were engaged in harassing women at your office, exposing yourself before them on more than one occasion, and generally behaving in a disgusting manner. A person in your position ought to be ashamed of himself, and I am afraid I have

59

no alternative but to administer the necessary punishment. You have been found guilty on four counts of molesting a particular typist, whom you threatened to dismiss unless she yielded to your lecherous advances.'

I went about this way and that and swished around to show the G-string. 'How would *you* like to be fondled and pestered by an ugly woman twice your age? Be quiet! There is nothing you can say that will spare you now.'

Acting the ugly woman, I advanced upon him and groped him as he stood to attention: his member was already hard. I undid his zip.

'Remove his trousers, girls.'

I played the role of the other girls.

'Tie him up.'

The girls tied him up, facing away, to a horizontal branch, an action simulated by his placing his hands on the branch.

'Huh, I'm going to enjoy this, you don't know how cruel we can be when roused. Bend over further, your body at right angles to your legs. And don't whimper before you're touched. You'll get 12 strokes, and every time you clench up, you'll get two extra. Where's the whip?'

I approached the victim, running my hands under his shirt, then ran my hands over his rump. 'Your organ is hard even in these lamentable circumstances, I see, how disgusting, can't you control yourself? Know that you are completely in my power, nothing can save you now, you must take your punishment.'

I slowly slid down his mustard coloured underpants, to reveal his snow-white buttocks. I saw the marks of a previous flagellation, but made no comment.

Then at last I stood back, and he rasped: 'Oh no, please no!'

I snarled: 'Come on, take it like a man, you coward. The first stroke is coming, look girls, see him flinch.'

I took careful aim and as luck would have it the thong hit him dead centre. He did not move and I delivered another, harder. The whip, some two centimetres thick, went 'wheep-slap' as I delivered four more.

'Oh girls,' I said, 'you don't know how this is exciting me.'

Giraud removed his right hand from the branch and began masturbating. After four more strokes, I saw his middle heaving rhythmically. I broke off to feel his wounds.

'Go on,' he ordered (this being outside the game), and I continued more slowly. I think I got two more strokes in before he jerked forward and I left him to it, for there was no point in going on.

He recovered slowly, and I watched. I was quite unroused myself, but made noises pretending I was. His bottom was streaked with pink weals, and I wondered how long it would be before they turned purple, before they disappeared eventually. I hoped I had whipped him hard enough, though not too hard.

After a minute or so he stood wincing and faced me. He said: 'Thankyou, *chérie*, it was marvellous.' We became our normal selves and he took me in his arms and hugged me, bestowing a brief kiss on my forehead.

At that instant we heard a shout. The vague shape of a man, no doubt a farmer or a gamekeeper of some kind, was heading for us 50 yards off.

'Shit,' Giraud snapped. 'Decoy him, I'll fetch you later, decoy him!'

His underpants were up in a flash, he grabbed his trousers and went flying for the car, leaping madly over the undergrowth with his yellow-clad rear aglow.

I spun round, somehow found the bag and ran off in the other direction. I threw myself into some brambles, ripping the tights for sure. The intruder was now shouting, but I was outdistancing him. I ran frantically for a while, then halted and listened. All was quiet, and I went cautiously back to the clearing. There was no trace of the farmer, and I prudently went to find my ordinary clothes. They were all there, except the panties! I changed and put the black gear into the bag. Still no noises, and it was creepy. The erstwhile golden sun now seemed sickly, the good earth damp and chilly. The tractor was still whining away. I thought it best to quit the scene of the crime and found the stump of a felled tree not far away. And there I sat, feeling cold without my vital garment, now utterly vulnerable with the farmer or whatever lurking around.

I reasoned Marcel Giraud would leave the car on the road and walk back eventually along the cart track to fetch me. So I shifted my position so that I could see him. It was now half past ten, it would take at least an hour to get back to the Club des Ducs, but the big question was: when should I stop waiting and set out?

My partner was true to his word, and 15 minutes later I heard the car stop.

In the car, with the leather upholstery warming me, he enquired as to my fortunes and then turned to me: 'Laure-Anne, you were terrific, what an actress you'd make! We must do it again some time. Now listen, Laure-Anne, I want to be really serious. I can't discuss it now, but could you possibly come into Evian tomorrow at nine? I have a proposition to put to you.'

'Why not?'

We fixed the venue.

'Can you tell me what this is all about?'

'It's a job offer, but I really prefer to discuss it properly tomorrow.'

And discuss it we did, at a table in a bistro. He was waiting for me with an expression that told me he had eaten broken bottles for breakfast.

'Right,' he declared, 'this is strictly business, take this pad and here's a ballpoint pen. You may care to make some notes, I'm a believer in taking notes. In fact my only regret about yesterday is that I did not ask you to wear glasses like my secretary; a real beauty but I daren't try anything because she loves her husband who's a schoolteacher and she's a stuckup bitch. Still, never mind that. Why I asked you to render that little service yesterday, apart from the pleasure it afforded me, was this: I wanted to test you, to test your amenability. You will understand in a minute. Now this is a copy of *Pariscope* and you'll see from these pages here that Paris is full of clubs and bars where people go to meet people, and there are live porn shows and the rest. Well, I am a committee member of the most elite of these clubs, the Rolls Royce of private clubs. We don't advertise. Would you be interested in being a hostess at our club? I am authorized by the committee to make you an offer.'

'Me?' I said gaping at him.

'We really need girls like you. But before you say anything I would like to ask you some rather personal questions. Could you please give me a quick summary of your life so far, not the intimate side, but the main lines?'

He was sticking his neck out, so I had no hesitation in responding to his open request. It took about two minutes.

'Good, excellent. It does seem to me that club work would suit you very well. I need not stress that you have all the attributes: class, pleasant personality, a good body, willingness to try anything, feminine appeal, discretion, compliance and so on. Tell me, Laure-Anne, about your plans for the future, what you want to do with your life?'

I let out a short laugh and said: 'You've caught me out there, I'm just trying to enjoy life after a sheltered start, I have no special plans. I like it at the Club des Ducs, and was intending to stay there a bit, to get some experience, find my bearings. I am only just beginning to feel alive.'

'Good. I do envy you your youth. Our club is patronized by the cream of France, and there are a few foreign guests from time to time. We have ministers, army people, church people – remember that bishop who dropped dead in the Rue Saint Denis? He was one of ours – and diplomats and those sort of people. We are a kind of freemasonry of the good life, of pleasure, if I can put it that way. Epicureans, hedonists and to hell with the rest. I must stress that we look after our girls, they earn good money, and in return they keep their mouths shut for the rest of their lives.' He paused to let the information sink in. 'We are fond of our girls, hold them in high esteem, as equals. You may be interested to know that one of our ladies met her future husband there, a permanent secretary at one of the ministries; poor fellow, his wife went off with some Vietnamese refugee. Anyhow, are you interested, that's the point? It's not hard work, but the one thing we must have is total loyalty, adherence to the spirit of the thing, and silence. Personally, if I were a young lady like you I'd jump at the chance, you never know what it can lead to.'

I put a fingernail in the gap in my teeth and began snapping it, a habit I have when under great anguish. I said: 'Thankyou for considering me, and for being so frank. It seems to me that you are running a kind of upper crust call-girl business.'

'Not at all. Any professional activity outside the premises is strictly forbidden. That may sound harsh, but I can only repeat that we look after you as if you were our own daughters.'

I said: 'All right, I'm interested. I think you've got a deal. Why not? I'm telling myself, but I want some more details.'

We conversed in subdued tones for a while longer, and then he handed me an envelope, fees for the whipping.

A week later I took a few days off to see the family, which was tricky, because I had to conceal so much about the recent past and my plans for the future.

I also spent part of a day and part of a night at the club in question.

I officially started my new job on November 1, 1976. A vote for hedonism if ever there was one.

Chapter Seven

The Top Club was not listed anywhere, and its two phone numbers were ex-directory.

It had a dull grey door easy to overlook, a street number, a bell and a small brass plate stating simply *Top*. It was in a busy street lined with shops and bars some 300 yards from the Arc de Triomphe.

Nobody noticed the door. Inside was the real entrance, a sort of conservatory with French windows, and it was guarded by a doorman while another man was presiding over a vestibule where the guests' coats were collected. You went through the French windows into a quiet lounge with a bar and a dozen tables. This room looked like a good hotel foyer. Immediately to the left on entering was a notice board, a letter rack and a shield reading: *Dignitas – Pietas – Fidelitas*. In due course I was given to understand that the cloak room usher had an automatic pistol under the counter, but also that it had never been used.

On any day shortly after noon you could expect to see a few of France's top people call in for a drink and a chat with the three or four girls present. But sometimes they simply got their drinks from the bar and took them to a table. Hot and cold snacks were served at the bar but no meals could be had. These refreshments were available at normal going prices as a safeguard against abuse, and members paid for them. The girls were strictly forbidden to drink alcohol, and no licentious behaviour of any kind was permitted in the lounge. Thus the tone was set. If a member desired the company of a girl or more than one, he took her or them into the other part of the establishment through a revolving door adjacent to the bar. No money was allowed to pass between the member and any of the personnel. Thus all pecuniary embarrassment was avoided. We were at the service of members, with whom no familiarity was allowed in the usual sense of the term. We employed the formal '*vous*'.

The atmosphere at this time of the day bordered on the austere, and this was deliberate. There was no music. After all, some people simply dropped in for a drink and a sit-down. Those who required sexual relief had four soundproof rooms in the recreational area at their disposal each containing two easy chairs, reproductions of paintings, a bed draped with a single sheet and an extra half-sheet that was changed after each member had finished his business. There was also a small washbasin.

The club closed at 2 p.m. and reopened at 6 p.m. The pace was far swifter after that, and at least six girls were on duty, while canned music was played. From 6 p.m. until 2 a.m. members had the use of a heated swimming pool, a room for group sex furnished with sofas and a huge round playbed, a large eliptical sunken bath with all necessary accessories, and a cellar equipped with instruments of torture – not forgetting the four cubicles. The other rooms consisted of the office, the retiring room for the girls including cupboards for clothes, and a store.

I have used the term 'top people' for a very simple reason. There were women as well as men among the members. Membership totalled roughly 500, taking one year with another. Ours was a non-profit-making establishment whose officers were elected once every two years. The resulting committee was in complete charge, and it appointed the manager and assistant manager who attended to staff and other administrative matters. The committee vetted every application for membership, the annual subscription being roughly six times the minimum guaranteed monthly wage in France.

As to the members, Marcel Giraud's description was about right. Several ministers were members, along with business magnates and others from influential walks of life. There was little we girls did not learn about sooner or later concerning their characters, but no word was breathed off the premises, and the comment within was negligible. One cabinet minister was strongly rumoured in the press to be a paedophile, but he personally swore to me that this was untrue, and his conduct at the club certainly indicated he was highly girlophile. The wife of one minister was said to have had a child by another minister and to have later teamed up with yet another minister; all I know is that she behaved like a maniac among us girls. A highly placed lady in the State was said to extract money from her husband when he wanted her to accompany him abroad on official functions; she was keener on a night out at the club, and had a penchant for our feminine company. A noted catholic priest of the traditionalist branch used to dress up in our clothes and have himself flagellated with a coat-hanger, but he was eventually asked to leave the club on the grounds that he was a disturbing influence.

It would be completely wrong to suppose that we were ravaged by the 500 members day in, day out. It was not like that at all. I estimate that on average I hosted about two or three men a day, no more, although once I week I was roped in for group activities on the playbed. We were universally known as '*les girls*' and were often asked to play with each other and perform in the torture room. In this connection I must amend a previous statement about free services; a bonus scale was applicable to any girl who allowed herself to be physically whipped or caned.

Having explained all this, I should like to backtrack to the day I first entered the Top Club, to sample the atmosphere and discuss contract terms.

Marcel Giraud took me in that evening round 8 p.m. and introduced me to the manager, whose name was Gérard. More about him later. We had just walked around and emerged from Gérard's office when I came face to face with none other than an old acquaintance!

'Is it really you, Blandine?' I cried. 'What are you doing here? You're the last person I expected to see in this place. What have you done to your hair?'

'*Mon chou*, what are *you* doing here?'

Neither of us answered, we simply embraced and the two men left us. We sat at one of the tables staring at each other and breathing in the leather smell.

Mademoiselle Groult, whom I later called Blandine at her request, was (or rather had been) a mathematics teacher at our school, and she gave private tuition during the run-up to the *Bac* at her nearby studio flat. She was also assistant captain of our girl guide section. A week after the *Bac* about two dozen of us went by coach to the Ardeche region of Central France, where a lorry had taken our tents and equipment. Mademmoiselle Groult shared a tent with the number two girl. I was allotted a tent with a girl called Delphine, who seemed just as withdrawn in character as myself. Both of us were late starters. Mademoiselle Groult requisitioned Delphine and I for stores duty. This consisted mainly of going to the nearest village and to a farm for things like bread, milk, eggs and poultry, trundling a two-wheeled cart behind us.

One afternoon after lunch, Mademoiselle Groult asked me to go with her to the village where she had to make a phone call. We looked smart in our uniforms as we strode out on the two kilometre journey. We had dark blue knee-length skirts and matching socks, dove grey shirts and red and white scarves with a woggle, plus berets. About 200 metres from the camp, my superior took my hand and we walked along gaily in step, telling ourselves how lovely the countryside was looking, pointing out birds and wild flowers. I decided I really quite liked 'La Groult' as she was tagged by the guides. She obviously sought friendship because, after I had been calling her mademoiselle for a while, she said: 'Do call me Blandine.'

I smiled and repeated 'Blandine'. She asked me to say it again as I had 'a charming low-pitched voice, with clear diction'. I said it again twice to please her, more than flattered by her comment.

The village was having its siesta in that early part of the afternoon, and we were able to admire the old houses and shops at our leisure and without noise and fumes, many of the buildings dating from centuries back. Even the post office was ancient, for it was merely a counter in the grocer shop that looked out onto the village green. I sat on a bench there while my companion made her phone call.

On the way back we decided to take a short rest on a tree trunk that lay back from the path in the shade of a small beech tree. We tarried, breathing deeply and saying, 'Ah, fresh air' once or twice.

Then suddenly Mademoiselle Groult, who was on my right, took both my hands in hers, and said: 'Laure-Anne, what a romantic name!' To which I naturally replied: 'Blandine's enchanting.'

She said: 'I'd like to kiss you, Laure-Anne.' And she acted on this wish without delay, placing one arm round my waist, turning me towards her and slipping the other arm round my neck. Her mouth was on mine and I was completely flabbergasted. I felt her lips opening, trying to force mine apart, and I was so unprepared that they yielded. She increased the pressure and kneaded my neck. I gawped at her and she stared into my eyes with a sort of frightened look, then kissed me on the lips again and this time my mouth was already receptive and the kiss spread. I felt her tongue enter my mouth and touch my own. She moved her head about so that her lips took mine with hers. She drew apart. The daringness of what we had done left me with a scampering heartbeat. There was no doubt as to her inclinations now, but the implications for me were tremendous.

'Oh Laure-Anne, how sweet you are, I am quite overcome. You don't mind do you? Kiss me back, please, at once!' I did so and found that I was quaking.

In the depths of the next night, my feet were gently shaken.

'It's only me – Blandine,' I heard as I struggled to my senses. She was unzipping my sleeping bag and taking my arm. I followed her out of my tent in a daze.

She led me into her tent, where I found her companion, a girl of about 18 called Christine who was setting out three beakers. We sat down, and the light of a torch shed a dull gleam on us from one corner. I told myself I was safe from La Groult's attentions with another person present.

'We're just having a little drink, and we thought you'd like to join us,' said La Groult. They both smiled at me, and produced a bottle of Johnny Walker, along with a packet of cheese biscuits.

Leaning back on the sleeping bags, we shifted from one elbow to the other and felt pleased with ourselves. All three of us were in pyjamas of course: mine were white with small flowers, Christine's pale blue and La Groult's wine-coloured and edged with white piping.

At length Blandine, as I was calling her now, drew nearer to me and placed an arm round my shoulders, ostensibly to give me support. I stiffened, knowing in a flash that I was the victim of a conspiracy, that Christine and she were both lesbians! Before I could arrange my thoughts, Blandine was playing with my hair and saying: 'She's lovely, isn't she Chris?' Christine was stroking my legs, and I saw she was pressing her nipples with the other hand. I was 'so warm and cuddly like a kitten,' La Groult observed, moving to my own bosom. I knew I ought to get up and rush away, but the whisky and the intimate radiance of the torch, the very secrecy of the gathering and the congenial nature of my conspirators, all had their effect. An appealing animal fragrance was issuing from Blandine, and this made me more susceptible. Her soft kneading of my breasts was breathtaking, and I could feel my nipples projecting under my pyjama top. Christine had ample breasts and I wanted now to feel what they were like. In a gesture of commitment, I put out a hand and touched them, lifting them and pinching them, as Blandine was doing to me.

Christine shuffled forward and kissed me on the mouth and I kissed her back. She had generous lips like my own.

Blandine withdrew, leaving us to play with each other. We sucked each others' orbs, and I looked round for Blandine, wanting to bring her into the game, and saw that she had a hand inside her trousers and was caressing herself with her legs apart. I found Christine's moist aperture, and did it for her, faster and faster until she suddenly opened her mouth and her eyes grew enormous. She stuffed a bit of a sleeping bag into her mouth and squeaked, her hair flaying around. I was absolutely fascinated and kept going. Then we were all on our knees and Christine heaved and heaved before falling back. Blandine clamped her lips to mine, her tongue slipping back and forth like a jelly, her fingers on my clitoris. She withdrew them to stroke my tummy and then returned to my orifice. A wave of heat poured into my lower body and I forced her hand against me, squeezing my legs together as the pleasure unfurled. A tidal wave engulfed me and I completely lost control, blowing into her mouth and biting at her cheek and neck.

We ended all three enfolding one another in a flurry of joy.

And now in the Top Club she was telling me: 'This isn't my hair, dear, it's a wig. Are you a member here, Laure-Anne? It can't be true.'

'It isn't. I've been offered a job and I'm taking a peek at the place. How wonderful to meet up with you again. To tell you the truth I'm lost, you can fill me in on the details here. But tell me about yourself first.'

She said: 'Well, after they threw me out of that school, I did various jobs connected with accountancy, but they never lasted. Except one at the offices of an art dealer next to the Rue Drouot auction rooms. And to cut a long story short I married the boss. But save your breath, this is a *mariage de convenance*. You see, he's a pederast and it wasn't doing him any good in the business, being so old a bachelor, so, as I don't go with men, he married me as a cover. We go our own ways, and I nominally act as his wife, in return for which I am handsomely provided for. He's tremendously rich and I get the spin-off. It means attending dinners and whatnot with him, and we sometimes entertain. It works out very well.'

I was blinking, and she went on: 'Now, as to this outfit, they call me in occasionally when they have a woman who wants to go with another woman. I'll name no names of course, you'll find out. I'm over age of course but I still count as a girl, especially with this wig. Rather nice, all these blonde curls, don't you think? Gets damned hot though sometimes. They also like me to come in for the Thursday passion parade.'

'Huh?'

'Yes, passion parade. Once a week we all take part in a kind of orgy on the playbed, and sometimes members like to watch girls having it off together, which believe me or not is pretty hard to simulate, so they ask me in to give it the authentic touch. Some members like to watch two of us in action privately too. Gaston got me the job, he's my husband.'

I whistled low in astonishment, and she told me she was at that moment waiting for a lady member to arrive from the *Assemblée Nationale*. We had gone on for a few minutes discussing the *mores* of the club and what the female staff were expected to do, when her client arrived. The lady knocked back a quick drink at the bar and came straight over. She was plump and fiftyish, and of course I did not know her as I took no direct interest in politics. She wore a wedding ring.

There was no time for introductions, as she said: 'Sorry I'm late, Blandine *chérie*, and I haven't got much time. My man's due back home at eleven. Let's go, shall we?' And they went through the revolving door.

Not wishing to draw attention to myself, I moved to another table in one corner, and sipped my orange juice. I decided I would hang around perhaps for another hour.

Within moments, a short man with a squashy face but impeccably dressed, slipped of a stool at the bar and headed for me.

'Good evening,' he declared sitting down. 'I'm Picard and I

72

understand you are Laure-Anne, I don't think we've met. You're new here.'

I felt awkward. The girls on duty were all in cocktail dresses, some of them slinky or with fringes, and they all wore expensive-looking jewellery which I assumed was imitation. I however was wearing an ordinary day suit of wool as if I had just left the office somewhere. I considered this man should have noticed the difference. Either he did not or cared not, but it seemed evident that he was going to seek my services, unless he happened to know someone in the family.

'Yes I'm Laure-Anne,' I smiled, breathing-in his discreet toilet water aroma. 'I have just been taken on, but I am not actually – er – on duty, as it were.'

He adopted an exaggerated expression of disappointment, and jerked his head about like a monkey's: 'Oh, so you're just starting, well I'm sure you are a delightful addition to the community. I would buy you a drink if it was allowed. Look, all the others seem busy, I wonder if we couldn't . . . ' He nodded his head at the revolving door. 'You don't mind, do you?'

I did not mind. He was friendly, smelled nice, looked like his clothes had just been delivered, and had one of those eager cruel faces that intrigue women.

I beamed at him: 'Of course not, Monsieur Picard.' I looked at him as if he was the Dalai Lama, determined to set out on the right foot.

He rose instantly, seized my elbow and propelled me into the recreation zone. He seemed in a hurry, and picked out number four cubicle. I thought it wise to start concentrating on the subject at once, otherwise I would not be ready for him and the Marcel Giraud semi-rape might be repeated.

Throwing my handbag onto the bed, I tried to be as nonchalant as possible, and turned towards him as he locked the door and switched on the only table lamp. He flung off his jacket and I did the same with mine, telling myself to smile, smile, smile. That at least could not be wrong. His shoes came off next and he guided my hand to his trouser belt. I lowered his trousers as he whisked off his shirt, I then removed my skirt, whereupon he leaned with his hands against one of the walls and indicated I should let down his underpants. This I did, to reveal one of the largest organs I have ever seen in repose. The flaccid shaft oscillated left and right over a huge round scrotum that probably contained Outspan oranges. All this in silence, and I wondered if I ought to say something, ask him if he had a good holiday, how his wife was, who the hell was he anyway? Even his genitalia were ugly, the penis almost purple, and this excited me.

He ordered: 'Off you go, I adore women with juicy lips and your top one is slightly longer than the other which I like too. If you only knew how I need this. What a day I've had!'

Kneeling down in my underclothes and stockinged legs, I bestowed a light kiss on his organ and took the heavy scrotum in one cupped hand, gently lifting it and clenching it several times. He sighed with contentment, and I ran the other hand over his hairy thighs and round the back. His shaft swelled and I deemed it appropriate to make some remark: 'You have a fine body, Monsieur Picard, full of energy and power, mmmm!'

'I've had an awful day,' was his response. 'Meeting after meeting at the *Assemblée*, phone calls by the dozen, everything going wrong, damned communists sabotaging us all along the line.' I flipped my tongue round his penis, and it jutted out horizontally then stiffened like a rod of steel at 45 degrees. 'I'm giving up hope of the Left ever getting to power, I think we should go it alone, what do you think?' I caressed his legs and buttocks, and drew away from his organ: 'No, you need the communist vote and you know it; it's too early to dump them.' Perhaps I had said too much, but he merely grunted; I hoped we knew what we were talking about, and decided to keep quiet and get on with my task. 'You have a superb thing, Monsieur Picard.'

A demure smile, and I resumed my stimulation. Picard, Picard? I knew the name. Wasn't he the young socialist who made a name for himself in May '68? I had seen his picture in the magazines, knew he was being tipped as the next but one

Head of State. My lips flattered the ballooning extremity of the presidential chopper and it seemed unlikely I could accommodate much of it wherever he sought to penetrate. Pending further instructions, I whirled my tongue saucily round the gland, then advanced my mouth, withdrew, advanced further and so on until I could take no more without choking. I tightened my lips and pulled back, then seized him again and again as if determined to snatch the protuberance from its fixture. His hands buried themselves in my hair and I pinched each bulbous testicle in turn, stroked the hairs and worked my way back to those behind. 'That's good, that's good,' he panted. I changed to a chewing movement, and my teeth lightly nipped him. He suddenly withdrew, and we waited until his desire subsided, before we continued. I pumped away, stopping repeatedly whenever he tapped my head as a signal. At last after many minutes I heard him holding his breath. 'Keep that, don't move, now!' I squeezed his scrotum hard, his organ seemed to vibrate and his warm seed burst into my mouth. It kept coming and coming, the fluid filled my cheeks and I was snorting frantically, willing myself to stay the course. I swallowed some, surprised that it had no taste, and then a little more, my stomach protesting as I ran out of breath. Then suddenly it was over, and I whirled round to let the remainder slop from my mouth on the half-sheet. He too grabbed a corner of the fabric for his own use. He moved over to the washbasin, and after that I was able to wash out my mouth.

He dressed, the pop-up politician gave me a curt thankyou, said I would make a wonderful geisha and planted a kiss on my forehead. I was to learn that this peck was the customary farewell gesture to one's hostess.

When Picard closed the door behind him, I sank onto the bed. I had of course done this previously for men, starting years ago with the boys, but had never actually taken their fluid into my mouth. With his hippo's genitalia still before my eyes, I perspired uncomfortably and was still breathing deeply. I was filled with apprehension, for at that point I had no idea how many men we were each likely to entertain per day, whether there was the risk of a particular girl being called upon again and again, whether we could 'go sick' if the pace wore us out.

In reality, as I have already said, the physical demands on us proved to be quite tolerable and most of our time was spent being pleasant, listening to members' woes and otherwise providing repose for the nation's quaking rulers. And my goodness how some of them arrived! They were akin to prima donnas collapsing after a gruelling performance on stage. We were geishas of a kind, but at that moment I really felt like a whore from the Bois de Boulogne, and had to fight back my tears. I was within an inch of storming into the manager's office and telling him I did not want the job after all.

I pulled myself together, repaired my make-up and crept out. The cubicles ran along the side of the large room containing the playbed, and I now counted a dozen men and some girls chatting and holding hands no more scandalously than at some village dance. The cubicles were all unoccupied now, and this gave me reassurance. I strolled into the room where the bath was and found it empty, then descended to the torture room which was in darkness. I sought out Blandine, told her I had entertained Picard and questioned her at some length. She confirmed I had no cause to worry about overwork. The members were courtesy itself and would be horrified to think they were treating the girls as prostitutes. She also mentioned that, owing to the absence of direct payment, there was no bickering among the staff.

One of the first gentlemen I took in hand is worthy of special mention. He was a sad ex-Foreign Ministry official called Dombre who knew Blandine's father. He admitted to 59 but was probably five years older. His most striking characteristic was a dreadful laugh, like a seagull's natter.

He arrived that day prompt at noon and I was the only girl actually in the lounge, seated at the bar on my own sipping a tomato juice. I must have drunk gallons of the stuff weekly but it was better than orange juice. He ordered a whisky, swallowed it in one gulp, and took me by the hand.

'Laure-Anne,' he said dolefully, 'may I have the pleasure of your company?'

I preceded him meekly to a cubicle, and stood before him awaiting his command in an olive green linen dress with three-quarter length sleeves and a V-neck. I chose to adopt conventional clothes, as others rather tended to play the vamp. I had unremarkable features and was feeling my way.

He took my hands and I almost thought he would kiss me, which was contrary to the accepted etiquette. But he immediately ran his hands all over my body, ending with my knees. He dwelt a while in this area, stroking my legs and exploring the shadows under my dress. He removed his clothing down to his underpants and ordered: 'Lift your dress slowly.' He kissed the exposed parts as they appeared, finally reaching my pubis. He moved me, still fully dressed, to the bed and pushed me on my back, then took off his underpants. I was totally unroused and called up the fantasy of a typist being forced, having found this useful at times already. *A chacune sa technique*! Dombre slipped off my panties with deliberation and lay upon me, his member quite limp. He rubbed against me but was unable to achieve an erection. This was not surprising, and it was clear that trouble lay ahead.

'*Merde*,' he muttered uncivilly, and to help him, I delicately ran my hands over his face, adopting my gaze of adoration.

'Let's take it slowly,' I said conspiratorially. 'We've got two whole hours ahead of us.' After a minute of fruitless effort, I suggested he sit in one of the easy chairs and I curled up at his feet, manipulating his venerable though slack parts.

'It's hopeless,' he growled at last. 'I'm so sorry to have wasted your time.'

'Don't be silly,' I scolded, 'we all have our ups and downs, don't you know that we girls have trouble sometimes? Let me just sit on your lap and hear what's troubling you.' I hitched up my dress and spread my thighs on his, one arm round his neck.

And it all came out. His wife had died of cancer a couple of years earlier, their sex life had been uninspiring anyway, and for some weeks he had sought to enjoy a new lease of life by going to a couple of call-girls who had terrified him and shrugged their shoulders at his failure to perform, finally he had joined the club and this was his first attempt to solve his difficulties. A colleague had kindly advised him to seek membership, and had recommended me as a partner. I said that was extremely kind of him, and he ended his account by saying: 'Perhaps I'm simply past it, we can't go on for ever, I suppose.'

He came to a halt lamely, and said in a thick voice: 'I am ashamed of myself.' He blinked, rubbed his eyes and the tears welled up. He buried his head in my bosom.

1 ran my hand through his depleted hair: 'Now don't dramatize, you are out of practice, Monsieur Dombre, and there is nothing to worry about. You have been through a grim experience and having to start out again on your own is dreadful. You're lonely perhaps, but you'll see it will all come right. I tell you what we'll do.' He raised his head with a sniff. 'Why don't you come in whenever you like, and I'll take you in hand? Half the men in Paris go through a bad patch, and you're not the first one I've helped out of a jam.' It was untrue, but I almost believed it myself, and I could not bear to let his appeal go unheeded.

He nodded, not daring to speak, and I boasted: 'We're expert at this sort of thing, believe me.' I had no idea if this was true, but it was all I could think of. I am not good with off-the-cuff zingers.

He perked up and said: 'What a charming personality you have. It's the atmosphere that's wrong, these pokey little rooms and everything. I have an idea, why don't you come away with me somewhere, to a country hotel. This could take weeks!'

'I doubt it, really I do. The idea's good but I'm afraid we are strictly forbidden to take members on our own, I'd get the sack.'

Quick to answer, he said: 'But you have time off, and what you do then is none of their business, surely. Look, the next time you have two days off we could go away and nobody would be any the wiser.'

I concealed my misgivings. A 40-hour week at the Top Club was a passable way of earning a living, but I had no intention of doing overtime. I was about to repeat that it was impossible, when he gripped my forearm like a drowning man seizing a lifebelt. Here was this destitute man, a human being who had lost all confidence, all pride in himself, pleading for assistance. He might even commit suicide! Worse, he might turn nasty and file a complaint against me! He was desperate enough.

'Well, if you like, I agree just this once. It's most irregular and you mustn't breathe a word. You promise me, Monsieur Dombre. Say you promise.'

One morning in late March 1977 I slammed the door of my taxi and clip-clopped up the steps at Gare Saint Lazare. He was waiting with a scowl at the departure indicator. We embraced on the cheeks and I put on my best smile for him.

'We've only got four minutes,' he said.

'Traffic,' I panted. 'No need to panic, we're not the last, I never miss trains. How do I look?' I lifted up my coat flaps to give him a glimpse of the shiny virginal white dress with vertical tone-on-tone stripes I had bought the previous morning, displaying the golden-brown stockings expensive enough to kindle the lust of any man.

'Ravishing,' he declared, dragging me onto the platform. I was starting to think he was more interested in trains than the matter in hand. I had decided to be the trusting bright little woman, and resolved to flatter his ego from start to finish, building him up as a big-shot, giving no orders, accepting any insults and upsets that came my way. I would keep purring come what may.

With my chatter I endeavoured to put some sparkle into his dismal existence, and I bubbled: 'This is fun, what a wonderful idea you had, and going to Deauville of all places, and the sun's shining too, I'm so thrilled!' I avoided mentioning that I had gone to sleep at 3 a.m. and woken five hours later, and had doubts as to whether I would last the day beyond teatime.

Around 3 p.m. we finished a late meal at the hotel and moved onto the glazed *terrasse* for coffee. I had maintained a stream of questions about his career during lunch, enabling him to drop plenty of big international names, most of which meant nothing to me, but it made him feel successful, took his mind off his wife, her death and their relationship.

Then I gradually guided him onto the subject that concerned us: 'Did we honestly drink a whole bottle of Fleury? They say wine has more effect on women than men, it's more aphrodisiac for them. Anyhow it is for me.' I wriggled my tiny hand under his. 'I love talking about sex too. Let me tell you some of the things that turn me on.'

Most of what followed, I invented; it had to do with trousers, donkeys, older men, bathing trunks and male odour. I invited him to reciprocate, and he slowly fell into my mood; his declared fantasies involved schoolgirls running and giggling, backs of knees, slit skirts (I wish I'd known), dogs, cows' udders (!) and wet one-piece swimsuits. I encouraged him to be poetic on these themes and then declared: 'My goodness I feel quite flushed. Let's go up to our room.'

Hoping to catch him on the up-beat, I helped him off with his clothes, and myself embarked on a striptease, doing it with a straight face as naturally as I could. I then slid between the sheets and lay there waiting for him to join me. Close scrutiny of his underpants told me I had made a mistake, I should have let him disrobe me, urged me to take the initiative. Damn! He lay next to me propped up on an elbow just looking down at me. I tried another tack, calmly pushing back the bedclothes and kneeling in front of him, my mammary orbs a short distance from his lips.

'You can do anything you like with me, Monsieur Dombre, please do it to me,' I whispered huskily with lowered eyes. Now we were both sitting back on our ankles and at last he gave a little kiss on both nipples. Big deal! What the devil did they do together, he and his wife?

'That's nice,' I said, 'do go on.'

Hardly daring to move, I took his hands and placed them under my armpits, he got the idea and began running his hands over me. A couple of discreet glances at his underpants told me we were going to need every minute of the 24 hours at this rate of progress. He was worse than a eunuch, but I let out a few 'mmmm's' and laid my hand modestly on my quim. Oh bliss, he removed it and began exploring my pussy. I felt like meowing but instead enclosed his hand and pushed his fingers into the entrance. He played around there and I deemed it none too soon to slip a hand into his remaining garment. Alas, we were still at square one; it did not matter to me, but his crestfallen expression informed me we had another lesson in theory ahead of us.

I insisted that he did not lay back and listen passively to my lecture. We were propped up on our elbows and holding hands as I assured him it was only a question of time, advised him to think about sex all the time, clearing everything else from his mind, told him his problem was all imaginary, recounted how Picasso and Charlie Chaplin and others were horny well into old age, said I actually preferred mature sensitive men, especially those who had achieved so much in their careers, finally said there were all sorts of ways of having 'it' and that some people hardly ever went as far as penetration and were perfectly happy. I then removed his pants, and toddled off next door to run the bath.

I stepped in and soaped myself sensually as he stood in the doorway: 'Like to join me? It's fun together, I used to bath with my brother when I was a little girl.'

That did it! He jumped in and we soaped each other for minutes, and to my delight his member responded. I giggled delightedly, kept up a flow of kittenish chat, constantly saying how it was like with my brothers, kept kissing him all over, telling him how huge he was growing.

Suddenly he ordered: 'Come on!' And we splashed out of the bath. 'On the floor!' and I was down in a jiffy, knees up and guiding him in.

'Oh, lovely, take me, rape me, I do love masterful men!' I was nowhere near coming myself but I thrashed about pretending. He ejaculated with a roar and I cried out and panted until at last he collapsed on top of me and lay completely stiff. So did I, wondering whether he had passed out, but he moved eventually. I clung to him as he rose, and then released him.

I was every bit as overjoyed as he was, and I nestled into his arms as we stood and he fondled my hair saying 'my little girl, my little girl, my little girl.'

We went for a long walk along the magnificent Deauville beach. He had a new lightness to his step and I kept looking up at him. It was misty in patches and the lights of the town and adjacent Trouville made it all like fairyland. I was so contented that I let Monsieur Dombre kiss me twice among the beach huts.

He imagined he was going to spend the night sleeping, but I had further treatment in store for him. I duly snuggled up to the patient and he dozed off happily while I awaited my chance. I knew my zizis and, as expected, this one grew stiff about an hour after. He was snoring contentedly, and I set to work. The point was that if I allowed him to slip back into indifference he might never get over his inhibitions: it was vital that he should have another 'success' at once. With infinite care I turned back the bedclothes and slid over him so that I straddled him, making sure I kept most of the weight off. I was nicely moist, and I inserted his ramrod, which I fancy was bigger this time. I would have liked to see it, but suppressed the urge to pull the lamp cord.

Monsieur Dombre mumbled and I said: 'Please again, let me do it this time, I can't resist the temptation, please, please.'

I rode his slippery digit for a while and, to my horror, felt him going slack, but immediately pulled him up and rolled him onto me. Still half asleep, he got going, I clenched up to increase the friction for him, and behold he had his second orgasm in seven hours.

I can't even remember him withdrawing, I was so tired.

Which goes to show that sex is most likely to work if you catch it unawares, a thought I give to the experts free of charge.

Talking of money, my *concierge* handed me an envelope three days later. Inside were 30 hundred-franc notes and a message on a sheet of plain paper.

'With heartfelt thanks,' was all it said.

I knew I had to take special care of him at the club. It was the thought that counted – and the money.

Chapter Eight

No man ever entered our studio-flat. I mean the one I was now sharing with my colleague Annick. Later, there were a couple of exceptions, but at that time it was our bolt-hole. And it was home for me.

When I started at the Top Club, I took a rather scruffy hotel room while I looked for a place for myself. Naturally one of the first things I did was to ask around the club, and Blandine quickly said I should move in with her until I got my own pad. I agreed, provided she didn't start anything with me, as it might ruin my new career.

'Nonsense,' she scoffed. 'When you've been here a few weeks you'll be so starved of affection you'll be crying out for a soft feminine hand on your brow, and anywhere else it hurts.'

'Let's wait until it happens,' I retorted.

'Why do you imagine *putes* have it off together after work?'

'We are not *putes* and I didn't know they did.'

'Well they do, because work's work and play's play, and the last thing you want when you get home is another stinking prick waving in your face.'

'Blandine, don't be so crude,' I said. 'You take all the poetry out of living.'

'Still, you needn't blush, my pretty maid. I'll keep my hands off you, even if you have to tie me down in bed. Huh, bondage chastity, I never thought of that.'

As will be seen, Blandine had developed quite a metallic line in humour. I don't think any of us liked it much; one does rather hope to retain one's illusion of being a lady, and indeed the Top Club constantly reminded me of a somewhat snooty finishing school, although that was perhaps a personal fantasy of mine. Meanwhile, we were afraid Blandine's caustic vulgarity might scare away some of the members, not least her own special clients. Politicians and whatnot are so susceptible.

Anyhow, she very decently took me in and kept her hands to herself, staring at me sometimes from her bed as I undressed and got into mine. Normally her husband lived elsewhere but he had a bed in the flat, which I used for a time. The atmosphere was embarrassing and highly charged, but she never broke her pledge.

I eventually found a diminutive studio not far from the club, and tried to make it cosy, though with little success. It was a short but unsatisfying interlude for me. The new job kept me in a permanently het-up state, as I was trying to learn the work, and also because only sometimes did we have full sexual pleasure ourselves. Thus, I was lonely, without a real home, feeling strange. I supposed this odd floating feeling would disappear, but even so I did not feel confident enough to re-activate the numerous friends I had made as a student, or to contact the people Uncle Xavier had listed for me in case of need. In the meantime, it was too early to make any special friends among my club colleagues.

Arid days they were. I spent a fair time window shopping and buying clothes, wandering in and out of the shopping arcades in the Champs-Elysées and scouring the environs. I was accosted frequently and, in view of the exalted nature of my calling, I found this molestation during my time off very impertinent. I knew I had nothing to fear; one phone call to the manager of the Top Club, and the romeos buzzing round me would be whipped off in a police van within minutes. For male readers to understand my attitude, let them imagine themselves to be, say, responsible executives in some big company being leered at and bombarded with suggestive remarks by wrinkled old women 20 years their senior in years, every time they walk along a street. They will then begin to comprehend the exasperation I felt. To be sure I was stuck-up, but it was hardly my fault if I found a 40-hour week administering to the nation's elite enough. The boulevard Romeos had just picked the wrong sex-object.

Having said that, I do remember that I felt especially sorry for the male population one wet January afternoon when I wanted to use the phones in the post office near Fouquet's. Ah, those Champs-Elysées phones! If some enterprising telephone engineer ever tapped them and recorded the conversation, he would have a best-seller on his hands. When I entered the building they were occupied entirely by men, as at all other instants, it seems! Men anxiously fixing up with call girls or trying to salvage an affair that has gone sour; sneaky men lying to their wives; wittering men so timid and frustrated they have to obtain their thrills phoning film stars for their pre-recorded promotional messages; bumptious men making a shot in the dark or importuning switchboard operators. All these pigs helping to make the world go round with one thought in mind, to bridge the gender gap. I stood there with the raindrops sliding down my transparent mac. I knew that rain turned a lot of men on, and I felt vulnerable among these particular specimens.

One of the great unwashed fraternity barged his way past me into a booth, somehow managing to slide his hand across my bottom; he said he was sorry and closed the door, staring out hungrily at me from behind the window. Three other men were waiting their turn, and they too examined me with bright eyes. So did at least two others confined in their cubicles. I was clearly the centre of attraction, which was flattering in its way, as I stood transferring my weight from one leg to the other with an unattainable look on my face, rotating a stockinged foot on a high heel, and then turning round a bit. All those pairs of eyes were leering at me, their owners wondering whether to try their luck with me, whether I had anything on underneath the fragile linen dress I wore beneath the mac and what colour. Some women hate to be drooled at, but I have never had any objection. Either one is a sex-object or one is not, and I like to think I am.

Warming to the men's hang-dog inspection, I lowered my eyelids and was lost in fantasies of my own. I reflected that the veneer of righteousness was as thin as the linen and plastic materials concealing my naked form. If we were locked in, just the nine or ten of us with myself the sole female, how many minutes would it be before a hand stretched out to fondle my hair, gliding down the silvery wet plastic to press at my curves? How long before the immaculate untouchable little *bourgeoise* they took me for was held in an arm-lock or pressed into wimpering submission in the hope that her composure could be shattered and she changed into a yelling she-cat? How long before two hulks had me on the floor with my dress and mac riding up and inflaming their desire? Would even five minutes have passed before the first thick unwashed sinewy penis forced its way deep inside me, its bulging pair of testicles banging against my bottom?

I love to pass the time of day in fantasies of this kind. My prudish sisters make me sad rather than angry, when I think that their moral principles forbid them the right to such poetic flights. Let them reflect that without lust there can be no life.

As for the pigs, I loved them all in a theoretical sort of way. The puritans can say what they like against the panting males in raincoats, those sinister brutes ever eager to sate their lust on us fragile flowers. But these are human beings, dependent upon us, they must have us, and most important, they are our last desperate *raison d'être* if all else fails. Any woman might be only too glad to have their attentions one day. Such men possess a natural passion, which is not to be despised in this passive electronic age. Do the pedagogues really think they are advancing the cause of Mankind by teaching kids to tap keyboards instead of watching the birds and the bees? How can they plunge us into economic depression with their new technology, and simultaneously complain about the low birth rate? Nature knows best – I *know*.

Enjoying my power, my nostrils twitching, I drew back my shoulders and took a deep breath now and again, letting the boys have an eyeful while they had the chance. The attendant eyes never failed to widen, rivetting on the two erect nipples straining through the linen and the plastic. The long wait continued. Then the booth at the end became free, and although it was not strictly my turn, with one accord the men waved me into it. What true gentlemen! I smiled sweetly, telling them they were *vraiment trop gentils*, and felt close to inviting them all back for a drink.

A shock awaited me. As I began delving into my coins, the door opened again and someone pushed in behind me. I turned to see the man who had just vacated the cubicle, a thick-set fellow of around 30 in a blue leather blouson and jeans. The blouson smelled new and blended with the curiously dank odour of our confined space.

'I'll buy them off you,' he said.

'What?'

'I'll buy them off you – your panties.' He was French, well-spoken, apparently clean, probably intelligent, and a good-looker with bright eyes that glistened at me so blatantly that I wondered if I ought to know him. He put me in the mind of the actor Jack Nicholson.

In my best English I said: 'Sorry, I don't think I know you. You have made a mistake, I'm an American and I don't understand.'

He replied in far better English: 'How fortunate, I spent three years in Minneapolis and New York, but I never met an American girl with a French accent.' Continuing in French he added: 'Just take off your panties and give them to me, you stuck-up little bitch. Bet you live in Versailles.'

'Fortunately, that is not Minneapolis and they are not for sale. Now get out and leave me alone. Whatever gave you the idea I was a pick-up; you seem to have a problem!'

'Yes, it's sticking out a mile.' We were still facing the same way and his middle was hard against my bottom. I could not have turned round had I wanted to. He went on: 'You have nice hair and you're no pick-up girl, you are too vulnerable and that's what's turning me on.'

I tried another tack: 'Molesting a policewoman in the course of duty will put you away for years. So be a good boy and disappear before I summon my colleagues.'

'Let's see your card,' he demanded, thrusting a hand towards my bag.

'You don't imagine we carry them on us. I warn you we're training in unarmed combat. And keep your hands off my bag.'

He twisted a strand of my hair: 'If I thought you were telling the truth I'd strangle you on the spot. I hate the decoy game, it's despicable, worse than cock-teasing.'

To my horror I felt his hand creeping up as he gathered the skirts of my dress and mac, then he got underneath. I tried to force his hand down but of course it was hopeless.

'If you don't stop I'll scream the place down,' I snapped.

The hand reached my gusset and he said excitedly: 'You're wet already, you certainly come on fast.'

'It's blood,' I said levelly. 'Why do you imagine I'm phoning?'

The expression on his face as he whipped his hand out and looked at his fingers was so funny I shook with suppressed glee. It changed everything, for I had revealed a hint of complicity. The people that laugh together fuck together! I made a last bid to evoke his better nature, if any: 'Now listen, you've had your grope so be a nice chap and let me make my call and get back to work, I'm late already.'

His palms lifted my bosom. 'Where's that?'

'A fashion boutique across the avenue.'

'Don't they have phones?' A foot pushed between my two shoes and his knee was between mine: 'Don't say a word, just let me take your panties down. I want them.'

I flung my head round the other way to appeal to whoever was in the next cubicle, and saw two other men glaring at me. 'My friends,' he added.

His two hands were under my clothes and he tugged at my

underwear. I was wearing an old pair of yellow briefs I had bought ages ago in Prisunic.

'All right, take them and get out,' I said, stepping out of them as he got them swiftly to my ankles. He observed the damp patch, sniffed at it and, to my astonishment, handed the briefs out to one of the two *voyeurs*. But he stayed inside the booth and thrust his fingers between my legs.

'My word, you're soaking. Admit it, you're simply pretending you don't like it. Why don't you just turn round and let's get together? I'll leave you alone afterwards.'

I said: 'No, and don't touch my bag! There's about 10 francs in it and you can have that.'

He forced me round and stood me in a corner: 'Do me a favour. Take the things out of the bag one by one and lay them on the shelf there, so that the boys can see. They have a fetish about lipsticks and things. Go on, it's only for fun.'

He was grinning and I liked what I saw. He had white teeth, a well-shaped mouth and his smile was attractive. I began unloading the bag: powder compact, lipstick, two safety pins, a tampax, a small hankie, a comb, a separate mirror, a nail file, crumpled Kleenex, cheque book . . .

'Now put them back,' he said.

I did so, glancing outside to see that the main ceiling light had gone out. I said: 'You'll have to stop now, the light's gone out and someone's coming to put it right.'

'That's what you think,' he said, now frigging me quite fast. 'My mate just undid the lamp a bit, so people will keep away.'

I sighed: 'Tell me, do you hang around here all the time doing this sort of thing?'

'Only when it's raining. Fluids excite us. We take it in turns to chat up the girl and the others watch. Telephone cubicles are the best, especially the ones out on the pavement, they're more public, but it's more dangerous.' He twisted his head to one side. 'Look, we've gone this far, all I want is a screw.' He undid the front flap of his jeans, held by two press-studs, and his stiff prick sprang out. He pulled his balls over the elastic of

90

his underpants, so that the skin was tight and said: 'How's that for a cock?'

I surrendered: 'All right, you win. But do it quick.'

'No,' he said, releasing his grip on my arms. 'I don't want to hurt you, we'll take it slowly and I promise I won't come until you have. Once I kept it hard for a whole half hour. Here, take hold of it.'

He placed my hand on his organ, and a strong whiff of his sweat reached my nostrils. I let him unbutton my mac and then my dress top. He pushed up my bra.

'Let the boys see your tits. They're beauties, you're wasted in that shop of yours. My God, you respond fast, the nipples are as hard as pips. Relax now, enjoy it, I want you to enjoy it, aren't you liking it?'

'Yes, but don't twist them, they're not transister knobs, they're tender. Take it easy, I'm cooperating, can't you see.'

He seemed a little crestfallen, saying: 'Of course, sorry. You've a superb body, what's your name?'

'Annick,' I said.

'Mine's André. I'll hold your dress up, Annick, don't be afraid, there's a good girl. Love your dress like that. Now guide me. I can feel your stockings – lovely!'

He moved his pelvis forward and his gland missed the target. I pulled back, parted my knees and went forward again, slipping his rod under me. There was little need for words now. My partner bent his knees and I rubbed myself shamelessly against his wand.

'Like it?' he said softly.

'Yes,' I breathed, 'it's big, are you sure it won't burst? That's good Gérard. No, no kissing please, take my breast.' He did so, sucking in a huge amount of flesh. 'Suck harder, yes that's good, keep going. Oh you're getting bigger, I can feel it, it's a whopper.'

He released the breast: 'My Rolls Royce. Nice and straight, eh? By the way, the name's André.'

My left hand glided between the fabric of his jeans and his bottom, and I found his testicles, gathering them. I whispered:

'I can't wait, I'm putting it in.' He squeezed my breast and protested: 'It's too soon, you must wait, it'll be better.'

Ignoring him, I lowered myself onto his thick tool, squirming until he fitted perfectly, then closed my eyes as I pushed down and enveloped him entirely. There was no stopping me now, an immense vacancy fuelled my longing, my loins were aching. I used him to stimulate myself, clamping the inside of my thighs against his hip bones each time I rose. A surge of warmth coursed through me and I stopped for a moment, my inside seeming to undulate around his penis. Then I resumed, flinging my arms round his neck, with him now thrusting to meet me as I speeded up. My hand was squeezing his balls and he groaned.

'Can you hold back?' I gasped, becoming frantic.

'As long as you like.'

'Oh God it's marvellous,' I croaked, 'it's burning, rippling, you too, I'm getting close.'

I clutched him and he began rotating so that I felt him widening me. His leather jacket was swishing with regularity against my mac and the phone booth made jumping noises.

'Push, I want the length,' I uttered. 'Go straight now, the length!'

Through glazed eyes I caught sight of one of his friends jacking off, his arm like a piston and his cheeks and lips blown out. Something went fizzy in my head and I climaxed, going to heaven and back.

Only then did I notice there were people outside looking in on us, men kind of panting and an oldish woman with a disbelieving hole for a mouth. But my hunky male had suddenly changed into a vicious animal, and let out an 'Aaah', taking his pleasure. I bit into his bull neck as he tensed up.

It was over. We slid to the floor and recovered our breath, neither speaking. I found I was shaking, my legs ached and my left knee hurt where it had scraped something.

'What do you do, André,' I asked, 'when you're not hanging around phone booths and touching up shopgirls?'

'I do voice-overs, dubbing films, TV ads and all that.'

'And my real name's Catherine Deneuve,' I quipped back.

He produced a handkerchief and we wiped ourselves. A puddle of fluid had formed on the floor and I looked away. We stood up, I said 'Bye-bye' in English and lunged for the door, walking off past the small band of onlookers.

It had all been his idea, and he could face the consequences. The last I saw was André trying to mop up the evidence of our tryst with a paper handkerchief.

To return to my accommodation problem, about four months after I reached Paris, Annick told me her flat-mate was leaving for Lyons, and she asked me if I was interested in moving in with her and sharing expenses.

Annick was a dizzy blonde doll a few years older than I was, and blessed with a cheerful disposition. I jumped at the chance. I myself tended to the contemplative and secretive, while she was a straightforward extrovert bordering on the vulgar but in some mysterious way avoiding the plunge. We got on wonderfully together and for my part I found her delightful. She was 100% female but rather dim, and I helped her in little ways like filling out her tax returns, finding out train times, advising her on what shoes to buy and so on.

Annick always seemed to be short of money, as she sent much of her income to Normandy where her mother had a hard job with her alcoholic husband who worked as a railway porter. The rest of Annick's money disappeared like magic, her aim apparently being to spend the franc before it fell through the 'floor', as the monetary jargon has it.

One night well after midnight on one of my days off, Annick arrived sobbing her heart out, saying she could stand the club no longer. This did not entirely surprise me, because everyone gets fed up with their job occasionally. But she had a more immediate reason.

'I'll give my notice in tomorrow,' she snuffled, her blonde curls bouncing in disarray. 'I can't stand these beatings any longer. Look!'

She upped her dress, pulled down her orange cami-knicks

and presented her bottom to me. Horrifying weals traversed her comely cheeks. I let out a gasp, for the skin was broken in some places.

'He's an absolute sadist,' she said, 'and that's the last time.'

'I should jolly well think so, you must be crazy. Who was it?'

'That brute Cattel from Radio-France. How was I to know he would hold nothing back? My God it hurts, it still hurts!'

'Of course it does, my kitten.'

She boo-hooed a bit and I told her to lie down on the bed while I massaged her rump for 20 minutes using a healing cream. It was a bottom simply asking to be whipped, huge and firm, wobbly but high and undrooping, a full moon jutting out from a still-narrow waist. In all the land there could not have been more than a dozen posteriors that had reached such perfection in maturity.

I made her promise she would not consent to chastisement again for a month.

'He got some water and splashed it on my panties knowing it would hurt more, but *I* didn't know it would,' she went on. It took me an hour to calm her down and get her to sleep.

We became close friends and I financed some redecorating, glad to be the loser on the accounts as I was so happy to have a place I could virtually call my own in the company of someone with a soothing character, and similar tastes to my own in some ways.

We found we both had a keen interest in painting, and would do the rounds of the galleries and flea markets. I encouraged her to try her hand at some canvasses, countering her self-deprecation as I persisted with my own efforts. Another pastime was music and I remember we once went to a recital of works by Palestrina, Monteverdi *e tutti quanti* at a church near the Luxembourg Gardens, emerging with the tears streaming down our faces, neither of us quite knowing why.

It was on a dark wet windy night when we were both off duty – it happened sometimes – that I had a phone call from my mother.

We had spent quite a time on routine maintenance to our physical attributes, a necessity in our job. The rain spattered on the window panes, we were warm and steamy and cosy in our nest, and decided to do a complete overhaul and wash our hair together. Mine presented no difficulty, there was really not much of it; but Annick had hers in a lion's mane style that ought to have been done by a hairdresser, only we were trying to economize for a hi-fi system. We had no bath, just a shower, and were doing our best with that. We were swapping tips as women do, I had a towel fixed as a turban round my skull and was untangling the lion's mane. We were looking forward to lying back on one of the beds reading our magazines, Annick flipping through *Paris-Match* and some photo-story things, myself ploughing through my business journals.

The phone rang and I abandoned my flat-mate. Mother said in so many words: 'I thought you ought to know, your Uncle Xavier has been taken ill . . . Well it began some time back but now they have diagonsed cancer, they've got him in hospital for treatment . . . Isn't it dreadful, I can't believe it, you simply must phone Aunt Ginette . . . I am afraid it's generalized . . . So sudden, he never even suspected . . . '

It was still wet and windy a few days later when I pushed open the heavy iron main door at the hospital.

I found Uncle Xavier under a drip-feed with a face as grey as old putty and bags under his staring eyes. I was shocked and in spite of my resolution during the journey I burst into tears immediately I saw him.

We discussed the illness, its origins, the treatment. Then he said: 'When you go back tonight, Ginette will give you a packet of bearer bonds, they're not worth much, about 50,000 francs, but we have agreed you should have them. No don't interrupt. She is also going to arrange the transfer to you of some shares in American and British companies; these should be worth something in years to come. We insist you have them.'

I protested and we argued about it for a while. He said: 'I

wanted to settle that before I go, so I won't take any nonsense from you.' He ended, coughing.

What can you say to a cancer patient who asserts that he will be dead in three months? Everyone knows he could be right. And what if it's someone you are deeply fond of, a light in your world, someone always there in the background in case anything goes wrong? How can you bear to think that soon he simply won't be there any more? I had never had a close bereavement, and like an idiot I wondered if the first time was the worst, as if time meant anything! I attempted a smile and told him 'all this' was nonsense, and we argued some more about the latest treatments and the cure statistics and how they were gradually gaining on cancer.

He wagged his head slowly: 'No, my little rabbit, I've no illusions. To tell you the truth I'd just as well prefer to go, I'd be a wreck and a burden anyway. It's better this way than dragging on. I only hope it's quick, I'm not afraid to die, but I don't like suffering, nobody does. They say they use morphine towards the end.'

'Oh stop it,' I said. We were silent for a while, my tears coming in waves as the idea of him dying took hold of me.

'When I was born Wall Street crashed,' he said, repeating a joke he'd told me before at the Club Des Ducs, but he added: 'God knows what'll happen when I die!'

He moved stiffly, indicating the cupboard door next to the bed. I opened it and he said: 'Take a swig from that bottle, my girl, it's a Chateau Barrateau '76, one of the best they ever produced. I drank half of it last night, the rest's for dinner tonight.'

'Do you think you ought to?'

'Of course I should, what else do you think now it's open? Good red wine cures everything, if anything can.'

Uncle Xavier told me about the funeral arrangements he had given, and how he didn't care either way about a religious service. If Ginette wanted one, that was all right by him. 'Huh,' he croaked, 'it's a lot of hypocrisy, handing over all that money to the funeral gang. But it keeps a lot of people happy.'

'It's out of respect, but let's stop all this, you ought to be praying to God for a recovery, making your peace with Him however it goes.' He chuckled. 'There may be a God after all, nobody can be sure,' I ended weakly.

'I can. There isn't one, if there was we'd know about it, there wouldn't be any doubt. We still can't make up our minds after thousands of years of recorded history, so I reckon we can forget the whole thing. At any rate I've never had a single religious experience in my whole life; no, there's nothing after this lot, plenty of evidence for that, I'd say.'

'Sshh,' I scolded. 'There must be something, otherwise it's ridiculous.'

'It's even sillier that we don't know. L'Anne, just you enjoy yourself. You're my favourite niece, the only decent female in the family, except old Ginette. What colour are you wearing today?'

'Well really! Peach. No, yellow – pale yellow.'

'Let's have a look.'

'No!' I stopped his hand snaking towards my hem.

'Come on, I don't get much fun these days. I still get a hard on though, I suppose it's the pain stimulating the nerves, or the drugs more likely going the wrong way. Come on pull your dress up, that's right, they're tight ones, never mind.' He tugged at my dress and I allowed myself to be pulled over. 'You're lovely and warm, doing me more good this is than a gallon of injections and stuff. Tell me about the Top Club.'

I told him and he ran his hand up and down my thighs. The poignancy overwhelmed me and I wanted to rush from the room. Then I realized there was no point in making an issue of it, and opened my legs to make it easier for him.

'Take your coat off and let's see your boobies. I've become a randy old man at last, got a big hard on. You're more sexy than ever and smell terrific. Here, lift the blanket, how about coming inside? You can slide out again if anyone comes. No wait, put that chair under the door handle.'

'Oh dear I couldn't, it wouldn't work, I'd be scared of doing some harm. Behave yourself.'

'You're as bad as Ginette. Fix the chair, go on, that's an order.'

'You're crazy, we'll bring down the drip-feed thing. I won't do it. Wait till you come out and we'll have a real good session. Honest.'

He pleaded: 'The chair! Please L'Anne, it'll start the recovery, you'll see. I'm feeling better already.'

I fixed the chair. 'You're the only person in the world I'd do this for.' I stuffed my panties into my bag, slid on to him and kept the weight off.

I gained not the slighest enjoyment from what followed, but faked it a bit and Uncle achieved an orgasm inside me, his emaciated frame coughing just after the climax. I took the chair away, knelt on the floor and cooled him off with a flannel. It might be the last time for him, I thought grimly, and was pleased I'd done it for him after all. Especially as it was for real love.

An orderly came in with his lunch tray, sniffed and opened the window slightly. I watched Uncle pick at the food, then said *au revoir*.

I smiled, trying to be radiant: 'I'll be down again soon, I'll phone anyhow before that. Don't forget we've got a date.'

I did phone, but that was the last I saw of him. He died suddenly just after New Year 1978 and nobody bothered to warn me the end was near. I was livid with anger, protesting that I had a right to know, but Aunt Ginette merely said the hospital 'helped him go and it was all too sudden.'

I was inconsolable from start to finish when we laid him to rest in the village where he was born near Aix-en-Provence.

Adieu Xavier, I'll love you for ever.

A few mornings later I heard funny noises coming from the bathroom.

As usual Annick and I had woken around 10 a.m. but I had been more than usually slow and she had finished her breakfast before I sat down to mine. It looked to be one of those blurry days when nature seems to drug the body and impose a rest.

98

Drizzle had coated the windows and it was dark grey outside.

A sort of woo-woo sound tickled my eardrums and, as we never really closed the bathroom door, I downed the last of my coffee and crept towards it, listening carefully. There was no doubt about it, the noise was of human origin, unless she had installed some kind of ape in there without telling me.

With infinite caution I eased the door open, and beheld my flat-mate sitting with her back to me on a stool in her strawberry bathrobe. Her left elbow was jerking rhythmically, the sleeve was flapping and she was rolling her head about moaning and snorting. I froze with my fingers on the door handle, not believing my eyes. Annick was masturbating!!! After years in the Top Club she was having to do that!

My jaw dropped and for perhaps 10 seconds I watched fascinated by the billowing bathrobe and the sway of her bottom on the stool.

'Annick?' I whispered, not wishing to frighten her. She stopped instantly and spun round, her face growing crimson.

'I was massaging my knee,' she said. 'I thought you were still in bed.'

I smiled fondly upon her, amused at her look of guilt: 'Oh Annick, you're telling stories. You're having it off on your own!'

'I – er – ,' she began, and then to my astonishment she burst into tears.

Running to her I cried: 'Annick, whatever's the matter?'

She blubbed: 'I feel so ashamed. I didn't want to, but I can't get excited any more at the club. I've got a sort of blockage and I'm all tensed up. I have to fake it all the time at the club, and I thought . . . Believe me I tried to sort it out, hoping it would go away, but this morning I felt like screaming, I couldn't stop myself. I feel awful, honestly I swear it's the first time, I'm so ashamed, I feel so randy in the morning these days.'

She sobbed some more and I smothered her with kisses: 'There's nothing wrong with that, Annick, *ma petite, ma chérie, mon amour, mon enfant.*' It was true, she had a mental age of about 15. As I mentioned earlier, she was a little dotty and I had rather taken her in hand. She also happened to be a super-dooper sexy blonde piece with a luscious hour-glass figure, Germaine Greer's 'female eunuch' in the flesh! And – in her present distraught condition – entirely at my disposal. More than once I had wondered what it would be like to make love with her.

Obviously I had to help her now, and I crooned: 'You should have told me, *ma chérie*, I'd love to play with you. I'm sure it's much better than doing it all alone.'

'But you're not a lesbian,' she exclaimed. 'Have you done it with other girls?'

'Once, years and years ago. It was fun. Have you ever done it?'

'Oh no, I've always been scared to.'

Adopting a confidential tone I said: 'Let's do it now. Who knows, we might be a.c./d.c.'

'What does that mean?'

'Lesbian as well as hetero.'

'I didn't know you could be both.'

'Of course you can, silly.' I gazed at her through my eyelashes, adding: 'You're very beautiful, Annick, and I'm already getting clammy just looking at you. Please let me play with you.'

She was clutching her garment round her, and my hand glided down to open the folds again, stroking her satin-smooth abdomen. 'Let me caress you,' I pleaded. 'I *desire* you.' Gently parting her legs, I slipped three fingers over her fleece, then moved them against her grotto. She gaped at me with a blend of innocence, gratitude and renewed shame as I entered her sloppy cleft. She buried her head in my neck, her lower torso thus moved forward and she said: 'Oooh, it's divine!'

It then dawned upon me that, if I gave her an orgasm at once, the fun would be over for me as well as her. And I was fast becoming as worked up as she was.

'Why don't we make it last, build up to it?' I cried. 'We can have a bath together! Oh yes, Annick, it'll be so good, you'll

100

see.' Already my eyes were darting about, looking for phallic objects we could use.

She wailed: 'Don't stop now.' But I left her and twisted the taps, poured in some bathsalts and filled the bath with about 10 centimetres of water. I asked her to choose a 'penis'; it was like teaching a child! Shyly she pointed to a shampoo bottle with a longish neck and a purple knob on the end.

In view of Annick's plumpness relative to mine, it was no easy task fitting us into the bath. We began by standing together in the shallow water and I was able to assess the full voluptuousness of my lady in distress. Her haunches and thighs were tight with solid flesh below a still-slender waist and her magnificent breasts pressed against mine, overpowering me. They were lolling outwards and I squashed them together once or twice and made them tremble like big jellies. Her mouth hung loose and she breathed fast between cushiony lips. Her mane of hair completed the wondrous picture. Annick was sensuality itself and I was hot for her. She would have let me do anything with her in the state she was in, and that fact stirred my cauldron of desire. I really wanted her now. We knelt down and I placed my own full lips against hers, while stroking her repeatedly from nape to thigh-top with a finger. This excited her, I knew, for she was unable to keep her mouth still. I opened my own mouth and explored hers with my tongue; as expected she copied me and I sucked greedily at her. We spent a long time in this manner, enjoying the tiny thrills.

I set out to lead this innocent through the paths of pleasure, soaping her all over, stroking her smooth sud-coated skin around the neck, her nipples, over her belly and between her legs, subsequently turning her round and massaging the cheeks of her gorgeous pink bottom. Everything I did she repeated for me and we ended up squeezing nipples in turn while clawing at the other's crotch.

Eventually we sat down so that our quims touched. Annick said she thought she was going to faint, and I said it would pass.

101

'Laure-Anne, you are laughing at me,' she said. 'I feel terribly wicked, but I still haven't come. Please give me an orgasm now, Laure-Anne, please!'

Stretching out for the shampoo bottle I told her: 'Place your legs on the sides.' This she did so that her crack was indecently exposed, and I bent down to lick her. I did not let up for a full five minutes, worrying her clitoris even when it retracted, lapping and lapping her splayed vulva scores of times until she was writhing and gasping, her hands pulling at my hair.

It was time for the ultimate baby talk: 'Now Annick, be a good girl and I'll insert this lovely purple penis. See how long it is, sticking out with the big round knob on the end. Shall I put it up?'

'Yes, oh please!'

'Keep saying it.'

'Please, please, please . . .'

'Say "I want it".'

'I want it.'

'Say you worship me and you're frantic.'

'I worship you, I'm frantic.'

'You're going to die with pleasure.'

'With pleasure.'

'Plead with me.'

This disconcerted her as she had to think, but I rubbed the bright coloured knob along the lips now swollen with blood and she cried: 'Oh yes, put it in now, please oh please! I worship you.' I began inserting it, pushing it in the full length and withdrawing it. She panted: 'Oh God, again, that's it, faster, oh faster, don't stop!!!'

Annick's belly rose and fell mountainously as she sought to draw in more of the makeshift dildo. With my other hand I found her anus and pushed a finger in, then two, such that her two openings were ravished. At last her countenance twisted horribly and she let out a series of shrieks, heaving again and again. My arms ached, but finally she gave an enormous sigh and splashed down into the water.

I was at fever pitch, my feet and thighs tingling as if I had rheumatism. Quickly I placed her limp ankles on the sides of the bath, then lay upon her so that our raw pussies were slithering together. I kneaded her big tits until she yowled in pain, but I was past caring, jigging rapidly until my whole body stiffened and I came – thrashing Phallic Woman in a savage paroxysm beyond all control.

In months to come she was to suffer from my hands perhaps a dozen times, usually when I was feeling nasty and had drunk too much. She never refused me and I treated her outrageously.

Whatever else my dream cottage of the future may contain, I vow that it will have a bath to my own specification!

Chapter Nine

Around that time Gérard called me into the office one evening at 7 p.m.

'Got a rush job for you – big stuff, orders from the Elysée. When did you last have a decent meal?'

I stuttered: 'Why, about two days ago, I don't bother too much about food.'

'You'll have to tonight. In about two minutes a police car will pick you up and take you to La Celle St Cloud where you will be fed by the Republic, in return for which you will cement Franco-West German relations by entertaining a VIP from Bonn until such time as he turfs you out. Get your coat on.'

'Oh yes, of course. Who is it?'

'You may recognize him, but you will forget him the moment you leave him. As far as you are concerned his name is Herr Bundesminister Schmidt.'

I gulped: 'Not the real Schmidt – Chancellor Helmut, I mean.'

'Don't be dumb. No, this one's more important. You will accept no money, you understand. And don't mess it up, otherwise they'll be reopening the Siegfried Line or whatever. And his name's not Schmidt. Believe me, the Club doesn't get a *sous* out of this, but we've had the Elysée on our side up till now, so let's keep it that way.'

Our doorman entered and said: 'Ready, Gérard.' I got my coat, scarf and bag and within a minute was charging across

104

the pavement, dodging between a pair of parked cars and diving through an open car door held by a plainclothes gent in a blue suit. He followed me in, slammed the door and we surged forward. I just had time to notice that the vehicle was an anonymous dull beige in colour with no markings or flashing light. But it had a strident pam-pom-pam-pom siren which the driver began using as we entered the maelstrom round the Arc de Triomphe.

After about five minutes of this I ventured to ask: 'Does he have to work that thing all the time?'

'Peak traffic, we're in a hurry. Just hold on to that strap and keep quiet, we may throw about a bit.'

I felt like throwing up, but held my tongue. The cop looked real mean.

Our journey took about 20 minutes. We bullied our way past the snarl-ups, sometimes on the wrong side of the road. I closed my eyes and thought about our manager.

Gérard was nice, which is about the best compliment a woman can give any man. I put his age at 35, knew he was getting a divorce and that it was taking ages. I had learned nothing else about him. Of average height with an open face, sandy hair and dark blue eyes, he dressed in good quality casual clothes most of the time, and was a born diplomat. I liked him from the moment I set eyes on him and fell in with his clipped style of handling people. He was fun, we all agreed, with a ready flow of wisecracks that took the tension out of any situation. All the girls loved him, and so did I. More than I was prepared to admit.

We hurtled to a halt outside a pair of gates with gold knobs on. The cop flicked his wallet open, and the CRS guard opened the portals. We scrunched up a drive to a large white house where the door was flung open.

'Here, carry this, it's your alibi – documents,' the cop said, handing me an attaché case. It felt empty and I got the message.

I was taken in, passed to another man who hustled me up a grandiose staircase and along a passage to a door. My escort knocked, the door was opened slowly, and I found myself looking at a white shirt collar and shiny yellow striped tie. Around these was a large blue double-breasted suit containing an elegant grey-headed man nearly 6 feet tall and about half as wide.

I raised my eyes and a smiling mouth between chubby cheeks declared: 'Bonshour, ma-de-moi-sselle.'

'Herr Bundesminister,' I replied. He waved me in and the door closed behind us.

The meal was ambrosial: following a glass of champagne, we had *consommé froid, toast de crevettes à la Rothschild, gratin de fruits de mer, salade Niçoise, plateau de fromages* (which neither of us touched), *ile flottante*. The wines, white and red, were in decanters and neither of us knew what they were. I had no say in the menu and could only suppose he had opted for fish because he was fed up with *wurst*.

Our conversation mainly concerned food, the language barrier, Japanese exports and his wartime experience in the Lyons area. There was an awkward moment when I asked if he ever came across a fellow called Klaus Barbie, but he said he drove a tank and my gaffe was lost in reminiscences of the good times his unit had had with the local population. I refrained from mentioning that Barbie and his mates had tortured and killed a relation of mine, a Dominican priest, and we moved on to an intricate discussion of the difference between a Federal Europe and a Confederation of Europe. By then we were having coffee on the settee.

He then said he was surprised I was not married.

In an attempt to provide him with some local colour I explained: 'We French women are very independent. In Paris, the latest census showed 430,000 women who were married, 310,000 spinsters, 150,000 widows and 80,000 single and divorced women.

'Really.'

'I think we like variety. I find all sorts of men fascinating, for example, I kind of flutter in their presence.'

It was true, I really find men a most absorbing subject for

study. In the case of my present companion, what gripped me were his hand movements; his fingers were chunky to match his bulk, and they were grouped like a bunch of miniature bananas, yet they were graceful.

The said fingers were soon warming my right knee. As to what followed, Herr Bundesminister displayed a voracious sexual appetite, but was gentle and considerate despite his size and the homeric proportions of his *schwere artillerie*.

We made love classically and lengthily, though only twice, ending the night giggling at 3 a.m. with champagne bottles strewn on the carpet, and he crowning me Empress of Europe with a curtain loop.

All in all, it was one of the most enjoyable duties that came my way. My client accompanied me to the hall and kissed my hand. Nobody had ever done that to me and I rode home in a pro-German haze. I cannot remember what happened to the attaché case.

Herr Schmidt asked for me when he visited Paris thereafter, and I saw him three more times if my memory serves me right.

Chapter Ten

Our wardrobe included a lady traffic warden's dark blue outfit, acquired by a club member, an Israeli women's army uniform acquired Lord knows how, the whipping set I had used on Marcel Giraud, a traditional Breton dress, some schoolgirl's gear, a nurse's white overall, G-strings and whatnot.

But pride of place went to our 10 drum-majorette outfits. Actually, the only reason they were so named is that we used to march in them round the playbed on Thursday nights. They consisted of a short white cotton tunic with shiny gold edging, gold belt, coronet and sandals, white speckled stockings and black underwear, plus red/gold/black garters with two ribbons trailing down a few centimetres.

The playbed was available for everyday use, of course, but was more specially intended for Thursday, which was gala night. Organized by the management, this was also termed ladies' night, because it was understood that the gentlemen would lay aside some of their arrogance and try to make the girls enjoy it too.

One of the gala performances has stuck in my memory because it was the night the mouse got loose and Nadège forgot an article of clothing.

Gérard the manager had switched on the Yankee-Doodle music and checked us over. He had given us our iced lollies to suck when he cried: 'Hold on a second. Nadège, you've got nothing on underneath.'

'My word, so I haven't,' she gasped and Gérard helped her into her puffy black knickers. Nadège, waving her frozen lolly about, promptly stuck it to the garment and we had to rip it off and some of the material went with it.

The lollies were Gérard's idea. He said they made us march without rolling our heads about and forced us to move our shoulders back and forth, which looked alluring. It also excited the clients when we pushed them in and out of our mouths. Anyhow, that's what Gérard said.

Off we went, lifting our knees so that our thighs were alternately 45° to the ground and the audience could get little peeks at our bloomers. Everyone clapped rhythmically and chanted '*les girls – les girls*' when we circled the king-size bed, our lips protruding round the lollies. The third time round, we offered the juicy things to our public, and allowed ourselves to smile. It was warm and the large oval-shaped room was thick with some special perfume the committee got hold of. I always found it rather sickening, but it was not my place to say so.

We did another turn, edging near the unlookers sitting on their couches, but some standing.

I think we were quite professional and at one stage there was talk about offering our services to the Elysée Palace. That night there were eight of us. Here is the roll-call in ascending order of height:

1) Bernadette. A curly-haired chubby red-head (dyed) who got away with her excess fat solely because she was small. Slightly bow-legged but nobody mentioned it.
2) Mauricette. Neatly turned out, mid-brown hair above a triangular face. Tendency to scowl, shaved between eyebrows. Much in demand from masochists for cellar duty. Enjoyed her work. Some men called her the nutcracker.
3) Annick the lion-headed blonde. Cheeks aglow, haunches asway. Highly popular with everyone.
4) Myself.
5) Blandine in her squirrel wig. A bit horsy, good nostrils.
6) Marie-Odile. Our oldest girl and worried about it. Her best feature were her perfect legs. She kept on about getting a

job at the *Lido*. Engaged to be married, so unlikely to be with us much longer. Her calves and ankles were thickening.

7) Nadège. Raven-haired and milk-white complexion, little breasts. A dreamy miss in the 1920's style. Lots of class, though.

8) Ghislaine. A copper-haired droopie, claimed she was related to the Comte de Paris, Pretender to the French throne. No pretence about *her*; she would make love with a warthog rather than wait.

There must have been about 25 or 30 people in the room. Half of them were there to look, and one or two of the women seemed to be members' wives. Gérard was scared of these, as he was certain they would one day sabotage us and the place would have to be closed down. The press like nothing so much as a 'secret orgy' scandal involving their betters.

The electrician was dimming the halogen lights and we were high-stepping round the apple-green playbed on its honey-hued wall-to-wall carpeting, when Mauricette released an almighty scream and jumped backwards.

A mouse was running towards us, and we scattered instantly, some of us jumping onto the bed and folding our tunics round our legs.

Everyone got up from the couches and someone said: 'It's a toy, it's a toy mouse!' And it was.

We all looked round to see who was responsible, and we saw a huge fat pink-faced white-haired man in a blue suit against the wall, shaking with noiseless laughter. The entire company frowned at the culprit and he scrabbled under a couch to retrieve the animal.

This man turned out to be a Swedish giant called Thage Something-or-other, who had come along as a Ministerial guest. He spoke no French apparently and there was nothing much anyone could do about him. Quite unaware of the poor taste of his joke, he gabbled away in English as we crowded wide-eyed round him.

Then he said: 'I tell you joke. What happen when the streaker ran through a convent? Come now what happen?' We

110

looked blank. 'I tell you: all the nuns have a stroke! Ha-ha-ha!'

None of us got the point, and he went on laboriously to explain what a streaker was, which most of us knew, and tried to show how 'stroke' was a play on words. Since nobody knew the French medical term for a stroke, we got into a shocking mess. But we had to be polite to this guest of the nation, a dealer in perfumery raw materials.

Gérard eventually stormed over and got us marching again. We could see he was furious, as the interruption had ruined the piety of the occasion.

The routine finished, we joined our admirers on the couches, at which point the music changed to a West Indian rhythm, and Blandine and Ghislaine moved towards the playbed. We all fidgeted, sharing our lollies, some of us sat on laps, and I landed on the knobbly knees of my old friend Monsieur Dombre. He was well into his new lease of life and he tickled my waist, making me wriggle. I had to stop him exploring further and slapped his hand.

Our electrician brought the lights up a little as the two women knelt facing each other on the bed. They kissed to start with and began running their hands over each others' bodies. This was the unvarying commencement to the evening's show. Blandine caressed Ghislaine's bosom and she responded likewise. They slowly took off their tunics, then the sandals, bloomers and bras, leaving just the stockings and garters. Alternately they licked and sucked each others' breasts while rubbing themselves between their legs. Blandine gently pushed her friend onto her back and buried her face in her quim. We all got nearer and watched them speed up, in due course Ghislaine spread her legs very wide and we could see Blandine's tongue at work. I never failed to appreciate this, as Ghislaine's sex was absolutely prodigious, a deep red gash with her clitoris the size of a hazelnut. Blandine's bottom jutted up in the air and she was still rousing herself as she operated. Eventually she pushed her partner's legs right back so that she lay there like a baby waiting for a new nappy. Then Blandine slid onto her so that their vulvas met and she went

111

back and forth, her rear moving up and down most erotically. They continued for ages and both got it off. Suddenly Blandine shifted so that she was operating diagonally to Ghislaine, whose arms and head thrashed around. My word how they wriggled! Ghislaine seemed to be in ecstasy but Blandine found it hard, finally letting out a deafening shriek and rolling away in a state of exhaustion. She once asserted to me that the most difficulty you have the better it is in the end. We applauded and they were helped away by Bernadette and Mauricette.

The six of us who were left climbed onto the bed and assumed artistic positions. As Gérard instructed we widened our eyes and looked innocent.

The men began undressing, leaving their clothes in disorder on the carpeted floor. We girls sat like mermaids holding hands, and a couple of men took Annick and Marie-Odile. Others joined us and I was soon being fondled all over by two of them. My bloomers were off within seconds and I said: 'Take your time, gentlemen, we have hours ahead of us.' A bony man with a long organ was the first to take his pleasure from me, and I could have wished for a less abrupt initiation, he had me in position 1. Another covered me at once, it was Marcel 'missile' Giraud.

A pause, during which I got up and walked round the bed, the music being softer now, a piano playing quiet swing. Another client took hold of my shoulders from the back and, working his way down my front, unclipped my bra. I was of course still in my tunic, for we endeavoured to keep these on as long as possible. I could not see who was trying to mount me, but he was using delicate movements and for the first time I felt the familiar tingling inside. The two buck rabbits had not worried me; on these occasions there were always a couple of cock-happy guys who could not or dared not wait, which is flattering for a girl but no substitute for pleasure. And pleasure was our right on gala night. Normally it was our duty to sacrifice our enjoyment if need be, but on Thursday the whole rhythm was more easy-going and everyone knows that women

112

do not like to be rushed. But I am digressing. My suitor massaged my neck, which is something I adore, and then readied me in an all-fours position with my bottom trembling at the brink of the bed. He spread my legs a wee bit and arched my back, then stood behind and moistened me generously with my own juices. I was enjoying this, arching further when his phallus replaced his fingers. He entered me warmly and slowly and I caught my breath. When he had finished I turned to discover a youngish man, quite good looking, someone else from the Foreign Ministry. I thanked him sweetly and we parted company.

A crescendo of noises was coming from the adjacent bathroom, and a group of us gathered to watch Thage the Swede taking a bubble bath and yapping away in English. His head and shoulders rose pinkly above the froth and he was trying to induce Mauricette to join him. But, small as she was, the idea was absurd in view of his bulk as there was hardly room for her, so she refused. This was always a dangerous thing to do but Thage seemed none the sadder having got his audience. 'I give you surprise,' he announced, rising to reveal a diminutive zizi stuck under his enormous belly. Without drying himself, this Falstaff from the Frozen North stepped out and rumbled over to his pile of clothes in the main room. He extracted a plastic bag of powdered stuff, which he raised high and described as 'one of my penises'.

Everybody came near and he went on: 'This bag contains freeze-dried penis of reindeer. A new order by Chinese in Singapore. Five tons they are ordering. Each penis weigh 10 pounds so we freeze-dry them. I take orders if you want.'

Marie-Odile said: 'What do the Chinese do with them?'

Thage: 'Eat them, so they have erection.'

Bernadette: 'How cruel.'

Thage: 'No, is nice for Chinese.'

Bernadette: 'I mean for the reindeer.'

Thage: 'Chinese eat dogs, don't they? Now they eat reindeer bits too, like you French eat pigs' balls.'

There was really no answer to that, and I asked: 'You mean this powder is aphrodisiac, is that it?'

Thage: 'Exact. We cut off one thousand penis for order, then freeze-dry. You like to taste?'

Me: 'How do we know this powder is the real thing? You could sell us anything.'

Thage: 'Not so. My Chinese came over and inspected reindeer. Chinese are very little and they walk under animal to find the best.'

Marie-Odile: 'How do you know it works, this powder?'

Thage: 'Not my problem that is, they pay 200,000 dollars, I am a very rich man.'

Me: 'Have you tried the powder yourself?'

Thage: 'No. We persons of Scandinavia have no need of powder. I now show you my North Pole ha-ha-ha!'

Before we could wonder what he meant, he made a grab at me, I skipped away and he caught his beloved Mauricette by the arm. He was overpowered by some of the men and dragged to the bed, where Ghislaine lowered her ever-willing pussy onto the fat man's now fully extended Polar region.

I did not stay for this, finding the whole episode distasteful and I specially wanted to avoid the Swede's attentions. Annick and I sneaked out for a snack at a bar near the club.

When we returned at midnight, he was gone. It appeared that he had drunk deep before arriving and considered it his role to liven up the party. Sadly he did not fit in with us at all, and we were glad to see the back of him.

Annick and I must have looked bad-tempered when we entered the room, which was unusual for us.

'Cheer up, girls,' one man said. 'Why so sad?' another asked. We were again fully dressed in our majorette outfits, and we contrasted with the general company, most of whom were in the nude.

Willing hands undressed us and we found ourselves, just the two of us, on the bed with several of the members. We knew what was coming and so did everyone else. A hush fell on the

114

room, and the men fell upon us, eventually separating us so that we were each adopted by two distinct groups of men. I was stood on my feet and assailed from all quarters: one set to work on my breasts and another sat under me with his mouth busy, a third spread my bottom and advanced his tongue there, others caressed my legs and neck, under my arms and all over my loins. We were a mass of bodies squirming in silence to the delight of the onlookers.

Needless to say, I was quickly roused to a state of white-heat and prepared to be sired repeatedly. I lost count as they took me as they liked: on my back, on all fours, standing, on my side. I became a raving, jerking object of the men's lubricity, a bitch on heat being ravaged by a pack of wolves. I lost all control and screamed and screamed as successive waves of frenzy swept through me.

The audience loved it, I am sure, and I vaguely heard their aahs and oohs. My heart was racing and I was streaming with sweat but the males kept on assaulting me *con passione e forza*. They gave me no respite, I could hardly breathe and cried out, 'It's too much, let me go.'

They paid no attention of course, and the large detachable sheet that covered the bed was wet with perspiration and spent semen, so that my body slithered about. I gave myself up to the voluptuous orgy. I had protested, but now I wanted the festival to go on for ever and they knew it! I wept in bliss, though I was now incapable of muscular effort and could hardly see.

I was lowered onto a penis, my hair was pulled back and another entered my mouth, my legs were thrust forward and I was bent double. The last ounce of vitality had been wrung from me, I offered no resistance as a phallus carefully entered my back passage, advancing within me and swelling as its owner vent himself. This hurt and I became aware again. Two others ripped into my vagina and pumped away like wild beasts. I wondered if I would split in two or pass out with a heart attack.

At last they released me, delicately laying me on my back to recover, the Foreign Ministry man soothing my brow with a kiss. The ordeal was over, the male pack left the bed and I stayed there trembling. Annick lay prostrate a yard or so from me on her side.

Applause reached our ears: 'Bravo Annick, bravo Laure-Anne, superb, superb, *les petites chattes*!'

If bouquets of flowers had landed at our feet we would not have been surprised.

I could hardly walk for a while. Fortunately we took it in turns to be the finale each week.

Chapter Eleven

The apartment occupied by Blandine and her art dealer husband Gaston contained some fine pieces of antique furniture including a *Directoire* secretaire and sofa and an Empire piano, the three items alone being worth a third of the flat's value. They also had a 17th century astronomer's armillary, a clock thing called a *nocturlabe*, some ship's compasses and some early toy steam engines. And it hardly needs mentioning that the premises boasted valuable old prints, paintings and china. It was all worth a fortune and they had a steel door with three locks plus an electronic surveillance system.

I was no different from the nation's other 50 million or so inhabitants, none of whom actually dislike acquiring wealth (no-one is on record as throwing banknotes into the Seine) and all of whom clutch it to their hearts when they have got it, as shown by the existence of a vast tax evasion industry.

I got to thinking more positively about money when Blandine invited me round to tea on one of the Sunday afternoons that Gaston was there.

'We prefer to invest in valuables rather than a country cottage or something,' Gaston explained. 'As the crisis deepens, the Government will be taxing everything they can lay their hands on, especially second homes, but they can't lay their grubby paws on all this. Meanwhile we have the pleasure of living with it.'

I nodded knowingly. He held forth in his professional

machine-gun style: 'Provided you stick to authentic items, you are certain to keep your head above water, by which I mean you keep ahead of inflation, maintaining buying power, which is of course why most things we deal in generate enormous added value with the crisis.' He went on to give some examples and added: 'Do you go in for this sort of thing, investment in valuables I mean?'

I answered honestly: 'Not so far, I'm afraid, I have no money and no place to put furniture. My uncle once told me that fortunes are made in the bad times, when the rich buy cheap and wait for it to go up.'

'Obviously,' he said, 'you have to be prepared to wait, sometimes 10 or 15 years or more. Your uncle sounds like a wise man, because the wealthy can indeed afford to sit out the crises, waiting to get wealthier. It's no good buying an Egyptian ushabty in September and hoping to make a profit for Christmas. May I be nosey and ask if you have much in the way of reserves at the moment? I might be able to do something for you.'

'That's very kind,' I said with sincerity, 'but I'm no Baronness Rothschild. I was left some bonds and American shares, worth about 120,000 francs the lot, the bank tells me. Plus a bit in the savings bank like most wage-earners.'

Gaston regarded me as if I'd stuffed my money behind the bidet: 'Savings bank! I hope it's not too much. My poor girl, how much is it?' He was really aggressive about it.

'I don't know, about 60,000 francs I think.'

'My angel!' he whooped. 'We must have that out straight away. They're giving you 7% with inflation at 12. Every year you're losing 3,000 francs, giving it to the government, who are the last people to be trusted with money. Those bonds, what do they yield?' I looked vague. 'Huh, 9% at best, I bet. How much is the principal?' I looked even more perplexed. 'The face value, if you like?'

'Fifty thousand francs.'

Bossy-boots leaned forward, clenching his fists: 'You can keep a wee bit in the savings bank, keep the shares but give me

118

100,000 francs and I'll make it 300,000 in five years. We'll restrict ourselves to articles that can be moved with your own two hands, the smaller the better.'

Blandine interjected: 'Don't bully her, perhaps she's got other ideas, buying a flat for example.'

'Rubbish,' he boomed. 'Property will stop dead soon. There are only two ways of winning now, currency speculation in some form, and antiques.' He shrugged as if to say it was none of his business how I threw my money away. 'Any more of that cake 'Dine? Blandine is fabulously rich of course thanks to me, in return for which she does my shoes and keeps the place clean.'

His wife snarled: 'Don't take any notice, he's a chauvinist pig.'

My visit drew to an end with further remarks along these lines. But, through Blandine, I was on to Gaston two days later.

'That you Gaston? It's Laure-Anne. You were serious on Sunday when you said you might be able to help me? I know you're busy but . . .'

'Of course, my dear. Blandine's friends are my friends; we need them these days.'

I tittered for no particular reason and said politely: 'There's just one proviso, Gaston. You're very kind, but could you possibly let me watch an auction first at the Hotel Drouot? I want to see what's involved. Perhaps one of your nice young men could walk me round and explain things.'

'No problem. We have lots of nice young men there, but unless there's an earthquake I shall insist on showing you the ropes myself. I shall personally handle your account. You see, I'm a nice *old* man, so much more reliable.'

A glow of amity surged within me: 'You're a real friend, Gaston, I'll never forget this gesture, I know I'll be safe in your hands.' He hooted with laughter and I said confusedly: 'Oh, I didn't mean it that way.'

We became serious and fixed a date.

We were serious too on the day. I was anxious to learn, and to learn fast. Gaston was a first class teacher, neither showing off nor making jokes, but simply giving me a basic run-down on the proceedings, what to look for and to avoid.

'Antiques,' he said, 'used to be defined as anything more than 100 years old, but now it covers Art Nouveau and Art Deco as well. But you don't need to worry about that. Half the transactions go through dealers and the rest through the auctions, except for some clandestine trade. Drouot is the safest because, as Renoir said, it's the real barometer. Some of our clients don't even see the stuff, they phone and say what they want and leave it to us. You could do that too. Now, as to added value, you can roughly count on doubling every 5 years, although we shall do better; if times are good people have money to spend, if times are bad they seek secure placements. A two-way certainty, provided you deal in good stuff. It can't help going up, what with money depreciation and museums and collectors grabbing everything, and people who are prepared to wait for decades, and the big houses needing good items. Quality always pays off but you should buy what appeals to you. Peter Wilson, ex-boss of Sotheby's once told me: "Never buy unless you really like it yourself, never buy solely as an investment, and you can be sure of making an excellent long-term investment." He was so right.

'Incidentally, if you ever do business with an antique dealer, never say "What's the price of that?" because he'll know you are a mug; always say "How much are you hoping for?" You see, when he realizes immediately that you are going to buy and know what you want, he has to bring down his price. Curious, but that's how it works. I'll do the buying for you, dear, don't worry. There's commission of course, usually 16% up to 6,000 francs, 11.5% up to 20,000 francs, 10% above that. So if only for that reason you should buy a few good pieces only.'

We watched an auction going on, and he instructed me: 'When you are buying, stand like this along the side of the hall, and you can see all the bidders. Over there, see that chap in the blue shirt, he's working a cartel with the fellow with the moustache, trying to force the price up; then they'll suddenly stop bidding, you'll see. Got to keep a cool head when you are bidding, and a good tip is to repeat your bid if you mean to go no higher because you stand a good chance of acquiring the lot in question. Keep still now, keep your hands down, because if they knock something down to you it'll cost you money even if you put it up for sale again and it reaches the same price; the charges, you see.'

And so it went on. I was infinitely grateful to him, and we parted on the understanding that I would concentrate on vases and statuettes. I came away scared about committing myself, and the fear heightened when I gave instructions to my bank to sell my bonds and to transfer most of my savings to the bank, in readiness to pay Gaston.

So it was that in the Spring of 1979 I became the owner of: a Roman terra cotta perfume vase from Apulia, 3rd century BC, supported by a crowned woman's head (colours sand, light turquoise and chalk); a clay Egyptian funeral statuette of a priest, 26th dynasty, with hieroglyphics on the shroud (colours chocolate, deep mauve and chalk); a Greek pottery vase, 4th century BC depicting Herakles and others with friezes (colours dark brown, red figures but white for Herakles). These beauties were photographed, certified, insured and stowed in a safe deposit at my bank, where I could go and look at them when I wanted to.

My assets thus comprised some 100,000 francs in antiques, 70,000 francs in foreign shares, 10,000 francs in savings. Total 180,000 francs. I was adding 3,000 francs a month from wages to the savings, so I would be buying more antiques in due course. Assuming Gaston's added value boast was right, and if the shares rose 50% in the five years due to a U.S. recovery, then my 180,000 francs of assets would become around 600,000 francs, I reckoned – the price of a fair-sized apartment! With capital jumping at that rate, it seemed too good to be true, and I sensed it probably was. I resolved to keep my own counsel in case events turned out to be less than a triumph.

So much for money. I was as much concerned at that time with the ravages caused by my job. I had been at the Top Club well over two years, during which, allowing for holidays of various kinds, I had experienced an average of 3 orgasms a day, according to my very rough estimate. This in itself is not in the nature of exhausting labour, but the fluctuations in pace and the late hours wrought havoc on the nerves.

I was striding along Rue Ponthieu one day shortly after midday late for work, when I bumped into Gérard.

'Why, hullo Gérard, given you the push, have they?' I chirruped.

'Not yet,' he laughed. 'Fred's in charge, I felt like a break.' Fred was the assistant manager. 'How about a drink? I was heading for that corner bar.'

'I'm late already, I've been for a health check,' I panted.

'Don't worry about being late, you're with me now.'

'Well fine, I could do with a beer, I'm parched. Isn't it warm!'

Gérard intrigued me. He was a quiet type, polite, easy to get along with, tons of self-control, but tight-lipped too. Aged now about 35-38 and single, he had a flat tummy, was of average height and had one of those pale, bland, flattish faces with a pointed nose stuck on it. Easy to overlook, he would have got by as a waiter, a supermarket manager, even a bank clerk. At the Top Club he behaved impeccably towards us girls, never suggesting anything, simply doing a self-effacing job of work.

With him so diffident and myself a mite bashful, I knew practically nothing about him as it was always strictly business between us. Professionally I respected him and he was always well-mannered towards me.

'You seem pale,' he said at the table inside. 'Check-up go all right?'

'Yes, except that I've got too many triglycerides and the

122

pressure's on the high side for my age. He gave me some tablets and I'm to cut down on salt and fats.'

'Anyhow your eyes are clear enough, so there can't be much wrong with you. The eyes tell everything.'

'That's nice of you Gérard.'

He rumbled: 'I still say you look pale, though. You need fattening up. Not overworking, I hope, let me know if you are.'

'Yes boss. It's general fatigue, it gets everyone at this time of the year. I do tend to walk into furniture at the moment. We should have a special rate for unsocial hours, like nightshift workers.'

'Are you drinking plenty of mint tea like I said? Keeping up the magnesium tablets and the yeast?'

'I'm fine really. You seem a bit green yourself.'

'I'm work-weary too, in a bit of a rut. We're down on the membership figures, austerity is starting to bite, and that's naturally worrying. And there's Ghislaine as well. You know she was taken off screaming last night?' He shook his head in dismay.

'Some imbecile brought in a vibrator. They're strictly banned of course but it got into the playroom. She got going with it and wouldn't stop. I was busy on the phone and the next thing I knew Nadège and Marie-Odile were dragging her into one of the cubicles half crazy. I knew she was virtually a nympho, but this time it looks like she flipped completely. She was raving for an hour afterwards, well half an hour perhaps, and Marie-Odile took her home. I've given her a week off sick but it's not certain she'll be back.'

He changed the subject suddenly: 'My God it's hot. I came out for air but there's not much about.'

We went on about pollution for a while, with me doing most of the talking, until I wondered if I was boring him.

Then he said in a clipped voice: 'I have an urge to breathe some clean air from over the Atlantic. I have relations on the Ile de Ré, little village called Loix. Like to come?'

He said it in a rush, averted his eyes a second, and then looked at me expectantly. How timid he was, in his own way.

I took an instant decision: 'I'd love to.'

He smiled broadly: 'There won't be a soul about at this time of year, it'll do us both good. I'll fix it up for next week.'

Yes, I liked Gérard. He was straight, I felt instinctively. But I found that things were not that simple.

Chapter Twelve

We did have clean air. It blew cold and damp all the way there
and carried on blowing. But we had sun after that.

I lay next to Gérard on the deserted stretch of sand. The
slight breeze at 20°C entered my nostrils, cooled the roof of
my mouth, slid through my throat and inflated my chest. I
held it for a while, then let it out slowly. I repeated the cycle
again and again: 10 seconds in, 5 seconds hold, 10 seconds out.
And it felt like a miracle every time. Neither of us had spoken
for ages, there was no need to.

We were mad the way we rushed out of Paris. At 2 a.m.
Gérard jostled the stragglers out and closed the club ten
minutes later. We picked up his Alfa-Sud from the George V
underground car park, lost 5 minutes collecting my travel bag
from the flat, but were through the Place de la Concorde and
heading for the Porte de Saint Cloud at 2.30 a.m. In continuous rain, we drove at breakneck speed to La Rochelle, rolled
onto the ferry at 7 a.m. and reached Loix at 8 a.m. There, our
hosts gave us breakfast and we slept all day. We ate at 7 p.m.,
noticed it was still raining and went back to our beds until the
next day. Beds in the plural, of course.

Gérard had ordered this programme, and I had to agree he
was right. We slept deeply, almost non-stop, for 14 hours, and
woke feeling it was great to be alive.

And now it was mid-morning and we were gazing up into a
deep blue heaven listening to the wavelets splashing against
the shore and the birds skimming them, twittering the while.

The sun grew warm enough for us to bathe a few minutes, but no longer, and I shivered afterwards. Gérard engulfed me in a huge towel and we ran up and down the beach.

'Feel good?' he said as we returned to the straw mats.

'Wonderful. All the fatigue has flown away. Imagine two whole days more, not counting today! You're a good organiser, Gérard.'

He grinned: 'You look stunning in that swimsuit.' It was a bright red one-piece, he was wearing black trunks. 'You realize how lucky you are, I suppose. About one woman in 50,000 is perfectly proportioned and you are one of them.'

I looked carefully to see if there was a catch: 'Nice of you to notice,' I said tartly, making a joke of it. 'You're well-preserved too.'

'Preserved! I'm only 37, for pity's sake.'

Avoiding the *gaffe* I responded: ' . . . And still highly eligible. Why don't you get married? I hope you don't mind me asking.'

'Of course I don't. I tried it once, didn't work. She kept ordering me about and after 18 months I quit. Why do so many women try to eat their men? How about you?'

'No, I don't eat men. Oh you mean . . . Well, what *do* you mean?'

'How about you and marriage? Or are you waiting to see how menopause goes first?'

I stared at him: 'Do I look that old in the daylight?'

'I'm saying you are ripe for marriage, sweetie. Any boy friend, fiancé, or what not? You started this, can I help it if I'm inquisitive too?'

'We-e-ell. I can't for the moment see myself coping with babies, so I guess that lets me out. Millions of women don't get married these days. I've never thought about it really. Let's go for a drive, you can show me the island. In any case I have my career!'

We saw the lot. The salt marshes, oyster beds, winegroves, the lighthouse, villages, sailing harbours, the sheep. Stopping now and again for a drink and finishing up at the house in the early evening. Oh bliss, they had a log fire going!

We were staying with an uncle and aunt in a U-shaped house right in the village. I learned that all the houses had to be low, white and with green shutters, and they all had masses of flowers in the patios, for the simple reason that simply anything grew on the Ile de Ré. Like many men on the island, Uncle Louis did a variety of jobs; nominally an oyster raiser, he collected other shellfish, helped in the fields and with the wine harvest, had his own plot of land, repaired roofs, moved furniture, fixed boilers and sinks, and so on. He drew the line at housework and we argued good-naturedly about women's liberation, which he was barely aware of.

Aunt Jocelyne gave us sea trout, followed by a charlotte she made. I praised the meal, she praised my appetite and Gérard said I never had a proper meal and lived on hamburgers, and I said I had pizzas too and he was making it all up. We were off on another friendly conversation involving fast food in which I agreed with everything the old folks said. We drank a fair quantity of local red wine which tasted like cough syrup and also *pineau* which is a local blend of wine or red wine and brandy. We sat by the fire and we townies revelled in its warmth. I grew hazy and Uncle Louis chided me for it. He and Gérard eventually gave me up for lost and they engaged in a rambling conversation about the new mayor, the future covered market, the sewage system, the tourists and the forthcoming bridge that would replace the ferry.

This took us to about 10 p.m. when, with one accord, we threw our arms round one another and mounted the stairs to our respective rooms. Blessed be Aunt Jocelyne who said we could do the washing up next morning; I said I would do it with Uncle Louis, whereupon we all went off into fits of laughter again. It was really a hilarious evening in an old-fashioned sort of way.

Not so the night. It was dreadful. I think it must have been the *pineau*, for the deep sleep this time ended around 2 a.m. when I had to go the toilet. This was next to Gérard's room and I

127

creaked by like a mouse. Returning to the bed, I was dismayed to find I had not covered it over and had let out all the warmth. On advice from my escort I had brought pyjamas, thin shiny black ones, and I now commenced to shiver. Burying myself within, I set about heating up the premises by breathing out *pineau*. It was not much good and I vaulted out to grab my raincoat which I laid on top.

Dozing fitfully until dawn, I had nightmare after nightmare. I waded in the sea trying to get away from some unidentifiable danger, I re-lived the trip round the island until I was sick of it, I re-entered the church we had visited at Ars-en-Ré and woke sweating in the certain knowledge that there was a God after all and I would be punished for my sins, I saw men around a woman nymphomaniac with a white face and thought it was Ghislaine but realized it was me. All this I remembered vividly. I also dreamed that an auctioneer banged his gavel and I bought a Gallé vase I could not pay for; this impinged itself on my mind when I finally woke, and I saw it as a sign that I should sell my shares and buy one of these vases. Which I did on return to Paris, reasoning that the crisis was getting worse rather than better.

When I edged down to breakfast, Gérard informed me that a 'force 5' was blowing outside and that we would walk along the dike to Loix harbour, then visit Louis Suire's studio near the lighthouse. Exhausted, we returned that evening to the relations and took them out to dinner at Saint Martin, capital of the isle, where I insisted on paying. We ate piles of oysters and then crab and drank Muscadet.

Alas, another cold and restless night it was, with the force 5 howling to get through my door. I felt miserable and apprehensive. There was a good reason for this. One of the things about sex is that the more you have the more you want. Three days without it for the first time in months threw me off-balance. They say that women's sexuality is diffuse, whereas men are direct; they have something to get hold of, so to speak. But tonight my claims were perfectly straightforward, and Gérard's face repeatedly hovered over me as I tossed and turned. Every time I stirred I found my knees up and my hand between my legs. My sex was splayed to the extent that if I had been a man my erection would have formed a tent of the bed-clothes. Yet I remained chaste, telling myself this was good for me. I would prove that I would not go the way of Ghislaine, that I was still in control, although in the prime of my sexual life.

The next day we looked for shellfish called *palourdes* buried under the sand and slime and theoretically detected thanks to two tiny breathing holes. I think the beasts must have heard me coming, for I hardly ever found any holes and in the end I left Gérard to carry on, contenting myself with a massive crop of winkles. Later, to cheer me up, he drove me back to Suire's studio where we each bought a painting, he a still life and I an Ile de Ré patio with a well in it.

The final night loomed. I was still cold. Gérard and I had an awkward relationship, and it occurred to me that he was not a bedjumper! Possibly he saw too much of it at the club, possibly he was impotent or homosexual. The desire to find out nagged me and I was tempted to knock on the wall and invite him to warm me up. But he was my employer and could render me jobless at the stroke of the pen if I made a wrong move, for he had the power to hire and fire. He was a strange man, we had swung along the dike holding hands, he had helped me across some seaweed-strewn rocks. But in the car, when I deliberately turned my knees towards him and allowed my skirt to ride up to see what would happen he had brushed my knees with his hand, only to apologize and shift the gear lever! Of course, maybe I did not appeal to him, and I sighed thinking I had never actually slept a night in the arms of a man; I would have loved just this once to have a muscled thigh between mine, just for affection's sake – so much more satis-fying than a bolster or a pillow. Romantically, I imagined the house on fire, the ferry sinking and Gérard rescuing me, holding me tight in his strong arms. Outside every woman is a little girl trying to get back in. I turned over, tucked the bed-clothes round me and went to sleep.

Nothing happened on the journey back either, and I was starting to wonder if I had body odour or bad breath or some other ghastly defect. My libido was in full cry and I felt every tremor from the road surface, but I had to sit still being bright and interested in his past. I extracted the information that his parents had both been killed in an air crash and that he had been in advertising and managed a cinema in Lyons before a local councillor got him the job at the Top Club, also that his ex-wife eventually found the brainy lecturer in archeology she had been yearning for. I never discovered what Gérard yearned for. I just sat there in the passenger seat, on the boil.

Body beautiful I may have been for him, but he dropped me unceremoniously on the pavement, we embraced on the cheeks and he refused to come up for a drink. He drove off to his flat in the 17th *arrondissement*, a lost opportunity slinking into the night.

In the mail Annick had stacked for me was a 5-liner from Terry the Englishman, redirected from Evian.

It said:

> My dear Laure-Anne,
> I am writing to you on the off-chance, hoping this
> reaches you if you have left Evian's quiet waters. I have
> never forgotten our stimulating encounter. Do you ever
> come to Paris? Do get in touch. Are you married? My
> plans were cancelled.
> > Best wishes,
> > Terry

A stupid letter to have written. I would have looked pretty foolish opening that up in the presence of my husband.

Resisting the urge to phone him at this late hour and tell him to get a grip on himself, I drank a double scotch in two gulps, and collapsed like a burst balloon on our settee. The apartment seemed stark.

I did not hear Annick come in, but she told me in the morning that I hugged her with all my might, though I did not wake.

* * *

For weeks I was disorientated, uneasy. More nightmares came and Gérard was in a lot of them. My existence had been revolving with the precision of a Viennese waltz, and suddenly, the music had switched to Abba and 'Gimme, gimme, gimme a man after midnight; won't somebody help me chase the shadows away.'

One of my problems was that, for obvious reasons, I could not confide in Annick or in anyone else at the club. Of *course* business was business and he was the manager and could not display any special liking for an individual on the staff. And of *course* there was no evidence that he harboured any special leaning towards me at all; for all I knew he gave all the girls a whirl in his Alfa-Sud in turn, perhaps he had a princess or a Chicago lady meat-packer stored away somewhere. On the other hand, for some complicated reason of his own, he might be playing hard to get, because his timidity was unconvincing in the light of his poise; Gérard was so charming and unassuming I sometimes wondered whether he was a KGB agent and would blackmail us all with photos! All right, business was business, but if he *did* want a special relationship, he alone could make the first move.

Meanwhile there was no denying the soft beat of passion's wings within my breast. I went back years, developing a schoolgirlish crush on him, listening for him, making sure I would look nice always in case he caught me unawares, observing his clothes, mood, timekeeping, hand movements. I was behaving like an idiot, a possessive shrew who meant to get her man, a scheming harpie. All to no purpose, for my yearning was unrequited and I was left nursing my thoughts, one of which was that if I missed this chance of a lifetime I would drift along until I finished up an old hag on a stool chatting up the drunks in some Marseilles bar.

I kicked myself. What would have happened, I wondered more than once, if in the car on the way back from Ré I had meekly slipped my hand into his? I may have changed the world for both of us. But it was too late now.

Eventually I pulled myself together in the Champs-Elysées

Prisunic one afternoon when I found myself weeping into a shelf of tinned cassoulet while taped sugar oozed out of the loudspeakers.

All that remained was the feeling that I was playing hide-and-seek on my own, that I had been stood-up in some indefinable way. Unless of course I had been found wanting, had been given the old-fashioned brush-off and failed to see it.

Chapter Thirteen

My problems were as nothing compared with those of the nation's leaders. And it was in the club cellar, more than anywhere else, that they let off steam and shed their inhibitions.

This large grotto-like place with imitation granite wallpaper and a low red ceiling was more in the nature of a surrealist's gymnasium.

We popsies were called upon to provide a variety of special services and it was mainly for these that we were paid our high salaries. The members could do what they liked with us, as long as we were not to be beaten, a service requiring our individual consent and earning us a 'merit bonus', as it was termed on our pay sheets.

Both client and girl were actors, and we had no choice but to act the role handed out to us. A kind of play was sketched out between us before we started; once the lines were rehearsed we were bound to stick to them as much as we could, not switching fantasies in mid-performance, unless of course the client altered his mind.

Some of the games were really quite fun and they broke the monotony of the quickies in the cubicles, which when all is said gave us the impression we were simply prostitutes. The dialogue was simple and there was no need to think too hard, which was a relief for us and the big-wigs too who were in a constant state of stress at the office and often at home.

One of our regulars was a highly-influential ex-army officer with a chip on his shoulder about lady traffic wardens. He had even provided the cellar with a car seat and a warden's uniform. He usually preferred Marie-Odile ('the legs') but I was asked to stand in one day.

'Did you put that ticket on my windscreen, Madame?' was his opening line.

'Yes sir, you are 15 minutes past your time and in any case you have been feeding the meter,' I scolded, fiddling with pen and pad. I walked back and forth on the imaginary pavement in my tight blue uniform and black stockings.

'Don't you look down your nose at me, young woman,' said the motorist. 'After all, as a taxpayer, I provide your wages! I've had enough of you women, you've taken 600 francs off me in the last month alone.' He grabbed my wrist and I struggled.

'Ow, there's no calling for insults. Kindly take your hand away or I shall report you for assaulting a sworn functionary,' said I in a schoolmistressy voice.

'Oh no you won't, it's more than your job is worth to upset me. I happen to know your boss, Durand.' He released his grip and ran his hand over my rump, adding: 'But you're very attractive and I won't say anything if you're nice to me. Take the ticket back, there's a dear.'

'I'm afraid I can't, the carbon copies are numbered and what's written is written.' He lowered his hand to my hem and squeezed a knee. I jumped aside, making a business of it.

He glowered: 'The street is completely empty, no-one can give evidence, so you've got what's coming to you. Into the car! Into the car, I say!'

At this point he mimed opening the door, and I struggled as he forced me inside the car. I plumped down on the seat, my skirt riding up, and he joined me and pawed me. He interfered with my clothing and I resisted. I screamed and waved my legs as his hand crept nearer its ultimate goal. He removed a key article of clothing and slipped off his own nether garments. The man's offensive missile was exposed, pink and unwrinkled like his face.

'Oh no,' I cried in mock-fright. 'No, please let me go, you're taking advantage of me, please please.'

'Aha, I have you at my mercy, you cannot escape my clutches. I'm getting my revenge on you. Obey me, or I'll get you into trouble, lose you your job, with one word from the Commissioner . . .'

'I need this job! You have me in your clutches!'

He undid my dress top, fondled and pawed my breasts and laid me along the seat as I continued pleading and whimpering. The irate motorist then exposed the lady traffic warden to the full extent of his wrath. I could only squeal, 'Oh-no-please, I am so ashamed, please stop, my husband will never forgive me,' as the rascal consummated his offence.

Afterwards he said, 'Thanks Laure-Anne you're as good as Marie-Odile,' gave me the statutory peck on the forehead, and we got dressed. I never asked him what he had against traffic wardens, but it certainly switched on his traffic lights, his pensioned-off penis exceeding the speed limit in a most impressive fashion for a man well over 60.

There was no accounting for taste, and some of my roles would have earned me an Oscar. At various moments I was a girl student in a bus, a girl coolie at the battle of Dien Bien Phu, a captured Kabyl belly dancer, and even one man's own wife! This particular client took the trouble to bring in some of her clothes, and I had to wear a black wig and special harsh make-up, We were usually taken lying face up or face down across a settee, or in a corner of the room, or simply on the floor, but often tied to some horizontal wall bars where the assailant could conveniently climb up as well and adjust his position relative to the victim.

Occasionally we were asked to do a striptease 'under protest'. My protégé Monsieur Dombre was one of our regulars too. As he was fond of me he was reluctant to humiliate me in this way, but would engage me in mutual arousal on the car seat as he issued his orders to his favourite stripper Nadège.

He stood her in the middle of the room dressed as the young

Duchess of Windsor and explained: 'Now milady, I know this is hateful for you but unless you do as I say I shall tell the Queen about you and your maids.' As will be seen, Monsieur Dombre was making up for lost time as enthusiastically as the rest of them. I was very proud of him.

Nadège twirled slightly left and right with her hands crossed meekly in front, eyes cast down and biting her lower lip. The pleated 1930's dress with its low-slung belt swirled prettily round her thighs and her fox fur wriggled round her neck.

'Undo your buttons at the top, all of them,' Dombre ordered. 'Pull the bodice apart, that's right. Take off your bra, feel your breasts, flick your nipples. Lift the dress a little, higher, higher, turn round, now face front.' Nadège clenched her knees. 'Lift the skirt right up so I can see everything. Take the dress off. Stand there until I tell you to move.' The Duchess crossed her arms modestly to conceal her bosom. 'Pull your knickers down slowly, no leave the shoes and stockings on, open your legs, squat down . . . ' And so it continued until he delivered himself into my hands, or thrust his organ into me at the last moment, still gaping at Nadège.

Some requests were really surprising. A man wanted to be examined by me as I acted the part of a nurse. Another made me wear the Israeli Women's Army uniform while he wore a woman's dress, and I had to seduce him in a forceful manner while he acted coyly. There was a steady flow of downright masochists; men whose signatures governed the lives of thousands if not millions would have me bind them to the wall bars and interfere with them, Monsieur Giraud and his chums would bare their posteriors and submit to whipping or caning, subsequently climbing down to lead the country further along the road to disaster. I used to wonder how they explained their weals to their spouses, although these good ladies may perhaps have been deprived of the sight for weeks on end.

I must emphasize that we of the personnel never laughed, ever. One morning Gérard had us all sitting in the lounge, just

136

before noon, wondering what dreadful news he was going to impart.

He smiled pleasantly and said: 'Thanks girls, for coming in ahead of time, and some on your time off, I really do appreciate it.' We relaxed visibly. 'We have had a complaint and I cannot ignore it as it's from one of our most influential members. He seems to think that some of you have been laughing behind his back at his *desiderata*' (Gérard was always coming out with these Latin expressions) 'I'm sure that is not true, but it may do no harm to remind ourselves that we are here to serve and not to judge. Who indeed has the right to judge others in a domain where there is no borderline between normal and eccentric? As you know, sex is the most delicate, complex and personal of all Man's experiences. We all have the right to our fantasies, the right to seek our pleasure as we wish provided we do not make others suffer against their will, the right to give our hormones free rein – and we all, all of us, have male and female hormones in various proportions, remember.'

We all had a quick peek at Blandine, and I wondered what Gérard's hormone mix was.

'The phallocrat of today,' he continued, 'can be tomorrow's transvestite, the frustrations of the past can spill over into uncontrolled passion. Girls, you are loyal and sincere in the services you render and, believe me, I am constantly receiving praise for the immense task you are performing so marvellously. Never forget that each time you are raped on the wall-bars you may be saving the life of a housewife or a little girl in Poitiers; every time you raise your skirts you may be safeguarding the career of a brilliant man who would otherwise be jailed for harassing a typist in the lift; when you provide a minister with the therapy he needs to urgently you bring peace to his body and soul and prevent a major calamity for the nation.' Annick was quite overcome, and blinked fast. 'The soft touch of a woman's hand, her lips against the burning protuberance of the male, the silken welcome within her body – these are a glimpse of paradise in the here-below, they restore

137

the faith of the faint-hearted, kindle hope in those who despair, bring solace to the trouble brow. Live up to your noble calling, the future of France is in your hands!'

We all cheered and broke into applause.

But the cellar was not always the sanctuary it was intended to be. Some of our best friends were husbands, and the incident we all feared occurred one evening early in 1980 when a spouse infiltrated past the doorman and was inexplicably not taken in hand. No doubt she was assumed to be a member as she prowled round the premises.

I was busy with one of our gentlemen, whom I had lashed to the wall-bars, and was about to climb up with him when a deep-throated 'Ah, there you are!' caused him to jump perilously in his elevated position.

'*Merde*, Marie-Chantal!' the man breathed.

I descended, leaving the elevated member fully vulnerable and rapidly deflating. The wife, dressed in a pink tweed suit and clasping a dangerously large handbag, stalked across, pushed me aside and with her gloved fingers captured the poor fellow's weapon and gave it a tweak.

'I can explain everything,' he declared with remarkable aplomb.

'You alley cat,' she spat at me, 'release this man at once. I intend to get an injunction and have this place closed down at once.'

I wondered whether the best tactic might be to tie her up as well, but wiser counsel prevailed. 'My name is Laure-Anne you must be Madame . . . '

'Get him down! You should be exposed!' Most of me was.

I began undoing the leather straps and ventured: 'I'm sure this is all a most unfortunate misunderstanding, I didn't know you were a member. This is a private part of the premises, so perhaps we can discuss it in more suitable surroundings upstairs.' I was saying the first thing I could think of, in a hopeless bid to placate her ire.

'And who's property do you think *he* is? Whose private part is *that*?' said the wife. She was most disagreeable.

Nadège had also been in the cellar with someone, and was now nowhere to be seen. The spouse scowled at my client while he climbed into his trousers meekly like the naughty boy he was.

Gérard appeared with Nadège, announced his name and status. He extended his hand which was refused.

He said: 'Well, this is most embarrassing, especially as your husband has only just become a member.' (this was a lie). 'I'm afraid we all had a little too much to drink (also untrue), and we are not quite ourselves. Entirely our fault, he bears no blame. *Licitum est omne quod non expliciter prohibitur,* as we all know, and if we might perhaps take a broad view . . . '

Gradually he smoothed her down and eventually took her upstairs. As far as that particular member was concerned, it was a case of *alea jacta est.*

On another occasion, it was I who was in his office, for a reprimand.

The person involved was a senior executive from a news agency which I suppose must remain anonymous, although I still haven't got over the way he treated me. He was a veritable pile-driver of a man.

I disliked Monsieur Poteau from the moment he set eyes on me and growled: 'I'd like a few moments with you if you don't mind.'

'Of course,' I purred, wondering who had rattled *his* cage that morning. He was a wiry, ginger-headed, pale-faced brute with mothballs in place of testicles. In a word, he smelt awful. He was also in a hurry on this occasion, but when you're on duty there is no arguing. All I could do was try to loosen up when he jumped me. His red hot poker scalded me, but he seemed delighted, so much so that he came back for more two days later. I should have been warned when he declared in his one-note robotic voice 'you're so submissive,' a lecherous gleam in his eyes. On the third engagement he invited me to see over his agency, and I accepted.

That afternoon after the 2 o'clock closure at the club, I

stepped through the automatic doors and we were whisked up to the main editorial room. He guided me through some swing doors and into an enormous place the size of a freighter's deck.

I recognised the smell, this was where the fetid odour came from. I said: 'Goodness, it's hot in here.'

'It is rather. We have a most unusual air-conditioning system. When it's cold outside it's cold inside, when it's hot out it's hot in.'

'Then what are all those people over there doing with their overcoats on?'

'That's the English service. The air's about −10 there, it's the only oxygen we have, so it's a matter of life and death.'

We strolled along the vast room where everyone was crouched over lines of hooded contrivances. 'They look as if they're sewing shirts,' I said. 'Oh they are screens and keyboards, but why is there so much paper? I thought electronics were supposed to do away with paper.'

'Well, we've got far more now. We are one of the world's major news agencies, a new story comes in every 5 seconds.'

'How exciting. But these journalists don't seem very stimulated, they don't rush about like Cary Grant.'

'That's all over now. Here we have the wonder of the new technology, everyone's half asleep. But production has leapt, we send out 300,000 words a day.'

'Really. Is that good?'

'Oh yes, we've taken on more journalists for the computer-ization, and sacked the teleprinter operators.'

We wandered about looking in screens and yards of paper. I tried to look intelligent, but was out of place, and felt everyone was looking at me. Several men were, and I smiled wanly at them, and they smiled wanlier back.

'And what do *you* do?' I asked.

'Oh I'm chief editor.'

'But that's most impressive. I didn't realize.'

'Not so wonderful, there are 64 chief editors, it's only a rank. My job is to run the English desk.'

'You must get cold,' I said.

'I keep away, just turn up twice a day to give them their orders.'

'You must be bilingual, then.'

'Actually I can't understand half what they're saying. They are mostly Asians these days, funny English they have, but they are less trouble, cheaper too.'

'I expect with all these words going out, the agency makes a lot of money.'

'Never made a profit. The Government provides most of our revenue. They pay 70% of our wages.'

'Ah, so you have to keep in with them. You're a sort of government department, like Tass.'

He winced: 'Oh no, we're independent.'

'Who owns the agency?'

'Nobody does. It was set up by the Government, which controls it.'

The anomaly shook me: 'But why should other countries want information from the French Government? Surely they get it officially anyway.'

'We carry other material too. Anyhow, we improve France's prestige abroad and provide a bulwark against communism.'

I was in a teasing mood, and laughed: 'It doesn't seem to work out too well: since your agency was set up the Reds have doubled their territory from one-sixth of the world in 1939 to one-third now. So you are wasting my money as a taxpayer.'

I thought he would laugh too, but his face clouded over and I knew I had overdone things. If you give a girl with an IQ of 138 the chance to join in a verbal to-and-fro, she may well seize the chance. I bit my lip and the walkabout came to a dreary end. I had the distinct feeling that I would pay for this later.

And so it turned out. Poiteau was more than usually ape-like on his next visit, and wanted to sodomize me. I refused.

This was how I came to be in Gérard's office.

'But he would do anyone an injury,' I complained weakly. 'He scares me.'

'He says you are a bit of a bolshie.'

'Well, if we're calling each other names, he's a nazi. A

micro-nazi but a nazi just the same. A punk with no brains or vision, a sadist with a cock like a bayonet. I can't consent to buggery, there is this AIDS disease about, and you can die from it. Please Gérard, I can't.' I was flushed and my throat thick.

Gérard took my hand across the desk: 'This puts me in an awkard position. We have to be so careful about journalists, but I understand your reluctance.'

I had a brainwave: 'Can't you lay down a rule, requiring our consent? After all, it's a matter of our health, we might all get infected with something.'

He released my hand: 'If it's humanly possible I'll fix it.'

He did fix it and I was off the hook. The only time I had been sodomized was in the heat of the gala night orgy that time. Some of the girls said they got to like the practice in due course, but I resolved it was not for me.

Nor had I consented to be whipped or caned, but in this case there was no health hazard.

After this incident, Poteau and I hated each other. He obtained a special thrill forcing me to go with him, but he dared not contravene the rule on consent.

Chapter Fourteen

On another occasion, still to do with the cellar games, Gérard waylaid me on the way to the lounge, telling me he had something he wanted to show me.

I had done a 20-minute stint in one of the cubicles with a gentleman voyeur. Not for the first time, he had blindfolded me with a white silk scarf, ordered me into the bed and instructed me to pleasure myself while he sat on the chair and watched. Oddly, this was something I found extremely difficult to do, the activity failing to excite me. However, by concentrating and contorting my face I usually managed to become turgid, though I invariably had to fake the orgasm. The result was that I was flushed and frustrated when Gérard bid me enter his office. His shirtsleeves were rolled up and more than anything in the world at that moment I yearned to stroke the hairs on his forearm.

In a rush of agitation I said: 'I thought your little speech to us the other day was sublime, uplifting. I was most moved.'

Gérard smiled wrily: 'I'm glad you liked it.' He changed the subject quickly. 'Now I suppose you've heard of the Iron Lady, Mrs Thatcher, Prime Minister of Great Britain and Northern Ireland or something like that. Well, this blond wig is supposed to be her hair. Try it on please.' I did so. 'Perfect. Now this little device is an iron chastity belt, of excellent quality I may say, made in Japan, it cost us 1,500 francs. It has been made specially for you, to your dimensions.'

143

'How did you . . . ?'

'I have all your measurements, remember. Madame Chavaux, one of our oldest members hates Mrs Thatcher, and you've got to beat her.'

'I didn't know Mrs Thatcher was a member of ours.'

'She's not, I mean you've got to beat Madame Chavaux.'

I stared at him: 'Gérard, dear Gérard, don't let's fall out over this. As I understand it, Madame Chavaux hates Mrs Thatcher and wants Mrs Thatcher, dressed in a chastity belt, to beat her. I *don't* understand it.'

He spread his hands: 'Laure-Anne, we've come a long way since Adam and Eve and you can't ask me to explain what goes on in Madame Chavaux's convoluted mind. It was not I who created Woman! So be sweet and just try it on and let me know if you can do the job.'

I grabbed the object, which seemed more like an attachment for cows: 'Yes Gérard, as you say.'

I turned and he said: 'Oh, I was forgetting, I had a phone call from your cousin. He wants to know your private address, but I told him to write here.'

'Cousin? Oh yes, my cousin. Ha ha, of course. Well thanks.'

Gérard's eyes narrowed: 'I didn't know you had an English cousin, unless he was an American.'

The blush started in my solar plexus and rose to my neck: 'Oh, he's English all right, I mean I haven't . . . Well, you know what I mean.'

I fled, trailing the Thatcher gear and clouds of disaster behind me. This time Terry had really put his foot in it, and mine too.

To cut a long story short, Terry wrote, I phoned him and we agreed on a trip to Scotland together. It was one of the silliest things I had ever consented to, a deliberate gesture of defiance so that I could get Gérard out of my hair. Otherwise I would go mad.

In any case the atmosphere at the club was thickening. We were in first half 1981 by now, and before and after the election of Socialist François Mitterrand to the Presidency our members were too busy gabbling about politics to think of pleasure.

We were anxious. Ghislaine had not yet been replaced and Marie-Odile had left us to get married. Annick and Mauricette had a dreadful row one evening in the lounge, tearing at each others' hair and overturning a table loaded with drinks. In my own case my voice took on a shriller, wobblier note, rather like Judy Garland's, a sure sign of impending disaster. I could feel it coming, and decided to head it off with some tape recorder lessons. People fail to realize the importance of vocal expression; if odour is the X factor in sex, voice is undoubtedly the Y component.

On election night itself, the place was like a morgue, empty of members, and we sat round the television for the results. Gérard made an important concession and allowed us girls to drink several bottles of Fleury, and we had pancakes at the bar. He himself popped in from a reception, appearing in a super dark blue suit, white shirt and expensive red and gold tie. He resembled one of those sexy G-men who ride on the US President's limousine. Alas he left within minutes for a date with his Chicago meat woman or whoever.

The premises were forlorn. Notable exceptions in the gloom were Picard and the lesbian after they became ministers. One of our members in the police was promoted.

Hence, for want of Gérard, it had to be Terry.

We must have made an impression as we leaned over the rail of the channel ferry in a blustery wind. Terry had matured physically, his cheeks a little fatter and his chest and belly showing the beginnings of a paunch. He wore dark blue trousers, a russet blazer, white shirt and blue neck scarf, and his shoes were reddish brown with braided uppers. He might have been a successful croupier, anyhow I pretended to myself that he was. I of course was a racehorse owner, quietly dignified in my lightweight suit which was white with vertical navy stripes and blue edging round the pockets and collar; I had a light blue jumper and a string of multicoloured reddish beads.

A lad of about 10 ran up to us: ' 'Scuse me mister, are you Ted Barridge?'

'No – er – I'm his brother Tom.'

'Funny, you don't look like him.'

'I'm sure you've never seen Ted's brother before.'

This was too much for the boy and he scampered off. Terry said: 'I've no idea who Ted Barridge is, but by the look of the kid I'd say he played football.'

For the first time I admired the white cliffs of Dover, on orders from Terry, because of course they meant nothing to me, and in fact looked rather scruffy. We stayed at a hotel in Sevenoaks, and then went straight for Gretna Green where we slept our second night at a farm, recommended by a colleague of Terry's.

It was gloriously quiet, and we spent long minutes at the open bedroom window in the dark looking out over a meadow where a dozen cows ambled about. The late June air was crisp and a slight mist hung over the field.

In fact it was distinctly chilly, and we jumped into bed like children and played footsies in the warm.

'I'm so happy,' said my companion, as I snuggled up to him, sighed as he fingered my hair. 'You know, for the first time I feel I'd like to be really wealthy and travel, just the two of us, to other places like Ireland, Norway, Iceland perhaps. Italy would be fun, Venice especially, yes we must do Venice on a long weekend. Come to think of it I rather like hotels and inns, nice ones of course off-season, provided you need spare no expense. . . . '

This was the last I remembered, for I fell asleep exhausted.

We had asked to be called as late as possible and the sight that greeted us in the little room set aside for meals was breathtaking. One of the tables had been cleared but the rest were loaded with expensive ware. All of it was silver, except the plates and cups and saucers.

'It must be worth thousands of pounds,' I gasped. 'How can

they dare leave it out for anybody to steal?'

'This is Great Britain,' Terry said huffily.

We ate with appetite in the good old British way. Fruit juice, shredded wheat, egg-bacon-sausages-fried bread, toast and butter and marmalade, excellent coffee.

'How do they do it?'

'Britain is now a poor country, especially Scotland,' my chap said drily. 'All this is part of four centuries' capital they're living on and as foreign visitors we are exploiting them.'

We were soon outside in our macs and stepping out along the route the farmer's wife had recommended. Two hours later in a country pub, I was introduced to best bitter, which we had with a meat pie. We lingered, ordering more beer for Terry, and then walked back another way, delighting in all we saw. The sun never came out but we did not care. Everything was new to me, and Terry answered my stream of questions as best he could. European agriculture was not his subject. We got to talking about his childhood.

'Since you ask,' he said, 'it was pretty enjoyable really. My mother was widowed early and I was brought up by my grandparents. Grandma was a subject woman, which was the general rule then, and it worked out well enough especially when the men, were like my grandad. He was a market gardener, raising pigs into the bargain. I used to love market days, walking about in the muck and straw, watching the auctioneers, shooing pigs into the lorry. We would sit up in the cab singing all the songs of his youth: "Daisy", "Under The Old Apple Tree", "Any Old Iron". They mean nothing to you of course, but they're part of my culture. He had a comical turn of phrase, had Grandad. "Only dirty people wash", he used to say, and of others: "Never washed his face since he got it." He himself never took a bath, even in the Great War when he practically won the war single-handed against the "Jerries". A great character, he dropped dead 10 minutes after winning a game of darts in the pub. A fine way to go.'

I interrupted: 'Were you poor in those times?'

'We had all his produce including pork and poultry. His big

enemy was a local "two-legged rat" who pinched chickens from time to time. Still, we managed to get through. They say he never signed or accepted a cheque in his life, always dealt in cash. Had two sets of books, everything entered in pencil so he could change it if need be. Ah, happy days.'

'Aren't these happy too?'

'Of course. Even so those times were good, and I'm glad to have known them. I saw the last of the steam lorries, the trams, the old puffer trains . . . ' And Terry embarked on an imitation of a train leaving a station.

The Scottish trip took in Stirling where I had to plead with him to come away from the military museum at the castle, as well as Fort William, Glen Coe, Glen Nevis, Pitlochry and Edinburgh. We were thrilled with the beauty of the vast rolling hills, the mauves and blues and pinks and greens, the astonishingly smooth lochs. The absence of advertisement hoardings was like clean air after a night in a bawdy house. Everywhere we stayed our hosts were honest and pleasant, the food commendable.

From Edinburgh we shot down a motorway to around Chester and meandered as far as Nantwich and the cosy old-style Lamb Hotel. That evening we did a pub crawl, and I learned to play darts and bar billiards.

'We must do all this again,' I said with enthusiasm. We had rejoined the car after a call at a gallery to see the paintings Terry referred to when we first met in Evian.

'I'll take you to East Anglia, Cornwall and the rest of the West Country, and Rye and Winchelsea, too. There are still lots of places worth living in.'

I told him: 'It's been a real holiday, the nicest I've ever had. Don't you wish you still lived here?'

'I'm not sure. It's my own country, but it's changed. You need money to get the best out of it.' He switched the subject quickly.

Yes, it was a marvellous holiday. But throughout the nine days we made love only once, for Terry should have drunk more water with his whisky. I don't know how the Scots get on, but my English companion suffered from low performance despite increased desire. Mr Shakespeare has already explained this in *Macbeth*, and we were in Scotland after all.

I kidded myself, imagining that this trip might be the start of a firm friendship, possibly something more exciting. As a result I committed a major error on return to Paris. I was to uncover his underlying weakness, and I suppose I was lucky I found out.

He did seem to be grinning rather excessively when I went to collect him from the doorman at the Top Club. Terry had asked to be sneaked into the club one evening. About half an hour later I began to observe that he was only just managing to perch on the barstool. He was definitely the worse for drink, but I could not argue about it as I had to leave him and get on with my work. The next thing I knew was that Gérard was rushing me into his office with a scared look on his face. He did not ask me to sit down.

'That man you were with, or are with, is he a friend of yours?' His tone was icy.

'Yes, he's called Terry.'

'Well, he's on the rampage in the cellar, and I've got to get him out quick or there'll be blood everywhere – yours and mine. He's completely drunk and climbing all over the walls yodelling like Tarzan. Did you let him in?'

'Yes, I'm sorry, it was a mistake. I don't know him very well. I didn't know he was drunk.'

'But you knew this is a members-only club. The dreaded Poteau's been in and bawled me out. Come on, there's no time to lose, let's ease him out gently, you should never have invited him, or at least asked me first.'

'Sorry Gérard, stupid of me.'

Terry had found a stick and was thrashing wildly at an imaginary victim when we arrived. I gained his attention and told him we have to leave now, then Gérard said it would be best to call it a day and the cellar had to be locked up. Terry came quietly until we reached the lounge and Gérard took one of his arms as he staggered. This and the sight of the door warned him he was being chucked out, for he suddenly drew himself to a great height and looked round threateningly at the company.

'I have decided to stay,' he announced in English. 'I will not be frog-marched out of here. You just try, froggie. You haven't got the guts, all French men are cowards, unless proved to the contrary. Always hit a man when he's down, that's your trouble. You know what a *gendarme* once told me? The only thing to do with a French man is to hit him before he hits you, and the only thing to do with a French woman is to kiss her.' He put up his fists: 'Come on, who's first, I'll take on the lot of you.' Nobody moved as he ambled round in the space between the tables and the bar, where assistant manager Fred was anxiously eyeing the glassware. 'There you are, what did I say, all rabbits the lot of you. They seek him here . . . that demned elusive Pimpernel . . . '

Terry was completely incoherent. This was going to cost me and Gérard our jobs if we didn't get him out quick. I took his hand gently when he was pointing at the exit and we strode over together as the doorman smiled politely, holding the inner door open.

'Good night, sir, thankyou sir, hope you had a pleasant evening sir,' he recited with admirable *sang-froid*. I shot him a glance of gratitude.

The unruly guest's coat was on in a flash and we were out on the pavement, just the two of us, before he knew what had happened to him. The doorman and Gérard watched over me as I walked him to the end of the street and despatched him in a taxi.

I then ran back to the club in my flimsy dress, while he disappeared from my life for ever. Or so I thought.

Now I had two strange men in elliptical orbit round my world, neither of whom seemed likely to splash down.

Chapter Fifteen

As a result of this incident, I was as popular thereafter as a bread crumb in a pair of pyjamas. Nothing was said but I felt the need to make reparation. I was steering a course for the rocks and had to act.

Hence the bravest decision I had made so far. For my 'penance', I consented to be whipped, something I had never had the courage to envisage hitherto. To use Terry's expression, I had no more guts than the average person, but I was once told by a member close to our worthy Gaullist Mayor of Paris Jacques Chirac that 'the glory that was Europe stemmed from the courage to leap in the dark.' The Englishman's jibes had raised my hackles, and in a great wave of resentment I sensed that someone had to accept the challenge.

But to be honest the motivations were numerous, and as everyone knows psychology is not one of the more accurate sciences. So I can summarise the factors as: shame at Terry's rampage, suspicion that I had to make some gesture to safeguard a job that seemed vulnerable anyway, frustration at Gérard's indifference, overall irritation at the pseudo-socialist regime, and finally a willingness to try anything once. I also have to confess a dash of masochism in my make-up.

However that may be, I deemed that I had to get 'it' out of my system, whatever 'it' might be. Consequently when Monsieur Cattel asked me for the tenth time whether I was interested in earning myself a bonus, I said I was.

151

Just as the Thursday night galas were elevated to the status of special events, so these whippings and/or canings were formal in the extreme. From the moment a girl signified her assent, an elaborate protocol got under way. It was not unknown for two or more girls to take part, and for more than one member to wield the instrument of torture. In this case, Monsieur Poteau was to join Monsieur Cattel. Grimly I judged that if I could take their combined punishment, I would be able to face any crisis that fortune might throw at me during the rest of my life, much as a man who has sailed the world single-handed is never afraid again, and walks tall for the rest of *his* life.

At 6 p.m. that day I sat alone in our apartment eating a large steak and chips, green salad, some ripe Camembert and a couple of pears, the whole washed down with most of a bottle of Cahors. After this fare, I walked for a while in the Tuileries Gardens and reported at the club, still a little woozy, at 8 p.m.

The cellar was in darkness when I was led in on the dot of 9 p.m., except that a ceiling spotlight beamed at the centre of the room, where a kind of cushioned altar had been set up with a stool placed before it. Around the walls I glimpsed the shadows of about 30 members standing or seated.

I was wearing a toga, rope belt and sandals over apricot panties and white elastic-topped stockings. My wrists were held together in front of me with an iron chain. Two colleagues, Annick and Nadège, dressed similarly to myself, led me in, pulling at the two loose ends of the chain. Someone was tapping on a drum: a roll followed by three slow muffled beats. Annick had said I need worry only about the first 4 lashes, after which it was plain sailing. But as she and Nadège paraded me slowly anti-clockwise round the room, I became really frightened for the first time.

I was to be given 8 strokes by Cattel and 8 by Poteau. Sixteen lashes! For this I would earn 1,200 francs, plus anything the bystanders cared to add – possibly 3,000 francs in all, the price of a superb new dress.

Round and round they led me. It was understood that the

victim would remain unsmiling, as if the stipulation was necessary! I was shaking with terror and everyone could see it. As I walked round, a hand or two touched my arms and other parts of my body. Somebody felt my hair.

Then my colleagues brought me to a halt in front of the altar and I was forced to stand there for a minute or two with the chain dangling on my thighs as the drum beat gradually speeded up. The door was ceremoniously locked and Cattel stepped out of the shadows wearing only a pair of shiny red shorts and tennis shoes, and bearing the whip. I had turned to glance at him and he tapped me on one cheek with the whip in a mute instruction to keep my eyes to the front. Cattel, a well-built man of average height and firm biceps, inspected me with his lantern-jawed face but did not touch me. When he peered into my eyes I had a hysterical desire to put my tongue out at him. I felt like giggling.

Controlling my nerves I lowered my eyes to discover that Cattel's big prick was already straining at the thin synthetic material of his only garment. My own vaginal muscles instantly contracted on seeing his manifest excitement. He now transferred the whip to his left hand and ran the other very slowly over my lips, neck and shoulders, then over each breast in turn, my pelvis, my bottom, finally by pubis. All this took a minute or so and the crowd looked on in silence except for their breathing. The drum kept beating and I knew that the punishment was imminent.

Cattel moved away behind me and cracked the whip, then again, and a third time. The drum stopped and, presumably on a signal from Cattel, the girls approached and removed the chain, took the hem of the toga and removed it, finally tying my wrists again with the chain. Thirty pairs of eyes leered at my bare quivering breasts and my silk pale apricot panties. I realized I was moist between my legs and forced my knees tight in embarrassment, afraid they would see the wetness. Cattel took the chain and led me to the stool, forcing me to kneel on it, and pulling the chain ends across the altar. He attached these to the far side so that my torso lay flat on the altar and my bottom was ready for the chastisement. I felt the silk straining against my buttocks and an erotic surge caused me to clench my bottom. At least I wanted to, but the angle I was bent over at prevented me; I was tethered like a sow for the boar, totally without defence.

I could not stop fidgeting, a honeyed warmth flowed through my loins and I knew that only when the first lash hit me would the itching ease. I *wanted* to be beaten! Cattel applied his hand to my back so that my bottom was rounded to his satisfaction, he ran a hand over my stockings, my posterior and the oh-so-thin material over my humid vulva. Everyone could see it was wet, I knew. I was so humiliated and I could hear them making remarks.

All eyes were on my ballooning rump, that was certain, and the gloating onlookers knew that nothing could save me now, that Gérard would halt the ceremony only if I fainted or bled. I closed my eyes thinking that, even though I was at Cattel's mercy, I had the entire cellar in my power.

Cattel cracked the whip just once, I held my breath, heard the 'whoip' sound as it flew through the air, and exhaled with pain as the whip hit me with a wicked sting. The other strokes would be worse as he judged the effect. It was 10 seconds before the second stroke came and it hurt terribly, then at 10 seconds' interval two others cut into me. I imagined my tormentor's stiff organ and his oncoming desire, and I myself derived a twin lasciviousness as both master and slave in my mind's eye. I squirmed with the pain and humiliation.

I knew the crowd was hoping Cattell would make it hurt more and with the final four strokes a groan escaped my lips as his wrath bit into my flesh. Then he stopped, and I felt his hands at the elastic round my waist. He slowly pulled my panties down so that they were round my thighs, a hand caressed the scalding weals across my bottom and then with his two hands he seized my waist and his hard penis entered my now gaping sex. Three or four thrusts, I felt him surge and a great yell broke from his throat.

Now it was Poteau's turn. He detested me and I knew he

would inflict the maximum pain upon me. My panties were pulled up and I smelt his nauseous odour behind me. Poteau slid a hand under the material and massaged my flesh, making it smart. I was now trembling with impatience and ached for him to thrash me. He lost no time and the whip struck with immense force so that I gasped. Oh how it stung! He took much longer than Cattel between the lashes and I lost count. Once he hit low and I cried out when the leather struck the back of my thighs just above my stockings. My abdomen began to jerk and my hair flew about as I threw my head left and right seeking to take off the pain. I was biting my lips and crying real tears. Men came round to the front to watch my facial expression, to enjoy the writhing of my lips. Spittle ran down my chin, I remember. I screamed again and again and Poteau himself gasped as he wielded each stroke. Then suddenly he was penetrating me and I shrieked as a volcanic orgasm erupted within me. It was one of the longest I had ever experienced, I kept tugging at the anchored chain, my wrists seemed raw.

Then it was over, my bottom on fire it seemed. Nobody moved for a while, and I was left heaving and quivering with my panties down. The chain was undone, and Annick and Nadège were helping me up and vesting me. They supported me under the elbows and I walked stiffly away amid prolonged applause and cries of '*bravo petite Laure-Anne*' and '*tu étais sensationelle*'.

Ten minutes later when my colleagues were patting ointment into the wounds as I lay prone in cubicle 4 Cattel and Poteau entered with a huge golden teddy bear. They planted kisses on my forehead and I pursed my lips so that they kissed me on the mouth.

I made 3 new friends that evening: Cattel, Poteau and the teddy bear.

My chafed bottom was driving me mad, and I craved for another last orgasm. Cattel said he could not oblige and left the room, but Poteau, I'll give him his due, valiantly took me again, standing against the wall. I kept telling him the weals still hurt and this excited him, but he was a long time coming

155

and I reached my peak before him, which is wonderful for a woman. I have always found that the second time is better.

What startled me was the thrill I derived from this experience: the poignant degradation that snatched at my entrails in the days that followed when I thought back to the whipping. I would quiver deliciously and almost faint. At last I fully comprehended the exquisite scene in the film *Night Porter* in which Charlotte Rampling, playing opposite the mouth-watering Dirk Bogarde, was a slave in leg-irons on a length of chain that slid about the room. Oh the perverseness of it! A whole new future spread out before me, and to be honest I was frightened. It was like a glimpse into heaven!

Chapter Sixteen

The axe fell early in 1983.

Profitable though flagellation might be for us girls, I submitted to it only that once. But it made me popular, and from then on I was much in demand as an outlet for the nation's ruling passions.

Meanwhile my own appetite was more voracious with maturity, and the members realized this. I became easy meat, and was subjected to a faster pace of work. This suited many, for whom I snaked and writhed more frenetically than in the past. But others took a jaundiced view of the refurbished Laure-Anne D.

As is well known, when a man is tired he goes to sleep, whereas a woman at the end of her tether just talks more. It was this over-confidence that led to my disgrace.

Eventually Gérard called me in before opening time one morning. I had a presentiment of calamity as I entered the building; suddenly I could not bear to face him, to hear him patiently explain anything, to submit to his suave voice and manner. I just wanted him to put me in irons like Charlotte Rampling, but that was the last thing he would do! Clearly, I was neurotically in love with the guy, but at that juncture I failed to recognize the symptoms.

As an opener he said: 'How are you getting on, still happy with us?'

'Of course.' I identified the gambit, and was already

wondering whether modelling was as overcrowded a profession as people said it was.

'We're living in tough times, it's difficult for us all,' he mumbled. 'The members are getting touchy, you know. There aren't so many now, and we have to trim costs. I have had one or two – er – complaints. People are still annoyed about your perfidious Englishman, and there have been remarks about your lateness and taking time off. Your political commentaries do not always score a laugh either, to put it mildly.'

My jaw dropped and I rasped: 'Whatever does that mean?'

He coughed: 'Well, it seems that it took 20 centuries to build Paris and 20 years for the quote Gaullist property mafia unquote to destroy it. Our esteemed Head of State is a part-time lawyer turned politician and therefore, in your view, an *ipso facto* menace to the nation. You also doubt that the economic crisis can possibly be solved by the bunch of wet schoolteachers occupying the Socialist benches. Finance Minister Jacques Delors comes within your firing range for his attempts to turn even the sackcloth into ashes. These are your views I am citing and they are regrettably out of place here.'

He paused and blew his nose. I felt a griping somewhere at the base of my torso, as he resumed: 'I didn't want to bother you with these things as they came up, and naturally I minimized them. But I must say that your personal remarks concerning Madame Chavaux and Monsieur Picard were going too far. In fact you now refuse to go with them, and that is unforgiveable.'

I pouted: 'She's got vaginal discharge, and Picard's prick has gone blotchy. Someone should take him to the vet.'

The manager sucked in his breath: 'That's quite enough of that. The fact is that I am afraid I – er – have bad news for you. At the last Committee Meeting it was decided that we shall have to divest ourselves of five staff including two hostesses, Stephanie because she was last in as the replacement for Ghislaine, and yourself.'

The blood drained from my face: 'I wasn't the next last in. I've been here 6 years, more even.'

'Yes, you were. Marie-Odile was not replaced. Besides the Committee is adamant. They consider you are no longer suitable as a hostess at this club, and no longer require your services. I'm sorry but there was nothing I could do. I even suggested a reprimand and a couple of months off without pay, but they would not hear of it. Believe me, Laure-Anne, this is not funny for me either. If it's any consolation, the whole outfit might be closed down soon. That's between ourselves, of course.'

The old familiar Gérard had gone, to be replaced by this tight-arsed mandarin. I was close to swooning. He assured me I would have a good written reference from the Committee and himself, and said he would personally ask around for openings for me. I said I was too upset to say anything. It was agreed I would continue for a couple of months, and I said I would not let him down and would be nice to everyone. The truth was I wanted to claw his eyes out. He said the accountant would fix up the best possible severance pay within the law and tell me how to obtain unemployment benefit. Finally he ordered me to take the day off.

It was still only 11.30 a.m. and I went back to the flat, crossing the avenue in a daze, with car horns blaring at me. Annick had not left, and I sobbed in her arms for a while, so that she was late for work. I felt brighter by early evening, and took a taxi to a cosy restaurant called the *Croque au Sel* near the Eiffel Tower and had a huge old-fashioned stew with a bottle of wine on my own. Then I walked back to the apartment under a windy cloudy sky.

Leaning over the Pont d'Alma I concluded that, since I was anything but destitute, I would give myself a couple of months off when I left. This prospect cheered me, and I was almost my usual serene self by the time I reached home. I spent a long evening in bed listening to France Musique with the teddy bear, I called him Xavier. I began to grow quite heady at the thought of leaving the Top Club. I had had a good run there, all good things must come to an end, it was no bad thing that we should part company.

I went next day to the Club Méditerranée headquarters for a catalogue and a chat, seeking a destination far from dark grey Europe: Tahiti perhaps, Senegal, Egypt, Guadeloupe. It had to be in the sun and I had to go alone.

That same week Gérard invited me out to lunch. I had reported for duty as usual at noon, and he was waiting for me in the lounge.

'You can't refuse, I've booked a table,' he said, 'and I'm still your boss. It's near the Opera.'

The expedition proved a disaster. He had become my enemy, and I sat sullen and silent in the taxi until he simply had to say something.

'Look, Laure-Anne, don't take it personally. It's not the end of the world. Believe me, I couldn't do a thing about it.'

It was raining again. The windscreen wipers slicked back and forth with gloomy regularity. We were stuck in a traffic snarl round the Place de la Concorde. I glowered at the back of the front passenger seat, slumped down and with my knees against the imitation leather.

'Well I don't believe you.'

'Oh come on, it's time you got away from the club anyway. There's no future in it for a decent intelligent girl like you.'

'You don't have to rescue me. It's my living, and you could have signalled what was coming, told me about the rings under my eyes.'

'I didn't want to worry you unnecessarily, it might have passed off. I told you the moment the decision was made.'

'You're a liar. You sacked me, cynically and without caring a damn. I could have walked out of that building and thrown myself over a bridge for all you cared. I nearly did.'

'That's ridiculous, not at 27 you wouldn't.'

'*You* wouldn't,' I snapped. 'Two months off without pay, that's your solution. You're too smart Gérard, smug and smooth and always two steps ahead. No, you wouldn't jump off a bridge, but others could do something dramatic. You haven't even spoken to me since you gave me the chop.'

'Rubbish. And don't keep pushing your hem down, your knees are one of your best features and the dress is too short anyway.'

'You sod! Is that all you think about, the right length for working in? As a B-film actor you're no better than Ronald Reagan, and you have the same soulless delivery. Doesn't it matter to you if a person gives her whole – her – her undying loyalty to you and the club, plays to the rules, sacrifices her private life? None of that counts, I suppose!'

The taxi drew up outside a Japanese establishment restaurant.

At the table we looked at each other tensely.

Gérard said: 'You are overwrought. All I want is to convey to you that this is your chance to make a new start. There's no reason for you to be a slave to the nation's rapacious bourgeoisie.'

'Is that your idea of a wisecrack? I told you, you don't have to rescue me. I don't want saving, it's affection and loyalty I need. If I want to go on the streets I will, whether you . . . '

'OK, I put it badly. And you're saying the first thing that comes into your head. You're starting to squawk.'

I hissed: 'And if this was a Hollywood studio I'd slap your face. I might still do it!'

A glance at the menu, which was in Japanese with French explanations, and then: 'Have you got a better suggestion than the streets? What do your friends say, have they got a job for me – a midnight live porn show maybe?'

'Like I told you, Laure-Anne, the entire club could be closed down any time. It's like the end of the Roman Empire. I'm already thinking of setting up my own thing, and there'll be openings.'

I snarled at him: 'There's no future in it, like you said. I wouldn't work for you now to save my life. You're not straight Gérard. Once I thought you were.'

We scowled at our menus. I was so angry I could scarcely focus. A waiter came up with an aperitif.

At length I said: 'I'm sorry, I couldn't eat a thing. Look at

161

this stuff: Toriwasa – raw chicken with lemon and green mustard; Kawa – chicken skin; Oshinka – salty vegetables. For pity's sake, I feel sick already. What's wrong with European food?'

We got up and left. I could see Gérard clenching his teeth in impatience. We took another taxi and I asked him to drop me at the flat.

'I'm in no condition to work today, take it off my wages, you lousy accountant. I just have to get away and think.'

We did not shake hands, because I charged indoors. The rain was a useful pretext, and I didn't care if I never saw him again.

A minute later I was looking sadly down from our window, unseeing, and was unable to tell whether it was through the rain or my tears, whether they were tears of anger or of self-pity.

To think that one kind word out of Gérard and I would have thrown my arms round his neck.

Instead of which I had burned my boats.

Chapter Seventeen

The Japanese foray had stolen my optimism. I worked for another month, but in a mood of complete indifference. I no longer wanted to take that holiday.

Then my mother phoned the date of my brother Yves' wedding. It was to be on the weekend before I left my employment. Normally I may well have shied away from this event, pleading duty, but I was in such a low key that I thanked her, confirming that I would be there. Then I asked for the Saturday off. This would perhaps snap me out of my dismal preoccupations, and in any case it would please the family. If you can't help yourself, you can help others, I argued, chuckling at this sanctimonious flash of inspiration.

The wedding was at Vezelay in the Yonne. The reception, arranged for some 200 guests, was to be at the home of my sister Sylvie who lived close by. She had married a wealthy farm equipment distributor, and their house which was a kind of small farm was a natural choice. Sylvie, now seven months pregnant with her second child, was spared any work.

My other brother, Jean-Marie, was still unmarried and he collected me by car at the crack of dawn. We spent the journey discussing everyone in the family, with the exception of myself. In answer to his questioning, I merely said I had worked as a kind of high-class waitress, and was looking around for something better. I told him I would rather forget work for that day.

Weddings tend to be erotic occasions in themselves, but I was unprepared for the effect on Sylvie's husband.

It was not far from midnight when Martin asked me to dance. I was hazy in the extreme, my legs as weak as if I had marched 30 kilometres, which was no doubt the distance I had moved since noon. He was puffing like a bull and his face was aflame. As we rotated, he had the annoying habit of holding his right leg in place longer than was strictly necessary, so that my loins pressed into his vast thigh each time. Endeavouring to maintain a deportment appropriate to the occasion, I sought to counter his ploy by slowing down too, so that several times we came to a halt, and he had to stop me falling. Martin was another of those dreadful men who pull their partners dresses up at the back for the benefit of their cronies, and I removed his hand more than once. He was trying to make me and I was annoyed, longing simply to enjoy the simple rustic delights as much as I could without being 'on duty'. Alcoholically, he was close to the point of no return, but I did not become aware of his true intentions until one of his hands started roaming over my superstructure and his cock dug into me like a pitchfork. Oh Lord, I thought, how wonderful if I could simply go for a stroll on my own to the top of some hill and sit and dream, breathing in the fresh air and gazing at the moon.

'You follow well when you dance,' he ventured. 'You have an amazingly supple figure. I bet you're a real pin-up in the nude.' At this 'daring' remark I gaped like a 15-year-old, pretending outrage, and he went on: 'How about a little trip to the hayloft? Just a kiss from those two hot lips is all I ask.' I still said nothing. 'Come on, Laure-Anne, a bit of a cuddle won't do you any harm. After all, I'm your host and you're a hostess. I've paid you in champagne twice over already.'

I was in a predicament. My personal wish was to knock his teeth in with a fencing hammer, but I could neither do this nor refuse his overtures for fear of creating bad feeling that could have gone on for decades. So when I was thinking of saying 'but what about your wife?' I blurted out: 'All right, but I want to leave soon if I can find Jean-Marie. No funny business, I'm

164

having – er – women's trouble at the moment.' It was almost true and I had hopes that Martin would abandon his project there and then. Alas, Martin danced me over to one of the buildings, and guided me swiftly to the foot of the ladder. With any luck he would break his neck on it.

'You go up first,' he ordered, and I was suddenly coy. There was just enough light from the dance area for him to get a cheap thrill as he stood below looking up at me if I went first. However, I was in no teasing mood, and in fact was in two minds whether to force the kissing game at once, and then escape.

'You first, hurry up,' he said, almost throwing me at the first rung of the ladder. So up I went, modestly holding my dress close around me.

In the loft he grabbed me without preamble and kissed me violently. Before I realized what was happening, he had me down in the hay. Finding myself there proved such a shock that I could only register a click noise in my backbone somewhere. How odd of him to use that kind of violence, I reflected; after all, we were not 15-year-olds. It then dawned upon me that he meant to take me by force, thinking that I was going to resist. He began tugging at my clothes.

'You pig, Martin,' I moaned, 'this is rape, control yourself!'

His hands groping around my thighs, he hissed: 'Come on, Laure-Anne, you've done this a thousand times, I can guess what you do for a living.' At this insult I struggled to get up, but his ox-like shoulders held me down. With one fist he had my two wrists neutralized and with his other hand he guided his fat penis to its destination, ramming it into my still-arid vagina. I squealed with pain, he jammed a paw over my mouth, and his trunk heaved and fell massively as he ripped into my passage again and again. I kept uttering little squeaks, bile rose in my throat and I flung my head to the side, spewing into the hay.

'Stop it,' I managed to gasp, 'can't you see I'm not ready?'

He cackled: 'I can wait. Badgers can do it a whole hour non-stop, and I'm going to make you come first, you tigress. I've wanted you ever since I saw you that afternoon at Marnes about four years ago.'

'You're hurting . . . ' But even as I complained, my labia responded of their own volition, in a conditioned reflex. My juices wept on their own as he slowly pushed in and withdrew again and again. He was huge and I could feel him all the way, his obdurate flesh stretching within me. But I was unaroused, felt myself as an observer looking at the procedure from a distance. He moved clumsily and his feet jabbed one of my shinbones, causing a searing pain there. At that point he ejaculated, so he had succumbed after all before the tigress.

I lay stunned under his weight. After a while he said: 'Thanks, Laure-Anne, I'll never forget that, my first for weeks.'

'Huh. I don't think I'll forget it either. Next time I'll wear a suit of armour.'

Such was rape. Not only had it been cheap and nasty with no enjoyment for me, but I had actually been defiled just when I was preparing to turn over a new leaf.

So henceforth I was to be the family whore!

Vowing to repay the bastard one day, I pushed the disgusting incident out of my mind.

Back in Paris, my forthcoming holiday was bobbing on the horizon like a life-raft. Finally, I wanted to go, it was the off-peak season and I could leave for anywhere whenever I chose to. The sooner the better, as I suddenly loathed France and my fellow-citizens. I was terribly lonely, and was willing to beg for an ounce of affection. But somewhere else.

It was not to be. A week after saying goodbye to the Top Club, I had an accident. I was caught up in a group of people on some steps at Porte Maillot Metro station, and slipped.

To my astonishment, I found I could not get up. And the pain was awful when I moved. The system sprang into action and in no time at all I was loaded into an ambulance for Ambroise Paré hospital in Boulogne-Billancourt.

I gave them Annick's phone number at the club. She arrived at once. Gérard came too.

X-rays disclosed a slipped disc.

Following which, I was virtually lost to the world for a week under the effect of painkillers.

It's hard being one's usual pleasant and cheerful self when one has been dismissed, raped and thrown down a dozen concrete steps. And I certainly was not my sprightly self.

I will spare the reader the close detail of my physical discomfort during the three weeks in hospital. Things such as having to keep still, my right leg dragging in macabre fashion when I went to the toilet, headaches from the drugs, constipation, the permanent unwashed feeling, my ashen face, my witch-like hair with the rinse growing out – all these things on top of the pain and the listlessness.

Uncertainty as to how long the affair would continue added to my woes. Nobody seemed to know, and a nurse told me encouragingly: 'Usually it's three days or three weeks or three months.' Of course, the only person I wanted to see was Annick; she was an absolute angel, bringing in my personal objects, filling in forms, negotiating with nurses, fixing up the phone in my room, and so on.

Actually the phone was a mixed blessing, because the whole family rang and I had to repeat the story over and over in my sepulchral voice; only to go through the entire rigmarole a week later when they called for a progress report. They meant well, as did the folk who put in a physical appearance. But frankly I was not one to be fussed over, and would have preferred to be left alone until I felt less foul.

The staff at the Top Club sent an enormous funny card, with all their signatures on it headed by Gérard's. They would visit me when I got home, they said. Annick had told them that that would be better.

In due course Annick fetched me by taxi and I prepared to sit out the siege, or more accurately to lie it out. I lurched about the apartment, but was under strict orders not to force

167

things, so Annick was kept busy. I told her she was a saint and she said I would be doing the same for her, and we were both overcome with emotion. In a fit of gloom I said she was my sole true friend.

This was an incorrect statement, especially as Gérard came in two days after I got home. He charged through the door with an enormous plank of timber. 'This should speed up things,' he said, stripping the bed and installing the plank under the mattress.

He was helpful in another practical way. As I said, the hospital was evasive about the cure period, and even about what to do. Gérard phoned the club's doctor and grilled her, interrogated the hospital staff and showed my X-rays to people he knew in the profession. The result was a four-page typed report which he presented to me. This confirmed that there was nothing to be done but kill the pain with drugs while I waited; that injections of various kinds would be necessary for months; that surgery must be avoided because it held out a risk of paralysis; that I would later do exercises; that an exercise bike was advisable in due course; that in a few weeks I would be up and about; that a year hence I would be back to normal.

'I'm so grateful,' I told him. 'You've spent hours going into all this.'

He shrugged: 'I know a lot of people, it was easier than you think. In any case I have more time now, we're closing down.'

'What! Annick didn't tell me!'

'She doesn't know. The club's closing at the end of the year. The Committee decided two days ago, and I'm telling the staff today.'

'But why?'

'Several reasons. Someone, we think a member's wife, has spilled everything to the press, and there have been all sorts of mean hints in the media. The Government has been putting pressure on us meanwhile, and we're throwing in the towel. Our lords and masters are incapable of solving the economic crisis so the little pork-bellies in pinstripe suits take it out on those of us making the most fun we can. In other words there's

168

a big clamp-down on the way, and a stack of places will be swept away in the next couple of years. We at the Top Club will all receive six months' notice from July 1. Normally we might have put up a fight just for the heck of it, start up in another form perhaps, but the membership has plunged.'

'It's a shock. Such a worthy cause. But Gérard, my dear friend, what are you going to do?'

'Same as you, think of something! But my first objective is to see you on your feet.'

'And the second?'

'To take you on the best holiday you've ever had in your life. We'll go to Crete, see a bit of Old Europe! I have to make up for that Japanese raw pork!'

'That would be marvellous, G – Ouch. Looks like it'll be some time yet.'

One afternoon in July the full bevy of club girls called to see me. They arrived with two large cakes and 5 bottles of bubbly to drink my health. I had a thimbleful of the champagne, because there is nothing worse than that for sciatica, said Gérard. I asked them what they were all proposing to do.

Bernadette: 'I am going to raise mink with daddy.'

Mauricette: 'I'm planning a mailing shot, Gérard's given me the membership list.'

Annick: 'I've no idea, but I intend to hibernate for months, I'm so tired.'

Blandine: 'I may invest in fast food. One has to survive and there's no need to eat the stuff.'

Nadège: 'I'm in touch with a Japanese film company exporting films to the Middle East.'

Solange: 'I propose to lease an airliner and run a Flying Eros Center between Paris and Miami. I have a backer in Switzerland, and I'll be needing hostesses.'

'Oo,' said Annick, 'you can call it the Flying Bed, or the Flying Nurse as it's therapy.'

Blandine growled: 'That's better than the Flying Bordel, I suppose.'

Chapter Eighteen

I had another visitor.

Annick went away for two weeks in August, leaving me on my own. I could now walk or stand or lie, but still found sitting uncomfortable.

On the night after she went away, the doorbell roused me from sleep. It kept ringing and ringing and I stretched out for the bedside lamp. One thirty five – a.m.! Whoever it was started banging on the door.

Burglars!

Scared stiff, I dialled the Top Club number.

'This is Laure-Anne, give me Gérard at once, it's an emergency!'

'He's on the other line.'

'Get him off, Fred, it's urgent, someone's breaking in here.'

Within seconds I was yelling this information to Gérard, but as I did so I heard the assailant bawl: 'Hey Laure-Anne, open up, it's me Terry, I've got some champers.'

I passed this news on in French to Gérard, adding that Terry was drunk. Gérard said he would be right over.

Terry kept ringing and banging. I rolled over and limped to the door.

'All right all right, stop making that row. You'll wake everyone up.'

As I reached the door, the phone rang. I hesitated, unlatched the door notwithstanding, and saw Terry grinning and waving

170

a couple of paper-wrapped bottles. I hastened as best I could to the phone.

'This is Monsieur Jourdan underneath. If you people don't stop making that noise at once, I'll have to call the police!'

I bawled back: 'What at two a.m.? That's ridiculous, go back to sleep!' I disliked the man, he was something high up in Gaz de France and went off to work at dawn.

I crashed the phone down and turned to find Terry falling towards me, the bottles threatening to hit the phone table. I pushed him back, and he released a bottle which smashed and foamed on the carpet. He put down the other bottle with exaggerated care.

Stepping clear, I said: 'Stay right where you are, Terry, don't you touch me!'

In hardly distinguishable French he drawled: 'Don't be absurd, you're my girl for the night.' Then in a curious English: 'Me lonely, we do fuckee-fuckee, me like Lauree-Lauree. You already get in bed!'

I was in fact standing next to the bed, and as he advanced upon me I spat out: 'I'm ill, you idiot, I'm in no state . . .'

I got no further. He upped my nightdress and threw himself upon me, so that I was forced back on the bed. I let out an almighty shriek of pain.

At that precise moment came two sharp raps on the door. Gérard, thank God.

Terry froze and Gérard shouted: 'I'm here, Laure-Anne, let me in!'

'Who's that?' Terry whispered.

'The downstairs neighbour. He just phoned. I'll send him away.'

'No, I'll send him away.'

'No, please let me speak to him, I don't want any trouble.'

He consented to this, released me, and once more I staggered to the door. I opened it, Gérard walked in, and I fell into his arms.

I heard Terry declare: 'You're the chap who tried to throw me out of that club.'

'Yes, and I'm throwing you out of here too.'

Gérard moved me to one side just as Terry lunged at him, landing a light blow on his cheek. My rescuer parried a second blow, forced Terry's arms into a lock and propelled him out of the apartment and into the lift. He took him down and I crept gratefully into bed, leaving the door open.

Ten minutes went by, maybe 15, maybe 20. I lay there riding the pain.

The phone went again exactly as Gérard returned and closed the door.

He took it and plunged into a conversation: 'Yes I'll be some time, Fred, everything's fine now, but I'll have to clear up here. Can you stay an extra hour? No, I've a better idea, just lock up and I'll collect the keys tomorrow morning early.'

Gérard sat on the edge of the bed.

'How is it? Must have shaken you up.'

'Better now, might have clicked me back into shape. What did you do with him?'

He took my hand, pushing my hair back with the other from my brow: 'I told the cab driver to dump him outside Charenton railyard.'

'Where's that?'

'In the general direction of Geneva. He'll take a while to find his way back.'

We looked at each other an instant and I said: 'Kiss me, Gérard.'

My eyes were damp when I opened them again.

Together we said: 'I love you.'

I wept for joy, my head on his neck: 'Oh Gérard, if you only knew how I have longed for that kiss.'

His deep blue eyes filled with mystery: 'I love you Laure-Anne. Fell for you when you walked into my office all those years ago, but couldn't do a thing about it. You see, the day before that my wife and I had separated, and we hadn't even discussed divorce proceedings. I'd just found a man in her bed and at that moment I could no more have dared shown my

172

interest than a boxer thrown out of the ring. I was done for, all the fight had left me. Later, when the divorce went through, it seemed too late. The moment had passed. In any case you appeared uninterested.'

I stared at him: '*I* was uninterested. Oh, you should have tried me. How absurd!' Instantly the truth of the situation struck me; it was I who had been absurd, expecting him to take his place in the queue, as it were, with the club members. I flushed hot at the thought.

He said: 'Everything's different now, it's all over.'

I said: 'Tell me again you love me. Keep saying it in case it goes away.'

'I love you, I love you, I love you . . . '

I pulled at his arms and brought his face within centimetres of mine: 'And I love you with everything I've got. And I feel so damned frustrated because if I wasn't like this I'd be climbing all over you and you'd be swinging me round and we'd go out and celebrate 'till dawn. Oh isn't it marvellous Gérard, I've got stars in front of my eyes, I'm such a fool crying like this but I can't help it because I've never been in love before and it is the most super thing that has ever happened to me. We're together at last, do you realize? Isn't it great we're made for each other, isn't it the most wonderful feeling?'

'The only one. I feel an immense peace.'

'Oh so do I? I love just being with you, every gesture you make I love, everything you say. Oh tell me, tell me how desirable I am, and let me tell you what bliss it is loving you. I'm so thrilled!!! You once said I had a perfect body, it'll be perfect again soon, won't it Gérard?'

Levelly he said: 'It's a sculptor's dream, lithe in movement, with a graceful lilt to the shoulders. You have a natural poise. I adore the way your hand pauses for a split second before you touch things.'

'Go on, it's delicious.'

'Your eyes are captivating.'

'And my big mouth and wonky teeth?'

He touched them: 'You have the hungriest lips I ever saw,

and now at last I have kissed them. A million times I shall sink my cruel mouth into your luscious cushions, seek repose there and inhale your musky fragrance. And that little gap in your teeth tells me we are going to enjoy ourselves wherever we go and whatever we do. You know too that your teeth are slightly pointed? That's for piquancy, for concupiscence.'

I stopped him: 'Gérard, I think I'm wetting the sheet. I've been chaste for ages, and now I'm flowing with love. But we'll wait, won't we *mon adoré*, it'll be even lovelier when we make love at last. Promise.'

'Promise.'

He rose and selected a cassette on the player. Brahm's violin concerto. Mysteriously he brought over a small table and found a glass. He then picked up the unopened bottle from the floor, popped the cork and filled the glass which he handed to me.

'To us,' we said. I drank and passed the glass to him and he turned it round and drank where I had.

After that we lay hand in hand, waiting for the dawn in the silence.

STAR BOOKS ADULT READS

FICTION

BEATRICE	Anonymous	£2.25*
EVELINE	Anonymous	£2.25*
MORE EVELINE	Anonymous	£2.25*
FRANK & I	Anonymous	£2.25*
A MAN WITH A MAID	Anonymous	£2.25*
A MAN WITH A MAID 2	Anonymous	£2.25*
A MAN WITH A MAID 3	Anonymous	£2.25*
OH WICKED COUNTRY	Anonymous	£2.25*
ROMANCE OF LUST VOL 1	Anonymous	£2.25*
ROMANCE OF LUST VOL 2	Anonymous	£2.25*
SURBURBAN SOULS VOL 1	Anonymous	£2.25*
SURBURBAN SOULS VOL 2	Anonymous	£2.25*
DELTA OF VENUS	Anais Nin	£1.60*
LITTLE BIRDS	Anais Nin	£1.60*
PLAISIR D'AMOUR	A.M.Villefranche	£2.25
JOIE D'AMOUR	A.M.Villefranche	£2.25

STAR Books are obtainable from many booksellers and newsagents. If you have any difficulty tick the titles you want and fill in the form below.

Name _____

Address _____

_____

Send to: Star Books Cash Sales, P.O. Box 11, Falmouth, Cornwall, TR10 9EN.

Please send a cheque or postal order to the value of the cover price plus: UK: 55p for the first book, 22p for the second book and 14p for each additional book ordered to the maximum charge of £1.75.

BFPO and EIRE: 55p for the first book, 22p for the second book, 14p per copy for the next 7 books, thereafter 8p per book.

OVERSEAS: £1.00 for the first book and 25p per copy for each additional book.

While every effort is made to keep prices low, it is sometimes necessary to increase prices at short notice. Star Books reserve the right to show new retail prices on covers which may differ from those advertised in the text or elsewhere.

*NOT FOR SALE IN CANADA

STAR BOOKS ADULT READS

FICTION

ADVENTURES OF A SCHOOLBOY	Anonymous	£2.25
AUTOBIOGRAPHY OF A FLEA	Anonymous	£2.25*
MEMOIRES OF DOLLY MORTEN	Anonymous	£2.25
LAURA MIDDLETON	Anonymous	£2.25
THREE TIMES A WOMAN	Anonymous	£2.25*
THE BOUDOIR	Anonymous	£2.25*
THE LUSTFUL TURK	Anonymous	£2.25*
MAUDIE	Anonymous	£2.25*
RANDIANA	Anonymous	£2.25*
ROSA FIELDING	Anonymous	£2.25*
JOY	Joy Laurey	£2.25
JOY AND JOAN	Joy Laurey	£2.25
INSTRUMENT OF PLEASURE	Celeste Piano	£2.25
OPUS PISTORUM	Henry Miller	£2.25*

STAR Books are obtainable from many booksellers and newsagents. If you have any difficulty tick the titles you want and fill in the form below.

Name _____

Address _____

_____

Send to: Star Books Cash Sales, P.O. Box 11, Falmouth, Cornwall, TR10 9EN.

Please send a cheque or postal order to the value of the cover price plus: UK: 55p for the first book, 22p for the second book and 14p for each additional book ordered to the maximum charge of £1.75.

BFPO and EIRE: 55p for the first book, 22p for the second book, 14p per copy for the next 7 books, thereafter 8p per book.

OVERSEAS: £1.00 for the first book and 25p per copy for each additional book.

While every effort is made to keep prices low, it is sometimes necessary to increase prices at short notice. Star Books reserve the right to show new retail prices on covers which may differ from those advertised in the text or elsewhere.

*NOT FOR SALE IN CANADA